CONCISE ENCYCLOPEDIA OF

POLYMER PROCESSING & APPLICATIONS

ADVANCES IN MATERIALS SCIENCE AND ENGINEERING

This is a new series of Pergamon scientific reference works, each volume providing comprehensive, self-contained and up-to-date coverage of a selected area in the field of materials science and engineering. The series is being developed primarily from the highly acclaimed *Encyclopedia of Materials Science and Engineering*, published in 1986. Other titles in the series are listed below.

BEVER (ed.)
Concise Encyclopedia of Materials Economics, Policy & Management

BROOK (ed.)
Concise Encyclopedia of Advanced Ceramic Materials

CAHN & LIFSHIN (eds.)
Concise Encyclopedia of Materials Characterization

CARR & HERZ (eds.)
Concise Encyclopedia of Mineral Resources

EVETTS (ed.)
Concise Encyclopedia of Magnetic & Superconducting Materials

KELLY (ed.)
Concise Encyclopedia of Composite Materials

MAHAJAN & KIMERLING (eds.)
Concise Encyclopedia of Semiconducting Materials & Related Technologies

MOAVENZADEH (ed.)
Concise Encyclopedia of Building & Construction Materials

SCHNIEWIND (ed.)
Concise Encyclopedia of Wood & Wood-Based Materials

WILLIAMS (ed.)
Concise Encyclopedia of Medical & Dental Materials

NOTICE TO READERS

Dear Reader
If your library is not already a standing order/continuation order customer to the series **Advances in Materials Science and Engineering**, may we recommend that you place a standing/continuation order to receive immediately upon publication all new volumes. Should you find that these volumes no longer serve your needs, your order can be cancelled at any time without notice.

CONCISE ENCYCLOPEDIA OF

POLYMER PROCESSING & APPLICATIONS

Editor

PATRICK J CORISH

Stratford-on-Avon, UK

Executive Editor

ROBERT W CAHN FRS

University of Cambridge, UK

Senior Advisory Editor

MICHAEL B BEVER

MIT, Cambridge, MA, USA

PERGAMON PRESS

OXFORD • NEW YORK • SEOUL • TOKYO

UK	Pergamon Press plc, Headington Hill Hall, Oxford OX3 0BW, England
USA	Pergamon Press, Inc., 395 Sawmill River Road, Elmsford, New York 10523, U.S.A.
KOREA	Pergamon Press Korea, KPO Box 315, Seoul 110-603, Korea
JAPAN	Pergamon Press Japan, Tsunashima Building Annex, 3-20-12 Yushima, Bunkyo-ku, Tokyo 113, Japan

---

First edition 1992

Library of Congress Cataloging in Publication Data
Concise encyclopedia of polymer processing & applications / editor, P. J. Corish. — 1st ed.
p. cm. — (Advances in materials science and engineering)
Includes index.
1. Polymers—Encyclopedias. I. Corish, Patrick J., 1921– .
II. Series.
TP1087.C65 1991
668.9'03—dc20 91–26436

British Library Cataloguing in Publication Data
Concise encyclopedia of polymer processing and applications.
— (Advances in materials science and engineering)
I. Corish, P. J.
668.9

ISBN 0–08–037064–0

™ The paper used in this publication meets the minimum requirements of the American National Standard for Information Sciences—Permanence of Paper for Printed Library Materials, ANSI Z39.48–1984.

Printed and bound in Great Britain by BPCC Wheatons Ltd, Exeter

CONTENTS

HONORARY EDITORIAL ADVISORY BOARD

FOREWORD

In the short time since its publication, the *Encyclopedia of Materials Science and Engineering* has been accepted throughout the world as the standard reference about all aspects of materials. This is a well-deserved tribute to the scholarship and dedication of the Editor-in-Chief, Professor Michael Bever, the Subject Editors and the numerous contributors.

During its preparation, it soon became clear that change in some areas is so rapid that publication would have to be a continuing activity if the Encyclopedia were to retain its position as an authoritative and up-to-date systematic compilation of our knowledge and understanding of materials in all their diversity and complexity. Thus, the need for some form of supplementary publication was recognized at the outset. The Publisher has met this challenge most handsomely: both a continuing series of Supplementary Volumes to the main work and a number of smaller encyclopedias, each covering a selected area of materials science and engineering, will be published in the next few years.

Professor Robert Cahn, the Executive Editor, was previously the editor of an important subject area of the main work and many other people associated with the Encyclopedia will contribute to its Supplementary Volumes and derived Concise Encyclopedias. Thus, continuity of style and respect for the high standards set by the *Encyclopedia of Materials Science and Engineering* are assured. They have been joined by some new editors and contributors with knowledge and experience of important subject areas of particular interest at the present time. Thus, the Advisory Board is confident that the new publications will significantly add to the understanding of emerging topics wherever they may appear in the vast tapestry of knowledge about materials.

The appearance of Supplementary Volumes and the new series *Advances in Materials Science and Engineering* is an event which will be welcomed by scientists and engineers throughout the world. We are sure that it will add still more luster to a most important enterprise.

Walter S Owen
Chairman
Honorary Editorial Advisory Board

EXECUTIVE EDITOR'S PREFACE

As the publication of the *Encyclopedia of Materials Science and Engineering* approached, Pergamon Press resolved to build upon the immense volume of work which had gone into its creation by embarking on a follow-up project. This project had two components. The first was the creation of a series of Supplementary Volumes to the Encyclopedia itself. The second component of the new project was the creation of a series of Concise Encyclopedias on individual subject areas included in the Main Encyclopedia to be called *Advances in Materials Science and Engineering*.

These Concise Encyclopedias are intended, as their name implies, to be compact and relatively inexpensive volumes (typically 300–600 pages in length) based on the relevant articles in the Encyclopedia (revised where need be) together with some newly commissioned articles, including appropriate ones from the Supplementary Volumes. Some Concise Encyclopedias will offer combined treatments of two subject fields which were the responsibility of separate Subject Editors during the preparation of the parent Encyclopedia (e.g., dental and medical materials).

Eleven Concise Encyclopedias have been contracted. These and their editors are listed below.

Concise Encyclopedia of Advanced Ceramic Materials	Prof. Richard J Brook
Concise Encyclopedia of Building & Construction Materials	Prof. Fred Moavenzadeh
Concise Encyclopedia of Composite Materials	Prof. Anthony Kelly CBE, FRS
Concise Encyclopedia of Magnetic & Superconducting Materials	Dr Jan E Evetts
Concise Encyclopedia of Materials Characterization	Prof. Robert W Cahn FRS & Dr Eric Lifshin
Concise Encyclopedia of Materials Economics, Policy & Management	Prof. Michael B Bever
Concise Encyclopedia of Medical & Dental Materials	Prof. David Williams
Concise Encyclopedia of Mineral Resources	Dr Donald D Carr & Prof. Norman Herz
Concise Encyclopedia of Polymer Processing & Applications	Mr Patrick J Corish
Concise Encyclopedia of Semiconducting Materials & Related Technologies	Prof. Subhash Mahajan & Dr Lionel C Kimerling
Concise Encyclopedia of Wood & Wood-Based Materials	Prof. Arno P Schniewind

Most of the new or substantially revised articles in the Concise Encyclopedias have been or will be published in one of the three Supplementary Volumes, which are designed to be used in conjunction with the Main Encyclopedia. The Concise Encyclopedias themselves, however, are "free-standing" and are designed to be used without reference to the parent Encyclopedia.

The Executive Editor is personally responsible for the selection of topics and authors of articles for the Supplementary Volumes. In this task, he has the benefit of the advice of the Senior Advisory Editor and of other members of the Honorary Editorial Advisory Board. The Executive Editor is responsible for appointing the Editors of the various Concise Encyclopedias and for supervising the progress of these volumes.

Robert W Cahn FRS
Executive Editor

ACKNOWLEDGEMENTS

I would like to acknowledge the general guidance and help given by Professor Robert Cahn, Executive Editor of the Advances in Materials Science and Engineering series of Concise Encyclopedias. Additionally, the editorial staff of Pergamon Press are thanked for their willing cooperation: formerly, Dr Colin Drayton and, latterly, Mr Peter Frank. I would like to thank the authors of the articles from the *Encyclopedia of Materials Science and Engineering* and its Supplementary Volumes for their valuable contributions. Finally, I should like to express my gratitude to authors and coauthors of the various new and revised articles which have served to give this volume more complete coverage of a broad, expanding materials area.

Patrick J Corish
Editor

GUIDE TO USE OF THE ENCYCLOPEDIA

This Concise Encyclopedia is a comprehensive reference work covering all aspects of polymer processing and applications. Information is presented in a series of alphabetically arranged articles which deal concisely with individual topics in a self-contained manner. This guide outlines the main features and organization of the Encyclopedia, and is intended to help the reader to locate the maximum amount of information on a given topic.

Accessibility of material is of vital importance in a reference work of this kind and article titles have therefore been selected not only on the basis of article content but also with the most probable needs of the reader in mind. An alphabetical list of all the articles contained in this Encyclopedia is to be found on pp. xv and xvi.

Articles are linked by an extensive cross-referencing system. Cross-references to other articles in the Encyclopedia are of two types: in text and end of text. Those in the body of the text are designed to refer the reader to articles that present in greater detail material on the specific topic under discussion at that point. They generally take one of the following forms:

> ...which is fully described in the article *Design of Elastomer-Based Engineering Products.*

> ...other applications of ceramic materials (see *Offshore Engineering Applications of Polymers*).

The cross-references listed at the end of an article serve to identify broad background reading and to direct the reader to articles that cover different aspects of the same topic.

The nature of an encyclopedia demands a higher degree of uniformity in terminology and notation than many other scientific works. The widespread use of the International System of Units has determined that such units be used in this Encyclopedia. It has been recognized, however, that in some fields Imperial units are more generally used. Where this is the case, Imperial units are given with their SI equivalent quantity and unit following in parentheses. Where possible the symbols defined in *Quantities, Units, and Symbols*, published by the Royal Society of London, have been used.

All articles in the Encyclopedia include a bibliography giving sources of further information. Each bibliography consists of general items for further reading and/or references which cover specific aspects of the text. Where appropriate, authors are cited in the text using a name/date system as follows:

> ...as was recently reported (Smith 1988).

> Jones (1984) describes...

The contributor's name and the organization to which they are affiliated appear at the end of each article. All contributors can be found in the alphabetical List of Contributors, along with their full postal address and the titles of the articles of which they are authors or coauthors.

The article *Polymer Processing and Applications: An Overview* provides an overview of polymer processing and applications and discusses in brief the coverage and organization of the Encyclopedia and how these issues are covered by the articles in the body of the work.

The most important information source for locating a particular topic in the Encyclopedia is the multilevel Subject Index, which has been made as complete and fully self-consistent as possible.

ALPHABETICAL LIST OF ARTICLES

AN INTRODUCTION TO POLYMER PROCESSING AND APPLICATIONS

by Patrick J Corish

Polymer usage is widespread in everyday life throughout the world, at home and in commerce. Under the three headings of elastomers (rubbers), plastics and fibers (textiles), singly and in combination—often with other materials—polymers are ubiquitous in their application. Since the 1970s, growth in polymer usage has exceeded that of metals.

With such wide-ranging end-use applications, it is not possible in a single volume to cover all polymer applications: the aim of this Concise Encyclopedia is to include all the mainstream areas of application and their associated processing techniques. The unique properties of elastomers, plastics, fibers and hybrid materials demand special processing techniques and equipment and these are reviewed. Variations in these processing techniques and the conditions under which processing is carried out can significantly alter the end-use properties of the materials and coverage of these areas is included.

Articles covering all aspects of polymer processing and the wide range of end-use applications have been chosen from the *Encyclopedia of Materials Science and Engineering* and updated. Emphasis has been placed on the practical handling and production techniques, assuming only a basic knowledge of polymers. Relevant articles from Supplementary Volumes 1 and 2 have been added and new contributions from leading experts have been commissioned to ensure that comprehensive coverage of the subject matter is obtained.

Some specialized applications areas of polymers are included in other Concise Encyclopedias in the Advances in Materials Science and Engineering series, covering specific topics such as building and construction materials, medical and dental materials and wood and wood-based materials. High- technology usage of polymers in combination with other materials is also covered in depth in a separate Concise Enyclopedia. The articles in the *Concise Encyclopedia of Composite Materials* and the *Concise Encyclopedia of Medical & Dental Materials* which discuss the specialized uses of polymers within these two important fields are listed in Tables 1 and 2.

The articles in this Concise Encyclopedia can be broken down into a number of broad categories, beginning with those covering the general aspects of polymers and the history of polymer production and use and moving on to those that discuss the three classes of polymers with their specific properties, applications and processing requirements. Where processing techniques are applicable to more than one category of polymer, articles reflect this in their coverage. Composite materials are dealt with in depth and specific applications are covered separately.

Table 1
Articles covering polymer usage in the *Concise Encyclopedia of Composite Materials*

Artificial Bone
Fiber-Reinforced Polymer Systems: Extrusion
Fibrous Composites: Thermomechanical Properties
Friction and Wear Applications of Composites
Helicopter Applications of Composites
In Situ Polymer Composites
Natural-Fiber-Based Composites
Paper and Paperboard
Polymer–Polymer Composites
Solid Fiber Composites as Biomedical Materials
Strength of Composites
Structure Performance Maps
Thermosetting Resin Matrices
Toughness of Fibrous Composites
Wood–Polymer Composites

Table 2
Articles covering polymer usage in the *Concise Encyclopedia of Medical & Dental Materials*

Acrylic Dental Polymers
Acrylics for Implantation
Adhesives from Protein–Polymer Grafts
Adhesives in Medicine
Arteries, Synthetic
Biocompatibility: An Overview
Biocompatibility of Dental Materials
Biodegradation of Medical Polymers
Chemical Adhesion in Dental Restoratives
Composite Materials
Denture Base Resins
Drugs: Attachment to Polymers
Elastomers for Dental Use
Heart-Valve Replacement Materials
Hydrogels
Maxillofacial Prostheses
Polyesters and Polyamides
Polyethylene
Polymers for Controlled Drug Delivery
Polysiloxanes
Polytetrafluoroethylene
Polyurethanes
Suture Materials
Wound Dressings Materials

What follows is a brief discussion of the contents of the Encyclopedia based on this classification.

1. Review Articles

General articles review the history of polymer science and technology, emphasizing processing, properties, engineering aspects and end-use applications (*Polymer Science and Engineering: Historical Development*; *Polymers: Classification by Use and Properties*; *Polymers: Technical Properties and Applications*; *Processing: Classification*; *Processing: Historical Survey* and *Processing: Relation to Product Properties*). Overview articles in the following subject areas are also included: elastomers, fibers and textiles, composites and adhesive joining.

2. Processing and Manufacturing Techniques

Product manufacturing systems for various end-use applications have an appreciable degree of commonality for elastomeric- and thermoplastic-type compositions. Thus, articles covering mixing and processing contain descriptions of both types of compositions and these are listed together. Specific aspects of elastomers and thermoplastics are covered separately. Mixing of both batch and continuous types is covered in a series of articles (*Mixing: Basic Concepts*; *Mixing: Internal Mixers*; *Mixing: Mills for Elastomers*; *Mixing: Rheological Behavior and Dispersion of Two-phase Systems*; *Mixing: Single-Screw Extruders* and *Mixing: Twin-Screw Extruders*). Melt delivery systems, essential to continuous mixing, are reviewed in another series of articles (*Melt Delivery Systems: Corotating Disk Processors*; *Melt Delivery Systems: Normal Stress Pumps*; *Melt Delivery Systems, Single-Screw* and *Melt Delivery Systems, Twin-Screw*).

In addition to the general articles on processing, specific areas are covered in more detail in the articles *Calendering*; *Coextrusion*; *Die Extrusion*; *Polymer Sheet: Manufacture and Applications*; *Processing: Rheological Properties and Fluid Mechanics*; *Processing: Structure Development* and *Rheological Characterization of Molten Polymers*.

Novel processing methods resulting in improved product physical properties are described in *Processing: Shear-Controlled Orientation* and the technique of *Reaction Injection Molding*, currently being used more widely, is discussed in some detail. The article *Radiation Processing* discusses the technique of producing crosslinking in elastomeric- or thermoplastic-based compositions. Additionally, the use of computers in process control and planning is discussed in some detail in the articles *Computer Control in Polymer Processing* and *Injection Molding of Polymers: Computer Simulation*. As an adjunct to successful processing, *Plasticization of Polymers* is reviewed. The article *Polymer Blends* covers the use of these materials for both processing and end-use applications and this theme is continued in the article *Thermoplastic Elastomers* which discusses a class of polymers that can be processed like thermoplastics but exhibit elastomeric properties in end-use applications. This class of material is an example of "tailor-making" polymers in both microstructure and macrostructure.

3. Elastomers (Rubbers)

Aspects more specific to elastomers are the basis of articles which include general topics (*Elastomers: An Overview* and *Elastomers: Applications Overview*). The processing of elastomers (rubbers) normally consists of mixing, then processing or shaping by sheeting, calendering or extruding, followed by vulcanization. Articles on all these topics are included (*Rubber Processing*; *Mixing: Internal Mixers*; *Mixing: Mills for Elastomers*; *Vulcanization; Vulcanization by Silanes*; *Vulcanization, Continuous* and *Polymer Sheet: Manufacture and Applications*). Other more specialized techniques for processing are described in the articles *Dipped Polymer Products*; *Elastomeric Products from Liquid Polymers*; *Powdered and Particulate Elastomer Processing* and *Radiation Processing*.

Compounding ingredients have an important role in the successful processing and vulcanization of elastomers for sometimes stringent end-use applications. The articles *Elastomers: Additives* and *Fillers in Elastomers* cover this aspect, emphasizing the necessity for good dispersion, especially of small particle size reinforcing fillers, to maximize properties. The use of plasticizers is referred to in the article *Plasticization of Polymers*: the conversion of a thermoplastic material to an elastomeric one by plasticization is also discussed in the article *Plastized PVC Elastomers*. The novel area of *Thermoplastic Elastomers* has already been highlighted.

Another aspect of elastomer processing reviewed in detail is that of quality (*Quality Requirements in Elastomer Processing*), an area of increasing importance in product liability. A different but equally important topic, with more stringent government requirements in many countries, is that of health and safety, covered in the article *Health Hazards in Elastomer Processing*. Another area of increasing interest in materials economy is recycling which is also reviewed in the articles *Reclaiming of Elastomers* and *Recycling*.

Some specialized elastomeric materials which have particular end-use applications are discussed in the articles *Elastomeric Coatings*; *Elastomeric Fibers*; *Elastomeric Foams* and *Hard-Fiber Elastomers*.

One general area of elastomeric products is reviewed in *Design of Elastomer-Based Engineering*

Products (see also *Design of Composite Structures*). The thematic use of elastomer–metal and elastomer–polymer bonded structures is also covered in five articles (*Elastomer–Metal Adhesion Processes*; *Elastomer–Wire Adhesion Processes*; *Elastomer–Elastomer Adhesion Processes*; *Elasomer–Textile Adhesion Processes* and *Elasomer–Thermoplastics Adhesion Proecsses*). Automobile tires are everyday examples of the use of elastomeric materials in conjunction with other polymeric and metallic components—really compositelike structures (see *Automobile Tires*). Other automotive applications of elastomers have also been included in the articles *Automotive Applications of Polymers* and *Automotive Composite Components: Fabrication*. Other applications of elastomers are discussed in the article *Sports Equipment Manufacture*, which includes information on many uses within this field including tennis, squash and golf balls, and in *Offshore Engineering Applications of Polymers*.

Additionally, some important aspects of the general testing of elastomeric materials and assessment of their environmental resistance are reviewed in *Aging of Elastomers*; *Elastomers: Tests for Mechanical Properties*; *Elastomers: Dynamic Mechanical Properties* and *Elastomers: Strength and Extensibility in Simple Tension*. Finally, two important properties of elastomers are reviewed that are doubly important in end-use applications such as cables for defense, offshore and other hazardous areas—those of *Electrical Properties of Elastomers* and *Flammability of Elastomers*).

4. Plastics

In addition to the common aspects of elastomer and thermoplastics processing mentioned in Sect. 2, more specific aspects for thermoplastics and thermosetting materials are included in *Blow Molding of Thermoplastics*; *Glass-Reinforced Plastics: Thermoplastic Resins*; *Glass-Reinforced Plastics: Thermosetting Resins*; *Injection molding of Thermoplastics: Mold Filling*; *Injection molding of Thermoplastics: Structural Foams*; *Injection molding of Thermoplastics: Structure Development*; *Polymer Sheet: Manufacture and Applications* (which predominantly covers thermoplastics); *Sandwich Molding of Plastics*; *Tubular Film Extrusion: Dynamics and Heat Transfer* and *Tubular Film Extrusion: Structure Development*.

Some processing and reprocessing techniques for thermoplastics are reviewed in *Carbon-Fiber-Reinforced Plastics*; *Recycling* and *Multilayer Polymer Films*. An aspect which has a cross-connection with fibers and textiles is covered in *Polyethylene Processing to Produce Ultimate Properties* (see also *High-Modulus High-Strength Organic Fibers*).

End-use applications of plastics are reviewed in a series of articles (*Appliances: Applications of Engineered Plastics*; *Bottles and Other Rigid Containers*; *Plastics and Composites as Building Materials* and *Plastics for Packaging*). Design aspects for product applications are also covered in the articles *Design with Plastics*; *Design of Composite Structures* and *Decorative Laminates*.

Product applications which involve both elastomers and plastics appear in the articles *Construction Materials, Cellular*; *Roofing Materials*; *Sports Equipment Manufacture*; *Novel Applications of Polymers* and also *Automotive Applications of Polymers* and *Automotive Composite Components: Fabrication*). Special properties of polymers with an emphasis on end-use applications include electrical properties (*Capacitor Materials: Solid Dielectrics*; *Electrical Insulation: Solid Polymeric Materials*; *Electrical Properties of Polymers*; *Electrically Conducting Polymers*; *Wire and Cable Systems for Submarine Transmission* and *Wire and Cable Systems for Underground Transmission*). Flammability of polymers is also of vital importance in many applications and is covered in the articles *Flammability of Polymers*; *Flammability of Polymers: Test Methods* and *Flammability of Structural Materials and Furnishings*.

Other properties of polymers that are important in end-use applications are reviewed in the articles *Polymers: Mechanical Properties*; *Polymer: Tests for Mechanical Properties*; *Impact Strength* and *Fatigue* and problems associated with stress are outlined in the article *Residual Stresses*. Finally, the environmental resistance of polymers is discussed in *Degradation and Stabilization, Thermal*; *Degradation and Stabilization, Ultraviolet* and *Degradation and Stabilization: Test Methods*) and radiation effects on polymers are reviewed in *Radiation Effects on Polymers* and *Radiation Effects on Polytetrafluoroethylene*).

5. Fibers and Textiles

Two general articles on fibers and textiles outline the various aspects of these types of polymers (*Fibers and Textiles: An Overview* and *Textiles: Structures and Processes*). Aspects of the production and processing of fibers which impinge on properties and applications are described in several articles (*Fibers: Drawing*; *Fibers: Formation and Processing*; *Fibers: Texturing and Draw Texturing*; *Melt Spinning: Dynamics* and *Melt Spinning of Fibers: Structure Development*). An important processing aspect of textile fabrics is described in *Fabrics: Finishing*, while the important safety topic of flame-resistance is covered under *Flammability of Structural Materials and Furnishings*. The natural fiber cellulose is described in the general article *Cellulose: Nature and Applications* while its widespread use in papermaking, together with other synthetic fibers, is reviewed in *Papers of Natural and Synthetic Fibers*. Interesting novel developments leading to various end-use applications are covered in the article *High-Modulus High-Strength*

Organic Fibers.

Fibers or fibrous structures in hybrid compositions are described in *Elastomeric Fibers*; *Hard-Fiber Elastomers* and *Hybrid Fiber–Resin Composites* and a processing aspect is described in *Extrusion of Fiber-Reinforced Polymer Systems*. Finally, some applications of fibers are covered in *Molded Fiber Products* and *Geotextiles*, the latter involving an interesting adhesion aspect (see also *Spunbonding of Polymeric Fibers*). Novel materials with fiberlike properties are described in the article *Liquid Crystal Processing*.

6. Composites

Composite materials are reviewed in two articles, the first relating to manufacture (*Composites: Manufacturing Overview*) and the second to end usage (*Composites: Applications Overview*). Articles on materials and processing of composites (*High-Performance Composites with Thermoplastic Matrices*; *Hybrid Fiber–Resin Composites* and *Automotive Composite Components: Fabrication*) are complemented by other articles on particular applications (*Aircraft and Aerospace Applications of Composites*; *Armor, Composite*; *Automotive Applications of Polymers* and *Plastics and Composites as Building Materials*). A review article *Design of Composite Structures* is also included (see also *Joining of Composites*). Only a selection of articles on composites appear here: a full cross reference list to the *Concise Encyclopedia of Composite Materials* is given in Table 1. It should be pointed out that many products, for example automobile tires, are really composites involving polymers of various types plus metals and fibers in various forms.

7. Adhesives and Bonding

Fundamentals and an overview of adhesion and adhesives are presented in three articles (*Adhesion: Fundamental Principles*; *Adhesive Joining: An Overview* and *Joining of Composites*). Different types of adhesives coupled with their areas of application are given in various reviews (*Adhesives for Wood*; *Adhesives in Building*; *Adhesives in Civil Engineering*; *Adhesives in Medicine*; *Acrylate Adhesives*; *Hot-Melt Adhesives*; *Pressure-Sensitive Adhesives*; *Structural Adhesives*; *Synthetic Adhesives*; *Sealants as Building Materials* and *Structural Sealant Glazing*). Various processes for adhesive joining or bonding including surface preparation are somewhat specific to the polymer substrates and end-use applications (*Bonding Nonwovens*; *Bonding: Preparation of Metallic Surfaces*; *Bonding: Preparation of Polymer Surfaces* *Elastomer–Elastomer Adhesion Processes*; *Elastomer–Metal Adhesion Processes*; *Elastomer– Thermoplastics Adhesion Processes*; *Elastomer– Textile Adhesion Processes*; *Elastomer–Wire Adhesion Processes*; *Spunbonding of Polymeric Fibers* and *Welding of Plastics*). Additionally, aspects of testing and durability of adhesive joints especially important in end-use applications are considered in *Adhesives: Mechanical Test Methods*; *Bond Durability*; *Bond Failure: Fracture* and *Bond Strength: Peel*.

8. Coatings

An allied area of polymeric coatings is covered in various articles: an important general area is reviewed in *Paints, Varnishes and Lacquers*. Other articles relate to the type of coating (*Elastomeric Coatings*) or to the coating process (*Dipped Organic Coatings*; *Electrodeposited Polymer Coatings* and *Protective Tape Coatings*). The final broad area relates to end-use applications (*Maintenance Coatings*; *Marine Coatings* and *Roll and Coil Coatings*).

A

Acrylate Adhesives

The term "acrylate adhesive" covers an extremely broad group of adhesive systems of varying degrees of chemical complexity. One generalization which may be used as a definition is that all acrylate adhesives are ultimately based on derivatives of acrylic or substituted acrylic acids, particularly esters, in monomeric, prepolymeric or polymeric form. It should be recognized that contained within this simplistic definition is an enormous variety of bonding and application capabilities spanning a range from relatively simple paper adhesives to adhesives formulated for high-performance bonding of highly engineered metal assemblies.

Perhaps the most familiar types of acrylate adhesives are those based on preformed polymers or copolymers of acrylic monomers. Typically, they find use as paper-converting, laminating, flocking, building and pressure-sensitive adhesives.

The term "engineering adhesives" has begun to find general acceptance as a generic designation for reactive adhesives capable of forming tough, permanent bonds between durable substrates. Included in this grouping are the epoxies, urethanes, anaerobics, cyanoacrylates and modified acrylic adhesives. Further subdivision has produced the term "convenience engineering adhesives" to denote those engineering adhesives that will cure at room temperature without recourse to metering, mixing, prerefrigeration or heat. Three classes of the convenience engineering adhesives fall within the definition of acrylate adhesives and are considered here: the anaerobics, cyanoacrylates and modified acrylics. The major point of differentiation between these three systems and the previously described acrylic adhesives is the fact that they are mostly based on monomers rather than preformed polymers. Conversion to a polymeric form occurs during the cure process.

The cyanoacrylate adhesives, based on esters of cyanoacrylic acid, are probably best known as the consumer-oriented "instant adhesives," but they also find numerous industrial applications. They are single-component adhesives whose cure is catalyzed by the trace amounts of moisture or alkaline substances found on almost any surface. In general, they are best suited for the ultrahigh-speed assembly of small, closely fitted parts, both metallic and nonmetallic.

There are basically two types of anaerobic adhesives. Formulations based on dimethacrylates of polyglycols or hydroxyalkyl methacrylates are frequently characterized as "machinery adhesives" and are used to improve the strength and reliability of threaded fasteners, to assemble cylindrical parts and to seal leakage in threaded, flanged and porous metal parts. These are essentially single-component systems, although cure may be accelerated by the use of surface activators. The second class of anaerobics, based on urethane prepolymers terminated with methacrylate esters, are formulated to develop extremely high performance on clean and/or prepared metal, ceramic and glass substrates. Although anaerobic in the sense of polymerization being inhibited by oxygen, these adhesives generally require surface activators to achieve an acceptable rate of cure.

The modified acrylic adhesives, based on a monomethacrylate monomer containing a dissolved polymeric rubber, find utility in bonding a broad variety of metallic and nonmetallic substrates and are particularly useful for unprepared or minimally prepared parts. They are also quasianaerobic in terms of oxygen stabilization, but they too require a surface activator to accelerate cure.

1. Chemistry and Performance

The acrylate adhesives, mainly the esters of acrylic or methacrylic acid, are available in polymerized [$\text{-(-}CH_2CRCO_2R'\text{-)-}$], prepolymerized or monomeric ($CH_2{=}CRCO_2R'$) liquid forms. If the acrylate is already in polymerized form it is likely to be dispersed or dissolved in a solvent, and procedures are necessary to drive off the solvent after application of the adhesive. These systems are in "prereacted" form and therefore are used in a fashion similar to contact cements.

When the acrylates are in a monomeric, unreacted form, procedures are implemented for the polymerization of the monomers to a cured, durable bond. Most of the monomer-based systems are cured by a free-radical mechanism involving the (meth)acrylate double bond.

Heat-activated initiators such as peroxides are used in systems where heat is tolerated by the substrates to be bonded. Radiation-sensitive initiators can be used to commence the curing cycles at lower or ambient temperature ranges in order to save energy, to lessen thermal shock and bondline stresses, and to reduce the venting of noxious solvents or volatile monomers. Radiation curing includes using radiant energy from any portion of the electromagnetic spectrum for curing. Commercial radiation curing processes utilize γ rays, electron beams and ultraviolet photoinitiation. The photoinitiated systems are of use where one of the substrates to be bonded is transparent to the particular ultraviolet wavelength of the initiator.

Another method for low-energy, ambient-temperature curing of acrylates is the use of anaerobic cure formulations. As mentioned earlier, there are two basic categories of anaerobic adhesives—machinery and high-performance engineering systems. The machinery anaerobics, discovered and developed by Krieble (1959, 1962a,b, 1965), are mainly based on the general composition of a dimethacrylate ester of a polyether glycol, a hydroperoxide such as cumene hydroperoxide, stabilizers such as quinones and accelerator combinations such as an aromatic amine and saccharin. These adhesive compositions are stable when stored in air-permeable containers, but once placed between two substrates from which air is excluded, polymerization takes place. Hence the term anaerobic—active in the absence of air. This phenomenon and the speed of the polymerization step are dependent on the activity of the surface of the substrate, that is, on the concentration of transition metal ions which can greatly accelerate the formation of the free-radical fragments from the hydroperoxide initiator.

The polymerizations can occur readily at ambient temperatures but if a surface is somewhat inactive then accelerators, which contain organometallic compounds, can be applied to the substrates to speed curing of the adhesive.

Other additives such as plasticizers, thickeners, adhesivity aids and comonomers may be present in the anaerobic machinery adhesives to impart further functional properties to either the liquid or cured state. Many formulation techniques to obtain further improvements such as better low-temperature properties (e.g., lowering the brittle point via the use of long-chain acrylate ester reactive diluents) or greater solvent resistance (the use of branched-chain esters) can be performed.

The high-performance engineering adhesives use basically the same cure systems. However, the key monomer prepolymers here differ in structure from the polyethylene glycol dimethacrylates—they are urethane prepolymers terminated with (meth)acrylate groups (Gorman and Toback 1965). These urethane prepolymers, once cured, exhibit high-strength, tensile and impact properties along with good elongation (5–50%). As these cured materials are crosslinked, good thermal and solvent resistance properties are also exhibited. The polyether glycol dimethacrylates and hydroxyalkyl (meth)acrylates, along with styrene, acrylonitrile, vinyl acetate and other vinyl monomers, are used as reactive diluents for the viscous urethane (meth)acrylate prepolymers. These diluents will cocure with the prepolymers and, hence, these adhesives are virtually 100% reactive systems.

The general advantages of the anaerobic adhesives, of either the machinery or high-performance engineering variety, are as follows.

Table 1
Properties of uncured acrylic engineering adhesives

Property	Anaerobic	Volatile acrylic	Nonvolatile acrylic
Shelf life (months)	6	6	6
Toxicity	very low	moderate	low
Flashpoint (K)	>365	285	>365
Viscosity ($mN\,s\,m^{-1}$)	0.8–2000	200–paste	2–8000
Solvents	none	none	none

(a) They are single-component systems and, therefore, there are no mixing or pot-life problems.

(b) By using surface activators, rapid room-temperature cures are possible—thermal cures are not needed.

(c) They are 100% reactive systems, with no volatile or nonreactive solvents.

(d) The anaerobic property eliminates curing outside the bondline, allowing easy cleanup.

The modified acrylic adhesives, which consist of monomers plus a dissolved rubber, use cure systems similar to the anaerobic acrylic systems, but the polymerization mode is more complex. It appears that polymerization of the (meth)acrylate occurs along with chemical grafting of these prepolymers onto the backbone of the predissolved rubber. The most common rubbers used are chlorosulfonated polyethylenes and the most common monomer is a volatile one, methyl methacrylate. Recently a nonvolatile version has been introduced. This system possesses many of the attractive final cured physical properties of the volatile system, but none of its disadvantages such as noxious odor and flammability.

The acrylic engineering adhesives can be compared from three standpoints: (a) the uncured liquid properties (see Table 1), (b) the curing characteristics (see Table 2) and (c) the cured physical properties of the bond (see Table 3).

2. Applications

Anaerobic machinery adhesives find use in a variety of industrial applications. Special, but readily available, application equipment is necessary because standard valves, tubing and pumps cause gelation of anaerobic systems since air is no longer in contact with the adhesive to stabilize it. The anaerobic character does, however, permit a wide variety of

Table 2
Curing characteristics of acrylic engineering adhesives

Property	Anaerobic	Volatile acrylic	Nonvolatile acrylic
Heat cure required?	no	no	no
Surface activator required?	yes	yes	yes
Fixture time (min)			
(0.05 mm)	1–5	5–12	1–2
(0.50 mm)	5–10	15–20	4–5
50% strength (h)	4–8	4–8	4–8
100% strength (h)	24	24	24
Maximum gap tolerance (mm)	0.75–1.0	0.75–1.0	0.75–1.0
Recommended surfaces	clean/prepared metals, glass, ceramics		oily and "as-received" metals, most plastics, wood, concrete

application techniques, some of which are uncommon in adhesive usage. For example, tracing, stencilling and silk screening can be carried out with only slight modification to the equipment.

The thread-sealing machinery anaerobics provide a tough, positive seal that closes off thread leak paths and prevents leakage from vibration loosening, temperature cycling and extreme pressures, while also protecting against rust and corrosion.

Typical applications include the sealing of fluid power system connections, stainless steel and Monel threaded pipe and fittings, sealing microporosity in brazed, soldered and welded joints and die castings, and sealing threaded connections against refrigerants, chemicals and gases.

One important application is the sealing and retaining of cup or core plugs in automotive engines. Specific equipment has been successfully developed for this application. The material is rotosprayed on the bore of the core hole and the cup is pushed through the ring of liquid adhesive when placed into the core hole.

The same technology, appropriately modified, provides formed-in-place anaerobic adhesive–sealant gaskets capable of replacing conventional cut gaskets in pump couplings, fuel and water pumps, engine thermostats, vacuum pump flanges, axle covers, power takeoff covers, compressors, outboard-engine water jacket covers and gearbox covers.

Table 3
Bond properties of cured acrylic engineering adhesives

Property	Anaerobic	Volatile acrylic	Nonvolatile acrylic
Tensile shear strength—steel (MPa)			
0.05 mm	31.0	34.5	18.6
0.50 mm	20.7	22.1	6.9
Tensile shear strength—aluminum (MPa)			
0.05 mm	22.8	26.7	17.9
0.50 mm	15.2	20.7	6.9
Impact strength—steel ($J\,cm^{-2}$)	5.25	2.5	2.3
T peel—steel ($N\,cm^{-1}$)	104.9	66.7	49.0
Thermal operating range (K)	220–475	235–420	235–395
Hot strength (%RT)			
365 K	56	38	30
420 K	49	12	11
Solvent resistance (% retention—360 K, 14 d)			
motor oil	⩾100	79	66
gasoline	55	51	22
water	55	46	61
phosphate esters	98	93	75

One further class of anaerobic machinery adhesives has been formulated to retain cylindrical parts, such as gears, rotors, pulleys, bearings, bushings and collars, reducing the expensive machining associated with the use of press fits, knurls, keyways, splines and tapers.

As previously mentioned, the anaerobic structural adhesives provide extremely high cured performance on metal, ceramic and glass parts which are clean and/or have been specifically prepared for bonding operations. As such, their applications span a spectrum of industries. A few typical applications include the bonding of speaker magnets, firearms parts, parts to auto glass, electronic components for high-temperature exposure, circuit-board components and honing stones to brackets, and the assembly of stemware components.

The broad variety of substrates bondable by modified acrylic adhesives and their ability to bond unprepared parts make acrylic adhesives highly attractive in an increasing variety of applications. The range of these applications is illustrated by the following examples: nameplate and decorative trim bonding; the assembly of wood, plastic and metal furniture; assembly of golf clubs, tennis rackets and fishing rods; the mounting of wear strips; the assembly of tools, appliances and machinery; and the bonding of glass to window and door frames.

See also: Adhesion: Fundamental Principles; Adhesive Joining: An Overview; Synthetic Adhesives

Bibliography

Allen K W 1990 Toughened adhesives—A conspectus. *Prog. Rubber Plast. Technol.* 6(2): 159–73

Gorman J W, Toback A S 1965 Polyurethane polyacrylate sealant compositions. US Patent No. 3,425,988 (2 April 1969)

Krieble V K 1959 Compositions containing hydroperoxide polymerization catalyst and acrylate acid diester. US Patent No. 2,895,950 (21 July 1959)

Krieble V K 1962a Anaerobic curing compositions containing acrylic acid diesters. US Patent No. 3,041,322 (26 June 1962)

Krieble V K 1962b Accelerated anaerobic curing compositions. US Patent No. 3,046,262 (24 July 1962)

Krieble V K 1965 Accelerated anaerobic compositions and method of using same. US Patent No. 3,218,305 (16 November 1965)

Loctite Corporation 1977 *Loctite Speedbonder TM Structural Adhesives,* Technical Data Sheet 52. Loctite Corporation, Newington, CT

Martin F R 1986 Acrylic-based adhesives. In: Kinloch A J (ed.) 1986 *Structural Adhesives, Developments in Resins and Primers.* Elsevier, Amsterdam, Chap. 2

Martin F R 1986 Cyanoacrylate adhesives. In: Bever M B (ed.) 1986 *Encyclopedia of Materials Science and Engineering.* Pergamon, Oxford, pp. 1012–15

Skeist I (ed.) 1977 *Handbook of Adhesives*, 2nd edn. Van Nostrand Reinhold, New York

Stamper D J 1987 Acrylate adhesives. In: Wake W C (ed.) 1987 *Synthetic Adhesives and Sealants.* Wiley, New York, Chap. 3

Toback A S, Cass W E 1967 Anaerobic composition and process for bonding nonporous surfaces, said composition comprising an acrylic monomer, a peroxy initiator, and a bonding accelerator. US Patent No. 3,625,930 (12 July 1971)

L. J. Baccei and M. Hauser
[Loctite Corporation, Newington, Connecticut, USA]

Adhesion: Fundamental Principles

The article outlines the fundamental principles of adhesion and, therefore, is largely concerned with the surface and interfacial aspects of adhesion science. The mechanisms of adhesion are considered, including those involved in attaining intimate contact between adhesive and substrate.

1. Interfacial Contact

In order to establish intimate molecular contact, the adhesive must be able to be spread over the solid substrate surface and to displace air and any other contaminants that may be present. Generally, this requires that the adhesive exhibits a zero or near zero contact angle and, at some time during the bonding operation, has a sufficiently low viscosity. Furthermore, it is essential that the adhesive and substrate are brought together at a rate and in a manner such that trapped air is displaced.

In order to assess the ability of a given adhesive–substrate combination to meet these requirements, it is necessary to consider wetting equilibria, to ascertain values of the surface free energies of the adhesive and substrate and the free energy of the adhesive–substrate interface, and to examine the kinetics of the wetting process.

1.1 Wetting Equilibria

Surface tension is a direct measurement of intermolecular forces which include dispersion (or London) and polar (or Keesom) forces. Dispersion forces are usually weaker than polar forces but they are universal and all materials exhibit them. Another type of force that may operate is hydrogen bonding.

Wetting may be quantitatively defined by a liquid drop resting in equilibrium on a solid surface as shown in Fig. 1. The tensions at the three-phase contact point are indicated and the Young equation relating these tensions to the equilibrium contact angle θ may be written as

$$\gamma_{SV} = \gamma_{SL} + \gamma_{LV} \cos \theta \tag{1}$$

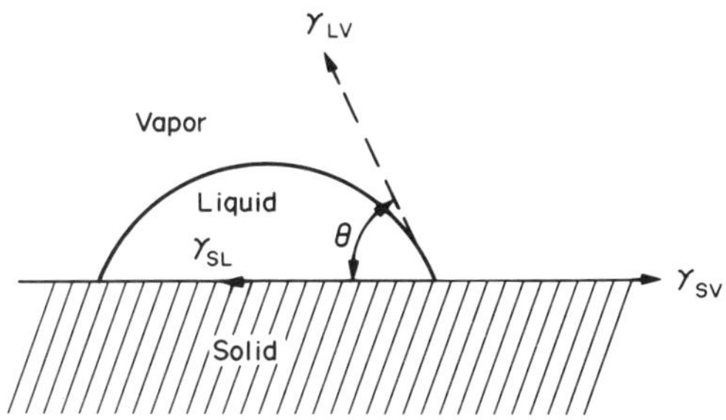

Figure 1
A liquid drop resting at equilibrium on a solid surface

The term γ_{SL} represents the free energy of the adhesive–substrate interface and γ_{LV} represents the surface free energy of the liquid adhesive; γ_{SV} represents the surface free energy of the solid substrate resulting from adsorption of vapor from the liquid and may be considerably lower in value than the surface free energy of the solid in vacuo γ_S. This reduction in the surface free energy of the solid when covered by a layer of vapor has been defined by the concept of equilibrium spreading pressure π_S, such that when the adsorbed layer obeys the ideal, two-dimensional gas law,

$$\pi_S = \gamma_S - \gamma_{SV} \tag{2}$$

Now when $\theta > 0°$ the liquid is nonspreading but when $\theta = 0°$ the liquid wets the solid completely and spontaneously, and spreads freely over the surface at a rate depending on the liquid viscosity and solid surface roughness. Thus, for spontaneous wetting to occur

$$\gamma_{SV} \geqslant \gamma_{SL} + \gamma_{LV} \tag{3}$$

It is also possible for a liquid to spread and wet a solid surface even when $\theta > 0°$, but this requires the application of a pressure or force to the liquid to spread it forcibly over the solid surface.

1.2 Surface and Interfacial Free Energies

A distinction may be made between low-energy and high-energy solid surfaces. Organic compounds, such as polymers, belong to the first group and their surface free energies are usually less than $100\,mJ\,m^{-2}$. Metals, metal oxides and ceramics belong to the second group and their surface free energies are typically greater than $500\,mJ\,m^{-2}$.

Zisman established that for low-energy solids and a series of liquids a linear relationship frequently existed between $\cos\theta$ and γ_{LV} of the wetting liquid. These plots defined a critical surface tension of wetting γ_c by the value to which γ_{LV} extrapolated as $\cos\theta$ tends to unity; that is, as θ tends to zero degrees. Thus γ_c is the surface tension of a liquid that will just spread on the surface. Some values of γ_c for polymeric solids are shown in Table 1.

Another approach was later adopted by Good and Girifalco who expressed the interfacial free energy in terms of the surface free energies of the pure phases by an analogy to the Berthelot relation. Thus, for the interfacial free energy γ_{SL}

$$\gamma_{SL} = \gamma_S + \gamma_{LV} - 2\Phi(\gamma_S\gamma_{LV})^{1/2} \tag{4}$$

where Φ possesses a maximum value of unity and is defined in terms of molecular geometrics and forces of attraction across the adjacent phases. The rela-

Table 1
Values of critical surface tension and surface free energies of various materials

Solid surface	Critical surface tension γ_c ($mN\,m^{-1}$)	Surface free energy ($mJ\,m^{-2}$)		
		γ_S^D	γ_S^P	γ_S
Polytetrafluoroethylene	18.5	14.5	1.0	15.5
Polychlorotrifluoroethylene	31.0	31.4	2.1	33.5
Polyethylene	31.0	31.3	1.1	32.4
Polystyrene	32.8	38.4	2.2	40.6
Polychloroprene	38.0			
Nylon-66	42.5	33.6	7.8	41.4
Typical amine-cured epoxy		41.2	5.0	46.2
Urea–formaldehyde	61.0			
Ferric oxide		105	1250	1355
Aluminum oxide		100	540	640
Silica		80	210	290

tionship between γ_c and γ_S is

$$\gamma_c = \Phi^2 \gamma_S \tag{5}$$

Fowkes has developed a similar argument in which he proposed that the surface free energy of a pure phase could be represented by the sum of the contributions arising from different types of force components. Recent work suggests that γ_S or γ_{LV} may be generally expressed by two terms, for example

$$\gamma_S = \gamma_S^D + \gamma_S^P \tag{6}$$

where γ_S^D is the dispersion force component and γ_S^P is the polar force component. Fowkes then proposed that the geometric mean of the dispersion force components is a reliable prediction of the interaction energies at the interface caused by dispersion forces. Kaelble (1971) and Owen and Wendt extended this proposal to take into account the polar force interactions. Thus,

$$\gamma_{SL} = \gamma_S + \gamma_{LV} - 2(\gamma_S^D \gamma_{LV}^D)^{1/2} - 2(\gamma_S^P \gamma_{LV}^P)^{1/2} \tag{7}$$

Combining Eqns. (1), (2) and (7) yields, for low-energy surfaces where $\pi_S \simeq 0$,

$$1 + \cos\theta = \frac{2(\gamma_S^D)^{1/2}(\gamma_{LV}^D)^{1/2}}{\gamma_{LV}} + \frac{2(\gamma_S^P)^{1/2}(\gamma_{LV}^P)^{1/2}}{\gamma_{LV}} \tag{8}$$

Values of γ_{LV}, γ_{LV}^D and γ_{LV}^P for many liquids are to be found in the literature. Consequently, if the contact angles of two liquids on a solid surface are measured, simultaneous equations may be formed from Eqn. (8) and solved to obtain values of γ_S^D and γ_S^P. Values of γ_S, γ_S^D and γ_S^P for various low-energy surfaces are given in Table 1.

Sharpe and Schonhorn have emphasized the importance of wetting and proposed that the single most important factor influencing adhesive joint strength is the ability of the adhesive to spread spontaneously on the substrate and, by ignoring the interfacial free energy and using Zisman's critical surface tension concepts, proposed that the γ_c of the adhesive must be less than that of the substrate. Indeed, the requirement of the increase of the low surface free energy of polyethylene and similar substrates is a major reason why surface modification of such materials is often necessary prior to bonding. Thus, the Sharpe and Schonhorn criterion essentially proposes that for effective bonding, a mobile, liquid adhesive with a small or zero contact angle is needed; this will spread readily, flow into crevices and achieve true contact with little opportunity for the voids which may act as stress concentrations.

A variety of techniques have been employed to ascertain the surface free energy of high-energy solids, but the results are often approximations at best. Some values are given in Table 1.

1.3 Kinetics of Wetting

The topography of the substrate surface may influence the kinetics of wetting. First, a liquid forming a contact angle of less than 90° with a solid surface may spread along fine pores, scratches and other inhomogeneities by capillary action. Second, surface roughness changes the apparent contact angle θ' compared with the angle θ observed on the smooth surface. This change can be expressed as

$$\cos\theta' = r\cos\theta \tag{9}$$

where r is the roughness factor or the ratio of the actual area to the projection area of the solid. On a smooth surface, if θ is less than 90°, roughening the surface will result in θ' being even smaller. However, if θ is greater than 90° roughening the surface will increase the apparent contact angle θ' still further and decrease the degree of wetting. Figure 2 shows the capillary penetration to be expected, under atmospheric pressure and against the back pressure of trapped air, as a function of the contact angle made on the solid by the penetrating liquid. The limited penetration into the "ink-bottle" pits is especially marked and this effect will be heightened if the liquid has a viscosity greater than a few tens of centipoises and is forced rapidly across the substrate surface. Under such circumstances the liquid near the surface cannot keep up with the

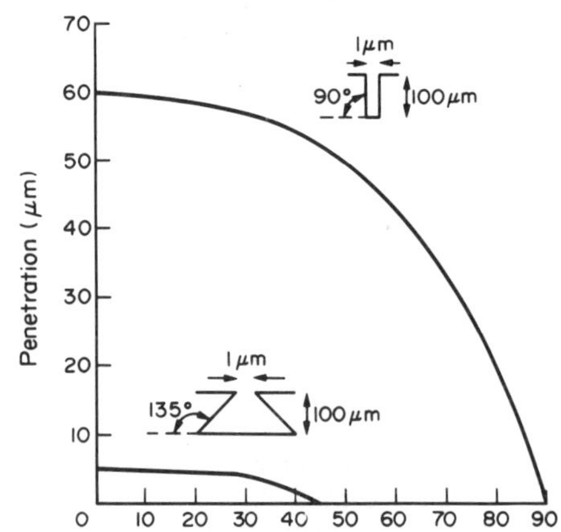

Figure 2
Comparison of the penetration of a film into cylindrical and ink-bottle pits

advancing front so that a higher dynamic contact angle develops.

Various workers have examined the kinetics of wetting of molten polymers on substrates by following the change in the contact angle of molten polymers as a function of time at different temperatures. It has been proposed that the spreading of a molten polymer across a metal surface could be described in terms of an activated rate process in which the activation energy barriers were the same as those which oppose flow, and the driving force for this derives from the action of surface forces during the replacement of a solid–vapor interface by the solid–liquid interface.

If the wetting or spreading conditions result in air entrapment, air voids may obviously act as stress raisers; increases in the peel strength of aluminum–epoxy adhesive joints of up to 30% have been reported upon complete void removal.

1.4 Surface Contamination

An important aspect of the bonding operation is the environmental conditions employed, especially when bonding metallic substrates which are usually covered by an oxide layer. Metals and metal oxides are high-energy solids and, hence, it would seem at first that such a surface would be readily wet by organic adhesives having low surface free energies once any machine oil, protective greases or other contaminants had been removed. However, even in the absence of these impurities, a high-energy surface adsorbs water vapor from the atmosphere and other contaminants such as nitrogen and organic adsorbates, such as hydrocarbons, which lower the surface free energy γ_{SV} of the substrate and may prevent spreading of the adhesive. Therefore, it is important that in the bonding environment contamination is kept to a minimum. However, some adhesives, such as the epoxies, are able to dissolve some organic contamination or possess a sufficiently high surface free energy to displace such contamination.

1.5 Orientation at Interfaces

It has been shown that oriented monolayers of amphipathic molecules, such as stearic acid, may be employed as extremely effective adhesives in the bonding of polyethylene to aluminum; indeed, so effective are these adhesives that joint strengths may exceed the cohesive strength of the polyethylene. Multilayer adsorption lowers the joint strengths because the relatively thick layer possesses low cohesive strength and thus behaves as a weak boundary layer. While chemisorption was thought to occur for the stearic acid on the aluminum oxide, it has been speculated that some diffusion occurs across the interface between the alkane chains of the acid and the polyethylene.

2. Mechanisms of Adhesion

The four main theories of adhesion which have been proposed are (a) the mechanical interlocking theory, (b) the diffusion theory, (c) the electronic theory and (d) the adsorption theory.

2.1 Mechanical Interlocking Theory

The mechanical interlocking theory proposes that mechanical keying of the adhesive into the irregularities of the substrate surface is the major source of intrinsic adhesion. The attainment of good adhesion between smooth surfaces exposes this theory as not being of general applicability and the results shown in Fig. 2 must raise doubts as to whether any significant penetration of adhesive into some configurations of irregular cavities suitable for forming a mechanical key, for example an ink-bottle-type cavity, would normally occur. However, there are some instances where mechanical interlocking has been demonstrated to contribute significantly to the intrinsic adhesive forces. The most notable example is the adhesion of polymers to textiles. Also, recent work on the adhesion of polyethylene to metallic substrates showed that high peel strengths were obtained when a very rough, fibrous-type, oxide surface was formed on the substrate. Actually, an increase in joint strength from increasing the roughness of the substrate need not necessarily arise from a mechanical interlocking mechanism, an increase in surface area for bonding, or from improved kinetics of wetting. It is quite possible that the surface irregularities promote greater viscoelastic energy dissipation than when the interface is smooth.

2.2 Diffusion Theory

Voyutskii (1963) is the chief advocate of the diffusion theory of adhesion which states that the intrinsic adhesion of high polymers to themselves (autohesion), and to each other, is due to mutual diffusion of polymer molecules across the interface. This requires that the macromolecules or chain segments of the polymers (adhesive and substrate) possess sufficient mobility and are mutually soluble; that is, they possess similar values of the solubility parameter. These conditions are usually met in the autohesion of elastomers and in the solvent welding of compatible, amorphous plastics. However, where the solubility parameters of the materials are not similar, or interdiffusion is not probable, this mechanism is not important.

2.3 Electronic Theory

If the adhesive and substrate have different electronic-band structures there is likely to be some electron transfer on contact to balance Fermi levels which will result in the formation of a double layer of electrical charge at the interface. The electronic theory of adhesion is due primarily to Deryaguin and co-workers and they have suggested that the electrostatic forces arising from such contact or junction potentials may contribute significantly to the intrinsic adhesion. The controversy this theory has caused is due to this final statement that electrostatic forces are an important cause, rather than merely a result, of high joint strength.

2.4 Adsorption Theory

The adsorption theory of adhesion is the most generally accepted theory and proposes that, provided sufficiently intimate intermolecular contact is achieved at the interface, the materials will adhere because of the surface forces acting between the atoms in the two surfaces. The most common such forces are van der Waals forces and are referred to as secondary bonds. In addition, chemisorption may well occur and thus ionic, covalent and metallic bonds may operate across the interface; these types of bonds are referred to as primary bonds. The terms primary and secondary are in a sense a measure of the relative strength of the bonds, as may be seen from Table 2 where the types of bonds are shown with estimates of the range of magnitude of their respective bond energies.

(a) Secondary force interactions. It has been calculated that the attractive forces between two planar bulk phases due solely to dispersion forces would be approximately 100 MPa at a separation of one nanometer. This is considerably higher than the experimental strength of most joints. This discrepancy between theoretical and experimental joint strengths is attributed to air voids, cracks, defects or geometric features acting as stress raisers when the joint is loaded. However, this calculation does indicate that high joint strengths may result from the intrinsic adhesion arising from solely dispersion force interactions.

The thermodynamic work of adhesion required to separate a unit area of two phases forming an interface may be related to the surface free energies by the Dupre equation. In the absence of chemisorption and interdiffusion the reversible work of adhesion W_A in an inert medium may be expressed by

$$W_A = \gamma_S + \gamma_{LV} - \gamma_{SL} \quad (10)$$

This equation only applies strictly to a solid–liquid or liquid–liquid interface but, by assuming that the surface energy of a liquid does not change significantly when it solidifies isothermally and ignoring any shrinkage stresses, it may be applied to solid-adhesive–substrate interfaces. By using Eqn. (7) to eliminate the interfacial free energy γ_{SL}, then

$$W_A = 2(\gamma_S^D \gamma_{LV}^D)^{1/2} + 2(\gamma_S^P \gamma_{LV}^P)^{1/2} \quad (11)$$

where the subscripts S and LV refer to the substrate and adhesive, respectively.

It is a recurrent hypothesis in the literature that the measured strengths of adhesive joints may be correlated with the thermodynamic work of adhesion W_A. Recent data suggests that when (i) the energy dissipated viscoelastically is taken into account, (ii) solely interfacial failure occurs, and (iii) only secondary forces are acting across the interface, then this is indeed possible.

(b) The role of primary interfacial bonding. Although it is evident that intrinsic adhesion arising from secondary bonding forces alone may result in adequate and high joint strengths, many adhesion scientists believe that the additional presence of primary bonding may often increase the measured joint strength and is certainly a necessary requirement for securing environmentally stable interfaces. However, although there is considerable indirect evidence emphasizing the importance of interfacial chemical bonding, studies which directly confirm its role are scarce. The use of sophisticated, surface-specific, analytical techniques such as laser-Raman spectroscopy, x-ray photoelectron spectroscopy, inelastic electron tunnelling spectroscopy and secondary-ion mass spectroscopy has produced definitive evidence that primary interfacial bonding may occur in certain circumstances and may make a significant, indeed often vital, contribution to the intrinsic adhesion.

Table 2
Bond types and typical bond energies

Type	Bond energy ($kJ\ mol^{-1}$)
Ionic	590–1050
Covalent	63–710
Metallic	113–347
Permanent dipole–dipole interactions	
hydrogen bonds involving fluorine	≤42
hydrogen bonds excluding fluorine	10–26
other dipole–dipole (excluding hydrogen bonds)	4–21
Dipole-induced dipole	<2
Dispersion (London) forces	0.08–42

2.5 Weak Boundary Layers

The topic of weak boundary layers is included here since it has been the subject of considerable controversy in discussions of the mechanisms of adhesion, especially with respect to low-energy substrates which are usually subjected to various surface pretreatments prior to bonding in order to improve their adhesion.

Bikerman (1968) initiated the debate when he stated that even though failure might appear to be interfacial, with the separation appearing to be exactly along the adhesive–substrate interface, a cohesive break of a weak boundry layer is the real event. More recently, Bikerman's ideas on weak boundary layers have been revitalized by the conclusions drawn by Schonhorn and co-workers. Schonhorn concluded that wettability was not the sole criterion for the formation of an adhesive joint and that the enhancement of the cohesive strength of a weak surface might be a factor of greater importance.

However, much theoretical and experimental evidence exists to show that these conclusions are not generally valid. For example, the new techniques of Auger and x-ray photoelectron spectroscopy have confirmed instances of interfacial failure beyond any doubt. Further, even if the locus of joint failure is cohesive in the adhesive or substrate, close to the interface, this does not necessarily imply the presence of a weak boundary layer. Several authors have recently suggested that the stress distribution in the joint and around the tip of a crack propagating close to the interface causes "mechanical focusing" of the failure path close the interface. Furthermore, many workers have established that nearly all pretreatments usually increase the surface free energy of the substrate and/or its rugosity. Thus, surface energetics and the production of chemically active species to which grafting may occur can account for the observed increased adhesion.

Therefore, while certain classes of weak boundary layers are well established, for example surface contamination, their presence in joints where low strengths are recorded should not be automatically assumed.

See also: Adhesive Joining: An Overview

Bibliography

Bascom W D, Patrick R L 1974 Surface chemistry of bonding metals with polymer adhesives. *Adhes. Age* 17(10): 25–32

Bikerman J J 1968 *The Science of Adhesive Joints*, 2nd edn. Academic Press, New York

Kaelble D H 1971 *Physical Chemistry of Adhesion*. Wiley, New York

Kinloch A J 1980 Review: The science of adhesion, Pt. 1: Surface and interfacial aspects. *J. Mater. Sci.* 15: 2141–66

Kinloch A J, Dukes W A 1977 Preparing low-energy surfaces for bonding. In: Wake W C (ed.) 1977 *Developments in Adhesives*, Vol. 1. Applied Science, London, pp. 251–82

Patrick R L (ed.) 1967 *Treatise on Adhesion and Adhesives,* Vol. 1. Dekker, New York

Shields J 1984 *Adhesives Handbook*, 3rd edn. Butterworth, London

Skeist I (ed.) 1977 *Handbook of Adhesives*, 2nd edn. Van Nostrand Reinhold, New York

Voyutskii S S 1963 *Autohesion and Adhesion of High Polymers*. Interscience, New York

Wake W C 1982 *Adhesion and the Formulation of Adhesives*, 2nd edn. Applied Science, London

A. J. Kinloch
[Imperial College of Science and Technology, London, UK]

Adhesive Joining: An Overview

The art of adhesive bonding can be traced into antiquity. For most of this history the adhesives were based on natural materials both organic and inorganic. Hide gelatin, blood, milk derivatives (albumin and casein) and vegetable derivatives (starch) formed the basis for many early "glues." Inorganic salts, primarily complex silicate salts, were also widely used. Their primary use was in bonding wood and, in fact, extended the usefulness of wood as a construction material over what could be achieved with mechanical fasteners, plywood being a notable example. A few of the natural adhesives are still in use. The advent in the early 1930s of synthetic polymers brought into existence new, more versatile adhesives, the first being based on phenol–formaldehyde polymers. In the ensuing 50 years, nearly every type of synthetic polymer has been used as an adhesive. The article *Synthetic Adhesives* reviews the major types of polymeric adhesives. The advantage of adhesives based on synthetic polymers is the versatility they offer in flow characteristics and curing or setting time, a property called tack which can be best described as the grabbing power of an adhesive when two adherends are brought together.

Since the first synthetic adhesives made their appearance there has been an explosion of new adhesive products for the home, office, industry and medicine. Examples include paste, aqueous and solvent drying glues, tapes with all manner of backings including double-sided tapes and a variety of preapplied adhesives which require moistening or simply the removal of a peel-ply covering. In industry the use of adhesives is less labor intensive and offers definite advantages over mechanical fastening; rapid production rates, better fatigue resistance and lower materials costs. The interior panels of modern automobiles and airplanes are literally glued on. Even before the advent of synthetic polymer adhesives, furniture manufacture relied heavily on "carpenters glues." The paper products industry uses large quantities of adhesives. Acrylate and cyanoacrylate adhesives find special utilization in industrial operations, notably for thread sealing to give mechanical fasteners additional holding power. These materials are described in *Acrylate Adhesives* and by Martin (1986). Adhesives are also put to novel uses in construction (see *Adhesives in Civil Engineering*).

In medicine, the use of adhesives ranges from the ubiquitous surgical tape to the bonding of implant components to themselves and to human tissue. In the article *Adhesives in Medicine* some of the challenges of using adhesives in oral or internal medicine are discussed.

Much of the diversity in adhesives and adhesive bonding that has developed since the early 1930s has come from imaginative innovation aided by a substantial amount of scientific investigation of adhesion and adhesives. Indeed, a large body of the literature in the field of surface science is devoted to the basic reasons why two different substances adhere to each other. The surface chemistry and physics of adhesion are discussed in *Adhesion: Fundamental Principles*.

At this point the important distinction between adhesion and bond strength must be made. Adhesion refers to the thermodynamic work of adhesion which can be related to certain surface chemical parameters as explained in the article *Adhesion: Fundamental Principles*. Too often, the word adhesion is used to mean bond strength or joint strength. With very few exceptions, the adhesive and the adherend surface preparation have been designed to have sufficient interfacial interaction, both chemical and mechanical, so that the cohesive strength of the adhesive determines the bond strength.

The testing of adhesives falls into two broad catagories: determining the chemical and physical properties of the adhesive itself and determining the ultimate bond strength. Properties such as tack, flow behavior and curing or setting times are poorly understood in physical chemical terms and so empirical techniques have evolved over the years and differ from one type of adhesive to another. This fact is quite evident from the articles *Hot-Melt Adhesives* and *Pressure-Sensitive Adhesives*. Testing for bond strength is a little better standardized and some of the more common test methods are discussed in *Adhesives: Mechanical Test Methods*. Two specific areas of testing are covered in *Bond Failure: Fracture* and *Bond Strength: Peel*.

Many of the advances that have been made and continue to be made in adhesive bonding have come from a better scientific understanding of the cohesive properties of the adhesive. The rheological, especially viscoelastic, properties of adhesives play a large part in their performance. This fact is made very evident in the article *Bond Strength: Peel*. The rheology of adhesives is a very active area of investigation, especially as tools for studying the time–temperature dependence of polymer properties become more sophisticated and are used to map out structure–property relationships.

It is axiomatic that good joint strength requires proper preparation of the adherend surface prior to the bonding operation. Failure to remove surface contaminants and to treat the surface to obtain sufficient mechanical or chemical interaction with the adhesive often results in the joint strength being limited by the strength along the adhesive–adherend interface. The subject of surface preparation is discussed in *Bonding: Preparation of Metallic Surfaces* and *Bonding: Preparation of Polymer Surfaces*. The importance of surface preparation is illustrated by the fact that the subject is at least touched upon in nearly all of the articles concerned with adhesion.

A new era in joining by adhesives began in the early 1970s when adhesives began to be seriously considered for "structural" application where adhesive bond(s) would be the sole load-bearing connection and that bond failure would mean the total failure of the structure. This definition differs from that given in the article *Structural Adhesives*. Nonetheless, both definitions are correct. The term has come to be used to describe any use of adhesives to join load-bearing components, notably in aircraft. To the author's knowledge, no critical load-bearing components have been built using adhesives without some backup mechanical fastener, the combination of adhesives and rivets being a typical example. Although many design studies have been made of pure structural adhesive joints, confidence in adhesives has not been good enough for their use in critical joints in any modern military or commercial aircraft. Engineering confidence in any joint, or any material for that matter, depends on having adequate design criteria. In the case of structural adhesives, the two most important criteria are fracture toughness and joint stiffness. The methodology and present state of adhesive bond fracture mechanics is discussed in *Bond Failure: Fracture*. The measurement of joint stiffness depends on the measurement of microstrains of adhesive layer. Reliable methodology for bond line microstrain measurement has only recently become available.

The lack of confidence in structural adhesives is not based on inadequate fracture resistance or stiffness. The problem is poor bond durability in wet environments. Under adverse conditions (i.e., tropical environments), total bond strength can be lost spontaneously. This problem for bonds to metals and other polar solids is discussed in the article *Bond Durability*. The importance of surface preparation for bond strength and moisture resistance is covered in *Bonding: Preparation of Metallic Surfaces*.

Before closing, it is worthwhile noting some of the areas of research and development that would benefit both the manufacturer and user of adhesives.

(a) The rheological properties of adhesives and the components of adhesives need to be more thoroughly studied. The articles *Pressure-Sensitive Adhesives* and *Hot-Melt Adhesives* suggest that there has been some progress along these lines. Nonetheless, adhesive formulation is still largely an art.

(b) Some progress has been made in understanding how water degrades bond strength as outlined in *Bond Durability*. However, much remains to be done in understanding the basic phenomena and then in developing adhesives and/or adherend surface preparations that prevent moisture attack.

(c) Methodology exists for developing design criteria for structural adhesives but actual data are rarely available from the adhesive manufacturer. There is a need to generate these data, and for the structural engineering community to develop methodology that can use the data in the design of structural components.

(d) The curing reactions of thermosetting polymers such as epoxies, polyurethanes and the acrylates have been rather thoroughly studied for model systems. However, much of this information is of marginal value in trying to understand the complex mixtures that constitute commercial adhesives. The time has come to ask how curing reactions are affected by such components as plasticizers, reactive diluents and fillers.

Additionally, legislation in the USA, the UK and Europe is restricting the use of solvents by various regulations which limit the exposure of workers to fumes and control the emission of waste solvent from factories. Also, the use of chlorinated solvents, which deplete the ozone layer, is to be phased out by 2005.

As solvent-based adhesives are usually elastomeric, there will be a general trend away from such systems to water-based adhesives, based on latices, and to 100% adhesives such as hot melts.

Bibliography

Allen K W (ed.) 1982 *Adhesion.* Applied Science, London

Alner D J (ed.) 1965–71 *Aspects of Adhesion*, Vols. 1–6. University of London Press, London

Anderson G P, Bennett S J, De Vries K L 1977 *Analysis and Testing of Adhesive Bonds.* Academic Press, London

Bikerman J J 1968 *The Science of Adhesive Joints.* Academic Press, New York

Broutman L J, Krock R H 1974 *Composite Materials*, Vols. 1–6. Academic Press, New York

Cagle C V (ed.) 1973 *Handbook of Adhesive Bonding.* McGraw-Hill, New York

DeLollis N J 1970 *Adhesives for Metals: Theory and Technology.* Industrial Press, New York

Eley D D (ed.) 1961 *Adhesion.* Oxford University Press, London

Herman B S 1976 *Adhesives—Recent Developments*, Chemical Technology Review No. 65. Noyes Data, Park Ridge, NJ

Houwink R, Salomon G (eds.) 1965 *Adhesion and Adhesives.* Elsevier, Amsterdam

Kaelble D H 1971 *Physical Chemistry of Adhesion.* Wiley, New York

Lee L-H (ed.) 1973 *Recent Advances in Adhesion.* Gordon and Breach, London

Lee L-H (ed.) 1975 *Adhesion Science and Technology.* Plenum Press, New York

Manly R S (ed.) 1970 *Adhesion in Biological Systems.* Academic Press, New York

Martin F R 1986 Cyanoacrylate adhesives. In: Bever M B (ed.) 1986 *Encyclopedia of Materials Science and Engineering.* Pergamon, Oxford, pp. 1012–15

Shields J 1970 *Adhesives Handbook.* Butterworth, London

Skeist I (ed.) 1977 *Handbook of Adhesives*, 2nd edn. Van Nostrand Reinhold, New York

Snogren R C 1974 *Handbook of Surface Preparation.* Palmerton, New York

Weiss P (ed.) 1962 *Adhesion and Cohesion.* Elsevier, Amsterdam

White L 1990a Elastomeric adhesives. *Eur. Rubber J.* 172(9): 28–32

White L 1990b Rubber/metal bonding. *Eur. Rubber J.* 172(9): 38–40

Wu S 1982 *Polymer Interfaces and Adhesion.* Dekker, New York

In addition to the above texts, there are a number of journals devoted to adhesion and adhesives. These include *The Journal of Adhesion, Adhesives Age, International Journal of Adhesion and Adhesives* and *Adhesion.*

Other journals that publish on adhesion include *Journal of Applied Polymer Science, Journal of Materials Science, Journal of Colloid and Interface Science, Sampe Quarterly, Polymer Engineering and Composites, Kunstoffe, Rubber Chemistry and Technology, Surface Science, European Polymer Journal* and *Macromolecules.*

W. D. Bascom
[Hercules Incorporated, Magna, Utah, USA]

Adhesives for Wood

The first evidence of adhesive-bonded wood products can be traced back over 3500 years, when the joining of wood was undertaken primarily for decorative purposes. The discovery of improved adhesives led to more sophisticated developments in wood composites and greater utilization of these products in structural applications. Until the twentieth century, most of these adhesives were derived from animal, vegetable or mineral sources.

Projected limitations in the available harvest of wood resources and the overall decrease in wood quality emphasize the importance of adhesives and the science of adhesion in improving wood utilization by allowing smaller trees and present waste wood to be restructured into useful wood products. By proper choice of wood component, wood orientation, adhesive and manufacturing conditions, a variety of engineered composites are possible; these can have greater mechanical strength, decreased variability and improved stress-distributing properties compared to conventional solid wood.

This article outlines the mechanism of wood bonding and pertinent physical and chemical factors involved in achieving good bonds. Both wood-related and adhesive-related characteristics influencing the formation and performance of bonds in products such as laminated lumber, plywood, flakeboard and particleboard are considered.

1. Nature of Adhesion

Bond formation depends on the development of physical and chemical interactions both within the bulk adhesive polymer and at the interface between adhesive and wood. Interactions within the adhesive accumulate to give cohesive strength while the forces between adhesive and wood provide adhesive strength. Both should exceed the strength of the wood, allowing substantial wood failure during destructive testing of high-quality bonds.

Most wood adhesives are liquids consisting of either 100% polymer material or, more frequently, an aqueous solution, emulsion or dispersion, wherein one-half to one-third of the mixture is polymer. Other solvents such as volatile alcohols, esters or aromatics may sometimes be used to replace all or part of the water. For more specialized bonding requirements, adhesives in a solid state (i.e., powders, hot melts) may be utilized with heat applied to transform the polymer into a flowable material.

The adhesive must accomplish three distinct stages during bonding processes: it must wet the wood substrate, flow in a controllable manner during pressing and finally set to a solid phase. Failure of the adhesive to accomplish any of these stages often results in reduced bond quality. The extent of wetting depends on the physical–chemical nature of both adhesive and the wood surface. Solvent molecules and low-molecular-weight polymer molecules, being smaller and more mobile, tend to wet and penetrate into the wood substrate most rapidly. Adsorption and diffusion of these liquids into the cell wall often leads to swelling of the lignocellulose wood substance. The porous structure of wood and defect areas created by machining processes facilitate the transport of these materials into the wood. In contrast, higher-molecular-weight polymer material would wet and penetrate the wood substrate at a significantly slower rate, remaining largely as a concentrated polymer on the wood surface. Ultimately, the bonding area should consist of a steadily decreasing concentration of adhesive polymer as one proceeds from the surface to a few micrometers within the wood. Achieving this condition will depend on polymer physical properties, thermodynamic considerations and wood-surface properties. With solid adhesives, an intermediate heating stage which melts or softens the adhesive is necessary to accomplish wetting of the wood surface.

Optimum bond formation requires intimate contact between adhesive and wood substrates to ensure molecular interaction over a large area. This is accomplished using pressure and heat which causes viscous adhesive components to transfer and flow throughout the bonding sites while deforming the wood to achieve better contact between wood surfaces of different texture. Adhesive penetration into the wood structure must not be so great that inadequate amounts of adhesive remain in the glueline (starved glueline) or too little so that fibers and microdefects just below the wood surface are insufficiently bonded together.

Solidification and cure of the wood–adhesive system occurs either by solvent loss, polymerization and crosslinking reactions, or both. This process is time, temperature and pressure dependent with bond quality being a reflection of how well the final solidification stage is accomplished in the particular adhesive and wood substrate present.

2. Mechanism of Adhesion

A number of theories have been proposed to explain bonding mechanisms in materials science. The unique heterogeneous characteristics of wood limit the relevance of any one particular theory, leading instead to wood bonding being rationalized in terms of a combination of mechanisms involving mechanical, physical and chemical parameters.

The ability of mechanical interlocking to occur readily is a function of the porous nature of most wood surfaces and the ease with which the adhesive can flow into these microcavities under the action of heat and external pressure. The mechanical anchoring of adhesive into wood occurs when the adhesive polymer hardens. This bonding mechanism is of obvious importance in hot melts and does increase the contact area available for bond formation. It cannot, however, account for the documented phenomena of bond-strength reduction with aging of wood surfaces and the limited correlation between adhesive penetration depth and bond strength development.

Adsorption and diffusion mechanisms allow for intimate molecular contact between adhesive and wood substance over a large area. The accumulative physical interaction arising from attractive van der Waals, hydrogen bonding and dispersion forces between adhesive polymer and wood components can result in strong bond formation which many consider a primary bonding mechanism. Complete wetting of the adhered surface enhances bonding potential. Both adhesive surface tension and wood surface tension have some bearing on wetting effectiveness. Adhesive spreading methods are additional factors of this mechanism since the mechanical energy input during glue application may overcome some surface force resistance to wetting.

Both adhesive and wood reactive functional groups allow for the high probability of covalent bond formation. In crosslinkable systems these chemical interactions can produce a strongly bonded wood–adhesive matrix capable of maintaining long-term stability under a variety of heat, moisture and stress conditions. While indications of covalent bonding have been found in model systems, conclusive evidence of their presence in wood-bonded products has not yet been made.

3. *Wood-Related Factors Affecting Adhesion*

Wood, being a complex, three-dimensional tissue, has a distinct but variable structural and chemical makeup which must be allowed for in adhesive formulation. The heterogeneous character of wood is reflected in both its bulk and surface properties.

3.1 Bulk Features

Anatomical characteristics, permeability, density and moisture content are among the more important interrelated factors influencing wood-bond formation.

Fiber orientation influences the ease with which the adhesive polymer can wet wood while at the same time retaining sufficient polymer on the bonding surface to form a glueline interface. Rapid adhesive penetration is possible along the grain because of open lumen structures, while penetration across the grain is highly restricted by fiber walls. Anatomical differences between species will affect the permeability of wood to liquids and gases, thus influencing the degree to which adhesive can successfully intermix with wood fibers. In softwoods, fiber lumina are laterally interconnected by small capillaries which allow the ready passage of air and liquid to various depths below the wood surface. In hardwoods, many of these interconnecting routes are blocked or absent, resulting in more restricted flow of the adhesive into bonding sites. Further variations such as knots or grain distortions associated with tree growth patterns can provide additional impediments to achieving bonding potential. Other structural features of importance are variations imposed by the presence of juvenile and mature wood and earlywood and latewood effects within one piece of wood. These features are interrelated to wood density.

The density of wood relates directly to its strength and has primarily a physical influence on adhesion. The stronger the wood, the stronger the adhesive must be in order to utilize the full strength of the wood. Density variations are associated with the growth ring structure of wood, particularly in the ratio of fiber-wall to fiber-lumen volume. In dense wood adhesive penetration is restricted, while in lower density woods care must be taken to prevent overpenetration. Where large differences in density can occur (i.e., earlywood–latewood zones), the adhesive must be designed to adapt adequately to both of these conditions. In general, woods with densities in the range 250–450 kg m^{-3} (i.e., spruce, pine, fir, aspen) are easier to bond than woods with densities greater than 600 kg m^{-3} (i.e., birch, oak, maple).

Wood physical properties are highly dependent on the interaction of wood density and moisture content, which in turn affect gluing behavior. High-moisture-content conditions often result in increased adhesive flow and penetration while retarding polymer cure by limiting solvent loss into the wood or restricting condensation polymerization reactions. Excessive water at the bonding interface can also act as a barrier to bond formation, especially with hot-melt systems. Typical moisture content ranges for bonding are 2–16% although successful bond formation has been reported at up to 40% moisture content, with some adhesives curing under ambient or slightly elevated temperature conditions. The likelihood of steam formation at higher moisture contents, with its destructive possibilities in the glueline, necessitates wood to be dried to 2–6% moisture content when using high-temperature-curing adhesives. Wood volumetric changes occurring between fiber saturation and oven-dry conditions result in directional shrinkage and swelling forces, directly proportional to density, which can place considerable stress on the glueline. Cured gluelines should partially maintain bond integrity with dimensional-change-induced stresses, even during continual cycling between wet and dry conditions, because they may weaken the wood–adhesive interface considerably in high-density woods. Moisture and density also affect wood deformational properties requiring some adjustments in bonding pressure (i.e., lower density or higher moisture content requires less bonding pressure).

3.2 Surface Features

Both physical and chemical properties of the wood surface can profoundly affect bond-formation potential. Their primary influence impacts on the accessibility of adhesives to wood-fiber bonding sites. From a microscopic perspective, wood surfaces have a rough physical appearance due to anatomy, in particular grain orientation and fiber size. Surfaces produced from transverse cuts (i.e., cut across the tracheids) expose end-grain areas whose large lumen volume often results in substantial overpenetration of adhesive. Consequently, direct bonding of end-grain surfaces is difficult and not commonly attempted. Cutting along tangential and radial surfaces inevitably excises some fiber walls, exposing different amounts and parts of fiber lumina. Surface topography may vary appreciably between species since lumen diameters may range from as small as 1 μm in fine-grained woods to as large as 200 μm in coarse-grained woods. The presence of resin ducts, vessel elements or knots gives added variations to the wood surface. Machine operations during surface formation often result in further irregularities which can be associated with anatomical features. These can include saw-tooth abrasions, indentations from planer or shaper knives, lathe checks and roughness resulting from the peeling of veneer, torn and broken fibers caused during flake cutting and other splits or checks. Although these features may increase available bonding area, they hinder good bond formation by weakening the wood surface. Heat and pressure can be utilized to compress and plasticize the interface to improve contact in lower-density species, but high-density woods must rely on smooth machining operations to achieve the intimate contact necessary for bonding. Mechanically weakened wood surfaces can be repaired during gluing by the action of adhesive penetration between and into the wood fiber.

Wood surfaces are chemically heterogeneous. Their composition does not necessarily correspond directly to the chemical makeup of bulk wood but rather is more a function of the conditions and methods of surface formation. Since wood fibers are composed of strands of oriented cellulose molecules intertwined with hemicellulose and lignin–polymer matrices, cleavage of a fiber wall may expose varying amounts of each chemical component at the wood surface layer. The carbohydrate components are highly polar, with fractions existing in both crystalline and amorphous forms and containing numerous functional groups available for reaction. Lignin, an amorphous polymer, offers additional reactive sites on its phenolic components. Both carbohydrate and lignin fractions are present in a range of molecular sizes and polarities which allow for bond formation with polar or nonpolar adhesives.

Extractives, although normally a minor component of the wood, can play a significant role in bond formation. These relatively low-molecular-weight materials are often highly mobile, nonpolar substances of limited reactivity which inhibit bond formation by blocking adhesive accessibility to reactive wood sites. This is accomplished by restricting wetting, altering the wood surface pH or forming complexes with adhesive or wood-bonding sites.

Wood-surface history is an additional chemical feature affecting bonding potential. A smooth, freshly cut wood surface dried at moderate temperatures offers the optimum conditions for bonding. As aging occurs, the surface is chemically modified at a rate depending on temperature and atmospheric conditions. Oxidation or contamination of the wood surface by nonpolar, mobile extractives and airborne pollutants are the primary phenomena occurring during this time. Oxidation reduces the number of active hydroxyl groups available for bonding. Under high-temperature drying conditions the chance of wood-surface oxidative degradation becomes highly probable. Sometimes chemical and thermal treatments can be utilized to change wood-surface chemistry and enhance bonding potential. Strong acids and oxidizing agents are common activators. They create new functional groups through depolymerization and ring cleavage reactions of lignin and carbohydrate molecules or generate free radicals which can lead to bond formation by oxidative coupling. Resulting gluelines tend, however, to be brittle and have highly variable strength because wood can be degraded by many of the chemicals used.

4. Wood Adhesives

Adhesives can be classified as either natural or synthetic. Typical natural adhesives are carbohydrate or protein based, derived from either animal or vegetable sources, or may be an inorganic material. Synthetic adhesives are produced by the controlled polymerization of various monomeric organic molecules. These synthetics can be further divided into thermosetting (most common for wood adhesives) and thermoplastic categories. They can function to produce structural or nonstructural bonds. This discussion deals only with major adhesives used in the wood industry. The majority of these adhesives are formed by condensation polymerization processes involving the use of formaldehyde as a crosslinking agent. Details of their chemistry are given by Skeist (1977).

Phenol–formaldehyde (PF) resin is the primary wood adhesive used in the production of strong, durable exterior products. These adhesives are thermosetting resins of low-to-moderate molecular weight, manufactured in aqueous solutions with acid or alkaline catalysts. By varying reaction time and temperature, catalyst type and amounts and ratio of reactants, a variety of different adhesives can be prepared. Resoles are a class of PF resins produced under alkaline conditions with a molar excess of formaldehyde to phenol. These reactions produce highly branched polymeric structures while remaining soluble in strongly alkaline solutions. Sufficient formaldehyde is present to allow a highly cross-linked network to be formed during cure. Novolacs are PF resins synthesized under mild acid or neutral conditions with a molar excess of phenol to formaldehyde. These polymers have a linear structure, relatively low molecular weight and maintain thermoplastic behavior unless additional formaldehyde is added when curing is undertaken. These are referred to as two-stage resin systems. Plywood, oriented strandboard, particleboard and fiberboard intended for exterior service are manufactured using aqueous resoles tailored to give particular application, flow and cure properties for a specific product. A typical aqueous resole has a molar ratio of phenol to formaldehyde to sodium hydroxide of 1:1.8–2.2:0.1–0.8, with a resin-solids range from 35–50% and averaging 5–40 phenolic units linked together. These alkaline resoles require cure temperatures greater than 100 °C to attain adequate crosslinking. Fast cure at ambient temperatures can be achieved by the addition of strong acids. Equipment corrosion and long-term acid degradation of the wood interface are, however, deterrents to this procedure.

For waferboard applications, the need for good adhesive distribution on large wafers and the limitations on moisture level in the glueline during high-temperature pressing require the use of powder PF resins. These are produced by spray drying or other solidification techniques.

Significant increases in phenolic resin reactivity are possible by copolymerizing with resorcinol to produce a phenol–resorcinol–formaldehyde (PRF) system. These are often used in specialized applications (i.e., glued laminated timbers, fingerjointed structures) where durable bonds and ambient cure conditions are needed. To ensure adequate pot life, the PRF is synthesized as a novolac system. Additional formaldehyde needed for cure is mixed in just prior to adhesive application. The high cost of resorcinol, relative to phenol, restricts most PRF applications to higher-value large-structure products.

Aqueous urea–formaldehyde (UF) polymers are low-cost, light-color, fast-curing adhesives with poor weather resistance. Their use is limited to products such as decorative plywood panels, particleboard subfloors or fiberboard for furniture or interior applications. Resins are prepared in molar ratios of formaldehyde to urea of 1–2:1 with the higher ratios usually utilized in solid wood or veneer bonding where higher strengths are required. Cure takes place at moderate acid conditions either by the addition of acid salts or by use of the inherent acidity of wood species such as oak or cedar. Heat accelerates cure speed; polymer crosslinking, however, can be achieved at ambient temperatures. Although a thermosetting polymer, UF is sensitive to heat and moisture with formaldehyde slowly released as decomposition occurs. Requirements exist restricting formaldehyde emissions in wood products to low concentrations in air. To help control these emissions, most particleboard is manufactured with formaldehyde-to-urea molar ratios less than 1.2:1. Alternatively, fortification of UF to improve its durability and strength is possible by copolymerizing with melamine. This modification adds increased chemical costs while requiring higher cure temperatures.

Polyvinyl acetates (PVAs) are thermoplastic, aqueous emulsions used primarily for furniture assembly and other nonstructural applications. Bond strength develops from water loss into the wood from the adhesive, with a resultant increase in polymer concentration in the glueline. While providing excellent dry adhesion strength and good gap-filling properties, the tendency of PVAs to coldflow or creep when subjected to sustained loads is a primary disadvantage. Modification of PVA with thermosetting phenolic resin results in crosslinkable PVAs with improved heat and moisture resistance but only moderate alterations in creep behavior.

Other adhesive systems such as epoxy, polyurethanes and polymeric isocyanates have limited applications to wood bonding at the present time. While these adhesives provide improved bond strength with no formaldehyde emissions, they have the disadvantages of greater cost and incompatibility with water, which requires the use of organic solvents for applicator cleanup. Applications of polymeric isocyanate in particleboard, waferboard and oriented strandboard have recently gained more favor due to their faster cure times and greater tolerance to moisture compared with conventional thermosets. Demand for durable, fireproof wood panels has also resulted in the application of inorganic cement adhesives in particleboard, although cure sensitivity to certain wood species and substantial increases in panel density limits their usage.

5. *Adhesive Factors Influencing Bond Formation*

Adhesive physical and chemical properties play as important a role as wood properties in determining bond performance. Strength, durability, working properties and cost are primary considerations governing adhesive selection. Interrelated polymer parameters which influence the overall performance of an adhesive include reactive functionality, molecular weight distribution, solubility, rheology and cure rate.

The number and reactivity of adhesive functional groups determine how many active sites are available for further polymerization. These affect directly the reaction speed and degree of cure possible during bond formation. The functionality of the adhesive components must be greater than 2.0 to achieve a crosslinkable system.

Molecular weight and molecular weight distribution are polymer properties affecting viscosity, flow and cure speed of the adhesive. Provided sufficient functional-group reactivity remains, higher-molecular-weight components can more rapidly attain cure but exhibit significantly greater solution viscosities and poorer flow properties than low-molecular-weight components. The optimum adhesive molecular weight distribution, however, depends on the wood composite. Solid wood products such as glued laminated lumber and plywood utilize a major portion of higher-molecular-weight components (i.e., $\bar{M}_N$ of 1500–3500) to avoid overpenetration of polymer into wood. Adhesives for waferboard and particleboard have their molecular weight distribution shifted to lower values (i.e., $\bar{M}_N$ of 500–1200). This allows efficient and homogeneous resin application while improving adhesive mobility to facilitate better resin transfer between wood particles during blending operations.

Polymer solubility or dispersibility in solution is an important factor affecting uniform application and cleanup of the adhesive. Improved solubility allows more concentrated adhesive solutions to be used which reduces the amount of solvent to be driven off during cure. Adhesive rheology and cure development, while related to polymer molecular weight and reactive groups present, are highly responsive to heat, pressure and wood moisture content. Various segments of the polymer molecule achieve mobility as plasticization occurs with moisture, or temperature increases to the glass-transition T_g or softening temperature. Cure results in an increase of T_g. Often, compromises are needed to ensure that heat and pressure are applied in a controlled sequence such that the adhesive adequately wets the wood surface and cures before resin overpenetration occurs.

6. Bond Requirements for Composites

Performance criteria for wood composites relate directly to product end use. This requires consideration of engineering strength needs, safety and short- and long-term response of the material to the service environment. Structural, exterior-grade products have the most demanding bond quality requirements, since glueline failure could be catastrophic to these structures. For these situations, glueline strength, durability and reliability must be assured, usually by extensive bond quality testing programs. While conventional bond strength and short-term durability evaluation methodology is well established, there is at present only a limited understanding of the effect of long-term aging on bond integrity. Insight on how product dimensional change and sustained loading affect the time-dependent performance properties of wood composites is developed by cyclic testing through simulated environments using variables of heat, moisture and loading stress. The nonlinear deterioration behavior of many adhesive bonds and our lack of knowledge of the aging process are serious obstacles to assessing long-term performance of new wood adhesives.

For conditions where bond performance is less crucial, cost may be the overriding factor—not only in terms of adhesive price but also of the impact adhesive costs have on overall production expenses. With more automated wood-composite manufacturing systems, compatibility of the adhesive with the assembly system is often the major consideration.

7. Future Direction

Continuing reductions in available harvest volumes and lower wood quality will lead to increased reliance on wood composites and the adhesives used to manufacture them. For commodity products, the importance of manufacturing efficiency will place greater emphasis on adhesives capable of tolerating variable wood properties and moisture contents while having rapid cure speed and reasonable cost. Adhesives will gain prominence in higher-valued wood composites by allowing products of varying shapes, improved engineering strength and greater aesthetic value to be more easily manufactured. Expanded use of continuous pressing techniques and heating methods such as steam pressing or microwave use will create added challenges for adhesive formulation development and wood-handling concepts. Improved knowledge of chemical factors influencing the reactivity of the wood-bonding surface would help enhance functional group reactivity, perhaps leading to reductions in the amounts of adhesive required or possibly to improved binderless wood composites. Advanced, high-strength composites are increasingly likely to utilize wood fiber or cellulose fiber as a directional, reinforcing matrix surrounded by adhesive polymers which may comprise up to 50% of the composite weight.

Future adhesive improvements will be concentrated in areas of durable, faster curing systems, capable of enduring moderate to long assembly times. Copolymer systems will increase in prominence with emphasis being directed towards multicomponent, cold-setting adhesives possibly applicable to on-site assembly. Isocyanates, because of their extensive reactivity and ability to bond at higher moisture contents, are likely to be a component of these newer wood-bonding systems. Reactive hot melts, capable of rapidly curing to thermosetting polymers, will also be a development having a significant impact on future bonding systems.

Bibliography

Collett B M 1972 A review of surface and interfacial adhesion in wood science and related fields. *Wood Sci. Technol.* 6: 1–42

Koch G S, Klareich F, Exstrum B 1987 *Adhesives for the Composite Wood Panel Industry*. Noyes Data Corporation, Park Ridge, NJ

Kollmann F F 1975 *Wood Based Materials*, Principles of Wood Science and Technology, Vol. 2. Springer, New York

Marian J E 1967 Wood, reconstituted wood and glued laminated structures. In: Houwink R, Salomon G (eds.) 1967 *Applications, Adhesion and Adhesives*, Vol. 2. Elsevier, New York

Oliver J F 1981 *Adhesion in Cellulosic and Wood-Based Composites*. Plenum, New York

Pizzi A 1983 *Wood Adhesives Chemistry and Technology*. Dekker, New York
Skeist I 1977 *Handbook of Adhesives*, 2nd edn. Van Nostrand Reinhold, New York
Subramanian R V 1984 Chemistry of adhesion. In: Rowell R (ed.) 1984 *The Chemistry of Solid Wood*, Advances in Chemistry Series 207. American Chemical Society, Washington, DC
Wellons J D 1983 The adherends and their preparation for bonding. In: Blomquist R F, Christiansen A W, Gillespie R H, Myers G E (eds.) 1981 *Adhesive Bonding of Wood and Other Structural Materials*, Vol. 3. Clark C Heritage Series on Wood, Pennsylvania State University, University Park, PA

P. R. Steiner
[University of British Columbia,
Vancouver, Canada]

Adhesives in Building

The process of adhesion, by which a substance makes intimate contact with two surfaces and undergoes physical and chemical changes to bond them together, is complex and only partly understood. Nevertheless, since the mid-1940s, age-old natural adhesives have been improved and numerous new adhesives have been chemically synthesized and now dominate the marketplace.

The construction industry is probably the largest user of adhesives. Adhesives increase the strength and stiffness of building components, effectively transfer and uniformly distribute stresses, reduce or eliminate mechanical fasteners and combine dissimilar materials, thin films, fibers and particles that could not be joined otherwise. Applications vary widely in structural capability, from the huge glued–laminated beams that support roof loads to nonstructural applications where adhesives only support wall coverings, accessories and trim.

Considerably greater amounts of adhesives are used in the manufacture of building materials such as structural softwood plywood, particleboard, hardboard, decorative hardwood plywood, finger-jointed lumber, siding, glassfiber insulation, doors, sandwich panels and other factory-laminated products. For the most part, they are bonded with phenol–formaldehyde and urea–formaldehyde resin adhesives, although small amounts of melamine–urea–formaldehyde and melamine–formaldehyde are used. Millions of kilograms of these adhesives are used annually in the building products industry.

Although used in lesser amounts, adhesives employed to construct industrial and residential buildings constitute a very significant portion of the adhesives market. This group of adhesives is used to assemble building materials either at the construction site (where bonding conditions are sometimes difficult to control) or in assembly shops. Generally, construction adhesives differ in chemical composition and methods of bonding from adhesives used to manufacture building materials. It is this relatively large and diverse group of construction adhesives which is the primary focus of this article.

1. Adhesion

Adhesion is the state in which two surfaces are held together by interfacial forces which may be valence forces, interlocking action or both. For adhesion to take place, the adhesive must be liquid and make intimate contact with the substrate. The more intimate the interfacial contact, where atoms and molecules approach electrostatic interaction, the more tenacious the bond. For strong bond formation, the surface must be free of contaminants and the adhesive must displace air and water molecules from the surface.

Wetting of the surface by the adhesive readily occurs when the contact angle between the edge of a drop of adhesive and the solid surface is low. The contact angle approaches zero when the following conditions exist: the substrate possesses a relatively high surface attraction energy, the adhesive has a real affinity for the substrate and the surface tension of the adhesive is low. If a drop of adhesive spreads to a thin film approaching zero contact angle, the adhesive will spread well and make intimate contact with the surface.

For maximum strength development on wood surfaces (the most common building material), an adhesive must penetrate to the most firmly attached fibers and interlock with the wood (see *Adhesives for Wood*). Low contact angle and viscosity are important in getting good capillary flow into crevices and pores. For the most part, the bonding surfaces of building materials are highly contaminated and replete with cracks and crevices. Such surface conditions cause gas pockets and blockages, which prevent complete wetting and introduce points for stress concentration. Furthermore, stresses develop in joints from dimensional changes brought about by the solvent, moisture or gain and loss of heat during bond formation, and from external environmental service conditions. Even if building adhesives were capable of high levels of adhesion by valence forces, it seems likely that the many sources for interfacial blockages would nullify the potential for that kind of adhesion. In bonding building materials, perhaps the greatest contribution to joint strength can be made by preparing smooth and clean surfaces and using an adhesive with a high affinity for the particular substrate.

The process of adhesion is essentially completed in the transition from a liquid to a solid film. This transition can be a physical change, as in thermoplastic adhesives, or it can be a chemical change, as in thermosetting adhesives. In thermoplastic adhesives, the physical change to a solid film occurs by the cooling of molten liquid on a cooler surface or by loss of solvent through evaporation and diffusion. With thermosetting adhesives, the solid state is achieved by polymerization. Resin intermediates in partly reacted states may react with other intermediates or complete their reaction when heated, to form highly crosslinked structures.

2. *Advantages of Adhesives*

Nails, screws and bolts are still used more than adhesives to assemble building materials. However, adhesives offer advantages that fasteners cannot provide. For example, adhesives distribute stresses over the entire bonded area, thereby avoiding concentration of stresses at the fasteners. When all members are bonded, all share the load. A laminated wood beam thus has much greater strength and stiffness than the combined laminates even when they are fastened together. Not only are these assemblies stronger and stiffer, but the designer can meet design requirements at a lower cost by using fewer pieces of lumber in smaller sizes and lower grades.

Adhesives reduce the number of fasteners that would otherwise be required. For example, 50–75% of the nails can be eliminated from stressed-skin panels by using adhesive. This reduces the cost of nails and can also drastically reduce nailing time and equipment needs.

Adhesives are advantageous when the appearance of fasteners is objectionable. Bonding of high-pressure laminates to kitchen counter tops is an example. Nail popping mars the appearance of smooth surfaces such as gypsum wallboard or resilient tile floors. Adhesives eliminate most, if not all, fasteners.

Adhesives make it possible to eliminate defects in lumber, thereby increasing strength, grade and value. Short pieces of lumber can be finger jointed into longer pieces of any length. Adhesives also permit randomizing of defects within a component, thereby avoiding a concentration of defects at one point, which could have disastrous effects on strength and stiffness.

Flexible adhesives have the ability to yield and distribute stresses during short-term dynamic loading. When used to bond modular or mobile homes for transporting, these adhesives can prevent damage to weaker materials such as gypsum wallboard and door and window frames.

3. *Structural Performance*

Adhesives in building construction have been categorized according to levels of expected structural performance. Five categories are presented here in descending order of structural integrity, with accompanying examples of applications and adhesives.

Prime structural is the most critical category because the adhesive is expected to contribute strength and stiffness for the life of the structure. The application is critical in the sense that failure could result in loss of life or serious damage to the structure. Examples of prime structural applications are glued–laminated beams, plywood and lumber I-beams and stressed-skin panels. The adhesives used are rigid types such as resorcinol, phenol–resorcinol and casein.

In semistructural applications, the adhesive is expected to impart stiffness to an assembly for the life of the structure. Failure of the adhesive would not cause any serious loss of structural integrity, because the load would be carried primarily by mechanical fasteners. An example of this category is the nailed–glued plywood floor system, where elastomeric construction adhesives are typically used.

Temporary structural adhesives contribute to the strength and stiffness of the structure, but the contribution is expected to be shorter than the life of the structure. If the adhesive should fail, damage might be significant but is repairable. Mobile and modular housing are examples where adhesives are used to resist severe dynamic racking, twisting and bending stresses that develop when transporting these large units over the highway. Poly(vinyl acetate) (PVA) resin emulsions are used for this purpose in mobile homes and elastomerics in modular units.

The secondary structural category comprises adhesives contributing strength and stiffness to an assembly but whose failure under loading would cause no significant damage to the structure. The failure could be easily recognized and repaired. An example is the bonding of decorative panelling to gypsum board or furring strips with elastomeric adhesives.

The last critical category is nonstructural. Here, adhesives are expected to support the deadweight of lightweight substrates such as accessories and trim. If failure should occur, damage would be slight and repair would be inexpensive. Examples are bonding of counter tops, floor coverings, ceiling and wall tiles and moldings. The adhesives are usually contact-bond and elastomeric types.

4. *Types of Adhesives*

4.1 *Phenolic Resins*

Resorcinol–formaldehyde and phenol–resorcinol–formaldehyde resin adhesives are used for critical, prime structural applications because they are exceptionally strong and durable and do not creep under high and long-term stresses. They are well suited for the slow laminating process because they tolerate long assembly times. They set at room temperature, thereby eliminating hot pressing of large, unwieldy beams and arches.

Resorcinol–formaldehyde resin is the product of a condensation reaction between resorcinol and formaldehyde. The reaction is carefully controlled to maintain a deficiency of formaldehyde to resorcinol, typically a 0.6–0.65 :1 molar ratio. The reaction is completed at the time of use by adding formaldehyde. Resorcinol is very expensive, so approximately 15% of phenolic resin may be substituted and copolymerized to make a lower-cost phenol–resorcinol resin. This adhesive still sets at room temperature with no sacrifice of strength and durability.

Resorcinolic adhesives are marketed as two-component systems; that is, resin and hardener are separate and mixed just before use. In laminating plants, they are applied by powered roller–spreaders or extruders. In the field, a brush or hand roller is adequate. Open and closed assembly time varies according to reactivity, air and wood temperature, humidity, moisture content of wood and spread rate. Even though lumber is planed to clean and smooth both bonding surfaces, a pressure of 0.7–1.4 MPa is needed to ensure intimate contact between adhesive and substrate. The time for which a laminate remains under pressure may vary from 4 h to 16 h.

Phenol–resorcinol adhesives with gap-filling capability are not commercially available, although there is a need for them in the construction industry. Conventional phenol–resorcinols are too thin to bridge gaps between rough, uneven plywood and lumber surfaces. Therefore, large presses, clamps or extensive nailing must be used to apply enough pressure to achieve intimate contact between adhesive and substrate. Recent studies show that certain thickening agents can be incorporated into a phenol–resorcinol resin to make an extrudable, nonsagging adhesive that will bridge gaps of 1.5 mm or more. Instead of 100–150 mm, nails can be spaced 300–400 mm apart. Nails are needed only to hold the assembly together until the adhesive sets. The phenol–resorcinol has exceptional strength and durability. With added fibrous ingredients, its strength, rigidity and durability, even in thick bonds, exceeds that of plywood substrates.

A few years ago, the desire to build houses rapidly spurred development of a fast-curing phenolic resin with gap-filling capability. Such a product was developed and used to assemble a few houses on a pilot scale but was never commercialized. The adhesive was made by reacting phenol–formaldehyde condensation products of low molecular weight with *m*-aminophenol (and other disubstituted and trisubstituted benzenes) to form a relatively stable phenol–formaldehyde *m*-aminophenol novolak-type polymer. This polymer was reacted with formaldehyde at the time of use. The cure rate could be varied between 20 s and 5 min. When the two components were mixed, the adhesive foamed and expanded to fill gaps between substrates. Sophisticated metering, mixing and application equipment was required, because the two components were highly reactive.

4.2 Polyurethanes

Polyurethane adhesives have also been developed for fast bonding of building components. They can be formulated for thin bondlines in laminating or for bridging gaps in construction joints. Two-component polyurethane systems are prepared from a prepolymer of a polyol and a multifunctional polyisocyanate. When the two components are mixed, the crosslinking reaction can be quite rapid. By altering the molecular structure, that is, by providing different degrees of chain flexibility and controlling the number of sites for subsequent urethane formation, a wide variety of mechanical properties can be obtained, ranging from flexible to rigid. Isocyanates are very reactive with hydroxyl groups and develop excellent bonds on a variety of surfaces. These systems also require sophisticated metering, mixing and extruding equipment because of their reactivity.

4.3 Epoxy Resins

Epoxy resin adhesives are used very little in building construction. However, their strength capability on wood, glass, metal and masonry makes them highly versatile. They are capable of prime structural applications, but are actually used only for repairs, patching of voids in materials and accessory and trim attachment. Epoxies have excellent gap-filling capability. They set at room temperature and do not shrink on setting.

Epoxy resin is made by alkaline condensation of epichlorohydrin and bisphenol A. At the time of use, the resin is mixed with a hardener, usually a polyamine, and the reaction proceeds exothermically. Depending on formulation, the pot lifetime may be a few minutes to a few hours. The substrates must be held together until the adhesive sets, but close fitting of substrates is not essential because of the gap-filling capability.

4.4 Casein

Casein was the most important moisture-resistant structural laminating adhesive before the introduction of synthetic resins such as resorcinol. A proteinaceous substance manufactured from skim milk, the casein molecule comprises a great number of different amino acids in its natural state. Its high molecular weight accounts for its colloidal properties and value as an adhesive. Water-resistant caseins are formulated as casein, calcium hydroxide and sodium salt. The proportions of these ingredients control the viscosity, working lifetime and water resistance. The casein–lime–sodium salt dry mix is readily dispersed in water at the time of use and sets at temperatures above 4.5 °C with a clamp time of 4–8 h. It is easy to use in factories and at job sites because of its long pot lifetime, short assembly time and fair gap-filling capability. However, even the most water-resistant caseins are not suitable for exterior use. Nowadays, casein is principally used to laminate structural beams and smaller components for interior service.

4.5 Vinyl Acetate Polymers

Poly(vinyl acetate) (PVA) emulsions, or white glues, are used in semistructural and temporary structural applications such as floors, walls and roofs of mobile homes. Although they have high strength, they are thermoplastic and yield under prolonged stress, high moisture and heat. Recently developed crosslinking PVAs partly overcome these structural disadvantages.

PVAs are formed by free-radical addition polymerization of the monomer. They may also be copolymerized with other monomer species. Polymer chains can be tied together in two- or three-dimensional networks by crosslinking agents that lead to a higher degree of resistance to heat and moisture.

Most PVA adhesives that set at room temperature are sold as ready-to-use, one-component systems. Catalyzed systems consist of resin and catalyst which must be mixed before use. These also set at room temperature or they may be set much faster with radio-frequency heating. PVA may be applied by brush or roller, but in mobile home construction it is applied in a bead from a squeeze bottle. PVA has some gap-filling capability, so reasonably strong joints can be made with only pressure from nailing. A bond strong enough for handling can be developed in less than 2 h on wood surfaces.

4.6 Elastomers

Elastomeric construction adhesives are a relatively new and important class of adhesives for job site and factory construction. They find use in practically all construction applications except the prime structural category. They are used extensively to bond plywood floors to joists and panelling to studs. Bonded floors are stiffer and stronger, and appeal to homeowners because they feel firm and do not squeak underfoot.

Elastomeric construction adhesives are not true elastomers; however, they are compounded with elastomers such as neoprene, styrene–butadiene, polyurethane, and reclaimed and natural rubbers. Some compounds may be blends or copolymers. The elastomer usually constitutes 30–50% of the total solid ingredients. A neoprene construction adhesive may consist of neoprene, antioxidant, magnesium oxide, zinc oxide, phenolic resin, fillers and organic solvents. High loading with various fillers and extenders imparts mastic consistency for gap-filling capability. Most adhesives in this class set by loss of organic solvent or water. An exception is the polyurethane system, which sets chemically on reaction with moisture in the air. The setting time is generally 30 d for most organic solvent systems, although the polyurethane sets within 3 d.

Elastomerics are marketed as ready-to-use, one-part systems, packaged in caulking cartridges for hand extrusion or in bulk-size vessels for automatic extrusion in factories. An extruded bead adheres to vertical surfaces without sagging and bridges gaps between rough and poorly fitting lumber and plywood with only a minimum of pressure from nailing. These adhesives bond to wet and frozen lumber surfaces although the temperature must rise to 4.5 °C soon after bonding to allow the solvent to volatilize. They develop stong bonds, but usually not strong enough to make wood fail. Elastomerics are unusually tolerant of field conditions and have proved to be an excellent solution to many on-site bonding problems.

4.7 Contact-Bond Adhesives

Contact-bond adhesives are used in secondary structural and nonstructural applications, particularly where sheet materials are laminated to rigid substrates and close-fitting joints can be constructed. Examples are laminated high-density counter tops in kitchens and bathrooms and double-laminated gypsum wallboard.

Most contact-bond adhesives are made from neoprene, although styrene–butadiene, nitrile and natural rubbers are used. In making a typical neoprene contact, neoprene is mill mixed with magnesium oxide, antioxidant and zinc oxide and then churn mixed with *t*-butyl phenolic resin and an appropriate mixture of organic solvents to give 15–30% solids. The use of organic solvents in contact adhesives is diminishing because of the fire hazard and toxic fumes. Though an interest is developing in water-based systems, water-based contacts are slower drying and lack the rapid bond strength development and high ultimate strength that characterize solvent-based systems.

Contact adhesives are low-viscosity liquids that are spread in one or two coats on both substrates. They are allowed to air dry to a tack-free state. The substrates are then pressed together, typically by roller pressure. Once in contact, the surfaces are stuck and cannot be repositioned. Setting takes place by loss of solvent, which is rapid on porous substrates but may take several days on nonporous substrates.

4.8 Mastics

A relatively large group of adhesives is used to bond wall and floor coverings. Generally, these are mastics able to bridge gaps between the flat covering materials and rough floor or wall surfaces. Adhesives for ceramic, metal and plastic wall tiles are rubber-based mastics in organic solvent or water, oil-based and resin-based mastics, and Portland cement with resin binder. Mastics with organic solvent are not used with plastic tile. The adhesives may be applied to almost any sound, dry and clean surface by trowelling or buttering techniques. The tiles are simply pressed into the mastic while the adhesive is wet and within a few hours the bond develops by loss of solvent.

Adhesives for rubber, vinyl and cork tile flooring are rubber-based mastics with a water vehicle and resin-based mastics with alcohol. These adhesives are suitable for linoleum and vinyl sheet flooring. The flooring materials must be applied while the adhesive is still wet. Rigid asphalt and vinyl–asbestos tiles are bonded with asphalt emulsion, asphalt cutbacks and rubber-based systems with a water vehicle. The asphalt emulsion and asphalt cutback must be dry before the tiles are placed; however, the tiles must be placed before the rubber–water system dries.

4.9 Animal Products

For the most part, animal adhesives have been replaced by synthetics. A few craftsmen still use them to bond kitchen cabinets and builtins. Animal adhesives are based on collagen—the fibrous protein found in bones, hides and the connective tissue of animals. The adhesive is sold as a solid that must be dispersed in water and heated before use. On cooling, it sets to a hard film that makes a strong bond with some gap-filling capability. However, the bond has limited durability and will weaken on exposure to moisture and heat.

See also: Adhesion: Fundamental Principles; Adhesive Joining: An Overview; Adhesives in Civil Engineering; Structural Adhesives

Bibliography

Bikerman J J 1961 *The Science of Adhesive Joints.* Academic Press, New York, pp. 1–258

Blomquist R F, Christiansen A W, Gillespie R H, Myers G E (eds.) 1983 *Adhesive Bonding of Wood and Other Structural Materials*, Vol. III. Educational Modules for Materials Science and Engineering (EMMSE) Project, Materials Research Laboratory, Pennsylvania State University, University Park, PA, pp. 1–436

Gillespie R H, Countryman D, Blomquist R F 1978 *Adhesives in Building Construction*, Agricultural Handbook 516. US Department of Agriculture, Forest Service, Washington, DC. pp. 1–160

Lees W A 1989 *Adhesives and the Engineer*. Mechanical Engineering Publications, London

Pizzi A (ed.) 1983 *Wood Adhesives Chemistry and Technology*. Dekker, New York, pp. 1–364

Selbo M L 1975 *Adhesive Bonding to Wood,* Technical Bulletin 1512. US Department of Agriculture, Forest Service, Washington, DC, pp. 1–122

Skeist I (ed.) 1977 *Handbook of Adhesives*. Van Nostrand Reinhold, New York, pp. 1–921

US Department of Agriculture Forest Service 1974 *Wood Handbook: Wood as an Engineering Material,* Agricultural Handbook 72. US Department of Agriculture, Forest Service, Washington, DC, Chap. 9, pp. 1–14

Wake W C 1982 *Adhesion and the Formulation of Adhesives*, 2nd edn. Applied Science, London, pp. 1–332

C. B. Vick
[US Department of Agriculture Forest Service, Athens, Georgia, USA]

Adhesives in Civil Engineering

Civil engineering concerns that branch of technology associated with large-scale structures and works: dams, roads, railways, harbor and coastal installations, large buildings and structures, earthworks and foundations. Such works have been carried out by man for centuries and we have much evidence of the success and ingenuity of our forefathers. Initially, they used stone or primitive bricks (sometimes with mortar for bonding and sealing), wood and earth. More recently, Portland-cement concrete (usually called simply concrete) and structural steel have become the principle materials used, although advantage is taken of natural rock and earth infill where possible for cost reasons alone.

Since concrete is less strong in tension than in compression (by a factor of about ten), it is normally used so that the tensile loading is small, or is reinforced with steel bar or mesh (see Fig. 1a). A bond

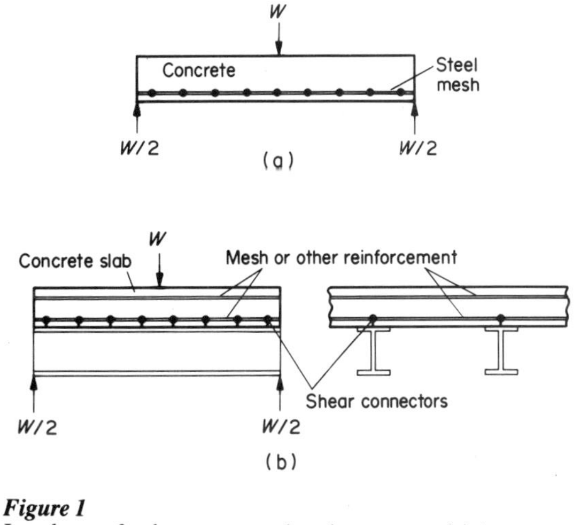

Figure 1
Load transfer between steel and concrete: (a) bar or mesh reinforcement for a simple slab: and (b) shear stud reinforcement

or adhesion between the steel and concrete is essential for the two materials to act as a composite. While shrinkage of the concrete on setting and hardening contributes some grip, the load transfer is enhanced by using twisted or specially deformed bars. In steel beam–concrete slab construction, composite action is achieved by welding or otherwise attaching steel studs to the steel section in the form of shear connectors as shown in Fig. 1b. However, in neither of these applications is there any attempt to use *structural adhesives* for the load transfer mechanism. It is the intention of this article to show what is available in structural adhesives for civil engineering applications, to discuss what problems are likely to be faced and to illustrate these by some actual applications.

It should be pointed out that there are two topics which will not be covered here. First, polymeric materials, together with pitches and tars, have been used as sealants, but these are not directly load-bearing applications. Second, there has recently been considerable interest in polymer concretes. In these, the aggregates are bound together with synthetic resins or polymer-modified cements. These polymer concretes can show a twofold increase in strength (compressive, tensile or flexural) over ordinary concrete. The polymer used is often a polyester resin. For further information, the reader is referred to the excellent article by Hewlett and Shaw (1977).

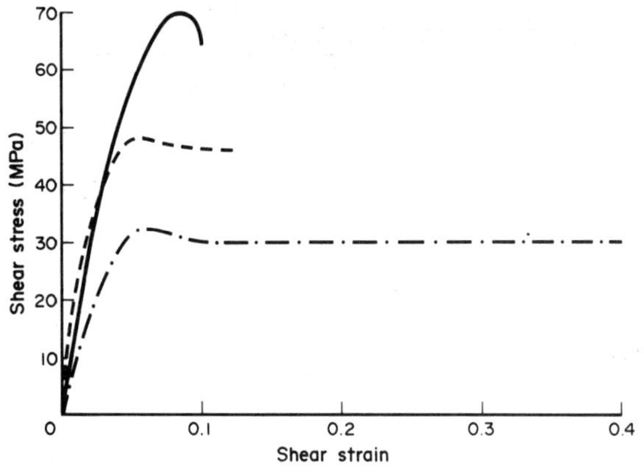

Figure 2
Shear stress–strain curves for some epoxy adhesives: ———, unmodified; – – –, filled, toughened; — · —, rubber-modified (failure strain 0.6)

1. Adhesive Materials Available

The chemistry of structural adhesives is complex and there are many variations on apparently similar formulations; the engineer who intends to use a structural adhesive is therefore well advised to consult the adhesives manufacturer about the chemical compatibility of the different types. Fortunately, these many variants fall into two basic forms: polyesters and epoxies. Generally speaking, epoxies are about three or four times as expensive as polyesters, but provide significantly improved performance. Because of the problems of heat provision and dissipation and shrinkage stresses arising during curing and cooling of hot-cured resins, cold-curing adhesives are often preferred in civil engineering applications. In special cases, heating strips might be laid in the adhesive, but the expense involved would rarely be justified.

Both epoxies and polyesters may be formulated so that they can be used for laminating glass or carbon fiber composites. In this form, the resins need to flow so as to wet the fibers completely. Thus, these resins would have very poor gap-filling properties for applications where two ill-fitting surfaces have to be bonded and where gravity would cause the adhesive to drain out. Fortunately, it is possible to use various thixotropic agents to overcome this problem. Also, fillers such as china clay, talc and powdered metals can be used to provide the necessary gap-filling properties. In addition, these fillers can reduce the cost of the product, since they are often cheaper than the resins. However, fillers, although providing some stiffening action, sometimes weaken the cured resin.

Resins, particularly epoxies, can be flexibilized by various polymer additives. This generally reduces the strength and modulus slightly and considerably increases the strain to failure, but reduces the creep resistance significantly. Of these effects, the gain in strain to failure usually outweighs the slight losses in the other mechanical properties.

Typical shear stress–strain curves of structural adhesives are shown in Fig. 2. These are meant to illustrate the range of available properties rather than to identify proprietary materials. It should be pointed out that the strength of adhesives is usually markedly poorer in tension than in compression or shear. (In this respect, they bear some similarity with concrete.) Typically, Young's modulus is in the range 1–7 GPa and Poisson's ratio is approximately 0.4. The shear modulus is typically 0.3–2 GPa.

2. Mechanics of Adhesive Joints

A considerable amount of mathematical analysis has been carried out on the type of overlap joints commonly used in aircraft and similar construction where thin sheets are bonded. Figure 3 shows that the shear stress is nonuniform, peaking at the ends of the joint, and that there are also considerable transverse tensile stresses at the ends. These transverse tensions are the usual cause of failure in this type of joint. Even when monolithic adherends are

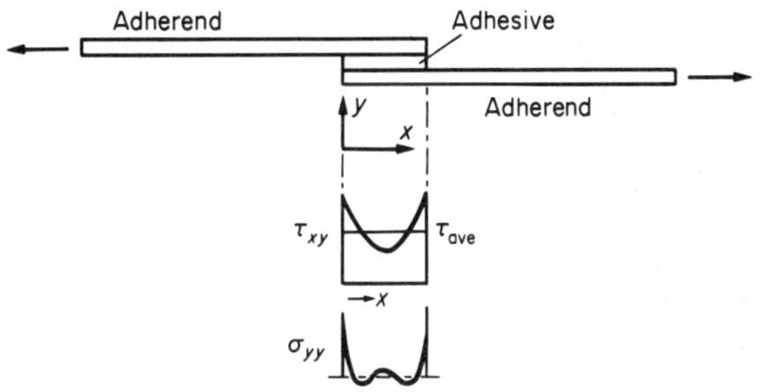

Figure 3
Stress distributions in a simple overlap joint

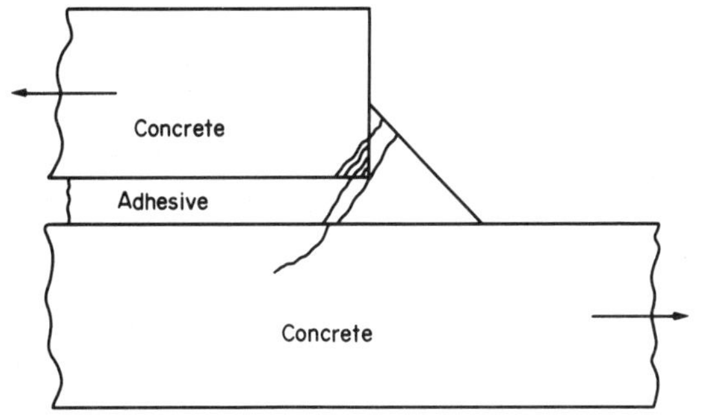

Figure 5
Failure in concrete bonded to concrete, loaded in shear

used, these stress concentrations still occur, particularly in the transverse direction. Adams and Peppiatt (1974) showed that at the ends of the joint a tensile stress of the order of 3.6 times the average applied shear stress exists and that this stress concentration is due to the high stress gradients at the joint ends.

It is generally recommended that load-bearing adhesives be used in shear rather than tension (they are rarely likely to fail in compression), owing to their low tensile strength. The same applies to concrete. But consider the case of a steel plate bonded to concrete and loaded in shear, as shown in Fig. 4; this is a common structural application of adhesives. Fracture will be initiated at the corner A and the crack will extend to the free surface of the adhesive at B and to the adhesive–concrete interface at C. There will now exist a large transverse tensile stress at C. If the lower adherend were also steel (or similar material with a high tensile strength), the crack would propagate through the adhesive near the steel–adhesive interface, as shown in Fig. 4b, to D. However, with a concrete adherend, the concrete itself may fail in tension, the crack now propagating to E (Fig. 4c). Conversely, if the upper adherend were concrete, as shown in Fig. 5, it is likely that this too would crack under the tensile loading, leading to a series of microcracks before the adhesive fails and the cracks again run into the lower adherend.

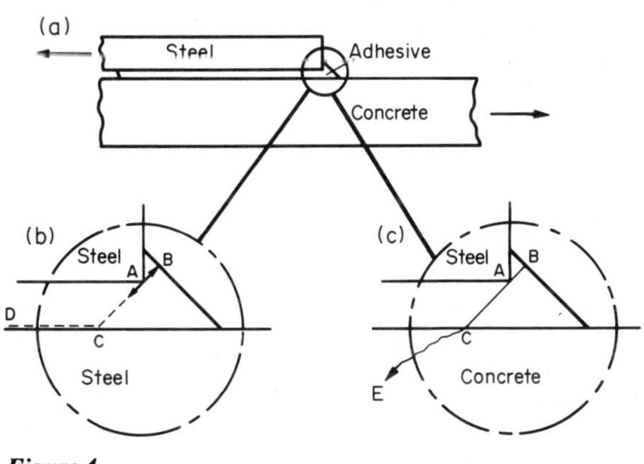

Figure 4
Failure in steel plate bonded to concrete, loaded in shear

One method of overcoming this problem is to use a flexibilized adhesive which yields considerably before it fractures. In this way, the stresses are redistributed, leading to a reduction in the stress concentrations in the concrete. Unfortunately, such flexibilized adhesives tend to creep readily and may not find favor in long-term applications, especially in hot locations.

3. Technical Problems

To the technical difficulties which face the mechanical or aeronautical engineer in using structural adhesives must be added a number of others which are specific to civil engineering. Principally, these are related to the large scale of the works. An excellent document which details the various precautions and tests and covers many aspects of using adhesives in civil engineering site work has been issued by the Fédération Internationale de la Précontrainte (1982). It is essential for engineers intending to use adhesives on site.

3.1 Surface Preparation

Good bonding requires good surface preparation. Adhesives will not normally give a satisfactory bond strength if the adherend surface is dirty, oily, lichen covered or otherwise contaminated. Grit blasting or bush hammering of both steel and concrete is usually satisfactory for removal of surface contamination and laitance. However, long-term protection of the steel is recommended to avoid corrosion at the adhesive–adherend interface. Any protective layer (such as is formed by galvanizing) must have a bond to the steel and adhesive at least as good as that of the untreated steel to the adhesive, otherwise it becomes a weak link in the chain. Also, it is important that the surfaces are not wet. If there is any moisture present, this should be dabbed away with a clean cloth.

In the aircraft industry, a primer is often applied to the surfaces prior to bonding. This acts as an intermediate compound, providing good adherence to both the adhesive and the substrate and protecting the latter from further contamination. This practice does not appear to be used in civil engineering, but it might be advantageous in allowing the use of different adhesive systems, such as one which does not bond well to concrete but otherwise has excellent mechanical and environmental properties. This could be a profitable idea for both adhesive manufacturers and civil engineers to follow.

3.2 Environmental Effects

The adhesive must not be unduly affected by the presence of ambient humidity or likely service temperatures. Fire is an additional hazard which must be given serious consideration where adhesives carry critical structural loads. In addition, the presence of chemicals, such as those of the alkaline environment associated with concrete or due to oil and similar hydrocarbon spillages, must not cause degradation of the adhesive.

3.3 Duration of Application

The components to be bonded are large, requiring a substantial amount of adhesive which has to remain active until the whole surface has been covered. Premature cure or skinning would be disastrous as it would severely reduce the ultimate bond strength.

3.4 Mixing

It is recommended that where two-component adhesives are used these should be of two distinct colors so that good mixing can be *seen* to have taken place. Power mixing is recommended and a 350 W electric drill on slow speed should be adequate for a 5 kg mix. High mixing speeds can lead to entrainment of air and may also cause the mix to overheat and cure prematurely.

3.5 Temperature of Application

In cold environments (5–20 °C), a fast-reacting adhesive system should be used while, in hot environments (>40 °C), a slow-reacting system is essential. It is best to consult the adhesive manufacturer to ensure that a properly cured bond is achieved.

3.6 Bondline Thickness and Gap Filling

In aerospace applications, bondlines down to 0.02 mm are possible but, in civil engineering, the thinnest practical bondline is in the region of 1 mm, owing to lack of fit of the components and the difficulty of removing the excess adhesive (squeezing out to the edges). Normally, the excess adhesive squeezed out is removed by a scraper before it has gelled.

The gap is often vertical, such as when bonding precast segmental bridge sections, and it is essential that the adhesive does not run out. There are various tests for thixotropy, a property which indicates the ability of a liquid to hold to a vertical surface without running. Such a test is described in ASTM D 2730. This property is a function not only of the adhesive but also of the ambient temperature. If problems are encountered, it may be necessary to caulk the edges of the gap or otherwise prevent the adhesive from running out.

3.7 Application

The adhesive may be applied to either or both of the surfaces to be joined. The use of a trowel or stiff brush is usually all that is required, although special tools are sometimes used.

4. Examples of Use

The following examples of the use of adhesives in civil engineering are not meant to be exhaustive; indeed, this technology is finding ever wider applications in construction.

4.1 Segmental Constructions

Since around the mid-1960s, there has been increasing interest in the technique of segmental construction, using precast concrete units made under factory-controlled conditions, possibly remote from the construction site. The sections are then erected and post-tensioned on the site. To achieve a good match, each segment is match cast against the end of its neighbor, using a release agent to aid separation. This technique has been widely used not only for bridge building but also for many other structures, such as tunnels, stadia and lighthouse foundations.

Initially, the gaps between the sections were large (>250 mm), using concrete cast *in situ* to make the joint. Stressing of the segments had to be delayed until the mortar had hardened sufficiently. In 1963, at Choisy-le-Roi in Paris, a bridge over the Seine was constructed using match-cast segments in which the joint was made using an epoxy resin. Match casting allowed for very small glue lines (about 2 mm thick in this case). This was the first use of adhesives in this way in a major construction. Since then, many other bridges have been built in this fashion in Europe and the Americas. The functions of an adhesive in such constructions are

(a) to act as a continuum to transmit shear, compressive and (in some cases) tensile stresses between the adjacent concrete sections;

(b) to develop strength (i.e., to cure) quickly so as to speed the erection of the segments;

(c) to act as a lubricant to facilitate the location of the mating surfaces; and

(d) to provide a watertight seal to protect the prestressing bars.

Two notable recent bridges in the UK in which adhesive-bonded segments have been used are the Byker viaduct, which is S-shaped in plan and some 800 m long (part of the Metro scheme at Newcastle-upon-Tyne), and the M180 motorway bridge at Scunthorpe. The Byker viaduct has the distinction of being the first to be built in the UK using cantilever construction and precast segments with epoxy-resin-bonded joints. The bonded construction allowed the use of woven fiberglass to provide joint thickening where needed, so as to correct for minor misalignments. The M180 bridge at Scunthorpe is interesting in that adhesives were successfully used even during severe (for the UK) wintry conditions. It was found necessary to warm the mating surfaces before applying the adhesive so as to remove all traces of frost. Also, space heaters were installed inside the hollow sections to help cure the adhesive. In an excellent technical account of the use of epoxy adhesives in the M180 bridge, Moreton (1981) recommends that heating must be used, at least to keep the concrete above 0 °C, and warns that allowance must be made for the possibility that the adhesive may not fully cure. It may be possible to insert a series of resistance wires to provide direct electrical heating of the adhesive and local concrete. These wires could either be withdrawn before cure or left in place; in the latter case, they will not significantly weaken the bond if they are nonreactive with the adhesive or the concrete.

Other notable structures in which adhesive-bonded segmental construction has been used include the roof of the Sydney Opera House and the large sugar silos in Durban, South Africa. A breakwater was formed around the Ekofisk caisson (North Sea Oil installation) by bonding concrete segments with epoxy adhesive.

4.2 Externally Bonded Reinforcement

A technique which has found substantial application in recent years has been the strengthening and stiffening of reinforced concrete structures by externally bonded steel plates. In particular, this has been applied to bridges in many parts of the world. Several reasons have prompted the use of this technique:

(a) to correct faulty design;

(b) to correct faulty workmanship;

(c) to increase the bridge capacity to cope with increased traffic loads; and

(d) to counter deterioration of the steel reinforcement, especially that subjected to salt-spray exposure.

This technique was pioneered in France and South Africa during the period 1965–1975. A particularly interesting example in France is on the Autoroute du Sud (Bresson 1972). The Japanese were quick to recognize the possibility of uprating many bridges for which the traffic loading (especially heavy commercial vehicles) had caused a serious problem. By 1975, at least 240 bridges had been strengthened in this way. More recently, several motorway bridges in the UK have been similarly strengthened.

The reinforcing plates may be applied to the underside of bridges, where the concrete is in tension, both to stiffen and to strengthen the structure. They can also be applied to the webs of concrete beams to increase the shear stiffness. The benefits of using bonded steel plates instead of additional beams or props are that the application can be carried out with a minimum loss of headroom and a minimum disruption to other traffic. However, the long-term durability of bonded joints used in this type of construction has not been established.

The Quinton interchange on the M5 motorway in the UK was strengthened during 1975 and the work has been monitored (Raithby 1980). Over 1000 steel plates, 250 mm wide, 6.5 mm thick and up to 3.5 m long, were bonded to the underside of the bridge. In some cases, two or three plates were bonded together to increase the thickness. Each plate was suspended from bolts located in the lower surface of the deck slab and then wedged into position so that the excess adhesive was squeezed out. To allay fears of detachment of the plates, the bolts remain, although they take virtually none of the structural loads, which are all carried in shear by the adhesive. In addition, the free ends of the plates are secured by small clamps bolted into the concrete; this is used to counteract the most likely cause of failure, local peeling at these free ends.

In 1977, two bridges on the M25 motorway at Swanley, UK, were strengthened by adding steel plates to the underside of the deck. Additional plates were bonded to the top of the deck over the supports to counteract tensile loads.

External plating has also been used to strengthen floors in buildings (again headroom advantages are obvious). An eight storey prefabricated building has been strengthened by bonding steel plates to the structure so as to withstand statistically rare, but possible, high wind loads (Shaw 1971, Hewlett and Shaw 1977).

Although bonded external reinforcement, used wisely, can be a very useful maintenance tool, there are as yet no codes of practice or standards (Mander 1981). Caution should therefore be exercised and the designer should consult those who already have experience. Also, the long-term effects should be carefully monitored.

4.3 Fixings in Masonry and Concrete

Resin-bonded anchors can be used in masonry, rock or concrete. These consist of a glass tube which contains a cold-setting polyester or epoxy resin, together with a separate internal tube containing the hardener. After drilling, the tube is placed in the hole and a rotary percussion drill is attached to the bolt. The bolt is then quickly driven into the tube so that the glass components are broken and the resin and hardener are mixed. Excellent adhesion is thus obtained between the bolt and the substrate when the resin has cured. Internally threaded sockets can also be bonded in by this method, thus leaving a demountable fixing. A variety of proprietary fixings are available for tensile loads of up to 52 kN (12 000 lbf) in K350 (35 MPa, 5000 lbf in^{-2}) concrete or shear loads of 39 kN (9000 lbf). Fast curing is essential if overhead work is contemplated. For this reason, polyester resins are often favored and a strong bond can be achieved in a matter of minutes.

4.4 Structural Repairs to Concrete

Internal repairs to concrete are carried out by injecting adhesives into cracks (Fuller and Kriegh 1973). Generally, the adhesive must have a low viscosity to penetrate narrow cracks; a viscosity in the region of 1 Pa s is often recommended, so filled adhesives are impractical. To avoid running of the resin, it is usual to close the crack either by taping or by caulking with tar or a hot thermoplastic polymer. The function of the adhesive is partly to bond the cracked surfaces together, thus restoring the monolith (provided that a good bond can be achieved), and partly to act as a sealant which prevents environmental attack on the steel reinforcement.

External repairs to spalled concrete structures tend to be partly cosmetic as well as structural. Typically, the defective concrete is cut away and the area is grit blasted. Any steel reinforcement thus exposed is treated with a rust inhibitor. The surface is then built up using a resin–cement grout cast on to the structure behind a thin, preformed section of glass-reinforced plastic.

4.5 Resin-Bonded Nonskid Surfaces

Since the mid-1960s, there have been many successful applications of nonskid surfaces, especially for roads. These consist of an epoxy resin, sometimes mixed with coal tar, overlaid with finely divided sand. Such surfaces have been used for pedestrian areas as well as bridges, highway intersections and pedestrian road crossings.

4.6 Cable Anchors

Cables are widely used in civil engineering for carrying substantial tensile loads. The decks of suspension bridges are usually carried by cables and, in offshore technology, there is an increasing use of cable-stabilized structures. The end attachments for these cables have to be treated carefully, for it is here that the maximum stress concentrations occur. Commonly, the ends of the cable are splayed into a conical socket which is then filled with molten zinc for large (structural) ropes or white metal for stranded wire ropes. When the cable is put under tension, the conically formed end is pulled into the conical socket, producing a compressive load between the socket and the formed end by wedge action. This compressive load results in a large interfacial compressive stress between the wires and the zinc or white metal. It is argued that, even in the absence of a good interfacial bond, this frictional force is sufficient to retain the cable end in the socket.

Recently, the molten metal has been replaced in some applications by a thermosetting polymer. By using a cold-setting resin which has a relatively low viscosity, excellent filling can be achieved without having to heat the socket and cable end. Polyester resins are normally used, since these tend to give a quicker cure, although the use of epoxy systems has also been reported. The rapid-curing polyesters exhibit an exothermic reaction. To reduce this effect, which can lead to overheating of the resin and to thermal shrinkage cracks, fillers such as silica can be used which both reduce the volume of resin and increase the heat capacity of the mix. Although epoxies generally give a better bond to steel than do polyesters, it has been argued that the system will work with only a small bond strength, owing to the cure shrinkage and to the wedging action under load. However, there must be some bond strength and action, otherwise the wires would simply pull out before the wedging action could begin.

Cables for which this system has been used range from 2 mm to 150 mm in diameter (2000 t breaking load). The big advantage of using resins compared with molten metals is that the application may be carried out on site with little danger and modest equipment. This is particularly useful when operating offshore with large diameter cables.

4.7 Steelwork

When an adhesive is introduced between the faces of a bolted steel connection, it is reported (Mander 1981) that both the static and the dynamic (fatigue) load-carrying capacities are increased.

Prevention of buckling failure by welding ribs and stiffeners to beams, plates and diaphragms can lead to distortion and fatigue. It is possible that bonding could replace welding. This would also have the advantage that dissimilar materials, such as aluminum or carbon-fiber-reinforced plastics, could be used which might provide useful weight savings—particularly if the stiffening had to be added as an afterthought.

See also: Adhesion: Fundamental Principles; Adhesive Joining: An Overview; Adhesives in Building; Bond Durability; Bonding: Preparation of Metallic Surfaces; Structural Adhesives

Bibliography

Adams R D, Peppiatt N A 1974 Stress analysis of adhesive-bonded lap joints. *J. Strain Anal.* 9: 185–96

Bresson J 1972 Renforcement par collage d'armatures du passage inferieur du CD126 sous l'Autoroute du Sud. *Ann. Inst. Tech. Batim. Trav. Publics.* Suppl. 297

Fédération Internationale de la Précontrainte 1982 *Proposal for a Standard for Acceptance Tests and Verification of Epoxy Bonding Agents for Segmental Construction*

Fuller J D, Kriegh J D 1973 *A Guide for Pressure Grouting Cracked Concrete and Masonry Structures with Epoxy Resin.* Army Construction Engineering Research Laboratory Technical Report AD-755 926. National Technical Information Service, Springfield, VA

Hewlett P C, Shaw J D N 1977 Structural adhesives used in civil engineering. In: Wake W D (ed.) 1977 *Developments in Adhesives—I.* Applied Science, London, Chap. 2, pp. 25–75

Lees W A 1989 *Adhesives and the Engineer.* Mechanical Engineering Publications, London

Mander R F 1981 Use of resins in road and bridge construction and repair. *Int. J. Cem. Compos. Lightweight Concr.* 3: 27–39

Moreton A J 1981 Epoxy glue joints in precast concrete segmental bridge construction. *Proc. Inst. Civ. Eng.* Part 1, 70: 163–77

Raithby K D 1980 *External Strengthening of Concrete Bridges with Bonded Steel Plates.* Transport and Road Research Laboratory Report No. 612. TRRL, Crowthorne, UK

Shaw J D N 1971 The use of epoxy resin impregnation for deteriorated concrete structures. In: Guy I N (ed.) 1971 *Advances in Concrete.* Concrete Society, London

R. D. Adams
[University of Bristol, Bristol, UK]

Adhesives in Medicine

At present, only a limited range of adhesives is used routinely in medical practice. Methyl silicones are employed as temporary bonding aids for the attachment of diagnostic or prosthetic devices to human skin, some cyanoacrylates are used in dental applications, and common acrylics such as poly(methyl methacrylate) find applications as cements in both dental and orthopedic reconstructions. Moisture-curing and ultraviolet-curing polyurethanes are now achieving some success in their use as "pit and fissure" sealants for children's teeth, aiding in the prevention of dental cavities.

Problems abound with these and all other adhesive substances proposed for medical use. For example, the adhesive bond strengths of the silicones are usually too low, the cyanoacrylates degrade slowly, releasing potentially toxic by-products, and the acrylics often contain migratable monomeric species that can be taken up by the body tissues and fluids with adverse secondary effects. In addition, the heats of polymerization of the acrylics are often so high as to kill adjacent living cells. Polyurethanes and many of the polycarboxylates used in dental applications can break down quickly due to the high-wear features of the oral environment.

1. Challenges of the Medical Environment

The fixation of biomedical or dental prosthetic devices in humans provides a severe challenge to adhesive science and technology. Custom-fitted strong (usually metallic) devices called subperiosteal implants provide a good example of the challenges to be met and the extreme complexity of the problems posed by contrasting adhesive requirements over very short distances along the same structures. These implants are expected to adhere to the remaining bone structures in a patient whose teeth have been lost as well as adhering to the remaining tissue flaps used to cover them. They must provide a post which protrudes with a bacterial-tight infection-free adhesive seal through the tissue into the nonsterile intraoral region to provide a mounting site for dentures or other load-bearing functional prosthetic restorations. The portion of the implant buried in the tissue must be firmly adherent in order to provide the immobile support required. The integrity of the tissue bond immediately adjacent to the permucosal post must be nearly perfect to prevent bacterial seepage, infection and resulting inflammatory responses. The structural parts residing in the saliva-bathed oral cavity should resist adhesion and colonization by biological materials, especially plaque-forming bacteria. The distance over which the adhesive properties must change so drastically can be as little as a few micrometers along the surface of such implants. Furthermore, the requisite conditions must be achieved and maintained in environments that begin as surgically prepared, often bloody and usually nonsterile, potentially contaminant-rich sites.

As may be expected, even though medical and dental implants of many configurations were being used clinically before the mid-1970s, the results obtained have been at best ambiguous. Many failures have been noted as a result of poor tissue bonding to the implanted portions, serious cellular and debris accumulation on the exposed segments and very poor healing around posts protruding through external tissue. It is not uncommon for the implants to extrude completely from their placement sites within only a few weeks. It is obvious that trials in animals, and premature applications of these otherwise promising devices in humans, have proceeded too rapidly in the absence of both necessary and sufficient knowledge of the critical properties of the materials that promote or inhibit biological adhesion.

One major difficulty with studies of biomedical adhesives for any use, be it in dental, orthopedic or cardiovascular environments, has been the tendency of investigators to focus on the longer-term behavior of the materials—from weeks to months to years after application—without proper characterization or even an elementary knowledge of the original surface states of the materials. Little knowledge has been gained of the interfacial conditions at the time of device placement or of the early bioadhesive sequelae (over minutes, hours and days) which change the original surface properties of the adhesive and tissue in ways that promote or inhibit adhesion. In more traditional engineering studies, such as those dealing with the rate and consequences of biological fouling of heat-exchange materials (as in power-plant condenser tubes), the same area of ignorance of the early adhesive or "conditioning" events is usually dismissed with the label "induction period."

There is much to be learned from the adhesion of marine fouling organisms. The biofouling of ship hulls, propellers and cooling-water piping includes the spontaneous adhesion of microfouling films and macrofouling aggregations of barnacles, tube worms, algae, bryozoa and hydroid communities that can diminish ship operating efficiency by 40% in as little as six months in tropical waters—an impressive demonstration of strong adhesion in wet, salty, biochemically active circumstances.

2. *Requirements of Good Medical Adhesives*

Assessing the surface properties of adhesive materials in the biological environment must be the subject of extensive and continuing research and development. The magnitude and difficulty of the subject can be appreciated by considering the number and types of different requirements.

The problem at hand requires the synthesis and study of selected materials for use as encapsulants, sealants and lead insulators exposed to biological environments, especially *in vivo* environments. Topics of interest in this study are dimensional stability, adhesion, permeability, ion migration, possibly electrical and radio-frequency interactions and physical fatigue.

Useful candidate adhesive systems should have a minimum sterile shelf life of 18 months, have a minimum adhesive force to oily or wet biological surfaces of 35 MN m^{-2}, cure effectively at temperatures from 5 °C to 40 °C, support a weight of at least a few kilograms within 30 s of application with an adhesive area of only 5 cm^2 or so when employed at 37 °C and function effectively on rough or irregular surfaces with heterogeneities as large as 0.25 cm. In addition, they should be able to maintain their attachment for periods in excess of 1 year without compromising the viability of sensitive adjacent tissues.

Assessment of candidate adhesive systems requires knowledge of their characteristics and performance in various environments:

(a) response to water and water vapor with respect to dimensional stability and permeability;
(b) response to inorganic ions typically found in biological fluids;
(c) methods and conditions of application that minimize failures in bonding;
(d) adherence to typical substrates in a saline environment;
(e) fatigue resistance to repeated flexure, rotation and elongation;
(f) ability to protect the electrical characteristics of semiconductor and interconnect circuits;
(g) attenuating effects over the entire electromagnetic frequency range);
(h) relative compatibilities and stabilities as passive layers in contact with sensitive living tissues; and
(i) flex life in physiological surroundings.

Similar information on composite adhesives or layered materials (i.e., reinforced composites) must also be compiled.

3. *Surface Effects on Bonding*

Recognizing that the use of medical adhesives to bond living tissue to solid substrates must bear some relationship to the original surface properties of the solid and tissue phases, studies have been made of the influence of some relevant surface-energy parameters in the presence and absence of significant variations of surface charge and texture (roughness). At least one empirical parameter of surface free energy, the "critical surface tension" defined in the mid-1940s and determined from simple contact-angle measurements with a variety of pure liquids (Baier et al. 1968), provides a useful "label" for the apparently bioadhesive character of solid or semi-solid (hydrogels, tissues) substrates. Materials with critical surface tensions between 20 mN m^{-1} and 30 mN m^{-1} exhibit minimal underwater adhesion, whereas materials with critical surface tensions above this range (including most common engineering polymers) or below this range (primarily fluorocarbons) support adhesion in saline media to greater degrees.

Examining the actual initial adhesive events at biological interfaces, it has been found that attachment is preceded by adsorption of macromolecular conditioning films on the nonbiological surfaces. These films are largely composed of amphipathic biopolymers, depositing from aqueous media at about the same rates but in different configurations, on substrates of differing surface properties. Initial rates of subsequent cellular or bacterial adhesion to such adsorbed films are also comparable, but the tenacity of adhesion varies with the degree of apparent "denaturation" of the conditioning film of macromolecules. The critical-surface-tension parameter is a useful descriptor of the bioadhesive properties of these conditioned substrates, revealing changes in surface properties as the initial biopolymer layer forms and as the cellular or bacterial material is deposited. Specific examples of adhesion to such devices as the artifical heart, substitute blood vessels and dental implants support the generalization of these findings to adhesion in most natural circumstances, and imply that practical and inexpensive approaches, such as contact-angle measurements, can be used in researching and controlling biological adhesion.

Particular attention has been given to the experimental requirements for making reliable determinations of surface-energy parameters from contact-angle measurements on living hydrated complex biological materials. These studies demonstrate the inadequacy of using water or water-miscible fluids alone. Although water droplets will quickly penetrate, swell or adsorb surface-active components when placed on biological substrates, exhibiting low contact angles or complete spreading, this complicated event cannot be taken to indicate an operationally high surface free energy for those substrates in contact with other environmental agents. In fact, contact angles measured with water-immiscible fluids of a variety of types on biological surfaces in equilibrium with their surroundings reveal the apparent surface free energy to be that of the "biologically stable" surface state, characterized by a critical surface tension of 20–30 $mN\,m^{-1}$, as opposed to greater than 70 $mN\,m^{-1}$ (normally expected for an aqueous surface). Desiccated biological surfaces, as exemplified by isolated cell-wall preparations and by extracted proteins, polysaccharides and their complex conjugates, typically display critical surface tensions of above 30 $mN\,m^{-1}$. The practical result of these findings is that medical adhesives and/or priming agents must, to be efficacious on living substrates, have surface-active ingredients capable of depressing the interfacial tensions between adhesive and tissue to above 20 $mN\,m^{-1}$.

4. *Role of Adsorbed Films*

One of the most important observations about adhesion in relevant biological circumstances is that it never occurs without the involvement of preadsorbed films of biological macromolecules, primarily glycoproteins. These films are spontaneously adsorbed from the aqueous biological media in which the adhesive is placed. It has become obvious that all bulk materials allow spontaneous deposition and continuing buildup of proteinaceous material from the biological medium in the same relative amounts and at the same rates. The immediate supposition could be that differences in the surface chemistries of these various materials are thus effectively masked or obscured by an overcoating of essentially the same biological macromolecules, accumulated in the same amounts over the same exposure times. If this supposition were correct, subsequent cell contacts, obviously limited to the new exterior face of each protein-coated substrate, could no longer exhibit differential degrees of cell adhesion or cellular spreading. However, the empirical observation is that striking differences in adhesive behavior are noted.

Returning to the parameter of critical surface, tension as one indicator of continuing substrate differences (even after conditioning-film coverage), the changes of the critical surface tension have been measured for different test materials vs incubation time at physiological conditions. Such tests revealed that the highest-surface-energy substrate used, plasma-cleaned (glow-discharge-cleaned) glass, underwent dramatic decreases in critical surface tension within seconds of immersion into the living media.

On the other hand, various glass surfaces treated with different "siliconizing" compounds to lower their surface energy remained at their original critical surface tensions (i.e., 20–30 $mN\,m^{-1}$ within experimental error) in spite of the independently demonstrated accumulation of about the same amount of proteinaceous material.

Using an improved discriminator, based on measures of the spread areas of adherent living cells, different spread areas were found for the same cell types in the same tissue culture media. The cells exhibited 50% greater spread areas on the high-energy (plasma-cleaned) glass than on the siliconized specimens. The differentiation was even more striking when assessing the number of pseudopods, or irregular projections, on cells adherent to substrates varying in surface energy. The initial substrate surface properties (e.g., surface energy) appear to influence cell adhesion at a considerable distance from the original interface, that is, through the conditioning layer. It is crucial to note that all such interactions occur through, or are mediated by, the initially adsorbed glycoproteinaceous layers. It is necessary to understand, predict and control the qualities of this biological coating in order to achieve the desired improvements in medical adhesives.

5. *Observed Failure Mechanisms*

There are at least three basic reasons why medical adhesives fail: rejection by the host tissue or some element of it; failure of the physical form as a result of structural or load-bearing forces; and chemical attack on the adhesive by some part of the biological environment. Each of these alone is a possible route to failure, but all three often act simultaneously.

The phenomenon known as biorejection has at its root the mismatched contact with a biological system. The manifestation of rejection may take the form of the biosystem rapidly encapsulating or "walling off" the adhesive itself, thereby rendering it ineffective. This may occur by the simple deposition of biological materials as in the formation of a lesion around an implanted device, thereby causing a widespread inflammatory or immunological reaction. Both of these reactions have been characterized by a number of investigators.

Interaction of biological fluids or other living systems with an adhesive phase may also be chemical in nature. Encapsulating metallic devices with adhesive polymers usually helps to extend the serviceable lifetime of the device, but problems often occur with the adhesion of the encapsulant to the metal, leading to blistering, loss of encapsulating film integrity and device leakage. The basis for this mode of attack lies in the loss in dimensional stability of the encapsulating agent due to swelling, the permeation of water and ions through the film and the lack of sufficient mechancial strength to withstand forces in the biological environment.

Water and water vapor are known to penetrate certain organic films to different degrees. This penetration can lead to failure of a protected device by promoting corrosion or other secondary effects. The rate of penetrant invasion, a thermally activated process, depends on the nature of the penetrant, the nature of the material to be penetrated and temperature.

Biological fluids contain several species of inorganic ions that are known to cause damaging effects to sensitive solid-state electronic devices. These ions are able to penetrate adhesive layers and either contribute to the breakdown of bonding between the adhesive and the substrate (thereby leading to rapid corrosion and possible catastrophic failure) or enter the electronic device and interfere with its function. This frequently occurred in early models of heart pacemakers.

6. Current Research Needs

At present, many aspects of medical adhesives and their use require investigation. These may be summarized as follows.

(a) An attempt should be made to resolve the relative influence of surface energy vs surface texture of the medical substrates as controlling factors in obtaining long-term adhesion. Further, resolution of the relative roles of the solid–liquid interfacial tension vs the critical surface tension of the cured adhesives should be attempted.

(b) A program of tensile testing should be developed to overcome the remaining difficulties in obtaining both valid and useful results. It should be aimed at defining the potential utility of a variety of surface treatments, including application of various "coupling agents," in establishing superior adhesion. These tests should also be addressed to determining the influence of water, of salt solutions and of protein solutions or biological fluid simulants on sealant and adhesive bond strengths.

(c) An attempt should be made to establish and validate routine and inexpensive quality control checks for adhesive formulations.

(d) The critical surface properties of human tissues *in vivo*, after various time lags from their initial exposure for bonding, should be determined on children and adults of both sexes. The measurement techniques must be safe and relatively simple.

(e) Investigations should be made of the reactivity of the candidate adhesive monomers and polymers with substrates precoated with different protein films, to model more closely situations encountered clinically.

(f) Studies should be made of the failure surfaces produced when medical adhesives are separated from substrates of different character. These should include consideration both before and after exposure to salt solutions, blood, saliva or tissue fluids.

(g) Experimentation with the use of tailor-made interface modifiers, also known as coupling agents, should be encouraged. This approach has given excellent results in other systems where strong adhesive bonds to moist surfaces were required, such as glass-fiber-reinforced plastics.

(h) Surface chemical characterization should be undertaken for all monomers, polymers and resins now being considered for medical adhesive applications. Legislative mandates require knowledge of such properties before the approval of materials for human applications.

See also: Adhesive Joining: An Overview

Bibliography

Baier R E (ed.) 1975 *Applied Chemistry at Protein Interfaces*, Advances in Chemistry, Vol. 145. American Chemical Society, Washington, DC

Baier R E 1984 The basics of biomedical polymers: Interfacial factors. In: Gebelein C G (ed.) 1984

Polymeric Materials and Artificial Organs, ACS Symposium Series 256. American Chemical Society, Washington, DC, pp. 39–44
Baier R E, Meyer A E 1988 Implant surface preparation. *Int. J. Oral Maxillofac. Implants* 3: 9–20
Baier R E, Meyer A E, Natiella J R, Natiella R R, Carter J M 1984 Surface properties determine bioadhesive outcomes: Methods and results. *J. Biomed. Mater. Res.* 18: 337–55
Baier R E, Shafrin E G, Zisman W A 1968 Adhesion: Mechanisms that assist or impede it. *Science* 162: 1360–8
Causton B E 1990 Chemical adhesion in dental restoratives. In: Williams D (ed.) 1990 *Concise Encyclopedia of Medical and Dental Materials*. Pergamon, Oxford, pp. 101–6
Coombe E C 1990 Acrylic dental polymers. In: Williams D (ed.) 1990 *Concise Encyclopedia of Medical and Dental Materials*. Pergamon, Oxford, pp. 4–14
Fowkes F M 1987 Role of acid–base interfacial bonding in adhesion. *J. Adhes. Sci. Technol.* 1: 7–27
Glantz P-O, Baier R E 1986 Recent studies on nonspecific aspects of intraoral adhesion. *J. Adhes.* 20: 227–44
Kronenthal R L, Oser Z, Martin E (eds.) 1975 *Polymers in Medicine and Surgery*, Polymer Science and Technology, Vol. 18. Plenum, New York
Manly R S (ed.) 1970 *Adhesion in Biological Systems*. Academic Press, New York

R. E. Baier
[State University of New York, Buffalo, New York, USA]

Adhesives: Mechanical Test Methods

Adhesive bonding for joining materials began when the Egyptians used casein adhesives in fabricating mummy cases. However, adhesive bonding for load-bearing structures was developed only several decades ago; bonding of the plywood laminae of the British wooden Mosquito bomber of World War II may have been the first aviation application. In the USA, adhesives were developed in the early 1940s for the Consolidated Vultee B-36 bomber (Vinson 1980). Since that time, the use of adhesive bonding has increased steadily.

Since the early usage of adhesives for joining structural components, research has concentrated on the improvement of adherend surface preparations and tougher/stronger adhesives. Significant achievements have been accomplished in these areas, with processes and bonding materials now available that provide bonded joints which equal or exceed (on a joint overlap per unit area basis) the shear loading strength of commonly used mechanical fasteners. In recent years, there has been the development of computer routines which are used to analyze, design and optimize joints of many configurations, subject to mechanical, thermal and hygroscopic loads. However, utilization of these improved materials, processes and new analytical tools is limited since the stress–strain properties of adhesives under environmental and load conditions are generally ill defined. This technological gap exists because suitable test techniques for generating this type of data have not been developed.

Current test-specimen configurations either (a) introduce strain concentrations which are not characteristic of the bond applications, (b) require extremely precise measurements of very small strains or (c) attempt to solve points (a) and (b) by testing thick adhesive castings, which have properties significantly different from those of the adhesive as a thin film (Hughes and Rutherford 1968, Zabora et al. 1971). This is especially true in relation to the surface chemistry at the adhesive–adherend interface, due to the residual strains locked in a bonded joint during an elevated-temperature cure and the constraining Poisson's effect of the stiffer adherend on the relatively soft adhesive. Thus, these bulk adhesive specimens can be excluded from consideration as candidate test specimens, being unrepresentative of real structural application environments.

1. Adhesive Characterization

The complete characterization of an adhesive requires that its response to various load, time and environmental conditions be ascertained. The specific adhesive-property data required to predict the response of the adhesive in a bonded joint under typical structural loads and environments are

(a) shear and tensile moduli;

(b) ultimate shear, tensile and combined shear and tensile strength;

(c) ultimate shear and tensile strain;

(d) creep-compliance master curves in shear and tension;

(e) fatigue shear and tensile moduli that reflect the wearout and/or stability of the adhesive just prior to catastrophic failure;

(f) fatigue tensile and shear strength (strain) versus cycles of failure; and

(g) coefficients of thermal and hygroscopic expansion.

For an isotropic, homogeneous adhesive layer, the Poisson's ratio may be determined in place of the shear or tensile modulus.

2. Criteria for Test-Specimen Evaluation

To evaluate objectively the adequacy of the various test specimens available, from which adhesive static-, creep- or fatigue-property data can be obtained for use in design of bonded structures, the following criteria need to be considered:

(a) The specimen loading mode must be dominant and controllable—tension and shear modes should be separable and their magnitude easily controlled. A biaxial or triaxial stress state should be present in the adhesive.

(b) The specimen must be capable of providing quantitative results—modulus and strength data in representative environments should be obtainable.

(c) The test specimen must be representative of materials and processing in production—it should reflect the chemical, physical and mechanical response of an adhesive bonded between two relatively rigid adherends.

(d) A realistic analysis of the biaxial or triaxial stress state in the adhesive-bonded test specimen must be possible—in closed form or by finite-element analysis.

(e) Multiple-specimen preparation should be relatively easy and of nominal cost—uniform control of adhesive-bondline thickness is necessary for proper test-data interpretation.

(f) The bondline of the specimen should be easily inspectable by nondestructive evaluation (NDE) means for debonds, voids and air bubbles.

(g) Other considerations—in addition to the above criteria, the specimen should be readily usable to obtain fatigue and viscoelastic properties of the adhesive. A brief summary of their importance in the design process follows.

3. *Fatigue Response*

Adhesives fail in fatigue. They do so, by analogy with the better-studied metals, because of a cycle-by-cycle accumulation of damage resulting from repeated nonelastic strain. This strain can be further influenced by frequency of loading, temperature, humidity and strain-rate (viscoelastic) effects. One of the items required in the design process is the stress–strain behavior of the adhesive during its lifetime, at a prescribed fatigue load, frequency and in a controlled environment. Master curves are constructed of the fatigue stress–strain relationship for a particular adhesive, for a definite strain ratio and various combinations of environment and load frequency.

4. *Viscoelastic Response*

Under sufficiently high strain rates and/or elevated temperature and relative-humidity levels, most adhesives exhibit a viscoelastic response. This can result in a creep-rupture failure of the bonded structure, so it is necessary to obtain viscoelastic-response data. This requires a joint geometry within which the adhesive sees an approximately uniform stress state.

5. *Candidate Test Specimens*

A summary of the various candidate-test-specimen geometries already in existence versus the evaluation criteria is given in Table 1. From this table, one can obtain the list of test specimens best suited to obtain desired adhesive mechanical properties shown in Table 2.

6. *Single Lap Joint*

It is unfortunate that the primary standard for adhesive strength is the ASTM Test Standard D1002-72, "Strength Properties of Adhesives in Shear Tension Loading." The specimen is a single lap joint (see Fig. 1), comprised of two thin adherends, usually steel or aluminum, bonded together by an adhesive layer. The specimen is pulled in tension and the mean value of shear strength is computed to be the maximum tensile load P divided by the shear area ($L \times$ unit width).

Because of the thin adherends, the tensile in-plane load causes considerable bending in the joint, as seen in Fig. 1. This results in shear-stress peaks near the ends of the overlap, which go to zero at the free surfaces. It is also noted that the normal or peel stresses peak at the ends and are compressive over a sizeable region in the joint interior.

The adhesive shear stress of the ASTM test specimen is far from constant, thus negating the determination of a true shear strength. Even worse is the fact that the maximum tensile or peel stress in the adhesive can be greater than the maximum shear stress. So, failure may occur in tensile rupture rather than in shear. Thus, the specimen is not recommended for determining adhesive mechanical properties.

7. *Thick-Adherend Symmetric Lap Joint*

The thick-adherend symmetric lap joint (see Fig. 2) has been proposed (Renton 1976, 1978) as a viable specimen with which to obtain the effective thin-film-adhesive shear modulus, proportional limit stress and ultimate shear strength, as well as viscoelastic and fatigue shear properties. Adhesive properties from this specimen provide data readily usable in the numerous methods of analysis currently in use for determining stresses and deformations in both the adhesive and the adherends. Additionally, it can be used to determine the effect of adhesive thickness, surface roughness, tempera-

Table 1
Candidate test specimens

Evaluation criteria	Thick-adherend specimen	Scarf joint	Butt joint	Single lap joint	Napkin-ring specimen	Iosipescu specimen	Double lap joint	Peel specimen
Loading mode	tension	tension, shear	tension	shear	tension, shear	shear	shear	tension
Specimen can provide quantitative results	static viscoelastic fatigue	static viscoelastic fatigue	static viscoelastic fatigue	none	static viscoelastic fatigue	static viscoelastic	none	none
Representative of real structural applications	yes	yes	yes	yes	no	yes	yes	yes
Biaxial or triaxial stress state in adhesive	yes	yes	yes	yes	yes	yes	yes	yes
Specimen analysis performed	yes	yes	yes	yes	no	yes	yes	no
Adhesive thickness								
controllable	easy	modestly difficult	difficult	modestly difficult	modestly difficult	easy	easy	easy
measurable	yes	yes	yes	yes	yes—one side only	yes	yes	yes
Specimen fabrication								
ease of replicating specimen	easy	easy	modestly difficult	easy	difficult	modestly difficult	easy	easy
relative cost	low	low	modestly expensive	low	high	modestly expensive	low	low
NDE inspectability	easy	difficult	modestly difficult	easy	modestly difficult	modestly difficult	easy	easy

ture, strain rate, residual strains, corrosive environment and cyclic loading. Accurate measurement of adhesive deformation is a challenge which can be met by attaching a measurement device at the adhesive bondline (Hughes and Rutherford 1968, Renton 1976, 1978). The cost of specimen fabrication and testing is quite reasonable.

Table 2
Test specimens suited to obtain desired adhesive properties

Specimen	Type load	Mechanical properties obtainable
Thick-adherend lap joint	shear	static, viscoelastic, fatigue
Butt joint napkin ring	tension	static, viscoelastic, fatigue
Scarf joint	tension plus shear	static, viscoelastic, fatigue

Figure 1
Deflections induced in a thin adherend lap joint

Figure 2 presents the optimum dimensions of the adherend recommended to obtain a near-uniform adhesive shear stress over the bond length and assumes that adhesive deformation is measured near the bond–adherend interface at the centerline (points A and B).

8. *Butt Joint*

The butt joint specimen illustrated in Fig. 3 has been used to obtain adhesive tensile properties under static, fatigue and viscoelastic loadings. The specimen has been rigorously analyzed using elastic and

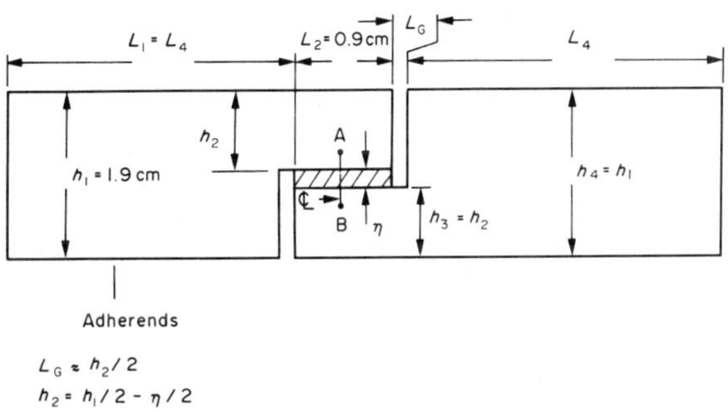

Figure 2
Geometry of thick adherend optimum adhesive test specimen

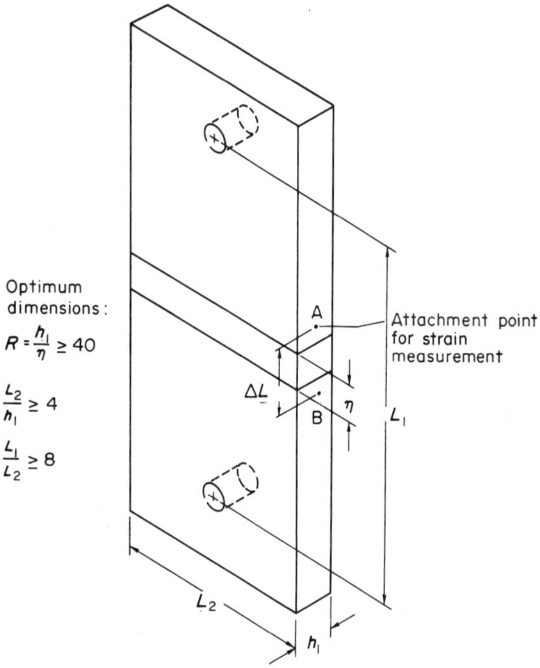

Figure 3
Geometry of tensile specimen optimum adhesive test specimen

photoelastic methods and has been verified by finite-element analysis (Renton 1978). A triaxial stress field is approximated in the center of the specimen when bending of the rigid adherends is negligible, an aspect ratio $(h_1/\eta) \geqslant 40$ is maintained and the Poisson's ratio of the adhesive is approximately 0.50. Optimum specimen dimensions are shown in Fig. 3.

Overall, the potential problems in using this specimen are adhesive thickness control, accurate adhesive deformation measurement of the bondline under load and the significance of eccentric load on the adhesive stress distribution. The cost is modest.

9. Scarf Joint

The scarf joint (see Fig. 4) has been proposed (Renton 1978) as a viable specimen with which to obtain the effective thin-film biaxial properties of

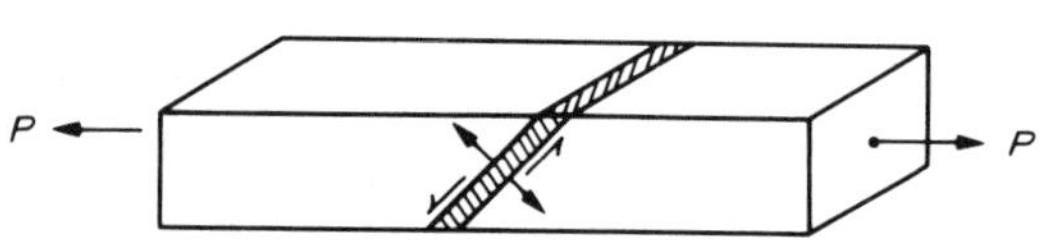

Figure 4
Scarf joint

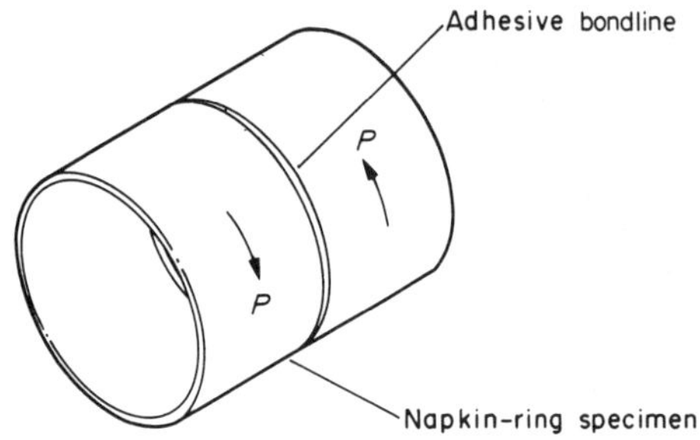

Figure 5
Napkin-ring test specimen

adhesives, as it can provide a controllable tension and shear-loading mode for static, viscoelastic and fatigue testing. This control is achieved by the scarf angle. While this joint is relatively easy to fabricate and of modest cost, there are potential problems in measuring adhesive bondline deformation accurately, and in bondline inspectability. However, the relatively new NDE technique of neutron radiography offers a realistic approach to resolving this problem. These specimens can also pose bondline thickness control and alignment problems.

Analytical and photoelastic studies (Renton 1978) have shown that the stress in the middle plane of the specimen is triaxial, with a superimposed shear stress.

In summary, by careful selection of the test-specimen geometry, one can obtain a near-uniform triaxial tensile-stress state and impart to the adhesive a controlled shear stress. An aspect ratio of 40 (up to 100 if the Poisson's ratio of the adhesive is $\geqslant 0.480$) is suggested to keep the edge effect to within one or two adhesive thicknesses of the bond edge. As the mechanical properties obtained are average properties of the adhesive constrained between the much stiffer elastic adherends, this minor edge effect should not appreciably affect the test results. Thus, the uniformity of the stress distribution over the bonded surface will enable the characterization of the linear and nonlinear viscoelastic and fatigue response of the adhesive.

10. Napkin-Ring Specimen

Many of the instrumented adhesive characterization tests performed to date have used the napkin-ring specimen (see Fig. 5). One of its principal attributes is that it can be used to determine the stress–strain response of an adhesive material in shear, tension or a combination thereof, for static, fatigue and viscoelastic loads. The use of the napkin-ring test piece can provide adhesive-shear-property data for typical bonded-joint thicknesses, but requires extreme care in loading to ensure pure torque. Additionally, the cost per test item is quite high and the specimen configuration is notably different from the geometrical configuration (Hughes and Rutherford 1968) in which structural adhesives are normally used. An accurate analysis of the stress (strain) distribution in the adhesive of this specimen has not been reported; it is believed to be biaxial or triaxial.

The specimen itself suffers from being difficult to fabricate to prescribed adhesive thicknesses, adhesive deformation measurement is difficult and control of the biaxial load into the specimen requires extreme care.

11. *Iosipescu Specimen*

The Iosipescu specimen (see Fig. 6) can be used to measure adhesive shear strength (Kutscha and Hofer 1969). The specimen has been analyzed by finite-element methods and the adhesive is in almost pure shear. However, to date displacements have not been measured; hence, it has not been used to measure the elastic modulus, the proportional limit or the strain to failure. Kutscha and Hofer recognized the potential of the Iosipescu specimen for measuring the shear strength of adhesives. They noted the difficulty in fabrication, but remarked that the difficulty is less than fabricating tubular torsion (i.e., napkin-ring) specimens and, with proper instrumentation, the specimen could be used to measure the shear modulus.

12. *Other Specimens*

In principle, the balanced double lap joint could be used in a shear mode with modestly thick adherends, but with relatively high fabrication and NDE inspection costs. The selection of the geometry of the joint would be extremely important in obtaining a uniform-shear-stress condition.

Several other test specimens to characterize the structural response of adhesives are available. The inadequacies of the block shear (ASTM-D905), crosstab tension and glueline cleavage tests have been reported (Renton 1978). It is important to note that the various peel tests that presently exist are representative of gross combinations of shear and tension. The main difficulty with these peel tests is that the adhesive properties under shear plus tension are grossly affected by the adherend. This factor makes the present peel tests unusable from the standpoint of obtaining precise quantitative mechanical-property data applicable to bonded-joint analysis.

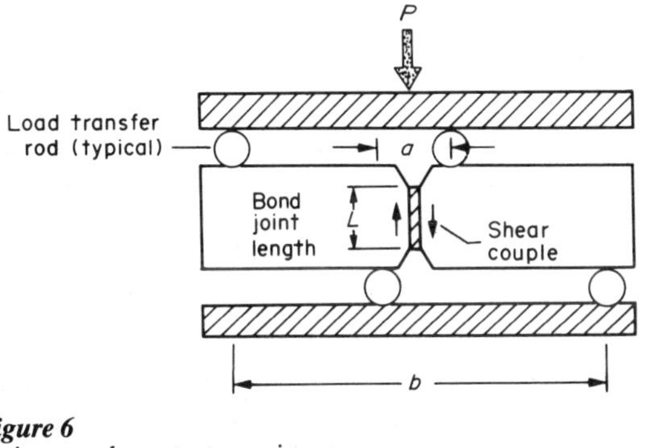

Figure 6
Iosipescu shear-test specimen

13. *Test-Methodology Criteria*

Once the test specimen has been selected, it is mandatory that simple, reliable test procedures be employed to minimize data scatter and cost. Moreover, proper conditioning of test specimens should ensure equilibrium of the bondline adhesive with the environment. Means to measure temperatures from −15 °C to 175 °C and relative humidities from 10% to 100% accurately before and during the performance of mechanical tests should be available. Sensitivity requirements for the adhesive deformation measurement device should be established, to determine the adhesive properties to within acceptable accuracy limits. Moreover, the deformation measurement device should be easy to attach to the test specimen, simple, stable with respect to time and environment, durable and relatively inexpensive.

14. *Test-Data Reporting*

A major deficiency in adhesive characterization results from the arbitrary test and report procedures followed by many investigators. Incomplete information makes data difficult to interpret and data comparisons impossible. This may be due in part to the lack of ASTM test specifications in this area.

15. *Adhesive Deformation Measurement*

The small dimensional changes that occur make the measurement of adhesive deformation during loading very difficult. For example, in a tensile test, 1% strain in a 125 μm bondline is 1.25 μm. Careful workers in the field (Renton 1978) express the need for a minimum accuracy in measurement of bondline dimensional changes of 25 μm (0.2%). This accuracy must be maintained over a very wide temperature and humidity range and for long periods of time. These conditions impose very strict requirements on the stability of the strain sensor.

Studies to date indicate that LVDTs, extensometers, Tuckerman gauges, air-gap capacitance gauges and strain gauges are the sensors most commonly used; of these, LVDTs and air-gap capacitance gauges show the most promise.

See also: Adhesion: Fundamental Principles; Adhesive Joining: An Overview

Bibliography

Brown R P 1986 Adhesion corrosion and staining. *Physical Testing of Rubber*, 2nd edn. Elsevier, Amsterdam, Chap. 18

Hughes E J, Rutherford J L 1968 *Study of Micromechanical Properties of Adhesive Bonded Joints*. Technical Report No. 3744. General Precision, Little Falls, NJ

Kutscha D, Hofer K E Jr 1969 *Feasibility of Joining Advanced Composite Flight Vehicles*, Report AFML-TR-68-391. US Air Force Materials Laboratory, Dayton, OH

Lehman G M, Hawley A V 1969 *Investigation of Joints in Advanced Fibrous Composites for Aircraft Structures*, Vol. 1, AFFDL Report No. TR-69-43. US Air Force Flight Dynamics Laboratory, Wright-Patterson Air Force Base, Dayton, OH

Renton W J 1976 The symmetric lap joint—What good is it? *Exp. Mech.* November: 409–16

Renton W J 1978 *Structural Properties of Adhesives*, Vol. 1, Final Report AFML-TR-78-127. US Air Force Materials Laboratory, Dayton, OH

Rose J L 1986 Adhesive bonds: Nondestructive evaluation. In: Bever M B (ed.) 1986 *Encyclopedia of Materials Science and Engineering*. Pergamon, Oxford, pp. 70–3

Vinson J R 1980 State of technology in adhesively bonded joints in composite structures. *Emerging Technologies in Aerospace Structures Design, Structural Dynamics and Materials*. American Society of Mechanical Engineers, New York.

Zabora R F, Clinton W W, Bell J E 1971 *Adhesive Property Phenomena and Test Techniques*, AFFDL Report No. AD-729873. US Air Force Flight Dynamics Laboratory, Wright-Patterson Air Force Base, Dayton, OH

W. J. Renton
[Vought Corporation, Dallas, Texas, USA]

Aging of Elastomers

Aging may be defined as unwanted deterioration in the physical properties of an elastomer resulting from the action of heat, radiation, chemical agents or mechanical work, either separately or in combination. The chemical agents that may affect elastomer properties include oxygen, ozone, oxides of nitrogen or sulfur, water, acids and bases.

The term degradation in its broadest sense is synonymous with aging. In the present context, however, it refers only to reduction in the molecular weight of polymer chains and scission of crosslinks. Thus, degradation is regarded as one manifestation of aging, the other being crosslinking during storage or service.

1. Thermal Aging

Heat accelerates the aging effect of chemical agents. Since elastomers are commonly used in the presence of atmospheric or dissolved oxygen, purely thermal aging is generally not a cause for concern. Thermal aging is controlled by the stability of the bonds in a polymer. Hydrocarbon-based elastomers such as polyisoprene or polybutadiene pyrolyze above 300 °C. If hydrogen atoms are replaced by fluorine, heat resistance is increased, since the carbon–fluorine bonds are stronger than carbon–hydrogen bonds. Many polymers have been designed for resistance to high temperatures (Cassidy 1980). These include inorganic elastomers such as polyphosphazenes which are stable to 310–410 °C in argon and tenboron cage siloxane polymers which are stable to 570–700 °C in argon.

The center of a large rubber article is substantially free of oxygen and aging in this location is controlled by the stability of the crosslinks. Carbon–carbon crosslinks resulting from peroxide or radiation vulcanization and carbon–sulfur–carbon crosslinks form so-called "efficient" vulcanization are thermally stable. Polysulfidic crosslinks mature on heating and produce further crosslinking which results in increased hardness and permanent deformation (set) in articles under compression.

The breakdown of elastomers at high temperatures is not always undesirable. Elastomers such as nitrile rubber filled with asbestos fiber and reinforced with a thermosetting phenolic resin, and also silicone rubber, have been used as ablative materials to protect rockets on reentry into the earth's atmosphere. The ideal breakdown yields a residual char structure to protect the rocket and evolves gases which promote cooling.

2. Oxidative Aging

Oxygen in conjunction with thermal, radiant or mechanical energy is the principal cause of elastomer aging. It is well established that, in hydrocarbon-based elastomers, oxidation is a free-radical process involving the cleavage of carbon–hydrogen bonds. In solution the following kinetic scheme applies:

Initiation:	production of $R\cdot$ or $RO_2\cdot$ radicals		(1)
Propagation:	$R\cdot + O_2 \rightarrow RO_2\cdot$		(2)
	$RO_2\cdot + RH \rightarrow ROOH + R\cdot$		(3)
Termination:	$2R\cdot \rightarrow$	nonradical products	(4)
	$R\cdot + RO_2\cdot \rightarrow$	nonradical products	(5)
	$2RO_2\cdot \rightarrow$	nonradical products	(6)

In unsaturated elastomers aging does not involve addition of oxygen to a double bond. It arises from hydrogen abstraction from the adjacent (allylic) methylene group. At atmospheric concentrations of oxygen, the reaction given in Eqn. (2) is rapid and the hydrogen extraction step (Eqn. (3)) is rate controlling. The rate depends both on the reactivity of the $RO_2\cdot$ radical and the ease of hydrogen abstraction. It is greater in natural rubber than in polybutadiene because of the electron-donating effect of the methyl group. It is greater in polybutadiene than in polychloroprene (poly(2-chloro-1,3-butadiene)) because of the electron-withdrawing effect of chlorine. Ease of hydrogen abstraction decreases in the order

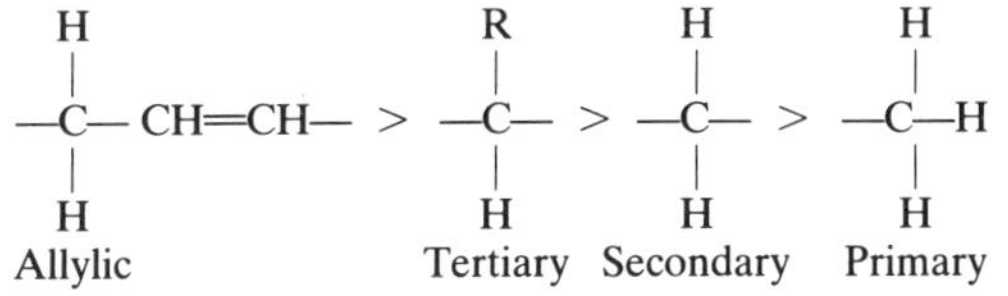

The pendant vinyl structure arising from 1,2 addition of butadiene in styrene–butadiene rubber (SBR) or nitrile–butadiene rubber (NBR) (see later) is particularly subject to abstraction of hydrogen, which is both tertiary and allylic:

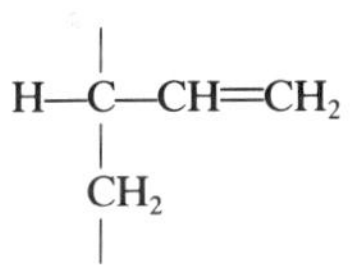

In highly purified elastomers, initiation (Eqn. (1)) proceeds very slowly but, since propagation (Eqn. (3)) regenerates a radical and produces hydroperoxide, the rate of initiation increases rapidly by

$$ROOH \rightarrow RO\cdot + \cdot OH \quad (7)$$

or

$$2ROOH \rightarrow RO_2\cdot + RO\cdot + H_2O \quad (8)$$

Thus, the overall reaction rate increases sharply with time. The origin of initiators in highly purified elastomers has largely been ignored. Direct interaction of molecular oxygen with hydrocarbon has been postulated, that is

$$RH + O_2 \rightarrow R\cdot + HO_2\cdot \quad (9)$$

This is unlikely since it would be endothermic to the extent of 125–190 kJ mol^{-1}.

Addition of oxygen to double bonds to give polyperoxides has also been postulated. A further possibility is reaction with singlet oxygen, which has been shown to greatly increase the rate of the oxidation of *cis*-polybutadiene. In processing or storage, singlet oxygen may arise from the interaction of ozone at double bonds:

$$\sim CH{=}CH \sim + O_3 \rightarrow \sim \underset{\diagdown O \diagup}{CH{-}CH} \sim + {}^1O_2 \quad (10)$$

In urban atmospheres singlet oxygen may be present in polluted air:

$$NO_2 \xrightarrow{\Delta} NO + O \quad (11)$$

$$NO + O_3 \rightarrow NO_2^* + O_2 \rightarrow NO_2 + {}^1O_2 \quad (12)$$

It has been postulated that, in the solid phase, the termination reaction (Eqn. (6)), which is the principal termination reaction in solution, could not occur. In any case, those elastomers which are subject to oxidation always contain stabilizers that are added during finishing or are present in the natural product. In this case the termination step is

$$RO_2\cdot + AH \rightarrow RO_2H + A\cdot \quad (13)$$

If the chain-breaking stabilizer (added during finishing) or the antioxidant AH (added during vulcanization) is efficient, the kinetic chain length will be reduced to one but hydroperoxide will be regenerated. For complete protection, it appears that a peroxide-decomposing additive should be used. In practice, hydroperoxide-decomposing antioxidants are too short lived to be used alone although some are believed to act by catalytic decomposition. Synergistic combinations of peroxide decomposers, such as organic phosphites, and chain-breaking additives, such as amines or phenols, are frequently used in elastomer stabilization. Combinations must be chosen with care since some are antagonistic. Antioxidants may also be incompatible with blacks and with vulcanization products.

Antioxidants may take part in other stages of the oxidation process with deleterious effect. Oxidation of the antioxidant may initiate chains:

$$\cdot O{-}O\cdot + AH \rightarrow HOO\cdot + A\cdot \quad (14)$$

Those compounds which most easily donate hydrogen to $RO_2\cdot$ also donate it most easily to oxygen. Since antioxidants may give rise to initiation, there is an optimum level of antioxidant for maximum efficiency.

Ideally the $A\cdot$ radical in Eqn. (13) should be inactive. Frequently it initiates new chains:

$$A\cdot + RH \rightarrow AH + R\cdot \quad (15)$$

The reactions discussed previously describe the oxidation of elastomers but do not explain the change in physical properties on aging. In practice some rubbers (such as natural, butyl and ethylene–propylene rubbers) degrade and soften; others (such as styrene–butadiene, polybutadiene, polychloroprene and nitrile) crosslink and harden. Generally, both degradation and crosslinking occur at the same time. In vulcanizates, the amount of degradation can be determined from the loss in stress in a specimen aged under constant strain (continuous stress relaxation), while the balance between degradation and crosslinking is indicated by the change in stress of a specimen which is aged unstrained and periodically extended to a constant elongation. Oxidative aging is dependent on oxygen concentration. Thus, the physical properties of a heat-resistant NBR vulcanizate deteriorate within 7 d in air at 150 °C, but they are still acceptable after 28 d in oil at 150 °C and after 150 d at this temperature if the oil has been purged with nitrogen.

The oxidative crosslinking of unsaturated elastomers is believed to occur through addition of $RO_2\cdot$ radicals to double bonds:

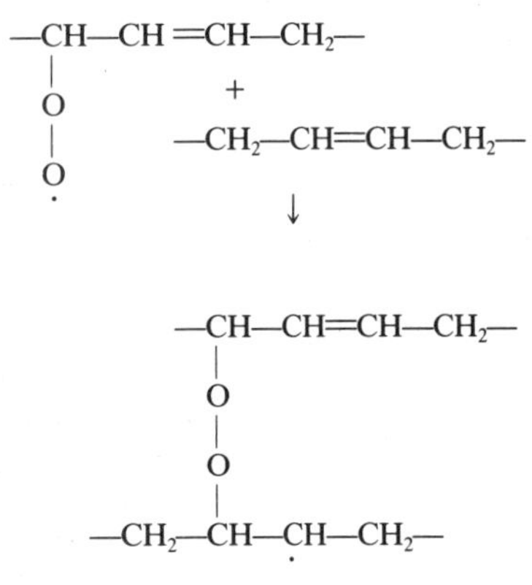

(16)

The involvement of $RO_2\cdot$ radicals in this process is indicated by the observation that radical-trapping antioxidants prevent crosslinking more effectively than those which decompose peroxides. The addition of $RO_2\cdot$ produces a new radical which can continue a chain reaction. At pressures below atmospheric, $R\cdot$ radicals may add to double bonds before they react with oxygen, and $R\cdot$ and/or $RO\cdot$ radicals may combine directly. Although the radicals produced during the oxidation of saturated polymers, such as ethylene–propylene rubber, have no double bonds to which to add, they may react with each other. In butadiene-based polymers, scission could occur following hydroperoxide formation at a pendant vinyl group:

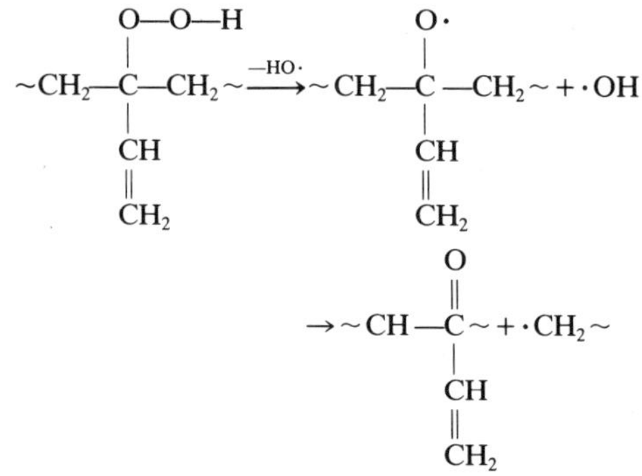

(17)

Several mechanisms have been proposed for the scission reactions in natural or synthetic polyisoprene that produce levulinaldehyde, formic acid, acetic acid and carbon dioxide. The mechanism proposed by Bevilacqua (Conley 1970) is shown in Scheme 1.

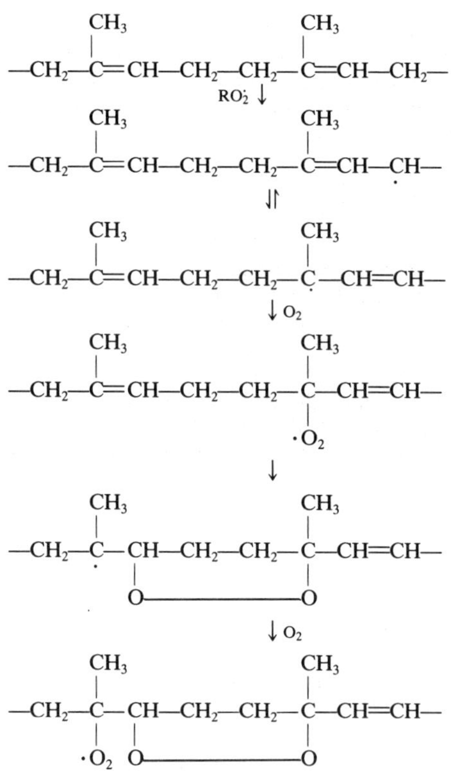

Scheme 1

3. *Metal–Ion Catalysis of Oxidative Aging*

Metal–ion catalysis of aging as it applies to rubber has been reviewed by Dunn (1977, 1978). The effect of metal ions is to catalyze the breakdown of hydroperoxides:

$$ROOH + M^{n+} \rightarrow RO\cdot + OH^- + M^{(n+1)+} \quad (18)$$

$$ROOH + M^{(n+1)+} \rightarrow ROO\cdot + H^+ + M^{n+} \quad (19)$$

This is equivalent to a bimolecular breakdown of hydroperoxide:

$$2ROOH \rightarrow RO\cdot + ROO\cdot + H_2O \quad (20)$$

According to this mechanism, the relative activity of a series of metals (cobalt, manganese, zinc, lead) should depend on their redox potentials and be independent of the substrate. In fact, comparisons of the activity of a number of metals in *cis*-polyisoprene, polybutadiene (high *cis*), polybutadiene (low *cis*), solution SBR, ethylene–propylene terpolymer, poly(propylene oxide) and butyl and nitrile rubbers have shown a different order of activity in each substrate. In addition, different workers have ranked the catalytic activity of metals differently in the same substrate. Of the metals tested, only cobalt always exhibits a strong pro-oxidative effect. Copper, iron and manganese, which are well-known aging catalysts, do not always act in this manner, and lead and zinc, traditionally inactive as catalysts, are not always inert in elastomer aging. The true catalyst may be a complex between the metal and the elastomer or some impurity in the elastomer. Synergistic effects between metals complicate the situation further since commercial elastomers generally contain traces of several metals. Cobalt appears to act as a positive synergist in combination with any other metal while calcium always has a zero or negative synergistic effect. Copper, iron and manganese have been found to enhance the catalytic effect of some metals and to reduce that of others. It should be noted that certain transition metal complexes prevent aging. Copper and iron dialkyl dithiocarbamates are powerful antioxidants, while nickel dialkyl dithiocarbamates are used to prevent photochemical degradation.

4. *Photochemical Aging*

The uv-initiated aging of elastomers is generally not a matter for serious concern since many vulcanizates contain high levels of carbon black, which serves as a very effective light screen. Light-colored vulcanizates and uncrosslinked thermoplastic elastomers exposed to uv light undergo surface degradation. Purified elastomers do not absorb strongly in the near uv but the absorption by hydroperoxides is 300–500 times greater. Thus, once a trace of oxidation has occurred, light is readily absorbed. In addition, carbonyl groups formed as a result of degradation also absorb at about 290 μm.

The net effect of uv light is to produce initiating radicals:

$$ROOH \xrightarrow{h\nu} RO\cdot + \cdot OH \quad (21)$$

In the presence of oxygen, propagation then proceeds as shown in Eqns. (2) and (3), and the overall effect on the elastomer is similar to that resulting from thermal aging. In the absence of oxygen, isomerization and degradation occur in unsaturated elastomers such as *cis*-polyisoprene or 1,4-polybutadiene.

Certain colorless additives such as *o*-hydroxybenzophenone, salicylic acid esters and *o*-hydroxyphenylbenzotriazole may be used to act as light absorbers, which strongly absorb the light and dissipate the energy harmlessly. Alternatively, nickel chelates of thiobisphenols and certain oximes serve to quench excitation energy in the elastomer after it has been absorbed. Since the photochemical oxidation of elastomers proceeds by a radical-chain reaction, conventional antioxidants act synergistically with uv absorbers once the reaction has been initiated.

5. *Aging Due to Nuclear Radiation*

The effect of nucelar radiation on elastomers is similar to that of uv light but it is much more severe and is capable of breaking bonds in the molecule. In the absence of oxygen, radiation crosslinks most rubbers but degrades polysulfide and butyl rubbers and vinylidene fluoride–perfluoropropylene copolymer.

6. *Degradation by Atmospheric Pollutants*

All elastomers containing unsaturation in the main chain, as well as polysulfide rubbers and certain polyurethanes, tend to crack when stressed and exposed to ozone. Ozone is present in the atmosphere and is particularly prevalent in areas subject to smog since it can be formed by interaction of NO_2 and peroxyacyl radicals. The product of ozonolysis is believed to be an ozonide which is produced by

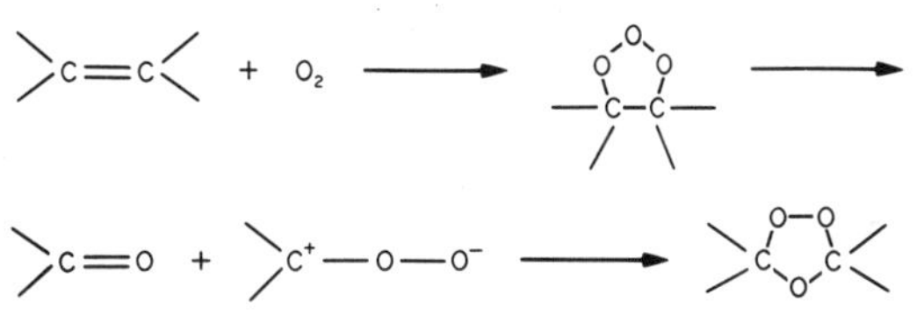

(22)

Ozone is known to attack unstretched rubbers but cracking occurs only when the rubber is under stress and the ozonide breaks down. In any elastomer there is a critical stress below which cracks will not grow. This depends on the size of deliberate or accidental cuts in the sample and on the square root of its Young's modulus. It does not depend on sample thickness, ozone concentration or temperature. On the other hand, the rate of crack growth does depend on ozone concentration, temperature and plasticizer content.

Although polychloroprene is as prone to ozone attack as natural rubber, polybutadiene or styrene–butadiene rubber, it does not crack because ozone reacts at the surface to form a soft flexible film. Butyl rubber is fairly resistant to ozone attack. This resistance is believed to arise from chain immobility rather than a shortage of unsaturated sites since butyl degrades as rapidly as natural rubber if it is exposed to ozone at elevated temperature. Ethylene–propylene terpolymers, being free of main-chain unsaturation, are not degraded by ozone.

Since ozone does not penetrate deeply, samples in storage may be protected simply by wrapping. Addition of of microcrystalline wax which blooms to the surface serves to protect elastomers in static service. Under dynamic conditions, arylenediamines are most effective in reducing the rate of crack growth, but unfortunately these are highly colored and staining. Light-colored materials, such as certain substituted thioureas and 6-ethoxy-2,2,4-trimethyl-1,2-quinoline, are much less effective. Dialkyl-*p*-phenylenediamines have been found especially useful in increasing critical stress.

Nitrogen dioxide may attack hydrocarbon elastomers as follows:

$$RH + NO_2 \rightarrow R\cdot + HNO_2 \qquad (23)$$

$$R\cdot + NO_2 \rightarrow RNO_2 \qquad (24)$$

Such reactions may be followed by scission and Eqn. (23) serves to initiate oxidative aging. In unsaturated elastomers, NO_2 can also add to double bonds with the production of free radicals:

$$\underset{\substack{|\\R}}{-C}=\underset{\substack{|\\H}}{C}- + NO_2 \rightarrow \underset{\substack{|\\R}}{-\dot{C}}-\overset{\substack{NO_2\\|}}{\underset{\substack{|\\H}}{C}}- \qquad (25)$$

Natural gas fumes contain NO_2, and latex foam based on unsaturated rubber is subject to degradation by them. Sulfur dioxide also attacks elastomers when it is activated by light, or by heavy-metal catalysts.

7. *Mechanical Degradation*

This complex subject has been reviewed by Dunn (1977) and Casale and Porter (1978). The mechanical degradation of unvulcanized rubber (mastication) is generally advantageous and it is used to render high-molecular-weight polymers more workable. The mechanical degradation of vulcanizates to produce reclaimed rubber is also advantageous but other mechanical effects in vulcanizates, such as flex cracking and fatigue, are usually undesirable.

In thermally oxidizable polymers, such as natural rubber, the amount of degradation on an open mill decreases with increased temperature to a minimum value, then increases at still higher temperatures, reflecting the balance between decreased shear and increased oxidation as temperature is increased. The temperature of least breakdown is about 115 °C for natural rubber and about 160°C for the less oxidizable ethylene–propylene–diene rubber. Mastication is found to have the greatest effect on high-molecular-weight components. *Cis*-polybutadiene breaks down less readily than natural rubber, an effect that has been attributed to the absence of a high-molecular-weight fraction and also to the absence of stiffening by stress crystallization. Nitrile rubber breaks down more readily at higher acrylonitrile content because of stiffening. Breakdown of nitrile is decreased by the considerable heat generated when it is mixed. Butadiene-based rubbers tend to gel when masticated in the absence of air because the radicals produced are not resonance stabilized and tend to add to double bonds. Fatigue results when, after repeated deformation, the elongation locally exceeds the elongation at break of the material.

Vulcanizates having polysulfidic or other shear labile crosslinks have superior fatigue life to those containing only carbon–carbon or monosulfide links. The better fatigue resistance results from reformation of the labile crosslinks after they have been severed. Since the radicals formed during mechanical breakdown react with oxygen and promote oxidative breakdown, fatigue is reduced by free-radical-trapping antioxidants. The most effective additives are arylalkyl- and dialkyl-*p*-phenylenediamines.

See also: Elastomers: An Overview

Bibliography

Bamford C H, Tipper C F H 1975 *Degradation of Polymers*. Elsevier, Amsterdam

Barnard D, Lewis P M 1988 Oxidative aging. In: Roberts A D (ed.) 1988 *Natural Rubber Science and Technology*. Oxford University Press, Oxford, Chap. 13

Casale A, Porter R S 1978 *Polymer Stress Reactions*. Academic Press, New York

Cassidy P E 1980 *Thermally Stable Polymers: Syntheses and Properties*. Dekker, New York

Conley R T (ed.) 1970 *Thermal Stability of Polymers*. Dekker, New York

Dunn J R 1977 Aging and degradation. In: Saltman W M (ed.) 1977 *The Stereo Rubbers*. Wiley, New York, pp. 511–82

Dunn J R 1978 Unsolved problems in polymer degradation. *Rubber Chem. Technol.* 51: 686–703
Dunn J R 1981 Stabilisation of vulcanised NBR, butyl and halobutyl rubbers. In: Scott G (ed.) 1981 *Developments in Polymer Stabilisation,* Vol. 4. Applied Science, London, pp. 223–45
Hawkins W K 1984 *Polymer Degradation and Stabilization.* Springer, Berlin
Hawkins W L (ed.) 1972 *Polymer Stabilization.* Wiley, New York
Jellinek H H G 1978 *Aspects of Degradation and Stabilization of Polymers.* Elsevier, Amsterdam
Ranby B G, Rabek J F 1975 *Photodegradation, Photo-Oxidation and Photostabilization of Polymers: Principles and Applications.* Wiley, New York
Reich L, Stivala S S 1971 *Elements of Polymer Degradation.* McGraw-Hill, New York
Sieron J K, Spain R G 1968 Elastomers. In: Rosato D V, Schwartz, R T (eds.) 1968 *Environmental Effects in Polymeric Materials.* Interscience, New York, pp. 1617–75
Winslow F H 1977 Environmental degradation. In: Schultz J M (ed.) 1977 *Treatise on Materials Science and Technology 10: Properties of Solid Polymeric Materials,* Pt. B. Academic Press, New York, pp. 741–76

J. R. Dunn
[Polysar, Sarnia, Canada]

Aircraft and Aerospace Applications of Composites

Fibrous polymer composites have found applications in aircraft from the first flight of the Wright Brothers' *Flyer 1*, in North Carolina on December 17th 1903, to the plethora of uses now enjoyed by them on both military and civil aircraft, in addition to the more exotic applications on space launcher vehicles and satellites. Their growing use has arisen from their high specific strength and stiffness, when compared to the more conventional existing and developing materials, and the ability to shape and tailor their structure to produce more aerodynamically efficient structures.

While fibrous polymer composites can, and will in the future, contribute up to 40–50% of the structural mass of an aircraft, the development of conventional alloys, aluminum–lithium alloys and the emerging metal and ceramic matrix composites will find increasing use, the latter in the more aggressive environments to which reentry vehicles and vertical takeoff aircraft will be subjected.

1. Composites For Aircraft Applications

Glass-fiber-reinforced composites have been used on military aircraft dating from 1940, but their poor relative specific stiffness has prevented them from extending the foothold they have found on fairings, doors and so on to the primary structural applications of wings, stabilizers and major fuselage sections. Aramid fibers introduced in the 1960s found parallel applications with glass fibers, but their lack of specific stiffness and poor compressive strength limited their use, despite the tolerance to damage that composites utilizing these fibers can afford.

The adoption of composite materials as a major contribution to aircraft structures followed on from the discovery of carbon fiber at the Royal Aircraft Establishment at Farnborough, UK, in 1964. However, not until the late 1960s did these new composites start to be applied, on a demonstrator basis, to military aircraft. Examples of such demonstrators were trim tabs, spoilers, rudders and doors. With increasing application and experience of their use came improved fibers and matrix materials resulting in composites with improved properties, allowing them to displace the more conventional materials—aluminum and titanium alloys—from primary structures.

1.1 Structural Properties of Composites

Whereas composite strength is primarily a function of fiber properties, the ability of the matrix to both support the fibers and provide out-of-plane strength is, in many load situations, equally important. The aim of the material supplier is to provide a system with a balanced set of properties. While improvements in fiber and matrix properties can lead to improved laminate properties, the all-important field of fiber–matrix interface must not be neglected.

All laminate properties, with the exception of those relying on interlaminar strength and stiffness, are almost directly proportional to the basic strength of the fiber. As laminate stiffness also follows this relationship, then clearly any improvement in fiber properties will almost inevitably lead to a product having improved laminate properties with a consequential improvement in their application to aerospace structures.

(*a*) *Effects of fiber properties.* The early carbon fibers, following on from the Farnborough discovery, exhibited properties moderate by current standards and improvements have occurred both in stiffness, strength and a combination of the two. Figure 1 shows a carpet plot of fiber strength against modulus with an indication of the years in which fibers were developed. While it is difficult to be specific as to the direction in which developments should best take place, the trend of development required for military aircraft would be one in which both properties were increased simultaneously along the "ideal" direction indicated. Satellite applications, in contrast, benefit from the use of high fiber modulus, improving stability and stiffness for reflector dishes, antennas and their supporting structure.

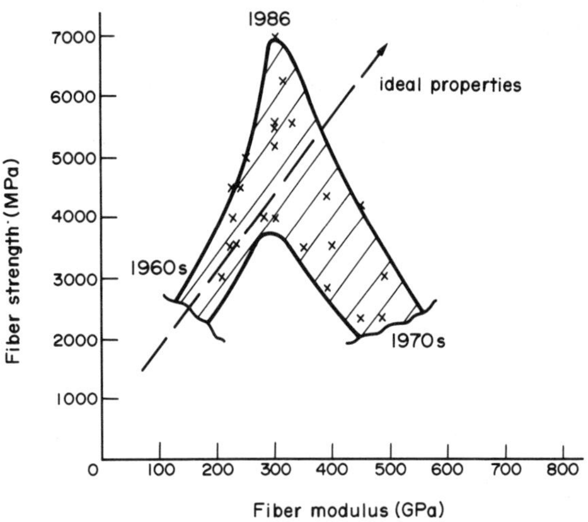

Figure 1
Tensile strength vs tensile modulus for carbon fiber

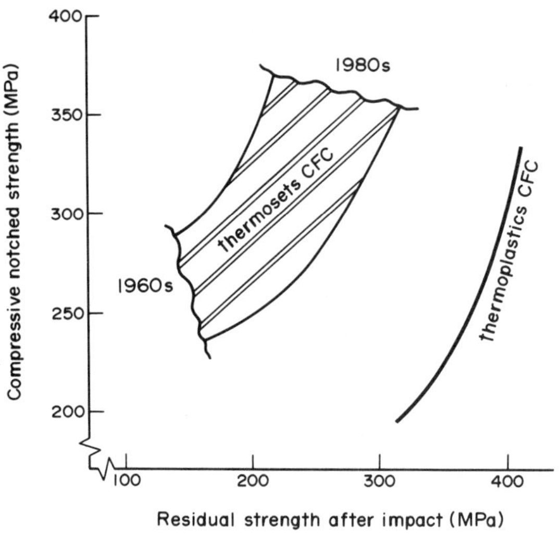

Figure 2
Comparison of notched compression and residual strength after impact of carbon-fiber composite (CFC)

(*b*) *Effects of matrix properties*. In order to produce a laminated structural element the fibers have to be bonded one to the other and, while thermoplastic materials are becoming available, the more conventional matrix material is a thermosetting epoxy. The matrix material is the Achilles' heel of the system and limits the fiber from exhibiting its full potential in terms of laminate properties. The matrix performs a number of functions, among which are stabilizing the fiber in compression, translating the fiber properties into the laminate, minimizing damage due to impact by exhibiting plastic deformation and providing out-of-plane properties to the laminate.

Matrix-dominated properties are reduced when the glass transition temperature is exceeded and, whereas with a dry laminate this is close to the cure temperature, the inevitable absorption of moisture reduces this temperature and hence limits the application of most high-temperature-cure thermoset epoxy composites to less than 120 °C.

The first generation of aerospace composites introduced in the 1960s and 1970s utilized brittle-matrix systems leading to laminates with a poor tolerance to low-energy impact. These impacts can be caused by runway debris thrown up by the aircraft wheels or the impacts occurring during manufacture and subsequent servicing operation. Although the emerging toughened epoxy systems provide improvements in this respect, they fall far short of the damage tolerance provided by the new solvent-resistant thermoplastic materials. A measure of damage tolerance is the laminate compression strength after impact and this property is plotted on the abscissa of Fig. 2 with the laminate notched compression strength on the ordinate. The ideal solution is to provide a material exhibiting equal properties and it can be seen that whereas the thermoplastic systems are tougher they have not capitalized on this by yielding higher notched compression properties.

(*c*) *The fiber–matrix interface*. This, the third factor governing strength, is considered by many to be crucial in maximizing the degree to which the fiber properties can be translated into the laminate, be the matrix thermosetting or thermoplastic. The interface between the fiber and matrix may be less than 1 μm yet excessive bonding between the fiber and matrix will result in poor notched tensile strength while an inadequate bond will result in poor interlaminar properties. Conflict therefore exists and the designer must select the material most nearly meeting his/her requirements.

2. *Alternative Fabrication Materials*

Of two materials with identical mechanical properties, the one with the lowest density will produce the minimum mass structure if all or part of the material is stability designed. As a consequence, carbon-fiber composites have an advantage over the more conventional materials used in aircraft fabrication. Composite materials are displacing the 2000 and 7000 series aluminum alloys from their dominant position in aerospace, but emerging metallic materials have the potential to challenge the growing composite applications. Aluminum-lithium alloys, again a development from the Royal Aircraft Establishment, Farnborough, UK, are offering improvements in strength with a parallel reduction

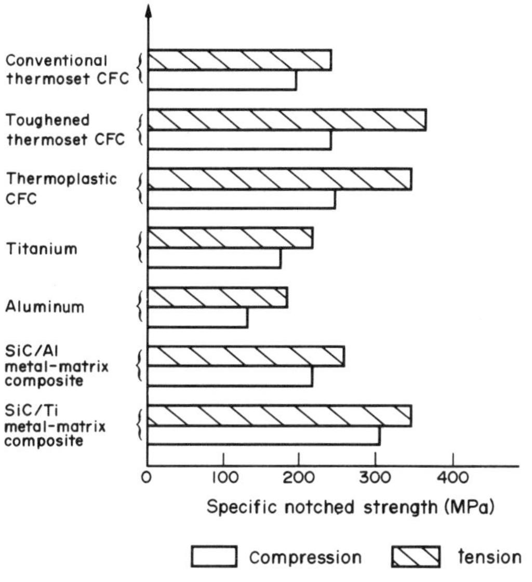

Figure 3
Structural material strength comparison

in density. These, in their own right, offer a challenge but when reinforced by ceramic particles, such as silicon carbide, yield specific properties approaching those of modern composites. However, these are prohibitively expensive and difficult to manipulate and will only find application in niche areas in the near future. A comparison has been made, and shown on Fig. 3, of the specific notched tensile strength of various aerospace materials including both thermoset and thermoplastic materials.

3. *Implications of Composites on Design*

Aircraft design from the 1940s has been based primarily on the use of aluminum alloys and as such a plethora of data and experience exists to facilitate the design process. With the advent of laminated composites exhibiting anisotropic properties the methodology of design had to be reviewed and, in many areas, replaced. It is accepted that designs in composites should not merely replace the metallic alloy but should take advantage of exceptional composite properties if the most efficient designs are to evolve.

Hitherto the material properties for direct use by the designer had been available and his/her function was, based on a limited number of properties, to ensure that the structure offered sufficient strength and stiffness. With composite materials, however, the extra function of designing the material from a set of laminae properties has to be undertaken. Although in many instances laminae properties are available and documented for strength, stiffness and stability, stacking sequences have to be compromised to ensure continuity and compatibility of structure. An example of the complexity of the process is the derivation of the bending stiffness of an anisotropic laminate which involves the manipulation of a 6×6 matrix, a task which can only be peformed by computer. Although these complexities lengthen the design process, they are more than compensated for by the mass savings and improvements in aerodynamic efficiency that result.

3.1 Aerodynamic Improvements

Because of the limitation of metal-manipulation techniques, the majority of aircraft control–lift surfaces produced have a single degree of curvature. Improvements in aerodynamic efficiency can be obtained by moving to double curvature allowing, for example, the production of variable-camber twisted wings. Composites allow the shape to be tailored to meet the required performance targets at various points in the maneuver envelope.

A further and equally important benefit is the ability to tailor the aeroelastics of the surface to further improve the aerodynamic performance. This tailoring can involve adopting laminate configurations which allow the crosscoupling of flexure and torsion such that wing twist can result from bending and vice versa. Modern computational techniques of structural optimization, using finite-element analysis techniques, allow this process of aeroelastic tailoring, along with strength and dynamic stiffness (flutter) requirements, to be performed automatically with a minimum of postanalysis engineering yielding a minimum mass solution.

3.2 Airworthiness Procedures

High material variability and the deleterious effects of temperature and moisture preclude the use of the conventional airworthiness route required for the clearance of aircraft structures. The composite design allowables, particularly in compression, can be as low as 60% of the mean test result obtained from a test conducted at room temperature. As both material variability and degradation are failure mode dependent, a room-temperature test cannot practically demonstrate the true strength of a structure. This is exacerbated if the structure is hybrid, containing metallic components, as the variability and thermal and moisture degradation of these materials can be negligible.

As the most critical condition, the production of minimum design values, generally takes place at elevated temperature when the laminate has absorbed moisture, then the obvious method would be to undertake the testing under these conditions. However, testing in this manner on full-scale flight vehicles can be prohibitive in terms of both cost and timescales and, as a consequence, alternative routes have been found to ensure the competence of the structure.

These routes have as their foundation the collection of data from many thousands of specimens, giving strength for the particular property along with variability and environmental degradation factors: from this, design data are produced. Verification of these data on elements containing structural features is then obtained by undertaking some of the tests under ambient conditions and other elements are degraded to confirm the degradation factor. Paralleling this is the verification of strain distribution by comparing strain measurements during the test with those obtained from a finite-element analysis of the specimen. The full-scale test is then performed at ambient conditions with each of the critical cases taken to the ultimate design load with the results from the extensively instrumented structure extrapolated to verify that, had the test proceeded to higher loads, both variability and degradation would have been accommodated. Should failure not have occurred at ultimate load then the test would be taken to failure, with this occurring in the area with lowest variability and degradation effects. A metallic component usually shows this property. Figure 4 has been constructed in an attempt to demonstrate graphically the way in which a full-scale static test undertaken at ambient conditions can be read across to provide the effective ambient condition design allowable.

3.3 Test Instrumentation

Instrumentation, particularly strain measurement, is essential to the airworthiness clearance of composite structures. The two major agents for understanding the strain distribution are electrical resistance strain gauges and photoelastic coatings, the former giving field strains while the latter are more useful for indicating the distribution around stress concentrations.

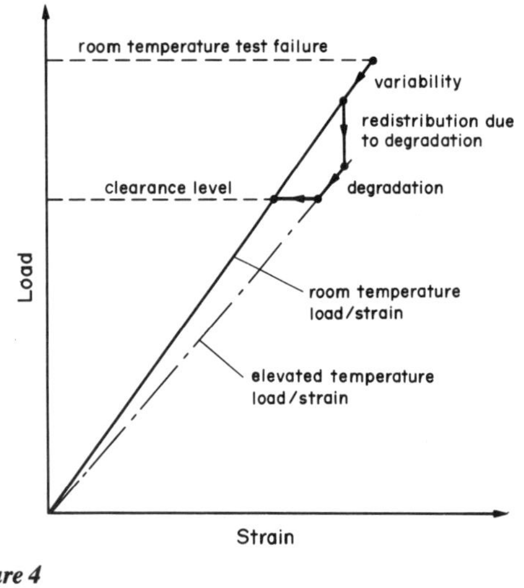

Figure 4
Airworthiness clearance of composite structures

Work is ongoing, taking advantage of the laminated form of composite structures, to embed strain-measuring fiber optics into the laminate during manufacture, producing a matrix of multipoint data measurements leading to an "intelligent" structure. Currently, however, only single-point strain measurement along a fiber is possible.

More direct methods of failure prediction have been investigated by many researchers and prominent among these methods is acoustic emission (AE). This relies on the ultrasonic emission from structures caused by matrix cracking, fiber failure, joint slip and so on. AE has been used as an inspection tool for such items as pressure vessels where a characteristic pressure–emission profile can be obtained from acceptable cylinders and then used to clear production components. However, on structures where this characteristic has not been determined, the current capability to predict failure is limited. Emission characterization is being developed to identify the impending failure mode and it is considered that this may lead to a more accurate prediction capability.

3.4 Electromagnetic Compatibility

All aircraft contain electronic equipment and with modern "fly-by-wire" aircraft the use of such equipment is increasing rapidly in the critical flight safety area of the systems. On metallic aircraft these delicate systems are in the main adequately protected from electromagnetic interference by the inherent shielding provided by the fuselage skins. Fibrous polymer composites unfortunately do not provide this protection and, as a consequence, special action has to be taken to ensure that this deficiency is overcome.

Electromagnetic interference can result from natural phenomena (e.g., lightning strike, radio transmitters) or in the case of military aircraft from malicious intent. The effect of the traditional aluminum structure can be replicated by lining the composite structure with foil and ensuring electrical continuity. The alternative is to provide adequate shielding to all the equipment and the interconnecting cabling. The use of optical fibers for communication will minimize the mass increase associated with this shielding requirement.

While a lightning strike can produce electromagnetic pulses affecting the electronic equipment, its direct effect can be catastrophic, producing significant structural damage. Arcing in fuel tanks, for example, can cause explosive disintegration of these tanks with potential loss of the vehicle. The inherent low conductivity of composite structures leads to the poor dissipation of arc current, producing hot spots, mechanical damage and preferential attachment to metallic items, leading to the internal arcing discussed. Protection against the direct effects of lightning strike are therefore aimed at limiting the physical damage to an acceptable level, particularly over fuel.

The aforementioned internal foil is inadequate for system protection, and external foils or meshes have to be provided on laminates in the more critical zones of the aircraft. Meshes can be cocured along with the laminate, as can foils, but an attractive protection capable of easy repair is that of flame or plasma spraying metallic coatings.

A side issue that must be addressed is the conflicting requirements of adequate electrical continuity and corrosion. In the presence of aluminum, carbon-fiber composites act as noble metals with the former corroding in their presence. Careful design is therefore essential to prevent the creation of galvanic cells and the resulting corrosion.

4. *Manufacture*

The aerospace industry has, over the past decades, geared itself to the batch manufacture of metallic parts for aircraft. This has led to sophisticated methods of manipulating these metals with numerically controlled machines for metal removal, routing, forming and so on. Further improvements, with the introduction of flexible manufacturing systems, ensure the optimum availability of part, tool and machine, producing a system capable of manufacturing a number of different components with minimal interference to efficiency.

With the extensive introduction of fibrous composites in the 1970s, new skills had to be developed in both the production engineering of the design and the subsequent manufacture of the component.

Mold tools capable of operating at the cure and postcure temperatures (200 °C) while remaining stable and durable had to be designed and, while this can be readily effected in steel, the mismatch in coefficient of thermal expansion between the two materials requires careful consideration. Early composite mold tools were neither durable nor stable but modern tooling materials, usually in preimpregnated form, are proving more reliable. Although the steel thermal mismatch can be overcome on laminate molding tools, the resolution of the problem is more difficult if this material is to be used in assembly tools requiring thermal cycling.

4.1 Laminate Manufacture

The largest proportion of carbon-fiber composites used on primary class-one structures is fabricated by placing layer upon layer of unidirectional material to the designers requirement in terms of ply profile and fiber orientation. On less critical items, woven fabrics very often replace the prime unidirectional form.

A number of techniques have been developed for the accurate placement of the material, ranging from labor-intensive hand layup techniques to those requiring high capital investment in automated tape layers. These latter machines can require, in certain instances, investments of up to US$3 million (at 1988 prices).

Processes adopted by the aerospace industry range from hand layup using foil-transfer techniques, through the use of numerically controlled ply profiling machines to numerically controlled tape laying or filament winding. The first of these is an obsolescent process and has, in general, been replaced by the latter two. It involves the hand profiling of plies and their subsequent location on the mold tool, again by hand, using draughting foils for their location.

Use has been made in both the tailoring and shoe industries of automated profiling equipment and out of these has arisen the broadgoods system in which numerically controlled machines under two-axis control profile the plies to the appropriate shape. Three cutting media have been used on these machines; high-pressure water, reciprocating knife cutting and laser cutting, with reciprocating knife becoming prominent because of the moisture uptake with the first method and charring of ply ends with the last. Although the use of broadgoods systems reduces the effort in profiling the plies, their manual laying remains a costly aspect of the laminate production. A number of attempts have been made to automate this ply-placement process, with success only being found on small components with limited complexity and curvature.

An obvious development is to combine the ply profiling and laying into one operation and this has been accomplished with limited success in automatic tape-laying machines operating under numerical control. These machines are currently limited in production applications to flat layup and significant effort is being directed by machine manufacturers at overcoming the many problems associated with laying on contoured surfaces.

Cost reductions in the manufacture of carbon-fiber composite components have been obtained by the progression from hand layup to the use of tape-laying machines. Table 1 shows the corresponding increase in productivity.

Table 1
Carbon-fiber composite layup rate comparison

Method of manufacture	Deposition rate ($kg\,h^{-1}$)
Foil transfer	0.1–0.25
Broadgoods	0.7–1.25
Tape layer	5
Filament winding	20

4.2 Assembly of Aerospace Components

Early composite designs were replicas of those which employed metallic materials and, as a consequence, the high material cost and labor-intensive laminate production jeopardized their acceptance. This was compounded by the increase in assembly costs due to the initial difficulties of machining and hole production. Evidence exists to support the argument that cost is directly proportional to the number of parts in the assembly and, as a consequence, design and manufacturing techniques had to be modified to integrate parts, thereby reducing the number of associated fasteners.

A number of avenues are available for reducing the parts count, among which are the use of honeycomb sandwich panels—sometimes frowned upon in metallic structures due to corrosion—the use of integrally stiffened structures and, more ambitiously, the cocuring or cobonding of substructures onto lift surfaces such as wings and stabilizers. Figure 5 has been constructed to demonstrate the cost of manufacture of aircraft structure in terms of units of cost per kilogram of structure using the conventional aluminum structure as the datum. Hand layup techniques and conventional assembly results in manufacturing costs 60% higher than the datum and only with the progressive introduction of automated layup and advanced assembly techniques can composites compete with their metallic predecessors.

4.3 Nondestructive Examination

All aerospace materials are quality assured to a high standard with the bulk of metallic materials being examined by the suppliers using ultrasonics, x-ray and other appropriate processes. The carbon-fiber preimpregnated material supplier quality-assures the product by various techniques, but the verification of a sound cured structural element is now the responsibility of the aircraft fabricator.

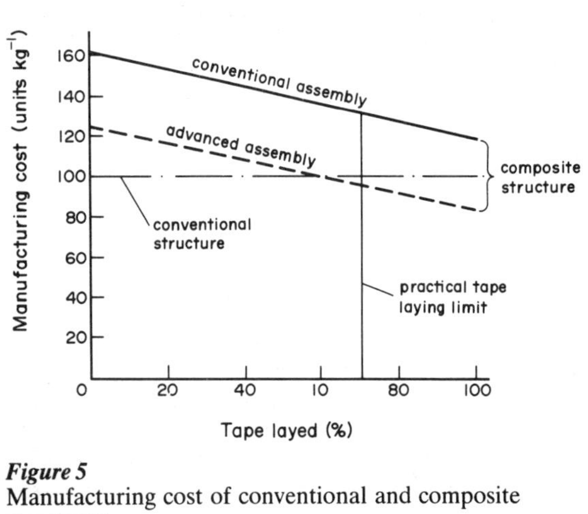

Figure 5
Manufacturing cost of conventional and composite structures

Composites attract a high level of both conventional inspection and the more sophisticated nondestructive examination (NDE) and these combined can account for up to 25% of the direct production effort. Of this figure, the proportion required for NDE is by far the largest. The main tool to ensure that any defects above those defined by the designer as tolerable are detected is ultrasonic examination, invariably on 100% of the structure. This, in many instances, can be supplemented by 100% radiographic examination.

A number of avenues are open for reducing this burden, by relaxing standards, based on defect investigation programs, by zoning of components through correlating allowable defect to stress–strain levels or by developing existing and new techniques. Current techniques based on single probes or multiprobes are time consuming and attention is being directed at phased arrays, thermography and so on.

5. *Aircraft and Aerospace Applications*

From the cautious application of composites in the early 1970s has developed the more confident use of these materials in the primary load-carrying structures of both military and civil aircraft. Although military applications are far in excess of civil aircraft applications, the latter are growing rapidly.

5.1 Civil Aircraft Applications

Civil aircraft applications have concentrated on replacing the secondary structure, bounding major structural elements, with fibrous composites where the reinforcing media has either been carbon, glass, Kevlar or hybrids of these. The matrix material, a thermosetting epoxy system, is either a 125 °C or 175 °C curing system with the latter becoming dominant because of its greater tolerance to environmental degradation. Typical examples of the extensive application of composites in this manner are the later aircraft from Boeing (the 757 and 767) and, from Europe, the aircraft produced by Airbus Industrie. This consortium of aircraft companies is currently producing the A320 which, like the later versions of the A310, carries a vertical stabilizer, a primary aerodynamic and structural member, fabricated almost in its entirety from carbon composite. In addition, the A320 has extended the use of composites to the horizontal stabilizer in addition to the plethora of panels and secondary control surfaces.

In two small transport aircraft attempts have been made to apply composite materials to almost the entire aircraft; these are the Lear Fan fabricated in Belfast, Northern Ireland, and the US Beech

Starship. Of these ambitious projects the former failed for technical and commercial reasons although the latter is showing every prospect of entering service.

5.2 Military Aircraft Applications

Military Aircraft applications throughout the world, initially funded from national research and development budgets, have enabled composite technology to be developed to its current level. This includes material manufacture, design techniques and product manufacture, as well as in-service experience and maintenance.

Without exception all agile fighter aircraft currently being designed throughout the world contain in the region of 40% of composites in the structural mass, covering some 80% of the surface area of the aircraft. Without this degree of compositization, the essential agility of the aircraft would be lost because of the consequential mass increase.

The US advanced tactical fighter (ATF) and the European Fighter Aircraft (EFA) are examples of the proposed degree of compositization for aircraft entering service in the 1990s. Examples of current applications in which the structural mass of composites is 25% of the total can be found in the McDonnell-Douglas AV8B (Harrier Mk. II) and the European Experimental Aircraft Programme (EAP).

5.3 Space Applications

Satellites and space platforms clearly require ultralight structures to minimize the launch costs and, to this end, exotic materials have been developed. These include ultrathin lamina and high-modulus materials, the former to enable precise tailoring to the structural requirement and the latter to provide stability for the antennas and reflectors. With the demand for observation and communication satellites ever increasing, the developments for this niche market will inevitably continue.

Current launch-vehicle availability in terms of tonnes per annum is incapable of meeting demand and the available vehicles, the US Space Shuttle and the European Arianne rocket, are prohibitively expensive. Worldwide studies are determining the feasibility of fully reusable launch vehicles requiring minimum turnaround activity, thereby reducing launch costs significantly. Amongst these are the US National Aerospaceplane (NASP) and the British Horizontal Take-Off Launch Vehicle (HOTOL). The latter experiences structural temperatures ranging from that of liquid hydrogen (20 K) to those caused by reentry (1750 K). Although materials are available that will perform structurally at these temperatures, their performance, particularly towards the upper temperature limit, is inadequate to meet the demanding specific-strength requirements. These inadequacies in available materials provide the next major challenge to material suppliers, designers and fabricators involved in the introduction of fibrous composites with polymer, ceramic or metal matrixes.

See also: Composites: Applications Overview

Bibliography

Ashton J E, Halpin J C, Petit P H 1969 *Primer on Composite Materials Analysis*. Technomic, Stanford, CT

Curtiss P T (ed.) 1984 Crag test methods for the measurement of the engineering properties of fiber reinforced composites. Royal Aircraft Establishment Report RAE TR84102. RAE, Farnborough, UK

Haresceugh R I 1987 Composites—The way ahead. *Proc. 6th Int. Conf. Composite Materials, 2nd European Conf. Composite Materials,* Vol. 5. Elsevier, London

Jones R M 1975 *Mechanics of Composite Materials*. McGraw-Hill, New York

Lechnitski S G 1963 *Theory of Elasticity of an Anisotropic Elastic Body*. Holden-Day, San Francisco, CA

Royal Aeronautical Society 1986 *Materials In Aerospace,* Vols. 1, 2. RAS, London

Thomas D K 1988 Advanced composites in aerospace engineering. *Prog. Rubber. Plast. Technol.* 4(4):1–19

R. I. Haresceugh
[British Aerospace Military Aircraft Division, Preston, UK]

Appliances: Applications of Engineered Plastics

Since the 1950s, technology in engineered plastics has undergone a revolution, and application of this technology to appliances has had a consequent effect on the appliance industry. As a result, products are affordable over a broader consumer market and there has been an improvement in real and perceived quality. The unique properties and processability of plastics compared to other materials have made this possible.

1. Growth of Polymers in Appliances

The growth in the use of polymers in appliances can be attributed to technological advances in engineered polymeric materials and fabrication processes, and the impact of business and marketplace driving forces. Continued growth will require the development and use of a creative design–material–process systems approach.

The use of plastics has grown at a phenomenal rate since the 1950s. Consumption of plastics for appliance applications alone has grown from 4.85 kt in 1959 to over 317 kt in 1981. This explosive growth has resulted from the invention of new materials, the modification of existing materials, the invention of new processes and the modification or refinement of existing processes. Particularly in appliance products, this growth has come from the creative coupling of the technological advances with design and manufacturing methods to produce a product which fulfilled a need. For example, plastic refrigerator inner doors added appreciable storage volume; urethane foam insulation reduced refrigerator energy consumption and expanded the interior volume; plastic clothes-washer and dishwasher pumps improved efficiency and product reliability; and plastic air conditioner fans improved air flow and reduced noise.

Plastics were first used in appliances only when no other material could meet the application requirements. An example of this is the breaker strips used in refrigerators, which connect the outer flanges of the cabinet to the inner liner. This part has to have low thermal conductivity to prevent excessive heat leakage from the exterior metal surface of the appliance to the refrigerated interior. Polymers, which inherently have low thermal conductivity, were ideally suited for this application. Other early applications of polymers in appliances, where no other materials could meet the requirements, included applications needing good dielectric properties (e.g., wiring and motor insulation) and gaskets to seal case–door assemblies where compressibility under load is a requirement.

Following this period when plastics were utilized in appliances because of their unique properties, it was found that plastics had to be used to maintain cost-competitive products. Most of the plastics applications introduced in the 1950s were direct substitutes for metal parts for cost-reduction purposes. In 1961, 95% of the polymers used in appliances cost less than US\$1 kg^{-1} and 90% cost less than US\$0.7 kg^{-1}. In many instances, the metal designs were copied and the plastic material was selected on a trial and error basis. Applications where direct substitutions were made included activators and pumps in laundry products, heat deflectors and hood fans in cooker products, spray arms and detergent dispensers in dishwashers, vegetable pans and ice-cube trays in refrigerators, and evaporator grilles in air conditioners.

Materials development has played a key role in the application of polymers to appliances. Apart from the creative technical work that was done in developing new polymers, it was the combining of polymers and additives that resulted in useful materials. The unmodified polymers have found limited use in appliances. Extensive work has been done in altering polymer properties to meet the needs of specific applications. Indeed, these modified materials often found applications in numerous other parts that had similar property requirements.

Polystyrene is a good example of polymer–polymer modification. Polystyrene is a brittle material that has low impact and low chemical resistance and, hence, limited usefulness in appliance applications. The impact resistance of the polymer is greatly enhanced by the incorporation of butadiene rubber. This produces a material that is called high-impact polystyrene (HIPS). Combination of acrylonitrile and styrene results in a copolymer material (polystyrene–acrylonitrile, SAN) that has better chemical and stress-crack resistance than polystyrene. By combining the styrene, butadiene and acrylonitrile, a blended terpolymer material, acrylonitrile–butadiene–styrene (ABS), is produced that has both improved impact and chemical resistance. These modified polystyrene materials have found numerous applications in appliances such as refrigerator inner doors and liners, television cabinets, air conditioner fans and grilles, coffee-maker bases and housings, radio cabinets and refrigerator crisper pans and trays.

Polystyrene was one of the first materials to be modified with other polymers. Since then the blending and copolymerization of polymers to give materials with unique properties has become commonplace in the polymer industry. Virtually every polymer has been modified as a copolymer or blended with other polymers and evaluated to determine the potential usefulness of the new materials. The result of this vast effort has been the development of new materials with properties that satisfy numerous applications and have sufficient potential production capacity to justify the commercialization of the new materials.

Improvements in material properties made it possible for the appliance industry to offer features and reliability not previously available. Refrigerator inner doors, thermoformed from a HIPS sheet, provided styling and formed-in features not available on metal doors. Air conditioner grilles, injection molded out of HIPS and ABS, were dent resistant, lightweight and eliminated sweating problems frequently found with metal grilles.

Of all the major appliances, however, the quality of dishwashers and clothes washers showed the greatest improvement. Corrosion resistance, inherent in plastics, gave long-term protection from high-pH detergent solutions, bleaches and rinse aids used in these appliances. In clothes washers, phenolic activators not only increased the reliability through better corrosion resistance over aluminum and zinc die-cast parts but also allowed the introduction of spiral designs for improved washability. These new designs, coupled with polypropylene automatic bleach and rinse-aid dispensers, and ABS lint removal systems, set new standards for washing performances and customer appeal.

In diswashers, glass-reinforced polyester pumps provided the means to introduce new water distribution techniques for efficient soil removal. Melamine and acrylic dials and escutcheons led to new appearance trends and plastic cutlery baskets replaced plastisol-coated wire baskets which were prone to knife cuts and subsequent rusting.

2. *Materials and Process Improvements*

In the early 1960s, plastic materials were identified as a key technology opportunity area for the appliance industry. Two of the factors affecting this were the successful implementation of plastics in the 1950s and the favorable impact of plastic on costs of appliance manufacture. A business philosophy evolved at that time that set the stage for the growth of plastics in appliances over the next two decades. Polymer and appliance manufacturers recognized that sustained growth could not be accomplished merely by replacing metal parts with plastic parts. To expand the use of plastics beyond the metal replacement market, it was necessary to establish a strong technical base: further development of tailored plastics materials to meet specific property–application requirements, technological advances in fabrication processes and a better understanding and control of the processing–properties interactions. It was also realized at that time that material, processing and design could not be considered independently, but must be viewed as a total system.

During this period further polymer alloying resulted in numerous new combinations of polymers. Two of the more important alloys are modified poly(phenylene oxide) resin, used in television cabinets, dishwasher control panels, air conditioner grilles and smoke-alarm covers; and modified polypropylene, used in dishwasher tubs and doors, refrigerator breaker strips and air conditioner outer cases.

Of equal importance at this time was the work expended in the development of polymer additives. This development effort required the coupling of a broad spectrum of technologies. A partial list of additives that are used to tailor polymers, producing a specific balance of properties, includes reinforcements and fillers, coupling agents for the fillers and reinforcements, heat and uv stabilizers, internal lubricants and pigments.

Screw-injection-molding equipment development advanced and its use was strengthened by the application of melt rheology and large, thin-wall injection-molding technologies. This, coupled with advancements in flame retardancy and weather resistance of polycarbonates, led to the introduction of the highly styled, corrosion-resistant, plastic air conditioner case. Promoted as a plastic case, this unit carried the first long-term warranty in major appliances for a structural plastic part.

In refrigeration, new insight into environment–stress interactions and the application of stress–temperature superposition principles led the industry to introduce ABS inner doors and foodliners. When combined with improved low-density, low-thermal-conductivity urethane foams, these plastic parts provided the means to design higher-efficiency, larger-capacity, thin-wall refrigerators and freezers and, at the same time, reduced the cost to the consumer.

For the dishwasher industry, polypropylene stabilization developments provided the opportunities for new product innovations and performance improvements. Introduced commercially in 1958, polypropylene, with its ease of molding, inherent chemical resistance, high heat-distortion and excellent appearance, found immediate acceptance in dishwasher applications. However, there was concern over the long-term resistance of polypropylene to failure by oxidative degradation. During the 1960s, cooperative programs between stabilizer manufacturers, polymer producers, universities and end users resulted in the development of highly extraction-resistant antioxidants that essentially eliminated this concern.

Along with the stabilization developments, significant changes were also occurring with the polypropylene polymer. Copolymers and rubber-modified grades improved impact and low-temperature performance. Mineral-filled grades added better dimensional stability, reduced mold shrinkage, reduced stress sensitivity at elevated temperatures and increased stiffness. New melt-compounding techniques, the development of direct processing from powder–glass-fiber blends and the introduction of chemically coupled systems led to the introduction of new glass-reinforced polypropylene compounds that competed with the engineering thcmoplastics. From all of these technologies evolved structural plastic tubs and baskets for clothes washers, coffee-maker housings and filters, modern spray and steam-iron housings and hair-setter bases.

Another class of materials that has also played a key role in appliances and that has undergone considerable development along with thermoplastics is the thermoset materials: polyesters, polyurethanes, phenolics and epoxies. On a weight basis, the polyesters and polyurethanes have been used more extensively than the phenolics and epoxies. Thermosets have gained increased use potential due to technological innovations introduced in the late 1960s and early 1970s.

The polyesters can be formulated to give high strength, high modulus, flame retardancy, chemical resistance, high surface hardness and good electrical properties. For appearance applications, these materials are pigmentable or paintable. Polyesters are used in one-piece air conditioner barrier and base pans, air conditioner cases, clothes-washer gear case covers, printed circuit boards and dishwasher pumps.

Large quantities of foamed polyurethane are used for refrigerator thermal insulation. Polyurethane has a decided processing advantage over other insulating materials in that the liquid mixture can be poured between the walls of a cabinet and it will foam and completely fill the void between the outer case and inner liner. The foam gives structural integrity to the cabinet and low thermal conductivity. The low conductivity results from the unique ability of the polymer to trap and retain the low-conductivity Refrigerant 11 gas within the tiny closed cells of the foam.

Phenolics in modern appliances are generally confined to those applications where good electrical or high-temperature properties are required, such as skirts on electric irons, oven handles, relay and switch housings and motor covers.

Epoxy resins find even more limited use in appliances than do the phenolics. Apart from their use for joining and sealing refrigerant tubing and a few adhesive applications, the major use of epoxides is for printed circuit boards and potting or casting of electrical and electronic components used in appliances.

3. *Effect of Oil Crisis on Growth*

The oil crisis in 1973–1974 broke the technology introduction momentum that had been so effective throughout the appliance industry. The focus changed from innovating to maintaining production. Material costs rose rapidly. Some plastic materials were temporarily removed from the market and some custom molders went out of business. In many instances, sales suffered and, in a few instances, quality standards were temporarily relaxed. Many major development programs were shelved and manpower was directed toward meeting these new challenges. Old tooling was modified to reduce part thickness in applications which were clearly overdesigned. Density reduction through the incorporation of chemical blowing agents, material substitution, and polymer–polymer blends were investigated as an approach to extending the use of polymers in short supply.

Through these efforts, the industry survived this difficult period and saw plastic materials supply, but not prices, restored to previous levels. However, new factors, common to most industries, began to add to the list of challenges. In addition to the rising material costs, and unprecedented cost increases in labor, energy and equipment, regulatory constraints added significant new cost burdens and made long-range planning uncertain. Also, increasing consumerism raised the level of expectations on the performance and quality of appliances. Coping with these challenges drained resources and funnelled efforts toward improvements and extensions of current applications, continuation of material substitution programs and new feature offerings.

The appliance industry is a leader in advancing injection-molding and thermoforming technologies. These two processes account for the major consumption of plastics by the industry. Injection molding is the more versatile process and, hence, is used to fabricate the majority of appliance parts, ranging from cams and gears weighing a few grams to dishwasher tubs weighing 6 kg. Thermoforming, although far less versatile, accounts for a large volume of the plastics used in the industry because of the size and weight of the parts made by this process (e.g., refrigerator food holders and doorliners).

4. *New Materials and Processes*

Since about 1970, a number of new materials and processes have emerged. The appliance industry has been quick to evaluate new materials that fit existing processes, but slow to evaluate and develop new processes. However, the automotive industry, driven by the need to reduce the weight of its products, has been exploring both new materials and processing techniques. Although the appliance industry is not driven by weight reduction, it is driven by other forces which are common to the two industries: rising labor and material costs and increasing capital equipment and processing energy costs. These driving forces are causing the appliance industry to give more serious attention to new materials and fabrication processes.

A new process, which the automotive industry has spent considerable effort developing, is reaction injection molding (RIM) (see *Reaction Injection Molding*). The process consists of injecting low-viscosity, reactive chemicals into a part mold and polymerizing the chemicals in the mold to a solid part. In a sense, the mold is the polymerization vessel. When this process is compared with conventional injection molding, it has the advantages of lower capital investment and considerably lower processing energy costs. The low viscosity of the reactants permits parts to be designed to minimum thicknesses and thus reduce material costs.

Until recently, RIM materials have been based solely on polyurethane materials. This chemistry has produced a whole family of products that results in parts ranging in properties from flexible to rigid and from low density to high density. The basic urethane systems have been further modified by additives technology to enhance the processability and the physical properties of the polymers.

More recently, there has been an explosive effort in the development of nonurethane RIM materials.

Some of the new materials are pigmentable and light stable, thus overcoming the major shortcomings of the standard urethane systems. Some of the materials being developed for commercialization include nylon, epoxy, polyolefin, polyester and a wide variety of hybrid polymers.

The use of polymers in external structural applications offers not only the largest, but also the most challenging growth opportunity for plastics in appliances. These structures must not only be cost competitive with painted steel, have good surface finish and durability and be flame resistant, but, equally importantly, must present a perceived quality image to the consumers. Of the emerging processing technologies it would appear that RIM is possibily the fabrication process closest to meeting the external structural requirements of appliances. The process has been successfully applied to automotive parts, business-machine housings and medical equipment. These industries have developed the materials and processing technologies that can be readily transferred to appliances.

5. *Application of New Technologies*

With the vast array of materials available to the appliance design engineer, the selection of the right material for a given application has been greatly complicated. The task has been made somewhat simpler by advances in the fundamental understanding of polymer properties and how they relate to engineering performance. Material selection has been further simplified, in recent years, by the application of computer-aided analysis (CAA) to part design.

The appliance industry, like all industries, is in urgent need of greatly improved productivity. To address these needs there has to be a better coupling of scientific advances with engineering application. For the appliance industry, the application of polymer technology to the products it produces is certainly an important method for improving productivity.

Technologies that are impacting quality and productivity of plastic parts in appliances include CAA of part design, mold design, polymer flow, mold cooling, robotics and microprocessors. The application of CAA to plastic part design makes it possible, by using finite-element methods, to predict the structural behavior of a part. By selecting node points and connecting them to form a structure of finite elements and by applying loads and boundary conditions to the structure, stress and strain can be determined from calculated nodal displacements at any location. In this way structural behavior can be predicted. These analyses make the selection of materials and the development of part design much easier tasks.

Computer programs assist in the mold design concept, which provides the basic framework from which the detailed mold can be fabricated. These programs also analyze mold economics and some aspects of polymer flow and part cooling. The programs answer questions related to part processability, runner and gate design, dimensional tolerances and surface finish. The results of the analyses indicate the optimum machine size, the number of cavities in the mold and the finished part cost. An in-depth computer analysis of polymer flow assists in the selection of molding conditions; gate locations, size and number; sprue and runner sizes; and part dimensions. Software is being developed to assist in part-cooling analysis. These programs will select coolant type, temperature, flow rate and cooling channel size and location. Improved productivity will result from this analysis by reducing part cycle time and improving part quality.

Productivity improvements are being realized through the use of robots in the manufacture of plastic parts for appliances. They are not only used to remove parts from the molds, but also to perform secondary operations such as hot stamping or assembly operations such as ultrasonic and vibration welding.

New technologies in microprocessors and other electronics are leading to sophisticated, on-line, closed-loop controls for improved process efficiency in injection molding. It is almost routine to have pressure transducers in molds to monitor and/or feedback cavity pressure for adaptive control to produce high-quality plastic parts. Closed-loop controls accommodate fluctuations in material processability and external conditions. Parts of consistently excellent quality are produced through the control of injection velocity, cavity pressure, holding pressure, feeding pressure and shot size.

Appliance products are in an intensely competitive market. The characteristics of this industry are high volume, high quality, low margins and value received. To remain viable, it will be necessary to take advantage of all pertinent technologies. In the past, the judicious use of polymers in appliances has helped to maintain the high quality, significantly contributed to the profitability of the industry, and proved adaptable to high-production methods. In the future, if plastic materials are to continue to have a major impact on this industry, then a systems approach which better integrates material, design, processing and control will have to be implemented.

See also: Polymers: Classification by Use and Properties; Polymers: Technical Properties and Applications

Bibliography

Hagan R S 1981 Durable consumer goods and engineering plastics. *Proc. Natl. Symp. Polymers in the Service of Man.* American Chemical Society, Washington, DC

Richards D A 1982 Computer aided analysis, design, and manufacture of plastic parts and molds. *Proc. Polycon '82 Conf.*, Noordwijkerhoot, The Netherlands, May 24–28

Thomas J R 1980 Opportunities for plastics in large appliances. *Proc. 1979 SPI Natl. Plastics Exposition Conf.* Society of the Plastics Industry, New York

R. W. Barito
[General Electric Company, Louisville, Kentucky, USA]

Armor, Composite

A list of possible composite armor materials may include traditional materials such as aluminum and steel or less traditional ones such as silk and leather. In the past, many combinations have been applied and found satisfactory, usually over a narrow range of applicability. Recent efforts to increase ballistic protection and reduce the weight necessary for this protection have led to a new focus in the armor community. Two classes of materials have emerged as having great potential over a wide range: fibrous composites and ceramics. However, a vital aspect of armor design is an understanding of the properties which control ballistic performance. Unfortunately, for these new materials, the fundamental research necessary for this understanding has only recently received attention. This is mainly a consequence of the fact that the processes occurring during the projectile penetration event occur at extremely high strain rates. Few materials have been fully characterized under these conditions, least of all ceramics. In this article the material properties important in armor design are examined, briefly for fibrous composites, and in greater detail for ceramics. Ceramics will receive more focused attention because of the lack of published information on their behavior at high strain rates.

Before proceeding further, it is necessary to describe how armors are rated for performance. Two fundamental quantities always considered in describing the performance of any armor system are weight and ballistic protection. Ballistic protection is characterized in terms of the projectile used and the maximum projectile velocity withstood by the armor. A popular method is the v_{50} method, in which v_{50} is defined as the velocity at which 50% of the projectiles are defeated and 50% penetrate completely through the armor. Many other methods are also used to determine when the armor has been defeated. Most often, a thin aluminum plate is placed behind the armor, and when either the projectile or debris from the armor penetrate the plate, the armor is considered to have been defeated. Weight efficiency is determined relative to an accepted standard material. The efficiency is expressed as the ratio of standard material weight necessary to defeat the projectile to the required weight of the new armor required to defeat the same projectile. The weight is usually computed per unit area or areal density of the armor. The standard material used by the US military is rolled homogeneous armor (RHA), a high strength steel defined by US specification MIL-A-12560.

The important point to be emphasized here is that many methods are used in armor peformance characterization. One should be certain of the method used before making direct comparisons.

1. Fibrous Composite Armors

The use of fibers or fabrics as armor dates from the sixth century AD when the Chinese used silk as padding for heavy metal outer armors. It soon became apparent, however, that such metal armor was so encumbering that a lightly armored, more mobile adversary had a tactical advantage. An efficient armor composed entirely of lightweight materials was needed. The use of silk as protection against handguns was first introduced in Russia in around 1914. Unfortunately, widespread use of composites for body armor did not occur until the Korean War when the US Army introduced a protective vest composed of hardened nylon plates for use by ground troops. This vest was only effective against low-velocity fragmentation and could not stop projectiles fired from rifles. The Vietnam War brought about a new vest capable of defeating high-energy rifle bullets. This vest consisted of ceramic plates backed by a fiberglass laminate. Its 60 kg m^{-2} weight was soon found to be an unacceptable burden for the average infantryman, so a lighter vest composed entirely of Kevlar was procured in the late 1970s and is still in use today. In addition to body armor for the military, many law enforcement and private security agencies use fibrous composites for protection. In this application, the threat is usually limited to low-energy handgun projectiles. The vest has the additional requirements of concealment and comfort. The role of fibrous composites as armor will be expanded in years to come.

As armored-vehicle designers are forced to reduce weight and increase protection, fibers will be combined with ceramics, metals and plastics to create more effective armors. These new armor systems will come in two forms, either as structural load-bearing components or as nonstructural parasitic appliqués.

The principal advantage of fibers is their high strength-to-weight ratio. Kevlar, for example, approaches a tensile strength of 2.75 GPa. Other properties such as tensile modulus, elongation, tenacity, energy absorption and sound speed have proven to be important in characterizing impact. Impact, however, is a complex problem and no single property should be relied on as a predictor; often, combinations of properties are the key. For example, glass fibers, with relatively low impact strength, and polyester resins, with poor tenacity, can be combined to produce a laminate of high impact strength. In addition, properties can be severely affected by strain rate. Most fibers show increasing strength and modulus with increasing strain rate, but elongation to break varies. Examination of fiber properties at the strain rate expected is important in ballistic design.

The application of fibers to ballistic protection generally takes two forms. Fabric or soft armor is constructed from multiple layers of woven fabric without a resin binder. The layers are sewn together using a common sewing thread in a straight or zigzag pattern. Composite laminate armors consist of multilayered fabrics combined with a resin binder. The application of the resin greatly affects the performance of the fabric and must be chosen to meet the given application. Depending on fiber and resin content, the laminate may be classified as either structural or nonstructural. Structural laminates usually contain a greater percentage of resin than of fiber.

Most ballistic threats which fiber composites are designed to resist are either soft lead projectiles or fragment-simulating projectiles (FSP). Ball ammunition common to handguns is typically fabricated from lead and easily "mushrooms" when striking a composite target. FSPs, on the other hand, are machined from hard steel and are used to simulate the effects of blunt-ended fragments propelled from larger exploding warheads. Fabric construction is used for the soft lead projectiles, whereas fragmentation is generally stopped using a laminate. Neither construction has proven particularly effective against the sharp point typical of armor-piercing projectiles. The sharp point can easily push aside the woven fibers and penetrate the composite.

1.1 Failure of Fibrous Composites

As a first step in the study of failure mechanisms in composites, it is instructive to examine the response of a single fiber to impact. In this discussion, it is assumed the fiber is oriented such that its longitudinal axis is perpendicular to the path of the projectile. At the moment of impact, a compressive wave is propagated outward along the longitudinal axis of the fiber. The speed c of this wave is defined by

$$c=\left(\frac{E}{\rho}\right)^{1/2} \tag{1}$$

where E is the fiber modulus and ρ is the density. The wave reaches the end of the fiber and is reflected as a tensile wave. Reflection of the wave as a tensile wave is necessary in order to satisfy the boundary condition of zero strain at the fiber ends. The resulting tensile strain is now experienced by the fiber in contact with the projectile and the magnitude of strain is dictated by the impact velocity of the projectile. Continued wave reflections intensify the tensile strain, with energy absorbed by the fiber being proportional to this tensile strain. At the same time, a second wave propagates along the transverse axis (across the diameter) of the fiber moving parallel to the projectile and at the same velocity. Because of this transverse wave, the fiber deflects in the direction of the projectile path and as a result increases the energy absorbed by the fiber. The fiber continues to deflect and absorb energy until the projectile is decelerated to a stop or the fiber strains past its dynamic yield point and breaks. If the impact velocity is sufficiently high, the fiber cannot respond fast enough to exhibit a strain. The lowest velocity for which this occurs is the critical velocity and is defined as the velocity for which the fiber will break without straining. Critical velocity is a function of wave speed and, in turn, a function of fiber modulus and density. Increasing the wave speed of a fiber will allow quicker strain response and thus increase its critical velocity.

The mode by which the fiber separates is also important in determining its resistance to penetration. Three principal modes have been identified: melting, brittle fracture and plastic deformation with longitudinal splitting. Nylon fibers melt but exhibit large deformations before breaking. High values of elongation equate to large work-to-rupture values and, thus, more energy absorption by the fiber. Brittle fracture with low elongation to break is typical of glass fibers. Kevlar fibers do not show as high a value of elongation as nylon, but have proven to be ballistically superior. This may seem to be a contradiction until the fiber breaking mode is examined. Kevlar fibers show some degree of plastic deformation before rupturing. More important, however, is the ability of the fibers to split along the longitudinal axis. Longitudinal splitting not only absorbs energy but acts as a crack arrester as well.

Failure in composite laminates is a far more complex process. The process is three dimensional in nature, characterized by interactions occurring in the individual fabric planes and across those planes to adjacent fabric layers. Additional strain-wave reflections at fiber crossovers serve to distribute impact energy over a larger area, thus increasing ballistic resistance. Because of the numerous crossovers found in woven fabrics, the energy is quickly dissipated as the wave encounters more and more fiber crossovers and the impact damage is usually restricted to a small area adjacent to the point of impact. Energy transfer to adjacent layers (through the thickness) is facilitated by use of a resin binder. Strain waves are transmitted from the fabric layer to the resin matrix and then to adjacent fabric layers.

At the same time, waves are reflected back into the fabric or matrix at each interface. The amplitudes of these waves are determined by a quantity called mechanical impedance I, defined as

$$I = \rho c \tag{2}$$

where ρ is the density and c is the wave speed defined in Eqn. (1). Waves may be reflected or transmitted according to the following relationships:

$$P_t = \left(\frac{2I_t}{I_t + I_0}\right) P_0 \tag{3}$$

$$P_r = \left(\frac{I_t - I_0}{I_t + I_0}\right) P_0 \tag{4}$$

where P_0, P_r and P_t are the pressure amplitudes of the incident wave, reflected wave and transmitted wave, respectively. The terms I_0 and I_t are the impedance of the incident and acceptor materials, respectively. As an example, suppose that Kevlar with an impedance of 1784 g cm^{-3} s^{-1} is used with an epoxy resin with an impedance of 320 g cm^{-3} s^{-1}. Further suppose that the wave is travelling from the Kevlar to the epoxy. At the interface, a wave of amplitude $-0.696\,P_0$ is reflected back into the glass and a wave of amplitude $0.304\,P_0$ is transmitted to the epoxy matrix (a negative sign indicates a tensile wave).

Transmission of waves to the resin matrix increases the energy-absorption capability of the laminate. Cracks are formed in the matrix which propagate parallel to the fabric layers. This process is called delamination and is characterized by separation of the fabric layers. Impact energy is dissipated in the formation and advancement of these cracks and in the deflection caused by separation of the individual fabric layers.

1.2 Designing with Fibrous Composites

No armor design will be applicable to all situations. The designer is often faced with compromises and must tailor the armor to the specific application. The ballistic performance of the system often depends on the interaction of its various components. It is difficult to separate out the characteristics important in each case but some generalizations can be made. The following paragraphs will examine some of the properties important in selecting components in a fibrous armor design. The properties to be examined include fiber selection, yarn denier, weave, resin binder and environmental susceptibility.

As was seen earlier, ballistic efficiency resulting from individual fiber properties is difficult to characterize. Modulus, wave speed and elongation to break, however, tend to be the predominating factors involved. Increasing these properties will result in a greater energy-absorption capability of the armor and, thus, an increase in protection.

Yarn denier is a measure of the diameter of the spun yarn used to produce the fabric. Denier numbers increase with increasing yarn diameters. Fabrics woven from high-denier yarns will have fewer fiber crossovers and less ability to distribute energy. Ballistic protection will usually decrease with increasing yarn denier with all other factors held constant.

Following the same logic, a close weave with many crossovers will be more efficient than a loose weave. In fabric armors, this seems to hold for handgun projectiles, but fragmentation-type projectiles show a lesser dependence, the difference being the area involved in the impact. Armors made with resin binders do not show a strong dependence on weave construction. The effect of weave is probably masked by the resin–fiber bonding which distributes the impact energy more efficiently.

Ductile resins such as vinyl esters perform better than more brittle resins like epoxy. Resins showing increased ductility will absorb more energy both in crack initiation and propagation. Evidence suggests that, for the best ballistic efficiency, the resin content should be 20–25 wt%. Laminates with less resin content show increased ballistic protection, but the deformations are severe and unacceptable for most applications. Some applications require the laminate to serve as a structural component as well as to give ballistic protection. The best mechanical properties occur at much higher resin contents, so a compromise must be reached between mechanical and ballistic performance.

Because of environmental factors, the fiber with the best ballistic properties may not make the best armor. Many fibers degrade when exposed to sunlight, moisture, seawater, temperature extremes and petroleum products. Military, and to a lesser extent civilian, armor designs, must take environmental factors into account. Fibers must be protected, or resistant fibers selected, when designing any such armor. In addition, cost, weight and space requirements may determine the design. No military commander will opt for an armor that is too heavy or costly when a lighter, less-expensive material will suffice. An armor whose cost does not justify the additional weight saving will not be implemented. For maximum ballistic efficiency, fibrous armors must be allowed to deflect or delaminate. For this reason, fibrous armors must never be combined with rigid components in a way which restricts deflection. When minimal space is allowed for the armor, fibrous armors are not the best choice.

2. *Ceramic Composite Armors*

If the definition of ceramics is limited to refractory materials, the history of their use as armor materials is a short one. A broader definition, however, extends their use to several thousand years. Stone, and more recently concrete, have long been recognized for their effectiveness as defensive materials aginst fragmentation, projectiles and blast. In the defensive posture, weight is not an issue. Offensive operations, on the other hand, require a high degree of mobility. Materials selected for this kind of mission must be light and effective. This requirement led the US Navy to evaluate ceramics for use on warships during World War II. It was found that plate glass backed by Doron (a glass-reinforced polyester) was an efficient combination against fragmenting warheads and resulted in weight savings over the existing steel armor. The question of structural integrity, however, prevented its acceptance. During the Vietnam War, the search for lightweight armors again turned to ceramics. In the early 1960s, Goodyear Aerospace Corporation and the US Army Materials and Mechanics Research Center investigated the use of ceramics for aircraft and river patrol boats, and for protection of ground combat personnel and aircrews. In 1970, Goodyear was issued a patent for an armor design consisting of a ceramic front face backed by a resin-impregnated glass fabric. The bonding agent between the ceramic and glass fabric was identified as an elastomeric adhesive such as polysulfide.

The principal advantages of ceramics over conventional steel are their high compressive strength and hardness. Hardened and sharp-pointed armor-piercing projectiles are effective against softer materials. This effectiveness is lost in impacts with ceramics because the dynamic stress limits in ceramics are much greater than in the projectiles. The sharp point is quickly eroded away leaving a less-effective blunt cylinder. The loss of projectile mass reduces the energy of impact and increases the protection afforded by the ceramic system. The system aspect of ceramic armor design is important and must be emphasized. Because of the inherent brittleness of ceramics, they may never be suitable as a "stand alone" armor. Ceramics are often combined with other more ductile materials to achieve the required durability and structural integrity. The following section is a discussion of important aspects of ceramic armor design. Design aspects are the result of the authors' observations and experience and are not founded on strict theoretical principles. Much fundamental research is needed before ceramic armor design is reduced to an exact science.

2.1 Designing with Ceramics

A typical ceramic armor design will consist of a forward spall cover, ceramic, adhesive and backing plate. Many variations of this design are possible of course, but a description of this design will aid in illustrating the important aspects of ceramic armor design. Most industrial ceramics suitable for use in armors are available in a square or rectangular configuration. This shape will be assumed in the following discussion and will be referred to as ceramic tiles. Two important criteria not discussed previously apply to ceramic armors more than any other type of armor. These criteria are multihit capability and collateral damage Multihit requirements are usually specified in terms of the maximum number of projectile caliber diameters allowed between impact locations without adversely affecting the protection afforded by the design. Collateral damage is a phenomenon often seen in improperly designed ceramic armor. Tiles adjacent to the impacted tile are cracked or damaged by the transmission of shock waves or debris ejected from the impact. This effect greatly reduces the multihit capability and overall desirability of the design. The armor designer should strive to reduce susceptibility to collateral damage.

Forward spall describes the ejection of projectile erosion products and ceramic debris out of the armor opposite to the direction of impact. Often this is of no concern and is beneficial in that it reduces the amount of energy the armor must absorb. If, however, the armor is to be used in close proximity to personnel, forward spall must be reduced. This can most easily be done by covering the ceramic with a flexible material such as a glass-reinforced plastic.

Ceramic material properties important in armor design include compressive strength, impedence, density and tensile strength. Ballistic efficiency can be correlated with the average of confined and unconfined compressive strengths

$$(Y + l_{\mathrm{H}})/2 \tag{5}$$

where Y is the unconfined compressive strength and l_{H} is the Hugoniot elastic limit, a measure of the confined strength. High tensile strength is desirable to delay the onset of tensile failure on the tile rear face described earlier. Delaying this failure only a few microseconds will significantly increase the perforation resistance of the ceramic. Low density allows the use of thicker tiles for the same areal density and reduces the stress on the back plate. Impedance should be high, as this is a measure of distortion produced in the ceramic. In short, the desired ceramic should have high compressive and tensile strength, high impedance and low density.

It has been observed that a single tile performs better than a stack of multiple tiles of equivalent thickness. The reason is that the single tile has fewer free boundaries where tensile cracks can form, particularly with respect to the time for wave travel to the rear surface of the tile. This time difference is large enough to greatly influence penetration resistance. The lateral dimensions of the tile should be as large as is consistent with acceptable multihit criteria. The larger the distance from the impact point to the sides of the tile, the longer it takes for relief waves to reach the point under the impacting projectile. The ceramics is then no longer confined and will fail in compression more readily. This represents a paradox for the ceramic-armor designer. A tile with large lateral dimensions is ballistically more efficient but has less capability for multihit. The thickness needed is dependent on the type of projectile and velocity. Generally, since lateral dimensions are larger than the thickness, thickness will be the predominant design factor. The final design thickness is often dictated by the corners and intersections of adjacent tiles. At corners where four tiles intersect, the ballistic efficiency may be reduced by as much as 30% compared with center-of-tile impacts. This reduction is a function of the gap size, so gap size should be reduced to the minimum value that does not cause collateral damage.

As mentioned earlier, collateral damage is a tremendous problem in ceramic armors. It has been observed that this problem can virtually be eliminated by placing an elastomer in the lateral gaps between tiles. The elastomer helps to dissipate the shock transmitted between tiles, but more importantly it prevents debris from impacting into adjacent tiles. A thin layer of elastomer between tiles and the back plate will also aid in mitigating transmitted shock waves and further reduce collateral damage. This aspect is more important if a metallic back plate is used, and less so if fibrous composites or other soft materials are used.

Backing plates are necessary in ceramic armors because of the inability of ceramics to provide structural integrity. In addition to a structural role, the backing plate catches the residual debris that is pushed out of the rear of the fracture conoid. In order to perform this task, the back plate must be a ductile material. Popular choices are aluminum and fibrous composites. Care must be taken not to allow too much deflection in the backing plate. Severe distortion can affect the ability of the armor to sustain multiple impacts and remain structurally sound. The ratio of ceramic to back plate thickness should be kept between 50% and 60%.

Adhesives are often used to secure the ceramic to the backing plate. In general, the adhesive has little effect on the ballistic performance of thick tiles, but has a significant effect on thin tiles. In thin tiles, the adhesive thickness should be kept as thin as possible to still secure the tile to the backing plate. Typical thicknesses are of the order of 0.5 mm. Care should be taken to ensure the adhesive is not too spongy and that it adequately transfers stress from the ceramic to the back plate. In other words, the adhesive must not sacrifice the rigid support provided by the back plate.

Bibliography

Agarwal B D, Broutman L J 1980 *Analysis and Performance of Fiber Composites*. Wiley, New York

Grady D E 1982 Local inertial effects in dynamic fragmentation. *J. Appl. Phys.* 53:322–5

Grady D E, Lipkin J 1980 Criteria for impulsive rock fracture. *Geophys. Res. Lett.* 7:255–8

Kipp M E, Grady D E, Chen E P 1980 Strain-rate dependent fracture initiation. *Int. J. Fract.* 16:471–8

Kolsky H 1963 *Stress Waves in Solids*. Dover, New York

Lankford J 1987 Temperature, strain-rate, and fiber orientation effects in the compressive fracture of SiC fiber-reinforced glass-matrix composites. *Composites* 18: 145–52

Liable R C (ed.) 1980 *Methods and Phenomena 5: Ballistic Materials and Penetration Mechanics*. Elsevier, New York

Lindholm U S, Yeakley L M, Nagy A 1974 The dynamic strength and fracture properties of Dresser basalt. *Int. J. Rock. Mech. Min. Sci. Geomech. Abstr.* 11:181–91

Zukas J A, Nicholas T, Swift H F, Greszczk L B, Curran D R 1982 *Impact Dynamics*. Wiley, New York

J. Lankford and W. Gray
[Southwest Research Institute, San Antonio, Texas, USA]

Automobile Tires

Although automobile and truck tires are manufactured in vast quantities—over 200 000 000 per year—it is not generally recognized that the tire is a highly sophisticated composite structure. Layers of rubber and rubber-impregnated sheets of parallel fiber bundles (cords) are combined to make what is basically a toroidal laminate with different properties in each of its three principal directions. Indeed, it is this carefully chosen anisotropy that enables the tire to perform diverse and often conflicting functions. Tire mechanics, outlined here, deals with the appropriate choice of materials and designs in order to achieve five main goals:

(a) high sliding friction in wet and dry conditions, over a wide range of temperatures and with varied road surfaces, to provide adequate acceleration, braking and steering;

(b) low rolling resistance under the same circumstances;

(c) a low vertical stiffness to cushion the ride;

(d) high longitudinal and lateral stiffnesses to minimize sliding motions in the contact patch (see Sect. 3); and

(e) resistance to cutting, puncturing, abrasion and so on.

How these diverse requirements are met is discussed in the following, after a description of the materials of which the tire is composed.

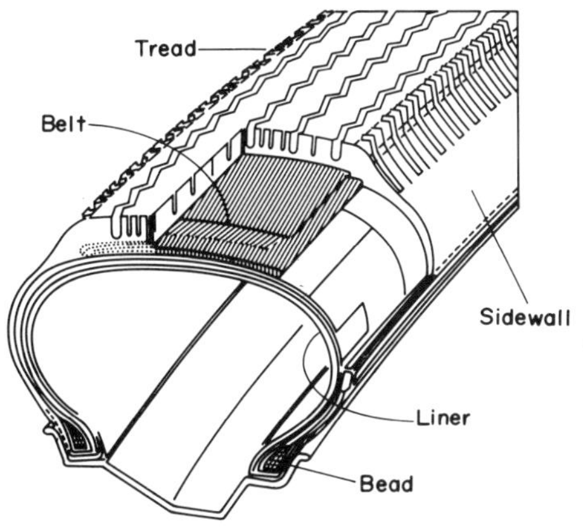

Figure 1
Structure of a radial tire

1. Tire Materials

R. W. Thomson, in 1845, was the first to employ air to provide pneumatic cushioning in a rubber tire, but the first practical use was developed in 1888 by Dr J. B. Dunlop. In modern tires, loss of air is minimized by a special rubber inner lining, usually of butyl rubber, chosen for its low permeability (see Fig. 1).

The tire must be reinforced by layers of relatively inextensible cords to contain the air pressure and to restrict longitudinal and lateral deformations and growth of the tire in service. Such cord layers have been made of a variety of materials over the years: cotton, rayon, nylon, polyester, glass, polyaramide and steel, the latter two materials being the dominant ones at the present time. The choice of cord materials is based on the need for high stiffness, good resistance to repeated flexing, high strength-to-weight ratio, good adhesion to rubber and low cost.

Rubber sidewalls of the tire (Fig. 1) must resist scraping, flexing and the attack of ozone in the air. Rubber treads must primarily resist abrasive wear. Indeed, this is one of the most severe applications of rubber and the continuous improvement in wear resistance by improved formulations has been a remarkable technological achievement. Typical formulations for tire sidewalls and treads are given in Table 1.

It must be emphasized, however, that rubber formulation is still more art than science. For example, the reason why a tread recipe based on polybutadiene is superior in abrasion resistance to one based on polyisoprene or on a butadiene–styrene copolymer is not known. It should also be pointed out that strong rubber compounds are themselves composite in nature. Typically, about 30% by volume of a fine particulate solid filler, usually carbon black, is included in the formulation. Without this ingredient, the abrasion resistance of rubber would be much lower. Indeed, the introduction of carbon black reinforcement in rubber around 1910 was a major factor in prolonging tire life, from about 8000 km to about 16 000 km but, again, the mechanism of reinforcement of rubber by fillers is still obscure.

2. Tire Structure

The most important feature of tire structures is the arrangement of the reinforcing cords, which determines the degree of anisotropy. At first they were placed to minimize distortion. For a straight tube, alternate layers would then be arranged at a cord (crown) angle to the tube axis of $\alpha = 57°$ (see Fig. 2) so that the line tensions along and across the tube, given by $nt\cos^2\alpha$ and $nt\sin^2\alpha$, would meet the inflation requirements, $PR/2$ and PR, respectively, where P is the inflating pressure, R is the tube radius, t is the cord tension and n is the number of cords per unit of width normal to the cord direction. In the early 1950s, however, Michelin introduced the radial tire (patented by Gray and Sloper in 1913), in which one set of cords is arranged to lie almost parallel to the tire circumference, forming a circumferential belt (Fig. 1). In practice, because the tire is a toroid rather than a cylindrical tube, the belt cords are made to lie at a crown angle of about 11° rather than at 0°. This arrangement maximizes the line tension in the circumferential direction for a given value of the inflation pressure and, hence,

Table 1
Representative formulations for passenger car tires in parts by weight. The mixes are converted into strong elastic solids by chemically interlinking the rubber molecules ("vulcanization"), on heating for about 20 min at a temperature of about 150 °C in a tire mold

	Compound	
Ingredient	Sidewall	Tread
Natural rubber	50	
Butadiene–styrene copolymer (75/25)	65	
Cis-1, 4 polybutadiene	50	35
Heavy aromatic oil		25
Carbon black (HAF)	50	65
Processing oil	10	15
Zinc oxide	4	4
Stearic acid	1	1
Sulfur	2	2
Vulcanization accelerator[a]	1	1
Antioxidant[b]	3	2
Protective wax	2	2

a Example: *N*-cyclohexyl-2-benzothiazolesulfenamide b Example: *N*-(1,3 dimethylbutyl)-*N'*-phenyl-paraphenylenediamine

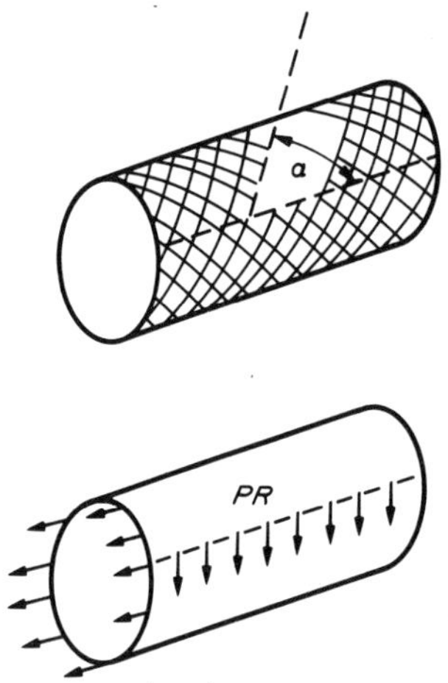

Figure 2
Cord angle α

gives greater resistance to lateral distortion of the tire on cornering. This feature provides improved steering and wear resistance (see Sect. 4).

3. Tire Traction

Consider a steadily applied traction or vehicle/maneuvering force, as in acceleration, braking or cornering. The tire footprint takes up a steady-state condition, in which tread rubber, entering and advancing through the footprint, is gripped fixedly by the ground while the interior of the tire body moves or "slips" with respect to the ground. Thus, the tread is progressively strained under the ground and wheel reactions. Figure 3 illustrates vehicle cornering, where the tread is displaced laterally from the plane of wheel rotation, set (by the driver) at a "slip" angle θ to the direction of vehicle motion. A point is reached, usually near the exit from the footprint, where the ground can no longer exert a sufficiently large stress to maintain the deformation because the local friction has been exceeded. The deformation of the tire then recovers by sliding of the surface over the ground, back to the plane of the wheel.

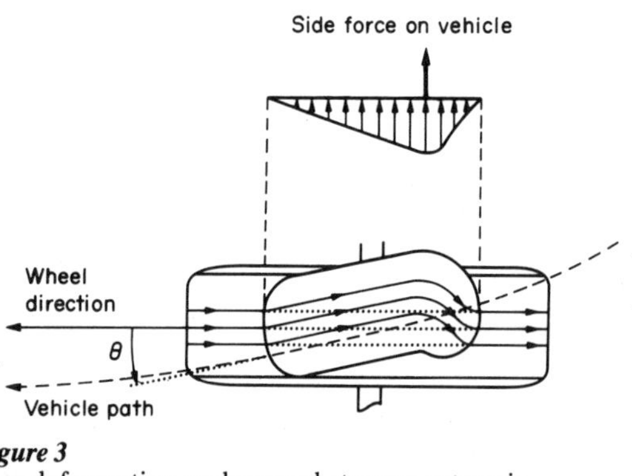

Figure 3
Tire deformation and ground stresses set up in cornering

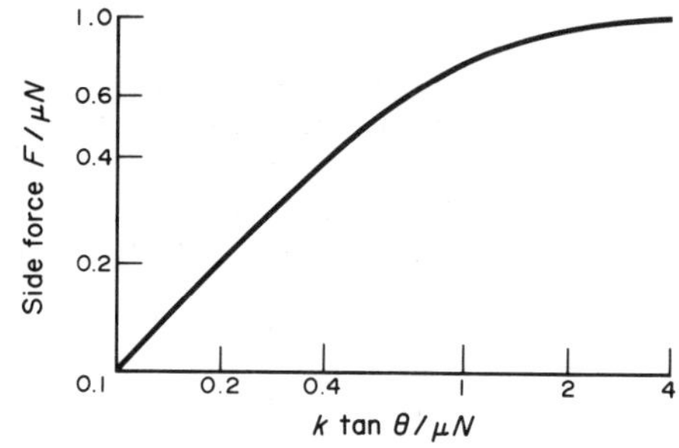

Figure 4
Side force F vs slip angle θ:N is the normal load on the tire, μ the coefficient of friction and k the lateral stiffness × the half-length of contact patch

The total lateral force exerted on the wheel is the summation of the ground stress distribution over the footprint. It varies with the slip angle θ typically as shown in Fig. 4. The initial slope is related to the lateral stiffness of the tire and the maximum value is governed by the coefficient of friction of tread rubber on the road, reached at a slip angle where the ground stress is everywhere equal to the friction limit. Above this value, the tire skids.

An analogous situation holds under longitudinal forces of braking and acceleration where the tire slips by "windup," producing a strain gradient in the wheel plane and a longitudinal stress distribution similar to that shown in Fig. 3. Here, the slip is the difference between vehicle and tire surface speeds, relative to vehicle speed. However, in braking, the tire gets so hot that the effective coefficient of friction of rubber against the road decreases and Fig. 4 becomes modified as shown in Fig. 5. A so-called friction peak is encountered, which antiskid devices try to maintain.

The initial slopes of curves like those shown in Figs. 4 and 5 are steeper for the stiffer radial construction so that more rapid responses and higher levels of tractive force are obtained for the same steering or braking (or accelerative) inputs.

Because the resultant of lateral and longitudinal forces cannot exceed the coefficient of friction times the weight on the tire, when both act together, as in a braked turn, the resistance to skidding is diminished.

Traction is maximized by the choice of tread materials having high hysteresis or energy dissipation, which raises the coefficient of friction of rubber. Addition of carbon black and the use of high styrene–butadiene copolymers have this effect.

4. Tire Wear

Life expectancy of tires has increased to approach 150 000 km because of advances in rubber compounding and tire design. Use of improved vulcanization systems and blending of elastomers, including the abrasion resistant *cis*-polybutadiene, has greatly improved the wear resistance of the tread. The almost universal adoption of the superior radial construction is another important factor.

However, radial tires are more prone than bias tires to uneven wear, in which parts of the tread pattern tend to wear away at a greater rate than their surroundings. When such concentrated wear reaches the vicinity of the belt, the tire must be replaced even though most of the tread remains. Uneven wear is minimized by empirical design of tread elements and by ensuring that the vehicle suspension does not impose severe camber or slip angles on the tire.

Wear proceeds by the mechanism mentioned in Sect. 3. At some point in the footprint the cumulative stress on the tread reaches the friction limit and the deformed tire slides back to its undistorted shape scrubbing the tread against the road. When wear occurs evenly over the tread face, the rate R ($m^3\,m^{-1}$) of material abraded from the tread in unit distance travelled is

$$R = \rho \gamma F^2 / k$$

where ρ is the tire resilience or fraction of energy not dissipated in a loading–unloading cycle and therefore available to produce abrasion; γ is the abradability ($m^3\,J^{-1}$), defined as the amount of material abraded for a unit of energy expended in sliding; F (N) is the root mean square value of the distortion force over the tire run, arising from accelerations imposed by the road configuration, by the necessity to stop and start, and by the vehicle speed and driver habit; and k is the appropriate stiffness coefficient in the ground plane, defined as the product of the half-length of the contact patch and the lateral or longitudinal force (depending on whether F refers to wear in cornering, or in braking and acceleration) for a unit of displacement of the contact patch with respect to the wheel rim.

Higher k means smaller excursions of the tread in slip and, therefore, shorter sliding paths and reduced wear. This explains the enhanced tread life of the stiffer radial over the softer bias construction.

See also: Elastomer–Elastomer Adhesion Processes; Elastomer–Textile Adhesion Processes; Elastomer–Wire Adhesion Processes

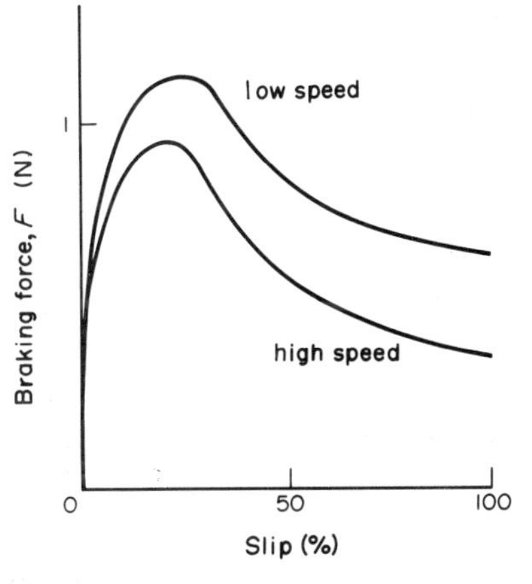

Figure 5
Braking force F vs slip (windup), expressed as the slip velocity relative to the circumferential velocity of the wheel

Bibliography

Anon 1990 Pneumatic tire centenary celebration 1988. *Prog. Rubber Plast. Technol.* 6(3): 201–98
Clark S K (ed.) 1981 *Mechanics of Pneumatic Tires,* 2nd edn. US Government Printing Office, Washington, DC
Eirich F R (ed.) 1978 *Science and Technology of Rubber*. Academic Press, New York
French T 1989 *Tire Technology*. Hilger, Bristol
Hays D F, Browne A L (eds.) 1974 *The Physics of Tire Traction.* Plenum, New York
Kovac F J 1978 Tire manufacture and engineering. In: Eirich 1978, Chap. 14
Morton G F, Quinton A R 1982 Manufacturing techniques: Pneumatic tires. In: Blow C M, Hepburn C (eds.) 1982 *Rubber Technology and Manufacture*. Butterworth, London, Chap. 10.1

A. N. Gent
[University of Akron, Akron, Ohio, USA]

D. J. Livingston
[Livington Associates, Akron, Ohio, USA]

Automotive Applications of Polymers

The range and diversity of materials used by the automotive industry is impressive and is listed in Table 1. Polymers play a critical role in the construction and performance of the automobile by virtue of their low weight, ease of fabrication, potential for parts integration, versatile shape and styling possibilities, and a range of mechanical properties that can be tailored to meet engineering needs. Polymers are used for bumpers, interior and exterior trim, fender extensions, front ends and instrument panels and are being evaluated for use in wheels, chassis members, driveshafts and springs. The US automobile industry consumes about 1.4 Mt of polymeric materials each year.

Table 1
Vehicle materials utilization as percentage of vehicle weight (1982)

Material	Percentage of vehicle by weight
Cast iron	14.0
Aluminum	5.0
Plastics	8.0
Steels	60.0
Glass	2.0
Others (e.g., elastomers, copper, lead)	11.0

Apart from physical and mechanical properties, there are unique social and economic forces that surround and govern the intense interest of the automobile industry in polymers:

(a) feedstock availability—from petroleum, natural gas, coal and biomass—is generally considered ensured;

(b) emphasis by the consumer on durability and abuse resistance highlights the advantages of corrosion-free and impact-resistant polymers and is a key factor in the gradually increasing use of elastomers;

(c) concern over resource conservation (both materials and fuel) has been important in emphasizing the weight-to-volume advantage of plastics and the importance of developing a recycling technology for these materials; and

(d) continued development by the chemical/plastics industry of new materials in conjunction with high-productivity fabricating processes offers to the automotive materials engineer a range of tailor-made systems from which to choose.

This article reviews the selection criteria employed by the automotive industry to optimize polymer usage in the automobile and indicates the important polymers and systems (excluding elastomers for tires) used for automotive applications.

1. Selection Criteria

The analysis, design and performance evaluation of potential plastic components and systems, compared with competing materials such as steel, aluminum and other plastics, is a complex process that presents continuing challenges to the automotive materials engineer. The selection criteria that lead to a particular system are very stringent. Major factors include availability from several sources to ensure a constant supply; cost; fabrication techniques that offer high productivity rates as well as being easily integrated into the automotive assembly line; suitability for painting and plating; performance on the assembly line, where vehicles may be exposed to metal-treating temperatures in excess of 200 °C; and performance of the finished part in the field, where creep, exterior durability and recyclability are major issues. The specific materials and applications that will be described illustrate these varied selection criteria.

2. Polyolefins

Polypropylene and polyethylene are relatively cheap compared with steel. Polypropylene is used in fender liners, dust shields, air conditioning and heating ducts, interior trim panels, battery cases and battery trays. It can be formulated to provide a range of properties of importance for automotive applications (see Table 2), molded and colored for novel styling effects, and electroplated to give durable chromeplated surfaces. Polypropylene can be easily fabricated by a number of techniques, including injection molding and sheet stamping—an efficient low-pressure forming process that is used to produce fender liners and air conditioning ducts.

These applications demand that the final part withstand the baking temperatures of the paint retouching step in vehicle manufacture, where temperatures above 120 °C are common. Additionally, temperatures in service may reach 100 °C. Proper formulation of polypropylene (with a high-modulus

Table 2
Mechanical properties of polypropylene (maximum values)

Propery	Homopolymer	Copolymer	40% Talc filled	40% Glass filled
Tensile strength to break (MPa)	41.4	37.9	34.5	103.4
Elongation to break (%)	600	700	8	4
Tensile modulus (GPa)	1.4	1.2	4.1	10.3
Flexural modulus (GPa)	1.7	1.4	4.3	6.9
Izod impact ($J\,cm^{-1}$)	0.54	10.85	0.33	1.08
Deflection temperature at 1.82 MPa (°C)	60	60	130	150

filler such as glass or mica, for example) yields a material with substantial stiffness at high temperatures that is suitable for instrument panels, trim panels and fan blades.

High-density polyethylene, which softens at somewhat lower temperatures than polypropylene, exhibits better impact strength at temperatures of about −30 °C and, hence, is used for fuel tanks. For such applications, the permeability of the polyolefin to gasoline, normally poor, must be improved on by coating the part with impermeable epoxy resin or fluorinating its inner surface to produce a polar, hydrocarbon-impermeable layer.

3. *Polypropylene Foams*

Foams are typically fabricated by injecting nitrogen gas into the polymer melt in the injection-molding operation. They possess a core density of about 800 kg m^{-3}, contain 20–50% air and exhibit excellent strength and stiffness. For most automotive applications, the relevant mechanical properties include the flexural modulus and the flexural, tensile, shear and compressive strengths. These properties reflect the properties of the base polymer. Although strength properties of foams are less than their solid counterpart, the rigidity of the finished part is a function of geometric factors and can be substantial. Further advantages of foams include absence of sink marks in large parts, dimensional stability (because the part is free from mold-induced stresses) and excellent sound damping characteristics. Since high stiffness-to-stiffness weight ratios are possible, thermoplastic foams find many applications for components where stiffness is the critical mechanical property (e.g., fender aprons, instrument panels and seat-back load floors).

4. *Polyurethanes*

The development of sophisticated urethane manufacturing systems—reaction injection molding mixing heads and metering devices—together with the versatile urethane polymers, which can be formulated to meet a variety of strength and stiffness needs, lies behind the use of urethanes for abuse-resistant front ends on passenger vehicles. Tensile strengths of up to 141 MPa, service temperatures of up to 100 °C and the ability to deform reversibly when subject to impact by a 1400 kg vehicle travelling at 8 km h^{-1} make polyurethanes well suited for applications that require high impact strength (e.g., bumpers, front ends, rear ends, fenders and doors).

5. *Engineering Thermoplastics*

The class of materials known as engineering thermoplastics includes the glass fiber-filled acetals, acrylonitrile butadiene styrene (ABS), nylons, polybutylene terephthalates (and alloys with other thermoplastics), polycarbonates and poly(phenylene oxide); newer materials under study include unfilled amorphous nylons (perhaps blended with other engineering thermoplastics). These systems exhibit mechanical and thermal properties substantially better than the polyolefins. The most important of these are their strength, stiffness, toughness, insensitivity to the multitude of solvents used in the automobile and stability up to about 120 °C. This combination of properties enables these materials to perform well in semistructural applications: gears, window mechanisms, instrument panels and fan blades; as well as in exterior small components such as fender extensions. The high-temperature stability of most of these materials permits the parts to be painted with the final top coat on the assembly line along with the metal components; this cost-saving feature is often a critical selection criterion, since off-line painting is costly and requires a different paint, resulting in problems with color matching. Properties of the major high performance automotive thermoplastics are given in Table 3.

Table 3
Mechanical properties of some glass-filled engineering thermoplastics

Material	Flexural modulus (GPa)	Tensile strength (MPa)	Deflection temperature at 1.82 MPa (°C)
ABS	4.8	75.8	200
Nylon	9.0	172.4	250
Polycarbonate	7.6	131.0	145
Poly(phenylene oxide)–elastomer blend	7.6	117.2	135
Acetal	7.6	127.6	165
Poly(butylene terephthalate)	9.0	158.6	225

6. *Unsaturated Polyesters*

The marked temperature sensitivity of the mechanical behavior of thermoplastics constitutes a significant difference between such plastics and metals. Although proper design and careful choice of materials can minimize this fundamental deficiency, applications involving exposure to temperatures in excess of around 100–125 °C require the greater stability of thermosets. The styrenated polyesters are highly crosslinked thermosets and can withstand the temperatures of the assembly-line metal cleaning and painting operations (>120 °C). This feature, together with the high modulus and low cost of these systems, has led to the development of sophisticated sheet molding processes (based on sheet molding compound (SMC)) for grille opening panels, fender extensions and truck tailgates. Flexural moduli of nearly 14 GPa are obtainable for mixtures of 30% random-chopped one-inch (2.54 cm) glass fiber, 30% calcium carbonate and 40% resin. The less desirable properties of these materials include difficulty in obtaining a perfectly smooth surface (because of solvent and gas liberation during cure), poor impact strength, and high density. Also, they tend to shrink in the mold, although this can be offset to a great extent by incorporating low-shrinkage additives (thermoplastics) into the molding formulation.

7. *High-Performance Thermosets*

Modifications of SMC systems, using both polyester and epoxy as the matrix resin with long glass fibers or even continuous glass filaments, result in high-strength formulations with excellent structural properties (see Table 4). Such materials comprise the category of newer, more exploratory automotive plastics and have the potential to replace steel parts such as transmission supports, energy-absorbing bumper parts, chassis members, road wheels and springs, with attendant weight savings of up to 50% over their metal counterpart. These materials are slowly finding their way onto vehicles for bumper beams and springs, although major developmental issues that still require resolution include a full understanding of mechanical properties in use (e.g., fatigue and creep) and the design of fabricating techniques to permit competitive manufacturing of the components. Filament winding, selective charge placement in the mold and pultrusion are promising fabrication processes under study.

Even longer-range developments indicate the versatility of high-performance resin–fiber composites and establish the fact that systems can be tailor-made to meet stringent structural automotive performance requirements (Table 5). Resin modification, such as the use of costlier epoxy resins, offers an improvement in the shear strength of the composite, and may be cost-effective for critical applications such as driveshafts. Carbon fiber and aramid fiber epoxy composites in many ways represent the ultimate composite system for automotive applications, offering potential weight reductions of up to 70% over steel, but with equivalent mechanical properties. However, a key issue for this new class of materials is their process development, that is, the requirement that such continuous-filament plastic composites be suitable for mass production. The continuous filament processes mentioned and the use of glass–graphite fiber hybrids may provide a unique combination of properties that may, for example, combine the desirable high extensibility of glass with the extraordinary stiffness of graphite fiber.

Table 4
Mechanical and physical properties of fiber-filled high-strength SMC (60% continuous and chopped glass fiber)

Tensile strength[a] (MPa)	551.6
Flexural strength[a] (MPa)	896.4
Flexural modulus[a] (GPa)	37.9
Density[b] (kg ,$^{-3}$)	1900

a In filament direction b For comparison, density of steel and aluminum are 7500 kg m^{-3} and 2500 kg m^{-3}, respectively

Table 5
Mechanical properties of carbon-filled epoxy composite (62 vol.% fiber, unidirectional orientation, measurements at room temperature)

Tensile strength (MPa)	1689.3
Compression strength (GPa)	1.2
Flexural modulus (GPa)	131.0

8. *Seals*

Seals are used extensively in automobiles. Typical examples are cylinder-liner seals, seals at rotating axles and seals in the cooling and fuel systems. These must work in different media such as air, coolant, brake fluid, grease, oil, diesel fluids, gasoline and mixtures of gasoline and alcohols in a temperature range of $-40\,°C$ to $150\,°C$.

Table 6 lists the most common rubber materials used. Only the main types are mentioned but relatively large modifications of the properties can be attained by altering the manufacturing processes such as the curing system, type of fillers and stabilizers.

Table 6
Seals: materials, applications

Material	Typical usage
Ethylene propylene rubber (EPDM)	coolant, air
Nitrile rubber (NBR)	fuels, oil, grease, temperatures <10 °C
Nitrile rubber/poly(vinyl chloride) (NBR/PVC)	fuels
Fluororubber (FPM)	fuels, oil, grease, temperatures <200 °C
Silicone rubber (MVQ)	oil, grease, temperatures <175 °C

In the engine, the cylinder-liner oil seal is often made of fluororubber (FPM) because of its high temperature resistance and compatibility with engine oils. The seal in contact with coolant is usually ethylene propylene rubber (EPDM), a material with a good resistance to water, glycols and coolant additives.

A typical seal for a rotating axle is the crank-shaft seal that is made in silicone rubber (MVQ) or FPM because of the high-temperature requirements.

Within the cooling system there are different sealing parts, for example the seal between plastic radiator tank and radiator package. The rubber material that can fulfill all the requirements, that is, aging resistance, compression set and ozone resistance, is EPDM.

The main requirement for the material in seals in the fuel system is, of course, resistance to different fuels. At lower temperatures, nitrile rubber (NBR) is commonly used while the much more expensive FPM is recommended at temperatures above 100 °C and when special requirements for a low compression set are stated.

9. Other Materials

In addition to the newer polymers described, the automobile industry utilizes a variety of other, more conventional, polymeric systems: plasticized poly(vinyl chloride) in seating, roof coverings, interior "soft" and "padded" trim and headliners; nylon fiber for carpeting and seating covers; and specialty elastomers, such as acrylates, epichlorohydrins, silicones and neoprenes for hoses, gaskets and engine mounts.

See also: Automobile Tires; Automotive Composite Components: Fabrication; Composites: Applications Overview

Bibliography

American Chemical Society 1977 *Impact of the Automobile on the US Polymer Industry*, ACS Symposium Series 72. American Chemical Society, Washington, DC

Avery S 1985 Advanced composites. *Mod. Plast.* 62:114

McCann M K 1979 Light trucks—Drive toward efficiency. *Automat. Ind.* 159(11):38–52

Wood A S 1982 All of a sudden, there are more materials for automotive jobs—And other markets too. *Mod. Plast.* 59(4):54–7

Wood A S 1984 Technological breakthroughs in RIM and SMC. *Mod. Plast.* 61:48–55

W. Burlant
[General Electric Company, Columbus, Ohio, USA]

Automotive Composite Components: Fabrication

There is a fundamental difference in the strategy of application of composites between the aerospace industry and the automobile industry. This is primarily due to the volume requirements of the two businesses. In aerospace and defense, the design of the structure is optimized to provide the required functionality and performance, and the manufacturing process (and associated cost) is subsequently selected on the basis that the process is capable of achieving the desired design. In direct contrast, in high-volume production industries such as the automotive industry, the rate of manufacture is critical to satisfying the economics of this consumer industry. Thus, manufacturing processes which are capable of satisfying production output are the primary consideration and design of a component or structure must be within the boundary constraints of the selected fabrication process. It is extremely important to appreciate this philosophical difference of approach in order to understand the difficulty of translating the extensive design and fabrication experience in aerospace and defense into the automotive industry. Perhaps the clearest example is the use of prepreg materials and hand layup procedures, common in aerospace and amenable to optimal design, which is clearly unacceptable in an industry requiring high manufacturing output but with a quality level of the same degree.

The use of plastic-based composite materials in automotive applications has gradually evolved during the 1970s and 1980s. Virtually all uses of these materials in high-volume vehicles are limited to decorative or semistructural applications. Sheet-molding component (SMC) materials are the highest-performance composites in general automotive use and the most widely used SMC materials consist of approximately 25 wt% chopped glass fibers in a polyester matrix and so cannot really be classified as high-performance composites. Typically, SMC materials are used for grille opening panels on many car lines and closures panels (hoods, deck lids and doors) on a few select models. A characteristic molding time for SMC is of the order of 2 min, which is on the borderline of viability for automotive production rates. A recent breakthrough by one manufacturer has reduced this cycle time to 1 min.

The next major step for composites in the automotive business is the extension of usage into truly structural applications such as primary body structure and chassis/suspension systems. These structures have to sustain the major road load inputs and crash loads and must deliver an acceptable level of vehicle dynamics so that passengers enjoy a comfortable ride. These functional requirements must be totally satisfied for any new material to achieve

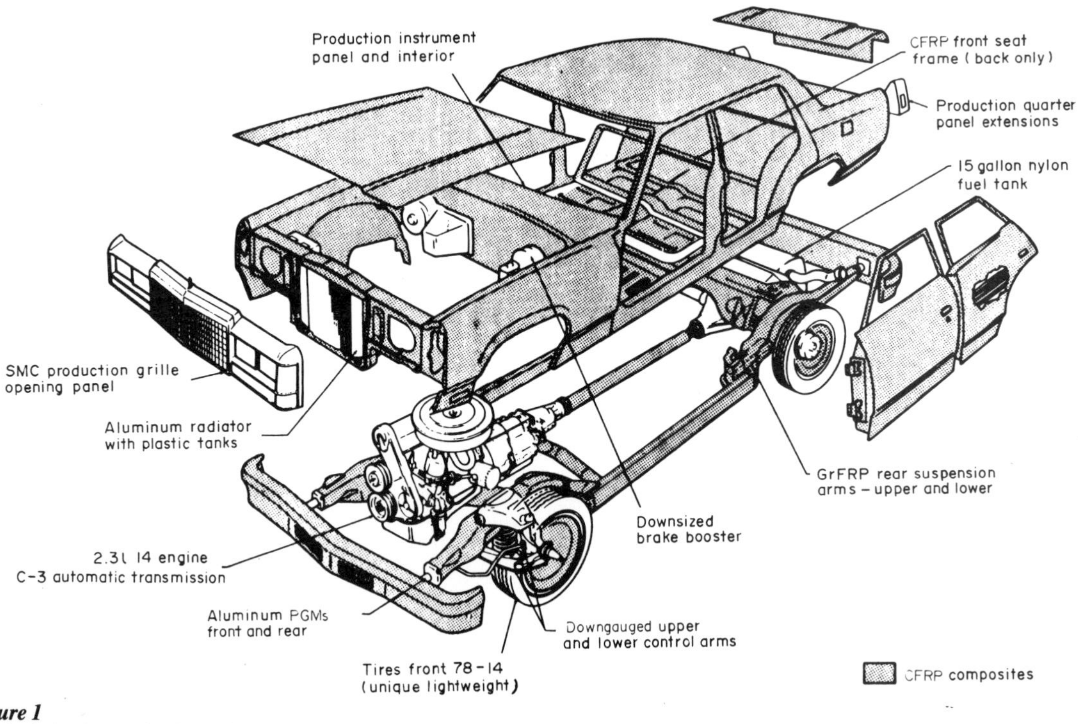

Figure 1
Exploded schematic view of a composite vehicle, with delineation of the composite components (GrFRP, graphite-fiber-reinforced plastic)

extensive application in body structures and must do so in a cost-effective manner. Appropriate composite fabrication procedures must be applied or developed which satisfy high production rates but still maintain the critical control of fiber placement and distribution.

1. Composite Materials for Automotive Usage

By far the most comprehensive property data have been developed on aerospace composites, in particular carbon-fiber-reinforced epoxy designed for fabrication by hand layup from prepreg materials. Relatively extensive databases are available on these materials and it would be very convenient to be able to build from this database for less esoteric applications such as automotive structures. Carbon fibers are the preferred material in aerospace applications because of their superior combination of stiffness, strength and fatigue resistance. Unfortunately for cost-sensitive mass-production industries, these properties are only attained at significant expense (typically carbon fibers cost US$55 or more per kilogram at 1988 prices). Intensive research efforts are being devoted to reducing these costs by utilizing a pitch-based precursor for production of these fibers, but the most optimistic cost predictions are around US$22 kg^{-1} which would severely limit the potential of these fibers for use in consumer-oriented industries. To illustrate the potential that carbon fiber could offer the automobile industry if a breakthrough ever occurred in the reduction of cost, it is interesting to summarize the data on the prototype carbon fiber LTD built by Ford to compare directly with a production steel vehicle (Beardmore et al. 1980, Kulkarni and Beardmore 1980). Although the vehicle was fabricated by hand layup procedures, several interesting features were evaluated. An exploded schematic showing the composite parts of the carbon-fiber composite car, made of carbon-fiber-reinforced plastic (CFRP), is shown in Fig. 1. Virtually all structures were designed using aerospace techniques, utilizing 0°/90° and ±45° prepreg materials. The weight savings for the various structures are given in Table 1. While these weight savings (of the order of 55–65%) might be considered optimal because of the use of carbon fibers, other more cost-effective fibers can achieve a major proportion of these weight savings (see Sect. 2.3). Although the CFRP vehicle weighed 1138 kg compared to a similar production steel vehicle of 1705 kg, vehicle evaluation tests indicated no perceptible differences between the vehicles. Ride quality and vehicle dynamics were judged at least equal to top-quality production steel LTD cars. Thus, on a

Table 1
Major weight savings in a CFRP vehicle

Component	Steel (kg)	Steel (lb)	CFRP (kg)	CFRP (lb)	Reduction (kg)	Reduction (lb)
Body-in-white	192.3	(423.0)	72.7	(160.0)	115.0	(253.0)
Front end	43.2	(95.0)	13.6	(30.0)	29.5	(65.0)
Frame	128.6	(283.0)	93.6	(206.0)	35.0	(77.0)
Wheels (5)	41.7	(91.7)	22.3	(49.0)	19.4	(42.7)
Hood	22.3	(49.0)	7.8	(17.2)	14.7	(32.3)
Decklid	19.5	(42.8)	6.5	(14.3)	13.1	(28.9)
Doors (4)	64.1	(141.0)	25.2	(55.5)	38.9	(85.5)
Bumpers (2)	55.9	(123.0)	20.0	(44.0)	35.9	(79.0)
Driveshaft	9.6	(21.1)	6.8	(14.9)	2.8	(6.2)
Total vehicle	1705	(3750)	1138	(2504)	566	(1246)

direct comparison basis, a vehicle with a structure made entirely of fiber-reinforced plastic (FRP) was proven at least equivalent to a steel vehicle from a vehicle dynamics viewpoint, at a weight of only 67% of that of the steel vehicle.

The CFRP car clearly showed that high-cost fibers (carbon) and high-cost fabrication techniques (hand layup) can yield a perfectly acceptable vehicle based on handling, performance and vehicle dynamics criteria. However, crash and durability performance were not demonstrated and these will need serious development work. An even bigger challenge is to translate that performance into realistic economics by the use of cost-effective fibers, resins and fabrication procedures.

The fiber with the greatest potential for automobile structural applications, based on optimal combination of cost and performance, is E-glass fiber (costing approximately US\$1.80 kg^{-1} at 1988 prices). Likewise, the resin systems likely to dominate, at least in the near term, are polyester and vinyl–ester resins based primarily on a cost–processibility trade-off. High-performance resins will find only specialized applications (in much the same way as carbon fibers) even though their ultimate properties may be somewhat superior.

The form of the glass fiber will be very application specific and both chopped and continuous glass fibers will find extensive use. It is expected that most of the structural applications involving significant load inputs will utilize a combination of both chopped and continuous glass fiber with the particular proportions of each depending on the component or structure. Since all the fabrication processes anticipated to play a significant role in automotive production are capable of handling mixtures of continuous and chopped glass, this requirement should not present major restrictions. One potential development which is likely if glass fiber composites come to occupy a significant portion of the structural content of an automobile is the tailoring of glass fiber and corresponding specialty-resin development. Approximately 35×10^6 vehicles per year are produced worldwide and consequently each kilogram of composite per vehicle implies the usage of 35×10^6 kg in the industry as a whole. This dictates that it should be economically feasible to have fiber and resin production tailored exclusively for the automobile industry. The advantage of such an approach is that these developments will lead to incremental improvements in specific composite materials, which in turn can promote usage.

Glass-fiber-reinforced composites must be capable of satisfying the functional requirements of the various elements of the structure in an automobile. The three primary criteria are fatigue (durability), energy absorption and ride quality (vehicle dynamics). Provided sufficient data are available on all three characteristics, it should then be possible to utilize these to satisfy the design requirements of the automobile. For example, unidirectional glass-fiber-reinforced plastic materials typically have a well-defined fatigue limit of the order of 35–40% of the ultimate strength (Dharan 1975). By contrast the chopped-glass composite would have a fatigue limit closer to 25% of the ultimate strength (Smith and Owen 1969, Aotem and Hashin 1976) and would exhibit much greater scatter in properties. Clearly, components and structures composed of a mixture of continuous and chopped fibers would have fatigue resistance specifically related to the proportions of each type of fiber. There is clear evidence that glass-fiber-reinforced composites can be designed to withstand the rigorous fatigue loads experienced under vehicle operating conditions. It must be emphasized, however, that fatigue design data are a critical function of the relevant manufacturing procedure.

In a similar manner, evidence is accumulating that fiber-reinforced-plastic composites can be efficient energy absorbing materials. Relative data from the

Table 2
Energy absorption (typical properties)

Material	Relative energy absorption per unit weight
High-performance composites	100
Commercial composites	60–75
Mild steel	40

collapse of tubes is given in Table 2 to illustrate this point and is based on the extensive work of Thornton (1979), Thornton and Edwards (1982) and Thornton *et al.* (1985). The fragmentation–fracture mechanism of energy absorption typical of glass-fiber-reinforced composites, as compared to the plastic deformation mechanism in materials, is actually very weight effective. The real issue is the translation of this effective fracture mechanism into complex structures.

The third and less quantifiable requirement for composites in vehicle applications is vehicle dynamics. Glass-fiber-reinforced composites are inherently less stiff than steel, typically by factors of up to ten (see Table 3). However, there are two offsetting factors to compensate for these apparent deficiencies. First, an increase in section thickness can be used to partially offset the decreased material stiffness; also, the flexibility of composite fabrication processs allows the thickening of local areas as is required to optimize properties. Since the composite has a density approximately one-third that of steel, a significant increase in thickness can be achieved while maintaining an appreciable weight reduction. The second, and perhaps major, compensating factor is the additional stiffness attained in composite structures by virtue of part integration. This integration leads directly to the elimination of joints, which results in a significant increase in effective stiffness. It is becoming increasingly evident that this synergism is such that structures of acceptable stiffness and considerably reduced weight are feasible in glass-fiber-reinforced composites. As a rule of thumb, a glass-fiber-reinforced plastic structure with significant part integration relative to the steel structure being replaced can be designed for a nominal stiffness of 50–60% of that of the steel structure. Such a design procedure should lead to adequate stiffness and typical weight reductions of 30–50%.

Table 3
Typical stiffness of composites

Material	Modulus (KPa)
Unidirectional CFRP	137.8
Unidirectional GFRP	41.3
Unidirectional Kevlar	75.8
XMC[a]	31.0
SMC-R50[b]	15.8
SMC-25	9.0

a Sheet-molding compound containing oriented continuous glass fiber
b Sheet-molding compound containing 50% chopped glass fiber

2. *Potential Applications*

The potential use of composites in structural applications in automobiles can be designated into two categories: the direct replacement of existing components and the integration of multiple steel components into one composite component or structure. The second category, involving part integration, is by far the most cost effective and ultimately will predominate. In the shorter term, however, much experience and confidence in the capabilities of composites is being generated by the singular component programs currently in production.

2.1 One-On-One Component Substitution

The significant differences in material costs between glass-fiber-reinforced plastic and steel (approximately US\$1.80 kg^{-1} versus US\$0.55 kg^{-1} at 1988 prices) means that, even with significant weight saving, direct substitution in a singular component can rarely be cost effective. There are two examples currently in production which can be cited in the direct substitution category, namely driveshafts and leaf springs.

Composite driveshafts have been utilized in production on at least two vehicles in North America. One example is in the special drive line configuration of the Ford Econoline Van. These vehicles would normally utilize a two-piece steel driveshaft incorporating a connecting center bearing. The total length of the drive line dictates that a two-piece steel shaft must be used because unacceptable vibrations occur in a one-piece steel driveshaft. In contrast, a one-piece composite driveshaft will provide satisfactory vibratory characteristics since the lower weight combined with high stiffness satisfies the bending-frequency requirements. The driveshaft is fabricated by filament winding on a continuous machine operating at speeds up to 2 m min^{-1}. The longitudinal fibers (0°) are carbon (220 Gpa modulus), to generate the required bending stiffness, and the ±45° fibers are E-glass, to provide the torsional strength. The resin used is a vinyl ester which gives the appropriate combination of properties, processability and cost. This particular driveshaft is economically feasible only because it results in the elimination of the center bearing—such cost-effective usage is impossible in the case of single-piece steel (or aluminum) driveshafts.

The leaf spring is an example of a singular component utilizing glass-fiber-reinforced epoxy which has

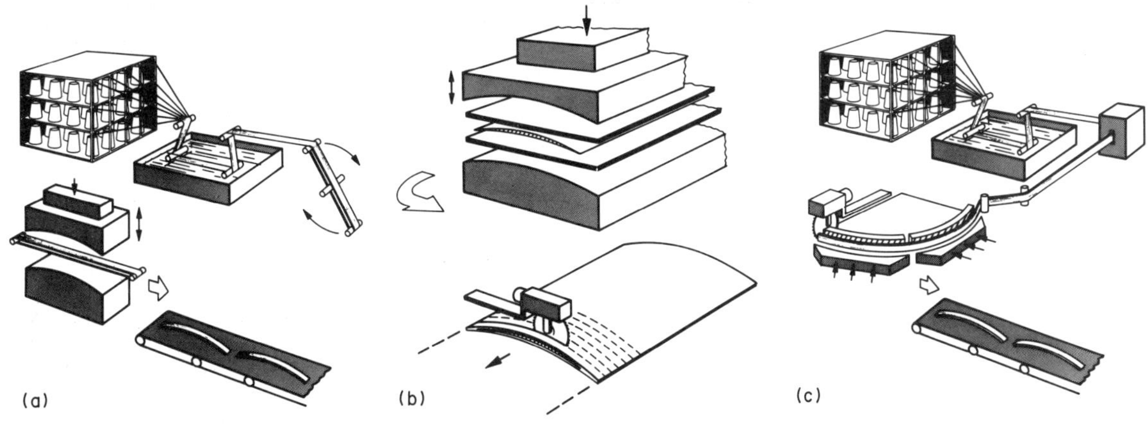

Figure 2
Different fabrication techniques for FRP leaf springs: (a) filament winding–compression molding, (b) compression molding and (c) pultrusion

found a niche in the North American vehicle market. Current production in the USA is greater than 500 000 FRP springs per year and their use to date has been a reliable demonstration of the feasibility and durability of composites. FRP leaf springs can be mass produced by three basic techniques, although there may be several variants on each particular process. The basic materials are always the same, namely continuous E-glass fiber and epoxy resin. (Attempts to substitute more economical resins have all been unsuccessful because of creep requirements.) The three potential techniques are illustrated in Fig. 2. The compression-molding technique involves development of a preform (usually by a filament-winding procedure) followed by molding and is the process currently used in the USA. An alternative procedure involves the automated layup of a wide slab of material, followed by molding under pressure and subsequent slitting of the slab to form springs. The third, least developed, technique is the pulforming procedure which has perhaps the greatest long-term potential for minimizing cost.

One for-one replacements of steel components with FRP components will be exceedingly rare, primarily because of economic considerations but also because of complex attachment considerations. As a general principle, therefore, multiple-part integration to reduce assembly costs will be necessary to make FRP structures cost effective relative to steel.

A simple example of this principle is demonstrated by the prototype composite integrated rear suspension for a Ford Escort (Morris 1986). The basic steel rear suspension and the replacement FRP integrated suspension are shown in Fig. 3. A simple extension of FRP leaf-spring design methodology allowed substitution for the complete steel rear suspension by allowing the FRP spring to perform the dual function of spring and suspension arms. The function and part integration resulted in a weight saving of approximately 4 kg (about 50% weight reduction). In addition to the weight saving and cost-effectiveness of this type of synergism, there are obvious improvements in package space, which is a major consideration in vehicle design.

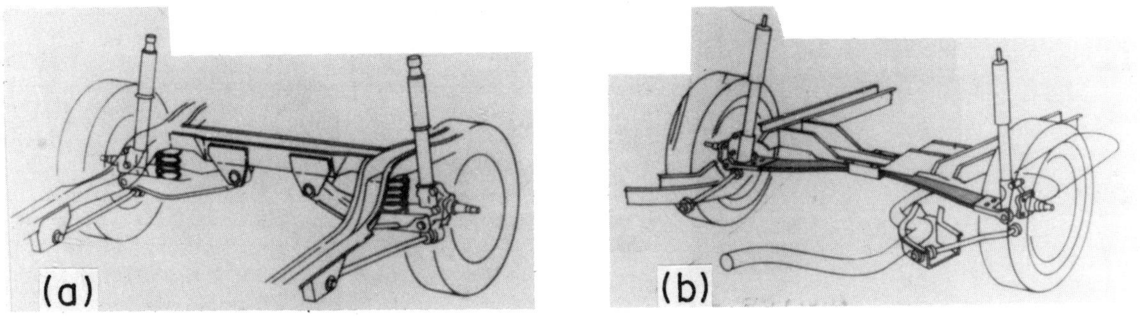

Figure 3
Comparison of (a) production Ford Escort steel rear suspension and (b) corresponding integrated composite suspension

2.2 Large Integrated Structures

The successful application of structural composites to large integrated automotive structures is more dependent on the ability to use rapid and economic fabrication processes than on any other single factor. The fabrication process must also be capable of close control of composite properties to achieve lightweight, efficient structures. Currently, the only commercial process which comes close to satisfying these requirements is compression molding of sheet molding components (SMC) or some variant of the process. There are, however, processs still at the development stage which hold distinct potential for the future in terms of combining high production rates, precise fiber control and high degrees of part integration. In particular, high-speed resin-transfer molding (HSRTM) offers these potential benefits, provided technical developments can be achieved. The following examples demonstrate these fabrication approaches and part-integration concepts which can be utilized in automotive production and compare the utilization of both SMC and HSRTM procedures. It is not the intent here to describe these two processes in detail but for illustrative purposes the techniques are summarized in Fig. 4. Further details on these and other fabrication procedures may be obtained from the *ASM Composite Engineered Materials Handbook* (Johnson 1988).

2.3 Primary Body Structure

The primary body structure of an automobile consists of approximately 250–350 major steel parts and utilizes approximately 300 assembly robots. For comparison of the two alternate composite construction techniques, consider the body side assembly. In Fig. 5a, a typical SMC approach is illustrated in which the complete body side consists of two moldings which would be bonded together. The HSRTM procedure for the same structure is shown in Fig. 5b and the major difference is the elimination of the adhesive bonding and the incorporation of a foam core. Utilizing either of these composite fabrication

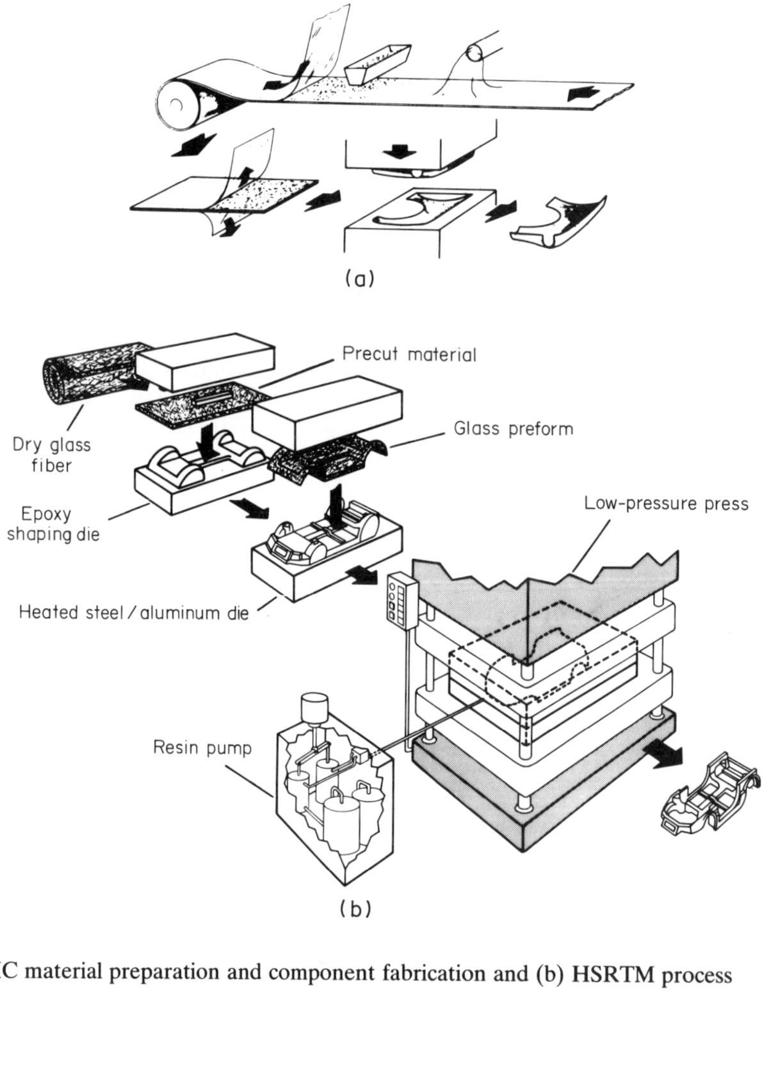

Figure 4
Depiction of (a) SMC material preparation and component fabrication and (b) HSRTM process

techniques for the other major segments of the body shell would lead to a typical body construction assembly as illustrated in Fig. 6a. In terms of reduction in parts, the molded SMC body structure could consist of 10–20 major parts (compared to approximately 300 major steel parts) and the HSRTM structure could be composed of 2–10 major parts. Note that the degree of integration is higher for HSRTM, reflecting the greater versatility of this procedure. Clearly the lower limit of just two major parts for the HSRTM technique would require assembly as illustrated in Fig. 6b.

While the all-out application of composites to the total body shell, as discussed previously, is appealing from the viewpoint of the materials connoisseur, reality dictates that the usage of composites in automobile structures will be evolutionary and progressive rather than revolutionary in nature. Therefore, it is to be expected that structural segments of vehicles will be first to utilize these procedures and there are several examples of prototype structures where these techniques have been used. Complex, high-

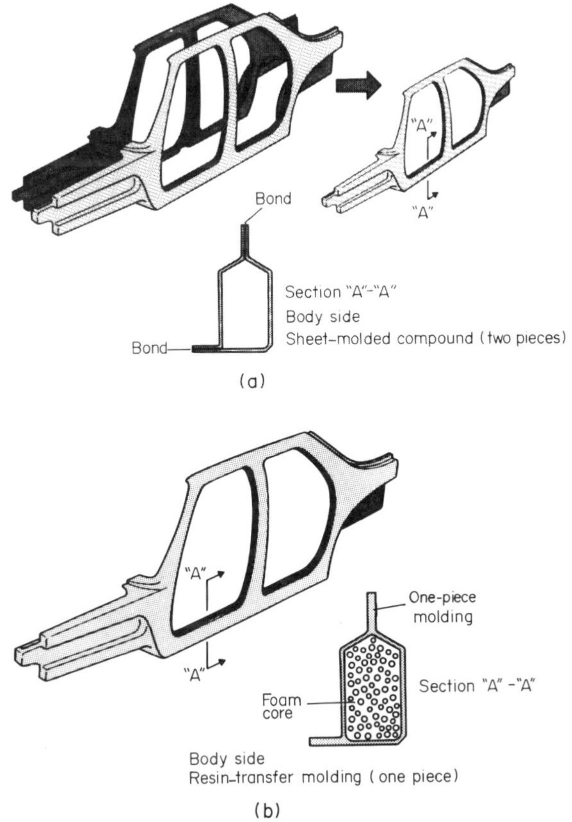

Figure 5
Typical assembly of composite body side panel by (a) adhesive bonding of inner and outer SMC molded panels and (b) HSRTM panels with foam core

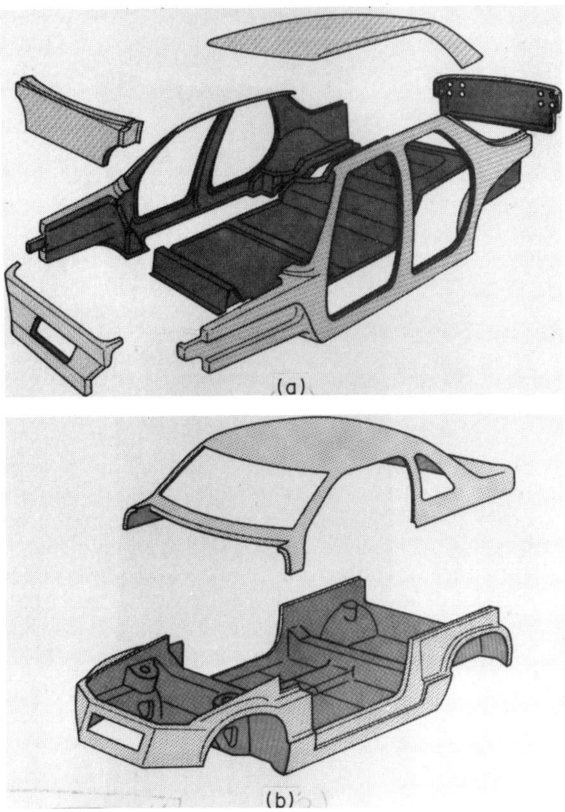

Figure 6
Composite body shell: (a) typical assembly and (b) assembly using two major moldings

load cross members have been developed for evaluation by both Chrysler (Farris 1987) and Ford. The cross member is fabricated from glass-fiber-reinforced vinyl ester utilizing the resin-transfer molding (RTM) process. The unique cross member is projected to save approximately 30% in weight relative to the comparable steel member. The fabrication and testing of these high-load, critical members is part of the gathering of detailed information on the realistic expectations that glass-fiber-reinforced plastic composites can satisfy the functionality required. Concurrently, the fabrication technique (RTM) has to be further developed to satisfy the economic constraints of the industry. The success of these prototypes is providing the impetus to drive the fabrication developments towards the required goal. A similar example concerning the fabrication of a highly integrated composite Escort front structure by RTM (Johnson et al. 1985) has been well documented and reported. In this prototype, 42 steel parts were replaced by one RTM molding resulting in a 30% reduction in weight and structural stiffness in excess of that of the steel structure. This structure is currently being utilized to develop energy absorption capability in a typical vehicle design.

3. Summary

The extension of composites use to automotive integrated structures will require an expanded knowledge of the design parameters for glass-fiber-reinforced plastic materials together with major innovations in fabrication techniques. There is accumulating evidence, both in the laboratory and in prototype vehicles, which strongly indicates that glass-fiber-reinforced composites are capable of meeting the functional requirements of highly loaded automotive structures. The more imperative requirement is the cost-effective fabrication advancements that appear necessary to justify such increased use of composites. High-volume, less stringent performance components can be manufactured by variations on compression-molding techniques. However, the high-volume, high-performance manufacturing techniques still need development and improved SMC materials and processes, and the HSRTM process holds promise in these areas.

See also: Automotive Applications of Polymers; Composites: Manufacturing Overview

Bibliography

Aotem A, Hashin Z 1976 Fatigue failure of angle ply laminates. *AIAA J.* 14: 868–72

Beardmore P, Harwood J J, Horton E J 1980 Design and manufacture of a GrFRP concept automobile. *Proc. Int. Conf. on Composite Materials*, Paris, August 1980

Dharan C K H 1975 Fatigue failure in graphite fiber and glass fiber–polymer composites. *J. Mater. Sci.* 10: 1665–70

Farris R D 1987 Composite front crossmember for the Chrysler T-115 mini-van. *Proc. 3rd Annual Conf. Advanced Composites*. American Society for Metals, Metals Park, OH, pp. 63–73

Johnson C F 1988 Compression molding; Resin transfer molding. In: *ASM Composite Engineered Materials Handbook,* Vol. 1. American Society for Metals, Metals Park, OH, pp. 559–68

Johnson C F, Chavka N G, Jeryan R J 1985 Resin transfer molding of complex automotive structures. *Proc. 41st Annual Conf. SPI,* Paper 12a. Society of the Plastics Industry, New York

Kulkarni H T, Beardmore P 1980 Design methodology for automotive components using continuous fiber reinforced materials. *Composites* 12: 225–35

Morris C J 1986 Composite integrated rear suspension. *Compos. Struct.* 5: 233–42

Smith T R, Owen M J 1969 Fatigue properties of RP. *Mod. Plast.* 46(4): 124–9

Thornton P H 1979 Energy absorption in composite structures. *J. Compos. Mater.* 13: 262–74

Thornton P H, Edwards P J 1982 Energy absorption in composite tubes. *J. Compos. Mater.* 16: 521–45

Thornton P H, Harwood J J, Beardmore P 1985 Fiber reinforced plastic composites for energy absorption purposes. *Compos. Sci. Technol.* 24: 275–98

P. Beardmore
[Ford Motor Company, Dearborn, Michigan, USA]

B

Blow Molding of Thermoplastics

Plastics blow molding is the process by means of which articles are formed by inflation of a molten resin to fill a mold cavity having the desired shape and dimensions. The mold cavity is usually cooled so as to cause the solidification of the article before the mold is opened and the product ejected. The technique of manufacturing hollow articles by the use of internal fluid pressure owes its initial development to the glass-blowing industry, which dates back to early Egyptian culture. The suggestion that containers could be molded from polymeric resins by forming and inflating a preformed molten cylinder or parison appeared for the first time in 1880, as Celluloid Novelty Co. of New York filed the first patent application covering the extrusion blow-molding process.

1. Extrusion Blow Molding

The most commonly used commercial process for the production of blow-molded articles is based on extrusion blow molding. This process can be divided into two main steps: plastication and blow molding. In the plastication stage, the polymeric resin, in solid pellet, powder or bead form, is conveyed from the hopper by means of an extruder screw. As it travels through the extruder, the resin is plasticated by heating and melting with the aid of mechanical friction and/or external heating. As it leaves the metering section of the extruder screw, the molten resin is pumped through an annular die to form a molten tube or parison with well-defined and controlled dimensions.

The blow-molding stage starts after the complete length of the parison has been extruded. At this point, the parison is clamped between the two mold halves and inflated by internal fluid pressure to take the shape of the mold cavity, which is usually cooled. Finally, the formed article solidifies as a result of cooling and the mold is opened to eject the article without damage.

This cycle is repeated to produce subsequent articles. A sketch of the extrusion blow-molding sequence is given in Fig. 1. A typical cycle and the distribution of cycle times during the various steps in the extrusion blow-molding sequence are illustrated in Fig. 2.

Depending on the thermal stability of the material, the extrusion process can be continuous or intermittent. The former technique is employed with materials having lower stability. Continuous extrusion may be employed in conjunction with one die feeding one mold or a multiplicity of molds rotating on a turntable. Alternatively, but at a higher cost and with the potential risk of inferior product quality, higher production rates might be obtained by using multiple die heads in conjunction with a larger extruder.

A variety of techniques may be employed to achieve the intermittent extrusion of the parison. In the reciprocating screw system, the extruder screw operates as a ram, in the same manner as in injection molding. Other systems employ a ram and an accumulator die head system, as in two-stage injection molding systems. Such an arrangement helps to overcome many of the shortcomings of intermittent-extrusion blow molding and particularly contributes to parison uniformity and thermal stability.

Commercial extrusion blow-molding systems cover a broad range of operating conditions and sizes. The ranges covered by typical commercial extrusion blow-molding machines are listed in Table 1.

2. Parison Programming

The dimensions of the ensuing parison are fundamentally influenced by the design of the die and several types of dies are commonly utilized. Typical designs include dies with cylindrically parallel annu-

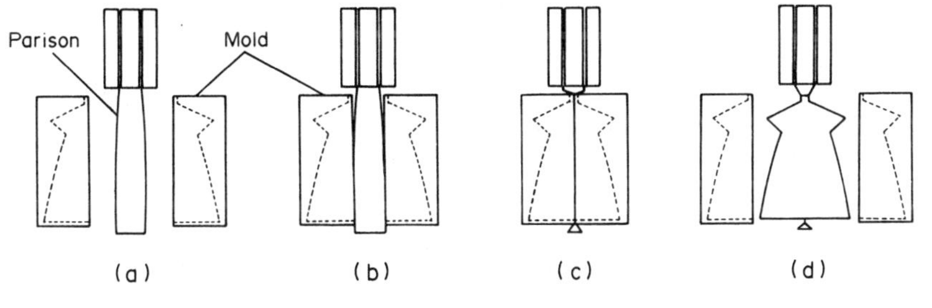

Figure 1
Sketch of the extrusion blow-molding sequence: (a) parison extrusion, (b) mold closing and parison pinch-off, (c) parison inflation and (d) mold opening

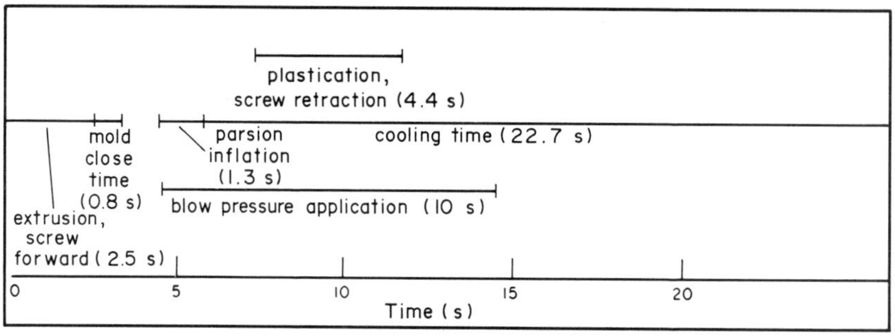

Figure 2
Diagram of a typical blow-molding cycle sequence

lar gaps and a compression section in the inflow zone. Tulip and dished or trumpet-shaped dies and dies with expansion inserts are commonly used to facilitate the adjustment of the die gap during extrusion.

It is of vital importance to control the dimensions of the parison, because the dimensions and hence the weight distribution of the parison determine the final thickness distribution of the blow-molded article. Thus, many extrusion blow-molding machines are equipped with parison programmers, which generate a parison wall thickness distribution and also control the overall article weight and its ultimate mechanical and optical properties. A typical thickness distribution in a small blow-molded high-density polyethylene bottle is shown in Fig. 3. Here, the bottle has been split vertically along its axis and effectively flattened to enhance graphical visualization.

There are essentially two major methods of programming the parison dimensions as a function of time or length. The first employs a constant die opening in conjunction with variation of the material flow rate as a function of time. In the second programming method, the extrusion speed is held constant and the die opening is varied as a function of time to generate the desired profile. In addition to the conventional solid-state controllers for parison programming, microprocessor-based programmable controls have been effectively employed for parison programming.

Table 1
Operating ranges of commercial extrusion blow-molding systems

Operating variable	Continuous	Intermittent
Extruder capacity ($kg\,h^{-1}$)	2.5–500	2.5–2500
Shot capacity (kg)		0.1–150
Clamp force (t)	1.4–30	3.0–300
Number of molds	1–18	1–12
Number of die heads	1–12	1–12
Number of blowing stations	1–18	1–12
Platen area (length × width, mm)	150 × 200 to 2000 × 300	150 × 200 to 2900 × 2000

3. *Parison Dynamics*

The dimensions of the parison are influenced to a large extent by phenomena that occur after the parison leaves the die. Normally, unless the extrusion stage is too long, cooling is minimal and the

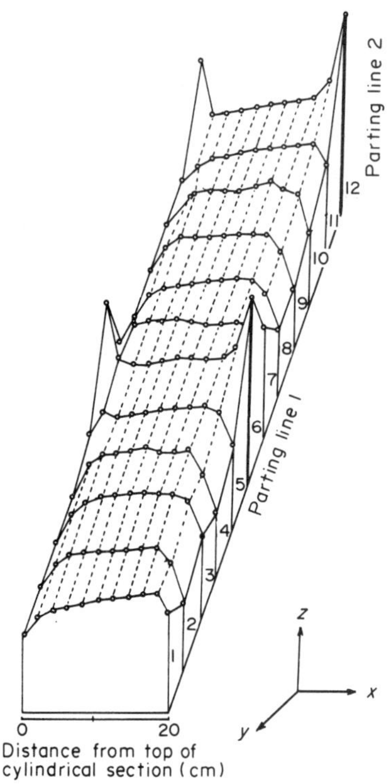

Figure 3
Typical three-dimensional thickness distribution in blow-molded sample (after Kamal et al. 1982)

parison maintains the original temperature distribution in the melt, except for a very thin zone near the outside surface, which is in contact with ambient air. Under these conditions, two important phenomena affect parison dimensions. These phenomena are parison swell and sag.

Parison sag or drawdown is due to gravitational forces acting on the parison. This phenomenon is associated with the stretching of the parison and the resulting reduction in its diameter and cross-sectional area. In a vertical parison, the minimum effect of sag is experienced by the leading (bottom) end, while the maximum effect is observed at the top, near the die exit.

Parison swell is associated with stress relaxation and melt recovery as the molten polymer leaves the die. It is a result of the viscoelasticity and fading memory of these melts, which are also characterized by the ability to snap back when the external deforming forces are removed. The net effect of these phenomena in extrusion blow molding is to cause the diameter and thickness of the parison to be larger than those of the extrusion die and to cause a springback of the parison, leading to an effective shortening of its length. Thus, the phenomena of parison swell and sag act in opposite directions.

Various techniques have been devised to measure parison swell and sag as functions of operating conditions and to relate them to more fundamental properties such as capillary die swell and other rheological parameters. Some of these techniques involve high-speed cinematography and the conventional pinch-off mold. Although high-speed photography provides more accurate data, the pinch-off mold is used more commonly, due to its simplicity and case of data analysis.

The pinch-off mold, which is normally water cooled, closes onto the parison and pinches it accurately into a number of segments. The mold remains closed until the segments solidify, at which point the mold is opened. By cutting the cooled parison on the pinch lines and weighing the separated segments, the weight and diameter distribution of the material is determined as a function of parison length. Three swell parameters are usually defined:

$$\text{weight swell, } S_w = \frac{4W_i}{\pi(D_o^2 - D_i^2)\rho\Delta L}$$

$$\text{diameter swell, } S_d = \frac{D_2}{D_o}$$

$$\text{thickness swell, } S_t = \frac{D_2 - D_1}{D_o - D_i}$$

where W_i is the weight of a parison segment with length ΔL; D_o and D_i are the outside and inside diameters, respectively, of the extrusion die; D_2 and D_1 are the undeformed outside and inside diameters, respectively, of the parison before pinch-off; and ρ is the density of the polymer melt.

In general, the weight and diameter swell parameters are increased as the shear rate in the die is increased or as the melt temperature is lowered. Weight swell decreases, up to a certain point, with increasing land length. Both diameter and weight swell tend to be greater for extrusion through converging dies than through straight or diverging dies.

4. *Parison Clamping and Inflation*

Blowing systems employ needle blow, which introduces air through a needle inserted in the side wall producing a relatively small hole, or the blow-pin method, which introduces the blowing device into the top or bottom of the parison. Calibration mandrel blow is employed to compression mold and finish the neck in addition to blowing the container. Sometimes the blow pin is rotated to eliminate some finishing operations.

The inflation of the parison involves the unsteady-state planar extension of a viscoelastic polymeric parison exhibiting outside diameter and thickness variation over its entire length. End effects are significant and the strain rates vary during the inflation. High-speed cinematography has been employed to study the inflation behavior during typical blow-molding conditions. The rate of inflation and the magnitude of stresses experienced by the material depend on the blowing pressure and the temperature of the melt during inflation.

In general, the circular parison is deformed during the clamping stage and takes on an oval or ellipsoidal shape, especially in the regions near the ends where it is clamped. Moreover, the segments near the middle part of the parison contact the mold wall before the other parts. The corners at the bottom and the top undergo complex deformations and they fill the cavity at the end of the inflation stage.

5. *Cooling and Solidification*

The optimization of the cooling system is an important aspect of the blow-molding process, since it greatly affects production rates, production costs and product quality. High cooling rates are required to increase production rates, especially since the cooling step tends to be the longest component of the blow-molding cycle. On the other hand, high cooling rates could lead to frozen stresses and deficiencies in the product. Refrigerated fluids have been used to cool the mold, and, in some instances, such fluids are blown inside the molding to enhance cooling rates from both sides. Also, molds are designed with allowance for zone cooling to compensate for thicker sections and higher cooling loads near pinch-off areas.

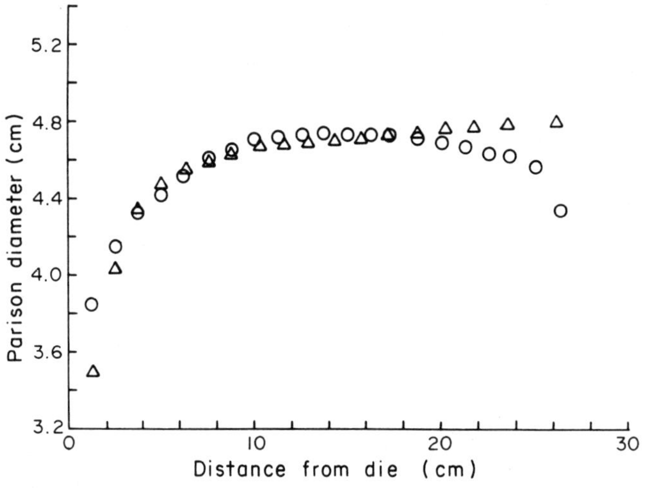

Figure 4
Comparison of calculated (△) and experimental (○) parison diameter distributions for a polyethylene resin (after Kamal et al. 1981)

6. Process Analysis and Product Microstructure

As in other plastics transformation processes, it is possible to employ mathematical modelling and computer simulation in order to gain some insight into the dynamics of the blow-molding process and the interactions between resin properties, machine and mold design and operating conditions, on the one hand, and the moldability of the material, process economics and the ultimate properties of the molded articles, on the other hand. These models are generally based on solving the equations of continuity, motion and energy for the system, in conjunction with appropriate initial and boundary conditions and suitable rheological constitutive equations. Since viscoelastic phenomena play an important role in the blow-molding process and since the available constitutive equations do not describe fully and accurately the rheological behavior of polymer melts, such models are complex and somewhat limited in their representation of the process and the product.

Nevertheless, there are some simple models which have been useful in evaluating some of the important interactions. These models permit the estimation of parison dimensions throughout the extrusion and inflation stages and therefore lead to an estimation of the dimensions of the molded article. Typical comparisons between model predictions and experimental data for parison dimensions and circumferential bottle distributions are shown in Figs. 4 and 5, respectively.

Heat transfer analysis of the cooling stage makes it possible to estimate cooling rates and their distributions in the molding. In crystallizing polymers, cooling rates control the distribution of both crystallinity and orientation in the material. The microstructure (crystallinity and orientation of the material) has a profound effect on the ultimate optical and mechanical properties of the molded

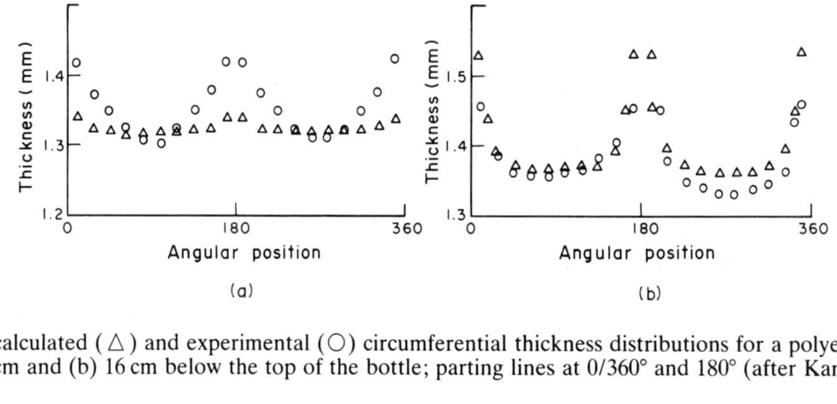

Figure 5
Comparison of calculated (△) and experimental (○) circumferential thickness distributions for a polyethylene resin bottle at (a) 12 cm and (b) 16 cm below the top of the bottle; parting lines at 0/360° and 180° (after Kamal et al. 1981)

article. Generally, it is found that crystallinity is lowest and frozen stresses are highest near the wall contacting the cold mold wall. Moreover, thick sections exhibit higher average crystallinity than thin sections.

7. *Modified Blow-Molding Techniques*

In recent years, the demands for higher production rates and improved precision in product dimensions and the desirability of molding new materials such as poly(ethylene terephthalate) (PET) have given impetus to the development and growth of novel blow-molding techniques. These techniques include coextrusion blow molding, injection blow molding and stretch blow molding.

7.1 Coextrusion Blow Molding

Coextrusion blow molding is employed to produce multilayered articles that meet a variety of specifications which could not be met by only one material. Two or three layers are usually employed in such articles. Typical products include carbonated beverage bottles (PET plus poly(vinyl alcohol) or nylon), fuel tanks and chemical drums. An adhesive layer is sometimes added to ensure good bonding between the components.

Coextrusion is a useful technique to improve the barrier properties of containers and to enhance the chemical or solvent resistance of the inner surface. For example, the shelf life of products containing high percentages of aromatics and volatiles (such as toiletries, paints and thinners) may be increased by as much as 150% by incorporating a barrier layer of poly(vinyl acetate) between two polyethylene and polypropylene layers. Usually, an accumulator die head is used for coextrusion systems and a separate extruder is used to feed each component to the die head.

7.2 Injection Blow Molding

Injection blow molding offers the advantages of accurate dimensional control, elimination of scrap and the molding in of threads before blowing. A preform is injection molded over a rod under standard injection-molding conditions, then the preform is introduced to the blow-molding machine, where it is blown to the desired shape. Sometimes the preforms may be vacuum formed. Nitrile resins, PVC, polypropylene and PET have been molded using this process. Besides soft-drink bottles, cosmetics and food containers, Christmas tree light bulbs and several noncontainer products have been manufactured by this process. However, extrusion blow molding is the preferred process for containers with high length–diameter ratios and products requiring handles.

Commercial machines differ in the number of stations available and the function of each station. Some of these functions include injection, blowing, stripping, conditioning and printing. The number of stations ranges between two and seven, and systems employing molds having as many as 20 cavities have been employed.

7.3 Stretch Blow Molding

Stretch blow molding is a process whereby biaxial orientation (in the axial and transverse directions) is optimized to yield desirable mechanical, optical, barrier and other physicochemical properties. An extruded parison or molded (by injection or transfer molding) preform is stretched axially and blown, either simultaneously or in two different stages, to yield the final product with the desired dimensions. Extrusion systems require low capital cost and simple tooling. On the other hand, injection molding is more versatile and yields precision products, though at higher capital costs.

Available commercial systems differ in the mode of producing the preform or parison, the temperature of the preform and parison at the time of introduction into the stretching apparatus, the time of stretching and the mode of stretching. The ultimate properties of the product are closely linked to the orientation temperature, the orientation ratios both axially and transversely, the rates of stretching and blowing and the rate of cooling. Therefore, very elaborate considerations are required when this process is used. The main area of application of stretch blow molding has been carbonated soft-drink bottles and containers for cosmetics and aerosols from PET and possibly containers for mineral waters and edible oils. Parallel activity with polyacrylonitrile (PAN) for containers for soft drinks, lager beers and paper-correcting fluid, PVC for soft-drink bottles and polypropylene for pharmaceuticals containers has been reported.

8. *Molds and Materials*

In blow molding, the mold is subjected to relatively low pressures in the region of 1 MPa. As a result, the most commonly used material in the construction of blow molds has been aluminum. This material has the advantages of low density, high thermal conductivity for cooling efficiency and ease of machining. Since the pinch-off region is exposed to higher loads and wear, the mold is usually reinforced with steel or beryllium–copper alloy inserts in these areas. Materials other than aluminum that have been used for mold construction include zinc alloys, steel, beryllium–copper alloys, cast bronze and metal-filled epoxy castings.

9. *Blow-Molding Materials and Products*

Traditionally, high density polyethylene, low-density polyethylene, poly(vinyl chloride), polystyrene and rubber-modified polystyrene and acrylics (e.g., ABS and XT polymers) have been the most commonly used materials for blow molding. Recently, polypropylene, high-molecular-weight polyethylene (HMWPE), nitrile resins, poly(ethylene terephthalate), polyester, poly(phenylene oxide) (PPO), fiber-reinforced resins and polycarbonates have found increased utilization in blow molding. One of the important uses of HMWPE is in the manufacture of large articles like 210 l drums. Nitrile resins and PET have been moving rapidly into the soft-drink bottle market. The development of new blow-molding techniques such as injection blow molding, stretch–blow molding and coextrusion blow molding have promoted recent growth and expansion of blow-molding products.

Typical products of the reciprocating screw blow-molding process include 2–20 l containers, acid bottles, labware, milk bottles, automobile horns, wax dispensers, electric coffee pots, special disposal containers and hospital syringes. Larger articles obtained with accumulator systems include 20 l containers to 1.1 m^3 drums, riding toys, streetlight globes, water coolers, displays, lawn ornaments, vaporizers, waste baskets, boots, underfender skirts, automotive gas tanks and large figurines.

See also: Bottles and Other Rigid Containers; Computer Control in Polymer Processing; Injection Molding of Thermoplastics: Mold Filling; Injection Molding of Thermoplastics: Structure Development; Processing: Classification

Bibliography

Belcher S L 1989 Blow molding of polymers. In: Aggarwal S L (ed.) 1989 *Comprehensive Polymer Science*, Vol. 7. Pergamon, Oxford, Chap. 15

Fisher E G 1971 *Blow Molding of Plastics*. Iliffe Books, London

Junk P B 1982 *Considerations on Melt Behavior in Continuous Extrusion Blow Molding*. Institut für Kunststoffverarbeitung, Aachen, Germany

Kamal M R, Kalyon D, Tan V 1982 Anisotropy and dimensions of blow-molded polyethylene bottles. *Polym. Eng. Sci.* 22: 287–91

Kamal M R, Tan V, Kalyon D 1981 Measurement and calculation of parison dimensions and bottle thickness distribution during blow molding. *Polym. Eng. Sci.* 21: 331–8

Morton-Jones D H 1989 Blow moulding. *Polymer Processing*. Chapman and Hall, London, Chap. 6

Plastics and Rubber Institute 1981 *Advances in Blow Molding II*. Plastics and Rubber Institute, London

Rosato D V, Rosato D V 1989 *Blow Molding Handbook*. Hanser, Munich, Germany

Ryan M E, Dutta A 1982 Mathematical modeling of the blow molding process. *Polym. Eng. Sci.* 22: 1076–85

M. R. Kamal
[McGill University, Montreal, Canada]

Bond Durability

Deterioration of bond strength on exposure to water is the Achilles' heel of adhesive bonds to polar substrates such as metals, glasses and ceramics. Moisture seriously reduces joint strength to as low as one-tenth of the dry strength, especially when the bond is under stress and at elevated (>25 °C) temperatures. In extreme cases, the adhesive joint spontaneously debonds in wet environments. Two examples illustrate the point: a major airplane manufacturer decided not to employ adhesive bonding in place of riveted construction because of the apparent unreliability of adhesive bonds despite potential weight savings and other advantages; during the Vietnam war many aircraft were disabled by the hot humid environment causing the aluminum wing skin to debond from the underlying honeycomb structure.

The methodology for testing adhesive-bond durability is reasonably well established. Test methods exist for ranking adhesives, surface treatments or different environments and for quantitatively characterizing the environmental cracking process. However, knowledge of the fundamental processes involved is less complete. It has been established that there is a thermodynamic driving force for the water to find its way to the adhesive–adherend interface, but the subsequent events which cause debonding are not well understood. The practical solutions to the problem are always complex, never simple. Indeed, it is difficult to understand why some of the empirically derived methods for adherend preparation actually give environmentally resistant adhesive joints.

1. Mechanisms of Environmental Failure

1.1 Stability of the Adhesive

It is obvious that if the service environment physically or chemically attacks the adhesive to any significant extent the joint may well be appreciably weakened. However, assuming that a degree of common sense has been exercised in adhesive selection, loss of strength in the adhesive is not usually a major mechanism of attack in water. This is evident from the frequent observation that, although failure of well-prepared joints is invariably caused by cohesive fracture in the adhesive layer, after environmental attack it usually occurs at or very close to the adhesive–adherend interface.

1.2 Stability of the Interface

The above observation highlights the importance of the interface when considering environmental-failure mechanisms. The thermodynamic work of adhesion required to separate unit area of two phases forming an interface may be related to the surface free energies by the Dupré equation; that is, in the absence of chemisorption, interdiffusion and mechanical interlocking, the reversible work of adhesion W_A in an inert medium is given by

$$W_A = \gamma_a + \gamma_b - \gamma_{ab} \quad (1)$$

where γ_a and γ_b are the surface free energies of the two phases and γ_{ab} is the interfacial free energy. In the presence of a liquid, the work of adhesion is

$$W_{AL} = \gamma_{aL} + \gamma_{bL} - \gamma_{ab} \quad (2)$$

For an adhesive–adherend interface the work of adhesion W_A in an inert atmosphere (e.g., dry air) usually has a positive value, indicating thermodynamic stability of the interface. However, in the presence of a liquid the work of adhesion W_{AL} may well have a negative value, indicating that the interface is unstable and will dissociate. Thus, calculation of the terms W_A and W_{AL} may enable the environmental stability of the interface to be predicted.

Kinloch and co-workers (Gledhill and Kinloch 1974, Kinloch 1982, 1983) have shown that values of W_A and W_{AL} may be calculated from

$$W_A = 2(\gamma_a^D \gamma_b^D)^{1/2} + 2(\gamma_a^P \gamma_b^P)^{1/2} \quad (3)$$

$$W_{AL} = 2[\gamma_L - (\gamma_a^D \gamma_L^D)^{1/2} - (\gamma_a^P \gamma_L^P)^{1/2} - (\gamma_b^D \gamma_L^D)^{1/2} - (\gamma_b^P \gamma_L^P)^{1/2} + (\gamma_a^D \gamma_b^D)^{1/2} + (\gamma_a^P \gamma_b^P)^{1/2}] \quad (4)$$

where γ^D and γ^P are the dispersion and polar force components, respectively, of the surface free energy.

Some values of W_A and W_{AL} for various interfaces and environments are shown in Table 1. The arguments presented in this section and much of the disucssion that follows apply to all polar (inorganic) adherends and, indeed, to any system where W_{AL} in Eqn. (2) is negative.

For those interfaces where there is a negative work of adhesion there is a driving force for the displacement of adhesive on the adherend surface by the liquid. It is to be expected, therefore, that if the joint is subjected to such an environment there will be a progressive development of debonded interface. This will have the effect of reducing the joint strength and causing the locus of failure to be interfacial between the adhesive and adherend. This is exactly what has been observed in practice.

1.3 Stability of the Adherend

Obviously, environmental attack on the adherend material (i.e. corrosion) may well cause a loss in joint strength. However, corrosion of the surface of a metallic adherend is often observed but is a post-failure phenomenon occurring after the displacement of adhesive on the metal oxide by water. Only in special circumstances—for example, with clad aluminum alloy or in a salt-water environment—is gross corrosion of the adherend an important failure mechanism.

A more subtle change induced in the nature of the surface of the adherend by an active environment is that water may enter adhesive joints and cause hydration of the oxide present on metallic adherends (Kinloch 1982, 1983). This, in turn, may cause weakening of the oxide layer and result in loss of joint strength. This mechanism is thought to be particularly relevant to environmental attack on aluminum-alloy bonded joints. However, the detailed reasons why certain surface pretreatments coupled with specific grades of aluminum alloy lead

Table 1
Values of W_A and W_{AL} for various interfaces and environments

Interface	Work of adhesion ($mJ\ m^{-2}$) W_A	W_{AL}	Evidence of interfacial bonding after immersion of unstressed joints
Epoxy adhesive/ferric oxide (mild steel)	291	ethanol:22	no
		formamide:−166	yes
		water:−255	yes
Epoxy adhesive/aluminum oxide	232	water:−137	yes
Epoxy adhesive/silica	178	water:−57	yes
Epoxy adhesive/carbon-fiber-reinforced plastic	88–90	water:22–24	no
Vinylidene chloride–methyl acrylate copolymer/polypropylene	88	water:37	no
		sodium *n*-octyl sulfate (3.5 wt% solution): 1.4	no
		sodium *n*-dodecyl sulfate (0.5 wt% solution): −0.9	yes
		sodium *n*-hexadecyl sulfate (0.05 wt% solution): −0.8	yes

to oxides of different surface topographies, microstructure and chemistry, which in turn result in adhesive joints possessing vastly different resistances to environmental attack, are far from being clearly understood. Some researchers have suggested that it is the accumulation of certain elements, such as copper and magnesium, at the oxide–metal interface or in the oxide layer which is detrimental to oxide stability and joint durability.

2. *Kinetics of Environmental Attack*

Comyn (Kinloch 1981 p. 279) has shown that the kinetics of the environmental-failure mechanism may often be governed by the rate of diffusion of water in the joint. Fortunately, water uptake by a crosslinked structural adhesive usually behaves according to Fick's law and, thus, from measuring the diffusion constant using bulk adhesive film samples, the water concentration profile within the joint as a function of geometry, temperature and water activity may be predicted. Such predictions have been shown to be reasonably accurate, certainly around the periphery of the joint where the initial debonding occurs.

It is also noteworthy that recent evidence suggests a critical concentration of water may be needed to initiate the environmental-attack mechanism, the actual concentration depending on the materials and surface pretreatment. Also, the diffusion data may be combined with a fracture-energy approach (see Sect. 3) to quantitatively predict the loss of joint strength.

3. *Environmental Failure of Adhesive Bonds to Metals*

The extreme vulnerability of metal-adhesive bonds to moisture is illustrated by a simple test in which aluminum rods were bonded end to end (butt joint) with an epoxy resin (Bascom et al. 1977). These specimens were exposed to ambient conditions (25 °C, 50% RH) and the time-to-failure noted for different static loads. The results are given in Fig. 1. Despite the large data scatter the combined effect of stress and moisture is evident. If, however, the bondline is protected by a desiccant the time to failure is orders of magnitude greater. Moreover, failure initiated near the metal–polymer interface for the specimens exposed to humidity, but the protected specimens failed by cohesive fracture of the adhesive.

The most common methods of evaluating the resistance of adhesive joints to degradation by the combined action of moisture, stress and temperature are time-to-failure (Fig. 1) or residual-strength tests. The simplest procedure is to expose lap-shear specimens in an environmental cabinet and record time to failure or measure the joint strength after specific time intervals. Some examples taken from the extensive work of Minford (1981) are presented in Fig. 2. The results clearly distinguish the relative environmental resistance of different adherend pretreatments.

Time-to-failure or residual-strength tests of bond permanence provide simple means of characterizing the effect of the type of environment, adherend surface condition, and adhesive composition on bond durability. However, in practice moisture attack of adhesive bonds is a dynamic environmental cracking process which these tests do not address directly. Consequently, a test methodology has been developed to characterize bondline cracking in aggressive environments. Basically, some measure is made of the stressed environmental cracking rate or the limiting load beyond which the cracking rate is

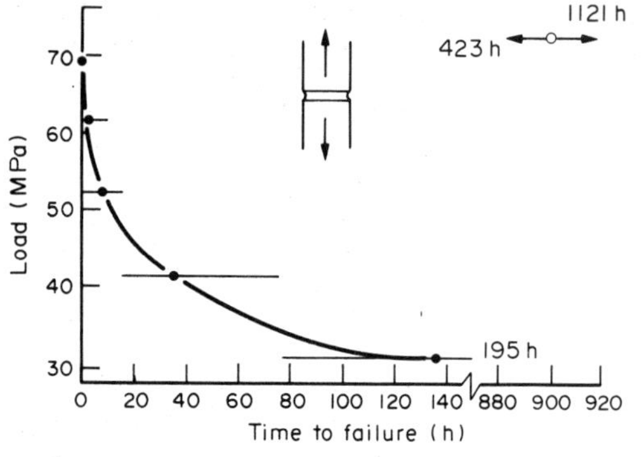

Figure 1
Time to failure of aluminum–epoxy butt joints in ambient humidity (●) and a very low humidity (○) (after Bascom et al. 1977)

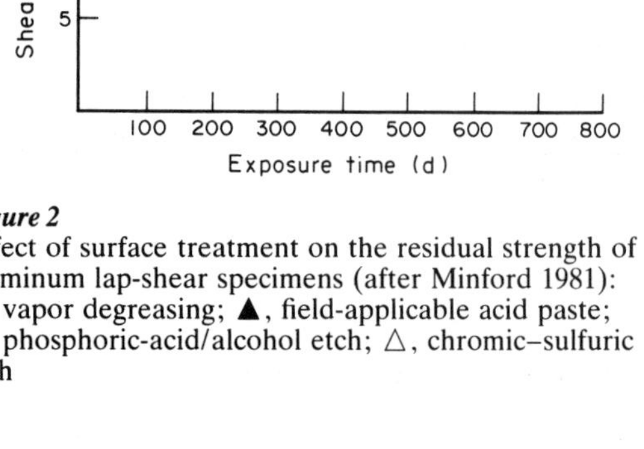

Figure 2
Effect of surface treatment on the residual strength of aluminum lap-shear specimens (after Minford 1981): ○, vapor degreasing; ▲, field-applicable acid paste; ■, phosphoric-acid/alcohol etch; △, chromic–sulfuric etch

essentially zero. The most sophisticated of these tests involves static loading of tapered double cantilever beams as illustrated in Fig. 3. The specimen consists of two tapered adherend beams bonded together by the adhesive. Details of this test are given elsewhere (see *Bond Failure: Fracture*). The essential feature is that the specimen is tapered for constant change in compliance with crack length and as a result the fracture energy is independent of crack length as the crack extends (see Fig. 4). Beam analysis gives the fracture energy $\mathcal{G}_{Ic}$ as a function of the load to drive the crack P_c, the bending modulus of the adherend beams E_B, and factors relating to the geometry of the specimen:

$$\mathcal{G}_{Ic} = \frac{4P_c^2}{b^2 E_B}\left(\frac{3a^2}{h^3} + \frac{1}{h}\right) = \frac{4P_c^2 m}{b^2 E_B} \qquad (5)$$

Here, b is the specimen thickness and m is a constant that characterizes the taper. To study environmental cracking, the specimen is static loaded in the desired hydrothermal environment and the crack rate determined as a function of load (applied $\mathcal{G}$). Figure 5 shows test data for an aluminum beam bonded with a commercial aerospace structural adhesive. There appears to be a limiting $\mathcal{G}$ below which the cracking rate is essentially undetectable. Mostovoy and Ripling (1972) designated this limit as $\mathcal{G}_{Iscc}$ by analogy with metals mode I, stress-corrosion cracking, and suggested that it is a design criterion for the resistance of adhesive bonds to stress-accelerated environmental debonding.

The existence of a true limiting $\mathcal{G}_{Iscc}$ has been questioned. Kinloch (1981 p. 82) has pointed out that a plot of $\mathcal{G}$ (or P) against log a gives a straight line with no evidence of a minimum. He argued that in many cases there is no fundamental basis for the crack rate reaching zero at a finite stress. Nevertheless, from an engineering point of view $\mathcal{G}_{Iscc}$ is a useful parameter in that it identifies a stress level below which environmental cracking will not sensibly affect the durability or reliability of the bonded structure.

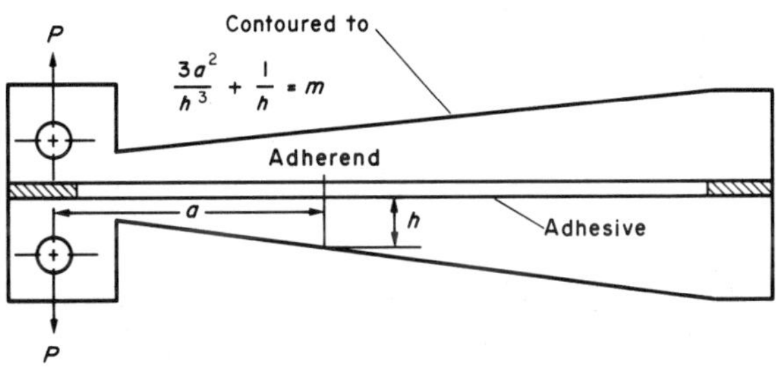

Figure 3
Tapered double cantilever beam (TDCB) specimen

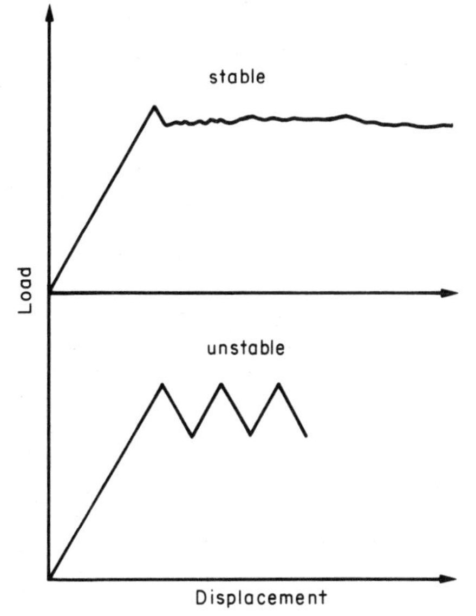

Figure 4
Load–displacement diagram for crack propagation in TDCB specimens

Variations of the tapered double cantilever beam specimen have been developed for large-scale stressed environmental cracking test programs. (The tapered double cantilever beam is too costly to machine and fabricate for large test programs.) Two of these simplified test specimens are illustrated in Fig.6. Both are designed for testing under constant-displacement conditions rather than constant load. The displacement is set by inserting a wedge

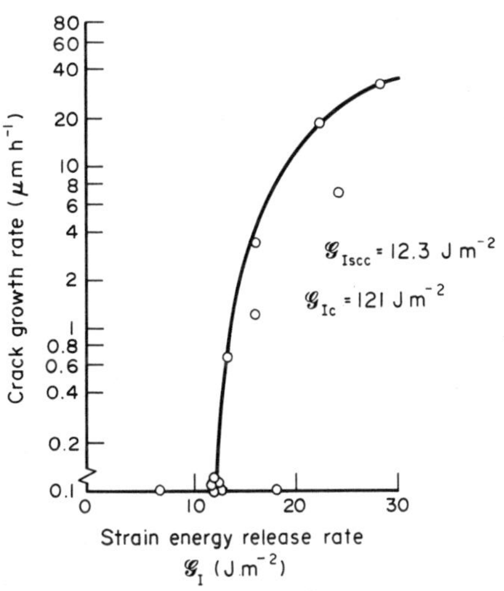

Figure 5
Crack growth rate vs load for an epoxy adhesive TDCB specimen immersed in water (25 °C)

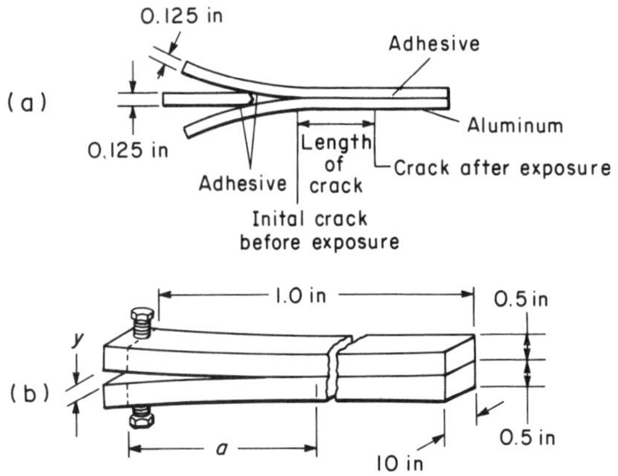

Figure 6
Constant displacement specimens for environmental testing of adhesive bond cracking resistance

(Fig. 6a) or adjusting a bolt (Fig. 6b) and immersing the specimen in an appropriate hydrothermal environment and periodically observing the specimen until the crack growth has sensibly ceased; the crack will grow until the applied $\mathscr{G}$ has reached $\mathscr{G}_{\mathrm{Isocc}}$. In the case of the specimen in Fig. 6b, $\mathscr{G}_{\mathrm{Iscc}}$ can be calculated from the following equation:

$$\mathscr{G}_{\mathrm{Iscc}}=\frac{y^2Mh^3[3(a+0.6\,h)^2]+h^2}{16[(a+0.6h)^3+ah^2]^2} \tag{6}$$

where $\mathscr{G}_{\mathrm{Iscc}}$ is the crack extension force (in lb in^{-2}, with 1 in lb in^{-2} = 175 J m^{-2}), y(=0.25 in, 1 in = 25.4 mm) is the displacement at the load points, a is the distance (in) from load point to crack tip, h (=0.5 in) is the height of one beam and m (=10 200 000 psi, 1 psi = 6.89×10^3 Pa) is the modulus of the adherend. The specimen in Fig. 6a cannot give $\mathscr{G}_{\mathrm{Iscc}}$ values; the thin adherends usually undergo inelastic deformation which invalidates a simple elastic analysis. The wedge specimen can be used to obtain a relative measure of corrosion-cracking resistance and is best suited to quality-control testing. Some results from the work by Shannon and Thrall (1977) using the wedge specimen are given in Fig. 7 where the crack growth is plotted as a function of time. Again, there is an apparent limit to the crack growth which may not be strictly true, but the expected bond durability may be long compared with the useful lifetime of the structure.

4. *Inhibiting Moisture Attack on Adhesive Bonds*

There have been extensive efforts to reduce the susceptibility of adhesive joints to degradation by moisture. In the case of metal adherends, various treatments have been devised to make the oxide environmentally resistant: anodization, chemical etching and conversion coatings. It is beyond the scope of this article to discuss the multitude of surface treatments that have been devised for metals to improve adhesive bond durability (see *Bonding: Preparation of Metallic Surfaces*). Snogren (1974) catalogs surface treatments for all commercially important metals and gives some ranking of the different procedures but makes no attempt to explain why they work. In fact, in most instances, the treatments were arrived at empirically. Where fundamental studies have been made, it has been found that the treatment produces a relatively thick oxide film: a few tens to a few hundred nanometers. Generally, these films have a porous structure which presumably allows penetration and mechanical locking of the adhesive. There is also evidence that these treatments tend to chemically stabilize the surface oxide.

Various adhesion promoters (sometimes called coupling agents) have been developed which significantly improve the moisture resistance of bonds when using glass and other adherends including metals. These agents include organic chromium complexes, organofunctional silanes and organotitanates, the silanes being the most widely used. These chemicals have the general structural formula $R'Si(OR'')_3$ where R' is an organic moiety such as $NH_2(CH_2)_n$ and R'' is either CH_2CH_3 or CH_3. In selecting a silane adhesion promoter, R' is chosen to have a reactive capability with the adhesive, for example, the amine group to react with epoxies. The alkoxy group OR'' is presumed to hydrolyze and react with the silanol (SiOH) groups on the glass surface to form siloxane bonds. This picture is an oversimplification since most adhesion promoters react to form polymeric films: the silanes form polysiloxane networks. Presumably the adhesive penetrates this film to form a thin, interpenetrating network at the interface. The mechanism by which

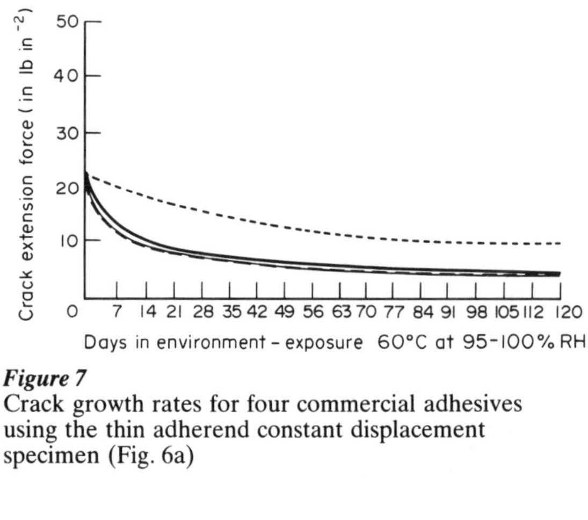

Figure 7
Crack growth rates for four commercial adhesives using the thin adherend constant displacement specimen (Fig. 6a)

these adhesion promoters reduce the rate of moisture attack is unclear; it has yet to be determined whether the interpenetrating network between the adhesive and the adherend plays a role, and also the extent and importance of the silane reacting with the adherend surface. Nonetheless, the simple concept of the adhesion promoter coupling the adhesive to the adherend is useful in selecting adhesion promoters for specific applications.

See also: Adhesion: Fundamental Principles; Adhesive Joining: An Overview; Adhesives: Mechanical Test Methods

Bibliography

Bascom W D, Gadomski S T, Henderson C M, Jones R L 1977 Fracture markings on stress-corroded epoxy/aluminium butt joints. *J. Adhesion* 8: 213–22

Gledhill R A, Kinloch A J 1974 Environmental failure of structural adhesive joints. *J. Adhesion* 6: 315–30

Kinloch A J (ed.) 1981 *Developments in Adhesives 2*. Applied Science, London

Kinloch A J 1982 The science of adhesion, Part 2: Mechanics and mechanisms of failure. *J. Mater. Sci.* 17: 617–51

Kinloch A J (ed.) 1983 *Durability of Structural Adhesives*. Applied Science, London

Minford J D 1981 Aluminum adhesive bond performance. In: Patrick R L (ed.) 1981 *Treatise on Adhesion and Adhesives*. Vol. 5. Dekker, New York, pp. 45–137

Mostovoy S, Ripling E 1972 Effect of temperature on the fracture toughness and stress corrosion cracking of adhesives. *J. Appl. Polym. Sci., Appl. Polym. Symp.* 19: 395–408

Shannon R W, Thrall E W 1977 PABST environmental testing of adhesive-bonded joints. *J. Appl. Polym. Sci., Appl. Polym. Symp.* 32: 131–44

Snogren R C 1974 *Handbook of Surface Preparation*. Palmerton, New York

A. J. Kinloch
[Imperial College of Science and Technology, London, UK]

W. D. Bascom
[Hercules Incorporated, Magna, Utah, USA]

Bond Failure: Fracture

Since the mid-1960s, adhesive bonding has been increasingly used in structural (load-bearing) applications. Aerospace construction is the most important example in terms of both technical advances and the amount of structural adhesives used. Structural bonding is also utilized in automobile manufacture, housing and shipbuilding, and is expected to be used in these and other applications more extensively in future years.

It is important to distinguish between structural bonding, where the bondline is expected to sustain a known stress (and strain), and nonstructural applications such as holding decorative panels, where neither the adhesive nor the panel is part of the load-carrying framework.

In order to predict the reliability of a structure or substructural element, such as an airplane wing or an automobile chassis, design criteria must be established for the materials used in fabrication, including the fasteners. Thus, when structural adhesives were being considered for military aircraft, rational design methodology for adhesive bonds needed to be developed. One criterion is fracture resistance: how sensitive is the bondline to the growth of flaws under sustained or fluctuating loads? Also, there needs to be some criteria for stress-corrosion cracking under hygrothermal (hot–wet) conditions.

The US Navy pioneered the development of adhesive bond fracture testing. In the early 1960s, with the urging of Dr George Irwin of the Naval Research Laboratory, Washington, DC, Mostovoy and Ripling (1966) of the Material Research Laboratory, Chicago, developed the methodology for adhesive fracture testing and produced a database of adhesive fracture data. In subsequent years, much has been added to this database and in-depth studies have been made of the micromechanics of adhesive fracture.

1. Test Methods

Fracture testing involves introducing a flaw of known dimensions into a specimen and determining the stress or energy necessary to cause the flaw to grow. There are three basic failure modes, as shown in Fig. 1: opening mode (I), in-plane shear (II) and torsional shear (III). Test methods are designed to determine the stress-intensity factor K_c or the strain energy release rate (fracture energy $\mathcal{G}_c$) for one of these failure modes, for example, K_{ic} or $\mathcal{G}_{ic}$ where i is I, II or III. These quantities are usually independent of the test method as long as the test conforms to the restrictions of linear elastic fracture mechanics (LEFM). The most critical assumption of LEFM is that the nonlinear (plastic) deformation at the crack tip is small compared with the overall specimen dimension. This constraint is easily met by brittle ceramics and some high-strength metals but, in the

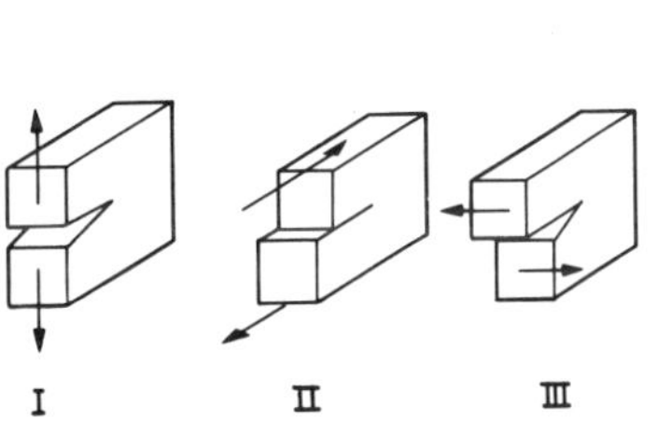

Figure 1
Three fracture modes

case of organic polymers, it sometimes takes heroic efforts to meet the required conditions.

Some of the more important relations pertaining to this discussion are

$$\mathcal{G}_{\mathrm{I}} = K_{\mathrm{I}}^2(1-\nu^2)/E \tag{1}$$

$$\mathcal{G}_{\mathrm{c}} = (\sigma^2/2b)(dC/da) \tag{2}$$

if the assumptions of LEFM are met. In these equations, E is the elastic modulus, ν is Poisson's ratio, σ is the applied load, C is the compliance and a is the crack length. For a tensile specimen with a machine notch in one side (single edge notch),

$$\sigma_{\mathrm{c}} = (E\mathcal{G}_{\mathrm{c}}/\pi a)^{0.5} \tag{3}$$

If $\mathcal{G}_{\mathrm{c}}$ (or K_{c}) is a material property then an expression can be developed to determine the load at which a flaw will begin to grow from $\mathcal{G}_{\mathrm{c}}$ and flaw size a. This is the basis of fracture mechanics design, albeit oversimplified.

Except for the most brittle materials, a deformation zone develops immediately ahead of the crack tip as the material is stressed. Just as the flaw becomes critical—it begins to grow slowly or move catastrophically—the radius of this zone reaches a critical value r_{c}. This critical damage zone size is related to the fracture energy for the case of a material that fails in an elastic–plastic manner by

$$\mathcal{G}_{\mathrm{c}} \simeq 6\pi\sigma_{\mathrm{y}}\varepsilon_{\mathrm{y}}r_{\mathrm{c}} \tag{4}$$

where σ_{y} and ε_{y} are the yield stress and strain, respectively.

Mostovoy and Ripling (1966) took advantage of Eqn. (2) and used the tapered double cantilever beam (TDCB) configuration illustrated in Fig. 2. The specimen is tapered so that aC/da is constant and equal to m. For this case,

$$\mathcal{G}_{\mathrm{Ic}} = \frac{4P_{\mathrm{c}}^2}{b^2E_{\mathrm{B}}}\left(\frac{3a^2}{h^3}+\frac{1}{h}\right) = \frac{4P_{\mathrm{c}}^2 m}{b^2E_{\mathrm{B}}} \tag{5}$$

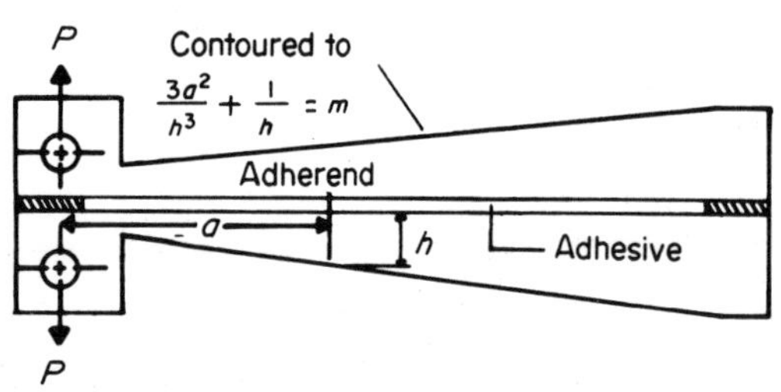

Figure 2
Tapered double cantilevered beam specimen

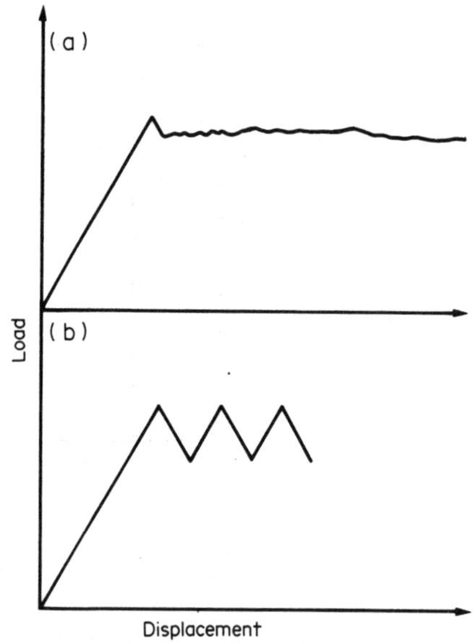

Figure 3
Load–displacement diagram for TDCB specimen: (a) continuous crack propagation driven by the test machine; (b) unstable, slip–stick crack advance

The advantage of this specimen is that the fracture energy is a function only of the fracture load P_{c}, the bending modulus of the beam material E_{B} and the specimen dimensions. The measurement is independent of the crack length, which can be difficult to determine. Also, because the TDCB specimen is tapered for constant change of compliance with crack length, the crack is driven at a constant load, as shown in Fig. 3a, or in a "slip–stick" fashion, as shown in Fig. 3b. The latter condition occurs when the crack speed is higher than the test machine loading rate. It is important to design engineers to know whether a crack will advance steadily and can be detected or whether failure will be catastrophic.

Mostovoy used the specimen in Fig. 2 to evaluate the stress-corrosion cracking characteristics of adhesive bonds by applying a constant load to a specimen submerged in water and determining the rate of crack advance. The subject of adhesive bond stress-corrosion cracking is discussed in the article *Bond Durability*.

Adhesive bonds are never deliberately designed so that they are under a purely mode I (cleavage) loading. Consequently, it is of interest to test adhesive fracture in other loading modes, especially mixed mode I and mode II. Various test specimens have been devised. In Fig. 4, two such specimens are illustrated. Figure 4a is a scarf joint with which, by varying the bond angle, different ratios of $\mathcal{G}_{\mathrm{I}}/\mathcal{G}_{\mathrm{II}}$ can be tested. In Fig. 4b, shear and cleavage loads are applied independently and can be varied to give

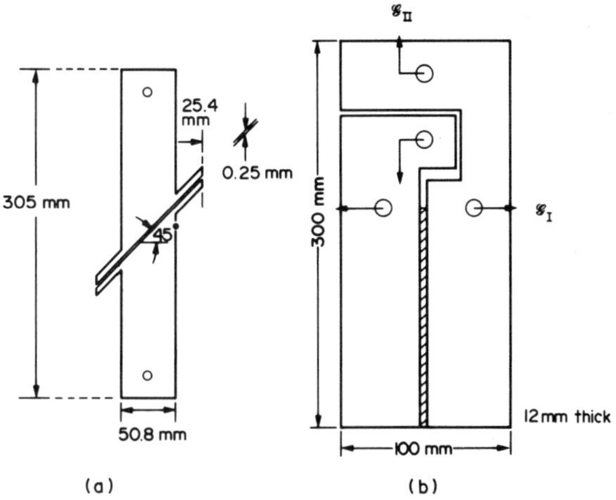

Figure 4
Specimens for mixed mode adhesive fracture testing

essentially any $\mathscr{G}_I/\mathscr{G}_{II}$ ratio. (It is problematical whether a pure mode II test is possible, owing to alignment difficulties because the bondline has a finite width). Another specimen which has been used to characterize mixed mode adhesive fracture is that for the cracked lap shear test, illustrated in Fig. 5.

2. *Adhesive Fracture Data*

The adhesive $\mathscr{G}_{Ic}$ values for some "model" adhesives and some commercial structural adhesives are given in Table 1. These data were obtained (Bascom et al. 1975) at loading rates of 1.25–1.5 mm min^{-1}; no significant strain rate effects have been reported within this small range. In all cases the adherends were aluminum beams. However, since failure was always a center-of-bond cohesive fracture of the adhesive, it is generally assumed that the adherend does not influence the results.

Note in Table 1 the very high toughness of the "rubber" modified epoxy adhesives. A large number of commercial structural adhesives are based on this type of composition and rival in toughness the metals they are used to bond; $\mathscr{G}_{Ic}$ for a high-strength aluminum is approximately 10 kJ m^{-2}. In Table 2 and Fig. 6, some results are given for the fatigue fracture of adhesives. All these results were obtained using the TDCB specimen. The effect of temperature on the fracture toughness of commercial adhesives is illustrated in Fig. 7. The bell shape of the curves is a result of the combined effect of temperature and bond thickness; as with most polymers, toughness increases with temperature, with an accompanying increase in the damage zone radius r. However, r can become comparable in size to the bondline thickness. When this occurs, the damage zone is restrained by the bondline, which according to Eqn. (4) has the effect of suppressing toughness; in other words, adhesive $\mathscr{G}_{Ic}$ is not as large as it would be for the adhesive resin tested in bulk.

There are relatively few data available for mixed mode adhesive fracture. In Table 3, some results are given for mixed, $\mathscr{G}_{Ic}$ and $\mathscr{G}_{IIc}$ adhesive fracture of low-toughness epoxy resins. As is generally the case

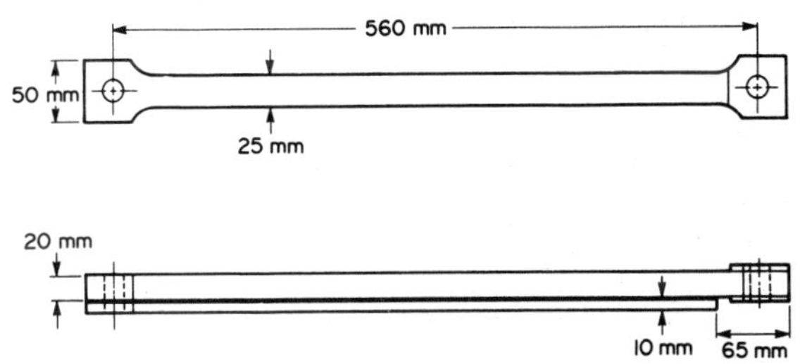

Figure 5
Cracked lap shear adhesive fracture specimen for mixed mode testing (after Romanko 1979a)

Table 1
Fracture energy of model and commercial adhesives

Adhesive	$\mathscr{G}_{Ic}$ (kJ m^{-2})
Bisphenol A diglycidyl ether–hexahydrophthatic anhydride	0.14
Bisphenol A diglycidyl ether–tetraethylpentamine	0.09
CTBN[a]–epoxy	3.5
Commercial elastomer–epoxy	2.3
Commercial nylon–epoxy	6.1

a Carboxy-terminated butadiene–acrylonitrile

Table 2
Fatigue fracture data for model and commercial adhesives

Adhesive[a] (chemical type/support)	Power law parameters[b] A	n
Model		
CTBN–DGEBA/none	4.85×10^{-17}	3.56
DGEBA–MPDA/none	8.87×10^{-13}	2.76
Commercial		
nitrile–epoxy/Dacron	6.28×10^{-14}	3.78
nylon–epoxy/none	4.56×10^{-14}	2.99
nitrile–phenolic/none	2.42×10^{-22}	5.80
modified epoxy/glass	1.20×10^{-16}	3.90
modified epoxy/none	2.05×10^{-16}	3.58

a CTBN, carboxy-terminated butadiene–acrylonitrile; DGEBA, diglycidyl ether of bisphenol A; MPDA, *m*-phenylenediamine, 14.5 wt% of epoxy b In the expression $da/dN = A\Delta\mathscr{G}_{Ic}$ (with units of m per cycle and J m^{-2})

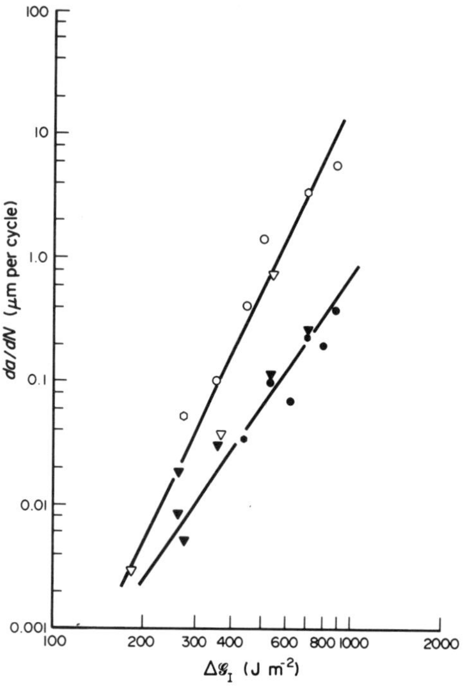

Figure 6
Fatigue crack growth for a model elastomer–epoxy adhesive

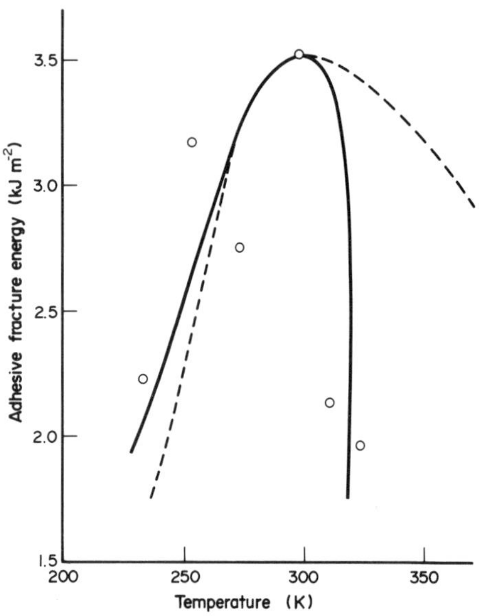

Figure 7
Effect of temperature on model and commercial elastomer–epoxy adhesives: ———, model; - - - -, commercial (after Mostovoy and Ripling 1972); ○, model (after Bascom et al. 1975)

Table 3
Effect of fracture mode on adhesive fracture energy of epoxy resins[a]

Resin[b]	Fracture energy ($J\,m^{-2}$) $\mathcal{G}_{Ic}$	$\mathcal{G}_{IIc}$	$\mathcal{G}_{(I,II)c45°}$
TEPA–DGEBA	70	1450	130
HHPA–DGEBA	116		140
Piperidine–DGEBA	121		255

a Mostovoy and Ripling (1966), Bascom et al. (1975) b TEPA, tetraethyl pentamine; DGEBA, diglycidyl ether of bisphenol A; HHPA, hexahydrophthalic anhydride

for isotropic materials, mixed mode $\mathcal{G}_c$ is greater than $\mathcal{G}_{Ic}$. Accordingly, the mode I toughness can be considered a lower limit for fracture resistance for these adhesives.

In the mixed mode testing of rubber-modified epoxy adhesives (both model formulations and commercial materials), Bascom et al. (1977) found $\mathcal{G}_{(I,IIc)}$ to be generally lower than $\mathcal{G}_{Ic}$. Some examples are given in Table 4. However, using the cracked lap shear test specimen, Mostovoy was unable to confirm the anomalously low mixed mode adhesive fracture values and, in fact, found a progressive increase in fracture energy with increasing mode II component, as would be expected from experience with isotropic materials. The discrepancy has not been resolved. However, it is a matter of some consequence, since if the anomaly does exist then the mode I fracture energy is not the criterion to

Table 4
Mode I vs mixed mode I and II adhesive fracture energy

Adhesive	Fracture energy ($J\,m^{-2}$) Cleavage $\mathcal{G}_{Ic}$	Cleavage + shear $\mathcal{G}_{(I,II)c45°}$
Unmodified epoxy[a]	116	140
CTBN–epoxy[b]		
10% CTBN[c]	3500	110
30% CTBN	2200	110
Commercial		
elastomer–epoxy	2300	870
nylon–epoxy	6100	750
ME 329[d]	630	55
AF 243[e]	860	220

a Hexahydrophthalic anhydride–DGEBA b Piperidine cured–DGEBA c Carboxy-terminated butadiene–acrylonitrile d Narmco Materials Inc. e 3M Company be used for conservative design with these tough adhesives.

3. Conclusion

The methods for mode I testing of adhesives are well established and a database has been amassed along with some understanding of the micromechanics of fracture. In some instances, the quasistatic (but not fatigue) data indicate a toughness comparable to that of the metal adherends, which suggests that the bondline is not especially flaw sensitive under these loading conditions. However, as new formulations are developed, their fracture resistance cannot be assumed, so it is necessary to maintain and expand the database for commercial adhesives.

Research activity continues on adhesive fracture, especially in the areas of fatigue (Brussat and Chiu 1979) and fracture design criteria (Chow et al. 1979). There is a major US Air Force program at General Dynamics/Fort Worth Division (Romanko 1979a) which includes adhesive fracture and fatigue fracture under mixed mode loading. Kinloch and Shaw (1981) have investigated the micromechanics of adhesive fracture.

See also: Adhesive Joining: An Overview; Adhesives: Mechanical Test Methods; Bond Strength: Peel

Bibliography

Bascom W D, Cottington R L, Jones R L, Peyser P 1975 The fracture of epoxy- and elastomer-modified epoxy polymers in bulk and as adhesives. *J. Appl. Polym. Sci.* 19: 2545–62

Bascom W D, Cottington R L, Timmons C O 1977 Fracture reliability of structural adhesives. *J. Appl. Polym. Sci., Appl. Polym. Symp.* 32: 165–88

Brussat T R, Chiu S T 1979 Fatigue crack growth of bondline cracks in structural bonded joints. *J. Eng. Mater. Technol.* 100: 39

Chow C L, Woo C W, Sykes J L 1979 On the determination and application of COD to epoxy-bonded aluminum joints. *J. Strain Anal.* 14: 37–42

Kinloch A J, Shaw S J 1981 The fracture resistance of a toughened epoxy adhesive. *J. Adhes.* 12:59–77

Mostovoy S, Crosley P B, Ripling E J 1967 Use of crack-line loaded specimens for measuring plane-strain fracture toughness. *J. Mater.* 2: 661–81

Mostovoy S, Ripling E J 1966 Fracture toughness of an epoxy system. *J. Appl. Polym. Sci.* 10: 1351–71

Mostovoy S, Ripling E J 1972 Effect of temperature on the fracture toughness and stress corrosion cracking of adhesives. *J. Appl. Polym. Sci., Appl. Polym. Symp.* 19: 395–408

Romanko J 1979a A behavior of adhesively bonded joint under cyclic loading. *Bonded Joints and Preparation for Bonding*, AGARD Lecture Series No. 102. NASA, Langley Field, VA, p. 4.1

Romanko J 1979b Integrated methodology for adhesion bonded joint life predictions. US Air Force Fort Worth Division Technical Report FZM-6879. USAF, Fort Worth, TX

W. D. Bascom
[Hercules Incorporated, Magna, Utah, USA]

Bond Strength: Peel

A peel test is often employed to assess the strength of an adhesive bond. In order to use this geometry, at least one of the two bonded components (adherends) must be sufficiently flexible so that the moment arm m of the peel force about the advancing peel front (see Fig. 1a) remains constant as peeling proceeds. This feature distinguishes a peel test from a cantilever-beam cleavage experiment, in which the moment arm m continually increases as the separation front progresses (Fig. 1b). Typical peel specimens consist of a pressure-sensitive adhesive tape adhering to a rigid substrate or two thin films bonded together. There are generally three components: the two adherends and a layer of adhesive between them. The mechanical properties of all three layers affect the resistance to peeling apart, as described later.

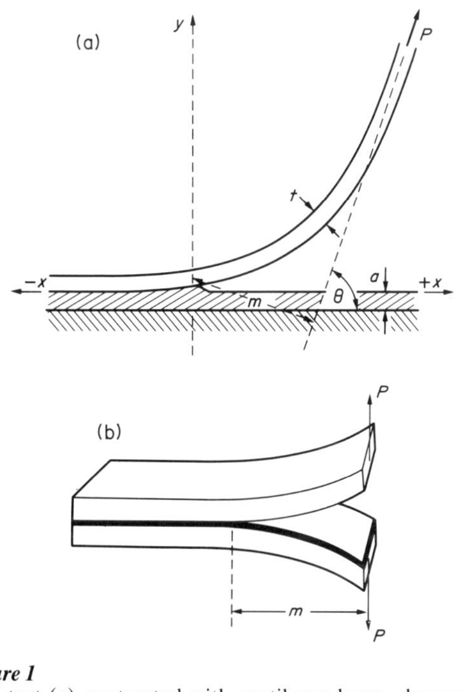

Figure 1
Peel test (a) contrasted with cantilever-beam cleavage test (b)

1. Peel Tests

The strength of a suitable specimen is usually determined at peel angles of 90° (see Fig. 2a) or 180° (Fig. 2b) or in a T-peel configuration (Fig. 2c). The bond strength is characterized by the peel force P per unit width of bonded surface when a constant stripping velocity v is imparted to the free ends. Once failure has started, the peel force generally fluctuates about an average value as peeling proceeds. Failure may take place by separation at either interface or within one of the three component layers if the layer itself tears apart instead of detaching. It is quite important to note the site of failure when testing an adhesively bonded joint. For example, if failure occurs within the adhesive layer, it is misleading to associate the measured peel strength with interfacial properties of the adhesive and adherends.

Scientifically, there are two simplifying advantages of the peel test compared with other methods, such as butt tensile or lap shear tests (Gent and Hamed 1977a). Failure proceeds at a controlled rate and the peel force is a direct measure of the work of detachment W_a (for interfacial failure) or the work of cohesive fracture W_c (for cohesive tearing). The values of W_a and W_c are, respectively, the energy which must be expended to detach a unit area of interface and the energy expended per unit area of material torn.

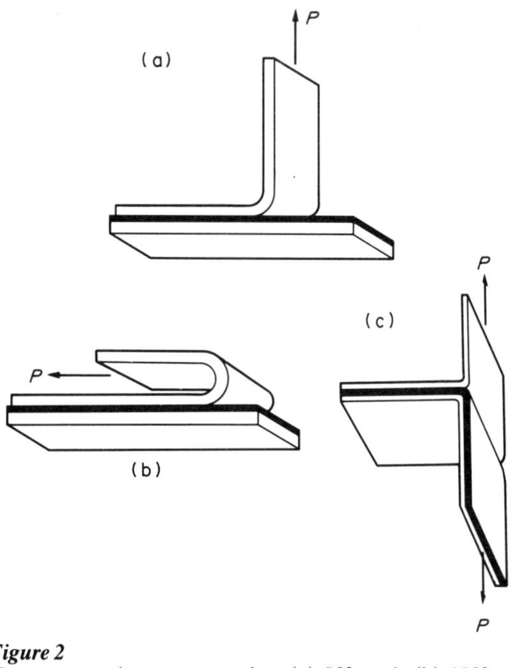

Figure 2
Common peel test geometries: (a) 90° peel, (b) 180° peel and (c) T peel

Two quite different approaches to the analysis of peeling rupture have been employed. The first sets up an energy balance for peeling an adhering strip (Lindley 1971). The second involves calculations of the stresses set up in the adhesive layer during peeling (Kaelble 1959, 1960).

2. Energy Considerations in Peeling

During steady-state peeling, conservation of energy requires that

work done by peel force = work of detachment + strain energy imparted to newly detached strip (1)

For peeling a unit length of bonded interface apart, Eqn. (1) becomes

$$P(1+\varepsilon-\cos\theta)=W_a+tE_{\text{strain}} \quad (2)$$

where P is the peel force per unit width, t is the thickness of the peeled strip and ε and E_{strain} are, respectively, the tensile strain and energy density imparted to the detached strip. Adhering strips are generally thick enough to detach without stretching significantly, at least at large peel angles, so that ε and E_{strain} are relatively small. In this case, Eqn. (2) can be simplified and the peel force becomes

$$P=W_a/(1-\cos\theta) \quad (3)$$

Thus, the peel force P provides a direct measure of the work W_a of detachment.

There are several contributions to W_a. The first is the intrinsic work $W_{a,0}$ required to break the interfacial bond. This is generally quite small, ranging from about $0.1\,\text{J m}^{-2}$ for van der Waals bonding between nonpolar materials up to about $100\,\text{J m}^{-2}$ for covalently bonded polymeric materials. Other contributions come from energy dissipated within the adhesive and adherend layers as they are deformed during peeling. If these layers were deformed reversibly, without dissipating any mechanical energy, the peel force would be small, reflecting only the intrinsic interfacial strength $W_{a,0}$. However, for typical adhesives this criterion of perfectly elastic deformations is not met and W_a is correspondingly much greater than $W_{a,0}$, being of the order of $10^4\,\text{J m}^{-2}$ or more for tough, ductile adhesive and adherend layers.

Because energy dissipated in deforming the adhesive and adherends and in bending the layers apart on separation depends generally upon the peel angle θ, the thickness of the various layers, the rate of peeling, the test temperature and the dissipative properties of the various components, the value deduced for W_a is found to depend upon all these factors also, even though the intrinsic interfacial strength $W_{a,0}$ is in principle independent of them.

The exact way in which the various components of W_a add together has not yet been worked out. They are clearly not independent of each other. If there is no intrinsic adhesion, no deformation energy is required to bring about separation. A factorization principle has been suggested (Gent and Schultz 1972):

$$W_a = W_{a,0}(1 + H) \tag{4}$$

where H is a numerical factor representing the dissipative properties of the adhesive and adherends. However, no method of calculating H from first principles has yet been devised.

3. *Stress Analysis*

An elastic beam (adhered layer) on an elastic foundation (adhesive) has been employed as a model of a peeling strip. Simple bending relations are used to calculate the peel force. It is assumed that the tensile stress σ in the adhesive layer is constant across the width of the bond and through its thickness. The stress at a distance $-x$ into the bond is obtained (Kaelble 1959, 1960) as

$$\sigma = \sigma_0(\cos Bx + K \sin Bx) \exp Bx \tag{5}$$

where

$$B = \left(\frac{Yw}{4EIa}\right)^{0.25} \tag{6}$$

and

$$K = \frac{Bm}{Bm + \sin\theta} \tag{7}$$

in which σ_0 is the maximum tensile stress at $x = 0$, Y is Young's modulus of the adhesive, E is Young's modulus of the adherend, a is the adhesive layer thickness, w is the bond width, I is the moment of inertia of the adherend layer cross section, m is the moment arm of the peel force and θ is the peel angle. Equation (5) reveals that the cleavage stress in the adhesive layer is a highly damped harmonic function of distance $-x$ from the peel front, involving alternating regions of tension and compression. However, it is strictly valid only for adhesive joints made of linearly elastic materials and for values of $K \simeq 1$ (i.e., $Bm \gg \sin\theta$), since the derivation is based on small-bending theory. Under these circumstances,

$$P = a\sigma_0^2 2Y(1 - \cos\theta) \tag{8}$$

It is instructive to compare Eqn. (8) with the result obtained previously from energy considerations, Eqn. (3). The predicted dependence of peel force on peel angle is identical in both equations.

4. *Effect of Thickness of the Adhesive Layer*

By combining Eqns. (3) and (8), it is seen that

$$\sigma_0 = (2YW_a/a)^{0.5} \tag{9}$$

Thus, the maximum tensile stress set up in a linearly elastic adhesive is predicted to be inversely proportional to the square root of the adhesive layer thickness. The stress at the point of failure would be expected to be rather independent of the adhesive layer thickness, but it should be noted that Eqn. (9) applies only to the mean tensile stress throughout the adhesive layer thickness and not to the local stress set up at the line of detachment.

In practice, it is found that the peel force increases as the thickness of the adhesive layer is increased, as shown in Fig. 3. This is due to energy dissipation within the bulk of the adhesive. As larger volumes of adhesive are subjected to deformation per unit area of detachment, so the total work expended in peeling increases. At sufficiently large thicknesses, however, the peel force become independent of thickness. The highly stressed region around the line of detachment is now small in comparison with the total thickness of the adhesive layer and the major dissipation process no longer involves the entire layer of adhesive.

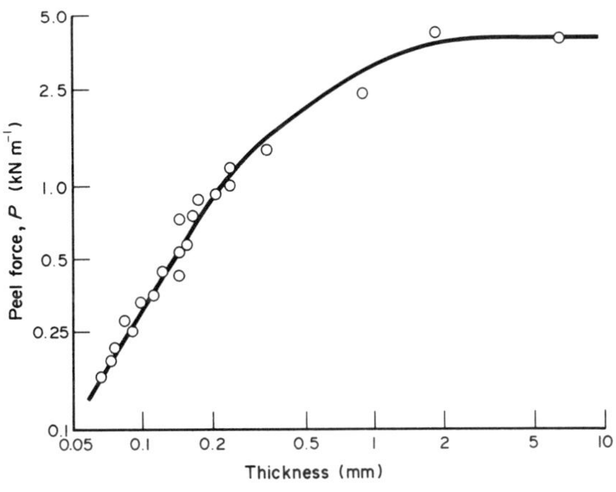

Figure 3
Dependence of peel force upon adhesive layer thickness for an elastomer detaching from a polyester substrate (after Gent and Hamed 1978)

5. *Effect of Thickness of an Elastic–Plastic Peeling Layer*

Generally, the peel force passes through a maximum with thickness of the peeling layer (see Fig. 4). This is due to energy dissipated in bending the layer away from the interface (Gent and Hamed 1977b). For a sufficiently thick peeling strip, the bending stresses set up are small and the strip remains substantially elastic as peeling proceeds. Thus, little energy is dissipated within it. However, if the strip is sufficiently thin, the bending stresses may be large enough to cause plastic deformation during peeling and W_a is then augmented by the work expended plastically in the strip. For an elastic–plastic adherend, peeled at 180°, the maximum contribution to the peel force due to yielding $P_{y,max}$, when the entire strip undergoes plastic yielding, is calculated to be

$$P_{y,max} = t\sigma_y/4 \tag{10}$$

where σ_y is the yield stress.

The critical thickness t_c of the adherend, above which no plastic yielding occurs, is calculated as

$$t_c = 12EP_0/\sigma_y^2 \tag{11}$$

where P_0 denotes the peel force per unit width in the absence of plastic yielding.

Thus, sufficiently thick strips do not undergo plastic yielding and, hence, there is no contribution to the observed peel strength from this cause. At the other extreme, a very thin strip undergoes complete plastic yielding; however, the total energy dissipated in this way is small because the thickness t is small. Hence, the additional peel force contribution due to plastic yielding is small (Eqn. (10)). As the thickness of the strip is increased, more energy is dissipated in yielding and the peel force rises proportionately. Eventually, at a large enough thickness, the bending stresses set up in the strip are insufficient to cause the entire strip to undergo yielding and the contribution to the peel force from this mechanism diminishes, eventually reaching zero again when the strip thickness exceeds t_c. Thus, the peel force for both infinitesimally thin and thick adhering layers is equal to P_0.

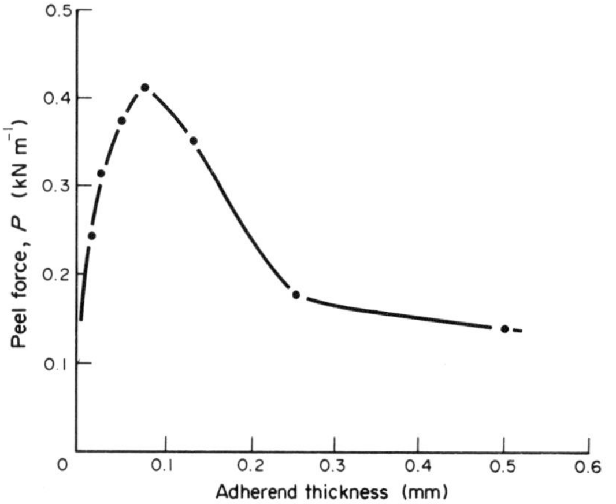

Figure 4
Effect of thickness of backing on peel force for a pressure-sensitive adhesive tape stripped from a rigid substrate (after Satas and Egan 1966)

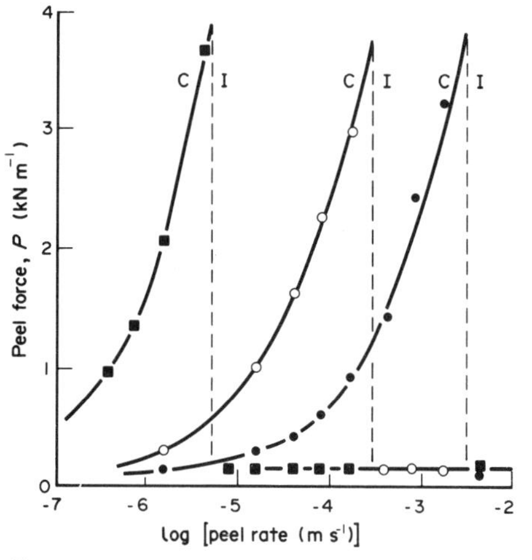

Figure 5
Peel force as a function of peel rate and test temperature for an elastomer adhering to a polyester substrate: ■, −5 °C; ○, 10 °C; ●, 23 °C; C and I denote cohesive and interfacial failure modes (after Gent and Petrich 1969)

6. *Rate and Temperature Effects in Pressure-Sensitive Adhesives*

Pressure-sensitive adhesives consist of soft elastomeric semisolids. Their peel strength depends strongly upon the rate of peel and the test temperature, as shown for a model system in Fig. 5. At low rates, the peel force increases with rate and failure takes place entirely within the adhesive layer, which fails by flowing apart. At a critical rate of peel (which depends on the test temperature), an abrupt transition takes place to interfacial separation, at much smaller peel forces. This transition occurs when the rate of deformation at the peel front is so high that the adhesive molecules are unable to disentangle and flow apart like a liquid but remain together as a coherent elastic solid. Although the local stress required to disentangle the molecules at low rates is relatively small, the work expended in ductile flow is large and the peel force (which measures the work of separation) is correspondingly high.

In the elastic state, the work of separation is mainly expended at the interface and is then relatively small.

The rate (and temperature) at which the abrupt transition occurs depend on the molecular weight of the polymeric adhesive and its segmental mobility. The peel strength above the critical rate depends upon the factors discussed previously. Below the critical rate, the peel strength is primarily a measure of the work of extending a viscoelastic liquid to the point of rupture.

See also: Adhesive Joining: An Overview; Adhesives: Mechanical Test Methods; Bond Durability

Bibliography

Gent A N, Hamed G R 1977a Peel mechanics of adhesive joints. *Polym. Eng. Sci.* 17: 462–6

Gent A N, Hamed G R 1977b Peel mechanics for an elastic–plastic adherend. *J. Appl. Polym. Sci.* 21: 2817–31

Gent A N, Hamed G R 1978 Adhesion of elastomers with special reference to triblock copolymers. *Plast. Rubber Mater. Appl.* 3: 17–22

Gent A N, Petrich R P 1969 Adhesion of viscoelastic materials to rigid substrates. *Proc. R. Soc. London, Ser. A* 310: 433–48

Gent A N, Schultz J 1972 Effect of wetting liquids on the strength of adhesion of viscoelastic materials. *J. Adhes.* 3: 281–94

Kaelble D H 1959 Theory and analysis of peel adhesion: Mechanisms and mechanics. *Trans. Soc. Rheol.* 3: 161–80

Kaelble D H 1960 Theory and analysis of peel adhesion: Bond stresses and distributions. *Trans Soc. Rheol.* 4: 45–73

Lindley P B 1971 Ozone attack at a rubber–metal bond. *J. Inst. Rubber Ind.* 5: 243–8

Satas D, Egan F 1966 Peel adhesion and pressure sensitive tapes. *Adhes. Age* 9: 22–5

A. N. Gent and G. R. Hamed
[University of Akron, Akron, Ohio, USA]

Bonding Nonwovens

An important distinction among types of fibrous webs is their primary mechanism of bonding. Wet-laid paper webs mainly rely on hydrogen bonding among cellulose fibers, textiles utilize frictional crossover forces for web integrity, while nonwovens, which are generally more porous and less dense than both papers and textiles, typically require additional bonding after forming. The most frequent bonding methods for nonwovens are chemical or adhesive bonding, mechanical entanglement, thermal bonding and inherent bonding, the first method being the most common. In the special case of continuous-filament nonwovens, self-bonding of fibers is achieved during web forming and consolidating using heat and/or pressure, sometimes augmented by the presence of reactive gases or more easily melted fibers.

1. Inherently Bonded Nonwovens

Inherently bonded nonwovens are those in which the fibrous structure is held together by its own internal polymer structure. This bonding method accounts for about 1.2% of all nonwovens produced. Polymers suitable for inherent bonding include polyethylene, polypropylene, polyester, polyurethane and nylon (see *Fibers and Textiles: An Overview*). One manufacturing method involves either slitting or mechanically fiberizing an extruded highly linear polymer sheet by passing the film over rolls which split the film into a network of interconnected fibers. Other examples of commercial inherently bonded nonwovens are extruded scrims such as Delnet, Sharnet and Conwed, which can be utilized as reinforcing scrims, surgical bandages, protective wrapping for equipment or parts and even as a melt adhesive. The major advantage of inherently bonded nonwovens is that they are inexpensive.

2. Thermally Bonded Nonwovens

Thermally bonded nonwovens rely on the presence of some thermoplastic powders, fiber or scrim to bond the fibrous web. This method accounts for about 1.2% of all nonwovens produced. If powders are used, the powder is generally sprinkled or otherwise metered onto the web and is then fused with heat. When fibers are utilized they may be in the form of bicomponent fiber, binder fiber or fiber binder as defined below.

Bicomponent fiber consists of high- and low-melting-point components in the same fiber, either extruded as a core and sheath or in a side-by-side configuration. Often products are made consisting of 100% bicomponent fiber.

Binder fiber consists of polymers that can be melted and remelted, whereas fiber binder can be melted or used only once. When either is used, 15–50% of this fiber must be incorporated in the nonwoven to obtain a suitably bonded product.

Thermal bonding may be accomplished using a heated oven or calender rolls. When calender rolls are used (see *Calendering*) they can produce either (a) an area or overall bonded nonwoven, or (b) a spot-bonded nonwoven, depending on whether the rolls are smooth or have a grooved or patterned effect, respectively. The spot-bonded products have greater bulk and flexibility than smooth calender-bonded nonwovens.

Some thermoplastic nonwovens are bonded using ultrasonics. One method is the Pinsonic process, which utilizes a roll having raised blunt pins positioned in a decorative pattern. In this method, the nonwoven to be bonded or laminated with other materials is passed betwen the rotating (feeding) pin roll and stationary metallic vibrating horns or tips (20–40 kHz) which cause the thermoplastic molecules to be excited, heated and then fused together.

The advantage of thermal bonding or ultrasonic bonding is that relatively little energy is required for the bonding step compared with the use of aqueous binder systems. In addition to being used to make various forms of spunbondeds, other applications of thermal and ultrasonically bonded products are in bedding, geotechnic fabrics, primary carpet backing, diaper cover stock, filter fabrics, furniture construction and apparel products.

3. Mechanically Bonded Nonwovens

About 40% of all nonwovens produced are bonded mechanically. This may be accomplished by the needle punch, stitchthrough or jet entanglement methods.

3.1 Needle Punch Method

The needle punch process generally utilizes thousands of barbed needles to entangle the fibrous web mechanically. The needles are about 8 cm long. They have a bend or crank at one end which is about 1 cm long and the needle has a triangular blade cross section. The blade usually has three barbs on each edge, making a total of nine. The three barbs on each edge are typically spaced about 6 mm apart; however, a spacing of 3 mm is sometimes used. For initial compaction of the nonwoven web, needles having only one barb on each edge of the needle are also used. One or more needle boards, each with thousands of needles, are installed in the needle loom.

Machine speeds range up to 1000 needle board lifts per minute. It is difficult to specify even an "average" corresponding production speed, since the numbers of needle board lifts and process steps vary for different designs of even the same product, but it may be helpful to note that an average blanket requires a total of about 8 min to produce. The width of the needle loom ranges from about 0.3 m for sample machines to 10 m for machines producing papermakers' felt.

During the needling process, the barbs of the needles carry the upper fibers of the nonwoven to a lower level to entangle the fibers mechanically and thus give strength stability and integrity to the product. The pattern spacing and density of needles, amount of barb kickup (the distance the barb is raised from the blade), size of needles, amount of penetration, draft and other variables must be considered when determining optimum needling conditions.

Virtually any fiber can be made into a nonwoven using the needle punch technique. One exception may be carbon fiber, which is very brittle and therefore cannot be needled successfully.

Sometimes, to give added dimensional stability, strength or improved dewatering properties in end use, for example, in papermaker's felt, a reinforcing scrim or other yarn structure is included in the nonwoven. Scrims are also used in needled blankets and indoor–outdoor carpeting.

Around 1972 a new type needle and needle loom was introduced which used a forked or U-shape tipped needle. This type of needle pushes the fibers all the way through the base material, causing them to form a raised pile on the other side. The pile may either be produced in a pattern or have a continuous uniform height. The latter is used extensively in automotive trunk liners and carpeting.

3.2 Stitchthrough Method

Stitchthrough nonwovens are made by another form of mechanical bonding, of which there are two broad categories. The Arachne, Maliwatt and VP systems stitch-bond fibrous webs, while the Malimo and Malipol systems stitch through a yarn system to form the nonwoven structure (Krcma 1971). The various Mali machines use an open compound needle system (latching outside the needle), whereas the Arachne machine uses a closed or hollow needle system (the latching device works inside the needle).

Initially, fabrics produced on either the Arachne or Mali machines tended to be inferior to either woven or knitted fabrics. Today, the stitchthrough nonwovens for apparel and household and industrial fabrics are competitive aesthetically, functionally and especially economically with conventional wovens and knitted fabrics. The state of the art now allows nonwovens to be used as high-quality terrycloth, imitation fur, upholstery, drapery, carpeting and substrate fabric for artificial leathers and other fabrics.

3.3 Jet Entanglement

The products obtained by this method, which have become known as spun lace, are produced by using high-pressure jets of water which are sprayed against the nonwoven web of fibers. The initial web may be formed by both dry- and wet-laid methods. The type pattern in the conveying medium which supports the web governs the pattern that is eventually created in the finished nonwoven. Because the jets entangle the fibers, a binder is not necessary for most products and the finished nonwoven has excellent softness, hand and draping properties. Applications include drapery, apparel, surgical sponges and wiping cloths.

4. Chemically Bonded Nonwovens

Chemical bonding is the method used for about 57% of the nonwovens produced and is achieved with solvents or adhesives, usually in the form of aqueous latex.

One method of solvent bonding is the Sulfalane process, in which the solvent is applied to acrylic fibers during the carding process. Subsequent heating of the nonwoven in an oven causes the outer shell of the fiber first to soften and then to bond at the fiber crossover points. Another example of a solvent-bonded nonwoven is Cerex, produced by the Monsanto Corporation. The Cerex process uses an acid to soften a nylon fibrous web, followed by pressure rolls to assist in the bonding at the fiber crossover points. The material is then neutralized, washed and dried to produce a nonwoven with excellent strength properties.

Bonding of wood fiber (cellulosic) nonwovens with latex involves the interplay of flexural moduli of the adhesive and the fiber adherend. Bonding of cellulosics with fiber-reactive dyes involves activating reactive sites, a degree of penetration of cross-linking molecules and polyelectrolyte entrapment (Allan 1972).

The majority of chemically bonded nonwovens are bonded using binders in the form of solvents, solutions and dispersions. The binders may be applied by printing, spraying, impregnating, rolling, splashing or foam methods. Application of binder by printing allows selective areas to be bonded. Unbonded areas between the bonded areas permit excellent softness and drape to be obtained. Spray bonding is used where maximum bulk and softness is required, such as in thermal linings, sleeping bags and air filtration fabrics. Impregnation completely saturates the nonwoven with binder and maximizes bonding. The advantage of foam bonding is that higher concentrations of binder can be applied, hence less energy is required to remove the water carrier.

Many materials have been used as binders. Examples include acrylics, vinyl acetate, vinyl acrylics, ethylene vinyl acetate, vinyl chloride, styrene–butadienes, carboxymethyl celluloses, epoxies and phenolics. Because the properties of the various binders vary extensively, nonwovens with a range of properties can be produced. The latitude of properties is further broadened by the range of fibers available and the different processing methods that can be used.

See also: Geotextiles

Bibliography

Allan G G 1972 Binder–fiber interactions in nonwovens. In: Jayne B A (ed.) 1972 *Theory of Design of Wood Fiber Composite Materials*. Syracuse University Press, New York, pp. 229–326

Anon 1981 The world of nonwovens. *Paper (London)* 196(10):30–2

Herlinger H, Bechter D, Kurz G 1981 Action of the binder in nonwoven stabilization, 1: Importance of the bonding mechanism for the strength of polyester nonwovens. *Chemiefasern/Text.-Ind.* 31(2): 940–2

Krcma R 1971 *Manual of Nonwovens*. Textile Trade Press, Manchester, UK

J. R. Wagner
[Philadelphia College of Textiles and Science, Philadelphia, Pennsylvania, USA]

Bonding: Preparation of Metallic Surfaces

The literature on the prebonding surface treatment of metal alloys is extensive. Until the comparatively recent work on the characterization by electron microscopic and surface analytical techniques of the oxides on aluminum alloys utilized in aircraft construction, surface preparation relied on bond testing to demonstrate adequacy of the surface for bonding. While bond performance is the ultimate test, advances in instrumental chemical analysis of metal surfaces have contributed to the controllability and reliability of the entire bonding process. Also, until relatively recently, there was no adequate mechanical test to discriminate between preparations. In general, the need for and the level of surface preparation depend on a number of factors, in particular the type of adherend metal, its surface condition, the type of adhesive material, the stress level, the type of loading on the bond and environmental exposure requirements. The importance of these factors depends on the reliability required and cost considerations.

1. General Aspects

In contrast to the mechanical joining of metals, adhesive bonding requires a higher level of technology if reasonably satisfactory and troublefree performance is to be achieved. Bonding has inherent advantages over mechanical attachments in many applications, and it is helpful to identify four different levels of surface preparation that may be selected for given applications, which provide:

(a) low adhesion and low environmental resistance,

(b) moderate adhesion with moderate environmental resistance,

(c) good adhesion with moderate environmental resistance, and

(d) optimum adhesion and maximum environmental resistance.

Many instances have been reported where excellent bonding is achieved to, say, automotive grade sheet steel with absolutely no surface pretreatment of the metal. What is not apparent, however, is that the steel was free of surface scale and that the adhesives utilized were excellent solvents for the thin oil film used as a lubricant on the metal. The durability of such bonds when stressed in a corrosive environment, however, is somewhat less than satisfactory.

For most metals, the above surface preparation levels are obtained, respectively, by using processes such as the following:

(a) descaling and degreasing,

(b) surface abrasion and degreasing,

(c) degreasing plus chemical etching, and

(d) degreasing, etching, anodizing and application of corrosion-resistant primer.

The requisite level of surface preparation must be chosen in relation to the performance and durability levels needed.

Examination of the literature on the various surface treatments available reveals a multitude of chemical etch formulations, many differing only in minor quantities of chemicals or recommended immersion temperatures and time. Almost none of the references considers the problems involved in the production quality control of the chemical baths, bath and rinse water purity, sensitivity of the bath to misformulation of ingredients, contamination, off-specification temperatures or allowable metallic ion buildup prior to replenishment or replacement of the bath, nor did they consider the effects of any of these factors on the long-term durability of the bond.

It is well known that certain adhesives such as the phenolic nitriles and phenolic butyrals are extremely tolerant to variations in metal surface preparations and provide excellent bond durability in spite of lack of attention to many of the factors mentioned, but in other instances utilizing epoxy-base adhesives, very serious trouble has developed even with what was thought to be excellent surface preparation control. Therefore, it is strongly recommended that the procedures and formulations described in the following be thoroughly explored by the interested user. The previously mentioned factors and others should be carefully investigated with the adherends and adhesives of interest, to develop a controllable, reproducible and reliable surface preparation process. Each situation and combination of materials should be thoroughly assessed and an adequate level of surface preparation selected and verified by a comprehensive and thorough evaluation program. The view taken here is that, in the overall bonding process, adherend surface preparation is the single most important factor in obtaining strong reliable adhesive bonds.

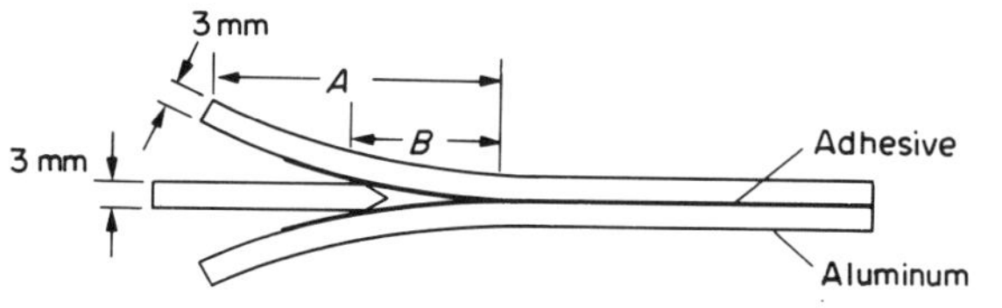

Figure 1
Wedge crack specimen: A, crack length (Douglas); B, crack length (Air Force Materials Laboratory)

The wedge crack specimen (see Fig. 1) introduced by the Boeing Corporation in 1972 has probably become the most widely used procedure for the rapid evaluation of the adequacy of metal surface preparation. The test is simple, rapid and inexpensive and the data obtained correlate well with those from more complicated and expensive tests such as the wet peel test. At present there are no non-destructive tests available to determine either the strength of a bonded joint or the quality of the surface of a metal pretreated for bonding. Quality and durability must be built into the bond by strict in-process control and the best end-item inspection possible.

The surface characteristics of a metal are determined largely by the properties of its oxide film or of the chemical conversion coating which may have been formed during prebond preparation. Pertinent properties are crystal structure, degree of hydration, cohesive strength, adhesion to the base metal, thermodynamic stability of the lattice structure, porosity, thickness and wettability. Only in rare instances, such as the noble metals, are bonds made to what may be considered a metallic surface. Once the surface of a metal has been properly pretreated for the formation of an adhesive bond, it should not be exposed to physical handling or to atmospheric environments that may damage or contaminate the surface. Dust, soil, grease and other contaminants are to be avoided.

The present state of knowledge about the surface aspects of adhesion is incomplete. Although new analysis techniques are providing much information, interpretation and analysis of this material will require strenuous effort. Meanwhile, adhesive bonding technology remains more of an art than a science.

2. *Aluminum and its Alloys*

Prebond surface preparation of aluminum and aluminum alloys is perhaps the most advanced of all metal prebond process techniques. This is due primarily to the past intensive efforts in the aerospace industry to produce long-lived bonded aluminum components. At least 25 procedures have been utilized to pretreat various aluminum alloys. They include hydrochloric acid etches, chromic acid etches, various anodizes and even sodium hydroxide etches. Relatively few, however, have been successful in the industry or have consistently resulted in environmentally resistant bonds. Only three such procedures are considered here: (a) optimized Forest Products Laboratory (FPL) etch, (b) phosphoric acid anodize, and (c) chromic acid anodize. Obviously the pollution control problems involved in properly disposing of large quantities of complex chromate solutions must be recognized.

2.1 Optimized FPL Etch

The steps involved in a typical FPL etch prebond treatment process are degreasing (solvent cleaning), alkaline cleaning and etching. These steps are separated by rinse procedures to prevent carryover and downstream contamination. Removal of oils and greases is accomplished in the first step usually by degreasing the metal in a condensing vapor of an industrial solvent such as Stoddard solvent of trichloroethylene. After evaporation of the solvent, the degreased metal is immersed in a solution of a highly buffered alkaline cleaner, usually maintained at 70–80 °C. After thorough rinsing, the metal is immersed in sulfochrome etch liquor, the composition of which differs from one organization to another. A common mix is

demineralized water:	30 parts
sulfuric acid (conc.):	10 parts
sodium dichromate:	1 part

Immersion in the bath at 70 ± 3 °C and spray rinsing immediately on withdrawal from the etch are critical to the durability of the bonds subsequently formed. Since bonds produced when freshly mixed FPL etch solutions are used are not as durable as those obtained with solutions that have been used for some time, the fresh bath is "sweetened" by the addition of several grams of aluminum alloy per cubic meter of etch to build up the amounts of dissolved aluminum and copper in the system.

The FPL etch surface preparation for aluminum has sometimes given highly variable results, in particular with the so-called second generation 120 °C curing epoxy resin adhesive systems. The service history of the FPL etch with 175 °C curing adhesives and their primers has been very satisfactory. The chemistry of the FPL etch is complex and the effects of its many variables are not yet well understood. It is known that certain contaminants, fluorine compounds for instance, can cause a bond of very poor durability to be formed.

Pretreated aluminum alloys are normally primed with a corrosion-inhibiting primer compatible with the adhesive to be used, as soon as the accompanying wedge specimens reveal that surface preparation is adequate. Once the surface is protected by the dried and baked primer, bonding may be effected or delayed for many months with no loss in bond performance. Critical process variables for the optimized FPL etch are the etch time, etch bath temperatures, rinse delay time, bath contamination and bath chemistry. These process variables must be kept under strict control in order to obtain satisfactory bond durability.

2.2 Boeing Phosphoric Acid Anodize

Introduced by the Boeing Company in 1974, Boeing phosphoric acid anodize, known as the "phos" anodize, has been shown by exhaustive testing to be the most durable prebond surface preparation available at present for aluminum alloys. It is characterized by relatively simple chemistry (one active component), modest temperature and electrical requirements and very few critical process requirements and, so far, it seems to be acceptable from most environmental aspects.

The basic steps are solvent degreasing, alkaline washing, deoxidizing and phosphoric acid anodizing. The anodizing solution is typically composed of a solution of about 100 g of concentrated phosphoric acid per liter of distilled water. Anodizing conditions are typically 10 V dc for about 22.5 min in a bath held at about 20 °C.

The "phos" process produces an oxide many times thicker and much more fragile than that produced by the FPL etch process; it must be protected as soon as possible after formation by a properly formulated and compatible corrosion-inhibiting primer. The process is characterized by having relatively few critical process variables, very near room temperature operation, highly reproducible results and is compatible with most commercially available primers and adhesives. Those process variables which it is important to control include temperature, voltage and the minimum time in the bath.

2.3 Chromic Acid Anodize

There is extensive literature on the chromic acid anodizing of aluminum alloys. Much of it seems confusing and conflicting. Yet the process has been used commercially for many years and has given extremely satisfactory service performance. Most of the controversy seems to be concerned with the various processes of sealing the prepared surface. Several authors consider that low-temperature sealed surfaces produce the highest bond strength, while others indicate that a more thorough, hotter seal produces more durable bonds.

A typical chromic acid anodizing process involves solvent degreasing, alkaline washing, deoxidizing, chronic acid anodizing and sealing. The anodizing bath is typically composed of a solution containing about 50–60 ml l^{-1} chromic acid. Anodizing conditions are about 20 V dc at 25 °C for about 35 min. The sealing step uses a bath containing 5% potassium dichromate for 8–12 min at about 88 °C. The sealing bath is made up with deionized water and the pH is adjusted to about 4.0 with chromic acid. Extensive studies of the process variables were conducted under the USAF/Douglas PABST program. The results indicated that the chromic acid anodize process must be controlled relatively closely in order to achieve reliable and reproducible bond performance.

3. Titanium and its Alloys

The surface treatment of titanium alloys, in particular 6% Al–4%V and 8%Al–1%Mo–1%V, is covered extensively in the literature. There are perhaps 20 methods for the treatment of titanium materials. Every organization that has had a need to bond titanium alloys seems to have developed its own procedure. These processes include phosphate etches, modified phosphate etches, phosphate–fluoride etches, sodium sulfate modified phosphate–fluoride etches, alkaline etches, chromic acid anodizes, sulfuric acid–hydrogen peroxide etches and several chemical conversion coatings. Some authors specify that surface abrasion such as grit blasting or vapor or liquid honing is mandatory for bonding titanium. Others do not mention this at all. The acid and alkaline etch treatments provide a pitted surface for bonding, while the anodize and conversion coatings provide a built-up oxide which may contain phosphates, fluorides or borides. Satisfactory commercial usage has been achieved with a modified cold etch developed by several aircraft companies and with several commercially available formulations.

In two treatments chosen for illustation here, the steps are typically vapor degreasing, liquid/vapor honing, acid or alkaline pickling and stabilizing. In nitric–hydrofluoric acid pickling, the metal is immersed in a solution of nitric acid (70%) 13 vol.% and hydrofluoric acid (50%) 4 vol.% for 45 s at room temperature, rinsed and then immersed in a solution of the following composition for 3–5 min at room temperature:

sodium sulfate: 5 wt%
sodium flouride: 1 wt%
hydrofluoric acid: 1 wt%

The second process, which has given good service particularly with adhesives cured at the 175 °C or above uses Pasa Jell 107C (product of Semco Corporation). The precleaned metal is immersed in a 1:2 (by volume) solution of Pasa Jell 107C and deionized water for 10–15 min at room temperature. Both treatments utilize extremely strong oxidizing acids, which should be handled and stored with proper safety precautions. The solutions should be replaced when the ratio of metallic ion to hydrofluoric acid exceeds about 1:2.

4. Magnesium Alloys

The reactivity of magnesium alloys and the serious corrosion problems encountered with them in service have led to their abandonment by most aerospace corporations. Owing to extremely poor service, both the US Air Force and the Navy have banned these materials from use in military aircraft. The recommended procedure for surface preparation of magnesium where bonding must be performed is the Dow 17 process. With this anodize treatment and the new corrosion inhibiting primers, adhesive bonds with reasonable environmental durability can be obtained. Since the treatment must be applied under very precisely controlled conditions of bath composition (voltage and current, temperature and time), few organizations are properly equipped to apply this conversion coating. A qualified operator should be sought for surface preparation of magnesium alloys for bonding.

5. Steels

Strong adhesively bonded joints with all steels can be obtained simply by solvent cleaning and light abrasion. However, in order to obtain adhesive bonds having maximum durability under conditions of stressed exposure to aggressive environmental conditions (e.g., industrial fumes, humidity, salt), more sophisticated methods of surface preparation must be used. Prior to chemical surface preparation, all contaminants and oxide scales should be removed by vapor degreasing, grit blasting (if necessary) and cleaning in a caustic detergent bath.

Many steels form a thin rust (oxide) coating even before they can be dried. Special procedures must be used with these materials. After chemical surface preparation and the final rinse, it may be necessary to dip into a methanol or methyl ethyl ketone bath and dry in moving heated air. The dry parts should be immediately primed with a corrosion-inhibiting primer in order to preclude rust formation.

For carbon and low-alloy steels a hydrochloric acid etch is recommended. The cleaned descaled metal should be immersed for 3–5 min in a solution of hydrochloric acid (30 vol.%) in deionized water at room temperature. Alternative methods that have been used extensively and have given good service performance involve the electrodeposition of a tin coating on the steel component, followed by conversion of the tin to the chromate or phosphate by one of several commercially available methods. The chromate or phosphate conversion coatings are highly resistant to corrosion and provide surfaces which can be handled readily.

For stainless or corrosion resistant steels there is extensive technical and patent literature on prebond surface preparations. These preparations cover processes such as chromic, dichromate, phosphoric, sulfuric, nitric and caustic anodizes, and chemical and alkaline etches. Many of these surface preparations have shown excellent performance in the wedge opening test in humid and salt fog environments. The following treatments have resulted in bonds of excellent durability: H_2SO_4–dichromate anodize; HNO_3–dichromate etch; HNO_3 anodize; H_2SO_4–dichromate etch; and NaOH etch. Anodizing in 50 vol.% HNO_3 at about 1 V and 5 mA cm^{-2} has given the best bond strength and wedge opening durability. The H_2SO_4–dichromate etch is performed by immersing the cleaned metal in a solution of 30 vol.% H_2SO_4 and 40 wt% sodium dichromate in 100 parts of water for 1 h at 80°C.

See also: Adhesion: Fundamental Principles; Adhesive Joining: An Overview; Elastomer–Metal Adhesion Processes; Elastomer–Wire Adhesion Processes

Bibliography

Air Force Flight Dynamics Laboratory 1977 *Primary Adhesively Bonded Structure Technology*, AFFDL-TR-77-107. Aik Force Flight Dynamics Laboratory, Dayton, OH

Air Force Materials Laboratory 1977 *Titanium Surface Preparations*, AFML-77-146. Air Force Materials Laboratory, Dayton, OH

Beck W, Bockris J P, McBreen J, Nanis L 1966 Hydrogen permeation in metals as a function of stress, temperature and dissolved hydrogen concentration. *Proc. R. Soc. London, Ser. A* 290: 220–35

Bikales N (ed.) 1971 *Adhesion and Bonding*. Wiley, New York

Bruno E J (ed.) 1970 *Adhesives in Modern Manufacturing*. Society of Manufacturing Engineers, Dearborn, MI

Cagle C V 1968 *Adhesive Bonding*. McGraw-Hill, New York

Evans T E, Sutton W H 1975 Control of electrolytic coloring of chromium-containing alloys. US Patent No. 3,804,730

Hochberg M S 1967 *Methods of Bonding Cryogenic Structures*, Report No. F-743. McDonnell Douglas Corporation, St Louis, MO

Houwink R, Salmon G 1967 *Adhesion and Adhesives*, 2nd edn., Vol. 2. Elsevier, Amsterdam

Hudson W R 1977 Ion beam texturing. *J. Vac. Sci. Technol.* 14: 286–9

Keith R E 1965 *Adhesive Bonding of Titanium and its Alloys*, NASA TM-X-53313. National Aeronautics and Space Administration, Washington, DC

McDonnell Douglas Corporation 1970a *Fretting Resistance Coatings for Titanium Alloys*, Department of Air Force Contract No. F33615-70-1538. McDonnell Douglas Corporation, St Louis, MO

McDonnell Douglas Corporation 1970b *Nondestructive Test Methods*, MCAIR Engineering Study Authorization Interim Report No. A0121. McDonnell Douglas Corporation, St Louis, MO

McDonnell Douglas Corporation 1975 *Investigation of Metalbond 329 Strength Reduction in Salt Spray Environment when Bonding Titanium*, MCAIR Technical Memorandum 256.756. McDonnell Douglas Corporation, St Louis, MO

Moji Y 1983 Preparing an environmentally stable stainless surface for bonding. US Patent No. 4,064,020

Monroe R E 1973 *Joining of Titanium*, TR DMIC240. Defense Metals Information Center, Battelle Memorial Institute, Columbus, OH

O'Brien R W 1968 *Strength Tests of Titanium Honeycomb Panels*, Report No. 604-175. National Aeronautics and Space Administration, Washington, DC

Pochily T M 1966 *Process for Anodizing Titanium*, Technical Report WVT-6605. Benet Laboratories, Watervliet Arsenal, Watervliet, New York

Reinhart T J 1982 *Evolution of Structural Adhesives*. Society for the Advancement of Material and Process Engineering, Azusa, CA

Smith T 1975 Photoelectron emission from aluminum and nickel measured in air. *J. Appl. Phys.* 46: 1553–8

Smith T 1976 Auger electron spectroscopy and ion sputter profiles of oxides on aluminum. *Surf. Sci.* 55: 601–24

Smith T 1977 Surface treatment for aluminum bonding (STAB). *J. Adhes.* 8: 313–20

Snogren R C 1974 *Handbook of Surface Preparation*. Palmerton, New York

Voss D L 1970 *Adhesive Bonding of Titanium*, Report No. A0449. Air Force Materials Laboratory, Dayton, OH

Walter R E 1970 *Titanium Processing Technology—Adhesive Bonding*, Report No. A0180-2. Boeing Commercial Airplane Company, Seattle, WA

Wegman R F 1989 *Surface Preparation Techniques for Adhesive Bonding*. Noyes Publications, Park Ridge, NJ

Winter J, Mechelburg C 1980 *Steel Bonding*. Van Nostrand, New York

T. J. Reinhart Jr
[Air Force Materials Laboratory, Dayton, Ohio, USA]

Bonding: Preparation of Polymer Surfaces

Polymers usually require pretreatment before they can be bonded successfully to other substrates. With some polymers, solvent degreasing or abrasion is sufficient, but with others, especially polyolefins and polytetrafluoroethylene, chemical modification may be necessary. There is much controversy regarding the reasons for the success of some of the important pretreatments; the importance of weak boundary layers, surface energy, specific interactions and surface topography are discussed in this article.

It is sometimes possible to avoid a surface pretreatment by using alternative joining methods: these include mechanical fastening, heated-tool welding, induction heating, ultrasonics and solvent cementing. Also, with polyolefins a pretreatment may sometimes be rendered unnecessary by using a melt adhesive. However, some of these alternatives are only suitable for joining two pieces of the same polymer and, except for mechanical fastening, they are not suitable for thermosets. When it is necessary to join a polymer to a dissimilar polymer or another material there may be no satisfactory alternative to a pretreatment followed by adhesive bonding.

Whether these alternatives are preferable to a pretreatment followed by adhesive bonding depends on cost, joint performance and any undesirable side effects such as crazing.

The most widely used synthetic polymers are the polyolefins. Without a pretreatment, only poor adhesion is obtainable, and it is therefore not surprising that these polymers present the most frequent adhesion problems. These problems are manifested not only in bonding problems, but also in printing, coating, heat sealing and metal plating. Like polyolefins, polytetrafluoroethylene (PTFE) normally requires a pretreatment to give satisfactory adhesion. Although PTFE is a low-production-tonnage polymer, it is of great technological importance. Because of the inherent difficulty in bonding polyolefins and PTFE and because of their technological importance, this article concentrates on these polymers; however, treatments for other important polymers are also discussed.

1. Physical Pretreatments

With some polymers (e.g., poly(vinyl chloride)), it may be sufficient simply to wipe the surface with a cloth moistened with solvent to remove most or all of the plasticizer and other contaminants from the surface. In other cases (e.g., Nylon 66), the following scheme is usually satisfactory: solvent wipe, abrade with emery cloth and, finally, remove dust by wiping with a cloth moistened by solvent. Mechanical abrasion will increase the potential bonding area and provide the possibility of mechanical keying; however, if the adhesive does not fill the crevices created by the abrasion, stress concentrations due to voids may result.

Solvents can cause severe etching of a polymer surface. For example, if polypropylene (PP) is exposed to trichloroethylene vapor for a few seconds, extensive roughening of the surface occurs; the adhesion strength to an epoxide adhesive then increases about sixfold (Garnish and Haskins 1969). However, after longer exposure to the vapor the adhesion strength falls sharply. The increased adhesion is presumably due to mechanical keying, but excessive treatment apparently results in a very weak surface. Solvent etching has three main disadvantages:

(a) it is critically dependent on the correct treatment time,

(b) damage to the polymer is likely, and

(c) practical difficulties stand in the way of high production rates.

2. Polyolefins

Although it is possible to achieve large increases in adhesion by means of a physical treatment, in practice chemical treatments are generally used prior to bonding polyolefins. The three most widely used pretreatments are corona discharge, flame treatment and etching with chromic acid. These methods were introduced in the mid-1950s for low-density polyethylene (LDPE), but they are also generally effective with high-density polyethylene (HDPE) and PP. Various other pretreatments have been proposed, of which two are in use to some extent. These are suppressed-spark treatment (Ayres and Shofner 1972) and a method which involves coating the polyolefin with benzophenone, followed by uv irradiation.

The corona-discharge method is almost invariably used for polyolefin films. The equipment consists of a high-frequency spark oscillator connected to one or more electrodes. Each electrode is placed about 1.5 mm from a roller, which has an external layer of a dielectric and a metal core. The plastic film is passed over the roller continuously and when the current is switched on the air between the electrodes and the film is made conductive.

Flame treatment is used for thicker sections such as bottles. It is the main method for treating blow molded and injected-molded parts. Chromic acid is used for treating complex shapes and is especially used for small production runs where specialized equipment is not available.

The changes in surface chemistry caused by the pretreatments have been studied by various spectroscopic techniques. The two most useful methods are x-ray photoelectron spectroscopy (XPS; alternatively known as electron spectroscopy for chemical analysis (ESCA)) and multiple internal reflectance infrared spectroscopy (MIR-IR). Both techniques can provide quantitative information about the degree and type of chemical modification and a combination of the two methods can be used to study acid etching and corona-discharge treatment.

The sampling depth of XPS is down to a few nanometers, while MIR-IR has a sampling depth of the order of 1 μm. Corona treatment of PE and PP produces surface changes of sufficient depth to be detected by MIR-IR, whereas the depth of the changes produced by flame pretreatment of LDPE has been calculated using XPS to be between 4 nm and 9 nm (Brewis and Briggs 1981). Studies of the chromic etching of PP using MIR-IR have failed to detect any changes in surface chemistry (Blais et al. 1974). It has been shown that the depth of treatment of PP with chromic acid is limited to less than 10 nm; the polymer chains on the surface are broken into fragments which pass into solution. However, with LDPE and HDPE this does not occur (at least, not to the same extent) and the depth of oxidation is sufficient for detection by MIR-IR.

The effects of the various parameters of these pretreatments on the resultant adhesion have been investigated by various workers. In a study of flame treatment (Ayres and Shofner 1972), the effects of the air-to-gas ratio, the distance between flame and polymer, and treatment time and temperature were examined. Also, it has been shown (Brewis and Briggs 1981) that a wide range of compositions of chromic acid are effective, including much milder treatments than normally recommended. In two separate studies (Briggs and Kendall 1979, Spell and Christenson 1979), the relationship between surface chemistry and discharge-treated PE and printability was demonstrated. In one study (Briggs and Kendall 1979), the importance of specific interaction was demonstrated by "blocking" the key functional groups.

In addition to chemical changes, the pretreatments can cause changes in the topography of the polyolefins, which, as noted earlier, can affect the resultant adhesion. Chromic acid and corona treatments generally cause considerable roughening, although the extent depends on the severity of the treatment, the type of polyolefin and its thermal history.

3. *Polytetrafluoroethylene*

To obtain satisfactory joint strengths with PTFE, it is usually necessary to pretreat the surface. The most widely used pretreatment involves a solution of naphthalenide ions in tetrahydrofuran (THF). The alternative treatment sometimes recommended is immersion in a solution of sodium in liquid ammonia. However, this treatment is clearly hazardous and it is also inconvenient to use a liquid with a boiling point of −33 °C. The sodium–liquid ammonia treatment is more rapid, but as the naphthalenide treatment only requires about 1 min, any saving in time is not likely to outweigh the disadvantages. Both these treatments result in the formation of a black carbonaceous layer which is readily bondable. It has been shown in XPS studies that this layer contains a substantial quantity of oxygen. The black layer can be removed with chromic acid, sodium hypochlorite or exposure to light.

In addition to the above methods there are several possible alternatives. These include electrical discharge and electrochemical methods, as well as alkali-metal amalgams.

In spite of its reputation as an inert material, PTFE is thermodynamically unstable in contact with nearly all metals (i.e., it may be reduced to carbon and fluoride ions). Reactions between PTFE and metals are not usually observed probably due to the formation of passive layers. However, under tribological conditions, such layers are likely to be broken down and the formation of metallic fluorides with steel and nickel has been observed. PTFE can be reduced electrochemically if placed in contact with a cathodically polarized electrode in an aprotic solvent such as dimethylformamide (DMF) containing an inert support electrolyte (such as 0.1 mol tetrabutylammonium tetrafluoroborate); the potential should be more cathodic than −1.5 V with respect to a saturated calomel electrode (SCE) and, preferably, near to −3 V SCE, which is the limit for this solvent.

The ease with which PTFE is reducible is highly dependent on the processing history of the polymer (Dahm et al. 1981). Thus, skived tape and ram-extruded rod, with their highly sheared atypical layers, are much more readily reduced than samples without such layers, such as dispersion-cast films or samples from which such layers have been removed chemically (i.e., with a pretreatment followed by removal of the carbonaceous layer with a reagent such as chromic acid). The lower reduction potential of naphthalene (−2.6 V SCE) compared with the formation potential of solvated electrons (about −3 V SCE) confirms that sodium naphthalenide in THF is a milder reducing agent than sodium in liquid ammonia. Sodium naphthalenide in THF can only be used to treat PTFE possessing an atypical layer. If this layer is removed, as described earlier, treatment with sodium naphthalenide is only possible if the surface is first reactivated either thermally, electrochemically, mechanically or with sodium–liquid ammonia.

The adverse effect of uv radiation on joints of PTFE (etched with sodium–liquid ammonia) with polyurethane has been demonstrated (Meier and Petrie 1973). The loss in peel strength is less severe if carbon black is incorporated into the PTFE or if hydroquinone (an antioxidant) is incorporated into the adhesive.

4. *Other Polymers*

For polymers other than polyolefins and PTFE, the most common chemical treatments are the corona-discharge method and chromic acid etching. The former treatment is used for plastic films (e.g., poly(ethylene terephthalate)), whereas chromic acid is used to pretreat molded parts.

Chromic acid etching of acrylonitrile–butadiene–styrene (ABS) is used to obtain good adhesion between metal plating and the polymer. Electron microscopy has provided evidence to show that mechanical keying makes an important contribution to the improved adhesion. However, the oxidation of the polymer is also advantageous to bonding, because improved wetting and stronger interaction with the adhesive are likely. Chromic acid may be used to improve adhesion with other polymers, including poly(phenylene oxide), acetal copolymer and polystyrene; the optimum conditions for the treatment vary with the polymer.

Corona-discharge treatment is used to improve the adhesion characteristics of poly(ethylene terephthalate). The chemical changes caused by the treatment have been studied using XPS and contact-angle measurements (Briggs et al. 1980). The main changes found are the introduction of phenolic–OH and carboxylic acid groups; improved wetting is also observed. Substantial chain scission occurs, the low-molecular-weight material migrating into the film on aging. This material can also be removed by washing with water.

5. Bonding Mechanisms

Usually, it is fairly easy to pretreat successfully any polymer, although additives can cause severe difficulties. Physical treatments are often sufficiently effective to give satisfactory joint strengths, except in the cases of polyolefins and fluorinated polymers. Much controversy has centered on the bonding problems of polymers and, in particular, polyolefins. Until 1959, it was generally thought that the poor adhesion to polyolefins was due to a lack of polar groups. However, in 1959, Bikerman proposed his theory of weak boundary layers, according to which the poor adhesion of polyethylene is due to low-molecular-weight material which separates from the melt on cooling. Zisman, on the other hand, argued strongly that the poor adhesion of polyethylene and similar polymers was due to their low surface energies and the consequential poor wetting by the adhesive.

Both theories have received support from other workers. For example, Levine et al. (1964) found a fairly good relationship between the joint strength obtained with a polymer and its critical surface tension. On the other hand, Schonhorn and coworkers have provided evidence in favor of the weak-boundary-layer theory: for example, they bombarded polyethylene and other low-energy polymers with activated inert-gas ions and obtained large increases in adhesion to an epoxide adhesive (Hansen and Schonhorn 1966). Using attenuated total reflectance (ATR)–infrared spectroscopy, they found no evidence of any chemical changes in the surface and they concluded that the small molecules on the polymer surface had been crosslinked to the long polymer chains in the bulk. In other work, PE was melted against aluminum and, after the polymer cooled, the aluminum was dissolved. The adhesion of an epoxide adhesive to the resultant polymer surface was much increased, but again no chemical changes were observed using ATR. Schonhorn and Ryan (1968) attributed the improved adhesion to a highly crystalline region which forced the small molecules into the bulk of the polymer. However, with the much more sensitive XPS, it was shown by Brewis and Briggs (1981) that substantial oxidation had occurred, thereby weakening the evidence in favor of the weak-boundary theory. Likewise, XPS weakened the hypothesis that chromic acid treatment of PP is effective by removing a weak boundary layer; MIR-IR had again failed to detect chemical changes, whereas XPS showed that substantial oxidation had occurred.

In recent years, the balance of evidence has shifted against the weak-boundary theory in favor of the need for functional groups to improve wetting and increase the magnitude of interactions. Some evidence has been given to show that small concentrations of functional groups can result in good adhesion, especially if specific interaction occurs (Brewis and Briggs 1981).

It is important to realize that the existence of small molecules on a polymer surface is probably common, but they do not necessarily constitute a weak boundary layer. The small molecules may originate from the migration of additives to the surface, from processing aids, from contamination or from degradation of polymer chains during a pretreatment. If the small molecules can be displaced by the adhesive then high joint strengths can still be obtained.

The importance of a displacement mechanism has been demonstrated by deliberately adding a layer of hydrocarbon grease ($\sim 2\times 10^3$ nm thick) to an HDPE surface which has been oxidized with chromic acid. The pretreatment increased the lap shear strength from 1.8 MPa to 17.3 MPa; the addition of the hydrocarbon only reduced the bond strength to 12.5 MPa. A potential weak boundary layer may actually be created by a pretreatment, as with excessive corona treatment of polyolefins or in corona treatment of poly(ethylene terephthalate). Whether a region of low strength is displaced depends on the nature of the adhesive, the nature of the polymer surface, the temperature and the time before the adhesive hardens.

Methods are now available to pretreate successfully most if not all polymers. However, migration of additives prior to pretreatment can sometimes have an adverse effect: for example, slip and antistatic agents can have an adverse effect on the corona treatment of PP. Other complications with pretreatments occur. For example, the corona-discharge treatment of polyolefins will usually much enhance bonding operations and also the adhesion of printing inks, but may seriously impair heat sealing.

See also: Adhesion: Fundamental Principles; Adhesive Joining: An Overview; Elastomer–Textile Adhesion Processes; Elastomer–Thermoplastic Adhesion Processes

Bibliography

Ayres R L, Shofner D L 1972 Preparing polyolefin surfaces for inks and adhesives. *Soc. Plast. Eng., Tech. Pap.* 28: 51–5

Blais P, Carlsson D J, Csullog G W, Wiles D M 1974 The chromic acid etching of polyolefin surfaces and adhesive bonding. *J. Colloid Interface Sci.* 47: 636–49

Brewis D M (ed.) 1982 *Surface Analysis and Surface Pretreatment of Plastics and Metals.* Applied Science, London

Brewis D M, Briggs D 1981 Adhesion to polyethylene and polypropylene. *Polymer* 22: 7–16
Briggs D, Kendall C R 1979 The chemical basis of adhesion to electrical discharge treated polyethylene. *Polymer* 20: 1053–4
Briggs D, Rance D G, Kendall C R, Blythe A R 1980 Surface modification of poly(ethylene terephthalate) by electrical discharge treatment. *Polymer* 21: 895–900
Dahm R H, Brewis D M, Gribbin J D, Barker D J, Hoy L R J 1981 Surface pretreatments of polytetrafluoroethylene. *J. Adhes.* 13: 67–76
Dukes W A, Kinloch A J 1977 Preparing low energy surfaces for bonding. In: Wake W C (ed.) 1977 *Developments in Adhesives—1.* Applied Science, London, pp. 251–82
Dwight D W, Riggs W M 1974 Fluoropolymer surface studies. *J. Colloid Interface Sci.* 47: 650–60
Garnish E W, Haskins C G 1969 The effect of surface conditions when bonding with epoxy adhesives. In: Alner D J (ed.) 1969 *Aspects of Adhesion—5.* University of London Press, London, pp. 259–76
Hansen R H, Schonhorn H 1966 A new technique for preparing low surface energy polymers for adhesive bonding. *J. Polym. Sci., Polym. Lett. Ed.* 4: 203–9
Levine M, Ilkka G A, Weiss P 1964 relation of the critical surface tension of polymers to adhesion. *J. Polym. Sci., Polym. Lett. Ed.* 2: 915–19
Meier J F, Petrie E M 1973 The effect of ultraviolet radiation on sodium-etched polytetrafluoroethylene bonded to polyurethane elastomer. *J. Appl. Polym. Sci.* 17: 1007–17
Petrie E M 1980 Joining the engineering plastics. *Adhes. Age* August: 14–23
Schonhorn H, Ryan F W 1968 Effect of morphology in the surface region of polymers on adhesion and joint strength. *J. Polym. Sci., Polym. Phys. Ed.* 6: 231–40
Spell H L, Christenson C P 1979 Surface analysis of corona-treated polyethylene. *Tappi* 62: 77–81
Wegman R F 1989 *Surface Preparation Techniques for Adhesive Bonding.* Noyes Publications, Park Ridge, NJ

D. M. Brewis
[Leicester Polytechnic, Leicester, UK]

Bottles and Other Rigid Containers

Rigid containers produced from polymeric materials are widely used by the packaging industry. Included in this product category are bottles, jars, tubes, drums, tubs, cups, blister packs and skin packs. Bottles are probably the most important of these categories. In the USA alone, over 725 kt of resin were consumed in 1979 to produce more than 15×10^9 bottles. Worldwide production is estimated as close to 25×10^9 bottles.

Dating back to the early commercial uses of polyethylene squeeze bottles in the late 1940s, plastic containers have enjoyed strong consumer acceptance. They provided drop-impact safety, package function and were lightweight. As other polymeric materials became available for packaging applications, they brought improved barrier properties, clarity, broader design capabilities and the potential for even lighter weights and better economics.

1. Applications of Plastic Packaging

Plastic bottles were first marketed in large quantities in the late 1940s for squeeze and spray uses. These bottles took advantage of the flexibility and recovery properties of low-density polyethylene. The material in widest usage for bottles today is high-density polyethylene, which accounted for about 75% of the resin consumed in 1979.

One advantage of the high-density polyethylene is its greater stiffness. This makes it possible to design bottles with thinner walls, which require less polymer weight and are thus cheaper to produce. Polyethylenes have excellent resistance to attack by most chemicals at normal temperatures with the exception of strong oxidizing acids. However, many hydrocarbons and some organic solvents and essential oils can permeate polyethylene and pass through the bottle walls. Water, being quite polar, does not dissolve into polyethylene, which is essentially nonpolar, and therefore the water permeability rate is low. The oxygen permeability of polyethylene is relatively high and this has limited its usefulness for the packaging of many oxygen-sensitive products, especially food items.

In general, polyethylenes, because of their low water permeability, are favored for packaging of dry products (e.g., powders, tablets) and for water-based liquid products. Major applications include containers for household products, such as detergents, bleach and fabric softeners; toiletries and cosmetics, such as shampoos, conditioners and lotions; pharmaceutical products, both liquid and dry; some foods and beverages, with milk the largest single use; automotive and marine products; and industrial chemicals.

Polypropylene has chemical resistance and permeability properties similar to the polyethylenes and is used for packaging of some items in these same product categories where better clarity and higher heat resistance are desired. These same properties are important in the medical area, where containers for injectable solutions are required to withstand autoclaving and permit visual inspection of the solution.

For those products which cannot be adequately packaged in polyethylene or polypropylene, the use of poly(vinyl chloride) (PVC) formulations or poly(ethylene terephthalate) (PET) often provides an answer. While neither has the superior water barrier properties of the polyolefins, they have been found suitable for many water-containing materials and they provide much better retention of hydrocarbons, oils, perfumes, flavorants and some solvents. They are also better barriers to gas transmission and, therefore, have been used for foods and beverages which need protection from the effects of atmospheric oxygen.

PVC has been used for the packaging of hair care products, toiletries, cosmetics, mouthwashes, household chemicals and vegetable oils. A major market for PVC bottles in Europe is for the packaging of mineral waters and, to a lesser extent, wines. An extended market test of liquors in rigid PVC bottles was terminated in 1973 when vinyl chloride monomer was implicated as a human carcinogen. The US Food and Drug Administration (FDA) determined that minute quantities of the monomer migrated from the containers to the packaged product and proposed a ban on further liquor packaging uses. Although the proposed ban was never finalized, it had the effect of terminating the market test and inhibiting the wider use of PVC for other products. Some producers of PVC resins are now able to reduce the amount of residual vinyl chloride monomer in their polymer to levels at which it cannot reasonably be expected to migrate into packaged products. It is expected that the FDA will amend its proposed regulations to permit food and beverage contact uses for these new PVC materials under strict monomer controls. This could result in much broader packaging applications for PVC.

PET bottles have found their major application in the packaging of carbonated soft drinks because, when oriented properly, they have the tensile properties to hold the internal pressure of carbonation and the barrier properties to retain the carbon dioxide. Other potential uses include the packaging of vegetable oils, salad dressings, liquors, mouthwashes and heavy-duty cleaners. Another specialized plastic bottle application is the returnable–refillable milk container made from polycarbonate (PC). PC provides excellent clarity, toughness and high heat resistance. It can be washed repeatedly without significant degradation or distortion. The relatively poor barrier properties of PC are not a handicap since the storage period for milk is so short. Other uses of PC include baby bottles and laboratory ware.

Copolymers of acrylonitrile have been used for consumer product packaging because of their superior oxygen and carbon dioxide barrier and their specific solvent resistance. The largest use by far was as a container for carbonated soft drinks. This use was discontinued when the FDA ruled that acrylonitrile monomer could be expected to migrate into the soft drink product and its safety as a food additive had not been demonstrated. As a consequence of a US Court of Appeals decision in 1979 in favor of the polymer producers, the FDA in 1984 issued a revised regulation permitting the use of a special low monomer acrylonitrile copolymer for the packaging of soft drinks.

Aside from the large consumer packaging markets for plastic bottles already mentioned, there are many other commercial applications. Bottles, drums, drum liners and carboys, usually of polyethylene, have been accepted for the packaging and transportation of many industrial chemicals. In addition to consideration of the compatibility of package and contents, the shipper must also be aware of regulations concerning the safety aspects during shipment. These larger size containers are also sold empty for storage of a variety of liquid products including gasoline and other petroleum derivatives.

The producers of laboratory equipment offer a wide range of products (e.g., bottles, containers, beakers, graduates) made from polyethylene, polypropylene, polycarbonate, polysulfone, polymethylpentene, fluorocarbon polymers, nylons and other plastic resins for specific job requirements.

Among other applications of polymeric materials to packaging, plastic tubes and laminated tubes are in prominent use. Dentifrice products, pharmaceuticals, hair products, artists colors and shaving creams are typical of the items packaged in tubes. All-plastic tubes are generally produced from polyethylene or polypropylene, although some PVC has been used for clear tubes. External coatings are sometimes applied to improve barrier properties. The laminated tube normally has an injection-molded polyethylene head with the body constructed from a lamination of plastic, aluminum foil and paper.

There is another large family of plastic packaging produced from sheet materials, which includes tubs, cups, trays and blister and skin packs. Tubs and cups are often used for dairy products, such as cottage cheese and yogurt, and for margarine. Since these items are usually refrigerated and their shelf life is short, water and oxygen barrier properties are usually not critical. Impact polystyrene and acrylonitrile–butadiene–styrene are often employed for ease of processing and for their stiffness, which permits the use of thin walls and light weights, while retaining sufficient package rigidity for handling and use. High-density polyethylene is also used, especially with soft margarines.

In producing trays for baked goods, clear oriented polystyrene is usually preferred. Meat trays may be formed from the oriented polystyrene or foamed polystyrene. The foamed material is also in common use for egg cartons.

Blister and skin packs are thermoformed packages utilized widely for the merchandising of notions, hardware, unit-dose drugs and unit-portion foods. They are usually combined with a backing card which provides additional support and serves to carry artwork, printed information, promotion and legal copy. The term blister pack identifies a part formed in a mold, the part subsequently being used to contain another product. Skin packs, on the other hand, use the packaged product as the mold. In other words, the plastic sheet material is softened by heat and then formed to shape over the item being packaged.

Clarity is usually important in blister packaging, so the clearer polymers such as PVC or, where greater toughness may be required, cellulose butyrate are commonly used. For skin packs the most common materials are polyethylene, PVC and ionomer.

Properties of common plastic packaging materials are summarized in Table 1.

2. *Production Processes*

The production of packaging items from polymers normally involves high-output, well-automated processes.

The most widely used method for plastic bottle production is extrusion blow molding. In this process, a molten tubelike shape called a parison is formed by an extruder. A split-cavity mold closes around the parison, pinching one or both ends. Compressed air is introduced to inflate the parison against the mold surfaces, where it is cooled and assumes the shape of the mold. There is frequently a secondary operation to trim excess material at the neck and base. Technology is available to program the extrusion of the parison for optimum control of thickness depending on the function and configuration of the final container.

Next in common usage for bottle production is the injection blow molding process. This is essentially a two-stage process in which the polymer is first injected into a preform mold. In the second stage, the hot preform is transferred into a blow mold, where compressed air forms it to the mold shape. There is no need for a trimming operation after injection blow molding.

There are many variations on these basic processes. One blowing technique being practised for special use polyethylene containers utilizes a dilute mix of fluorine in nitrogen as the blowing medium in place of compressed air. The fluorine reacts with polyethylene to produce an inner surface with different chemical affinity and polarity, resulting in better barrier properties and extending the types of product which can be packaged.

The bottle-forming process which has received most attention in the past few years is stretch blow molding. This method is used basically to produce orientation in the walls of the finished container, thus improving the barrier properties, tensile properties and impact resistance. The most easily recognized example of stretch blow molding is the two-liter PET carbonated-beverage bottle. The process consists of injection molding a test-tube-style preform under closely controlled conditions. This preform is brought to orientation temperature, stretched axially by mechancial means and blown radially with compressed air to produce the final container form. To hold the pressure of soft drink carbonation (~410 kPa) without excessive creep or distortion, many molders produce the bottle with a hemispherical bottom and assemble this with an injection-molded base cup. A specially designed one-piece unit is, however, also available.

For some of the larger container sizes, such as drums and carboys, rotational molding may be used as an alternative to extrusion blow molding. In this process, the polymer charge is introduced into the mold and the mold and charge are rotated on two axes while heat is applied. The mold is then cooled to set the container shape, which is then ready for removal.

Molded containers formed from sheet materials are relatively simple designs amenable to high-speed, multicavity production. The sheet stock is heated to forming temperature, clamped in a frame then pulled into a cavity or draped over a male mold. The pressure for forming can be atmospheric (vacuum forming), compressed air, mechanical or a combination. The formed parts are trimmed from the web in place or in a secondary operation. The sheet used can be roll fed or taken in line from the extruder. Many food items use an in-line form–fill–seal operation.

3. *Design Considerations*

In designing a plastic bottle or other rigid container, it should be recognized that despite the connotation of the word rigid there will probably be some flexibility in the final product. For instance, flat panels on a container of square or rectangular cross section are difficult to mold, may warp on cooling and may bulge when the container is filled.

Packaged products which produce an internal pressure will do best in a container of round cross section and products which produce an internal vacuum should use a design which will compensate for, or conceal the effects of, the vacuum. The effects of hot or cold storage temperatures should also be taken into account. Simple designs without sharp angles or features will generally provide better impact resistance. Reentrant angles should be avoided to permit removal of the part from the mold without scuffing.

4. *Environmental Considerations*

Although polymeric materials are derived essentially from petroleum and natural gas, studies have shown that most plastic packaging actually consumes no more energy, and often significantly less energy, then the conventional packages they replace. This low energy requirement is due mainly to the small amounts of material required and the lower temperatures of processing. After use, discarded plastic

Table 1
Typical properties of common plastic packaging materials

Material	Resin density ($kg\ m^{-3}$)	Flexural modulus (GPa)	Oxygen permeability[a] ($10^{-16}\ m^2\ s^{-1}$)	Water vapor permeability[b] ($10^{-14}\ kg\ m^{-1}\ s^{-1}$)	Chemical resistance/ permeability	Drop impact resistance	Appearance	Typical uses
Acrylic multipolymer	1090–1140	2.07–2.41	13.7	4.6–6.8	good grease/oil resistance; poor to good with acids and alkalis; poor with solvents	fair to good	clear	bottles, jars for food and drugs
Ionomer	930–960	0.24	103–228	0.6–1.0	good grease/oil resistance	excellent	clear to slight haze	blister, skin packs
Nitriles	1120–1170	3.10–3.38	0.5–0.9	2.3–2.7	good barrier for grease, oils, flavorants, odorants; excellent oxygen and carbon dioxide barrier	poor to good	clear, slightly yellow	bottles for soft drinks, ink; blister packs for food
Polycarbonate	1200	2.21	103–114	4.6–6.4	good grease/oil resistance; fair to good with alkalis and solvents	excellent	clear	bottles for milk, water, mineral oil
Poly(ethylene tetrephthalate)	1340–1380	2.34	2.3–3.7	0.9–1.4	good resistance to grease/ oils; good barrier for oxygen, carbon dioxide, flavorants, odorants, some solvents; poor with alkalis; fair to good with acids	excellent	clear	carbonated beverage bottles, drug packages
Polyethylene					good resistance to chemical attack except oxidizing acids; good barrier for polar chemicals, poor for nonpolar; poor to fair grease/oil resistance			bottles, jars, tubs, lids, drums, pails
low-density	910–925	0.17	205.2	0.5		excellent	hazy	
high-density	940–965	0.93	50.2	0.2		very good	hazy, translucent	

Polypropylene	890–910	⩾1.38	68.4	0.2	similar to high-density polyethylene	good	slight haze	bottles, boxes, closures, thermoforms
Polystyrene								
general-purpose	1040–1080	3.38	159.6	4.6	fair to good resistance to grease/oil; poor solvent resistance especially to aromatic solvents	poor to good	clear	trays, closures, bottles, vials, jars
impact	1030–1100	2.62	54.7	5.9		fair to good	hazy	tubs, cups, trays, closures
Polysulfone	1240	2.69	77.5	6.8	very good grease/oil resistance; fair to good with acids, alkalis	good to excellent	clear	laboratory ware, microwave cooking
Styrene–acrylonitrile	1080	3.10	31.9	5.9	similar to polystyrene	poor to good	clear to slight tint	cups, trays, jars, bottles
Poly(vinyl chloride) (impact moderate)	1350–1370	2.21–2.41	3.6–6.8	0.9–1.4	good resistance to grease/oil, acids, alkalis; poor with solvents except good aliphatics	fair to good	clear to slight tint	bottles, jars, blister and skin packs
Polymethylpentane	830	1.10	1.37–1824	2.7	fair to good grease/oil resistance; very good with alkalis	poor to good	moderate haze	laboratory ware, microwave cooking
Acrylonitrile–butadiene–styrene	1010–1110	2.41–2.62	54.7	4.6–6.8	good resistance to oils, greases, weak acids and bases; not suitable for aromatic or halogenated hydrocarbons	fair to good	opaque	tubs, trays, closures

a At 22.8 °C and 1 atm b At 37.8 °C and 100% relative humidity packages can usually be handled without problems in normal waste disposal systems, whether incinerator or sanitary landfill.

See also: Blow Molding of Thermoplastics; Novel Applications of Polymers; Plastics for Packaging; Recycling

Bibliography

Current Industrial Reports: Plastics Bottles (monthly). US Department of Commerce, Bureau of the Censors, Washington, DC

Frados J (ed.) 1976 *Plastics Engineering Handbook of the SPI,* 4th edn. Van Nostrand Reinhold, Cincinnati, OH

Jones G 1967 *Packaging Information Sources,* Management Information Guide Series No. 10. Gale, Detroit, MI

Simms W C (ed.) 1981 *Packaging Encyclopedia,* Vol. 26, No. 4. Cahners, Chicago, IL

E. J. Temple
[Monsanto Chemical Co., Bloomfield, Connecticut, USA]

C

Calendering

Calendering is the pressing of materials such as cloth, rubber, thermoplastics or paper between rollers or plates in order to produce a smooth and glossy finish or to thin the materials into sheets. The calendering of polymeric materials, like thermoplastics and rubbers, is an operation for the production of continuous sheet or film of uniform thickness. This is accomplished by squeezing the molten material between a pair of heated driven rolls.

1. General

The modern technological developments in the calendering of thermoplastic materials are offshoots of fabric and rubber calendering, which dates back to the early 1800s. Elden and Swan (1971) present a short history and a detailed account of the major technological developments up to about 1968.

Industrial calenders consist usually of 3–6 rotating rolls in a Z or L arrangement. The basic forming operation is completed by the calender and is normally followed by additional treatment of the plastic sheet or film produced. A typical modern calendering layout is shown in Fig. 1. The molten material is fed to the calender rolls from a Banbury mixer and two-roll mill system, or from a large extruder. A feedback control loop between the thickness gauge and the calender gap adjustment mechanism can make sheet thickness corrections automatically.

The major thermoplastic material that is calendered is poly(vinyl chloride) (PVC). There are over 150 calendering lines in the USA and these process about one-fifth of the US PVC production. This includes production of all film of thickness less than 0.25 mm and sheet of thickness ranging usually from 0.25 mm to about 1 mm. Products range from wall coverings and upholstery fabrics to reservoir linings and agricultural mulching materials.

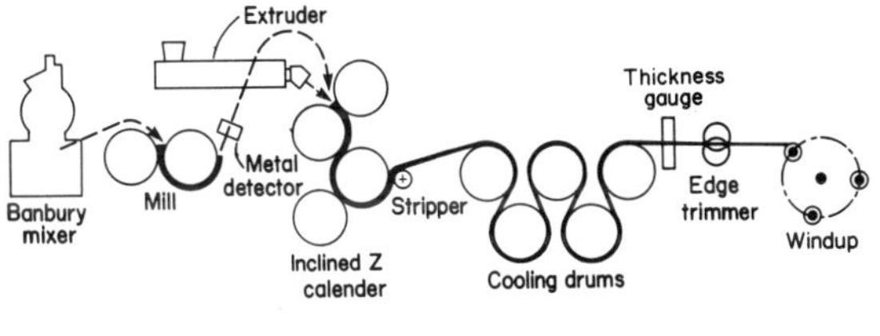

Figure 1
A typical calendering layout for the production of plastic film

2. Flow and Heat-Transfer Analysis in the Calender Gap

The operational characteristics of a calender are primarily determined by the flow and heat-transfer phenomena in the gap between two calender rolls (see Fig. 2). The mathematical formulation is based on the equations of conservation of mass, momentum and energy for creeping flow in a passage of variable width (lubrication approximation). Middleman (1977) and Tadmor and Gogos (1979) detail the various models that have been proposed.

For the rectangular coordinate system of Fig. 2, the lubrication approximation equations are

$$Q = 2\int_0^h u\,\mathrm{d}y \tag{1}$$

$$\frac{\partial p}{\partial x} = \frac{\partial \tau_{xy}}{\partial y} \tag{2}$$

$$\rho C_p u \frac{\partial T}{\partial x} = k\frac{\partial^2 T}{\partial y^2} + \tau_{xy}\frac{\partial u}{\partial y} \tag{3}$$

where Q is the flow rate, u is the velocity in the x direction, p is the pressure, T is the temperature and ρ and C_p are the density and the specific heat capacity of the material, respectively. The shear stress τ_{xy} for thermoplastic materials can be reasonably well represented by the power-law model

$$\tau_{xy} = m\left(\frac{\partial u}{\partial y}\right)^n \tag{4}$$

The boundary conditions are

(a) at the entry to the gap, $p=0$ and $T=T_0$;

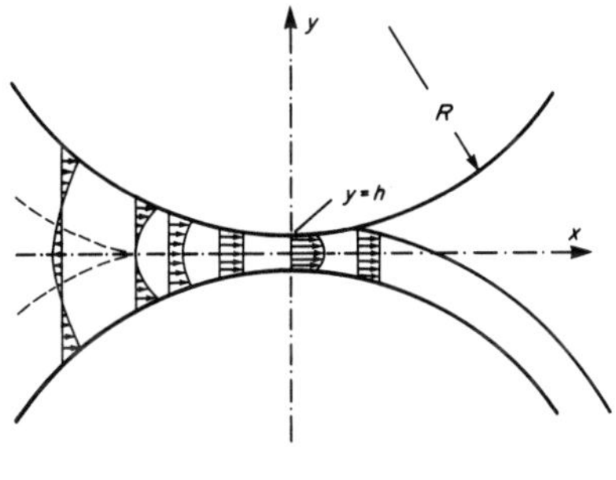

Figure 2
Flow configuration and velocity profiles in the gap between two equal-diameter rolls

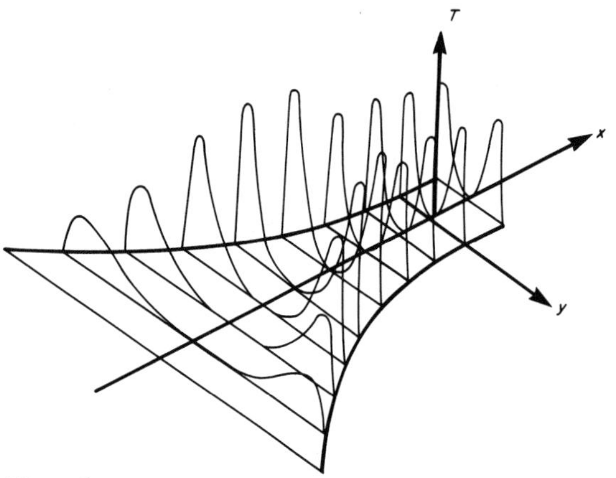

Figure 3
Schematic representation of temperature distribution due to viscous dissipation

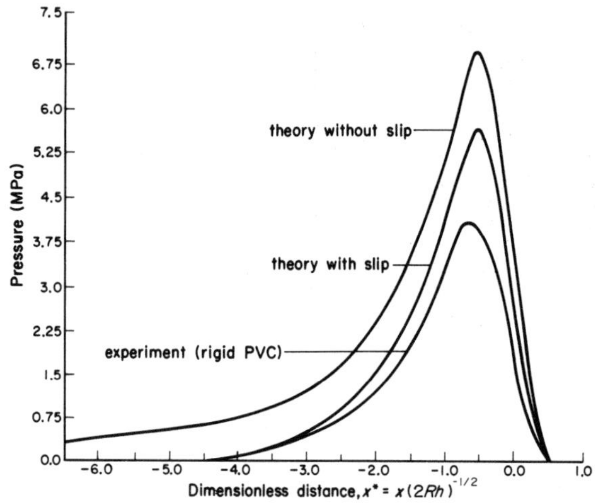

Figure 4
Pressure distribution

(b) at the leave-off point, $p=0$ and $dp/dx=0$;

(c) at the upper cylinder surface, $u=U_1$ and $T=T_{w1}$; and

(d) at the lower cylinder surface, $u=U_2$ and $T=T_{w2}$.

These equations have been solved numerically by a number of investigators for both isothermal and nonisothermal conditions. The velocity and temperature profiles are shown schematically in Figs. 2 and 3. Experimental data on calendering are scant and the only data that have been obtained on a rheologically characterized PVC sample are those of Vlachopoulos and Hyrmak (1980). It should be noted, however, that the possibility of slip exists, especially with PVC, and, consequently, the velocities at the cylinder surfaces should be lowered by the amount of slip velocity. Figure 4 shows that the introduction of slip improves the agreement between theory and experiment.

Calculations of sheet thickness and roll torques are in relatively good agreement with experimental values. Elasticity effects have not been included in any of the analyses presented thus far. These effects are expected to be of importance in the entry region where flow recirculation occurs. It is possible that certain surface defects on the resultant sheet originate in this region of rapidly accelerating flow. Other defects, such as the appearance of small bubbles, may be due to the peculiar temperature distribution (Fig. 3) arising from frictional heating (viscous dissipation).

The complete two-dimensional flow equations without the lubrication approximation have been solved by the finite-element method (Mitsoulis et al. 1985). Figure 5 shows the streamlines in the melt bank for a rigid PVC formulation under nonisothermal conditions with slip at the wall.

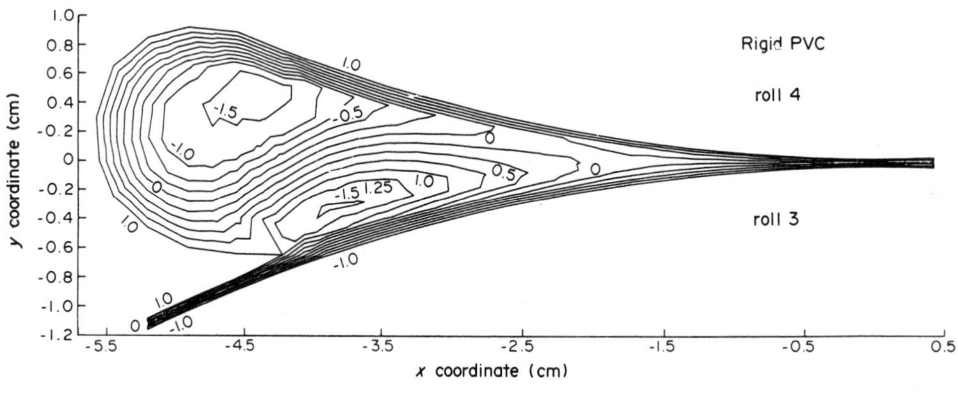

Figure 5
Streamline pattern in the melt bank formed between rolls

3. Machinery Considerations

Calenders are made in a wide variety of designs depending on the materials and the products. For floor coverings, two-roll calenders are generally used. For rigid PVC processing, the four-roll L calender is common. Other designs are also available (Perlberg and Burnett 1977). In modern calender installations, multimotor drives are used which provide the capability of fine adjustments.

The rolls are heated by means of circulating fluids (superheated steam, pressurized water or oil). At low speeds, heat is transferred from the rolls to the calendered material, whereas at high speeds, the reverse is true because of frictional heating.

Owing to the large separating forces developed in the calender gap, the rolls tend to bend. This may result in undesirable thickness variations in the finished product. Compensations for roll deflections are provided by using crowned rolls having a larger diameter in the middle than at the ends or by roll bending or roll skewing (Meinecke 1965). Bearings of special design are used to keep the roll separation constant.

Calender installations require large initial capital investment. Film and sheet extrusion are competitive processes because the capital investment for an extruder is only a fraction of the cost of a calender. However, the high quality and volume capabilities of calendering lines make them far superior for many products.

See also: Polymer Sheet: Manufacture and Applications; Processing: Classification

Bibliography

Elden R A, Sean A D 1971 *Calendering of Plastics.* Elsevier, New York

Freakley P K 1985 Calendering and milling. *Rubber Processing and Production Organization.* Plenum, London

Meinecke E 1965 Calendering. *Plastic, Resins, Rubbers, Fibers,* Encyclopedia of Polymer Science and Technology, Vol. 2. Wiley, New York, pp. 802–19

Middleman S 1977 *Fundamentals of Polymer Processing.* McGraw-Hill, New York

Mitsoulis E, Vlachopoulos J, Mirza F A 1985 Calendering analysis without the lubrication approximation. *Polym. Eng. Sci.* 25: 1–13

Morton-Jones D H 1989 Polymers in the rubbery state. *Polymer Processing.* Chapman and Hall, London, Chap. 10

Perlberg S E, Burnett P P A 1977 Calendering and calender laminating. In: Nass L I (ed.) 1977 *Encyclopedia of PVC,* Vol. 3. Dekker, New York, pp. 1361–414

Tadmor Z, Gogos C G 1979 *Principles of Polymer Processing.* Wiley, New York

Vlachopoulos J, Hrymak A N 1980 Calendering poly(vinyl chloride): Theory and experiments. *Polym. Eng. Sci.* 20: 725–31

J. Vlachopoulos
[McMaster University, Hamilton, Canada]

Capacitor Materials: Solid Dielectrics

Although it is true that, since the energy density of a capacitor is proportional to $\varepsilon_r E^2$, the primary technical qualities sought for in a capacitor dielectric are high dielectric constant ε_r and high electric field strength E, the diversity of modern applications has placed many other demands on capacitor materials. The adoption of a new generation of impregnant fluids introduced to replace the outlawed polychlorobiphenyls imposes compatibility constraints on the solid dielectric. Furthermore, the solid is often required to operate at elevated temperatures, thus limiting the use of many traditional materials. In addition, the dielectric loss of the material has become an overriding factor for some applications involving elevated-frequency, elevated-temperature or low-loss situations. Other requirements such as processibility and assured material quality and stability have also placed important demands on dielectrics, which may have to withstand electrical stresses as great as 80 V μm^{-1} over a 40 year life span. These exacting requirements have led to changes in the use of the traditional materials. Synthetic polymers and ceramic formulations have emerged as preferred dielectrics for most nonelectrolytic capacitors.

1. The Changing Material Emphasis

Impregnated paper has been widely used but can be very variable in thickness and its working electric stress (electric field strength) is limited to about 20 V μm^{-1} because of the inherent limitations of the paper and also the migration of ions in the impregnant, enhancing the stress at the paper surface. Impregnated with mineral oil, kraft paper exhibits a loss tangent of about 0.003 but, since many polymer films yield values which are an order of magnitude better, there has been a gradual switch to polymeric materials for many applications, such as power capacitors, previously served by paper. The temperature limit for paper capacitors is dependent on the working electric stress but is generally about 100 °C. High-temperature operation results in a sharp reduction in insulation resistance and in lifetime (Mullin 1978). Generally, paper capacitors exhibit a positive temperature coefficient of capacitance of typically 150 ppm °C^{-1}.

Naturally occurring muscovite mica is nonporous, relatively nonhygroscopic and substantially free of electrolytic impurities. With a dielectric constant between 6 and 7, loss tangent in the range 10^{-3}–10^{-4} and an electric field strength in excess of $100\ \mathrm{V\,\mu m^{-1}}$, mica exhibits an attractive range of properties for use in low-loss, high-frequency capacitors. It is also stable up to about 550 °C, where the hydroxyl groups decompose. Precision mica capacitors with silver electrodes can also yield an excellent stability (1%) over long periods of time, but the variation in properties of the natural dielectric from one batch to the next and the nature of the material has made it difficult to manufacture competitively to close tolerances. Furthermore, the trend to lower voltages with the introduction of semiconductors has left the high dielectric strength of mica underutilized. For these reasons, the silvered-mica unit is being replaced by low-loss polymer films (see Sect. 2) and high-permittivity ceramics (see Sect. 3), except for specialized applications such as radio transmitting equipment (Reed 1981, Components Technology Institute 1982).

Although high-voltage impregnated units are invariably hermetically sealed in aluminum cans, plastic encapsulation is attractive for automated production of low-voltage capacitors. This may be achieved by thixotropic-resin dipping or by suitable molding of polypropylene or alkyd materials (Whitby et al. 1968).

2. *Synthnetic Polymer-Film Dielectrics*

Before any film can be considered as a commercial dielectric it must be developed as a pinhole-free electrical-grade film in the thickness range 1–10 μm. Table 1 indicates a range of candidate polymeric materials. The parameters are dependent on the grade and processing (e.g., biaxial orientation) and thus are only approximate. In particular, the loss tangent often varies considerably with both temperature and frequency. It is clear from Table 1 that polypropylene provides a low-cost, low-loss material with a service temperature which is acceptable for most general applications. Its low dielectric constant is more than offset by high inherent dielectric strength and, in combination with a synthetic impregnant, it is very widely used. Polystyrene, although having a somewhat higher dielectric constant, has a severe temperature limitation and a tensile strength of 55 MPa which is only a third that of polypropylene. For applications requiring high-temperature operation, polyimide or polysulfone films are often employed, but the capability for higher operating temperature carries a large cost penalty. Selection of materials must also take account of the operating duty. For example, since the dissipation factor of poly(vinylidene fluoride) is 0.1% at 60 Hz at a temperature of 120 °C, it is unacceptable as a capacitor material for sustained high-temperature ac operation. However, for high-energy-density capacitors required for intermittent discharge duty, it is an attractive material on account of the high relative permittivity resulting from its polar molecular structure.

For applications requiring a high current-carrying capacity (e.g., power capacitors), the film is wound onto a roll sandwiched between aluminum foil typically 5–10 μm in thickness. The foil edges represent an area of electric-stress concentration which limits the voltage if the unit is to be operated without large partial-discharge magnitudes. Some stress relief may be obtained by folding the foil edges or by coating or discharge treating. Although the foil edges are easily impregnated due to their position on the outside of the roll, interior regions of a tightly wound capacitor are difficult to impregnate. For film/foil capacitors, this has been overcome by the use of paper in combination with polypropylene film and, more recently, by the use of textured film and embossed foil.

For lower-voltage applications, volume efficiencies may be raised from about $0.2\ \mathrm{\mu F\,cm^{-3}}$ for film/foil units to $1.5\ \mathrm{\mu F\,cm^{-3}}$ by the use of metallized polymer film introduced in the 1960s (Apps 1970). The electrodes are composed of 30–50 nm of vapor-

Table 1
Representative properties of polymeric materials for capacitor dielectrics

Material	Dielectric constant	Dissipation factor (%)	Service temperature (°C)	Cost[a]
Polycarbonate	3.0	0.1–1.0	110	7
Poly(ethylene terephthalate)	3.2	0.2–2.0	85	3.6
Polyimide	3.5	0.2	240	68
Polypropylene	2.2	0.01–0.03	90	1.8
Polystyrene	2.5	0.01–0.1	75	1.3
Polytetrafluoroethylene	2.0	0.003–0.025	250	21
Polysulfone	3.1	0.1–0.4	150	23
Poly(vinylidene fluoride)[b]	11	0.01–0.12	160	

a Relative to low-density polyethylene (=1) b Not presently available commercially as a capacitor-grade film deposited aluminum and the metallized film is tightly wound onto a cylindrical core. Metallized units also have the desirable quality of being self-healing: any flaws in the dielectric film result in a breakdown, but the resulting current feeding the fault rapidly causes vaporization of the local metallization. This relieves the stress in the vicinity of the defect within a few microseconds, thus permitting recovery. The success of this clearing process is sensitive to the resistance of the metal layer and resistances of 1–5 Ω (per square) are found to be optimum. Failure of a device to clear is often associated with a deposition of carbon from the structure of the film or impregnant to form resistive faults. For example, the aromatic structure of polystyrene creates greater problems in obtaining clearing than is evident with semicrystalline polypropylene (Sillars 1973), unless it is used in combination with other material such as cellulose acetate lacquer. Although metallized technology permits high reliability, it does suffer from capacitance losses due to removal of metal and electrode corrosion.

Another metallized construction involves the sandwiching of a polypropylene film between layers of doubly metallized paper (Preissinger 1981). Although the metallization of both sides of the paper means that it is unstressed and thus not contributing to the capacitance directly, the construction does allow the benefits of complete impregnation and clearing to be combined, thus permitting working electric stresses in the film to approach 80 V μm^{-1}.

In an effort to maximize the volume efficiency, technology has also been developed to fabricate stacked-film capacitors in large quantities and at low cost. In this process, metallized film is wound on a large wheel and then subsequently slit into long strips which can be cut to the requisite lengths. In this way, a rectangular stack of many layers of metallized film connected in parallel is formed.

3. *Ceramic Technology*

A mixture of inorganic nonmetallic materials, shaped in the powder state or slip cast, forms an integral polycrystalline structure when sintered at temperatures of the order of 1200–1500 °C. Such ceramics can be realized in two different classes of dielectric:

(a) formulations which offer low losses and high stability and which can provide a predetermined temperature coefficient; and

(b) high-permittivity ferroelectric ceramics which, although less stable and often lossy, can exhibit relative permittivities up to 15 000.

In the first category, low-loss ceramic materials have been traditionally fabricated from synthetic magnesium silicate ($MgSiO_3$), which is not susceptible to humidity in the same way as naturally occurring porous steatite. Small amounts of barium and calcium carbonates are sometimes added to provide a bonding agent for the $MgSiO_3$. Such a low-loss material exhibits a dielectric constant of about 7, a loss tangent of 0.001 and a positive temperature coefficient of about 100 ppm °C^{-1}, making it suitable for capacitors in resonant-circuit applications. Medium-permittivity dielectrics ($\varepsilon_r = 15$–800) based on TiO_2 or its derivatives, the alkaline-earth titanates, are also used for this application since the formulation may be adjusted to give temperature coefficients ranging from about 100 ppm °C^{-1} to −1500 ppm °C^{-1} to compensate for a particular inductance characteristic. Barium, calcium and magnesium oxides are often used for admixing.

The major constituent of many high-permittivity formulations is $BaTiO_3$. This material has a relative permittivity of about 1500 at 20 °C but exhibits ferroelectric hysteresis and is thus characterized by a high dissipation factor (2×10^{-2}) and a voltage-dependent dielectric constant. It is also very sensitive to temperature in the range of the Curie point at 120 °C. Substitutions of lead, strontium, calcium and cadmium for the barium, and similar substitutions of tin, zinc and cerium for the titanium, may be used to provide more stable formulations (Nicker 1974). Considerable success has also been achieved with multiphase ceramics such as lanthanum-modified $PbZrO_3$–$PbTiO_3$ (PLZT), which has a reduced dielectric-constant dependence on electric field strength, and with glass ceramics based on the niobates of barium, strontium or lead, which can provide 200 °C operation.

Ceramic capacitors are often fabricated as multilayer devices. In one process, the ceramic slip is drawn out onto a belt and dried prior to printing of the electrodes, which consist of precious metal ink or paste such as silver–palladium. After stacking, units are fired to form a monolithic structure with integral electrodes. Although the electric strength of these dielectrics is ultimately limited by their porosity, failure most usually occurs because of cracks, voids and electrode delamination induced during the manufacturing process. The tendency for crack formation makes them vulnerable in pulse-discharge applications. An important failure mode which results in a drastic reduction in insulation resistance is thought to result from ionic migration in the dielectric layers, compounded by the ingress of moisture into the defects.

4. *Emerging Trends*

New polymers, such as a recently announced polyetherimide, are being evaluated for capacitor duty and specialist applications are attracting novel approaches. The possibilities of synthesizing thin-film polymeric materials using plasma and radiation sources could radically change traditional capacitor manufacture if such processes could be made commercially viable for volume production.

Since the working stress is usually limited by dielectric imperfections rather than by the intrinsic strength of the material, emphasis is also placed on improved fabrication technology and quality control. The use of postfiring electrode insertion in multilayer ceramics to avoid delamination problems and laser-scanning inspection systems are examples of these initiatives.

The use of the Helmholtz double layer as a capacitor is also a concept which is now being commercialized (Sanada and Hosokawa 1979). The principle utilizes the extremely small dimensions (measured in angstroms) of the electric double layer separating charges at the interface of a polarizable electrode and an electrolyte to generate very large capacitances. In a practical embodiment of this principle, activated carbon particles are used as electrodes separated by suitably porous material through which a sulfuric acid electrolyte can migrate. The electrolyte decomposition voltage limits the working potential to about 1 V, necessitating series connection of several cells. However, capacitance values can be made very high indeed, providing considerable space savings over standard electrolytic devices. The major drawback at present is the very high loss; a typical 5 V, 1 F unit may have an equivalent series resistance of about 5 Ω. The major application is for battery substitutes, especially for standby power for volatile semiconductor memories.

See also: Electrical Properties of Polymers

Bibliography

Apps L T 1970 Metallised plastics capacitors for electronics. *Electron. Power* 16: 369–72

Components Technology Institute 1982 *Proc. 2nd Capacitor and Resistor Technology Symp.* CTI, Huntsville, AL

Mullin W F 1978 *ABC's of Capacitors.* Howard Sams, Indianapolis, IN

Nicker D A 1974 High voltage ceramic capacitors. *Electrocomponent Sci. Technol.* 1: 113–20

Preissinger H 1981 Breakdown field strength of thin insulating oil films in MKV capacitors. *Siemens Rev.* 38: 235–8

Reed C W (ed.) 1981 *Proc. Symp. High-Energy-Density Capacitors and Dielectric Materials.* National Academy of Science, Washington, DC

Sanada K, Hosokawa M 1979 Electric double layer capacitor "Super Capacitor." *NEC Res. Dev.* 55: 21–8

Sillars R W 1973 *Electrical Insulating Materials and Their Applications.* Peregrinus, Stevenage, UK

Whitby F J, Simpson W E, Pettican K A 1968 Plastics encapsulation. *Electron. Commun.* 43: 220–7

J. K. Nelson
[Rensselaer Polytechnic Institute, Troy, New York, USA]

Carbon-Fiber-Reinforced Plastics

To utilize the high stiffness and strength that carbon fibers possess it is necessary to combine them with a matrix material that bonds well to the fiber surface and which transfers stress efficiently between the fibers. Fiber alignment, fiber content and the strength of the fiber–matrix interface all influence the performance of the composite. Furthermore, highly specialized processing techniques are necessary that take account of the handling characteristics of carbon fibers, which are different from those of other reinforcements. It is the processibility of polymers that makes them particularly attractive for use in a whole range of composite materials. For example, one form of carbon-fiber-reinforced plastic (CFRP) contains continuous carbon fibers, possibly in the form of a unidirectional or woven mat, which are impregnated by a thermosetting resin such as an epoxy. Preimpregnated (prepreg) material of this type can then be layed up to ensure the appropriate orientation of the fiber before curing of the resin is undertaken. Alternatively, if fiber orientation does not have to be so rigorously defined, then the processibility of thermoplastic polymers means that, with the reinforcement in the form of short fiber (typically 0.2–10 μm), a molding compound of fiber and thermoplastic can be injection molded to produce directly finished components.

From the perspective of materials performance, the potential of CFRP is enormous: a civil aircraft constructed, where possible, entirely of CFRP instead of aluminum alloy would have a total weight reduction of 40%; a communications satellite built from CFRP can carry more telecommunications channels than permitted by previously specified materials; and a new generation of tactical aircraft can be planned, which will have a versatility regarded as impossible prior to the advent of CFRP (e.g., forward-swept wings for low stalling speeds, and very small turning-circle radii). However, the high cost of the fiber and the lack of cost-effective manufacturing processes are still inhibiting the more general use of CFRP, as is the newness of the art of composites design.

1. Matrix Materials

1.1 Thermosets

Examples of thermoset resins commonly used for carbon-fiber composites are listed in Table 1, together with their advantages and disadvantages. Long shelf life is an important advantage for prepregs, while low viscosity is vital to processing techniques such as pultrusion, filament winding and rapid resin injection molding (RRIM). Cure of the resin should not take long times or require extreme temperatures, while a high value of the glass transition temperature T_g (the temperature of the glassy–

rubbery transition) is necessary if the composite is to be used at elevated temperatures. Carbon-fiber–epoxy composites represent about 90% of CFRP production. As can be seen in Table 1 the properties of an epoxy are dependent on the curing agent used and the functionality of the epoxy. The attractions of epoxy resins are that

(a) they polymerize without the generation of condensation products which can cause porosity;

(b) they exhibit little volumetric shrinkage during cure, which reduces internal stresses; and

(c) they are resistant to most chemical environments.

The major disadvantages of epoxy resins are their flammability and their vulnerability to degradation in agressive media. Phenolic resins are, however, quite stable in both respects, although as structural composite materials, phenolic CFRPs are quite uninteresting as their fatigue and impact properties are very poor. As indicated earlier, the prospect of using CFRPs as a substitute for light metal alloys is particularly attractive but there are problems to be overcome at high temperatures. Using polyimides, peak service temperatures of 400 °C can be reached. However, conventional polyimides present significant processing problems. Toxic solvents have to be evaporated and, during curing, condensation products have to be vented from the tooling. Cure cycles are also prohibitively long. PMR 15 is a polyimide developed to surmount the above problems. Three monomer reactants are mixed in an alcohol, the solution being used to impregnate the fiber. The alcohol is then evaporated, during which time a thermoplastic prepolymer is formed. This prepreg can then be cured using tooling identical to that used with epoxies, although much higher temperatures (~380 °C) are required.

Another compromise using polyimide chemistry to enhance CFRP performance is to form copolymers with epoxy resins. In this way, flame retardancy and reduced moisture absorption are realized, but there is no substantial increase in high-temperature performance. Polybismaleimides represent a very good compromise in many respects and should command the applications arena between the epoxies and the polyimides.

1.2 Thermoplastics

A considerable number of thermoplastics are currently being used in CFRP, they may be either semicrystalline or amorphous. Semicrystalline thermoplastics offer greater solvent resistance, but are generally more difficult to process.

In load-bearing applications, carbon fibers are most widely used in combination with Nylon 66 which gives the highest strength and modulus over a fairly wide temperature range. A thermoplastic that is currently of considerable interest to fabricators looking to mass-produce articles is poly-1,4 butanediol terephthalate (PBT). It has a melting point of 224 °C and, if a low-molecular-weight grade is used, it can be injection molded at 250 °C. It crystallizes very rapidly, so very short cycle times can be employed. CFRP based on PBT exhibits a heat distortion temperature of 200 °C with adequate resistance to creep up to 120 °C.

The prospect of using high-temperature engineering thermoplastics, examples of which are shown in

Table 1
Examples of thermosetting resins used for CFRP

Resin	Advantages	Disadvantages
Epoxides		
DGEBA[a] (aliphatic amine curing agent)	very low viscosity; room-temperature cure	very low T_g; short shelf life; flammability
DGEBA[a] (aromatic diamine curing agent)	medium viscosity; long shelf life	medium T_g; difficult cure
TGMDA[b] (aromatic diamine curing agent)	high T_g; long shelf life; toughness	high viscosity; difficult cure; moisture uptake; flammability
Novalac (phenolic)	nonflammability; high T_g	difficult cure; shrinkage; brittleness
PMR 15 (polyimide)	very high T_g; low moisture uptake; high strength	difficult cure; thermal fatigue

a Diglycidylether of bisphenol A b Tetraglycidylmethylenedianiline

Table 2
High-temperature engineering thermoplastics

Generic name (trade name; manufacturer)	Heat distortion temperature at 1.82 MPa (°C)
Polycarbonate (Lexan, Merlon; GE, Mobay)	133
Polyphenylene sulfide (Ryton; Phillips)	137
Polyether ether ketone (Victrex PEEK; ICI)	148
Polysulfone (Udel; Union Carbide)	174
Polyetherimide (Ultem; GE)	204
Polyethersulfone (Victrex PES; ICI)	204

Table 2, with continuous-fiber reinforcements instead of thermosetting resins is particularly exciting. The advantages of thermoplastics are lower production costs, improved toughness and impact resistance and lower susceptibility to moisture uptake, which is found to reduce the value of T_g for epoxy matrices. Additional benefits are the ease of repair of the composites and the recovery of waste water or used material.

In Table 3 the properties of injection-molded Nylon 66, polyethersulfone (PES) and polyether ether ketone (PEEK), each with 40 wt% carbon fiber, are compared. The Nylon 66 compound exhibits superior properties at normal ambient temperature. It is with properties such as creep modulus (stress to induce 0.1% strain within 100 s) and tensile strength at elevated temperatures that the high-temperature thermoplastics reveal their merit.

2. Fiber–Matrix Interphase

The properties of the material between the fiber reinforcement and the plastic matrix have a strong influence on the mechanical properties of the CFRP. To produce a high-modulus composite requires maximizing adhesion between the fiber and matrix. However, if the polymer matrix is brittle (e.g., epoxies) there is a corresponding reduction in impact strength as mechanisms of energy dissipation such as debonding and fiber pullout become suppressed. If a tough polymer matrix is to be used (e.g., PEEK), where under impact the principal mechanism of energy dissipation is yielding within the polymer, good adhesion between the fiber and matrix is required.

Unfortunately, this region of the composite, sometimes known as the interphase, is extremely complex. Constituents of the interphase might include a modified fiber surface, the treatment being undertaken to promote better adhesion, or a coating of polymer present to ensure compatibility with the matrix polymer. In addition, the presence of the fiber surface may affect the cure of thermosets, resulting in chemically different polymer, or may affect the morphology of semicrystalline thermoplastics.

2.1 Surface Treatment

Surface-treatment processes are now established as an integral part of fiber manufacture. Most are either wet oxidative treatments using solutions of, for example, sodium hypochlorite, sodium bicarbon-

Table 3
Properties of three thermoplastics reinforced with carbon fiber ("Grafil" product, reference RG40)

Property	Material		
	40% CF–Nylon 66	40% CF–PES	40% CF–PEEK
Density (kg m^{-3})	1340	1520	1450
Tensile strength (MPa)			
at −55 °C			275
24 °C	246	176	227
50 °C	168	168	208
100 °C	108	140	165
140 °C	75	112	129
180 °C			72
Tensile elongation (%)	1.65	1.1	1.35
Tensile creep modulus (GPa)			
at 24 °C	26.2	22.1	26.5
120 °C	11.5	21.4	23.9
160 °C	9.0	21.1	10.5
200 °C		17.5	6.6
Flexural strength (MPa)	413	244	338
Flexural modulus (GPa)	23.4	16.8	20.8
Compressive strength (MPa)	240	216	
Compressive modulus (GPa)	20.2	17.2	

ate or chromic acid, or dry treatments in which the oxidizing agent is ozone. The precise action of the treatment is undoubtedly complex, but is currently thought to include the introduction of chemically active sites onto which the polymer can graft, as well as chemical cleaning, an increase in rugosity of the fiber surface and the removal of any weak surface layer on the fiber.

A plethora of techniques has been proposed to measure fiber–polymer adhesion. Such techniques are experimentally tedious but, more significantly, the assumptions necessary to deduce adhesive strengths from the various types of measurements make them only comparative. The techniques include measurement of the composite short-beam interlaminar shear strength, the distribution of fiber lengths in a fully fragmented single fiber, the stress required to pull out a fiber embedded in the polymer and the stress required for a bead of polymer to become detached from a single fiber.

2.2 Surface Coating and Sizing

Polymer coatings and sizes are frequently applied to fibers immediately after manufacture. The light application of resin binds individual filaments together and maintains a compact bundle for processing into an intermediate product or final composite form. Size compositions applied to the surface can afford protection against filament damage and provide lubrication for a smooth delivery of fibers in continuous processing. The sizing or coating may be bulk polymer or resin specially functionalized to improve fiber wetting and adhesion while still retaining compatibility with the bulk polymer.

3. Properties of CFRP

The main points to appreciate when considering a carbon-fiber composite as compared with a conventional material such as a metal alloy are that the composite is heterogeneous and anisotropic. By heterogeneous it is meant that the properties vary from point to point within the material and by anisotropic it is meant that the properties depend on the direction in which they are measured. These two factors are important when considering not only mechanical properties (e.g., stiffness and strength) but also the physical, electrical and thermal characteristics of the composite.

The physical and mechanical properties of a CFRP are strongly dependent on the properties of the particular fiber used. There are a variety of fiber types available and the properties of fibers are being advanced continually by the development of new grades (see Table 4).

The section below outlines some of the main features of the mechanical, thermal and electrical properties of CFRP.

Table 4
Types of commercially available carbon fibers

Carbon fiber type	Tensile strength (GPa)	Elastic modulus (GPa)	Breaking strain (%)
High strength (PAN)	3.5	240	1.5
High strain (PAN)	5.5	240	2.2
Intermediate modulus (PAN)	4.5	285	1.6
High modulus (PAN)	2.5	350	0.7
Ultrahigh modulus (PAN)	2.0	450	0.5
Isotropic (pitch)	1.0	35	2.5
Ultrahigh modulus (pitch)	2.0	550	0.4

3.1 Mechanical Properties

CFRP containing continuous aligned fibers (which would typically have a volume fraction of 60%) are essentially elastic to failure (when loaded in tension parallel to the fibers) and exhibit no yield or plasticity region. There is, however, a slight nonlinearity in the stress–strain curve, with the stiffness increasing slightly with applied strain, as a result of improved orientation of the graphitic planes within the fibers. Although the failure strains in tension for CFRP may seem small when compared with those of other structural materials, the predictably elastic behavior under load allows a high proportion of the ultimate strength to be utilized in practice. Consequently, strain levels at useful working stresses are comparable with those of metals, alloys and glass-fiber composites. The properties of unidirectional CFRP are compared with similar performance characteristics of other commonly used structural materials in Table 5. A range of material properties typical of those required for design calculations is given in Table 6. These data confirm the anisotropy of the unidirectional material (which is much more marked in CFRP than in glass or aramid composite); the stiffness and strength of the materials are very much lower when they are loaded perpendicular to the fibers than when they are loaded parallel to the fibers.

It should also be noted that the unidirectional compression strength of the lamina is only about two-thirds of the tensile strength. In early carbon-fiber composite materials (from the early 1970s) the tensile strength and compression strength were approximately equal, with the failure mode in compression being shear failure. Improved fiber tensile properties have been achieved with a reduction in fiber diameter and, during this time, more ductile resins have come into favor. These factors have tended to promote fiber instability or microbuckling failure when unidirectional materials are loaded in compression. Hence the compression strength of the material is not a true measure of the fiber strength,

Table 5
Some properties of fiber-reinforced composites (55–70 vol.% fiber) and metals

Material	Density (kg m^{-3})	Ultimate tensile strength (GPa)	Young's modulus (GPa)	Specific tensile strength (GPa)	Specific Young's modulus (GPa)
CF–Epoxy					
high strength	1500	1.9	130	1.27	87
high modulus	1600	1.2	210	0.94	119
E-glass–epoxy	2000	1.0	42	0.50	21
Aramid–epoxy	1400	1.8	77	1.30	56
Steel	7800	1.0	210	0.13	27
Titanium	4500	1.0	110	0.21	25
Aluminum L65	2800	0.5	75	0.17	26

but more of the ability of the matrix to support the fiber against buckling and the integrity of the fiber–matrix interface. As a result of these developments, compressive performance of CFRP can be a limiting design condition.

There are very few practical applications for CFRP which require only unidirectional reinforcement. Laminates are assembled from unidirectional layers stacked at various orientations to one another. Layers are included at 90° to give adequate transverse properties and at 45° to give good shear properties. A laminate construction of 0°/90°/±45° plies has negligible anisotropy when loaded in-plane and is known as quasiisotropic. Typical data for such a laminate are shown in Table 7. The mechanics of laminated composites are quite complicated and the interested reader is referred to Chou (1989).

As a result of the marked anisotropy, CFRP can experience large stress concentration factors at cutouts or mechanically fastened joints. Given that the material is predominantly elastic to failure this might be expected to present serious practical problems. Fortunately, certain damage mechanisms (e.g., matrix cracking parallel to the fibers and delamination cracking between layers) can occur in laminates so that the stress concentrations do not have the deleterious effects that might be expected. Even so, the designer must take care when dealing with cutouts or joints. Bonded joints must also be designed carefully to avoid failure of the laminate by interlaminar (through-thickness) failure.

The range of basic mechanical properties of CFRP is enhanced by excellent resistance to creep rupture and fatigue. Creep rupture times at equivalent stress levels are several orders of magnitude greater for CFRPs compared with those for glass-fiber-reinforced composites (GFRP) and also significantly better than for aluminum. CFRP is also capable of operating well under conditions of alternating stress. For high strength fibers, S–N fatigue curves are much flatter for CFRP compared with those for GFRP or aluminum, and a higher ratio of working stress to ultimate stress can be utilized. However, the newer fiber-matrix systems are not yet characterized fully and there are signs that, with these materials, fatigue may be more of a consideration than in the past. In passing it should be noted that fatigue failure in composite materials is not caused by the initiation and growth of a dominant crack as it is in metallic materials; instead there is an accumulation of damage (matrix cracks, delamination, fiber breaks) which may result in a lowering of the stiffness and the residual strength. Much more work is needed to develop damage-tolerant design for CFRP composites.

Table 6
Typical performance data for unidirectional laminates

Property	High strength	High modulus
0° Tensile strength (GPa)	1.90	1.15
0° Tensile modulus (GPa)	137	200
Elongation (%)	1.4	0.6
Interlaminar shear strength (MPa)	90	70
0° Flexural strength (GPa)	1.80	1.10
0° Flexural modulus (GPa)	130	190
0° Compression strength (GPa)	1.25	0.70
0° Compression modulus (GPa)	140	210
Density (kg m^{-3})	1550	1630
90° Tensile strength (MPa)	50	35
90° Tensile modulus (GPa)	10	7

Table 7
Performance data for multidirectional laminates with 60 vol.% fiber and 0°/90°/±45° laminate construction

Laminate	Tensile strength (GPa)	Tensile modulus (GPa)
High strength	0.5	50
High modulus	0.35	75

Table 8
Typical thermal expansion data for CFRP with 60 vol.% fiber

CFRP	Coefficient of thermal expansion[a] (10^{-6} K^{-1}) Longitudinal	Transverse
High modulus–epoxy	−0.25 to −0.60	20–65
High strength–epoxy	0.30 to −0.30	20–65

a Over the temperature range 100–400 K

The impact resistance of CFRP is generally considered low and, in fact, the post-impact compression strength is considered to be a design-limiting factor. There is, however, scope for the modification of impact performance and fracture toughness by fiber hybridization with glass or aramid reinforcements leading to structures with much greater damage tolerance. Moreover, impact performance will improve as strain-to-failure of the fiber continues to increase.

3.2 Thermal Properties

CFRP is thermally conductive along the direction of the fibers. Conductivity perpendicular to the fibers is much less as it is dominated by the polymer matrix. The ability to dissipate heat is thought to contribute to the very good fatigue properties of CFRP. The conductivity of fibers increases with graphite content and highly graphitized fibers, such as high-modulus fibers, have thermal conductivity values of 700 W m^{-1} K^{-1} in the longitudinal direction which surpass the value for steel (50 W m^{-1} K^{-1}).

Coefficients of thermal expansion for carbon-fiber composites are dependent on the fiber type and on temperature. The fibers themselves have small negative coefficients of thermal expansion and this leads to the possibility of designing composites with almost zero overall thermal expansion. This dimensional stability, which can be available over extremes of temperature, is very attractive in applications such as antennas for aerospace vehicles. As with the thermal conductivity, the coefficient of thermal expansion is anisotropic—the value measured perpendicular to the fibers is dominated by the polymer matrix and is usually much higher (see Table 8) than the value measured parallel to the fibers. This can have important practical consequences in that residual stresses are left trapped in a laminate after it has cooled from the processing temperature. These stresses can be large enough to promote matrix cracking, especially in systems such as PMR-15, which, as described previously, is processed at high temperatures. For such systems to achieve wider use, this problem has to be overcome by either toughening the resin further or further development of a damage-tolerant approach to design.

3.3 Electrical Properties

CFRP is electrically conductive along the direction of the fiber. As with thermal conductivity, electrical conductivity increases with the level of graphitization of the fiber.

There are practical applications which make use of the electrical properties of carbon fiber, notably to provide resistance heating effects (e.g., in nonmetallic tooling) or static charge dissipation in composite structures.

The electrochemical behavior of graphite can lead to problems when carbon-fiber composites are in contact with metal alloys, as they may be in an aircraft structure, for example. Graphite is cathodic to most metal alloys and this can lead to appreciable corrosion (especially in mechanically fastened assemblies) unless adequate precautions are taken.

4. Forms of Carbon Fibers

Continuous-filament fibers represent the primary production route for manufacture and other products are derived from this form by secondary processes. The conversion of fibers into some conveniently handleable intermediate product plays an important part in the effective utilization of the fibers as a reinforcement. Specific products are needed to meet the individual processing requirements of different fabrication techniques.

4.1 Continuous-Filament Tow

Continuous fibers are grouped together in tow bundles, each containing a specified number of individual parallel filaments. The tows are produced in different bundle sizes, ranging in discrete steps from 400 to 320 000 filaments per tow. The main sizes used are in the 3000–12 000 range; finer tows are only employed for specialized applications. Heavier tows offer some advantages for rapid volume conversion to composite and can lead to economies of sale for both fiber production and composite manufacture by certain processes.

Twisted tows, containing typically 15 turns per meter, arise from processes employing twisted precursor fibers. The presence of twist reduces the efficiency of the fiber reinforcement but can enhance the handleability of the tow, aiding conversion to secondary products such as woven fabrics and braids.

4.2 Woven Fabrics

Carbon fibers can be woven into a variety of unidirectional and bidirectional fabrics capable of being processed efficiently into composite materials. The weave construction (e.g., plain, satin, twill) determines the handle of the fabric and controls its ability to conform to contoured shapes.

4.3 Preimpregnated Products

Forms of preimpregnated materials (prepregs) include continuous tows, unidirectional tapes and sheets, and woven fabrics. A resin is applied to the fibers using a solvent, melt or film impregnation technique. The resins used are specially formulated with a latent cure, so that at room temperature the cure reaction proceeds only very slowly and the material can be handled and stored at ambient temperature for a reasonable length of time. Curing temperatures are typically between 100 °C and 180 °C. A choice can be made from the wide range of commercial prepregs to suit the end use and fabrication technique employed.

4.4 Nonwoven Fabrics

Carbon-fiber mats, felts and papers manufactured by conventional nonwoven processes from randomly distributed short fibers (about 6 mm in length) provide multidirectional reinforcement but convert inefficiently into composite form. New products containing longer fibers (about 25 mm) are now being developed, exhibiting good handling characteristics and generating improved composite performance.

4.5 Molding Materials

Short-length fibers chopped from continuous tows are widely used for the manufacture of molding compounds in both thermoplastic and thermosetting matrices. The application of special resin sizes and coatings facilitates processing into molding materials and other products such as papers and felts.

Thermoplastic molding compounds are supplied as granules containing short-length (0.2 mm) fiber reinforcements for processing on conventional injection-molding equipment.

Sheet molding compounds (SMCs) and dough molding compounds (DMCs) are already well established with glass fibers. The use of a small specific quantity of carbon-fiber reinforcement is envisaged in SMC with careful placement of aligned continuous fibers for optimum reinforcing effect at minimum additional cost. Carbon-fiber-reinforced DMCs in epoxy and polyester systems achieve excellent stiffness but strength levels achieved so far are disappointing.

Aligned short fibers in the form of prepreg sheets (ASSM) are available from special processes developed in the UK and Germany. The products can be molded to generate high performance in composites with highly curved or contoured shapes by virtue of the aligned fiber reinforcement coupled with the ability to promote controlled fiber movement for mold conformity.

5. Fabrication Techniques

Fabrication techniques have an important bearing on the properties obtained from carbon-fiber composites, and efficient processes must be devised for handling the fibers to ensure the maximum contribution to mechanical performance. The manufacture of GRP provides some useful experience although two major points of difference arise. First, the handling characteristics of the two fibers are different; second, the aim in the manufacture of carbon-fiber composites is to obtain the best possible mechanical performance, requiring the fiber to be incorporated in the precise direction and in the exact amount necessary to meet the stresses applied.

The basic principles of the manufacture of carbon-fiber composites are similar to those used for GFRP (see *Glass-Reinforced Plastics: Thermosetting Resins*), although changes in the actual techniques are necessary in most cases. Processes unique to CFRP production are also being developed.

5.1 Compression Molding

Three suitable material forms containing carbon-fiber reinforcement are available: DMCs, SMCs and preimpregnated tapes, sheets and fabrics.

Certain considerations which affect plant and mold tool design and process procedures are necessary for the manufacture of CFRPs. The thermal conductivity and low thermal-expansion characteristics of carbon fibers introduce the need for carefully controlled heating and cooling cycles and special attention to molding shrinkage (and expansion) to achieve dimensional accuracy in the final part. SMCs usually comprise hybrid reinforcements, mainly glass, with a small amount of carbon included for specific stiffening. The effectiveness of the carbon fiber is influenced by careful placement for maximum reinforcing effect and for minimum risk of misalignment caused by resin and glass-fiber flow during molding. Alignment accuracy, parallelism of mold-tool platens and additional considerations arise with the use of prepreg materials.

5.2 Autoclave and Vacuum-Bag Molding

The production of high-quality composites from prepreg materials (tapes and fabrics) is achieved by the consolidation on a single-sided tool of a laminated preform. The method makes use of a flexible bag or blanket through which pressure is applied to the prepreg layup on the mold. The mold is supported on a platen inside an autoclave which is pressurized with gas heated to the required cure temperature.

Unidirectional and woven materials can be molded equally successfully and the choice depends on the end use requirements and the layup techniques employed. Highly automated procedures for the precise placements of prepreg layers in laminate preforms are now well developed and are being introduced on a production basis in aircraft component manufacture. Computer-controlled tape-laying and ply-profiling machines are a major step towards a fully automated integrated composite-manufacturing scheme.

Current trends are towards prepregs with simplified cure cycles and zero resin-bleed characteristics. The careful use of viscosity modifiers in resin formulation offers greater flow control which is important for large or thick composite moldings in which significant temperature variations may exist.

Vacuum molding represents a simplified version of the autoclave technique where the pressure chamber is substituted by an oven and consolidation is achieved by virtue of the vacuum drawn inside the flexible membrane. The relatively low net pressure applied requires the resin to have easy flow characteristics.

5.3 *Filament Winding*

The filament-winding process is appropriate for the production of hollow carbon-fiber components which have an axis of rotation. Excellent mechanical properties are achieved by virtue of the accurate positioning of the fibers and the high volume fractions possible in the composite. The process is highly automated and can be programmed to lay down specific reinforcement patterns to comply with complex design requirements in multistressed components.

Carbon fibers can be processed successfully by both wet-winding and prepreg-winding techniques. Special attention to the gradual buildup of winding tensions and to the choice of smooth-surfaced materials for guide points is necessary in order to promote even, high-speed delivery of carbon fibers with minimum risk of abrasion and filamentation.

5.4 *Mandrel Wrapping*

Prepreg wrapping is used widely in the manufacture of sports goods, for the production of parallel and taped tubular shapes. Special purpose-built equipment is available for semiautomated production. Mechanical performance is determined by the layup pattern of the tube, although simple configurations are required for successful volume throughput. Unidirectional tapes and woven fabrics are equally applicable. Good material handleability is essential; unidirectional prepregs are often combined with a light scrim support to improve transverse properties.

5.5 *Pultrusion*

Pultrusion is a well-established technique for the production of profiles of glass-reinforced polyester composites and offers scope for CFRP manufacture. The process is simple, continuous, can be highly automated and does not require a high level of financial or technological investment. The use of carbon-fiber reinforcement with high-performance systems, such as epoxies, presents a completely new set of considerations for a continuous molding operation. The interacting effects of thermal expansion and resin shrinkage impose strict requirements for the tool design both in terms of dimensional accuracy and material of construction.

High-quality, high-performance pultruded stock is commercially available. Hybrid reinforcement configurations, new matrix materials (e.g., thermoplastics offering postformability) and methods for introducing off-axis reinforcement are being developed to extend the capabilities of the process.

5.6 *Wet Layup*

The contact molding method is simple and is a low-cost method, but is prone to variability in quality and performance. The method is therefore not attractive for the widespread use of carbon fibers. However, the addition of a small proportion of carbon fibers as a specific reinforcement for GFRP structures can be both cost and performance effective. For example, unidirectional woven tapes strategically incorporated into GFRP laminates promote maximum stiffness for minimum added cost, and nonwoven fabrics applied as surfacing tissues for GFRP structures provide chemical resistance and electrical conductivity.

5.7 *Other Methods*

Thermoplastic molding compounds are processed by injection molding. Long-fiber reinforcements, combined with PES, PEEK and other polymers can be converted to composites by thermoforming operations (see *Glass-Reinforced Plastics: Thermoplastic Resins*).

6. *Applications*

Applications for CFRPs fall generally into three broad categories: aerospace (aircraft, space and satellite structures), sports goods and engineering and industrial equipment. The overriding application is that of metals substitution for improved mechanical performance and weight reduction. The stiffness-to-weight ratio for high-modulus CFRPs is the highest available for any structural material and hence carbon-fiber composites are applied where stiffness, strength and low weight are essential.

6.1 *Aerospace*

Military aircraft applications have been the main driving force behind much of the investment in advanced composites technology. New aircraft are being designed to take advantage of the superior strength- and stiffness-to-weight ratios of composites compared with metals. The General Dynamics F16 fighter commenced production in 1977 with CFRP parts representing approximately 90 kg of airframe weight and the McDonnell Douglas F18 is programmed to include almost 10% of its structural weight as advanced resin-based carbon composite. In Europe, the Panavia Tornado, Dassault Mirage and Dornier Alphajet are typical of military aircraft containing production CFRP parts.

Further prospects for continued applications growth are good; for example, the McDonnell Douglas AV8-B (developed from the British Aerospace Harrier) has an all-composite wing with CFRP components representing 26.3% of the structural weight of the aircraft. Based on this experience, combined with the availability of new high-strain carbon fiber grades, design predictions for 50 wt% composite-based aircraft are being made.

Civil aircraft production has been slower to take advantage of carbon-based composites, but the Boeing 767 and 757 programs are contributing significantly towards eliminating this imbalance. Other commercial aircraft set to contain major CFRP structures are the European Airbus A310 and A320 and the Learfan 2000. The successful experience in helicopters such as the Aerospatiale Dauphin, Puma and the Westland series is extending the use of CFRP from rotor blades into other rotational and fuselage structures.

In space applications, CFRP components are used to meet the ultralightweight requirements and to match the need for materials with dimensional stability over a wide range of temperatures. The payload-bay doors of the NASA Space Shuttle Orbiter utilize a carbon-fiber–epoxy composite construction and the payload manipulation arm relies on the use of very high modulus (VHM) carbon fiber for its capability to deploy a 15 t payload in space.

6.2 Sports Goods

The sports goods industry provided the dramatic advance for carbon fibers into commercial end use in 1973 with the development of the CFRP golf shaft. The principle of manufacture (i.e., mandrel wrapping) has been applied to many items of sports equipment, such as fishing rods, ski poles, sailing boat masts and racquet shafts. In addition, the concept of using advanced composite materials for enhancing competitive sports equipment performance has been extended to virtually the whole range of sports and leisure goods. Structural applications in the sports category now include sailing boat hulls, sailplane wing spars and racing car chassis.

6.3 Engineering and Industrial Uses

Textile machinery has served well to demonstrate the effective use of CFRP in engineering applications where considerations of strength, stiffness and inertia apply to high-speed reciprocating parts. Components such as picking sticks and rapier arms are supplied as original equipment by machinery manufacturers although growth has been severely restricted by the depressed state of the textile industry.

The potential for weight and energy saving has been recognized in the automotive industry, where extensive development has been directed towards achieving the targets set in the USA for improved fuel economy. Propeller shafts and leaf springs of hybrid carbon–glass construction are of primary interest, together with other components contributing to unsprung vehicle weight, and short-fiber composites for engine, ignition, fuel system and transmission parts.

Successful medical applications include prostheses, surgical implants and x-ray equipment.

7. Future Perspective

About 4000 t of carbon fiber per year are consumed worldwide. Compared with glass fiber used in composite materials, estimated at five to six million tonnes per year, this is indeed a modest market. The latter is now a mature product and advances at the approximate rate of world economic growth. Before carbon fiber arrives at this phase of its history, it must first establish a concrete role in the engineering consciousness and achieve an economic scale of production. Carbon fiber costs US\$30 kg^{-1} at 1988 prices. Potentially, using polyacrylonitrile (PAN) produced on a commercial scale, it could reduce in price to US\$10 kg^{-1}. On the basis of modulus, volume for volume, this would make CFRP less expensive than GFRP. If the early promise of using pitch as a precursor to carbon fiber should be fulfilled, the cost expectations would exceed this prediction by far.

The most important factor influencing the future of CFRP is the cost of energy. By the late 1980s, earlier anxieties about energy supplies had abated, but ultimately the driving force in the materials economy will be the energy cost of materials plus the energy cost of their use. CFRP offers considerable advantages in both respects compared with conventional materials.

See also: Composites: Applications Overview; Composities: Manufacturing Overview; Glass-Reinforced Plastics: Thermoplastic Resins; Hybrid Fiber–Resin Composites; Sports Equipment Manufacture

Bibliography

Anon 1981 *Processing and Uses of Carbon Fiber Reinforced Plastics*. VDI, Düsseldorf, Germany

Anon 1983 Fiber composites; Design, manufacture and performance. *Composites* 14(2): 87–139

Anon 1986 *Carbon Fibers, Uses and Prospects*. Noyes Data Corporation, Park Ridge, NJ

Chou T-W 1989 Laminates: Elastic properties. In: Kelly A (ed.) 1989 *Concise Encyclopedia of Composite Materials*. Pergamon, Oxford, pp. 159–65

Clegg D W 1986 *Mechanical Properties of Reinforced Thermoplastics*. Elsevier, London
Delmonte J 1981 *Technology of Carbon and Graphite Fiber Composites*. Van Nostrand Reinhold, New York
Goodman S 1986 *Handbook of Thermoset Plastics*. Noyes Data Corporation, Park Ridge, NJ
Kelly A, Mileiko S T 1983 *Fabrication of Composites*. North-Holland, Amsterdam
Lubin G (ed.) 1982 *Handbook of Composites*. Van Nostrand Reinhold, New York
Pamington D (ed.) 1981 *World-Wide Carbon Fibre Directory*. Pammac Directories, Slough, UK
Schwarz M M 1983 *Composite Materials Handbook*. McGraw-Hill, New York
Sittig M 1980 *Carbon and Graphite Fibers—Manufacture and Applications*. Noyes Data Corporation, Park Ridge, NJ
Tsai S W, Halpin H T 1980 *Introduction to Composite Materials*. Technomic, Westport, CT
VDI 1982 *Processing and Use of CFRP*. VDI, Düsseldorf, Germany

P. J. Mills and P. A. Smith
[University of Surrey, Guildford, UK]

E. M. Trewin
[Courtaulds, Coventry, UK]

Cellulose: Nature and Applications

Cellulose is the most abundant organic material on earth, comprising the principal component of plants. Over 100 billion metric tons grow annually. This article discusses the composition and structure, properties, sources and applications of cellulose.

1. Composition and Structure

Cellulose is a long-chain polymer of β-D-glucose in the pyranose form linked together by 1,4′ glycosidic bonds to form cellobiose residues which are the repeating units in the cellulose chain (see Fig. 1).

As a result of the β configuration the alternate glucose units are rotated through 180° as can be seen from the figure. This structure strongly favors the organization of the individual cellulose chains in bundles with crystalline order held together by hydrogen bonds. X-ray diffraction studies since the early twentieth century have established that the small crystallites are oriented parallel to the fiber axis and have a monoclinic unit cell with dimensions

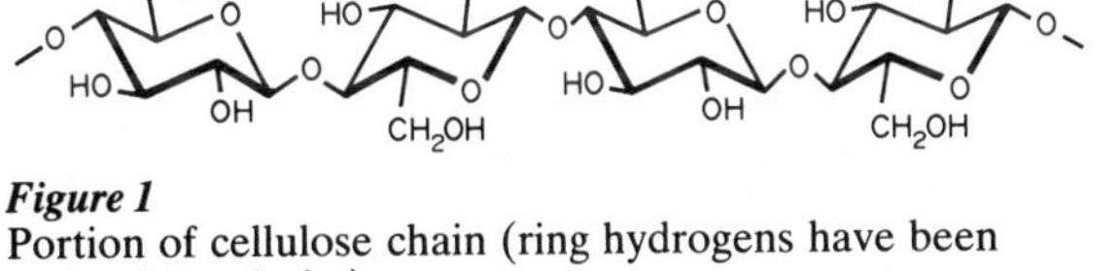

Figure 1
Portion of cellulose chain (ring hydrogens have been omitted for clarity)

of about 1.03 nm along the fiber axis (the cellobiose repeating unit), 0.82 nm and 0.79 nm with an angle of about 96°. These cells containing cellobiose residues of two chains are considered an adequate approximation of the structure of native cellulose, although some uncertainty still exists about fine structure details.

Cellulose is polymorphic and the structure of native cellulose (cellulose I) is changed by strong alkali as in mercerization of regeneration to cellulose II, which is also monoclinic with the same fiber axis repeating unit of 1.03 nm, but with other dimensions of 0.8 nm and 0.91 nm and angle of 117°. Other polymorphic forms of cellulose have been reported as cellulose III and cellulose IV. However, there is considerable opinion that these are not true polymorphic forms, but are rather disordered versions or mixtures of cellulose I and II.

Cellulose consists not only of these highly ordered crystalline regions, but contains disordered or amorphous regions as well. These different regions do not have clearly defined boundaries, but rather blend into each other. Cellulose samples of different origin and history contain differing relative amounts of ordered and disordered material. This degree of crystallinity decreases in the order cotton > wood pulp > mercerized cellulose > regenerated cellulose.

The chain length of cellulose in solutions can be determined by various techniques such as osmotic pressure, light scattering, ultracentrifugation and viscosity measurements. Not unexpectedly, celluloses of different origins and treatment histories show considerable variation. Values for the degree of polymerization (DP) can range from less than 1000 for regenerated cellulose to 7–10 000 for wood pulp and as high as 15 000 for cotton.

The macrostructure of cellulosic materials as defined by light microscopy is specific for the plant material concerned, for example cotton, softwoods, hardwoods and grasses. However, the native celluloses regardless of origin are made up of fibrils of only a few tenths of a micrometer in diameter, near the limit of resolution of the light microscope. Electron microscopy has shown the fibrils to be composed of still finer microfibrils. These range from 10 nm to 30 nm in diameter and thus may be presumed to contain several hundred cellulose chains.

Closely associated with cellulose in the plant cell walls are other carbohydrate polymers known collectively as hemicelluloses. They differ from cellulose in that they consist for the most part of sugars other than glucose, both pentoses and hexoses, are usually branched and have much lower degrees of polymerization, from less than 100 sugar units to about 200 sugar units. In softwoods hemicellulose mannan is the principal constituent, in hardwoods xylan, while in grasses arabinan occurs as well.

2. Properties

The most remarkable property of cellulose is its insolubility in water despite the fact that it is a polymer of glucose. Hydrogen bonding between cellulose chains is so intense that water cannot disrupt it by complexing with the hydroxyl groups. Other reagents, however, such as concentrated strong acids and bases, concentrated salt solutions and metal complexes do exert a solvent effect, especially on fractions of low molecular weight. Hemicelluloses are also water insoluble, but can be dissolved in alkali, permitting their separation from the total carbohydrate fraction, called holocellulose, leaving essentially pure or α-cellulose behind.

However, the presence of three hydroxyl groups on each anhydroglucose residue in the cellulose chain does make cellulose very hygroscopic and it readily adsorbs and desorbs water with changes in relative humidity. Many applications of cellulosic materials are influenced by this property, both beneficially and detrimentally.

The mechanical properties of cellulose, cellulosic materials and cellulose derivatives are high in relation to their weight. These desirable strength properties coupled with the abundance of cellulosic materials have led to their wide industrial application.

Chemical properties of cellulose are influenced by its structure and functional groups. Reagents which interact with the hydroxyl groups must first penetrate the fibrils, so accessibility or availability of the hydroxyl groups is an important factor in all cellulose reactions. Cellulose derivatives may be prepared by esterification, etherification, xanthation and grafting.

Cellulose esters of strong inorganic acids such as nitric, sulfuric and phosphoric may be prepared by direct reaction of the acids with cellulose to give the respective esters. Other strong acids such as perchloric acid and the halogen acids do not esterify cellulose. Formic acid is the only organic acid which can directly esterify cellulose to any appreciable extent. However, the anhydrides and acid chlorides of other organic acids can yield high degrees of substitution in the presence of suitable acidic catalysts (with anhydrides) and bases (with acid chlorides).

Ethers of cellulose are formed in the presence of alkali with organic halides, sulfates and alkene oxides. Some important ethers are methyl cellulose, ethyl cellulose, carboxymethylcellulose, hydroxyethylcellulose and cyanoethyl cellulose.

Although cellulose xanthate is the ester of dithiocarbonic acid it is prepared as in etherification by the reaction of alkali cellulose with carbon disulfide. The resultant derivative is water soluble and can be regenerated to cellulose by coagulation in an acid medium.

Graft copolymers in which chains of monomer units of a different kind are attached to a backbone of cellulose can be prepared by reaction with ethylene oxide or β-propiolactone, but more generally by allowing vinyl monomers to polymerize on active free-radical or ionic sites on the cellulose. The initiating sites may be formed by peroxides, irradiation, oxidation or mechanical shear. Properties of cellulose graft polymers can be controlled to enhance or decrease hygroscopicity, solubility and thermoplasticity.

Some reactions of cellulose involve degradation which can be desirable in controlling the viscosity and solubility of cellulose derivatives or producing sugars, or can be undesirable when the mechanical properties are adversely affected. Hydrolysis involves the scission of the acetal links between glucose units by acid with reduction of degree of polymerization (DP) and, when carried to completion, production of glucose. An initial rapid degradation to a levelling-off DP corresponds to hydrolysis of the amorphous cellulose. The hydrolysis rate then decreases to a new lower value as the crystalline regions are degraded.

Oxidative degradation of cellulose, especially by oxygen in the presence of alkali, involves a free-radical mechanism which is catalyzed by transition metals (cobalt, iron, manganese). Solution viscosity is decreased indicating reduction of DP and the introduction of aldehyde and carboxyl groups is significant. The exposure of cellulose to ionizing radiation also generates free radicals which can become involved in further oxidative degradation.

Since cellulose is a sugar polymer it serves as food for many microorganisms and the degradation of cellulosic materials by the enzymes secreted by these organisms is of great importance. Enzymatic degradation is similar to hydrolytic degradation by acids, but is more localized because of the much larger size of the enzymes. This biodegradability of cellulosic materials is an intrinsic part of the carbon cycle, causes billions of dollars of damage to structures and fabrics, and may yet serve to convert abundant cellulose to chemicals and fuels.

3. Sources

The most abundant source of cellulose is from woody plants in which 40–50% of the plant material consists of cellulose. So plentiful a material should be relatively inexpensive and, in fact, raw wood of low quality is cheap. The price of cellulose, however, depends on its purity and the extent to which it has been separated from the lignin and hemicelluloses with which it is naturally associated. Highly purified cellulose for use in the preparation of cellulose derivatives is known as chemical pulp or dissolving pulp, and results from extensive bleaching and extraction beyond that customary for ordinary paper pulps.

Seed hairs, of which the most important is cotton, also consist of cellulose. Of all the commercial raw materials for cellulose, the cotton fiber contains the lowest percentage of noncellulosic material, 3–15%, making purification simpler than for other cellulosic materials. The longer lint fibers are used in textile applications, while the shorter fibers removed from the seed are known as linters which are used for the production of chemical cellulose.

Bast fibers, the long fibers of the inner bark of various plants, include such cellulosic fibers as flax, hemp, jute and ramie. Many cellulosic fibers are obtained from the leaves of plants. Since cellulose is the major constituent of the cell walls of all plants, grasses, straws and agricultural residues have been extensively used as a source of cellulose.

4. Applications

Most of the applications of cellulosic materials are in the form of cotton, paper and wood. Regenerated cellulose products and cellulose derivatives which are based on chemical cellulose have a wide range of uses.

Regenerated cellulose refers to products that may have undergone chemical modification during processing to bring about changes in shape or form, but which consist of cellulose in the final product. Parchment paper prepared by the gelatinization of cellulose with concentrated sulfuric acid and subsequent washing as well as vulcanized fiber in which strong zinc chloride solutions are used to swell the cellulose are examples. Viscose rayon, fibers prepared by spinning cellulose xanthate into an acid bath, and cellophane, the analogous film product, are also regenerated cellulose.

In describing the applications of cellulose derivatives it is more convenient to categorize them by the use rather than the specific cellulose derivative involved. Important fiber applications in addition to viscose rayon use cellulose acetate and cellulose triacetate. Earlier cellulosic rayons such as cellulose nitrate and cuprammonium rayon, a regenerated cellulose, are no longer produced.

Protective coatings in the form of lacquers, textile and paper coatings and varnishes use cellulose nitrate, cellulose acetate, the acetate propionate, acetate butyrate and ethyl-cellulose either as the main film-forming material or in combination with synthetic resins. Films and foils, principally of cellulose acetate, are used in photographic applications and packaging.

The first synthetic plastic, celluloid, was made from cellulose nitrate. Organic cellulose esters, chiefly cellulose acetate, but also mixed esters with higher acids for water resistance, are important in molding applications. Cellulose ethers find application as gums, viscosity modifiers and thickening agents. Nitrocellulose is still used as an explosive.

Newer applications of cellulose derivatives are based on graft polymers with hydrophilic monomers which show greatly enhanced water-holding capacity and ionic substituents which are useful in ion-exchange applications.

See also: Paints, Varnishes and Lacquers; Papers of Natural and Synthetic Fibers

Bibliography

Attala R H (ed.) 1987 *The Structures of Cellulose,* ACS Symposium Series No. 340. American Chemical Society, Washington, DC

Bikales N M, Segal L (eds.) 1971 *Cellulose and Cellulose Derivatives,* Vols. 4, 5. Wiley, New York

Immergut E H 1963 Cellulose. In: Browning B L (ed.) *The Chemistry of Wood.* Interscience, New York, Chap. 4

Ott E, Spurlin H M, Grafflin M W (eds.) 1954 *Cellulose and Cellulose Derivatives*, Vols. I–III. Wiley, New York

I. S. Goldstein
[North Carolina State University, Raleigh, North Carolina, USA]

Coextrusion

It is often desirable to produce continuous multilayer products in sheet, film, tubing, filament or other forms. The process of forming such products is called coextrusion and is generally carried out by feeding separate polymer fluids, in most cases melt streams, from different pumps to a die feed block. The fluids are generally combined in the die and emerge in a combined form as a continuous extrudate.

1. Historical Development

The earliest applications of the coextrusion of polymeric phases may well have been in the rubber industry. Combined tread–sidewalls may have been coextruded before World War I. Another early application, dating from the 1940s, was the formation of so-called bicomponent fibers with the coextrusion and reaction wet spinning of two differently aged cellulose xanthate solutions (Sissons 1948). The coextrusion and dry spinning of solutions of acrylonitrile copolymers to produce bicomponent acrylic fibers was developed about 1960 (Ryan and Tichenor 1961, Hicks et al. 1967). More recently, bicomponent fibers have been developed by combining two melt streams in a spinneret (Buckley and Phillips 1969). The fibers produced in this manner have cross sections in which each phase is a semicircle. The unique characteristics of these filaments result from the differential shrinkage of the two polymeric phases, which buckles the filaments, producing the phenomenon of crimp (Brand and Backer 1962, Hicks et al. 1967, Buckley and Phillips 1969).

The coextrusion of molten plastics dates from the late 1950s, though earlier patents relating to the formation of plastic pencils may be so interpreted (Johnson 1946, Lorenian 1957). Patents on the coextrusion of multilayer coatings for pipe (Colombo 1958) and multilayer tubing (Lainson 1958) appeared in this period. The early 1960s saw the patenting of processes using coextrusion to produce spectacle frames (Rowland 1961), tubular blown multilayer film (Raley 1965) and additional multilayer tubing processes (Szipila 1962, Amsden 1964, Reynolds 1965, Schrenk 1967). Processes for coextruding film through slit dies and multilayer sheet appeared in succeding years (Corbett 1967, 1968, Squires 1969, Chisolm and Schrenk 1971, Nissel 1974). In the same period, multilayer films were developed with layer thicknesses of the order of the wavelength of visible light (Schrenk and Alfrey 1973a, b), producing striking optical effects (Alfrey et al. 1969).

2. Coextrusion Technology

Several basic technologies for producing multilayer extrudates have been developed. The simplest method for producing continuous multilayer products is to extrude separate layers of polymer melts from separate dies and then combine them. One of the earliest true coextrusion technologies for multilayer film used an annular multimanifold die coupled with tubular blowing to produce multilayer film (Raley 1965). More recently, there has been a trend for flat dies, often combined with chilled roll casting. Early flat die technology involved individual manifolds extending across the entire width of the die for each layer (see Fig. 1) (Corbett 1967, 1968, Schrenk 1974). The manifolds combine the completely spread melt streams just ahead of the die lips. A second method uses a conventional single manifold

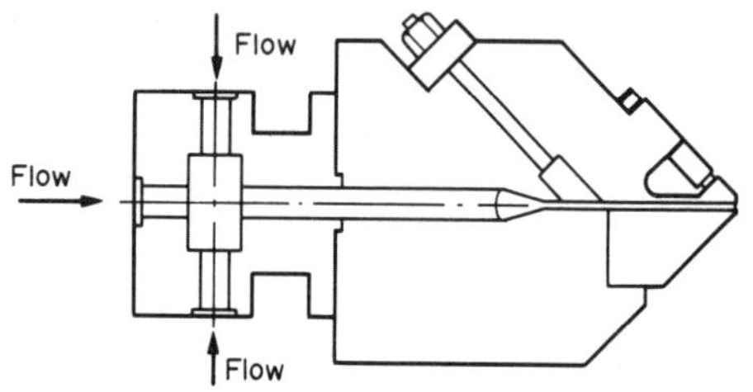

Figure 1
Coextrusion feedblock

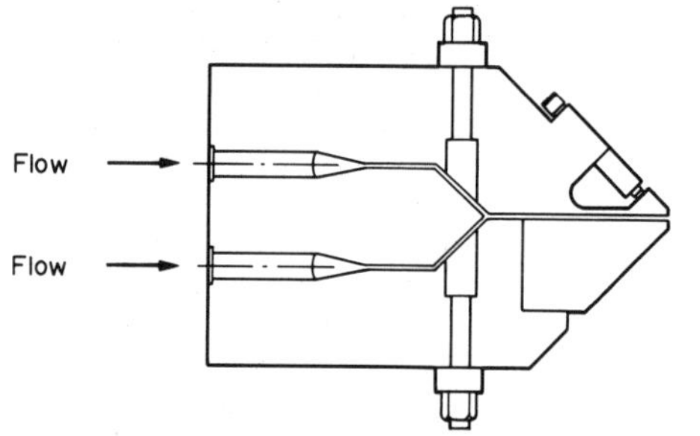

Figure 2
Multimanifold coextrusion feedblock (after Garton 1981–82 Extrusion dies. *Modern Plastics Encyclopedia*. McGraw-Hill, New York)

die combined with a feedblock at the die inlet. The feedblock introduces a prearranged multilayer stream into the die (Chisholm and Schrenk 1971, Schrenk 1974) as shown in Fig. 2. The multimanifold die system has the capability of using polymers of widely different viscosities. These dies, however, become very expensive and difficult to operate when more than three layers are involved. The feedblock method requires a moderately close match of viscosities but has greater versatility, since it may produce products with as many as several hundred layers. The layer thicknesses of the film or sheet are governed by the volumetric extrusion rate through the individual port.

Multilayer film with very thin layers has been developed by Schrenk and Alfrey (1973a, b). A multimanifold annular die is used in which the die mandrel and outer ring rotate, deform the layers emerging from the manifold and form them into long, thin spirals. Films with layers of thickness 90–200 nm have been prepared. The number of layers and their thickness can be computed.

The formation of bicomponent fibers involves other design problems. In a single spinneret there are numerous cylindrical holes, each of which must be fed by two manifolds.

3. Fluid Mechanics of Two-Phase Stratified Flow of Polymer Melts

Basic studies of two-phase stratified flow of polymer melts date from the 1970s. However, fundamental investigations of the fluid mechanics of two-phase flow have a much longer history. Perhaps the most important earlier investigations of two-phase flows were the studies of the Research Council of Alberta on the side-by-side of oil and water (Russell and Charles 1959, Charles and Redberger 1962). These studies were later extended by Yu and Sparrow (1969). Analyses of stratified two-phase laminar flow of Newtonian fluids in slit, circular and other cross-sectional dies were presented. These assumed flat interfaces. Profiles and pressure gradient–flow rate relationships were derived. Experimental studies generally indicated greater reductions in pressure gradients due to the presence of the water than were predicted from the analyses. Flow visualization studies by Yu and Sparrow (1969) suggested that the water tended to encapsulate the oil during flow.

Experimental and theoretical studies of the coextrusion of multilayer film have been reported by Schrenk and Alfrey (1969), Yu and Han (1973), Schrenk (1974), Han and Shetty (1976, 1978a, b) and Schrenk et al. (1978). Yu and Han (1973) and Han and Shetty (1976) analyzed flow through a slit die, assuming power law rheological behavior based upon

$$0=-\frac{\partial p}{\partial z}+\frac{\partial \sigma_{zy}}{\partial y}=-\frac{\partial p}{\partial z}+\frac{\partial}{\partial y}\left[K\left(\frac{\partial v_z}{\partial y}\right)^n\right]$$

where z is the direction of flow, y is the direction of shear, $\partial p/\partial z$ is the pressure gradient, σ_{zy} the shear stress, v_z the velocity along the die length, n the power law constant and K the consistency index. Velocity fields and extrusion rate–pressure gradient relationships were determined. Han and Shetty contrasted the latter relationships with experimental investigations of coextrusion through a slit die.

Several experimental studies of the stratified two-phase flow of polymer melts through dies of circular cross section have appeared. Southern and Ballman (1973) made the striking observation that the interface between two polymer melts in flow through a tube often distorts and one phase often encapsulates the other. Similar observations are described by Han (1973) and by Lee and White (1974) (see Fig. 3).

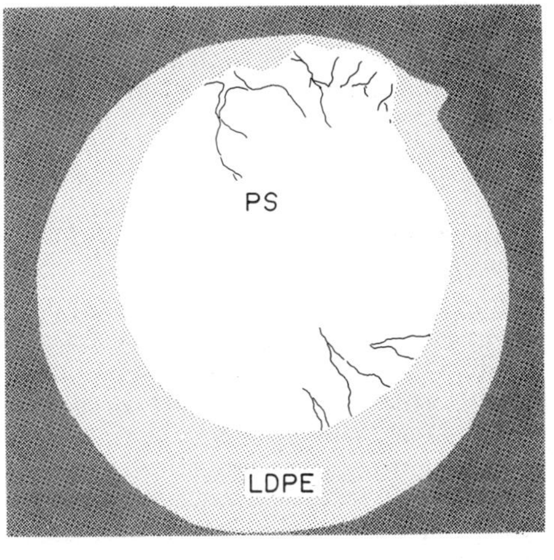

Figure 3
Interface for polystyrene (PS)–low density polyethylene (LDPE) extrudate

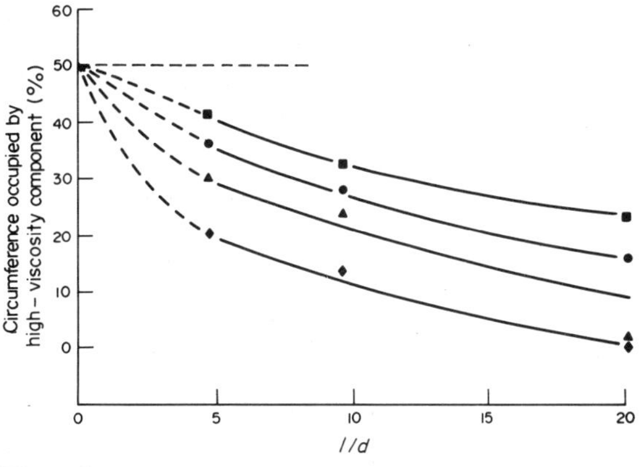

Figure 4
Interface configuration as a function of die length–diameter ratio for different viscosity ratios: ■, 3.6 (HDPE/LDPE); ●, 9.1 (PS/HDPE); ▲, 22.5 (PMMA/HDPE); ◆, 41.6 (PS/LDPE)

The same phenomenon has been observed in flow through dies with other cross sections (Minagawa and White 1975, Khan and Han 1976). Southern and Ballman and the later investigators found that the interface curvature development was correlated with the difference in viscosity between the two phases (see Fig. 4). The mechanism appears related to the problem of melts of different viscosities and specified extrusion rates maintaining the same pressure gradients in each phase. White and Lee (1975) analyzed this in terms of hydrodynamic lubrication theory.

Wavelike distortions at the interfaces of coextruded film were first noted by Schrenk, probably about 1970, but not published until later (Schrenk et al. 1978). The occurrence of these patterns was correlated by Schrenk et al. with a critical shear stress. This view is confirmed by a later study of this apparent interfacial instability by Han and Shetty (1978a) who also attempted an interpretation in terms of the non-Newtonian viscoelastic properties of the phases.

More recently (Belcher 1989), coextrusion has been used as a modification to free extrusion blow molding with the addition of precise technology that has 3–7 extruders feeding a die head. The multilayers of polymers form a parison in the die head in precise layers, each of which has a specific purpose.

See also: Die Extrusion; Extrusion of Fiber-Reinforced Polymer Systems; Processing: Classification; Tubular Film Extrusion: Structure Development

Bibliography

Alfrey T Jr, Gurnee E F, Schrenk W J 1969 Physical optics of iridescent multilayered plastic films. *Polym. Eng. Sci.* 9: 400–4

Amsden D L 1964 Apparatus for extruding tubing. US Patent No. 3,147,515
Belcher S L 1989 Blow molding of polymers. In: Aggarwal S L (ed.) 1989 *Comprehensive Polymer Science*, Vol. 7. Pergamon, Oxford, Chap. 15
Brand R H, Backer S 1962 Mechanical principles of natural crimp of fibers. *Text. Res. J.* 34: 39–49
Buckley R A, Phillips R J 1969 The development of bicomponent fibers. *Chem. Eng. Prog.* 65(10): 41–5
Charles M E, Redberger P J 1962 The reduction of pressure gradients in oil pipelines by the addition of water: Numerical analysis of stratified flow. *Can. J. Chem. Eng.* 40: 70–5
Chisolm D S, Schrenk W J 1971 Extrusion of plastic laminates. US Patent No. 3,557,265
Colombo R 1958 Apparatus for coating articles with multi-layer linings. US Patent No. 2,820,249
Corbett H 1967 Multiple flow diverter for thermoplastic materials. US Patent No. 3,320,636
Corbett H 1968 Laminating extrusion die. US Patent No. 3,398,431
Han C D 1973 A study of bicomponent coextrusion of molten polymers. *J. Appl. Polym. Sci.* 17: 1289–303
Han C D, Shetty R 1976 Studies on multilayer film coextrusion I. The rheology of flat film coextrusion. *Polym. Eng. Sci.* 16: 697–705
Han C D, Shetty R 1978a Studies on multilayer film coextrusion II. Interfacial instability in flat film coextrusion. *Polym. Eng. Sci.* 18: 180–6
Han C D, Shetty R 1978b Studies on multilayer film coextrusion III. The rheology of blown film coextrusion. *Polym. Eng. Sci.* 18: 187–99
Hicks E M, Tippetts E A, Hewett J V, Brand R H 1967 Conjugate fibers. In: Mark H F, Atlas S M, Cernia E (eds.) 1967 *Man-Made Fibers: Science and Technology*, Vol. 1. Wiley, New York, pp. 375–408
Johnson T L 1946 Multiple extrusion. US Patent No. 2,408,398
Khan A A, Han C D 1976 On the interface deformation in the stratified two-phase flow of viscoelastic fluids. *Trans. Soc. Rheol.* 20: 595–621
Lainson J J 1958 Extrusion apparatus. US Patent No. 2,859,476
Lee B L, White J L 1974 Rheological properties of polymer melts in laminar shear flow and of interface deformation and its mechanisms in two-phase stratified flow. *Trans. Soc. Rheol.* 18: 467–91
Lorenian Z 1957 Method and apparatus for manufacturing pencils. US Patent No. 2,790,202
Minagawa N, White J L 1975 Co-extrusion of unfilled and TiO_2-filled polyethylene: Influence of viscosity and die cross-section on interface shape. *Polym. Eng. Sci.* 15: 825–30
Nissel F R 1974 Method of extruding a plural layered sheet. US Patent No. 3,833,704
Raley G E 1965 Melt extrusion of multi-wall plastic tubing. US Patent No. 3,223,761
Reynolds M 1965 Dual extrusion apparatus. US Patent No. 3,222,721
Rowland W P 1961 Manufacture of spectacle frames. US Patent No. 2,985,556
Russell T W F, Charles M E 1959 The effect of the less viscous liquid in the laminar flow of two immiscible liquids. *Can. J. Chem. Eng.* 37: 18–24
Ryan J F, Tichenor R L 1961 Process for spinning polyacrylonitrile having low degree of crimp and crimp reversibility. US Patent No. 2,988,420
Schrenk W J 1967 Die. US Patent No. 3,308,508
Schrenk W J 1974 Multilayer film from a single die. *Plast. Eng.* 30: 65–8
Schrenk W J, Alfrey T Jr 1969 Some physical properties of multilayered films. *Polym. Eng. Sci.* 9: 393–9
Schrenk W J, Alfrey T Jr 1973a Coextruding multilayer blow film—Part 1. *SPE J.* 29(6): 38–42
Schrenk W J, Alfrey T Jr 1973b Coextruding multilayer blow film—Part 2. *SPE J.* 29(7): 43–7
Schrenk W J, Bradley N L, Alfrey T Jr, Maack H 1978 Interfacial flow instability in multilayer coextrusion. *Polym. Eng. Sci.* 18: 620–3
Sisson W A 1948 Method of manufacturing artificial filaments. US Patent No. 2,443,711
Southern J H, Ballman R L 1973 Stratified bicomponent flow of polymer melts in a tube. *Appl. Polym. Symp.* 20: 175–89
Squires P H 1969 Process for coextruding multiple-layered thermoplastic sheeting. US Patent No. 3,476,627
Szipila J S 1962 Plastic pipe extrusion head. US Patent No. 3,024,494
White J L, Lee B L 1975 Theory of interface distortion in stratified two phase flow. *Trans. Soc. Rheol.* 19: 457–79
Yu H S, Sparrow E M 1969 Experiments on two component stratified flow in a horizontal duct. *J. Heat Transfer* 91C: 51–8
Yu T C, Han C D 1973 Stratified two phase flow of molten polymers. *J. Appl. Polym. Sci.* 17: 1203–25

J. L. White
[University of Akron, Akron, Ohio, USA]

Composites: Applications Overview

Composite materials now exert an influence on the daily lives of most people in industrialized societies. Much personal transport, and indeed safety, depends on the cord or fiber-reinforced elastomer tire; many commercial airliners fly with some parts of the primary structure and much of the interior furnishings and trim made from composites; many types of sporting and leisure goods, such as boats, gliders, sailboards, skis and racquets, make extensive use of composite materials in their construction; and in the home, most plastic-bodied appliances incorporate reinforcement in the form of short, chopped fibers.

The term composite material can be defined in a number of ways. The most commonly held view is of a material in which strong, high-modulus man-made fibers, with diameters of the order 10–100 μm, are embedded in a matrix material which may be a polymer, a metal, a glass or a ceramic. The objective is to enhance the mechanical or physical properties of the host material. Alternatively, from a mechanical standpoint, the matrix may be regarded simply as a binder whose role is to transfer stress to the reinforcing fibers and ensure their cooperative

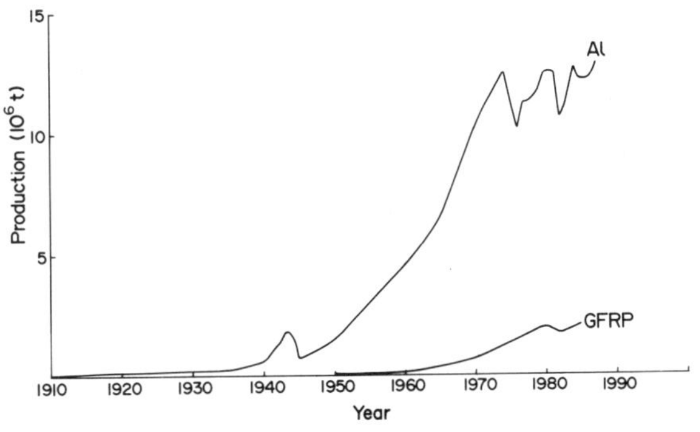

Figure 1
Comparative growth of aluminum and GRP consumption

interaction. The matrix may also fulfill other functions, such as protecting the fine reinforcing filaments from corrosion, oxidation or other forms of environmental degradation.

Reinforcing agents are not necessarily confined to man-made fibers, although continuous filaments of glass, carbon, ceramics and high-modulus polymers are the most widely used at present. However, vegetable fibers, inorganic whiskers, metal wires and particles of refractory solids all find a place in the spectrum of reinforcing materials. On a somewhat different scale, concrete reinforced with steel bars can be regarded as a composite with mechanical behavior analogous to that of fiber composites, but where application in tonnage terms dwarfs that of other constructional materials. However, this article is concerned with the families of materials known as fiber-reinforced plastics (FRPs), metal-matrix composites (MMCs) and ceramic-matrix composites (CMCs), which at least two of the three dimensions of the reinforcement are typically of the order of 100 μm or less.

Of these materials, the group that has reached the most advanced stage of both technical development and market penetration is FRPs. One of their principal attributes is low density and to indicate the scale of their application, world consumption of glass-fiber-reinforced plastic (GRP) is compared in Fig. 1 to that of aluminum, a material of roughly comparable cost and density, but established in the market place some 50 years earlier. The data for aluminum are plotted from the time (approximately 1890–1900) when commercial quantities became available following the development of the Hall process for the electrolysis of fused cryolite. Figures for GRP date from the mid-1940s when significant quantities of glass fiber first became commercially available. Both curves reflect the recession in world trade during the 1980s, and the aluminum plot also illustrates the boost and subsequent fall in demand associated with World War II and the influence of the 1973 oil crisis.

Although present consumption of GRP is small compared to that of aluminum, it is apparent from Fig. 1 that the rate of growth of consumption over the first 30 years of composites technology, averaging 10–15% per year, was considerably greater than in the first 30 years of aluminum technology. It is probably not meaningful to pursue the comparison further in view of the general increase in the pace of technological development that has occurred over the past few decades, except to note that at least part of the growth for GRP has resulted from metals substitution and that most market projections are for continuing growth in the applications of composites.

GRPs, in volume terms, represent the major part of the reinforced plastics market, mainly on account of their relatively low cost, good corrosion resistance, good specific strength and modulus, and the ease with which complex shapes can be molded or fabricated by a number of well-established processes. The same general advantages apply to plastics reinforced with even stronger or stiffer fibers, notably of carbon or of highly oriented polymers such as the aramids, and although the resulting composites are considerably more expensive than GRPs, they can nevertheless be highly cost effective in applications where weight reduction gives rise to operating economies, as in aerospace. For example, one source estimates the acceptable cost of weight saving in helicopter construction as US\$250 kg^{-1} and in a satellite as US\$650 kg^{-1} (Jardon and Costes 1987).

One of the major limitations of polymer-based composites is the decrease in their mechanical properties with increased temperature. A large number of polymer formulations, both thermosetting and thermoplastic, are used as matrices for composites and although a few, presently expensive materials are suitable for extended service above 300 °C, most reinforced plastics are limited to temperatures below around 200 °C for continuous operation. It is necessary to employ more refractory reinforcements with metal or ceramic matrices in order to exploit the advantages of composite materials for high-temperature applications, such as in heat engines. Table 1 indicates maximum temperatures for continuous operation of the three principal types of composite material considered in this article.

Reinforcements currently being explored with metal- and ceramic-matrix composites include continuous fibers of carbon, silicon carbide, boron, aluminum oxide and whiskers of silicon nitride and silicon carbide. Most of the resulting composites are still at the research and development stage (see, e.g., Anderson et al. 1988) and to give some indication of the relative scale of applications for polymer,

Table 1
Upper temperature limits for continuous operation of composite materials

Composite	Maximum operating temperature (°C)
Polymer matrix	400
Metal matrix	580
Ceramic matrix	1000

Source: Hancox and Phillips 1985

metal- and ceramic-matrix composites, Table 2 lists some estimates of approximate consumption levels for the principal reinforcing materials. Despite the fact that carbon fibers are used to reinforce some low-melting-point metals and alloys, it can be assumed that virtually all the glass, carbon and aramid fibers listed in Table 2 are used for plastics reinforcement at the present time. Although the quantities of carbon, aramid and ceramic fibers used in composites are small compared to glass fiber consumption, most market analysts predict growth rates up to around 20% per year for high-performance polymer-matrix composites, based largely on assessments of the continuing adoption of these materials by the aerospace industry.

In discussing applications for composite materials it is important to recognize that they are generally more expensive than other construction materials. Table 3 gives broad ranges of cost for reinforcements, matrix resins and some conventional construction materials. Exact costs will depend on the form in which the basic materials are to be supplied.

In the case of reinforced plastics, for example, cost will be determined by whether the fibers are in the form of rovings or woven fabric, whether preimpregnated with resin, and on the tow size or areal weight of the fabric. The costs of manufactured items will also depend on the nature of the production process and on the complexity of the component. For GRP, the cost of a finished component may be typically two or three times the materials cost, and the added value may be significantly higher if there is a substantial degree of proof testing and quality assurance involved in the manufacturing process, as with many aerospace parts.

Table 2
Approximate production levels for the principal reinforcements for composite materials

Reinforcement	Production capacity/consumption ($t\ year^{-1}$)
Continuous fibers	
glass[a]	1280×10^3
aramid[a]	3600
carbon[a]	3300
ceramics (B, SiC, Al_2O_3)[b]	few
Short ceramic fibers[b]	~40
Whiskers (SiC, Si_3N_4)[b]	~100

a Brehm and Sprenger 1985 b Feest et al. 1986

Table 3
Approximate price ranges for reinforcements, thermosetting resins and some traditional materials of construction

Material	Approximate price, 1988 (£ $10^3\ t^{-1}$)
Continuous fibers	
glass	1–2
aramid	20–75
carbon	25–100
SiC	200–400
Thermosetting resins	1–10
Whiskers	50–100
Unreinforced materials	
metals (e.g., steel, aluminum)	0.1–1
timber	0.3–1.5
engineering thermoplastics	2–10

1. *Applications of Fiber-Reinforced Plastics*

In view of the difference between the market sizes of GRP and of composites based on carbon and aramid fibers, it is convenient, for a broad description of applications, to discuss the two areas separately. There is, however, no clear-cut definition of what constitutes a high-performance composite in terms of its composition. Thus, although some types of GRP fulfill relatively undemanding uses such as decorative panels or cable ducting, other types are used in the primary structures of some aircraft designs (Lubin and Donohue 1980) and in the construction of deep-sea submersibles (Oliver 1980).

1.1 Applications of GRP

Table 4 shows a breakdown of the European market in terms of applications of GRP based mainly on the

Table 4
European market for glass-resin composites in 1986, breakdown by application (estimated volume 860 000 t)

Application	Distribution (%)
Transport	21
Electrical industry	19
Industrial and agricultural equipment	20
Building and public works	16
Sports and leisure	6
Consumer goods	6
Others	12

Source: Chevalier 1987

relatively low-priced polyester resins. Some attempt is made in the following to give an indication of the types of application in some of the larger sectors.

(*a*) *Transport*. In Western Europe, some 170 000 t of GRP in its various forms is used in land transportation applications and in the USA total consumption in this sector is substantially greater. The incentive to introduce reinforced plastics into land-based transport is mainly to take advantage of their corrosion resistance and of manufacturing economies. Weight reduction, although leading to improved fuel economy or performance, is unlikely to be an important factor in new designs unless it can be achieved without a cost penalty.

The use of reinforced plastic panels for car bodies is at present mainly restricted to the small-volume production of specialist vehicles for which the versatility of GRP fabrication methods enables design changes to be made fairly easily and for which the high capital cost of large metal press tools is not justified. The largest production run to date is probably the General Motors' Fiero car which employs body panels, produced by press molding a sheet molding compound (SMC) and by the reinforced reaction injection molding (RRIM) process, attached to a steel structural frame (Ferrarini et al. 1984). Similar schemes have been used for many years for the production of cabs for heavy commercial vehicles (Seamark 1981).

The adoption of molded reinforced plastics for a wide variety of panels, bumpers and interior fittings appears to be growing steadily, at least in terms of the average quantity per car (Waterman 1986). In the USA, reinforced plastic front and rear ends have almost completely superseded die-cast zinc because a single molding can replace an assembly of many individually cast items. In Europe, Renault were producing 9600 GRP bumpers per day in 1983, and the Citroen BX features an SMC bonnet (hood) and tailgate, produced at a rate of the order of 1000 per day (Buisson 1983). Attention has also been directed towards underbonnet components, of which one of the most ambitious has been the moding, by a lost-wax-type process, of a dough molding compound (DMC) inlet manifold (Suthurst and Rowbotham 1980).

The use of composites for structural automotive components has not been overlooked and there are numerous references in the literature to the development of stressed components such as drive shafts and GRP springs. Leaf springs make use of the excellent fatigue properties of fiber composites and can give extended lives in comparison to steel springs, particularly for commercial vehicles (Lea and Dimmock 1984, Jardon and Costes 1987). Figure 2 shows a GRP spring undergoing fatigue testing. Such springs provide a substantial reduction in mass over steel springs, as well as providing an improved ride and a decrease in the noise transmitted to the driving compartment.

The use of composites is not confined to road transport, however. Applications in rolling stock on European railways have been reviewed (Anon 1980) and the construction of the cabs on British Rail's Intercity vehicles has been described by Gotch and Plowman (1978). In the USA, a rail-borne grain transporter, 15 m long and 4.6 m in diameter, has been produced by filament winding (Ruhmann et al. 1982).

(*b*) *Electrical*. The most common type of glass fiber used for reinforcement purposes is E-glass, which was originally developed for electrical insulation purposes, so it is not surprising that, when combined with organic resins, a class of materials results which can readily be molded into complex shapes possessing excellent insulating properties and a high dielectric strength. Thus switch casings, junction boxes, cable and distribution cabinets, relay components and lamp housings, to name just a few examples, are produced from a variety of glass-fiber molding compounds. Cable ducts and standard cross-sectional shapes for transformer insulation are produced by the pultrusion process and reinforcing bands for rapidly rotating electrical machinery are often filament wound. Mathweb, a proprietary process for winding lattice-type structural beams (Preedy and Kelly 1982), is now used by British Rail to provide emergency support to overhead power lines because the low weight simplifies replacement of damaged gantries. Pultruded composites are also used in the construction of pantograph arms for current pickup on electric traction vehicles.

(*c*) *Industrial and agricultural*. Generally speaking, applications in agriculture, in chemical process plants and in tanks and pipes for industrial purposes rely on the combination of low density and good corrosion resistance of GRP. One of the advantages of the material is the relative ease with which large and complex-shaped components can be fabricated on a one-off basis. Typical applications include vats and silos for the storage of aggressive chemicals, pipelines for the transport of water and sewerage, water storage tanks, wine vats and the construction of certain types of chemical process plants as, for example, in the manufacture of chlorine. A particular advantage due to the low density is that the cost of transporting components is minimized and this particularly facilitates the installation of pipelines in remote or rugged terrain. It is also possible to transport and handle longer lengths of pipe, thus minimizing the number of joints and hence the assembly costs. Figure 3 illustrates a PVC-lined GRP storage vessel for hydrochloric acid, of dimensions 10.5 m in length and 3.5 m in diameter.

(*d*) *Building*. The corrosion resistance and lightness, and hence ease of transport and installation, of

Figure 2
Fatigue test of a GRP automotive leaf spring (courtesy of GKN Technology Ltd)

Figure 3
GRP tank lined with PVC for storage of HCl (courtesy of Plastics Design and Engineering Ltd)

GRP, coupled with the ability to mold surfaces with decorative textures and color, have led to extensive use of this material as cladding for prestige buildings (Jaafari et al. 1976). Considerable quantities of GRP in the form of flat and corrugated translucent sheeting are also used for roofing purposes, particularly in outdoor canopies of large structures such as stadia, where the reduced weight of the roofing panels enables economies to be made in the design of the roof support structure. In the USA, the development of prefabricated bathroom units comprising sanitary ware, furniture and partitioning has generated a significant market (Seymour and Tompkins 1975).

(*e*) *Marine*. This area provides a good example of how a number of attributes of GRP in combination have resulted in almost complete substitution for traditional materials, particularly timber. Low density, corrosion resistance and the ability to produce complete hull and deck moldings repetitively from a single set of molds have revolutionized the leisure and small workboat building industries, and reduced maintenance costs have played a large part in gaining widespread consumer acceptance of GRP. Of less general applicability, but of particular importance to naval mine-clearance operations, is the nonmagnetic character of GRP. The UK pioneered the introduction of GRP minesweepers with the launch, in 1972, of HMS Wilton with an overall length of 46 m (Dixon et al. 1973), and vessels up to 60 m in length and with displacements of approximately 600 t are now being built in a number of countries (Anon 1987a).

1.2 Applications of High-Performance Composites

As indicated earlier, the term high-performance composites is somewhat loosely employed to describe composites based primarily on the more expensive carbon and aramid reinforcements, although it also reflects the improved performance to be gained in the two major market sectors to have developed thus far—aerospace and sports goods.

(*a*) *Aerospace*. The use of lighter materials of construction means either that a greater payload can be carried or that operating costs can be reduced for the same payload. The outstanding example of the cost effectiveness of lightweight composites is in spacecraft technology where the cost of launching may often exceed the cost of designing and building the spacecraft (see, e.g., Zweben 1981, Burke 1986). Consequently, carbon- and aramid-fiber-reinforced plastics are frequently used for such items as the basic structure of a spacecraft, as supports for solar arrays, for the construction of dish antennas, in pressure vessels for propellant gases and in apogee motor cases and thrust cones. In addition to weight savings, the low axial thermal expansivity of these materials enables very high dimensional stability to be achieved in antennas, microwave resonant cavities and filters and in telescope support structures (see, e.g., Reibaldi 1985). Such components may experience temperature excursions in orbit within the range −100 °C to approximately 100 °C depending on whether the component is exposed to full sunlight or is in the shadow of the spacecraft.

In military aircraft, composites offer improved performance, and a number of aircraft design and construction programs are in progress involving major use of advanced composites (Hadcock 1986). For example, approximately 35% of the structural weight of the projected European Fighter Aircraft (EFA) is likely to be built from composite materials, involving the main wing, the forward fuselage and the fin and rudder. Figure 4 illustrates the use of composites in the construction of the AV-8B Mk. 5 Harrier II airframe giving rise to a 25% weight saving in the airframe compared to all-metal construction (Riley 1986). In commercial aircraft, composites have for some time been used to reduce weight in internal fittings such as cargo floors, galley furniture, luggage lockers and for internal trim. Seat frames are now beginning to be built in carbon-fiber-reinforced plastics (CFRP) since they not only lead to weight savings but occupy less space. With regard to the use of composites in primary aircraft structures, considerable caution has been, and still is, exercised in proving such materials before flight certification is granted, but nevertheless major structural items are now being flown on passenger aircraft. In the A310 Airbus, for example, the vertical stabilizer makes substantial use of composites construction involving glass, carbon and aramid fibers, with a total weight saving of 397 kg (Pinzelli 1983). The A320 Airbus will use CFRP in the fin and tailplane leading to a weight saving of 800 kg over aluminum-alloy skin construction (Anon 1986). As an indication of the benefit of such weight savings, it has been estimated that a 1 kg weight reduction on a DC-18 saves over 2900 l of fuel per year (Zweben 1981).

On a smaller scale, a number of designs of light aircraft and gliders (Riddell 1985), built primarily of GRP, have been flying for a number of years and attention is now being directed to the development of all-composite aircraft in the executive jet category. In this class, two prototypes of the Lear Fan 2100 aircraft with 70% of structural weight in CFRP were flown before the company went into liquidation for reasons unconnected with the use of composites (Noyes 1983). The Beech Starship, currently under development, is also constructed largely from composite materials and one phase of the program involves a comparison between hand-lay and filament-winding techniques for the construction of the fuselage (Wood 1986).

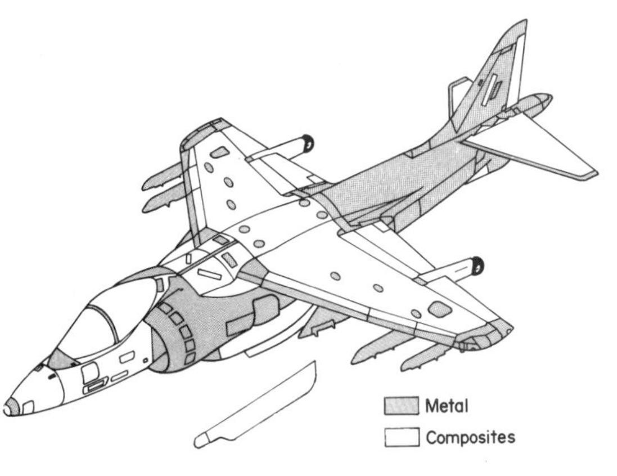

Figure 4
The extent of composites construction in the AV-8B Mk. 5 Harrier II aircraft (from Riley 1986)

As frequently happens when an existing technology is threatened by innovation, the aluminum industry has responded to the introduction of composites with the development of alloys, such as the aluminum–lithium series, with improved strength and stiffness (Clementson 1985). Although their specific properties do not match those of the advanced composites, they possess the advantage that established metal skin design and fabrication procedures can be applied, obviating the need for investment in new skills to handle design with anisotropic materials and the very different manufacturing techniques involved with composites. Nevertheless, whatever the outcome of the trade-off between metal or composite properties, one important economic advantage of composites remains, namely the ability to fabricate large and complex parts as a single molding, as opposed to assembly of a number of individual metal pressings.

Thus far, little has been said of the excellent fatigue behavior of composites in fiber-dominated directions. An outstanding example of a component whose life is governed by fatigue damage is the helicopter rotor blade which experiences complex dynamic stresses including torsion, together with bending in, and perpendicular to, the plane of the rotor (Goddard 1985). Composite blades are now fitted to most major helicopter designs on the basis that they give virtually unlimited fatigue life in relation to the lifetime of the aircraft, reduce manufacturing costs and enable novel and more efficient aerofoil structures to be created. Most helicopter manufacturers also plan to use composites extensively in future airframes. For example, the Boeing 360 demonstrator has an all-composite fuselage consisting of CFRP frames to which are bolted and bonded honeycomb sandwich panels with aramid-fiber-reinforced skins. This form of construction gives a 25% weight saving in the fuselage and substantial cost reductions related to parts integration and the shorter time required for assembly (Anon 1987b). As with helicopter rotor blades, aircraft propeller blades made from composites show excellent fatigue and corrosion resistance compared to metals and at reduced weight. McCarthy (1986) has described the design and construction of composite blades and has illustrated the weight savings compared to metal blades.

(*b*) *Leisure and sporting goods*. Generally speaking, the production of aerospace structures and components is a low-volume manufacturing operation measured typically in terms of a few tens or hundreds of parts per year. The second largest market sector for composites based on carbon and aramid fibers, and to some extent glass fibers, has in recent years been the leisure goods industry, where production rates have been measured in thousands or tens of thousands of units per year. However, in items such as tennis and squash racquets, fishing rods, skis and ski poles, golf shafts, windsurfers and other small sailing craft, the quantities of expensive reinforcements are small, but aggregate to approximately one-third of present carbon and aramid fiber consumption.

Reference to the use of composites in sailplanes and small boats has already been made in this article. A feature of the use of such materials in high-performance sailing craft is that the advanced materials and techniques developed for the aerospace industry are being adapted to produce racing vessels with strong rigid hulls constructed from honeycomb sandwich panels, giving substantial reductions in weight and increases in performance. For similar reasons, many racing cars now employ a composite chassis consisting of CFRP skins bonded to a honeycomb core (Clarke 1985).

2. Metal-Matrix Composites

Among the attributes of fiber- or whisker-reinforced metal-matrix composites (MMCs), in addition to improved strength and stiffness, are increased creep and fatigue resistance, giving rise to higher operating temperatures than those of the unreinforced metal (Feest 1986). The presence of ceramic reinforcements, whether in fibrous or particulate form, can also provide increased hardness, wear and abrasion resistance. These properties are potentially exploitable in many aspects of pump and engine technology including compressor bodies, vanes and rotors, piston sleeves and inserts, connecting rods, cylinder heads and clutch components. Many such components are under development (Feest et al. 1986) but few have yet entered commercial production. One exception is the use by the Toyota Motor Corporation (Donomoto et al. 1983) of aluminum-alloy diesel engine pistons incorporating a ceramic-fiber-reinforced top land and piston ring groove (see Fig. 5). Honda have also engaged in a limited production run of aluminum-alloy connecting rods reinforced with stainless steel wire for their 1.2 l City vehicle in Japan. On a much smaller scale of application, boron-fiber-reinforced aluminum-alloy struts are used in the midsection framework of the US space shuttle fuselages (Irving 1983). An application that exploits the thermal properties of aluminum reinforced with silicon carbide particles is the packaging of microwave circuits (Thaw et al. 1987). In this respect, the presence of the silicon carbide reduces the thermal expansion coefficient of the packaging material, providing a better match to the ceramic circuit substrate without significantly decreasing the high thermal conductivity required for good heat dissipation from the device.

3. Ceramic-Matrix Composites

The present picture for ceramic-matrix composites (CMCs) is similar in many ways to that of MMCs in

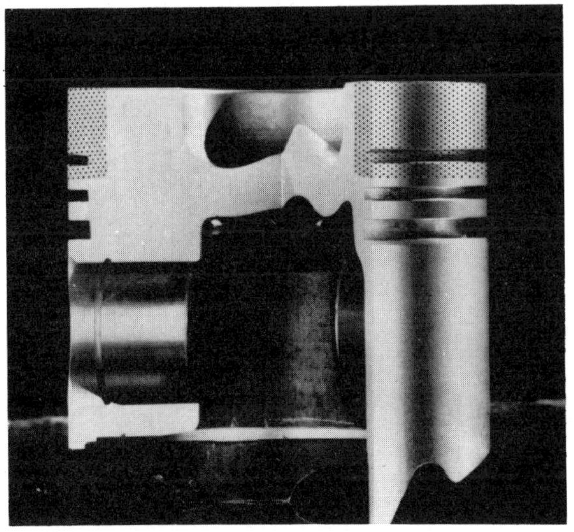

Figure 5
Toyota diesel engine piston: the MMC insert can be seen as the darker region incorporating the upper piston ring groove

that considerable research and development activity exists (Phillips 1985) but there are few established applications. Potential applications for CMCs include high-temperature seals, bearings, blades and rotors, particularly in gas turbine engines, piston crowns in reciprocating engines and wear-resistant duties such as valve bodies, cutting and forming tools. In the latter context, can-forming punches made from alumina reinforced with silicon carbide whiskers are reported to show better performance than tungsten carbide punches in machines that produce 400 cans per minute round the clock (Anon 1988).

The availability of strong, tough, lightweight materials with good thermal shock resistance, capable of continuous operation in air at temperatures above 1200 °C, would make possible significant increases in the efficiency of gas turbine engines. Phillips (1987) and Jamet (1987) have reviewed the prospects of developing CMCs to meet this type of requirement. Existing ceramic reinforcements suffer degradation in strength or undergo creep in the range 1000–1200 °C and the only material available in the late 1980s with attractive mechanical properties above 1200 °C was carbon-fiber-reinforced carbon, operating in a nonoxidizing atmosphere. (Hill et al. 1974). Carbon–carbon materials are currently used for ablative purposes on those surfaces of spacecraft and reentry vehicles subject to severe aerodynamic heating and considerable research is being devoted to identifying ways of increasing oxidation resistance, particularly for reusable space vehicles.

At lower temperatures and for short periods of time at temperature, carbon–carbon materials have proved suitable in disk brakes for aircraft and racing cars, where sufficiently large amounts of energy are generated in emergency braking to cause conventional steel disk brakes to fade. For a review of carbon–carbon technology, see Fitzer (1987).

See also: Aircraft and Aerospace Applications of Composites; Automobile Tires; Automotive Composite Components: Fabrication; Sports Equipment Manufacture

Bibliography

Anderson L I, Lilholt H, Pederson O B (eds.) 1988 Mechanical and physical behaviour of metallic and ceramic composites. *Proc. 9th Risø Int. Symp. Metallurgy and Materials Science*. Risø National Laboratory, Denmark

Anon 1980 GRP for rolling stock on European railways. *Vetrotex Fibreworld* 9: 11–18

Anon 1986 A320—Fly-by-wire airliner. *Flight Int.* 130(4026): 86–94

Anon 1987a Growth of composites in military naval construction: Slowly but surely. *Vetrotex Fibreworld* 24: 6–7

Anon 1987b Boeing 360—Helicopter hi-tech. *Flight Int.* 131(4058): 22–7

Anon 1988 ACMC's ceramic composites—Tooling up for a first in cans. *Mater. Edge* 7: 10

Brehm B, Sprenger K H 1985 Modern fibre reinforced materials. *Sprechsaal* 118(3): 253

Buisson M 1983 SMC and the car. *Vetrotex Fibreworld* 17: 10–12

Burke W R (ed.) 1986 *Proc. Workshop Composites Design for Space Applications*, ESA-SP-243. European Space Agency Publications Division, Noordwijk, The Netherlands

Chevalier A 1987 Glass/resin composite market. *Vetrotex Fibreworld* 24: 8

Clarke G P 1985 The use of composite materials in racing car design. *Proc. 3rd Int. Conf. Carbon Fibres—Uses and Prospects,* Paper 18. Plastics and Rubber Institute, London

Clementson A 1985 Materials and manufacturing in aerospace. *Proc. 2nd Conf. Materials Engineering*. Institute of Mechanical Engineers, London, pp. 189–94

Dixon R H, Ramsey B W, Usher P J 1973 Design and build of the GRP hull of HMS Wilton. *Proc. Symp. GRP Ship Construction*. Royal Institute of Naval Architects, London, pp. 1–32

Donomoto T, Funatani K, Miura N, Miyake N 1983 *Ceramic Fiber Reinforced Piston for High Performance Diesel Engines*, SAE Trans. 830252. Society of Automotive Engineers, Warrendale, PA

Feest E A 1986 Metal matrix composites for industrial application. *Mater. Des.* 7(2): 58–64

Feest E A, Ball M J, Begg A R, Biggs D A 1986 *Metal Matrix Composites Development in Japan*, Report on OSTEM visit to Japan, Oct. 1986. Harwell Laboratory, Didcot, UK

Ferrarini L J, Spence D H, Walker M G 1984 Broadening the limits of reinforced polyurethane RIM. *Proc 1st Int. Conf. Fibre Reinforced Composites*, Paper 25. Plastics and Rubber Institute, London

Fitzer E 1987 The future of carbon–carbon composites. *Carbon* 25(2): 163–90

Goddard P N 1985 The use of new materials in helicopter load bearing structures—Present trends and future predictions. *Proc. 2nd Conf. Materials Engineering.* Institute of Mechanical Engineers, London, pp. 243–51
Gotch T M, Plowman P E R 1978 Improved production processes for manufacture of GRP on British Rail. *Proc. Reinforced Plastics Congr.*, Paper 4. British Plastics Federation, London
Hadcock R N 1986 Design of advanced composite aircraft structures. In: Dorgham M A (ed.) 1986 *Designing with Plastics and Advanced Plastic Composites. Proc. Int. Association for Vehicle Design.* Interscience, Geneva
Hancox N L, Phillips D C 1985 Fibre composites for intermediate and high temperature applications. *Proc. 2nd Conf. Materials Engineering.* Institute of Mechanical Engineers, London, pp. 139–44
Hill J, Thomas C R, Walker E J 1974 Advanced carbon–carbon composites for structural applications. *Proc. Conf. Carbon Fibers, Their Place in Modern Technology,* Paper 19. Plastics Institute, London
Irving R R 1983 Metal matrix composites pose a big challenge to conventional alloys. *Iron Age* 226(2): 35–9
Jaafari A, Holloway L, Burstell M L 1976 Analysis of the use of glass reinforced plastics in the construction industry. *Proc. Reinforced Plastics Congr.*, Paper 27. British Plastics Federation, London
Jamet J F 1987 *Ceramic–Ceramic Composites for Use at High Temperature, New Materials and Their Applications,* Institute of Physics Conference Series No. 89. Institute of Physics, Bristol, pp. 63–75
Jardon A, Costes M 1987 Mass production composites. *Proc. 6th Int. Conf. Composite Materials*, Vol. 1. Elsevier, London, pp. 1.1–1.4
Lea M, Dimmock J 1984 The development of leaf springs for commercial vehicle applications. *Proc. 1st Int. Conf. Fibre Reinforced Composites,* Paper 10. Plastics and Rubber Institute, London
Lubin G, Donohue P 1980 Real life ageing properties of composites. *Proc. 35th Annual Conf. SPI,* Paper 17-E. Society for Plastics Industry, New York
McCarthy R 1986 *Manufacture of Composite Propellor Blades for Commuter Aircraft,* SAE Trans. 850875. Society of Automotive Engineers, Warrendale, PA, pp. 4.606–4.613
Noyes J V 1983 Composites in the construction of the Lear Fan 2100 aircraft. *Composites* 14(2): 129–39
Oliver P C 1980 Applications for composites in the offshore environment. *Proc. 3rd Int. Conf. Composite Materials.* Pergamon, Oxford, p. 2395
Phillips D C 1985 Fibre reinforced ceramics. In: Davidge R W (ed.) 1985 *Ceramic Composites for High Temperature Engineering Applications.* CEC Publication EUR 9565EN. Commission of the European Communities, Luxembourg
Phillips D C 1987 High temperature fibre composites. In: Matthews F L, Buskell N C R, Hodgkinson J M, Morton J (eds.) 1987 *Proc. 6th Int. Conf. Composite Materials,* Vol. 2. Elsevier, London, pp. 1–32
Pinzelli R 1983 Concept, benefits and applications of aramid fiber in hybrid composites. *Proc. 4th Int. Conf. SAMPE European Chapter.* Comité Pour la Développement des Matériaux Composites, Bordeaux, France, pp. 63–77
Preedy J, Kelly J F 1982 An application for Mathweb structures—Development of a temporary overhead electrical gantry by British Rail. *Proc. Reinforced Plastics Congr.*, Paper 23. British Plastics Federation, London
Riddell J C 1985 Composite materials and the sailplane market. *Proc. 2nd Conf. Materials Engineering.* Institute of Mechanical Engineers, London, pp. 253–8
Riebaldi G G 1985 Dimensional stability of CFRP tubes for space structures. *Proc. Workshop Composites Design for Space Applications,* ESA-SP243. European Space Agency Publications Division, Noordwijk, The Netherlands
Riley B L 1986 AV-8B/GR Mk. 5 airframe composite applications. *Proc. Inst. Mech. Eng.* 200(50): 1–17
Ruhmann D C, Mundlock J D, Britton R A 1982 Glasshopper—The fiberglass reinforced polyester covered hopper car. *Proc. 37th Annual Conf. SPI,* Paper 4-E. Society for Plastics Industry, New York
Seamark M J 1981 Facelifting the world's first all-SMC clad truck cab after 5 years—A unique case study. *Proc. 36th Annual Conf. SPI,* Paper 11-C. Society for Plastics Industry, New York
Seymour M W, Tompkins D D 1975 Design approaches to the fibrous glass reinforced polyester bathroom as related to market needs. *Proc. 30th Annual Conf. SPI,* Paper 3-A. Society for Plastics Industry, New York
Suthurst G D, Rowbotham E M 1980 Achieving the impossible—Plastic intake manifold. *Proc. Reinforced Plastics Congr.*, Paper 33. British Plastics Federation, London
Thaw C, Minet R, Zemany J, Zweben C 1987 Metal matrix composite microwave packaging components. *SAMPE J.* 23(6): 40–3
Waterman N A 1986 The economic case for plastics. In: Dorgham M A (ed.) 1986 *Designing with Plastics and Advanced Plastic Composites Proc. Int. Association for Vehicle Design.* Interscience, Geneva, pp. 1–15
Wood A S 1986 The majors are taking over in advanced composites. *Mod. Plast. Int.* 16(4): 40–3
Zweben C 1981 Advanced composites for aerospace applications. *Composites* 12(4): 235–40

D. H. Bowen
[UKAEA Harwell Laboratory, Didcot, UK]

Composites: Manufacturing Overview

Examples of composites involving natural fibers, such as clay bricks reinforced with straw and plaster reinforced with animal hair, can be traced back over many centuries. For significant engineering purposes, however, fiber composites can be said to originate with the development during the 1940s of processes for spinning glass fibers in commercial quantities. Subsequent development during the 1960s and 1970s of methods of manufacturing strong stiff fibers from carbon, aramid polymers and ceramics such as silicon carbide and aluminum oxide has now made feasible the production of a variety of composite materials in which the fibers are used to enhance the properties of metals and ceramics as well as polymers.

To illustrate the range of possibilities, Fig. 1 plots the strength and modulus of materials of construction in bulk form and as fibers and whiskers. It can be seen that conventional materials occupy the lower two decades of strength whereas most fibers display strengths greater by one or two orders of magnitude. Whiskers, as a result of their high degree of crystal perfection, show even higher strengths which in some cases approach theoretical values as calculated from interatomic binding energies.

The physical and mechanical properties of a composite can, to a first approximation, be represented by a simple mixtures law:

$$P_c = P_f V_f + P_m V_m$$

where P_c represents the composite property (for example tensile strength, Young's modulus or density) and P_f and P_m are the values of the corresponding property for the fibers and matrix, respectively. V_f and V_m are the fractions of the total volume occupied by the fibres and the matrix, respectively. The value of V_f typically lies in the range 0.25–0.65, depending on whether the fibers are aligned and closely packed or randomly oriented with respect to each other. Since most synthetic fibers are spun continuously as multifilament bundles or rovings it is relatively easy to achieve high packing fractions in components. Figure 1 indicates that there is then considerable potential to increase the strength, and to a lesser extent the modulus, of many bulk engineering and constructional materials provided suitable manufacturing methods are available to combine the fibers and the matrix. Unlike spun fibers, whiskers tend to be only a few millimeters in length and their organization into parallel arrays is a slow and expensive process. Consequently, they are more frequently used as randomly oriented mats with low packing fractions, and composite properties are correspondingly low despite the high intrinsic properties of the whiskers themselves.

At the present time only polymer-matrix composites have achieved significant market penetration and this is due partly to the relative ease with which fibers and liquid-thermosetting polymers can be combined and processed and partly due to the fact that suitable reinforcements for metal and ceramic matrices have become available only comparatively recently. Consequently, most of the manufacturing techniques described in this article for metal-matrix and ceramic-matrix composites are still at the research and development stage whereas most of the methods outlined for polymer-based composites have become established for industrial production.

It is important to recognize that the manufacture of composite components differs in several respects from, say, the manufacture of metal parts, where material of a specified or controlled composition is supplied by a primary producer and is then shaped,

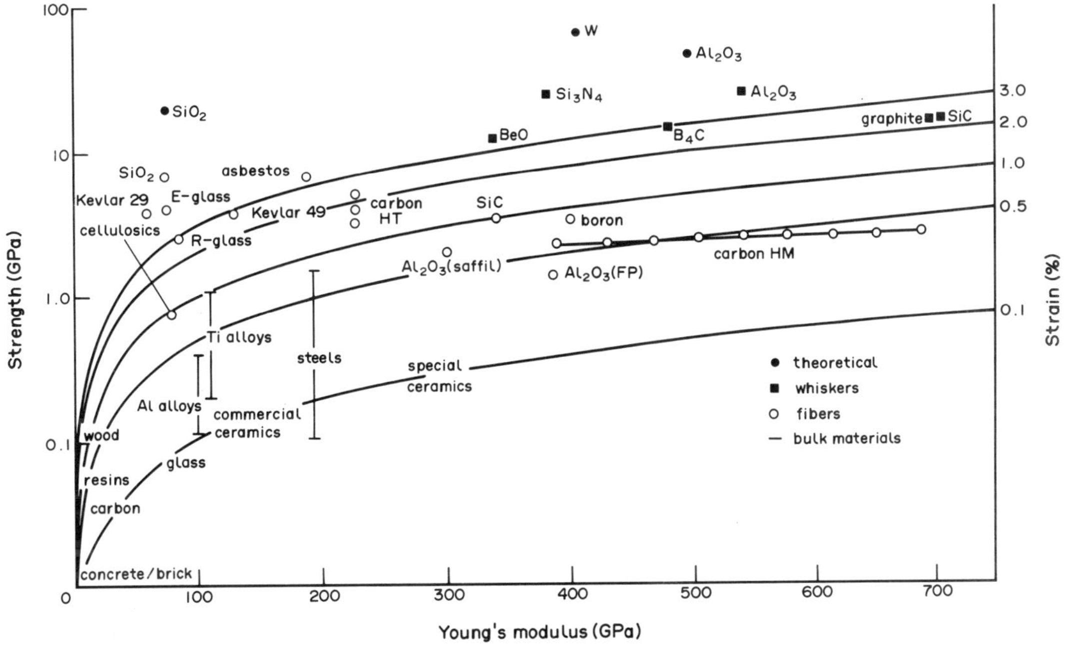

Figure 1
Mechanical properties of various material forms

joined and assembled by a fabricator. During this process, the basic materials properties either remain unaltered or, in the case of heat treatment, are changed in a predictable way. In composites manufacture, although some feedstocks of fiber and resin may be preassembled by a primary producer, the composite material itself is generally produced by the fabricator in the process of forming the component. As a result, the quality of the material may be a function of the type of fabrication process employed and the degree of quality control exercised by the fabricator.

Although the maximum reinforcing effect is achieved with arrays of continuous parallel fibers, the properties in the transverse direction are governed primarily by those of the matrix material and by the strength of the bond between the fiber surface and the matrix. Thus, unidirectionally reinforced materials are highly anisotropic in their properties and the anisotropy must be taken into account in the design of a component and may have an influence on its manufacturing conditions. Anisotropy may be reduced, of course, by randomizing fiber directions or by laminating sheets of parallel fibers with the fibers in each sheet oriented in different directions.

1. Reinforced Plastics

1.1 Feedstock Materials

Manufacturing processes for composites, particularly those with polymer matrices, are intimately bound up with the forms in which the fibers and matrices are available to the fabricator. The range of materials is very wide. For instance, there are at least three types of reinforcing fiber of commercial importance—glass, carbon and aramid—each available in different grades, and there are many types and formulations of resin, both thermosetting and thermoplastic. The fiber length, depending on the mechanical duty required of the component, may fall within the range 1–50 mm or may be effectively continuous; the fibers may all be aligned parallel or laminated in sheets at angles to each other or lie in a plane in random orientations. Thus an enormous number of combinations can be made, depending on the design and cost of the end result. The fabricator has the choice either to combine fibers and resin himself when making the component or to use various types of preassembled materials produced by specialist suppliers. A description of manufacturing processes can therefore usefully be prefaced by an account of the principal forms of feedstock materials. For more detailed descriptions of materials and manufacturing processes see Mohr (1973) and Lubin (1982).

(*a*) *Resins*. Apart from thermoplastics used for injection molding, which are often filled or reinforced with short (less than 1 mm) glass fibers, most reinforced plastics are based on thermosetting resins. A full description is beyond the scope of this article but polyesters (Dudgeon 1987), epoxies (Bauer 1980), phenolics (Knight 1980) and bismaleimides (Stenzenberger 1986) are among the principal types used for composites manufacture. From a manufacturing standpoint, resins of these types can be formulated to cure catalytically or thermocatalytically, and some polyester systems can be cured photolytically: curing can therefore be achieved over a range of temperatures from ambient up to about 200 °C. Choice of resin type depends on a variety of design and manufacturing considerations, not least of which are the viscosity and flow characteristics of the resin as a function of time and temperature. An important step in the production of some types of composite feedstocks is the ability to cure partially (B-stage) a thermosetting resin to convert it from a liquid to a tacky solid, and then to inhibit further curing, usually by refrigeration, until the component has been assembled, when completion of the cure can be achieved by raising the temperature.

Although in terms of volume, thermosetting resins are used predominantly for making composites, interest has developed in recent years in continuous-fiber-reinforced thermoplastics. Candidate materials range from commodity thermoplastics such as polypropylene and nylon to high-performance engineering materials including polysulfone (PS), polyether sulfone (PES), polyetherimide (PEI) and polyether ether ketone (PEEK) (McMahon 1984, Cogswell 1986). Thermoplastic-matrix composites possess certain potential manufacturing advantages over thermosets. For example, feedstocks possess unlimited shelf life at ambient temperature: the fabricator does not have to be concerned with proportioning and mixing resins, hardeners and accelerators as with thermosets; and the reversible thermal behavior of thermoplastics means that components can be fabricated more quickly because the lengthy cure schedules for thermosets, sometimes extending over several hours, are eliminated.

(*b*) *Fibers*. Rovings are the basic forms in which fibers are supplied, a roving being a number of strands or bundles of filaments wound into a package or creel, the length of the roving being up to several kilometers, depending on the package size. The term rovings is usually applied to glass fibers, whereas bundles of continuous carbon or aramid fibers are often referred to as tows, this name reflecting the textile fiber technology involved in their manufacture. A strand of glass fibers contains typically some 200 filaments, each approximately 10 μm in diameter, and the number of strands in the roving will depend on the type of application for which the glass is employed (Waring 1970). Carbon fibers, with similar diameters to glass fibers, are produced in a variety of tow sizes ranging from a few hundred filaments per tow to 320 000 per tow. The commonest sizes in most manufacturers' ranges are 3000, 6000, 10 000 and 12 000 ends (Lovell 1988). Fibers are usually sized during production, partly to protect the surface against mechanical damage during subsequent processing or handling, partly to promote wetting by matrix resins and partly to increase adhesion between the fiber and the matrix material. Glass fibers are usually sized with a complex organosilicon compound with other additives; carbon fibers are often treated with a dilute resin compatible with the matrix resin.

Rovings or tows can be woven into fabrics, and a range of fabric constructions is available commercially, such as plain weave, twills and various satin weave styles, woven with a choice of roving or tow size depending on the weight or areal density of fabric required. Fabrics can be woven with different kinds of fiber, for example, carbon in the weft and glass in the warp direction, and this increases the range of properties available to the designer. One advantage of fabrics for reinforcing purposes is their ability to drape or conform to curved surfaces without wrinkling. Special forms of woven rovings include braids and knitted constructions which can be used to reinforce tubular components and will readily accommodate bends or curves. Indeed it is now possible, with certain types of knitting machine, to produce fiber preforms tailored to the shape of the eventual component. Generally speaking, however, the more highly convoluted each filament becomes, as at crossover points in woven fabrics, or as loops in knitted fabrics, the lower its reinforcing efficiency.

Chopped fibers are used in lengths ranging from less than 1 mm as a reinforcement or filler in thermoplastics for injection-molding purposes, to approximately 50 mm randomly oriented in sheets or mats for the production of laminates with isotropic properties. For handling purposes the fibers in chopped strand mats are held together with a polymeric binder designed to be soluble in the matrix resin. Chopped strand mats composed of glass fibers are one of the most commonly used feedstocks for hand-placement manufacturing processes, being relatively cheap and easy to cut and drape over curved mold surfaces. Thin mats or tissues are often used in the surface layer of moldings to impart a smooth finish.

An alternative use for chopped fibers is in sprayed constructions. Guns are available that simultaneously spray liquid resin and fibers, the latter being derived from a continuous roving that is reduced to appropriate lengths by a chopper within the gun. Spray-up processes are used to build large shell structures on a suitable core and can be controlled robotically to eliminate exposure of operatives to the styrene vapor associated with the polyester resins widely used in glass-reinforced-plastics technology. Spraying of chopped fibers and resin is also sometimes used in conjunction with other manufacturing processes, such as the filament winding of pipes, to introduce a degree of isotropy in the properties of products.

To simplify the handling of fibers and resins for the fabricator, preimpregnated fibers (prepregs) are available from specialist suppliers. The simplest form, continuous warp sheet, consists of an array of parallel rovings or tows in which the filaments have been spread out laterally to produce a uniform distribution within the thickness of the sheet. The sheet is then impregnated with liquid resin to produce a known fiber–resin volume fraction. Prepregs, developed originally with thermosetting resins, are now also available with certain high-melting-point thermoplastics. In the case of thermosets, the resin is partly cured (B-staged) to a slightly tacky condition after impregnation and the sheet surfaces are protected with siliconized paper during storage. Fabrics can also be obtained in the preimpregnated condition and such materials are used typically to build up shell structures by stacking as a preliminary to press or autoclave molding operations. Unidirectional warp sheet is frequently stacked with adjacent layers oriented with respect to each other in such a way as to reduce anisotropy in the final product, rather like the orthogonal grain directions in adjacent veneers in plywood. Warp sheet is also produced in tape of various widths for use in the automated assembly of thin shell structures such as aircraft wing skins and helicopter blades.

Sheet and dough molding compounds are materials which generally comprise polyester resin, an inert particulate filler and chopped fibers, typically in roughly equal proportions by volume. Sheet molding compounds (SMC) are produced by sandwiching a layer of chopped-strand mat between layers of resin mixed with a filler such as talc or chalk, and rolling to ensure penetration of the mixture into the bed of fibers. A material of paste-like consistency results, additives being employed to obtain the appropriate rheological characteristics to ensure the flow of the mixture into mold cavities. Sheets are usually supplied in thicknesses between 3 mm and 10 mm and are protected on each surface until use by a polyethylene film to prevent evaporation of volatile constituents of the resin.

In standard sheet molding compounds containing 25–35% fiber, the fibers are generally short and are oriented randomly to give isotropic properties and good flow characteristics into deeply drawn sections. Mechanical properties are correspondingly poor but can be improved by increasing the fiber content at the expense of the filler. Fiber contents up to 65% can be achieved, in which case the materials are usually referred to as HMCs. Further improvements in mechanical strength can be attained with XMCs, which embody continuous fibers or rovings in two directions at the surface of the sheet. Clearly the presence of continuous fibers modifies the ability of such compounds to flow readily and to conform to complex surfaces.

Dough molding compounds (DMC) or bulk molding compounds (BMC) differ from SMCs principally in employing a shorter fiber length. The compounds are made by combining the constituents in a high-shear type mechanical mixer and the fibers, although short, may provide a greater degree of three-dimensional reinforcement than in sheet molding compounds.

It may sometimes be desirable to prefabricate the mass of reinforcing fibers to fit a shape to be molded, rather than to rely on flow processes to carry fibers to all parts of the mold. For example, in resin-transfer molding processes, the mold is first filled with dry reinforcing fibers and the resin is then injected, ideally to flow around the fibers to fill the mold. However, some fibers may be displaced by the flow of the resin. In these circumstances, the fiber mass may first be shaped to fit the mold cavity, for example by spraying on to a porous screen along with a dilute binder, whose function is simple to hold the fibers to the required shape while impregnation with the matrix resin and molding take place. Such a fiber mass is called a preform.

1.2 Manufacturing Methods

A basic aim in almost all reinforced-plastic manufacturing processes is to avoid the entrapment of air or vapor in the form of bubbles or voids. These represent flaws, often located at the fiber–matrix interface or between plies or sheets of material, and generally reduce the strength of the cured composite. A rule of thumb for glass-fiber-reinforced plastics (GRP) is that the interlaminar shear strength decreases by approximately 7% for every 1% of porosity. The use of low-viscosity resins, the vacuum outgassing of resins prior to application and pressure consolidation of the composite are among methods adopted to reduce porosity in the final product.

(*a*) *Hand and spray placement*. In the simplest form, chopped-strand mat or woven rovings are laid over a polished mold surface previously treated with a release or nonstick agent, and liquid thermosetting resin is worked into the reinforcement by hand with the aid of a brush or roller. Polyester resins are most commonly used with glass fibers because of their low cost and good chemical resistance. Resin and curing agent are mixed prior to application and most systems are formulated to cure at ambient temperature. The principal advantages of hand-lay processes are versatility—there is little limitation on the size and complexity of the molding—and low capital cost, the major investment usually being the cost of the tooling. The major disadvantages are low fiber volume fraction and hence low mechanical properties, partly due to the absence of positive consolidation measures. Spray-up methods offer greater rates of material lay-down in addition to the advantages previously outlined. Both methods are well suited to one-offs or short production runs and an outstanding example of the importance of this type of technology is the boat-building industry where GRP has virtually superseded traditional construction materials. Hand-lay methods have been used for the construction of quite large craft such as minesweepers up to approximately 60 m in length (Dixon et al. 1973).

(*b*) *Press molding*. Molding under pressure in heated matched male/female tools is the most widely used process for the volume production of reinforced plastic components. Most of the feedstocks described in Sect. 1.1 can be molded in this way, but the process is particularly appropriate for use with sheet and dough molding compounds. These materials are formulated first to undergo a rapid reduction in viscosity from their pastelike consistency as the charge of compound heats up by contact with the mold surfaces. This enables the compound to flow smoothly, under molding pressures of the order 3–7 MPa, into complex and deeply reentrant parts of the mold. A rapid gelation and cure then ensues, total cycle times ranging typically from seconds to minutes, depending on the size and thickness of the molding. The process is extensively employed for the production of domestic articles, cabinets and containers for electronic equipment and office machinery, and for automotive parts in sizes up to those of truck doors and cab panels. Figure 2 illustrates schematically a simple press-molding operation.

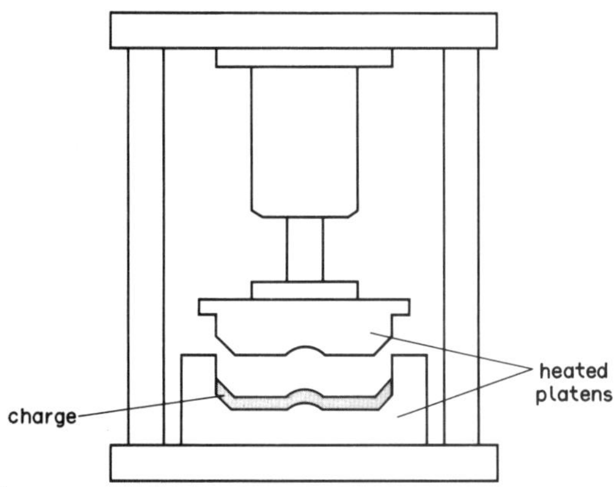

Figure 2
Simple press molding arrangement

Press molding is also used for components reinforced with continuous fibers, but the fibers inhibit flow in the mold and such materials are used primarily for the production of parts of constant wall-thickness and of limited curvature.

In production runs involving many thousands of components, steel or plated-steel tools are employed and for large complex parts can represent a substantial capital investment. For short production runs or the development of prototypes, molds can be made by casting a high-strength epoxide resin around a pattern at a relatively modest cost.

(*c*) *Vacuum molding*. This process makes use of atmospheric pressure to consolidate the material while curing, thereby obviating the need for a hydraulic press. The laminate, in the form of preimpregnated fibers or fabric, or as chopped-strand mat or woven rovings impregnated with liquid resin, is placed on a single mold surface and is overlaid by a flexible membrane which is sealed around the edges of the mold by a suitable clamping arrangement. The space between the mold and the membrane is then evacuated and the vacuum maintained until the resin has cured. Quite large, thin-shell moldings can be made in this way at low cost.

(*d*) *Autoclave molding*. If a higher density and lower void content are required, a consolidating pressure greater than atmospheric pressure may be necessary. The component, laid up on a mold, is enclosed in a flexible bag tailored approximately to the desired shape and the assembly is enclosed in an autoclave—a pressure vessel designed to contain a gas at pressures generally up to approximately 1.5 MPa and fitted with a means of raising the internal temperature to that required to cure the resin. The flexible bag is first evacuated, thereby removing trapped air and organic vapors from the composite, after which the chamber is pressurized to provide additional consolidation during cure. The process is used extensively in the aircraft industry for the production of thin-shell structures of high mechanical integrity, and large autoclaves capable of housing complete wing or tail sections have been installed.

(*e*) *Resin-transfer molding*. The molding processes thus far described start with an open mold or tool which is subsequently closed to surround and compress the charge of fibers and resin. In resin-transfer molding, the fibers only, sometimes as a preform, are enclosed in the mold and the resin is subsequently injected to infiltrate the fibers and fill the cavity. Low-viscosity resins are used to ensure low voidage and good penetration of resin to all parts of the mold, and in vacuum-assisted resin injection, a variant of the process, air in the mold is pumped out prior to injecting the resin. Resin-transfer molding

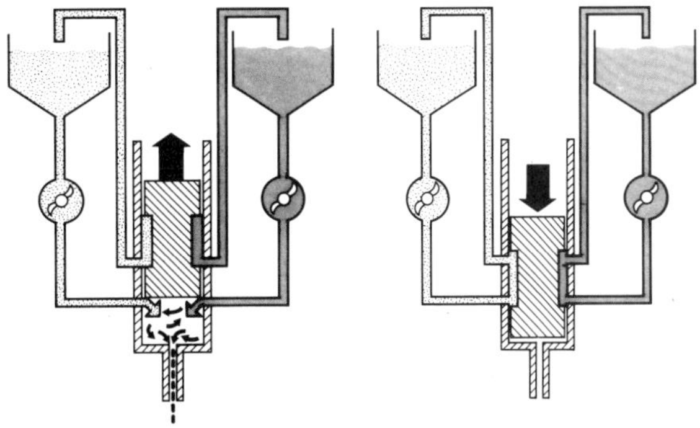

Figure 3
Reaction injection molding process: (a) mixing/injection stage, (b) mixed polymer ejection stage

can be used to produce quite large moldings, such as car body-shells, in small-scale production of a few hundred moldings per year. Compared with press-molding of similar-sized parts, the capital cost is low because the low pressures involved enable the molds themselves to be made from GRP.

(*f*) *Reaction injection molding*. Reaction injection molding (RIM) and reinforced reaction injection molding (RRIM) also use relatively low pressures (approximately 0.5–1 MPa) to produce large moldings, principally in polyurethane elastomers (Becker 1979). The principal difference from resin-transfer molding is that, instead of using a precatalyzed resin with a relatively slow cure, the RIM process brings two fast-reacting components together and mixes them just prior to injection into the mold. The mixed system can be tailored to cure in the mold within 30–60 s, thus giving rise to component cycle times of the order 1–2 min. The basis of the process is illustrated in Fig. 3. The equipment comprises two reservoirs storing, typically, a diisocyanate and a polyol, which polymerize when intimately mixed. Figure 3a shows the mixing/injection stage during which the two reactants are pumped through an impingement chamber in the mixing head, and from there to the mold cavity. The impingement chamber produces highly turbulent flow conditions and hence rapid mixing of the two constituents, and this is followed by rapid gelation and cure in the mold. In the second stage, represented by Fig. 3b, all the mixed polymer is ejected from the mixing head and the two reactants are separated and continuously recirculated ready for the next molding cycle. The mixing head is thus self-cleaning and any cured polymer in the connecting pipe from the mixing head to the mold cavity is arranged to be removed as a piece of flash attached to the molding.

A variety of urethane formulations is available to give polymers with a range of hardness from soft and

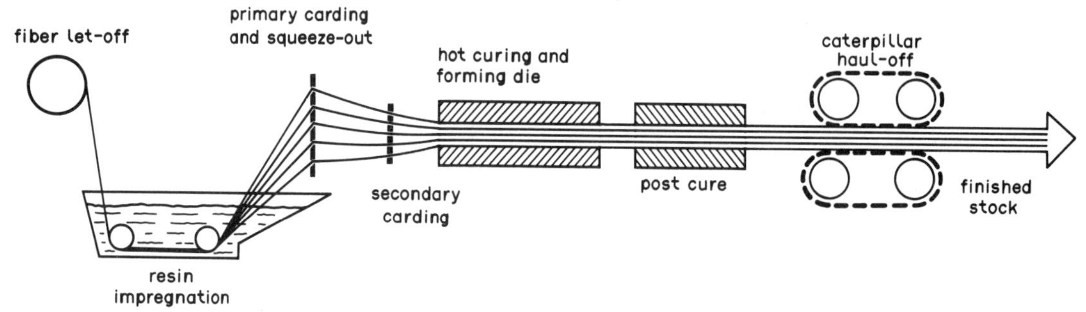

Figure 4
Schematic of pultrusion process

rubbery to hard and brittle. The moldings can be reinforced to some degree with short glass fibers by including them in one or both of the liquid reactants to form a slurry. The limitation on fiber length and hence reinforcing efficiency, and the quantity of fiber that can be incorporated, is set by the design of the mixing head and its ability to handle the fibers without risk of blockage or of undue agglomeration of the fibers, resulting in a nonuniform distribution in the molding. RIM and RRIM processes compete in many ways with press-molded sheet and dough molding compounds for markets in office and domestic equipment and for automotive products such as bumpers, body and trim panels, and both materials and equipment are the subject of intensive development at the present time.

(*g*) *Pultrusion*. The processes thus far described produce discrete moldings. Pultrusion enables profiles of constant cross section to be manufactured continuously. The process is analogous to the extrusion of plastics and nonferrous metals in that the profile is shaped by continuous passage of the feedstock through a forming die, but in the case of pultrusion the reinforcing fibers are used to pull the material through the die. In Fig. 4, continuous rovings are passed through a bath of resin followed by a series of carding plates to encourage complete penetration of resin into the fiber bundles and to remove excess resin. The impregnated fibers are then drawn through a heated die in which the resin first gels and then cures. The finished profile finally passes through a traction mechanism before being cut to length.

Glass fibers and polyester resins are the most commonly employed materials because of their relatively low cost and good corrosion resistance. Preheating by radio-frequency methods is often used to increase the throughput, and production rates of several meters per minute are achieved depending on section thickness. The process is ideally suited to the production of structural elements of high strength and stiffness owing to the presence of the high proportion of axial fibers necessary to sustain large tractive forces, particularly when high volume fractions of fiber (up to approximately 0.7) are required in the cross section. It is, however, possible to incorporate varying cross-sectional geometry although of constant cross-sectional area.

(*h*) *Filament winding*. In this process, strong stiff shell-structures can be made by winding bands of continuous fibers impregnated with resin on to a former or mandrel which can be withdrawn after the resin has been cured. The method is analogous to the technique of wire wrapping for the reinforcement of steel pressure-vessels and gun barrels and is extensively used for the manufacture of reinforced-plastic pipe, and for cylindrical tanks and vessels in GRP for the storage of aggressive chemicals and a wide variety of agricultural products. The size of component that can be manufactured is limited only by the capacity of the winding machine, and methods also exist for the building of large vertical silos *in situ*, involving the construction of a circular railway track upon which a carriage, dispensing fiber and resin, travels continuously around a vertical former.

Modern filament-winding machinery is capable of producing shapes that are much more complex than the simple axisymmetric structures indicated above, largely as a consequence of adopting robotics technology. Figure 5 shows schematically a gantry-type filament-winding machine with five axes or machine motions that are controlled independently by a microprocessor. Two such motions, rotation (θ_1) of a mandrel between head and tailstock, and translation (x) of a carriage backwards and forwards parallel to the axis of mandrel rotation, suffice for the winding of cylindrical shells at helix angles dictated by the relative rates of rotation and translation. The additional facility of movement of the feed eye along the y and z directions, however, enables a band of rovings to be laid in a controlled way on convex surfaces of any given shape, and rotation of the feed

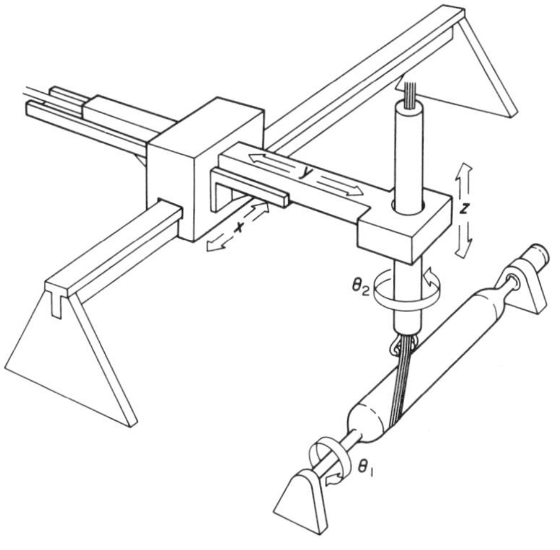

Figure 5
Basic motions of a 5-axis filament winding machine

eye (θ_2) enables the plane of the band to be matched to the inclination of the mandrel surface at any given point.

For a stable winding it is necessary to identify geodesic paths, defined as the shortest distance between two points on a surface, to avoid the possibility of the band of fibers slipping from the desired trajectory. This condition, coupled with the sheer magnitude of the task of programming a multiaxis robot to traverse a constantly changing and nonrepetitive path, has led to the development of computer-based methods aimed at integrating design and manufacturing procedures for filament-wound components by analogy with CADCAM processes for metal machining (Wells and McAnulty 1987). This type of technology opens up the possibility of winding components as complex as aircraft fuselages, steering wheels and bicycle frames.

2. *Reinforced Metals*

Reference to Fig. 1 shows that the mechanical properties of the nonferrous metals and alloys commonly used for structural purposes are considerably better than those of unreinforced polymers, and the improvements to be obtained by combining them with fibers or whiskers will therefore be less dramatic than with polymers. However, other worthwhile improvements can be achieved by incorporating fibers or even particulates into a metal matrix. For instance, creep resistance at a particular temperature may be improved or the resulting material may be more wear and abrasion resistant, or less susceptible to fatigue failure. Such improvements, even if modest, may be of great technological importance if, for example, the temperature limit for continuous operation of an engine component or an airframe structural part can be increased. Considerable research is now being devoted to both the development of reinforcements suitable for metal matrices, and of manufacturing methods. Much of the drive behind the upsurge in metal-matrix composites development derives from the recent development of processes for producing refractory fibers and whiskers in commercial quantities (Bunsell 1987).

2.1 *Feedstock Materials*

(*a*) *Matrix materials*. The most common matrices are the low-density metals, such as aluminum and aluminum alloys, and magnesium and its alloys. Some work has been carried out on lead–tin alloys, mainly for bearing applications, and there is interest in the reinforcement, for example, of titanium-, nickel- and iron-base alloys for higher-temperature performance. However, the problems encountered in achieving the thermodynamic stability of fibers in intimate contact with metals become more severe as the potential service temperature is raised, and the bulk of development work at present rests with the light alloys.

(*b*) *Reinforcements*. The principal reinforcements for metal matrices include continuous fibers of carbon, boron, aluminum oxide, silica, aluminosilicate compositions and silicon carbide. Some ceramic fibers are also available in short staple form, and whiskers of carbon, silicon carbide and silicon nitride can be obtained commercially in limited quantities. There is also interest in the use of refractory particles to modify alloy properties such as wear and abrasion resistance. In this case, particle sizes and volume fractions are greater than those developed metallurgically in conventional alloys, and incorporation of the particles into the metal is achieved mechanically rather than by precipitation as a consequence of heat treatment.

Most metal-matrix composites consist of a dispersed reinforcing phase of fibers, whiskers or particles, with each reinforcing element ideally separated from the next by a region of metal. An alternative approach is to reinforce the metal, generally in sheet form, with a fiber-reinforced polymer. An example is ARALL (Vogelesang and Gunnink 1983), a laminate consisting of alternate plies of aluminum alloy bonded to sheets of aramid-fiber-reinforced resin. This has the advantage of a simple low-temperature production process in which the continuous parallel fibers can display a high reinforcing efficiency. Applications are, however, restricted to thin shell-type structures.

Apart from sheet materials similar to ARALL, the production of feedstocks in which reinforcement and matrix can be purchased preassembled for the convenience of the fabricator is less well advanced than the production of polymer-matrix composites. Wires or tapes of aluminum containing up to 500 continuous silicon carbide filaments are available in development quantities and in some cases reinforcements can be obtained woven into fabrics or agglomerated into randomly oriented mats.

2.2 Manufacturing Methods

Many of the processes used for the production of unreinforced metal parts, or for the deposition of metal layers, have been explored or adapted with the aim of producing a continuous, void-free metal matrix with a controlled distribution of reinforcements. Fabrication methods can involve processing the metal in either the molten or the solid state, and components can be formed either by direct combination of matrix and reinforcement or by the production of a precursor composite which, in the form of composite wires, sheets or plies, is used to build up a component. In the latter case, the assemblage of plies must be consolidated and bonded in a subsequent process.

In liquid-metal techniques, composites can be prepared by infiltrating mats or fiber preforms with liquid metals or, under carefully controlled conditions, by physically mixing the reinforcement and the liquid metal together. A pseudoliquid route is offered by plasma or flame spraying in which metal-powder particles are heated above their melting point and are sprayed onto an array of fibers on a thin sheet of the same matrix metal. The resulting sheet of fiber-reinforced metal can then be stacked with other sheets and consolidated in a subsequent operation. An alternative approach is to build up a solid body by codepositing metal and particulate reinforcements by spraying techniques.

The simplest solid-state preparation route is to mix short fibers or particulates with metal powders. The mixture can then be processed by standard powder-metallurgical procedures. Alternatively, the metal may be coated onto the reinforcement by electrochemical or chemical-vapor deposition methods.

Many of these processes are still under development or evaluation and, as with polymer-matrix composites, the methods of manufacture that will eventually be adopted commercially will depend on production costs as well as technical suitability for particular components. For this reason, considerable attention is being focused on infiltration and casting processes because of their relatively low cost. For detailed reviews of manufacturing techniques see Mileiko (1983) and Chou et al. (1985).

(*a*) *Liquid-metal infiltration*. Direct impregnation of bundles of continuous filaments has been used to produce precursor wires or tapes that can be subsequently consolidated into component forms. A typical scheme involves drawing a bundle of filaments through a bath of molten metal and then through an orifice which shapes the impregnated bundle to a circular or rectangular cross section. For complete impregnation to take place it is important that the metal should wet the fiber surface, and considerable research has been devoted to coatings for fibers or additives to the metal that will reduce the contact angle of the molten metal at the fiber surface and promote wetting. Penetration of fiber bundles may be enhanced by infiltrating under vacuum or using inert-gas pressure over the molten metal (Masur et al. 1987).

(*b*) *Squeeze casting*. In this process, a mass or preform of fibers or whiskers is infiltrated with molten metal under a high pressure derived from a hydraulic press. The method is illustrated schematically in Fig. 6 and can result in a fully dense composite with good dispersion of the reinforcement. Whisker concentrations of up to 30% by volume have been achieved, resulting typically in increases in strength and Young's modulus of a factor of two over unreinforced aluminum alloys. The method can be used to produce billets that can then be shaped by conventional metal hot-working techniques such as extrusion or forging, or it can be used directly for near net-shaping of components such as automotive connecting rods. With some materials it may be desirable for the infiltration to be carried out under vacuum, to avoid either the risk of oxidation of the reinforcement or the possibility of entrainment of air bubbles in the matrix.

(*c*) *Stir casting and compo-casting*. In some cases it is possible to introduce particulate or short-fiber reinforcements simply by feeding into the vortex region produced by vigorously stirring the molten metal. Long stirring times may be necessary to develop a good bond between the two components, and the volume fraction of the reinforcement may be limited to about 10% by the rapid increase in the viscosity of the mixture with the further addition of reinforcement.

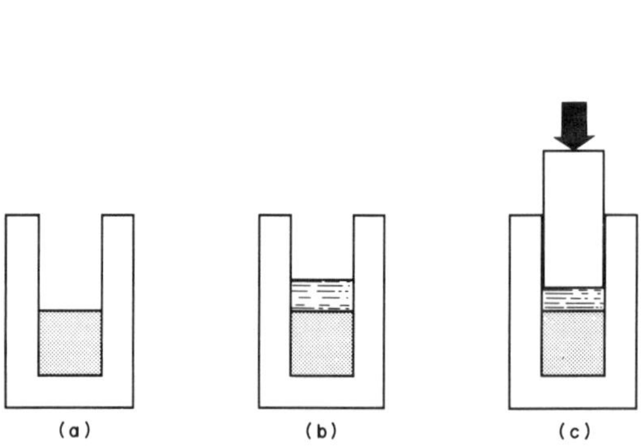

Figure 6
Schematic of squeeze casting process: (a) fiber preform placed in heated mold, (b) liquid metal poured into mold and (c) infiltration under pressure

Differences in density between molten metals and reinforcing fibers or whiskers may mean that separation of the two components will occur due to gravitational force if simple mixing of the fibers in molten metal is attempted. The tendency to separate can be reduced or eliminated in compo-casting whereby stirring takes place as the metal undergoes solidification. Under these conditions, mixed phases of solid and liquid metal can coexist in the form of a slurry in which reinforcement particles can be mixed and become trapped between the solid-phase regions with uniform dispersion.

(*d*) *Precursor production*. Preimpregnated sheet that can be used to build up more complicated structures can be made in a variety of ways. The most common method using continuous monofilament or bundles of filaments is to assemble them with the required spacing by winding on to a drum or around pegs to produce a parallel array on a flat surface. The metal is then coated onto the filaments by plasma spraying, electroplating or by chemical or chemical vapor deposition techniques. Alternatively, bundles or tows of filaments can be coated by electrochemical or plasma-activated deposition processes. These various precursor forms must then be shaped, stacked, consolidated and bonded to form a continuous matrix phase.

(*e*) *Consolidation and bonding methods*. Densification and bonding of sheets of precursor materials or of mixtures of metal powder and reinforcement are usually carried out in the solid state under conditions of temperature and pressure such that consolidation results from a combination of plastic deformation and diffusion in the metal phase. Methods of placing the material under the appropriate conditions of temperature and pressure include conventional hot-pressing, hot and cold isostatic pressing, hot rolling and explosive welding. As an example see Hack and Amateau (1982) who describe diffusion bonding of carbon-fiber-reinforced aluminum alloy wires to aluminum alloy foils in a press at a temperature between 554 °C and 562 °C and a pressure of 24 MPa for a dwell time of 20 min. An account of the technology of boron-reinforced aluminum composites, including details of diffusion bonding, has been given by Miller and Robertson (1977).

Many of the composites incorporating short-fiber or particulate reinforcements are produced in the form of billets by the methods described above and may require subsequent processing by standard metallurgical techniques such as extrusion, forging or drawing to achieve their final form.

3. Reinforced Ceramics

Research into the properties of fiber-reinforced ceramics has been in progress since the 1950s, but until the advent of refractory, oxidation-resistant fibers such as alumina and silicon carbide in commercial quantities, much of the work was confined to developing an understanding of the principles of reinforcement and mechanisms of failure (Phillips 1983). As with metal-matrix composites, there is currently an increase in research and development activity, although little commercial application of materials thus far. This description of manufacturing methods consequently relates to laboratory-type processes.

3.1. Feedstock Materials

(*a*) *Matrix materials*. The choice of matrix depends primarily on the mechanical and thermal duty required of the composite. For potential applications in aerospace and in heat engines, oxidation resistance is of prime importance and most refractory oxides, carbides and borides have been considered as matrices. The most extensively employed, however, have been aluminum oxide, silicon carbide and silicon nitride, mainly because the technology of producing well-consolidated artifacts in the unreinforced state is better established for these materials. Glasses and silica have also been explored because of the relative ease with which matrix consolidation can be accomplished by processing in the liquid state. Glass-ceramic materials that can be combined with reinforcements while in a liquid state and then devitrified by heat treatment have also been the subject of considerable investigation.

Although not a ceramic, carbon is one of the most refractory materials in nonoxidizing atmospheres, and Fig. 1 indicates that substantial improvement to the mechanical properties of bulk carbon is feasible by fiber reinforcement. In samples unidirectionally reinforced with carbon fibers, tensile strengths of up to 1 GPa have ben reported (Hill et al. 1974). For some applications, reinforcement in three dimensions is required and weaving or knitting processes have been developed for producing continuous-carbon-fiber preforms that can subsequently be invested with a carbon matrix.

(*b*) *Reinforcements*. Only a few of the fibers available commercially are suitable for ceramic-matrix composites by virtue of their compatibility and stability with the matrix during fabrication and at intended service temperatures. Continuous fibers include silicon carbide, boron, aluminum oxide, aluminosilicates and carbon (Bunsell 1987). In oxidizing atmospheres, only silicon carbide and the oxide fibers offer the prospect of high-temperature composites and, depending on the stress to which they are subjected, the upper limit of ceramic-matrix composites is currently in the region of 1200 °C.

(*c*) *Manufacturing methods*. Hot-pressing of sheet-material preforms is the route most widely adopted for ceramic-matrix composites, and hot-pressing conditions generally follow established procedures for consolidation and sintering of monolithic ceramics. Early work has been reviewed by Sambell (1970) and a method of producing sheet preforms is described by Sambell et al. (1974). More recently, chemical vapor deposition (CVD) methods have been adopted to infiltrate a fiber preform with, for example, silicon carbide derived from the gaseous-phase cracking of methyl trichlorosilane in hydrogen. Alternatively, the preform may be impregnated with a liquid organometallic compound which is then pyrolyzed to yield a refractory ceramic residue (Bernhart et al. 1985).

Similar methods are used for the manufacture of carbon-fiber-reinforced carbon materials (Fitzer 1987). Preforms are either coated with carbon produced in the vapor phase by reacting, say, methane and hydrogen, or are subjected to impregnation of a polymer possessing a high carbon yield when pyrolyzed. One of the major disadvantages of both CVD and pyrolysis routes is that successive depositions are necessary to produce a fully dense matrix. The processes are therefore slow and correspondingly expensive. Nevertheless, such processes are operated commercially, notably for the manufacture of carbon–carbon disk brakes for aircraft and for racing cars.

4. Conclusion

Within the scope of this article it has been possible to describe only briefly some of the methods used for manufacturing composite materials and components. In practically every case, however, the aim is to control the disperson of the reinforcement in the matrix in such a way as to maximize the properties of the composite while minimizing the manufacturing cost. The great variety of materials and material forms involved in composites technology provides considerable promise for the evolution of materials with improved performance, but the availability of cost-effective manufacturing methods capable of providing products with a high degree of consistent and reproducible performance is a key factor in determining the rate at which they can compete with more traditional materials of construction.

See also: Composites: Applications Overview; Sports Equipment Manufacture

Bibliography

Bauer R S 1980 The versatile epoxies. *Chem. Technol.* 9: 692–700

Becker W E 1979 *Reaction Injection Molding*. Van Nostrand Reinhold, New York

Bernhart G, Chateigner S, Heraud L 1985 Les composites ceramique–ceramique, du concept à la piece. In: Bunsell A R, Lamicq P, Massiah A (eds.) 1985 *Developments in the Science and Technology of Composite Materials*. European Association for Composite Materials, Bordeaux, France, pp. 475–81

Bunsell A R 1987 Fibre reinforcements—Past, present and future. In: Matthews F L, Baskell C R, Hodgkinson J M, Morton J (eds.) 1987 *Proc. 6th Int. Conf. Composite Materials*, Vol. 5. Elsevier, London, pp. 1–13

Chou T W, Kelly A, Okura A 1985 Fibre-reinforced metal-matrix composites. *Composites* 16(3): 187–206

Cogswell F N 1986 Continuous fibre reinforced thermoplastics. In: Collyer A A, Clegg D W (eds.) 1986 *Mechanical Properties of Reinforced Thermoplastics*. Elsevier, London, pp. 83–119

Dixon R H, Ramsay B W, Usher P J 1973 Design and build of the GRP hull of HMS Wilton. *Proc. Symp. GRP Ship Construction*. Royal Institution of Naval Architects, London, p. 1

Dudgeon C A 1987 Polyester resins. *Materials Engineering Handbook*, Vol. 1. American Society for Metals, Metals Park, OH, pp. 90–6

Ewald G 1981 Curved pulforming—A new manufacturing process for composite automobile springs. *Proc. 36th Ann. Conf. Reinf. Plastic/Composites Inst.*, Paper 16C. Society for Plastics Industry, New York

Fitzer E 1987 The future of carbon–carbon composites. *Carbon* 25(2): 163–90

Hack J E, Amateau M F 1982 Mechanical behavior of metal matrix composites. *Proc. AIME Conf. Dallas, Texas. 16–18 Feb. 1982*. American Institute of Mining Engineers, New York

Hill J, Thomas C R, Walker E J 1974 Advanced carbon–carbon composites for structural applications. *Carbon Fibers, their Place in Modern Technology*, Paper 19. The Plastics Institute, London, pp. 122–30

Knight G J 1980 High temperature properties of thermally stable resins. In: Pritchard G (ed.) 1980 *Developments in Reinforced Plastics*, Vol. 1. Applied Science, London, pp. 145–210

Lovell D R 1988 *Carbon and High Performance Fibres Directory*, 4th edn. Pammac Directories, High Wycombe, UK, pp. 45–74

Lubin G (ed.) 1982 *Handbook of Composites*. Van Nostrand Reinhold, New York

McAdams L V, Gannon J A 1986 Epoxy resins. *Encyclopedia of Polymer Science and Engineering*, Vol. 6, 2nd edn. Wiley, New York, pp. 322–82

McMahon P E 1984 Thermoplastic carbon fibre composites. In: Pritchard G (ed.) 1984 *Developments in Reinforced Plastics*, 4. Elsevier, London, pp. 1–30

Masur L J, Mortensen A, Cornie J A, Flemings M C 1987 Pressure casting of fiber-reinforced metals. In: Matthews F L, Buskell N C R, Hodgkinson J M, Morton J (eds.) 1987 *Proc. 6th Int. Conf. Composite Materials*, Vol. 2. Elsevier, London, pp. 320–9

Mileiko S T 1983 Fabrication of metal-matrix composites. In: Kelly A, Mileiko S T (eds.) 1983 *Handbook of Composites*, Vol. 4. North-Holland, Amsterdam, pp. 221–94

Miller M, Robertson A 1977 Boron/aluminum and borsic/aluminum analysis, design, application and fabrication. In: Renton W J (ed.) 1977 *Hybrid and Select Metal Matrix Composites: A State of the Art Review*. American Institute of Aeronautics and Astronautics, New York, pp. 99–157

Mohr J G (ed.) 1973 *SPI Handbook of Technology and Engineering of Reinforced Plastics/Composites*, 2nd edn.

Van Nostrand Reinhold, New York
Phillips D C 1983 Fibre reinforced ceramics. In: Kelly A, Mileiko S T (eds.) 1983 *Fabrication of Composites*, Vol. 4. Elsevier, Amsterdam, pp. 373–428
Sambell R A J 1970 The technology of ceramic-fibre ceramic-matrix composites. *Composites* 1(5): 276–85
Sambell R A J, Phillips D C, Bowen D H 1974 The technology of carbon fibre reinforced glasses and ceramics. In: *Carbon Fibres, their Place in Modern Technology*, Paper 16. The Plastics Institute, London, pp. 105–13
Stenzenberger H 1986 Bismaleimide resins. In: Kinloch A J (ed.) 1986 *Structural Adhesives—Developments in Resins and Primers*. Elsevier, London, pp. 77–126.
Vogelesang L B, Gunnink J W 1983 ARALL, a material for the next generation of aircraft—A state of the art. In: Jube G, Massiah A, Naslain R, Popot M (eds.) 1983 *High Performance Composites Materials: Proc. 4th Int. Conf. Society for the Advancement of Materials and Process Engineering, European Chapter*. SAMPE, Bordeaux, France, pp. 81–92
Waring L A R 1970 Reinforcement. In: Parkyn B (ed.) 1970 *Glass Reinforced Plastics*. Butterworth, London, pp. 121–45
Wells G M, McAnulty K F 1987 Computer aided filament winding using non-geodesic trajectories. In: Matthews F L, Buskell N C R, Hodgkinson J M, Morton J (eds.) 1987 *Proc. 6th Int. Conf. Composite Materials*, Vol. 1. Elsevier, London, pp. 161–73

D. H. Bowen
[UKAEA Harwell Laboratory, Didcot, UK]

Computer Control in Polymer Processing

The growing requirements placed on plastics processing in terms of materials utilization, manufacturing time and cost, and reproducible product quality, together with the continued penetration of microelectronics into all fields of technology, are bringing an ever-increasing degree of automation into production.

The application of microprocessors and microcomputers has made it possible and economically worthwhile to replace the conventional control and regulation systems of plastics-processing machines by processor systems (Dormeier 1977, Schwab et al. 1979, Bongardt et al. 1980). Development and practical application have concentrated both on special single-function processor systems, designed for tasks such as the direct digital control of one or more temperature control circuits, and on multiprocessor systems for the full-scale operation, control and regulation of plastics-processing machines.

This article covers a number of economically interesting and technically highly developed process control systems using microcomputers, taken as representative of the application of microelectronics in plastics processing.

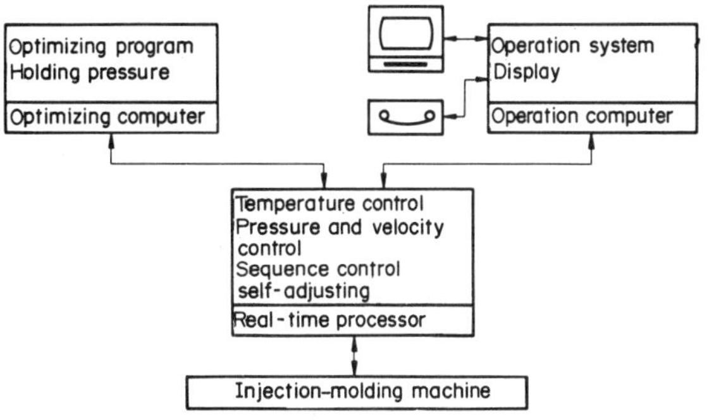

Figure 1
Computer control of injection-molding machine

1. Microcomputers in Injection Molding

A microcomputer system for the automation of injection-molding machines has been developed at the Institut für Kunststoffverarbeitung (IKV) at Aachen Technical University in conjunction with Lehrgebiet für Prozessdatenverarbeitung (PDV) at the same university and with industry (Bongardt et al. 1980). The microcomputers in the system assume the following individual functions (see Fig. 1):

(a) system command and display of chief process quantities (operation computer);

(b) machine sequence control, zone wall temperature regulation, injection velocity and internal mold pressure regulation, fully automatic self-adaptive control parameters (real-time computer); and

(c) optimization of pressure holding phase in process technology terms (optimizing computer).

The operator–machine dialog is effected through a video display unit with aid of the operation computer. When the machine needs to be reset, the operator is guided through the setting-up process by the microcomputer system. The necessary inputs are requested and erroneous inputs are ignored and notified to the operator. In this way, machine setting errors are generally eliminated. Once the machine has been set initially, any target value desired can be accessed via the free-setting facility. Target values and actual values are set out on sheets, suitably arranged on the basis of the process technology.

Target value profiles and actual value curves are given in graphical form as bar charts. Once the injection-molding machine has completed its test runs, the setting data obtained can be stored on a magnetic tape cassette and used subsequently in automatic machine setting.

The operation computer likewise organizes the data transfer between the operation computer and the real-time processor.

1.1 Sequence Control

The real-time processor itself monitors the machine control sequence. For control sequence purposes the injection-molding cycle is divided into characteristic subphases, that is, machine states. These machine states are: 0—mold closed, 1—injection unit forward, 2—injection and so on.

Each subphase is handled by a special basic block in the automatic cycle, which is dedicated to this phase alone. In this way the operator knows which control signals to expect in each phase during normal control sequencing and an erroneous machine sequence is immediately recognizable by the appearance of unexpected control signals. In this event, a clear-text output will follow, indicating the machine state in which the error occurred, the control signal that was expected but not received and the unexpected control signal that signalled the sequencing error. The control signals for the sequence control are generated by a program from the signals emitted by the displacement transducers on the clamping unit and screw, individual limit switches, meters, hydraulic pressure signals and internal mold pressure signals.

Once the correct signals have been generated during a specific machine state, the sequence control switches machine functions in and out in the correct order. This involves the switching of magnetic valves and the screw advance, holding pressure and back pressure controls. The subsequent state (the next basic block) with which the next control signals are to be processed is determined as a function of the current machine state.

1.2 Temperature Regulation

The temperatures of cylinder heating zones 1, 2 and 3 and the melt temperature in front of the screw are regulated and monitored to ensure that they are within upper and lower limits. The mold wall temperatures are controlled by two or more temperature control units, whose target values are controlled by the real-time processor. The target and limit values for the digital temperature control are preset by the operator. Temperature measurement follows solely by means of a precision amplifier with reference junction and an analog input into the microcomputer. The microcomputer selects the different measurement points via an electronic analog multiplexer.

1.3 Screw Advance Velocity

Screw advance velocity is controlled by the microcomputer using a servo valve and an inductive displacement transducer. A screw advance velocity profile can be preset, as a function of the screw path, through ten support points. By selection of the profile it is possible to set a desired flow front velocity for the melt, independent of molding geometry and outside disturbance. After the operator has selected the corresponding function type, the real-time processor will adjust the screw advance controller to its optimum setting within five to ten shots.

1.4 Maximum Pressure Regulation

In addition to the familiar time-, path- or pressure-dependent switching from the injection phase to the pressure holding phase, a new control strategy has been developed using the microcomputer system (Bongardt and Menges 1980). In this case, the switch from the injection phase to the pressure holding phase is no longer triggered by the arrival at a certain switching threshold but is continuously controlled.

It is thus possible to set the maximum inside mold pressure for each shot at a level determined by the operator, which is reproducible and independent of machine and environment-conditioned influences. This serves to avoid pressure peaks, which could damage the molding, and the formation of flash, and also ensures good filling of the mold. It is no longer necessary to seek the pressure switchover point at which the maximum desired inside mold pressure is just reached at the end of the compression phase. Temperature changes and intermittent machine stoppages no longer have any effect on the maximum pressure inside the mold. The same applies to fluctuations in injection velocity on machines without screw advance velocity control. On machines fitted with injection control there is no need to adjust the pressure switchover point when the advance velocity target value is changed.

1.5 Holding Pressure Setting

Holding pressure can be regulated through a servo valve as a function of pressure inside the mold, hydraulic pressure or locking pressure. It is also possible to switch from internal mold pressure-governed control to hydraulic pressure-governed control after a preset period. The operator can set a time-dependent pressure profile with ten support points for each form of control. Once the operator has made the selection, the holding pressure optimization program in the optimizing computer calculates the pressure profile inside the mold and transmits it to the holding pressure regulator through the real-time processor. In this case, it is not necessary for the operator to feed in the pressure profile.

1.6 Control Parameter Setting

Upon request, the real-time processor can proceed to an optimum adjustment of all the control parameters to changed process conditions. Different materials and working temperatures in the cylinder heating zone and in the mold can, under certain circumstances, lead to considerable changes in the path properties of the injection-molding machine during the injection and pressure holding phase. If the controller has been set at a fixed setting, this will modify the dynamic behavior of the corresponding control systems. In order to achieve optimum control procedures, with which the preset target value profile is actually observed, the control parameters need to be adjusted in the event of a change in production conditions. This is executed by the microcomputer system itself by a program. The program assesses the dynamic quality of the control process from the control difference profile that is measured between the target value and the actual value during each cycle. The absolute value of the control difference is thereby integrated over the screw path for the screw advance velocity control and over time for the pressure holding control. The control parameters are adjusted from shot to shot to give a better-running control process, up to the point where the optimum values have been found. This takes some six to eight shots.

1.7 Holding Pressure Optimization Program

The holding pressure phase can be said essentially to affect the mass and shrinkage quality features of the molded part and also on macromolecular orientation in the inner zone of the molding. Applying the *p*, *V*, *T* method developed at the IKV, the machine setting for the pressure holding phase is calculated in such a way as to give the desired molding shrinkage without any orientation in the center of the molding (Vargel 1974). In addition, long-term temperature fluctuations can be compensated through spontaneous pressure adjustment.

Once the optimum course has been established for the thermodynamic melt state variables during the pressure holding phase and, after calculation of the cooling conditions in the molding, the optimizing computer proceeds, as already indicated, to determine the pressure level inside the mold which will give the preset state course. This information is then transmitted to the real-time processor. The target point of the process in the state diagram is a mean specific molding volume at ambient pressure which will produce the desired molding volume shrinkage at ambient pressure. The optimum state course is also always established in such a way, with consideration of other secondary conditions like maximum permissible pressure inside the mold, that melt backflow during the pressure holding phase is avoided. The orientation that develops in conjunction with melt backflow, particularly just before the sealing point, is thereby also avoided. The data required for optimization are fed into the optimizing computer through a dialog or by a data carrier, or are called up for the process through the real-time processor. The real-time processor determines the mean melt temperature and the mean mold wall temperature for the cooling calculation from measurements recorded by the temperature probes in front of the screw and in the mold. Any changes in these temperatures, such as can result during machine start-up or during large-scale disturbances, and which cannot be offset quickly enough by the temperature controller, are smoothed out by this holding pressure optimization program on a shot-by-shot basis, thereby giving immediate compensation.

1.8 Mold Design

New developments in computer-aided mold design have been described by Menges and Recker (1989), concentrating chiefly on layouts of thermoplastic and elastomer injection molds. An overall analysis of the design process is first presented, followed by simulation of filling methods. An analysis and simulation of the holding pressure phase in thermoplastics injection molding is presented in the third section. The fourth section considers thermal analysis of the processing methods. Finally, mechanical analysis of molds is discussed.

2. Microcomputers in Blow Molding

A multiprocessor computer system has been developed at the IKV which allows additional functions to be fulfilled, over and above the standard functions that have been executed in extrusion blow molding to date (Schwab and Menges 1979). The system was designed for the following areas of application (see Fig. 2):

(a) central display and command system with optimization (microcomputer 1);

(b) DDC temperature control for ten zone wall temperatures, including three zones for heating and cooling, cascade control for melt temperature (microcomputer 2);

(c) adaptive parison length control with subordinated screw rotational velocity control (microcomputer 3);

(d) screw rotational velocity control with ac and dc shunt motors (microcomputer 4);

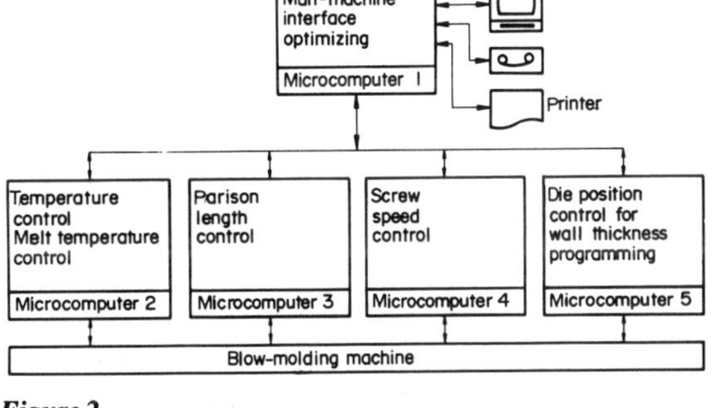

Figure 2
Computer control of blow-molding machine

(e) control of nozzle position for parison wall thickness regulation (microcomputer 5).

So as to give decentralized process control, the individual function areas are each resolved by means of their own individual microcomputers.

2.1 Central Computer

Microcomputer 1 serves for the input of target values and parameters and for logging the process course in all the function areas. In addition, it contains a program for optimizing cycle time on the blow-molding unit. A prerequisite for development of the system was the execution of a full process sequence analysis and the compilation of a process model (Junk et al. 1978).

2.2 Temperature Control Computer

Microcomputer 2 is used for the DDC temperature control with ten zone wall temperatures, including three zones for heating and cooling, and for the cascade control melt temperature regulation. The actual values of all the temperatures including melt temperature are logged by a multiplexer with reference point, linearization facility and amplifier, controlled from the computer, and these are transmitted to an analog computer input with an accuracy of 12 bits. Relays are used as control components for the heater bands and cooler fans. The setting values for the zones are calculated by the control algorithm as percentage switch-in times relative to the scanning time.

The zone wall temperature target values, the melt temperature target value and preset controller parameters are transferred to the control program by means of a dialog program through a visual display unit. A PID control algorithm with cutoff for the integral part of the controller starting value is used for values that are greater or smaller than the achievable setting values.

2.3 Parison Length Control

The adaptive parison length control which follows by means of microcomputer 3 works on the basis of logging the parison length and velocity together with the length of tail waste (Dormeier 1977). The actuating signals used are (a) the cycle time so as to give optimum parison length with the same cycle time, and (b) screw rotational velocity so as to eliminate long-term fluctuations in length. Parison length and velocity are logged by the first three light barriers (LS1, LS2, LS3) below the nozzle. The length of the tail waste is logged by light barriers LS4 and LS5 as the mold is closed. The length from the cutoff point to light barrier LS4 corresponds to the required parison processing length. The distance between LS4 and LS5 represents the tolerance range.

Apart from serving as start and stop signals in the machine cycle, light barrier states LS1 and LS5 also constitute digital input signals for the computer. In addition, the position of the mold and the time at which blowing commences are also logged. Digital output signals from the computer serve to control the movement of the mold, to cut off the parison and to terminate blowing. In addition, a target value is transmitted to the screw rotational velocity control through an analog output signal (microcomputer 4). Cycle time and vent time (and other variables) for the process are fed in through a visual display unit.

With a microcomputer system of this type it is possible to manufacture optimum hollow bodies both during the start-up of the blow molding machine and when disturbing factors occur in steady-state operation.

2.4 Wall Thickness Adjustment Computer

Control of the nozzle position for the regulation of wall thickness is carried out with the aid of microcomputer 5, which is equally suited for continuous operation and for accumulator head operation. Dialog mode is likewise obtained through the visual display unit, as are the input and output of the 32 program points of the basic gap. Up to ten wall thickness programs can be stored. Contrary to the case with traditional units, a special quadratic interpolation is carried out between the program points, which guarantees optimally smooth running at the bulges in the program curve. In addition, every 20 ms a new interpolation support value is calculated and driven towards. The control circuit consists of a proportional position control with subordinated proportional–integral velocity control in the computer. This gives improved control properties compared with the proportional control incorporated in current units.

3. Sheet and Film Production

Moving away from the standard process control used for film units to date, the IKV has developed a process model that no longer requires continuous thickness measurement of the product (Michaeli 1975, Bergweiler et al. 1980). Thickness measurement is needed only for purposes of machine calibration and start-up.

With this model, thickness distribution over the product is calculated from measurements taken by pressure and temperature probes placed in the die system. If these measurements are taken at several points over the width of the die, they provide information on local throughput distribution and thus on thickness distribution over the product width.

Both the logging of actual values and the calculation of the necessary actuating signals for choker bar or lip adjustment and their output are effected through a microcomputer system. It has been found that choker bar adjustment alone allows sufficiently sensitive influence on the thickness profile.

The stretching that occurs between die and take-up results in a change in thickness (even if only slight in most cases) over the product width. The process model is aware of this thickness modification and appropriate allowance is made when the choker bars are adjusted. It would be worthwhile to examine similar strategies for the thickness control of profiles.

Hence, at the time of writing, a pipe unit is being fitted out so that wall thickness likewise results from pressure and temperature measurements in the die gap. The corresponding measuring points are arranged at three points at 120° intervals around the die channel.

4. Extrusion Process

4.1 Cable Sheathing Units

Extrusion units for high-voltage cables have to be operated at varying speeds on account of the necessary drum changing. This leads to changes in output that are not proportional to rotational velocity. These are caused by modifications in viscosity due to the conversion of different proportions of drive energy into heat and due to the convective heat components.

Using a computer system, the unit is operated by means of an adaptive control in such a way that the desired insulation wall thickness is retained even with changing line velocities. To do this, the controller parameters are adjusted to the section behavior with the aid of data from a thickness logger located directly behind the cooling section, in such a way that the required tolerance range of <5% is reached and maintained as quickly as possible.

It has been seen that the insulation wall thicknesses come within the requisite tolerance range after a run of just a few meters and are maintained within this range even if material or take-up conditions are altered.

4.2 Control Problems

A review of the control problems of polymer extrusion processes, their identification and control has been given by Lee and Lee (1989). The multivariable process needs to be controlled to tight tolerance by manipulation of appropriate process variables such as screw speed, power to the die and barrier heaters, take-up speed and so on. Control of separate components should be followed by multivariable control. Improvements in measuring techniques and understanding of the process from steady-state and dynamic models will enable more effective and reliable control systems to be developed.

5. Other Applications

The hardware and software components of the injection-molding machine control (Sect. 1) are currently prepared in such a way that they can be used on pressure-setting machines and elastomer machines. The main feature here is the monitoring of the melt with regard to its viscosity and the degree of crosslinking. In addition, the injection process and the holding pressure program are controlled in such a way that overinjection and the flash formation that has been unavoidable to date are prevented (Buschhaus and Menges 1981).

Optimization programs for control of reaction injection-molding machines have already been developed (Schwesig 1978) and were being tried out on the machines at the time of writing.

See also: Blow Molding of Thermoplastics; Injection Molding of Polymers: Computer Simulation; Processing: Classification; Processing: Shear-Controlled Orientation; Reaction Injection Molding

Bibliography

Bergweiler E, Menges G, Dierkes A, Michaeli W 1980 A new method for thickness control in flat film and sheet extrusion. *Soc. Plast. Eng. Tech. Pap.* 26: 50–2

Bongardt W, Menges G 1980 Der Einsatz von Mikrorechnern beim Spritzgiessen. *Kunstst.-Plast.* 2: 10–13

Bongardt W, Plessmann K W, Menges G 1980 Systemtechnisches Gesamtkonzept der selbsteinstellenden Spritzgiessmaschine. *Kernforschungszentrum Karlsruhe, [Ber.] KfK Ext.* 80: 331–57

Buschhaus F, Menges G 1981 Grundlagen für die Prozessführung beim Spritzgiessen von Duromeren und Elastomeren. *Plastverarbeiter* 3: 322–5

Dormeier S 1977 Ein Beitrag zur Automatisierung des Extrusionblasformens. Dissertation, Technischen Hochschule Aachen

Junk P B, Schwab E, Dormeier S, Ramm F 1978 Von der Prozessanalyse zur Rechnersteuerung-dargestellt am Beispiel des Extrusionblasformens. *Kunstst.-Berat.* 4, 5, 6

Lee W-K, Lee L J 1989 Computer control of extrusion. In: Aggarwal S L (ed.) 1989 *Comprehensive Polymer Science*, Vol. 7. Pergamon, Oxford, Chap. 13

Menges G, Recker H 1989 Computer-aided engineering in injection molding. In: Aggarwal S L (ed.) 1989 *Comprehensive Polymer Science,* Vol. 7. Pergamon, Oxford, Chap. 14

Michaeli W 1975 Zur Analyse des Flachfolien- und Tafelex-trusionprozesses. Dissertation, Technischen Hochschule Aachen

Schwab E, Kiel H, Wiegand G 1979 Mikrorechner in der Kunststoffverarbeitung. *Elektrotechnik* 8: 14–18

Schwab E, Menges G 1979 Mikrorechnereinsatz zur prozessführung beim extrusionsblasformen. *Kunstst.-Berat.* 7: 324–7

Schwesig H 1978 Beitrag zur Abschätzung der PUR-Integral-Schaumteilbildung. Dissertation, Technischen Hochschule Aachen

Vargel J 1974 Entwicklung eines Prozessmodells zur Adaptierung des Spritzgiessprozesses. Dissertation, Technischen Hochschule Aachen

G. Menges and H. Recker
[Technische Hochschule Aachen,
Aachen, Germany]

Construction Materials, Cellular

Many materials have a cellular structure, composed of an interconnected network of struts or plates. Such cellular materials occur in nature as wood, cancellous bone and cork. Recently, mankind has begun making its own cellular materials, in the form of honeycomblike materials and foams. Although initially honeycombs and foams could only be made from a limited range of materials, both can now be made from a wide range of polymers, metals, ceramics and glasses. Micrographs of several natural and synthetic cellular materials are shown in Fig. 1.

The cellular structure of these materials gives rise to unique properties which make them useful in many applications. Since they can undergo large deformations at relatively low loads, they are widely used in the packaging industry for providing protection from impacts. Their low volume fraction of solids and small cell size give them a low thermal conductivity, making them ideal thermal insulators. Their low weight is exploited in structural sandwich panels, where they are used as a lightweight core to separate two stiff, strong faces; the resulting sandwich gives an efficient member for resisting bending and buckling loads.

In this article, the structure and mechanical behavior of cellular materials are reviewed and their use in construction is described.

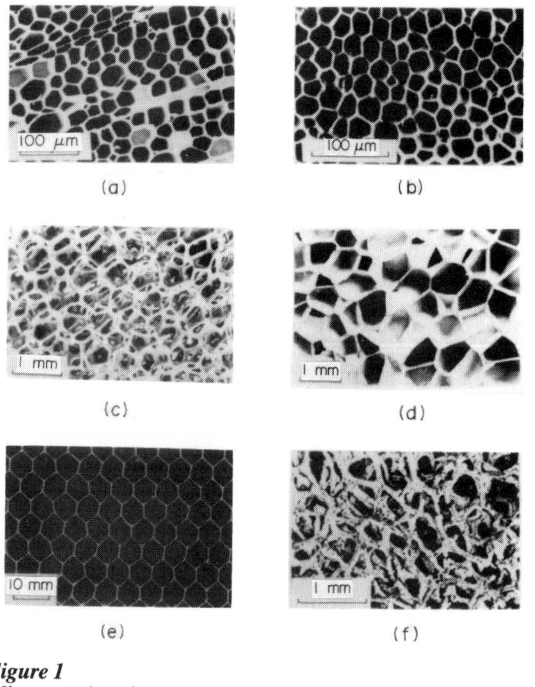

Figure 1
Micrographs of cellular materials: (a) balsa wood, (b) cork, (c) polyurethane foam, (d) polyethylene foam, (e) aluminum honeycomb and (f) copper foam

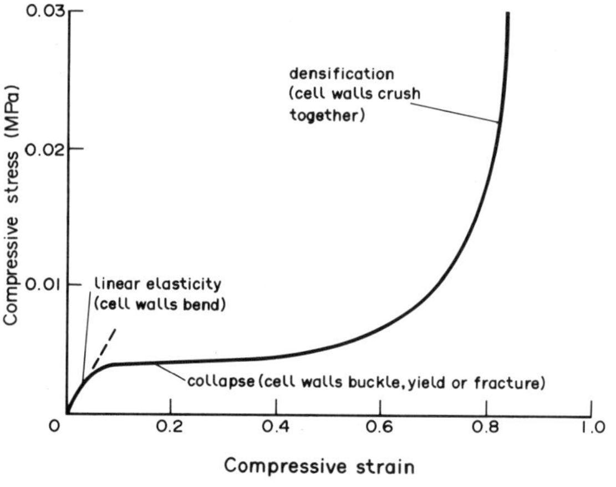

Figure 2
Typical stress–strain curve for a cellular material in uniaxial compression

1. Structure and Mechanical Behavior

Several features characterize the structure of a cellular material. The volume fraction of solids in the material can be described by its relative density, that is, the density of the cellular material divided by that of the solid cell wall material from which it is made. The cells may extend in one plane only (as they do in honeycombs), so that the normals to all of the cell walls lie in a single plane, producing a "two-dimensional" cellular structure. Alternatively, they may extend in any direction (as they do in foams), giving a less restricted, "three-dimensional" structure. The cells may be open or closed, depending on whether or not the faces of the cells are empty or covered by thin membranes. The cells may be geometrically anisotropic, with different mean intercept lengths in different directions. Anisotropy is often the result of the rise of the foaming gas in the liquid matrix which produces cells which are elongated in the rise direction and roughly equiaxed in the plane normal to it. The mechanical properties of foams are often anisotropic as a result of anisotropy in the cell geometry. Each of these structural features is shown in Fig. 1.

A typical stress–strain curve for a cellular material is shown in Fig. 2. There are three distinct regimes of behavior. At relatively low loads, the material is linear elastic. At some critical level of load, the cells begin to collapse by elastic buckling, plastic yielding or brittle fracture, depending on the nature of the cell wall properties. Cell collapse then progresses at a roughly constant load, producing a roughly horizontal stress plateau, until, at relatively large strains (typically about 0.8), the opposing cell walls begin to meet and touch. The stress then rises steeply as the material densifies (Maiti et al. 1984).

Each of these regimes of behavior is related to the mechanism by which the cells deform within that regime. Observation of model honeycomb materials and of foams in a scanning electron microscope reveals that the initial linear elasticity is produced by bending in the cell walls. Cell collapse is caused by elastic buckling in elastomeric honeycombs and foams, by plastic yielding in materials which yield and by brittle crushing in materials which fracture (Gibson and Ashby 1982, Maiti et al. 1984). The strain at which the honeycomb or foam begins to densify depends on its relative density.

The tensile behavior of cellular materials is similar to the compressive behavior with the following two exceptions. First, elastomeric honeycombs and foams do not buckle so that the stress plateau between the linear elastic regime and densification is eliminated in tension. Second, the tensile stress–strain curves of brittle cellular solids are terminated by brittle fracture at the end of the linear elastic regime.

This stress–strain behavior can be characterized by several mechanical properties. The linear elastic regime is described by a set of elastic moduli such as Young's modulus E, Poisson's ratio ν and the shear modulus G. The plateau stresses corresponding to elastic buckling, plastic yielding and brittle crushing are σ^*_{el}, σ^*_{pl} and σ^*_{cr}, the limiting strain for densification is ε_D and the tensile fracture toughness is K_{Ic}. By modelling the structure of cellular materials and the mechanisms by which they deform, relationships between the properties of the cellular material and its relative density, its solid cell wall properties and its geometry can be found. Table 1 contains a summary of these relationships for both honeycombs and foams. The constants of proportionality are different for each property and each case (foams, honeycomb in plane, honeycomb axial). The symbols E_s, σ_{ys} and σ_{fs} are the Young's modulus, yield strength and modulus of rupture, respectively, of the solid cell wall material. The characteristic length of the cell size is l. A more detailed analysis can be found in Maiti et al. (1984).

2. Applications of Cellular Materials in Construction

2.1 Lightweight Structural Sandwich Panels

Structural panels composed of two stiff, strong faces separated by a lightweight core are known as sandwich panels. The core acts to separate the faces, increasing the moment of inertia of the panel, with little increase in weight, producing an efficient structure for resisting bending and buckling loads. Usually, the core is made of a lightweight cellular material. In aerospace applications, aluminum or paper–resin honeycomb cores are preferred for their extremely low weight. In building applications, foam cores, which are typically somewhat heavier than honeycomb cores, are preferred as they offer low thermal conductivity in addition to relatively low weight. Typical face materials include fiber-reinforced composites, steel, aluminum and plywood.

The technology for producing sandwich panels was first developed in the aerospace industry, beginning with the balsa core–plywood face sandwiches of the World War II Mosquito aircraft (Allen 1969). Modern aerospace sandwich components are now made almost exclusively with honeycomb cores and fiber-composite faces bonded together with advanced adhesives. Sandwich panels are now made routinely and are reliable for components such as the rotor blades of helicopters, wing-tip sections, tail sections and flooring panels. This technology is now well developed and is being transferred to the construction industry.

The main use of sandwich components in the construction industry is in building panels. Several manufacturers are now producing wall, roof and door panels using sandwich construction. Different combinations of face and core materials are used; typically, the faces are made of steel, aluminum or waferboard and the cores are made of foamed polyurethane, foamed polystyrene beadboard or foamed glass. The foam core gives excellent thermal insulation while the faces provide the structural

Table 1
Property–density relationships for cellular materials

		Honeycombs	
Property	Open-celled foams	In plane	Axial
E	$C(\rho/\rho_s)^2E_s$	$C(\rho/\rho_s)^3E_s$	$C(\rho/\rho_s)E_s$
ν	C	C	C
G	$C(\rho/\rho_s)^2E_s$	$C(\rho/\rho_s)^3E_s$	$C(\rho/\rho_s)E_s$
σ^*_{el}	$C(\rho/\rho_s)^2E_s$	$C(\rho/\rho_s)^3E_s$	$C(\rho/\rho_s)^3E_s$
σ^*_{pl}	$C(\rho/\rho_s)^{3/2}\sigma_{ys}$	$C(\rho/\rho_s)^2\sigma_{ys}$	$C(\rho/\rho_s)^2\sigma_{ys}$
σ^*_{cr}	$C(\rho/\rho_s)^{3/2}\sigma_{fs}$	$C(\rho/\rho_s)^2\sigma_{fs}$	$C(\rho/\rho_s)\sigma_{fs}$
K_{Ic}	$C(\rho/\rho_s)^{3/2}\sigma_{fs}(\pi l)^{1/2}$	$C(\rho/\rho_s)^2\sigma_{fs}(\pi l)^{1/2}$	$C(\rho/\rho_s)\sigma_{fs}(\pi l)^{1/2}$

stiffness and strength of the panel. The panels are used in housing, in low-rise, nonresidential buildings and as fascia panels in commercial buildings.

The reduced weight of sandwich construction is also useful in applications where the component must be lifted into place by crane or helicopter. For example, a recent EPA ruling required the addition of roofing covers to exposed water distribution reservoirs. Domed covers made with fiberglass-reinforced plastic faces and a paper–resin honeycomb core were assembled adjacent to the reservoir sites and then lifted into place by crane. The low weight of the sandwich reduced the installation costs. In another application, modular sandwich lighthouse structures with fiber-reinforced plastic faces and balsa cores were installed using helicopters; again, the low weight of the units reduced installation costs.

Sandwich construction is also used for portable structures, such as the folding bridges being developed by the US Army. The deck of these bridges is doubly sandwiched, with a top face of high-strength aluminum and a bottom face of a graphite-epoxy–aluminum composite. Between these two faces is a webbed core which is itself a sandwich with aluminum faces and a honeycomb core. The low weight of the sandwich structure allows the bridge to be transported easily. The bridges are hinged so that they fold during transport and can be deployed by a hydraulic actuator for use.

The weight of sandwich panels can be minimized by appropriate optimization analyses. Ackers (1945), Wittrick (1945) and Kuenzi (1965) have developed minimum-weight analyses for pin-ended columns subject to an end load. Huang and Alspaugh (1974) and Ueng and Liu (1979) have minimized the weight of a sandwich panel subject to multiple constraints. More recently, Gibson (1984) and Demsetz and Gibson (1987) have found the core density as well as the face and core thickness which minimize the weight of a sandwich beam or plate for a given stiffness. These optimization analyses are useful in applications where the weight of the component is critical.

2.2 Wood

Wood is still one of the most widely used structural materials throughout the world. World production is about 10^9 t per year, roughly equal to that of iron and steel (Dinwoodie 1981). In North America more wood products are used for construction than all other construction materials combined (Bodig and Jayne 1982).

The cellular structure of wood is shown in Fig. 1a. Softwoods used in construction have two main types of cells: tracheids and parenchyma. The tracheids are the long, narrow cells (aspect ratio about 100 : 1) which make up the bulk of the tree. They conduct fluids and nutrients within the tree and give it its structural support. The parenchyma cells, found in the rays, store sugars.

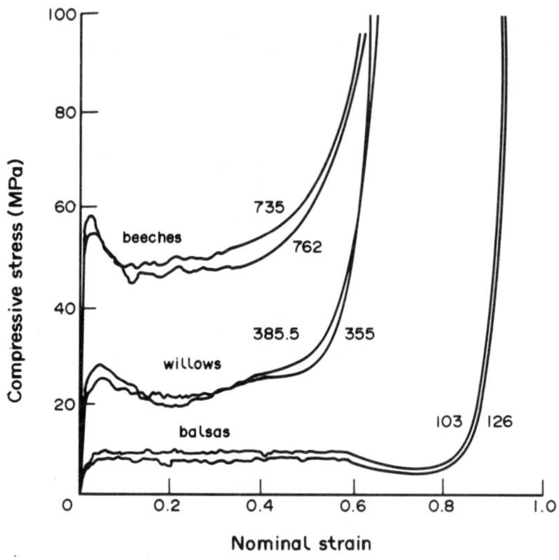

Figure 3
Stress–strain curves for wood loaded along the grain: figures by curves indicate density (kg m^{-3}); axial compression = 1.25–1.5 × 10^{-5} s^{-1}

At a finer level, the walls of the cells have a fiber-composite structure with partially crystallized cellulose molecules forming stiff, strong, fiberlike chains helically wound in a matrix of amorphous lignin and semicrystalline hemicellulose. Each cell wall is made up of four layers, each with a different orientation of cellulose fibers.

Mechanically, wood behaves very much like other cellular materials. Stress–strain curves for several types of wood in uniaxial compression along and across the grain are shown in Figs. 3 and 4. These curves show that wood is highly anisotropic; it is much stiffer and stronger when loaded along the grain than across. Part of this anisotropy results from the different mechanisms by which wood deforms when loaded along and across the grain and part comes from the anisotropy in the cell wall properties. The general shape of the curves resembles that of Fig. 2.

The mechanical behavior of wood can be modelled by analyzing the mechanisms by which it deforms (Easterling et al. 1982). When loaded across the grain, the cell walls bend in the same way that a honeycomb does when loaded in the plane of the honeycomb and, because of this, the Young's moduli and compressive strength of woods loaded across the grain depend on the cube and the square of the relative density of the wood (see Table 1). When loaded along the grain, the cell walls compress uniaxially, as a honeycomb does if loaded normal to the plane of the hexagonal cells and, because of this, the Young's modulus along the grain

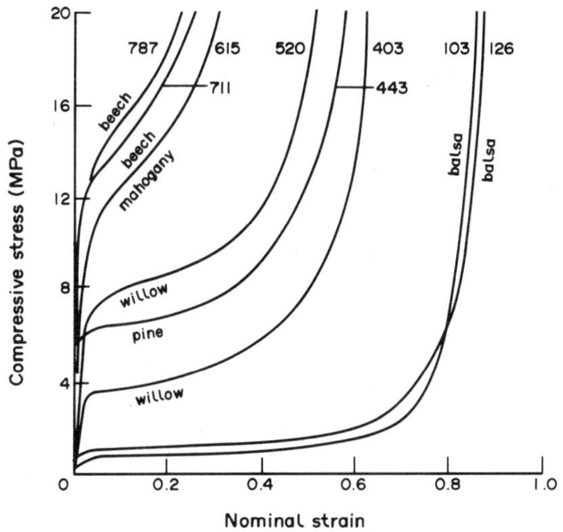

Figure 4
Stress–strain curves for wood loaded across the grain: figures by curves indicate density (kg m^{-3}); tangential compression = 1.6×10^{-3} s^{-1}

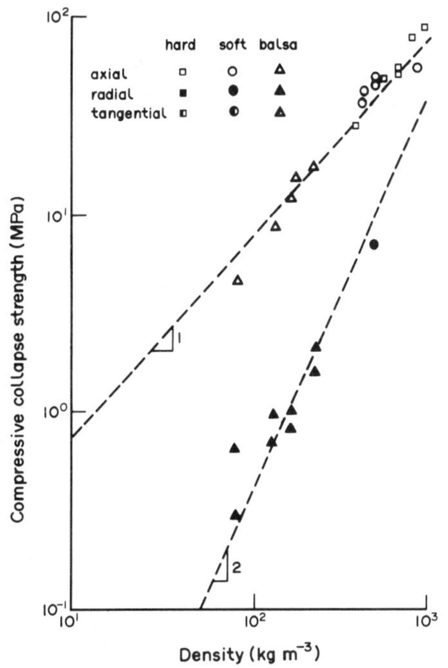

Figure 6
Compressive strength for woods, plotted against density: the lines indicate the theoretical dependencies

depends linearly on the relative density of the wood. Finally, the axial compressive strength of the wood is related to yielding in the cell walls, again giving a linear dependence on relative density. These results

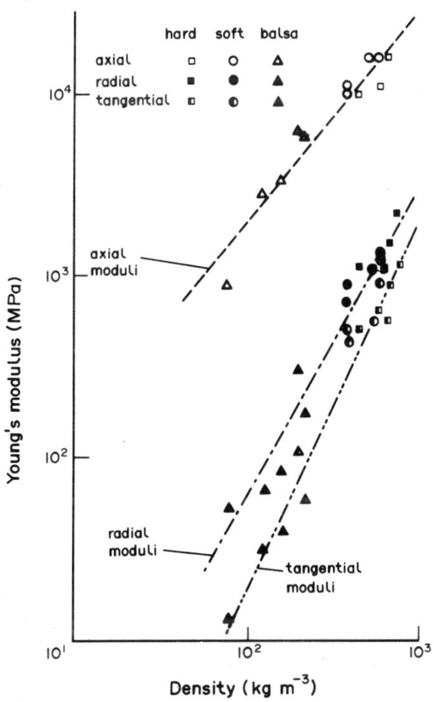

Figure 5
Young's modulus for woods, plotted against density: the lines indicate the theoretical dependencies

are plotted in Figs. 5 and 6, along with data for wood. This modelling begins to explain the dependence of the properties of wood on the loading direction and the relative density of the wood. The tensile fracture toughness of wood has been explained in a similar way; see, for example, Ashby et al. (1985).

The mechanical properties of wood tend to be variable, due to the presence of defects in the wood, for example, nonparallel grain and knots. Recently, glue-laminated wood members (glulam) have been developed to minimize the effect of such defects. Glue-laminated members are made by gluing together thin strips of wood from which the defects have been cut out. The properties of such members are more uniform than those of dimensioned lumber. Glulam sections also have the advantage that they can be curved and built up to any size. The efficiency of wood members can also be improved by using them in sandwich constructions. Typically, plywood faces are used in conjunction with honeycomb, foam or balsa cores.

2.3 Thermal Insulation

The thermal conductivity of cellular solids is typically low, partly because of their low volume fraction of solids (about 10%) and partly because of their small cell size (usually between about 0.04 mm and 1 mm). This makes them ideal thermal insulators; one of the most prevalent applications of foams is in thermal insulation.

In Arctic construction, icing can be a problem; foams are sometimes used as thermal insulation to reduce this. For example, polyethylene foam sheets have recently been used to insulate railroad tunnels outside Anchorage, Alaska, and in British Columbia, Canada. Near the portals of the tunnels, ice buildup on the crown and sidewalls of the tunnels and on the track caused severe icing problems. Insulation shields, made of polyethylene foam faced with galvanized steel plates, installed at the portals of the tunnels, reduced icing to a tolerable level.

Foam sheet has also been used to insulate foundations from permafrost in the Arctic. In this case, the foam is used to prevent the permafrost melting from the heat of the building since melted permafrost can cause excessive sliding and settlement of the foundations.

See also: Injection Molding of Thermoplastics: Structural Foams

Bibliography

Ackers P 1945 The efficiency of sandwich struts utilizing calcium alginate core, R & M 2015. Aeronautical Research Council, Farnborough, UK

Allen H G 1969 *Analysis and Design of Structural Sandwich Panels*. Pergamon, Oxford

Ashby M F, Easterling K E, Harrysson R 1985 The fracture toughness of woods. *Proc. R. Soc. London, Ser. A*. 398: 261–80

Bodig J, Jayne B A 1982 *Mechanics of Wood and Wood Composites*. Van Nostrand Reinhold, New York

Buist J M 1986 Cellular polymers. In: Buist J M, Grayson S J, Woolley W D (eds.) 1986 *Fire and Cellular Polymers*. Elsevier, Amsterdam, pp. 5–23

Demsetz L A, Gibson L J 1987 Minimum weight design for stiffness in sandwich plates with rigid foam cores. *Mater. Sci. Eng*. 85: 33–42

Dinwoodie J M 1981 *Timber: Its Nature and Behavior*. Van Nostrand Reinhold, New York

Easterling K E, Harrysson R, Gibson L J, Ashby M F 1982 On the mechanics of balsa and other woods. *Proc. R. Soc. London, Ser. A*. 383: 31–41

Gibson L J 1984 Optimization of stiffness in sandwich beams with rigid foam cores. *Mater. Sci. Eng*. 67: 125–35

Gibson L J, Ashby M F 1982 The mechanics of three-dimensional cellular materials. *Proc. R. Soc. London, Ser. A*. 382: 43–59

Huang S N, Alspaugh D W 1974 Minimum weight sandwich beam design. *AIAA J*. 12: 1617–18

Kuenzi E W 1965 Minimum weight structural sandwich. US Forest Service, Research Note FPL-086. Forest Products Laboratory, Madison, WI

Maiti S K, Gibson L J, Ashby M F 1984 Deformation and energy absorption diagrams for cellular solids. *Acta Metall*. 32: 1963–75

Richardson M O W 1987 Cellular polymers in building and construction. *Prog. Rubber Plast. Technol*. 3(1): 10–23

Ueng C E S, Liu T L 1979 Least weight of a sandwich panel. In: Craig R R (ed.) 1979 *Proc. ASCE Engineering Mechanics Division 3rd Specialty Conf*. University of Texas at Austin, pp. 41–4

Wittrick W H 1945 A theoretical analysis of the efficiency of sandwich construction under compressive end load, R & M 2016. Aeronautical Research Council, Farnborough, UK

L. J. Gibson
[Massachusetts Institute of Technology, Cambridge, Massachusetts, USA]

D

Decorative Laminates

A decorative laminate is a product resulting from (a) saturation or coating of a web of cellulosic material, such as paper or other carrying media, with a thermosetting type of synthetic resin, such as melamine, polyester or phenolic, and (b) consolidation of one or more of these webs under heat and pressure.

The history of decorative laminate development follows that of polymers. Initially, two basic polymers were used in decorative laminates: melamine–formaldehyde and phenol–formaldehyde. The phenolics, which typically give a dark color, were and are used for core stock, whereas the malamines, which can be clear, are employed for the printed and clear overlay face sheets in high-pressure decorative laminates. The development of decorative laminates was expanded and a variety of new polymers, such as poly(vinyl acetate), acrylics, methacrylates, ureas, polyesters and vinyls, were introduced for low-pressure decorative laminates and the new synthetic venteer or foil approach (Cannady 1981).

1. High-Pressure Decorative Laminates

A high-pressure laminate consists of multiple plies of resin-impregnated papers that have been consolidated under heat and pressure to a specific thickness. The visible outer ply of the sandwich is a decorative layer. There are three types of papers employed in these multiple plies.

(a) Overlay or face sheet consists of as blend of α-cellulose and bleached kraft fibers. The range of basis weights encountered is 20–29 lb (9.1–13.2 kg) per ream. Resin content in the overlay will vary in the range 64–68 wt%.

(b) Decorative sheet consists of pigments and a blend of bleached hardwood and softwood fibers. This ply can be used as a solid color or it can be calendered and printed. The range of basis weights encountered is 50–110 lb (22.7–49.9 kg) per ream. Resin content when used with an overlay on a print will vary in the range 40–44 wt% whereas, when the decorative sheet is used as a solid color without an overlay, resin content will vary from 48–51 wt%.

(c) Core sheet consists of unbleached kraft fibers. The range of basis weights encountered is 110–150 lb (49.9–68.0 kg) per ream. Resin content will normally be 28–32 wt%, the lower the pressure of consolidation, the higher the resin content. The number of core sheets in a particular laminate is adjusted to satisfy thickness and performance requirements.

The types of resins employed are all of the thermosetting class, requiring heat but no catalyst for cure.

The overlay or face sheet and the decorative sheet are impregnated with melamine–formaldehyde resins, which provide a tough, attractive surface that is heat, wear and stain resistant. The core sheets, impregnated with phenol–formaldehyde resins, are considered as the carrier for the decorative surface. Resin modifiers can be used to alter properties or lower manufacturing costs.

In the manufacture of high-pressure decorative laminates, each type of ply is saturated and partially cured as a continuous web. Individual sheets, cut from partially cured webs, are stacked according to the predetermined design and thickness requirements.

A horizontal, multiple-opening platen press is used to produce the final decorative laminate. The press must be supplied with (a) hot water or steam for heating the platens during the final resin cure, (b) cold water to return the platens to nearly room temperature after the cure and (c) sufficient hydraulic pressure to consolidate or fuse the laminate layers. A typical press opening can produce 10–16 laminates depending on laminate thickness and overall press opening. Stainless steel plates are used to separate adjacent laminates and to control surface roughness. The production capacity of a press is a function of the number of openings and cycles per day. A typical press cycle lasts about 75 min, with the platen reaching a maximum temperature of approximately 150 °C and pressures up to 1500 psi (10.5 MPa).

The physical properties of the decorative laminates, such as flexural strength, compression strength, impact strength and elastic modulus depend on the specific melamine and/or phenol–formaldehyde resins, the quality of the various papers and the laminate thickness.

The high-pressure decorative laminates are synthetic counterparts of natural products, such as wood veneers, slate and marble. Therefore, they must be combined through the use of adhesives with various substrate materials, such as plywood, particle board, flakeboard, hardboard or medium-density fiberboard, in order to produce a panel of functional construction (see *Adhesives for Wood*).

2. Low-Pressure Decorative Laminates

Low-pressure decorative laminates have surfaces of similar quality to those of high-pressure decorative products. The differentiating factors between low- and high-pressure products are: (a) makeup or construction of the laminate, (b) types of resins used in the decorative papers, (c) time–temperature cycles for curing the resins, (d) pressures utilized for the consolidation phase and (e) press equipment.

Low-pressure decorative laminates utilize a presaturated decorative paper, core material (chipboard, particle board, flakeboard, medium-density fiberboard or hardboard) and a presaturated paper to balance the decorative face.

For low-pressure laminate products, the types of resins employed in the decorative and balance papers are polyesters and modified melamine–formaldehydes. The polyester resins are of the addition polymer type and are activated by a catalyst through heat to bring about curing. The modified melamine–formaldehydes are heat activated for curing but may contain a catalyst to induce a faster rate of reaction. The resin content normally encountered for either resin system is approximately 56 wt% and is sufficient to produce a good bond between the paper and substrate.

The time–temperature cure cycles used for low-pressure decorative laminates vary from less than 1 min to 6 min at temperatures of 125–175 °C. The pressures employed in making low-pressure laminates range from 150 psi (1 MPa) to 340 psi (2.3 MPa). The lower pressure level is satisfactory for good resin flow and bonding with polyester resin systems, whereas the higher pressure level is required for proper flow and bonding with melamine–formaldehyde resin systems.

Press equipment used for the production of low-pressure laminates is either a single-opening, fast cure/fast cycle press, or multiopening horizontal platen press. The single-opening presses are generally used for the production of melamine-surfaced decorative laminates. Multiopening presses can be used for melamine resins; however, they are used principally for polyester resin laminates in North America. The procedure for either type of press requires loading or unloading while the press remains hot.

After the curing process the decorative laminates are considered semifinished. An edge treatment may be necessary to cover the core material after it is cut into component parts. Areas of application for low-pressure laminates are both competitive and complementary to those for high-pressure decorative laminates.

3. *Synthetic Veneer*

The subject of synthetic veneer is an adjunct to the two items previously discussed, namely high- and low-pressure decorative laminates. The development of synthetic veneer began in Europe a number of years ago when the cost and availability of high-quality wood veneer became nearly prohibitive.

The raw material for veneer is a pigmented paper (37–74 lb (16.8–33.6 kg) per ream) which has been printed with a decorative design as is done for high- and low-pressure decorative laminate papers. The printed paper is saturated or impregnated with various modified thermosetting resins, such as ureas, melamines, acrylics and poly(vinyl alcohol), and cured at high temperatures (e.g., 150 °C). The cured product, which retains a fair degree of flexibility as a continuous web, is then top coated with a lacquer or other type of wear-resistant coating and wound into rolls.

There are three principal methods—sheet, roll and board—for processing synthetic veneer. In the first method, the synthetic veneer is sheeted and stacked in a flat position next to a single-opening press. The substrate or core board is coated with a urea adhesive at the level of 10.25–15.3 lb Mft^{-2} (0.05–0.07 kg m^{-2}) per side. The synthetic veneer is placed on the adhesive-coated board and transferred into the single-opening press for approximately 30 s at a temperature of 170 °C and under a pressure of 280–350 psi (1.9–2.4 MPa).

In the second method, synthetic veneer is fed from a continuous roll onto a base material precoated with an adhesive and processed through a single-opening press, as is done with the sheet method. The laminated boards are separated into the individual pieces following exit from the press. The major advantage of the roll-fed method is that a shorter press cycle can be accommodated, resulting in a higher production rate.

In the third method, the synthetic veneer is fed through a hot nip-roll laminator onto the core, which has been coated with an adhesive, such as urea, poly(vinyl acetate) or hot melt. The process has the advantage of being essentially continuous, since the decorative panels are separated into the individual panels after exiting the laminator.

The surface physical properties of the decorative laminates produced by the veneer method are more than adequate for the intended applications; however, the surface quality is not as good as that produced with a high- or low-pressure decorative laminate.

4. *Material Properties*

Early research (Cox and Pepper 1944, Pepper and Barwell 1944) established the following principles for properly fabricated phenolic laminated boards.

(a) Above 30% resin content in a board, the air voids are substantially absent and bonding is good.

(b) Shear and compression strength increase up to 30% resin level and are essentially constant thereafter. The shear and compressive strengths parallel to the laminations, and the water absorption, approach those of the resin used.

(c) Tensile strength and Young's modulus are functions of the type, amount and orientation of fiber. Numerical estimates of these properties may be calculated from fiber strengths and stiffnesses assuming perfect adhesion.

(d) Impregnated paper should, under current technology, be dried to a level of 5–8% volatiles to have a good resistance to water.

(e) Nonabsorbent high-density paper pressed at low pressures will have high tensile strength, but because of voids will tend to be weak in shear and compression and have very poor water resistance. Paper properties are thus important in making quality board.

Further research has confirmed and expanded these findings; for example, Straka (1948) and Morris and Verdi (1959) explain the role of processing and moisture in dimensional stability and curl of laminates, respectively. The need for good absorbency, absence of impurities which can affect resin cure, proper fiber refining, uniform caliper and uniform basis weight in laminate-saturating grade paper is stressed. Morris and Verdi (1959) confirmed the need for a volatile content as low as 3% in order to have a dimensionally stable laminate.

Bibliography

Cannady D L 1981 Decorative laminates. *Mod. Plast.* 58(10A): 126–7

Cox H L, Pepper K W 1944 Paper-base plastics—I. The preparation of phenolic laminated boards. *J. Soc. Chem. Ind., London, Trans. Commun.* 63: 150–4

Deutsches Institut für Normung 1979 *Test Methods for Decorative Laminated Sheets on the Basis of Amino Plastic Resins*, DIN No. 53799. Deutsches Institut für Normung, Berlin

Glomb J W, Mulligan D D 1989 Paper and paperboard. In: Kelly A (ed.) 1989 *Concise Encyclopedia of Composite Materials*. Pergamon, Oxford, pp 217–27

International Organization for Standardization 1981 *Plastics-Decorative Laminated Sheets Based on Thermosetting Resins*, Pts. 1 and 2, ISO 4586/1 and 4582/2. ISO, Geneva

Kussman J E 1953 Paper base laminates. *Pap. Trade J.* 137(20): 254–6; 258; 260–1

Lair W M 1981 Industrial laminates. *Mod. Plast.* 58(10A): 131–6

Morris W E, Verdi J M 1959 A study of curl in decorative laminates. *Tappi* 42: 583; 585–7

National Electrical Manufacturers Association 1977 *Decorative Boards*, NEMA Standard Publication LQ-1977. NEMA, Washington, DC

National Electrical Manufacturers Association 1980 *High Pressure Decorative Laminates*, NEMA Standard Publication LD3-1980. NEMA, Washington, DC

Pepper K W, Barwell F T 1944 Paper-base plastics—II. Production at low pressure. *J. Soc. Chem. Ind., London, Trans. Commun.* 63: 321–9

Straka C J 1948 Causes of dimensional instability of paper base phenolic resin laminates on total immersion in water. *Pap. Trade J.* 126(14): 50–4

Technical Association of Pulp and Paper Industries 1980 *Plastic Laminate Seminar*, 27–31 October 1980. TAPPI, Atlanta, GA

J. F. Haskins
[JFH Associates, Brunswick, Maine, USA]

Degradation and Stabilization: Test Methods

Numerous tests have been developed for measuring the stability of polymers, taking into account the many different conditions responsible for polymer degradation. Some tests measure the stability of polymers to thermal degradation; others provide for evaluation of the adverse effects of weathering. There are also tests to measure the effect of specific chemical reactants or unique conditions on special classes of polymers (e.g., the test for ozone resistance of unsaturated elastomers under stress). As the stability of polymers increased, it became necessary to develop tests that measure stability over shorter time intervals. These procedures are referred to as accelerated tests.

Accelerated tests for polymer stability provide data which must be extrapolated down to conditions to be encountered in service. It is necessary to intensify test conditions in order to accelerate degradation, but extrapolation of the results from accelerated tests to service conditions is not always linear. Changes in morphology or in the relative rates of individual reactions in an overall mechanism could result in a nonlinear relationship as shown schematically in Fig. 1. As a general principle, the closer accelerated tests approximate service conditions, the more accurate will be the prediction of service life.

There are two different approaches to measuring the stability of polymers. In the development of a new polymer or in evaluating the effectiveness of stabilizers, a material-oriented test is used. As the name implies, this type of test is intended to measure the stability of a polymer or polymer composition apart from contact with other materials which might be used in a final product. Design-oriented tests, on the other hand, measure polymer stability in a complete device or article. The time to failure can be affected by other materials with which the polymer is in contact. For example, stabilized polyolefins lose most of their stability when extruded as insulation over copper conductors. When polymers are laminated onto metals, similar loss of stability can occur. Stabilizers may also be extracted from a stabilized polymer by other materials present in the final product. Design-oriented tests should be used to predict how long a polymer will retain its essential

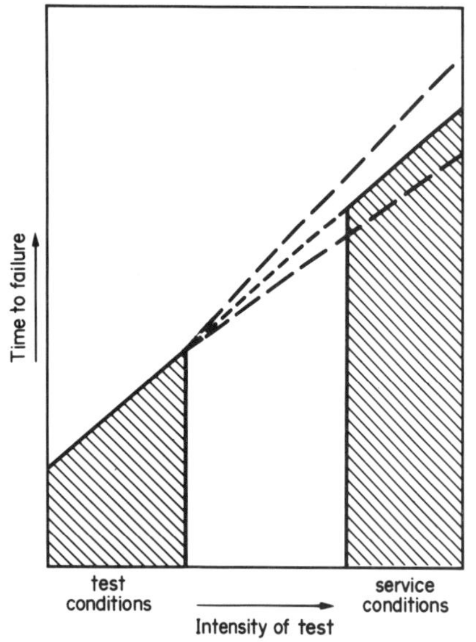

Figure 1
Schematic representation of the relationship between test conditions and service life

properties in each specific application. Improvements in the stability of polymer compositions are usually evaluated by materials-oriented tests before incorporation into a finished product.

1. Measurement of Thermal Stability

Early in the development of synthetic polymers, considerable research effort was directed towards their pyrolytic behavior. Tests to measure the decomposition temperature were based on weight loss during heating, either in vacuum or in an inert atmosphere. In an alternative procedure, the pressure developed by volatile fragments from the polymer was determined. Test samples in sealed, evacuated tubes were heated at programmed rates of temperature increase and changes in sample weight measured. Although these tests made important contributions to the structural analysis of polymers, the conditions used are not encountered in most applications. Thermal degradation of polymers in the presence of oxygen, however, is common to all polymers and several important tests have been used to measure this effect.

1.1 Milling Tests

The earliest tests for stability to thermal oxidation originated in the study of the aging of natural rubber. In milling tests, rubber samples were passed continuously between heated rolls, held at close clearance. The combination of mechanical energy and heat accelerate degradation. Sticking of the sample to the rolls was taken as the onset of degradation. This test was never precise, but did have the advantage of resembling processes used in making rubber formulations.

The torque rheometer is an adaptation of the milling test. In this device, test samples are heated and milled between rotating blades while simultaneously recording changes in the torque developed by the rotors. Degradation of the polymer is indicated by a rapid change in torque. Polymers which degrade by chain cleavage show a decrease in torque, whereas those in which crosslinking predominates show an increase.

1.2 Oven Aging Tests

Among the most widely used tests for measuring oxidative stability are those based on oven aging. In the simplest version of this test, samples are exposed to air while heated isothermally. Static and circulating ovens have been used, but the latter may cause excessive loss of stabilizers by volatilization. The progress of degradation is observed by removing samples at selected intervals and observing changes in appearance such as discoloration or surface cracking. More precise data are obtained by measuring changes which occur in mechanical or electrical properties. Modulus, elongation and tensile strength are among the more common mechanical measurements, and dielectric strength or dissipation factor are often measured to detect changes in electrical properties. Since some of these measurements are destructive, samples are required for each test interval. Multiple samples may also be tested to correct for the scatter in some of these results. Infrared spectroscopy can also be used to follow chemical changes in a polymer during oven aging. Measurement of the increase in concentration of carbonyl groups is often employed.

Oven aging tests have the advantage of simplicity, varying with the technique used for following degradation. However, they do not provide a continuous measurement. Accuracy of the method depends on the frequency with which samples are removed for evaluation and enough samples must be used at the start for every anticipated test interval. Despite this disadvantage, oven aging tests have the advantage of following degradation by observing changes in a property which could be critical in a particular application.

1.3 Oxygen Uptake Measurements

Measurement of the rate at which oxygen reacts with a polymer is an adaptation of oven aging, used extensively to follow thermal oxidation. In this test each sample is enclosed in a separate test chamber, usually a glass tube. Measurements are conducted in pure oxygen. This eliminates the variable concentrations of water and trace catalysts which are usually present in air. A desiccant is placed in the test chamber to absorb any water evolved as the sample oxidizes.

Small samples can be used in oxygen uptake measurements, but sample thickness must be carefully controlled. If the samples are too thick, diffusion of oxygen into the polymer bulk may become the controlling rate factor. Reproducible results can be obtained above the melting range of the polymer, provided that sample thickness does not change significantly.

Originally, ovens were used for this test but more uniform sample temperatures are obtained in an oil bath or in a metal block drilled to accommodate sample tubes and control devices and heated by external heaters. The rate of oxidation is measured either volumetrically or manometrically. With the former method, sample tubes are connected to a burette which is fitted with a levelling bulb to maintain atmospheric pressure. It is possible to follow oxygen uptake manometrically, but the rate of reaction is pressure dependent, decreasing sharply as the pressure falls in the reaction vessel.

Oxygen uptake measurements have several advantages over the basic oven aging test. The rate of reaction with oxygen is followed continuously and there is no interruption in the test to withdraw samples. Since sample size is very small, many tests can be run simultaneously. However, it is advisable to run each sample in duplicate to compensate for variations in the material. This method is widely used in evaluating the relative effectiveness of stabilizers. The point of failure is taken as the time at which oxidation becomes autocatalytic and is usually reported as the induction period (i.e., time interval between the start of the test and the intercept of the steady-state rate extrapolated back to the time axis).

Mechanical, dielectric or aesthetic failure do not always coincide with the induction period. A better correlation with actual failure can be obtained by first determining the level of oxidation at which a critical property exceeds the allowable limit, and then determining the time required to react with this amount of oxygen.

1.4 Thermal Analysis

Thermoanalytical techniques have been adapted to the measurement of thermal degradation, both in the absence and in the presence of oxygen. Commercial equipment is available for thermogravimetric analysis (TGA), differential thermal analysis (DTA) and differential scanning calorimetry (DSC). Each of these methods has been used in the measurement of oxidative degradation in polymers.

Modern TGA is an adaptation of early methods used to study the pyrolysis of polymers. The apparatus consists of a highly sensitive balance enclosed in a glass vessel. The test chamber may be evacuated or filled with an inert gas for measuring the temperature T_c at which polymer decomposition takes place. The test is programmed to a selected rate of temperature increase and sample weight is monitored continuously. Typical TGA data are shown in Fig. 2 for the thermal degradation of several different polymers.

TGA is also used to measure the rate of thermal oxidation. For these measurements the sample is heated isothermally usually in the 200–260 °C range. An inert atmosphere is maintained over the sample until the test temperature is reached and the inert gas is rapidly replaced with oxygen at this point. Weight changes are plotted against time. For stabilized polymers, there is no significant weight change until the stabilizer is exhausted. Oxidation then takes place and there is a rapid gain in weight. The period between the introduction of oxygen and the onset of rapid oxidation is taken as the induction period.

DTA is based on the principle of a difference in temperature response between the polymer and an inert control during heating. Thermocouples in contact with both polymer and control detect the temperature differential ΔT. A strong exotherm indicates the onset of the autocatalytic reaction and the time required to reach this point defines the induction period (see Fig. 3).

DSC is an adaptation of conventional adiabatic calorimetry. The test is not as accurate as conventional calorimetry, but is adequate for measuring the thermal degradation or oxidation of polymers. Oxidative degradation appears as a strong exotherm in the DSC measurement. Thermomechanical analysis is used to measure changes in a mechanical property during heating. Torsional braid and volatilization analyses have also been used to measure

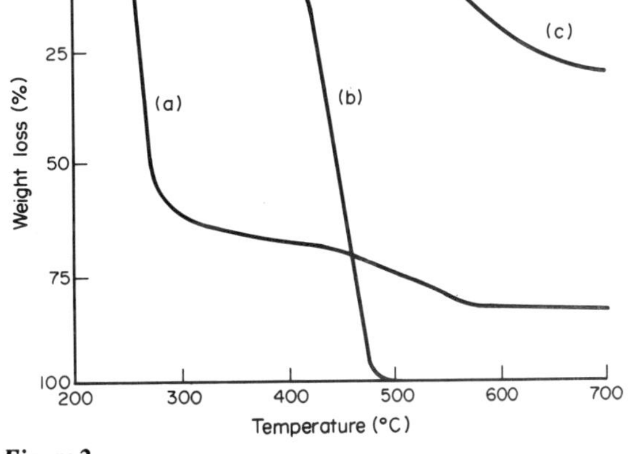

Figure 2
TGA scans for the thermal degradation of typical polymers: (a) poly(vinyl chloride), (b) polyethylene and (c) a polyamide

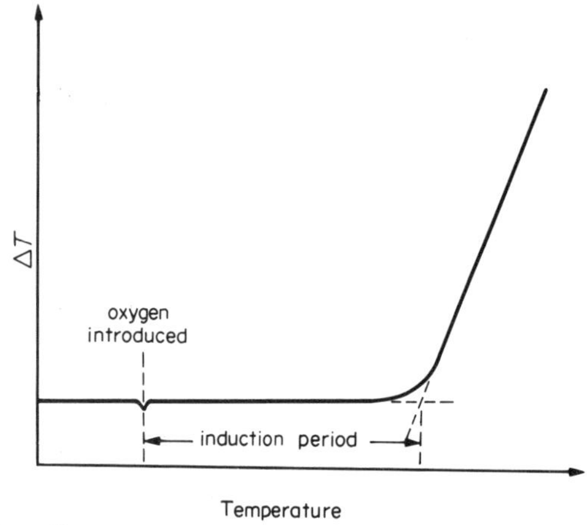

Figure 3
Typical DTA scan for the thermal oxidation of a stabilized polymer

thermal degradation, but these techniques have had only limited application in measuring polymer degradation.

1.5 Chemiluminescence Measurements

Light is emitted during the oxidation of many polymers producing the phenomenon of chemiluminescence. Light emission is attributed to radical-induced reactions in the polymer mass. Test samples are heated at constant temperature under a stream of oxygen. As degradation proceeds, emitted light is focused by a lens system onto a photomultiplier. The quantity of emitted light is plotted against time. This technique has been used to detect degradation in the latter stages of the reaction, but it may also be used to detect early reactions occurring during the induction period. Since the quantity of light emitted early in the reaction is very small, it is necessary to magnify the effect. This may be done either by rapidly replacing an inert atmosphere with oxygen or by briefly exposing the sample to ultraviolet (uv) radiation. This modification of the method could be useful in the early detection of oxidative degradation. Chemiluminescence can be used to evaluate both thermal and photooxidation in polymers.

2. *Test for Ozone Resistance*

Elastomers which contain ethylenic unsaturation in the backbone chain are degraded rapidly by ozone when held under stress. Resistance to ozone attack is usually measured by laboratory tests. A standard method has been established by the American Society for Testing and Materials (ASTM). In this test, samples are placed in an aluminum chamber and held at a prescribed level of stress. Temperature is maintained at 40–50 °C while a mixture of 500 ppb of ozone in oxygen is passed over the samples. Cracks, which always form perpendicular to the direction of applied stress, indicate the onset of degradation. These cracks may eventually extend through the test samples. Ozone concentration is monitored frequently by titration. This test is often used to rate antiozonants and waxes for their ability to inhibit ozone attack on elastomers.

3. *Weatherability Tests*

The degradation which occurs in polymers during outdoor exposure is measured by either outdoor or indoor tests. The simplest of the outdoor tests consists of exposing samples at locations selected for their high intensity of solar radiation. Samples are examined periodically for evidence of degradation such as discoloration or surface cracking. When more precise data are required, samples can be removed from the exposure racks and changes in mechanical, electrical or chemical properties measured.

Outdoor exposure conditions have been standardized by the ASTM. The angle at which samples are mounted has been established as 45° with samples facing due south. Material to be used in constructing the exposure racks and the nature of the underlying surface are also specified. Application of these standards provides a reasonable correlation between different locations, but the time to failure is usually only about half that encountered in service. A somewhat higher level of acceleration is obtained by continuously rotating samples to follow the sun's path. These solar-tracking devices are complicated and require considerable maintenance. Furthermore, there is little additional acceleration over fixed-angle exposure.

The highest level of acceleration in an outdoor test has been reported for devices which, in addition to solar tracking, use a series of polished metal mirrors to accumulate solar radiation and to direct it to the surface of test samples. In these devices, known as EMMA (equatorial mount with mirrors for acceleration) devices, samples may reach fairly high temperatures. An air stream or a water jet is used to control the temperature, but the contribution of thermal degradation may still exceed that which occurs outdoors. However, test results can be obtained in about one-tenth the time required for normal outdoor exposure.

Results obtained with any outdoor test are subject to weather variability. Furthermore, exposure is limited to daylight hours and this seriously restricts the level of acceleration. These factors have stimulated development of devices to measure weatherability in a laboratory environment. The devices for indoor testing are known as weatherometers.

The basic component of a weatherometer is an artificial source for uv radiation. Samples, mounted on racks, rotate slowly around the light source. It is very important to use a light source with a uv spectrum closely resembling that of the sun. Carbon and mercury arcs were used originally, but the xenon arc is gaining in popularity since its spectrum most closely resembles that of sunlight. Incandescent sunlamps and fluorescent uv tubes are also used. Filters screen out undesirable radiation peaks and combinations of lamps may be used to provide a closer match to the solar spectrum.

Fluorescent tubes have an important advantage in that they generate much less heat than most arcs or incandescent bulbs. It is important that excessive heat is not generated in a weatherometer in order to prevent thermal reactions from becoming the primary factor responsible for degradation. Sample temperature can be lowered by increasing the distance from the light source, but this also decreases the amount of uv radiation reaching the test samples. Baffles have been used to shield samples from radiation for a fraction of the rotation cycle. Water sprays have also been used as a means of cooling as well as to introduce the effect of moisture on test samples. Tap water, however, leaves a film on the surface which reduces the amount of uv reaching the samples. Condensation of water vapor on test samples is used in some weatherometers: this resembles dew formation without leaving a film on the surface.

Reasonably accurate correlations to outdoor weathering are obtained with modern weatherometers. The time required to reach polymer failure can be of the order of one-tenth that observed in service. By using an actimeter mounted alongside the samples, the actual level of uv radiation can be measured continuously and accurately: these devices correct for any change in output from the light source. When an actimeter is used, failure is based on energy units absorbed rather than on elapsed time. The actual time to failure can then be calculated from weather bureau reports which show the time interval over which the critical amount of radiation reaches any specific location.

See also: Degradation and Stabilization, Thermal; Degradation and Stabilization, Ultraviolet

Bibliography

Brown R P 1986 Effect of temperature and environmental resistance. *Physical Testing of Rubber*, 2nd edn. Elsevier, Amsterdam, Chaps. 15–16

George G A 1981 Use of chemiluminescence to study the kinetics of solid polymers. In: Grassie N (ed.) 1981 *Developments in Polymer Degradation*, Vol. 3. Applied Science, Barking, UK

Gray V E, Cadoff B C 1967 Survey of techniques for evaluating effects of weathering on plastics. *Appl. Polym. Symp.* 4: 85–96

Hawkins W L (ed.) 1972 *Polymer Stabilization*. Wiley, New York

Hirt R C, Searle N Z 1967 Energy characteristics of outdoor and indoor exposure sources and their relation to the weatherability of plastics. *Appl. Polym. Symp.* 4: 61–83

Kaufman F Jr 1967 A new technique for evaluating outdoor weathering properties of high density polyethylene. *Appl. Polym. Symp.* 4: 131–9

Madorsky S L 1964 *Thermal Degradation of Organic Polymers*, Polymer Reviews, Vol. 7. Interscience, New York

Quackenbos H M, Samuels H 1967 Practical problems in predicting weathering performance. *Appl. Polym. Symp.* 4: 155–65

Turi E (ed.) 1981 *Thermal Characterization of Polymeric Materials*. Academic Press, New York

W. L. Hawkins
[Plastics Institute of America, Hoboken, New Jersey, USA]

Degradation and Stabilization, Thermal

Thermal effects are a major cause of deterioration of physical properties of polymers. Relatively high temperatures are often encountered in polymer processing and fabrication, and may result in breaking chemical bonds. Volatile products are formed by elimination reactions and depolymerization. The relative thermal stability of a material is thus related to the chemical structure of the polymer. More moderate temperatures are encountered during long-term storage and in typical end-product uses. Oxidative degradation of polymers occurs in contact with air at temperatures encountered in processing, and also in the range typical for storage and service. Thermal and photoautoxidation together account for much of the loss in physical properties resulting in failure of polymer materials. This article first describes nonoxidative thermal degradation and stabilization problems, and then focuses primarily on thermal autoxidation and the stabilization of polymers by incorporation of antioxidants. Ultraviolet degradation and stabilization are covered in the article *Degradation and Stabilization, Ultraviolet*).

1. Nonoxidative Thermal Degradation

Thermal degradation is of practical importance only in situations where the availability of oxygen is limited or when oxidation-resistant polymers are used. Processing often involves mixing, extrusion and molding procedures that limit or exclude oxygen and, when high temperatures are used, nonoxidative thermal degradation can be a significant problem. When polymers are used at reduced pressures or in space vehicles outside the earth's atmosphere, thermal stability (rather than oxidative stability) becomes the more important consideration. Some polymeric materials are used at high temperatures in air and require the development of special polymers that are resistant to both oxygen and heat.

Chemical bonds in polymers are broken when sufficient energy is acquired from heat, light, ionizing radiation or chemical reaction. Elimination of volatile fragments from substituent groups usually occurs at lower temperatures than are required for cleavage of the polymer backbone. Rupture of the backbone chain frequently involves a weak bond and may result in the evolution of monomer. At high temperatures, random scission of the main chain bonds occurs and lower-molecular-weight polymer fragments are formed.

Two types of degradation effects result. Rapid loss of weight occurs as volatile products are formed either by elimination reactions from substituent groups or by unzipping of the main chain. Little or no change in the molecular weight of the remaining polymer is observed. Random chain scission produces the opposite effect, with little or no weight loss but with a rapid decrease in molecular weight.

Only a few polymers have substituent groups that readily undergo elimination reactions with activation energies lower than would be required to break bonds along the main chain. Random elimination of isobutylene is observed on heating poly(*t*-butyl methacrylate):

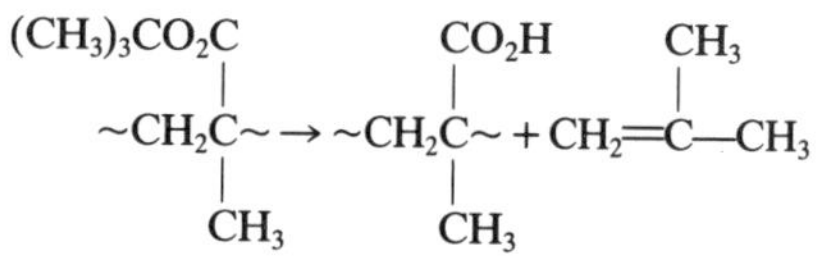

Poly(vinyl acetate) yields acetic acid, and poly(vinyl alcohol) (formed by hydrolysis of the acetate) undergoes dehydration on heating to form unsaturated units in the polymer chain:

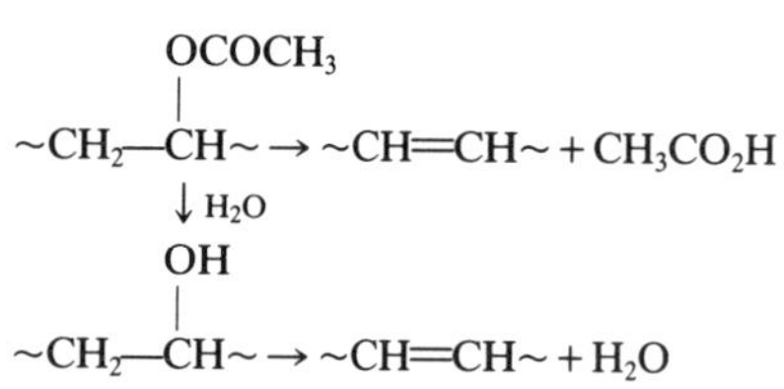

The most important example of degradation by an elimination reaction without monomer loss is that of poly(vinyl chloride). Temperatures in the range 160–200 °C are typically used in processing and stabilizers are required to prevent discoloration by loss of hydrogen chloride from adjacent polymer units to form a sequence of conjugated double bonds:

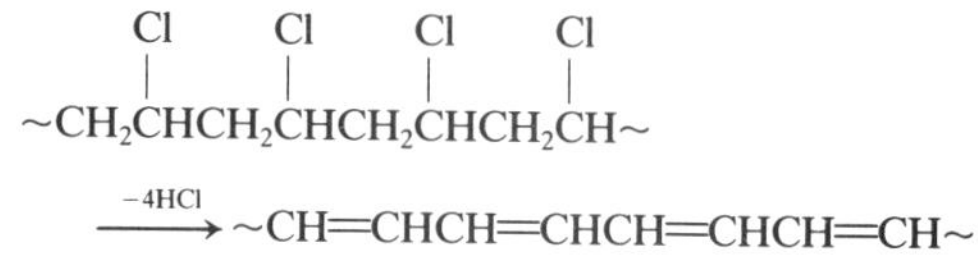

Poly(vinyl chloride) (PVC) is probably the least stable of the major commercial polymers. Since the saturated head-to-tail structure would be expected to be thermally stable, it is thought that initiation of dehydrohalogenation is due to labile defect structures. Various sites have been proposed: (a) chain ends substituted by initiator residues; (b) unsaturated chain ends; (c) branch points with tertiary chlorine; (d) random unsaturation with adjacent allylic chlorine; (e) oxygen present as, for example, peroxide or carbonyl; and (f) head-to-head units.

The mechanism of thermal dehydrohalogenation of PVC is still controversial in spite of extensive research. Various workers have proposed radical chain, ionic, concerted or some combined mechanisms. The rate of dechlorination has been shown to follow an inverse relation to molecular weight, which indicates substantial involvement of groups at or near the chain ends. A free-radical chain mechanism was widely accepted at one time, but support for an ionic mechanism has come from studies on model compounds and the observation that HCl is a catalyst for the reaction. The number of allylic chlorides has been estimated as 3–4 per 1000 monomer units, and the number of tertiary chlorides as 27–29. The mean length of polyene segments in degraded PVC is reported to be about 9–14 units, but some researchers have estimated that sequences of up to 25 conjugated double bonds may be formed. Defect structures containing allylic chlorides are predicted to have a more pronounced effect on dehydrochlorination than tertiary chlorides:

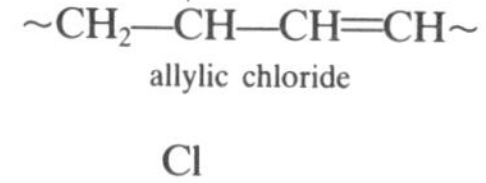

allylic chloride

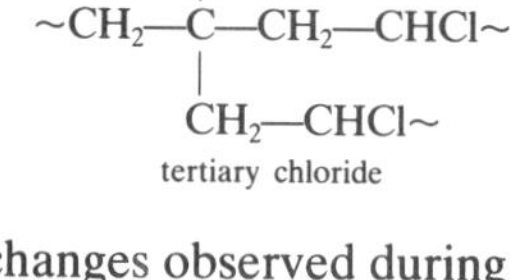

tertiary chloride

The main changes observed during thermal deterioration of PVC are evolution of hydrogen chloride, discoloration and embrittlement. In the absence of oxygen and stabilizers, the color changes rapidly from colorless to yellow, then red and ultimately black. Oxygen increases HCl evolution and decreases color development (yellow, rather than the red typical for PVC heated in an inert atmosphere). Stabilizers are used to decrease color formation both in the presence and absence of air.

Thermal degradation by depolymerization to monomer occurs in polymers with structures that favor the unzipping reaction. Most important commercially is polyformaldehyde. The hydroxy end group, $\sim CH_2—OH$, in the unmodified polymer permits depolymerization by a polar mechanism. This reaction is favored when the monomer unit contains a tertiary carbon atom as in the polymethacrylates:

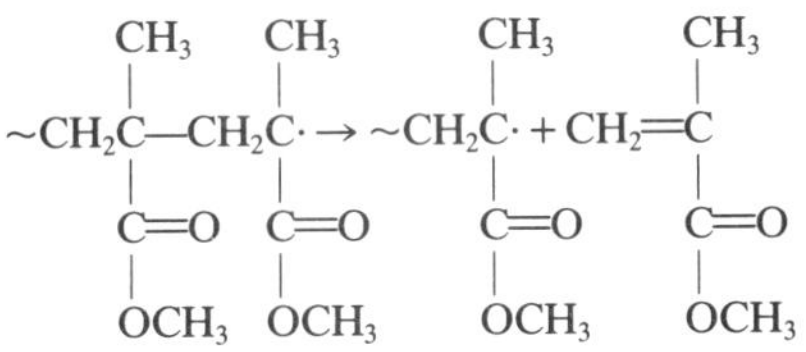

poly(methyl methacrylate)

Thermal decomposition of polymers that do not readily undergo elimination or depolymerization reactions occurs by random scission. The initial break occurs at a weak link in the polymer chain. Typical reactive sites include branched, unsaturated and oxygen-containing groups. The initial chain scission produces free radicals at the ends of the chain fragments, but intramolecular and intermolecular chain transfers produce a random distribution of radical sites. Subsequent β scission results in a random distribution of molecular weights in the degraded material. Intramolecular transfer via a six-membered cyclic mechanism and β scission proceeds as follows:

$$\sim(CH_2)_3CH_2(CH_2)_3CH_2(CH_2)_3\dot{C}H_2$$
$$\rightarrow\sim(CH_2)_3CH_2(CH_2)_3\dot{C}H(CH_2)_3CH_3$$
$$\rightarrow\sim(CH_2)_3\dot{C}H(CH_2)_3CH_2(CH_2)_3CH_3$$
$$\rightarrow\sim CH_2\dot{C}H_2 + CH_2{=}CH(CH_2)_7CH_3$$

2. *Methods of Improving Thermal Stability*

Two methods are available to obtain improved resistance to thermal degradation: the polymer structure may be modified to make it more difficult for degradation to occur, or additives may be used to improve stability. In the case of an important commercial polymer such as PVC, it was essential to find additives to suppress the harmful dehydrohalogenation and the resultant discoloration and embrittlement.

An obvious early approach to the stabilization of PVC was the use of additives to absorb HCl and thus prevent autocatalysis. Lead salts are commonly added to PVC for electrical applications. Other examples include basic lead carbonate, dibasic lead phosphite, dibasic lead phthalate and tribasic lead sulfate.

Another approach to stabilization of PVC is the removal of defect structures such as reactive allylic and tertiary chlorine atoms. Metal soaps based on barium, cadmium, zinc and other heavy metals react with active chlorine groups present in defect sites to replace them with carboxylic ester groups. The widely used organotin stabilizers react similarly to replace reactive chlorine with various ligands, including both thio and ester types. Typical examples are dibutyltin dilaurate and dibutyltin diisooctylthioglycolate:

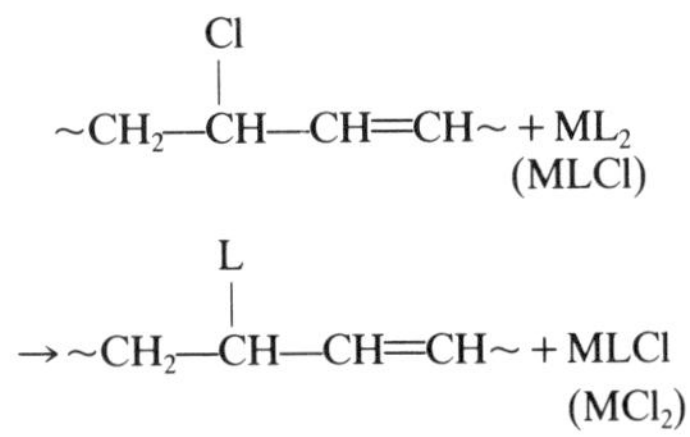

where M = Ba, Cd, Zn, Ca, Pb, LSb, R_2Sn, etc., and L = $—O_2CR$, —OR, —SR, etc.

Another important way to stabilize PVC is to decrease the formation of long polyene sequences (or alternatively to break up such sequences) that are responsible for the observed discoloration. In this function, mercaptides are particularly effective due to their tendency to add to double bonds. Another reaction that can interrupt conjugation in polyene sequences is a Diels–Alder reaction between a dieneophilic stabilizer (i.e., containing a maleate group) and a diene segment:

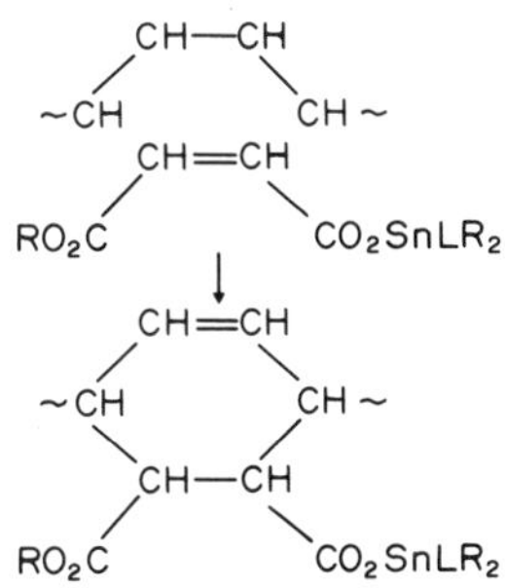

The stabilization of PVC is thus a complex combination of many mechanisms and interactions that are not yet fully understood.

Stabilization against degradation caused by unzipping can be obtained with appropriate blocking groups. For example, the terminal hydroxyl group on polyformaldehyde can be converted to a stable ester or ether group. In the case of radical chain ends on a polymer that readily depolymerizes, an inert blocking group can be inserted by copolymerization. A small amount of methyl acrylate can be used for this purpose as a comonomer with methyl methacrylate.

Radical scavengers and blocking groups are ineffective in polymers that degrade by random scission. Increased resistance to the effects of scission can be obtained by introduction of crosslinks. Synthesis of special heat-resistant polymers such as the polyimides is required for polymer service at temperatures in excess of 300 °C.

3. *Degradation by Thermal Autoxidation*

The degradation resulting from reaction with oxygen in air at temperatures encountered in normal use is particularly damaging to the properties of polymers. The reaction is autocatalytic in that hydroperoxides formed by oxidation are a source of free radicals which initiate the chain mechanism. The reaction of elemental oxygen with organic materials is thus called autoxidation:

Initiation	$ROOH \rightarrow RO\cdot + HO\cdot$
	$2ROOH \rightarrow RO\cdot + RO_2\cdot + H_2O$
Propagation	$RO_2\cdot + RH \rightarrow ROOH + R\cdot$
	$R\cdot + O_2 \rightarrow RO_2\cdot$
Termination	$2R\cdot \rightarrow R{-}R$
	$R\cdot + RO_2\cdot \rightarrow RO_2R$
	$2RO_2\cdot \rightarrow$ nonradical products $+ O_2$

Thermal decomposition of hydroperoxide can be either unimolecular or bimolecular. Higher concentrations and mobility favor association and the bimolecular process. In solid polymers with low initial extent of oxidation, first-order decomposition would be expected. If the polymer structure favors intramolecular propagation to place hydroperoxide groups on adjacent monomer units, hydrogen-bonded association will occur, favoring second-order decomposition. The tertiary C—H sites in adjacent units of polypropylene may react in this manner:

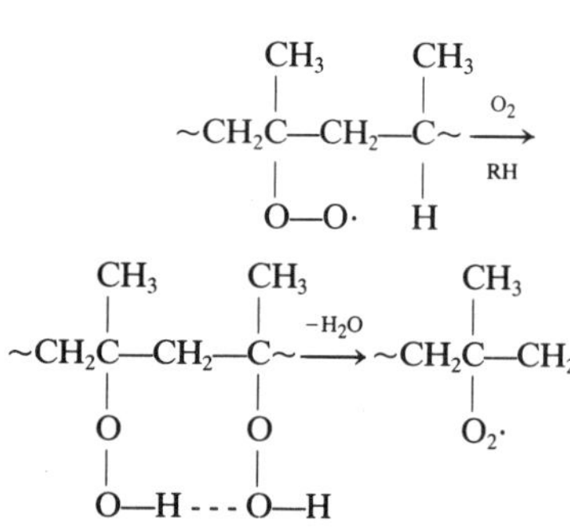

All of the oxy free radicals ($RO_2\cdot$, $RO\cdot$, $HO\cdot$) can abstract hydrogen atoms from reactive sites (tertiary, allylic, or adjacent to functional groups or heteroatoms) to form $R\cdot$, which reacts rapidly with O_2 to form $RO_2\cdot$ and thus initiates the propagation reactions. Many hydroperoxide groups can result from a single initiation, and as they decompose more initiations occur and the rate increases autocatalytically.

In the absence of antioxidants, termination occurs by combination of any two radicals to form nonradical products. If oxygen is readily available in the system, the concentration of $R\cdot$ will be negligible, and termination will result from bimolecular combination of two peroxy radicals. The tetroxide RO_4R will decompose to form O_2 and $2RO\cdot$ if R is tertiary. These tertiary alkoxy radicals may abstract hydrogen to reinitiate the oxidation mechanism, but they also readily undergo a β-scission reaction to form —CO— and $R\cdot$, which adds O_2. Primary and secondary $RO_2\cdot$ form tetroxides which decompose by a nonradical concerted intramolecular reaction to form stable oxy derivatives:

$$2R_3CO_2\cdot \longrightarrow R_3CO_4CR_3 \longrightarrow 2R_3CO\cdot + O_2$$

$$R_3CO\cdot \longrightarrow R_2C{=}O + R'CH_2\cdot \xrightarrow{O_2} R'CH_2O_2\cdot$$

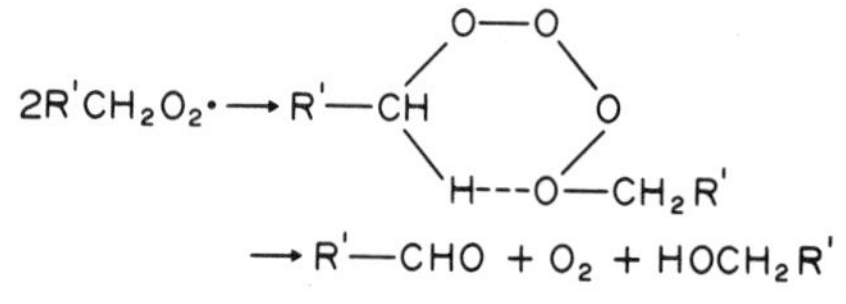

The observed deterioration of physical properties caused by thermal autoxidation results from the net effect of two degradation processes: (a) chain scission, which softens the polymer, and (b) crosslinking, which increases hardness. Thus, the net effect upon the modulus (stress required to produce a given elongation), or the hardness, may be either an increase or a decrease depending upon the ratio of chain scission to crosslinking. If oxygen is limited, as in processing by molding or extrusion, or by diffusion of air into thick sections of material, the carbon free radicals, $R\cdot$, will couple to form crosslinks.

Chain scission increases relative to crosslinking at higher temperatures, higher oxygen concentration, and with more bulky substituents on the chain which hinder the coupling reaction. Alkoxy radicals (which readily undergo β scission) are formed in termination by coupling of t-$RO_2\cdot$, as illustrated previously, and by ROOH decomposition in the initiation reactions. If a peroxy radical adds to a double bond (rather than the usual abstraction of allylic hydrogen), an epoxide is formed plus an $RO\cdot$ radical:

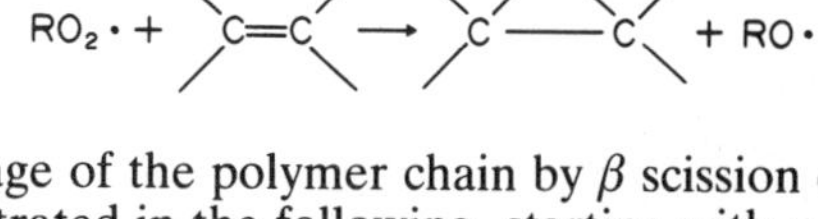

Cleavage of the polymer chain by β scission of $RO\cdot$ is illustrated in the following, starting with polypropylene hydroperoxide:

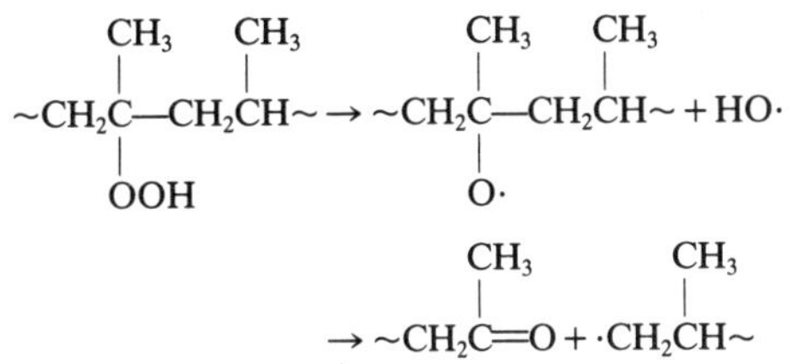

When peroxide is present and oxygen is limited or absent, crosslinking results from combination of R· radicals. The vulcanization of rubber by heating with dicumyl peroxide is nearly quantitative, with one crosslink formed for each peroxide molecule decomposed:

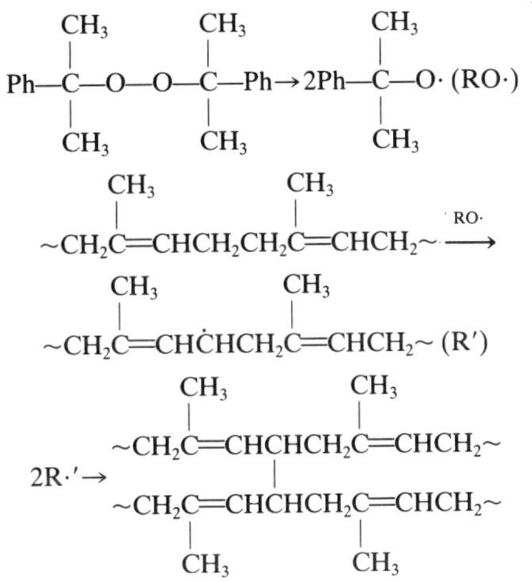

Crosslinking is also favored when pendent vinyl groups are present (as in emulsion styrene–butadiene copolymer), or if conjugated double bonds are introduced during autoxidation or vulcanization or by elimination reactions. Carbon radicals readily add to such structures:

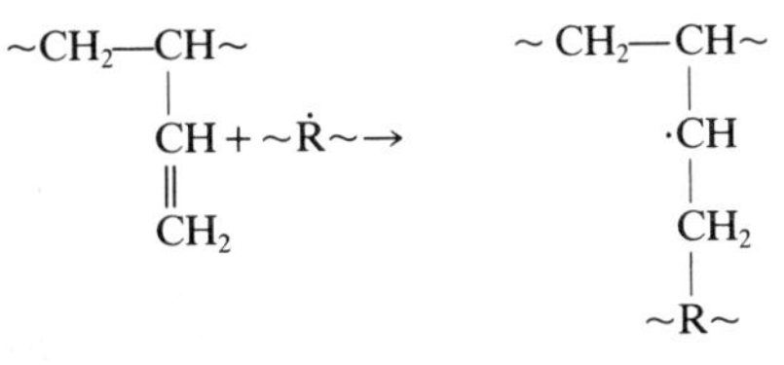

The relative resistance of polymers to autoxidation varies with the structure of the monomer units and also with the morphology or arrangement of the polymer chains. Some polymers are essentially unordered or amorphous, while others are largely oriented to form crystalline regions. The degree of orientation depends on the linearity of the main chain, regularity of repeat units and the nature of any substituents, functional groups or heteroatoms. Linear and branched polyethylene oxidize similarly at temperatures above the crystalline melting point, but at 100 °C the linear polymer oxidizes slower and levels off at a lower oxygen uptake. The cumulative amount of oxygen which can be absorbed varies inversely with the crystalline content of the polymers. Oxidation of these polyolefins is evidently limited to the amorphous regions and to the folded chains on the edges of crystalline regions.

4. Stabilization During Synthesis, Fabrication, Storage and Use

Polymers that are susceptible to thermal autoxidation must be protected by addition of stabilizer as soon as they are formed, and before they are exposed to air. The presence of unsaturation in natural rubber and synthetic elastomers makes them particularly susceptible to oxidative attack. Natural antioxidants are present in latex from rubber trees, but synthetic latex obtained by emulsion copolymerization requires an added antioxidant. A hindered phenol such as 2,6-di-*t*-butyl-4-methylphenol is often used. Only a small amount (less than 1% in the rubber) is required since additional antioxidant will be added during compounding before vulcanization.

Physical properties that are useful as a measure of polymer stability include: (a) plasticity, or its inverse, viscosity changes (e.g., Mooney viscosity), (b) solution viscosity of soluble polymers, or of the sol fraction, and (c) gel content of a partially soluble polymer (where an increase may indicate crosslinking resulting from thermal autoxidation). Processing temperatures may be fairly high for short periods of time, as in mixing or extrusion. Storage or shelf life will involve longer times at more moderate temperatures. Prolonged shelf life is not essential in most cases. Adequate stabilization is indicated in normal practice if there is no observed change in viscosity or gel content during shipment and storage before compounding.

Stabilization of the fabricated material to provide protection over the expected useful lifetime of the product requires additional amounts of stabilizers and possibly different types of antioxidants. A variety of accelerated testing procedures have been used to select suitable stabilizers for a particular formulation and use. Oxygen absorption measurements show the relative order of effectiveness of stabilizers in a given formulation. Stress relaxation and creep measurements show the effects of chain scission and crosslinking which result from thermal autoxidation. Testing can be accelerated to obtain results in shorter times by increasing temperature and oxygen concentration. However, the most reliable correlations with natural aging have been obtained with the individual test specimens isolated in test tubes and heated at atmospheric pressure in a circulating air oven at moderate temperatures (e.g., 70 °C).

Environmental factors are also important in the selection of appropriate stabilizers. Volatility of the antioxidant may result in significant losses at elevated temperatures or reduced pressure. Exposure to steam causes losses by a steam-distillation effect, and leaching with water and repeated laundering can also remove antioxidant. Dry-cleaning solvents are particularly harmful in extracting antioxidants. The loss of stabilizers by volatilization and extraction can be reduced by using higher-molecular-weight compounds and polymeric antioxidants. Polymer-bound antioxidants can be incorporated by reaction of appropriate functional groups on the antioxidant with reactive sites on the polymer chain. Monomers with antioxidant groups attached can be used in copolymerization. These chemically attached antioxidants appear to be as effective as the same concentration of stabilizer physically dispersed in the polymer.

5. *Stabilization by Chain-Breaking Antioxidants*

The free-radical chain mechanism of thermal autoxidation proceeds at a rate that is dependent on the ratio of the rates of initiation and termination, and also varies with the rate of propagation:

$$\text{rate} = k_p(R_i/2k_t)^{1/2}\,[\text{RH}]$$

where R_i is the rate of initiation of autoxidation of substrate RH and k_p and k_t are rate constants for propagation and termination. Antioxidants can be used to retard the rate of oxidation either by reducing the rate of initiation or increasing the rate of termination. Scott (1965) has classified antioxidants into two groups according to the mechanism by which they function: (a) preventive antioxidants, which reduce the rate of initiation in some way (e.g., by decomposing hydroperoxides to nonradical products), and (b) chain-breaking antioxidants, which intercept the free radicals that propagate the kinetic chain and thus provide an additional mechanism of termination.

Chain-breaking antioxidants may react with either R· or RO_2· to stop the kinetic chain. Since R· reacts rapidly with O_2 to form RO_2·, the concentration of carbon radical is much smaller than that of peroxy radical. Hindered phenols and aryl amines are excellent antioxidants which react readily with RO_2·, but they are not very reactive with carbon radicals and so are not very effective in protecting polymers against hardening by crosslinking.

Antioxidants may react with the propagating radical by addition or coupling, by one-electron transfer to convert the free radical to an anion, or more commonly by hydrogen transfer. Direct termination results if the initial reaction product cannot reinitiate the free-radical reaction. In most cases a new radical derived from the antioxidant is formed which terminates a second kinetic chain. In some cases it reacts either with oxygen or directly with the substrate by chain transfer, to reinitiate and autoxidation. A few substances, such as quinones and dialkyl or diaryl nitroxide free radicals, can trap alkyl free radicals, and a few stabilizers function by electron transfer. However, most commercially imporant antioxidants are either secondary aryl amines or hindered phenols which react by hydrogen transfer.

The reactive N—H of the amine and the O—H of the phenol enable them to compete with polymer RH for the RO_2· radical. Hydrogen transfer forms RO_2H and a radical derived from the antioxidant AH. The antioxidant A· typically serves as a radical trap and terminates a second kinetic chain:

$$RO_2\cdot + AH \rightarrow ROOH + A\cdot$$
$$RO_2\cdot + A\cdot \rightarrow RO_2A$$

The products of the antioxidant reaction may be capable of further reactions to terminate several free radicals per mole of original antioxidant. This mechanism was questioned when initial studies failed to show the expected kinetic deuterium isotope effect. Shelton (1972) has reported significant isotope effects in the retarded autoxidation of styrene–butadiene rubber and purified polyisoprene with both diaryl amines and hindered phenols in which the reactive hydrogen had been replaced by deuterium. Polymer containing the deuterated antioxidant oxidized faster than samples containing the normal N—H or O—H form of the antioxidant. At higher temperatures and higher antioxidant concentrations, an isotope effect was observed in the opposite direction, consistent with initiation by direct attack of O_2 on the antioxidant.

Most stabilizers exhibit prooxidant effects at higher concentrations and temperatures in addition to the expected antioxidant effect. The observed behavior thus depends on which effect predominates, and an optimum concentration for stabilization can be established for a particular antioxidant and polymer. It appears that direct attack of oxygen on the antioxidant is the most important source of initiating free radicals in the first stage of retarded autoxidation. Peroxide initiation becomes more important as hydroperoxide accumulates along with other oxy derivatives of polymer and antioxidant:

$$AH + O_2 \rightarrow HO_2\cdot + A\cdot \xrightarrow{O_2} AO_2\cdot$$
$$ROOH \rightarrow RO_2\cdot, \text{etc.}$$
$$\vdots$$
$$\text{oxy groups}$$

6. *Stabilization by Preventive Antioxidants*

Stabilizers that protect polymers by decreasing the rate of initiation of autoxidation are called preventive antioxidants. Initiation by decomposition of hydroperoxide is accelerated by heat, light and transition metal ions. Appropriate chelating agents can be used to complex metal ions and thus deactivate the metal as a catalyst of autoxidation. Similarly, ultraviolet absorbers and quenchers are used to reduce the rate of photoinitiation of autoxidation. Initiation by thermal decomposition of hydroperoxides can be reduced by compounds that convert them to stable nonradical products. Stabilizers that function in this way are commonly referred to as peroxide decomposers.

Many organic sulfur, nitrogen and phosphorus compounds are known to accelerate the decomposition of organic hydroperoxides to nonradical products. These include certain (but not all) sulfides, disulfides, sulfoxides, sulfonic acids and metal salts of dithiocarbamates and dithiophosphates. The catalytic nature of the reaction has been clearly shown for the decomposition of cumene hydroperoxide (CHP) by these metal derivatives. Complete decomposition of CHP was obtained with as little as one mol.% of these additives relative to the hydroperoxide. A polar mechanism of decomposition was suggested by the observation that phenol was the principal decomposition product, consistent with the known acid-catalyzed conversion of CHP to phenol and acetone.

The superior aging resistance of natural rubber vulcanized with tetramethylthiuram disulfide in the presence of zinc salts has been attributed to the formation of zinc *N,N′*-dimethyldithiocarbamate. Removal of this salt by acetone extraction destroyed the improved aging resistance. The activities of this and other metal salts of similar organosulfur compounds are thought to be due primarily to their ability to decompose hydroperoxides by a polar process.

The mechanism of oxidation inhibition by zinc dialkyl dithiophosphates has been reviewed by Burn (Gould 1968a p. 323). There is evidence that this material and the dithiocarbamate salts function as chain-breaking antioxidants in addition to their well-established activity as peroxide decomposers. The mechanisms are not yet established, but most workers agree that reaction products of the metal salts are the active hydroperoxide decomposers.

Alkyl and aryl phosphite esters are very efficient peroxide decomposers, and consume one mole of hydroperoxide per mole of stabilizer as the phosphite is converted to phosphate. This stoichiometric reaction is rapid and probably involves initial formation of a complex of hydroperoxide with phosphite:

$$(ArO)_3P + ROOH \longrightarrow (ArO)_3P\text{---}\overset{H}{O}\text{---}OR \longrightarrow (ArO)_3P{=}O + ROH$$

The phosphate esters are usually not peroxide decomposers, so a sufficient amount of phosphite stabilizer must be used to consume all the hydroperoxide that may be formed. In more complex derivatives, such as 2,6-di-*t*-butyl-4-methylphenyl phosphate and some cyclic phosphates, the ester undergoes reactions to form acidic products which decompose hydroperoxide catalytically.

Among the many types of organic sulfur compounds that show activity as peroxidce decomposers, the most widely used are the dialkyl thiodipropionate esters. The dilauryl ester (DLTDP) is commonly used along with a hindered phenolic antioxidant to stabilize polypropylene. Marshall (Gould 1968a p. 140) observed an initial slow reaction of ROOH with DLTDP at 30–90 °C followed by a faster main reaction. The initial reaction was first order in both reactants, consistent with oxidation of the sulfide to sulfoxide:

$$(H_{25}C_{12}O\overset{\displaystyle O}{\overset{\|}{C}}CH_2CH_2\text{—})_2S \xrightarrow{ROOH} (H_{25}C_{12}O\overset{\displaystyle O}{\overset{\|}{C}}CH_2CH_2\text{—})_2S{=}O$$

Addition of dilauryl β,β'-sulfinyldipropionate (DLTDP sulfoxide) did not eliminate the induction period. The formation of a more reactive species from decomposition and further oxidation is suggested to account for the catalytic nature of the main reaction, which consumed at least 20 moles of hydroperoxide per mole of sulfur compound.

The only organic sulfides and disulfides that show activity as preventive antioxidants are those that are oxidized by hydroperoxide to sulfides (—S(=O)—) and thiosulfinates (—S—S(=O)—), which are thermally unstable. The active hydroperoxide decomposers must be among the decomposition and oxidation products of these compounds. Studies in the author's laboratory have established the reactions that occur and the nature of the catalytic hydroperoxide decomposers. This work has been described in a recent review (Shelton 1981) and is illustrated below with a reaction sequence starting with di-*t*-butyl sulfide as a simpler example of stabilizers such as DLTDP:

$$(CH_3)_3CSC(CH_3)_3 \xrightarrow{ROOH} (CH_3)_3C\overset{\displaystyle O}{\overset{\|}{S}}C(CH_3)_3 \xrightarrow{heat} \underset{\text{sulfenic acid}}{(CH_3)_3CSOH} + (CH_3)_2C{=}CH_2$$

$$2(CH_3)_3CSOH \xrightarrow{-H_2O} \underset{\text{thiosulfinate ester}}{(CH_3)_3CS\overset{\displaystyle O}{\overset{\|}{S}}C(CH_3)_3}$$

$$(CH_3)_3C\overset{\overset{O}{\|}}{S}SC(CH_3)_3 \xrightarrow{\text{heat}} \underset{\text{thiosulfoxylic acid}}{(CH_3)_3CSSOH} + (CH_3)_2C{=}CH_2$$

$$(CH_3)_3CSOH + 2ROOH \rightarrow \underset{\text{sulfonic acid}}{(CH_3)_3CSO_3H} + 2ROH$$

$$(CH_3)_3CSSOH + 2ROOH \rightarrow \underset{\text{thiosulfuric acid}}{(CH_3)_3CSSO_3H} + 2ROH$$

The sulfide consumed one mole of ROOH to form sulfoxide, which had no effect on either *t*-butytl hydroperoxide or cumene hydroperoxide. Heating the sulfoxide to 80 °C in benzene gave sulfenic acid, RSOH, identified by NMR and ir spectroscopies and trapping with appropriately substituted alkenes and alkynes. In the absence of hydroperoxide or trapping agents, two moles of RSOH reacted to form thiolsulfinate (identical to the product formed from disulfide and hydroperoxide). The thiolsulfinate ester had no effect on ROOH at 25 °C in benzene until after an induction period during which it decomposed in part to form thiosulfoxylic acid, RSSOH. The sulfenic acid rapidly consumed two moles of ROOH per mole of RSOH to form sulfonic acid. The thiosulfoxylic acid is oxidized to thiosulfuric acid and other products.

The only compounds in this reaction sequence that catalytically decomposed hydroperoxide were RSOH and RSSOH, together with the stronger acids formed by their reaction with ROOH. The formation of phenol and acetone from CHP in the presence of these materials confirmed the polar nature of the mechanism. However, some products of free-radical reactions were also observed, consistent with prooxidant effects. There is also evidence that RSOH is an efficient free-radical chain stopper.

The greater activity of DLTDP compared with that of di-*t*-butyl sulfide as a stabilizer in polymers is due to the greater ease of decomposition of sulfoxide formed from the thiopropionate esters. The hydrogen on the carbon adjacent to the ester group is activated by the ability of the carbonyl to stabilize the developing anion in the transition state of the intramolecular elimination reaction:

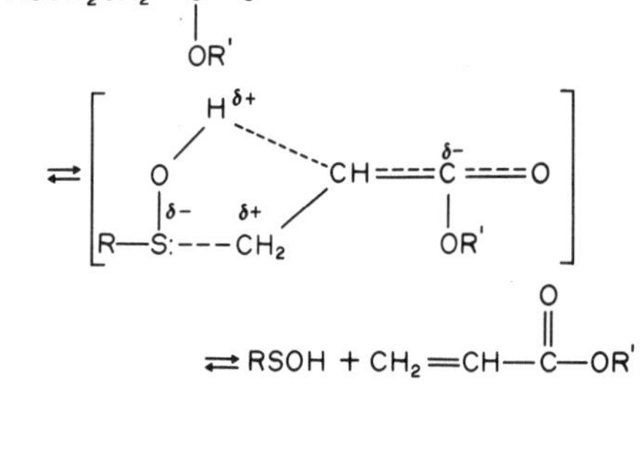

7. Use of Combinations of Stabilizers Against Thermal Degradation

Combinations of antioxidants in polymers frequently give better protection against oxidative degradation than the sum of the effects observed when the same amounts are used independently. Such a beneficial interaction is called synergism. The desired effect requires careful selection of the stabilizers and verification by actual testing. The wrong choice can in some cases result in antagonistic effects. Scott (1965) has suggested a classification into two mechanistically different types of synergism: (a) homosynergism, involving compounds of unequal activity that operate by the same mechanism; and (b) heterosynergism, involving two or more stabilizers that function by different mechanisms.

An example of homosynergism would be the use of two phenolic antioxidants that differ in the degree of hindrance by bulky *ortho* substituents. The less hindered phenol will be a more efficient radical scavenger and will be consumed more rapidly. However, it will not be depleted if the more hindered, but less efficient, antioxidant can serve as a reservoir of available hydrogen to regenerate the more reactive phenol.

The most obvious type of heterosynergism is the combination of a preventive antioxidant with a chain-breaking antioxidant. Many combinations are available that show marked synergistic effects. If an effective peroxide decomposer is present, there will be less hydroperoxide available for initiation and fewer oxidation chains will need to be terminated. Similarly, with an efficient free-radical chain stopper present the kinetic chain length will be shortened and less hydroperoxide will be formed. Thus, the two types of stabilizer protect each other by mutually reducing the demand on the other.

A widely used combination for stabilization of polyolefins is dilauryl thiodipropionate used with 2,6-di-*t*-butyl-4-methylphenol. This pair of stabilizers is of particular interest because both compounds are among the additives that have been approved for use in packaging materials for food products.

Autosynergism may account for the greater effectiveness of a stabilizer that can function by more than one mechanism. Many peroxide decomposers have some activity as chain-stopping antioxidants, and both phenolic and amine antioxidants have been observed to decompose some hydroperoxide. Organic phosphites and organic sulfur compounds can be combined with a hindered phenol. The thiobisphenols, such as 4,4′-thiobis(2-methyl-6-*t*-butylphenol), are well-known antioxidants.

The roles of stabilizers in autoxidation frequently include prooxidant effects. Increased rates of oxidation have been observed with sulfoxides and their decomposition products. The several ways in which an antioxidant may function under thermal autoxidation conditions are illustrated for the specific case of *t*-butanesulfenic acid (formed from di-*t*-butylsulfoxide):

Prooxidant

Induced decomposition

$$t\text{-BuSO}\overset{\text{H}}{\overset{|}{\text{H}}}\cdots\text{OOR} \rightarrow t\text{-BuSO}\cdot + \text{RO}\cdot + \text{H}_2\text{O}$$

Initiation

$$\text{RO}\cdot + \text{RH} \rightarrow \text{ROH} + \text{R}\cdot \xrightarrow{\text{O}_2} \text{RO}_2\cdot$$

Antioxidant

Chain stopper

$$t\text{-BuSOH} + \text{RO}_2\cdot \rightarrow \text{ROOH} + t\text{-BuSO}\cdot$$

Peroxide decomposer

$$t\text{-BuSOH} + 2\text{ROOH} \rightarrow t\text{-BuSO}_3\text{H} + 2\text{ROH}$$

$$t\text{-BuSO}\cdot + \text{O}_2 \rightarrow t\text{-Bu}\overset{\text{O}}{\overset{\|}{\text{S}}}\text{O}_2\cdot \rightarrow t\text{-BuO}\cdot + \text{SO}_2$$

$$\text{ROOH} \xrightarrow[\text{SO}_2]{t\text{-BuSO}_3\text{H}} \text{nonradical products}$$

The mechanism of retarded autoxidation in the presence of stabilizers of the chain-breaking type includes participation of the antioxidant in all three stages of the reaction sequence:

Initiation

Prooxidant

$$\text{AH} + \text{O}_2 \rightarrow \text{HO}_2\cdot + \text{A}\cdot \xrightarrow{\text{O}_2} \text{AO}_2\cdot$$

$$\text{HO}_2\cdot(\text{AO}_2\cdot) + \text{RH} \rightarrow \text{R}\cdot + \text{H}_2\text{O}_2(\text{AO}_2\text{H})$$

Propagation

Chain transfer

$$\text{RO}_2\cdot + \text{AH} \rightarrow \text{ROOH} + \text{A}\cdot \xrightarrow{\text{O}_2} \text{AO}_2\cdot$$

$$\text{AO}_2\cdot + \text{RH} \rightarrow \text{AO}_2\text{H} + \text{R}\cdot \xrightarrow{\text{O}_2} \text{RO}_2\cdot$$

Termination

Chain stopping

$$\text{RO}_2\cdot + \text{AH} \rightarrow \text{ROOH} + \text{A}\cdot$$
$$\text{RO}_2\cdot + \text{A}\cdot \rightarrow \text{RO}_2\text{A}$$
$$2\text{A}\cdot \rightarrow \text{A—A}$$

The preceding sets of reactions involving both preventive and chain-breaking antioxidants in thermal autoxidation will be superimposed on the free-radical chain-reaction sequence presented in Sect. 3 for the mechanism of uninhibited autoxidation.

Lists of antioxidants used in stabilization of polymer material are available in a review by Ambelang et al. (1963). Information on available commercial antioxidants, including both preventive and chain-breaking types, may be obtained from manufacturers and suppliers.

See also: Polymers: Technical Properties and Applications

Bibliography

Allara D L, Hawkins W L (eds.) 1978 *Stabilization and Degradation of Polymers*, Advances in Chemistry Series 169. American Chemical Society, Washington, DC

Ambelang J C, Kline R H, Lorenz O M, Parks C R, Wadelin C, Shelton J R 1963 Antioxidants and antiozonants for general purpose elastomers. *Rubber Chem. Technol.* 36: 1497–541

Barnard D, Lewis P M 1988 Oxidative aging. In: Roberts A D (ed.) 1988 *Natural Rubber Science and Technology*. Oxford University Press, Oxford, Chap. 13

Conley R T (ed.) 1970 *Thermal Stability of Polymers*, Vol. 1. Dekker, New York

Dweik H S, Scott G 1984 Mechanisms of antioxidant action. *Rubber Chem. Technol.* 57: 735–43; 908–17

Gould R F (ed.) 1968a *Oxidation of Organic Compounds*. Advances in Chemistry Series 75–1. American Chemical Society, Washington, DC

Gould R F (ed.) 1968b *Stabilization of Polymers and Stabilizer Processes*, Advances in Chemistry Series 85. American Chemical Society, Washington, DC

Grassie N 1956 *Chemistry of High Polymer Degradation Processes*. Butterworth, London

Grassie N 1987 *Developments in Polymer Degradation*, Vol. 7. Elsevier, Amsterdam

Hawkins W L (ed.) 1972 *Polymer Stabilization*. Wiley, New York

Lundberg W O (ed.) 1961–1962 *Autoxidation and Antioxidants*, Vols. 1, 2. Interscience, New York

Madorsky S L 1964 *Thermal Degradation of Organic Polymers*. Interscience, New York

Nethsinghe L P, Scott G 1984 Mechanisms of antioxidant action. *Rubber Chem. Technol.* 57: 779–91; 918–24

Reich L, Stivala S S 1969 *Autoxidation of Hydrocarbons and Polyolefins: Kinetics and Mechanisms*. Dekker, New York

Scott G 1965 *Atmospheric Oxidation and Antioxidants*. Elsevier, Amsterdam

Scott G 1983 Recent trends in the stabilization of polymers. *Br. Polym. J.* 15: 208–23

Scott G 1987 *Developments in Polymer Stabilization*, Vol. 8. Elsevier, Amsterdam

Scott G 1991 *Mechanisms of Polymer Degradation and Stabilization*. Elsevier, Amsterdam

Shelton J R 1972 Review of basic oxidation processes in elastomers. *Rubber Chem. Technol.* 45: 359–80

Shelton J R 1981 Organic sulfur compounds as preventive antioxidants. In: Scott G (ed.) 1981 *Developments in Polymer Stabilization*. Applied Science, London, pp. 23–69

Shelton J R 1983 Oxidation and stabilization of rubbers. *Rubber Chem. Technol.* 56: G71–86

J. R. Shelton
[Case Western Reserve University, Cleveland, Ohio, USA]

Degradation and Stabilization, Ultraviolet

Ultraviolet (uv) degradation of polymers is a natural phenomenon, involving mainly sunlight and oxygen, which causes deterioration of physical properties and appearance. Most commercial polymers would not have a useful life except for stabilizing additives which are present in small amounts, generally less than 1%. These additives permit the polymers to survive the severe conditions of processing and fabrication and to withstand the conditions of end use, whether as articles for use indoors or outdoors.

Various stabilizing additives are available for the protection of polymers. These may function either as chemical stabilizers or simply as light screens to restrict penetration of uv radiation into the polymer. Light screens, carbon black for example, give excellent protection against uv degradation, but their use is limited to those applications where pigmentation is acceptable. Chemical stabilizers are of several types—uv absorbers, hindered amines, nickel compounds, benzoate esters, trivalent phosphorus compounds, hindered phenols and thiosynergists. These additives stabilize polymers by different mechanisms and the selection of a stabilizer for a specific use is dependent on the polymer to be stabilized, other components of the formation, the article thickness, the use and the required lifetime.

Polymer uv stability should be evaluated by a method which will permit prediction of performance under conditions of end use. A number of exposure options, both indoors and outdoors, and a number of test methods are available for consideration.

1. Ultraviolet Degradation of Polymers

Applications involving sunlight exposure are the most challenging and most stressful for commercial thermoplastic polymers. With the exception of poly(tetrafluoroethylene), which has been reported to endure outdoor exposure for more than 20 years without significant modification of physical properties, all thermoplastic polymers undergo loss of physical properties or color changes to varying degrees when exposed to direct sunlight.

Sunlight is harmful to polymers because the near-uv portion of the sun's spectrum which reaches the surface of the earth, 290–400 nm, includes wavelengths which possess sufficient energy to induce photochemical reactions in polymers when absorbed by suitable chromophores in the polymers. Table 1 shows the relation between wavelength of light and energy. Table 2 lists the bond energies of some common bonds in organic molecules. The low energy of the oxygen–oxygen bond as in peroxides should be noted.

Table 1
Correlation of wavelength and energy of uv and visible radiation

Wavelength (nm)	Energy ($kJ\,mol^{-1}$)
290	405
400	295
600	195
800	150

Table 2
Bond energy of selected bonds in organic molecules

Bond	Energy ($kJ\,mol^{-1}$)
O—H	~420
C—H	~395
C—O	~375
C—C	~345
O—O	125–145

Most thermoplastic polymers undergo degradation on exposure to sunlight, either because they contain functional groups which are chromophores and can absorb light directly to cause photochemcial reactions or because they contain trace amounts of impurities introduced during manufacture or subsequent thermal processing, which behave as sensitizers of photodegradation on exposure to light. Some substances—dyes, pigments and some aromatic compounds in particular—can act as sensitizers for the photodegradation of susceptible polymers by the photochemical generation of singlet oxygen which adds to unsaturated groups to yield hydroperoxides are known to be common low-level impurities in hydrocarbon polymers which sensitize the photodecomposition of the polymers by absorption of near-uv light and subsequent generation of alkoxy and hydroxy free radicals, initiators of photooxidation.

The portion of the sun's spectrum between 290 nm and 400 nm shown in Fig. 1 is a small fraction of the total spectrum but is important for polymers because it includes the highest-energy region and is the only portion which can cause direct harm to unmodified polymers, that is, polymers without colorants, fillers, coatings and so on. The visible region of the

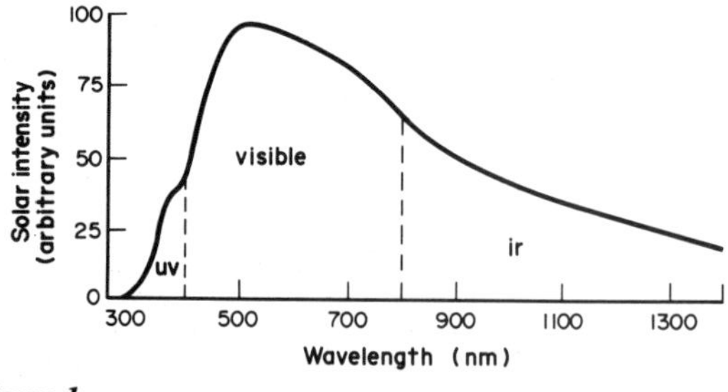

Figure 1
Typical midday solar energy distribution

solar spectrum, 400–800 nm, does not usually cause direct harm to polymers but can do so by interaction with sensitizing substances in the polymer. The infrared (ir) portion, above 800 nm, is generally considered harmless in a photochemical sense but may have a role in the thermal oxidative degradation of some polymers.

Direct sunlight in a weathering situation is most harmful to polymers and presents the greatest challenge to the stabilization technologist. Sunlight which reaches indoors has been filtered by glass which removes the shortest, most energetic wavelengths, 290–310 nm, attenuating the effect. Nevertheless, indoor sunlight can have an important effect on the indoor lifetimes of plastics articles as well as carpeting, upholstery and draperies. Usual artificial lighting, fluorescent and incandescent, plays a less important role in polymer photodegradation than sunlight, mainly because such lighting has only a very small fraction of wavelengths harmful to polymers in its emission spectrum. However, special attention needs to be paid to the stabilization of polymers used close to artificial light sources, as diffusers or lenses for example, to protect against discoloration and loss of physical properties.

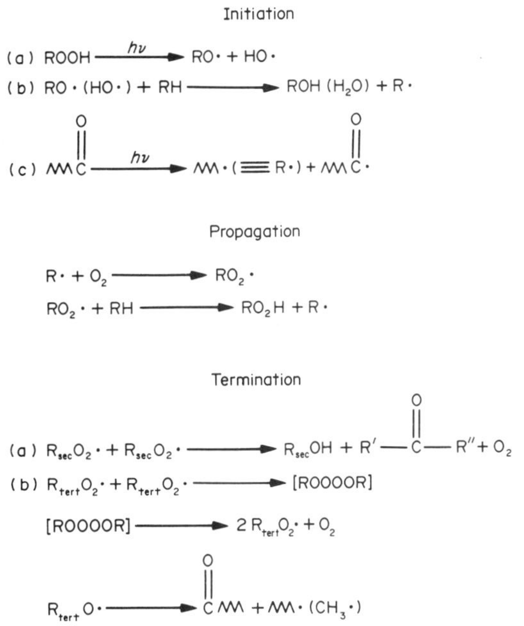

Figure 2
Reaction scheme for photooxidation of hydrocarbon polymers

2. *Mechanisms of Photodegradation of Polymers*

Many of the reactions and pathways of polymer photodegradation are identical with those of polymer thermal degradation. These mechanisms were elucidated by Bolland and Gee (1946) and later workers for the thermal oxidation of liquid hydrocarbons. They are the core reactions of hydrocarbon oxidation. Key elements of the photooxidative scheme for hydrocarbon polymer oxidation are shown in Fig. 2. The basic reactions in polymer photooxidation involve free radicals. The basic propagating reaction is a free radical chain reaction. A major difference between the reaction schemes for thermal oxidation and photooxidation lies in the relative instability of hydroperoxides in light. Whereas hydroperoxides are relatively stable at moderate temperatures in the dark, they are sensitive to near-uv light, especially the shorter wavelengths. Thus, the major product of thermal oxidation and photooxidation, hydroperoxide, is a major initiator of photooxidation by virtue of its facile cleavage into alkoxy and hydroxy radicals on absorption of a photon of uv light. Hydroperoxides have very low extinction coefficients for absorption of light but since they do not appear to have higher energy states which can be filled, an absorbed photon of light results directly in homolytic cleavage of the oxygen–oxygen bond.

Chromophores other than hydroperoxides—ketones, esters, aromatic rings—are generally able to fill excited states on absorption of a photon of light. Interrelationships between the ground state and excited singlet and triplet states are shown in Fig. 3. The absorption of a photon of requisite energy raises the chromophore from the ground state to an excited singlet state. There are several possibilities for the excited chromophore:

(a) it may emit the excitation energy as a photon of light (fluorescence) and return to the ground state;

(b) it may undergo photochemical change from the excited singlet state;

(c) with favorable energetics, it may undergo intersystem crossing and convert to the triplet state;

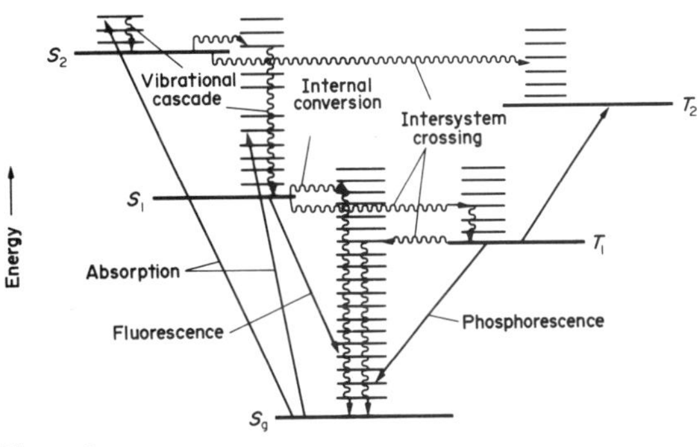

Figure 3
Photophysical processes in a typical organic molecule

(d) it may return to the ground state from the excited triplet by emitting a photon of light (phosphorescence);

(e) it may undergo photochemical change from the excited triplet; or

(f) it may return to the ground state by internal conversion from either the excited singlet or excited triplet by entering a higher vibrational level of the ground state and dissipating the excitation energy as heat to reach the lowest vibrational level of the ground state.

Photoexcited ketone chromophores usually undergo reaction by the Norrish I and II reactions shown in Fig. 4. Both paths lead to polymer backbone cleavage. By the former, radical intermediates are formed, but in the latter the cleavage takes place by means of a cyclic transition state without radical intermediates. Carboxylic esters photodegrade by undergoing reactions similar to the Norrish I and II reactions of ketones, as shown in Fig. 5.

Aliphatic hydrocarbon polymers, if perfect, should not in theory be subject to near-uv photoinduced reactions, since they would contain no chromophores responsive to sunlight. However, in practice they contain small amounts of hydroperoxide and carbonyl impurities which do respond to sunlight and are commonly considered to be the initiating sites for photooxidation of these polymers. The impurities arise from low levels of thermal oxidation occurring during manufacture, processing, and subsequent storage of those polymers.

Some substances, most notably dyes and pigments, incorporated into polymers can have the undesired effect of sensitizing the polymers to photodegration. Many dyes are known sensitizers of singlet oxygen (Zweig and Henderson 1975) and, if incorporated into a polymer with unsaturated carbon–carbon bonds, would result in the formation of hydroperoxides. The anatase form of titanium dioxide is known to a photochemically active pigment which catalyzes the photochemical oxidation of isopropanol to acetone (Irick 1972) and which contributes a surface-renewing effect to coatings through photooxidation of coating surfaces and subsequent chalking (surface powdering). Precautions should be taken to avoid adding photosensitizing additives to polymers.

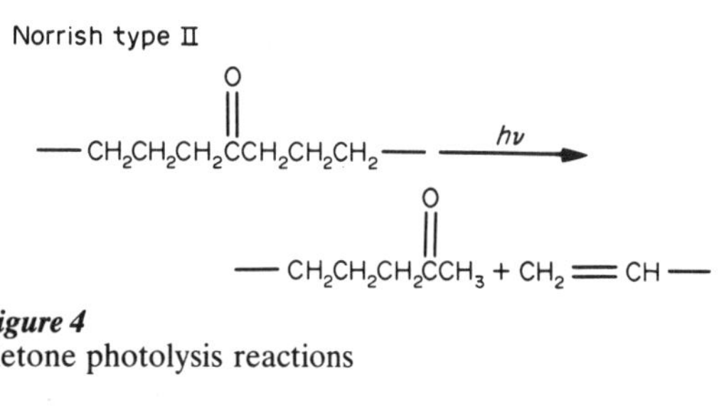

Figure 4
Ketone photolysis reactions

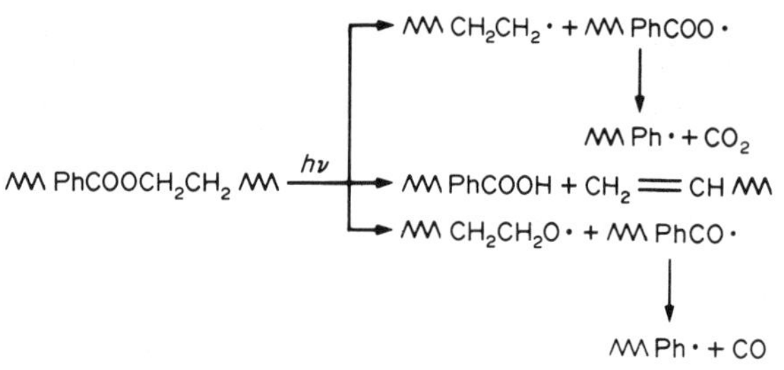

Figure 5
Ester photolysis reactions

3. Light Stabilizing Mechanisms for Polymers

Prevention of radiation from interacting with polymers has been achieved with opaque surface coatings which block the light from the polymer and with pigments such as carbon black and rutile (TiO_2) incorporated into the polymer, but both methods have obvious disadvantages. A more acceptable method is to incorporate stabilizers into the polymer.

A variety of stabilizer types is available to permit polymers to be used in light exposure applications. Since sunlight exposure is one of the most stressful situations for thermoplastic polymers, relatively high levels of light stabilizers (0.2–2%) may be needed to achieve the desired lifetimes. Generally, higher levels of stabilizers are required for effective photostabilization than for thermal stabilization. In addition, primary stabilizers for photostabilization are different in nature from those used for thermo-oxidative stabilization. Hindered phenols, which are commonly used as the primary stabilizers in thermal environments, play a secondary role in the light stabilization of polymers, where they function mainly to protect the polymers during thermal processing. The main classes of photostabilizing additives are uv absorbers, hindered amines, organonickel compounds, benzoates and phosphites. These function by a variety of mechanisms. Advantage may often be taken of different mechanisms, using a formulation of two or more classes of photostabilizer.

The light stabilization of polymers has been an art and empirical science since its inception. Understanding the mechanisms by which light stabilizers function has generally followed actual application for each specific class of compounds. A notable current example is the case of the hindered amines, which were discovered to form stable nitrooxyl radicals in the early 1960s in a search for stable radicals with antioxidant activity. Since then they have evolved to become a major class of light stabilizing additives for polymers. In the intervening years, several mechanisms by which hindered amines function as light stabilizers have been developed, but the complete story remains to be written.

3.1 Ultraviolet Absorbers

Several chemical classes (Fig. 6) are represented among commercially available uv absorbers: hydroxybenzophenones, hydroxyphenylbenzotriazoles, oxanilides, α-cyanoacrylates and phenyl salicylates. Phenyl salicylates are only moderately absorbing by themselves, but on exposure to uv light they undergo a photo-Fries rearrangement and are transformed into the hydroxybenzophenone uv absor-

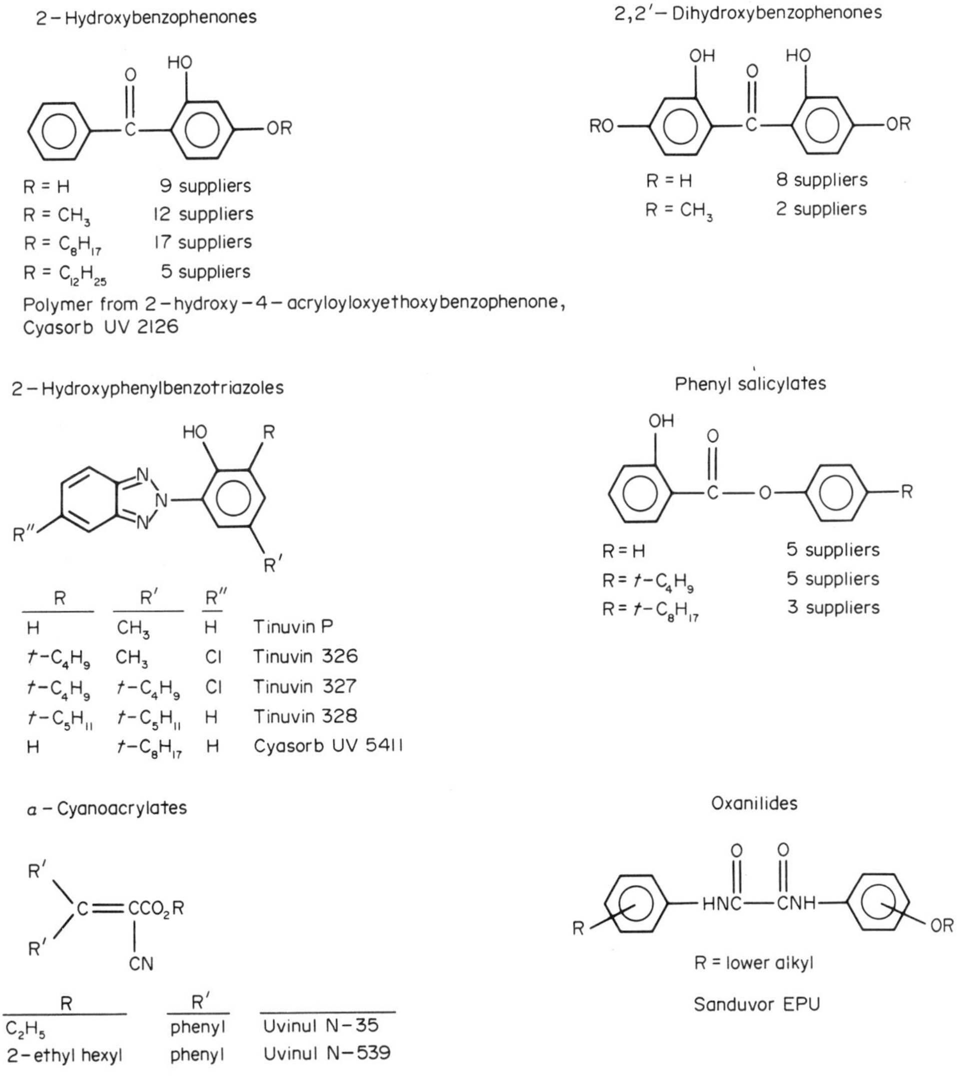

Figure 6
Chemical structures of commercially available uv absorbers

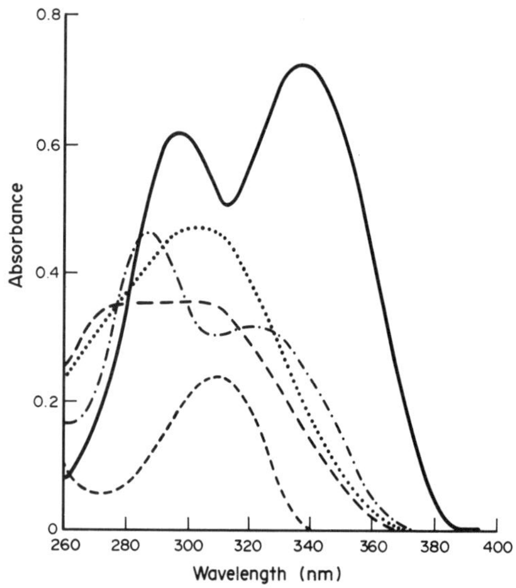

Figure 7
Absorbance of uv absorbers as a function of wavelength (at a concentration of 10 mg l^{-1} in $CHCl_3$, 1 cm cell): —— 2-(2′-hydroxy-5′-methylphenyl)-benzotriazole; · · · ethyl 2-cyano-3,3-diphenyl acreylate; – · – · 2 hydroxy-4-octyloxy-benzophenone; – – – oxanilide (Sanduvor EPU); - - - phenyl salicylate

bers, which are more strongly absorbing and more effectively stabilizing.

Figure 7 shows the absorption spectra of examples of the major classes of uv absdorbers. There are qualitative differences in the spectra, but it is apparent that they effectively absorb the harmful uv light. Four major requirements of uv absorbers are:

(a) they must show a sharp cutoff of absorption at 400 nm so that they do not absorb in the visible and contribute color but absorb strongly in the 290–400 nm region (Fig. 8).

(b) they must be photostable—able to undergo photoexcitation countless times without photochemical reactions;

(c) the photoexcited molecule must have a means of returning to the ground state and dissipating the photoexcitation energy in harmless ways, generally as heat (Otterstedt 1973, Werner 1979); and

(d) the thickness of the specimen in relation to the absorber concentration must be such as to permit complete absorption of harmful radiation—thin samples (<0.05 mm) and the surface of thicker samples are difficult to stabilize because the number of absorber molecules in the light path is too small for complete absorption of the uv light.

Ultraviolet absorbers are effective in virtually all polymers.

3.2 Hindered Amines

Hindered amine light stabilizers, based on 2,2,6,6-tetramethylpiperidine, are one of the most important developments in light stabilizers for polymers (Carlson et al. 1979, Gugumus 1979). Figure 9 shows the structures of some commercially available hindered amines. Like uv absorbers, they are useful in a large number of polymers, except where the basicity of these compounds may be a drawback. Hindered amines provide degrees of polymer stability which were previously unachievable and as a result are permitting polymers to be used successfully for more demanding applications. Very often, combinations of uv absorbers with hindered amines are the systems of choice for polymer stabilizing applications. The two classes of stabilizers function by different mechanisms which often supplement each other (Leu et al. 1977). The piperidine moiety of hindered amines does not absorb in the near uv and is generally acknowledged to function by trapping carbon or peroxy free radicals by means of nitroxyl free radicals as shown by Fig. 10.

The termination of photooxidative chains at low levels of conversion and especially by trapping carbon free radicals before they have an opportunity to react with oxygen makes hindered amine light stabilizers particularly effective inhibitors of photooxidation. Hindered phenols do not trap carbon radicals but usually trap peroxy free radicals, to form hydroperoxides and phenol oxidation products, such as substituted cyclohexadieneones (Fig. 11). Hydroperoxides and phenol oxidation products are potential initiators of photooxidation. A major difference between nitroxyl and hindered phenol radical trappers is the recyclable nature of the former and the sacrifical nature of the latter.

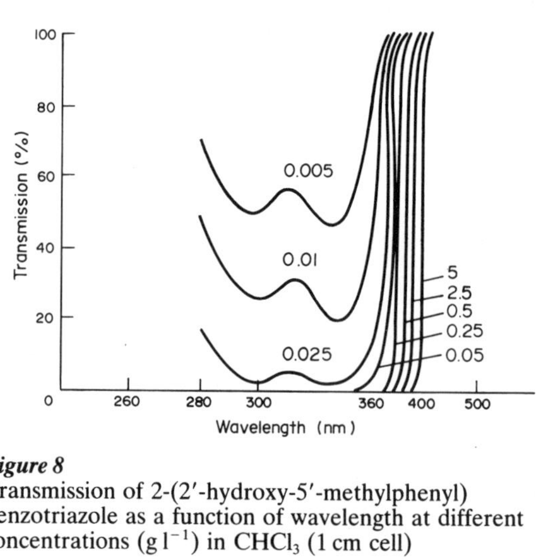

Figure 8
Transmission of 2-(2′-hydroxy-5′-methylphenyl) benzotriazole as a function of wavelength at different concentrations (g l^{-1}) in $CHCl_3$ (1 cm cell)

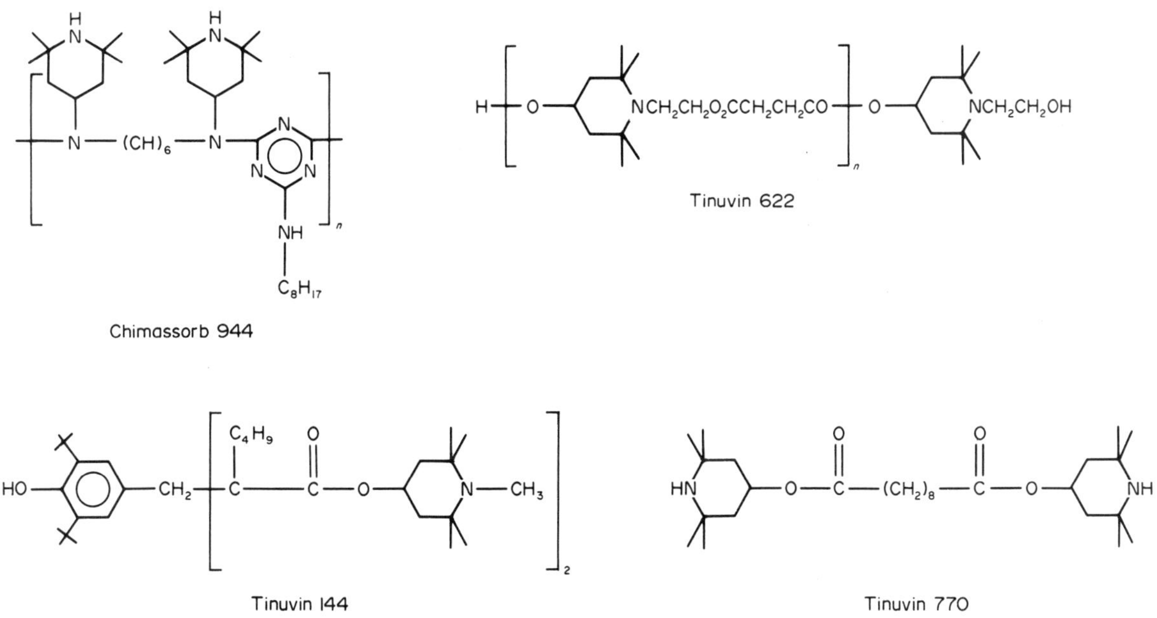

Figure 9
Chemical structures of commercially available hindered amine stabilizers

3.3 Nickel Compounds

Polyolefins respond to stabilization by certain nickel-containing compounds (Fig. 12). Nickel compounds have been shown to be effective quenchers of photoexcited chromophores including anthracene (Briggs and McKellar 1968), diethyl ketone (Chien and Conner 1968) and benzophenone (Harper et al. 1974). Some have also been found to reduce the lifetimes of polypropylene phosphorescence (Harper et al. 1974). This behavior of nickel compounds is similar to that of deactivation of photoexcited triplets by a large number of transition metal complexes. Once the excited chromophores have transferred their excitation energy to the metal complex, the excited metal complex dissipates the energy as heat. The effectiveness of nickel compounds as light stabilizers for polyolefins is attributed to the deactivation of photoexcited carbonyl chromophores in the polymer before they undergo photochemical reactions with adverse consequences for the polymer. However, some nickel-containing light stabilizers absorb significantly in the near uv and some decompose hydroperoxides; both effects may contribute to the efficiency of these compounds as light stabilizers.

3.4 Benzoate Esters

2,4-Di-*t*-butylphenyl 3,5-di-*t*-butyl-4-hydroxybenzoate and other esters, including aliphatic alcohol esters, are also effective light stabilizers for polyolefins (Fig. 13). The mechanisms by which these compounds function as light stabilizers are not completely known. The aromatic esters have been shown to be transformed via the photo-Fries rearrangement to 2-hydroxybenzophenone uv absorbers, but aliphatic esters, which cannot undergo the transformation to hydroxybenzophenones, are effective light stabilizers as well.

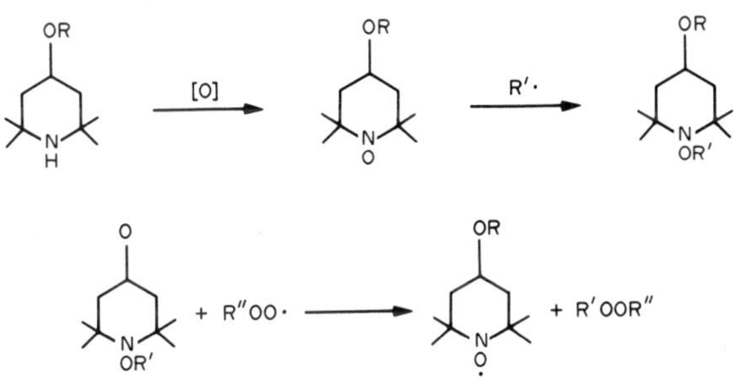

Figure 10
Free radical trapping mechanism of the piperidine moiety of hindered amine stabilizers

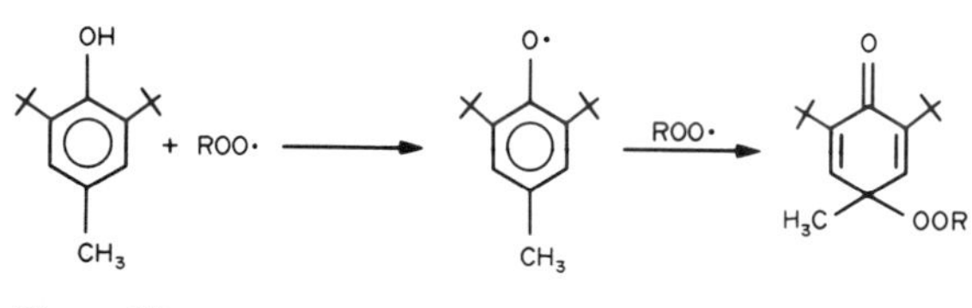

Figure 11
Free radical trapping mechanism of hindered phenols

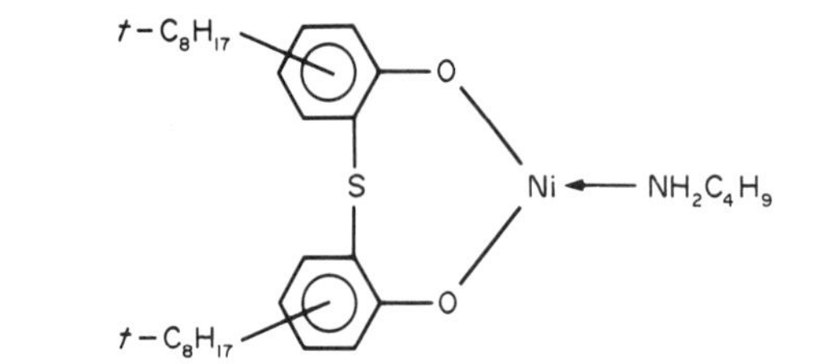

Cyasorb UV 1084

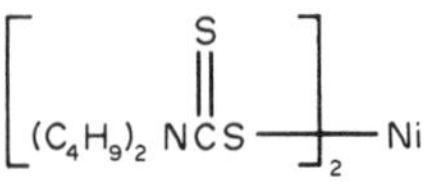

NBC

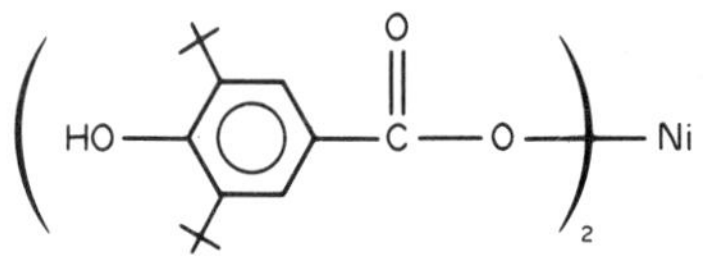

UV-Check AM 126

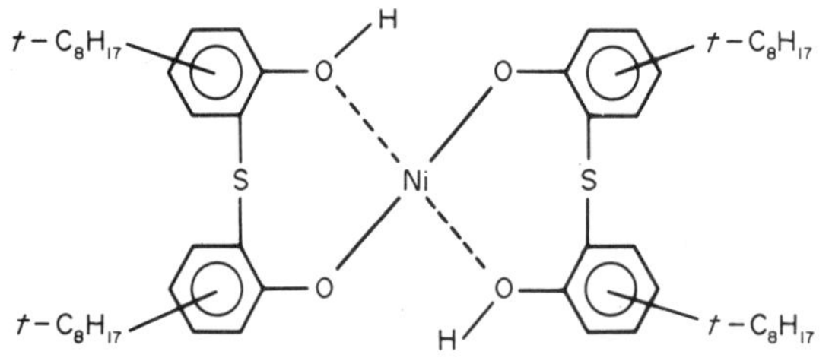

UV-Check AM 101

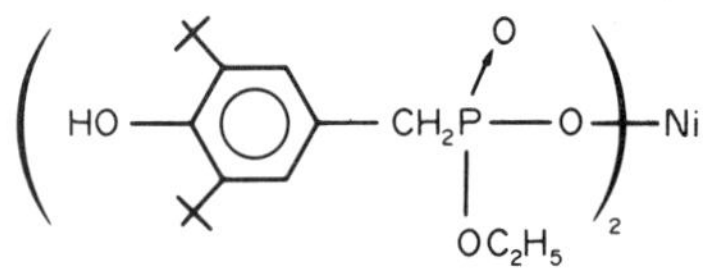

Irgastab 2002

Figure 12
Chemical structures of commercially available nickel stabilizers

3.5 Hindered Phenols

Hindered phenolic antioxidants play a secondary role in the light stabilization of polymers. When used alone in a polymer they are not anywhere as effective as primary light stabilizers—uv absorbers, hindered amines or nickel compounds. They are needed in formulations with primary light stabilizers mainly to provide protection against thermo-oxidative degradation, for example during high-temperature processing or for elevated-temperature-use conditions.

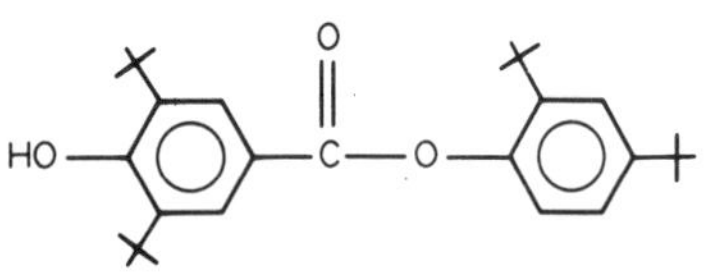

UV-Check AM 340
Tinuvin 120

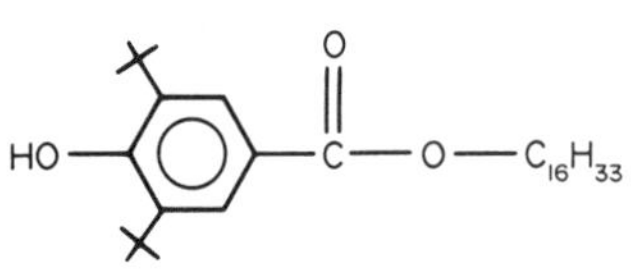

Cyasorb UV 2908

Figure 13
Chemical structures of commercially available benzoate stabilizers

3.6 Phosphorus Compounds

Trivalent phosphorus compounds contribute in two ways to the light stabilization of polymers. Some provide protection during processing, reducing thermooxidative degradation and thereby reducing potential initiators of photooxidation and helping to retain physical properties. Others contribute directly to the photostabilization by primary light stabilizers by a mechanism generally considered to involve deactivation of hydroperoxides and peroxides by abstraction of peroxygen with concomitant oxidation to pentavelent phosphorus:

$$(RO)_3P + ROOH \rightarrow (RO)_3PO + ROH$$

Few, if any, trivalent phosphorus compounds are highly effective in both stabilization modes. Some are effective processing stabilizers while others are effective in light stabilization.

4. *Important Light Stabilizer Properties*

In addition to the obvious light stabilizing requirement, additives should possess several additional properties:

(a) polymer compatibility,

(b) low volatility,

(c) thermal stability,

(d) photostability, and

(e) low discoloration tendency.

Most additives used for the light stabilization of polymers possess these properties. However, as polymers are used for more demanding applications, such as artificial turf, automotive coatings, outdoor furniture, fibers for many uses, glazing and solar applications, greater requirements are placed on the light stabilizing additives. Increased light stabilizing effectiveness and greater permanence are the major requirements. Hindered amine light stabilizers, alone and in conjunction with other light stabilizers, are making it possible for polymers to be used in more exacting situations. Oligomeric hindered amine light stabilizers are now available which are highly compatible with many polymers and have low migration rates and low vapor pressures, permitting them to meet the new requirements.

Some recent uv absorbers are less volatile than previously available materials, reflecting the recognition that more permanence is required to permit polymers to be used for long-term applications. There has been work recently on polymeric uv absorbers (Tirrell 1981), but the cost of providing them in that form is still too high to permit their use except in a limited number of applications.

5. *Evaluation of Light Stabilizers*

Light stabilizers are best evaluated in conditions and compositions as close to the end use as is consistent with obtaining the results in a reasonable time. If the polymer is to be used for outdoor applications, exposures to evaluate stability are best carried out at outdoor weathering sites such as are available in the USA in Florida, Arizona, Puerto Rico and Texas, where there is more intense sunlight and faster weathering than in temperate regions. Samples may be exposed at outdoor weathering sites in a number of modes depending on the intended application. In the northern hemisphere they are 45 °S, 5 °S, horizontal, vertical, in a so-called black box with horizontal orientation and supplemental heating, in a device at 45 °S which follows the sun from sunrise to sunset to maximize exposure, and in an apparatus which follows the sun and increases light intensity by means of ten mirrors which reflect sunlight onto the exposed samples. Direct outdoor exposure can be modulated by exposure behind glass to correct the quality of light for indoor exposure and still accelerate the rate of degradation.

A number of pieces of equipment are available throughout the world for accelerated light exposure of polymer samples. All present some compromise with regard to correlation with sunlight and natural weathering, but they do offer the advantage of convenience. The light sources for these in-house accelerating devices are varied. Among them are xenon arc lamps, carbon arcs, medium pressure mercury lamps, fluorescent sunlamps and combinations of fluorescent sunlamps and fluorescent black lights. All these light sources emit different amounts of uv, visible and ir radiation. None matches the quality of sunlight exactly, but in fact sunlight exhibits seasonal variations in quality and intensity, so there is not a fixed standard of summer sunlight, for example, for outdoor weathering. Many of the commercially available light exposure devices have provision for exposure to water as well as light. The water exposure is controllable to several ratios of wet–dry cycles.

The effectiveness of light stabilizing systems is evaluated in a number of ways. Exposed samples are evaluated for physical property retention—impact strength, tensile strength and elongation—and for appearance—color and gloss—as a function of exposure.

See also: Degradation and Stabilization, Thermal; Polymers: Technical Properties and Applications

Bibliography

Allen N S (ed.) 1983 *Degradation and Stabilization of Polymers*. Elsevier, Amsterdam

Bailey D, Vogl O 1976 Polymeric ultraviolet absorbers. *J. Macromol. Sci., Rev. Macromol. Chem.* 14B: 267–93

Bolland J L, Gee G 1946 Kinetic studies in the chemistry of rubber and related materials. *Trans. Faraday Soc.* 42: 236; 244

Briggs P J, McKellar J F 1968 Mechanism of photostabilization of polypropylene by nickel oxime chelates. *J. Appl. Polym. Sci.* 12: 1825–33

Carlsson D J, Garton A, Wiles D M 1979 The photostabilization of polyolefins. In: Scott G (ed.) 1979 *Developments in Polymer Stabilization—1*. Applied Science, London, Chap. 7, pp. 219–59

Chien J C W, Conner W P 1968 Quenching of excited states of diethyl ketone by paramagnetic metal chelates. *J. Am. Chem. Soc.* 90: 1001–6

Davis A, Sims D 1983 *Weathering of Polymers*. Elsevier, Amsterdam

Gugumus F 1979 Developments in the u.v.-stabilization of polymers. In: Scott G (ed.) 1979 *Development in Polymer Stabilization—1*. Applied Science, London, Chap. 8, pp. 261–308

Hardy W B 1982 Commercial aspects of polymer photostabilization. In: Allen N S (ed.) 1982 *Developments in Polymer Photochemistry—3*. Applied Science, London, Chap. 8, pp. 287–346

Harper D J, McKellar J F, Turner P H 1974 Photostabilizing effect of Ni(II) chelates in polymers II. Effect of diamagnetic chelates on polypropylene fluorescence. *J. Appl. Polym. Sci.* 18: 2805–7

Henman T J (ed.) 1982 *World Index of Polyolefine Stabilisers*. Kogan Page, London

Irick G Jr 1972 Determination of the photocatalytic activities of titanium dioxides and other white pigments. *J. Appl. Polym. Sci.* 16: 2387–95

Leu K W, Gugumus F, Linhart H, Wieber A 1977 Neue Stabilisieringssystems für Polyolefine in Ausseneinsatz. *Plaste Kautsch.* 24: 409–12

Otterstedt J-E A 1973 Photostability and molecular structure. *J. Chem. Phys.* 58: 5716–25
Owen E D (ed.) 1984 *Degradation and Stabilization of PVC*. Elsevier, Amsterdam
Pospíšil J 1981 Photooxidation reactions of phenolic antioxidants. In: Allen N S (ed.) 1981 *Developments in Polymer Photochemistry—2*. Applied Science, London, pp. 53–133
Tirrell D A 1981 Preparation of polymeric ultraviolet stabilizers. In: Pappas S P, Winslow F H (eds.) 1981 *Photodegradation and Photostabilization of Coatings*, ACS Symposium Series 151. American Chemical Society, Washington, DC, Chap. 4, pp. 43–9
Werner T 1979 Triplet deactivation in benzotriazole-type ultraviolet stabilizers. *J. Phys. Chem.* 83: 320–5
Zweig A, Henderson W A Jr 1975 Singlet oxygen and polymer photooxidations. I. Sensitizers, quenchers, and reactants *J. Polym. Sci.* 13: 717–36

P. P. Klemchuk
[Ciba Geigy Corporation, Ardsley, New York, USA]

Design of Composite Structures

The design approaches necessary for composites (*Advanced Composites Design Guide* 1973, Kulkarni and Beardmore 1980) are significantly different from those normally employed for metals. A composite design methodlogy is described here which outlines the steps in developing an efficient structural design using continuous-fiber-reinforced composites. In particular, the design of the initial laminate and the subsequent stress analysis are highlighted as the major conceptual steps, in comparison to design with isotropic materials. In addition, the importance of manufacturing feasibility and minimization of the number of joints cannot be overemphasized. The composite design methodology is generic in nature, and is applicable to any type of component or structure and to any type of continuous-fiber-reinforced material.

The approach taken here is to initially outline the overall design methodology for development of structural designs. In Sect. 1, the stages generic in nature to all materials are indicated and the stages of major differences in approach between composites and conventional (isotropic) materials are highlighted. Subsequently, those areas unique to composites are treated in some detail. Finally, a condensed example of designing with composites is given to illustrate the procedure.

1. Design Methodology

The overall design cycle for a composite structure consists of three phases, as depicted schematically in Fig. 1. Note that the overall design cycle is not unlike the typical design cycle for conventional (isotropic) materials, with the major difference lying in the second phase.

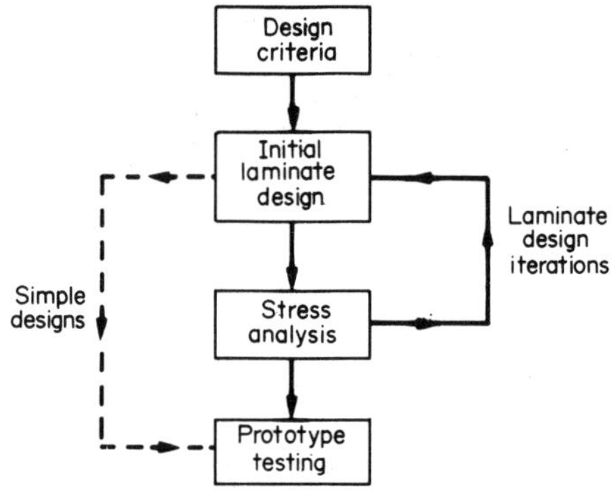

Figure 1
Design cycle for composite structures

1.1 Design Criteria

The initial phase of the design methodology consists of the development of the data which dictate the design of the component or structure. The design criteria may be broken into two stages, as depicted in Fig. 2. The structural criteria stage is merely the requirements imposed by the demands on the structure in order to perform efficiently and safely. The other stage consists of the constraints imposed on the design by nonfunctional considerations.

1.2 Design Phase

The primary design stage consists of the iterative procedure between initial laminate design and stress analysis (Fig. 1) until a final design is developed. The geometry of the component or structure and the composite laminate layup (or other material thickness) are developed. The integration of several parts into a reduced number of large fabricable composite pieces must also be considered in this phase. Such integration increases the overall stiffness (owing to elimination of joints) and reduces assembly costs. This initial design concept is subsequently analyzed for potential failure modes due to the imposed loads, using simple formulae. If the designs satisfies

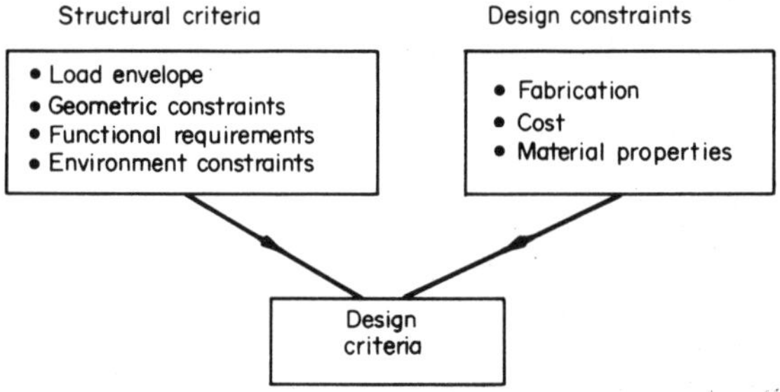

Figure 2
Inputs to the design criteria

this preliminary analysis, a detailed stress analysis may be undertaken.

The design stage described here is generic in nature. The major distinction in designing with composites lies in the laminate design and stress analysis, and these two specific areas of composite design are described separately in Sect. 2.

1.3 Prototype Testing

The last stage of the overall design cycle (Fig. 1) involves the building and testing of prototype components, based on the primary design. This is to verify the efficiency of the design and to allow feedback into the design loop if any problem occurs. When composites are used, tests for unique composite failure mechanisms must be included, but the prototype testing procedure is analogous to that for components fabricated in conventional materials.

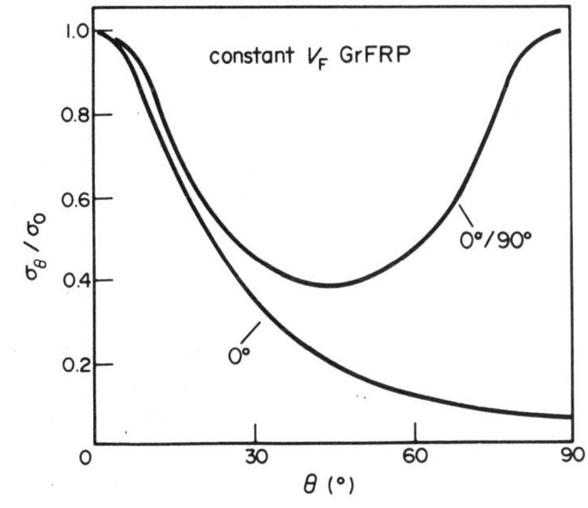

Figure 3
Typical relationship of fracture stress and angle θ between testing direction and fiber orientation for unidirectional tape and 0°/90° woven cloth

2. Design of Laminates

2.1 Laminate Design

The stacking of individual composite layers to form a part involves the selection of the individual layers and the use of lamination theory to yield the required laminate properties. Consequently, the basic design parameters going into the design of the laminate are the layer properties and the laminate parameters. These are

(a) lamina parameters—type and orientation of fibers, form of lamina (tape or woven cloth), thickness of individual layers and type of matrix; and

(b) laminate parameters—stacking sequence of the layers and symmetry of layers about centerline of thickness.

The mechanical properties of the individual lamina are functions of the type and orientation of the fiber in the matrix material, as illustrated in Fig. 3. The other significant point illustrated in Fig. 3 is the importance of appropriate fiber alignment in the resulting component, since the overriding concept in composites is to have all significant stresses supported by the fibers, not the matrix.

The design of the initial laminate must be based on a selected predominant design criterion, for example a specified stiffness or specified strength, or equivalence to an existing metal structure. Based on the predominant criterion selected, a laminate section is developed from specified individual layer properties, which will approximately satisfy the governing condition. This is followed by simple analysis to approximately verify the layup (e.g, using strength of material formulae). A reasonable laminate is developed by this iterative technique prior to full stress analysis. The process is shown schematically in Fig. 4.

Suppose, for example, a simple plate or beam is required with two basic criteria: (a) a specified bending stiffness (functional criterion) together with adequate torsional properties and (b) woven cloth must be used (fabrication requirement) as the individual layers. Let the required bending stiffness be 6.78 N m. For composite panels, the bending stiffness per unit width can be calculated from

$$\text{bending stiffness} = \sum_{i=1}^{i=n} E_i\left(\frac{t_i^3}{12} + t_i y_i^2\right) \qquad (1)$$

where t_i is the layer thickness, y_i is the distance from the neutral axis and n is the number of layers. With graphite-fiber-reinforced plastics (GrFRP) as the construction material, $E = 65.5$ GPa for 0°/90° cloth, $E = 24$ GPa for ±45° cloth and, with a layer (prepreg) thickness of 0.28 mm, bending stiffness = 6.99 N m for a layup consisting of four plies (namely, 0°/90°, ±45°, ±45°, 0°/90°).

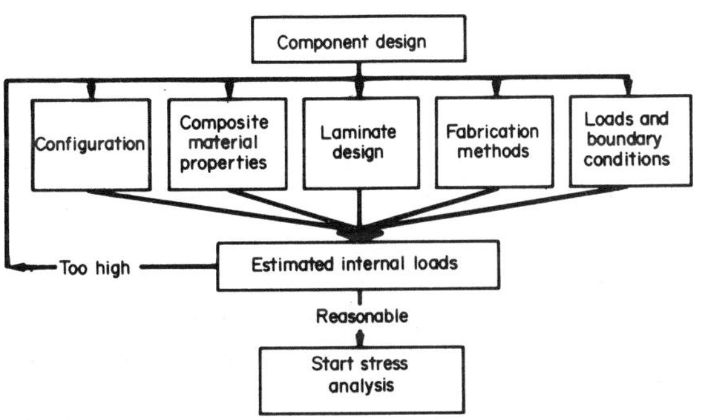

Figure 4
Flow chart of initial laminate design

One other common procedure for development of initial laminate design, particularly for complex components, is by the use of so-called carpet plots (*Advanced Composites Design Guide* 1973, Kulkarni and Beardmore 1980), generated by the application of lamination theory.

The initial laminate design must usually be verified by comprehensive stress analysis before prototype hardware is built. For simple structures, an abbreviated analysis based on simple formulae (e.g., beam formulae) can be utilized and indeed may be used for the initial iterations of complex structures. However, a full stress analysis to confirm the adequacy and efficiency of the structure is required for highly complex structures. The output of the stress analysis allows a detailed iterative design procedure to commence. The laminate is progressively modified to achieve structural adequacy relative to required functionality, to ensure that failure will not occur in any individual layer and to optimize cost–weight trade-offs.

2.2 *Stress Analysis*

The stress analysis for composites involves lamination theory (i.e., the ability to examine the stresses in the individual layers which make up the laminate). The normal criterion for failure is overstressing relative to the design allowables in any layer in the structure.

The analysis is greatly facilitated by generating the finite-element mesh in a manner conducive to stress analysis of composites. Basically, this consists of following two criteria: (a) local element axes should be coincident with principal material axes (commonality of element directions and major fiber orientations) and (b) elements with similar laminate properties should be grouped (definition and analysis are based on large groups of elements). Since a laminate of n individual layers requires up to n computations for a single calculation for isotropic materials, it is important to minimize computational time where possible.

The stress analysis consists of numerical computations involving static and/or dynamic analysis. The resulting nodal displacements and rotations are translated into stresses and strains. Most standard computer programs used for conventional materials (e.g., NASTRAN) do not have the capability of layer-by-layer analysis (laminate analysis) which is critical for composite analysis. Such laminate analysis requires the addition to the standard analytical program of preprocessing and postprocessing capability for the input and output data, respectively. This type of modification can be developed for standard programs such as NASTRAN, but is not yet generally available; many aerospace companies have developed in-house preprocessors and postprocessors specifically for composite analysis. Alternatively, several commercial stress-analysis computer programs for composites are available. Typically, the analytical programs contain, as part of the stress output analysis, a series of evaluations of the stress data based on appropriate failure criteria. Failure modes based on fiber failure, matrix failure and interlaminar-shear failure are normally examined. For example, fiber and matrix failure in an individual layer is usually evaluated by comparing the transformed stresses in the layer with the design-allowable strengths for that particular layer.

At least a few iterations are usually necessary to finalize the layup and ensure structural adequacy and efficiency. Any changes will usually consist in modifications of (a) the number of layers, (b) the stacking sequence, (c) the type of fibers or (d) the component configuration. In general, the iterations will involve progressive changes in the first three procedures.

3. *Design Example*

An example has been chosen of the application of composite design methodology based on experience in designing prototype automotive components in continuous-graphite-fiber-reinforced composites. This illustration should only be viewed as a condensed example of composite design. The example is a GrFRP front-end structure, that is, a stiffness-critical structure (Kulkarni et al. 1979). The basic material properties used in the design are listed in the example, but it should be emphasized that other composite materials could be substituted in the design by use of the appropriate mechanical properties.

Three major assemblies—the right fender, left fender and radiator support—are included in the front-end structure. Since the right and left fenders are symmetrically opposite, the design is divided into two major assemblies: the fenders and the radiator support.

3.1 *Design Criteria*

The dominant design factor for the front-end structure was selected as stiffness, based on the paramount need to meet critical ride-quality requirements. (Critical ride quality was the design criterion assumption for this concept vehicle, but it should be noted that for production vehicles additional criteria which may or may not be treatable analytically are necessary. Durability and crash, for example, are two necessary criteria which are not currently amenable to analytical treatment and require experimental verification.) Thus, the overall design philosophy was based on deflection requirements, with the modification that the deflection could be adjusted to allow for the lower input loads, since the overall weight of the composite vehicle was reduced by about one-third compared to the steel vehicle. Based on this philosophy, the individual components of the front-end

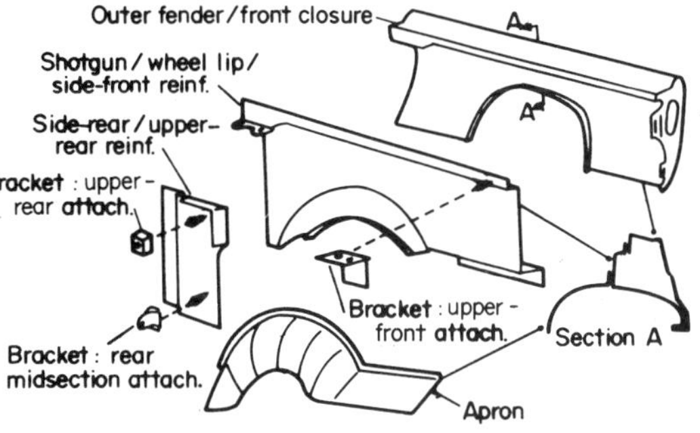

Figure 5
Schematic of 1979 Ford Motor Company composite integrated fender assembly

structure were divided into two different groups: bending-critical components and torsion-critical components. In addition, local areas such as the bolt attachment points were separately analyzed for strength-critical requirements to ensure the integrity of the joints.

Several other general design rules were adapted based on fabrication considerations:

(a) to use GrFRP cloth whenever possible for ease of fabrication;

(b) to minimize part warpage, maintain laminate centerline symmetry based on lamina orientation;

(c) to use primarily 0°/90° and ±45° orientations if possible, to simplify design and fabrication; and

(d) to integrate as many individual (steel) components into a single (composite) structure: in the fender, nine components were reduced to four major composite components (see Fig. 5) and, in the radiator support, fifteen steel components were replaced by four major composite components.

3.2 Laminate Design

The mechanical properties of the individual layers used in the design of the front-end structure are given in Table 1. As a consequence of choosing a critical stiffness parameter, the GrFRP components only matched the steel in either torsional or bending stiffness, but not both. This follows the basic design philosophy of composites, in that the maximum weight reduction is obtained by using the directionality of these materials to match the critical requirements and having lesser properties in other less-critical directions. The stiffness of the individual components of the front-end structure was matched by using the bending stiffness Eqn. (1) and for torsional stiffness

$$G_s K_s = G_g K_g (W_s / W_g) \quad (2)$$

where G is the shear modulus, W is the weight of the car, K is the torsional constant and subscripts g and s refer to GrFRP and steel, respectively.

Based on these calculations, the thickness of each individual composite part was computed. Only one of the two critical stiffnesses was matched in the composite material relative to the steel component. It should also be pointed out that whichever of the design criteria are used (i.e., bending or torsional stiffness) the GrFRP component only has to be approximately 67% of the steel stiffness to match the same deflection criteria.

All the bolted joints between the radiator and fender and between the fender and the body were examined for failure modes such as shear out, net tension and bearing strength. The baseline steel was analyzed to determine the ultimate load for these three failure modes. The composite attachment area was then designed to take the same ultimate load P as the steel, based on the following design criteria:

$$\text{shear out, } P_s = (\sigma_s 2et)$$

$$\text{bearing, } P_b = (\sigma_B td)$$

$$\text{net tension, } P_{uts} = \frac{\sigma_{uts} t W_{net}}{K_t}$$

where e is the edge distance, d is the hole diameter, $W_{net} = W - d$, σ_s is the shear strength, σ_{uts} is the ultimate tensile strength, σ_B is the bearing strength and K_t is the stress concentration factor.

3.3 Finite-Element Analysis

Once the layups were defined for each individual component on the front-end structure, further analysis was carried out on a finite-element model of the assembled front end. Loads were applied to the

Table 1
Mechanical properties of front-end structure layers

Material form	t (mm)	E_{11} (GPa)	E_{22} (GPa)	G_{12} (GPa)	ν_{12}	ρ (kg m^{-3})
GrFRP cloth (0°/90°)	0.28	71	71	6.9	0.09	1550
GrFRP tape	0.127	145	8.25	4.5	0.25	1550
GrFRP cloth (±45°)	0.28	24	24	34.5		1550

Table 2
Relative stiffness of front-end assemblies based on finite-element analysis

Total weight of car (kg)	Front-end fabricated with	Stiffness (%) Bending	Stiffness (%) Torsion
1700	steel	100	100
1700	GrFRP	80	81
1135[a]	GrFRP	121	121

a Input loads to the GrFRP front-end assembly will be two-thirds (i.e., $\frac{1135}{1700}$) of the input loads to the steel front-end assembly

assemblies at two frame-mount locations on the radiator-support lower extension. Models in both steel and composite were analyzed for both bending- and torsional-load cases. The static analyses were carried out using finite-element codes having the capability of both isotropic- and anisotropic-material analysis. The results of the analysis are summarized in Table 2. Note that based on the 1700 kg weight of the steel vehicle, the GrFRP front-end structure is only 80% as stiff in both bending and torsion as the steel structure. However, based on the 1135 kg weight of the composite vehicle, the relative stiffness of the GrFRP front-end structure is 20% greater than that of the steel structure in both bending and torsion. Thus, the analysis would predict that the composite front-end structure in the composite vehicle would act as a structure of increased stiffness relative to vehicle performance when compared with the steel vehicle; that is, the analysis predicted that no problem should result in the ride quality characteristics of the front end from a stiffness viewpoint.

See also: Aircraft and Aerospace Applications of Composites; Automotive Composite Components: Fabrication; Composites: Applications Overview; Composites: Manufacturing Overview; Sports Equipment Manufacture

Bibliography

Advanced Composites Design Guide, 1973: Vol. I, *Design;* Vol. II, *Analysis*; Vol. III, *Manufacturing*; Vol. IV, *Materials*; Vol. V, *Applications*. Air Force Materials Laboratory, Wright-Patterson Air Force Base, Dayton, OH

Kulkarni H T, Beardmore P 1980 Design methodology for automotive components using continuous fiber reinforced composite materials. *Composites* 12: 225–35

Kulkarni H T, Lim G, Knight R, Beardmore P 1979 *Design and Fabrication of Automotive Front-End in GrFRP Composites,* SAE Paper 790031. Society of Automotive Engineers, New York

P. Beardmore and H. T. Kulkarni
[Ford Motor Company, Dearborn, Michigan, USA]

Design of Elastomer-Based Engineering Products

Elastomers are a class of materials with very special properties. Because of their great extensibility and relative incompressibility, elastomers have to be thought about in different ways to more "normal" engineering materials such as steel or concrete. In spite of, or rather because of, their unusual characteristics, elastomers perform many vital functions in engineered products.

Natural rubber was the first elastomer and was used in antiquity by the indigenous peoples of America. It was not possible to exploit its unusual elastic characteristics for mechanical applications until Charles Goodyear discovered that the addition of sulfur, followed by heating, changed the rubber into a truly elastic material. This process, known as vulcanization or curing, results in crosslinks being formed between the long molecular chains of the base polymer, thus eliminating viscous flow.

After World War I, synthetic elastomers began to appear. Currently, there are more than 20 types of synthetic elastomer produced commercially, each with its own characteristics. Advances have also been made in the vulcanizing systems, antiozonants and antioxidants, fillers and extending oils used with the elastomer in the formulation of an engineering material. With modern formulation methods, elastomers are available to cater for a wide range of service temperatures and environmental conditions, singly or as blends. The abbreviations and some of the more important attributes and deficiencies of the available elastomers are given in Tables 1 and 2. A number of elastomer names in the tables are generic (NBR, BR, SBR and polyurethane are notable examples) and the properties of the final elastomeric material depend on details of the chemical structure of the elastomer among other factors. Thermoplastic elastomers (TPEs), a class of materials not mentioned in the tables, are reviewed by Rader and Stemper (1990). Tables 1 and 2 will be referred to repeatedly and the significance of the properties listed and how their values determine which elastomer is suitable for an application will be explained.

This article will focus on the designing of those engineering products that exploit the mechanical properties of elastomers.

1. *Elastic and Hysteretic Properties of Elastomers*

1.1 Elastic Properties

The elastic properties of elastomers are considered by Treloar (1975).

Elastomers have exceptionally low shear moduli (typically 0.3–3 MPa) and strain up to several hundred percent is recoverable. Both characteristics can be explained by the theory of rubber elasticity

Table 1
Common elastomers: basic properties

Name	Symbol[a]	Operating temperature[b] (°C)	Resistance[c] Ozone	Water (20 °C)	Mineral oil (20 °C)	Comments
Natural rubber	NR	−60/ +80	L	H	L	Strain crystallizes to give outstanding tear strength properties unfilled or filled. Can be formulated to give low hardness and low mechanical damping. Low-temperature crystallization most rapid at ∼−25 °C
Polyisoprenes (synthetic)	IR	−60/+80	L	H	L	Crystallizing grades are available. Properties can be similar to those of NR although both strength and creep rate are generally lower for IR
Butyl rubber halobutyl	IIR CIIR BIIR	−60/+120	M	H	L	Strain crystallizable, but unfilled strength properties do not approach those of NR. Filled elastomer has good strength properties. High damping over a wide temperature range. Lower gas permeability than most elastomers. Halogenated butyls are faster curing making them suitable for covulcanization with other rubbers.
Styrene butodiene rubbers carboxylated	SBR XSBR	−40/+80	L	H	L	Like many of the following elastomers, the basic polymer can vary widely in character. The styrene content profoundly influences final properties. Acceptable strength properties and good abrasion resistance when filled
Polybutadienes	BR	−70/+80	L	H	L	As for SBR, a generic title covering many structural variations. Damping can be even lower than for NR. Acceptable strength only when filled; good abrasion resistance obtainable
Nitrile rubbers hydrogenate carboxylated	NBR HNBR[d] XNBR[d]	−20/+110	L	H	H	Brittleness temperature, oil and high-temperature resistance increase with increasing acryonitrile content. Acceptable strength properties only when filled. High damping and low gas permeability. HNBR provides improved heat resistance and resistance to oxidized fuels (i.e., sour gasolines). XNBR has better strength and abrasion properties than NBR
Ethylene propylene rubbers terpolymers copolymers	EPDM EPM	−40/+130	H	H	L	Acceptable strength properties when filled. Can accept very large amounts of filler and oil
Polychloroprenes	CR	−30/+100	M	M	M	Two general types available: crystallizable and noncrystallizable. Strain crystallization gives good tear strength properties unfilled or filled. Low-temperature crystallization most rapid at ∼−10 °C.
Polyurethanes						Castable types can be made up to very high hardnesses with good strength properties. Damping adjustable from moderately low to high
polyester	AU	−30/+80	H	M	H	
polyether	EU	−30/+80	H	H	H	

a Unless otherwise indicated, ASTM D1418 or ISO 1629 symbols are used b Typical long-term continuous operating temperature ranges are given as indications only. Formulation details are often important c Intrinsic levels: L, low; M, medium; H, high. In practice, antiozonants are often added to enhance ozone resistance. M fluid resistance means either intrinsic medium resistance or special care needed in formulation d Not ASTM/ISO symbol

Table 2
Specialized elastomers: basic properties

Name	Symbol[a]	Operating temperature[b] (°C)	Resistance[c] Ozone	Water (20 °C)	Mineral oil (20 °C)	Comments
Polypropylene oxides	GPO	−50/+130	H	M	M	Low damping obtainable. Moderately low strength even when filled
Polynorbornenes	PNR[d]	−40/+80	L	H	L	Very high levels of plasticizer are necessary and very high levels of filler can be used: these largely determine vulcanizate properties. Very soft vulcanizates of acceptable strength can be obtained
Epoxidized natural rubbers	ENR[d]	−20/+80	L	H	M or H	Properties depend on degree of epoxidation. Strain crystallizes to give excellent tear strength. High damping and low gas permeability can be obtained
Epichlorohydrin homopolymers	CO	−10/+130	H	L	H	Fair strength properties. Low damping. CO has substantially lower gas permeability than IIR
copolymers	ECO	−20/+130	H	M	H	
Silicones	MQ, VMQ PMQ, PVMQ	−90/+230	H	H	L	Rather weak at room temperature even when filled but strength properties deteriorate only gradually with increasing temperature. Poor resistance to water at high temperatures. High gas permeability
Fluorosilicones	FVMQ	−90/+230	H	H	M	As MQ, etc.
Ethyl vinyl acetate copolymers	EVA[d]	−40/+120	H	H	M	With above ~30% vinyl acetate, essentially elastomeric. Can be heavily filled, e.g., with fire retardant types in cable covering
Chlorosulfonated polyethylenes	CSM	−20/+120	H	H	M	Temperature range depends on degree of chlorosulfonation. Moderate strength filled or unfilled. Hardness tends to be quite high. Brilliamt white vulcanizate can be produced
Chlorinated polyethylenes	CM	−20/+100	H	H	M	Less rubbery than CSM
Polyacrylics	ACM	−20/+140	H	L	H	Relatively hard, moderately low strength filled and unfilled
Ethylene methyl acrylate copolymers	AEM	−40/+150	H	H	M	Fair strength properties. High damping over a wide temperature range. Can accept substantial quantities, e.g., of fire retardant
Fluorocarbons	FKM or FPM	−20/+230	H	H	H	A relatively hard elastomer with fair strength properties and relatively low extensibility
Perfluorinated rubbers	FFKM	−20/+250	H	H	H	As FKM but resistant to a wider range of fluids
Tetrafluoroethylene propylene copolymers	TFE/P[d]	−20/+200	H	H	H	As FKM but resistant to a wider range of fluids
Polysulfides	OT, EOT	−40/+80	H	H	H	Rather weak even when filled

a Unless otherwise indicated, ASTM D1418 or ISO 1629 symbols are used b Typical long-term continuous operating temperature ranges are given as indications only. Formulation details are often important c Intrinsic levels: L, low; M, medium; H, high. In practice, antiozonants are often added to enhance ozone resistance. M fluid resistance means either intrinsic medium resistance or special care needed in formulation d Not ASTM/ISO symbol which ascribes the high extensibility to the ability of the polymer molecules to uncoil under the action of a force. Like the pressure in a gas, the restoring force arises from the low entropy of the uncoiled configuration, rather than from the distortion of chemical bonds, and is predicted to be proportional to absolute temperature. High extensibility and low modulus are thus interrelated, and materials which have a higher shear modulus than is typical for elastomers yield or break at relatively modest strains. Some polymeric materials (e.g., certain grades of polyurethane and polyethylene) have mechanical properties between those of typical elastomers and traditional engineering materials capable of strains of approximately 1% or less. Classical (linear) elasticity theory is fully appropriate only to infinitesimal strains but gives useful approximations for strains up to around 30%; a mathematical theory of finite-strain elasticity has also been developed to deal with large deformations.

Volume changes in elastomers do not occur by the uncoiling of polymer chains and are opposed, as in other materials, by the resistance of interatomic distances to change. The bulk moduli of elastomers (~2000 MPa) are comparable with those of liquids. The high ratio of bulk to extensional and shear modulus implies that, for most deformations, elastomers can be regarded as incompressible.

1.2 Elastomer–Metal Bonded Components

Elastomer–metal bonded components are discussed by Freakley and Payne (1978) and Lindley (1978).

Metals bonded to elastomers provide convenient attachment points and are often also necessary to give the required force–deformation behavior. Because of near incompressibility, uniaxial compression of an elastomer causes pronounced lateral expansion. Inextensible layers bonded onto or into the elastomer normal to the direction of compression restrict the tendency to expand and, thus, have a dramatic stiffening effect. In the simplest case, an elastomeric block bonded between plane parallel rigid plates, classical elasticity theory has been used to derive the formula for compression stiffness:

$$\mathrm{k_c} \simeq 3GA(1+2S^2)/h \tag{1a}$$

where G and h are the shear modulus and thickness of the elastomeric material, respectively, and S is the shape factor of the block defined as the ratio of the area A of one bonded face (the loaded face) to the force-free area. For a circular block,

$$S = \text{radius}/(2 \times \text{thickness}) \tag{1b}$$

The inextensible layers have little effect on the stiffness of the component in directions parallel to their plane. These effects are exploited in a wide range of components (see Table 3).

1.3 Dynamic Properties

The dynamic properties of elastomers are considered by Freakley and Payne (1978), Hepburn and Reynolds (1979) and Ferry (1980).

Dynamic stiffness and damping are important design parameters for many components (see Sects. 2.2–2.4). Damping expressed in terms of the loss tangent $\tan\delta$ may be defined in the equation relating the shear stress σ required to strain a linear viscoelastic material in simple shear with a sinusoidal amplitude γ_0 and angular frequency ω:

$$\left.\begin{aligned}\gamma(t) &= \gamma_0 \sin\omega t\\ \sigma(t) &= G'\gamma_0\sin\omega t + G''\gamma_0\cos\omega t \equiv \sigma_0\sin(\omega t+\delta)\end{aligned}\right\} \tag{2}$$

where G' is the storage modulus, G'' is the loss modulus, $\sigma_0 = \gamma_0(G'^2+G''^2)^{1/2}$ and $\tan\delta = G''/G'$. The energy dissipated per unit volume per cycle is given by

$$\pi\sigma_0\gamma_0\sin\delta = \pi\gamma_0^2 G'' = \frac{\pi\sigma_0^2 G''}{G'^2+G''^2} \tag{3}$$

Similarly, the dynamic stiffness has storage and loss components, which are proportional to G' and G'', respectively, in the case of linear deformations.

1.4 Frequency and Temperature Effects

Ferry (1980) considers frequency and temperature effects in elastomers.

The dynamic properties of elastomers depend more strongly on temperature and frequency than do those of materials such as metals. However, interrelationships between properties exist that are quantitative for unfilled elastomers and give guidelines for highly filled elastomers.

According to linear viscoelasticity theory the, often advantageous, loss tangent of elastomers implies frequency dependence in the storage modulus:

$$\frac{d\ln[G'(\omega)]}{d\ln(\omega)} \simeq \frac{2}{\pi}\tan\delta \tag{4}$$

As in other glass-forming materials, the viscoelastic properties of elastomers exhibit a particularly strong temperature dependence as the temperature is reduced towards the glass-transition temperature T_g. G' rises towards the glassy value (~ a thousand times the rubbery value); G'' also rises, but passes through a maximum just above T_g.

In the range T_g to $T_g + 100\,°\mathrm{C}$, the effects of temperature and frequency are interrelated in a manner which is approximately independent of molecular structure. Thus, for a given elastomer, the frequency dependence of the viscoelastic properties can be estimated from their temperature dependence over a restricted frequency range and vice versa. Choice of an elastomer with a high T_g (e.g., NBR, see Table 1) to achieve high damping results in strong temperature and frequency sensitivity.

1.5 Static Stiffness, Dynamic Stiffness and Filler Effects

Freakley and Payne (1978), Hepburn and Reynolds (1979) and Fuller et al. (1988) discuss static and dynamic stiffness and filler effects.

The shear stress–strain behavior of unfilled vulcanizates is reasonably linear (see Fig. 1a), as predicted by finite-strain elasticity theory. The material shown (NR) also has low damping, hence the small area enclosed by the loading and retraction curves.

Most engineering elastomeric materials do not consist of only crosslinked polymer, but also contain a specified loading of solid "filler" particles. The use of filler improves processing and increases modulus. "Reinforcing" fillers, typically fine-particle carbon blacks, are particularly effective at increasing modu-

Table 3
Formulae for stiffness of bonded elastomeric components

Component		Stiffness equation: Not affected by shape factor	Stiffness equation: Affected by shape factor
Elastomer pad bonded on flat parallel faces (e.g., bridge bearing)	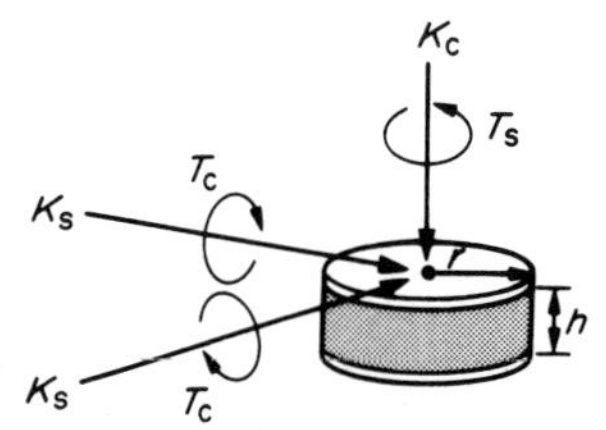	$K_S = \dfrac{\pi r^2 G}{h}$ $T_S = \dfrac{\pi G r^4}{2h}$	$K_C = \dfrac{3\pi r^2 G(1+2S^2)}{h}$ $T_C = \dfrac{3\pi r^4 G(1+2S^2/3)}{4h}$ where $S = r/2h$
Bush	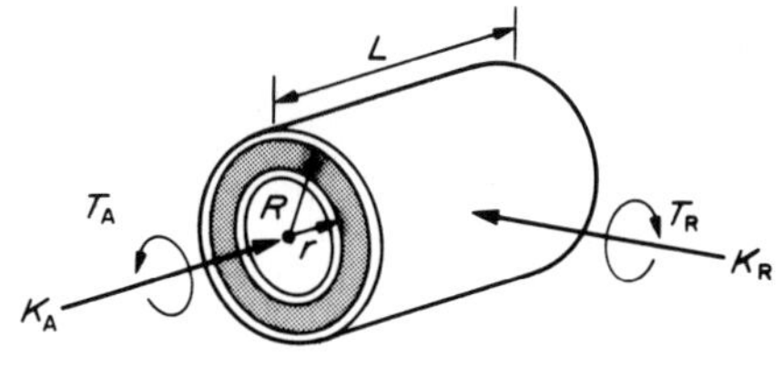	$K_A = \dfrac{2.73\,GL}{\log_{10}(R/r)}$ $T_A = \dfrac{4\pi GL}{1/r^2 - 1/R^2}$	for $L>2R$: $K_R \simeq \beta LG$ $T_R \simeq 1/12\beta L^3 G$ where $\beta = \beta_L$ or $\beta = \beta_S + (\beta_L - \beta_S)\left(\dfrac{0.04L}{R-r}\right)$ whichever is the smaller, where $\beta_S = \dfrac{80\pi(R^2+r^2)}{25(R^2+r^2)\ln(R/r) - 9(R^2-r^2)}$ $\beta_L = \dfrac{4\pi(R^2+r^2)}{(R^2+r^2)\ln(R/r) - (R^2-r^2)}$
Inclined bonded elastomer pads (as above) V configuration as used for transmission train mounts, etc.	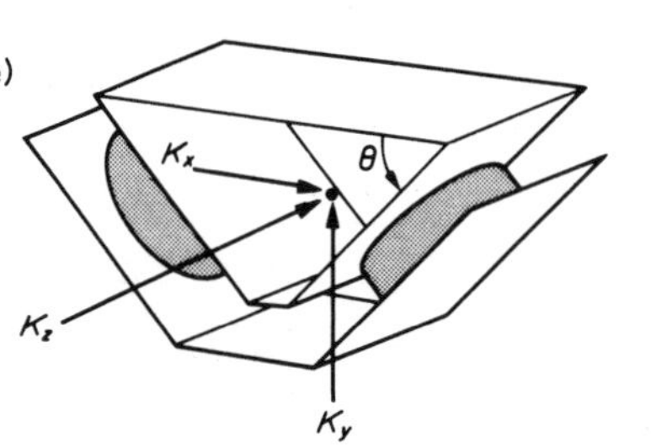	$K_z = 2K_S$	$K_y = 2(K_C\cos^2\theta + K_S\sin^2\theta)$ $K_x = 2(K_C\sin^2\theta + K_S\cos^2\theta)$ (see Sect. 2.5)

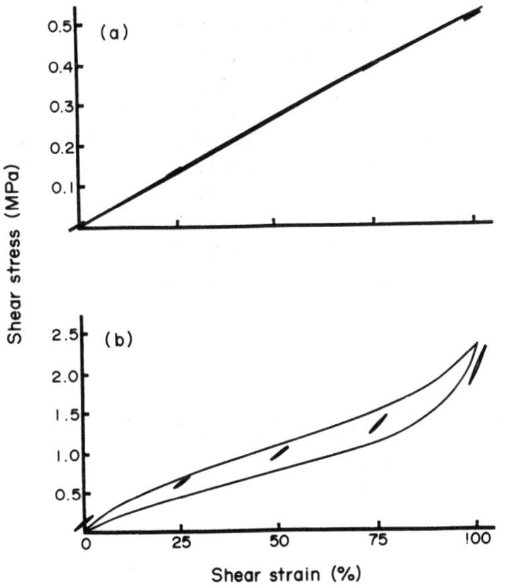

Figure 1
Tenth cycle quasistatic shear stress–strain curves with ±2% sinusoidal shear strain superimposed about static strains of 0%, 25%, 75% and 100%: (a) unfilled NR and (b) NR filled with 75 pphr (parts per hundred of rubber) of N330 carbon black

lus (cf. Figs. 1a and 1b, noting the changed ordinate scale). Appreciable levels of damping can also be achieved by using reinforcing filler in a low-damping elastomer (see Fig. 1b for NR with N330 (HAF) carbon black filler). The nonliner shear stress–strain behavior of this material raises the question of what value to use for the shear modulus G in the formulae given in Table 3; a reasonable choice is the value of G appropriate to a strain level representative of those prevailing in the component.

The dynamic modulus of a filled, low-T_g elastomer might be expected to approximate to the large-strain quasistatic value, because of the relative insensitivity to frequency. However, Fig. 1b shows that, instead of the frequency effect of high-T_g elastomers, such filled elastomers exhibit an amplitude effect. The smaller the dynamic amplitude the larger the dynamic stiffness, until a limiting value at strains of approximately 0.1% or less is reached. Aspects of this behavior can be represented by a model with friction pad elements. Figure 1b shows that the slope of the stress–strain curve of filled NR is steep at small strains; the dynamic modulus (for small dynamic strains) is comparable in magnitude.

The stress–strain gradient rises at high strains (~90% shear for the highly filled elastomer in Fig. 1b). A further and important aspect of the nonlinear behavior of filled elastomers is that dynamic properties can depend on strain history to a marked extent. Repeated deformation, especially at large amplitude (scragging), results in a reduction of the dynamic stiffness for a sinusoidal deformation of a given amplitude. In time, however, the dynamic stiffness does recover towards the value prior to scragging.

1.6 Creep and Stress Relaxation

Gent (1962) considers creep and stress relaxation in elastomers.

The stresses in an elastomeric component held at a constant deformation gradually decrease with time as the crosslinked network equilibrates. This is termed stress relaxation. Creep (i.e., change of deformation at constant load) is a consequence of the same molecular processes. The creep is usually taken to be the change of the strain e expressed as a fraction of the initial strain. The rate of stress relaxation S and the rate of creep C may be defined by

$$S = -\frac{1}{\sigma}\frac{\partial \sigma}{\partial t}\bigg|_e \qquad (5)$$

$$C = \frac{1}{e}\frac{\partial e}{\partial t}\bigg|_\sigma \qquad (6)$$

where σ is stress. These definitions lead to a relationship between creep and stress relaxation

$$C = \frac{\sigma}{e}\frac{S}{(\partial \sigma/\partial e)_t} \qquad (7)$$

So the creep rate of a spring depends not only on the stress-relaxation rate of the elastomer, but also on the degree of nonlinearity of the force–deformation behavior of the spring.

According to the theory of linear viscoelasticity, stress relaxation and loss tangent are interrelated:

$$tS \simeq \frac{2}{\pi}\tan\delta \qquad (8)$$

For filled rubbers, damping is largely nonviscoelastic and Eqn. (8) substantially overestimates S. Plots of deformation during creep or stress during relaxation against $\log t$ are approximately linear, thus the rates are reported in percent per decade of time.

At long times, stress relaxation may largely be caused by chemical changes in the elastomer rather than by viscoelastic or other physical processes. The "chemical" relaxation rate is approximately independent of time and has a different dependence on temperature than does physical relaxation. For a typical NR-based material at 10 °C, the rate of physical relaxation falls to about the same level as the anaerobic chemical relaxation rate after a year or two.

In sustained cold conditions, some elastomers crystallize, also resulting in stress relaxation; these effects, in contrast to other physical and chemical relaxation processes, can be reversed by sufficient warming.

1.7 Finite-Element Analysis

Finite-element (FE) analysis is increasingly being used to calculate the force–deformation behavior of, and stresses within, elastomeric components. For unfilled elastomers, there is little difficulty in describing the material properties needed for either small-strain or large-strain programs. Small-strain linear programs may be adequate in many cases where the strains are approximately 30% or less since the relative error will be of the order of the prevailing strain. The near incompressibility of elastomers does, however, present a difficulty for many FE packages.

The small-strain nonlinearity of filled elastomers (Sect. 1.5) can present a serious difficulty for small- or large-strain FE programs. Thus, FE methods cannot, as yet, readily yield precision force–deformation predictions for black-filled elastomers. However, they are especially useful for analyzing components of complex shape.

2. *Design of Functional Engineering Products*

2.1 Thermal Expansion Bearings and Joints

Elastomeric bearings and joints are discussed by Freakley and Payne (1978) and Fuller et al. (1988).

Modern bridges require bearings that allow thermal and other dimensional changes as well as articulation of the deck due to traffic loading while generating minimal forces in the superstructure and supports. Metallic roller bearings can be used but, being susceptible to corrosion and fretting wear, require maintenance. These difficulties are avoided with elastomeric bearings, which take up the relative displacements by deforming elastically.

The bearings support the weight of the bridge deck and traffic and, in order to keep the vertical deflections acceptably small, they must have high vertical stiffness. However, their horizontal stiffness must be sufficiently low to allow dimensional changes in the bridge deck without producing unacceptably large bending forces in the piers. Typically, a ratio of vertical to horizontal stiffness of about 100 is required; this can be achieved by a laminated bearing (alternate bonded layers of rubber and steel or glass fabric) with a shape factor of around 4 (see Eqn. (1)). If angular displacements of the bridge deck, due to changes in the traffic loading, need to be accommodated, the tilting stiffness of the bearings is also specified (Fig. 1). Another requirement is stability to high compressive loads and to horizontal displacements. Normally, steel surfaces are protected by elastomer to give a design life of the order of 50 years.

2.2 Isolation of Vibration

The isolation of vibration is considered by Snowdon (1968), Freakley and Payne (1978) and Hepburn and Reynolds (1979).

A spring will isolate an object from a vibrating structure if the dynamic force associated with deflection of the spring is small compared with the dynamic (inertial) force required to vibrate the object at the amplitude of the excitation. Stiffening of the spring at high frequency would impair isolation, so that if an elastomeric spring is used the damping level should be small (see Eqn. 4)). Isolators are sometimes referred to as "absorbing" the vibration but, as this might lead to the erroneous conclusion that the higher the damping the better, a preferable term is "attenuation" of the vibration.

Although isolators with low damping perform best over the principal frequency range, a modicum of damping is beneficial when performance at frequencies below and above this range is considered. At a certain (low) frequency, the force required to deflect the spring dynamically is equal to that required to accelerate the object, but of opposite phase: this is resonance and the vibration amplitude of the "isolated" object is greater than that of the structure. If the vibration source is a motor, for example, resonance will occur during runup to and rundown from full operating speed. For such situations, some damping is required to prevent excessive amplification of the vibration at resonance. The effect of damping is quantified in Fig. 2, which shows transmissibility against the normalized excitation frequency f/f_n (f_n is the resonance frequency). By definition,

$$\text{transmissibility} = \left|\frac{x_{\text{object}}}{x_{\text{source}}}\right| = \left|\frac{F_{\text{out}}}{F_{\text{in}}}\right| \tag{9}$$

The first formula is appropriate to the case of isolation of an object from a structure vibrating with amplitude x_{source} and the second to isolation of a structure from a machine causing an excitation force of amplitude F_{in} (see inserts in Fig. 2).

For very high frequencies, some damping is also beneficial since the spring has internal resonances, due to standing-wave effects. As for the primary resonance, these correspond to peak values of transmissibility; however, damping reduces both the peak and trough levels (Fig. 2). The inherent damping of elastomers thus gives good isolation at very high frequency. Metallic springs have low damping and although they can be supplemented by a damper (for primary resonance control) and an elastomeric

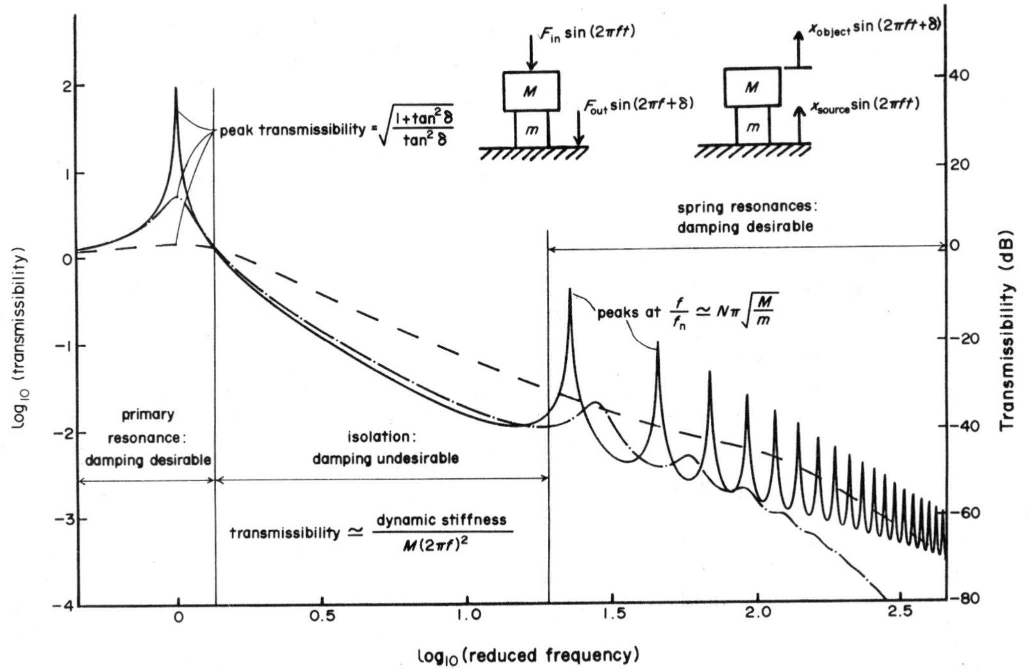

Figure 2
Transmissibility of a spring of mass m used to mount a mass M as a function of frequency f/f_n (where $f_n = (1/2\pi)(\text{storage stiffness}/M)^{1/2}$ is close to the natural frequency). In the graph $m/M = 0.02$; tan δ is independent of frequency taking the values—— 0.01,— · — · — 0.2; — — — 1.0; and the storage stiffness increases with frequency according to Eqn. (4) (so that the spring resonance peaks are progressively displaced to higher frequencies as the parameter tan δ is increased)

"noise-stop" pad (to improve high-frequency performance), it may be difficult to match the performance of a well-designed elastomeric spring.

Many isolators are load bearing (mounts) and static stiffness and susceptibility to creep are then important parameters, in addition to the dynamic properties. Creep is undesirable because extra clearance is required to accommodate it and differential creep can lead to redistribution of the loading of the mounts. Sometimes nonlinear mounts of the "buckling" or softening type of load–deflection behavior are used, to achieve a low dynamic stiffness for good isolation, with a low static deflection (see Fig. 3). However, such mounts are particularly prone to creep (Eqn. (7)).

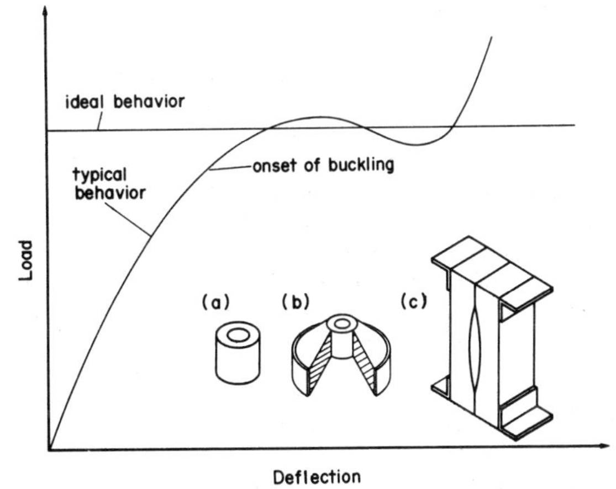

Figure 3
Load–deflection behavior of a typical buckling elastomeric component in comparison with the theoretical ideal rectangular behavior. Insert: elastomeric components that exhibit buckling on loading vertically: (a) hollow rubber cylinder; (b) conical shear bushing, cut away to show construction and (c) dock fender (Andre)

2.3 Vibration Absorbers

Snowdon (1968) and Freakley and Payne (1978) discuss vibration absorbers.

The high vibration amplitudes associated with undamped resonances in machines may result in noise generation or fatigue problems. An auxiliary spring–mass system, tuned to the resonant frequency and attached to the system, reduces the vibration amplitude for excitation at the original resonance frequency. However, two new resonances appear, one at a frequency below the original resonance and one at a frequency above it. If these new resonances are excited, they may be equally unacceptable—unless damping is introduced into the auxiliary spring. The damping attenuates the two new resonant peaks, so that the original high-amplitude peak is replaced by two peaks each of lower amplitude. Such auxiliary damped spring–mass systems, known as vibration (or dynamic) absorbers, make effective use of the inherent damping of elastomeric materials.

Optimization of absorber performance is mathematically complicated, but shows that the resonant frequency of the absorber should be near the original system resonance and the damping level of the absorber should be a substantial fraction of critical damping. For vibration absorbers tuned to optimum frequency and damping, the larger the ratio of auxiliary to primary mass, the more effective the absorber. Critical damping of a damped spring–mass system corresponds to $\tan \delta = 2$ (at the natural frequency) so high-damping elastomers are required for dynamic absorbers. Absorbers are typically tuned to high frequencies so that static deflections are small and creep, associated with high damping (see Eqns. (7) and (8)), is therefore not a problem. Moreover, because the storage modulus of high-damping elastomers tends to increase strongly with frequency, the absorber will be effective (i.e., close to optimum tuning) over a broader range of frequencies.

2.4 Shock and Seismic Isolation

Shock and seismic isolation are considered by Snowdon (1968), Freakley and Payne (1978) and Coveney et al. (1988).

A structure subjected to a violent transient displacement $x_s(t)$ may be protected (with its contents) by placing suitable compliant elements between it and the source of the shock. Isolation of vibration, (i.e., periodic displacements), can be regarded as a special case of shock isolation but mathematical analysis is considerably more complex for transients and numerical methods are often required.

The damage caused by a transient can depend on details of the fragile elements within the structure. If these are very softly suspended, the maximum (abrupt) displacement of the structure determines whether damage occurs—and, if so, its extent. However, the maximum acceleration of the structure determines the damage of relatively rigid structural elements.

Although some aspects of isolator design (e.g., stiffness) depend on details of the structure and the shock, adequate deflection capability and damping are general requirements. For a succession of transients, a buildup of oscillations at the isolation natural frequency can occur if damping is insufficient.

A common applicatrion of shock isolation is in packaging. On impact, the deceleration (in units of g) of the contents must be kept below the fragility factor a_f when the package is dropped from a height h. Packaging thickness, for fixed a_f and h, is minimized if the force is constant with deformation; buckling elastomeric components approach this ideal (Fig. 3). Similar considerations apply to dock fenders which ideally absorb the kinetic energy of a berthing ship by imposing an acceptable force (or deceleration) over a minimum distance. Head injuries depend on deceleration magnitude *and* duration so impact-softening elastomeric matting in playgrounds and so on is most effective when it has damping to reduce rebound.

Seismic isolation of large structures can be achieved with laminated elastomeric bearings. This technology evolved from those of bridge bearings and antivibration mounts for buildings. High damping, advantageous in reducing peak accelerations and displacements, is obtained from a supplementary device or the elastomer itself. Usually, horizontal isolation is satisfactory as the horizontal accelerations of the earthquake are generally the most damaging; the bearings can therefore be made vertically stiff (by high shape factor, see Sect. 1.2) to avoid rocking of the building and to minimize the effects of elastomer creep—a corollary of high damping. Earthquake bearings operate closer to safety limits than do bridge bearings, for example, and so require closer determination of these limits. The key advantage of seismic isolation of a building is the protection afforded to the contents.

2.5 Elastomeric Suspension Systems: Scaling Effects and Component Layout

Elastomeric mountings for machinery and suspension systems for vehicles often function as both vibration and shock isolators. An important feature is an adequate deflection limit; an abrupt onset of large forces at that limit is undesirable and can be avoided by using elastomeric snubbers or "bump stops" to cushion "bottoming out".

If an application requires a particular maximum displacement x_{max}, regardless of scale, a spring design may be suitable over only a restricted range of scale since x_{max} increases with L, the linear dimension of the spring, whereas the load capacity increases with L^2. Elastomeric bearings are suitable for high-load applications in civil engineering and heavy road and rail transport. Scaled down versions have too limited a deflection to be suitable for car springs, although a more complicated system using a lever arm can be used, as on the Mini.

Rotational and translational natural modes of vibration are uncoupled for the fail-safe V configuration (Table 3), if the lines of action of the reaction forces from the elastomeric mounts pass through the center of gravity of the supported body; this can be arranged by selecting the angle of the mounts and their compression–shear stiffness ratio.

2.6 Elastic Joints and Couplings for Articulation

Göbel (1974) and Freakley and Payne (1978) consider elastic joints and couplings for articulation.

Joints which take up relative movements elastically avoid many problems associated with rolling or sliding elements (wear, stick–slip behavior, noise and backlash). The extensibility and low modulus of elastomers make them suitable materials for elastic joints which, when correctly designed, have long fatigue lives.

In elastomeric bushes (Table 3), the choice of dimensions and the introduction of cavities or additional metal layers into the elastomer enable wide variation of the ratios of axial, radial, torsional and tilting stiffnesses. Their use in the joints between the metal segments of the tracks on tracked vehicles exploits the large torsional deflection capability, the elastomer being deformed essentially in simple shear. Bushings which connect the chassis (or bodies) of road vehicles to the suspension have the dual role of providing specified stiffnesses in all directions and attenuating vibration—especially in the horizontal plane where the suspension system is relatively rigid.

Manufacture of bushes bonded to metal or other materials poses a difficulty: thermal contraction relative to the metal tubes after vulcanization can result in unacceptably large tensile stresses in the elastomer. Such stresses are often removed or replaced by compressive stresses, by swaging one of the metal tubes (i.e., decreasing the diameter of the outer tube or expanding the inner tube). Alternatively, the elastomer can be unbonded at one or both metal surfaces but kept in position by frictional restraint produced by compressing the vulcanized elastomer during manufacture or installation. If the compressive stress is insufficient or relaxes away, slip leads to abrasion and premature failure. However, properly designed unbonded bushes are cheap and long lasting.

Flexibile couplings in torsional transmission systems cater for misalignment arising from geometric necessity, imperfection or dynamics (e.g., if the power source is mounted on antivibration springs). The key deflection capabilities and stiffnesses are axial, radial and angular, to take up misalignment, and torsional for shock and vibration isolation and satisfactory torque transmission. For elastomeric couplings of reasonable torque capacity, the fatigue life normally required ($\gtrsim 10^7$ rotations) is achievable for continuous shaft misalignments up to 4°.

A coupling can be a tube or rod of elastomer, but in demanding applications elastomeric couplings often exploit the full range of design features (from shape factor effects to snubbing devices).

2.7 Elastomeric Seals

Freakley and Payne (1978) discuss elastomeric seals.

The low modulus and high extensibility of elastomers allow wide dimensional tolerances in the joints to be sealed. For static applications, no leakage should occur—a requirement met if the seal contact stress is greater than the pressure difference across the seal. The contact stress depends on the sealing force, the modulus of the elastomer (the contact material being assumed rigid) and geometric details including surface roughness. Low-pressure seals, utilized in underground drains and sewers, usually rely on compressive strain at installation to maintain contact stress so a low stress-relaxation rate is required. Higher-pressure seals require a geometry such that the compression force is produced by the pressure differential. An O-ring seal in a rectangular groove provides such an example. Here, creep rate is important as the seal can extrude into the clearance gap over long periods of time.

For some static applications (e.g., pipe seals), it may be necessary to accommodate greater movement than is possible by elastic deformation. The seal may then be allowed either to roll (e.g., rolling O-ring or rolling diaphragm seals) or slide. Rolling will subject the elastomer to strain cycles which may result in fatigue, while sliding may result in damage of the seal surface by abrasion.

Reciprocating and rotary sliding seals for liquids are usually designed to permit a slight leakage to form a film of liquid which separates the elastomeric and rigid surfaces and greatly reduces friction and wear. The film thickness, which must be at least comparable with the surface roughness to achieve good lubrication, can be calculated from the geometry, sliding speed and liquid viscosity according to the theory of elastohydrodynamic lubrication. Resilience (i.e., low damping) of the elastomer is generally advantageous for dynamic applications so that the seal can respond quickly to movements of the counter surface.

As seals are not generally bulky, resistance of the elastomer to swelling by the fluid is crucial. NR is suitable for sealing water at low and moderate temperatures but EPDM is more appropriate at higher temperatures. Mineral oil is normally sealed with NBR at moderate temperatures, while at higher temperatures more specialized elastomers such as FKM must be used (Table 2).

3. Design of Enginering Products for Service Longevity

3.1 Mechanical Failure and Fatigue

Bulky elastomeric components undergoing high dynamic loading (e.g., solid tires) can become very hot internally, through conversion of some elastic energy to heat (Eqn. (3)) which cannot readily be removed owing to the low thermal conductivity of elastomers. Long-term exposure to high temperatures can affect the dynamic properties, increasing the loss tangent; eventually, runaway heat buildup may occur, resulting in pyrolysis of the polymer and possibly mechanical failure due to "blowout" by gaseous products. Choice of a low-damping elastomer and maximizing heat removal reduce operating temperatures. "Fatigue" is sometimes taken to include failure caused by heat buildup, but more usually refers to failure by growth of cracks in the elastomer, which is the mode of failure of many elastomeric engineering components.

According to fracture mechanics, the propagation rate of a crack into an elastomer is governed by the strain-energy release rate

$$T = -\left.\frac{\partial U}{\partial A}\right|_l \qquad (10)$$

where U is the total elastic energy and A is the area of one fracture surface; the partial derivative is evaluated with a fixed boundary to the elastomer so that the external forces do no work.

If T is less than a threshold value T_0, the crack will propagate only by ozone attack for susceptible elastomeric materials (Table 1). T_0 is of the order of $50\,\mathrm{J\,m^{-2}}$, but there are significant differences between elastomers, while compounding details such as crosslink density and filler loading have some effect. Unfortunately, values of T_0 are not available for many of the specialty elastomers, which are often less well characterized than general-purpose elastomers.

The existence of T_0 implies indefinite life for components (in the absence of attack by ozone, etc.) if the deformation is kept below a threshold level. Deformation in simple extension provides an illustration. For an edge crack in a tensile test piece of an elastomer with Young's modulus E, linear theory gives

$$T \simeq 3Ee^2c \qquad (11)$$

where e is the tensile strain and c is the crack length. For a high-quality well-mixed elastomeric material $c \simeq 25\,\mu\mathrm{m}$ is representative of flaw size so, for typical values of $E = 3$ MPa and $T_0 \simeq 50\,\mathrm{J\,m^{-2}}$, indefinite life is predicted if the strain is kept below about 50%. This does occur in the absence of ozone cracking.

If T exceeds T_0, mechanical crack growth proceeds at a rate which depends on T. A distinction must then be drawn between two classes of elastomer. Strain-crystallizing elastomers (notably NR, but also some grades of IR and CR) do not normally suffer mechanical crack growth unless the strain is periodically relaxed. The amount of crack growth depends on the number of cycles and the maximum value of T in the cycle. If the strain is only partially relaxed, the crack-growth rate is reduced by an amount depending on the minimum strain level. Nonstrain-crystallizing elastomers (e.g., EPDM, SBR) suffer mechanical crack growth even under noncyclic conditions ("time-dependent" crack growth, again a function of T) while, under cyclic conditions, there is an extra contribution to crack growth. For this reason, strain-crystallizing elastomers are more suitable for components that bear substantial static loads. Reinforcing filler can be used to reduce the rate of time-dependent crack growth (many noncrystallizing polymers would be useless for most engineering applications without it) but may not eliminate the problem.

To assure long fatigue life, T should be designed not to exceed the critical value throughout the component. Although T is known for some geometries, it is difficult to evaluate for complex shapes. Clearly, though, stress concentrations should be alleviated by, for example, filleting the free surface of an elastomer bonded to a rigid substrate and radiusing embedded metal edges. T should be minimized on the elastomer surface in particular, because the effective flaw size is likely to be increased there by abrasion, action of ozone and so on.

3.2 Environmental and Aging Effects

Anaerobic chemical changes may occur in elastomers in service. Some may harden (e.g., those based on CR) because crosslinking continues. Others may soften or "revert" (notably conventional NR vulcanizates at elevated temperatures) because crosslinks break down. Temperature determines the details and rates of the reactions and the best choice of polymer and vulcanizing system.

Specifications tend to concentrate on the irreversible effects of exposure to high temperatures rather than the strength properties at those temperatures. Noncrystallizing elastomers are inclined to weaken substantially as the temperature is raised, due to the fall in damping, whereas strain crystallization still occurs and promotes the strength of NR even at temperatures as high as 140 °C. For some applications, this retained strength may outweigh the disadvantage of an increased rate of deterioration, especially if the exposure to high temperature is sporadic rather than continuous.

Aerobic chemical changes may also occur. Unsaturated polymers (e.g., NR, IR, SBR, NBR and CR (Table 1)) are prone to degradation by oxygen and ozone, and normally require the addition of antidegradants. Oxidation rate increases with temperature and depends on the particular polymer (CR being less susceptible than the others). The depth to which oxidative degradation occurs is controlled by the diffusion and consumption of oxygen. The oxidative degradation of bulky components, generally restricted to the surface layers, can be unacceptable on cosmetic grounds but may not affect performance. Nevertheless, for engineering applications, antioxidants are generally used and can be essential for thin products. Ultraviolet radiation encourages oxidation but can be blocked by small loadings of carbon black in the elastomer.

Ozone reacts extremely rapidly with C=C bonds in unsaturated elastomers causing direct cleavage. Without strain only a very thin surface layer is affected, but cracking can occur for stretched rubber. Antiozonants are used with unsaturated elastomers to increase the threshold strain for cracking and to reduce the rate of crack growth. An alternative is an inert surface layer such as wax, polyurethane paint or a protective layer of (substantially) saturated polymer (e.g., EPDM or IIR). For dynamic applications, the cover layer must flex and extend without itself cracking.

Elastomers that have been heavily swollen by liquids are weak. The amount of swelling depends on chemical similarity with the liquid (Table 1). Generally, the rate of swelling can be predicted by the laws of diffusion and is inversely related to the liquid viscosity. Splashing of a bulky component (e.g., an engine mount) by a viscous swelling liquid is not, therefore, normally cause for concern, whereas immersion of a thin component (e.g., a seal) in the liquid would be. Again, a layer of resistant elastomer can be used to protect a susceptible elastomer (e.g., NBR lining to NR flexible piping for mineral oil).

Significant exposure to liquids can leach antidegradants from the elastomer. The more effective, mobile protectants are generally more susceptible to leaching; bulky components are less affected since the surface layers can be replenished by outward diffusion from the interior.

All elastomers undergo an almost instantaneous reversible increase in the storage modulus as the temperature is lowered towards T_g because molecular uncoiling motions become difficult (see Sect. 1.1). Below T_g, the material becomes a stiff, brittle "glass." T_g can be lowered by plasticizing the elastomer (i.e., incorporating a compatible liquid before vulcanizing). Low-temperature crystallization, a separate phenomenon exhibited notably by NR and CR, is time dependent and can substantially stiffen the material which may then show a pronounced yield stress. The crystallization disappears fully at relatively high temperatures (e.g., ~30 °C for NR). Low-temperature crystallization can be reduced or avoided by chemical modification of the polymer, for example by using a high-sulfur vulcanizing system with NR, while grades of suitably sulfur-modified CR are available.

See also: Elastomers: Applications Overview; Offshore Engineering Applications of Polymers

Bibliography

Bhowmick A K, Stephens H L (eds.) 1988 *Handbook of Elastomers—New Developments and Technology*, Dekker, New York

Brydson J A 1988 *Rubber Materials and Their Compounds*. Elsevier, London

Coveney V A, Derham C J, Fuller K N G, Thomas A G 1988 Vibration isolation and earthquake protection of buildings. In: Roberts A D (ed.) 1988 *Natural Rubber Science and Technology*. Oxford University Press, Oxford, pp. 938–57

Ferry J D 1980 *Viscoelastic Properties of Polymers*, 3rd edn. Wiley, New York

Freakley P K, Payne A R 1978 *Theory and Practice of Engineering with Rubber*. Applied Science, London

Fuller K N G, Gregory M J, Harris J A, Muhr A H, Roberts A D, Stevenson A 1988 Engineering use of natural rubber. In: Roberts A D (ed.) 1988 *Natural Rubber Science and Technology*. Oxford University Press, Oxford, pp. 892–937

Gent A N 1962 Relaxation processes in vulcanized rubber. Part I. *J. Appl. Polym. Sci.* 6: 433–41

Göbel E F 1974 *Rubber Springs Design*. Newnes-Butterworths, London

Hepburn C, Reynolds R J W (eds.) 1979 *Elastomers: Criteria for Engineering Design*. Applied Science, London

Lindley P B 1978 *Engineering Design with Natural Rubber*. Malaysian Rubber Producers' Research Association, Hertford, UK

Rader C P, Stemper J 1990 Thermoplastic elastomers—A major innovation in rubber. *Prog. Rubber Plast. Technol.* 6(1): 50–99

Snowdon J C 1968 *Vibration and Shock in Damped Mechanical Systems*. Wiley, New York

Treloar L R G 1975 *The Physics of Rubber Elasticity*, 3rd edn. Oxford University Press, Oxford

V. A. Coveney
[Bristol Polytechnic, Bristol, UK]

A. H. Muhr
[Malaysian Rubber Producers' Research Association, Hertford, UK]

Design with Plastics

Successful manufacture of plastics products requires a combination of sound judgement and experience. Experience and familiarity with polymers and their properties, and knowledge of mold design, mold construction methods and procedures subsequent to molding are essential. Early in the design process, there should be consultation and rapport between product designer, material supplier, mold builder and processor. Discussions should be continued throughout the design stage to identify the appropriate manufacturing processes, their variables and the influence on the product. This should be continued through material selection and prototype evaluation.

Often, products which are to incorporate plastics components are designed before final selection of the polymer materials. Because of this, compromises may be required and the processor should review the design for manufacturing technique and available equipment to be able to produce the plastic part within the required economic bounds. Consultation with specialists ensures that the most effective material and processes are selected, that processing limitations and variables have been identified and that a testing procedure for ensuring product quality has been developed. Participation by specialists in product assurance, maintenance, packaging, materials and environmental requirements will also contribute to a successful design.

1. Basic Requirements

It must first be determined what the material must do and environmental conditions to which it will be exposed. The design must also take into account economic factors, production rates and special feature requirements. For each product, an initial description of performance must be established. Since the resulting part design and material selection are generally arrived at through many design compromises, it is important to distinguish between true requirements and those which are only desirable. The general considerations in the design proess can be grouped as

(a) configuration:
  (i) basic use;
  (ii) size and shape;
  (iii) weight;

(b) structural or mechanical properties:
  (i) static, dynamic or cyclic forces;
  (ii) load magnitudes and duration;
  (iii) strength and stiffness;

(c) nonstructural or physical properties:
  (i) thermal, electrical, optical and other physical properties;
  (ii) color, surface finish and identification;
  (iii) fire resistance and material compatibility;
  (iv) permanence;

(d) environmental factors:
  (i) temperature and humidity range in storage and use;
  (ii) weathering and water resistance;
  (iii) chemical and biological resistance;
  (iv) service life and degradation characteristics;
  (v) recyclability;

(e) production and cost:
  (i) total quantities and production rates;
  (ii) secondary operation and special features;
  (iii) cost restrictions;
  (iv) standardization.

2. Materials Selection

There are hundreds of distinctly different plastics materials, the result of combining generic types, fillers, reinforcements and manufacturing processes. Designing with plastics demands knowledge of specific material behavioral factors which include creep, strength, stiffness, water absorption, dimensional stability, fatigue, toughness, chemical resistance and stess-crack resistance. For this reason, the selection process for a plastic material requires expertise in materials technology.

2.1 Initial Screening

The initial screening of materials is performed to select the type of plastics to be examined as possible final candidates. This choice is based on inherent materials properties not subject to enhancement by design. These "go or no go" requirements, which may include transparency, chemical resistance and/or temperature resistance, immediately limit the types and numbers of materials to be considered. The materials list may also be narrowed down by considering part shape and general strength requirements.

Screening should also consider processability, that is, ease of production. Limitations of the manufacturing process and its tooling cover such factors as parting line, draft, thicknesses, tolerances, undercuts, inserts and subsequent joining techniques.

2.2 Final Selection

Choosing the right material is probably the most important decision to be made in designing with plastics. The selection of candidate materials can range from a simplified choice based on a well-established precedent to a complex assessment of key properties for the full range of available plastics.

A key element in the selection process is the analysis of part design requirements and the translation of these into meaningful material property requirements. Be they simple or complex, the requirements must be described in terms of all important properties, namely, strength and resistance to impact, temperature, stress, time and chemicals.

A problem faced in the selection of materials resides in the property data that are available. It is difficult for the uninitiated to assimilate the large volume of available information and to adapt the data to a specific need, without a thorough understanding of polymer materials. A fundamental understanding is required of what properties are useful in design. In cases where there is no material property that relates directly to a part requirement, a qualitative judgement must be made, followed by testing to evaluate performance. The result of the selection process should be a list of candidate materials, with the associated cost factors, worth pursuing in the prototype evaluation phase.

3. Change in Material Properties during Processing

When designing any plastics product, a number of variables must be kept in mind. For example, the strength of high-impact plastics containing long glass fibers or rag stock fillers generally remains high when the materials are compression molded. On the other hand, values may drop by as much as 50% when the materials are transfer, plunger or injection molded, because gating can break up filler fibers. Certain sections become weaker per unit of thickness than other sections. Geometric design of the product may offset any advantage to be gained by compression molding. If a high-impact material is selected, design of the part may produce sections no

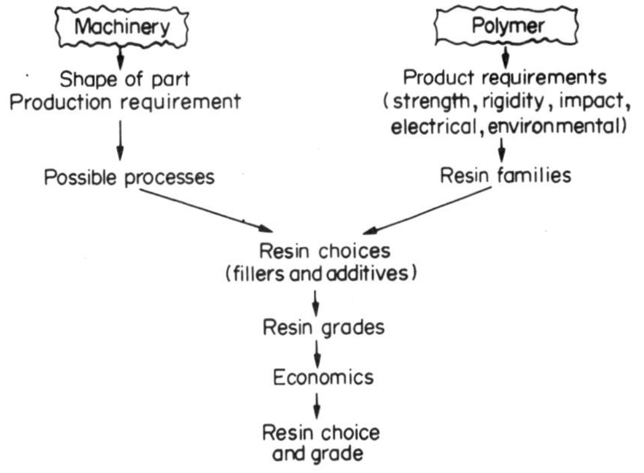

Figure 1
Choice of polymer and manufacturing process (courtesy of Dr R. Progelhof, New Jersey Institute of Technology)

Table 1
Processing limitations to part design: ○, process compatible with part type (courtesy of Dr R. Progelhof, New Jersey Institute of Technology)

Process	Shape limitation	Factor limiting maximum size	Complex shapes	Controlled wall thickness	Open hollow shapes	Closed hollow shapes	Very small items	Plane area > 1 m²	Inserts	Molded in holes	Threads
Blow molding	hollow thin wall	platten			○	○		○			○
Calendering	sheet	width of roll		○				○			
Casting		mold	○	○					○	○	
Compression molding		platten	○	○	○				○	○	○
Extrusion: film	sheet or tube	die		○				○			
profile		die		○							
sheet	sheet	width of roll		○				○			
Filament winding	surface of revolution			○		○		○			
Hand or spray layup	large thin wall	mold		○	○			○	○	○	
Injection molding: solid		platten	○	○	○		○		○	○	○
foam		platten	○		○			○	○	○	○
reactive		platten	○		○			○	○	○	○
Machining			○	○	○		○				○
Melt flow stamping		platten		○							
Pultrusion	linear	die		○							
Rotational molding	hollow				○	○		○	○	○	○
Thermoforming	thin wall	platten			○			○			
Transfer molding		platten	○	○	○		○		○	○	○

stronger than if they had been molded of the least expensive, low-impact material. Figure 1 and Tables 1–4 summarize various processing considerations and should assist in the selection of materials and processes.

4. *Shrinkage and Tolerance*

The actual dimensions of the final part will be different from those of the mold. The net shrinkage, or growth in some instances, depends on various factors, such as the stock temperature and the molding pressure. Two different forms of shrinkage must be considered when designing plastics products—the initial shrinkage which occurs while the part is forming, called "mold shrinkage," and that which occurs in the part after a period of time, called "postshrinkage" ("aftershrinkage").

Some materials are more stable than others during aging, regardless of their initial shrinkage. Commonly used thermoset plastics, such as a glass-filled phenolic compound, may have a mold shrinkage of 0.6% but no postshrinkage. On the other hand, an α-cellulose-filled melamine may have a mold shrinkage of 0.7–0.9% but show an aftershrinkage of as much as 0.6–0.8%; thus, the total shrinkage may be as high as 1.3–1.7%. In thermoplastic materials, there are large variations in shrinkage; for example, 30% glass-filled polycarbonate has 0.2% shrinkage, whereas the value is 2.5–3.5% for unmodified polyethylene.

The product design should allow for as broad a tolerance as possible. If it has been determined that the molded part must be postcured, stress relieved or baked, allowance must be made for the additional shrinkage that could occur during these treatments. Figures 2 and 3 indicate the tolerances required for various polymers as a function of dimension.

5. *Product Evaluation by Prototyping*

The prototype phase of a design project provides the first opportunity to evaluate product performance. Prototype evaluation presupposes the fabrication of a part or product by the manufacturing process that will be used in production. A prototype could be made by machining or other simplified modelmaking techniques. However, this practice should be avoided when detailed engineering tests are to be performed. Parts manufactured by a method other than the manufacturing process can provide very misleading results, since material properties can be greatly affected by the fabrication process.

Prototype evaluation should also be representative of the final design with respect to the specific material, the configuration and the use of special manufacturing techniques such as ultrasonic welding. For example, it is essential when testing colored parts for impact to have the dye incorporated, since the presence of the dye may have a significant effect on impact resistance or other mechanical properties.

Table 2
Process suitability for thermosetting resins: ●, primary choice; ○, secondary choice (courtesy of Dr R. Progelhof, New Jersey Institute of Technology)

Process	Alkyd	Allyl DAP	Amino melamine	Urea	Epoxy	Phenolic	Polyester	Polyurethane	Silicone
Blow molding									
Calendering									
Casting					●		●	○	●
Compression molding	●	○	●	●	○	●	●		○
Extrusion: film									
fiber									
profile									○
sheet									
wire and cable									○
Filament winding					○		●		
Hand or spray layup					○		●		
Injection molding: solid		○		○	○		●		○
foam									
reactive								●	
Lamination		○	○		○	○	○		○
Melt flow stamping									
Powder coating					●		●		
Pultrusion					○		●		
Rotational molding								○	
Thermoforming									
Transfer molding	●	●			●	○	○		●

Testing of the final product under actual end use conditions is the most meaningful form of evaluation. Although this may not be practicable, due to cost or induced hazards in the event of failure, all performance requirements are encountered and a fairly complete assessment can be made. When end use testing is not practicable, simulated service conditions must be employed in a manner which duplicates the end use conditions as closely as possible. Tests may include temperature cycling, load cycling, electrical performance, environmental exposure, static and impact strength tests and others.

Since testing for the service life of the item is often impracticable under conditions of actual service, accelerated testing can be used to evaluate performance. It has been found to be advantageous to initiate long-term testing as early as possible, concurrent with accelerated tests, and to continue the long-term tests indefinitely.

Another useful approach to plastics part evaluation is that of "overtesting" or testing to failure. Such testing may include temperature cycling beyond the required limits, drop tests from greater heights and at lower temperatures, application of induced stresses until part failure occurs and other

Table 3
Process suitablity for thermoplastics: ●, primary choice; ○, secondary choice (courtesy of Dr R. Progelhof, New Jersey Institute of Technology)

Process	ABS	Acetal	Acrylic	Fluoropolymers (melt proc.)	Nylon	PBT	PET	Polycarbonate	Polyethylene	Polyphenylene oxide based	Polypropylene	Polystyrene	PVC	SAN
Blow molding	○	○			○	○	●	○	●	○	●	○	●	○
Calendering													○	
Casting			●		○								●	
Compression molding														
Extrusion: film				●	●		●		●		●		●	
fiber					●		●				●			
profile	●		●		○				○	○	○	○	●	○
sheet	○		●					○	○	○	○	●	○	○
wire and cable					●				●		○		●	
Filament winding														
Hand or spray layup														
Injection molding: solid	●	●	●	○	●	●	●	●	●	●	●	●	○	●
foam	○					○		○	○	○	○	○		
reactive														
Lamination														
Melt flow stamping					○			○			○			
Powder coating				○	○				○					
Pultrusion														
Rotational molding	○				○			○	●				○	
Thermoforming	●		●				●	○	○	○	●	●	●	●
Transfer molding														

destructive tests. These tests tend to establish the ultimate performance limits of the product and provide an estimate of the factor of safety for the ultimate user.

In the final design phase, consideration is normally given to specifications to control production. These specifications usually include raw material identification, part dimensions, tolerances, color, finish and other pertinent appearance features. In many cases an engineering test is needed to assure product performance. This test would be based on critical product requirements and the means of measurement, and may involve the measurement of material properties. Generally, quality assurance procedures should be developed from the prototype testing phase.

The prototype phase affords the opportunity to assess the influence of the manufacturing process variables on product performance. Plastic fabrication processes subject the material to rather severe conditions involving elevated temperature, high pressure and high shear-flow rates. These conditions, coupled with variable cycle times, other machine variables and procedures subsequent to molding, can result in internal stresses which can lead to product failure. Knowledge of these variables and their influence on the material will contribute significantly to a successful design. The evaluation of process variables can also be used to determine the effectiveness of the product assurance procedures intended for use as a measure of product integrity.

6. Design Specifications

The completion of a design must be followed by development of appropriate engineering drawings. These drawings detail dimensions and tolerance limits, materials, surface finish, color and marking, as well as special requirements and features as needed for production control. Design finalization should also include technical documentation of the entire integrated product design process, describing the requirements, the design, material selection and product evaluation. Such reports can be the basis for future reference with respect to production, product improvement and maintenance.

Table 4
Processing cost indication (courtesy of Dr R. Progelhof, New Jersey Institute of Technology)

Cost per unit part	Process
Very low	calendering film extrusion RIM solid injection molding wire and cable extrusion
Low	flow molding[a] foam injection molding melt flow stamping profile extrusion rotational molding[a] sheet extrusion thermoforming transfer molding
Medium	blowmolding[a] compression molding powder coating pultrusion rotational molding[a]
High	casting filament winding hand or spray layup machining

a Varies with size

6.1 Considerations of Product Assurance

Hardware or products are produced chiefly on the basis of a requirements package. The package contains drawings and specifications detailing the requirements for each part and assembly. In essence, it contains the product assurance considerations, which include quality control and performance specifications. The emphasis placed on the design concept is to develop more effective performance specifications.

6.2 Standards and Specifications

Quality provisions are frequently cited by reference to specifications or standards, particularly for materials acceptance. Use of available documents must be scrutinized to ensure that conformance to the specifications also results in compliance with product requirements.

6.3 Quality Control

The term "quality control" is a general term implying the control of product quality, either through product measurement or in the control of a manufacturing process. The requirements package sets the general standards by which the product is accepted. This includes initial material purchase, dimensions, tolerances, surface finish, color and general workmanship. These requirements, although a measure of acceptability, may not completely reflect the true quality of the item. For instance, the dimensional stability cannot be assessed accurately unless the item is exposed to an elevated temperature as expected in use.

6.4 Performance Specifications

Performance specifications are defined as those methods used to evaluate performance, the basis of which is related to the function of the component.

Performance specifications are extremely important in designing with plastics, to ensure that the production techniques employed do not hinder performance. Testing can be either destructive or nondestructive. It may employ existing methods or novel ones unique to the product. Tests may be applied to individual components, subassemblies or

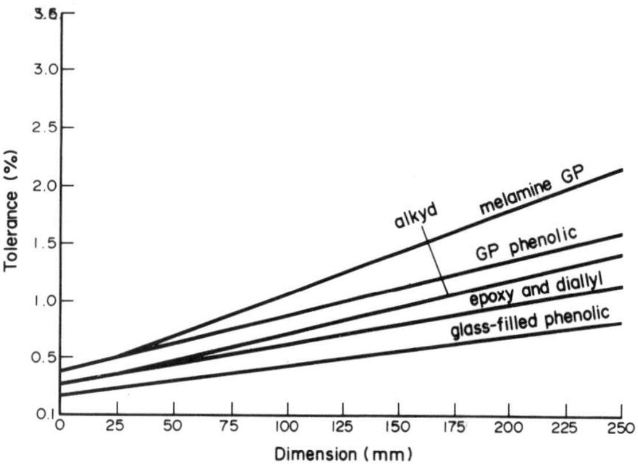

Figure 2
Dimensional tolerances for thermoplastics (no glass reinforcement)

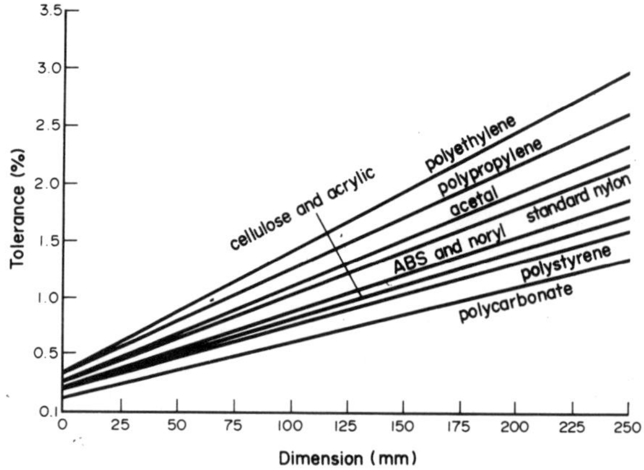

Figure 3
Dimensional tolerances for thermosetting resins

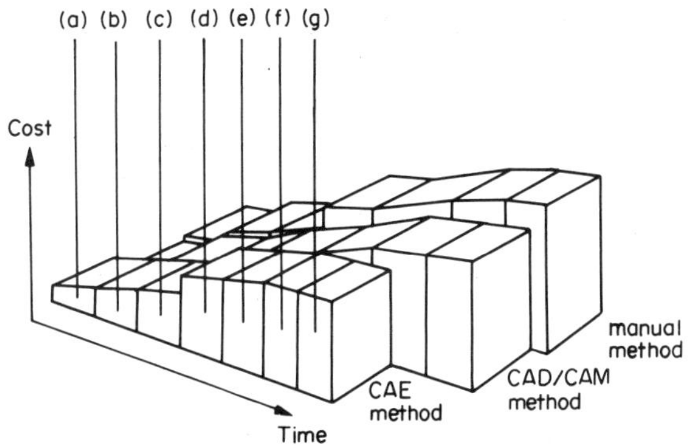

Figure 4
Illustration of the benefits of computer-aided engineering in part design: (a) system simulation/concept development, (b) design guided by CAE analysis, (c) CAE detail design, (d) first prototype built, (e) preproduction engineering, (f) production prototype and (g) release to manufacturing (courtesy of Structural Dynamics Research Corporation)

total assemblies as appropriate. In all, performance specifications are the most meaningful form of product evaluation.

7. *Secondary Process Operations in Designing*

There are numerous types of secondary processs operations, which fall into the categories of finishing, joining and decorating. Many are quite useful in achieving design features not readily possible with other materials. Their adaptability to design is important in controlling product uniformity.

7.1 Finishing

Finishing operations include degating, deflashing and machining, which are required after the part has been fabricated. Although degating and deflashing are very simple procedures, machining of plastics warrants greater attention. Techniques used to machine metals are generally applicable to plastics but must be modified for effective performance. Such modifications are necessary because plastics are of lower stiffness, lower thermal conductivity, higher thermal expansion and lower softening point. Each type of polymer has unique properties and as a result it can be assumed that each has different machining characteristics.

7.2 Joining

The joining of plastics is generally accomplished by mechanical methods, adhesive bonding or welding. Each requires different design configurations, shape and size, which must be considered early in product development, to ensure evaluation in prototype testing.

Mechanical means of joining include screw fasteners, inserts, snap fits, hinge mechanisms and staking. These may be used for joining plastics to plastics or to metals. Because of potential induced stress and thermal considerations, mechanical joining cannot be treated routinely and must be given careful consideration in design.

The field of adhesive bonding is a highly specialized area. Adhesives for plastics fall into five basic groups: solvent cements, bodied adhesives, monomeric cements, elastomeric adhesives and reactive adhesives. There is no universal adhesive or cement that will bond all types of plastics materials. In addition, it should be noted that certain plastics materials cannot be joined with adhesives, due to their chemical makeup. Although a very cost-effective means of joining plastics, adhesive bonding requires a good deal of knowhow for effective utilization.

Welding processes for plastics include ultrasonic assembly, hot gas welding and heat sealing. Each process is an art in it itself and employs specific design configurations for effective utilization.

The many techniques for joining plastics afford the designer great latitude in product development to facilitate assembly, incorporate unique features and reduce cost.

7.3 Decorating

Whether applied for functional or aesthetic reasons, decorating is frequently required in the manufacture of plastic parts. The methods include painting, printing, coloring, metallizing and hot stamping. Many of the basic molding procsses are also adaptable to decorating, in that raised or lowered letters, two-color molding and in-mold decorating are practicable. To meet product requirements effectively these secondary processes and their variations must be compatible with the material and the design.

8. *Use of Computers in Part Design*

The computer provides a useful tool both for improving original designs and for upgrading existing designs by incorporating different materials. Component designers are faced with designing the lightest possible component meeting performance specifications set by the systems designer and providing optimal cost–benefit balance. Development and evaluation of various design concepts in terms of cost and performance must be iterated until this "best" design is achieved. Faced with many parameters, designers need a technique for evaluating many alternatives.

8.1 Computer-Aided Engineering

Computer-aided engineering (CAE) is a systematic process of applying computer to each design application concern. CAE organizes design processes into unique elements (design, analysis, test, drafting, documentation and control, and related manufacturing), to which computer tools are applied to gain information and make better decisions. With CAE, performance of alternative materials and concepts is investigated via "computer prototypes." By using computer models rather than hardware prototypes, CAE can substantially shorten a design cycle, improve product and reduce cost. The CAE method is compared favorably with the CAD/CAM and manual methods in Fig. 4.

8.2 Materials Considerations

Past lightweight component design efforts reveal two characteristic materials problems. First, theoretical equations are used to evaluate relative material effectiveness without consideration of overall component design, an approach satisfactory only for simple structures or properties. The second problem occurs when alternative materials are substituted directly in a design.

See also: Bottles and Other Rigid Containers; Plastics for Packaging; Polymers: Technical Properties and Applications

Bibliography

Grafton P 1980 Design considerations for plastics components. *Plastics Mechanical Performance and Physical Testing*, PIA graduate course. Plastics Institute of America, Hoboken, NJ

Hagan R S, Thomas J R 1965 *Plastics Part Properties and Their Relationship to End-Use Performance*. American Society for Testing and Materials, Philadelphia, PA

Hagan R S, Thomas J R 1979 Environmental effects on plastics parts and properties. *Plastics Mechanical Performance and Physical Testing*, PIA graduate course. Plastics Institute of America, Hoboken, NJ

Maddux K C 1983 Use of computers in plastics part design. *Applications of Computer Aided Engineering in the Injection Molding Industry,* PIA graduate course. Plastics Institute of America, Hoboken, NJ

Maddux K C, Amgas G D 1984 *Lightweight Component Design via Computer-Aided Engineering*, SAE paper No. 790984. Society of Automotive Engineers, New York

Nardone J 1981 Engineering plastic products. *Plastics Product Design*, PIA graduate course. Plastics Institute of America, Hoboken, NJ

O'Toole J L 1981 Applications and design properties of plastics. *Plastic Mechanical Performance and Physical Testing*, PIA graduate course. Plastics Institute of America, Hoboken, NJ

Progelhof R C 1981 Process technology and product assurance. *Plastics Product Design*, PIA graduate course. Plastics Institute of America, Hoboken, NJ

Starita J 1978 Getting meaningful impact data for product design. *Plastics Mechanical Performance and Physical Testing,* PIA graduate course. Plastics Institute of America, Hoboken, NJ

A. Spaak
[Plastics Institute of America, Hoboken, New Jersey, USA]

Die Extrusion

Dies are used to produce a desired shape in continuous extrusion processes. Depending upon the die cross section, circular, annular or complex profiles may be produced. The products obtained from dies include fialments, tubing and pipes, and film and sheet. The origins of die extrusion of polymers date from patents of Brooman (1845) and Bewley (1845). In this article, a description is provided of various aspects of flow through dies including flow from pumping devices into dies, pressure losses and their calculation and the characteristics of emerging extrudates.

1. Feeding Melt into a Die

One of the basic problems of die design is feeding polymer melt into the die section, called the land, to produce the desired extrudate cross section. For simple die shapes such as a capillary or low-aspect-ratio slit (tape) die, the melt is generally fed from a reservoir. For the feeding of a more complex die, such as a high-aspect-ratio slit die to produce sheeting, more complex die designs are required.

There is extensive experimental literature on the flow of polymer melts through constrictions. Some polymer melts exhibit vortexlike patterns in a 180° angle into the die entrance (see Fig. 1). This has been found to be the case for low-density polyethylenes (Tordella 1957, Bagley and Birks 1960, Ballenger and White 1971, Cogswell 1972), polystyrene (Ballenger and White 1971, White and Kondo 1977a), and poly(methyl methacrylate)

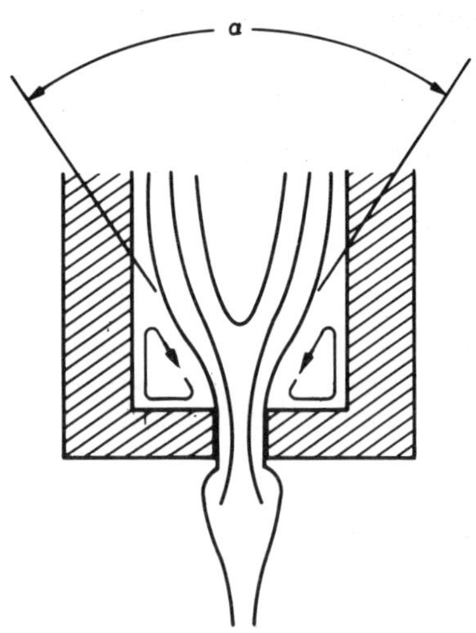

Figure 1
Flow from a reservoir into a slit or capillary die and associated vortex flow patterns; angle α determines the convergence of flow into the die

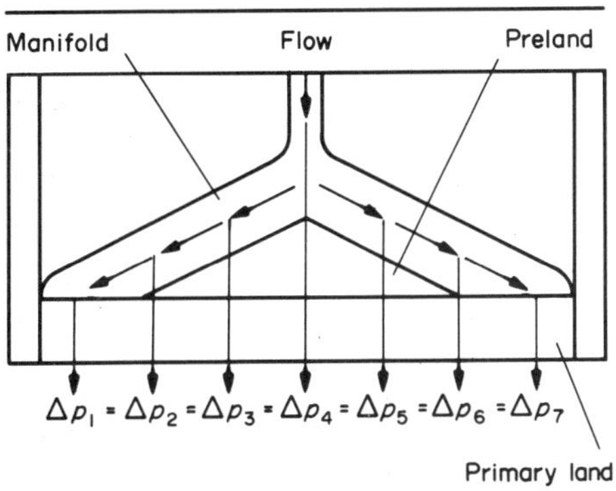

Figure 2
Polymer melt flow in coathanger film dies (after Garton 1982)

(Cogswell 1972). Similar vortices have been observed in polymer solutions (Giesekus 1968). However, certain molten plastics do not appear to exhibit vortices but rather flow in streamlines directly into the die. This has been found to be the case for commercial high-density polyethylene (Bagley and Birks 1960) and polypropylene (Ballenger and White 1971, Cogswell 1972). Studies of flow patterns in a polymer melt in a converging die have been reported by various investigators (Bagley and Schreiber 1961, White and Kondo 1977a). Generally, as the die entrance angle decreases, so does the vortex intensity; with small die entrance angles, vortices disappear (White 1973).

The major problem for dies with large complex cross sections is feeding the melt to give uniform flow. Sheeting dies generally have a coathanger design (Tadmor and Gagos 1979, Garton 1982) where melt is fed from the screw through a connector conduit to a die manifold. This feeds melt from the center of the longer die dimension to its extremities. The manifold may decrease in cross section in a programmed manner as does the length of the die preland (see Fig. 2). This is carried out in such a manner that the melt is fed uniformly from the manifold into the die land.

Annular dies used to produce film, tubing and parisons often have similar complex designs to produce flow uniformity around the circumference. Melt may be fed into the annular channel from two sides or from a spiral groove manifold on the die mandrel. Sometimes the die mandrel may rotate (Tadmor and Gogos 1979, Garton 1982).

2. Pressure Losses in Dies

Generally, the pressure drop necessary to extrude a polymer melt through a die must be considered as the sum of pressure losses associated with the entrance and exit, as well as with flow within the die itself. For flow from a reservoir through a constriction, this was first clearly shown in the research of Bagley (1957).

In most cases, the combined, rather than the separate, entrance and exit pressure losses for flow through a constriction are considered; these studies mainly reflect the characteristics of the entrance pressure loss, since it is the larger of the two (Han 1976). The relative magnitude of these pressure losses for different melts has been described by Ballenger and White (1971). The largest values of the ratio of this pressure loss to the die-wall shear stress were found with low-density polyethylene, where values of over 50 have been recorded. For a Newtonian oil, ratios of the order of unity were obtained. Many interpretations and correlations have been made relating the large entrance pressure losses to the principal normal stress difference in shear flow (Philippoff and Gaskins 1958, La Nieve and Bogue 1968, White and Kondo 1977b). An alternative approach has been to relate it to the elongational flow behavior occurring in the converging flow at the die entrance (Cogswell 1968, 1972, 1978).

For many die designs including coathanger dies, there are no abrupt constrictions, but instead there is a Poiseuille-like steady shearing motion. Calculations of pressure losses in manifolds may be made using hydrodynamic lubrication theory (Tadmor and Gogos 1979).

Flow within the die cross section, or land, is usually the primary source of pressure loss. In this region, the cross section is the same all along the die and the melt moves in fully developed laminar shear flow. The pressure drop may be calculated by balancing the shear stress field with the pressure gradient along the direction of flow. The shear stresses are equated with the products of the viscosity and velocity gradient perpendicular to the direction of flow. If z is the direction of flow and x and y are directions perpendicular to the die wall, this force balance has the form (Middleman 1965, Wheeler and Wissley 1965, Rothemeyer 1970)

$$0=\frac{\partial p}{\partial z}+\frac{\partial}{\partial x}\left(\eta\frac{\partial v}{\partial x}\right)+\frac{\partial}{\partial y}\left(\eta\frac{\partial v}{\partial y}\right)$$

where p is pressure, v is velocity and η is the shear viscosity, which depends on $\partial v/\partial x$ and $\partial v/\partial y$, and the fluid is presumed to adhere to the die walls. Calculations have been reported for circular, slit, rectangular, trapezoidal and complex cross sections. The theories for the circular and slit cross sections are so well established that they are both used in viscometers (Han 1974, 1976). Fewer comparisons exist between theory and experiment for more complex cross sections (White and Huang 1981).

If the pressure is measured at various positions along the direction of flow and extrapolated to the die exit, the result is a finite pressure. The most extensive studies of exit pressure losses in dies have been those of Han (1974, 1976) who has sought to use the measurements to determine normal stresses in polymer melts.

The pressure losses in more complex dies have also received some attention; Tadmor and Gogos (1979) discuss in detail pressure losses in coathanger dies including the interaction between the manifold and die land.

3. Extrudate Swell

Polymer melts emerging from a die generally swell to cross sections much greater than that of the die. The magnitude of the swell increases with extrusion rate and with decrease in length of the die. Most studies are for capillary dies (Bagley et al. 1963, White and Roman 1976, Tadmor and Gogos 1979). A lesser number of investigators consider slit dies (Kruse 1974, Huang and White 1979, 1980) and a few others more complex cross sections (Han and Park 1973, White and Huang 1981). Generally, the swell from capillary and slit dies increases with increasing extrusion rate. Its relationship to die-wall shear rate or shear stress is independent of die diameter.

A serious problem arises in dies of complex cross section where the emerging extrudates exhibit rather different shapes than the dies (Han and Park 1973). This is due to a combination of effects; one of these is the velocity-profile arrangement effect and is inherent in Newtonian fluids. A second problem arises with the nonuniform extrudate swell of the melt emerging from the different sections of the die.

4. Extrudate Distortion

At high extrusion rates, polymer extrudates emerging from a die exhibit distortion. This phenomenon was first studied by Spencer and Dillon (1949) for polystyrene. They found that a smooth uniform extrudate becomes helically distorted at a critical die-wall shear stress which varies inversely with average molecular weight. The phenomenon was shown to be more general in the later studies of Tordella (1958), occurring in essentially every type of polymer melt at a die-wall shear stress of the order 10^5 Pa. The distorted extrudate character, however, varies from polymer to polymer. In some cases, it continues with small surface roughness amplitudes down to low extrusion rates (Howells and Benbow 1962). Comparative characteristics of distorted extrudates are discussed by White (1973). Most studies of extrudate distortion are for capillary dies. Huang and White (1979) discuss extrudate distortion for slit dies.

Various mechanisms have been proposed for the inception of the unstable flows leading to extrudate distortion. Investigations of low-density polyethylene (LDPE) have centered on observations of the breakup of the "wine-glass-stem-vortex" structure at the die entrance shown in Fig. 1 (Tordella 1957, Bagley and Birks 1960, Ballenger and White 1971). Studies of high-density polyethylenes (HDPEs) have centered on the oscillating shear phenomenon in which the pressure drop or flow rate oscillates at the inception of the distortion. At higher extrusion rates, distorted extrudates continue to be produced on the low-pressure-drop branch (Tordella 1963, Metzger and Hamilton 1964, Vinogradov et al. 1972). These distortions are associated by many researchers with the initiation of slip at the die wall. Tordella (1963) supports this view with flow-birefringence investigations of LDPE and HDPE at the onset of extrudate-distortion-related flow instabilities. High fringe concentrations (stress levels) are found in LDPE but not HDPE at the die entrance. Studies on polystyrene melts (Ballenger and White 1971) suggest mechanisms similar to that occurring in LDPE, with the entrance flow breaking down to give a spiralling motion into the die entrance.

5. Extrusion Processes

A more general review of the extrusion of rubber and plastics has been presented by Stevenson (1989), including the geometry and operation of extrusion equipment, an analysis of operations by mathematical models and an organization and review of extrusion literature. It also contains details of die classifications and design criteria.

See also: Coextrusion; Processing: Classification

Bibliography

Bagley E B 1957 End corrections in the capillary flow of polyethylene. *J. Appl. Phys.* 28: 624–7

Bagley E B, Birks A M 1960 Flow of polyethylene into a capillary. *J. Appl. Phys.* 31: 556–61

Bagley E B, Schreiber H P 1961 Effect of die entry geometry on polymer melt fracture and extrudate distortion. *Trans. Soc. Rheol.* 5: 341–53

Bagley E B, Storey S H, West D C 1963 Post extrusion swelling of polyethylene. *J. Appl. Polym. Sci.* 7: 1661–2

Ballenger T F, White J L 1971 The development of the velocity field in polymer melts in a reservoir approaching a capillary die. *J. Appl. Polym. Sci.* 15: 1949–62

Benbow J J, Lamb P 1963 New aspects of melt fracture. *SPE Trans.* 3: 7–17

Bewley H 1845 Flexible syringes, tubes and bottles. English Patent No. 10,825 (4 September 1845)

Brooman R A 1845 Manufacture and application of gutta percha thread. English Patent No. 10,582 (27 March 1845)

Cogswell F N 1968 Tensile deformations in molten polymers. *Rheol. Acta* 8: 187–94
Cogswell F N 1972 Converging flow of polymer melts in extrusion dies. *Polym. Eng. Sci.* 12: 64–73
Cogswell F N 1978 Converging flow and stretching flow: A compilation. *J. Non-Newtonian Fluid Mech.* 4: 23–38
Garton D R 1982 Extrusion dies. In: Agranoff J (ed.) 1982 *Modern Plastics Encyclopedia*. McGraw-Hill, New York, pp. 412–17
Giesekus H 1968 Nicht-lineare Effekte beim Strömen viskoelastischer Flüssigkeiten durch Schlitz- und Lochdüsen. *Rheol. Acta* 7: 127–38
Han C D 1974 On slit- and capillary-die rheometry. *Trans. Soc. Rheol.* 18: 163–90
Han C D 1976 *Rheology in Polymer Processing*. Academic Press, New York
Han C D, Park J Y 1973 A study of shaped fiber formation. *J. Appl. Polym. Sci.* 17: 187–200
Howells E R, Benbow J J 1962 Flow defects in polymer melts. *Trans. J. Plastics Inst.* 30: 240–53
Huang D C, White J L 1979 Extrudate swell from slit and capillary dies: An experimental and theoretical study. *Polym. Eng. Sci.* 19: 609–16
Huang D C, White J L 1980 Experimental and theoretical investigation of extrudate swell of polymer melts from small (length)/(cross-section) ratio slit and capillary dies. *Polym. Eng. Sci.* 20: 182–9
Kruse R L 1964 Extrudate swelling of molten polypropylene from an extrusion rheometer. *Polym. Lett.* 2: 841–3
La Nieve H L III, Bogue D C 1968 Correlation of capillary entrance pressure drops with normal stress data. *J. Appl. Polym. Sci.* 12: 353–72
Metzger A P, Hamilton C W 1964 The oscillating shear phenomenon in high density polyethylenes. *SPE Trans.* 4: 107–12
Middleman S 1965 Flow of power law fluids in rectangular ducts. *Trans. Soc. Rheol.* 9: 83–193
Philippoff W, Gaskins F H 1958 The capillary experiment in rheology. *Trans. Soc. Rheol.* 2: 263–84
Rothemeyer F 1970 Das Druckströmungsfeld in zylindrischen Fliesskanälen mit beliebigen Querschnitt. *Rheol. Acta* 9: 259–63
Spencer R S, Dillon R S 1949 The viscous flow of molten polystyrene II. *J. Colloid Sci.* 4: 241–55
Stevenson J F 1989 Extrusion of rubber and plastics. In: Aggarwal S L (ed.) 1989 *Comprehensive Polymer Science*, Vol. 7. Pergamon, Oxford, Chap. 10
Tadmor Z, Gogos C G 1979 *Principles of Polymer Processing*. Wiley, New York
Tordella J P 1957 Capillary flow of molten polyethylene—A photographic study of melt fracture. *Trans. Soc. Rheol.* 1: 203–12
Tordella J P 1958 An instability in the flow of molten polymers. *Rheol. Acta* 1: 216–21
Tordella J P 1963 Unstable flow of molten polymers: A second site of melt fracture. *J. Appl. Polym. Sci.* 7: 215–29
Vinogradov G V, Insarova N I, Boiko B B, Borisenkova B B 1972 Critical regimes of shear in linear polymers. *Polym. Eng. Sci.* 12: 323–34
Wheeler J A, Wissley E H 1965 The friction factor—Reynolds number relation for the steady flow of pseudoplastics fluids through rectangular ducts. Part 1: Theory. *AIChE J.* 11: 207–12
White J L 1973 Critique on flow patterns in polymer fluids at the entrance of a die and instabilities leading to extrudate distortion. *Appl. Polym. Symp.* 20: 155–74
White J L, Huang D C 1981 Extrudate swell and extrusion pressure loss of polymer melts flowing through rectangular and trapezoidal dies. *Polym. Eng. Sci.* 21: 1101–7
White J L, Kondo A 1977a Flow patterns in polyethylene and polystyrene melts during extrusion through a die entry region: Measurement and interpretation. *J. Non-Newtonian Fluid Mech.* 3: 41–64
White J L, Kondo A 1977b Ends pressure losses in extrusion of polymer melts through dies. *J. Appl. Polym. Sci.* 21: 2289–92
White J L, Roman J F 1976 Extrudate swell during the melt spinning of fibers—Influence of rheological properties and take-up force. *J. Appl. Polym. Sci.* 20: 1005–23

J. L. White
[University of Akron, Akron, Ohio, USA]

Dipped Organic Coatings

Immersion, dip or free coating is a process in which the surface to be coated is immersed in the coating fluid and then withdrawn. A layer of liquid remains on the surface of the object, the amount depending on the viscosity and surface tension of the coating fluid and on the speed of withdrawal. Gravity causes the fluid to tend to drain off the object, but this is prevented by the drying or curing of the coating.

There are three forms of immersion coatings. The simplest case is when the object to be coated is a continuous web or belt of material (i.e., film or coil coating). Another is when the object is a cylindrical filament (i.e., fiber or wire coating). The third form is when a wet coating is applied to a vertical surface (i.e., dip coating).

Dipping is the most economical method of industrial finishing. It is fast, lends itself to a continuous conveyorized operation and all parts of the object, inside and outside, are coated. In a conveyorized dipping operation, all processes from the initial surface cleaning stage to the final curing of the coating can be carried out in a continuous fashion.

Despite its conceptual simplicity, dipping is a complex process. Many operating variables such as paint composition and viscosity, temperature, line speed, solvent evaporation rate, rate of withdrawal of the object and the effectiveness of drainage must be controlled to produce a coating of a desired thickness and uniformity.

1. Application

Dipping processes are used mainly at opposite ends of the painting scale:

(a) in the painting of large objects where protection rather than decoration is desired such as auto body parts; and

(b) for obtaining a high-class decorative finish on small objects.

Quick dipping is mainly used for large objects. The workpiece is transported by conveyor, dipped into a large stirred tank of paint, lifted out and allowed to drain while passing over a trough which collects the drainage and returns it to the tank. This is a rapid process, taking only 1–2 min, and requires the workpiece to be suitably designed to allow the paint to reach all parts of the surface and to allow the excess paint to drain cleanly and not be trapped in pockets. Examples of quick-dip application are the coating of constructional steelwork, agricultural equipment and electrical goods and the application of primer coats to large objects such as car bodies. In the auto primer application, the parts are mounted on a spigot and rotated both in the dip tank and during the drainage process (Rotodip process).

The slow-dip process is mainly used for decorative coating of small objects. For a high-grade decorative finish, high-solids, viscous paints which dry quickly by solvent evaporation are applied by the low-dip process.

Dip coating is especially useful for the application of thin coatings which are difficult to achieve by other means. Therefore by thinning the paint sufficiently, that is, reducing both viscosity and solids content, it should be possible to apply a dip coating routinely at a thickness of 2.5 μm or less with good accuracy.

The formulation of dipping paints must be carried out with great care. Dipping tanks are generally used for months or even years without being completely emptied and therefore require an unusually stable paint. Settling and viscosity buildup are major stability issues that can interfere with reproducibility of the coating operation. Usually dipping paints have low viscosity to ensure a thin film and promote rapid wetting and clean runoff. However, they must not settle and should have a sufficient and uniform solids content to leave an adequate paint layer on the workpiece. Furthermore, because of the nature of the dipping operation, the paint comes into contact with large amounts of air and must therefore be stable against oxidation and subsequent thickening due to oxidative crosslinking.

The solvent evaporation must also be kept to a minimum, since the loss of solvent not only raises the viscosity but also affects the drying and drainage behavior of the coating. The latter is due to the change in the solvent balance through the differential evaporation of lower-boiling constituents. Adding more of the original thinner can restore the proper viscosity but does not restore the original solvent balance. In industrial finishing, a practical solution to the evaporation problem is to maintain a nearly solvent-saturated atmosphere in the dipping room; however, this greatly increases the risk of fire and requires additional precautions against sparks and flame if flammable solvents are used.

To mitigate the paint stability problems, the dip tank should be no larger than is necessary to accommodate the object and should have a minimum cross-sectional area to minimize solvent evaporation and coating deterioration. An agitator should be provided for continual stirring of the paint. The tank should also be covered when not in use, to minimize evaporation and related fire hazards.

Another potential problem for dipping is clean runoffs. There is a natural tendency for lower edges to acquire a thicker layer of paint during drainage. The manner in which the object is hung (particularly if it has a complex shape) is important in achieving a uniform coating without unsightly drip beads or fat edges. Such fat edges and drops or "tears" may be removed at the end of the drainage process by passing the part over a highly charged grid which attracts the drip or tear electrostatically. This process is known as electrostatic detearing.

In general, the dip application is used whenever possible, because of its simplicity and economy. Other advantages of the dip coating techniques are excellent reproduction of results, no need for special skills in carrying out the operation and adaptability to nonplanar surfaces.

Two special cases of dip coating are plastisol dipping and fabric coating.

1.1 Plastisol Dipping

Wire goods, tool handles, bobby pins, electrical components, lamp sockets, surgical gloves, shoes, spark plug covers, springs, glass bottles, pump diaphragms and toys are coated by plastisol dipping. If the object to be coated can be heated, the hot-dipping process gives greater latitude in controlling coated weight pickup and thickness control. The parts to be coated are brought up to dipping temperature in a convection oven and then conveyed to a dip tank. They are submerged in the plastisol paste to a desired height and for a given time. To ensure a uniform coating height, the level of paste in the tank is kept constant. The object is removed at a controlled rate from the dip tank and conveyed to another tank where excess paste is allowed to drain by gravity or mechanical means—by spinning or vibration. The object may be rotated on its axis during this stage, to make the drip marks less conspicuous. After the proper drainage schedule, the coated object is conveyed through a fusion oven where the plastisol reaches the flux temperature and cures. The coated objects are cooled before removal and storage.

1.2 Fabric Coating

The dip method of coating fabrics is one of the oldest employed in the industry. In this process, the fabric to be coated is threaded over guide rolls and passes through the coating tank and vertically into a drying tower. The drying tower must be designed to dry the coating completely before it reaches the top roller. These rollers are usually water-cooled to prevent sticking of the coated fabric. Multiple coats can be applied continuously by allowing the web emerging from the first pass to dip into the next coating tank and continuing until the desired number of coats have been applied. This method of coating is particularly well adapted for relatively weak, open-weave materials, since the fabric is not subjected to a great deal of tension.

2. *Rheology*

The rheology of the dipping process is mainly defined by the viscosity of the paint and the speed of withdrawal of the part. These factors affect the levelling and runoff characteristics of the coating as well as its thickness and uniformity. Surface tension is the driving force that causes a coating to a level. Generally, the thicker the coating and the lower the viscosity, the faster a coating will level.

In general, increasing the viscosity of the paint results in a thicker coating at a given withdrawal rate. Furthermore, increasing the speed of withdrawal of the part also results in a thicker coating. Coatings with uniform thickness can be produced at relatively low withdrawal speeds. At higher withdrawal rates, the coating thickness becomes nonuniform and the lack of uniformity becomes progessively worse, the higher the paint viscosity.

A theoretical treatment of the relationship between paint viscosity, withdrawal rate of the part and the coating thickness has been given by Middleman. Briefly, the weight of the wet coating on the part is compensated by the frictional forces, which are in turn proportional to the viscosity and the rate of change of the velocity across the thickness of the coating:

$$\eta\frac{d^2v}{dx^2}=\rho g$$

where η is the viscosity, v the paint velocity in the direction of withdrawal, x the coordinate in the direction of coating thickness (perpendicular to the direction of withdrawal), ρ the paint density and g the acceleration due to gravity.

Therefore, both a higher velocity of withdrawal and a higher paint viscosity result in a larger mass of wet paint carried out of the dip tank and, thus, a thicker coating on the part. Eventually, when the part is completely out of the tank and the withdrawal velocity becomes zero, the coating tends to sag and run off under gravity; however, this is prevented by drying or curing, which results in a rapid increase in the viscosity of the coating.

3. *New Developments*

Recently, a chemically activated immersion coating process has been introduced. This process is called autodeposition and is used mainly for primer coating of steel parts, such as automotive parts. In the autodeposition process the metal is cleaned, rinsed, deionized water rinsed, immersion coated, water rinsed, chemically rinsed and baked in a continuous conveyorized fashion. The autodeposition bath comprises a polymer (e.g., latex) and pigment dispersed in an aqueous solution of metal activators (e.g., acid or peroxide). The bath solids content may vary widely, but for most commercial operations it is in the range of approximately 5–10%. The bath pH is in the range 2.5–3.5. The mechanism of autodeposition is explained by the dissolution of iron from the steel surface to form Fe^{2+} ions, which in turn destabilize the resin and pigment colloids at the metal surface, resulting in coagulation of the colloidal particles and deposition of an organic protective film. The film is later baked to achieve the final properties.

See also: Electrodeposited Polymer Coatings; Paints, Varnishes and Lacquers

Bibliography

Anon 1978 Autodeposition: A new approach to steel coatings. *Automat. Eng.* 86: 52–5

Haritani T 1979 Dipping or saturating coater. *Kami to Purasuchikku* 7: 10–16

Payne H F 1943 The dip coater. *Ind. Eng. Chem., Anal. Ed.* 15: 48

Payne H F 1944 Application of uniform films. *Off. Dig., Fed. Soc. Paint Technol.* 16: 400

Sato T 1979 Effect of turnover rate on solid content in dip-coating bath. *J. Coat Technol.* 51: 61–73

Takahashi T 1976 Development of dip coating of aluminum with aqueous coating material. *Aruminyumu Kenkyu Kaishi* 105: 11–22

B. Siadat
[W R Grace and Company, Columbia, Maryland, USA]

Dipped Polymer Products

The process of forming artifacts by dipping is among the earliest fabrication techniques known to have been used. The origins of dipping probably go back many thousands of years, although it was not until the exploration of equatorial South America by Juan de Torquemada and Charles de la Condamine, in the seventeenth and eighteenth centuries, respectively, that the first observations were recorded. South American Indians were seen to make a type of shoe by dipping clay molds into a milky liquid obtained by cutting into the bark of certain trees.

Drying between successive dippings using heat from a fire not only accelerated the process, but the smoke also had an antiseptic effect and conferred additional strength and durability on the shoes. The milky liquid we now know to have been latex from the natural rubber tree *Hevea brasiliensis*.

The early part of the nineteenth century saw the formation and rapid growth of the rubber industry. At that time, dipped products were made from dry, masticated natural rubber dissolved in hydrocarbon solvent. It was not until the late 1920s, when ammonia-preserved latex became commercially available, that a major step forward was made. Not only was natural rubber latex safer to use than the inflammable rubber solutions, but the higher solids content meant that thicker articles could be made more quickly. The adoption of latex was so broad that the only product it did not displace was surgeons' gloves, where the solution-dipped type continued to be regarded as offering greater suppleness.

At present, natural rubber latex is supplemented by a wide range of synthetic latices made by means of emulsion polymerization. Originally, these were confined to styrene-butadiene rubber (SBR) and polychloroprene (neoprene) elastomers. They have now been joined, however, by acrylonitrile-butadiene rubber (NBR), poly(vinyl acetate) (PVA) and copolymers, poly(vinyl chloride) (PVC), poly(vinylidene chloride), carboxylated-SBR types and polymers derived from esters of acrylic or methacrylic acid, including styrene-acrylate. Together with polyurethane and other artificial latices, PVC plastisols, liquid polymers and polymer solutions, the range of material types now available is considerable. Nevertheless, the bulk of dipped products is still derived from natural and synthetic rubber latices; hence, this article is mainly confined to latices and the latex dipping process.

1. The Nature of Latex

A latex is a two-phase system in which colloidal polymer particles are dispersed in an essentially aqueous medium. Generally, the particles may be regarded as spherical, although it should be noted that more complex geometrical forms are found in natural rubber. All latices comprise particles of a range of sizes. The nature and extent of the polydispersity may be expressed in several ways, the most common being to base it on the linear dimension of the particle. Synthetic latex particles are typically 80–140 nm in diameter, whereas the particles in natural rubber latex are an order of magnitude larger. Furthermore, the particle-size distribution is Gaussian compared with the skewed distribution towards the larger particle diameters that is found with natural rubber. Normally, latex that is used for dipping has a solids content in the range 50–60 wt%. Almost invariably, synthetic latex has a somewhat lower solids content than natural rubber latex, owing to its higher specific surface area.

Each particle of latex is contained within a protective, interfacial film that is solvated by water molecules and carries an electrical charge. The interfacial film comprises surfactant molecules, or other protective agents, on the outside of which is a layer of tightly bound water molecules that act as a mechanical barrier against coalescence. The electrical charge, which produces a coulombic repulsion between particles, can be either negative or positive. The corresponding anionic and cationic latices are known, but the former are more commonplace in practice.

The viscosity of all latices is quite temperature sensitive, particularly if they are allowed to cool to 5–10 °C. Since water is the continuous phase, latices normally exhibit Newtonian behavior; that is, the viscosity is a linear relationship between shear stress and shear rate. Compared with polymer solutions, which are very viscous even at concentrations as low as 10 wt% solids, polymer latices remain freely fluid up to concentrations as high as 60 wt%.

2. Destabilization of Latices

Whereas thin-walled articles may conveniently be made by a simple dipping process, the deposition of thick films is best facilitated by means of a coacervant. A coacervant destabilizes the latex by destroying the protective interfacial film and neutralizing the particle charge, resulting in clustering and aggregation.

Chemical coacervants may be classified into three groups: contact, heat sensitive and delayed action. Contact coacervants bring about immediate destabilization when added to a latex. In the case of anionic latices, the usual procedure is first to coat the exterior of the mold, which is called a former with either 10% solution of formic acid in an alcohol–water mixture or calcium chloride in alcohol. A coacervant that is especially useful when products of optimum clarity are required is cyclohexylammonium acetate. On drying the deposited film of latex, no residual coacervant remains, since cyclohexylamine itself is readily volatile.

Heat-sensitive coacervants have virtually no effect on the stability of the latex at room temperature but, when they are heated, coacervation occurs rapidly once a particular threshold temperature has been exceeded. Heat sensitizers are based on zinc ammine, poly(vinyl methyl ether) or poly(propylene glycol). In the case of zinc ammine, it is necessary for there to be a source of zinc ions, free ammonia and ammonium ions. Although the mechanism is obscure, it is believed to be associated with the thermolability of the zinc tetrammine ion, an effect which becomes particularly evident at temperatures above 50 °C. Poly(vinyl methyl ether) is a nonionizable polymer which is soluble in cold water but insoluble in hot water. When a latex containing a solution of poly(vinyl methyl ether) is gradually warmed to 35 °C, the latter precipitates and coacervation occurs. Poly(propylene glycol) behaves in a similar manner.

Delayed-action coacervants have little effect when initially added to a latex but bring about gradual coacervation after a certain time has elapsed. The most important examples of this group are the alkali metal salts of hydrofluorosilicic acid, commonly known as silicofluorides. The mechanism is believed to involve the slow, reversible hydrolysis of the silicofluoride anion to form an unstable intermediate orthosilicic acid which readily condenses to form a gelatinous precipitate of colloidal, hydrated silica.

Finally, brief mention should be made of the "anode process." Here, latex particles are caused to move, under the influence of an applied potential difference, towards an anode where they lose their charge and are deposited. As yet, the process has attracted little commercial attention because there are still a number of problems which need to be overcome, the most important of which is the evolution of oxygen at the same electrode where the deposition of the latex is occurring.

3. *Latex Compounding*

Compounding ingredients are added to latices to modify their processing characteristics, to bring about vulcanization or to effect economies. Such additives are seldom introduced in the dry state because of the risk of coagulation occurring, for example by dehydration. The usual procedure is to make an aqueous predispersion of the particulate or liquid additive, incorporating a surfactant to assist wetting and minimize reagglomeration, and to convert to a very fine particle size by ball milling, ultrasonics or colloid milling.

The additives that are typically incorporated into latices comprise stabilizers, viscosity modifiers (thickeners) and plasticizers, and fillers such as whiting and china clay to achieve cost reduction. Where necessary, accelerators, zinc oxide and sulfur are added for vulcanization.

The equipment required for mixing the finely divided dispersion of powders or liquid emulsion into a latex is relatively simple and inexpensive. Only low-shear stirring is essential and no particular order of addition is important, except that it is necessary to introduce any stabilizer which may be needed at the onset of the process. After mixing, the compound should be stored, preferably in bulk, under conditions of mild agitation and temperature control to minimize sedimentation.

4. *Latex Dipping*

As a process, latex dipping is ideally suited to the manufacture of seamless thin-walled products such as gloves, novelty balloons, prophylactics, catheters, baby feeder teats and soothers, meteorological balloons, rubber boots and overshoes. Indeed, such items would be either extremely difficult or impossible to make by any other means, especially from normal bale rubber.

In latex dipping, the product is deposited on the outside of a former. The strength of the material from which a former is made must be sufficient to withstand the necessary in-process handling and be chemically resistant to, and readily wetted by, the latex. Should uniform wetting by the latex not occur, air entrapment and an uneven film thickness will result. Formers must be made from materials which can be easily shaped. They should have a smooth line and provide a surface that is free from imperfections. Soda-lime glass fulfills most of these requirements and is used for prophylactics, but is fragile. Teats and thin gloves are normally made on glazed porcelain formers, whereas thick gloves, balloons and similar articles are made on aluminum.

In simple dipping, the latex is not destabilized and the thickness of the dry rubber film is determined by the amount of latex which adheres naturally to the former. An adequate wall thickness may require more than one dip to be made, in which case partial drying is necessary between dips. Wall thickness may also be increased from a single dip by incorporating cellulosic or acrylic thickening agents.

Coacervant dipping is a widely used process. The former, which preferably should be unglazed porcelain in order to absorb the coacervant, is dipped into a bath of coacervant solution and dried, after which it is lowered into the latex. As noted in Sect. 2, typical coacervants are calcium chloride or cyclohexylammonium acetate dissolved in alcohol solvent or an alcohol–water mixture. For a given coacervant, the wall thickness of the rubber which is deposited depends on the concentration of the coacervant solution and the length of time the former is immersed in the latex. For example, it may take 3 min to build a thickness of 0.5 mm and 9 min for 0.75 mm.

Following dipping, the former is withdrawn slowly from the latex and normally inverted to minimize blob formation and to ensure a uniform film thickness. The deposit is then dried at 55–65 °C, beaded, leached to remove water-soluble matter and, where appropriate, vulcanized in a hot-air oven in the region of 100 °C. Beading is usually necessary on dipped articles in order to strengthen the lower edge. Since thick sections have a tendency to unroll, it is preferable to form the bead by rolling a long, thin film. Furthermore, consolidation of the bead is best achieved if it is carried out prior to vulcanization.

There are many applications where latex-dipped articles are required to have special properties, for example oil and solvent resistance or antistatic properties. In the case of oil-resistant gloves, where only the surface comes into contact with the harmful conditions, the interior is made from natural latex and the surface from NBR latex. For antistatic goods, the exterior would be made from polychloroprene latex, because it has intrinsically higher electrical conductivity, and the interior from natural latex.

5. *Dipped Products from PVC*

The object being dipped is a suitably shaped metal former, probably treated with a release agent. In common with latex dipping, the coating becomes a hollow molding when it is stripped from the former after fusion and cooling.

Two processes are used: hot-dip molding and cold-dip molding. In the first, the formers are preheated in an oven and dipped into the plastisol at a controlled rate that is sufficient to avoid differential cooling and an uneven coating thickness. The immersed formers are allowed to remain in the plastisol for a time sufficient to build up the required thickness of deposit. They are then withdrawn at a slow, uniform rate, so as to avoid excess plastisol adhering to the coated former, and inverted. The PVC is then fused by passing through an oven at 175–200 °C.

By comparison, cold-dip molding has no preheating cycle. The former at room temperature is immersed into a PVC plastisol at the same temperature. In order to retain a sufficient quantity of the plastisol on the former, the rheology of the system must be very carefully controlled. This particular technique has found less favor that the hot-dip process.

See also: Processing: Classification; Processing: Relation to Product Properties

Bibliography

Blackley D C 1966 *High Polymer Latices, Their Science and Technology*, Vols 1, 2. Maclaren, London

Bratby D M 1982 Latex dipping. In: Calvert 1982, pp. 193–205

Calvert K O (ed.) 1982 *Polymer Lattices and Their Applications*. Applied Science, London, pp. 1–20

Esser H 1978 Production of dipper articles from synthetic latices. *Introd. Polym. Sci. Technol. SPE Textb.* 5(9): 67–70

Feast A A J 1982 Synthetic latices. In: Calvert 1982. pp. 21–46

Gorton A D T 1980 Developments in elastomeric latex technology. *Prog. Rubber Technol.* 43: 79–98

Stern H J 1967 *Rubber: Natural and Synthetic,* 2nd edn. Maclaren, London

Sutton S D 1953 Rubber dipped goods. In: Schidrowitz P, Dawson T R (eds.) 1953 *History of the Rubber Industry*. Heffer, Cambridge, pp. 303–9

Titow W V 1984 *PVC Technology*, 4th edn. Elsevier Applied Science, Barking, UK

P. J. Grey
[BTR Industries, Burton-on-Trent, UK]

E

Elastomer–Elastomer Adhesion Processes

In many applications employing elastomers, such as the manufacture of tires, oil-transporting hoses and conveyor belting, it is necessary to obtain strong, durable bonding between successive elastomeric layers. Although a degree of bonding can be achieved by applying certain adhesives to the surfaces of cured elastomers and then bringing these into contact, this level of adhesion is not sufficient for many applications. Best interelastomer bonding is obtained when elastomers are brought into contact at the uncured stage, so that durable bonds can form across the interface during vulcanization. Prior to the formation of these permanent bonds, some elastomers possess the ability to adhere together while still uncured: this ability facilitates the fabrication of elastomer composites.

Table 1
Measured autohesive tack data under optimum conditions

	Tack strength kN m^{-1}
Elastomeric polymer only	
natural rubber (NR)	12.7
cis-polyisoprene (*cis*-IR)	10.3
styrene–butadiene copolymer (SBR)	5.7
cis-polybutadiene (*cis*-BR)	5.4
butyl rubber (IIR)	3.7
ethylene propylene terpolymer (EPDM)	2.2
Elastomer bases (compound)	
typical NR engineering	15.6
NR containing 33% carbon black only	13.3
typical blended NR, IR and SBR containing black and oil	10.1
SBR containing 30% carbon black only	9.3
typical EPDMs containing black and oil	5–8

Source: Campion (1977)

1. Uncured–Uncured Adhesion

If two smooth sheets of certain unvulcanized elastomeric compounds are pressed together, the sheets adhere and remain attached when the contacting force is removed, the bond resisting any subsequent attempt at separation. This is the phenomenon of autohesive tack, more commonly termed elastomeric tack or building tack. The coalescence can occur within seconds of initial contact. In forming many rubber composite articles, different elastomeric compound layers are pressed into contact and combine together initially due to building tack so that, after vulcanization, the composite acts as an optimized integral unit.

Compounds from different elastomers possess different degrees of tack. Those elastomeric compounds that possess high tack do so largely because their base polymers display the same phenomenon. Hence, the basis of tack is autoadhesion or autohesion, the spontaneous adherence between two unvulcanized surfaces of the same unfilled elastomeric polymer. Under optimum conditions, natural rubber (NR) possesses a high degree of autohesion whereas ethylene propylene elastomeric polymers (EPDM) have little autohesion. Consequently, EPDM compounds are generally not tacky.

1.1 Tack Strengths

The optimum conditions mentioned previously refer to the optimizing of several factors that can influence the degree of tack, quantified as tack strength. In practice, it is necessary to employ a destructive technique to measure this property. Suitably contacted test pieces are separated after a chosen contact time, either in tensile or peel modes, and the associated forces and/or energies are measured. The peel mode is the more reliable. The wide variation in tack magnitudes can be seen in Table 1 for a number of elastomeric polymers and compounds, measured using a rotary tackmeter—a dynamic peel test. The compound data are in good agreement with building tack ratings expressed by those experienced in the product manufacture of, for example, tires (Campion 1977).

1.2 Factors Influencing Tack Strengths

The major factors influencing tack levels are molecular mass of the elastomer, contact force, contact time and temperature. The contact force should be sufficiently large for the "high spots" of the contacting uncured elastomer surfaces to flow in a viscoelastic manner bringing other regions of each surface together, resulting in a rapid mutual contact of the complete surfaces.

An inverse relationship occurs between tack strength and molecular mass M. Because bulk strength has a direct relationship with M, results are best expressed as plots of bulk and tack strengths vs M on one graph (see Fig. 1). As the molecular mass is reduced, the tack strength increases, meeting the decreasing bulk strength curve. An increase in contact time also causes an increase in tack level towards a maximum; however, for "tacky" rubbers, using contact times of several seconds only allows measured tack strengths to equal green (bulk)

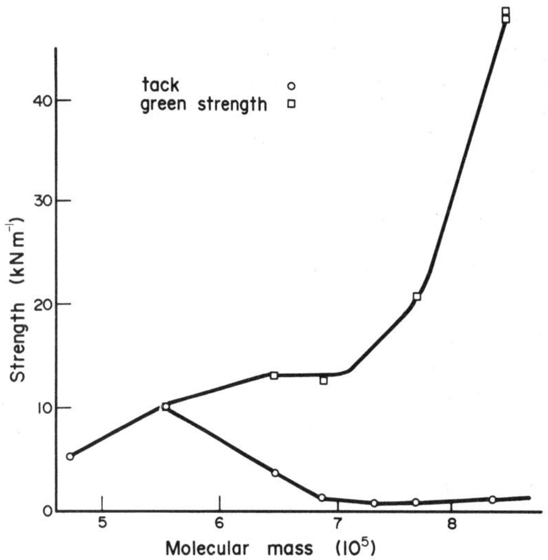

Figure 1
Relation of tack and green strength to molecular mass for natural rubber

strengths, making testing impracticable. The rotary tackmeter results in Table 1 employed a contact time of about 1 s, allowing discernible differences to be obtained for the various elastomers. This contact time makes practical sense also, as it is appropriate for production techniques in tire building.

Higher test temperatures also increase tack magnitudes, but testing techniques become difficult because green strength decreases with increasing temperature. Another factor that can increase measured tack levels is an increase in the speed of separation of the two surfaces during testing. Fracture mechanics approaches have been applied to relate data obtained at various test rates and temperatures; data at one set of conditions can be superimposed on those at another set using well-established shift factors.

From a compounding viewpoint, although the tack strength of an elastomer arises largely from base polymer autohesion, certain additives that facilitate the motion of polymer chains, such as tackifying resins, process oils and plasticizers, can be incorporated in the formulation to improve tack strengths. Excessive amounts of these additives could cause surface bloom or the deterioration of other elastomer properties: loadings should therefore be optimized (see *Elastomers: Additives*).

1.3 Mechanism of Autohesive Tack

Various authors have proposed different mechanisms to explain the phenomenon of autohesive tack. Interfacial adsorption has been favored by some, but the general opinion favors the diffusion approach of Voyutskii (1963) and others. Their concept is based on the interdiffusion of elastomer molecular chain segments between the two intimately contacting surfaces resulting in mixing of the chain segments and ultimate disappearance of the interface. Radiometric determinations (e.g., Skewis 1966) have supported the diffusion theory and resulted in self-diffusion coefficients of a size suggesting that chain segments can diffuse 5 nm into the opposite surface after 1 s and 50 nm after 100 s: clearly, considerable intermingling of chains can result quickly from the two-way interfacial diffusion.

Bueche (1962) has further proposed that, for each diffusive step, cooperative movements of adjoining segments in one surface occur to form a hole of sufficient (critical) size to allow penetration by a chain segment from the other surface. Campion (1975) has extended this approach to explain the large differences in autohesion between NR and EPDM elastomers, the model involving the influence of fine-scale polymer chain structural "cavities" on the location of free volume (an expression describing and quantifying the space which exists among the chains in an ever-changing pattern due to their continual thermal movements). Critical size hole formation, accompanying thermal movements, can be assisted for elastomers such as NR by contributions from these cavities. In contrast, for molecular structural reasons, the distribution of free volume around EPDM polymer chains is low, making critical hole formation more difficult. Wake (1982) and Kinloch (1987) have separately reviewed this model along with a complementary mathematical approach developed to indicate fundamental reasons why most of the factors outlined in Sect. 1.2 influence tack strength: these authors have also considered other adhesion theories.

The thermodynamics of autohesive tack occurs in a manner analogous to the mixing of liquids, but applied to surface regions only, as a consequence of the diffusion theory. Thus, mixing at temperature T will occur as a result of a negative free-energy change ΔG, where

$$\Delta G = \Delta H - T\Delta S \quad (1)$$

For identical materials there is no enthalpy change ΔH and the entropy change ΔS will be positive because the removal of an interface by intermixing represents an increase in disorder. Hence, ΔG will be negative, as required for "surface mixing": diffusion across the interface is thermodynamically favorable.

1.4 Interadhesion

Similar good levels of uncured–uncured bonding can occur in some cases between two different, tacky elastomers. However, this is not a universal happening: the overruling factor is whether the system is thermodynamically favorable or not. This condition depends on ΔH, which in turn is largely influenced by the values of the solubility parameter δ for the two elastomers; that is, if the difference in δ is large, ΔH will also be large and outweigh $T\Delta S$ so that ΔG will be positive (see Eqn. (1))—a thermodynamically unfavorable situation. The solubility parameter of an elastomer is a measure of its intermolecular attraction and allows compatibility of elastomers and their swelling/dissolution in solvents to be elucidated.

Differences in δ can be large as demonstrated by values for δ for quite dissimilar elastomers. Oil-resistant types such as butadiene acrylonitrile copolymers (NBR) and fluoroelastomers (FKM) possess δ values of 10–11 $cal^{1/2}\,cm^{-3/2}$ (20–22 $MJ^{1/2}\,m^{-3/2}$), depending on monomer proportions. Elastomers such as NR, EPDM, butyl (IIR) and *cis*-polybutadiene (*cis*-BR), compatible with oil in the physicochemical sense, possesses δ values of 8–9 $cal^{1/2}\,cm^{-3/2}$ (16–18 $MJ^{1/2}\,m^{-3/2}$). The maximum allowable difference in δ between contacting tacky elastomers to attain good uncured–uncured bonding has been estimated to be about 0.25 $cal^{1/2}\,cm^{-3/2}$ (0.5 $MJ^{1/2}\,m^{-3/2}$), although use of low-molecular-mass polymers could allow greater differences to be tolerated.

2. Permanent Elastomer–Elastomer Bonding Associated With Vulcanization

2.1 Similar Elastomers

In practice, multilayered elastomeric articles held together by autohesive tack are not yet ready for any useful engineering purpose: the vulcanization process is still required. The application of heat and pressure causes crosslinks between elastomeric chains to form chemically, so that each component layer becomes elastic. Moreover, chain segments from one layer which have previously diffused across an interface into the adjacent elastomer will be available for participation in the vulcanization reactions of the second layer and will then become chemically linked into it. Since these diffused segments are still part of molecular chains residing in the first elastomer, if sufficient interdiffusion has occurred, the two elastomeric layers become bonded in a manner which is as permanent as the crosslinks of vulcanized products. Consequently, the article achieves an integrity as a whole: this is often aided by diffusion of various compounding ingredients during the initial stages of vulcanization which will create gradients in stiffness between the different component layers. It follows that if the elastomer layers were initially nontacky, it is unlikely that any permanent rubber–rubber bond will form during vulcanization unless different approaches are employed.

One means of assessing the adhesion between two elastomeric strips placed in contact under pressure and then vulcanized together is by a 180° peel test. For complete bonding, the energy of peeling apart is in the order of the fracture (tearing) energy of the elastomer, measured for NR as 40 $kJ\,m^{-2}$. The permanent nature of these bonds is such that, after 30 d immersion in either sea water or oil, the mode of test failure remains cohesive (i.e., fracture of the bulk).

2.2 Dissimilar Elastomers

By comparison with Sect. 1.4, if the difference in δ values for two elastomers is less than 0.25 $cal^{1/2}\,cm^{-3/2}$ (0.5 $MJ^{1/2}\,m^{-3/2}$), then bonding can occur according to the method of Sect. 2.1. However, the differences in δ for many pairs of elastomers are considerably outside this limit. Successful permanent bonding between the dissimilar elastomers NR and NBR has been obtained using other approaches. In one of these, because of the unique position of polymers in having characteristics between those of liquid and solid, intimate interfacial contact was attained as a liquid/solid wetting phenomenon. The critical surface tension γ_c of a solid surface is an empirically based term pioneered by Zisman. Only liquids having a surface tension $\gamma<\gamma_c$ will wet that solid, that is, spontaneously spread over its surface indicating that the solid–liquid intermolecular forces are higher than the liquid–liquid forces. This phenomenon is well utilized in adhesives technology. The approach taken was that, for two incompatible elastomers, the elastomer of lower viscosity was termed a 'notional liquid' and that of higher viscosity a 'notional solid'; if the surface tension of the former was less than that of the latter, intimate contact by wetting was favored thermodynamically. In practice, the fluidity of the notional liquid was increased by incorporating a compatible low-molecular-mass polymer (i.e., a compatible true liquid). In this way, NR extended by liquid polybutene (PB) was bonded permanently to NBR. Formation of interfacial crosslinks has also been shown despite the absence in this system of significant interfacial polymer chain interdiffusion. Curing ingredients diffuse across the interface enabling crosslink networks to form across it. To bond successfully, a fast-curing formulation for the NBR was required together with a formulation that produced crosslinks containing polysulphides (i.e., long crosslinks) for the NR. Such bonds also withstood immersion in sea water and oil (Campion 1989).

Alternatively, chemical treatments of stiff, low-energy surfaces may also be employed to raise the surface energy of that elastomer, enabling the surface to be wetted by a second, dissimilar lower-viscosity elastomer possessing a surface energy higher than that of the untreated first elastomer but lower than the treated value. Typical treatments include hypochlorite solutions or similar proprietary liquids. The chemical groupings on the treated surface also provide sites for the formation of permanent bonds during cure.

Other approaches to the bonding during cure of dissimilar elastomers involve the application to surfaces to be bonded of adhesives which, after drying, subsequently cocure with the elastomer and thus form a permanent bond. Many of these adhesives are proprietary and are of the type mainly applied in elastomer–metal adhesion (see *Elastomer–Metal Adhesion Processes*). For instance, various Chemlok, Typly, Thixon or Cilbond adhesives might be applied, paying attention to the achievement of good wetting. The solvent in the adhesive contributes to wetting during its application to one elastomer surface to be bonded: after drying, the resulting film is often based on higher surface energy polymer–resin systems which are quite stiff. Hence, during molding at elevated temperatures and pressures, the second contacting elastomer should wet the adhesive film, flowing under the molding pressure and adopting the contours of the film. Again, curatives and additives in the dried adhesive will interdiffuse to form crosslinks across the interface and thus cause bonding: direct chemical coupling with some active groups on the polymer chain of the adhesive is another possibility.

In one particular approach to the bonding of dissimilar elastomers, Corish et al. (1980) used a novel particulate-type adhesive. The particles employed were of a reinforcing filler known to exhibit strong links with the elastomeric polymer when performing its reinforcing role. Low results for peel adhesion obtained from various pairs of standard carbon-black-filled compounds cured together under normal conditions were improved appreciably by using a 5% dispersion of high-abrasion furnace black in petroleum naphtha as an adhesive, brush coated on to the interfacial surfaces. Preparation of a typical adhesive of this type involved dispersing the black in a suitable solvent using a high-speed, high-shear mixer. Further development involving the use of low levels of polymeric binders to "fix" the coating while retaining its adhesive ability are described in the article *Elastomer–Thermoplastics Adhesion Processes*.

3. Uncured–Cured Adhesion

The molecular chains of a cured elastomer no longer possess the freedom to migrate across an interface into another surface; in addition, curatives have been used up and are not available to migrate similarly. However, segments of the crosslinked chains are thought to be still sufficiently mobile to form some critical holes (see Sect. 1.3). These are then visualized as acting at the cured elastomer surface as recipients to interdiffusing chain segments from another, uncured surface belonging to an elastomer of similar δ. The process might be considered as one-way tack. Similarly, curatives must interdiffuse from uncured to cured elastomer so that, with the application of further heat and pressure, the diffused segments can bind into the cured bulk.

Practical processes have been developed based on these considerations. Cured surfaces are usually abraded (buffed) to remove any curing additive which has migrated to the surface (bloomed) during cure and to provide rough surfaces for good mechanical keying. In addition, a "tie coat" adhesive, comprising some uncured elastomer with extra curatives dissolved in a suitable solvent, is brushed on to the abraded surface and allowed to dry before the bulk of the uncured elastomer is molded into place. The tire retreading industry is based on such procedures: strong bonding is achieved, although probably not to the magnitude of bonding arising from initial uncured–uncured contact. Proprietary adhesives are also capable of giving good levels of adhesion; buffing and/or chemical pretreatments of the cured elastomer surface are usually necessary prerequisites.

Dissimilar elastomers can be bonded by procedures of the same type. NR extended by liquid polybutene has formed durable bonds to cured nitrile (see Sect. 2.2), without the use of a tie coat.

4. Cured–Cured Adhesion

Because neither elastomer surface can provide polymer segments or curatives for interdiffusion, bonding between cured elastomers can only be performed through suitable adhesives. In addition, chemical treatments of abraded surfaces are necessary to achieve high surface energies for wetting, plus sites for bonding. Chemical treatments include strong hypochlorite solutions, concentrated sulfuric acid or proprietary liquids. After a few minutes treatment, the cured surfaces can be washed with water followed by acetone, for quick drying; the adhesive should then be applied immediately. Epoxy adhesives and cyanoacrylates have been used with some success in this area; however, bond strengths are often inferior to those obtained when the elastomer was initially uncured.

An alternative novel method for elastomeric compounds based on low-unsaturation polymers, especially EPDM, is based on fusion bonding, in which the molten surface of a plastics material is brought into contact with a chemically compatible elastomeric compound. EPDM and EPM precured compounds, which do not respond well to chemical pretreatment or to buffing, have been effectively bonded to other similar elastomers by means of polyethylene (or polypropylene) film. Use of reasonably high contact pressure is necessary, which should be maintained during cooling (see *Elastomer–Thermoplastics Adhesion Processes*).

See also: Bonding: Preparation of Polymer Surfaces; Elastomer–Textile Adhesion Processes; Elastomer–Thermoplastics Adhesion Processes; Elastomers: Additives

Bibliography

Bueche F 1962 *Physical Properties of Polymers*. Interscience, New York

Campion R P 1975 The influence of structure on autohesion and other forms of diffusion into polymers. *J. Adhes.* 7: 1–23

Campion R P 1977 Microstructure and autohesive tack. In: Allen K W (ed.) 1977 *Adhesion 1*. Applied Science, London, Chap. 5

Campion R P 1989 Elastomer composites for engineering applications—Spontaneous and other forms of bonding between components. *Mat. Sci. Technol.* 5: 209–21

Corish P J 1986 Some aspects of rubber/rubber adhesion. *Prog. Rubber Plast. Technol.* 2: 12–22

Corish P J, Osmant T H, Clarke D I 1980 Novel "particulate-type" adhesives. In: Allen K W (ed.) 1980 *Adhesion—4*. Applied Science, London, Chap. 10

Hamed G R, Shieh C-H 1983 Relationship between the cohesive strength and tack of elastomers, Part I *J. Polym. Sci., Polym. Phys. Ed.* 21: 1415

Hamed G R, Shieh C-H 1985 Relationship between the cohesive strength and tack of elastomers, Part II. *Rubber Chem. Technol.* 58: 1038

Hamed G R, Shieh C-H 1986 Relationship between the cohesive strength and tack of elastomers, Part III. *Rubber Chem. Technol.* 59: 883–95

Kinloch A J 1987 *Adhesion and Adhesives: Science and Technology*. Chapman and Hall, London

Mori K 1987 Cure adhesion in the interface of fluorinated and nitrile rubbers. *Rubber Chem. Technol.* 60: 822–36

Skewis J D 1966 Self-diffusion coefficients and tack of some rubbery polymers. *Rubber Chem. Technol.* 39: 217–25

Voyutskii S S 1963 *Autohesion and Adhesion of High Polymers*. Interscience, New York

Wake W C 1982 *Adhesion and the Formulation of Adhesives*, 2nd edn. Applied Science, London

Wragg R T, Yardley J F, Nightingale A F 1981 Fusion bonding *Kautsch. Gummi, Kunstst.* 34(8): 657

P. J. Corish
[Patrick J. Corish & Associates,
Stratford-on-Avon, UK]

R. P. Campion
[Materials Engineering Research Laboratory,
Hertford, UK]

Elastomer–Metal Adhesion Processes

Elastomers in engineering are frequently employed as the deformable component in composite structures, the rigid reinforcements being metals. Strong and durable adhesion bonding between elastomer and metal is crucial to the continued successful service of the composite. The levels of adhesion required for elastomer components vary considerably but, as a guide, an elastomer shear mount may be expected to deform up to 300% shear strain. The average shear stress acting on the rubber–metal bond might be as high as 10 MPa (or 0.65 ton in^{-2}). The bond strength developed between these two material types of widely different rigidity and chemical type must withstand such a stress throughout a service life that might also involve the exposure of the mount to hostile environments. However, if elastomeric compounds are pressed against clean smooth steel surfaces during vulcanization, only a low degree of bonding is observed afterwards. The bond is essentially physical, being the result of initial viscous flow into microscopic surface defects with the subsequent vulcanization resulting in mechanical keying. Separation at the original interface can easily be achieved. Indeed, because of the absence of a permanent bond to steel, it is advantageous to form elastomers under pressure to required shapes employing steel molds. Hence, use of appropriate adhesives is necessary to achieve good adhesion in bonded elastomer–metal composites. In fact, bond levels can be obtained from correctly chosen systems which are effectively higher than the strength of an elastomer so that, during service or testing, failure occurs in the elastomer bulk rather than at the bonding interface.

1. General Principles and Mechanism

In seeking to attain such quality of bonding, the adhering of vulcanized elastomer by conventional structural adhesives is possible, but is rarely practised. It is far more common for elastomer to be bonded to metallic substrates during the course of vulcanization using a proprietary bonding agent adhesive precoated on the appropriate metal surfaces. An application of primer to metal is often required first. There are several advantages to such a practice. At molding temperatures, unvulcanized elastomer is a viscous fluid rather than the elastic quasisolid it becomes after vulcanization. The greater mobility of the elastomer before vulcanization means that the application of molding pressure causes flow of elastomer until it becomes tightly constrained against the coated metal surface. With successful bonding systems, the elastomer will possess a lower surface energy than that of the dried film of adhesive on the metal, the correct requirements for thermodynamic wetting, as expressed by Zisman (1964), then being satisfied. Thus, molding pressure renders the elastomer–adhesive interface sufficiently intimate for molecular interactions to be experienced across the interface and the thermodynamic wetting means that these interactions are stronger than those between molecules in the elastomer bulk. Moreover, during the subsequent vulcanization stage, both elastomer and adhesive become cured and, because of interdiffusion of various constituent species across the elastomer–adhesive interface during the early stages, by the end of cure the interfacial region becomes permanently enmeshed and continuous, thus providing strong bonding. The primer–adhesive interface can be viewed in a similar way.

The bond between adhesive primer and metal is less understood but is thought to involve mechanical keying and induced molecular attractions. Because the surface energy γ of clean metal (or its oxide) is greater than γ of the applied liquid primer dispersion and, similarly, because γ of the dried primer film is greater than γ of the applied adhesive, all thermodynamic requirements for wetting are again satisfied. The metal (oxide) surface must be firm and every crevice in this surface must be wetted by the low-viscosity primer. The associated formation of permanent bonding across this interface is thought to involve the formation, for steel, of chelating bonds between iron oxides and high-molecular-weight condensation products (formed during vulcanization) from, for example, phenolic resins in the primer. At the end of the vulcanization stage, when bonding reactions at each interface have been completed, an added advantage for the good distribution of stresses in the bonded region during service has arisen because a modulus gradient has been developed across the region in the sequence

$$\text{metal} > \text{cured primer} > \text{cured adhesive} > \text{cured elastomer}$$

and curative diffusion has rounded off the modulus discontinuities at each interface.

For bonding elastomer to steel and many other metals, the proprietary low-viscosity adhesives usually employed are, for instance, the Chemloks and Typlys (Lord Corp., USA), the Thixons (Dayton Chemical Products, Whittaker Corp., USA), the Cilbonds (Compounding Ingredients, UK) and Desmodur R (Bayer Chemicals, Germany). Appropriate grades are recommended by the manufacturers for use with particular elastomers. Different bonding agents vary considerably in choice of active agent. Their base elastomers are often chlorinated or brominated; nitroso or isocyanate chemicals are often incorporated and lead oxide, phosphorus, silicon or titanium chemicals might also have been added. High resin loadings are also frequently included by the adhesive manufacturer. Good substrate preparation is essential and, as the bonding agents are often dispersions rather than solutions, they must be stirred extensively before application. Bonding agents are applied to the substrates by painting, spraying or dipping techniques. Complete evaporation of solvent must be allowed before application of another layer or contact with elastomers. Typical evaporation times at ambient temperatures are 0.5–2 h; the procedure can be quickened by employing slightly elevated temperatures (up to 40 °C). However, once dry, the resulting films should be quickly progressed to the next stage, protecting them from light and contamination. Optimum thickness values of the dried adhesive layers are recommended by manufacturers; these are usually in the order of 75 μm and can be measured by electrical resistance devices.

General descriptive literature in this area includes Pickup and Weatherstone (1969), Meier and Findley (1973), Sexsmith (1979) and Stevenson (1990).

2. *Metal Preparation*

The performance of any bonding agent is strongly influenced by the effectiveness of preparation of the metal surface. The basis is degreasing followed by either mechanical or chemical procedures for producing fresh surfaces, oxide layers and absorbed moisture being thus removed. Sand blasting, shot blasting, grit blasting and wire brush or emery cloth application are examples of the former; for good all-round performance, grit blasting using fairly fine particle-size alumina is known to be effective. The prepared surface is higher in area and energy due to the process, and sites for mechanical keying may be formed. Chemical treatment usually consists of immersion in acid, rinsing to remove traces of acid and drying. Phosphating of the metal surface may also be used to give a stable corrosion-resistant surface, and controlled reduction in hydrogen or treatment in a dilute alkali bath may convert an oxide layer into one containing reactive hydroxyl groups.

Regarding the initial degreasing step, a commonly used organic solvent is 1,1,1-trichloroethane, which is effective for dissolving greasy contaminations and is relatively safe to use. Some contaminations may be removed more easily by trichloroethylene or carbon tetrachloride; however, these solvents (especially the latter) are potentially greater hazards to health. The step is frequently performed with the articles to be cleaned being placed in the vapor phase of the solvent, using a reflux distillation process. Degreasing before chemical treatment may involve the use of hot alkaline aqueous solution. On occasions, an extra degreasing step, with organic solvent, is used after the surface treatment stage with the purpose of removing excess blasting grit particles and so on.

For more specific consideration of metallic surface preparations see *Bonding: Preparation of Metallic Surfaces*.

Once the complete metal surface has been prepared in this way, it should be maintained in a clean environment. The next stage, primer application, should be carried out as soon as possible.

3. *Elastomer Processing Factors*

Processing factors can also be relevant to the quality of elastomer–metal bonding obtained. Use of transfer or injection-molding techniques means that fresh uncured compound is presented to the adhesive film with much elastomer flow across the surface until high molding pressure is attained. When compression or plunger molding, however, little elastomer flow occurs, so that care is required to apply fresh elastomer surfaces to the adhesive film surface when filling the mold; plunger molding gives greater confidence that the applied high molding pressure is reaching the elastomer–metal interface. Care should also be taken in avoiding bloom, the appearing of supersaturated compounding ingredient which has migrated out of the elastomer and formed a film on the surface. This film will inhibit bond formation and should be wiped away with a pad soaked in suitable hydrocarbon solvent (e.g., naphtha) or, preferably, the elastomer should be remilled to provide a new, clean, surface for bonding.

4. *Design of Compound Formulations*

Traditionally, manufacturers of elastomer composite goods have developed compound formulations which are associated with successful elastomer–metal (especially mild steel) bonding. When designing these formulations, claims sometimes made are that sulfur-curable elastomers bond better than those employing other curing systems (e.g., peroxides) and that high sulfur loadings lead to higher bond strengths than do low sulfur loadings. However, under laboratory conditions, it is known that successful bonding can be achieved even with low sulfur or peroxide cure systems—indeed, some elastomers (e.g., highly saturated hydrogenated nitriles, fluoroelastomers) cannot be cured with sulfur and yet these materials can be bonded to steel—providing the correct choice of adhesive is made and correct procedures followed. It is easier to control laboratory conditions than plant manufacturing conditions in terms of cleanliness, metal preparation, viscosity and age of compound and adhesive, protection of prepared surfaces and control of molding pressure, application of adhesive and cure time and temperature. Nevertheless, manufacturing procedures now exist which rely on strict attention to detail and very good quality-control procedures throughout to achieve good bonding between many elastomers and metals.

The choice of both base elastomer and compound formulation is also dependent on other factors which will be required for reasons of manufacture or service, for example ease of processing, fatigue life, resistance to low-temperature crystallization and resistance to oil exposure. Hence, it is more usual to attempt to bond a compound of a given formulation by optimizing conditions and varying the choice of adhesive until successful; information produced by adhesive manufacturers tabulates both basic adhesive types for various elastomers and other high-activity adhesives for use where bonding or bond durability problems arise.

Weih et al. (1985) have described how these curable adhesives can be applied to cured elastomers, employing high contact pressures and temperatures.

5. *Assessment of Adhesion*

To assess the magnitude of elastomer–metal adhesion, in common with other forms of adhesion testing, it is usual to mold samples specially for the purpose and then measure the force necessary to test the bond to destruction (Brown 1979). In all cases, metal surfaces to be bonded are prepared as described previously and a chosen primer dispersion is applied carefully and allowed to dry. As soon as possible, the required adhesive is similarly applied and molding is performed, preferably by transfer or plunger molding techniques. After cure, samples should be allowed at least one day as a conditioning period before testing.

The main modes of testing are peel, tensile and shear: in peel testing, an energy of adhesion can be obtained from the test force. Geometries frequently tested are illustrated schematically in Fig. 1. Figures 1a–c all involve cylindrical specimens. The test in Fig. 1a complies with the USA standard test ASTM D429–81 method A; in this, the force required in tensile mode to remove one of two bonded metal buttons is measured. The test in Fig. 1b, which follows D429 method C, employs the same technique, but replacement of the buttons by metal cones means that the separation at an adhesive interface is now in shear, with a concentration of stresses at the cone apex. In both cases, failure occurs suddenly at that interface which, for reasons of slight variation in preparation, application, quality of adhesive locally and so on, is the weaker of the two. The "double-shear" sample in Fig. 1c comprises three metal pieces and two elastomer regions, that is, four adhesive interfaces. Testing in shear is performed as shown, considerable deformation of the elastomer first taking place; once again, failure occurs suddenly at the weakest interface.

Figs 1d and e involve flat rectangular metal plates with bonded elastomer suitably positioned as shown. Shear overlap testing is shown in Fig. 1d, failure

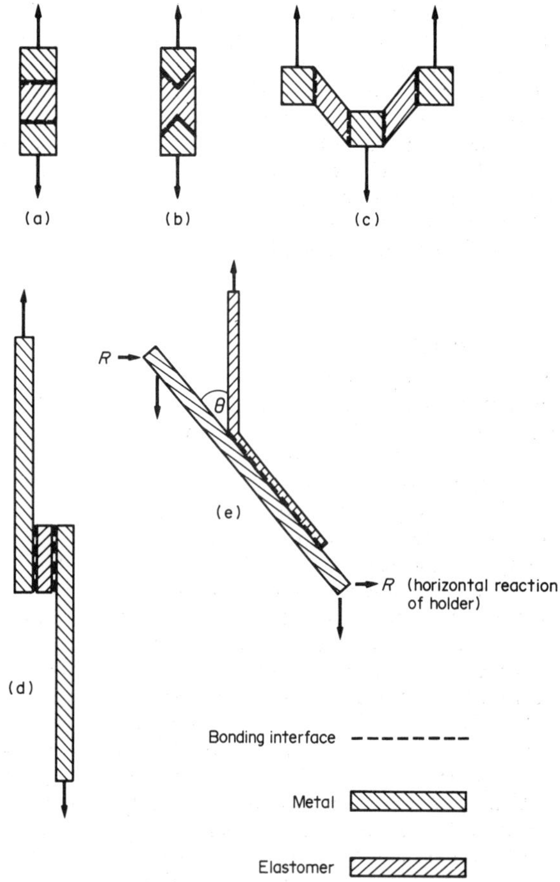

Figure 1
Schematic illustration of five different testing geometries (side view): (a)–(c) are cylindrical specimens; (d) and (e) involve flat rectangular metal plates with bonded elastomer suitably positioned

again occurring at the weaker of the two adhesive interfaces. In contrast, in peel testing (Fig. 1e which complies with ASTM standard D429 Method B when θ is 90°), as peeling occurs, one interface is progressively assessed, the line of separation gradually sweeping the complete bonded area. The test force will be altered if the peel angle θ (which is applied by use of suitable test grips) is changed. During testing, initially the unbonded tab becomes extended until peeling commences, after which each debonded portion of elastomer becomes extended to the same strain. These high extensions can be eliminated if, during molding, the elastomer is reinforced on its "upper" surface by a backing layer of inextensible, but flexible, rubberized fabric.

The choice of test employed could depend on many factors such as reliability, similarity to the mode of application of deformations in service, whether the test is for quality control (QC) or for development and design, and ease of molding samples. In general, although samples for the test in Fig. 1a can be molded in large numbers quite easily, the scatter in measured data is considerable. Also, as each result (given as MPa) is the mean of the whole bonded area, this average is diminished by local low-adhesion areas, whereas for peel testing such a deficiency will be represented as a trough in the force–extension curve and can be discarded after suitable inspection of the tested surfaces. The cone test in Fig. 1b requires that the apex of the metal cone is near perfect in shape (not always the case when cones have been previously used); under these conditions only, the scatter is usually less than that for the test in Fig. 1a. As failure always initiates at the cone apex, these test pieces are not suitable for durability work in which samples are exposed to fluids for a set period, after which the bond strength is assessed. The shear overlap test in Fig. 1d is often unwieldy for use with elastomers, being more useful for testing metal–adhesive–metal systems (for epoxy resins, etc.). The test in Fig. 1c is normally employed as a QC check on, for instance, elastomer shear mounts and thus may not always be continued to failure—instead, the confirmation is obtained that the bond will withstand the greatest deformation (plus safety factor) to be met in service. It may be relevant to perform an extended-time test by clamping a deformed sample followed, perhaps, by exposure to relevant fluids.

If no strong case exists for using other tests, the peel test in Fig. 1e is to be preferred for reasons already given and because a single large sample can be used to give an initial peeling force vs extension graph, using a few millimeters only of the bonded area, and then the same sample can be exposed to fluids before further testing. This procedure can be repeated several times. In addition, the representation of bond failure in energy terms assists the designer who is employing fracture mechanics to assess the fatigue life of the elastomeric component of the composite. Crack growth rate in the elastomer under service conditions can be compared with rate of bond failure for the same conditions. The designer seeks to ensure that bond failure will not occur first.

To assess the peeling energy of adhesion P, Lindley (1971) considered the energy balance before and after the peeling of a previously bonded elemental length of elastomer away from the metal strip at angle θ. Arising from these considerations, for nonlinear extension

$$P = (F/w)(\lambda - \cos\theta) - Wh \qquad (1)$$

where F is the peeling force, λ is the extension ratio of the elastomer, w and h are the width and thickness of the elastomer tab, respectively, and W is the strain energy density of the elastomer. For linear elasticity (Hata et al. 1965)

$$P = (F/w)[(1+\lambda)/2 - \cos\theta] \qquad (2)$$

Because λ is $(e+l)/l$, where l is the original elastomer tab length and e is the elongation at force F, Eqn. (2) can be modified to

$$P = (F/w)(1 - \cos\theta + e/2l) \qquad (3)$$

Inspection of typical peel force vs extension traces shows that it is not unreasonable to assume linear elasticity over this range, to a first approximation. Hence, Eqn. (3) gives a convenient means of obtaining peel energy from θ, test-piece dimensions and F and e which are read from the test chart. In practice, F will be a representative value obtained by averaging peaks and troughs obtained at different times on the chart, due to the "stick-slip" patterns produced when testing high-adhesion interfaces.

From Eqn. (3), when testing at 90°

$$P = (F/w)(1 + e/2l) \qquad (4)$$

and when the sample is reinforced with fabric so that e is zero

$$P = (F/w)(1 - \cos\theta) \qquad (5)$$

Hence, for reinforced samples at $\theta = 90°$,

$$P = F/w \qquad (6)$$

These equations do not allow for energy concerned with bending the elastomer (more important at high angles) or with tearing into the elastomer. However, successful bonding systems usually provide P values above the tearing energies of the elastomers, which is the required situation. In practice, it is easier during testing to maintain θ when at an angle of 180° (when P is theoretically $2F/w$ for reinforced test pieces) rather than when at 90° for reasons of geometry of the test rig. However, if possible, testing at 90° is preferable, especially for thick samples, as energy losses, for instance due to bending, are less.

As a final general point concerning these "static" tests, besides the bond magnitudes, the appearance of failed samples also gives an indication of the quality of adhesion. Providing the elastomer is of good quality, a 100% fracture in the elastomer (sometimes conveniently represented as 100%R) is desirable. Low bond strengths are associated with 100% adhesive failure (100%A). Intermediate levels can be assessed visually (e.g., as 50%R/50%A). Some coding systems (e.g., ASTM D429–81) employ other symbols to describe, when appropriate, at which interface failure occurs.

Much information can also be obtained from dynamic tests, which can test bonded elastomer–metal composites in a manner simulating the action of either the product in service or that of any of the static tests already described. In the latter case, a force less than that obtained during static testing is applied in a cyclic fashion; rate of failure can be obtained from either the rate of growth of fracture at the interface during testing or the number of cycles required to produce complete failure.

6. *Durability*

The durability of bonded test pieces is, as indicated in Sect. 5, best performed qualitatively with double-shear testpieces (Fig. 1c) and quantitatively by peel testing. In the latter, samples are tested initially and then after various exposure times in relevant fluids, which are frequently held at elevated temperatures to accelerate any relevant effects. Plots of peel energy vs the square root of time are often linear because the deterioration in bond magnitude is dependent on the amount of fluid which has entered the adhesive by diffusion. It is sometimes possible to extrapolate from the gradients of these lines at various temperatures to other temperatures of interest, for example service temperature, using Arrhenius-type relationships (i.e., log(rate) is plotted against reciprocal absolute temperature).

One particular aspect of durability which is important, for instance in offshore applications of elastomer–metal articles, is whether the bonding will survive the application of cathodic potential employed to protect the metal itself against corrosion from the sea water. Stevenson (1985) showed that a well-established bonding system was affected adversely by cathodic protection. Other adhesives are affected less. In addition, if the elastomer is molded to encompass completely the bonding interface, preventing direct contact with sea water, the deterioration is avoided.

7. *Quality Control and Assurance Testing*

In manufacturing industries, the QC/QA testing which will be applied to the individual components of elastomer–metal composites should be reinforced by appropriate adhesion testing. Individual batches can be subjected to routine peel testing, with set peel force magnitudes to be achieved; appropriate durability tests can be included and qualitative double-shear testing can be performed if necessary. In addition, for some elastomer–metal laminated products, one product per batch can be sacrificed for QA testing, with peel adhesion test pieces being cut out of the article to give bond magnitudes under realistic manufacturing conditions.

8. Other Bonding Systems

In the early days of the rubber industry, metal parts were brass plated prior to contacting the elastomer, these materials bonding together spontaneously during vulcanization. This metal pretreatment is time consuming and expensive and has now been superseded by the use of adhesives as already described; the only remaining important area of use is described in the article *Elastomer–Wire Adhesion Processes*.

On occasions, as an anticorrosion measure, steel parts are provided with a protective coating of zinc, either as plating ("galvanizing") or by "sherardizing" (the tumbling of the steel part in zinc powder). Unlike brass, zinc can be difficult to bond to elastomer, although it becomes firmly attached to the steel; certain adhesives of the curable types already discussed are necessary to bring about good bonding of the whole composite.

One system which has traditionally utilized another form of bonding is the use of ebonite—elastomer (normally natural rubber) with very high sulfur loadings and little or no accelerator—as the adhesive for bonding tank linings. The relatively inert elastomers used as linings are cured at low temperatures for long times, which are conditions also suited to forming the hard ebonites for producing good bonding. Finally, a degree of bonding is possible between metals and cured elastomer treated with hypochlorite solution and so on (see *Elastomer–Elastomer Adhesion Processes*) using cold-curing adhesives such as epoxy resins. However, the bond magnitudes are low compared with those obtained from the curable adhesives described here and the bonds are not flexible and are therefore unsuitable for dynamic applications.

See also: Bonding: Preparation of Metallic Surfaces; Elastomer–Elastomer Adhesion Processes; Elastomer–Wire Adhesion Processes

Bibliography

American Society for Testing and Materials 1991 *Rubber, Natural and Synthetic—General Test Methods,* Annual Book of ASTM Standards, P. 37. ASTM, Philadelphia, PA

Brown R P 1979 *Physical Testing of Rubbers.* Applied Science, London, Chap. 18

Hata T, Gamo M, Doi Y 1965 Mechanics of peeling, I: Peel equation. *Kobunshi Kagaku* 22(239): 152–9

Lindley P B 1971 Ozone attack at a rubber–metal bond. *J. Inst. Rubber Ind.* 5: 243–8

Meier J L, Findley H J 1973 Current trends in rubber to metal bonding. *Proc. Conf. ACS (Rubber Division), 104th Meeting.* American Chemical Society, Washington, DC, Paper 10

Pickup B, Weatherstone E 1969 Some factors influencing the bonding of metal to rubber. *J. Inst. Rubber Ind.* 3(6): 1–8

Sexsmith F H 1979 Milestones in elastomer-to-metal bonding. *Adhes. Age* May 21–6; June 31–6

Stevenson A 1985 On the durability of rubber/metal bonds in seawater. *Int. J. Adhesion Adhes.* 5(2): 81–91

Stevenson A 1990 *Rubber in Engineering, Kempe's Engineering Year Book* 1 (A1–F10). Morgan-Grampian, London, Chap. C3

Weih M A, Silverling C E, Sexsmith F H 1985 Effect of compounding ingredients on elastomer to metal adhesion. *Proc. Conf. ACS (Rubber Division), 128th Meeting.* American Chemical Society, Washington, DC, Paper 97

Zisman W A 1964 *Commemorative Volume on the Work of W A Zisman*, Advances in Chemistry Series, 43. American Chemical Society, Washington, DC

R. P. Campion
[Materials Engineering Research Laboratory, Hertford, UK]

P. J. Corish
[Patrick J. Corish & Associates, Stratford-on-Avon, UK]

Elastomer–Textile Adhesion Processes

In the polymer industry, polymeric textile materials are used to reinforce elastomers (and plastics). The strength and modulus of the composite article as well as its dimensional stability are provided by the textile. The enveloping elastomer provides flexibility and holds the composite together as well as providing protection of the textile from abrasion and wear. Good elastomer–textile adhesion is a necessary prerequisite for the achievement of optimum properties.

Factors influencing textile choice include strength, modulus, stability, durability, flexibility and cost, as well as adhesion. The first fibers used for elastomer reinforcement were natural products, cotton, the most important type, being incorporated in the original pneumatic tires. Synthetic fibers were produced and used later, including rayon, nylon, polyester, poly(vinyl alcohol) and aromatic polyamides (aramids). General properties of these fibers are reviewed elsewhere in the Encyclopedia.

It should be recognized, however, that although textiles based on organic fibers have lower strength and modulus values than denser glass and steel materials, their tenacity (strength per unit weight) values can be equal or higher.

1. Development of Adhesive Systems

The originally used textile fiber, cotton, does not require adhesive pretreatment to achieve adequate adhesion to elastomers after molding together at

Table 1
Typical RFL adhesive dip formulation

Constituent	Parts	
	Wet	Dry
Resorcinol	9.4	9.4
Formaldehyde (37%)	13.8	5.1
Sodium hydroxide (10%)	7.0	0.7
Latex (40%)	212.0	84.8
Water	257.8	
Total	500.0	100.0

high temperature and pressure. As cotton is a hairy fiber, adhesion is achieved mechanically by embedment of the fiber ends into the elastomer and by elastomer penetration into the textile structure.

The first widely used synthetic fiber, rayon, required the use of adhesive pretreatment for adequate fiber–elastomer adhesion. The originally used natural rubber latex–casein mixture was later modified to obtain improved adhesion, replacing the casein with resorcinol–formaldehyde resins.

Further development during World War II (1939–1945) led to replacement of the natural rubber part of the dip system with emulsion SBR latex. Introduction of high tenacity grades of rayon yarn and of various types of nylon caused further changes in which a latex based on a terpolymer of styrene, butadiene and vinyl pyridine (originally Gentac) replaced part or all of the SBR latex.

Normally, the adhesive pretreatment is performed by dipping the textile, in the form of yarn, cord or fabric, in a continuous process through a liquid mixture of a latex and a resole resin, formed by an alkali catalyst. Two methods of preparation may be employed: a two-stage process, in which the resin is formed before the latex is added or a one-stage process, in which all ingredients are mixed and an alkali added.

A typical resorcinol–formaldehyde–latex (RFL) adhesive dip formulation of resole type is given in Table 1.

A less popular dip system is the novolac resin type, in which a precondensed linear novolac resin is prepared with an acid catalyst and additional formaldehyde is added to form a three-dimensional resin network during cure of the dip.

Yarn (or fabric) is normally impregnated with the dip by passing through a dip bath, removing excess by, for example, air jets or squeeze rollers. After removal of water at 100–120 °C, a short higher-temperature cycle is employed to cure the dip film. Time and temperatures vary according to the fiber type: 1 min at 150 °C may be used for rayon but higher temperatures (170–200 °C) are needed for nylon, as dipping is normally part of the heat-setting process.

Table 2
Du Pont D417 system

Constituent	Parts	
	Wet	Dry
Predip		
Phenol-masked isocyanate (40% dispersion)	180.0	72.0
Epoxy resin	27.2	27.2
Gum tragacanth (2%)	25.0	0.5
Dispersant (50%)	0.6	0.3
Water	1767.2	
Total	2000.0	100.0
Second-stage dip		
Standard RFL formulation		

Polyester and polyaramid fibers present problems due to the chemical inertness of the high-modulus surfaces. The standard RFL dips do not give rise to good adhesion to elastomers and special pretreatments or alternative systems have been developed. For both types of fiber, pretreatment with solvent-based isocyanates followed by RFL dips can give reasonable success. However, use of "free" isocyanates and the presence of organic solvent, lead to problems in production. Alternative materials have led to success with polyester fibers, including Du Pont's D417 system which uses a phenol-masked isocyanate together with a water-soluble epoxy resin. Predipping polyester fabric in an aqueous solution typified by a formula shown in Table 2 is followed by heat activation at about 230 °C.

The predipping must be followed by a second-stage dip with a standard RFL subsequently to achieve good levels of adhesion. Another system developed by ICI uses Pexul (Vulcabond E), a trimer obtained by reacting resorcinol with *p*-chlorophenol. A typical formula is shown in Table 3. A high-temperature treatment (>200 °C) is needed to achieve the subsequent good adhesion. Aromatic polyamide (Kevlar) fibers also respond favorably to a predip based on epoxy resin before RFL dipping; alternatively, polyester-type systems can be used.

2. *Surface Activation of Textiles*

Alternative methods of attaining good adhesion to elastomers for polyester and aramid fibers without recourse to use of special predips, which sometimes give rise to problems in service performance, involve surface activation of yarns or cords.

Polyester yarns that are chemically treated at the spinning stage of manufacture with epoxy materials, such as reaction products of ethylene glycol with

Table 3
Pexul (Vulcabond E) system

Constituent	Parts Wet	Parts Dry
Vulcabond E (20%)	197.0	39.4
Resorcinol	9.8	9.8
Formaldehyde (37%)	5.9	2.2
V P latex (41%)	121.5	48.6
Water	165.8	
Total	500.0	100.0

epichlorohydrin, are termed spin-finished textiles. Subsequent dipping with a standard RFL dip results in good levels of adhesion to a range of elastomeric compounds.

Other methods of surface activation before composite manufacture which can be used for polyester or aramid fibers include ultraviolet (uv) irradiation and exposure to glow discharge. Both these methods result in chemical modification of the surface, enabling subsequent RFL dipping to give improved adhesion levels.

3. *Dry-Bonding Systems*

The adhesive systems and pretreatments described relate to processing of the textile before it is coated with an elastomeric compound. An alternative system has been developed in which resin-forming ingredients are added to the elastomeric compound. Curing of such dry-bonding compounds gives acceptable adhesion to undipped textiles, although spin-finishing or uv treatment is still necessary for polyester and aramid fibers. The resin-forming ingredients are resorcinol and a methylene donor in place of formaldehyde, which is too volatile for such use. Suitable donors are hexamethylene tetramine (HMT) and hexamethoxy methylmelamine (HMMM). Use of a fine-particle hydrated silica in the compound enables higher levels of adhesion to be obtained.

The mechanism of the bonding process is thought to be the production of a resorcinol–formaldehyde (RF)-type resin at the interface between elastomer and textile. The role of the silica is believed to be that of retarding the cure of the resin, allowing low-molecular-weight resin to migrate to the surface for subsequent bonding to the textile. The high surface area of the active silica may also be conducive to resin formation.

For such a process to occur, an adequate loading of additives in the compound is necessary, for example, resorcinol at 2.5 parts per hundred (pph) elastomer, HMT at 1.5 pph. A reasonable thickness of dry-bonding compound is also required to minimize effects of back migration of resin-forming ingredients into adjacent compounds.

Release of amine residues from HMT can cause problems of degradation with polyester and aramid fibers; thus HMMM is preferred as the methylene donor for these fibers.

Dry or *in situ* bonding gives acceptable levels of adhesion in various elastomers, from high-unsaturation types such as natural rubber to low-unsaturation types such as ethylene–propylene terpolymers.

4. *Polar and Low-Unsaturation Elastomers*

Adhesive formulations for polar and low-unsaturation elastomers need some modifications from those used for general purpose high-unsaturation elastomers such as natural rubber, polybutadiene and styrene–butadiene copolymers. For elastomers such as polychloroprenes (CR) or butadiene–acrylonitrile copolymers (NBR), special RFL systems are needed, replacing the latex in the dip partly or wholly by CR or NBR latices.

For butyl and ethylene propylene terpolymer (EPDM) type elastomers, emulsions (not latices) are available for use, however, subsequent adhesion levels attained with RFL dips incorporating these are not as high as those with highly unsaturated elastomers.

An interesting extension of the use of RFL dips for bonding is provided by plasticized poly(vinyl chloride) (PVC), a thermoplastic with elastomeric characteristics. If PVC latices are used in the dips, it is usually necessary to incorporate a plasticizer as an emulsion to reduce stiffness of the dip film. Alternatively, a dip incorporating NBR latex gives satisfactory adhesion, as NBR forms compatible blends with PVC.

5. *Elastomer Compounding Effects*

In addition to the resin-forming ingredients such as resorcinol and HMT, plus silica, already discussed under dry bonding, several other types of compounding ingredients can effect adhesion.

Materials such as plasticizers, oils, waxes or antidegradants can adversely affect adhesion. They can bloom to the surface, causing a weak boundary layer and also diffuse into the dip film on the textile, reducing its cohesive strength. Removal of such a layer is recommended before plying up the textile–elastomer composite.

However, the most significant effects on adhesion are produced by curative type and level. In natural rubber, high-sulfur/low-accelerator systems, such as sulfur 2.5/accelerator 0.5–0.6, give the highest adhesion levels. As sulfur content is reduced in semiefficient vulcanization (semi-EV) and efficient vulcanization (EV) systems, such as sulfur 0.5/accelerator 4.0, adhesion levels are halved. With truly sulfurless vulcanization systems, such as tetramethyl thiuram disulfide 3.0, adhesion levels are reduced to very low values (all numbers refer to pph).

6. Adhesion Mechanism

The mechanism of the adhesion of elastomer to fiber via an RFL aqueous dip is not completely understood. However, it can be considered in a three-part way.

The dip film adheres to the textile fibers by mechanical penetration of dip into the yarn or fabric structure (15–20% contribution) and also by diffusion into the surface of the fiber (30% for rayon, 5% for nylon). Chemical bonding is responsible for the rest of the adhesion, either primary covalent bonds (60% for nylon, 25% for rayon) or secondary hydrogen bonding (20% for nylon, 25% for rayon).

Adhesion between baked dip film and elastomer is largely associated with cocrosslinking of the elastomer (latex) content of the dip to the elastomer in the main polymer phase. The major effect of curative systems on adhesion levels is readily explained, as migration of sulfur and accelerator into the dip film is necessary for cocrosslinking to occur. In sulfurless systems, no sulfur migration can occur, thus adhesion is low. A minor contribution to the total adhesive bond is thought to be due to reaction of the RF resin with the elastomer, leading to formation of chroman rings. This is obviously a major factor in dry-bonding systems.

Finally, the dip film itself is probably a type of interpenetrating network (IPN), where the elastomer reptates through the RF resin network. It provides an intermediate layer between the relatively high-modulus, low-extensibility fiber and the low-modulus, high-extensibility elastomer and has intermediate properties in view of its biphasic character. The system is thus an example of a composite with a modulus graded bond.

See also: Bonding: Preparation of Polymer Surfaces; Elastomer–Elastomer Adhesion Processes; Elastomer–Thermoplastics Adhesion Processes

Bibliography

Dunlop Company 1968 Treating polyester materials with ultraviolet radiation. British Patent No. 1,228,173

Dunlop Company 1968 Polyester filaments exposed to plasma/gas discharge. British Patent No. 1,230,982

Skeist I 1977 *Handbook of Adhesives*. Van Nostrand Reinhold, New York

Wake W C 1982 *Adhesion and the Formulation of Adhesives*, 2nd edn. Applied Science, London

Wootton D B 1982 Heat-setting and adhesive treatment. In: Wake W C, Wootton D B (eds.) 1982 *Textile Reinforcement of Elastomers*. Applied Science, London, Chap. 4

P. J. Corish
[Patrick J. Corish & Associates,
Stratford-on-Avon, UK]

R. P. Campion
[Materials Engineering Research Laboratory,
Hertford, UK]

Elastomer–Thermoplastics Adhesion Processes

Many composite articles in current use are based on elastomer and plastics layers. The plastics component provides reasonably high bulk rigidity/modulus and the elastomer provides flexibility, resilience or damping to the composite. The service life and performance of the composites are very dependent on the attainment of good bonding: environmental resistance of the bonded composite is especially sensitive to this factor. Elastomer–plastics composites offer weight savings compared with elastomer–metal composites and their end uses reflect this advantage.

Several different bonding methods can be used, some involving surface treatments of the plastics substrate.

1. Adhesive Bonding

Proprietary bonding agents are available from several suppliers and are employed as one-coat or two-coat systems. Variants are available depending on the nature of the plastics and elastomeric components. The primer coating is designed to have good adhesion to the plastics substrate and usually consists of polar materials, for example chlorinated natural rubber dissolved in a suitable organic solvent. The cover coating is formulated to give good adhesion to the elastomeric composition by formation of chemical bridges or crosslinks.

Application of these adhesives coatings can be by brushing, spraying, dipping or roller coating, depending on whether production processes are discontinuous or automated. Normally, for in-mold bonding, the adhesive-coated plastics component is positioned in the mold and the elastomeric composition is introduced by compression, transfer or injection molding. The combination of heat and pressure of the molding operation completes the curing of the elastomeric composition and its bonding in the elastomer–plastics assembly.

2. *Preparation of Plastics Substrates*

Many plastics components can be prepared for adhesive bonding merely by solvent degreasing followed by application of the primer coat. Initial light abrasion by emery cloth, wire wool or silica grit before degreasing improves the wettability of the primer onto the plastics surface. These surface preparations are sufficient for polar plastics such as nylons, polycarbonates and even cured thermosets (e.g., phenolics and epoxides). However, some plastics have difficult bonding surfaces and require special pretreatment methods (see *Bonding: Preparation of Polymer Surfaces*).

Polyethylene and polypropylene may be subjected to etching either with a chlorinated organic solvent such as trichloroethylene or by chromic acid. Alternative methods include corona discharge, in which the surface is treated with an electric corona discharge process, or flame treatment, using an oxidizing gas flame.

Fluorinated plastics, such as polytetrafluoroethylene (PTFE), need special etching processes to provide a suitable bonding surface. Most involve reduction of the polymer surface with one of a number of reducing agents, for example sodium in liquid ammonia or sodium naphthalenide in tetrahydrofuran. Alternatively, electrochemical pretreatment may be used or pretreated PTFE may be obtained from suppliers. After these special pretreatments, immediate application of a cover coat of adhesive is necessary.

3. *Postvulcanization Bonding*

In some products involving intricate designs or a multiplicity of components, (e.g., sandwich constructions), it is preferable to vulcanize the elastomeric material(s) in an initial operation. Subsequent assembly of the vulcanized elastomeric and plastics components enables postvulcanization bonding to take place using heat and pressure.

The component assembly is normally heated to give a bondline temperature between 100 °C and 160 °C for 15–60 min. Clamping or compression is normally necessary during this bonding operation. Compression ratios of 5–10% normally give optimum bonding but this is dependent on the initial dimensions of the elastomeric component. This compression must be maintained throughout the vulcanization and until the assembly has cooled down to below 50 °C.

The vulcanized elastomer surface needs to be thoroughly cleaned and degreased prior to adhesive application, for example light abrasion with a wire brush followed by use of a chlorinated solvent, such as trichloroethylene. Nonpolar elastomers such as butyl and ethylene propylene copolymers and terpolymers, natural rubber and styrene butadiene copolymers require chlorination or acid cyclization treatment of the surface to achieve good adhesion. Preparation of the plastics surface is described in Sect. 2.

Primer coatings should be applied as necessary to the pretreated elastomer and plastics surfaces immediately after cleaning and chemical activation. Assembly of the adhesive-coated components (one- or two-coat adhesive systems may be involved) should be carried out when they are free from solvent. Handling should be minimized.

4. *Novel Particulate-Type Adhesives*

Recent novel developments (Corish et al. 1980) have resulted from the discovery that a dispersion of a fine-particle-size carbon black in solvent, applied at the interface, promotes good bonding between plastics and elastomers when molded together at or above the melting point of the plastics. The molding times were sufficient to vulcanize the elastomeric composition.

Some disadvantages occurred using carbon black alone as an adhesive system, including difficulty in controlling the molding of the elastomeric composition under pressure when the plastics component was fluid. Additionally, the high temperatures required to melt the plastics materials were detrimental to the properties of the elastomer and the coating was dirty, and thus contaminated other materials and could be easily disrupted during handling.

Use of low levels of polymer binder, dissolved in the solvent containing the black dispersion, obviated these disadvantages. Typically, adhesive systems consist of a 5% total solids high-abrasion furnace (HAF) black dispersion in 1,1,1-trichloroethane. A black–binder ratio of 80:20 was preferred, the binder being a butadiene acrylonitrile copolymer of medium nitrile content (37%) and fairly high Mooney viscosity level (ML-4 at 100 °C = 95).

This allowed a technique to be developed for bonding elastomeric compositions to plastics, without the need for high curing temperatures and long heating and cooling cycles. The particulate adhesive described is applied first to the surface of a mold by air spraying, the coating drying within seconds. Injection molding of plastics materials into the mold produces an opaque black film on the plastics. Vulcanization of elastomeric compositions by compression molding against the coated plastics surface produces assemblies with good levels of adhesion. A wide range of plastics was found to give good bonding, both static and dynamic, to many different types of elastomers. No pretreatment of the plastics was carried out.

5. *Fusion Bonding*

Another new elastomer–plastics bonding option has been recently described. Under certain conditions, when the molten surface of a plastics material is brought into contact with a compatible elastomeric composition, either during vulcanization or afterwards, spontaneous adhesion occurs giving a strongly bonded composite. Compatible plastics–elastomer pairs have been identified. Polyethylene, polypropylene and ethylene vinyl acetate copolymers can be fusion bonded to elastomeric compositions based on ethylene propylene elastomers and their blends with styrene butadiene copolymers, polybutadienes or polyisoprenes. Polyesters and nylons can be fusion bonded to polyepichlorohydrin and polyacrylate elastomers.

The presentation of the plastics component to the elastomer component can be achieved in several ways. These include compression molding, during or after cure, injection molding, extrusion bonding and friction bonding. This last process is similar to the concept of friction and vibration welding used for bonding plastics to plastics.

See also: Bonding: Preparation of Polymer Surfaces; Elastomer–Elastomer Adhesion Processes; Elastomer–Textile Adhesion Processes; Welding of Plastics

Bibliography

Corish P J, Osmant T H, Clarke D I 1980 Novel "particulate-type" adhesives. In: Allen K W (ed.) 1979 *Adhesion—4*. Applied Science, London, Chap. 10

Lovatt C H, Aitchinson G D 1976 The preparation of substrate surfaces in modern adhesive technology. In: Jackson B S (ed.) 1976 *Industrial Adhesives and Sealants*. Hutchinson Benham, London, Chap. 4

Skeist I 1977 *Handbook of Adhesives*. Van Nostrand Reinhold, New York

Wegman R F 1989 *Surface Preparation for Adhesive Bonding*. Noyes Publications, Park Ridge, NJ, Chap. 8

Wragg R T, Yardley J F, Nightingale A F 1981 Fusion bonding. *Kautsch. Gummi, Kunstst.* 34(8): 657–60

P. J. Corish
[Patrick J. Corish & Associates, Stratford-on-Avon, UK]

R. P. Campion
[Materials Engineering Research Laboratory, Hertford, UK]

Elastomer–Wire Adhesion Processes

Since the early 1960s, the means of reinforcing tires has changed both in design, from cross ply to radial ply, and material, from textile to largely steel wire cord. However, one factor which has remained unchanged is the need for strong durable bonding to form between the reinforcing cords and the elastomer of the tire during vulcanization. Steel surfaces do not bond directly to elastomers during vulcanization (see *Elastomer–Metal Adhesion Processes*), this comment applying to wires as well as to flat surfaces. However, one metallic species well known for its ability to bond to rubber is brass, especially with copper–zinc ratios of around 70:30. In the early part of the twentieth century, production of most elastomer–metal composite structures involved a first stage of brass plating of the metal parts, to provide bonding. Although this procedure has been superseded by use of adhesives for general elastomer–metal bonding, one important use of brass plating, the coating of steel wires, has increased over recent years. These wires are twisted or laid into strands and/or cords to produce the steel cords used in steel-braced tires. This article is therefore mainly concerned with elastomer–brass adhesion.

As well as their use in the tire industry, steel wires (often brass plated) are also employed for reinforcement purposes in the manufacture of hoses and belting.

1. General Pointers and Background

When producing steel wires, the brass-plating process is performed before the final drawing stage in which coarse wires are pulled out to their final diameter; consequently, the brass plating becomes stretched during drawing, with formation of plating defects and unevenness (van Ooij 1979) and an increase in its level of surface energy. Both factors enhance subsequent rubber–brass adhesion, by providing sites for the elastomer to key into and by increasing the brass surface potential for chemical bonding reactions.

In tire manufacture, the term steel fabric is used to describe a thin composite layer comprising parallel steel cords, all close together, sandwiched by veneers of elastomeric topping compound. This compound is formulated to produce optimum levels of bonding and bond durability. In the production of steel fabrics, using a calendering process for instance, extra compound is forced in between cords and partially between the strands making up the cords (i.e., elastomer penetration occurs "into" the cords). It is essential for the following operations that steel fabrics are formed to tight tolerances. After cutting to suitable shapes, so that the cords will ultimately take up the correct angles in the finished product, the rubberized steel fabric pieces are fabricated into the tire construction, initially relying on building tack (see *Elastomer–Elastomer Adhesion Processes*). The chemistry of the bonding between rubber and brass plating then occurs during the subsequent vulcanization at high temperatures and molding pressures.

The elastomer–brass bonding requirement for a tire is that a high level of adhesion is formed during cure and is retained throughout its service life in the carcass of the tire; for a truck tire, because several retreads are possible, this service life can relate to distances of 160 000–320 000 km. The achievement of these aims is dependent on tire design, on the choice of recipe for the topping compound and on various characteristics of the brass plating. The durability of the bond will depend, in the first instance, on elastomer stiffness or modulus, as this property will influence actual strains at the bond interface when a force is applied; later aspects of durability include the resistance of the system to fatigue and hostile fluids.

2. *Bonding mechanism*

The elastomer–brass bonding mechanism has been discussed extensively in the literature, particularly by van Ooij (1978, 1979, 1983), Haemers (1980) and van Ooij et al. (1981). In separate investigations, these workers have employed fine-scale analytical techniques such as x-ray photoelectron spectroscopy (XPS or ESCA) and Auger electron spectroscopy (AES) to examine the elastomer–brass interface. Quantity versus distance profiles across the interface were established for copper, zinc, sulfur and other relevant chemical elements and compounds. In addition, these workers have simulated bonding reactions by immersing brass plates in suitable liquids which model the repeating chemical units of natural rubber, adding vulcanization ingredients, heating to curing temperature and examining various reaction stages by AES and x-ray fluorescence.

To summarize the detailed conclusions reached by these researchers, the chemistry of bond development, occurring concurrently with vulcanization, is initially associated with attack of the brass surface ("sulfidization") by sulfur or sulfur donors present as vulcanization ingredients in the elastomer. A mixed sulfide layer then forms, with copper sulfide tending to be on the surface and zinc sulfide beneath, bonding to the brass perhaps via zinc oxide (initially present as a result of atmospheric oxidation during plating). If the copper–zinc ratio is too small (below ~60:40) an excess of zinc sulfide and/or change in brass alloy type inhibits the bondability of the copper sulfide layer which otherwise has been suggested to be an *in situ* adhesive to elastomer. The stability of the bond during product service is then dependent on this sulfide layer and the brass beneath. Excessive sulfide formation (e.g., because of incorrect sulfur/curative levels, overlong vulcanization times, or high copper content in the brass) causes excessive layer thickness, embrittlement, easy fracture and lower bond strengths. Resistance to exposure to hostile environments is affected by diffusive phenomena involving additives, reaction products and hostile fluids, and electrochemical interactions with the brass surface components. The presence of cobalt at the interface is thought to affect these interactions, resulting in improved bonding and bond durability; the cobalt is usually added as a chemical compound to the elastomer mix, but has also been incorporated as a trace element into the brass alloy.

An alternative point of view (Haemers 1980) is that copper sulfide forms as a by-product of the bonding reaction, with direct brass–rubber links forming in between domains of sulfide.

The views mentioned thus far have mainly stemmed from observing bonding from the brass side. The most probable means of linkage to an elastomer molecular chain is through an attached pendent chemical grouping which, in the elastomer bulk, would react with another chain later in the vulcanization process to form a crosslink. (In the 1950s and 1960s, Bateman et al. (1963) carried out extensive experimentation on accelerated sulfur vulcanization chemistry and produced the mechanisms now generally accepted for crosslinking, culminating in a so-called sulfurating species which attacks the chain to give first this pendent group, then the full crosslink.) Campion (1989) provided kinetic and other experimental evidence to support the view that brass bonding reactions follow similar chemical paths to vulcanization reactions, but occur more quickly.

3. *Steel Cord Aspects*

Many variations in wire diameter and steel cord construction exist, to enable cords of the necessary strength and flexibility to be produced for many different sizes of tire, hose and so on. The magnitude of adhesion is influenced mainly by the overall diameter of the cord and embedded length (Campion and Redmond 1975). Secondary improvements in adhesion can arise if extra bondable brass surface is available in the cord surface region in the interstices between neighboring strands or between wires if these are of large diameter. Elastomer penetration between strands affects service performance and dynamic adhesion magnitudes through the load-carrying ability of penetrated elastomer rather than basic adhesion (Campion and Corish 1977). Cords constructed from ultrafine wires experience little penetration and adhesion levels consequently are lower.

3.1 Brass-Plating Aspects

Adhesion tests (see Sect. 5) have been used to measure levels of adhesion for various combinations of brass-plating thickness and copper content in the brass. Bertrand (1970) reported optimum adhesion at a copper content of 70% for many different elastomer recipes in which the main variant was the accelerator. When an organic salt of cobalt (a common adhesion promoter) is also in the formulation, adhesion is less sensitive to decreases in copper content, especially when measured after environmental aging. With 70% copper plated cords, Campion and Corish (1977) have shown that there is an optimum plating thickness of 0.2–0.3 μm associated with a standard natural rubber topping compound, obtaining data both from a static test and from a dynamic test simulating tires in service under accelerated conditions.

4. *Compound Formulation Aspects*

Particular attention has been paid for many years to the effect on adhesion of variations in the elastomeric compound formulation. Many papers have been published, a few of which are mentioned here. A general criticism is that, although over recent years the quality of test methods has improved considerably, most of these publications have not employed high-quality tests, which makes correlations with other properties difficult to assess. Buchan (1959) reported, among other observations, that successful bonding could be attained by employing both high sulfur loadings in unaccelerated recipes and low sulfur loadings when accelerating and activating ingredients were present (although the sulfur level present in the "low" loadings was still twice that employed in typical modern recipes for vulcanizing elastomers). Bertrand (1970) showed that a particular class of chemical accelerator, the sulfenamides, are associated with high bond strengths. The optimum sulfur–accelerator ratio is around five. The cure time and temperature employed can also have an effect on adhesion magnitudes. Compound recipes would ideally be chosen to produce maximum adhesion at the optimum cure time for other elastomer properties; however, measurements at various cure times have indicated that optimum adhesion develops more quickly than do optimum elastomeric properties (van Ooij 1979, Haemers 1980). It is also necessary to design recipes so that good adhesion levels are maintained during exposure of the bond interface to aging or hostile environments; to this end, the addition of some organic cobalt compounds or other adhesion promoters has often been used successfully (van Ooij et al. 1981).

Other adhesion promoters employed are the so-called dry bonding ingredients hexamethylene tetramine and resorcinol (see *Elastomer–Textile Adhesion Processes*) which are added together to the compound; fine particle hydrated silica may also be included (Wagner and Derringer 1971). Other systems with modifications to the basic dry bonding recipe include the use of proprietary partially cured resins instead of resorcinol (Nichols and Ohm 1976). Unlike the elastomer–textile case, the mechanism of bonding with these promoters is probably not directly chemical; however, one feature which they always bring about is an increase in cured elastomer modulus, which perhaps lessens the strain at the bonded interface when a disruptive force is applied during service or testing. One reason for including these promoters is to inhibit adhesion failure at the cut, brass-free, ends of cords at the edges of the steel fabric when fabricated into, for example, the tire.

The numerous publications on compounding considerations have been reviewed together with many other aspects of rubber–brass bonding by van Ooij (1978, 1979, 1983) and van Ooij et al. (1981)—much experimentation is also described. Some typical topping compound recipes (van Ooij 1979) are given in Table 1.

5. *Assessment of Adhesion*

Because steel cords are usually employed in dynamic service applications, resistance to adhesion failure is best assessed by a dynamic test which simulates the service situation. However, dynamic testing tends to be time consuming, so that when performing large-scale comparisons of effects of compound, brass or wire variations on adhesion magnitudes, screening static tests are first carried out, to indicate which systems should be progressed to dynamic testing. In addition, in production, static tests are necessary for monitoring successive batches of elastomer compound and steel cord during quality-control testing.

5.1 Static Adhesion

Basically, adhesion is quantified statically by measuring the force to fracture the bond when pulling a cord out of a block of cured elastomer. The blocks are produced containing many cords passing through them, so that a mean value and standard deviation associated with the removal force can be obtained by testing them all. It is desirable that high molding pressures are used when molding the uncured elastomer around the cords laid in suitable grooves in the mold. However, tests in common usage until the 1970s did not always comply with this requirement. Furthermore, the usual means of extracting the cord with the early tests, by supporting the elastomer block in a holder containing a slot and noting the force to pull the cord out through the slot, was shown by Eccher and Canevari (1969) to cause stress concentrations at one point on the bond interface, thus giving erroneously low measurements. Failure occurred first at this point and then progressed over a period of 1 s or so until the cord was removed, often with a chunk of elastomer, torn from the bulk, still attached. Obviously, this was not a true assessment of adhesion level. Hence, these workers developed the EC test, using photoelastic modelling techniques to give a cylindrical test sample, laterally

Table 1
Some typical steel cord topping compound formulations

Formulation	Parts by weight (phr)[a]				
	1	2	3	4	5
Natural rubber	100	100	100	100	100
Filler: carbon black (type HAF[b])	60	55	55	55	55
Activators					
zinc oxide	8	10	10	10	10
stearic acid	0.5	2	2	2	2
Process oil	5	5	5	5	5
Antioxidant	1.5	1	1	1	1
Sulfur	6	4	4	4	4
Curatives					
accelerator DCBS[c]	1				
accelerator MBS[d]		0.8	0.8	0.8	0.8
Prevulcanization inhibitor	0.2				
Adhesion promoters					
cobalt naphthenate			3		
cobalt stearate				3	
cobalt octoate					3

Source: van Ooij (1979)
a Parts per hundred of rubber b ASTM designation N330: mean particle diameter 29 nm c *N,N*-dicyclohexyl-benzthiazole-2-sulfenamide d *N*-oxydiethylene-benzthiazole-2-sulfenamide

reinforced by bonded shaped steel plates (the test cord running axially through the cylinder). Molding was performed at high pressures. During testing, the sample was supported by the lateral plates only, so that a more uniform shear stress was applied to the bonded interface. The success of this method is reflected by high levels of measurement and by cords being extracted almost instantaneously with a clean snap; in addition, the standard deviation between replicate samples was low.

Unfortunately, the EC method possessed two large disadvantages—the complexity of the mold required and the small number of cords (in their cylindrical samples) possible per single molded test piece. Hence, Campion and Redmond (1975) developed the DMP test, with test pieces comprising wide rectangular elastomer blocks, laterally reinforced with steel fabric, which contain many test cords per single molding, as an equivalent test to the EC test in terms of magnitude, low standard deviation and nature of bond fracture (Campion 1979). Uniform shear at the bonded interface was achieved by extracting the cord through a circular hole in the test jig during testing, and optimizing test-piece dimensions and hole size. High molding pressure was obtained with a semiplunger mold. This test now complies with method 1 in the International Organization for Standardization (ISO) International Standard 5603. Method 2 in this standard, which is also the American Society for Testing and Materials (ASTM) D2229-73 standard test, is again a hole test and, hence, is also preferable to earlier slot tests. However, reservations have been expressed about the contribution of friction to test magnitudes with this test, as the test piece (which is unreinforced) can partially enter the test hole when deformed by strain during testing (Campion and Redmond 1975)—test magnitudes from the two methods do not agree.

An alternative form of cord pull out test (TCAT) developed by Fielding-Russell et al. (1980) employed a cord which enters the elastomer block from one side only—it does not pass right through. This geometry is of a suitable form for a fracture analysis to produce the energy of adhesion (cf. *Elastomer–Metal Adhesion Processes*), the analysis being subsequently performed by Gent et al. (1981). Separation involves three stages—initial separation at the cord end, failure propagation and eventual catastrophic rupture. Skill is required in locating the unsupported cord end when producing the test piece if high molding pressures are to be achieved. These factors increase the chance of raising the variation between replicate measurements. More recently, a fracture analysis has also been performed on a cylindrical test-piece configuration, in which a stiff steel rod replaces the steel cord (Ellul and Emerson 1988).

5.2 Dynamic Adhesion

Many dynamic tests have been developed, but few have demonstrated correlation with appropriate service performance. In some of these tests, test pieces from static tests have been cycled to strains well below those associated with static bond failure, and the number of cycles until dynamic failure occurred were recorded. Additionally, Langevin et al. (1974) and Vorachek et al. (1975) have described different means of applying cyclic shear strains to the bonded interface until failure occurs. Campion and Corish (1977) described the multicord test which simulates tire performance, unlike other tests, in three ways:

(a) interfacial bond failure occurs inside the elastomer, not at an outer surface;

(b) maximum shear stress operates over less than 5 mm of bonded cord; and

(c) constant load conditions apply.

Test pieces possessed six accurately molded parallel steel cords along the center plane of an elastomeric block having a tread bar on each of its large outer surfaces, running normally to the cord direction. When testing, these tread bars were deformed and released repeatedly by a cycling compressive ramrod at 3 Hz. The number of cycles was recorded at the initiation of failure at the elastomer–brass interface, which always occurred under one edge of the tread bar—the position of maximum stress. Failure was detected by an associated change in electrical resistivity of the steel cords (due to local heat buildup) which were previously wired into suitable circuitry. The applied load could be altered according to the type of cord under test, and factors such as compound cure, wire and brass variations were assessed. It was found that the resistance to adhesion failure was governed by basic adhesion (i.e., static magnitudes from the DMP test), compression modulus (due to compound penetration into the cord), brass thickness and type, and cord construction and so on. Good correlation with tire testing data was illustrated.

6. *Durability*

During tire service, it is necessary for rubber–brass bonds to resist failure by fatigue due to repeated flexing. Besides the mechanical aspects, the bonded layer and the adjacent elastomer might become altered somewhat with the passage of time, as the high temperatures associated with the running of tires (especially large truck tires in which the elastomer–brass interface can be at 100 °C for long periods) means that vulcanization and bonding chemistry can continue—the crosslinks and the bonded brass surface can change in character. The brass surface can also be attacked if hostile fluids, for instance salt water, penetrate to the region—this can happen perhaps if a puncture is experienced during the winter months when salt has been scattered on the roads. The tire manufacturers seek to avoid the problem by compounding procedures, developing very stable formulations and maximizing elastomer penetration into the cords during molding (by varying cord construction, etc.) to minimize the possibility of wicking of hostile fluids along the cords.

Much laboratory work has been performed in exposing test pieces to hostile fluids such as salt water, air and hot water before adhesion testing, to assess the quality of systems developed to resist hostile fluids (van Ooij 1979).

7. *Adhesion to Zinc*

Zinc-plated steel cords find limited usage in hoses and belting where the resistance to possible corrosive attack (e.g., from seawater) is more important than attaining the highest levels of adhesion. Van Ooij (1979) discusses this property—the mechanism of bonding resembles that for brass, but certain stages occur much more slowly. Reported results on compounding formulation aspects show that high levels of sulfur, addition of dry bonding ingredients and addition of cobalt salts can all improve the adhesion of elastomer to zinc; at best, a level of about 80% of the elastomer–brass bond level can be obtained.

See also: Bonding: Preparation of Metallic Surfaces; Elastomer–Metal Adhesion Processes; Elastomer–Textile Adhesion Processes

Bibliography

Bateman L, Moore C G, Porter M, Saville B 1963 Chemistry of vulcanization. In: Bateman L (ed.) 1963 *The Chemistry and Physics of Rubber-like Substances.* Maclaren, London, Chap. 15

Bertrand G 1970 Adhesion du caoutchouc naturel a des fils d'acier laitonne—Influence de certains facteurs chimiques. *Proc. Conf. Int. Caoutchouc,* Paris, Sessions 3 and 4, pp. 109–14

Buchan S 1959 *Rubber to Metal Bonding.* Crosby Lockwood, London

Campion R P 1979 Observations on DMP rubber/wire adhesion testing. *Eur. Rubber J.* September: 46–8

Campion R P 1989 Elastomer composites for engineering applications—Spontaneous and other forms of bonding between components. *Mater. Sci. Technol.* 5: 209–21

Campion R P, Corish P J 1977 Rubber/wire-cord adhesion tests, part 2—Dynamic: The Dunlop multicord test. *Plast. Rubber Mater. Appl.* 2: 65–70

Campion R P, Redmond G B 1975 Rubber/wire-cord adhesion tests, part 1—Static: The DMP hole test. *Rubber Ind.* 9: 19–24

Eccher S, Canevari C 1969 Improvement of the testing method for adhesion of vulcanized rubber to steel wire cords. *Kautsch. Gummi, Kunstst.* 22: 228–32

Ellul M D, Emerson R J 1988 A new pull-out test for tire cord adhesion. *Rubber Chem. Technol.* 61(2): 293–308; 309–23

Fielding-Russell G S, Livingston D I, Nicholson D W 1980 Physical factors in cord-to-rubber adhesion by a new tire cord adhesion test. *Rubber Chem. Technol.* 53: 950
Gent A N, Fielding-Russell G S, Livingston D I, Nicholson D W 1981 *J. Mater. Sci.* 16: 949
Haemers G 1980 The influence of the interface reaction on rubber-to-brass adhesion and adhesion retention. In: Allen K W (ed.) 1980 *Adhesion 4.* Applied Science, London, Chap. 9
International Organization for Standardization 1986 *Rubber, Vulcanized—Determination of Adhesion to Wire Cord,* International Standard ISO 5603. ISO, Geneva
Langevin P P, Payne R J, Shephard C S 1974 A dynamic rubber–cord composite adhesion and fatigue test. *Rubber Chem. Technol.* 47: 171–87
Nichols M J, Ohm R F 1976 Compounding for rubber adhesion in the HRH adhesion system. *Adhes. Age* June: 31–9; July: 25–30
van Ooij W J 1978 Mechanism of rubber-to-brass adhesion: Effect of rubber composition on the adhesion. *Rubber Chem. Technol.* 51: 52–71
van Ooij W J 1979 Fundamental aspects of rubber adhesion to brass-plated steel tire cords. *Rubber Chem. Technol.* 52: 605–75
van Ooij W J 1983 Mechanism and theories of rubber adhesion to steel tire cords—An overview. *Proc. Conf. ACS (Rubber Division), 124th Meeting.* American Chemical Society, Washington, DC, Paper 59
van Ooij W J, Weening W E, Murray P F 1981 Rubber adhesion of brass-plated steel cords: Fundamental study of the effects of compound formulation variations on adhesive properties. *Rubber Chem. Technol.* 54: 227–54
Vorachek J J, Causa A G, Fleming R A 1975 A steel tire cord oscillatory shear dynamic adhesion test. *Proc. Conf. ACS (Rubber Division), 108th Meeting.* American Chemical Society, Washington, DC, Paper 60
Wagner M P, Derringer G 1971 Improved rubber–steel cord adhesion: A kinetic study. *Proc. Conf. ACS (Rubber Division), 100th Meeting.* American Chemical Society, Washington, DC

R. P. Campion
[Materials Engineering Research Laboratory, Hertford, UK]

P. J. Corish
[Patrick J Corish & Associates, Stratford-on-Avon, UK]

Elastomeric Coatings

Synthetic elastomeric coatings possess excellent resistance to vibration and to debonding caused by thermal expansion. They find wide use in environments where erosive and corrosive conditions are severe (e.g., in chemical processing equipment, storage tanks, dock structures, cement mixers, oil refinery equipment and ship propeller shafts). Properties of some important elastomeric coatings are given in Table 1.

1. Butyl Rubber

Butyl rubber coatings are used as a lining for chemical tanks, pumps and hoses because of their excellent resistance to concentrated acids, alkalis, corrosive salts, and fresh and salt water. They are also resistant to esters, animal fats, vegetable oils, alcohols and ketones, but are attacked by chlorinated solvents, petroleum and aromatic hydrocarbons.

An unusual feature of butyl rubber coatings is that they remain flexible over long periods at temperatures down to about −240 °C. Because of this feature, they find wide use in cryogenic applications. They can also be used at temperatures up to 100 °C. In addition, butyl rubber is a good electrical insulator and is, therefore, used as a wire coating.

2. Fluoroelastomers

At the other end of the temperature scale, fluoroelastomers can be used at about 300 °C for short periods and almost indefinitely at about 250 °C. These coatings retain their flexibility at temperatures down to −125 °C.

An outstanding characteristic of fluoroelastomers is that they resist nearly all corrosive chemicals: the only materials that attack them are molten alkalis and highly halogenated compounds. Their uses include linings for reaction vessels and protective coatings for tanks, drums and other equipment exposed to harsh chemical environments. Because of their slippery nature, fluoroelastomeric coatings are used in machine parts that cannot be lubricated. Other characteristics include excellent resistance to weathering and aging. They are also nonflammable.

3. Polyurethanes

Polyurethane elastomeric coatings possess excellent abrasion resistance and so are used in applications such as bowling pins, flooring shoes and cement mixers. They also exhibit excellent resistance to oxygen, ozone and ultraviolet light, and have high tensile and tear strength together with excellent resilience. The elastomer weakens, however, when subjected to high temperatures (e.g., hot water or steam) and hardens at temperatures below about −20 °C.

A relatively new use for elastomeric urethane coatings is on foamed plastics used as insulation in aircraft cabins and air-conditioning equipment. Flotation devices such as life preservers, buoys and dock floats are examples of other uses.

4. Chloroprene

Chloroprene coatings provide above-average resistance to oils, weathering, abrasion, heat and alkalis. Introduced in the early 1930s, chloroprene finds use in coal chutes, pump and blower blades, ship propeller shafts, storage tanks, heater and brake hose, conveyors, ductwork and pipelines. Although it withstands exposure to sulfuric and phosphoric acids, it is decomposed by nitric and chromic acids, hydrogen peroxide and sulfur trioxide. Aromatic and halogenated hydrocarbons, ketones, esters, carbon disulfide, turpentine and aromatic alcohols cause the material to swell and soften. Chloroprene retains its desirable properties at temperatures ranging from −20 °C to 65 °C. It stands up well to long exposure to sunlight and weathering.

5. *Chlorosulfonated Polyethylene*

Chlorosulfonated polyethylene coatings have excellent resistance to nitric and chromic acids, hydrogen peroxide, ozone and hypochlorite. They can be used at service temperatures up to about 100 °C. Their uses are similar to those of chloroprene, including coatings for textiles, leather and rubber, for such items as heater and brake hose conveyors, industrial gloves and belts.

6. *Nitriles*

Nitrile elastomeric coatings possess high resistance to heat, sunlight, oils, greases, waxes and most solvents. They find their widest use as coatings and impregnants for paper, fabric and leather products. Vulcanization is achieved with sulfur and sulfur-bearing materials.

7. *Polysulfides*

High solvent resistance is the outstanding feature of polysulfide elastomeric coatings. Thus, they find wide use as solvent-resistant linings for chemical

Table 1
Properties of elastomeric coatings

Property	Chloroprene	Polysulfides	Fluoroelastomers	Nitriles	Silicones	Chlorosulfonated polyethylene
Mechanical properties						
hardness[a]	30A–95A	20A–80A	55A–95A	30A–100A	20A–90A	40A–95A
tensile strength (MPa)	3.4–24.0	3.4–10.3	13.7	6.8–24.0	10.3	3.4–24.0
modulus (100% elongation) (MPa)	0.69–2.1		1.3–13.7	3.3		0.69–2.1
elongation (%)	100–800	210–450	150–450	400–600	100–800	100–700
flex cracking resistance	very good		good		fair–excellent	very good
tear resistance	good	poor–fair	fair–very good	good	fair	fair
abrasion resistance	excellent	poor–fair	good	excellent	poor	excellent
impact resistance	excellent	poor–fair	good	good	poor–good	very good
Thermal properties						
service temperatures (K)						
min for continuous use	222	227	233	222	156	227
max for continuous use	380	>394	>560	422	588	408
Environmental resistance						
ozone	very good	excellent	outstanding	poor	excellent	outstanding
oxidation	very good	excellent	outstanding	fair–good	excellent	outstanding
weathering	very good	excellent	excellent	good	excellent	
water	good	good	good	excellent	excellent	good
radiation	good	fair	fair–good	fair–good	fair–good	very good
alkalis	excellent	good	fair–good	fair–good	poor–fair	excellent
aliphatic hydrocarbons	good	excellent	excellent	excellent	poor–good	good
aromatic hydrocarbons	fair	excellent	excellent	good	poor–good	fair
halogenated hydrocarbons	poor	fair–good	good	poor	fair	poor
alcohol	good	very good	fair–good	very good	fair	good
animal, vegetable oils	good	excellent	excellent	excellent	good	good
acids						
dilute	excellent	good	fair–excellent	excellent	very good	excellent
concentrated	fair–good	poor	fair–excellent	excellent	good	very good
synthetic lubricants (diester)	poor	good	good–excellent	fair–good	good	poor
hydraulic fluids						
silicates	poor–good	poor–good	excellent	fair	poor	poor
phosphates	poor	poor–fair	excellent	poor	good	poor
permeability to gases	low–medium	very low	low	very low	high	low
limiting oxygen index	38–45		50–100	17–20	20–30	30–36

a Tested using a Shore D durometer

tanks and equipment and as a solvent-resistant coatings for fabrics, wood and leather. Polysulfide coatings also have outstanding resistance to sunlight, ozone, oxidation and weathering. Their service temperature range is −55 °C to 120 °C.

A water-dispersed polysulfide rubber is used as a liner for concrete fuel-storage tanks, its advantage being that it forms a film without cracking. Thick films up to about 2.5 mm are possible with this coating.

8. Silicones

Silicone elastomeric coatings can be used over a wide temperature range (−70 °C to 260 °C) and are unaffected in continuous service at 150 °C. They also are very resistant to ozone and hot oils. However, they have poor resistance to hydrocarbon solvents, acids, strong alkalis and steam.

Silicones have superior electrical properties. The coatings are supplied as solutions, pastes and water dispersions. Application methods include spreading, dip coating, spraying and brushing. Typical uses include coatings for fabrics, paper, ceramics, metals and wire.

See also: Elastomers: An Overview; Elastomers: Applications Overview; Marine Coatings

Bibliography

Federation of Societies for Coatings Technology 1978 *Paint Coatings Dictionary*. Federation of Societies for Coatings Technology, Philadelphia, PA

Martens C R 1981 *Waterborne Coatings: Emulsion and Water-Soluble Paints*. Van Nostrand Reinhold, New York

Technomic 1979 *Urethanes in Elastomers and Coatings*, Vol. 2. Technomic, Westport, CT

US Department of Commerce, National Technical Information Service 1981 *Elastomer and Plastic Protective Coatings*. National Technical Information Service, Springfield, VA

Walker B M (ed.) 1979 *Handbook of Thermoplastic Elastomers*. Van Nostrand Reinhold, New York

J. A. Mock
[Society of Plastics Engineers, Brookfield Center, Connecticut, USA]

Elastomeric Fibers

An elastomeric fiber must be capable of being stretched to several times its original length and of returning rapidly to its original state after the removal of the stretching force. This behavior is related to the molecular structure of the fiber-forming polymer. The elastomeric fiber is composed of long, coiled, flexible chains which are uncoiled under stress if the ambient temperature is higher than the polymer glass-transition temperature. Chemical or physical interconnections act as crosslinkages and create a network structure. This structure restricts the imposable strain and provides physical stability for the fiber. The definition excludes texturized fibers of normally hard, inelastic polymers gaining their elasticity only from the physical crimps.

Elastomeric properties can be created by two main principles: first, by chemically crosslinking flexible polymer molecules as encountered in the sulfur vulcanization of rubber and, second, by building a two-phase, multiblock or segmented polymer. Nonpolar, soft chain constituents are bonded to polar, hard moieties capable of strong interaction, which means that the linear polymer molecules are connected by physically crosslinking domains. This structure is obtained in segmented polyurethanes, block copolyolefins and special polycondensates.

Only a few of all the known, inherently elastic materials can be successfully converted into elastomeric fibers and, in this respect, the nature of the polymer and the manufacturing process are both important factors.

The very first polymer used as an elastomer was natural rubber and the development of articles with the unique rubber elasticity only became possible through Goodyear's discovery of vulcanization in 1839. For nearly a hundred years, rubber was the only material available for elastomeric fibers. Rubber fibers have fairly good elastic properties but some disadvantages in textile applications, such as poor dyeability, abrasion resistance and chemical stability. Furthermore, these threads are relatively coarse with respect to other fibers.

Segmented polyurethanes, prepared according to Bayer's diisocyanate polyaddition process, have been found to be very useful elastomers which have successfully entered the field of elastomeric fibers after intensive research and development work on fiber synthesis and manufacture. Among the great variety of polyurethanes, systems especially tailored for use as fibers are obtained by combining appropriate components such as long-chain polyols with short-chain urethane and urea moieties via isocyanate reactions. The polymers are converted into elastomer fibers by conventional spinning techniques. Important improvements over rubber are achieved notably in tensile strength and in textile processibility. The fibers' generic name is *spandex*, provided at least 85 wt% consists of segmented polyurethane. The first commercial spandex fiber was introduced to the market in 1956. Since then, the spandex market has grown continuously and reached a worldwide production of about 14 000 t in 1979.

Since the mid-1960s, many attempts have been made to use various, chemically different elastomeric systems in fiber manufacture. Fiber-forming

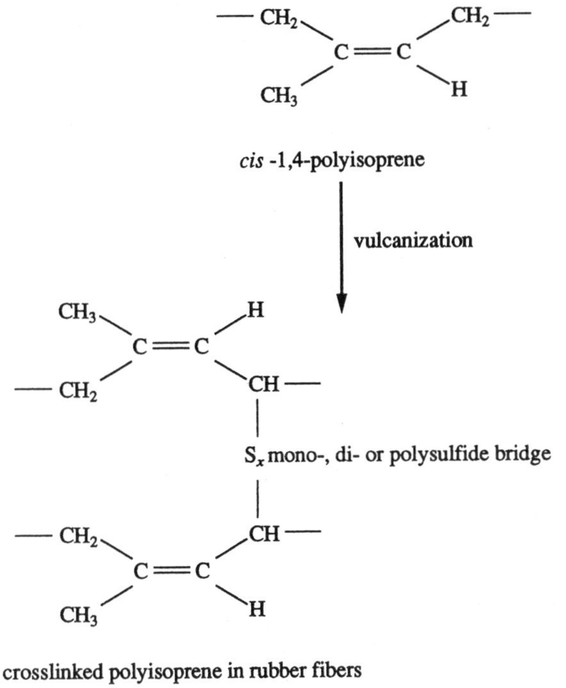

Figure 1
Rubber fiber chemistry

elastomeric polyacrylates with the generic name *anidex* have been marketed for some time. Anidex fibers are manufactured from block copolymer emulsions of butyl acrylate as a soft segment and vinylidene chloride as a hard segment. After spinning, the fibers are crosslinked. Their elastic and textile properties are reported to be acceptable compared with rubber and spandex.

Other systems, such as segmented polyamides, polyether esters, polycarbonates and graft copolymers of cellulose and alkyl acrylates, have been converted into elastomeric fibers but have not gained any commercial importance.

1. Rubber Fibers

Still the most important raw material for rubber fibers is natural rubber. It is a *cis*-polyisoprene latex, obtained from rubber trees, notably *Hevea brasiliensis*, in tropical regions. The latex has a solids content of 30–35 wt%. For making shaped rubber articles, the latex is coagulated and compounded with the agents necessary for vulcanization, usually sulfur or sulfur-releasing derivatives. In vulcanization, the allylic carbon atoms adjacent to olefinic double bonds of different polyisoprene chains react with the sulfur to form S_x bridges and crosslink the individual chains. The chemistry of rubber fiber production is shown in Fig. 1.

Synthetic rubbers have been used to a limited extent to produce elastomeric fibers. The best-known product is a fiber made of a sulfur-vulcanized acrylonitrile butadiene rubber.

1.1 Rubber Fiber Chemistry and Manufacture

There are two completely different methods of producing rubber thread. The older one is the cut-thread process, a technique otherwise not used in man-made fiber technology. Raw rubber is plasticized in heavy mixing equipment (e.g., roll mills), compounded with sulfur, accelerators, antioxidants, pigments and lubricants, and calendered to a thin sheet of carefully controlled dimensions. The sheet is heat vulcanized and cut into narrow strips. These are desulfurized to avoid further stiffening by post-curing during use. The cut threads are dusted with talc to prevent tacking. To obtain long threads, the rubber sheet may be wrapped on a drum and cut in a spiral. Cut threads have square or rectangular cross sections which cause problems during subsequent textile processing. Figure 2a shows a microscopic view of a cut-rubber thread.

The second, more recently developed method of producing rubber thread is the latex thread process. In some respects, this resembles the wet-spinning techniques of conventional fiber manufacture. Only top quality natural rubber latex can be used. The latex is concentrated to about 60 wt% solids, precured by heat treatment, mixed with the vulcanization agents and additives, and degassed *in vacuo* to remove air bubbles and to avoid filament rupture in subsequent spinning. Extrusion is performed through glass spinnerets into an acidic coagulation bath, usually aqueous acetic acid. The emerging latex strands coagulate to filaments which, while being mechanically supported, are washed, dried, vulcanized and finished before final take-up. Precise control of temperature, latex viscosity, pH, drying and vulcanization conditions and process rate is essential to obtain good-quality rubber threads. Production rates are in the range 50–100 m min^{-1}. Latex rubber threads are round and allow lower linear densities than cut threads.

Figure 2
Microscopic view of (a) cut and (b) latex rubber threads

Table 1
Mechanical properties of rubber fibers

	Cut thread		Latex thread	
Modulus	low	high	low	high
Density (kg m^{-3})	960	980	1010	1010
Linear density (counts)	5	85[a]	30	160
(tex)[b]	25 800	88	563	20
Elongation at break (%)	900	600	800	600
Tensile properties (cN tex^{-1})				
breaking strength	1.6	2.1	3.6	3.7
modulus at				
200% elongation	0.06	0.09	0.14	0.16
400% elongation	0.11	0.17	0.33	0.40

a Square cross section b Assuming a density of 1000 kg m^{-3}

The vulcanization of the natural rubber is one of the most important property-determining steps in fiber manufacture. Normally, 1–5 wt% sulfur is used. Higher concentrations lead to high degrees of crosslinking and, consequently, reduce the elongation at break and increase the modulus of the fibers. Obviously, rubber-thread production benefits from various improvements in rubber vulcanization. Less sulfur, more efficient compounding and the use of zinc oxide reduce the amount of unvulcanized sulfur and give fibers with better aging and nonstaining properties. The use of thiuram derivatives instead of sulfur leads to heat-resistant fibers. Efficient antioxidants inhibit discoloration and degradation of the rubber threads in end products. Figure 2b shows a microscopic view of a latex rubber thread.

1.2 Rubber Fiber Properties

The mechanical and elastic properties of rubber threads vary considerably according to different manufacturing conditions and product demands. Rubber threads are commercially available in various types which can be roughly divided into low- and high-modulus fibers. The linear density of rubber fibers is expressed in the count system. The number of fibers placed side by side to form a 1 in (25.4 mm) ribbon is quoted as the linear density. For example, if 100 fibers placed side by side make up a 1 in width, then the fibers are 100s count. The count values can be transferred into the tex system used to characterize the linear densities of textile fibers. The number in tex is the weight of 1000 m of the fiber. Assuming a density of 1000 kg m^{-3} for rubber, a 100s count circular rubber filament has a weight of 0.051 kg per 1000 m and thus a linear density of 51 tex.

A generalized set of mechanical data is given in Table 1. As may be seen from the curve labelled R in Fig. 3, rubber has a rather flat stress–strain curve. Consequently, high retractive forces can only be achieved with comparatively thick fibers. This is sometimes a disadvantage in the construction of fine, lightweight fabrics. Cut-rubber threads show somewhat lower breaking strength and modulus compared with correspondingly compounded latex threads, probably due to higher orientation of the molecular chains by spin-draw effects during latex extrusion.

Rubber fibers are chemically stable in dilute inorganic, nonoxidizing acids and concentrated alkalis, no hydrolytic degradation taking place. Organic solvents, oils and dry-cleaning fluids may cause swelling and often the destruction of rubber fibers. Strong oxidizing agents, especially ozone, chlorine and peracids, damage the rubber severely. The effect of normal oxidizing agents, such as hydrogen peroxide in textile bleaching, can be controlled by using suitable antioxidant systems. Dry heat above 100 °C and prolonged heat exposure destroy rubber. Light resistance is a further problem which must be overcome through the use of suitable light stabilizers. Rubber is also attacked by human perspiration, a severe effect in foundation garments and swimming costumes. This attack is even more pronounced in the presence of light and oil-containing suntan or cosmetic lotions.

2. *Spandex Fibers*

The elastomeric properties of spandex fibers depend on the chain architecture of the segmented polyurethane molecules. Flexible amorphous polyol moieties, which are completely molten at use temperatures due to their low glass-transition temperatures, stretch easily and regain their coil form after stress relief. These soft segments are connected via diisocyanate moieties with rigid, high-melting urea groups formed between isocyanate and amino functions. In these hard segments, strong intermolecular interactions are possible, notably as hydrogen bonds and as conformational symmetry. The elastomeric system can be described as a physically crosslinked network between the usually linear polyurethane molecules with a domain structure of soft and hard

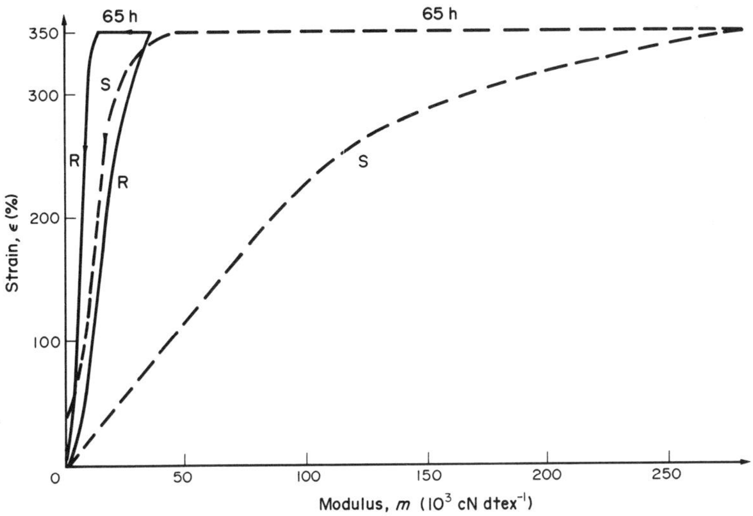

Figure 3
Typical example of the viscoelastic behavior (relaxation hysteresis) of a spandex filament (S) and a rubber thread (R): m is the modulus for go and return, and ε is the elongation; the relaxation time is 65 h and the extension velocity 400% min^{-1}

segments. Linear polyurethanes can be shaped from solutions and therefore most commercial spandex types are linear. It is possible, however, to produce spandex fibers with additional chemical crosslinkages in their molecular structure. A list of spandex manufacturers and products is given in Table 2.

2.1 Spandex Fiber Chemistry and Manufacture

Three classes of raw materials are necessary in spandex polyurethane synthesis. The first consists of soft, segment-building aliphatic polyols with a molecular weight in the range of 800–3000 (number average), preferably 2000, and a hydroxyl functionality of at least (and preferably) two. Suitable hydroxy-terminated compounds are either polyethers, namely poly(tetramethylene ether) diol, made by cationic ring-opening polymerization of tetrahydrofuran, or polyesters, for example, polycaprolactone and preferably cocondensates of adipic acid with a slight excess of diol mixtures, for

Table 2
Commercial spandex fibers

Trade name	Manufacturer	Composition	Spinning technique
Cleerspan (Glospan)	Globe Manufacturing, USA	polyester/TDI/diamine	reactive
Dorlastan	Bayer AG, Germany	polyester/MDI/diamine	dry
Elaston	Chemitex Cellviskoza, Poland		
Espa	Toyobo, Japan	polyether/MDI/diamine	dry
Fujibo Spandex	Fuji Spinning, Japan	polyester/polyether/ MDI/diamine	wet
Lycra	Du Pont, USA, Brazil Netherlands, Canada, UK	polyether/MDI/diamine polyester/MDI/diamine	dry dry
Lynel	Fillatice, Italy	polyester/MDI/diamine	reactive
Mobilon	Nisshin Spinning, Japan		
Opelon	Toyo Products, Japan	polyether/MDI/diamine or hydrazine	dry
Roica	Asahi Chemical Industries, Japan	polyester/polyether/ MDI/diamine	dry
Vispan	Elastofibre, Italy	polyester urethane	reactive

example ethylene glycol, propanediol, butanediol or hexanediol.

The second class of compounds consists of diisocyanates. Among the aromatic diisocyanates methylene-bis-(4,4′-diphenylenediisocyanate) (MDI) is required for high-performance spandex. Isomeric toluene diisocyanates (TDI) play only a modest role, because TDI and, moreover, aliphatic diisocyanates give spandex fibers with thermal and mechanical stability levels which are often too low for further textile processing.

The third class of compounds, chain-extension agents, are commonly NCO-reactive, bifunctional, low-molecular-weight species bearing OH or NH_2 groups. Spandex fibers having adequate properties can only be obtained with NH_2 derivatives. Although hydrazine, dihydrazides and semicarbazides were formerly used as chain extenders, the modern high-performance spandex requires aliphatic diamines, mostly ethylene diamine. It creates sterically well-ordered, crystalline, high-melting urea groups with the diisocyanate-tipped polyol.

The segmented polyurethane is generally prepared by first reacting one mole of polyol with an excess (usually two moles) of diisocyanate to yield an NCO-terminated prepolymer (see Fig. 4). Prepolymerization is possible either continuously or in batches; solvents like toluene, chlorobenzene or acid amides can be used optionally. Reaction temperatures vary from 40 °C for solvent processes to 90 °C for melt processes. The prepolymer is dissolved in either dimethylformamide or dimethylacetamide, prior to chain extension, which is carried out by reacting stoichiometric amounts of diamine with the NCO prepolymer. Because of the extremely fast rates of urea formation by the NCO–NH_2 reaction, chain extension calls for very efficient mixing and, possibly, cooling to prevent gel formation.

As an alternative, solid carbon dioxide can be added to the amine solution to yield a suspension of carbamate salt which has a considerably slower reaction rate during chain extension.

For convenient spinning and filament properties, the molecular weights of the segmented polyurethanes are adjusted with a few mole percent of a monoamine (e.g., diethylamine) as a chain-stopping agent. The molecular weights range from 50 000 to 100 000 and lead to spinning dopes with solids contents of 18–30 wt% and a bulk viscosity of 50–150 Pa s. Various additives, such as rutile-type titanium dioxide as a delustring agent, antioxidants, stabilizers (e.g., sterically hindered phenols), lubricants (e.g., polysiloxanes) antitacking agents (e.g.,

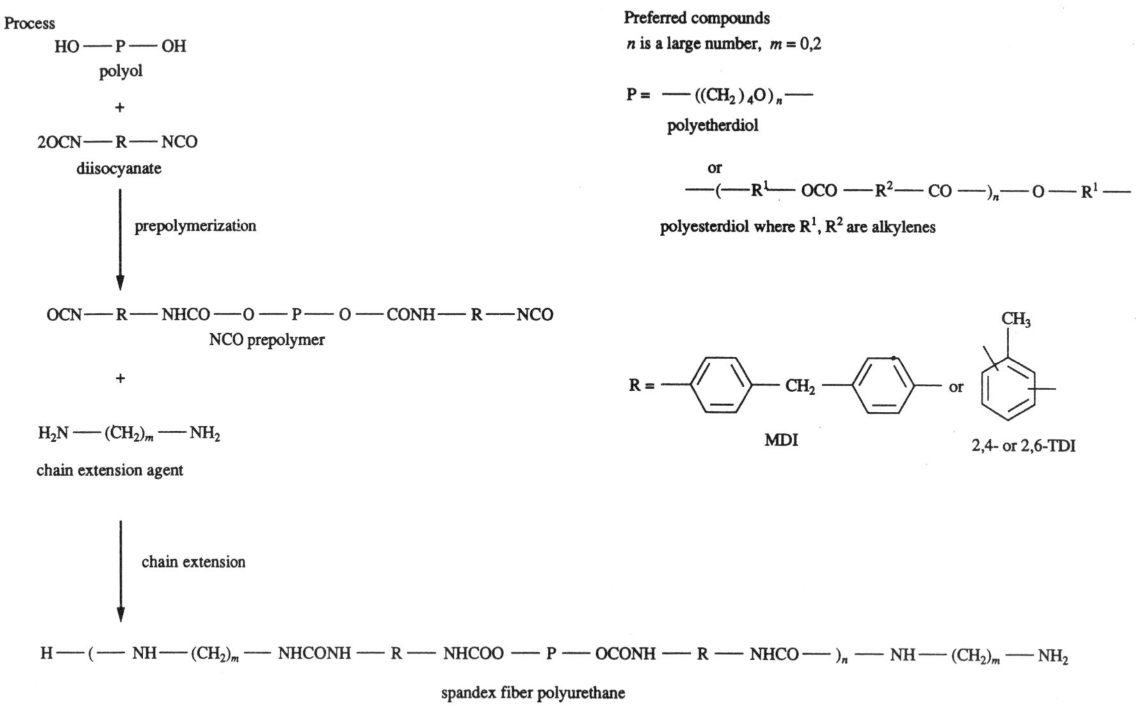

Figure 4
Spandex fiber chemistry

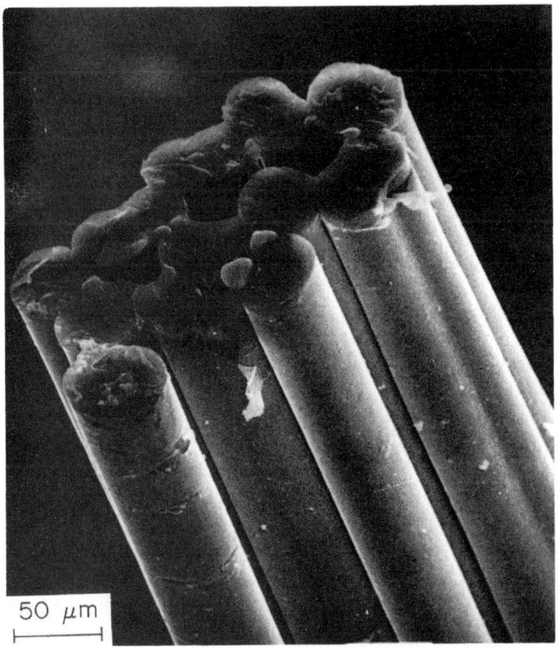

Figure 5
Microscopic view of a dry-spun spandex filament yarn

metal salts of fatty acids) and dyestuff-receptive compounds are incorporated into the spinning dope prior to spinning.

Dye receptors usually bear tertiary amino groups (e.g., *t*-alkylaminoalkylacrylate polymers). Alternatively, such functions may be incorporated in the polyurethane by using NCO-reactive, bifunctional tertiary amino derivatives (e.g., *N*-*t*-alkyl dialkanolamines) in polymer synthesis.

The filament formation is performed by dry-, wet- or chemical spinning processes. About 80% of all spandex fibers are dry spun. In this process, the spinning dope is extruded by a metering gear pump through metal spinnerets in a heated duct, where the solvent is removed by a flow of hot gas. The single filaments coalesce to form a fused multifilament by applying a mechanical or aerodynamic false twist at the end of the duct. Spinning rates range between 200 m min^{-1} and 800 m min^{-1}. A microscopic view of a dry spun spandex fiber is shown in Fig. 5.

In wet spinning, the solvent is removed by coagulation of the extruded dope to filaments in a nonsolvent bath, usually water. The production rates (5–50 m min^{-1}) are smaller than dry-spinning rates; productivity, however, can be equally high by using more spinnerets in the production line.

Some spandex fibers are produced by chemical or reactive spinning methods. Starting with a prepolymer with branched molecules (e.g., by using a trifunctional polyol), a covalently crosslinked, insoluble, segmented polyurethane is produced with diamines or triamines in the reaction spinning bath after extrusion of the prepolymer simultaneously with filament formation. The fiber is cured by heat treatment in water or amine solution.

The spinning process is followed by finishing treatments to optimize certain fiber properties. A brief heat treatment at or near 150 °C increases the tensile strength of the prestretched spandex fibers as a result of the increasing orientation of the hard segments in the fiber axis. Although spandex cannot be melted without degradation, the degree of thermoplasticity inherent in polyurethanes makes it possible to produce fibers with low linear densities from coarser ones by a heat stretching and relaxation process. As spandex fibers are tacky, they are finished with talc or, more recently, with polysiloxanes and/or white oils.

2.2 Spandex Fiber Properties

The mechanical properties of commercial spandex fibers are summarized in Table 3. A characteristic stress–strain diagram of spandex is shown in Fig. 3 in the curve labelled S. Compared with rubber, spandex fibers exhibit higher tensile strength at equal elongation. Their breaking strength, if based on the linear density at break, is similar to the corresponding data of partly crystalline fibers such as polyamides or polyesters. The permanent sets of spandex fibers are higher than those of rubber. Repeated stress–strain cycles or prolonged extensions under load lower the tensile strength and the stress relaxation, respectively. It should be noted that the amount and type of hard and soft segments and the manufacturing conditions have a decisive effect on the mechanical properties of spandex fibers. Adjustments to various product requirements must be made in each case.

Among the physical properties, the thermal behavior of spandex is most important for textile processing. The fibers are rigid below the polyol glass-transition temperature, (i.e., −40 °C to −60 °C for polyether and −20 °C to −40 °C for polyester

Table 3
Mechanical properties of spandex fibers

Linear density (tex)	2–500
Elongation at break (%)	
normal types	420–570
low-modulus types	600–800
Breaking strength (cN tex^{-1})	
based on initial linear density	4.5–12
based on linear density at break	30–70
Modulus in stress–strain cycles of 300% maximum elongation (cN tex^{-1})	
at 150% elongation	0.4^a–1.0^b
at 300% elongation	1.0–2.5
Permanent set (%)	10–30

a Low-modulus types b Normal types

spandex). At ambient temperatures, the completely mobilized, amorphous polyols give the fiber entropic elasticity. The high-temperature behavior is mainly governed by the hard segments and the higher amounts and increased symmetry of these groups (e.g., with the MDI/ethylene diamine system) increase the heat distortion temperature. Fabrics containing spandex fiber can be heat treated to achieve thermosetting of given dimensions. Water acts as a plasticizer for spandex, decreasing the stress and increasing the elongation compared with the initial values imposed in the dry state. Prolonged exposure to temperatures above 100 °C in air weakens the spandex fibers.

The chemical stability of spandex fibers is generally acceptable. Polyester spandex is hydrolytically unstable and degraded by mineral acids and alkalis. Normal laundering or swimming pool water has little influence on either type. Commercial spandex fibers are stabilized against thermal and/or photooxidative aging by means of suitable additives. Stabilization is necessary, in particular where sensitized ultraviolet degradation is to be expected. Ultraviolet sensitizers are certain unsaturated compounds in suntan lotions and human skin perspirants. Stabilizers should not be extractable by water or dry-cleaning fluids. Industrial waste gases, especially nitrogen oxides, cause a more or less pronounced yellowing of spandex fibers, which is a serious drawback in white foundation garments. This effect is suppressed by special additives (e.g., tertiary alkylamino derivatives or organophosphorous compounds). Spandex fibers withstand normal textile bleaching and dyeing procedures without any appreciable deterioration.

In general, spandex fibers are to be regarded as technical auxiliary fibers used to impart elasticity to textile goods. This means that their properties must be considered within the limits and possibilities of the other fiber components in elastic textile fabrics.

3. Applications of Elastomeric Fibers

Elastomeric yarns are used to impart high extensibility and an efficient retractive force to textile fabrics. To achieve this in fabrics constructed with conventional yarns, elastic threads must be incorporated in an elongated state. Constant stress during processing, which is necessary to obtain fabrics having reproducible elastic properties, is achieved either by suitably designed yarn-braking devices or by defined prestraining of the elastic yarn.

Spandex fibers are converted to fabrics either in bare form or covered with other textile fibers (e.g., as covered yarn, core-spun yarn or elastic twisted yarn). Rubber threads are mostly used as covered yarns and bare only if the high rubber friction is especially needed for rendering adhering properties (e.g., in foundation garments).

3.1 Elastic Yarns

Bare elastic yarns show typically uniform elastic properties, poor friction behavior and a nontextile surface. Sheathing the elastic core with other textile fibers greatly improves its attractiveness, as well as its processability and abrasion resistance. The extensibility of covered yarns can be adapted to product requirements within certain limits.

The most important elastomeric composite yarns are the covered yarns. They are produced by winding one or two textile yarns in opposite directions around the highly strained elastomeric core yarn (see Fig. 6). Core-spun yarns are made on specialized staple fiber spinning machines by twisting the staple fibers around the elastic core. Core-spun yarns exhibit the softest textile hand of all elastomeric yarns.

Spandex composite yarns can be heat set at about 80 °C which causes them to temporarily lose their elasticity. This gives them a processing performance similar to other, normal textile fibers. The fabrics regain the desired elasticity during dyeing in boiling water and without any tension.

3.2 Fabrics with Elastomeric Yarns

The most important applications of elastomeric fabrics are summarized in Table 4. Knitted, warp-knitted and braided fabrics are preferred depending on the requirements made on the article with respect to extensibility, elastic recovery and textile characteristics. Spandex fibers dominate in elegant, lightweight articles with high retractive forces. Rubber yarns give fabrics for which high extensibility and minimal permanent set are specified.

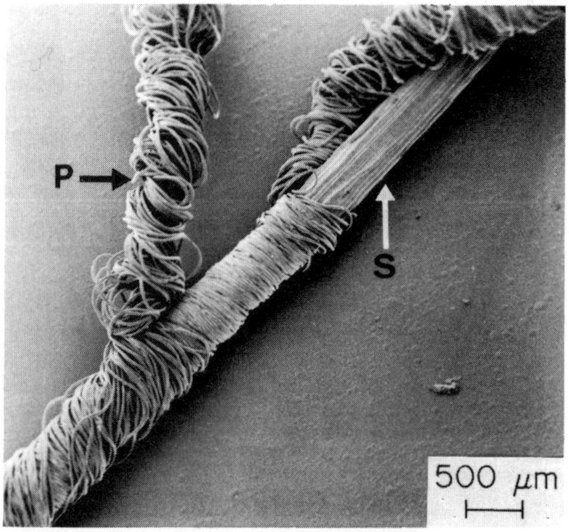

Figure 6
Crosswise covered elastic yarn: S, spandex; P, polyamide filament

Table 4
Applications and fabric manufacturing techniques for elastic textile articles

Article	Fabric manufacture	Processing	Type[a]	Form	Linear densities (tex)
Underwear (except foundation garments)	circular knitting	overhead	S	bare or covered	2.2–8
	warp knitting	rolling[b]	S	bare	2.2–8
Foundation garments	warp knitting	rolling	S	bare	4.5–80
	raschel knitting	rolling	R	bare	80–300
Bathing fashions	warp knitting	rolling	S	bare	2.2–8
	circular knitting	overhead[c]	S	bare or covered	4.5–16
Stocking waistbands	circular knitting	overhead	S	bare	13.5–40
			R	covered	40–80
Support hosiery, socks	circular knitting	overhead	S	corespun, covered	2.2–48
			R	corespun, covered	20–100
Fabrics for trousers and anoraks	weaving	overhead	S	elastic twists, corespun	4.5–16
Ribbons, belts	ribbon weaving	overhead/	S	bare/covered	31–500
	raschel knitting	rolling	R	bare/covered	20–25 800
Elastic webbing	ribbon weaving, braiding, crochetting	overhead/ rolling	R	bare, covered or core knitted	20–500
Nets	special raschel knitting	rolling	R	covered	800–1400

a S denotes spandex, R denotes rubber b Driven yarn spool, predeterminable in-run strain c Fixed yarn spool, predeterminable in-run stress

In most of the knitwear fabrics, the elastomeric yarn is located on the inner side of the knit structure. Accordingly, the article surface shows the characteristics of the inelastic fiber component, (e.g., polyamide in power-net fabrics). Bare elastomeric yarns can be used in this field.

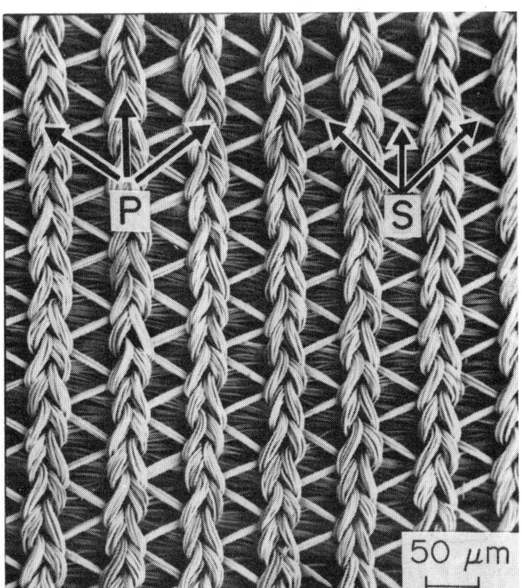

Figure 7
Warp-knitted elastic fabric with spandex (Dorlastan): 4.5 tex; S, spandex; P, polyamide filament

Covered elastomeric yarns dominate in the manufacture of support hosiery because they wear better and are more easily processed. Fine denier fabrics for cosmetic, form-persuasive support pantyhose are produced mainly with covered spandex whereas coarse denier articles for medical purposes with higher stretch and little permanent set are made, at least partially, with covered rubber threads.

Spandex filaments are used, usually bare, in warp knitting and raschel knitting to give fabrics for foundation garments and swimwear. Tight, nontransparent knitted fabrics are obtained in which the inside spandex filament is almost completely hidden. Figure 7 shows the spandex side of a typical warp-knitted fabric for bathing suits. The spandex is only visible because the fabric is stretched considerably.

Raschel-knitted elastic nets (e.g., for packaging applications) usually consist of covered rubber threads and show high extensibilities of about 300–400%.

Classical woven fabrics for suiting have recently been elasticized by incorporating fine denier covered or core-spun spandex yarns. Fabrics with elasticity in the weft as well as in the warp direction can be produced. In weft elasticizing, neither stress nor strain can be held definitely constant and, therefore, heat-set combination yarns must be used exclusively. Elasticity is regained during subsequent fabric dyeing.

Elastic bandages, ribbons or belts, manufactured on special looms, show warpwise elasticity. Covered spandex or rubber yarns (only seldom bare yarns) meet these purposes. Webbings for elastic articles such as foundation garments and bathing fashions usually contain spandex; such items as elastic bandages and belts for trousers and suitcases are made with rubber threads, often by braiding techniques.

Elastomeric fibers and yarns provide textile fabrics and articles with predetermined extensibilities between 20% and 350% and exactly defined elastic, retractive, form-fitting and support forces. In the textile industry where, in addition, aesthetic viewpoints play an important role, spandex yarns are in widespread use and have replaced the formerly used rubber yarns.

See also: Elastomers: An Overview; Hard-Fiber Elastomers

Bibliography

Allport D C, Mohajer A H 1973 Spandex fibers. In: Allport D C, Janes W H (eds.) 1973 *Block Copolymers*. Applied Science, London, pp. 240–64

Cuthbertson G R, Kelly R J, Logan L R, Porter D S, Brass P D 1968 Elastomeric (spandex) fibers. In: Mark H F, Atlas S M, Cernia E (eds.) 1968 *Man-Made Fibers, Science and Technology*, Vol. 3. Interscience, New York, pp. 401–24

McDonald M 1970 *Spandex Manufacture, Chemical Process Reviews No. 48*. Noyes Data Corporation, Park Ridge, NJ

Meredith R 1971 *Elastomeric Fibers*. Merrow, Watford, UK

Oertel H 1981 Polyurethanfasern und sonstige Elastomerfasern. In: von Falkai B (ed.) 1981 *Synthese fasern Grundlagen, Technologie, Verarbeitung und Anwendung*. Verlag Chemie, Weinheim, Germany, pp. 179–90

Saunders J H, Frisch K C 1964 *Polyurethanes, Chemistry and Technology*, High Polymers Series Vol. 16–II. Interscience, New York, pp. 694–733

K. H. Wolf and H. Gall
[Bayer A G, Leverkusen, Germany]

Elastomeric Foams

Elastomeric foams are flexible cellular materials of low density with relatively small cells distributed uniformly throughout; the cells may be either closed pore, open pore or a combination of both. A flexible material is defined in this context as one that will not rupture within 60 s when a specimen measuring 200 × 25 × 25 mm is bent around a 25 mm diameter mandrel at a uniform rate to produce one lap in 5 s in the form of a helix at a temperature of 18–19 °C. Products that fit this definition include elastomeric foams made from rubber, rubber latex, urethane, polyolefins, poly(vinyl chloride) and silicone.

1. Expanded Rubber

Expanded rubber is made by one of two processes—a continuous low-pressure method or a high-pressure method—using a nitrogen-releasing blowing agent such as azodicarbonamide, dinitrosopentamethylenetetramine and benzenesulfonyl hydrazide.

Table 1
Expanded rubber physical properties

Property	Natural	Styrene–butadiene	Polychloroprene
Cure (°C min^{-1})	163/5	153/12.5	149/15
Cell structure	open	closed	closed
Density ($kg\ m^{-3}$)	497	400	272
Compression deflection at 25% (MPa)	0.086	0.138	0.069

The continuous low-pressure process involves extruding a fast-curing rubber compound through a die into either a hot-air oven, a liquid curing medium, a fluidized bed or a microwave oven where the product expands and is quickly vulcanized to the desired density.

The high-pressure process uses a two-step procedure in which a mold cavity is completely filled with a rubber compound and heated under pressure to partially vulcanize the stock and activate the blowing agent. The mold is then opened, allowing the rubber to expand and vulcanization to be completed. Natural rubber, and polymers of styrene–butadiene, acrylonitrile–butadiene and polychloroprene are used in making these products (see Table 1).

2. Latex Foam Rubber

Latex foam rubber (see Table 2), primarily used for cushioning applications, is made by mixing air into a suitably compounded and concentrated (62–68% total solids) rubber latex. In one process, a delayed gelling agent, such as a water dispersion of Na_2SiF_6, is added and the fluid froth is poured into a mold or onto a surface such as cloth or carpet. Following gelation, which takes only a few minutes, the gelled foam is vulcanized with heat and dried.

In a modification of this method, the partially foamed latex is put into a closed mold and a vacuum is applied to expand the foam, thus causing it to fill the mold. The mass is frozen and then CO_2 gas is passed through the frozen mass to gel the foam. Subsequent heat vulcanization and drying operations complete the process.

No-gel foam used for carpet and fabric backing was first used in 1927. The development of super-high-solids latices (68–70%) which were able to accept large quantities of pigment loading and the use of improved synthetic froth stabilizers has

Table 2
Physical properties of latex foams

	Gelling agent			
	Ammonium acetate	Sodium silicofluoride	No-gel	Sodium silicofluoride
Styrene–butadiene/natural rubber	80/20	100/0	100/0	100/0
Density (kg m^{-3})	282	281	279	64
Filler type	$CaCO_3$	$CaCO_3$	$CaCO_3$	clay
Filler (phr[a])	125	125	125	45
25% ILD[b] compression resistance (MPa)	0.0317	0.0558	0.0655	0.0030
Tensile strength (MPa)	0.119	0.128	0.115	0.0483
Elongation (%)	170	208	213	200
Compression set (%)	8.1	10.1	14.2	7.7
Delamination strength (kg m^{-1})	21	32	46	
Application	undercarpet	undercarpet	undercarpet	automobile cushioning

a phr = parts per hundred rubber b ILD = indentation load deflection

aroused renewed interest. The major advantage of this type of foam is ease of processing and, because of high pigment loading, lower compound cost on a weight basis.

3. *Flexible Urethane Foam*

Flexible urethane foam represents the major volume cushioning foam used worldwide mainly because of its overall economy in manufacture and its relatively low density. The two major components used in making this foam are a hydroxyl-terminated resin, called a polyol, and a polyisocyanate such as toluene diisocynate. An example is

$$HOROH + O{=}C{=}NR'N{=}C{=}O \rightarrow \left[ORO_2CNHR'NH\text{–}C(O)\right]_n$$

with the polymer being characterized by the urethane linkage $-O_2CNH-$. A typical formulation for urethane foam would, in addition, require one or more activators, an organosilicone for foam stabilization, water, and a fluorocarbon for density control. These ingredients are accurately metered into a mixing head in two or more streams, giving rise to an almost instantaneous exothermic reaction. In commercial practice, this reacting liquid is quickly deposited onto a moving belt for the purpose of making slab- or block-type foam. It is poured into a mold cavity and an engineered lid is secured on top to shape the foam. When stripped from the mold, the cushion is pressed between rollers to rupture any closed cells that may adversely affect cushioning.

Polyethers have largely replaced polyesters as the polyols used because of their better hydrolytic stability and lower cost. Slab foam with a density of 28.8–32.0 kg m^{-3} is used for seating applications and foam with a density of 16.0 kg m^{-3} is used for pillow-back chairs. Mattresses are made from both slab and molded types. Urethane foam is easily molded around metal reinforcing members for efficient installation in automobiles and furniture.

Both cold foam, also called HR (high resilient) foam, and hot foam are manufactured. Hot foam production employs molds heated to 66 °C or higher and the formulation is adjusted accordingly. HR foam utilizes polyethers of higher molecular weight and is a little higher in density compared with hot foam (see Table 3).

Polyurethane foams in their large-volume applications need attention to be paid to flame retardancy. This is achieved by compounding with additives and/or by polymer design (see *Flammability of Elastomers*).

4. *Polyolefin Foams*

Low-density polyethylene (LDPE), high-density polyethylene (HDPE) and polypropylene (PP) are used to make polyolefin foams with essentially closed cells, good chemical and oil resistance, low water pickup and densities in the range 32–640 kg m^{-3} (see Table 4). Applications include buoyant products, athletic equipment, expansion joints, footwear and packaging. High-density (160–640 kg m^{-3}) foam applications include insulation, cables, tools and furniture.

Polyethylene foam may be made by mixing the polymer in an extruder with a blowing agent such as azodicarbonamide and a peroxide crosslinker, transferring this product as chips or sheet to a mold where it is heated under pressure which first triggers the crosslinker and then the blowing agent, causing the mass to expand and fill the mold.

Low-density foam is made by first mixing polyethylene with dichlorotetrafluoroethane blowing

Table 3
Physical properties of polyurethane foam

	Polyurethane type			
Property	Molded hot	Molded high resilient	Polyether slab	Polyester slab
Density ($kg\,m^{-3}$)	35.24	26.59	19.22	32.0
Tensile strength (MPa)	0.18	0.12	0.08	0.24
Elongation (%)	225	170	200	380
Block tear ($kg\,cm^{-1}$)	0.38	0.27	0.36	1.01
Flammability[a] ($cm\,min^{-1}$)	7.1	4.3		
Ball rebound (%)	35	60	50	
Breathability ($m^3\,s^{-1}$)	0.97	4.07	6.80	

a According to MUSS 302

agent in a cooled extruder. Expansion takes place on extrusion into the atmosphere.

5. Poly(Vinyl Chloride) Foams

Poly(vinyl chloride) foams consist of polymers and copolymers of vinyl chloride mixed with one or more plasticizers such as di-2-ethylhexylphthalate, tricresylphosphate, or a polymeric plasticizer. Improved aging is obtained by incorporating compounds of barium, cadmium and zinc which also contribute to good color retention and cost efficiency. The chemically blown vinyl foam is made by increasing the temperature to its hot-melt state prior to the triggering of the blowing agent, such as azodicarbonamide.

In the plastisol method, resin and plasticizer are mixed to a highly viscous mass with a low-temperature blowing agent such as *N,N'*-dinitrosoterephthalamide. Foams made by this process have fine cell structure, most of which is open pore (see Table 5).

Flexible foam has also been made from poly(vinyl

Table 4
Physical properties of polyolefin foam

	Ethafoam (Dow Chemical)	Foamace (Furakawa Elec)	Soft light (Gilman Bros)	Toray PEF (Toyo Rayon)	Minicel (Haveg)
Polymer	LDPE	LDPE	Ionomer	LDPE	PP
Blowing agent	volatile liquid	decomposing solid	volatile liquid	decomposing solid	decomposing solid
Crosslinking	none	chemical	ionic	radiation	chemical
Thickness (cm)	0.32	1.59	0.32	0.64	0.32
Density ($kg\,m^{-3}$)	43.2	43.2	43.2	35.2	80.0
Cell size (mm)	1.2	1.0	1.7	0.7	fine
Compressive strength at 25% (MPa)	0.029	0.054	0.027	0.047	0.828
Compression set (% of original deflection)	51.3	50.6	86	73	85
		10.3	30	5.3	
Open cell					
Tensile strength (MPa)	0.38–0.94	0.35–0.38	0.70–0.93	0.30–0.40	1.10
Elongation (%)	40–60	125	40–50	113–217	25
K-factor (10^{-4} cal $s^1\,cm^2\,°C\,cm^{-1}$)	0.966	1.000	0.828	0.828	0.932
H_2O absorption (24 h immersion, % volume change)	1.8	0.03	0.33	0.01	

Table 5
Physical properties of PVC foam

Property	Flexible	Rigid	Flexible
Cell structure	closed	closed	open
Density (kg m^{-3})	49.7–99.3	49.7–80	>160
Tensile strength (MPa)	0.345–1.03	>6.9	0.069–1.379

Table 6
Physical properties of silicone foam

	Cell structure	
Property	Open	Closed
Density (kg m^{-3})	160.2	400.5–544.7
Tensile strength (MPa)	0.310	0.552–0.690
Maximum service temperature (°C)	260	232

chloride) copolymer latex blended with an emulsion of plasticizers, pigment extender, froth stabilizer, and foamed by beating air into the liquid, pouring onto plate, drying and then fusing at 163°C.

6. Silicone Foam

Silicone foam (see Table 6) is made by blending a high-molecular-weight linear siloxane polymer with blowing agent, catalysts and fillers and heating in the presence of an organic peroxide such as dibenzoyl peroxide. Ethylene links between the siloxane chains are thus formed by abstraction of hydrogen from the methyl groups. Silicone rubber is masticated on a mill or internal mixing machine and the necessary compounding ingredients such as coloring agent(s), blowing agent and peroxide are added and uniformly dispersed.

Free-blown foam is made by blowing extruded or calendered sheets of this stock in a hot-air, vulcanizing unit or a steam autoclave. Press-blown stock employs a mold in which sheets of the mix are placed with sufficient space being allowed for expansion. The extent of blowing and vulcanization depends on thickness and temperature required for a given curing agent.

See also: Construction Materials, Cellular; Elastomers: An Overview; Flammability of Elastomers

Bibliography

American Society for Testing and Materials 1978 Standard specifications for flexible cellular materials: Latex foam, ASTM D1055-69. *Annual Book of ASTM Standards*, Pt. 38. ASTM, Philadelphia, PA, pp. 213–20

American Society for Testing and Materials 1978 Standard specification for flexible cellular materials: Sponge or expanded rubber, ASTM D1056–77. *Annual Book of ASTM Standards*, Pt. 38. ASTM, Philadelphia, PA, pp. 221–32

Bobear W J 1973 Silicone rubber. In: Morton M (ed.) 1973 *Rubber Technology*, 2nd edn. Van Nostrand Reinhold, New York, pp. 368–406

Buist J M 1986 Cellular polymers. In: Buist J M, Grayson S J, Woolley W D (eds.) 1986 *Fire and Cellular Polymers*. Elsevier, Amsterdam

Byrne P S, Pole E G, Fitzgerald P L 1978 Latices in carpet applications, *Latex Symposium, ACS Rubber Division Meeting*. Polysar, Sarnia, Canada

Calvert K O 1982 *Polymer Latices and Their Applications*. Elsevier, Amsterdam

Dunn E R 1972 Process of making solid foams from polymer emulsions. US Patent Reissue No. 27,266

Gazeley K F, Gorton A D T, Pendle T D 1988 Technological processing of natural rubber latex. *Natural Rubber Science and Technology*. Oxford University Press, Oxford, Chap. 4

Otterstedt C W 1978 Closed cell sponge rubber. *Vanderbilt Rubber Handbook*. Vanderbilt, Norwalk, CT

Rogers T H, Sassaman W O 1978 Elastomeric products made with low energy systems. *J. Elastomers Plast.* 10: 17–28

Sundquist D J 1972 Polyolefin Foams. Dow Chemical Company, Midland, MI

Technology of Celogen Blowing Agents. Naugatuck Chemical, Division of Uniroyal, Naugatuck, CT

Wells W V, Strecker L A, LaRose B A 1971 Properties of free-blown cellular-expanded rubber. *Rubber World*, Ridgefield, CT

T. H. Rogers and A. F. Finelli
[Goodyear Tire and Rubber Company, Akron, Ohio, USA]

Elastomeric Products from Liquid Polymers

The use of low-molecular-weight liquid polymers with functional groups to effect chain extension and crosslinking was perceived in the 1950s and 1960s as a major advance in the production of elastomeric products. For example, the prospect of producing tires by low-pressure molding, or even by casting, excited an industry accustomed to using almost intractable solid elastomers in high-energy mixing and molding processes. In actuality, liquid elastomers play a relatively small part in the modern rubber industry, occupying a number of specialized niches. General elastomeric moldings made in large batches are rarely based on liquid polymers and the prospect of producing car or truck tires from liquid elastomeric compounds has yet to be realized.

Nevertheless, functional liquid polymers are playing an increasing role in a number of industries. In addition to the rubber industry, liquid elastomers are used in the manufacture of adhesives, sealants and coatings and as a basis for electronic encapsulants. In all these areas the important features of liquid polymers are exploited—flow at ambient temperature, ease of processing and versatility in their conversion into products via vulcanization over a wide temperature range at low or extremely high rates. Specific properties relating to the chemical structure of the liquid polymers are applied to the end use, for example heat and chemical resistance, low-temperature flexibility and good electrical performance.

The shortcoming of liquid elastomer technology is the relatively poor strength of the vulcanizates compared with products based on their solid, millable elastomer counterparts. This appears to be an insoluble problem, related to the chemical conversion of the functional groups and imperfections in the vulcanizate network. Only in systems based on urethane chemistry, where there is internal reinforcement through hard block—soft block phase separation and through hydrogen bonding, do properties approach these of carbon-black-reinforced synthetic elastomer and natural rubber vulcanizates.

1. *Liquid Polymers*

Liquid polymers are true polymers with a low number average molecular weight (<10 000) when compared with solid elastomers (>100 000). That is, their molecular structure consists of a number of identical repeat units, the number being small in comparison with solid polymers. Liquid means liquid under normal ambient conditions. Liquid elastomers are liquid polymers whose glass-transition temperatures are well below normal ambient temperature. When chain extended to a solid form they exhibit normal elastomeric characteristics.

The liquid elastomers described in this article are low-molecular-weight analogs of well-known solid elastomers. They have reactive functional groups (usually at the ends of the polymer chains) for chain extension and crosslinking. They are known as telechelic polymers. Not included in this article are liquid elastomers formed by depolymerization of high-molecular-weight polymers, for example liquid natural rubber. Also excluded are liquid polymers where the reactivity is along or pendent to the backbone of the polymer, for example liquid ethylene propylene terpolymers. These latter two types of liquid polymer are more likely to be used as reactive plasticizers or modifiers for solid elastomers than for the direct production of elastomeric products. They are chemically bound to the host elastomer through their double bonds and the chemical processes of vulcanization. Though polyether and polyester polyols are telechelic polymers, polyurethane chemistry is also excluded from this article.

The telechelic liquid polymers under review fall into three categories:

(a) the functionally terminated liquid polybutadienes and copolymers of polybutadiene,

(b) the liquid silicone elastomers, and

(c) functionally terminated sulfur-containing polymers such as liquid polysulfides.

All these polymers are available commercially in a range of molecular weights, and hence viscosities, and with a wide variety of functional groups. These are hydroxyl, carboxylic acid, amino, thiol, halogen and epoxy. Silicone elastomers are polydialkylsiloxane polymers, where the alkyl group is generally methyl. They too can be designed to have a number of types of reactive groups, including hydroxyl, vinyl and amino.

2. *Conversion of Liquid Elastomers to Solid Products*

The chain extension and crosslinking of telechelic polymers and their conversion to solid elastomers with useful mechanical properties (curing) is brought about by the reaction between the functional groups on the polymer and a catalyst or initiator, which directly or indirectly couples the low-molecular-weight molecules into an infinite network (reticulation). The use of polymers with a degree of functionality greater than two or the presence of multifunctional chain-extending molecules ensures the formation of the three-dimensional polymer network. The ultimate strength and modulus of the product elastomer will depend on the efficiency of the chain-extension chemistry (i.e., the degree of chemical conversion). The inability of these liquid polymer systems to approach 100% conversion is a contributing factor to their generally inferior properties.

Where the chain extension is by the stepwise addition of a functional polymer to a functional chain extender, complete conversion depends on the following.

(a) The use of the exact stoichiometry for the ingredients (i.e., the chemical balance of the reacting ingredients) must be exact. An excess of either polymer or curative will reduce the extent of network formation.

(b) No side reactions should occur.

(c) There must be an absence of monofunctional polymer. The degree of chain extension is significantly reduced by the presence of a few percent monofunctionality. This is akin to chain stopping in the polymerization process. Monofunctional monomer can deliberately be added in small amounts to a liquid elastomeric compound to control the modulus of the cured product.

As stated, using the correct stoichiometry is essential to optimizing the physical properties of products based on liquid elastomer. This places the onus on the processor of liquid polymer systems to measure the separate constituents accurately prior to mixing. To achieve this accuracy, when dispensing the liquid elastomer components from machinery, requires precise engineering.

2.1 Chemical Processes

The variety of functional groups available on the liquid polymers offers considerable scope to the formulator and product designer. Room-temperature, elevated-temperature and water-activated processes are used in the fabrication of products from liquid elastomers.

Hydroxyl-terminated polybutadiene-type polymers are generally chain extended using isocyanates, thus becoming a branch of conventional polyurethane chemistry. The principles developed for polyurethanes apply (e.g., the use of tin and tertiary amine catalysts, prepolymer techniques in combination with polyols and water-activated single-component systems). The relatively low strength obtained with polybutadiene-based polyurethane chemistry indicates the presence of structural defects in the final network and the absence of hydrogen bonding.

Carboxylic-acid-terminated polymers are chain extended with polyfunctional epoxides using basic catalysts such as tertiary amines. Similarly, amine-terminated polymers use epoxy compounds for chain extension, while epoxy-terminated liquid polymers can use polyfunctional amines or carboxylic acids. In all cases, tertiary amines or metal soaps and chelates accelerate the network formation. Halogen-terminated polymers are chain extended with polyfunctional amines.

Mercaptan-terminated polymers (thiol-terminated polymers) are chain extended through the use of isocyanates, epoxides or amines. Another method is the oxidation of the mercaptan groups, producing disulfide linkages in the process of chain extension. Inorganic dioxides can be used (e.g., manganese(IV) dioxide and calcium peroxide), plus hydrogen peroxide precursors (e.g., sodium perborate monohydrate). Organic peroxy compounds such as cumene hydroperoxide are also effective oxidizing agents. The inorganic peroxides are unique in that they are used in stoichiometric excess. The chemical mechanism allows an excess to be used without detriment to the conversion of the mercaptan groups or the formation of a complete network. This nondependency on stoichiometry is an advantage in the control of mixing processes at the conversion stage in a factory.

A common method of chain extending and crosslinking room-temperature-vulcanizing (RTV) silicone polymers is through the process of hydrosilylation, 01in which a vinyl functional silicone polymer is reacted through its double bonds with a silicone hydride functional group on a siloxane or other silicone polymer. Catalysts, usually platinum compounds, are used. In another example, a silanol-terminated silicone polymer is reacted with an alkoxy functional silicone curing agent in the presence of a tin catalyst to produce chain extension.

With some formulation adjustment, RTV silicone systems can be made to react and cure very rapidly by raising the temperature in excess of 100 °C. Production cycles of about 10 s are feasible, for example in liquid silicone injection molding. A major advantage of most liquid polymer systems is their ability to cure at room or elevated temperature.

The potential for crossreaction between telechelic liquid polymers has not been fully exploited. One example, found in adhesives technology, is when a liquid epoxy functional resin is reacted with carboxylic acid functional liquid polybutadiene elastomer. A toughened adhesive with outstanding performance is produced.

There is potential in both the rubber and plastics industries to produce alloys (e.g., through the interpenetrating network route), block copolymers and two-phase toughened systems. Such combinations of polymers are expected to offer a unique balance of properties.

2.2 Physical Properties

Without the use of reinforcing fillers, the strength achieved with liquid elastomers is relatively low and, hence, unfilled compounds are rarely used in the manufacture of general elastomeric products.

Typical tensile properties for unfilled liquid telechelic polymers are tensile strength 1–2 MPa, elongation at break (EB) 200–600% and Shore A hardness 35–50. Tear strength is always low. Better figures can be obtained by adding polyfunctional bridging compounds, which are selected to add structure to the cured system. For example, when short-chain diols or diamines are used with a hydroxyl-terminated polybutadiene, using an isocyanate prepolymer technique, tensile strengths in excess of 15 MPa are possible. However, the elastomeric characteristics of these systems have moved towards polyurethane type, with Shore A hardness over 90 and reduced EB. Similarly, unfilled mercaptan-terminated polysulfide liquid polymers, chain extended with epoxy-terminated resins, give rise to tougher products than those obtained by straight oxidation.

Silicone polymers also require reinforcement to reach a useful level of properties. Without fillers, polydimethyl-siloxane-based (PDMS-based) liquid systems achieve tensile strengths between 0.5 MPa and 2.0 MPa.

Telechelic liquid polymers have applications

where reinforcing fillers are not used, but not usually in molded products. Examples are found in the encapsulation of electronic components, fabric coatings and some specialized adhesives.

The majority of applications of liquid polymers rely on reinforcement by particulate fillers; some use fillers as extenders or as pigments. In all cases, the free-flowing liquids increase in viscosity as a result of the addition of the fillers and, in the extreme, become viscous pastes. This trade-off between strength and processability is an important design consideration. The more reinforcing the filler the more it influences the viscosity change. The polymer–filler physical and chemical interactions which determine the strength of the cured product similarly influence the base rheology prior to cure. Even at an equal volume loading, a precipitated calcium carbonate will be less reinforcing and thickening than either a fumed silica or a furnace-type carbon black. However, the viscous, filled liquid elastomers are shear thinning, so that the semisolid pastes can be pumped, extruded and molded under the influence of the applied shear. Other systems have a viscosity–temperature relationship, with viscosity falling as temperature rises, which can be a useful attribute for castable and sprayable liquid elastomeric systems.

With reinforcement, properties become more acceptable to the rubber industry, but generally still fall short of those obtained by conventional reinforced millable elastomers. The addition of 20 parts per hundred (pph) of a furnace-type black (HAF) to an isocyanate-cured hydroxyl-terminated polybutadiene increases the tensile strength threefold to 4 MPa; 50 pph gives a tensile strength of 17 MPa. The latter compound is very viscous and its cured Shore A hardness is 80. The addition of 50 parts of a low-structure furnace black to a mercaptan-terminated polysulfide polymer increases its tensile strength from less than 1 Mpa to 7 MPa. In all cases, the reinforcement brings about an improved tear strength. In liquid silicone elastomer technology, it is usual to reinforce with fumed silicas, thus retaining a colorless, translucent form suitable for pigmenting. Tensile strengths are typically in the range 4–10 MPa.

Other mechanical properties are improved by the presence of reinforcing filler, for example abrasion and cut growth resistance. The dynamic properties will depend on polymer–filler interaction, the chemical structure of the polymer and the nature of the chain-extension and crosslinking chemistry. The dynamic performance is usually inferior for systems based on liquid polymers, which reflects the relatively short polymer chain length between crosslinks. Isocyanate chain-extended hydroxyl-terminated polymers have a dynamic performance similar to conventional polyurethane elastomers.

2.3 Other Properties

When designing products to be made from liquid elastomers the selection criteria must relate to the end requirements for the finished products. Apart from basic strength, other parameters such as compression set resistance, chemical, solvent and fuel resistance, heat resistance, ozone and ultraviolet (uv) resistance, and gas and water vapor permeability, as well as electrical properties, will be considered. The range of liquid polymers available on the market covers these requirements and so selection of a specific type is a question of balance and compromise. The characteristics of commercial telechelic liquid polymers are listed in Table 1.

As with millable elastomers, the overall balance of properties is determined not only by the basic polymer structure but also by the method chosen for vulcanization and by the level and type of other ingredients, fillers, plasticizers, antidegradants and so on. Considerable compound development precedes selection and processing into finished products. For these liquids and paste systems, the rheology of the base compound and the curing system are important parameters controlled by formulation. The methods chosen for mixing the components and processing the compound into a cured product will also be determinants. A castable system differs in rheology from an extrusion compound. In addition, when the two components are mixed the cure chemistry commences, even at ambient temperature. This implies that there is a finite time available for further processing before the compound becomes too viscous. This time is known as the work life or pot life.

Table 1
Characteristics of commercial telechelic liquid polymers

Silicones	flexible to −55 °C, heat resistant to 230 °C, resilient, resistant to ozone and uv, good electrical properties, resistant to alcohols, flame resistant
Polysulfides	flexible to −50 °C, resistant to ozone and uv, low permeability to gases and water vapor, highly resistant to fuels and oils including aromatic fuels and alcohol-modified fuels
Polybutadienes	flexible to −55 °C, good electrical properties, hydrolytically stable (unlike conventional polyester-type polyurethanes), low permeability to water vapor
Butadiene–acrylonitrile copolymers	flexible to −50 °C, resistant to oils, heat resistant to 130 °C, low permeability to gases

Controlling the work life is an important design factor and may involve the use of accelerators, retarders or blocked, temperature-activated curing agents.

3. Processing of Liquid Elastomer Compounds into Products

Two-component liquid elastomer systems are based on a fully compounded base compound mixed with a curing agent, the latter being formulated to provide both the rheology to match the base and a simple mix ratio. After mixing the two components, the compound is pumped to its conversion line—involving extruder, injection-molding machine, spreader and so on. One potential attraction of liquid elastomer technology is the possibility for automated production, with raw materials being fed into continuous mixers and the resulting compounds being pumped to dedicated product lines. However, the relatively low volumes of liquid elastomers used do not justify the engineering design effort and the capital investment needed for such a project. Currently, the industry uses batch processes.

3.1 Compounding of Liquid Elastomers

The part of the two-component system containing the polymer is the base. As well as the polymer, the base may contain reinforcing filler (carbon black, fumed silica), plasticizers, cure accelerator, antioxidant and pigments. Thus, compounding the base involves mixing powders with liquids and successful performance in the finished product will depend on the degree of dispersion. This is particularly true of the reinforcing filler. Reinforcement relies on a uniform dispersion, the breakdown of agglomerates and "wetting out" of the filler. A rating in the range A-1 to A-4 on the Cabot carbon black dispersion scale is the usual goal.

One of the early assumed advantages for liquid elastomer technology was the ease with which reinforcing fillers could be mixed into the polymer, when compared with the large rubber mills and internal mixers used with conventional elastomers. This proved to be a fallacy as ordinary paddle mixers are unable to break down agglomerates or apply sufficient shear through the liquid polymer medium to effect dispersion. High-shear paint mixing equipment is required—three-roll mills, high-speed dissolvers or continuous extruder mixers. The energy input to achieve good dispersion is similar for millable and liquid elastomers.

There are handling advantages in liquid technology in that raw materials and compounds can be pumped in and out of storage containers.

The most commonly used mixers are change-pan planetary mixers equipped with high-speed dissolver blades and vacuum. The latter is important to exclude moisture and air which may be detrimental to subsequent processes and to the cured product. A final refinement of the filler dispersion is sometimes achieved using a three-roll paint mill, using a nitrogen blanket where necessary.

The order of addition of the ingredients to the mixer is important to the efficiency of mixing and to mimimize excessive heat generation which may damage the more heat-sensitive materials.

The manufacture of the second part of the two-component mixture, the cure paste, uses the same techniques. As the cure paste contains the active curing agent, special care is taken to avoid excess heat or moisture ingress. Cooling jackets, vacuum and dry nitrogen are used.

3.2 Mixing of Base and Cure Paste

For all two-component liquid elastomer systems, the mixing of base and cure paste is an important stage and one which must be controlled accurately. While it is possible to simply pour the cure paste into the base and hand mix, possibly with the aid of a geared-down drill with paddle, it is more usual to use meter-mix machines. These are pneumatic or hydraulic in operation and consist of two drum pumps, one for the base and one for the cure paste, coupled to a proportioning device. This is most often based on a leverage system, which is adjusted to give a range of mix ratios by varying the relative lengths of the pump pistons strokes for the base and cure paste. Prevention of mix ratio drift with time is important and this is checked for frequently. Wear of critical components is minimized by the use of hardened steel.

The metered components are pumped to the mixing zone. Mixing is either static or dynamic. In the static system, the two components are pumped through a tube containing a series of fins and spirals which divide, invert and recombine the components. The internal design and length of the static mixer tube is determined by the viscosity of the components being mixed and the pressure drop between the inlet and the outlet. Dynamic mixers consist of series of interlocking blades rotating in a mixing chamber. The volume of the chamber is kept small to minimize waste. Meter-mix machines are equipped with the means to flush the mixing zone and beyond, either with solvent or with the base component.

The heat of mixing in the dynamic mixer is controlled as it will influence both the rheology and the work life of the mixed compound. For some applications, a continuous flow of mixed material is required (i.e., without pulsing) and this is achieved through the use of reservoirs and gear pumps.

3.3 Manufacture of Products from Liquid Elastomers

All the techniques used to convert millable elastomers into products apply to liquid elastomers.

Some modification to the conversion machinery is necessary in order to take the mixed liquid compound directly from the meter-mix machine. Liquid systems lend themselves naturally to some processes, for example casting and encapsulation. In processes such as fabric spreading or spray coating, solventless liquid polymer compounds replace the solvent slurries and solutions of millable compounds normally used.

(*a*) *Injection molding*. The most successful application for liquid polymers in the rubber industry is injection molding. The technology was developed and refined around liquid silicones, but is applicable to most liquid polymers including polysulfides and polybutadienes.

The low injection pressure required with liquid elastomers allows the use of small injection-molding machines, low clamping forces and complex, multicavity, runnerless molds. The low pressure in the mold eliminates flash and consequently any trimming step. Very fast high-temperature cures give short cycle times (10 s for silicone elastomers at 200 °C, 30 s for polysulfide elastomer at 150 °C). Full development of properties may require a postcure. The short cycle times, low capital costs, significantly lower energy costs and automated production lead to high productivity and low unit costs.

Conventional elastomer injection-molding machines are used with the input from the meter-mix machine replacing the feed hopper. The work life for the mixed elastomer is long enough to cope with overnight or longer shutdowns. The barrel is not heated as melting is not required; it may be cooled to extend work life. The mold is preheated to the cure temperature. Demolding depends on the elastomeric system and the surface finish. Mold release agents are used. When designing products, linear shrinkage is taken into account (~3–4% for liquid silicone elastomer cured at 200 °C or more).

The products made by liquid injection molding are predominantly based on silicone elastomers. The products tend to reflect the properties of silicone polymers—high temperature resistance (e.g., for heat-sterilizable medical products) and good electrical properties. For fuel-resistant products, polysulfide liquid polymers can be used.

Injection-molded products based on liquid silicone polymers include baby bottle teats, funnels and tips for catheters, electrical connectors, light bulb holders, spark plug boots, computer key pads, valves, diving masks, o-rings, gaskets, diaphragms and rollers.

Compression molding of liquid elastomers is possible although the economics are not as attractive as injection molding. After the mixing stage, the materials are pumped via suitable valving directly into the preheated mold, which is already clamped in the press under pressure. Pressure from the meter mix is applied to assist with effective filling of the mold and consolidation of the material. Use of a manifold or rotary press system enables several presses to be served by one meter-mix machine. The principles for mold design are similar to those for injection molding.

(*b*) *Extrusion*. The rheological characteristics of liquid elastomeric compounds makes unsupported extrusion difficult. The shearing forces applied during transportation of materials to the die, and at the die, lower the viscosity to a level where profile shapes flow out. Nevertheless, with the use of high-structure fillers and thixotropic agents, the extrusion of simple profiles is possible. Unlike the extrusion of conventional elastomers, heated barrels and dies are not necessary and this helps to maintain a high viscosity. The extrudate must be supported immediately after extrusion on a release surface such as polyethylene film. This support surface will be on the haul-off conveyor, if it is a horizontal extrusion process. The weight of the extrudate will lead to a flattening of the surface in contact with the support. This should be taken into account in the design of extrusions produced by this method. Delivery of the mixed material to the die must be continuous and not pulsed. This requirement is built into the meter-mix machine.

Supported extrusions of liquid polymers are more common, but limited to simple profiles such as wire coating. A conventional extrusion crosshead die arrangement is used. The thickness of the coating is controlled by the die diameter.

The cure of liquid elastomer extrusions can be batch or continuous. It is possible to cure supported extrudates at high temperatures and fast rates. In wire and cable coating, this takes place in a vertical hot-air oven. The throughput rate and temperature and speed of cure are matched to the output rate of the meter-mix machine and the length of the oven, such that the crosshead is not starved of material and the coated product exits the oven sufficiently cured for windup. Where high-temperature cures would lead to flow and loss of profile, a room-temperature batch cure is used. While this need not be slow (cure times at room temperature can be as short as 10 min) by definition, there is a limit to the length of the extrudate.

Supported and unsupported extrusion are being manufactured from liquid polysulfides and silicone polymers. Profile strips used in the sealing of joints in buildings are made up to 180 mm width. Wire and cable coating, polymer-coated fabric tape and glass-reinforced sleeving are examples of supported liquid elastomer extrusions.

(*c*) *Fabric coating*. The spreading of liquid-polymer-based compounds onto fabric is a natural extension of the current practice of using elastomer–solvent cements. There are several advantages. First, the absence of solvent has implications for health and safety and the environment. Second, with solvent-based systems, the final depth of the elastomeric coating is built up gradually over many passes through the spreading machine and on each pass the solvent is removed. With the liquid elastomer system the final required thickness is put down in a single pass, from a fraction of millimeter to 1 mm or more. Finally, the cure cycle can be designed such that with the application of heat the coating has reached a sufficient state of cure to be wound up at the end of the line (i.e., subsequent vulcanization of the coating in an autoclave is not needed). Problems with liquid elastomer fabric coating are encountered with the rheology at the spreading stage and with work life/cure-rate control.

The two-component system is fed directly from the meter-mix machine onto the front end of the spreader to form a rolling bank. The stability of the rolling bank is important and depends on the feed rate from the meter-mix machine, on the viscosity and viscosity–shear rate–temperature relationship for the compound, on the size of the gap under the doctor blade and on the web speed. The work life of the system must be such that, for a known rate of consumption, none of the material in the rolling bank gels.

After spreading, the coated material passes through ovens normally used for solvent drying. The temperature is raised to the maximum compatible with a fast cure and the thermal stability of the elastomer and the fabric in air. Heat transfer into the bulk of the elastomer is important and the surface of the elastomer must be tack free at wind-up. Adhesion between the fabric and the elastomer is achieved either through the use of a bonding agent or through a tie coat of the same elastomer forced into the interstices of the fabric.

The economic advantages of spreading with solvent-free liquid elastomeric compounds have not been fully exploited, partly because of the large capital investment already put into conventional spreaders with solvent removal ovens. In addition to polyurethane systems, liquid silicone and polysulfide elastomers have been coated onto fabrics. The products fabricated from such materials include hospital blankets, fire screens, hot-air ducting, gas meter diaphragms, printing blankets, flexible fuel and water tanks.

(*d*) *Casting*. Casting is a widely used application for liquid elastomers. For small moldings, it is possible to use single-component, moisture curing systems. Two-component chemical curing systems are mostly used. Vulcanization is usually at room temperature (RTV).

The requirements are for a free-flowing, air-free compound. It will be capable of filling complex molds and replicating precise surface details. The two-component systems may be hand mixed (followed by degassing) or dispensed from a meter-mix machine. Where high strength or hardness is required, reinforcing fillers will be used in the compound. This means finding a compromise between the mold-filling characteristics and the cured properties. Liquid polysulfide and silicone elastomers are used in casting systems. The liquid elastomers are also used in rotational casting.

Examples of the use of casting systems are flexible molds, rollers for printing and conveying, linings for pipes and pumps, doctor blades, industrial wheels, seals and gaskets, large moldings such as dock fenders and dental impression compounds.

(*e*) *Other applications for liquid elastomers*. The main usage of liquid elastomers is outside the rubber industry, where they are used as the basis for sealants, adhesives, coatings and electronic encapsulants.

Formed-in-place gasketing consumes several tens of thousands of tonnes of liquid polysulfide, silicone and polybutadiene elastomers annually. The edge seal in double glazing units, factory applied windscreen sealants, seals for integral fuel tanks in aircraft, elastomeric sealants for joints in structures and roads and gaskets in structural glazing are examples. The ability of these systems to adhere to a wide range of surfaces and to perform under a variety of adverse conditions has led to their specification for such products.

See also: Sealants as Building Materials; Structural Sealant Glazing

Bibliography

Flory P J 1953 *Principles of Polymer Chemistry*. Cornell University Press, Ithaca, NY

Goethals E J (ed) 1989 *Telechelic Polymers: Synthesis and Applications*. CRC Press, Boca Raton, FL

Isayev A I H 1989 Injection molding of Rubbers. In: Aggarwal S L (ed.) 1989 *Comprehensive Polymer Science*, Vol. 7. Pergamon, Oxford, pp. 355–78

Mark H F et al. (eds.) 1989 *Encyclopedia of Polymer Science and Engineering*. Wiley, New York

Stevenson J F 1989 Extrusion of rubber and plastics. In: Aggarwal S L (ed.) 1989 *Comprehensive Polymer Science*, Vol. 7. Pergamon, Oxford, pp. 303–54

Willoughby B G 1985 Liquid rubbers. In: Goodman I (ed.) 1985 *Developments in Block Copolymers*. Elsevier, Amsterdam, pp. 239–92

T. C. P. Lee
[Morton International, Coventry, UK]

Elastomers: Additives

The earliest vulcanization processes involved only one kind of rubber (natural rubber) and one additive (sulfur, the vulcanizing agent). Although such processes overcame the sticky-when-hot/brittle-when-cold characteristics of unvulcanized (uncured) rubber, its mixing, curing and long-term aging characteristics were still unsatisfactory for many applications. Continuing efforts to simplify processing procedures, speed up vulcanization and provide increased protection against the environment have led to increasingly more complex rubber formulations; there also has been a proliferation of polymer types, fillers, curatives, processing aids and other additives. This article briefly explains the functions of the more commonly used additives, except for carbon black, fillers and vulcanization system chemicals which are covered elsewhere (see *Fillers in Elastomers*; *Vulcanization by Silanes*; *Vulcanization, Continuous*).

1. Processing Aids

Processing aids are typically used to alter the properties of uncured rubber so that it can be processed into manufactured articles quickly and economically. They may, for example, reduce viscosity for rapid flow, or improve tackiness for making built-up articles. They can also affect properties of finished goods. Although processing aids are generally classified as extenders, softeners or plasticizers, the differences between the classifications are not well defined. For example, a hydrocarbon oil, used at a low level (<15 parts per hundred parts of rubber polymer (phr)) acts as a softener while, at a higher level of use (>15 phr), it is thought of as an extender, since it is generally lower in cost than the rubber polymer.

The compatibility of a processing aid with a given rubber must be considered, otherwise blooming, exudation, or diffusion losses, may occur. However, since processing aids differ widely in chemical nature, by using the rule that "like dissolves like," a compatible processing aid can generally be found.

Probably the most common processing aids are paraffinic, naphthenic and aromatic hydrocarbon oils. Paraffin oils (straight-chain hydrocarbons) are relatively incompatible with many elastomers such as styrene–butadiene rubber (SBR). They tend to exude or "bleed" from the rubber stocks and are typically used at low levels. They do, however, impart low-temperature flexibility without imparting color.

Naphthenic process oils are generally more compatible than paraffins with most elastomers. They provide satisfactory low-temperature flexibility with only moderate staining (darkening) under light.

Aromatic oils, the most widely used processing aids, show excellent compatibility with most elastomers and provide optimum carbon black dispersion. Thus, compared with paraffins and naphthenics, the aromatics yield the best processing and mechanical properties (including lower hysteresis). They do, however, show significant staining under light, give poorer low-temperature flexibility and give compositions less resistant to certain types of degradation.

The hydrocarbon oils, which are products of petroleum distillation, are generally classified according to the major hydrocarbon constituent; however, most are mixtures of all three (paraffin, naphthenes and aromatics). In addition to the hydrocarbon oils, vegetable oils, coal-tar products or pine products are used as softeners.

A plasticizer is defined as a material that is mixed with a polymer to improve flexibility, processibility, extensibility or resistance to low-temperature embrittlement. The difference between a plasticizer and softener (process oil) is somewhat semantic. However, the term plasticizer is generally applied to materials which are produced synthetically, and the extender oils are generally used in larger proportions.

Phthalates and other esters are typical plasticizers used in the rubber industry. Although they provide the same general properties as process oils, they can be selected or even tailored to give better compatibility and low-temperature flexibility.

Other materials that are used as processing aids include fatty acids and their salts (e.g., stearic acid, zinc stearate) which act not only as plasticizers, but also as processing lubricants which aid in reducing friction from wetting between the uncured rubber and metal during its passage through extruders and calenders and during mold filling. Fatty acids and their zinc salts are also essential for vulcanization activation. Finally, dusts and dipping materials are used to prevent cohesion of rubber during storage and to provide fresh surfaces for subsequent assembly steps.

2. Resins

Resins are incorporated into elastomers to improve building tack, for reinforcement, to promote adhesion and to enhance processibility. They are used in rubber formulations largely because of their characteristically high glass transition temperatures. At processing and curing temperatures above their glass transition temperatures, the resins can act as plasticizing processing aids, improving processing and mold flow. Vulcanizate properties are not necessarily grossly affected at use temperatures, which are generally below the resin glass transition temperature.

Table 1 gives a brief list of resins commonly used in rubber formulations. Aliphatic hydrocarbon resins impart tackiness and give increases in the strength of uncured rubber (i.e., increased green strength). The combination of these effects leads to improvements in building tack.

Phenolic and other aromatic resins provide reinforcement. The addition of a methylene donor

Table 1
Resins used as elastomer additives

Type	Source	Major effect
Aliphatic resins	petroleum or turpentine fractionation to give mainly alicyclic resins	tackification
Cyclopentadiene resins	low-cost refinery streams containing cyclopentadiene	tackification, reinforcement
Polynuclear aromatic resins	polymerization of oil-cracking unsaturates	reinforcement
Coumarone–indene aromatic resins	polymerization of light oil fractions from coal coking and oil cracking (containing coumarone, indene, methylindene, styrene, methylstyrenes, etc.)	reinforcement
Alkylphenol and resorcinol (phenolic) resins	reactions of alkylphenols or resorcinol with formaldehyde or other aldehydes	reinforcement tackification, adhesion promotion (e.g., to tire cords and metal surfaces), vulcanizing agent

(e.g., hexamethylenetetramine and melamine–formaldehyde reaction products) leads to the formation of an "extra network" crosslinked structure at cure temperatures. This results in increased tensile strength, ultimate elongation, or adhesion. Certain phenolic resins (2,6-dimethylol-4-alkylphenols and their oligomers) are also curatives, especially for butyl rubber, as well as being methylene donors.

The selection of a specific resin is not only based on its function, but also on its compatibility. For example, it is generally thought that a limited amount of incompatibility between elastomer and resin is required to achieve optimum building tack.

3. *Protective Materials*

Oxygen, light, ozone and mechanical flexing are extremely hostile to vulcanized rubber. These degradative forces can render an unprotected rubber compound useless within hours after the curing process is complete. Continuing studies of these destructive processes have, however, led to many types of chemicals which delay or prevent degradation, resulting in rubber products of extended service life.

Oxidation usually softens natural or butyl rubber because of polymer main-chain scission. Most synthetic rubbers such as SBR and nitrile–butadiene rubber (NBR) harden when oxidized since, in these cases, oxidation generally leads to crosslinking. Although many reactions can and do occur during oxidation, chain scission and crosslinking are the most prevalent. They usually occur simultaneously with one or the other predominating. As expected, increased temperatures accelerate the oxidation reactions.

Metal ions (e.g., copper ions) can accelerate oxidation. While some additives can counteract this effect, the general approach is to avoid contact with certain pro-oxidant metal salts.

Light can also induce the degradation of rubber, causing surface crazing and failures in thin cross sections. It is a serious problem only in unpigmented and light-colored compounds (e.g., rubber bands).

Rubber is frequently used because of its flexibility. The mechanical stress due to flexing, in combination with chemical stresses in the environment (e.g., due to O_2 or O_3) can cause a breakdown in the molecular structure which can lead to the premature failure of rubber articles. Fatigue life or resistance to repeated deformations can be improved through proper engineering design and by the use of chemical additives.

Ozone reacts primarily at rubber surfaces with the unsaturated sites in the polymer backbone. In an unstrained state, the buildup of ozonation products on the surface prevents further contact between rubber and ozone and thus prevents serious damage. In a strained state, however, cracks develop which continually expose fresh surfaces for further ozone attack.

Nondiscoloring antioxidants are used in light-colored stocks. They are generally less effective than the amine antioxidants. Most commercially available nondiscoloring antioxidants may be classified as phenolics, bisphenols or substituted hydroquinones.

The phenolics (see Fig. 1) are generally used in latex and in light-colored mechanical goods. The size and placement of the substituted groups vary

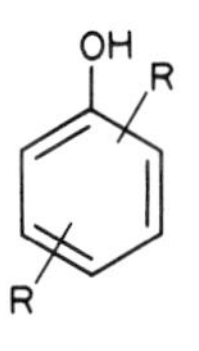

Figure 1
General structure of phenolics

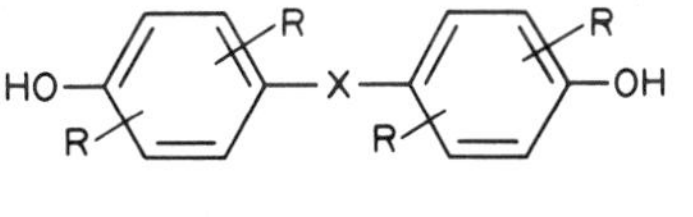

Figure 2
General structure of bisphenols

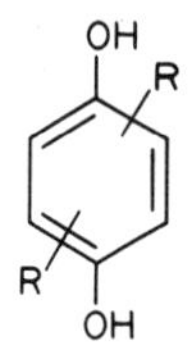

Figure 3
General structure of substituted hydroquinones

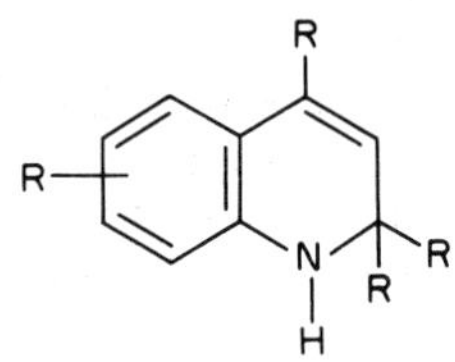

Figure 5
General structure of dihydroquinolines

widely among the large number of available materials. Products of this class often give moderate protection at low-to-moderate cost. They tend to be somewhat volatile relative to other classes and may not perform as well in high-temperature applications.

Bisphenols are the most potent, persistent and expensive class of nondiscoloring antioxidants. Their general structure is shown in Fig. 2 where X may be a sulfur or an alkylidene bridge.

Substituted hydroquinones (see Fig. 3) are a rather more specialized class of products. They are used primarily for tack retention in uncured rubber films and adhesives.

Amine-based antioxidants are chosen for severe service applications where color stability is not important. Polymerized dihydroquinolines are among the most potent and widely used general purpose amine-based antioxidants. They offer oxidation resistance, high-temperature stability and reduced frosting (ozone crazing). The polymerized dihydroquinolines also protect rubber against oxidation, especially when it is accelerated by ions such as copper, manganese or nickel. Although these antidegradants show little effect on other key performance properties in natural and most synthetic rubber compounds, they are not often used in neoprene because they activate vulcanization to an uncontrollable extent.

Another class of amine antidegradants, the diphenylamines, give slightly less antioxidant protection than the polymerized dihydroquinolines. *N*-phenyl-β-naphthylamine (see Fig. 4) is one of the oldest of the amine antioxidants. Its use is declining rapidly, however, due to suspected health problems.

The most effective antiozonants are those which, unfortunately, become highly colored (stained); only two classes are in general use, dihydroquinolines and *p*-phenylenediamines (PPDs). The dihydroquinolines (see Fig. 5) are the oldest, but they are still used because they are relatively inexpensive. PPDs (see Fig. 6) are by far the most common antidegradants used by the rubber industry. These materials protect elastomers from both ozone and oxygen, as well as from fatigue and pro-oxidant metal ions. They are relatively expensive but their ability to protect against all the common degradative forces justifies their use. There are three classes of commercially available PPDs, differing in the nature of the substituent groups.

N, *N'*-dialkyl-PPDs give superior ozone protection in a static environment (i.e., in the absence of motions such as flexing or stretching). They tend to be more sensitive to oxidation, however, and may suffer from a lack of persistence in antiozonant activity. *N*-alkyl-*N'*-aryl-PPDs offer superior ozone and fatigue resistance in a dynamic (cyclic-flexing) environment. When combined with wax, they also provide excellent static protection and are normally free of bloom problems.

N,*N'*-diaryl-PPDs are the least expensive of the PPDs. However, they exhibit less antidegradant activity and may cause bloom problems at effective use concentrations.

Properly chosen, waxes can offer a significant degree of protection against ozone attack under nonflexing conditions. Being relatively insoluble in rubber, they bloom to the surface to form a physical barrier preventing contact between ozone and elastomer.

The ability of waxes to bloom to the surface of rubber is affected significantly by the ambient storage temperature. In general, it is best to use a blend of various waxes of different molecular weights to provide optimum protection over a wide range of temperatures.

Flame retarders, fungicides, germicides, uv-light absorbers and chemical stabilizers are specialized antidegradants that provide protection against hostile environments to which most elastomers are not commonly exposed.

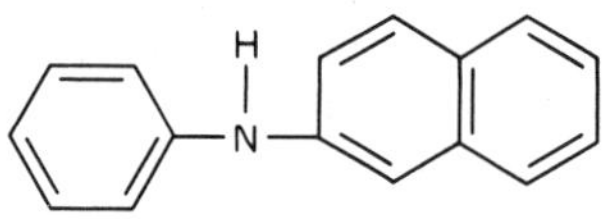

Figure 4
Structure of *N*-phenyl-β-naphthylamine

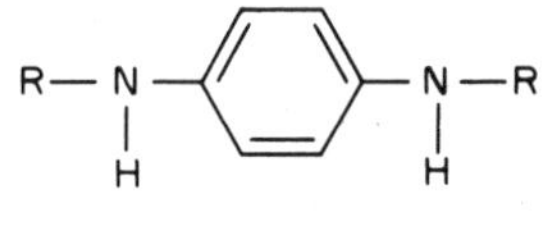

Figure 6
General structure of PPDs

4. Specialty Additives

Because of the seemingly infinite number of intended uses for elastomers, a wide range of specialty chemicals have been developed to impart additional functional and cosmetic characteristics to rubber compositions.

Finishes, odorants, antistaining agents and optical brighteners are available to enhance the appearance of a finished product. Antiblocking, antislip and antistatic agents provide highly specialized surfaces on rubber articles intended for sensitive industrial applications (e.g., belts and conveyors). Blowing-agent systems are used in applications where an elastomeric foam is required.

As with processing aids, the selection of a specialty additive is in part based on its compatibility with a particular elastomer, as well as on its functionality and effectiveness.

See also: Degradation and Stabilization, Thermal; Degradation and Stabilization, Ultraviolet; Elastomer–Elastomer Adhesion Processes; Elastomers: An Overview; Fillers in Elastomers; Plasticization of Polymers; Polymer Blends; Vulcanization

Bibliography

Brown G L 1969 *Resins in Rubber: Or, What Every Young Compounder Should Know about the Use of Hydrocarbon Resins in Rubber*. Pennsylvania Industrial Chemical Corporation, Clairton, PA

Eirich F R (ed.) 1978 *Science and Technology of Rubber*. Academic Press, New York

Morton M (ed.) 1973 *Rubber Technology*, 2nd edn. Van Nostrand Reinhold, New York

Sears J K, Darby J R 1982 *The Technology of Plasticizers*. Wiley, New York

Studebaker M L, Beatty J R 1978 The rubber compound and its composition. In: Eirich 1978, Chap. 9

Van Alphen J 1973 *Rubber Chemicals*. Reidel, Dordrecht, Holland

A. Y. Coran and D. A. Lederer
[Monsanto Company, Akron, Ohio, USA]

Elastomers: An Overview

Elastomers share the property of rubberlike elasticity, which may be operationally defined as very large deformability with essentially complete recoverability. In order for a material to exhibit this type of elasticity, three molecular conditions must be met:

(a) the material must consist of polymeric chains,

(b) the chains must be joined into a network structure, and

(c) the chains must have a high degree of flexibility.

The first requirement arises from the fact that the molecules in a rubber or elastomeric material must be able to alter their arrangements and extensions in space dramatically in response to an imposed stress, and only a long-chain molecule has the required extremely large number of spatial arrangements of very different extensions.

The second characteristic is required in order to obtain the elastomeric recoverability. This is achieved by joining together, or crosslinking, pairs of segments, approximately one per hundred, thereby preventing the extended polymer chains from irreversibly sliding by one another. These crosslinks may be either chemical bonds, as would occur in sulfur-vulcanized natural rubber, or physical aggregates (e.g., the small crystallites in a partially crystalline polymer or the glassy domains in a multiphase block copolymer).

The third characteristic specifies that the different spatial arrangements be accessible; that is, changes in these arrangements should not be hindered by constraints resulting from the inherent rigidity of the chains, extensive chain crystallization or the very high viscosity of the glassy state.

There are a wide variety of polymers which exhibit rubberlike elasticity under conditions suitable for applications requiring this long-range, reversible deformability. They include relatively simple hydrocarbon polymers, semi-inorganic polymers, ionomers, block copolymers with thermoplastic properties and polymer blends. Many of the polymers are produced industrially by means of standard polymerization techniques, while others are produced in living systems, such as the "natural rubber" polyisoprenes produced in plants and the bioelastomeric proteins synthesized in mammals and in some types of insects.

Polymers to be used in the preparation of commercially useful elastomers are first blended with a variety of additives, some for the crosslinking (curing or vulcanization) mentioned and some to give better properties (such as abrasion resistance or thermal stability) in the anticipated applications. The processing steps involved in their incorporation and utilization are complex, but of crucial importance in the rubber industry. The final materials are generally solid, molded objects, but may also be in the form of a coating or foam.

Elastomers thus prepared are generally subjected to a range of tests to evaluate their properties, including their structural characteristics, transition temperatures, equilibrium, dynamic and ultimate properties, flammability and environmental resistance. Measurements of their mechanical properties at equilibrium are particularly useful for the evaluation of the theories of rubberlike elasticity, both molecular and phenomenological.

The applications of elastomeric materials are extremely diverse. They include universally important industrial items such as rubber tires and conveyor belts, garments, biomedical devices for implantation in the body and soft contact lenses. Growing concerns with limitations in nonrenewable resources and with environmental problems have added a final important consideration regarding materials of any type. This is their reclaiming or reclamation, a subject of increasing importance with regard to the efficient utilization of rubberlike materials.

See also: Elastomers: Applications Overview; Elastomers: Dynamic Mechanical Properties; Elastomers: Tests for Mechanical Properties

Bibliography

Chompff A J, Newman S (eds.) 1971 *Polymer Networks: Structure and Mechanical Properties. American Chemical Society Symp. Highly Cross-Linked Polymer Networks.* Plenum, New York

Dunn A S (ed.) 1974 Rubber and rubber elasticity. *J. Polym. Sci., Polym. Symp.* 48

Dušek K, Prins W 1969 Structure and elasticity of noncrystalline polymer networks. *Adv. Polym. Sci.* 6: 1–102

Eirich F R (ed.) 1978 *Science and Technology of Rubber.* Academic Press, New York

Flory P J 1953 *Principles of Polymer Chemistry.* Cornell University Press, Ithaca, NY

Labana S S 1977 *Chemistry and Properties of Cross-Linked Polymers.* Academic Press, New York

Mark J E 1981 Rubber elasticity. *J. Chem. Educ.* 58: 898–903

Mark J E, Lal J (eds.) 1982 *Elastomers and Rubber Elasticity.* American Chemical Society, Washington, DC

Meares P 1965 *Polymers: Structure and Bulk Properties.* Van Nostrand, New York

Morton M 1973 *Rubber Technology*, 2nd edn. Van Nostrand Reinhold, New York

Nash L K 1979 Entropy and rubbery elasticity. *J. Chem. Educ.* 56: 363–8

Smith K J Jr 1972 Theories of chain coiling, elasticity, and viscoelasticity. In: Jenkins A D (ed.) 1972 *Polymer Science.* North-Holland, Amsterdam, pp. 323–441

Treloar L R G 1975 *The Physics of Rubber Elasticity*, 3rd edn. Clarendon Press, Oxford

Wall F T 1974 *Chemical Thermodynamics: A Course of Study*, 3rd edn. Freeman, San Francisco, CA

J. E. Mark
[University of Cincinnati, Cincinnati, Ohio, USA]

Elastomers: Applications Overview

Rubber products cover a wide range of articles used by industry and consumers. The development of synthetic polymers, in addition to natural rubber, has provided the flexibility of matching the characteristics of these new elastomers to meet the needs of specific applications. Polymers can be selected for abrasion resistance, low-temperature flexibility, hardness, resilience, high-temperature strength, and oil, fuel and ozone resistance, or any combination of these properties.

Most rubber products require the application of heat to accomplish vulcanization, whether applied directly by steam in a pressure vessel or indirectly by a press. This process cures the product into its final form so that it is unaffected by normal temperature changes. There are many products, however, which use rubber in its uncured or unvulcanized state. Typical products include friction tape, insulating tape and coated fabrics. Another product group includes adhesives, rubber solutions and dispersions, which can be used as caulks, sealants, paints and surface protectants, as well as for bonding similar and dissimilar materials. One special application for uncured rubber is as the base for chewing gum and bubble gum.

In general, rubber products are manufactured using molding, extruding or calendering processes, followed by vulcanization or crosslinking.

1. Molded Products

Molded rubber goods are products manufactured by compression, transfer or injection-molding processes to meet the requirements of a specific application. The uses for molded goods are extremely diverse, with the products taking a great variety of forms, such as cable connectors, vibration insulators, ball mill liners, champagne stoppers, hydraulic brake cups and rubber seals.

The most familiar molded rubber products are tires. The world rubber consumption for 1990 was approximately 15 Mt, of which almost half was used by tire manufacturers, with another 12–15% in tire-related products, such as inner tubes, valves and retreads.

The automobile itself is one of the largest users of rubber based on an estimated annual usage of 0.75 Mt for the USA. Each new car requires about 100–130 kg of rubber in 600 different rubber products, with tires, as one product, accounting for 40–75 kg per car.

Styrene–butadiene, *cis*-1,4-polybutadiene and *cis*-1,4-polyisoprene can be formulated to give high abrasion resistance, low heat buildup, low compression set and excellent cut and chip resistance for transportation applications.

Aircraft requirements provide a different environment to challenge the versatility of rubber. For example, the eight main tires of the Concorde jet airliner have to support an overall weight of 23 t, land at 140 km h^{-1}, take off at 170 km h^{-1} and withstand exposure to temperatures ranging from −40 °C to 100 °C. There are also over 500 different types of rubber seals, ranging from door and hatch seals to tail-fin seals.

The one important characteristic of most rubbers that no other natural material has is resilience—the ability to return to its original form after being stretched, twisted or distorted. Resilience and good fuel and oil resistance are critical for applications where rubber is in contact with aggressive fluids or environments.

Polychloroprene, acrylonitrile–butadiene, butyl, chlorosulfonated polyethylene and epichlorohydrin polymers are used in the manufacture of articles which require good fuel and oil resistance, low-temperature flexibility, high-temperature stability and/or good fire-resistance properties. In products where weathering and outdoor exposure are critical, these polymers can be formulated to give good property retention after aging or they can be covered with polymers such as ethylene–propylene rubber, silicone and fluorocarbons which are not subject to the destructive action of oxygen, ozone and weathering in general. The service life for these rubbers can be in excess of 20 years.

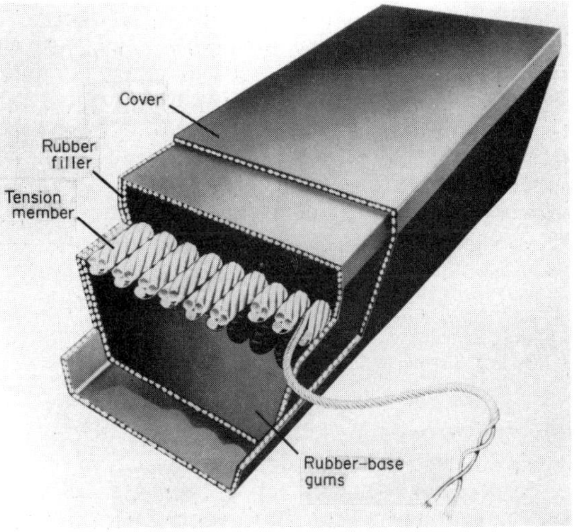

Figure 1
Basic V-belt construction

2. *Extruded Products*

Hoses are the major example of specialty elastomers manufactured by an extrusion process and their applications (and hence construction and selection of materials) are extremely diverse. They can be used whenever a fluid or a gas is moved from one area to another, with fluids ranging from the most active corrosive acid to waste sludge. Hoses range widely in size, from tubing to replace blood vessels to high-pressure flotation hoses reaching from a supertanker to the shore. A hose usually has three basic parts: the tube, the reinforcement and the cover. The essential characteristics of a given hose are determined by its intended use in service. Special properties may be built into any or all of the three parts through proper design and compounding. The tube may be extra thick; resistant to oil, abrasion, heat, chemicals and solvents; static conducting; or a combination of these qualities. The reinforcement may be resistant to high pressures and nonkinking, noncollapsing, crush resisting or designed to hold the coupling. The cover may be extra thick; resistant to oil, grease, chemicals and abrasion; weatherproofed; and treated to resist mildew.

3. *Calendered Products*

Sheet rubber products are made by the calendering process. Typical products include conveyor belts, protective liners and flooring tiles. A major rubber use in industry is flat belting, which is divided into conveyor belting for the transport of materials and transmission belting for power-transmission applications.

A conveyor belt is constructed with an inner load-support member (usually plies of rubber-coated fabric) and a protective cover designed to withstand the physical conditions to which the belt is exposed. The surface of the belt may be smooth for carrying material on a level or slightly inclined plane, have a molded design to carry material on a sharp incline or have buckets attached to carry material straight up. A conveyor belt may be used for short distances or combined with other conveyor belts to carry materials for kilometers over all types of terrain. As with hoses, belts are made from a variety of polymers and in various widths and lengths.

The most common conveyor belt is the general-service type designed to convey materials such as coal, rocks and other mining elements over long distances from where they are mined to the transportation or processing points. Special types of belts are made for a great variety of uses, such as food handling, agricultural services, package and baggage handling and for light-industrial automation lines.

Power transmission has been a major growth area for V belts, where power is efficiently transferred from a motor to drive another machine. V belts, like the conveyor belt, consist of rubber-coated load-bearing reinforcement plies surrounded by a surface ply and are molded or cut into a V shape fitting tightly into the groove of a pulley (see Fig. 1). The most common applications are the fan, air-conditioning and power-steering belts in automobiles. As with hoses and conveyor belts, V belts are made in different sizes depending on specific operating conditions.

Calendered sheet rubber is the next largest product application. The sheets are usually a single ply of rubber of various thicknesses used as a sheet or fabricated into various shapes. The major uses are for flooring, protective linings and roofing membranes. Floorings are nonslip, weatherproof, resilient, resistant to bacteria, chemicals and cigarette burns, long wearing and puncture proof. A more resilient application has been its use as a synthetic turf in sports areas such as tennis courts, running tracks and swimming pool borders.

For protective linings and membranes, the rubber must have good water and fuel resistance and, of course, longevity. Rubber tank linings allow the basic holding tank to be built from porous materials or from lightweight metals. The rubber lining is thus needed to protect the tank structure from corrosive attack of the contained fluids and allows a variety of aggressive materials to be stored or transported safely. Large sheets of rubber are used to help control the conservation of water through use as pit and pond liners.

Sheet rubber can also be cut and bonded together to form different shapes and sizes. Typical products are oil-spill containment booms in harbors and inland waterways, inflatable dams, aircraft fuel cells, collapsible containers and tarpaulins.

4. *Specialized and Consumer Applications*

Being relatively impremeable to gases, rubber is essential in the manufacture of such pneumatic articles as inner tubes, tubeless tires, air hoses, inflatable boats and hot-air balloons.

Consumer products cover a wide range of rubber applications, each utilizing a specific characteristic of the polymer. Household products include the rubber coverings of most electrical wiring, garden hoses, backings on carpets and underlay, rubber gloves, latex paint and adhesives.

Various rubbers have special properties to meet very specific needs. Elastomers are essentially nonconductive and those such as polychloroprene, butyl and EPDM (polymerized from ethylene, propylene and diene monomers) are used for electrical insulation. Yet, these and other elastomers can readily be made conductive by the use of special compound additives. Odor and taste properties are important in the food and drug industries and in other applications where the product comes into contact with people. Rubber can also be chemically blown into a foamed structure which can be used in many applications where solid rubber does not permit adequate compression, such as weatherstripping of all types, household sponges, mattresses and seat cushions.

See also: Automobile Tires; Design of Elastomer-Based Engineering Products; Elastomers: An Overview; Offshore Engineering Applications of Polymers; Sports Equipment Manufacture; Synthetic Adhesives; Wire and Cable Systems for Underground Transmission

Bibliography

Dannenbaum A K, Lovell M R Jr, Broadbent R A, Beretta D, Riley R A, Forlenza G A 1978 1977 Review; 1978 preview. *Elastomerics* 110(1): 17–24

Greener G B 1979 Fabric-reinforced rubber meets dynamic field requirements. *Rubber World* 179(3): 45–9

Murtland W O 1978 Elastomers in Detroit's future. *Elastomerics* 110(10): 35–40

Murtland W O 1979 Elastomers in Detroit: State of the art, 1980. *Elastomerics* 111(10): 41–8

Oxley C E 1979 Elastomers in sealants. *Elastomerics* 111(10): 29–32

Schultz D L 1979 Elastomers and the auto challenge. *Rubber World* 180(4): 38–41

J. H. Sudekum
[B. F. Goodrich Company, Breckesville, Ohio, USA]

R. S. Varga
[B. F. Goodrich Company, Akron, Ohio, USA]

Elastomers: Dynamic Mechanical Properties

Elastomers are frequently used for applications in which they undergo rapid cyclic deformations at a certain frequency or over a range of frequencies. Examples are engine mounts, where use is made of the flexibility, friction, cushioning and damping properties of rubber, and the sidewall or tread of a rotating tire. The dynamic mechanical properties are strongly dependent on temperature, frequency, the presence of fillers and the extent of deformation if it is large. For very small deformations, the properties are found to be independent of the magnitude of the deformation (linear viscoelasticity) and temperature–frequency superposition can be applied in many cases, a feature that greatly facilitates obtaining or predicting the dynamic response over a wide frequency range.

1. *Definitions of Dynamic Mechanical Properties and Viscoelastic Functions*

"Dynamic" mechanical properties refer generally to responses to periodically varying strains or stresses. They are most simply defined for a small sinusoidally varying strain or stress, whose response is a small sinusoidally varying stress or strain, respectively, with the same frequency but generally out of phase. Large and/or nonsinusoidal (e.g., square, triangular or pulsing) periodic stresses are, of course, also of great practical importance.

For small deformations in simple shear, if the strain γ varies sinusoidally as

$$\gamma(t) = \gamma_0 \sin \omega t \qquad (1)$$

where γ_0 is the peak strain, ω is the radian frequency

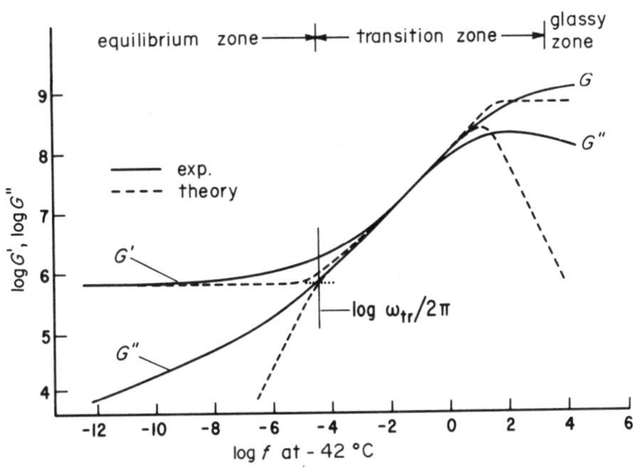

Figure 1
Zones of dynamic viscoelastic behavior for a crosslinked elastomer (polyurethane rubber) illustrated by logarithmic plots of G' and G'' (Pa) against frequency (Hz). Dashed curves drawn for Rouse–Mooney theory (after Schwarzl et al. 1972)

(2π times the frequency f in Hz) and t is the time, the stress σ will be

$$\sigma(t)=\sigma_0 \sin(\omega t+\delta) \tag{2}$$

Here the stress σ is the tangential force per unit area producing the deformation and the strain γ is the tangent of the angle by which a cubical element is deformed. The stress can be decomposed into a component in phase with the strain proportional to $\sin\omega t$ and another component 90° out of phase proportional to $\cos\omega t$; the total stress can then be expressed as

$$\sigma=\gamma_0(G' \sin\omega t+G'' \cos\omega t) \tag{3}$$

thus defining two shear moduli, $G'(\omega)$ and $G''(\omega)$. The storage modulus G' is a measure of elastic energy stored and recovered in cyclic deformation; the loss modulus G'' is a measure of energy dissipated as heat. The ratio G''/G' is $\tan\delta$, the loss tangent.

The two moduli are functions of frequency as illustrated in Fig. 1 for a typical crosslinked gum stock rubber (Schwarzl et al. 1972). At low frequencies, G' approaches the equilibrium shear modulus G_e and G'' becomes very small; the rubber is nearly perfectly elastic in small deformations. With increasing frequency, G' and G'' increase and become roughly equal in magnitude (at room temperature around 10^6 Hz). This change is related to local molecular mobility as will be explained below. Eventually, G' approaches a high value characteristic of a hard glass rather than a rubber, but at room temperature this would require exceedingly high frequencies. The curves shown are constructed by combining data obtained at various temperatures, and at 25 °C the zero on the logarithmic frequency scale would lie eight decades to the left (see Sect. 3).

For small deformations in simple extension (elongations less than a few percent, so that Hookean behavior is approached) the corresponding Young's moduli, E' and E'', are defined in an analogous manner as the ratios of the in-phase and out-of-phase tensile stresses to the practical tensile strain ε ($\varepsilon=\lambda-1$, where λ is the stretch ratio, i.e., the stretched length divided by the unstretched length). Unless the frequency is so high that the moduli approach glasslike magnitudes ($G'>10^8$ Pa), the ratios E'/G' and E''/G'' can be well approximated by 3, so dynamic shear and extensional experiments give equivalent information.

In an alternative formalism, the strain can be expressed in terms of two components, one in phase and one 90° out of phase with the stress. Ratios of strain to stress are called compliances, and the storage and loss compliances $J'(\omega)$ and $J''(\omega)$ are defined by the equation

$$\gamma=\sigma_0(J' \sin\omega t-J'' \cos\omega t) \tag{4}$$

Although J' and G' are both measures of energy storage, they differ in that comparisons of J' are comparisons at corresponding stresses, whereas G' compares at corresponding strains. They are interrelated as follows:

$$J'=G'/(G'^2+G''^2) \tag{5}$$

$$J''=G''/(G'^2+G''^2) \tag{6}$$

and the same equations hold with G and J interchanged. The loss tangent, a measure of "lossiness" or relative dissipation vs storage of energy, is $\tan\delta=G''/G'=J''/J'$. The compliances for simple extension are D' and D'', smaller than J' and J'', respectively, by a factor of 3 (except at very high frequencies).

The quantities G', G'', $\tan\delta$, J' and J'', all functions of frequency, are called linear viscoelastic functions; linear because, although dependent on frequency, they are independent of the magnitude of peak stress as long as the latter is sufficiently small. For certain other well-known linear viscoelastic functions, time t rather than frequency is the independent variable, for example the creep compliance $J(t)$ and the relaxation modulus $G(t)$. These are related to the dynamic functions defined in the preceding paragraphs by standard interconversion formulae.

2. Zones of Dynamic Viscoelastic Behavior

The frequency scale in Fig. 1 is divided into three characteristic zones. In the equilibrium zone, the period of oscillation is long compared with the relaxation times associated with configurational changes of the strands of the crosslinked network. In the transition zone, the period is shorter and some modes of configurational motion cannot be accomplished within it. Analysis of these modes, for example by the Rouse–Mooney theory, permits a prediction of the frequency dependence of G' and G'' in the transition zone which agrees moderately well with experiment. In the glassy zone, the period is short compared with all relaxation times except those associated with very short-range motions. Only in the equilibrium zone is the mechanical behavior really elastomeric. The boundary frequency ω_{tr} (rad s^{-1}) as drawn between equilibrium and transition zones is close to the reciprocal of τ_{tr}, the longest relaxation time of a network strand.

A similar plot of the frequency dependence of G' and G'' for an uncrosslinked elastomer of high molecular weight is shown in Fig. 2. The transition zone is very similar but the equilibrium zone of Fig. 1 is replaced by the plateau and terminal zones. This behavior reflects strong topological constraints of the tangled molecules which may be modelled by entanglements acting somewhat like temporary crosslinks or by restraint of each molecule to move primarily along its contour as though confined in a tube. In the plateau zone, the period of oscillation is long compared with relaxation times for rearrangements between entanglements (or within the tube diameter) but short compared with the time required for escape from the entanglements or the tube. In the terminal zone, the period is long enough to permit escape of a molecule from its topological restraints, primarily by diffusion along its contour.

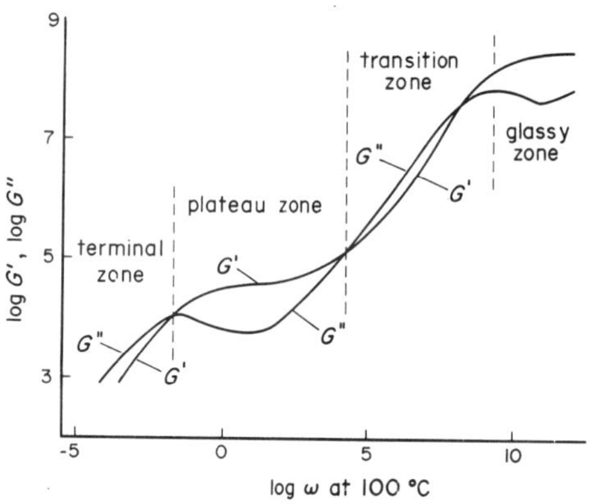

Figure 2
Zones of dynamic viscoelastic behavior for an uncrosslinked elastomer (poly(n-octyl methacrylate)) illustrated by logarithmic plots of G' and G'' (Pa) against radian frequency

3. The Transition Zone: Dependence of Frequency Scale on Temperature and Other Variables

The longest relaxation time τ_{tr} in the transition zone, and all the other relaxation times, are predicted by the Rouse–Mooney theory and other theories to be proportional to the monomeric friction coefficient ζ_0, the force per unit velocity per monomer unit of a segment of the macromolecule as it moves through the local environment with random thermal motion. Thus an increase in ζ_0 will cause ω_{tr} ($\simeq\tau_{tr}^{-1}$) and most of the transition zone to move to lower frequencies without changing the form of the frequency dependence in Fig. 1 or Fig. 2. The effect of changing temperature, pressure and diluent concentration on dynamic viscoelastic properties in the transition zone is primarily a shift in the frequency scale through change in ζ_0. This can be approximately formulated in terms of the fractional free volume f, which has the qualitative significance of relative free space but can be considered as defined by the equation

$$\log \zeta_0 = C + \frac{B}{2.303 f} \tag{7}$$

where C and B are constants with $B \simeq 1$.

For temperature dependence, the factor a_T is defined as the ratio of any relaxation time at temperature T to that at a reference temperature T_0; $\log a_T$ is the shift of the frequency scale for a change from T_0 to T. If $df/dT = \alpha_f$ then a_T is given by the Williams–Landel–Ferry (WLF) equation

$$\log a_T = -\left(\frac{B}{2.303 f_0}\right)(T - T_0)\left(\frac{f_0}{\alpha_f} + T - T_0\right)^{-1} \tag{8}$$

where f_0 is the value of f at T_0. It is found that α_f is usually about $5\times10^{-4}\,K^{-1}$ and f is about 0.025 at the glass-transition temperature, although the values differ somewhat from one polymer to another. A plot of a_T against temperature (for the same elastomer as Fig. 1) is compared with the WLF equation in Fig. 3. Figure 1 was actually constructed by combining data obtained over a much more limited frequency range at different temperatures by the use of Fig. 3 and Eqn. (8).

For pressure dependence, the factor a_p is defined as the ratio of any relaxation time at pressure p to that at a reference pressure, usually atmospheric. There are several formulations relating f to p; one provides in conjunction with Eqn. (7)

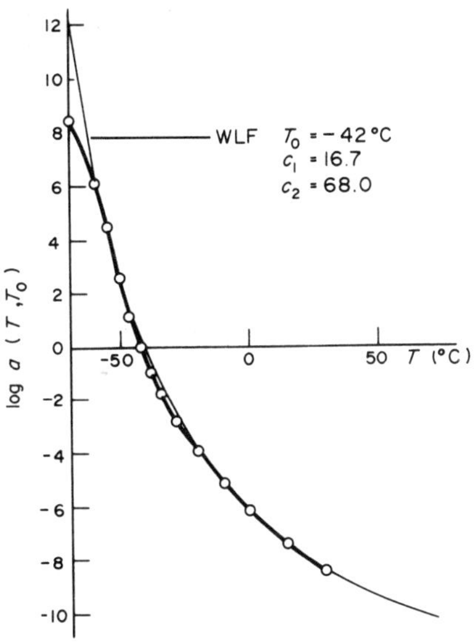

Figure 3
Temperature shift factor a_T plotted logarithmically against temperature for the rubber of Fig. 1, $T_0 = 231$ K. WLF curve drawn from Eqn. (8) with $\alpha_f = 3.8 \times 10^{-4}$, $f_0 = 0.026$

$$\log a_p = C'(p - p_0) \tag{9}$$

but at high pressures better agreement with experiment is obtained with more complicated relations. At moderate pressures, the derivative $(\partial T/\partial p)_f$ is about 0.020 ± 0.003 K atm^{-1} (2×10^{-7} K Pa^{-1}).

For dependence on diluent concentration, the factor a_c is defined as the ratio of any relaxation time at polymer concentration c (in g ml^{-1}) to that in the undiluted polymer. The volume fraction of diluent is $v_1 = 1 - c/\rho_2$, where ρ_2 is polymer density. If $df/dv_1 = \beta'$ and $B = 1$, then a_c is given by the Fujita equation

$$\log a_c = v_1\left[2.303 f_2\left(\frac{f_2}{\beta'} + v_1\right)\right]^{-1} \tag{10}$$

where f_2 is the value of f for the undiluted polymer. For typical diluents (plasticizers), β' is found to be in the range 0.1–0.3.

In addition to the shifts in frequency scale described by Eqns. (8–10), small changes in the magnitudes of G' and G'' are caused by changes in polymer mass per unit volume accompanying thermal expansion, bulk compression or dilution with solvent (plasticizer) and, in the case of temperature dependence, direct proportionality to absolute temperature T. The frequency shifts are strictly applicable only in the transition zone. In the terminal zone of an uncrosslinked elastomer, Eqn. (8) is usually applicable, but the effect of diluent is more complicated than described by Eqn. (10) because of dilution of the topological constraints.

4. *Practical Aspects of Dynamic Mechanical Properties*

The energy dissipated per second in small sinusoidally oscillating deformations with a prescribed peak strain γ_0 is $\dot{\varepsilon} = \omega G''\gamma^2/2$; for deformations with a prescribed peak stress σ_0 it is $\dot{\varepsilon} = \omega J''\sigma_0^2/2$. The ratio of energy dissipated to average energy stored per cycle is proportional to $\tan\delta$. These quantities clearly increase rapidly with increasing frequency as the transition zone is entered. Even in the equilibrium zone, G'' is significant as seen in Fig. 1 and is often considerably larger than predicted by the Rouse–Mooney theory. Its magnitude and frequency dependence are not completely understood but are probably associated with network imperfections such as a distribution of strand lengths, trapped entanglements and dangling branched structures incompletely attached to the network. These features are important in the performance of vehicle tires and other applications.

Damped oscillations, damping of propagated shear waves, resilience and rebound and rolling friction are primarily related to $\tan\delta$; these properties are important in many applications of elastomeric materials.

5. *Dynamic Mechanical Properties in Large Deformations*

In oscillating deformations with large amplitudes, the stress–strain relations are more complicated than described by Eqn. (3); terms may be added with higher odd powers of γ_0 at the fundamental frequency as well as at higher harmonics. The onset of such nonlinear behavior may be at quite large strains for unloaded (gum) crosslinked elastomers but at very low peak strains ($\gamma_0 \simeq 0.001$) for heavily loaded elastomers. For the latter, the properties depend strongly on stressing history. Small oscillating deformations superposed on large static strains show a complicated behavior which is related to the shape of the equilibrium stress–strain curve and, in the case of loaded elastomers, strongly dependent on stressing history.

See also: Elastomers: Strength and Extensibility in Simple Tension; Elastomers: Tests for Mechanical Properties

Bibliography

Ferry J D 1980 *Viscoelastic Properties of Polymers*. Wiley, New York

Graessley W W 1982 Entangled linear, branched and network polymer systems. Molecular theories. *Adv. Polym. Sci.* 47: 67–117

Kramer O, Ferry J D 1978 Dynamic mechanical properties. In: Eirich F R (ed.) 1978 *Science and Technology of Rubber*. Academic Press, New York, pp. 180–221
Medalia A I 1978 Effect of carbon black on dynamic properties of rubber vulcanizates. *Rubber Chem. Technol.* 51: 437–523
Nielsen L E 1974 *Mechanical Properties of Polymers and Composites*. Dekker, New York
Payne A R 1958 Temperature–frequency relationships of dielectric and mechanical properties of polymers. In: Mason P, Wookey N (eds.) 1958 *Rheology of Elastomers*. Pergamon, London, pp. 86–112
Schwarzl F R, van der Wal C W, Bree H W 1972 On the shear behavior of a polyurethane rubber in the rubber-like and in the glass–rubber transition region. *Chim. Ind. (Milan)* 54: 51–8
Snowdon J C 1968 *Vibration and Shock in Damped Mechanical Systems*. Wiley, New York

J. D. Ferry
[University of Wisconsin, Madison, Wisconsin, USA]

Elastomers: Strength and Extensibility in Simple Tension

When an elastomer is continuously stretched in simple tension, it fractures at a nominal stress σ_b, termed the tensile strength, which equals the tensile force per unit cross-sectional area of the undeformed specimen. The extension ratio λ_b is the length of the specimen, when fracture occurs, per unit initial length; the strain ε_b is thus $\lambda_b - 1$. Because the volume of a specimen increases insignificantly during a test, the true stress at fracture equals $\lambda_b\sigma_b$, which is the tensile force per unit cross-sectional area of the deformed specimen.

Although fracture in pure shear, equibiaxial tension and triaxial tension have also been studied, this discussion is limited to strength and extensibility in simple tension. When both of these quantities—sometimes termed the ultimate tensile properties—are large, considerable energy is required to effect fracture and the elastomer is said to be tough. Another measure of toughness is the fracture (or tearing) energy G. This is the energy required to advance a crack or cut by unit area and is commonly obtained from the force required to propagate a tear at a constant rate.

The strength and extensibility of an elastomer depend strongly on its molecular and supermolecular structures and on the temperature, extension rate and deformation history generally. To delineate the sources of toughness and its wide variability, elastomers can be divided into four categories. These consist of those elastomers that (a) are homogeneous and noncrystallizable; (b) contain a reinforcing particulate filler; (c) show strain-induced crystallization; and (d) are block copolymers that contain plastic domains in a rubbery matrix.

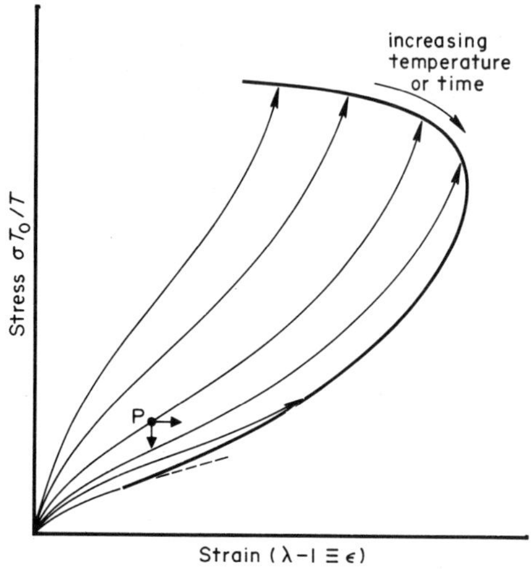

Figure 1
Schematic representation of stress–strain curves measured at different temperatures and extension rates; with an increase in either temperature or time (equal to $(\lambda\text{-}1)/\dot{\lambda}$), the fracture point moves clockwise around the failure envelope, depicted by the heavy curve

Although the homogeneous and noncrystallizable elastomers lack toughness under ordinary test conditions and, thus, are of minor technological importance, an understanding of their properties is basic to an understanding of toughness in other types of elastomers.

1. Phenomenological Aspects of Fracture of Homogeneous Noncrystallizable Elastomers

A schematic set of stress–strain curves, measured at a constant rate of extension $\dot{\lambda}$ over a wide temperature range, is shown in Fig. 1. At a particular temperature, an increase in $\dot{\lambda}$ is equivalent to a decrease in temperature. The lowest curve is independent of the extension rate or time (equal to $(\lambda-1)/\dot{\lambda}$). Such an equilibrium curve results whenever the molecular segments are highly mobile and all molecular relaxation processes occur rapidly compared with the extension rate. At a fixed strain, the equilibrium stress is nearly proportional to the absolute temperature. Therefore, the ordinate in Fig. 1 is $\sigma T_0/T$, where T_0 is an arbitrary reference temperature and T is the test temperature, both in kelvin. The nonequilibrium, or viscoelastic, stress–strain curves reflect molecular rearrangements whose type and extent depend on the strain, time, network topology and the mobility of molecular segments. In turn, the mobility depends on the temperature and chemical structure of the segments.

The ends of the curves in Fig. 1, indicated by arrows, define the heavy curve known as the failure envelope. When either the temperature is increased or the extension rate is decreased, corresponding to an increase in time, the fracture point moves clockwise around the envelope. The envelope continues down the equilibrium stress–strain curve and ends at the threshold values of $\sigma_b T_0/T$ and ε_b (see Sect. 2.1).

Except along the equilibrium curve, the stress and strain cannot simultaneously be held constant. Suppose that point P is reached during a test and thereafter the strain is held constant. The stress now relaxes along a vertical line. However, the anticipated equilibrium stress is unstable, as indicated by the failure envelope. Therefore, the specimen breaks either during or after stress decay. Similarly, if the stress is held constant after reaching point P, creep occurs and leads to fracture at a deformation near the failure envelope.

Data determined at numerous extension rates and temperatures can be interrelated as illustrated in Fig. 2. The abscissa is $\log \dot{\lambda}\, a_T$ where a_T is a temperature function, given by the WLF equation (see *Elastomers: Dynamic Mechanical Properties*) that represents the temperature dependence of segmental mobility. Moving left along the abscissa represents either an increase in $\dot{\lambda}$ or a decrease in temperature. This interrelation implies that the marked dependence of the ultimate tensile properties on temperature and extension rate results from the viscous dissipation of strain energy. The dissipation rate depends in a complex way on the extension rate and the rate of molecular relaxation processes.

2. Fracture Process and Sources of Strength

Fracture is a multistep process that effects the progressive breakage of the molecular chains that cross the fracture plane. The process involves the formation of a microcrack that grows slowly until an instability develops; thereupon the crack propagates catastrophically. In certain instances, many microcracks undoubtedly form and grow either sporadically or continuously until some crack propagates at high speed.

The evolution of a microcrack clearly involves dilation or cavitation. In theory, an absolutely homogeneous elastomer can cavitate when the hydrostatic tension ($\lambda\sigma/3$ for a specimen in simple tension) exceeds a critical value. In reality, a crack usually, if not always, forms near a heterogeneity, such as a foreign particle, gel microparticle, microvoid or surface defect, that gives a stress concentration equivalent to that from a sharp surface crack, in a rigid solid, 40 ± 20 μm in depth. Commonly, crack initiation occurs on a surface or near an edge of a specimen but sometimes within the specimen.

2.1 Homogeneous Noncrystallizable Elastomers

In a homogeneous noncrystallizable elastomer at a temperature far above its glass-transition temperature T_g, the molecular segments are highly mobile and allow cracks to form readily. More importantly, the mobile segments offer little resistance to crack growth. When the mobility is sufficiently high so that no energy is dissipated viscously near the tip of a growing crack (where the deformation rate can be very high), the fracture energy equals its threshold value G_0. According to theory and experiment, G_0 varies inversely with the concentration of elastically effective network chains (Gent 1978).

The threshold values of $\lambda_b\sigma_b$ and ε_b can be approximated by invoking the Griffith theory. Accordingly, $(\lambda_b\sigma_b)_0 = (EG_0/\pi c)^{1/2}$ and $\varepsilon_{b,0} = (\lambda_b\sigma_b)_0/E$ where E is the equilibrium tensile modulus and c is the crack length. For the SBR vulcanizate (Fig. 2), E reduced to 263 K is 0.76 MPa. Equating G_0 to 50 J m^{-2} and C to 40 μm, which are representative values, gives

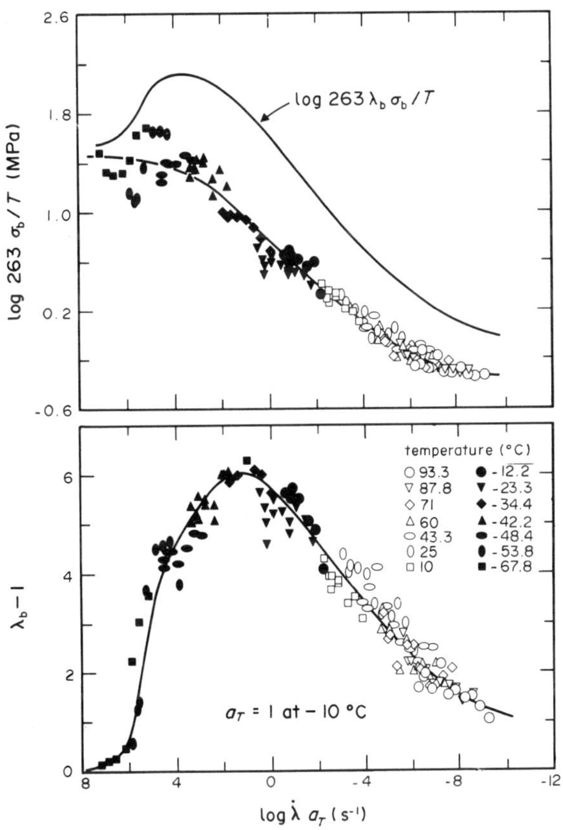

Figure 2
Ultimate tensile properties measured at various extension rates and temperatures on an unfilled styrene–butadiene rubber (SBR) vulcanizate; the composite curves result from superposing data at different temperatures

0.55 MPa and 0.72 for $(\lambda_b\sigma_b)_0$ and $\varepsilon_{b,0}$, respectively. At the right-hand end of the curves in Fig. 2, 263 $\lambda_b\sigma_b/T$ and ε_b are about 1.0 MPa and 1.0, respectively. These values, though clearly not minima, are only somewhat larger than those calculated from the Griffith equations which were derived for a linearly elastic solid containing a sharp edge crack and therefore are not strictly applicable to elastomers.

Although the curves in Fig. 2 have shapes typical of those for other unfilled noncrystallizable elastomers, ε_b does decrease rather rapidly with the concentration of effective network chains. Usually, $\lambda_b\sigma_b$ depends weakly on chain concentration and chemical structure provided data on different elastomers are compared at test conditions such that the rates of energy dissipation are similar. Observed differences in the tensile strengths of many elastomers arise primarily from differences in their viscoelastic properties. Even when a specimen is stretched such that its response is elastic (i.e., time independent), the deformation rate near a slowly developing crack is usually sufficiently high that viscoelastic effects are operative locally and retard its growth.

2.2 Filler Reinforcement

Incorporation of carbon black or other finely divided fillers (see *Fillers in Elastomers*) in a noncrystallizable elastomer increases its strength and toughness, except at low temperatures, without affecting its T_g significantly, if at all. Reinforcement results primarily from the broadening of the relaxation spectrum (see *Elastomers: Dynamic Mechanical Properties*) of the elastomer by the filler. Therefore, the capacity for the dissipation of strain energy is increased and maintained at temperatures above those for the corresponding unfilled elastomer. Secondarily, the filler particles may retard or stop a slowly growing crack, and also cause a crack to follow a tortuous path and possibly to bifurcate. Clusters of particles may promote cavitation and formation of microcracks. Such events decrease the strain energy locally and thus the driving force for crack growth.

Filler particles larger than a few micrometers in size are nonreinforcing. When a specimen is deformed, cavities usually form near or at the interfaces between particles and the rubbery matrix. Gross dewetting occurs thereafter and vacuoles form around the particles.

2.3 Crystallizable Elastomers

Certain elastomers, exemplified by a vulcanizate of natural rubber, crystallize rapidly when stretched two or three hundred percent. Reinforcing crystalline domains are thus generated, possibly beginning in highly deformed regions near incipient or actual microcracks. Crystallization not only reduces the strain energy but, more importantly, leads to domains that deflect, retard or block growing microcracks. When the strain energy becomes sufficiently high, the domains can deform plastically, thereby dissipating considerable energy and imparting high strength and toughness. Within the temperature range in which crystallization occurs readily, the strength and extensibility are large and relatively insensitive to temperature and extension rate. Typically, $\lambda_b\sigma_b$ is about 200 MPa and ε_b is between 6 and 10.

When strain-induced crystallization occurs only at temperatures substantially below that at which chemical degradation becomes significant (an example is butyl rubber), the incorporation of a reinforcing filler has a twofold effect. First, the filler increases the highest temperature at which strain-induced crystallization can occur and therefore it broadens the temperature range in which high strength and toughness are observed. Second, at temperatures above that at which crystallization can occur, the filler provides reinforcement (see Sect. 2.2).

2.4 Elastomeric Block Copolymers

Examples of elastomeric block copolymers (see *Thermoplastic Elastomers*) are segmented polyurethane and poly(urea–urethane) elastomers and the triblock copolymers such as SBS where S and B denote relatively long polystyrene and polybutadiene segments, the relative lengths of which are typically such that the volume fraction of polystyrene is between 0.20 and 0.30. These elastomers contain a rubbery phase and plastic domains that serve as multifunctional physical crosslinks. More importantly, the domains are highly effective reinforcing agents provided their volume fraction exceeds 0.20, their size is less than 0.1 μm and their softening temperature is substantially above the test temperature. The domains impart toughness more effectively than particulate fillers because they are deformable and usually dispersed uniformly. Their deformability serves to reduce stress concentrations, to delay formation of microcracks and to blunt crack tips. For a microcrack to form and grow sufficiently to become unstable, domains must be disrupted. This process necessitates considerable strain energy and is the primary source of strength and toughness. Although the strength and extensibility of these elastomeric materials depend strongly on temperature, $\lambda_b\sigma_b$ commonly exceeds 300 MPa and ε_b is typically beween 4 and 10.

See also: Elastomers: An Overview; Elastomers: Dynamic Mechanical Properties; Fillers in Elastomers; Thermoplastic Elastomers

Bibliography

Andrews E H 1968 *Fracture in Polymers*. Elsevier, New York

Gent A N 1978 Strength of elastomers. In: Eirich F R (ed.) 1978 *Science and Technology of Rubber*. Academic Press, New York, pp. 419–54

Halpin J C 1965 Molecular view of fracture in amorphous elastomers. *Rubber Chem. Technol.* 38: 1007–38
Lake G J, Thomas A G 1988 Strength properties of rubber. In: Roberts A D (ed.) 1988 *Natural Rubber Science and Technology*. Oxford University Press, Oxford, Chap. 15
Smith T L 1969 Strength and extensibility of elastomers. In: Eirich F R (ed.) 1969 *Rheology*, Vol. 5. Academic Press, New York, pp.127–221
Smith T L 1978 Strength of elastomers. A perspective. *Rubber Chem. Technol.* 51:225–52

T. L. Smith
[IBM Research Division, San Jose, California, USA]

Elastomers: Tests for Mechanical Properties

The scope of materials science involved and the types of tests made to determine physical properties of elastomers are illustrated by the following general topics (Brown 1986): density and dimensions; short-term stress and strain properties; friction and wear; creep, relaxation and set; fatigue; electrical tests; thermal properties; effect of temperature; environmental resistance; permeability; and adhesion, corrosion and staining. This article presents a rationale for these varied tests rather than attempting to summarize such a large field.

1. *Reasons for Mechanical Property Tests on Elastomers*

Physical properties of elastomers are determined for quality control, for compliance with specifications or for research and development.

Tests used for other solid materials are seldom directly applicable to elastomers because of their many unique properties. Stresses, for example, are very dependent on time and temperature and are nonlinear with strain. Because of the ultrahigh deformability and low moduli of elastomers, their test equipment must be accurate at higher displacements and lower forces than are needed for most engineering materials. Good reproducibility of results can be obtained only from thoroughly developed methods, which is one of the reasons for the many standardized tests that exist. Elastomers may also have unique values of Poisson's ratio, frictional properties and the Gough–Joule effect, and have very high energy storage. Hysteresis is also high and there is no metallic-type yield point. Many of these differences from other engineering materials relate directly to the nature of the force of retraction from a deformation, which is mostly due to entropy in elastomers but to internal energy in more rigid materials (Brown and Read 1984).

2. *Nature of Tests*

Elastomers and their products are tested for initial and continued performance in the unique applications that are enabled by their viscoelastic properties. Specimens are, therefore, almost always deformed either statically or dynamically, or both, in the test procedures (Conant 1985).

Static or low-deformation-rate tests for the common stress–strain properties of modulus, tensile strength and elongation are usually made only for quality control since elastomers are seldom elongated in service. Low-rate tests for creep, stress relaxation and set, however, reflect concern for limitations of service because of the time-dependent properties that are peculiar to elastomers. Examples are motor mounts, gasket seals and bridge bearings. The amount of static deformation also affects other properties such as gas permeability, friction, electrical characteristics and crystallization that are important, for example to tire inner liners, tire treads, antistatic applications and to some polymers when used under arctic conditions. Flexibility rather than continuous deformation is involved directly or indirectly in such applications as coatings, diaphragms and wearing apparel. Hardness (the resistance to indentation without puncture) is largely a function of modulus and requires different instruments for each of three ranges of hardness.

Elastomers are usually deformed dynamically in service, thereby transforming mechanical into heat energy. Deformation is thus resisted by both elastic and damping forces. The principal purposes of dynamic mechanical tests are to evaluate these forces and the attendant loss of useful energy. A bewildering number of tests for storage, loss and combination properties has been developed from the viewpoints of material properties, engineering properties and technological properties. This relation of physical test to the property involved has been treated in detail by Freakley and Payne (1978) and by Hepburn and Reynolds (1979).

3. *Quality Control*

Tests for controlling the quality of elastomeric materials and products start with those made by the manufacturer on the raw materials. Both physical and chemical properties of even the nonpolymeric components must meet such specifications as those for particle size, density, viscosity, purity, surface area, pH and stability. Tests for elastomeric properties, however, start with the polymer and continue through the various stages of preparation of first the uncured compound, then of the vulcanizate. The primary requirement during fabrication is that of having and maintaining suitable plasticity, which is the case of producing nonrecoverable deformation. For example, if the product is obtained by extrusion, it must also be well formed. Plasticity can be quantified by compression, rotation or extrusion plastimeters, which are often simplified versions of processing equipment. Although plasticity of the uncured material can affect product performance, its most important effect is on processibility in such operations as extrusion, milling, incorporation of compounding ingredients and shaping of products. Power consumption in each of these operations can be estimated from torque measurments on internal mixers.

Processibility is changed by mastication, addition of compounding ingredients, temperature, storage time and scorch. Scorch, which is a prevulcanization, is measured by curemeters or viscometers which can also be used for continuous evaluation of the state and rate of cure while vulcanization proceeds.

Factory batches are sampled at critical stages of their fabrication and their processing then completed in a laboratory. Quality control tests on the resulting product often merely determine whether or not tensile, modulus and elongation are within prescribed limits. Such tests are made to ensure that components having correct properties have been added in the correct proportions and that the specified mixing procedure has been followed. More tests on elastomers are probably made for control than for any other reason. However, quality control at one stage, say polymer manufacture, becomes acceptance criteria at another stage, say the manufacture of rubber sheeting.

4. Specifications

Specifications state the requirements that must be met by a material or product. They dictate the tests to be made and acceptable test results. Specifications may be made by a design engineer who requires certain performances, by a customer to ensure a uniform product or adequate quality, by the manufacturer to detect irregularities during fabrication or by a regulatory agency to maintain safety standards. Duplicate tests are often made by the final inspection department of a producer and by the acceptance laboratory of a consumer or merchandiser. This requires that the equipment and procedures used by each group should give results that differ by no more than an agreed amount. Systems for classifying rubbers by line call out given in ASTM Standards D2000 and D1207 are useful in writing specifications, especially for automotive applications.

Tests made for conformance to specifications, like those for quality control, have results which either permit or do not permit acceptance. Tests discussed in the following sections are more comprehensive in that they can be used to characterize the subject materials and are useful for research and development purposes.

5. Suitability for Particular Applications

Comprehensive tests on elastomers are required for compound development, for engineering applications and for validation of theoretical studies. Often standardized, such tests may range in complexity from simple determination of trends in the level of some property to thorough, designed experiments for determining practical levels of manufacturing variables for optimum combinations of properties.

A designed experiment consists of a statistical procedure that requires systematic determination of multiple properties (responses) on a multivaried system. It usually requires computer-aided analysis. Elastomers are particularly suited to this type of optimization of the wide range of properties and combinations thereof that can be achieved. A motor mount, for example, may require that static modulus, dynamic modulus and energy absorption remain within stated limits over a stated range of temperature, load, frequency, amplitude and concentration of ozone and certain oils. The property profile specified by the engineer must take into account the effects of variable component levels and their interactions. Since a particular profile may be achievable by more than one combination of manufacturing variables, the test system must also provide for evaluations of varied component levels and processing methods and of all possible interactions.

These complex requirements, supplemented by those for theoretical studies, have led to the development of very versatile test equipment. Not only can a given property, say dynamic modulus, be evaluated on a single specimen over a wide range of test conditions in a single apparatus, but the same apparatus can often be used for related properties, for example, energy absorption, stress relaxation or failure conditions. Such versatility is particularly important for testing elastomers which, because of high and often nonuniform deformability, have uncertain shape factors. Thus the results obtained from different instruments, which usually have differently shaped specimens, are generally very inconsistent.

6. Special Considerations for Testing Elastomers

Reproducibility of results, especially between laboratories, is very poor for elastomers unless detailed instructions are given and followed. Those for pretest procedures must include sampling, mixing, type of mold, mold geometry, vulcanizing time and temperature and rest time before testing; forming test pieces by cutting, buffing or plying; thermal and mechanical conditioning; and avoidance of specific interferences such as surface contamination, irradiation, ozone or other chemical attack, strains, high temperature, crystallization or undue aging time.

Equally detailed instructions are necessary for the interpretation of test results. These include scope, significance and applicability of results, precision and validity, accuracy checks, calculation and presentation of results. Tests on elastomers are often made to rank a series of samples; the results are relative with no necessity to meet specified values. Corrections for day-to-day variability are often made by inclusion of a control compound with each of a series of experimental compounds.

Some tests yield basic properties such as thermal conductivity, specific heat, thermal expansion, density and refractive index. These are independent of the measuring instrument and method. Values of these constants for typical vulcanizates of some common polymers have been published by Wood (1975). Usually, however, results are highly dependent on both the test method and the chosen conditions within that method. For example, the statement that a vulcanizate has a modulus of $5.0\,MN\,m^{-2}$ is meaningless unless the state and rate of deformation are given along with the temperature. Furthermore, the value must be identified as Young's modulus or rigidity modulus, as equilibrium or transient, as tangent or secant, whether based on original or attained cross section and whether determined on the first or subsequent deformations. These are some of the many important details that are included in standardized tests.

7. *Standards*

Tests may be standardized at international, national or industrial level. They represent a consensus of interested parties including producer, consumer, sales manager, regulator and theoretician. They are usually outgrowths of tests developed for company use, agreement between producer and consumer, or made-to-order tests for particular research or development work. Only by such broad-based input can the inclusion of all necessary considerations be assured.

Some 300 standards or recommendations on rubber have been developed by the International Standards Organization (ISO). Various national bodies such as ANSI in the USA, BSI in the UK, GOST in the USSR, AFNOR in France and DIN in Germany can supply ISO standards on request. These groups also have comprehensive standards of their own which usually conform essentially to corresponding ISO standards if they exist. Most of the ANSI standards are included in the ASTM *Annual Book of Standards*. A widely accepted list of standard test temperatures and other intrastandard agreements such as rest time between curing and testing are very helpful for operational efficiency and for correlating results from different tests.

The various types of standards include test methods, specifications, terminology, classification, practice, guide and information. All of these contribute to communication among interested parties, which includes those that make ordinances and set tariff rates.

See also: Elastomers: An Overview; Elastomers: Dynamic Mechanical Properties

Bibliography

American Society for Testing and Materials *Annual Book of Standards*, Pts. 9.01, 9.02. American Society for Testing and Materials, Philadelphia, PA

Brown R P 1986 *Physical Testing of Rubbers*, 2nd edn. Applied Science, London

Brown R P, Read B E (eds.) 1984 *Measurement Techniques for Polymeric Solids*. Elsevier, New York

Conant F S 1985 Physical testing of vulcanizates. In: Morton M (ed.) 1985 *Rubber Technology*, 3rd edn. Van Nostrand Reinhold, New York

Freakley P K, Payne A R 1978 *Theory and Practice of Engineering with Rubber*. Applied Science, London

Hepburn C, Reynolds R J W 1979 *Elastomers: Criteria for Engineering Design*. Applied Science, London

Wood L A 1975 Tables of physical constants of rubber. In: Brandrup J, Immergut E H (eds.) 1975 *Polymer Handbook*, 2nd edn., Pt. V. Interscience, New York, pp. 7–12

F. S. Conant
[Firestone Tire and Rubber Company, Akron, Ohio, USA]

Electrical Insulation: Solid Polymeric Materials

Electrical insulation, either solid, liquid or gaseous, is defined as a medium offering very high resistance to the conduction of electrical current. It is used in electrical systems to confine the current flow within the conductors and to isolate conductors in high-voltage systems from ground.

Solid insulation (also known as dielectric), in contrast to liquid and gaseous insulation, is not self-healing and must be repaired or replaced once it has failed. It is generally applied using a large safety factor to accommodate operational and environmental deterioration, as well as that of aging. Solid insulation for power equipment is generally built up using a combination of materials often applied in tape or wrapper form. The tapes are made up of a matrix of fibrous materials, such as glass fibers, synthetic fibers (e.g., Nomex) or natural fibers (e.g., paper), high-dielectric-strength materials (e.g., mica) and binders such as polyesters and silicones. Following the wrapping process, the entire insulation structure is filled and consolidated with liquid resin, often a monomeric solution which cures to a crosslinked polymer when heated *in situ*. Thus, primary solid insulation often depends on secondary insulation in the form of tapes, wrappings, varnish, support members and possibly even liquid and gaseous insulants to preserve long-term integrity.

1. Evaluation of Materials

Test methods for solid insulation fall into two categories: nondestructive testing (NDT) and destructive testing (measurement of dielectric strength). NDT involves the following dielectric parameters: dielectric constant, dissipation factor and resistivity or insulation resistance. Dielectric strength involves insulation thickness, density and composition.

1.1 Nondestructive Testing

For NDT, the electrodes are disk foil pieces attached to the flat specimen, one on each side with a guard ring around the ungrounded disk. If the specimen is an insulated conductor, a foil wrapper electrode is applied to the outer surface and two foil strips are applied beyond the edges of the test electrode (guard rings). In this case the conductor is grounded. Low-voltage capacitance bridges are used when measuring thin specimens, whereas high-voltage capacitance bridges are used for measuring properties of built-up solid insulation, especially if it is to be used in high-voltage power apparatus (Hague 1946).

The basic dielectric loss parameters which form the bases for comparison of various types of insulation material concern leakage current through the material and displacement current related to capacitive charge.

The dielectric constant K' is the ratio of the capacitance of the insulation to the capacitance of the same geometrical arrangement of electrodes with the solid insulation replaced by a vacuum. It is thus a relative factor, depending on density, molecular arrangement, polarizability of the molecules, presence of polar contamination such as water, ambient heat and frequency and other factors. Dielectric constants range from 1 for a vacuum to between 2 and 8 for common solid insulators and from 1000 to 100 000 for titanate ceramics and ferroelectric solid insulation.

The dissipation factor is an expression of the imperfection of a capacitor and, therefore, of its dielectric. In a perfect capacitor (using a vacuum between the plates) the current vector leads the voltage vector by 90° (see Fig. 1). The displacement current for a vacuum is given by

$$A_c = V2\pi fF \tag{1}$$

where f is the frequency (Hz), V is the voltage (V) and F is the capacitance (F).

For the solid insulation case, there is a leakage current A_g which changes the total current A_t to a

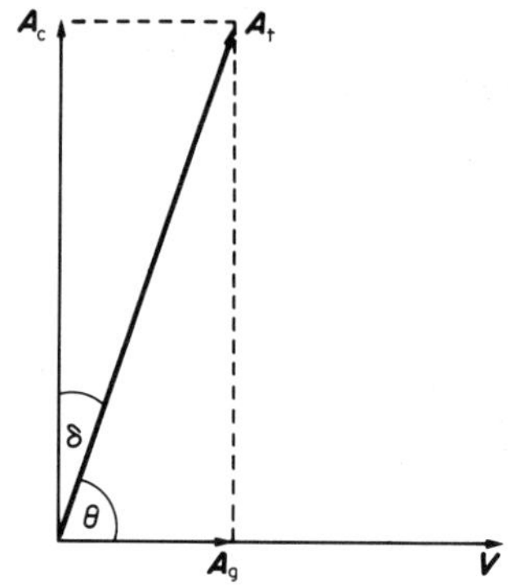

Figure 1
Vector diagram for a capacitor

new position with phase angle θ less than 90°, thus creating the loss angle δ:

$$A_g = V2\pi fF \tan \delta \tag{2}$$

where $\tan\delta$ is known as the dissipation factor. Therefore, $\tan\delta = A_g/A_c$. As with the dielectric constant, the dissipation factor is affected by factors such as heat, moisture, frequency, molecular arrangement and ionic content.

Many engineers erroneously use the term "power factor" in place of the term "dissipation factor." The power factor is really the cosine of the angle θ in Fig. 1. Thus,

$$\cos\theta = A_g/A_t = \tan\delta/(\tan^2\delta + 1)^{1/2} \tag{3}$$

For dissipation factor values at or below 0.10 (10%), the power factor is nearly its equal. However, many insulations in useful devices have dissipation factors greater than 10%, especially at operating temperatures. Therefore, the two parameters should not, in general, be considered interchangeable.

Multinetwork bridges can be set up to measure both the dissipation factor and the capacitance (Hague 1946). From the geometrical calculation of the cell factor (i.e., the capacitance with a vacuum between the electrodes) and dividing this into the measured capacitance, the dielectric constant is evaluated.

Resistivity, in ac use (60 Hz), is expressed by

$$\rho_{ac} = (3.00 \times 10^4)/K' \tan\delta \quad (\text{M}\Omega\,\text{cm}) \tag{4}$$

For the dc case (most commonly listed in property tables), the resistivity (for m meters of cable) is given by

$$\rho_{dc} = (2.73 \times 10^8)Rm/\log(D/d) \quad (\text{M}\Omega\,\text{cm}) \tag{5}$$

where R is the measured resistance (MΩ), D is the outer diameter (cm) and d is the conductor diameter (cm).

The power loss in m meters of cable at 60 Hz is given by

$$W = (9.12 \times 10^{-9}) V^2 K' \tan\delta\, m/\log(D/d) \quad \text{(W)} \qquad (6)$$

1.2 Destructive Testing (Dielectric Strength)

Dielectric strength, as defined by the American Standards Association, is the maximum voltage gradient (volts per unit thickness) that a solid insulator can withstand without breaking down. Often "dielectric strength" is listed erroneously as the actual measured breakdown voltage under specified conditions for a given solid insulation. In this case, thickness measurements are needed since dielectric strength varies markedly with thickness of the dielectric.

To calculate the dielectric strength, well-rounded electrodes are applied to the sample under oil to determine breakdown voltage. Several tests may be run on solid insulation: (a) impulse wave (microsecond rise to peak), (b) 60 Hz straight rise (at 500 V s^{-1} or more) and (c) 60 Hz 1 min hold. Values obtained are in the range 100–150 kV cm^{-1} at 5 cm thickness, 600–750 kV cm^{-1} at 1 mm thickness and 1500–1900 kV cm^{-1} for 0.1 mm thick samples. This function, modified by the nature of the solid insulation, is roughly represented by the empirical formula

$$G_s = 250t^{-0.385} \qquad (7)$$

where G_s is the dielectric strength (kV cm^{-1}) and t is the thickness (cm). G_s for plastics is found to be approximately 14 times greater than the dielectric strength of air in comparable spacings.

There are many theories of breakdown mechanisms. These are discussed by Eichberger (1966).

2. Solid Insulation Types

Solid insulation falls into four main categories: inorganics, organic polymers, elastomers and fibers (Fink and Beaty 1978).

2.1 Inorganics

Inorganic solid insulation includes ceramics, glass, mica and asbestos. More information on these is provided by Bever (1986).

2.2 Organic Polymers

Organic polymers make up the largest group of solid insulators in the industry. The chemical field has expanded in this area with the search for more heat-resistant polymers. Over 35 polymers are in use for bulk solid insulation and another 35 polymers are used in elastomers (see Sect. 2.3), fibers (see Sect. 2.4) and magnet wire. Many of the new polymers have good thermal endurance properties, some have improved electrical properties and others have special properties such as solderability without prior removal of insulation.

Polymer insulators are divided into two general classifications: thermoplastic and thermosetting resins. Thermoplastics are useful in flexible insulation, extrudable forms, potting and molding areas, and in sealing. Thermoplastics usually have a definite melting point, which is always higher than the use temperature.

Thermosetting polymers, with molecular crosslinks between chains, have no definite melting point and are useful wherever solid insulation must maintain its shape throughout the life of an electrical device; for example, magnet wire coating must continually separate the turns in a winding over the operating-temperature range of a coil. Thermosetting resins are often applied in monomeric or thermoplastic form to enhance filling and/or impregnation. Then, after application, the resin is heated or otherwise activated with or without a catalyst to cause crosslinking and the formation of a thermosetting solid insulation. Curing methods include heat, nuclear radiation, ultraviolet radiation, vulcanization and catalytic reaction.

Polymers are available in a series of heat-stable categories. Table 1 lists representative polymers in each category. The temperature quoted for each category is the limiting temperature for expected "long-life" use. These temperatures may be exceeded if shorter lifetimes are acceptable. At present, long life is set at 20 000 h continuous full-load use. This allows for normal operation in intermittent use at less than full load, which can lead to lifetimes as long as 10 years in the field.

The dielectric properties of polymers (Harper 1975) are governed by, among other factors, their molecular structure. The presence of polar radicals such as halogens, dipolar molecules and ionic impurities produces poor electrical properties in general. Thus, the chlorinated and fluorinated polymers tend to show higher losses but have better thermal stabilities. Noted exceptions to this trend are the fluorocarbon polymers, polytetrafluoroethylene and polychlorotrifluoroethylene, both of which have excellent electrical properties as well as high thermal stability. The nonpolar polymers such as polyethylene have excellent dielectric properties but are largely in the 105 °C category. They have a very wide range of applications not requiring high thermal performance. There are exceptions, as seen in Table 1. Crosslinking of some polymers, either by chemical means or by exposure to some type of radiation, usually raises their temperature index. The best polymers thermally, apart from the fluorocarbons,

Table 1
Polymer insulation shown by temperature category

Temperature (°C)	Polymer insulation
90	Cellulose acetate, cellulose acetate–butyrate, poly(vinyl chloride)
105	Acrylonitrile–butadiene–styrene, polystyrene, poly(vinyl formaldehyde), glycerol phthalate, methyl methacrylate, urea–formaldehyde, phenol–formaldehyde, chlorinated polyether, melamine–formaldehyde, ionomer polymer, methyl pentene olefin, polyallomer, polyamide, poly(phenylene oxide), polyarylether, polycarbonate, poly(ethylene maleate), polyethylene, polypropylene, polyurethane
130	Epoxy, fluorinated ethylene, propylene, polyethylene–polytetrafluoroethylene polythylene–polychlorotrifluoroethylene, phenoxy, crosslinked polyethylene
155	Poly(ethylene terephthalate), poly(phenylene sulfide), poly(aryl sulfone), polysulfone
180	Poly(vinyl fluoride), poly(vinylidene fluoride)
200	Polyesterimide, polyesteramideimide, poly(methyl siloxane)
220	Polyamideimide, aromatic polyamide, diallyl phthalate, diallyl isophthalate, polychlorotrifluoroethylene
250	Polytetrafluoroethylene, polyimide

are those based on the imide structure. Polyimide in tape or sheet form is capable of 250 °C operation. Other imidized polymers appear in the 200 °C and 220 °C sections of Table 1.

Factors that affect the electrical properties of polymers are heat, moisture, aging, radiation, frequency, voltage and thickness (see *Electrical Properties of Polymers*). As the temperature is raised, little change is seen in loss and dielectric strength until the softening temperature (glass-transition point) is reached. This temperature, different for each polymer, is the point at which molecules become free to move within the polymer structure and thus absorb energy by friction. The result of any further raising of the temperature is a sharp rise of the dissipation factor and a sharp fall in the dielectric strength. It is important that the chosen operational level for the polymer is not greatly in excess of its glass-transition point. The higher-heat-stable polymers do not have sharply defined transition points and some do not soften at all prior to pyrolysis.

Moisture affects some polymers such as polyamide and cellulose derivates, but many are not seriously affected unless they contain fillers or are in built-up insulation structures where moisture can seep in to lower the combined electrical properties. Surface moisture promotes leakage across the surface, resulting in "tracking." Tracking occurs when surface leakage current leads to heating of the insulation surface which, over a period of time, is sufficient to cause carbonization of the organic insulation. The final result is a surface breakdown by conduction and/or flashover.

Aging early in service life actually improves properties since the polymer hardens and the insulation loses volatiles. Thereafter, however, aging has the effect of increasing losses and lowering dielectric strength. Finally, the dielectric strength reaches levels comparable to that of air. In addition, aging may cause the polymer to crack or become carbonized resulting in conduction to ground. Thermal endurance is the term used to describe the resistance of a material to prolonged aging in heated air.

Nuclear radiation changes many molecular radicals to free ions. These ions carry leakage current through the polymer wall, thus contributing to high dissipation, low resistivity and low dielectric strength. Polymers need to be protected or used in conjunction with inorganic insulation such as mica to remain in high-radiation fields. The aforementioned effects of nuclear radiation can be removed (after exposure) by heating the polymer for a given time period. Simultaneous exposure of a polymer to both heat and radiation produces less severe changes in losses and dielectric strength. Radiation also causes surface erosion in the polymer similar to that caused by corona attack.

The frequency of line voltage can affect the electrical properties of a polymer. As the frequency is raised, the dissipation factor rises to peaks at resonant frequencies corresponding with, first, molecular movement and, at a higher level, electronic movement. Dielectric heating occurs at the electronic peak.

Voltage alters the dissipation factor for high-voltage testing of polymer insulation. The dissipation factor remains low in the low-voltage region until ionization occurs in or around the insulation, when the dissipation factor rises sharply with voltage. The location of this point gives information about consolidation and internal voids. Oil is often used to surround the test piece to minimize external ionization.

2.3 Elastomers

Elastomers are a special type of polymer insulation (Harper 1975). Sufficient freedom of movement of the molecular chains gives these polymers great elasticity and flexibility. Rubber, both natural and synthetic, is important in electrical insulation for cable coating, potting compounds, coatings for cloth and tape, adhesives and protective articles such as boots, hoods, gloves and mats. Much of the research on rubber has been directed towards improving ultraviolet, oil, corona and ozone resistance. One exception is the development of silicone rubbers, which are designed for, among other things, high-temperature use.

Some of the elastomers currently in use are natural rubber (NR), polyisoprene (IR), butyl rubber (BR), chlorobutyl rubber (CBR), chlorosulfonated polyethylene (CSM), crosslinked polyethylene (CLPE), polysulfide elastomer (Thiokol, PSE), polyurethane rubber (AU, EU), acrylic rubber (ACM, ANM), butadiene–acrylonitrile–poly(vinyl chloride) (NBR–PVC), acrylonitile–butadiene rubber (ABR), Buna-N (NBR), styrene–butadiene (Buna-S, SBR), polychloroprene (Neoprene, CR), epichlorohydrin (CO), polyethylene–polypropylene–butadiene (EPDM), ethylene–propylene terpolymer (EPT), polyethylene–polypropylene (EPM), polyfluoro elastomer (FKM), flowable rubber (depolymerized rubber, DPR), silicone rubber (RTV) and polyurethane rubber (PUR). Of these, Neoprene, ABR, Buna-N, NBR–PVC, CO and FKM have poor electrical properties but, because of their other properties, they are useful as protective members such as cable jackets. This group also has good mechanical strength and resistance to oil, ozone, water, corona and radiation—properties not shared by many of the elastomers with good electrical properties. The better electrical rubbers are Buna-S, EPDM, EPT, butyl rubber, CLPE, IR, PSE, RTV and PUR. The silicone rubber family is extensive and they are all room-temperature vulcanizable (RTV), as are all the other products in the group. They are listed as general purpose elastomers resisting high-heat conditions. Methyl–vinyl–phenyl (MVP) siloxane, although having greater strength, is for lower-temperature uses; its major feature is excellent radiation resistance. Methyl–phenyl siloxane is also used for lower temperatures and has similar radiation resistance to MVP silicone rubber. Methyl–fluoro siloxane has oil, fuel and solvent resistance (Saums and Pendleton 1973).

2.4 Fibers

Fibers in solid electrical insulation cover a wide range from natural fibers to glass fibers to synthetic organic polymer fibers. Fibers take the form of thin serving strands for magnet wire, roving for tie cords, woven tapes and cloths, paper, chopped fillers and sleeving structures.

In the natural form, wood is used with wax or resin impregnation for support. However, wood has been largely replaced by composites of compressed glass fibers, quartz or minerals impregnated with various binders, or plastic laminates. Wood products such as paper chopped fibers for resin fillers, pressboard for barriers and the many cellulose derivatives in fiber form have long been used as solid insulation. Paper is formed in a wet process from wood pulp fibers. The sheets are modified with various binders and often calendered to increase their density. Sometimes cotton fibers are added to improve mechanical strength. Paper is used in small coils, in slots of small motors, impregnated with oil in power cables and power transformers, and in laminates with polymer binders. The thinness of paper, combined with its ability to absorb oil or some other liquid insulator, makes it an ideal high-dielectric-strength insulation when wound in multiple layers on a conductor.

Cotton, once used extensively in a number of solid electrical insulation applications, has now been replaced by synthetic yarns and woven products, but is still used as a filler in resin moldings.

Polyamide, nylon fiber, the first synthetic fiber to be produced, has excellent mechanical strength but, owing to its affinity for water, has limited applications as electrical insulation, although it is used in sleeving to some extent.

Poly(ethylene terephthalate), Dacron fiber, has proved to be an important synthetic yarn in electrical insulation use (Saums and Pendleton 1973). Dacron is used in the form of paper, mat, serving thread, roving, woven tapes and cloth, sleeving and laminates. It is capable of 155 °C use.

Aromatic polyamide (aromide), Nomex fiber, is a high-temperature synthetic fiber. It can be made into paper, cloth and roving. Nomex has good resistance to the effects of heat and radiation and has thermal capabilities in the 220 °C range. It can be used in dry-type transformers for power applications and is also useful in power motors, laminates and high-temperature coils. The most widely used form is paper. Although it is not easily wet by liquids, Nomex can be impregnated using vacuum techniques. Nomex is used with glass fibers, plastic films and mica for a number of applications.

Glass fibers, used since the 1940s, are still important in solid electrical insulation. Fiberglass can be served on magnet wire, used in the woven form for tape, sheeting and sleeving, and in the roving form as tie cord. The woven sheeting is used extensively in laminates. In combination with silicone or polyimide binders, fiberglass laminates are useful for operation at 220 °C or higher.

Asbestos, an inorganic fiber, was considered until recently to be too hydrophilic to be useful in many solid insulation structures. However, modern techniques have produced resin-treated asbestos papers and tapes well suited to dry-type apparatus such as transformers operating at temperatures in the range 155–220 °C.

3. Application and Use of Solid Insulation

Solid insulators, more than liquid and gaseous insulators, must be applied with concern for the following factors: elimination of defects, matching of various components (compatibility), environmental effects, homogeneity of processed material, and mechanical and thermal stresses apart from thermal aging.

The broad range of use of solid insulation is in (a) power equipment (motors, transformers, capacitors, generators, transmission lines, cables and switch gear), (b) electronic, communications and aerospace, and (c) home and commercial uses such as appliances, farm equipment and air conditioning.

Power equipment must be insulated to withstand overvoltages caused by power switching, lightning strokes and line disturbances. The insulation must also withstand long-time voltage application effects. Voltage endurance experiments indicate that dielectric strength is adversely affected by extended time on voltage when the possibility arises of corona occurring in the vicinity of the insulation. The insulation, if organic, is gradually eroded by corona attack, eventually leading to a decrease in dielectric strength. This function can be represented (over a limited range of voltage gradient) by an inverse power law:

$$s = kG^{-n} \tag{7}$$

where s is the time, G is the voltage gradient and k and n are constants.

Power devices also must be built with high resistance to mechanical forces, which can be large as, for example, in the mechanical crushing action in a transformer when a short circuit occurs.

Electronic equipment used in high-frequency applications must, in addition to having adequate dielectric strength, be insulated with low-loss (low-dissipation-factor) and low-dielectric-constant materials, such as polystyrene and polyethylene. Special problems exist for some electronic insulators. In integrated circuits, the ceramic substrates need to be single-crystal types such as sapphire and magnesium aluminate spinels to permit epitaxial deposition of the single-crystal semiconductor circuits (Saums and Pendleton 1973). In potting of electronic units, special catalysts are needed to ensure high volume resistivity after humidifying. In inertial instruments, such as gyros, care must be taken to eliminate foreign particles, closely match expansion coefficients and keep outgassing to a minimum.

In domestic and office appliances the most demanding application is in hermetically sealed motors for air conditioning. These motors operate in a refrigerant atmosphere. The insulation is subject to cyclic temperature and cyclic pressure changes while exposed to both liquid and gaseous fluorocarbons. The polymers tend to soften and be plasticized by the refrigerant. Special magnet wire coatings have been developed to minimize these effects.

In summary, the problems of concern to the user of solid insulation are

(a) thermal—thermal endurance, shock and conductivity;

(b) mechanical—including system and material strength and stability, resistance to shock, vibration fatigue, creep, flow;

(c) chemical—for example, resistance to water, solvents, refrigants, aging by-products;

(d) electrical—dielectric strength (initial, after aging, after corona attack, in humidity), surface tracking, dielectric constant, dissipation factor, resistivity; and

(e) environmental—radiation effects, ozone, infrared, ultraviolet, salt spray, contamination.

The product designer has to take data from the insulation producers, select system components, build prototypes and run system tests with thermal aging and any other operational and/or environmental test conditions. Thus, knowledge of the integrity of the system in the proposed use may be built up. This is known as functional-life testing.

As solid inorganic insulation ages it becomes more brittle and shrinks in size. Under the influence of mechanical forces, an aged polymer is vulnerable to cracking and scaling. The dielectric strength falls with time according to an inverse power law similar to that of the corona-effects law (Eqn. (7)). When about half the initial dielectric strength has been lost, organic and paper form solid insulators are at the brittle point, where failure could occur by vibration. Therefore, half dielectric strength is chosen as the end-point criterion for these materials. However, for solid insulators in fixed positions, such as magnet wire coatings, it is usual to take about 7% of the initial dielectric strength as an end point, since these materials last longer in actual machines. The principal factor responsible is that the polymer coatings are more tightly held between adjacent conductors. Another factor is that turn-to-turn voltages are very small and the air gap can sustain the stress until metallic contact is made.

Good thermal dissipation in electrical equipment, especially by the solid insulation itself, is important as it lowers the temperature of the conductors, thereby reducing losses and improving the efficiency of a device. Many organic polymers have poor thermal conductivities. They are greatly improved by use with higher-thermal-conductivity insulators such as mica, glass fibers and asbestos fibers. One combination found to be particularly good is glass fibers bonded with triaryl cyanurate polyester resin. Alumina, in the ceramics class, is one of the best solid insulators in terms of its thermal conductivity.

See also: Electrical Properties of Elastomers; Electrical Properties of Polymers

Bibliography

Bahder G, Garrity T, Sosnowski M, Eaton R, Katz C 1982 Physical model of electrical aging and breakdown. *IEEE Trans. Power Appar. Syst.* 101: 1379–90

Bamji S, Bulinski A, Densley J, Garton A 1983 Etching and the morphology of cross-linked PE. *IEEE Trans. Electr. Insul.* 18: 32–41
Bever M B (ed.) 1986 *Encyclopedia of Materials Science and Engineering*. Pergamon, Oxford
Dakin T W 1981 *Electrical Effects*, Final EPRI Report Technical Planning Study TPS 79–723. Institute of Electrical and Electronics Engineers, New York
Eichberger J E 1966 Electrical failure in solids. *Electro-Technology (Beverly Shores, Indiana)* 77(5)
Field A W, McAllister D 1982 Polymeric insulation and sheathing materials. In: McAllister (ed.) 1982 *Electric Cables Handbook*. Granada, London, pp. 41–59
Fink D G, Beaty H W 1978 *Standard Handbook for Electrical Engineers*. McGraw-Hill, New York
Hague B 1946 *Alternating Current Bridge Methods for the Measurement of Inductance, Capacitance and Effective Resistance at Low and Telephonic Frequencies*. Pitman, London
Harper C A 1975 *Handbook of Plastics and Elastomers*. McGraw-Hill, New York
Porter C H, Boyd R H 1972 *Dielectric Properties of Polymers*. Plenum, New York
Saums H L, Pendleton W W 1973 *Electrical Insulating and Dielectric Functions*. Hayden, New York
Struik L C E 1978 *Physical Aging of Amorphous Polymers and Other Materials*. Elsevier, Amsterdam

W. W. Pendleton
[Muskegon, Michigan, USA]

Electrical Properties of Elastomers

Elastomers, like other polymeric materials, are widely used for electrical insulation in electrical and electronic systems. Their elastomeric physical properties, coupled with the wide range of electrical insulation properties available, offer product opportunities not possible with many other polymeric materials. This is particularly the case for electrical and electronic applications requiring flexible insulating materials, a major example of which is wiring and cabling. Thermoplastic nonelastomers can be used for small-diameter wiring and cabling, but only elastomers are sufficiently flexible for larger diameters. There are also, of course, many other electrical and electronic applications requiring elastomeric insulating materials.

The important electrical properties requiring consideration are resistance-related properties such as volume resistivity, surface resistivity and insulation resistance; bulk-voltage-related properties such as dielectric strength, voltage breakdown, voltage endurance and internal corona; surface-voltage-related properties such as track resistance and surface or external corona; and electrical loss properties such as dielectric constant (permittivity) and dissipation factor (tan δ). The general electrical properties of elastomers are shown in Table 1. It should be mentioned that since wide variations in formulation and processing are possible, so wide variations in electrical properties are possible. However, some guideline statements can be made.

(a) *Resistance-related properties*. Most elastomers are basically good insulators at room temperatures and humidities. For all elastomers, resistance properties decrease rapidly with increasing temperatures or humidities. Silicones and fluoroelastomers have highest stability in this respect.

(b) *Bulk-voltage-related properties*. Most elastomers have basically good dielectric strength and voltage-breakdown properties. These properties decrease with increasing temperature

Table 1
General electrical properties of elastomers

Elastomer	Volume resistivity (Ω cm)	Dielectric strength, short-time V μm^{-1})	Dielectric constant (at 1 kHz)	Dissipation factor at (1 kHz)
NBR	10^{10}		13.0	0.055
SBR	10^{15}		2.9	0.0030
IIR (butyl)	10^{17}	23.6	2.1–2.4	0.0030
CSM (chlorosulfonated polyethylene)	10^{14}	19.7	7–10	0.03–0.07
EPR (EPM)	10^{15}–10^{17}	35.4	2.7–3.4	0.0065–0.0075
EPT (EPDM)	10^{15}–10^{17}	35.4–39.4	3.0–3.5	0.004 (at 60 Hz)
FPM (hexafluor)	10^{14}	23.6	5.9	0.055
FPM (chlorotrifluor)	10^{13}	9.8–29.5		0.03–0.04
FSI (fluorosilicone)	10^{13}–10^{14}	11.8–13.8	6.9–7.4 (at 100 Hz)	0.03–0.07 (at 100 Hz)
CR (Neoprene)	10^{11}	5.9–23.6	9.0	0.030
NR or IR (natural or synthetic)	10^{15}–10^{17}		2.3–3.0	0.0025–0.0030
Polysulfide	10^{12}	9.8–23.6	7.0–9.5	0.001–0.005
Urethane	10^{11}–10^{14}	13.8–20.7	5.0–8.0	0.015–0.010
SI (silicone)	10^{11}–10^{17}	9.8–25.6	3.0–3.5	0.001–0.010

and electrical operating frequency. Corona (ionization of air in entrapped voids) decays most elastomers in sufficient time.

(c) *Surface-voltage-related properties*. Under high-voltage arcing conditions, tracking (formation of carbon paths) occurs in most elastomers due to their carbon structure. Most resistant are silicones and fluoroelastomers, due to the presence of silicon and fluorine atoms in the structure.

(d) *Electrical-loss properties*. The dielectric constant and the dissipation factor increase with increasing temperature and electrical operating frequency for all elastomers. Silicones and fluoroelastomers exhibit very good stability in this respect, when compared with other elastomers.

See also: Electrical Insulation: Solid Polymeric Materials; Electrical Properties of Polymers

Bibliography

Harper C A (ed.) 1975 *Handbook of Plastics and Elastomers*. McGraw-Hill, New York

C. A. Harper
[Westinghouse Electric Corporation, Baltimore, Maryland, USA]

Electrical Properties of Polymers

The most familiar electrical property of polymers is their resistance, $R=\rho s/A$, where A is the electrode area, s is the electrode separation and ρ is the resistivity. The resistivity is an intrinsic material property which chiefly depends on temperature, humidity and the nature of the polymer and its impurities. The mode of conduction is usually ionic and is enhanced by humidity by several orders of magnitude.

Under ac electric fields, polymers store electrical energy proportional to their capacitance, $C=\varepsilon\varepsilon_0 A/s$, where ε(or ε') is an intrinsic material property termed the dielectric constant or relative permittivity and ε_0 is the permittivity of free space. Like ρ, ε also depends on conditions and the nature of the material; it is typically in the range 2–8. In addition to charge storage, polymers dissipate part of the ac electrical energy as heat. The intrinsic material property that describes this process is the loss factor ε'' (related quantities are $\tan\delta=1/Q=\varepsilon''/\varepsilon'$).

The ac and dc properties are related in that the Fourier transform of the transient current gives the complex permittivity, $\varepsilon^*=\varepsilon'-j\varepsilon''$. Because of the complex nature of ε^*, the change of ε' with frequency ω is related to the integral of ε'' with respect to $\ln\omega$, the Kramers–Kronig relation. At very high (optical) frequencies, $\varepsilon^*=\eta^2$, where η is the (complex) refractive index.

1. Material Concepts

1.1 Electric Field Coupling

The electrical response of polymers is caused by coupling of the polymer motion to an electric field. This coupling occurs in three principal ways.

(a) *Polarization*. The polarizability α measures the ability of atoms to distort their normal electron configuration in response to an applied field. It depends on the nature of the atoms present, increasing with increasing atom size, and is manifest in the optical and all lower-frequency responses.

(b) *Dipole interactions*. Dipoles result from an asymmetric distribution of charge between nearby atoms. Their moment μ measures the product of charge and its separation. Dipole motion in polymers is determined by the viscous motion of the polymer chains in which the dipoles occur. Together, these two effects determine the relaxed dielectric constant ε_R according to

$$\frac{\varepsilon_R-1}{\varepsilon_R+2}=\frac{N}{\varepsilon_0 M}\left(\frac{\alpha+\mu^2}{3kT}\right) \tag{1}$$

where N is Avogadro's number, M is the molecular weight and T is temperature.

(c) *Ionic interactions*. Ions, both free and attached to a polymer chain, also couple their motion to an applied field. Free ions contribute to dc conduction in proportion to their concentration and mobility. Attached ions reflect the motion of the polymer chains to which they are bound.

The relation between the relaxing orientation of one molecule and its relaxing environment is accounted for in the Onsager model:

$$\frac{(\varepsilon_R-\varepsilon_u)(2\varepsilon_R+\varepsilon_u)}{3\varepsilon_u}=\frac{N\mu^2}{3kT} \tag{2}$$

where ε_u is the "unrelaxed" dielectric constant. This may be extended to the time-dependent response (Nee and Zwanig 1970).

1.2 Polymer Motions

As a result of the couplings described previously, the response to electric probes reflects the motion of polymer chains (e.g., by the temperature or frequency dependence of ε''). Polymer motion results from thermally activated, viscous motions or polymer-chain segments. The simplest case, the Cole–Cole model, is that in which all moving units are alike. This model produces the result

$$\varepsilon^* - \varepsilon'_\infty = \frac{\varepsilon_0 - \varepsilon'_\infty}{1 + (j\omega\tau)^\alpha} \qquad (3)$$

For $\alpha = 1$, a single relaxation time τ is characterized, but the model accounts for a distribution of relaxation times for $0 \leqslant \alpha < 1$. The chain motion is thermally activated and so τ is normally described by the Arrhenius equation:

$$\tau = \tau_0 \exp(-H/kT) \qquad (4)$$

The case for partially crystalline polymers is more complicated, as different contributions of amorphous and crystalline regions to the total relaxation process may result in different values of τ.

The special case of a phase transition has a profound effect on local polymer motions and, hence, on electrical response. Because polymers are rarely used above their melting points, this transition is usually of less interest than the second-order, rubber–glass transition at the glass-transition temperature T_g. Near T_g, the WLF equation describes τ much better than the Arrhenius equation:

$$\frac{\ln \tau}{\tau_g} = \frac{a(T - T_g)}{b + T - T_g} \qquad (5)$$

where $a \simeq 40$ and $b \simeq 52$ for typical polymers.

The various dispersions are conventionally denoted by Greek letters (i.e., α, β, γ and δ) according to their occurrence in frequency or inverse temperature space. Hence, this letter designation does not, per se, connote the relaxation mechanism. For example, the glass–rubber transition in an amorphous polymer is associated with its α relaxation. By contrast, the glass–rubber transition in a polycrystalline polymer is associated with its β mechanism.

2. *Electrical Properties of Commercial Polymers*

Most plastics in commercial use are mixtures of polymers with various additives. The additives, which may constitute a small (<1% for antioxidants in PE) or a large proportion (>50% for plasticizers in poly(vinyl chloride)) of the plastic, affect the electrical properties. Mixtures and blends of polymers which retain their individual identity tend to have ac properties like those of their constituents. However, copolymers are likely to have properties different from those of the constituent homopolymers.

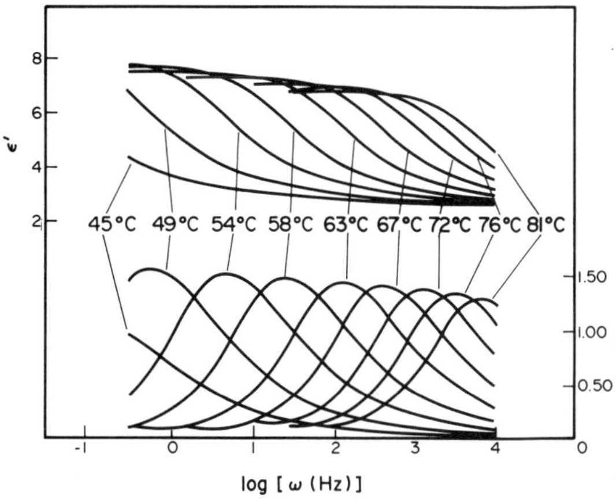

Figure 1
Complex permittivity of poly(vinyl acetate) for the glass–rubber (α) transition

Two dielectric processes are commonly seen in amorphous polymers. Poly(vinyl acetate) is a typical example, showing α and β transitions. The mechanism (see Fig. 1) follows a WLF equation with regard to the frequency of ε'' maxima. The mechanism's strength, $\varepsilon_0 - \varepsilon_\infty$, decreases with increasing temperature, approximately as T^{-1}. The secondary mechanism, β, is considerably broader and has a much lower strength; it, however, follows an Arrhenius law with regard to the frequency of ε'' maxima. Both α and β behavior are shown in Fig. 2.

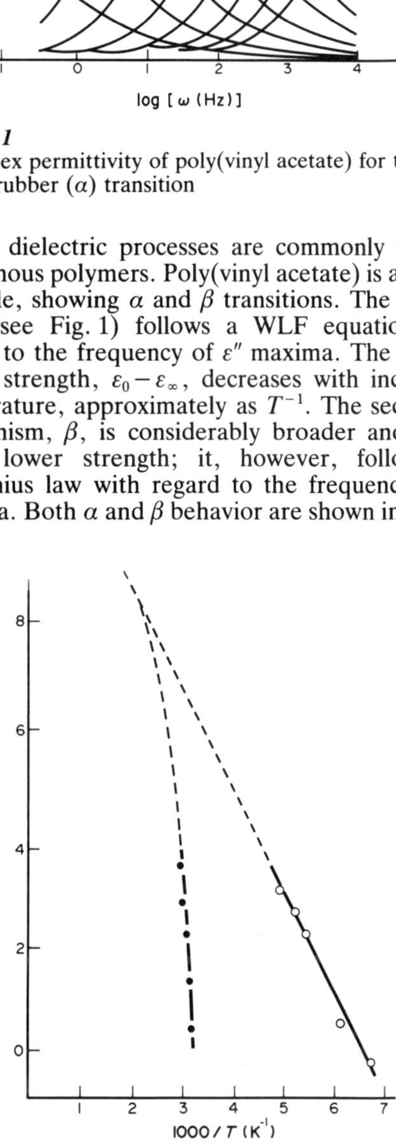

Figure 2
Frequency of ε'' (f) maxima against reciprocal temperature for poly(vinyl acetate): ●, α transition; ○, β transition

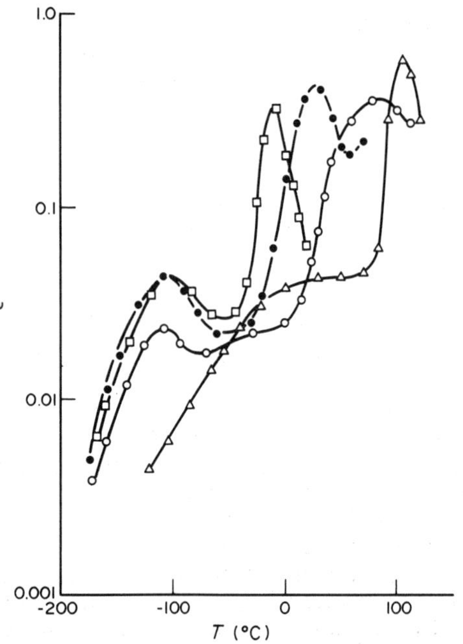

Figure 4
Temperature dependence of dielectric loss at 1 kHz for ABS/PVC bends: □, 100% ABS; ●, 60% ABS–40%PVC; ○, 20% ABS–80%PVC; △, 100% PVC

Polycrystalline materials often show at least three relaxation mechanisms. For example, branched (low-density) polyethylene exhibits three distinct loss maxima at 10 kHz as a function of temperature. Figure 3 shows oxidized polyethylene with α, β and γ mechanisms giving distinct peaks. In simple terms, the α mechanism is related to dipole rotation in the crystalline region, whereas β represents a glass–rubber mechanism associated with the amorphous region. The γ mechanism is often attributed to local motion in the main chain. Alternative explanations attribute all or part of γ to crystalline defects.

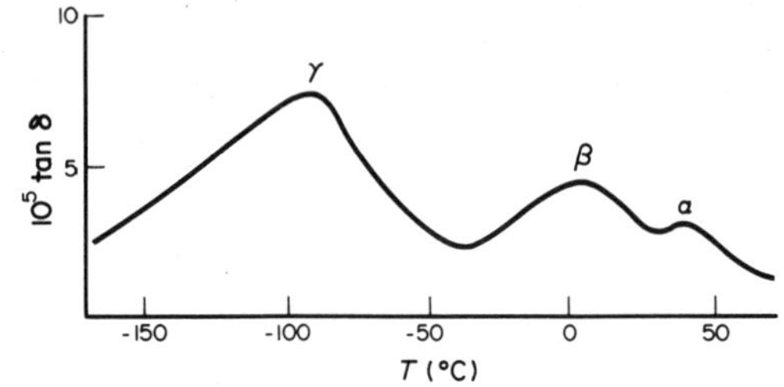

Figure 3
Dielectric loss at 10 kHz as a function of temperature for low-density polyethylene

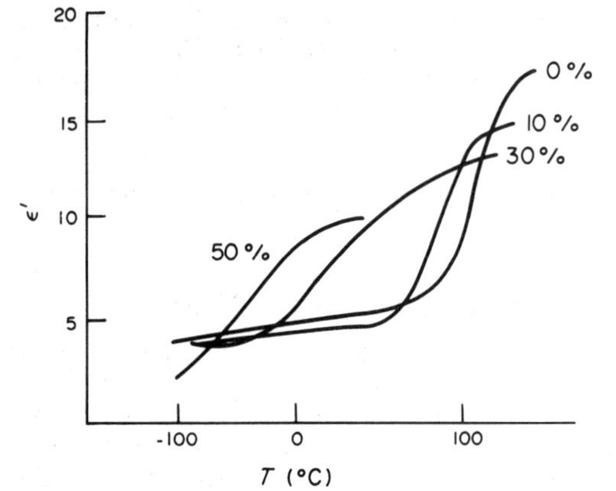

Figure 5
Temperature dependence of permittivity at 30 Hz for a diisooctyl phthalate/poly(vinyl chloride) system as a function of plasticizer content

Significant crosslinking will affect the dielectric behavior of a polymer. This can be illustrated by the changes in the behavior of epoxy resins caused by extended curing; the additional crosslinking raises the sample T_g, affecting the α mechanism and eventually significantly lowering the transition strength as the crosslink density bcomes high.

Block copolymers show the distinct dielectric behavior of their constituents, whereas random copolymers show intermediate behavior. The same dual type of behavior is noted in those blends which are compatible and show one glass-transition mechanism. An acrylonitrile–butadiene–styrene/poly(vinyl chloride) (ABS/PVC) blend is shown in different proportions in Fig. 4. Only one T_g is noted in all proportions, though intermediate proportions broaden the α transition. The secondary transition in ABS resins is not significantly affected, indicating that the localized motion of the ABS chain is primarily intramolecular.

The effect of plasticizer is similar to that observed in compatible blends. Figure 5 shows the temperature-dependent permittivity at (30 Hz) of PVC plasticized with diisooctyl phthalate as a function of plasticizer content. The broadening of the glass–rubber transition is also similar to that described for the ABS/PVC blend.

See also: Electrically Conducting Polymers

Bibliography

Hedvig P 1977 *Dielectric Spectroscopy of Polymers*. Wiley, New York

Hill N E et al. 1969 *Dielectric Properties and Molecular Behaviour*. Van Nostrand Reinhold, London

McCrum N G, Read B E, Williams G 1969 *Anelastic and Dielectric Effects in Polymeric Solids*. Wiley, New York
Nee T-W, Zwanzig R 1970 Theory of dielectric relaxation in polar liquids. *J. Chem. Phys.* 52: 6353

G. L. Link and G. E. Johnson
[Bell Laboratories, Murray Hill, New Jersey, USA]

Electrically Conducting Polymers

Organic conjugated polymers, with a delocalized π-electron system along the polymer chain, can show high electronic mobilities and behave as semiconductors or, when suitably doped, as metals. Extensive studies of semiconductor and metallic behavior in chemically doped polyacetylene are described by Heeger and MacDiarmid (1986). Since the initial interest in their high electronic conductivities and the reversible electrochemical doping/undoping which can be exploited by using these polymers as electrodes in secondary batteries, there has been considerable progress in the fundamental understanding of electronic phenomena in this class of materials and, more recently, in the synthesis and processing of polyacetylene and other conjugated polymers. There is now a wide range of polymers available with differing band gaps, air reactivities and morphologies. The three topics selected in this article cover some of these recent advances. General references to this field include Chien (1984) and Skotheim (1986).

1. *Materials Synthesis and Processing*

A problem common to all conjugated polymers is that the strong π bonding along the polymer chain generally confers rigidity to the chain and a loss of the generally desirable processing properties of conventional polymers (solubility, melt processibility). Polyacetylene prepared by the direct polymerization of acetylene gas by the method developed by Shirakawa (Heeger and MacDiarmid 1986) is usually in the form of a fibrillar mat of low density. An important improvement to this process has been achieved by Badische Anilüs und Sodafabrik (BASF) (Basescu et al. 1987); the Ziegler–Natta catalyst is dissolved in a silicone oil and the polymerization is carried out at room temperature. The material still retains its fibrillar morphology but, when stretch oriented and subsequently doped with iodine, the conductivity along the stretch direction may reach a value as high as 145 kS cm^{-1} at room temperature. This value is only a little lower than that of copper (though when expressed as conductivity per unit mass is greater than that of copper). Other routes for the synthesis of conjugated polymers of recent interest are summarized in the following sections.

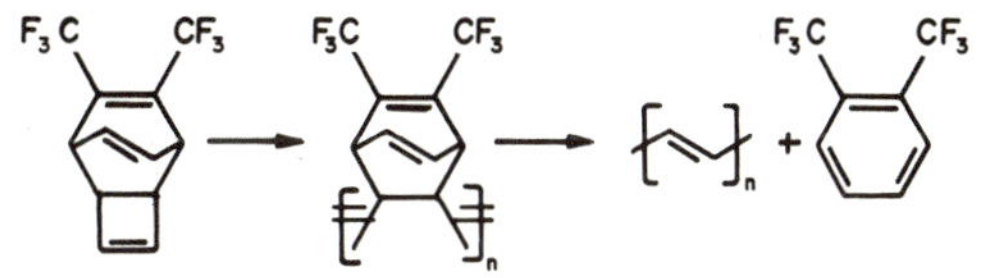

Figure 1
Durham precursor route to polyacetylene

1.1 Precursor Polymer Routes

An effective means of conferring solution processibility on an insoluble polymer is to synthesize, as an intermediate, a solution-processible "precursor polymer" which can be converted, usually by a thermal elimination reaction, to the conjugated polymer after the desired processing. The first and most studied of these routes is the "Durham" route to polyacetylene due to Edwards et al. (1984). The monomer, precursor polymer and elimination reaction are summarized in Fig. 1. Conditions for the thermal elimination reaction to give polyacetylene and hexafluoroorthoxylene (typically 60 °C for 1 h) are discussed by Bott et al. (1986) and the polyacetylene obtained from a solution-cast film of the precursor is fully dense, although retaining the disordered, amorphous structure of the precursor polymer film. Kahlert and Leising (1985) found that stretch orientation during the transformation reaction can give a remarkably high degree of orientation and crystallization. These oriented films have greatly extended the scope for experimental measurements of the anisotropy of the electronic properties (Friend et al. 1987).

The precursor polymer route shown in Fig. 2 invovles a water-soluble sulfonium polyelectrolyte salt, which is fully converted to polyphenylenevinylene when heated to 300 °C (Murase et al. 1984, Karasz et al. 1985). This polymer has a larger band gap than polyacetylene, with its absorption edge at 2.5 eV. It degrades slowly (if at all) in air and it can still be chemically doped to high conductivities. As with the Durham route to polyacetylene, highly ordered films can be produced by stretch orientation during the thermal conversion. The synthesis can be adapted to produce a series of polyheteroaromatic vinylenes. This route is reviewed by Bradley (1987).

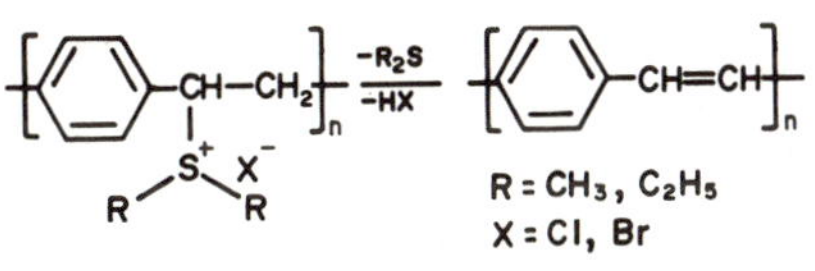

Figure 2
Sulfonium polyelectrolyte precursor route to polyphenylenevinylene

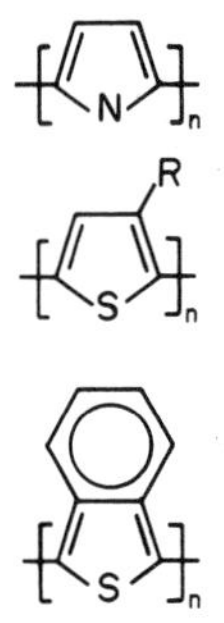

Figure 3
Various conjugated polymers commonly prepared by electrochemical polymerization: (a) polypyrrole, (b) poly 3-alkylthiophene and (c) polyisothionaphthene

1.2 Electropolymerization

Electrochemical oxidation of several monomers, including pyrrole and (substituted) thiophenes, can achieve polymerization and the formation of highly conducting, doped films of the polymer adhering to the electrode surface. These films can subsequently be electrochemically reduced and reoxidized. Polypyrrole (see Fig. 3a) was first extensively studied by the IBM group (Street 1986). The as-made films can be grown to thicknesses of several millimeters and show air-stable metallic conductivities of the order of $100\ S\ cm^{-1}$. Polythiophene and alkyl-substituted polythiophenes can similarly be electropolymerized from suitable monomers (Tourillon 1986) and show readily reversible electrochemical reduction and oxidation. The associated changes in optical properties are discussed in Sect. 3. A derivative of polythiophene, polyisothionaphthene (see Fig. 3c), is also prepared by electrochemical oxidization of the monomer and is of particular interest in that it has the smallest optical gap yet reported, with a band edge at 1 eV (Kobayashi et al. 1985). The doped polymer is both highly conducting and also partially transparent in the visible part of the spectrum and may be of use as a transparent conductor. Electropolymerized aniline, known for many years as aniline black, is now recognized to be a further example of this class of polymer, with considerable potential as a secondary battery electron (Heeger and MacDiarmid 1986). The electrochemistry achievable with this system includes reversible protonation/deprotonation at the nitrogen sites in addition to the redox reactions found for the examples mentioned (MacDiarmid et al. 1987).

1.3 Soluble Polymers

The addition of appropriate side groups to the conjugated polymer backbone can confer solubility to the polymer. The most studied example is that of the alkyl-substituted thiophenes (Fig. 3); the polymers made with butyl and longer side chains are readily soluble and have been synthesized both electrochemically and chemically (Elsenbaumer et al. 1987, Hotta et al. 1987, Kaeriyama et al. 1987). In spite of the considerable volume fraction of alkyl chain, these polymers can still be doped to high conductivities ($100\ S\ cm^{-1}$). Of further interest is the fact that polymers doped in solution still retain high conductivities.

1.4 Polydiacetylenes

Polydiacetylenes are prepared by the solid-state polymerization of the diacetylene monomer, as shown in Fig. 4. With an appropriate choice of the side groups R and R′, the crystal structure of the monomer can be chosen so that polymerization requires only slight atomic rearrangement and single crystals of high quality have been fabricated for some time. Their preparation and electronic properties are reviewed in Bloor and Chance (1985). These crystalline materials are one-dimensional semiconductors, with the conjugated polymer chains well separated from one another by the often bulky side groups. They are not readily doped to give metallic conductivities and had been considered apart from the other polymers mentioned here. However, the increased interest in the linear and nonlinear optical properties of conjugated polymers (see Sect. 3) has removed this distinction.

2. Electronic Excitations of the Polymer Chain: Solitons, Polarons and Bipolarons

Charges added to the chains of conjugated polymers cause a rearrangement of the π-electron bonding on the chain in the vicinity of the charge. For the case of the *trans*-isomer of polyacetylene, the localized state thus produced is a bond alternation defect separating sections of chain on either side of it with opposite

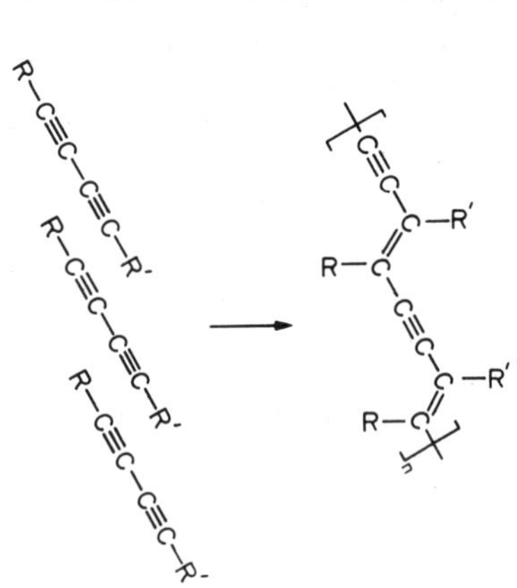

Figure 4
Schematic representation of the preparation of polydiacetylene by the solid-state polymerization of diacetylene monomers

senses of bond alternation. It has unusual topological and spin/charge properties which are characteristic of solitary waves or solitons (Heeger and MacDiarmid 1986). From general symmetry considerations the soliton has one nonbonding p_z level associated with it, which within one-electron models lies in the middle of the π–π^* band gap. Optical spectroscopy associated with transitions involving these levels is discussed in Sect. 3.

The particular symmetry with respect to the sense of bond alternation found in *trans*-polyacetylene is not found in other conjugated polymers. For these, the single bond alternation, or soliton, defect is not stable, since it would migrate rapidly to reduce the length of the higher-energy phase. This may be illustrated by the polyphenylene chain shown in Fig. 5. The "aromatic" conformation of the chains, with a "long" bond between the phenylene rings, is of lower energy than the "quinoid" conformation obtained by changing the sense of bond alternation. Nevertheless, the preferred site for a charge added to the chain is a region with quinoid conformation supported between two bond alternation defects. The defect shown in Fig. 5 is singly charged, has a spin of $\frac{1}{2}$ and is not a topological defect. this "charge + associated lattice distortion" is generally termed a polaron. The defect is capable of accepting a further charge, removing the electron spin and forming a bipolaron.

Calculations of the energetics of formation and the electronic structure of these defects have been carried out using the "continuum Hamiltonian" model to produce analytic solutions (Fesser et al. 1983) and also using "quantum chemistry" techniques (Brédas 1986). The polaron defect can be considered as a confined pair of solitons and its electronic structure reflects this, with two "bonding" and "antibonding" states equally displaced about midgap, as shown in Fig. 6. The occupation of the levels is shown for the positive polaron and bipolaron, and also for the neutral "polaron–exciton," which is expected to be produced following photoexcitation across the π–π^* gap and which can decay radiatively as indicated. The most direct evidence that charge is stored in polarons and bipolarons comes from the optical spectroscopy discussed in Sect. 3 and the success of these descriptions of the electronic excitations for this more general class of polymers has been an important advance for the field. Brédas (1986) has considered how the tendency to form polarons survives for high levels of doping. The polaron levels are calculated to broaden

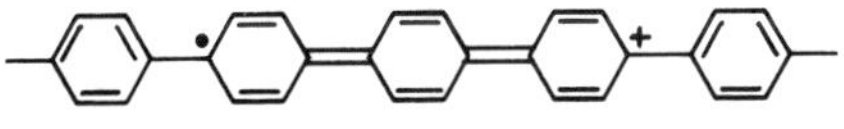

Figure 5
Schematic representation of a polaron defect on a polyphenylene chain

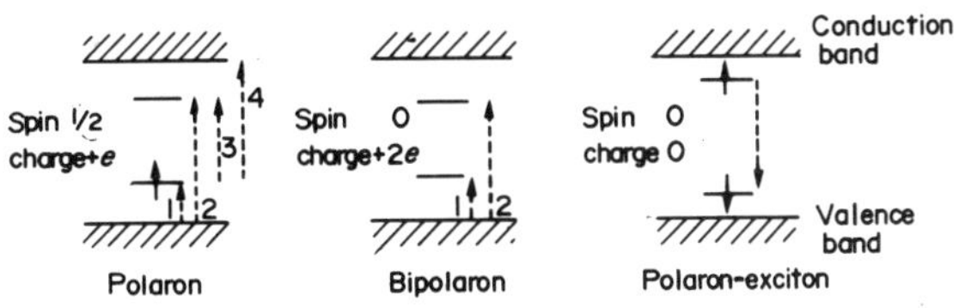

Figure 6
Energy levels and occupations for a positive polaron and bipolaron and a neutral polaron–exciton; allowed optical transitions are indicated

to form bands, but the gap between these and the band edges remains and the polymer, while highly conducting, does not have a high density of states at the Fermi level.

3. Linear and Nonlinear Optical Properties

The optical response of conjugated polymers is now recognized to be among their unusual and possibly exploitable properties. The absorption spectra of the doped polymers show direct evidence for the creation of soliton or polaron states. For *trans*-polyacetylene, optical transitions from band states to the midgap soliton level are strongly allowed and a midgap absorption feature is observed, peaking near 0.75 eV (Suzuki et al. 1980). For the other polymers that support polaron or bipolaron states, the absorption spectra show more structure below the gap; this is illustrated in Fig. 7 for polythiophene

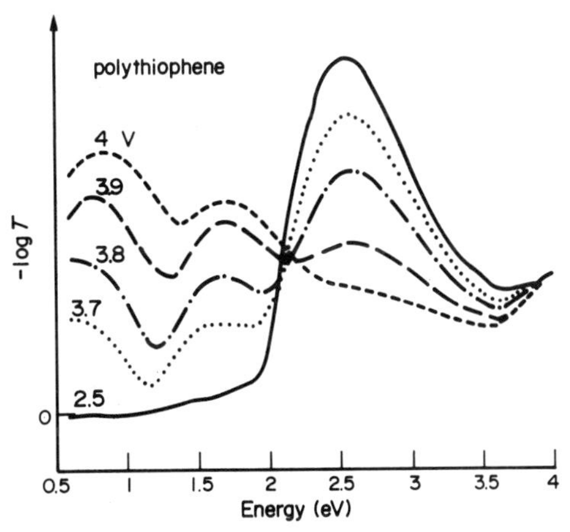

Figure 7
Optical absorption spectra for polythiophene at various levels of oxidative doping with perchlorate ions, achieved electrochemically with the cell potentials vs lithium as indicated: potentials of 2.5 V and 4.2 V give reduced and fully oxidized ($\approx +0.3e$ per thiophene ring) material, respectively (data from G. P. Evans and R. H. Friend)

(see also Chung et al. 1984). The undoped polymer shows a band edge at about 2 eV and the doped polymer shows loss of intensity from this absorption together with the appearance of two subgap absorption bands, peaking near 0.6 eV and 1.6 eV. These are identified with transitions from the valence band to the two polaron levels in the gap for the bipolaron defect, as shown in Fig. 6.

Su and Schrieffer (1980) showed from their simulations of the chain dynamics following photoexcitation of an electron–hole pair that, if the photoexcited charges are able to separate, they will form the localized defect states appropriate to that polymer. These are detected in measurements of photoinduced absorption, which show spectra very similar to those obtained following chemical doping (Orenstein 1986). Thus polyacetylene shows induced absorption below the gap, near 0.45 eV, characteristic of soliton–band transitions, and polaron-supporting polymers show the two absorption features associated with transitions between the band edges and the two bipolaron gap states (Friend et al. 1987).

An important prediction of the calculations of Su and Schrieffer (1980) is that the polymer chain can relax very quickly to the changed electronic density following photon absorption, within a time of 10^{-13} s for polyacetylene. This has been experimentally verified in a series of subpicosecond time-resolved experiments, reviewed by Orenstein (1986), and on these very short time scales the changes in optical constants both above and below the gap can be large. This demonstration of a large, rapid, nonlinear optical response shows that this class of polymeric semiconductors has considerable potential in optic or electrooptic applications. Measurements of nonlinear optical properties include those of Greene et al. (1987) and Kajzar et al. (1987).

See also: Electrical Properties of Polymers

Bibliography

Basescu N, Liu Z-X, Moses D, Heeger A J, Naarmann H, Theophilou N 1987 High electrical conductivity in doped polyacetylene. *Nature (London)* 327:403–5

Bloor D, Chance R R (eds.) 1985 *Polydiacetylenes: Synthesis, Structure and Electronic Properties*. Nijhoff, Dordrecht, The Netherlands

Bott D C, Brown C S, Chai C K, Walker N S, Feast W J, Foot P J S, Calvert P D, Billingham N C, Friend R H 1986 Durham polyacetylene: Preparation and properties of the unoriented material. *Synth. Met.* 14: 245–69

Bradley D D C 1987 Precursor route poly(phenylenevinylene): Polymer characterisation and control of electronic properties. *J. Phys. D* 20: 1389–410

Brédas J-L 1986 Electronic structure of highly conducting polymers. In: Skotheim 1986, pp. 859–913

Chien J C W 1984 *Polyacetylene, Chemistry, Physics and Materials Science*. Academic Press, Orlando, FL

Chung T-C, Kaufman J H, Heeger A J, Wudl F 1984 Charge storage in doped poly(thiophene): Optical and electrochemical studies. *Phys. Rev. B* 30: 702–10

Edwards J H, Feast W J, Bott D C 1984 New routes to conjugated polymers, 1: A two-step route to polyacetylene. *Polymer* 25: 395–8

Elsenbaumer R L, Jen K Y, Miller G G, Shacklette L W 1987 Processible, environmentally stable, highly conductive forms of polythiophene. *Synth. Met.* 18: 277–82

Fesser K, Bishop A R, Campbell D K 1983 Optical absorption from polarons in a model of polyacetylene. *Phys. Rev. B* 27: 4804–25

Friend R H, Bradley D D C, Townsend P D 1987 Photoexcitations in conducting polymers. *J. Phys. D* 20: 1367–84

Greene B I, Orenstein J, Millard R R, Williams L R 1987 Nonlinear optical response of excitons confined to one dimension. *Phys. Rev. Lett.* 58: 2750–3

Heeger A J, MacDiarmid A G 1986 Electrically conducting polymers. In: Bever M B (ed.) 1986 *Encyclopedia of Materials Science and Engineering*. Pergamon, Oxford, pp. 1399–1402

Hotta S, Rughooputh S D D V, Heeger A J, Wudl F 1987 Spectroscopic studies of soluble poly(3-alkylthienylenes). *Macromolecules* 20: 212–15

Kaeriyama K, Sato M, Tanaka S 1987 Electrochemical preparation of conducting polyalkylthiophene films. *Synth. Met.* 18: 233–6

Kahlert H, Leising G 1985 Properties of $(CH)_x$-fibers and films prepared from BTFM-TCDT. *Mol. Cryst. Liq. Cryst.* 117: 1–8

Kajzar F, Etemad S, Baker G L, Messier J 1987 $\chi^{(3)}$ of trans-$(CH)_x$: Experimental observation of $2A_g$ excited state. *Synth. Met.* 17: 563–7

Karasz F E, Capistran J D, Gagnon D R, Lenz R W 1985 High molecular weight polyphenylene vinylene. *Mol. Cryst. Liq. Cryst.* 118: 327–32

Kobayashi M, Colaneri N, Boysel M, Wudl F, Heeger A J 1985 The electronic and electrochemical properties of poly(isothianaphthene). *J. Chem. Phys.* 82: 5717–23

MacDiarmid A G, Chiang J C, Richter A F, Epstein A J 1987 Polyaniline: A new concept in conducting polymers. *Synth. Met.* 18: 285–90

Murase I, Ohnishi T, Noguchi T, Hirooka M 1984 Highly conducting poly(p-phenylene vinylene) prepared from a sulphonium salt. *Polym. Commun.* 25: 327–9

Orenstein J 1986 Photoexcitations of conjugated polymers. In: Skotheim 1986, pp. 1297–335

Skotheim T J (ed.) 1986 *Handbook on Conjugated Electrically Conducting Polymers*. Dekker, New York

Street G B 1986 Polypyrrole: From powders to plastics. In: Skotheim 1986, pp. 265–91

Su W P, Schrieffer J R 1980 Soliton dynamics in polyacetylene. *Proc. Natl. Acad. Sci. USA* 77: 5626–9

Suzuki N, Ozaki M, Etemad S, Heeger A J, MacDiarmid A G 1980 Solitons in polyacetylene: Effects of dilute doping on optical absorption spectra. *Phys. Rev. Lett.* 45: 1209–13; erratum *Phys. Rev. Lett.* 45: 1463

Tourillon G 1986 Polythiophene and its derivatives. In: Skotheim 1986, pp. 293–350

R. H. Friend
[University of Cambridge, Cambridge, UK]

Electrodeposited Polymer Coatings

Electrochemical electrode processes in the presence of monomeric or polymeric organic molecules in solution may lead to insoluble polymers forming a surface coating on an electrode. However, if nucleation and solid-phase formation do not occur on the surface, precipitation in solution, rather than the growth of a coherent polymer layer, proceeds at the phase boundary. The complicated colloid-chemical processes along these lines are not clearly understood at present. Throughout this article, all systems are represented by a sandwich structure according to

metal or semiconductor electrode	polymer layer	monomer or polymer and supporting electrolyte in solution
	(1) (2)	

While both outer phases are pure electronic and ionic conductors, the polymer layer may conduct ionically or electronically. In the former case, the electrode process is located at the phase boundary (1), and molecules (besides ions) must be transported through the intermediate phase. This should hamper the rate of the overall process. In the latter case, the electrode process is continuously shifted to the outer phase boundary (2).

The efficiency of the electrochemical coating process is governed by the electrochemical equivalent m_e, which is defined by Faraday's law:

$$m_e = \frac{\Delta m}{\Delta Q} = \frac{M}{zF} \qquad (1)$$

where Δm is the mass electrodeposited by charge ΔQ, M is the molecular weight of the polymer (or the polymer section), which needs z electrons for its formation, and F is the Faraday constant. The time t for the growth of a polymer layer of thickness L and density s is then easily derived to be

$$t = sL/(m_e j_{eff}) \qquad (2)$$

where j_{eff} is the effective current density, which is linked to the actual current density via the current efficiency γ ($\gamma < 1$):

$$j_{eff} = \gamma j \qquad (3)$$

Another important quantity for the electrochemical process is the voltage drop ΔU, driving the current density j through the film. ΔU is given by Ohm's law (ρ = resistivity):

$$\Delta U = \rho j L \qquad (4)$$

Under practical conditions, j may be $10\,mA\,cm^{-2}$ and L may be $10\,\mu m$. Then ΔU does not exceed 1 V, which is of the order of electrochemical potentials, if ρ remains below $10^5\,\Omega\,cm$. In the presence of low concentrations of charge carriers in the film, charge separation will occur, a space charge will be established and ΔU can be calculated by twofold integration of Poisson's equation:

$$\Delta U = C_0 F L^2/(2\varepsilon\varepsilon_0) \qquad (5)$$

where C_0 is the molar concentration of charged species, ε is the electric permittivity and ε_0 is the electric field constant.

Film growth may be limited by diffusion of components in the film. The time constant τ is represented by

$$\tau = L^2/4D \qquad (6)$$

The diffusion constant D may differ for molecular species from $10^{-20}\,cm^2\,s^{-1}$ (polyacetylene) up to $10^{-5}\,cm^2\,s^{-1}$ for free solutions. For $D = 10^{-10}\,cm^2\,s^{-1}$ (polypyrrole) and $L = 1$ mm to 1 nm, τ changes widely, from 100 d down to 0.025 ms.

The following description is classified according to the three mechanisms of electrodeposition.

1. Electrochemical Initiation

The majority of present plastic material produced industrially is based on the polymerization of monomers of the vinyl type $CH_2{=}CH{-}X$. This is a chemical chain reaction, started by a chemically generated radical or ionic initiator (starter). However, the initiator may be produced via an electrode process, for example radicals by cathodic proton discharge:

$$[H(H_2O)_x]^+ + e^- \rightarrow H^{\cdot}_{ad} + xH_2O \qquad (7)$$

or by anodic oxidation of solvent molecules:

$$RH \rightarrow R\cdot + H^+ + e^- \qquad (8)$$

If this is done in the presence of a vinyl monomer, these radicals may start a radical polymerization quite analogous to the classical process:

$$R\cdot + pCH_2{=}CH{-}X \rightarrow R{+}CH_2{-}\underset{\substack{|\\X}}{CH}{+}_{p-1}CH_2{-}\underset{\substack{|\\X}}{CH}\cdot \qquad (9)$$

Termination and chain branching follow the conventional mechanisms. The solvent–electrolyte system is mostly aqueous, for example dilute sulfuric acid. Organic cosolvents may improve the solubility of the monomer.

Radicals and radical ions are quite common intermediates in organic electrochemistry. From this point of view, the structure of the organic molecule present in the system determines whether the reaction products are molecules of low molecular weight (functionalized or dimerized), or whether polymerization as described by Eqns. (7–9) is the prevailing process.

The first successful electrochemical polymerizations, appearing about 1950, were of this type. They were reviewed later by Breitenbach and Srna (1962), Parravano (1973) and Beck (1974). The rapid development of conducting polymer layers since the mid-1970s (see Sect. 2) has led to a diminished interest in this area and, today, only a few laboratories are concerned with this topic.

A special case of electrochemical initiation is ring-opening polymerization, which was studied by Schulz and co-workers (Beck 1974). Tetrahydrofuran (THF) may serve as an example. It is initially oxidized at the anode to yield the radical cation which, by ring opening, starts an ionic polymerization to form a polyether (see Scheme 1).

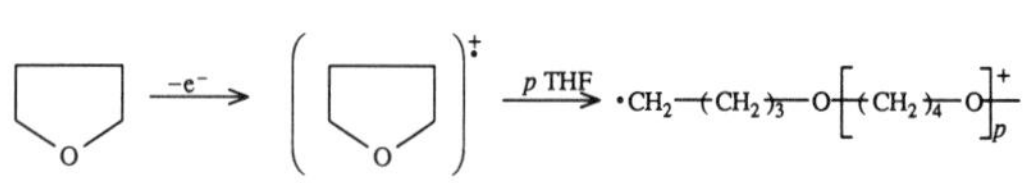

Scheme 1

The reaction proceeds under aprotic conditions (i.e., the solvent has no accessible protons) and film formation occurs only when careful optimization of the solubilizing and nucleophilic character of the solvent is carried out. Ionic polymerization may also be realized with vinyl monomers, forming radical ions by themselves or ions via solvent–electrolyte decomposition (see Eqn. (7)). Electrochemical initiation of the cationic type is found only in those cases where the relevant acids (HBF_4, HPF_6) are not normally stable in the anhydrous state, but nevertheless are readily formed *in situ* in the vicinity of the anode under anhydrous conditions. Thickness growth of the films is difficult owing to the fact that the polymers, carrying only single bonds along their backbone, are insulators. However, ionic charge transport through the film can be maintained by three different strategies:

(a) through electrolyte-filled pores;

(b) by modification of the monomer structure to allow a pronounced swelling of the polymer film with the electrolyte medium; and

(c) by modification of the solvent–electrolyte system in order to introduce ionic species via swelling in the film.

If the second strategy is followed, homogeneous ionic conductivity is readily induced. An example is in the polymerization of diacetone acrylamide, described in a patent of Grace Company (see Beck 1974). Using the third strategy, Grčev showed that the addition of 0.1 M $ZnCl_2$ to the electropolymerization mixture of acrylamide in aqueous electrolytes leads to polyacrylamide films capable of growing to thicknesses of approximately 20 μm.

If termination reactions are retarded effectively, molecular weight may reach the order of 10^5 amu. With a monomer of molecular weight 100 amu, this means a degree of polymerization as high as 1000. The electrochemical equivalent is very high for this case, namely $1\ g\ C^{-1}$. Appropriate layer thickness may be obtained with small amounts of electricity, even without further care to maintain ionic conductivity in the film. However, mechanical properties are poor, owing to brittleness and bad adherence; lower molecular weights (10^3–10^4 amu) are thus preferred, leading to elastic and well-adhering films.

The excellent insulating characters of the polymer films may be applied in providing nonconducting layers in electronic devices. The electrical deposition process greatly facilitates the automated fabrication of these layers even onto microstructures. The work of Bogenschütz et al. (see Beck 1974) is an example of this potential application. Coatings for general corrosion protection on iron may also be deposited. An advantage in comparison with electrodeposited paint (see Sect. 3) should be the elimination of the preceding steps of manufacturing the polymer itself.

2. *Electrochemical Propagation*

While the propagation process, which couples the monomers to the growing polymer, may be purely chemically *in situ* (Sect. 1) or *extra situ* (Sect. 3), it may also be due to an electrochemical process itself. This possibility has been realized with many aromatic or heteroaromatic monomers. Kanazawa et al. (1979) and Diaz (1981) were the first to demonstrate that well-adhering, homogeneous black films of polypyrrole could be obtained by anodic oxidation of pyrrole in acetonitrile with $LiClO_4$ or $N(C_4H_9)_4BF_4$ as an electrolyte. The current density was relatively low, usually below $1\ mA\ cm^{-2}$. The anode was platinum. Other studies have shown that similar polymer layers could be obtained from pyrrole solutions even in aqueous solutions, with diluted sulfuric acid or $NaBF_4$ as an electrolyte. Besides platinum, iron and other metals could be electrocoated; in this case KNO_3 was the appropriate electrolyte. The work of Diaz has been summarized to some extent (see Diaz 1981). An overview of more recent work has been provided by Skotheim (1986).

The electrodeposition starts with the formation of the monomer radical cation (see Scheme 2).

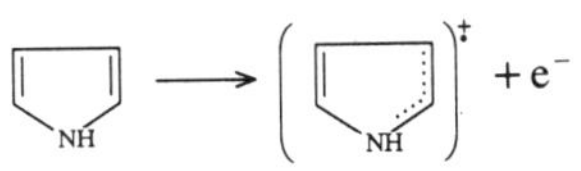

Scheme 2

The radical cation may react with a pyrrole molecule (see Scheme 3) or with a second pyrrole radical cation.

Scheme 3

The dimeric radical cation, linked in the 2-position, may be further oxidized and finally eliminates two protons to yield the dimer. If XH_2 denotes a pyrrole molecule, these processes may be written as follows.

$$XH_2 \xrightarrow{-e^-} (XH_2)\cdot^+ \xrightarrow{XH_2} (HXH\text{—}HXH)\cdot^+ \xrightarrow[-2H^+]{-e^-} H\text{—}X\text{—}X\text{—}H \qquad (10)$$

However, the dimer shown as Scheme 4 is more easily oxidized as the monomer.

H-X-X-H = 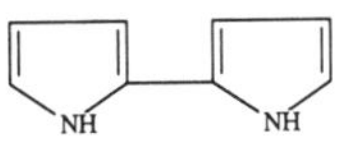

Scheme 4

It therefore undergoes an analogous electrooxidation sequence to yield the tetramer

$$2H\text{—}X\text{—}X\text{—}H \rightarrow H\text{—}X\text{—}X\text{—}X\text{—}X\text{—}H + 2H^+ + 2e^- \qquad (11)$$

The dimer may also couple with an oxidized monomer to form a trimer. As a general rule, the ease of oxidation increases as the degree of oligomerization increases. Finally, the solubility is exceeded and the insoluble oligomer precipitates onto the surface of the anode. This mechanism of formation, which has not been proved experimentally in all of its aspects, shows clearly:

(a) a limited degree of polymerization, of the order of ten;

(b) a low crystallinity of the polymer;

(c) the incompatibility with the chain growth mechanism of any vinyl polymerization;

(d) a high consumption of electricity for the average addition of one monomer, namely $2F\,mol^{-1}$, corresponding to an electrochemical equivalent (for $M=100$) of only $0.5\,mg\,C^{-1}$; and

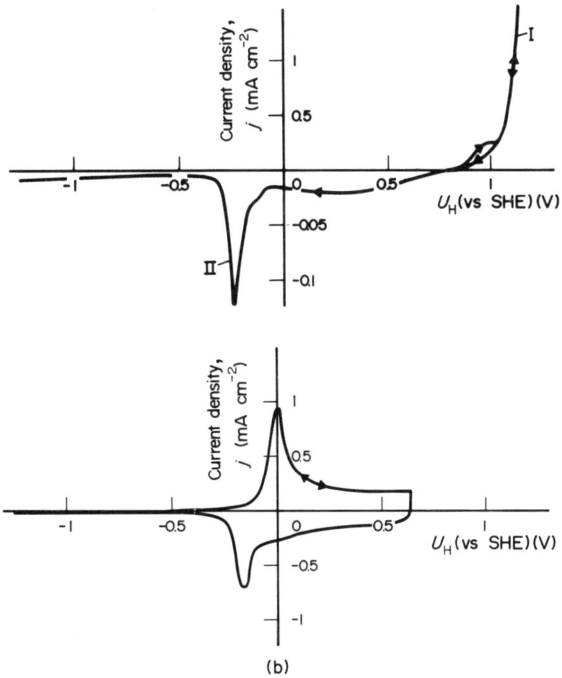

Figure 1
Potentiodynamic current–voltage curves for (a) the anodic electrode position (I) and first cathodic discharge (II), and (b) cyclic charge and discharge of polypyrrole: curves in (a) were measured with a voltage scan rate of $v_s = 2\,mV\,s^{-1}$, the curve in (b) with $v_s = 20\,mV\,s^{-1}$; $Q_0 = 116\,mC\,cm^{-2}$, $Q_{E,0} = 14\,mC\,cm^{-2}$; the electrolyte was 0.1 M pyrrole, 0.1 M NBu_4BVF_4, 1 mM NaBr in acetonitrile

(e) an important role for nucleation steps that cause the solid to form on the electrode rather than in the bulk of the electrolyte.

The current efficiency for polypyrrole formation was found to be nearly 100%. The potentiodynamic current–voltage curve for the anodic deposition of polypyrrole is shown in Fig. 1a as part I. The current rises exponentially at potentials of about 1.1 V vs standard hydrogen electrode (SHE). A prewave, which is controlled kinetically, is also observed. If the electrodeposited polypyrrole film is polarized in a negative direction, a "first discharge" is found with a current peak at about −0.25 V (Fig. 1a, part II). The discharged polymer can be cyclically charged and discharged again (Fig. 1b), but the corresponding discharge curves differ widely from II. This reversible redox process in the solid state is due to the anodic oxidation of the polymer backbone to the polymer radical cation and the simultaneous insertion of an anion A^- to compensate the positive charge for the charging process. In the discharge process, the polymer radical cation is reduced and the anions are released from the film (see Scheme 5).

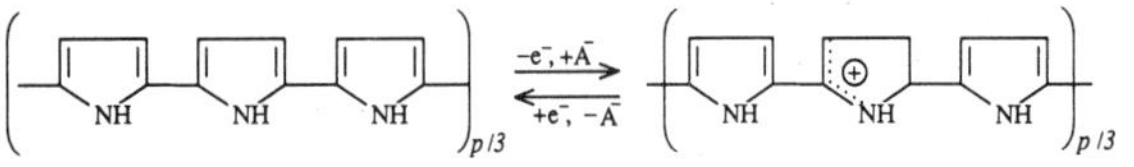

Scheme 5

The maximum degree of insertion y_{max} is 33 mol.% with reference to one monomer unit as schematically represented by Scheme 5. In physical language, these solid radical ion states are called polarons. A further oxidation of the conjugated chains may lead to spinless di-ions, named bipolarons in physics:

$$2R\cdot^{+} \rightarrow R + R^{2+} \text{ or } R\cdot^{+} \rightarrow R^{2+} + e^{-} \qquad (12)$$

The electronic conductivity of the polymer, which carries an extended system of conjugated double bonds, rises dramatically from the semiconducting state ($\kappa \sim 10^{-10}$ S cm^{-1}) to the metallic conductivity level (100 S cm^{-1}) in the course of the doping process according to Scheme 5. This was discovered for polyacetylene by Heeger and McDiarmid in 1977 (see Seymour 1981, Baughman et al. 1982).

Polyaniline, known as aniline black since 1862, can be deposited as a black film from a solution of aniline in aqueous sulfuric acid (Genies et al. 1987). Typical concentrations are 0.1 M. Potentiodynamic or potentiostatic conditions are necessary to obtain layers of a good quality. The real structure of the electrodeposited polymer is not perfectly understood; besides poly(*p*-phenyleneimine) units, it may be characterized by phenylphenazine elements or phenylindole elements. Redox transitions in polyaniline differ from those in polypyrrole in two main ways:

(a) electron transfer in two steps, and

(b) proton transfer as an alternative ion-exchange reaction.

Scheme 6 demonstrates this for a dianiline subunit in the polymer backbone.

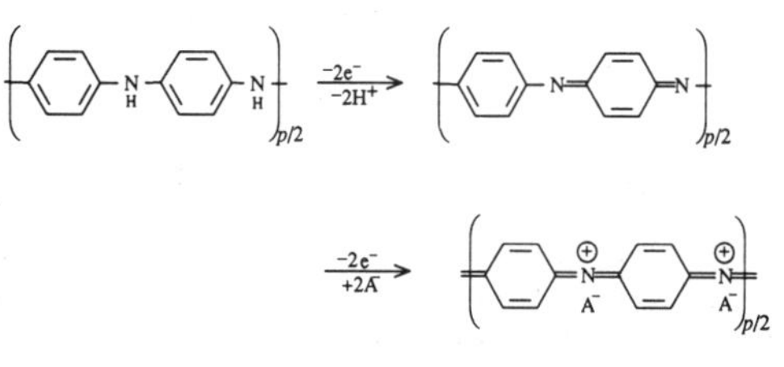

Scheme 6

However, the theoretical degree of insertion of $y = 2$ is never attained in practice. The maximum is $y_{max} = 0.6$.

Poly(*p*-phenylene) (PPP) has been electrodeposited from solutions of benzene in 70% hydrofluoric acid by Rubinstein in 1983. Contrary to the products of chemical polymerization with $CuCl_2/AlCl_3$, obtained by Kovačič (1963), the black films seem to contain not only *p*-products, but also crosslinked structures. Polythiophene was obtained as a green–black film from aprotic solutions of thiophene in acetonitrile. As a quasiaromatic system, the potential of deposition is, like PPP, much more positive with regard to polypyrrole or polyaniline. The current efficiencies for the formation of the 2,5-coupled polymer are found to be only 60–80%. The maximum degree of insertion is up to 50 mol.%.

Table 1 lists the characteristic electrical and electrochemical parameters for conducting polymer layers. For comparison, chemically polymerized polyacetylene and poorly conducting polyfurane and poly(*p*-phenylene sulfide) are included. Table 1 shows that most of these polymers have a good electronic conductivity well above 1 S cm^{-1}. Consequently, the thickness growth of the layer will never be limited by ohmic voltage drop. Typical voltages are low, of the order of a few volts. Thin layers below 1 μm are smooth and well adhering. However, thick layers up to some 100 μmm are rough and highly porous. This may be a great advantage for use in batteries.

Phenole and its derivatives can be electropolymerized in aqueous, alkaline electrolytes according to the same principles (Mengoli et al. 1980). The resulting poly(*p*-phenylene oxide) is poorly conducting, even in the doped state. Ionic conductivity is improved in the presence of certain amines.

Three strategies can be followed for further modification of the conducting polymer layers:

(a) physical modification with conventional polymers in terms of sandwich structures or as a binder with dispersed conducting polymer, (e.g., polypyrrole/PVC);

(b) chemical modification of the monomer (e.g., 3-methylthiophene); or

(c) chemical modification of the polymer via the inserted anion (e.g., poly(styrene sulfonate)).

The interesting electrical, electrochemical and optical properties of these types of polymer layer, together with their chemical stability, their nonsolubility and their high decomposition temperatures, lead to the following potential applications:

(a) corrosion protection and decoration of technical surfaces;

(b) contact layers in (micro)electronic devices;

(c) antistatic layers, electromagnetic shielding;

Table 1
Structural, electrical and electrochemical parameters for some important conducting polymer layers

Structure			Electrical specific conductivity ($S\,cm^{-1}$)		U_H/SHE[b]	
Polymer	Monomer unit	y^{a}_{max} (mol.%)	At $y=0$	At $y=0.01$	Reversible cycling	Electrode position
Polyacetylene	—CH=	0.25	10^{-5}	10^{3}	0.8	c
Poly(*p*-phenylene)		0.50	10^{-10}	5×10^{2}	1.4	2.2
Polypyrrole	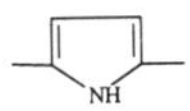	0.33	10^{-8}	10^{2}	0.0	1.2
Polythiophene	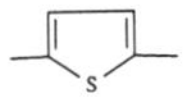	0.50	10^{-8}	50	1.15	1.9
Polyaniline	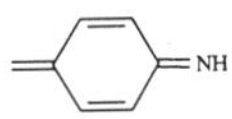	0.60	10^{-10}	10^{2}	0.4/0.9	1.15
Polyfuran	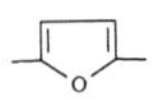			10^{-2}	1.8	>2.0
Poly(*p*-phenylenesulphide)	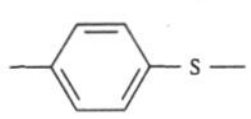	0.33	10^{-10}	1	1.5	>2.0

a y = degree of insertion b U_M = electrochemical potential/the standard hydrogen electrode c Chemical polymerization

(d) thin layers for physical and electrochemical solar cells;

(e) electrochemical and electronic displays;

(f) construction of light filters;

(g) active masses in rechargeable batteries;

(h) electrochemical injection and extraction of ionic materials; and

(i) electrocatalysis and new electrodes.

3. *Electrocoagulation of Polymers*

Contrary to the systems discussed hitherto, ionmeric species, that is, polymers with ionic groups, react at the electrode when dissolved in an aqueous medium. Anionic systems are deposited at the anode, while cationic polymers form cathodic layers. Typical polymer structures developed since the 1960s for large-scale industrial application are summarized in Table 2. These systems have found a worldwide application in the electrodeposition of paint (electrophoretic painting), providing the first protecting layer on metallic mass-produced ware such as automobile bodies. Two synthetic principles hold for the introduction of ionogenic groups into the polymer:

(a) polymerization of monomers which already carry these groups; and

(b) modification of nonionic polymers.

These polymers are rendered soluble in the aqueous bath by partial neutralization with amines or acetic acid. Typical concentrations are 10 wt%. The macro-ions are aggregated to micellar structures.

The mechanism of electrodeposition is due to electrolytic water decomposition, for example at the anode:

$$H_2O \rightarrow \tfrac{1}{2}O_2 + 2H^+ + 2e^- \qquad (13)$$

The ions, injected by this process, shift the pH in the diffusion layer. As monomeric counterions are not available, the formation of space charges cannot be ruled out *a priori*. However, the voltage increases only after the electrocoagulation of the polymer due to an acid–base reaction. For this reason, the pH shift is limited to 1–2 units around the neutral point:

$$p\text{—}COO^- + H^+ \rightleftharpoons p\text{—}COOH \qquad (14)$$

producing the insoluble form of the free acid and shifting back the original equilibrium, which has been established upon solubilization:

$$p\text{—}COOH + NR_3 \rightleftharpoons p\text{—}COO^- + NR_3H^+ \qquad (15)$$

Table 2
Typical polymer structures for electrocoagulation

Type	Synthesis	Ionomer	Structure
Anionic	I	acrylate	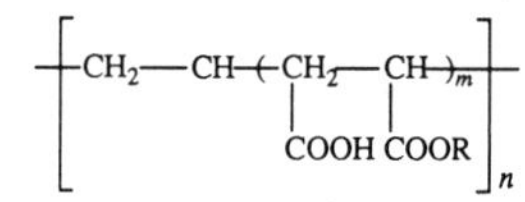
Anionic	II	maleinized oils	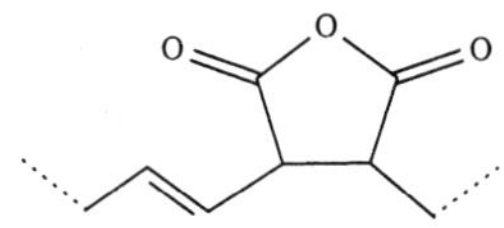
Cationic	I	dimethylaminoethyl-methacrylate	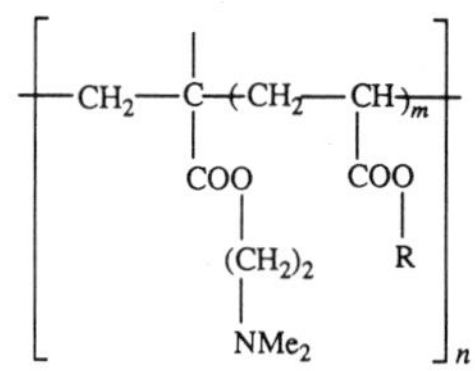
Cationic	II	epoxide systems	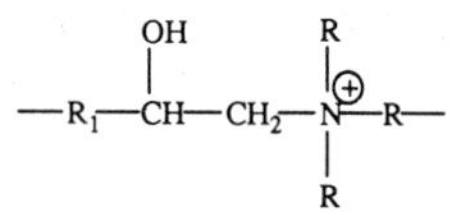

The free amine, diffusing into the diffusion layer, may also capture protons:

$$NR_3 + H^+ \rightleftharpoons NR_3H^+ \qquad (16)$$

This indirect mechanism has been proved by Beck (1966) with the help of experiments with a rotating-disk electrode. The present state is thoroughly described in various review articles (see Beck 1976, 1986). The situation at the cathodic phase boundary at the instant of electrocoagulation is well represented by Fig. 2. It shows that OH^- ions accumulate in the diffusion layer. Off-transport in the steady state is due to diffusion and migration; a steady-state concentration gradient is established. The polymer cations p^{z+} are mainly transported towards the cathode via electrophoresis. A concentration gradient is not needed in this case. At the point where a critical OH^- ion concentration $c^*_{OH^-}$ is attained, coagulation of the polymer analogous to Eqn. (14) proceeds and the concentration of the free ions falls to nearly zero. The formation of the polymer phase may be retarded due to nucleation phenomena.

Practical electrodeposition of paint is performed at a constant high voltage of 100–400 V. The electrocoagulation through the high electrolysis current I is

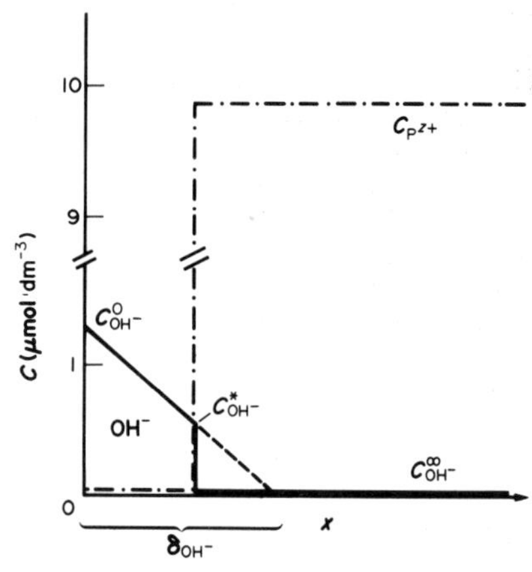

Figure 2
Concentration profiles for OH^- ions and polymeric cations p^{z+} in the steady state of cathodic electrodeposition of paint (plot of concentration C vs distance x from the cathode surface): $C^0_{OH^-}$ is the surface concentration of OH^- ions, $C^\infty_{OH^-}$ is the bulk concentration of OH^-, $C^*_{OH^-}$ is the critical concentration of OH^- ions, $C_{p^{z+}}$ is the concentration of polymeric cations and δ_{OH^-} the thickness of the Nernst diffusion layer

forced in a short time and as soon as the polymer has coagulated, I decreases rapidly. Finally, a residual current density below $1\ mA\ cm^{-2}$ is attained, and further deposition ceases because this is below the critical current density j^* which is needed to establish the critical ion concentration in the diffusion layer. The electrodeposited ionomer is a poor ionic conductor with a specific conductivity of the order of $10^{-8}\ S\ cm^{-1}$. This corresponds to an ion concentration in the solid as low as $1\ \mu mol\ dm^{-3}$. In the high electric field it leads to a charge separation in the polymer film of the order of $10^5\ V\ cm^{-1}$. The high voltage applied to the system is absorbed in the space charge within the film according to Eqn. (4). However, portions of the surface which have not yet been film coated draw a relatively high current density and electrodeposition is forced at these regions as well. In this way, the throwing power of the process is made excellent.

If the deposition is performed at constant current densities, the voltage rises steeply after an induction time τ, indicating the final electrodeposition of the polymer film. The induction time τ can be much longer than the time constant for the establishment of the steady-state concentration profiles in the diffusion layer according to Eqn. (5), especially on approaching the critical ion concentration. This means that τ may reflect a time constant for the nucleation process.

The electrochemical equivalent m_e attains a medium level with respect to the two preceding sections. Let us assume a copolymer of 1 mol acrylic acid and 10 mol acrylic acid butylester as a model (see Table 2). Then the molecular weight is 1472 and m_e (for the case of 100% neutralization) is $15\ mg\ C^{-1}$. The same order of magnitude is found for all other systems. The thickness of electropaint layers is up to $20\ \mu m$ and, for special cationic polymers, up to $50–100\ \mu m$. The residence time of an electrode in the bath is only 1–2 min.

Industrial systems contain up to 20 components in an aqueous medium, namely binders, pigments, cosolvents, solubilizers (amines or acids) and corrosion inhibitors. The electrodeposition is performed in stirred tanks, mainly to prevent the sedimentation of the pigments. The maintenance of a constant bath composition may become problematic due to the various distribution constants of the bath components (Brewer 1973, 1983). The presence of a low concentration (1–3 wt% with regard to the solid) of carbon black fillers leads to electronically conducting polymer layers after the stoving step. This even allows the electrodeposition of a second layer (Beck and Guder 1987).

It should be pointed out that, in practice, every electrodeposition of paint is preceded by a spray or dip pretreatment (phosphating) of the iron substrate. Finally, after thorough rinsing, a baking process is performed in an oven at 160–210 °C (Brewer 1973).

Possible alternatives to electrodeposition of paint are the following.

(a) Chemiphoresis or autophoresis—a process in which the iron substrate dips into an acid solution containing HCl, H_2O_2 and other components. Corrosion of the inner and outer surface of the metal body leads to an appropriate change of Fe^{3+} and H^+ concentration in the diffusion layer.

(b) Electrodeposition of colloids, latices or dispersions of nonionic polymer articles, stabilized by surfactants—this possibly has been known for a long time. The accumulation of the articles at the electrode due to electrophoresis plays a major role.

(c) The so-called EPLA process—solid particles of thermoplastic resins are dispersed in water and the dispersions are stabilized by anion active emulsifying agents. At $j = 80\ mA\ cm^{-2}$, enough heat is generated to smooth the thick layers. The deposition time is only 10–20 s.

4. Outlook

Electrodeposition of paint is the only process which is applied on a large scale all over the world. Cationic systems have almost totally replaced the former anionic systems since about 1980, owing to pronounced advantages in corrosion protection. Further improvements are mainly to be expected in paint preparation and chemical feedstock.

The electrodeposition of "synthetic" metals is a rapidly growing area. Further progress and first applications (batteries, displays) are expected in the near future.

Even use as a preparative method to yield the material as a continuous film after detachment from a moving electrode has been realized (Naarmann 1987). This resembles the classical electrodeposition of latex films to yield rubber sheets or hoses (see Beck 1976).

Sandwich structures, for example swollen PVC films with polypyrrole, open a new, interesting field with an enormous practical potential. "Dry" processes, like plasma polymerization (analagous to Sects. 1 and 2) and electrostatic spray or powder painting (analagous to Sect. 3) are outside the scope of this article.

See also: Marine Coatings

Bibliography

Baughman R H, Bredas J L, Chance R R, Elsenbaumer R L, Shacklette L W 1982 Structural basis for semiconducting and metallic polymer/dopant systems. *Chem. Rev.* 82: 209–22

Beck F 1966 Zum Mechanismus der elektrophoretischen Lackierung. *Farbe Lack* 72: 218–24
Beck F 1974 *Elektroorganische Chemie*. Verlag Chemie, Weinheim, Germany, Chap. 4.11
Beck F 1976 Fundamental aspects of electrodeposition of paint. *Prog. Org. Coat.* 4: 1–60
Beck F 1986 Electrical and chemical aspects of electrodeposition of paint. In: Wilson A D, Prosser H J (eds.) 1986 *Development in Ionic Polymers*. Elsevier, London, Chap. 8
Beck F, Guder H 1987 Electrodeposition of paint in carbon black filled systems. *J. Electrochem. Soc.* 134: 2416–24
Breitenbach J W, Srna C 1962 Electrochemical initiation of polymerization. *Pure Appl. Chem.* 4: 245
Brewer G E F (ed.) 1973 *Electrodeposition of Coatings*, Advances in Chemistry Series 119. American Chemical Society, Washington, DC
Brewer G E F 1983 Electrophoretic painting. *J. Appl. Electrochem.* 13: 269–75
Diaz A F 1981 Electrochemical preparation and characterization of conducting polymers. *Chem. Scr.* 17: 145–8
Genies E M, Lapkowski M, Santier C, Vieil E 1987 Polyaniline, spectroelectrochemistry, display and battery. *Synth. Met.* 18: 631–6
Grčev T, Cvetkovska M, Arsov L, Petrov G 1985 Electrochemically initiated polymerization of acrylamide. *Kem. Ind.* 34: 325–40
Kanazawa K K, Diaz A F, Gill W D, Grant P M, Street G B, Gardini G P, Kwak J F 1979 Polypyrrole: An electrochemically synthesized conducting polymer. *Synth. Met.* 1: 329–36
Kovacič P, Kyriakis A 1963 Polymerization of benzene to p-polyphenyl by aluminium chloride–cupric chlorides. *J. Am. Chem. Soc.* 85: 454–8
Mengoli G, Daolio S, Musiani M M 1980 The influence of amines on the anodic coupling of phenols to polyoxyphenylene films. *J. Appl. Electrochem.* 10: 459–71
Naarmann H 1987 New aspects on intrinsically conducting organic systems and their synthesis. *Makromol. Chem., Macromol. Symp.* 8: 1–15
Parravano G 1973 Electrochemical polymerization. In: Baizer M M (ed.) 1973 *Organic Electrochemistry*. Dekker, New York, Chap. 24
Rubinstein I 1983 Electrochemistry of polyphenylene films deposited anodically on platinum or glassy carbon electrodes in HF–benzene system. *J. Electrochem. Soc.* 130: 1506
Seymour R B 1981 *Conductive Polymers*. Plenum, New York
Skotheim T A (ed.) 1986 *Handbook on Conjugated Electrically Conducting Polymers*. Vol. 2. Dekker, New York

F. Beck
[Universität Gesamthochschule Duisburg, Duisburg, Germany]

Extrusion of Fiber-Reinforced Polymer Systems

The incorporation of discontinuous reinforcing fibers, such as chopped fiberglass or natural cellulose, into a polymer matrix produces a composite whose mechanical properties fall between the unreinforced polymer itself and high-performance composites reinforced with continuous fiber strands. An advantage of the short-fiber composite is that it can be fabricated into useful parts by the same high-output flow processes as for the unreinforced polymer (i.e., extrusion, molding and drawing). A longer fiber produces higher reinforcement. However, the effect of the fiber on the melt rheology and, conversely, the influence of the flow on the composite structure, especially fiber directionality, are then more severe.

1. Flow Characteristics

Compared with particulate additives, the 500% increase in shear viscosity at low deformation rate ($<0.01\,s^{-1}$), when 15 vol.% fibers of 40:1 length–diameter ratio (l/d) are added to a polymer melt, is dramatic. In an extrusion die, where the shear rate is higher ($\sim 100\,s^{-1}$), viscosity doubles. (The extruder head pressure usually rises further to produce an increased output.) One explanation for the reduced viscosity buildup in fast flows is a blunting of the velocity profile. At a high 50 vol.% loading of 250 l/d fibers, the composite flows as a plug through channels of uniform cross section. Shear is then limited to a localized region near the walls of the channel.

When the channel cross section varies, the entire flowing mass must deform. The extensional flow field produced as the area reduces and the flow accelerates in the entrance to an extrusion die generates very large tensile stresses. Each fiber must be pulled past its neighbor; this induces high shear stresses in the intervening matrix. It has been shown that the rheology of fiber-reinforced polymers is dominated by these large hydrodynamic interactions between fibers (Czarnecki and White 1980).

Despite the high stresses, very low extrudate swell is characteristic of fiber-reinforced systems. However, the high shear stresses between fibers may cause rough, torn surfaces on the extrudate.

2. Fiber-Orientation Control

In shear fields, fibers rotate end over end with an unsteady motion that places them in a time-averaged position nearly parallel to the flow direction. Fiber orientation near the surface of an extrudate is of this type. The structure in the interior is dominated by extensional strains. In a converging die, the positive velocity gradient dv_x/dx stretches the material elements and aligns the fibers parallel to the x or flow direction. An increasing cross section would turn the velocity gradient negative and rotate the fibers perpendicular to the flow.

The equations for fiber rotation in two special extensional flows (Goldsmith and Mason 1967) reduce to

$$\frac{d\theta}{dt} = -k\lambda \frac{dv_x}{dx} \sin 2\theta \qquad (1)$$

where θ is the mean fiber angle measured from the flow direction, λ is a coefficient measuring the orientability of the fibers and k is a constant taking a value of $\frac{3}{4}$ in uniaxial extension $(dv_y/dy = dv_z/dz = -\frac{1}{2}(dv_x/dx))$ or $\frac{1}{2}$ in planar extension $(dv_y/dy = -dv_x/dx;\ v_z = 0)$. Uniaxial extension applies to extrusion through a die orifice and planar extension to the special tube dies described in Sect. 3. Integrating along the flow between channel positions 1 and 2 yields

$$\frac{\tan \theta_2}{\tan \theta_1} = \left(\frac{A_2}{A_1}\right)^{2k\lambda} \qquad (2)$$

where A is the channel cross section and λ is unity for long, straight fibers rotating freely, but decreases to 0.4 is concentration is increased or fiber length is reduced.

The overall balance between surface and internal fiber orientation in the extruded part depends upon the channel geometry and the material composition, which affects both the value of λ and the extent of shear generation. Since fiber reinforcement varies strongly with directionality, these considerations are important to the design performance of the extruded part. A balanced orientation distribution is required to avoid warpage. Reviews of fiber-orientation effects in plastic composites are available in the literature (McNally 1977, Goettler 1984).

3. Applications

In a profile or rod extruded through a converging die, the fiber orientations due to shear and extension are each directed parallel to the flow. The extrudate has both high axial strength and stiffness, typically 3–5 times the crosswise properties. Production of a smooth surface requires a long, heated die.

When the reinforcement is needed in a transverse direction, the fibers may be rotated by using a special die geometry to establish the proper extensional strains in the flowing material. Types of fiber orientation produced by different expanding tube dies are shown in Fig. 1. The fiber pattern generated in die (b) provides the needed hoop strength to double the pressure rating of a short-fiber-reinforced hose. At the same time, flexibility and surface appearance are improved by the absence of axial orientation.

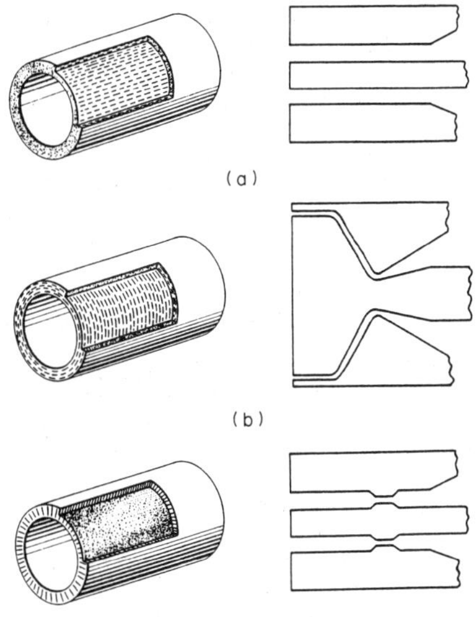

Figure 1
Tube extrusion dies for control of fiber-orientation patterns: (a) conventional converging, (b) expanding diameter and (c) expanding thickness (after Goettler et al. 1979)

See also: Glass-Reinforced Plastics: Thermoplastic Resins; Glass-Reinforced Plastics: Thermosetting Resins; Processing: Rheological Properties and Fluid Mechanics; Processing: Shear-Controlled Orientation

Bibliography

Czarnecki L, White J L 1980 Shear flow rheological properties, fiber damage, and mastication characteristics of aramid-, glass- and cellulose-fiber-reinforced polystyrene melts. *J. Appl. Polym. Sci.* 25: 1217–44

Goettler L A 1984 Mechanical property enhancement in short-fiber composites through the control of fiber orientation during fabrication. *Polym. Compos.* 5: 60–71

Goettler L A 1988 Short fiber–rubber composites. In: Bhowmick A K, Stephens H L (eds.) 1988 *Handbook of Elastomers*. Dekker, New York

Goettler L A, Leib R I, Lambright A J 1979 Short fiber reinforced hose—A new concept in production and performance. *Rubber Chem. Technol.* 52: 838–63

Goldsmith H L, Mason S G 1967 The microrheology of dispersions. In: Eirich F (ed.) 1967 *Rheology: Theory and Applications*, Vol. 4. Academic Press, New York, pp. 86–250

McNally D 1977 Short fiber orientation and its effects on the properties of thermoplastic composite materials. *Polym.-Plast. Technol. Eng.* 8(2): 101–54

L. A. Goettler
[Monsanto, Akron, Ohio, USA]

F

Fabrics: Finishing

The purpose of finishing, the last step in the treatment of fabrics, is to improve either the aesthetic or performance properties of the fabric or both. Such modification of properties often involves a chemical treatment (e.g., with softeners, whiteners or fire-retardant compounds), although mechanical finishing processes are also employed, notably to change the appearance of a fabric.

1. *Removal of Impurities*

In processing fibers and yarns to fabrics, sizes are usually added to the yarn, the removal of which is accomplished in a final water-washing step. In addition, oils added to the surface of the fibers during processing and soil accumulated during manufacture are removed in this washing. The process of washing fabrics after manufacture, called scouring, is accomplished with the use of various washing additives plus agitation.

2. *Changing the Hand*

There are many methods of changing the hand or handle of a fabric. Wool fabrics may be compacted in a finishing step called fulling. Singeing or passing a fabric over a gas flame or a heated metal plate may be used to burn off the fiber ends of a fabric made from staple yarns. If a fabric is limp, it may be stiffened or given added body by the application of various additives. Nonpermanent additives or sizes that are used include starches, gums, gelatins, poly(vinyl alcohol) and carboxymethylcellulose. Durable stiffening is accomplished by the application of resin finishes. The hand of fabrics may be made softer or more limp by the addition of a softener. Oils, waxes and long-chain cationic and nonionic materials are applied as softeners to fabrics, but they are usually not durable to washing or dry cleaning. Increased durability is obtained with polymeric softeners such as polyethylene and silicones. Softeners lubricate the fiber surfaces and reduce friction between fibers, so that in addition to softening the fabrics, improved abrasion resistance may also be provided.

3. *Changing the Appearance*

The appearance of fabrics may be changed in a variety of ways. Fabrics may be whitened by bleaching or by the application of fluorescent whiteners. Bleaching may be accomplished with the use of a chlorine bleach or other oxidizing agents capable of removing colored material from the fabric. Fluorescent whiteners or optical bleaches whiten fabrics by absorbing ultraviolet light and reemitting the light in the blue part of the spectrum, counteracting any yellow cast.

The luster of cotton fabrics may be increased by mercerizing, which involves immersion of the fabric in sodium hydroxide solution followed by washing. The mercerization process produces greater luster by swelling the cotton fiber and changing its shape from a collapsed to a cylindrical structure.

A glossy surface can be produced on a fabric by calendering or passing the fabric through rollers to produce a smooth finish by mechanical means. If a friction calender is used, a glazed finish results. With a Schreiner calender, which contains many fine diagonal lines, a soft luster and hand is produced. Moiré finishes are achieved by passing fabric through a calender with a pattern etched into the surface of the roller, which flattens one area more than another. Embossed designs are similarly produced by pressing a pattern onto a fabric. If the fabric is impregnated with a resin prior to calendering, a durable finish can be obtained provided the calendered fabric is subsequently cured.

Napped fabrics are produced by brushing with machines containing small wires, thereby raising the loose ends of fibers on the fabric surface. Napped fabrics are usually made of yarns with a fairly loose twist so that the yarns will brush easily. The brushed fabrics may be sheared to make the nap uniform in thickness. Napping changes both the appearance and the hand of the fabric and also increases the warmth by trapping air which provides an insulating effect.

An effect similar in appearance to a napped fabric can be obtained by flocking, in which short staple fibers are attached to a fabric surface by means of an adhesive. The adhesive may be uniformly coated on the fabric surface or applied in a printed pattern. The fibers are held on that part of the surface coated with the adhesive in a direction perpendicular to the surface of the fabric. The durability of a flocked fabric is dependent on the properties of the adhesive used.

4. *Control of Biological Degradation*

Synthetic fibers are not normally damaged by mildew and bacteria. However, the natural and regenerated natural fibers are susceptible to microbiological damage, cellulosic fibers being especially prone to damage. Indeed, cellulolytic fungi can destroy cellulosic fabrics that are subjected to moist, warm conditions. To resist biological degradation, cellulosic fabrics may be treated with antimildew agents such as chlorinated phenols, copper naphthenate and other organometallic compounds. Treatment of cellulosic fabrics with formaldehyde reactants such as melamine–formaldehyde will also protect the fabrics from biological degradation, as will acetylation and cyanoethylation of the cellulose.

5. *Durable Press*

Since cellulosic fibers do not recover well from deformation, formaldehyde reactants are used to treat these fabrics to improve this property and consequently the recovery of the fabric from wrinkling. If sufficient formaldehyde reactant is applied, the resulting fabric will have both wet and dry recovery from wrinkling and consequently will have durable press properties.

Historically, the treatment of fabrics with formaldehyde-containing reactants to provide recovery from wrinkling started with an invention of Foulds, Marsh and Wood in 1926, which involved the finishing of rayon fabrics with water-soluble urea–formaldehyde or phenol–formaldehyde resins. The first resin treatment was carried out on rayon which has greater elongation than cotton and is embrittled less by the resin treatment. Subsequently, urea–formaldehyde resins were applied to cotton fabric to give dimensional stability and recovery from wrinkling. Later, blends of polyester and cotton were resin treated to provide durable press fabrics with greater strength and resistance to wear. The resins used included not only urea–formaldehyde but also water-soluble reaction products of formaldehyde and triazine, triazone, uron, ethylene urea, propylene urea and carbamates. The most widely used product at present is dimethyloldihydroxyethyleneurea (DMDHEU) or its ethers. DMDHEU has the form

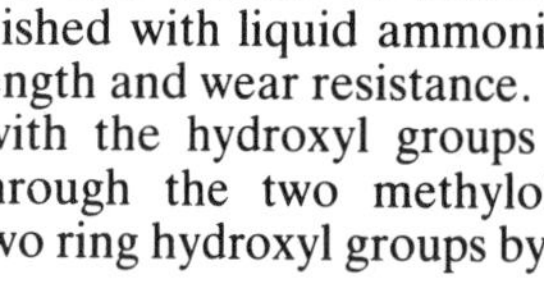

Treated fabrics also include all-cotton fabrics that have been finished with liquid ammonia to provide improved strength and wear resistance. Reactions of DMDHEU with the hydroxyl groups of cellulose can occur through the two methylol groups or through the two ring hydroxyl groups by the elimination of water.

Durable-press treatment involves immersion of the fabric in an aqueous pad bath containing the formaldehyde reactant and a metal-salt catalyst, squeezing the fabric through pad rolls, drying and curing. The generally accepted mechanism involves crosslinking of adjacent cellulosic chains by reaction of the formaldehyde-containing reactant with cellulose. This crosslinking prevents the slippage of adjacent cellulosic chains that may not be prevented by van der Waals and hydrogen bond forces when stress is applied to the fibers or fabric.

Formaldehyde release at high temperatures and humidities from durable-press fabric treated with formaldehyde reactants depends on many factors. With regard to reactant composition, those reactants having low reaction rates with cellulose such as DMDHEU release less formaldehyde. The rate of hydrolysis of the formaldehyde reactant bears a direct relation to its reaction with cellulose and to formaldehyde release. The leaving group has an important effect; earlier work indicated that, of the groups studied, the methyl leaving group resulted in the lowest release of formaldehyde from the treated fabric. Later studies showed that the polyols gave lower formaldehyde release values than the methyl group. Increasingly, therefore, the polyol ethers of DMDHEU, especially the diethylene glycol ether, have been used. Some finishers add the polyol to pad baths containing DMDHEU rather than using the polyol ether of DMDHEU. A strong catalyst such as zinc nitrate is preferable, along with a cure time and temperature which gives optimum crosslinking with a minimum of pendent NCH_2OH groups. Impurities such as urea in the glyoxal–urea reaction product from which DMDHEU is prepared can result in increased formaldehyde release from the treated fabric; the presence of impurities such as alkali in the fabric prior to treatment and the use of additives such as buffers in the pad bath can also have a deleterious effect. On the other hand, formaldehyde scavengers added to the pad bath can reduce the amount of formaldehyde released from the treated fabric. Formaldehyde release can also be decreased by afterwashing the treated fabric or aftertreating with a formaldehyde scavenger.

In the manufacture of garments from durable-press fabrics, the garment is often made before the curing operation is carried out. The garment is then cured with the creases in place. This operation is referred to as postcuring, as differentiated from precuring.

In order to avoid the problem of formaldehyde release, reactants have been developed which either do not contain formaldehyde or do not decompose and liberate formaldehyde. A product used to a limited extent in Japan, where there are stringent requirements for formaldehyde release, is *N,N'*-dimethyldihydroxyethyleneurea:

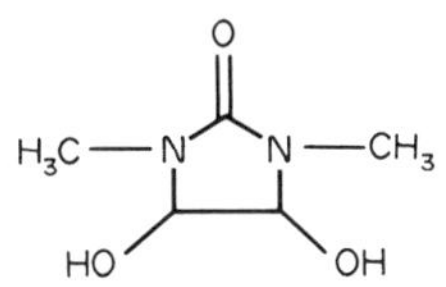

This product can react with cellulose through the ring OH groups, but the durable-press properties are not equal to those obtained with a difunctional *N*-methylol compound.

Although the main method of increasing the elastic recovery of cellulosic easy-care fabrics is by crosslinking with cellulose reactants inside the fiber, another method, which gives a lesser effect, is surface application. Since the cotton fiber surface has a high rugosity, it has a comparatively high coefficient of friction. Therefore, a topical application to decrease the coefficient of friction of the fiber and yarn will to some extent increase the recovery of the fabric from deformation. Emulsions of silicones and low-molecular-weight polyethylene are used extensively for this purpose as well as to provide a softer hand. If, in addition to the reduction in the coefficient of fiber and yarn friction, a surface application can produce an elastic fiber–fiber or yarn–yarn bond, a still further increase in the elastic recovery of the fabric may result. Therefore, lattices of elastomers, such as polyurethanes, are sometimes added to the formaldehyde reactant pad bath to impart additional recovery from deformation. Care must be taken, however, not to apply excessive quantities of elastomeric products to the fabric; otherwise an undesirable cold, clammy hand may result.

6. *Shrinkage Control*

Fabric shrinkage may be due to many causes. The shrinkage of cellulosic fabrics during washing is due to the relaxation of stresses imposed during manufacture of the fabric. This type of shrinkage, referred to as relaxation shrinkage, is usually greater in the warp direction of woven fabrics because the warp yarns are subjected to greater tension than the filling yarns. Knit fabrics tend to be stretched more than woven fabrics and therefore tend to shrink more during washing. Mechanical agitation in washing wool fabrics produces a compaction or felting of the fabric. This felting is due to the scalar surface of the wool fiber which causes the fiber to migrate preferentially in the direction of the fiber root since the overlapping scales point towards the fiber tip.

The tendency of fabrics to shrink during washing can be minimized by finishing treatments. The shrinkage of cellulosic fabrics can be reduced by a treatment that causes a compression or compaction in the plane of the fabric. An example of this type of treatment is Sanforizing. In this process, the fabric and a backing blanket of wool or rubber held under tension are fed between a feed roller and a curved braking shoe. The tension on the blanket is released after passing the fabric and blanket between the roller and braking shoe. The result of this action is a compaction of the fabric. Another method to reduce the shrinkage is aftertreatment with formaldehyde reactants such as those used for durable-press treatments. In fact, the durable-press treatment itself provides shrinkage control.

The felting of wool fabrics can be reduced by chemical treatment or by application of resins. The most commonly used chemical treatments involve chlorination, which partially dissolves the scalar structure of the wool fiber. As a result, the differential friction effect (the difference between the "antiscale" and "with scale" coefficients of friction) is reduced so that the preferential migration is minimized. Unless carefully controlled, chlorination may reduce the strength of the wool fiber and change the hand of the treated fabric. Various resin treatments have also been developed for controlling the felting of wool fabrics. These treatments have included (a) impregnation with melamine–formaldehyde, (b) interfacial polymerization primarily utilizing polyamide resins and (c) application of polyurethane resins. The resin treatments also reduce the differential friction effect and further reduce felting shrinkage by spot welding the fibers together.

In the case of fabrics made of thermoplastic synthetic fibers, shrinkage can usually be controlled by heat setting the fabrics. If the temperatures to which the fabric is exposed during service are below the temperature at which the fabric is heat set, shrinkage of the fabric is not usually a problem.

7. *Water and Oil Repellency*

There are a number of chemicals that can be applied to fabrics to impart water and/or oil repellent properties. One of the earliest and simplest water-repellent treatments involves the application of wax to a fabric from an aqueous emulsion to impart a hydrophobic surface. These wax treatments have no durability to dry cleaning and limited durability to washing. If aluminum or zirconium salts are applied in a second treatment, limited durability to dry cleaning as well as washing is obtained. Formaldehyde reactants are also coapplied to improve the durability.

The next type of water-repellent finish involved long-chain pyridinium compounds:

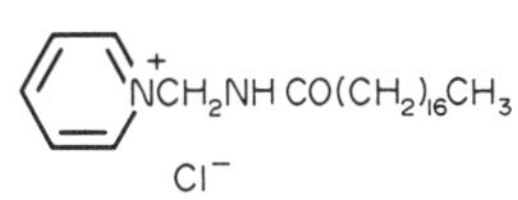

These were applied to cellulosic fabrics via an aqueous padding procedure followed by drying and curing. It is generally believed that during the curing operation, these compounds lose pyridine and hydrogen chloride and react with cellulose. The resulting finishes have fairly good durability to washing, but limited durability to dry cleaning. The coapplication of formaldehyde reactants such as melamine–formaldehyde increases the durability to washing.

A number of resin-based water repellents were then developed in which formaldehyde reactants such as melamine–formaldehyde were combined with long-chain hydrocarbon moieties to give durable finishes. Often, quaternized nitrogen moieties were incorporated to give improved emulsification properties.

Historically, the next water repellents developed were based on the following silicone polymers:

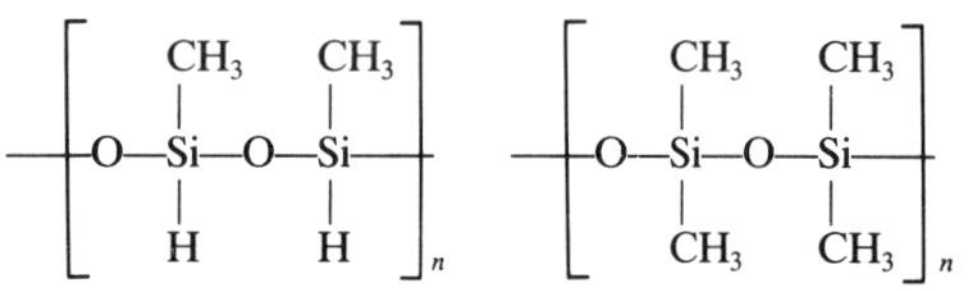

These imparted low surface energy properties to fabrics. The Si—H groups are reactive, whereas the Si—CH_3 groups are not. The silicone polymers are normally applied from an aqueous emulsion with a metal soap catalyst. They provide good durability to both washing and dry cleaning.

The most recently developed and most widely used products are based on fluorochemicals. These products are both water and oil repellents because of their very low surface energies. The compounds are usually provided as fluoroacrylate polymers with fluorine constituents present at the end of the polymer side chains as in

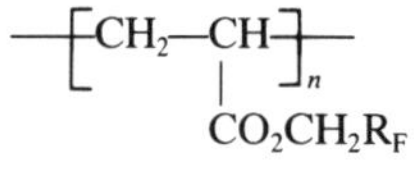

where R_F represents a perfluoroalkyl chain. The polymers are usually applied to the fabric in the form of an emulsion and the treated fabric is then dried and cured.

8. Soil Release

Even fabrics treated with oil-repellent fluorochemicals show a propensity to retain a stain when exposed to oils for a long period of time. For soil release, polymers with hydrophilic groups are needed. Fluorocarbon polymers with terminal fluorocarbon groups and modified with hydrophilic segment (containing carboxyl or other hydrophilic groups) when applied to fabrics as aqueous emulsions will provide both oil repellency and soil release. Many of the compounds developed as soil-releasing agents are polymers or copolymers of acrylic or methacrylic acid. They are usually provided as aqueous emulsions and can be applied alone or with formaldehyde reactants used to obtain durable-press properties. It is believed that the soil is held by the hydrophilic groups, perhaps by the acid moiety, and released during washing. The major disadvantages of these finishes are the change in hand of the treated fabric (stiffening effect) and their limited effectiveness as soil releasers.

9. Fire Retardancy

Although most organic fibers support combustion, their flammability can be reduced if they are treated with fire retardants. Some organic fibers such as the aramid fibers are thermally stable and withstand the high temperatures at which most organic fibers support combustion. In contrast, polyamide, polyolefin and polyester fibers usually melt before they burn and the molten polymer may be removed from the source of flame before combustion occurs. Inorganic fibers such as glass and asbestos do not burn.

The flammability of fibers can be reduced by (a) modification of the polymer used for the fiber, (b) addition of a fire retardant to the dope prior to extrusion of the fiber and (c) treatment of the fiber (usually in fabric form) with a fire retardant. An example of the first type of modified fiber is a fire-retardant acrylonitrile fiber made by the copolymerization of vinyl chloride with acrylonitrile. An example of the second type of fiber is a polyester fiber made from a dope containing brominated biphenyl as the fire-retardant additive. An example of the third treatment is the finishing of cellulosic fabric with tetrakis(hydroxymethyl)phosphonium chloride (THPC). The fire retardants that have been used are many and varied in composition. However, some of the common ones are halogenated compounds and compounds containing phosphorus and nitrogen.

Several theories have been advanced to explain the modes of action of fire retardants. These include (a) noncombustible gases are generated which dilute the combustible volatiles and the oxygen supply, (b) the fire retardant decomposes endothermally, (c) the fire retardant degrades into free radical acceptors which interfere with flame chain reactions, (d) a nonvolatile char or liquid coating is formed which protects the fibers and (e) a physical interaction occurs which alters the course of gas-phase reactions.

A great deal of research has been conducted on fire-retardant finishes for cellulosic fabrics. Some of the earlier nondurable finishes for cellulosic fabrics included inorganic salts such as borates and phosphates. Ammonium phosphate and diammonium phosphate are the most widely used nondurable flame retardants. A flame retardant with some durability that is commonly used for flameproofing of industrial and military fabrics is a mixture of antimony oxide and chlorinated paraffin. Durable flame retardants developed for cellulosic fabrics used in clothing include many phosphorus compounds capable of reacting (a) directly with cellulose or (b) with a formaldehyde reactant that reacts with cellulose. The most widely used of these fire retardants are THPC, its reaction product with sodium hydroxide (THPOH) and THPS where sulfate replaces chloride in THPC. These flame-retardant finishes are usually applied with formaldehyde reactants to give durable-press properties and greater durability to washing, but they may produce an undesirable hand and reduced wear characteristics.

Although wool is less combustible than cotton, its flammability can be reduced by many of the fire retardants that are used on cotton and by the application of zirconium or titanium salts. The THPC treatments for cellulose fabrics are also effective on both wool and silk fabrics. Acetate fibers can be made fire retardant by the addition of retardants such as haloalkylphosphonates to the dope prior to spinning. For polyester and polyamide fibers, a fire retardant can be added to the spinning bath or tris(2,3-dibromopropyl) phosphate (Tris) can be applied. However, the use of Tris has essentially ceased since it was found to be mutagenic and a potential carcinogen. Acrylic, polyamide and polyethylene fibers can be made fire retardant by adding an organic phosphorus or halogenated compound to the dope prior to spinning.

10. Antistatic Properties

Cellulosic fibers and most of the natural fibers are hydrophilic, having polar groups on the fiber surface, and are therefore capable of dissipating static charge to the atmosphere so that an antistatic finish is not needed. However, with most synthetic fibers, static charge can be built up on the fiber surface, especially under low humidity conditions, causing garments to cling and electric shocks when the wearer touches a conductor such as a metal door knob.

An antistatic finish or the incorporation of a conducting carbon or metallic fiber into the fabric can solve these problems. Antistatic materials can be applied to the fiber as a temporary finish or added to the spinning bath prior to fiber formation to give a more durable finish. These finishes can either coat the surface of the fiber with a conductive material or attract enough moisture to increase the conductivity of the fiber surface.

Antistatic materials applied to polyamides include hydrophilic polymers, polyethylene glycols and salts of amphoteric and cationic surface-active agents. Antistats used on polyester include lauryl phosphate, morpholine, polyethylene glycols, organosilicones and polyamines. Antistatic agents for polyacrylic fibers include polyglycol esters, fatty acids and hydroxyl and amino compounds that are crosslinked with aldehyde or epoxy compounds.

See also: Fibers and Textiles: An Overview; Textiles: Structures and Processes

Bibliography

Lewin M, Atlas S M, Pearce E M 1975 *Flame-Retardant Polymeric Materials*. Plenum, New York

Lewin M, Sello S B 1983 *Chemical Processing of Fibers and Fabrics*, Vols. 1, 2. Dekker, New York

Mark H F, Atlas S, Cernia E 1968 *Man-Made Fibers: Science and Technology*. Wiley, New York

Mark H F, Wooding N S, Atlas S 1971 *Chemical Aftertreatment of Textiles*. Wiley, New York

Needles H L 1981 *Handbook of Textile Fibers, Dyes and Finishes*. Garland STPM, New York

Tortora P G 1978 *Understanding Textiles*. Macmillan, New York

T. F. Cooke
[Textile Research Institute, Princeton, New Jersey, USA]

Fatigue

The term fatigue refers to the reaction of materials to alternating loading. Under such conditions, common in many engineering applications, polymeric components may fracture after many cycles of loading, even though the applied stress amplitude is well below the tensile strength of the material. The fatigue process involves a crack initiation phase and a crack propagation phase. Which phase dominates depends on the type of polymer and the test conditions. The fatigue resistance depends on many material variables, such as composition, molecular weight, extent of orientation and presence of additives, and on many external variables, such as frequency, temperature, stress and environment.

1. General Nature of Fatigue Fracture

Observation of polymer specimens that have fractured in service or in laboratory fatigue tests yields much information about the principal characteristics of fatigue fracture (Hertzberg and Manson 1980). There is little macroscopic plastic deformation and the fracture surface, generally perpendicular to the stress direction, has rather characteristic markings.

Part of a typical fatigue fracture surface for a glassy polymer, polystyrene, is shown in Fig. 1. The smooth region surrounding the source is indicative

Figure 1
Part of fatigue fracture surface of polystyrene sample tested at 17.2 MPa and 2 Hz

of crack propagation through a surface craze. The craze, consisting of oriented fibrillar elements interspersed with voids, forms after a certain number of alternating cycles, as a result of a high local stress concentration at a surface flaw. Hence, the resistance offered by different polymers to the development of a craze, or localized plastic zone, greatly affects their fatigue performance.

The series of concentric rings, called discontinuous growth bands, is a distinctive feature of many fatigue fracture surfaces. The crack is arrested near the end of each band, usually for hundreds of cycles. Meanwhile, the craze in front of the crack tip grows continuously until some critical length or some critical level of cyclic damage is reached. At this stage the crack jumps through about two-thirds of the existing craze and is again arrested (Döll 1983). The stability of the craze to cyclic breakdown is a determining factor in the resistance of different polymers to fatigue fracture. Molecular rearrangements can occur during fatigue cycling. Changes have been observed in infrared absorption bands, in mechanical loss spectra, in extent of local packing and in defect structure (Kausch 1978).

2. *Effects of Material Variables*

Crystalline polymers are more resistant to fatigue fracture than are the glassy amorphous polymers. Many amorphous polymers, such as polystyrene (PS) and poly(methyl methacrylate) (PMMA), have a fatigue strength that is only about one-fifth of the tensile strength of the material. In crystalline polymers, such as Nylon 66 and polyoxymethylene, this ratio is appreciably higher (0.3–0.5). Figure 2 shows the fatigue crack propagation (FCP) rate da/dn, where a is the crack length and n is the number of cycles, as a function of the stress-intensity-factor range ΔK for a number of thermoplastics. The crystalline nylon shows lower FCP rates than the amorphous polymers (Hertzberg et al. 1975). The greater resistance of the crystalline polymers to fatigue is attributed to the presence of the many small crystallites which hinder crack growth.

It is also evident from Fig. 2 that ductile amorphous polymers, such as poly(vinyl chloride) (PVC) and polycarbonate (PC), are superior in fatigue performance to the brittle glassy polymers such as PS and PMMA, where crazing is the predominant deformation mode.

Molecular weight is an important material variable. This is demonstrated by Fig. 3 for PS specimens. As the molecular weight increases from 1.6×10^5 to 2×10^6, the average lifetime is increased by more than two orders of magnitude and the fatigue

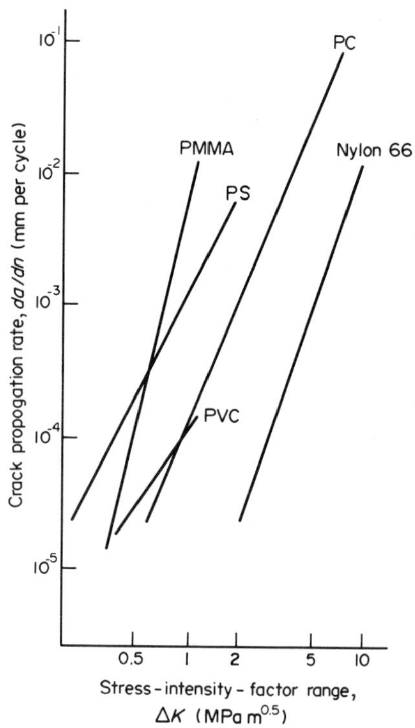

Figure 2
Fatigue crack propagation rate vs stress-intensity-factor range for various thermoplastics

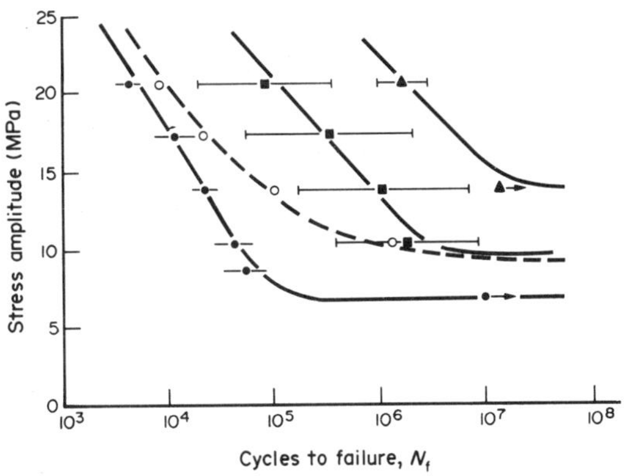

Figure 3
Stress amplitude vs number of cycles to failure for polystyrene samples of different molecular weight: ●, 1.6×10^5; ○, 2.74×10^5; ■, 8.6×10^5; ▲, 2×10^6; —— narrow, --- broad molecular weight distribution

strength is doubled. Reductions in FCP rate with increase of molecular weight have also been observed in other polymers, such as PVC, PMMA and polyethylene (PE). The beneficial effects of molecular weight are attributed to an increased craze stability as a result of an increase in chain entanglements and a reduction in the number of chain ends.

Fatigue resistance of polymers can be increased by glass-fiber reinforcement and by lamination. For many polymers thus modified, the ratio of fatigue strength at 10^7 cycles to tensile strength lies between 0.3 and 0.5, or considerably above that of the amorphous glassy homopolymers. The better fatigue performance arises largely because of the impediments to crack propagation produced by the reinforcement or by the laminations.

Orientation in polymers tends to hinder craze development and thus can have a beneficial influence on fatigue performance. The fatigue response of highly oriented polymers, such as fibers, is complex; different deformation modes may be encountered, depending on the polymer, extent of orientation, stress state, microstructure and frequency (Bunsell and Hearle 1974).

Rubber modification is a standard method for raising impact strength, but it has less effect on fatigue response. Tests on notched specimens, as well as fractographic studies of fractured fatigue samples, show that resistance to FCP is improved. However, unnotched rubber-modified polymers fracture sooner in fatigue than the unmodified polymer (Chen et al. 1981). This is an example of where the rubber modification, although enhancing FCP resistance, alters the crack initiation phase adversely by accelerating craze initiation.

Addition of diluents, such as plasticizers, generally decreases fatigue resistance. Plasticizers induce greater chain mobility; thus, there may be localized heating near the crack tip. This can lead to blunting of the crack tip and, hence, to a decrease in FCP rate. Both effects have been observed.

3. *Effects of External Variables*

When polymers are subject to alternating stress, the temperature may rise above ambient, owing to hysteresis. The seriousness of the resulting thermal softening effect depends on the magnitude of the applied stress, on the test frequency and on the viscoelastic nature of the polymer. For polymers having low internal friction, such as PS, the temperature rise is negligible. However, when more viscoelastic polymers, such as PE, are subject to cyclic stresses and high frequencies, appreciable temperature rises are observed and failure may occur, not by progressive growth of a fatigue crack but by thermal rupture or even melting (Sauer and Richardson 1980).

The effects of frequency on fatigue life and on FCP rate depend on the polymer, on its mechanical relaxation spectra and on the ambient temperature. In some polymers, such as PS, PMMA and PVC, an increase in test frequency tends to increase average fatigue lifetimes and to decrease FCP rates. In other polymers, such as Nylon 66, PC and polysulfone, no effect of frequency on FCP has been noted. It has been suggested that the frequency sensitivity of polymers is a maximum when the frequency of segmental jumps of chain segments associated with the principal secondary transition is in the range of test frequencies.

The effect of increasing the test temperature on fatigue performance is to shift the stress–lifetime curve to shorter fatigue lifetimes and to lower stresses, without altering its general shape.

The stress-intensity-factor range, as Fig. 2 shows, is an important factor controlling FCP rate. However, other factors, such as the mean stress or the mean stress intensity, are also important. In general, as these factors increase, the fatigue resistance decreases. Many different forms of rate equation have been proposed; their possible application to fatigue crack growth in polymers has been reviewed (Radon 1980).

Environment may also play a role in fatigue failure of polymers. Liquid coatings, for example, such as silicone oils for PS, can reduce fatigue lifetime even when the environment in the absence of stress is inert. Low-viscosity liquids, which more readily penetrate surface crazes and cracks, and environments that are more compatible with the polymer and thus readily plasticize can severely reduce fatigue performance. On the other hand, fatigue resistance can be enhanced by protecting the polymer surface against aggressive environments.

Other means of extending the fatigue lifetimes of polymers include improving surface finish, developing an oriented surface layer and applying thin adherent elastomeric films. All of these inhibit development of a localized damage zone such as a surface craze and, hence, enhance fatigue performance. More detailed information on these matters, together with fatigue data on many types of polymers, including rubbers, fibers, plastics and composites, is given in the references cited in the bibliography.

See also: Polymers: Mechanical Properties

Bibliography

American Society of Testing and Materials 1975 *Proc. Symp. Fatigue of Composite Materials*, ASTM Special Technical Publication No. 569. ASTM, Philadelphia, PA

Andrews E H 1969 Fatigue in polymers. In: Brown W E (ed.) 1969 *Testing of Polymers*, Vol. 4. Interscience, New York, pp. 237–96

Bunsell A R, Hearle J W S 1974 The fatigue of synthetic polymeric fibers. *J. Appl. Polym. Sci.* 18: 267–91

Chen C C, Chheda N, Sauer J A 1981 Craze and fatigue resistance of glassy polymers. *J. Macromol. Sci., Phys. B* 19: 565–88

Döll W 1983 Optical interference measurements and fracture mechanics analysis of crack tip craze zones. *Adv. Polym. Sci.* 52/53: 105–68

Hertzberg R W, Manson J A 1980 *Fatigue of Engineering Plastics*. Academic Press, New York

Hertzberg R W, Manson J A, Skibo M 1975 Frequency sensitivity of fatigue processes in polymeric solids. *Polym. Eng. Sci.* 15: 252–60

Kausch H H 1978 *Polymer Fracture*. Springer, Berlin

Radon J C 1980 Fatigue crack growth in polymers. *Int. J. Fract.* 16: 533–52

Sauer J A, Richardson G C 1980 Fatigue of polymers. *Int. J. Fract.* 16: 499–532

J. A. Sauer
[Rutgers University, Piscataway, New Jersey, USA]

Fibers and Textiles: An Overview

The first fibers used by man were natural fibers such as cotton, wool, silk, flax, hemp and sisal. The first man-made fiber was probably glass; the Egyptians made vessels out of coarse glass fibers. The first use of glass fibers in textiles took place in the eighteenth century. The idea that man could make an organic fiber is attributed to Hooke, who in 1664 suggested that a "glutinous substance" could be drawn into fibers. However, it was not until the latter part of the nineteenth century that Despassis was awarded a French patent for the production of rayon fibers by the cuprammonium process. Then in 1893 a British patent was granted to Cross, Bevan and Beadle for making rayon fibers by the viscose process.

It was only at the beginning of the twentieth century, however, that man-made fibers started supplementing and replacing natural fibers. The first important man-made fibers were the cellulose-derived fibers, rayon and cellulose acetate. Then came nylon, the first truly synthetic fiber, followed by polyesters, polyacrylics and polyolefins. Also man-made elastomeric, glass and aramid (aromatic polyamide) fibers became important commercial products. Finally, the high-temperature, high-performance inorganic fibers including carbon, alumina, boron, silicon carbide and silicon nitride, some in the form of whiskers, have been produced and are used primarily as reinforcement fibers in composites.

Man-made fibers are now available, ranging in properties from the high-elongation and low-modulus elastomeric fibers, through the medium-elongation and medium-modulus fibers such as polyamides and polyesters, to the low-elongation, high-modulus carbon, aramid and inorganic fibers. With such a wide variety of man-made fibers available, the volume of synthetic fibers consumed worldwide is now greater than that of natural fibers.

1. Classification

The major fibers used in textiles can be classified into the groups given in Fig. 1.

Although most man-made fibers are composed of one class of polymers only, some biconstituent or bicomponent fibers are manufactured; these fibers may be spun from different polymers side by side or in other configurations. One purpose of such fibers is to produce a crimp in the fibers by exploiting the differential thermal properties of the two polymers. Also, differential coloring effects may be produced in this manner.

In fabric manufacture, blends of fibers are often used to obtain the desired fabric properties. An example is a blend of polyester and cotton that is treated with a formaldehyde reactant to give a durable press fabric. The tensile strength and abrasion resistance of the polyester fibers counteract the loss of these properties in the treated cotton fibers. Another example is a blend of nylon and wool, which is often used for clothing fabrics where both warmth and wear resistance are important.

2. Physical Properties

Some of the more important physical properties of the most extensively used fibers are given in Table 1.

Fibers used in textile structures must have sufficient length to be made into twisted yarns, but can be as short as 1.3 cm in stable fibers. The diameter of fibers with the exception of some inorganic fibers

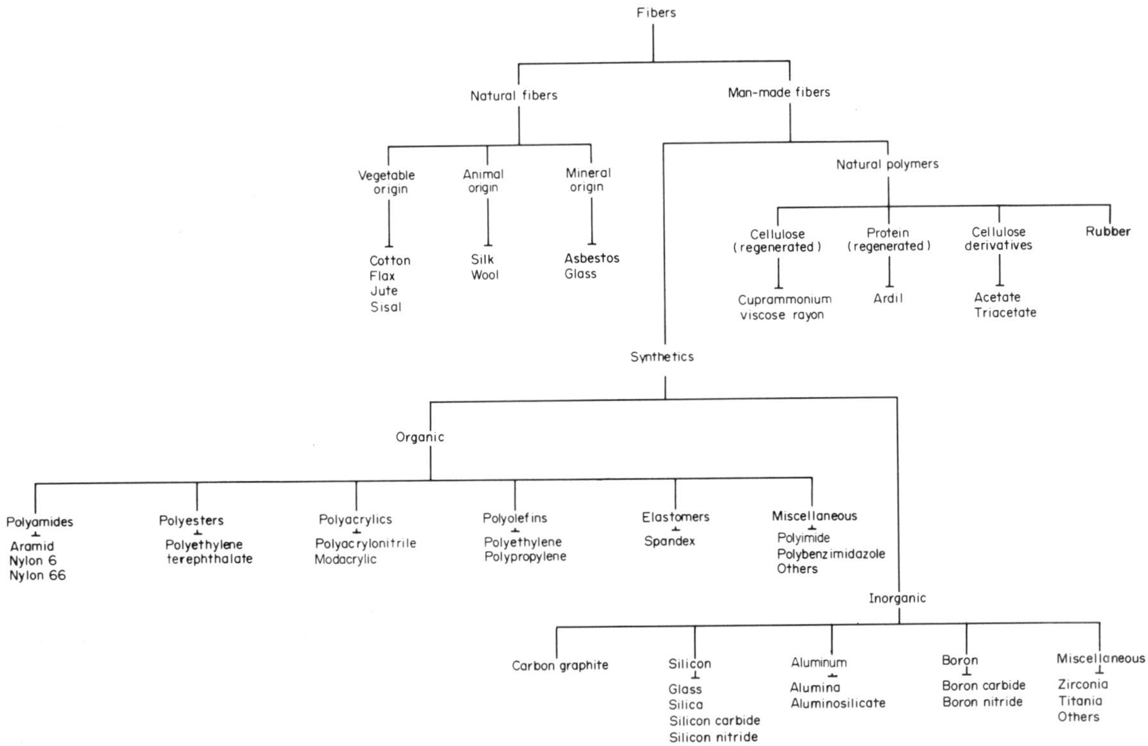

Figure 1
Classification of fibers used in textiles

generally varies from 5 μm to 45 μm. However, nylon, polyester and polyacrylic fibers are now being manufactured with diameters approaching 1 μm; these fibers give fabrics that are soft and have a good drape. Even smaller diameter fibers have been produced by solvent removal of one component of a bicomponent fiber.

The tensile strength of a fiber increases with decreasing fiber diameter as is illustrated in Table 2. Not only does the tensile strength increase with decreasing fiber diameter, but the modulus generally increases as well. The generally accepted explanation for these phenomena is that there are fewer imperfections and faults as the fiber diameter is reduced. Probably the ultimate in small fiber diameter and in perfection, with resulting high tensile and modulus values, are the inorganic whisker fibers. These high-performance fibers with their high-temperature resistance are used as reinforcing fibers in metallic and ceramic composites used in space vehicles.

The cross-sectional shape of fibers varies considerably. The cross section of a cotton fiber is a collapsed circle, which is somewhat flat and irregular. Silk has a triangular cross section, which accounts for its sheen and feel. Wool is circular in shape with scales on the surface. Rayon can be circular or serrated circular, and it often has a center or core that differs from the outer layer. Man-made fibers, particularly melt-spun fibers, can also be made oval, multilobal and even hollow circular. The shape of the fiber cross section affects the optical properties of the fiber, including luster and translucency. It also affects the feel or hand of fabric made from the fiber.

The surface of fibers also varies considerably. Cotton has a surface that has many irregularities or rugosities. Wool has a scalar structure, which has a tendency to make a wool fabric felt or shrink during washing. Most man-made fibers have relatively smooth surfaces; however, rayon may be serrated.

Some fibers, such as wool, have a natural undulating physical structure or crimp, whereas man-made fibers are frequently subjected to various mechanical and heat-setting processes to provide crimp. These processes include false twist, stuffer box, gear crimping, knife-edge, antitwisting, knit–deknit and air-jet crimping. Most of these processes are self-descriptive and all involve heating the fiber or yarn while it is being held in a crimped configuration. Fabrics made from crimped fibers tend to have increased stretch, bulk and warmth.

Table 1
Physical properties of textile fibers

Fiber type	Diameter (μm)	Density (kg m^{-3})	Tenacity (gf tex^{-1})	Breaking extension (%)	Initial modulus (gf tex^{-1})	Work of rupture (g tex^{-1})	Moisture regain[a] (%)	Melting point (°C)
Natural								
cotton	11–22	1520	35	7	500	1.3	7	decomposes
flax	5–40	1520	55	3	1840	0.8	7	decomposes
jute	8–30	1520	50	2	1750	0.5	12	decomposes
sisal	8–40	1520	40	2	2500	0.5	8	decomposes
wool	18–44	1310	12	40	250	3	14	decomposes
silk	10–15	1340	40	23	750	6	10	decomposes
glass	≥5	2540	76	2–5	3000	1	0	800
asbestos	0.01–0.30	2500			1300		1	1500
Regenerated								
viscose rayon	≥12	1450–1540	20	20	500	3	13	decomposes
acetate	≥15	1320	13	24	350	2	6	230
triacetate	≥15	1320	12	30	300	2	4	230
Synthetic								
Nylon 6	≥14	1140	32–65	30–55	250	6–7	2.8–5	225
Nylon 66	≥14	1140	32–65	16–66	250	6–7	2.8–5	250
Qiana	≥10	1030	25	26–36			2.5	274
Nomex	≥12	1380	36–50	22–32		7.5	6.5	decomposes above 380 °C
Kevlar	≥12	1440	200	2–4	4500–8000		4	decomposes above 500 °C
polyester	≥12	1340–1380	25–54	12–55	1000	2–9	0.4	250
acrylic	≥12	1160–1170	18–30	20–50	650	5	1.5	sticks at 235 °C
polypropylene		910	60	20	800	8	0.1	165
polyethylene		950	30–60	10–45		3	0	115
spandex		1210	6–8	444–555		18	1.3	230

Source: Goswami et al. 1977
a At 65% relative humidity

The density of fibers (see Table 1) varies from about 3900 kg m^{-3} for alumina to 900 kg m^{-3} for polypropylene, with polyester being about 1400 kg m^{-3} and cotton about 1500 kg m^{-3}. The low density of polypropylene makes it desirable for marine ropes, since they will float in water.

Fibers must have sufficient strength to be useful in textile structures. The strength of textile fibers is usually referred to as tenacity, which is defined as the force required to rupture or break the fiber per unit linear density. Tenacity is generally measured in grams-force per denier (where denier is the weight in grams of 9 km of fiber or yarn), or as grams-force or newtons per tex (where tex is the weight in grams per kilometer of fiber or yarn). The tenacity of fibers varies from slightly less than 6–8 gf tex^{-1} for the elastomeric spandex fibers to 76 gf tex^{-1} and even higher for some inorganic fibers, with nylons and polyesters generally being 25–65 gf tex^{-1}.

In addition to having sufficient strength for its various end uses, a fiber must also have sufficient extensibility to withstand the stresses to which it will be subjected in service. The breaking extension or elongation-at-break varies from less than 1% for some inorganic fibers to over 500% for some spandex fibers, with nylons and polyesters generally varying from 12% to 66%. For many end uses, an elongation of less than 2% is not sufficient, so the use of fibers with elongations of this order is limited. However, for some composite applications, fiber elongations of less than 1% can be tolerated.

Table 2
Influence of diameter on tensile strength of glass fibers; fiber length held constant

Diameter of test fibers (μm)	Ultimate tensile strength (GPa)
19.0	0.7
15.2	0.9
12.7	1.0
10.2	1.3
8.6	1.7
5.1	2.8
2.5	6.0

Source: Anderegg 1939

Another fiber property that is important is recovery from deformation. This property of the fiber is one that is also imparted to yarns and to fabric made from them; for example, the ability of a fabric to recover from wrinkling depends to a large extent on the recovery from deformation of the individual fibers. Recovery can be measured by imposing deformation in a cyclic manner and determining the elongation recovered. The initial modulus or slope of the first part of the stress–strain curve can also be used as a measure of this property. Generally, fibers with low elongation (e.g., glass fibers) have high modulus values and those with high elongation such as spandex fibers have low moduli. Nylons and polyester fibers generally have intermediate elongation and modulus values. The area under a fiber stress–strain curve represents the toughness of the fiber and, if measured to rupture, is called the work of rupture. Values of work of rupture are generally high for spandex, medium to high for nylons and polyesters and low for glass fibers.

Other important properties of fibers include abrasion resistance, water absorption, dimensional stability, thermal characteristics, flammability, density and resistance to attack (e.g., from chemicals, sunlight and mildew). Generally, fibers with a low work-of-rupture value, such as glass fibers, have poor abrasive resistance, whereas fibers such as nylon and polyester have good abrasion resistance. Water absorption is an important factor for comfort in clothes. Fibers with high water absorption such as cotton and rayon are usually considered more comfortable than polyamide, polyester or polyolefin fibers, which have low water absorption. Since it is important that fabrics neither shrink nor stretch excessively in an irreversible manner, man-made fibers are often heat set. This gives the fibers dimensional stability provided that they are not exposed to a temperature in excess of that of the heat-setting. Some cellulosic fabrics shrink during washing if they have been stretched during manufacture. Finishing of cellulosic fabrics with a formaldehyde reactant may impart dimensional stability as well as recovery from wrinkling.

While some fibers such as the cellulosics do not melt but will burn, others (e.g., nylons, polyesters and polyolefins) melt at relatively low temperatures, but burn with difficulty because the molten polymer falls away from the flame.

3. *Morphology*

The morphology, or spatial arrangement of the individual polymer molecules and groups of molecules, has a bearing on fiber properties. The naturally occurring fibers already have a built-in morphology, which can sometimes be altered. For example, cotton can be mercerized by treatment with caustic to swell the fiber and give it higher luster, and it can also be crosslinked by various formaldehyde reactants to enhance its recovery from wrinkling. Similarly, cellulose can be chemically modified and spun into rayon or acetate fibers, and wool can be chemically treated to decrease its felting properties.

4. *Production*

A large number of man-made fibers with a variety of properties has been produced from polymers by various spinning techniques, including melt, dry, wet and emulsion spinning. Polyamide, polyester and polyolefin fibers are formed by forcing the melted polymer through an orifice into air. Polyacrylonitrile fibers are spun from a solvent solution of polyacrylonitrile by either dry spinning (into air) or wet spinning (into a coagulating bath). Spandex fibers can be formed from polyurethanes by either dry or wet spinning.

Generally, after being spun, the fibers are mechanically stretched or drawn. In this process, the molecules are oriented in a direction parallel to the axis of the fiber. The resulting aligned molecules may become closely packed, so that crystallization can occur in some areas. This crystallization stabilizes the fiber and imparts improved properties of strength and recovery from deformation. However, some areas are left relatively unaligned and in these amorphous areas the molecules are loosely held together by secondary forces including hydrogen bonding, van der Waals forces and dipole–dipole interactions. These permit more flexibility of movement than in the crystalline regions. The amorphous regions also allow the penetration of dyes and finishes into the fiber.

Fibers are usually processed into yarns, and the yarns are woven or knitted into fabrics. Filament fibers can be made into yarns with little or no twisting, but staple fibers, being short, must be twisted together to form long continuous yarns. Often sizes, which are usually water-soluble polymers, are applied to the yarns to facilitate weaving. Fibers can also be randomly arranged in so-called nonwoven textile structures, without first being made into yarns. Nonwovens can be formed by spunbonding, needle punching or resin bonding. In spunbonding, bonds between the fibers are formed by heating the fibers to near their melting point. Fibers, especially the inorganic fibers, are used as reinforcing fibers in composites in which the matrix may be an organic resin, metal or ceramic. In order to obtain optimum reinforcement, the fibers are usually laid as nearly parallel to one another as possible.

5. *Dyes and Dyeing*

Fibers, yarns and fabrics may be colored with dyes (a) containing anionic groups (acid, direct and mordant dyes), (b) containing cationic groups (basic dyes) or (c) requiring chemical reaction during dyeing (azo, reactive, sulfur and vat dyes). Special colorants such as disperse dyes and pigments may also be used. All of the dye molecules have a chromophore group, which is responsible for the color, and usually a functional group, which increases the affinity of the color for the fiber.

Acid dyes are water-soluble anionic dyes that are applied to textiles from acid solutions. Fibers that develop a positive charge in acid solutions (e.g., wool, silk and nylon) can be dyed with acid dyes. Direct dyes frequently contain azo groups connecting aromatic chromophores and are mostly used on cellulosic fabrics. Mordant dyes are acid dyes that are chelated with metal ions such as aluminum, chromium, cobalt and copper, and are used chiefly for wool and silk. These dyes are normally applied to a fiber and then mordanted.

Basic dyes are dyes that yield colored cations in aqueous solutions. They can be applied in neutral or mildly acidic dye baths to acrylic, cellulosic, nylon and protein fibers. These dyes may be treated with complexing agents such as tannin to improve their fastness to washing.

Azo dyes are formed in the fiber by interaction between two separate dye components. They are chiefly applied to cotton, but can be used with man-made fibers. Reactive dyes contain groups that can react covalently with a chemical group in the fiber to make the color durable to washing and dry cleaning. They are chiefly applied to cellulosic fibers, with the reaction occurring with the hydroxyl group of the cellulose, but are also used with nylon and protein fibers. Sulfur dyes are products formed by the reaction of aromatic compounds with sodium polysulfide. They are reduced under alkaline conditions before application to a fiber and then oxidized after the fiber is dyed. They are chiefly applied to cellulosic fibers. Vat dyes are generally water insoluble. They are reduced under alkaline conditions to form a water-soluble, colorless leuco form, which is applied to the fiber and then oxidized to an insoluble colored state within the fiber. Vat dyes are chiefly used with cellulosic fibers, but can also be applied to acrylic and nylon fibers.

Disperse dyes are nonpolar, relatively insoluble molecules, usually containing azo or anthraquinone groups, and are applied to the fiber from an aqueous dispersion in water. They were developed originally to dye acetate fibers but are now also used in dyeing other man-made fibers. Pigments are insoluble colors that are normally added to the polymer from which the fiber is spun, or bonded to the fiber surface with a resin binder.

6. *Printing*

Fabrics may be printed with a design using any of the classes of colorants already described. The colors are dispersed in a thickened printing paste and applied to the fabric in a design. The printed fabric is then treated in some fashion such as by heating or an oxidation process to fix the colors in the fabric. With pigments, a resin binder is used to bind the pigment to the fabric.

7. *Finishing*

Fabrics are finished in a great variety of ways; the finishing may be to improve either the aesthetic or performance properties, or both.

7.1 Removal of Impurities

In processing fibers to fabrics, sizes are often applied to the warp yarns. These sizes may be removed in a final washing step. With natural fibers, impurities such as waxes, grease and vegetable matter may remain in the fabric and are removed in a final scouring and bleaching process.

7.2 Changing the Hand or Feel

In the case of woollen fabrics, a felting or compacting operation may be the finishing step. Singeing or passing a fabric over a gas flame or a heated metal plate may be used to burn off the fiber ends of a fabric made from staple yarns. If a fabric is limp, it may be stiffened by adding a nonpermanent starch or other sizing, or a permanent size may be obtained by applying a resin finish.

7.3 Changing the Appearance or Luster

Fabrics may be whitened by bleaching and/or by the application of a fluorescent whitener. A glossy fabric surface can be produced by calendering or passing it through rollers. If a friction calender is used, a glazed finish is obtained. With a Schreiner calender, a soft luster and hand are produced. Embossed designs can be produced with an embossing calender. With the addition of resins to the fabric, a durable calendered finish is produced. Napped fabrics are made by brushing the fibers on the surface of the fabric, and napped or pile fabrics may be sheared to make the nap uniform in thickness.

7.4 Control of Biological Degradation

Cellulosic fabrics are especially susceptible to damage by mildew and bacteria. Various antimildew and antibacterial chemicals including chlorinated phenols and organometallic compounds are applied to these fabrics to retard biological degradation. In addition, treatment of cellulosic fabrics with formaldehyde reactants such as melamine–formaldehyde will also protect the fabrics from biological degradation. Wool is susceptible to damage from moth larvae unless properly treated.

7.5 Durable Press

Since cellulosic fibers do not recover well from deformation, formaldehyde reactants are used to treat these fabrics to improve this property and consequently the recovery of the fabric from wrinkling. If sufficient formaldehyde reactant is applied, the resulting fabric will have both wet and dry recovery from wrinkling and consequently will have durable press properties.

The formaldehyde reactants used to produce a durable press have at least two functional groups capable of reacting with the hydroxyl groups of cellulose. Such reactants are normally applied in an aqueous solution with a metal salt catalyst by immersion of the fabric in a pad bath, followed by passage through a pad roll. The treated fabric is dried and then cured in an oven. The most frequently used formaldehyde reactant is dimethyloldihydroxyethyleneurea (DMDHEU). This agent can react with the hydroxyl groups of cellulose and crosslink adjacent cellulose chains. Other formaldehyde reactants that are used include the dimethylocarbamates and dimethylolethyleneurea. The methyl, ethylene glycol or diethylene glycol ethers of DMDHEU are often used because the release of formaldehyde from the etherified products is less than from DMDHEU.

Since the formaldehyde reactant tends to lower the tensile strength and abrasion resistance of cellulosic fibers, blends with polyester are commonly used to improve both of these properties. Also, ammonia treatment of the cotton in durable press fabric, either before or after treatment, is now being practised. With the ammonia treatment, less formaldehyde reactant is required to obtain durable press properties.

In the manufacture of garments from durable press fabrics, the garment is often fabricated before the curing operation is carried out, and then cured with the creases in place. This operation is referred to as postcuring.

7.6 Shrinkage Control

The tendency of some fabrics, especially wool and cellulosic, to shrink during washing can be minimized by finishing treatments. The shrinkage of wool fabrics caused by felting can be reduced by treatment with chlorine, by resin treatment or by a combination of the two. The shrinkage of cellulosic fabrics during washing can be reduced by compressive shrinkage of the fabric during finishing; an example of this kind of treatment is Sanforizing. Another method of controlling shrinkage is with formaldehyde reactants such as those used for durable press. In fact, durable press treatments will reduce the shrinkage of cellulosic fabrics during washing.

7.7 Water and/or Oil Repellency

There are a number of chemical compounds that will impart water and/or oil repellency to fabrics. Durable water repellency can be obtained with silicone polymers and with long hydrocarbon chain pyridinium chloride derivatives. Both water and oil repellency can be obtained if a fabric is treated with certain polymeric fluorocarbon-containing materials. With these compounds, it is important that the fluorine groups be present at the end of the polymer side chains to impart a low-energy surface to the fabric. The most effective compounds have a terminal trifluoromethyl group in the polymer.

7.8 Soil Release

Fabrics with a certain degree of soil release can be obtained by the application of hydrophilic polymers (e.g., polymers or copolymers of acrylic or methacrylic acid or hybrid fluorocarbon polymers). These finishes can be applied along with the formaldehyde reactants to give durable press fabrics with soil-release properties.

7.9 Fire Retardancy

Most textile fibers with the exception of the inorganic fibers are flammable to some degree. Some of the aramids are thermally stable. Some fibers such as wool and some modacrylics are inherently flame retardant. Others such as nylon, polyester and polyolefins will melt and the molten beads will drop away from the flame. Cellulosic fibers will burn readily unless chemically treated (e.g., with organic compounds containing halogen and/or phosphorus such as tetrakis (hydroxymethyl) phosphonium chloride). The flammability of man-made fibers can be reduced by the addition of flame retardants to the polymer before spinning, although only limited commercial success has been obtained with these fibers. The flame retardant can also be applied to the fabric, as was done in the unfortunate treatment of children's sleepwear with tris(2,3-dibromopropyl) phosphate, which was found to be a potential carcinogen.

7.10 Antistatic Properties

The buildup of static electricity on fibers, with resultant clinging and/or shock, is mainly a problem with hydrophobic man-made fibers. This problem can be minimized by the addition of antistatic agents, which discharge the electrostatic charges that tend to build up on fiber surfaces. These agents are surface active materials, and include ethylene oxide derivatives and polyamines.

See also: Elastomeric Fibers; Textiles: Structures and Processes

Bibliography

Anderegg F O 1939 Strength of glass fibers. *Ind. Eng. Chem.* 31: 290

Goswami B C, Martindale J G, Scardino F L 1977 *Textile Yarns: Technology, Structure and Applications*. Wiley, New York

Hearle J W S, Grosberg P, Backer S 1969 *Structural Mechanics of Fibers, Yarns and Fabrics*. Wiley, New York

Hearle J W S, Lomas B, Cooke W D, Duerden I J 1989 *Fibre Failure and Wear of Materials*. Ellis Horwood, Chichester, UK
Lewin M, Sello S B 1983 *Chemical Processing of Fibers and Fabrics*, Vols. 1, 2. Dekker, New York
Mark H F, Atlas S M, Cernia E 1967 *Man-Made Fibers: Science and Technology*. Wiley, New York
Mark H F, Wooding N S, Atlas S H 1971 *Chemical Aftertreatment of Textiles*. Wiley, New York
Needles H L 1981 *Handbook of Textile Fibers, Dyes and Finishes*. Garland, New York
Summerscales J 1987 High performance reinforcement fabrics. *Prog. Rubber Plast. Technol.* 3(3): 20–32
Tortora P G 1978 *Understanding Textiles*. Macmillan, New York

T. F. Cooke
[Textile Research Institute, Princeton, New Jersey, USA]

Fibers: Drawing

In fiber manufacture, newly spun fibers generally have very poor mechanical properties and need to be strengthened by stretching highly. This process is known as drawing. Thermoplastic fibers such as poly(ethylene terephthalate), nylon, polypropylene and polyethylene form a sharp neck when cold stretched. They immediately attain their natural draw ratio, numerically the ratio of the cross-sectional areas either side of the neck. In poly(ethylene terephthalate) and nylon this ratio is almost independent of temperature and rate of straining, whereas in polypropylene and polyethylene recent work has shown that the ratio may be dramatically altered not only by the conditions of drawing, but also by the molecular weight and morphology of the spun fiber. Since drawing with a sharp neck may produce an unstable process, industrial processes are usually carried out in a tension gradient at temperatures sufficiently high for adequate molecular mobility. In amorphous polymers this is conventionally above the glass-transition temperature T_g. In crystalline polymers it is in the range of the crystalline α-relaxation, so that the molecular chains can move through the crystalline regions. The drawing is performed continuously by leading the yarn to a "feed" roll which has a nip to stop slipping, by supplying heat to the yarn, usually by heating the feed roll, and then winding the yarn up at a higher speed on a "draw" roll. The ratio of the two roll speeds gives the nominal draw ratio; the actual draw ratio is slightly less because of some residual contraction after leaving the draw roll. To produce a coherent yarn bundle the filaments in the yarn are usually intermingled with an air jet. This method supplants the more costly technique of twisting during drawing on a "draw twister." The drawn yarn can subsequently be stabilized against shrinkage by some form of heat treatment (e.g., by use of a heater plate between the feed and draw rolls).

Nylon 6 and Nylon 66 are "cold" drawn 250–300% and the neck can be localized to reduce variability by wrapping the yarn round a metal or ceramic snubber pin placed after the feed roll. ("Cold" drawing is a misnomer since adiabatic heating occurs due to the work of drawing.) Very high tensile strengths can be obtained by a second hot-drawing stage. Poly(ethylene terephthalate) is drawn 200–400% at speeds of 400–800 m min^{-1} using heated feed rolls at 80–90 °C and a heater plate at a much higher temperature (e.g., 180 °C). Again, a second drawing stage can be used where high strength is required. For polypropylene the extensions may be higher, ranging from 300–900%, the higher value being used for high strength. The drawing process resembles that for nylon or poly(ethylene terephthalate). Acrylic fibers are also stretched 300–900% at 80–100 °C, using hot air after dry spinning and steam or hot water after wet spinning. Poly(vinyl alcohol) is stabilized against shrinkage by drawing up to 800% simultaneously with high-temperature heat treatment at 140–210 °C. Poly(vinyl chloride) fibers are drawn about 400% at their T_g of 75 °C.

The most important effect of drawing is orientation of the polymer molecules and the development of a highly anisotropic structure. As a consequence, desirable industrial properties, such as tensile strength and modulus, together with optical measurements of orientation such as birefringence, all increase monotonically with draw ratio. The extension during drawing can be viewed as consuming part, if not most, of the potential extensibility of the initial molecular network formed at spinning. Thus, the extensibility of the drawn fiber decreases with increasing draw ratio. Since orientation of the molecules increases their lateral packing, properties which depend on this packing are also changed. The density and heat of fusion are increased, while the rate and extent of dyeing and moisture absorption, the gas permeability and the degree of transverse swelling in certain liquids are reduced. Although, in general, longitudinal properties along the fiber are improved, transverse mechanical properties (e.g., abrasion resistance) may be impaired.

The initial state of the spun fiber is very important in determining the drawing behavior. The most critical factor is the magnitude of the preorientation produced by the spinning process. For example, the natural draw ratio of poly(ethylene terephthalate) fibers has been found to decrease steadily from about 5.0 to 2.5 as the spun birefringence increases from near zero to 0.01. This is because the total permissible deformation can be viewed as the sum of the deformation at spinning and drawing and, hence, the more that occurs at spinning, the less that can be obtained at drawing. The degree of crystallinity of some spun fibers is also important, since if it is too high, for example for poly(ethylene terephthalate) and poly(vinylidene chloride), the fibers are too brittle to draw.

The drawing conditions are also very important in controlling those properties which depend on the final degree of orientation. Although a high draw temperature is desirable to provide adequate mobility, if the temperature is too high the molecular network does not retain its continuity and the fiber decreases in diameter without an accompanying increase in molecular orientation, a condition known as flow drawing. Since the plastic deformation processes depend on strain rate as well as temperature, a suitable combination of strain rate and temperature is required for effective drawing.

Even after many years of commercial exploitation and academic study, the basic mechanisms underlying the molecular changes occurring on drawing are still not very well understood. Normal drawing is an adiabatic process and considerable heat can be generated, sufficient even to melt the fiber at high drawing speeds. This phenomenon has been used as the basis for a general viscoelastic theory of neck formation in polymers, which does not involve molecular models. It is claimed that the rise in temperature produces an incipient drop in the load with increasing strain, an unstable condition that leads to a rapid increase of strain. The main objection to this theory is that necking can occur under apparently isothermal conditions. For example, polyethylene necks when drawn at an extremely slow rate, and Nylon 66 necks when drawn in liquids where heat transfer is good. It has been shown that the conditions for a stable neck involve the strain-rate sensitivity of the flow stress, in addition to its dependence on plastic strain.

In crystalline polymers major structural changes occur in the neck, involving the breakdown of the original lamellar texture by crystal slip and twinning processes. The fibrillar structure formed consists of lamellar stacks with a new characteristic spacing (the long period), which has been shown in polyethylene and polypropylene to depend only on the temperature of drawing. In spite of the complex structural changes, the effects of drawing on the overall molecular orientation and other properties such as modulus, can be quantified quite well in some cases (e.g., poly(ethylene terephthalate) and low-density polyethylene) by viewing the deformation as the affine rotation of microelements, each of which possesses the properties of the fully oriented fibers. Such modelling does not work well for linear polyethylene and polypropylene, where very drastic structural changes occur. Moreover, the behavior at high draw ratios requires totally new concepts and it has been suggested that ultramodulus polyethylene is achieved because the lamellar stacks become increasingly linked by intercrystalline bridges at very high draw ratios. At this stage, the crystallite orientation is nearly perfect and there is also a substantial increase in the orientation of the noncrystalline material. The high modulus, however, relates primarily to the degree of crystalline continuity in terms of the intercrystalline bridges.

See also: Fibers: Formation and Processing; Fibers: Texturing and Draw Texturing; Melt Spinning of Fibers: Structure Development; Processing: Classification

Bibliography

Hearle J W S, Peters R H 1963 *Fibre Structure*. Butterworth, London

Hill R 1953 *Fibres from Synthetic Polymers*. Elsevier, Amsterdam

Ludewig H 1964 *Polyester Fibers: Chemistry and Technology*. Wiley, New York

McFarlane S B 1953 *Technology of Synthetic Fibers*. Fairchild, New York

Mark H F, Atlas S M, Cernia E 1968 *Man-Made Fibers*, Vols. 2, 3. Wiley, New York

Tanner D, Fitzgerald J A, Phillips B R 1989 The Kevlar story—An example of the innovative process. *Prog. Rubber Plast. Technol.* 5(3):229–51

Walczak 1977 *Formation of Synthetic Fibers*. Gordon and Breach, New York

Ward I M 1975 *Structure and Properties of Oriented Polymers*. Applied Science, London

Ward I M 1988 Processing, structure and properties of oriented polymers. *Prog. Rubber Plast. Technol.* 4(2): 1–13

Ziabicki A 1976 *Fundamentals of Fiber Formation: The Science of Fiber Spinning*. Wiley, New York

H. Brody
[ICI, Harrogate, UK]

I. M. Ward
[University of Leeds, Leeds, UK]

Fibers: Formation and Processing

Synthetic fibers are produced by a variety of commercial processes, all of which utilize suitable modifications of four or five basic steps. First, a polymer is provided in some fluid state (solution, melt or dispersion) and then is forced through capillaries (the spinneret) to provide thin streams of the fluid. These fluid streams are changed to solid filaments by solvent removal, by cooling the melt or by converting a dispersed phase into a continuous phase. The filaments are attenuated to reduce their diameters and to partially align polymer chains with the fiber axis. In most cases, an additional drawing step is included in the process to further align polymer chains with the fiber axis and to enhance crystallization.

In most commercial processes the polymers used have high molecular weights, are linear (or almost linear) and are not intended to undergo chemical reaction during spinning and drawing. In a few cases the molecular weight is purposely increased, and crosslinking may even be induced in special cases. Fibers may be produced in a continuous filament form, or in "staple" (i.e., cut lengths of about 3–15 cm).

1. Solution Spinning

The spinning of polymer solutions is preferred when the polymer does not melt or decomposes below its melting point. Thus, cellulose, cellulose derivatives (such as the acetate), acrylic polymers and copolymers, and high-melting aromatic polymers (e.g., aromatic polyamides, polyhydrazides, polyimides and polyoxadiazoles) are spun from solutions. Solution spinning is also usually preferred for inducing chemical crosslinking in the fiber after it is spun, as in the case of some elastic polyurethane fibers.

In general, the polymer is dissolved in a suitable solvent, usually at about 10–30% concentration. The solution (or "dope") is pumped through a filter to a spinneret containing from about 20 to several thousand capillaries, to provide the filaments. It is in the filament solidification step that two major modifications are introduced: solidification may be achieved by evaporation of solvent (dry spinning) or by extraction of the solvent from the fiber by a nonsolvent (wet spinning).

1.1 Dry Spinning

Acrylic copolymers (90–95% acrylonitrile and 5–10% of comonomers such as vinyl acetate, vinyl bromide or vinylidene chloride) are soluble in powerful solvents such as dimethylformamide (bp 153 °C). These solutions are spun commercially by the dry-spinning method (as is cellulose acetate). Streams of solution are extruded through capillaries into the top of a tower, where the solvent is evaporated in a stream of hot drying gas, usually air. The resulting filaments are combined into a yarn bundle, are washed to remove the last traces of solvent and are then dried.

This process produces fibers of continuous lengths, but in most cases it is sold as staple. The yarn obtained as described above is combined with other similar yarns to give a "tow." A finish is then applied and the tow is drawn over hot rolls to develop the desired morphology and strength. It may be cut into staple at this point or may be crimped first and then cut to staple.

1.2 Wet Spinning

In wet spinning, dope is pumped through a spinneret and the streams of polymer solution pass into a spin bath containing a liquid which is not a solvent for the polymer, but which is compatible with the solvent used in making the dope. For example, an acrylic copolymer dissolved in dimethyl acetamide may be spun into a water bath. Water extracts the solvent, coagulating the polymer into fibers.

The fibers are then washed further to remove traces of solvent, dried, finish is applied and yarns are combined to give a tow for processing as in the case of dry spinning.

Wet spinning has several advantages over dry spinning. Quenching filaments so that they do not stick together is easier when dope is spun into a liquid bath, so many more filaments can be spun from a single spinneret. In addition, a much wider range of solvents can be used, since the solvent is not evaporated from the filaments. The initial polymer can be converted to a derivative to aid solubility, as in the xanthate process for rayon spinning, followed by regeneration of the original polymer in the spin bath. Crosslinked fibers can be obtained also by spinning a solution of a reactive polymer into a spin bath containing a curing agent.

Commercial examples of wet spinning include acrylic fibers spun from dimethylacetamide solutions, rayon spun from caustic solutions to which carbon disulfide has ben added (viscose or xanthate process) and aromatic polyamides spun from concentrated sulfuric acid.

Dispersions or emulsions such as those of poly(vinyl chloride) have also been wet spun, with coagulation of the polymer phase taking place in the spin bath. This technique, however, is not widely used in industry.

2. Melt Spinning

Melt-spinning processes are usually preferred for economic reasons when the polymer can be processed as a melt without serious degradation. In practice, melt temperatures much in excess of 300 °C are not practical because thermal decomposition is usually too severe above this temperature. This limit means that polymers must melt below about 270–80 °C to be readily processable by this technique. Poly(ethylene terephthalate), Nylon 66, Nylon 6 and polypropylene are examples of polymers which are eminently successful in melt spinning.

In this process, molten polymer is provided by melting in an extruder or by heating on a metal grid (an early process for nylon spinning), or is obtained as a melt directly from a polymerizer. The processing temperature is usually chosen so that the melt viscosity is 1000 Pa s or less, for ease of polymer flow. The melt is delivered under pressure of 7–70 MPa through a filter to a meter pump, which accurately meters polymer to the pack, a combination of a small filter and spinneret. During these steps the polymer temperature is maintained as uniform as possible, to ensure uniform flow to each capillary in the spinneret and equal heat history for each filament.

On leaving the spinneret, the molten filaments (20–1000 in number) pass into a vertical chimney where they are cooled by a crosscurrent of air. At the same time they are attenuated to the desired diameter (usually expressed as denier per filament or denier per yarn bundle, grams per 9000 m of fiber or yarn length). Finish is applied and the yarn is wound on a bobbin.

Fiber properties depend greatly on the spinning speed (i.e., the speed of winding on the bobbin) and on the ease of crystallization of the polymer. Speeds may vary from about 500–5000 m min^{-1}, usually with relatively little, if any, crystallinity developing at the lowest speeds and essentially complete crystallinity forming at the highest speeds. Unless the yarn is spun at a sufficiently high speed that crystallinity is complete, a drawing step will be needed to develop the final morphology and full strength characteristics.

3. *Factors Affecting Fiber Properties*

Properties are influenced mainly by polymer composition, molecular weight, spinning speed, filament diameter and drawing conditions (temperature and degree of stretch or draw ratio). Some classes of polymers and copolymers, such as the acrylic copolymers, do not develop well-ordered crystallinity. To compensate for this, a higher molecular weight than with nylon or polyester is used. Even with a higher molecular weight, these polymers cannot readily be produced at as high a tenacity as the following two classes of polymers.

Polymers which crystallize readily in folded chain structure (such as Nylon 66, Nylon 6, poly(ethylene terephthalate), polypropylene and polyethylene) can be hot drawn to form well-developed crystalline structures, with both crystal and amorphous orientation nearly parallel to the fiber axis. Under these conditions very strong fibers ($\geqslant 12$ grams per denier) may be obtained.

The third principal class of polymers used for fibers is that group containing many cyclic structures in the main chain, such as cellulose, aromatic polyamides and oxadiazoles. With the right choice of structure, certain of these polymers may be spun from solutions in such a way that considerable strength is obtained in the as-spun fiber, especially if the polymer solution forms liquid crystals. Even with these, however, hot drawing leads to more ordered structures and higher tenacity and modulus.

The optimum choice of polymer and processing conditions will vary according to cost and the desired set of properties. For example, for polyester tire yarn a diameter of 5 denier per filament, maximum tenacity of 9–10 grams per denier and an elongation of only 10–15% may be preferred. For nylon carpet yarn the denier per filament may be about 8–20, with 4–5 grams per denier tenacity being adequate, and 40–60% elongation desired. For aromatic polyamide yarn for the belts of radial tires 20 grams per denier tenacity, 4% elongation and 800 grams per denier modulus may be desired. For nylon or polyester apparel yarn 2–5 denier per filament may be preferred for comfort, with 3–5 grams per denier tenacity and 25–35% elongation being suitable. Spinning and drawing conditions can be chosen to provide these properties.

See also: Fibers: Drawing; Fibers: Texturing and Draw Texturing; Melt Spinning of Fibers: Structure Development; Processing: Classification; Spunbonding of Polymeric Fibers

Bibliography

Mark H F, Atlas S M, Cernia E (eds.) 1967, 1968 *Man-Made Fibers: Science and Technology*, Vols. 1–3. Wiley, New York

Moncrieff R W 1975 *Man-Made Fibers*, 6th edn. Halsted, New York

Walczak Z K 1977 *Formation of Synthetic Fibers*. Gordon and Breach, New York

Ziabicki A 1976 *Fundamentals of Fiber Formation: The Science of Fiber Spinning and Drawing*. Wiley, New York

J. H. Saunders
[Monsanto Textiles Company, Pensacola, Florida, USA]

Fibers: Texturing and Draw Texturing

Textile yarns are linear polymer structures which can be either woven or knitted into planar fabrics for subsequent use as textile materials. These linear structures are generally fabricated from staple short fibers or continuous long filaments. With the advent and subsequent manufacturing of synthetic continuous filaments, techniques have been developed to enhance these materials with characteristics that are generally common to staple fiber yarns. One of these techniques is the texturing process. Fabrics made from textured yarns possess breathability and enhanced permeability, are generally more comfortable and provide bulk, cover and ease of care.

Texturing is a process of taking thermoplastic polymer yarns and changing their filamentous topological configuration to provide fabrics with the aforementioned properties. The general principle which is used to make most thermoplastic textured yarns involves the heat setting of these materials in a temporary deformed configuration. Thus, the texturing process involves: (a) deforming, (b) heating to relax stresses, (c) cooling, (d) undeforming, and (e) relaxing to allow redeformation.

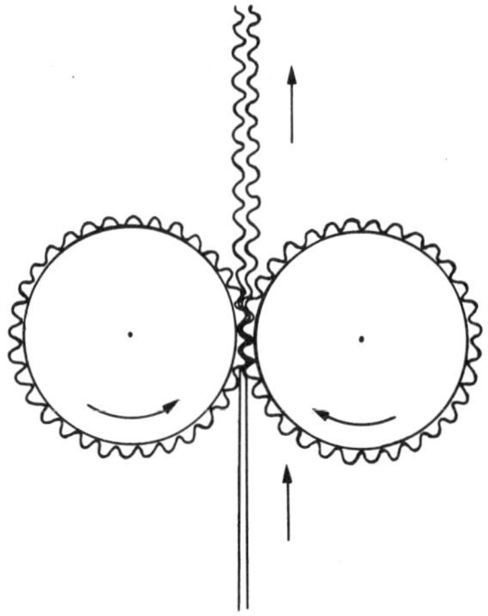

Figure 1
Gear crimping process

Texturing can be accomplished in a variety of ways. The simplest method involves gear crimping, whereby the fibers are embossed with the teeth of gears. The resulting bends are angular, in contrast to the rounded bends and crimps which are found in naturally crimped fibers. Further, the crimp is generally planar. A diagram of the gear crimping operation is shown in Fig. 1.

Another method of texturing yarns involves the stuffing box process, which is shown in Fig. 2. Here, thermoplastic fibers are continually deformed in a stuffer box and subjected to heat. The fibers relax and the crimp which is formed in the stuffer box is heat set in the yarn. The yarn is then withdrawn through a weighted tube and allowed to relax. Stuffer box yarns are usually plied to hold the filaments together. The stuffer box process is a relatively fast method and not very costly. Yarns crimped in this process are used in carpets, upholstery, sweaters and knitted materials.

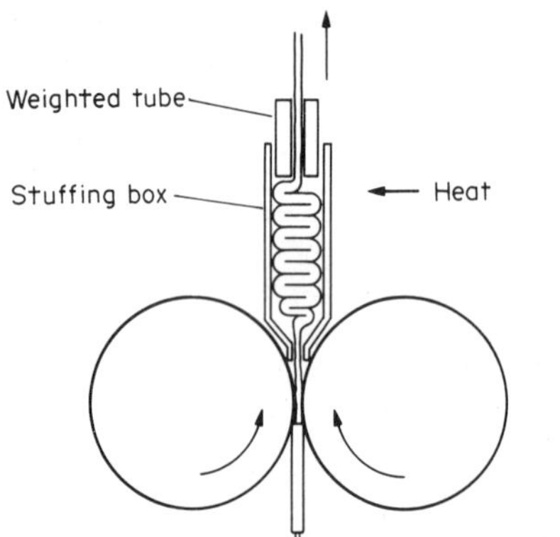

Figure 2
Stuffing box process

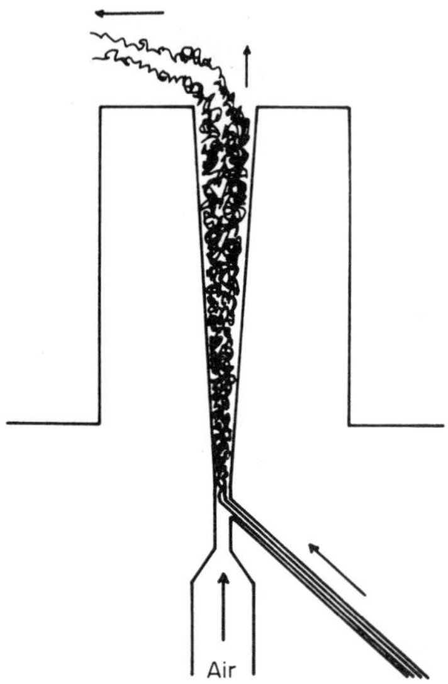

Figure 3
Air-jet process

A third process which can be used to texture yarns is the air-jet process, which is shown in Fig. 3. Here, thermoplastic filaments are passed over a turbulent air jet at a faster rate than they are drawn off. The blast of hot air forces the filaments to deform into very tiny loops, increasing the subsequent volume of the yarn by 50–150%. Not all of the fibers are looped, however, and thus air-jet textured yarns generally maintain their size and bulk when they are placed under tension, as the straight sections of the yarn will bear load and allow the looped sections to remain relatively unaffected. Accordingly the yarns possess very little stretch. Heat setting is generally not necessary in this process, which can therefore be used for non-heat-settable yarns.

Edge crimping is a process which can be used to impart to thermoplastic filaments a curled configuration much like that obtained when a ribbon is pulled over a pair of scissors. The filament cross section changes from round to flat and the filaments undergo a differential shrinkage which causes a nontorqued curl. The process, shown diagrammatically in Fig. 4, is relatively low cost and expedient.

The knit–deknit process is used to give fibers a crimped configuration which is similar to that found in knitted loops in fabrics. As shown in Fig. 5, the fabric is knitted, heat set and subsequently unravelled. Crimp size and frequency can be varied by

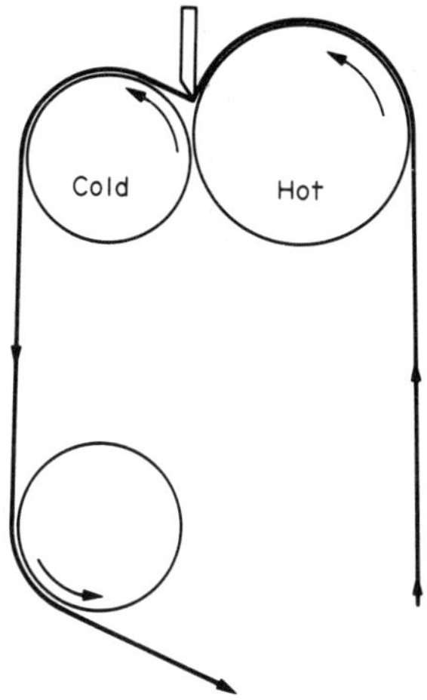

Figure 4
Edge crimping process

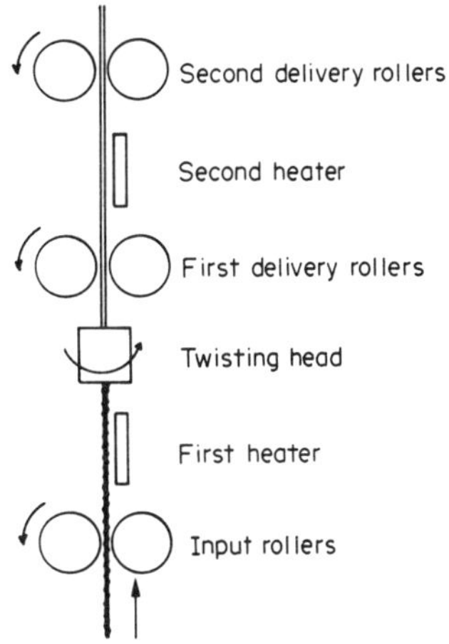

Figure 6
False-twist process

changing both the stitch size and the tension. It is important that the size of the stitch be different from that of the knitted stitch in the finished fabric, otherwise pinholes will form where the crimp gauge and knit gauge match. Knit–deknit yarns are used in upholstery, swimwear and outerwear.

The most common method employed to texture yarns involves false-twist texturing. In this process, thermoplastic filaments are temporarily twisted to a relatively high helical twist level, heat set in the twisted configuration, untwisted and then allowed to relax. Over 80% of all textured yarns are textured in this manner. A drawing operation can be combined with texturing to draw and texture thermoplastic fibers simultaneously; this process is called draw texturing.

A diagram of the false-twist process is shown in Fig. 6. Continuous filaments are fed into the system via input rollers. They are immediately twisted by a torque field which is supplied by the twisting head, which can be of the pin spindle, friction spindle or belt type, the first two of which are described later. The yarns are then passed over the first heater, which is generally at a temperature greater than the glass-transition temperature of the fibers. Immediately after passing over the first heater, the yarns are allowed to cool and thus the twisted configuration is now heat set. After progressing through the twisting head, the yarn immediately untwists and proceeds to the first delivery rollers. It is then passed over a second heater and through the second delivery rollers, which usually have a surface speed slightly less than the first delivery rollers, thus allowing the crimp to reform over the second heater. The fibers are then heat set again and any residual torque in the yarn due to heat setting between the input rollers and the first delivery rollers is usually relaxed. Sometimes the second heater is eliminated and the resulting yarns are very lively or stretchy. For draw texturing, the first delivery rollers have a surface speed significantly greater than that of the input rollers. Thus the material is drawn as it is being twisted.

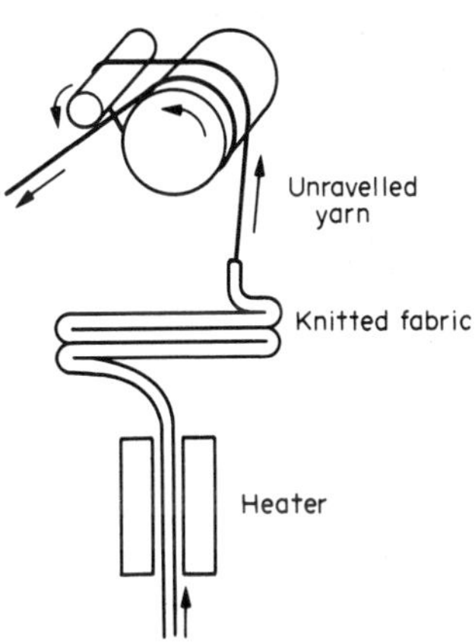

Figure 5
Knit–deknit process

The original method of imparting twist involves the use of a pin spindle with a diablo, shown in Fig. 7. The yarn is wrapped once around the pin spindle. Thus, every time the pin spindle rotates, one twist is imparted to the yarn upstream of the twisting head. The maximum rotational speed of the yarn between the twisting head and the input rollers is limited by the rotational speed of the pin spindle. The second method of imparting twist involves the use of a disk spindle or friction spindle, shown in Fig. 8. Here, the yarn is passed over a rotating disk, the rotational speed of the yarn being developed by the rotational speed of the disk. The surface speed of the disk is

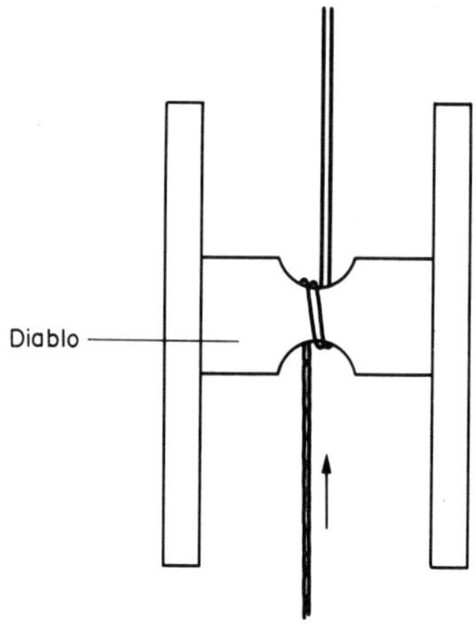

Figure 7
Pin spindle twisting method

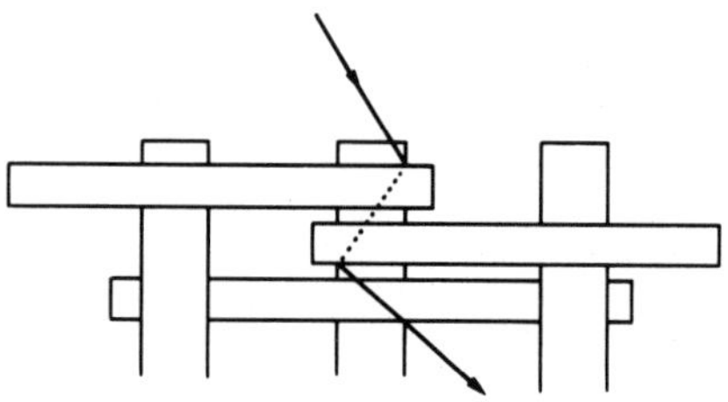

Figure 8
Threadline on a stacked-disk friction tester

generally greater than that of the yarn. Since the diameter of the disk is significantly greater than that of the yarn, one rotation of the disk results in several rotations of the yarn. Accordingly, the rotational speed of the yarn is not limited as it is with the pin spindle. This disk spindle is generally called a friction spindle.

See also: Fibers and Textiles: An Overview; Fibers: Formation and Processing; Textiles: Structures and Processes

Bibliography

Hearle J W S, Grasberg P, Backer S 1969 *Structural Mechanics of Fibers, Yarns and Fabrics*. Wiley, New York

Mark H F, Atlas S M, Cernia E (eds.) 1967, 1968 *Man-Made Fibers: Science and Technology*, Vols. 1–3. Wiley, New York

Redmond G B, Wood J O 1982 Textiles. In: Blow C M, Hepburn C (eds.) 1982 *Rubber Technology and Manufacture*, 2nd edn. Butterworth, Guildford, UK

Textile Institute 1976 The yarn revolution, new developments in the production of spun and textured yarns and their exploitation in fabric form. *Pap. 60th Annual Conf. Textile Institute*. Textile Institute, Manchester, UK

Zimmerman J 1989 Polymers for fibers. In: Aggarwal S L (ed.) 1989 *Comprehensive Polymer Science*, Vol. 7. Pergamon, Oxford, Chap. 8

D. S. Brookstein
[Albany International Research Company, Dedham, Massachusetts, USA]

Fillers in Elastomers

Finely divided solids are used almost universally in formulating elastomeric compositions. Their function may be to reinforce the rubber, to reduce compound cost, or both. The most important fillers used in rubber are carbon blacks, silica and silicates, clays and whiting (calcium carbonate), although many other materials have found limited application.

Reinforcing effects generally increase with decreasing filler particle size or increasing specific surface area up to a limit of about 300 m^2 cm^{-3}. The morphology of the filler particles is another important factor. In the most highly reinforcing fillers (carbon black and silica), the particles consist of more or less random aggregates of fused, roughly spherical subunits or nodules. The spatial arrangement of the nodules within the aggregates determines the filler structure.

1. Filler Characteristics

The most general technique for estimating specific surface areas is the low-temperature nitrogen-adsorption method. However, in some highly reinforcing carbon blacks and silicas, a portion of the surface area is contained in micropores. A more realistic measure of surface area, approximating the surface accessible to macromolecules, may then be obtained with a larger adsorbate molecule. Such measurements are most easily carried out in aqueous solution using a surfactant as the adsorbate; for example, in ASTM Method D3765, the adsorbate is hexadecyltrimethylammonium bromide, whose molecules are too large to penetrate the micropore space. Iodine number is used widely as a control test for carbon blacks, but it is not always an accurate reflection of the specific surface area, because the adsorption of iodine is also influenced by surface chemical characteristics. Because of the generally broad size and shape distributions of carbon-black aggregates, geometric parameters are not easily determined. Tinting strength, used as an empirical measure related to carbon-black aggregate size, is governed by an average aggregate volume. It generally increases with specific surface area and decreases with structure. Filler structure is estimated by "vehicle demand." The vehicle, usually dibutylphthalate (DBP), is added to the dry powder up

Table 1
Characteristics of some important fillers (typical values)

Filler	ASTM grade number	Specific surface area ($m^2 g^{-1}$)	DBP number ($cm^3 g^{-1}$)	Principal uses
Carbon black	N339	95	1.00[a]	tire treads
Carbon black	N330	83	0.88[a]	tire sidewalls, treads
Carbon black	N326	83	0.70[a]	tire breaker stocks
Carbon black	N660	35	0.76[a]	tire carcasses, extruded goods, wire and cable insulation
Carbon black	N990	7	0.33[a]	mechanical goods
Silica		150	1.93[b]	light-colored articles, polysiloxane rubbers
Hard clay		22	0.35[b]	footwear, mechanical goods
Precipitated $CaCO_3$		2–20	0.3–0.6[b]	mechanical goods

a ASTM D3493 b ASTM D2414

to the point of incipient formation of a coherent paste. The test (ASTM Method D2414) is performed in a small internal mixer fitted with a device sensing the torque on the rotor blades. The end point is marked by a sharp increase in torque. The DBP number of a carbon black (volume absorbed at end point) contains a contribution from interaggregate packing (Medalia 1970) which is related to the densification history of the material rather than the volume pervaded by the aggregates, which is the relevant descriptor of structure. For this reason, DBP numbers are often measured after a standard precompression treatment (ASTM Method D3493); DBP numbers are also used in the characterization of inorganic fillers.

Electron microscopy was used extensively in the past to measure nodule size distributions of carbon blacks, but the practice has lost favor in recent years. Computerized automatic image analysis methods have been applied with some success to characterize the details of filler morphology.

Common methods for the assessment of surface chemical characteristics of carbon blacks include measuring the slurry pH, diphenylguanidine adsorption or the content of volatiles. These quantities are related to the degree of surface oxidation of the black, which affects the rate of sulfur vulcanization. Toluene extractables, moisture, ash and grit are impurities in carbon black which are monitored regularly in manufacture. Carbon black is sold in the form of pellets (agglomerates) roughly 1–2 mm in diameter. Pellet size distribution, crushing strength, pack point and bulk density are parameters important in bulk handling and the dispersion of the material in rubber. Commonly specified properties of mineral fillers, in addition to specific surface area and vehicle demand, are moisture, sieve residue, chemical composition, pH, specific gravity and optical brightness.

Table 1 lists a few important fillers and some of their characteristic properties. Carbon blacks are available in many combinations of specific surface area (7–150 $m^2 g^{-1}$) and structure (DBP absorption: 0.30–1.30 $cm^3 g^{-1}$). The first digit in the ASTM grade designation is intended to follow the arithmetic mean nodule size; hence, in general, the larger the grade number, the smaller the specific surface area. Surface chemical properties in carbon blacks are rarely specified as they are similar throughout the ASTM N category of blacks. There is also an S category comprising oxidized and, hence, slow-curing blacks. The largely obsolete channel process carbons belong to this group, as do some oxidized furnace blacks.

2. *Mixing and Dispersion*

Dispersion of fillers in rubbers is usually accomplished in internal mixers of the Banbury type or on two-roll mills. The attainment of a technologically adequate filler dispersion is rendered more difficult the higher the specific surface area of the filler and the lower its structure. Whereas a low-structure carbon black (small DBP number) will be incorporated more rapidly than a high-structure black, the final dispersion of the latter will usually be better. Inhomogeneities of 1 μm or less are not harmful to vulcanizate properties, so that the attainment of a perfect dispersion is not necessary (Boonstra and Medalia 1963). The easiest, and usually sufficient, means of ascertaining the quality of a dispersion is by examining a razor-cut surface of the vulcanized compound under a low-power optical microscope. The method is rendered semiquantitative by establishing a set of standards. The examination of microtome sections at magnifications of about 400 is also used widely. Electron microscopy presents a serious sampling problem and is rarely employed except in research studies.

Carbon black–rubber master batches, prepared by the coagulation of mixtures of a rubber latex and a carbon-black slurry, are available commercially. Although the dispersion of the carbon black in a master batch is rarely adequate, the subsequent addition of vulcanizing agents and other compounding ingredients provides sufficient additional mixing to refine the carbon-black dispersion.

Typical filler concentrations are 20% by volume of the mix for highly reinforcing types and greater for mildly reinforcing or nonreinforcing fillers.

3. *Effects of Fillers in Unvulcanized Rubbers*

In uncrosslinked rubber compositions, fillers increase the batch viscosity; to partly overcome this effect, plasticizing hydrocarbon oils are used routinely in rubber compounding. Common mixture rules for viscosity, such as the modified Einstein equation

$$\eta = \eta_0(1 + 2.5c + 14.1\,c^2) \qquad (1)$$

where η is the viscosity of the mixture, η_0 that of the medium and c the volume concentration of filler, rarely hold for any but low-surface-area fillers, devoid of significant structure, for example N990 (medium thermal) black. Highly reinforcing fillers invariably produce viscosity increases greatly in excess of those predicted by Eqn. (1). These result from the occlusion of polymer in the internal void space of filler aggregates (thereby raising the effective filler concentration), adsorption of polymer segments on the filler surface with loss of segmental mobility and filler agglomeration effects. Invariably, fillers also accentuate non-Newtonian behavior.

A technologically important aspect of filler structure, especially with carbon blacks, the DBP value of which can be held within narrow specification limits in manufacture, is the control of extrusion shrinkage. The higher the DBP number the lower the shrinkage or degree of die swell. The decrease in recoverable strain results from inclusion in the mix of a rigid phase incapable of elastic recovery. The abundance of this phase increases with the ability of the filler to shield rubber from elastic deformation by occlusion in its internal void space.

4. *Physical Properties of Filler-Reinforced Vulcanized Elastomers*

Figure 1 shows, schematically, the effects of some 20% by volume of various fillers on the uniaxial stress–strain curve of an amorphous elastomer. Figure 1a represents the unfilled vulcanized rubber, Fig. 1b the same rubber filled with a weakly reinforcing filler of small specific surface area (e.g., $CaCO_3$), Fig. 1c the rubber filled with a highly reinforcing carbon black (surface area ~100 cm² g⁻¹) and Fig.

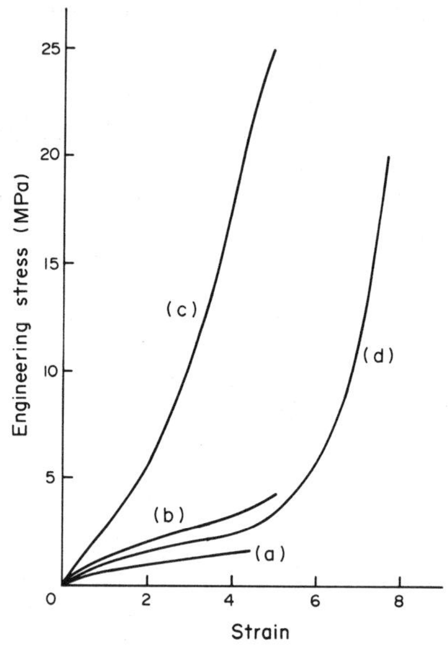

Figure 1
Schematic stress–strain curves in uniaxial tension: (a) unfilled rubber, (b) rubber filled with coarse filler of low activity, (c) rubber filled with highly reinforcing carbon black and (d) rubber filled with graphitized carbon black

1d the rubber filled with the same black after deactivation by heat treatment at 2700 °C (graphitization). Carbon blacks carry reactive sites capable of grafting rubber molecules onto their surface by several mechanisms; these sites are lost in graphitization. The result of the removal of active sites is a dramatic loss in modulus or stiffness (stress at a given strain), but only a moderate loss in tensile strength. This phenomenon is quite general. Silica, of 100 $m^2\,g^{-1}$ surface area, which does not graft to general-purpose rubbers, gives low-modulus vulcanizates of good tensile and tear strength, but using an organosilane coupling agent such as 3-mercaptopropyltrimethoxysilane, which provides surface grafts to the vulcanized rubber network, gives results similar to carbon black. The relatively coarse $CaCO_3$ filler (Fig. 1b) has neither the surface area for the development of high tensile strength, nor the ability to bond rubber to produce a vulcanizate of high stiffness at the moderate loading of the present example.

In structured fillers, the ability to occlude polymer leads to high effective filler concentrations; thus vulcanizate stiffness increases with the DBP number of the carbon black.

Many applications of rubber involve periodic deformations of modest magnitude, so that the role of fillers in the dynamic viscoelastic response of vulcanizates is of great importance. In general, fillers increase both the storage and loss components of the dynamic modulus in a rather complicated way which is strongly dependent on the deformation amplitude and the filler loading. The amplitude dependence, which is absent in unfilled vulcanizates, is illustrated in Fig. 2. It has been explained as the result of filler agglomeration by van der Waals forces, which is increasingly disrupted as the strain amplitude increases, the constitutes a major part of the energy losses occurring in the periodic deformation of filled rubbers. In a given compounding recipe and at fixed strain amplitude, the storage modulus is governed by filler structure, while the loss tangent is essentially a function of specific surface area; consequently, the loss modulus is a function of both morphological characteristics.

The agglomeration of carbon black in rubber has another important consequence. When the black concentration reaches a critical value, electrically conducting paths which pervade the entire volume of the rubber are formed, leading to a precipitous drop in electrical resistivity. Certain high-structure blacks of very large specific surface area can form conductive networks at volume concentrations as low as 5%.

The glass-transition temperature of rubbers is raised very little by carbon black and other fillers. However, the transition is broadened on the high-temperature side, this being the result of loss in segmental mobility of the polymer in the vicinity of the filler surface.

The ability of fillers, especially carbon black, to impart abrasion resistance to rubbers is of utmost importance in their use in automotive tire treads. Tread-grade carbon blacks have specific surface areas in excess of 75 $m^2 g^{-1}$ and DBP values above 0.90 $cm^3 g^{-1}$. Under all conditions of wear, abrasion resistance increases with specific surface area up to a limit of about 150 $m^2 g^{-1}$. The carbon-black structure has only a minor effect under mild wear conditions, but becomes a major contributor to abrasion resistance in severe service. The ability of fillers to graft rubber molecules to their surface is an important factor in wear, but it is not a major variable within the ASTM N series of carbon blacks (see Table 1). Graphitization of carbon blacks leads to significant loss in wear resistance. Conversely, the use of coupling agents with silica increases resistance to abrasion, albeit not to the level attainable with the most highly reinforcing carbon blacks.

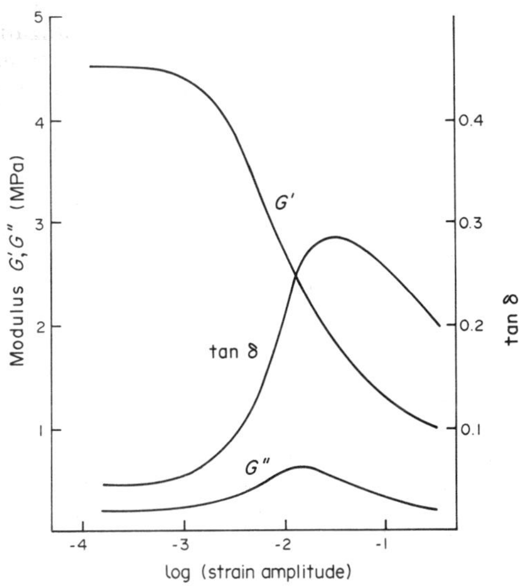

Figure 2
Schematic curves for dynamic storage modulus G', loss modulus G'' and loss tangent for carbon black (20 vol.%) reinforced vulcanizate (scales convey an idea of magnitude only)

There are some important distinctions in the response to filler reinforcement of amorphous and strain-crystallizing rubbers (natural rubber, high *cis*-polyisoprene). Strain-induced crystallization provides a reinforcing mechanism which is not simply added to the effects of filler reinforcement. Consequently, the tensile strength of most natural rubber vulcanizates, when measured at room temperature and modest strain rates, is hardly increased by fillers (and may even be lowered). On the other hand, in tire wear, where strain rates are large and temperatures higher so that strain-induced crystallization does not come into play, response to carbon black is similar to that of amorphous butadiene–styrene rubbers. Important differences between crystallizing and noncrystallizing rubbers exist in cut-growth and fatigue behavior.

5. *Mechanism of Reinforcement*

The mechanism of filler reinforcement of rubbers is not completely understood. It is generally agreed that dissipation of strain energy at the tip of growing cracks by viscoelastic processes is important in the mechanical failure of all polymers and that fillers certainly contribute additional loss mechanisms. The stiffening or "modulus reinforcement" of fillers is governed by morphological features of the filler and the formation of strong attachments of polymer molecules to the filler surface.

See also: Elastomers: Additives; Rubber Processing

Bibliography

Boonstra B B, Medalia A I 1963 Effect of carbon black dispersion on the mechanical properties of rubber vulcanizates. *Rubber Chem. Technol.* 36: 115–42

Dannenberg E M 1985 Reinforcement of rubber with carbon black and mineral fillers. *Prog. Rubber Plast. Technol.* 1(1): 13–26

Donnet J-B, Voet A 1976 *Carbon Black*. Dekker, New York

Eirich F R (ed.) 1978 *Science and Technology of Rubber*. Academic Press, New York
Horn J B 1982 In: Blow C M, Hepburn C (eds.) 1982 *Rubber Technology and Manufacture*, 2nd edn. Butterworth, London, pp. 202–63
Kraus G (ed.) 1965 *Reinforcement of Elastomers*. Wiley, New York
Kraus G 1971 Reinforcement of elastomers by carbon black. *Adv. Polym. Sci.* 8: 155–237
Kraus G 1978 Reinforcement of elastomers by particulate fillers. In: Eirich 1978, Chap. 8
Medalia A I 1970 Morphology of aggregates, VI. Effective volume of aggregates of carbon black from electron microscopy; Application to vehicle absorption and to die swell of filled rubber. *J. Colloid Interface Sci.* 32: 115–31
Voet A 1980 Reinforcement of elastomers by fillers: Review of period 1967–1976. *Macromol. Rev.* 15: 327–73

G. Kraus
[Phillips Petroleum, Bartlesville, Oklahoma, USA]

Flammability of Elastomers

The increasing use of polymers has caused considerable concern regarding the hazard these materials may present when brought into contact with an ignition source. Much of the interest originally was in plastics. However, sizable quantities of elastomers also are used in situations that require low flammability. Consequently, these materials have also recently received a great deal of attention.

Cellular elastomer compositions are used in seat cushions, rug backing, carpet underlay and as thermal insulation, while dense elastic products find use as conveyor belts in mine shafts, electrical wire and cable insulation, hose and ducting material and seals and gaskets. All of these applications call for elastic materials exhibiting some degree of flame-retardant behavior.

The flammability of a material is characterized by its response to a specific ignition stimulus. Typically, the flammability is expressed in terms of ease of ignition, surface flame spread and the amount and rate of generation of heat, combustible material, smoke and toxic gases. A wealth of laboratory tests has been designed to obtain relative numerical values for these characteristics under strictly specified test conditions ranging from simple exposures to ignition sources of varying energy (e.g., a burning cigarette or an electrically heated silicon carbide rod) to more elaborate measurements using instruments that permit the determination of more than one of the above parameters at the same time.

Among the more common procedures are the Motor Vehicle Safety Standard 302 (for automotive interiors), a vertical flame test (DMS 1510A) for aircraft interior materials, the Steiner tunnel test (ASTM E84) for carpet and rug underlay, as well as the radiant panel (ASTM E162) and heat-release-rate calorimeter tests (Smith 1972) which attempt a more realistic simulation of actual fire conditions by exposing the material sample to radiant heat fluxes in addition to ignition stimuli.

The formation of combustible volatile matter necessarily has to precede any burning process. The chemical structure of an elastomer will, therefore, greatly determine its flammability characteristics (Rogers and Fruzzetti 1975, Fabris and Sommer 1977). Hydrocarbon rubbers (i.e., natural rubber, ethylene–propylene rubbers, polybutadienes and styrene–butadiene copolymers) produce large amounts of combustible gases at relatively low temperatures (350–400 °C) and ignite and burn with relative ease. Elastomers containing halogen, phosphorus or silicon have a much greater resistance to ignition.

Generally, flame-retardant elastomers can be obtained by two fundamental methods: (a) the addition of flame retardants or (b) the preparation of elastomers in which the flame retardant is an inherent part of their structure. Chamberlain (1978) has reviewed the various mechanisms by which flame retardants operate.

The additive approach to flame retardance is widely used with the low-cost, commodity hydrocarbon rubbers. These rubbers account for approximately 70% of the total elastomer consumption (excluding polyurethanes—described in the following) in applications requiring some control of flammability. Antimony trioxide in combination with chlorine-containing compatible compounds (e.g., chlorinated paraffins) and specific fillers such as alumina trihydrate are frequently the chosen additives, although in certain cases phosphorus-containing plasticizers are used also (Fabris and Sommer 1977). Extensive research continues to develop improved additives and techniques for their incorporation into the final composite.

Polyurethanes (flexible foams and dense elastomers) represent the largest volume of elastomers used in applications where flame retardancy is a consideration. Flame retardancy in these polymer classes is achieved both by compounding with additives or by polymer design (Papa 1975, Fabris and Sommer 1977).

Polymers with built-in flame-retardant characteristics have been prepared by polymerization of, or copolymerization with, special monomers or by carrying out suitable postreactions on otherwise flammable polymers (Fabris and Sommer 1973). The polymerization of special monomers yields elastomers with various degrees of flame retardance. These include the well-known polychloroprene (Neoprene) and a group of specialty polymers which are used in a diversity of highly critical applications. Fluoroelastomers, nitroso rubbers, triazine elastomers and polyphosphazenes burn with difficulty even in oxygen-rich atmospheres and are used in spacecraft and in high-velocity aircraft to meet extreme requirements of flame retardance. Silicone elastomers, although they do ignite, burn only slowly and produce no flaming drip and relatively little smoke. Because of their outstanding flexibility at low temperatures, they have found use as seals and gaskets in commercial aircraft and as components in space suits. Polychloroprene has a chlorine content of 40% and, therefore, inherently burns less readily than the hydrocarbon rubbers. The choice of fillers is most critical for the retention of flame-retardant behavior in this polymer. While inert fillers show the normal dilution effect in combustible matter giving a slight improvement, calcium carbonate, a very common filler, has been found to reduce the flame retardance of polychloroprene (Fabris and Sommer 1977).

Appropriate modification of preformed polymers is an alternative way to arrive at elastomers with built-in flame retardance. In particular, the chlorination and chlorosulfonation of polyethylene have led to materials of commercial value (e.g., Dow Chemical Company's "CM" series and Du Pont's "Hypalons"). The chlorine content of these modified materials is generally high enough to pass all but the most severe flammability test requirements.

The copious amount of smoke produced by burning rubbers (particularly those derived from dienes and styrene) obscures vision and contributes to the creation of panic in a fire and, therefore, is a serious deterrent to their acceptability in many applications. Flame retardants added to a polymer generally prevent complete combustion and further tend to increase smoke evolution. Inert inorganic fillers lower the amount of smoke only to the value predicted from the dilution of organic matter in the total composition. Certain hydrated fillers (alumina trihydrate and magnesium hydroxide) appear to be exceptions to this rule. These two fillers are found to lower both the total light obscuration as well as the initial rate of smoke evolution caused by burning styrene–butadiene rubber (SBR) and natural rubber (Hecker et al. 1973, Lawson et al. 1975).

In addition to these flammability factors, increasing attention is being given to the toxicity of airborne matter formed by burning or thermally degrading polymers. Toxic gases (e.g., carbon monoxide, hydrogen cyanide and hydrogen chloride) evolved during burning of specific polymers or polymer compositions have to be considered not only by themselves but also in possibly synergistic combinations. Temperature and oxygen depletion and their effect on the human organism have to be taken into account. No adequate tests have been developed yet to assess fully the hazards related to this important characteristic of burning or pyrolyzing elastomer systems and, although research is being vigorously pursued, sufficient information on this subject will probably not be available in the immediate future.

The field of elastomer flammability is a very dynamic one and much research is directed toward the development of compositions that ignite with greater difficulty and also (at least in the early stages of a fire, when the heat flux is still relatively small) burn at slower rates. Much needs to be done to reduce the smoke and the toxic gases produced during burning. Test procedures will have to be developed that make it possible to predict the burning behavior of materials with sufficient accuracy to allow choosing the best elastomer for any specific application.

See also: Flammability of Polymers; Flammability of Polymers: Test Methods; Flammability of Structural Materials and Furnishings

Bibliography

Chamberlain D L 1978 Mechanism of fire retardance in polymers. In: Kuryla W C, Papa A J (eds.) 1978 *Flame Retardancy of Polymeric Materials,* Vol. 4. Dekker, New York, pp. 109–68

Fabris H J, Sommer J G 1973 Flame retardation of natural and synthetic rubbers. In: Kuryla W C, Papa A J (eds.) 1973 *Flame Retardancy of Polymeric Materials,* Vol. 2. Dekker, New York, pp. 135–99

Fabris H J, Sommer J G 1977 Flammability of elastomeric materials. *Rubber Chem. Technol.* 50: 523–69

Hecker K C, Fruzzetti D G, Sinclair E A 1973 Flammability and smoke properties. *Rubber Age* 105(4): 25–32

Lawson D F, Kay E L, Roberts D T Jr 1975 Mechanism of smoke inhibition by hydrated fillers. *Rubber Chem. Technol.* 48: 124–31

Papa A J 1975 Flame-retarding polyurethanes. In: Kuryla W C, Papa A J (eds.) 1975 *Flame Retardancy of Polymeric Materials,* Vol. 3. Dekker, New York, pp. 1–133

Rogers T H Jr, Fruzzetti R E 1975 Flame retardance in rubbers. In: Levin M, Atlas S M, Pearce E M (eds.) 1976 *Flame Retardant Polymeric Materials,* Vol. 1. Plenum, New York, pp. 223–37

Smith E E 1972 Measuring rate of heat, smoke and toxic gas release. *Fire Technol.* 8(3): 237–45

H. J. Fabris
[General Tire and Rubber Company,
Akron, Ohio, USA]

Flammability of Polymers

Polymeric materials are being increasingly used as construction materials in a number of areas such as home furnishings, domestic and industrial buildings, fabrics and vehicles. Selection or design of a flame-retardant system for a particular application is often difficult. It is usually desirable that a flame-retardant polymer has high resistance to ignition, low rate of combustion and smoke generation, low toxicity of product gases, retention of low flammability during use, acceptability in appearance and properties, no environmental or health impact, and little or no economic penalty. In order to select or design a polymeric material with desirable flammability properties, a familiarity with the principles of combustion, the relationship of flammability to polymer structure and the modes of flame inhibition is essential.

1. Basics of Polymer Flammability

The presence of fuel, heat and oxygen is necessary to initiate and sustain a fire. A schematic flammability cycle is shown in Fig. 1. When heated sufficiently by the external ignition source, the polymeric material reaches a characteristic temperature at which it begins to degrade. The extent to which the molecular oxygen plays a role in this surface decomposition depends on the type of polymer. Gaseous combustible products may then be formed, with a rate dependent on factors such as the intensity of external heat, temperature and rate of polymer decomposition.

Flammable gases (the fuel) thus produced diffuse to the flame front, where a series of heat-generating complex free-radical chain reactions take place in the presence of surrounding oxygen. Since the flaming of organic polymers simply represents the oxidation of hydrocarbons, the well-understood methane–oxygen combustion system can be used as a model for the more complex polymer flames. The mechanism of methane combustion is a complicated free-radical chain reaction consisting of propagation, chain branching and termination steps:

Propagation

$$CH_4 + HO\cdot \rightarrow CH_3\cdot + H_2O \tag{1}$$

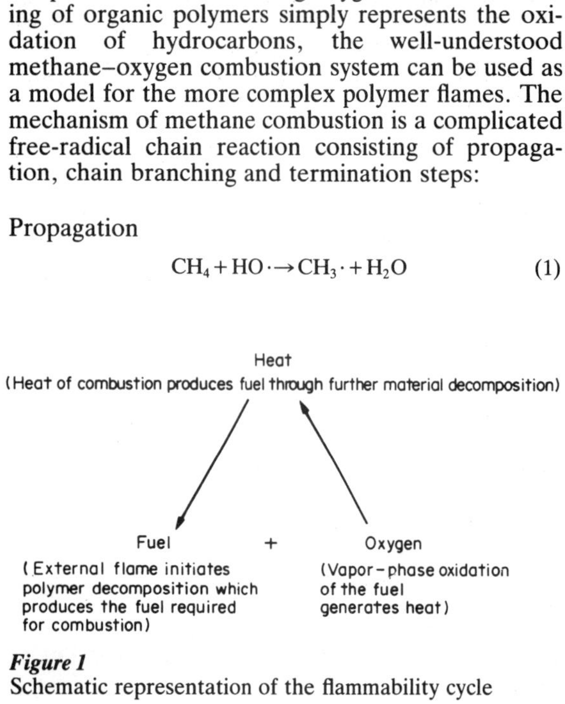

Figure 1
Schematic representation of the flammability cycle

$$CH_4 + H\cdot \rightarrow CH_3\cdot + H_2 \tag{2}$$

$$CH_3\cdot + O\cdot \rightarrow CH_2O + H\cdot \tag{3}$$

$$CH_2O + CH_3\cdot \rightarrow CHO\cdot + CH_4 \tag{4}$$

$$CH_2O + HO\cdot \rightarrow CHO\cdot + H_2O \tag{5}$$

$$CH_2O + H\cdot \rightarrow CHO\cdot + H_2 \tag{6}$$

$$CH_2O + O\cdot \rightarrow CHO\cdot + HO\cdot \tag{7}$$

$$CHO\cdot \rightarrow CO + H\cdot \tag{8}$$

$$CO + HO\cdot \rightarrow CO_2 + H\cdot \tag{9}$$

Chain branching

$$H\cdot + O_2 \rightarrow HO\cdot + O\cdot \tag{10}$$

Termination

$$H\cdot + R\cdot + M \rightarrow RH + M^* \tag{11}$$

It is interesting that oxygen is only found in one step, the major branching step. In the branching step, two chain-carrying radicals are produced for every one consumed and this accounts for the explosive nature of burning. The main chain-carrying radicals in the propagation steps are the hydrogen atom and hydroxyl radicals. This mechanism clearly indicates that hydrogen, and not carbon, is mainly responsible for the flammability of hydrocarbons.

After ignition and removal of the ignition source, combustion becomes self-propagating if sufficient heat is generated and radiated back to the material to continue the decomposition process. The combustion process is governed by such variables as the rate of heat generation, the rate of heat transfer to the surface, surface area and the rate of decomposition.

2. Approaches to Flame Retardation

Polymer combustion, a highly complex process, is believed to involve a vapor phase, in which the reactions responsible for the formation and propagation of the flame take place, and a condensed phase, in which fuel for the gas reactions is produced. Flame retardancy, therefore, can be improved by appropriately modifying either one or both of these phases. The approaches to reducing the flammability of polymer systems can be grouped into the following three categories.

2.1 Vapor Phase

In the vapor-phase approach, a flame retardant or the modified polymer unit releases a chemical agent on heat exposure that inhibits free-radical reactions involved in the flame formation and propagation. For example, the presence of HX(X = halogen), produced by the pyrolysis of a halogen-containing organic material in the polymer, acts as a radical scavenger and replaces less reactive halogen atoms for active chain carriers (Factor 1974):

$$H\cdot + HX \rightarrow H_2 + X\cdot \quad (12)$$

$$HO\cdot + HX \rightarrow H_2O + X\cdot \quad (13)$$

Halogenated flame retardants such as chlorinated paraffins and alicyclic compounds, and chloro- and bromo-aromatic additives, are postulated to function primarily by a vapor-phase flame inhibition mechanism. Flame retardation is implemented by incorporating fire-retardant additives, by impregnating the material with a flame-retardant substance or by using flame-retardant comonomers in the polymerization or grafting.

2.2 Condensed Phase

In condensed-phase modification, the flame retardant alters the decomposition chemistry such that the transformation of the polymer to a char residue is favored. This is achieved by the addition of additives which catalyze charring rather than flammable product formation or by designing polymer structures which favor char formation. Carbonization which occurs at the cost of flammable product formation also shields the residual substrate by interfering with the access of heat and oxygen. Phosphorus-based additives are typical examples of flame retardants which act by a condensed-phase mechanism.

2.3 Miscellaneous

Miscellaneous approaches include dilution of the polymer with nonflammable materials (e.g., inorganic fillers), incorporation of materials which decompose to give nonflammable gases such as carbon dioxide and formulation of products which decompose endothermically. A typical example of such a flame retardant is alumina trihydrate ($Al_2O_3.3H_2O$). This kind of material acts as a thermal sink to increase the heat capacity of the combustion system, lower the polymer surface temperature via endothermic events and dilute the oxygen supply to the flame. Thermoanalytical techniques can prove to be helpful in determining the mechanism of flame retardation (Pearce et al. 1981, Cullis and Hirschler 1983).

Of the several test methods for evaluating the burning behavior of different polymers (see *Flammability of Polymers: Test Methods* and Kishore and Mohandas 1982), the limiting oxygen index of Fenimore and Martin (1966) will be used here to illustrate the relative flammability of materials. This test measures the minimum concentration of oxygen in an oxygen–nitrogen atmosphere that is necessary to initiate and support a flame.

3. Polymer Structure and Flammability Properties

The flammability of a particular polymer mainly depends on its structure. The amounts of char and incombustible gases formed on thermal decomposition determine to a great extent the flame resistance of a polymeric material. A material with a limiting oxygen index of ≤26 should be considered flammable (van Krevelen 1975). Based on the polymer structure and limiting oxygen index, it appears that polymeric materials can be broadly classified into three categories. Table 1 presents the structures, thermal decomposition products and limiting oxygen index data for some representative polymers belonging to each of the three classes.

Thermal decomposition of class I polymers gives rise to little or no char residue. This absence of a pyrolysis residue is due to the high hydrogen content of these polymers, which allows the entire decomposition product to be disproportionated into volatiles. High yields of monomers or other hydrocarbons from the thermal decomposition of these systems constitute the fuel needed for the flammability cycle shown in Fig. 1. The CN-containing polyacrylonitrile and the OH-containing polymers (cellulose and poly(vinyl alcohol)) must be considered as exceptions in that they cyclize and therefore produce some char. However, at the same time, they give off a large amount of flammable products. The polymers based on structures similar to those included in class I have limiting oxygen indices lower than 26 and are considered to be flammable.

Class II covers those materials which are known to be high-temperature polymers. Structurally, they are characterized by the presence of either wholly aromatic or heterocyclic–aromatic functional units. An aromatic ring in the polymer backbone is not only relatively unreactive thermally, but also increases the chain rigidity. Stiffer chains resist thermally induced vibrations and, therefore, a higher temperature is required for their decomposition. These materials not only degrade at high temperatures, but also undergo extensive crosslinking and cyclization, as a result of which their pyrolysis generates high char residues; typically 20–70% at 700 °C. The formation of char residue is conveniently measured using thermogravimetric analysis (TGA) in an inert atmosphere (Pearce et al. 1981).

Since the formation of char takes place at the expense of flammable products (i.e., the elements are converted into carbonaceous involatile residues rather than volatile species), it should be expected to result in the reduction of flammability. In fact, it has been demonstrated that there is a significant correlation between the char residue (e.g., at 850 °C) and the oxygen index of polymers (van Krevelen 1975). As shown in Fig. 2, the greater the amount of char, the higher the oxygen index and, hence, the lower the flammability. The high temperature polymers of class II, such as those in Table 1, are considered

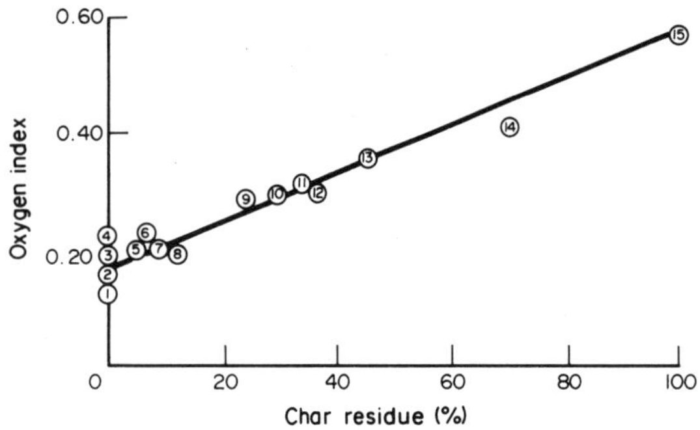

Figure 2
Correlation between oxygen index and char residue: 1, polyformaldehyde; 2, polyethylene, polypropylene; 3, polystyrene, polyisoprene; 4, nylon; 5, cellulose; 6, poly(vinyl alcohol); 7, poly(ethylene terephthalate); 8, polyacrylonitrile; 9, poly(phenylene oxide); 10, polycarbonate; 11, Nomex; 12, polysulfone; 13, Kynol; 14, polyimide; 15, carbon (after van Krevelen 1975. © IPC Business Press, London. Reproduced with permission)

flame retardant (limiting oxygen index >26). The lower flammability is attributed to factors such as reduction in the amount of combustible volatiles, high energy requirements for continuous fuel generation and insulating effects of the resultant char.

Class III consists of halogen-containing polymeric materials. Some of these form a small char residue, others none at all. These polymers are inherently flame retardant. This is a direct consequence of the fact that the halogen radicals act as radical scavengers in the vapor phase and therefore inhibit combustion as described earlier. Also, the formation of noncombustible gases such as HCl, HF and C_2F_4, which seal the polymer surface from the air, is partly responsible for the flame retardancy. All these factors thus influence the interaction between pyrolysis and ignition.

4. *Flame Retardation of Polymeric Materials*

The techniques of reducing the flammability of polymers are based on one or more of the three fundamental approaches described in Sects. 2.1–2.3. Whether a particular flame-retarding method is suitable or not is often determined by the nature of the polymer material (e.g., thermoplastic or thermoset).

4.1 Thermoplastics

Many flame-retardant chemicals have been developed for use in thermoplastics. The majority of these flame retardants are of the additive type, usually halogen- and/or phosphorus-based compounds. Some examples of fire-retardant additives for use in polyolefins (Kuryla and Papa 1973) are hexabromocyclododecane, octabromobiphenyl, hexabromobiphenyl, chlorowax, chlorinated triphenyl, chlorendic acid, tris(tribromophenyl) phosphite and tris(dibromopropyl) phosphate. The choice of flame retardants is dependent on the nature of the polymer, the method of processing, the proposed service conditions and economic considerations. Although the processing, service and economic factors are important, the flame-retardancy potential of an additive is of primary importance and this can be readily evaluated by thermal analysis. Einhorn (1971) has described the use of TGA in selecting the appropriate fire retardants. Figure 3 represents the TGA thermogram for a hypothetical polymer which exhibits a simple unimolecular degradation process. Point A represents the start of the decomposition process and point B represents the temperature of the maximum rate of degradation. Fire retardants are screened so as to select a material having thermal characteristics similar to those in Fig. 3. The efficiency of matching the degradation curve of the polymer with the volatilization or degradation characteristics of the fire retardant has been cited as the key to effective flame retardancy. If the flame-retardant additive possesses low thermal stability compared with that of the polymer, it will be lost before its function is needed; if the additive has greater stability, it may be intact at the time its function is needed. Depending on the thermal degradation characteristics of the flammable substrate, more than one fire-retardant additive could also be employed.

The concept of synergism is very important in fire retardation, since it can lead to efficient flame retardants with less expensive polymer systems and minimal effects on other desirable properties. One of the classical illustrations of synergism observed in flame retardation is the addition of Sb_2O_3 to halogen-containing polymers. For example, it is believed that the reaction of a chlorine source with Sb_2O_3 produces SbOCl as an intermediate with a lower energy barrier, which on thermal decomposition evolves $SbCl_3$, the actual flame retardant (Khanna and Pearce 1978). The existence of phosphorus–halogen and nitrogen–phosphorus synergisms has also been suggested. Compounds based on antimony–halogen, nitrogen–phosphorus and phosphorus–halogen combinations have been utilized as flame-retardant additives for thermoplastics.

4.2 Thermosets

Fire retardancy in thermosetting polymers is achieved largely by the use of reactive fire retardants, since the common fire-retarding additives lack permanence. The flammability of thermosetting materials can be reduced by the addition of inorganic fillers and/or reactive flame-retardant components. Flame-retardant vinyl monomers or other crosslinking agents are also frequently employed.

Table 1
Effect of structure on polymer flammability

Class	Name	Structure	Degradation products	Limiting oxygen index
I	Polyethylene	$+CH_2-CH_2+_n$	continuous spectrum of saturated and unsaturated hydrocarbons from C_2–C_{90}	17.4
	Polybutadiene	$+CH_2-CH=CH-CH_2+_n$	about 2% monomer in addition to saturated and unsaturated hydrocarbons	18.3
	Polystyrene	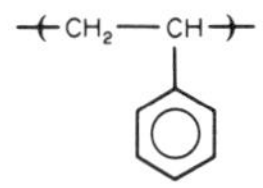	about 41% monomer, 2% toluene; remainder dimer, trimer and tetramer	18.3
	Cellulose	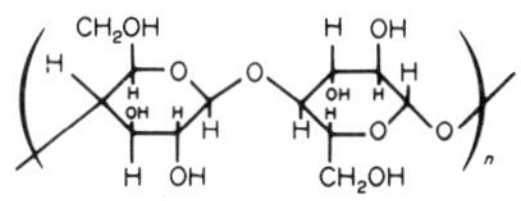	H_2O and small amounts of CO_2, CO and a tar containing principally levoglucosan	19.9
	Poly(ethylene terephthalate)	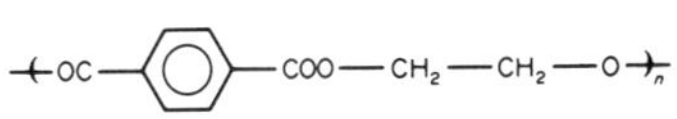	acetaldehyde major product with CO_2, CO, C_2H_4, H_2O, CH_4, C_6H_6, terephthalic acid and more complex chain fragments	20.6
	Nylon 66	$+H_2N-(CH_2)_6-NHCO-(CH_2)_4-CO+_n$	H_2O, CO_2, cyclopentanone, traces of saturated and unsaturated hydrocarbons	21.5
II	Polycarbonate	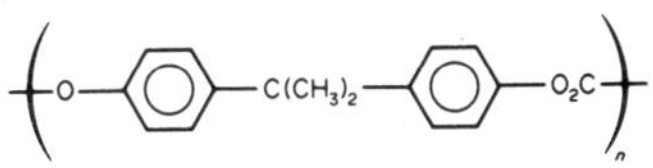	major products CO_2, bisphenol A; minor products CO, CH_4, 4-alkylphenols; much char	29.4
	Nomex	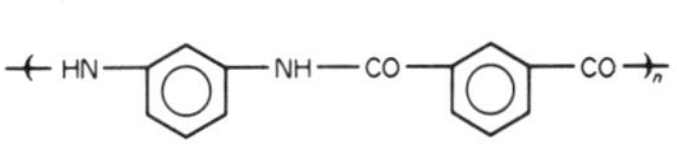	high char yield; volatiles contain CO_2, CO, H_2O, 1,3-dicyanobenzene, 3-cyanobenzoic acid	29.8
	Kynol	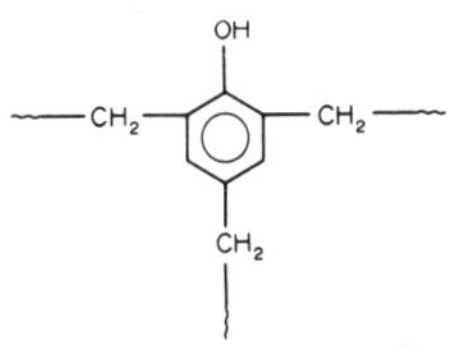	high char yield; volatiles comprise xylene (76%), traces of phenol, cresol, benzene	35.5
	Polybenzimidazole	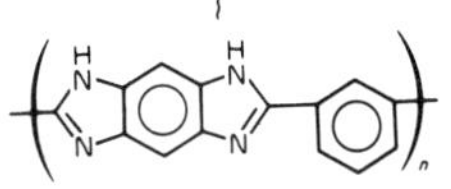	very high char yield, small volatiles may include NH_3 and H_2	41.5
III	Poly(vinylidene fluoride)	$+CH_2-CF_2+_n$	35% HF and high yields of products involatile at 25 °C; some carbonization	43.7
	Poly(vinyl chloride)	$+CH_2-CHCl+_n$	quantitative yields of HCl, aliphatic and aromatic hydrocarbons	47.0
	Poly(vinylidene chloride)	$+CH_2-CCl_2+_n$	high yields of HCl	60.0
	Polytetrafluoroethylene	$+CF_2-CF_2+_n$	>95% monomer, 2–3% C_3F_6, no larger fragments	95.0

A large number of halogen-containing reactive diols, polyols, anhydrides and other functional-group-containing intermediates have been used to produce flame-resistant unsaturated polyester resins. Flame-retardant polyester resins have been made by using bromostyrene as a partial replacement of styrene for crosslinking reactions.

As in polyester resins, reactive halogen-containing fire-retardant chemicals are most often used in epoxy materials. Tetrabromobisphenol A is perhaps the most widely used component for flame-retarding epoxy resins.

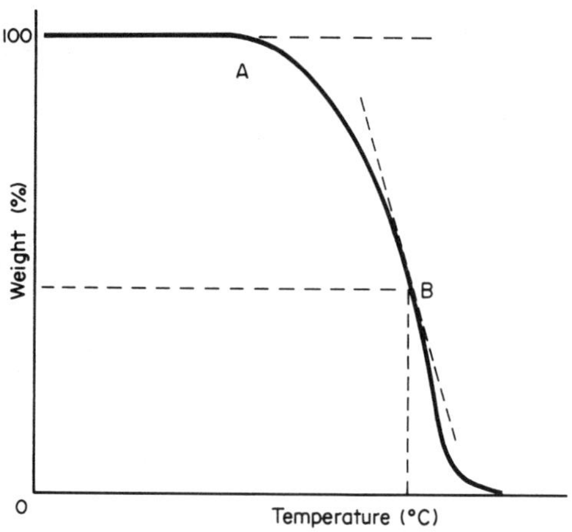

Figure 3
TGA thermogram for a hypothetical polymer (after Einhorn 1971. © National Academy of Sciences, Washington, DC. Reproduced with permission)

Polyurethane foams, however, are generally flame retarded by the incorporation of halogens, phosphorus and/or nitrogenous compounds. Conley and Quinn have reviewed the fire retardation of various resin systems (Lewin et al. 1975).

4.3 Elastomers

Many applications of rubbers (e.g., tires, gaskets and washers) normally do not require flame resistance. Where improvement in flammability is required, it can be achieved by the addition of halogen-containing materials, phosphorus compounds, oxides of antimony and combinations of these materials. Rubbers containing chlorine and silicon atoms (e.g., neoprene and silicone) have self-extinguishing properties. Rogers and Fruzzetti have described the flame retardance of elastomer systems (Lewin et al. 1975).

4.4 Fibers

The flame retardation of fiber-forming polymers is generally achieved by the incorporation of additives based on halogens, phosphorus and/or nitrogen. The use of antimony oxide as a synergist with halogen is also common.

The most common route for making cellulosic fibers flame retardant is through the use of catalysts (e.g., antimony trichloride, phosphoric acid) which enhance the formation of char (Kuryla and Papa 1979).

Retardation of combustion of polyamides (e.g., poly(ethylene terephthalate)) can be decreased by using bromocompounds in conjunction with an Sb_2O_3 synergist, red phosphorus and triphenylphosphine oxide (Lewin et al. 1975 pp. 193–221).

The flammability of high-temperature fiber-forming polymers can also be improved. Durette made by chlorination or oxychlorination of Nomex, a polymer of *m*-phenylene isophthalamide, exhibits much improvement in flammability properties over Nomex. Retardation of flammability by the use of halogen-substituted diamines has also been described for aromatic polyamides (Khanna and Pearce 1981). In these high-temperature materials, the resistance to burning is also achieved by thermal treatment (Bingham and Hill 1975). Hirsch and Holsten (1968) have reported that the fabric made from poly(*m*-phenylenebis(*m*-benzamido)-terephthalamide) can be stabilized thermally such that the product will not burn even if heated to red hot in a gas flame. Such treatments are believed to involve cyclodehydration of the original polymer.

See also: Flammability of Elastomers; Flammability of Polymers: Test Methods; Flammability of Structural Materials and Furnishings

Bibliography

Bingham M A, Hill B J 1975 A study of the thermal behaviour of flame-resistant fibres and fabrics. *J. Therm. Anal.* 7: 347–58

Cullis C F, Hirschler M M 1983 The significance of thermoanalytical measurements in the assessment of polymer flammability. *Polymer* 24: 834–40

Einhorn I N 1971 The chemistry of fire resistant materials and suppression. *Fire Res. Abstr. Rev.* 13(3): 236–67

Factor A 1974 The chemistry of polymer burning and flame retardance. *J. Chem. Educ.* 51: 453–6

Fenimore C P, Martin F J 1966 Candle-type test for flammability of polymers. *Mod. Plast.* 44(3): 141–92

Field A W 1982 Cables in fires—Special designs and finishes for optimum performance. In: McAllister D (ed.) 1982 *Electric Cables Handbook*. Granada, London, Chap. 6

Fristom R M, Westenberg A A 1965 *Flame Structure*. McGraw-Hill, New York

Hall C 1989 Combustion: Fire properties of polymeric material. *Polymer Materials,* 2nd edn. Macmillan, London, pp. 141–7

Hirsch S S, Holsten J R 1968 Enhancement of aromatic polyamide thermostability by heat treatment in air. *Polym. Prepr., Am. Chem. Soc., Div. Polym. Chem.* 9: 1240–1

Khanna Y P, Pearce E M 1978 Synergism and flame retardancy. In: Lewin et al. 1978, pp. 43–61

Khanna Y P, Pearce E M 1981 Aromatic polyamides III. Mechanistic studies on the role of substituent chlorine in flame retarding aromatic polyamides. *J. Polym. Sci., Polym. Chem. Ed.* 19: 2835–40

Kishore K, Mohandas K 1982 Laboratory test methods for the evaluation of flammability of polymers. *J. Macromol. Sci., Chem.* 18(3): 379–93

Kuryla W C, Papa A J (eds.) 1973, 1975, 1978, 1979 *Flame Retardancy of Polymeric Materials,* Vols. 1–5. Dekker, New York

Lewin M, Atlas S M, Pearce E M (eds.) 1975, 1978 *Flame-Retardant Polymeric Materials,* Vols. 1, 2. Plenum, New York

Pearce E M, Khanna Y P, Raucher D 1981 Thermal analysis in polymer flammability. In: Turi E A (ed.) 1981 *Thermal Characterization of Polymeric Materials*. Academic Press, New York, Chap. 8, pp. 793–841
van Krevelen D W 1975 Some basic aspects of flame resistance of polymeric materials. *Polymer* 16: 615–20

Y. P. Khanna
[Allied Chemical Corporation, Morristown, New Jersey, USA]

Flammability of Polymers: Test Methods

All polymers will burn, given sufficient heat and oxygen. Some, such as polyethylene and poly(methyl methacrylate), burn readily under normal ambient conditions, whereas others (e.g., polytetrafluoroethylene) burn only under extreme conditions of applied heat or oxygen content.

Typical characteristics measured by flammability tests are ignitability, surface flame spread, heat-release rate and smoke evolution. All of these flammability characteristics are extrinsic properties of a polymer. Unlike intrinsic properties, they are dependent on external factors such as geometry and orientation, as well as the amount of oxygen available for combustion. Because of this dependency on external conditions, it is difficult to predict "real fire" performance by extrapolation from fire test results.

1. *Fire Development Regimes*

The stages of a fire from ignition to inferno are characterized in Fig. 1 where fire intensity is plotted versus time. The intensity scale conceptually represents the fire growth parameters of mass burning rate, rate of heat release, temperature and degree of involvement of combustibles. Three regimes of fire intensity are shown: incipient, developing and fully developed.

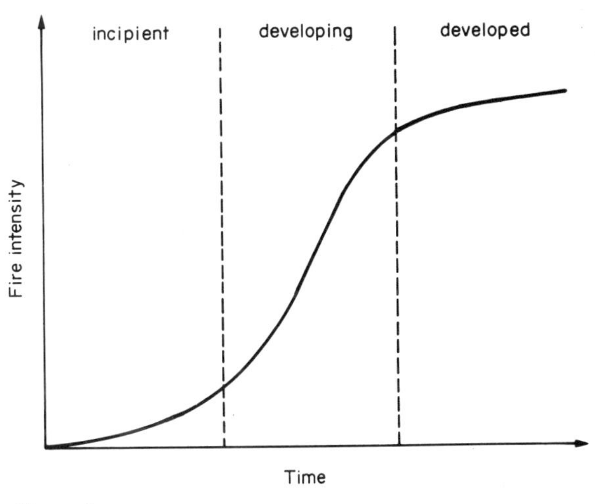

Figure 1
Idealized fire development curve showing the three regimes of fire growth (©Underwriters Laboratories, Melville, New York. Reproduced with permission)

In the incipient regime the fire is small with the materials involved in the fire being subjected to, and producing, relatively low levels of thermal energy. When the rate of heat evolution from the burning materials exceeds the rate of heat loss to the environment, the temperature rises, causing the polymers to burn faster. In the developing regime, the temperature and the heat-release rate increase. Energy feedback (convective and radiative) accelerates the rate of burning until the fire becomes limited by the availability of either fuel or oxygen. The fire is then in the fully developed regime.

Flammability tests for polymers operate in one or more of the three regimes. Most small-scale laboratory tests which utilize small ignition sources such as a Bunsen burner operate in the incipient regime. The performance of a polymer in small-scale tests may relate to its performance in an incipient fire, but such tests are not reliable predictors of performance in developing or fully developed fires because they do not simulate the thermal environment of these larger fires.

Flammability tests that operate in the developing regime typically have higher-energy ignition sources than incipient tests, are usually larger in scale and often provide containment so that heat loss is restricted to provide a severe thermal environment. Perhaps the most widely used test in this regime is the ASTM E84 surface flame spread test which utilizes a 1.4 m long ignition flame and a test sample 7.6 m long.

Flammability tests that operate in the fully developed regime typically have high-flux ignition sources and provide for containment of energy and full involvement of test samples. Generally, they are used to measure the fire endurance of structural elements such as walls.

In the section that follows, some of the widely used flammability tests applicable to polymers have been selected and grouped according to the regime of fire intensity where they are most applicable.

2. *Incipient Regime Tests*

In ASTM D1929 (Test for Ignition Properties of Plastics) a 3 g sample is exposed to flowing air (see Fig. 2), and the air temperature is increased until the plastic ignites. When tests are conducted with a pilot flame, the flash-ignition temperature is measured. Without a pilot light, the self-ignition temperature is determined.

The ASTM D2863 oxygen index test (see Fig. 3) determines the relative flammability of plastics by measuring the minimum amount of oxygen, in a

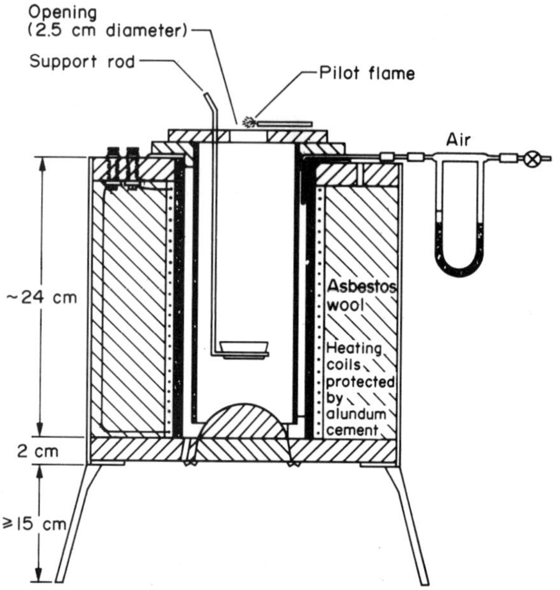

Figure 2
Apparatus for measuring ignition temperatures by ASTM D1929

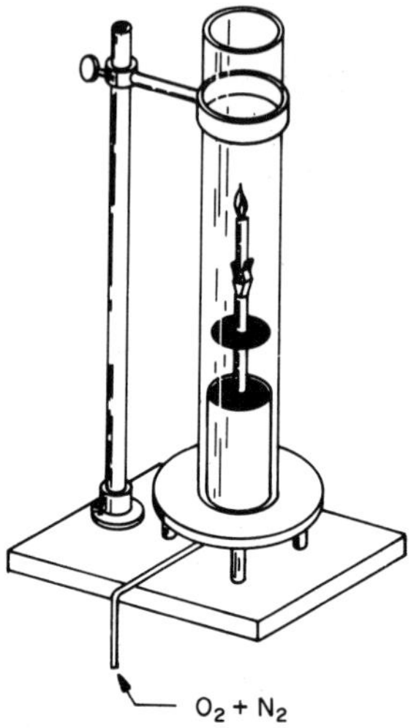

Figure 3
Apparatus for measuring oxygen index by ASTM D2863

mixture of oxygen and nitrogen, that is required to support combustion. Materials can then be rank ordered with the flammable materials having low oxygen indices (below the 21% O_2 in the air) and fire-resistant ones having indices considerably above 21%. It is common practice to specify a minimum oxygen index of 28% for fire-resistant materials.

The oxygen index test is one of the most widely used flammability tests. It is easy to run, reproducible and, most importantly, yields a quantitative result rather than a pass/fail result for assessing the flammability of materials. It is an excellent test for quality control and research, but provides insufficient information for predicting the performance of a material in a developing fire.

Several factors affecting the oxygen index test help explain the poor correlation between oxygen index test results and developing fire performance. Unlike a developing fire, there is minimal energy feedback from the flame to the polymer substrate—the oxygen level is adjusted so the flame is on the verge of extinguishing. This situation is in contrast to a large fire where radiative feedback is the dominant mode of energy transfer.

The oxygen index decreases with increasing sample temperature. For example, the oxygen index of a polychloroprene compound containing clay and carbon black as the principal additives is 33% at room temperature and 19% at 300 °C (Johnson 1974). The oxygen index also decreases with decreasing sample thickness. Another factor is that thermoplastic materials often drip and flow away from the flame. Since the melted thermoplastic carries heat away, a higher oxygen concentration is needed to maintain combustion and a higher oxygen index results. In a real fire, thermoplastics will often drip and burn in a molten pool.

In UL94, *Tests for Flammability of Plastic Materials for Parts in Devices and Appliances* (Underwriters Laboratories 1980), methods are provided for both vertical and horizontal testing. In the vertical test (see Fig. 4), a Bunsen burner with a 19 mm flame is applied to test specimens 13 cm long,

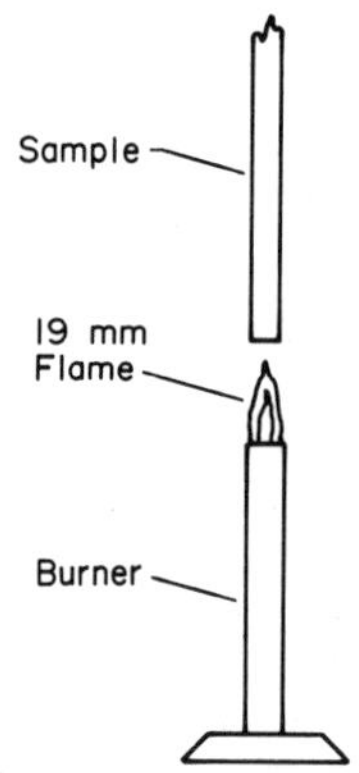

Figure 4
The UL94 vertical test setup

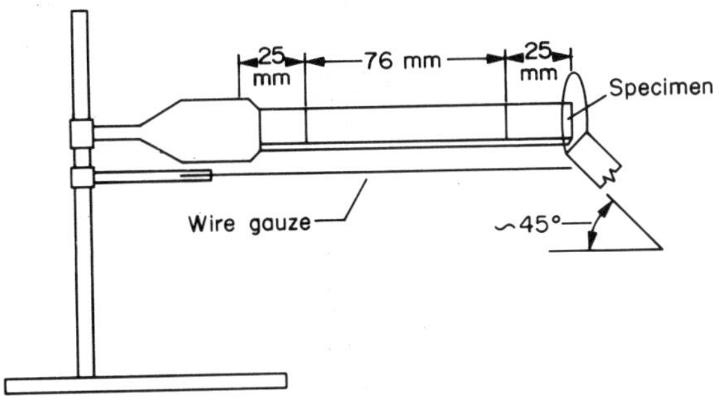

Figure 5
The UL94 horizontal test setup

13 mm in width and up to 13 mm in thickness. The burner is placed under the sample for 10 s, then withdrawn until flaming ceases and reapplied for an additional 10 s. The duration of flaming combustion after each flame application is used to rate materials as 94V-0 if flaming combustion averages 5 s or less without dripping flaming particles and 94V-1 if flaming combustion averages 25 s or less without dripping flaming particles. Materials that produce dripping flaming particles that ignite surgical cotton are rated 94V-2.

In the UL94 horizontal test (see Fig. 5), a Bunsen burner with a 25 mm flame is applied to test specimens 13 cm long and 13 mm in width and the rate of flame spread over the horizontal surface is measured. Materials that have flame spread rates less than specified maxima are rated 94HB. This test is nearly identical to ASTM D635 (*Test Method for Rate of Burning and/or Extent of Time of Burning of Self-Supporting Plastics in a Horizontal Position*).

3. Developing Regime Tests

ASTM E84 (*Test for Surface Burning Characteristics of Building Materials*) is the standard test to certify building materials for compliance with building codes. It utilizes the Steiner tunnel furnace (see Fig. 6) where a 7.3 m long by 51 cm wide test specimen forms the roof of the furnace. The lower surface of the specimen is ignited by a large (1.4 m) methane flame. A 1.22 m s^{-1} draft and an enclosure, which provides for conservation of heat energy, aid flame spread. The fire is observed through windows in the tunnel wall.

The tunnel is calibrated with red oak, which spreads flame 5.9 m from the end of the ignition flame to the tunnel end in 15.5 min, and with asbestos cement board, which does not burn. Asbestos cement board is assigned a flame spread classification of 0 and red oak 100.

A smoke value is determined by measuring the percent attenuation of a beam of light across the exhaust duct of the tunnel. The area under the percent attenuation versus time curve for red oak in a 10 min test is assigned a value of 100. The smoke values of other materials are reported relative to the red oak value of 100.

The combination of a large ignition flame, with a heat-release rate of 88 kW, and an insulated furnace

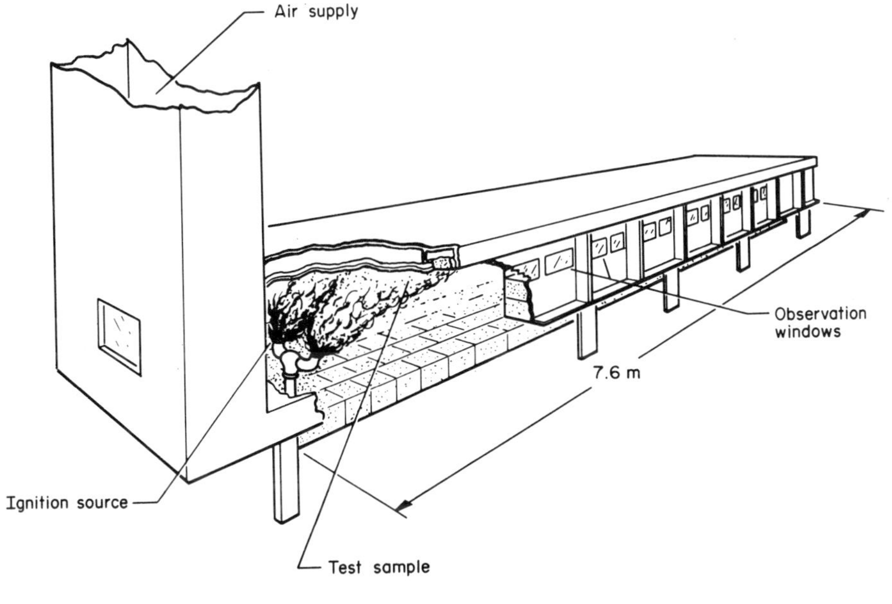

Figure 6
The Steiner tunnel furnace for measuring flame spread and smoke developed by ASTM E84

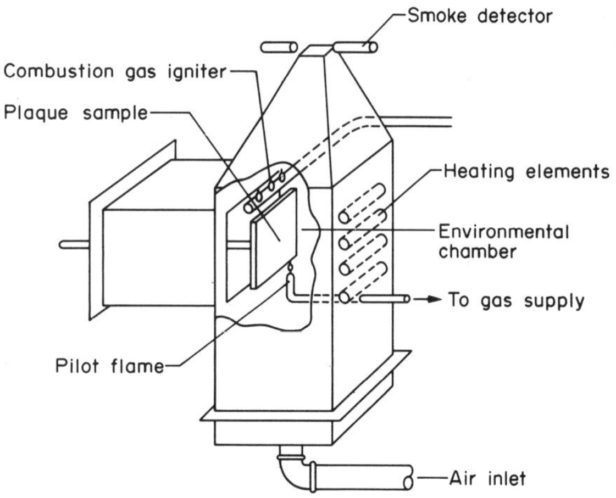

Figure 7
The Ohio State University heat-release-rate calorimeter (ASTM E906)

which reradiates heat once it is heated up provides a severe test environment. Measurements made by Parker (1977) indicate that the sample in the flame impingement area is subjected to 40–60 KW m^{-2} incident heat flux.

Materials with a flame spread classification of 25 or less (less than 1.5 m flame spread) are generally regarded by building codes as providing a high degree of fire safety. The correlation of a low flame-spread classification with good fire performance was established for cellulosic (wood) materials, but recent experiences with thermoplastic foams show that ASTM E84 should not be employed as the sole source of fire hazard rating for polymeric materials.

The rate at which a material burns (i.e., the rate at which it releases heat) is an important flammability property since it is a measure of fire intensity. Rate of heat release calorimeters measure rates of heat and smoke release of a material subjected to an externally applied heat flux. Unlike most other fire tests which utilize only one exposure condition, heat-release-rate calorimeters allow the incident heat flux to be varied from test to test and, hence, the influence of flux on release rates can be determined.

In the Ohio State University (OSU) calorimeter (see Fig. 7), which is used in ASTM E906 (*Test Method for Heat and Visible Smoke Release Rates of Materials and Products*), incident flux can be varied from a level characteristic of an incipient fire (0–10 kW m^{-2}) on through those characteristics of developing fires, up to values associated with fully developed fires (80–100 kW m^{-2}). Rowan and Lyons (1978) have published some experimental results that demonstrate the importance of externally imposed heat flux on the burning behavior of materials. Figure 8 shows peak heat-release rates vs incident applied flux for a series of fire retardant polyurethane and polyisocyanurate foams. The rate of burning generally increases with increasing incident flux and for flame retardant materials there is a minimum flux required to sustain combustion. The rank ordering of materials can change when flux is increased. The material represented by the dashed line is best (i.e., it has the lowest heat-release rate) below 40 kW m^{-2} incident flux, but it is the worst or next to worst material at incident fluxes above 45 kW m^{-2}.

Smoke evolution is also dependent on incident heat flux. Figure 9 shows the variation in total smoke output versus incident flux for several polymers tested in the OSU release rate calorimeter.

The most widely used instrument for measuring smoke is the National Bureau of Standards smoke chamber (see Fig. 10). Test conditions are specified in ASTM E622 (*Standard Test Method for Specific Optical Density of Smoke Generated by Solid Materials*). The specimen is subjected to a 25 kW m^{-2} incident flux. Tests are run either in the flaming mode where the sample is ignited by a pilot flame or in the smoldering mode where the pilot flame is not used. Smoke is accumulated within the smoke chamber box and the specific optical density of the smoke is reported. Specific optical density D_s is defined by

$$D_s \equiv (V/LA)\log(100/T)$$

where V is the chamber volume, L is the optical path length, A is the exposed specimen area and T is the percent transmittance of the light beam.

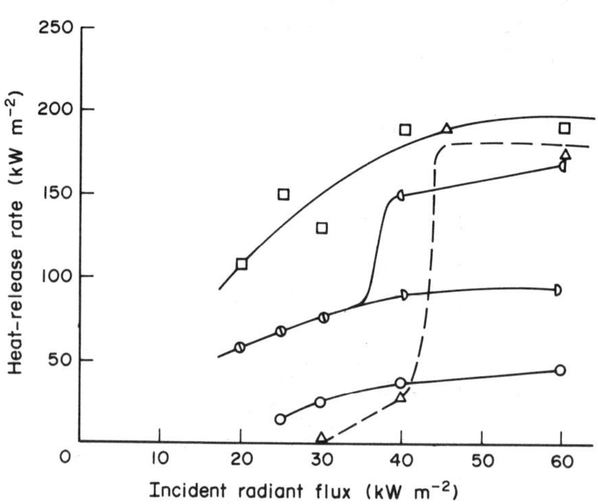

Figure 8
The influence of externally applied flux on peak heat-release rates for a series of polyurethane and polyisocyanurate foams (after Rowan and Lyons 1978. ©Technomic, Westport, Connecticut. Reproduced with permission)

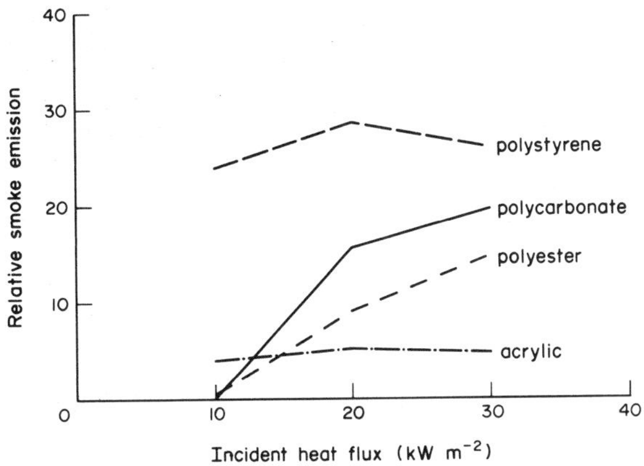

Figure 9
The influence of externally applied flux on total smoke emissions for various polymers (courtesy Professor E E Smith, Ohio State University)

There are several differences between smoke measurements in the NBS chamber and in the OSU release-rate calorimeter. The NBS chamber is normally run at only one standardized flux. Thus it cannot be related to a real fire situation where materials are exposed to a spectrum of fluxes. The NBS chamber measures total smoke by accumulating it within the box whereas the OSU calorimeter operates with an air flow through the combustion chamber and total smoke is determined by integration of the instantaneous smoke values. Both test methods report smoke in optical density, which is linearly related to human visibility through smoke. Both methods differ greatly from the ASTM E84 smoke measurement, where the amount of smoke generated is dependent on the flame spread value of the material tested.

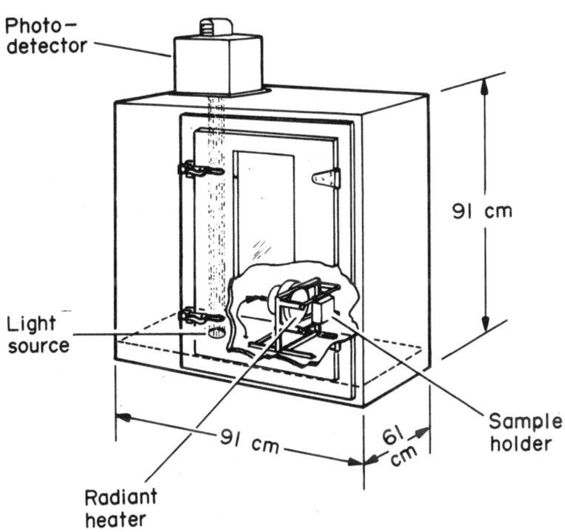

Figure 10
National Bureau of Standards smoke chamber (ASTM E662)

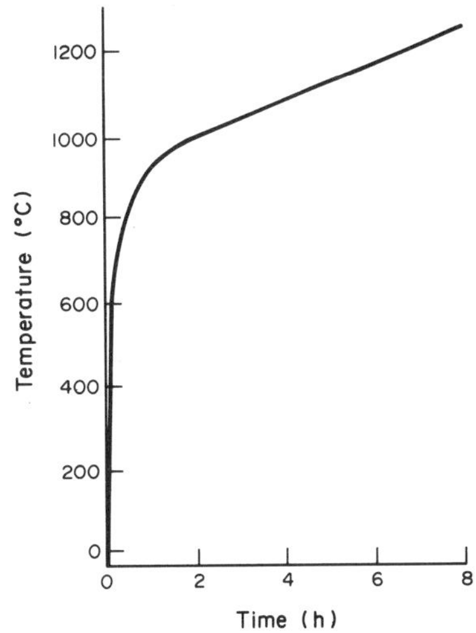

Figure 11
The ASTM E119 time–temperature curve for testing the fire endurance of structures

4. Fully Developed Regime Tests

ASTM E119 (*Fire Tests of Building Construction and Materials*) is used for rating the fire endurance of materials and structures. Large specimens, minimum exposed area 9.3 m^2, are exposed to a standard fire whose time–temperature curve (see Fig. 11) is characteristic of a postflashover fire. Assemblies are rated by the elapsed time to failure. Several failure criteria are provided, such as ignition of cotton waste on the unexposed side of the specimen or temperatures exceeding 139 °C on the unexposed side.

ASTM E814 (*Fire Test of Through Penetration Fire Stops*) is applicable to polymeric materials used in penetrations (e.g., cables, plastic ducts and firestop materials). It utilizes the ASTM E119 time–temperature exposures with a firestop and penetration in place to test the endurance of a firestop assembly.

See also: Flammability of Polymers; Flammability of Structural Materials and Furnishings

Bibliography

American Society for Testing and Materials 1966 *Symposium on Fire Test Methods—Restraint and Smoke 1966*, ASTM Special Technical Publication 422. ASTM, Philadelphia, PA

American Society for Testing and Materials 1972 *Ignition, Heat Release and Noncombustibility of Materials*, ASTM Special Technical Publication 502. ASTM, Philadelphia, PA

Hilado C J 1973 *Flammability Test Method Handbook*. Technomic, Westport, CT

Hilado C J (ed.) 1974 *Flammability Handbook for Plastics*, 2nd edn. Technomic, Westport, CT
Johnson P R 1974 A general correlation of the flammability of natural and synthetic polymers. *J. Appl. Polym. Sci.* 18: 491–504
Kanury A M, Alvares N J, Martin S B 1977 *Flammability Testing of Polymers*. Manufacturing Chemists Association, Washington, DC
National Materials Advisory Board 1979 *Fire Safety Aspects of Polymeric Materials—II: Test Methods, Specifications and Standards*, Publication NMAB 318-2. National Academy of Sciences, Washington, DC
Parker W J 1977 *An Investigation of the Fire Environment in the ASTM E84 Tunnel Test*, National Bureau of Standards Technical Note 945. US Government Printing Office, Washington, DC
Paul K T 1987 Fire testing in polymers. *Prog. Rubber Plast. Technol.* 3(2): 23–43
Rowan J W, Lyons J W 1978 The importance of externally imposed heat flux on the burning behavior of materials. *J. Cell. Plast.* 14: 25–31
Smith E E, Harmathy T Z (eds.) 1978 *Design of Buildings for Fire Safety*, ASTM Special Technical Publication 685. ASTM, Philadelphia, PA
Underwriters Laboratories 1980 *Tests for Flammability of Plastic Materials for Parts in Devices and Appliances*, UL94. Underwriters Laboratories, Melville, New York

S. Kaufman
[AT&T Bell Laboratories, Norcross, Georgia, USA]

Flammability of Structural Materials and Furnishings

The flammability hazards of structural materials and furnishings are of great importance: they have been significant factors in many catastrophic fires such as the Cocoanut Grove nightclub fire in 1942, the Beverly Hills Supper Club fire in 1977 and the MGM Grand Hotel fire in 1980, and they continue to be significant factors in less newsworthy fires which occur in residences every week if not every day.

1. Fire Response Characteristics

Flammability hazards of materials can be viewed in terms of nine fire response characteristics: (a) smolder susceptibility, (b) ignitability, (c) flash-fire propensity, (d) flame spread, (e) heat release, (f) fire endurance, (g) ease of extinguishment, (h) smoke evolution and (i) toxic gas evolution. Smolder susceptibility is the tendency to undergo progressive combustion in the solid state without flame. Ignitability is the ease of initiation of burning by a small flame or spark. Flash-fire propensity is the tendency of a heated material to emit flammable gases and vapors which can propagate flame with extreme rapidity. Flame spread measures the progress of flame over a combustible surface. Heat release relates to both the amount of heat released from a unit quantity of burning material and the rate at which that heat is released. Fire endurance is a measure of the elapsed time during which a material or product maintains its design integrity under exposure to heat and fire. Ease of extinguishment is the facility with which a fire can be extinguished in a burning material. Smoke evolution is the tendency of a burning or smoldering material to produce visible mixtures of gases, aerosols and particulates which obstruct light transmission and vision. Toxic gas evolution is the production of gases during burning or smoldering which are poisonous or destructive to body tissues.

2. Structural Materials

The primary functions of structural materials are to support and maintain the integrity of a structure. Fire endurance is the most important fire response characteristic for load-bearing and fire-barrier materials. Building codes recognize this importance in classifying types of construction Fireproof, fire-resistant and noncombustible types of construction generally involve steel, iron, concrete and masonry; the concrete and masonry are reinforced by the steel and iron, and the steel and iron are protected from direct fire contact by the concrete and masonry. Unprotected steel performs well in terms of ignitability, flame spread, heat release, smoke evolution and toxic-gas evolution, but is limited in fire endurance. Special recognition is given to heavy timber construction, because the characteristic charring of wood at the surface exposed to fire provides a measure of protection to the underlying material in large wood cross sections. Ordinary wood-frame construction involves a material which has general acceptance because its flammability hazards are widely understood and considered in the design and occupancy of a structure. Unreinforced plastics are generally not used as structural materials because of their tendency to creep under load and heat; substantial reinforcement in the form of glass fibers and carbon fibers is required for even relatively small structural members. The use of combustible materials such as wood and plastics in construction increases the importance of smolder susceptibility, ignitability, flash-fire propensity, flame spread, heat release, ease of extinguishment, smoke evolution and toxic gas evolution in assessing flammability hazards. Wood and wood-based composites (plywood, laminated beams) are preferred to plastics for structural applications because both performance characteristics and flammability hazards are better understood as a result of long experience and can be predicted with greater confidence in both design of the structure and control of its occupancy.

Thermal insulation materials are incorporated in structures to reduce the energy requirements for heating and air conditioning. Inorganic insulation materials such as glass fiber and mineral wool batts and blankets are essentially noncombustible, but foam plastics can represent a significant fire hazard if not properly applied. Rigid polyurethane and polyisocyanurate foams present the least hazard when protected by noncombustible materials, as when they are used as the core of sandwich panels or covered by plaster. Rigid polyurethane foam has had an unfavorable history of fires when spray applied on interior surfaces and left unprotected.

3. Furnishings

Furnishings are designed with a view to utility, comfort and aesthetic considerations. The majority of materials used in furnishings are combustible materials and all nine fire response characteristics are important, each to a degree that varies with the material. Smolder susceptibility is more important in cellulosic fabrics and cushioning materials and in polymeric and elastomeric foam materials than in most other materials. Flash-fire propensity appeared to be more relevant to some early polyurethane foams than to most other materials. Smoke evolution may be more of a hazard with poly(vinyl chloride), polystyrene, acrylonitrile–butadiene–styrene and foam rubber than with many other materials. Toxic gas evolution is a potential hazard from all combustible materials because of their ability to produce potential oxidation products, including carbon monoxide, in a fire. Certain plastics have aroused additional concern because of the possible production of other toxic gases such as hydrogen chloride from poly(vinyl chloride) and hydrogen cyanide and nitrogen oxides from certain nitrogen-containing plastics, wool and leather. The evolution of heat, smoke and toxic gases from furnishings was a significant factor in the Beverly Hills Supper Club fire and in the MGM Grand Hotel fire.

The particular furnishings in which materials are used determine to a large extent their potential exposure to fire and their potential for harm. Approximately 39% of residential fires originate in the bedroom and about 36% in the living room. Upholstered furniture and bedding are the furnishings most likely to be ignited in a fire and their smolder susceptibility, ignitability and flash-fire propensity are important. They are furnishings which represent a major source of heat and other products of combustion and so their flame spread, heat release, smoke evolution and toxic gas evolution are also important. The susceptibility of certain combinations of upholstery fabrics and cushioning materials to smoldering is a well-known fire hazard. Smoldering can continue for relatively long periods of time, during which it may or may not kindle into flame, but during which it produces smoke and toxic gases. Smolder susceptibility is not a function of the fabric or cushioning material alone, but of the specific fabric–cushion combination. Cellulosic fabrics have been observed to be more smolder susceptible than thermoplastic fabrics in furniture upholstery, but this susceptibility is greatly reduced when the underlying cushioning material is resinated polyester batting or neoprene foam interliner. Thermoplastic fabrics exhibit no smolder susceptibility except when untreated cotton batting is the cushioning material. Smoldering combustion generally requires a porous fuel and charring appears to be a prerequisite for smoldering. Cellulosic fabrics such as cotton and rayon are, like wood, prone to char and smolder. Thermoplastic fabrics such as nylon and polyolefin inhibit smoldering by not forming a porous char and by melting down to help seal off areas against smoldering.

In ignitability, as in smolder susceptibility, the physical characteristics of the material are important. Thermal diffusivity and absorptance characteristics are more decisive factors in the rate of increase of surface temperature than differences in chemical composition. The effect of density, which influences thermal conductivity, volume specific heat and thermal diffusivity, appears to override any differences in chemical composition. A dark-colored material would probably ignite more rapidly than a light-colored or clear material in a fire.

Flash-fire propensity is of the greatest concern in occupied compartments. Some early polyurethane flexible foams were noted for their flash-fire propensity. Polychloroprene and poly(vinyl chloride) appear to be less prone to flash fires, since they do not readily yield flammable volatiles on heating.

Flame spread is more relevant to floor coverings, wall coverings and ceiling linings. These materials, which can be classified as interior finish, are discussed in Sect. 4. Flame spread becomes more relevant in furniture when the objects have at least one long dimension, as for sofas and beds.

Heat release is of greatest concern with materials which burn rapidly. In furnishings, untreated polyurethane flexible foam and latex foam rubber have been noted for high heat release. Polystyrene structural foam used in kitchen cabinet doors and in baby cribs and bathtub–shower enclosures of reinforced thermosetting polyester have been notable cases of high heat release.

Smoke evolution from furnishings has been of greatest concern with materials such as poly(vinyl chloride), polystyrene, acrylonitrile–butadiene–styrene and foam rubber.

4. Interior Finish

Floor coverings, wall coverings and ceiling linings are furnishings in that they are primarily selected for utility, comfort and aesthetic reasons, but they may be considered as structural materials to the extent that they are attached to part of the structure. Because of the large surface area exposed by these materials, flame spread is their most important fire response characteristic. Floor coverings are more likely to be exposed to small ignition sources such as smoldering cigarettes and lighted matches than are wall coverings and ceiling linings. In the event of a substantial fire, however, the ceiling linings and wall coverings are more likely to become involved because of the tendency of the hot combustion gases to rise toward, and rapidly heat materials in, the upper portions of an enclosure. Wainscoting, for example, is not as likely to contribute to flame spread as the upper half of a wall.

Carpet may be the most widely used floor covering material in residential and commercial occupancies. Many carpets are installed with an integral cushion backing or on a separate cushion pad. The addition of a carpet cushion pad under the carpet in many, if not most, cases may have a greater effect on flame spread than differences between carpets. Since the rate of flame spread is dependent on the rate at which succeeding surface areas are heated to their ignition temperature, the thermal characteristics of the carpet–underlay system may prove to be more important than the chemical composition of the carpet.

Various materials are used in wall coverings and ceiling linings. The paints and coatings used on walls and ceilings are important factors in response to fire. The flammability hazards of these materials are generally judged on the basis of their flame spread classification according to the ASTM E84 tunnel test. A flame spread classification of 0–25 is considered the least hazardous, followed by a flame spread classification of 26–75. The smoke-developed classification by this same test is gaining importance.

5. *Legal and Regulatory Aspects*

Product-liability litigation is one of the major economic factors in flammability hazard evaluations. The increasing number of claims, the increasing size of settlements and awards, and the increasing cost of litigation have all multiplied the cost of flammability hazards to manufacturers and their insurance companies. Opponents of such litigation contend that the substantial resources expended in litigation could be better spent in making the product safer. Proponents of such litigation believe that only the economic penalty of prohibitive jury awards is effective in forcing manufacturers to improve the safety of their products.

Some materials which have been the subject of a series of lawsuits, substantial lawsuits or both are rigid polyurethane foam used as thermal insulation, flexible polyurethane foam used as cushioning and poly(vinyl chloride) used as electrical insulation.

Regulatory action has generally paralleled litigation. Misuse and misleading advertising of rigid polyurethane foam have led to action by the Federal Trade Commission, which resulted in a consent decree involving payment of US$5 million by polyurethane manufacturers. Hazardous flexible polyurethane foam has been banned from upholstered furniture sold in the state of California and more fire-safe polymeric foams have been developed to meet California state requirements. There has been considerable opposition to permitting the use of poly(vinyl chloride) in electrical conduits because of its potential for producing smoke and toxic gases.

See also: Construction Materials, Cellular; Elastomeric Foams; Plastics and Composites as Building Materials

Bibliography

Berl W G, Halpin B M 1979 *Human Fatalities from Unwanted Fires*, NBS GRC 79-168. National Bureau of Standards, Washington, DC

Best R L 1977 *Reconstruction of a Tragedy: The Beverly Hills Supper Club Fire.* National Fire Protection Association, Quincy, MA

Damant G H, Young M A 1977 Smoldering characteristics of fabrics used as upholstered furniture coverings. *J. Consum. Prod. Flammability* 4: 60–113

Hilado C J 1981 *Flammability Handbook for Plastics,* 3rd edn. Technomic, Westport, CT

Hilado C J, Huttlinger P A 1980 Toxicity of off-gases from furnishing materials: A review. *J. Consum. Prod. Flammability* 7: 189–99

Hilado C J, Murphy R M 1979 *Fire Response of Organic Polymeric Materials (Organic Materials in Fire: Combustibility)*, ASTM STP 685, American Society for Testing and Materials, Philadelphia, PA, pp. 76–105

Paul K T 1986 Fires, foams and furniture. In: Buist J M, Grayson S T, Woolley W D (eds.) 1986 *Fire and Cellular Polymers.* Elsevier, Amsterdam

C. J. Hilado and P. A. Huttlinger
[Product Safety Corporation, Sunnyvale, California, USA]

G

Geotextiles

The term geotextile is almost self-explanatory, meaning a textile material involved with the earth. The American Society for Testing and Materials (ASTM) defines a geotextile as "any permeable technical material used with foundation, soil, rock, earth or any geotechnical-related material as an integral part of a man-made project, structure or system." An even broader definition applied to geotextiles also includes impermeable textiles such as coated fabrics; impermeable coated fabrics and polymer films are called geomembranes.

The main functions of geotextiles are the following: (a) separation, (b) reinforcement, (c) filtration, (d) drainage, and (e) as a moisture barrier. To illustrate, geotextiles are used to separate different layers of earth, to protect drainage pipes and embankments, to consolidate damaged roads, to stabilize railroad beds and to protect waterways. Geomembranes are used as liners to contain either potable or contaminated water (landfills, toxic wastes and so on), or as covers to prevent evaporation of water.

The most commonly used fibers in geotextiles are (in decreasing order of use): (a) polypropylene, (b) polyester, (c) polyamide (nylon), (d) polyethylene and (e) other polymers. The fibers may be monofilament, staple or slit film, and are used in the form of woven, knit and nonwoven fabrics. The last category includes spunbonded materials.

1. Fibers

A major requirement of any fiber used in geotextiles is resistance to deterioration by microorganisms in the soil. All of the four most important fibers used in geotextiles satisfy this requirement. Also, all have sufficient strength and resistance to weathering to be satisfactory for most applications. All four fibers, however, have some weaknesses. Polypropylene has comparatively poor resistance to ultraviolet (uv) radiation and, for uses requiring exposure to uv light, the addition of a uv light absorber is usually required. Polypropylene and polyethylene have comparatively low melting points. Polyester fibers have comparatively poor resistance to hydrolysis in alkaline soils, whereas nylon fibers have relatively poor resistance to acid soils.

Most fibers used in geotextiles are melt spun; that is, the polymer is melted and then forced through an orifice. The fiber is solidified after extrusion by cooling and simultaneously or subsequently stretched by drawing. During stretching the molecules are oriented in the stretching direction, increasing the tensile strength and modulus and decreasing the elongation. The fibers can be produced as continuous filaments or cut into short staple fibers which may subsequently be twisted together to form spun yarns.

Another type of fiber used in geotextiles is called a slit-film fiber. It is produced by cutting or slitting films of polymer with a knife, an air jet or by other means, resulting in ribbonlike fibers.

It is important to choose a fiber with appropriate properties for the particular geotextile end use. Of course, the cost of the fiber and the fabric is an additional important factor in this choice.

2. Fabrics

Fabrics can be prepared from fibers or from yarns made from fibers by weaving or knitting, or they may be made in a nonwoven form. Nonwoven fabrics have historically been the commonest type of geotextile, followed by woven fabrics, with knitted fabrics used infrequently.

Nonwoven fabrics for geotextile use are usually manufactured by spunbonding, which includes melt-bonding, resin-bonding and needlepunch processes. Spunbonding is a continuous process in which fibers from one or more polymers are extruded from a spinneret or series of spinnerets and laid on a moving conveyor belt to form a continuous web. The desired fiber orientation is achieved aerodynamically, electrostatically or mechanically and the fibers are bonded together by thermal, resin or mechanical treatments before being wound into a finished roll. In the thermal treatment or melt-bonding process, the fibers are bonded together using heated rolls. The fibers may be a mixture of two fibers with different melting points or a bicomponent fiber with the sheath of the fiber having a lower melting point than its core. By the application of heat and pressure, the lower-melting material melts and fuses onto the higher-melting component. In resin bonding, a resin, such as an acrylic resin latex, is applied to the nonwoven web to bind fibers together at crossover points. In the mechanical treatment or needlepunch process, needles punch through a fibrous web held between a bed plate and a stripper plate, reorienting the fibers so that mechancial bonding is achieved by fiber entanglement.

Combination treatments are sometimes used in nonwoven fabric manufacture. For example, in order to achieve dimensional stability in a lightweight needlepunched fabric, an additional treatment by resin bonding or melt bonding may be given to the fabric. In another example, melt bonding may be followed by needlepunching. Melt-bonded fabrics are usually thin, typically 50–1000 μm, whereas needlepunched fabrics are thicker, typically 1000–5000 μm or more. Generally, 90% of the nonwoven fabrics used in geotextiles are in the 100–200 $g\,m^{-2}$ range, although some are as heavy as 2000 $g\,m^{-2}$.

Geogrids are another form of geotextile. They are relatively stiff, matlike materials with large open spaces (typically 10–50 mm) between the ribs that make up the structure.

Yet another form of geotextile is the geomembrane, which is usually a polymer film but may be a polymer coating over a textile fabric. North American sales of geomembranes in 1984 were estimated to be $8.3\times10^6\,m^2$ poly(vinyl chloride), $5.1\times10^6\,m^2$ Hypalon (chlorosulfonated polyethylene) and $3.7\times10^6\,m^2$ other polymers, which include polyolefins, nylon and various rubbers and combinations of polymers.

Geotextiles may be composites of various geotextile structures to obtain additive performance properties. Such composites may consist of a geomembrane and a geogrid, for example, or a sandwich of a geogrid between two fabrics so as to enhance the strength of the structure while retaining soil separation and water filtration properties. A further example is lamination of a geomembrane with a fabric, giving added resistance to puncture, tear propagation, abrasion and tensile failure. There are an infinite number of combinations of composite structures that can be produced for geotextile use.

Geotextiles can be used in most areas of geotechnical engineering: they are used when they bring about performance and/or cost enhancement. The functions that geotextiles can perform are separation, reinforcement, filtration and drainage. They can also be used as moisture barriers. Although the last three could be lumped together as hydraulic functions, since specific types of fabrics are often required for these functions, it seems appropriate to discuss them separately. It should be emphasized, however, that in many applications geotextiles serve a combination of functions.

3. *Separation*

Geotextile separation involves the introduction of a fabric barrier between dissimilar materials. Separation is important in nearly all uses of geotextiles. Examples of the use of geotextiles for separation purposes are: between subgrade and stone base in paved and unpaved roads and airfields, between subgrade and ballast in railroads, between foundation and embankment soils for earth and rock dams, between foundation soils and retaining walls and beneath sidewalk slabs, curb areas, parking lots and athletics fields.

A typical example is the separation of stone aggregate from subsoil in an unpaved road. Two simultaneous events tend to occur when stones are placed on soil:

(a) the soil fines attempt to enter the voids of the stone, thereby reducing its drainage capability; and

(b) the stone attempts to intrude into the soil, diminishing the strength of the stone.

To hinder both events by using a geotextile to separate stone and subsoil, therefore, the geotextile must fulfill several requirements. One is that the size of the openings in the geotextile should be small enough to prevent the subsoil from penetrating into the voids of the stones placed over it and, of course, prevent the overlaid stones from penetrating the subsoil. Another requirement is that the geotextile must have sufficient permeability to prevent pressure buildup under the separation layer if the ground water level should rise. Also, any water in the stone layer should be able to drain out through the geotextile. Other requirements of geotextiles used for the separation function are sufficient burst strength, tensile strength, puncture resistance and impact resistance for the various conditions to which the geotextile is exposed. There are, of course, minimum requirements for all these properties for any function or functions for which a geotextile is used.

4. *Reinforcement*

Geotextile reinforcement involves the use of a fabric to distribute a load over a larger area of soil, rock and so on in order to avoid local overloading. Examples of applications in which geotextiles are used to reinforce weak soils and other materials include strengthening soft soils for unpaved roads, airfields, railroads, landfills and athletics fields; constructing fabric-reinforced walls; reinforcing embankments; aiding in the construction of steep slopes; reinforcing earth and rock dams; and preventing the puncture of geomembranes by subsoils, landfill materials and stone bases.

An unpaved road provides a good example of the use of a geotextile placed between the subsoil and the rock aggregate for the reinforcement function. Calculations can show the thickness of stone required over a subsoil without a geotextile and with a geotextile between the stone and subsoil. From these calculations, it can be determined whether use of a geotextile will effect a saving. High-modulus fabrics provide greater rock aggregate reduction, less rutting and better system performance than low-modulus fabrics. Generally there is a greater saving in stone aggregate as the subsoil becomes softer; also, heavier and stronger fabrics usually result in greater savings. Properties of the geotextile important for the reinforcement function are soil-to-geotextile friction, creep and stress relaxation.

In addition to their use in unpaved roads, geotextiles are used for reinforcement purposes over the subsoil in the construction of new paved roads. Where the road is to be paved with asphalt or concrete, the geotextile cannot be placed in excessively yielding soil subgrade or the road surface will crack after being subjected to a few loads. The use of geotextiles in paved roads may serve the functions of separation and/or filtration in addition to reinforcement.

Geotextiles are also used as overlays in resurfacing asphalt roads that have excessive cracks (called reflective cracking). Such resurfacing is usually done with bituminous (asphalt cement) overlays ranging in thickness from 2 cm to 10 cm. First, the cracks are cleaned and filled. Then the road surface is coated with an asphalt sealant and the fabric is placed over the sealant. Finally, a hot mix overlay is placed over the fabric. With this use of geotextiles, the moisture barrier function of the asphalt-sealed fabric appears to be an important function in addition to the reinforcement function. The moisture barrier prevents the water from seeping up into the road surface, thereby causing damage. For reflective cracking, needled nonwoven fabrics have predominantly been used; these are mainly composed of polyester, polypropylene and glass. If the moisture barrier function is predominant, it can be seen why fabrics composed of these fibers are used since they saturate very well with sealant. However, if reinforcement is the dominant function, high-modulus fabrics give superior results.

A special case of the reinforcement function for geotextiles is the stabilization function, which is important in wall embankments, foundations and slopes. Geotextiles used to stabilize embankments are normally laid horizontally with the exposed soil surface not covered. In most applications of geotextiles in embankments it is impossible to use a single roll width, so that overlap or seaming must be used. Although either method is satisfactory, the labor cost of sewing with a portable hand sewer is low, so this method is commonly used. Geotextiles are used in foundation stabilization by placing horizontal layers of fabrics in the soil.

5. *Filtration*

Filtration, in connection with a fabric-to-soil system, means that water can move across the plane of the geotextile, while soil is retained indefinitely on the upstream side without clogging the fabric. Examples of the use of geotextiles for filtration are beneath stone base for unpaved and paved roads and airfields, beneath ballast under railroads, around crushed stone surrounding underdrains and in tile fields, as a snow fence, as protection for chimney drain material, between backfill soil and voids in retaining walls and as a filter beneath stone riprap and precast blocks.

The voids in a filtering geotextile must be large enough to allow water to flow through but sufficiently small to prevent "soil piping" or passage of finer soil particles through the fabric along with the flowing water and ultimate collapse of the soil structure. In order to assess the suitability of fabrics for filtration, therefore, it is necessary to be able to measure the open area of the fabric and its crossplane permeability. Since many geotextiles, especially nonwoven fabrics, compress under loads, permittivity, which is the ratio of crossplane permeability (Darcy's coefficient of permeability) to fabric thickness, is used in evaluating the hydraulic properties of geotextiles. Permittivity is defined by the ASTM as "the volumetric flow rate of water per unit cross-sectional area per unit head under laminar flow conditions in the normal direction through a geotextile." Thus,

$$\psi = k_n / t$$

where ψ is the permittivity, k_n is the permeability coefficient normal to the fabric and t is the fabric thickness. Using Darcy's law,

$$\psi = q / \Delta h A$$

where q is the flow rate, Δh is the head lost (difference between heads at beginning and end of test) and A is the area of fabric under test. Typical permittivity values for geotextiles range from $0.02\ s^{-1}$ to $2.2\ s^{-1}$ and permeability ranges from $0.0008\ cm\ s^{-1}$ to $0.23\ cm\ s^{-1}$.

One of the important uses of geotextiles involving filtration is behind retaining walls. Behind conventional reinforced-concrete retaining walls there should be a vertical layer of sand serving as a flow path that allows water from the soil to escape into an underdrain system or weep holes. Without such a sand layer, hydrostatic pressures will build up and, together with the horizontal soil pressure, could easily cause overturning or movement of the wall. The sand must stay free-draining or it will become useless. Therefore, it must be protected either by a graded soil filter which is expensive and difficult to replace, or by a geotextile.

Another important use of geotextiles involving filtration is around underdrains; in particular, they make excellent replacements for graded soil filters around perforated pipe underdrains. Major application areas are highways, airfields and railroads, Another application is the use of geotextiles beneath erosion control structures.

6. Drainage

Some examples of the use of geotextiles in drainage situations are as a chimney drain or a drainage gallery in an earth dam, as a drain behind a retaining wall or beneath railroad ballast, as a water drain beneath athletics fields, as a replacement for sand drains, as a capillary break in frost-sensitive areas and for salt migration in arid areas.

The variation in drainage of different fabrics is great. Fabrics may be listed in order of increasing in-plane drainage capability as follows: woven slit film, woven monofilament, nonwoven–heat set or resin bonded, nonwoven–needled and thick hybrid systems, particularly geocomposites.

An important hydraulic property of geotextiles used for drainage is transmissivity, the product of the in-plane coefficient of permeability and the thickness of the fabric. Thus,

$$\theta = k_p t$$

where θ is the transmissivity, k_p is the in-plane permeability coefficient (hydraulic gradient) and t is the thickness of the geotextile.

7. Moisture Barriers

A geotextile used as a moisture barrier, as the term implies, is one that prevents or slows down the passage of water either in the liquid or vapor state. Such geotextiles, called geomembranes, may be either polymer sheets or fabric-reinforced polymer sheets. Some of the major uses of these geotextiles are as moisture barriers in pond liners, underneath landfills and in covers for ponds. Other uses include caps and closures for landfills, liners in canals and for damming the flow of water.

Probably the largest single use of geomembranes is in the area of liquid containment as pond liners. Ideally, a hole is dug, a geomembrane liner is placed in it and then it is filled with liquid. However, under most conditions the liner would need to be shielded from ozone, uv light, high temperatures, vandalism and the like, so that a covering layer of soil at least 30 cm deep is required. The soil layer extends up over the pit and over the liner anchorage area. The friction between the liner and cover soil must be evaluated and proper design features planned. A geotextile beneath the geomembrane, which may be placed directly on the prepared soil subgrade before the liner is put in, has many advantages. It provides a clean working area for making field seams and added puncture resistance when loads (either during construction or from the contained material) are applied. It can add frictional resistance to the liner–soil interface, thereby preventing excessive stresses on the liner. A well-designed geotextile allows lateral and upward escape of subsurface water and gases that rise up beneath the geomembrane during its service life.

The choice of geomembrane material is an important one. For containment of water, poly(vinyl chloride) has been found to be satisfactory and is economical; however, unless it is protected by covering with soil, it can become brittle from loss of plasticizer and action of the sun. An additional consideration is the resistance of the adhesive, if one is used, for seams. If the seams are made by welding (melting of the polymer), then an additional adhesive need not be used; however, it is prudent to test the resistance of the seams to the liquid to be contained. High-density polyethylene is often the polymer of choice for geomembranes used to contain liquids due to its relatively good resistance to chemicals. Since, with polyethylene, seaming is usually accomplished by extrusion welding or heat fusion, no other materials such as adhesives are added.

Geomembranes are also added as liners for landfills and surface impoundments. When used for the disposal of hazardous wastes, such facilities are required by the US Environmental Protection Agency (EPA) to have two or more liners and a leachate collection system above (in the case of landfill) and between such liners.

Another use of geomembranes is for pond covers. There are a number of important reasons why liquid containment structures should be covered. These include losses due to evaporation of water and chemicals such as chlorine used in water treatment, reduced air pollution from reservoirs holding chemicals, increased safety against accidental drowning and protecting reservoirs from such natural pollution as animal excretions.

Geomembranes are used to cap and close landfills. Often these geomembranes are used in conjunction with soil covers; the EPA requires the use of clay to cap landfills containing hazardous materials.

When geomembranes are used as liners for canals, they are usually covered with 300–600 mm of soil. Poly(vinyl chloride) is the most widely used geomembrane for canal liners because of its low cost, but it may be unsuitable in hot desert areas.

8. Physical and Mechanical Test Methods

Many test methods for geotextiles are borrowed from existing clothing and industrial textile test methods, which are standardized as ASTM methods. Some tests, however, are not yet standardized. Work towards standard tests is being conducted worldwide; in the USA, ASTM Committee D35 has been set up to work on geotextiles.

The mass per unit area of a fabric is measured by

ASTM Method D3776. The thickness of a fabric is measured by ASTM Method D1777; compressibility of a fabric is also measured under various pressures described in ASTM Method D1777.

Tensile strength measurements are made by ASTM Method D1682. There are a number of variations permitted including wide-width fabric samples, which are often used especially with those fabrics that exhibit a "necking down" or a serious Poisson's ratio effect under tensile stress. A round-robin testing program of ASTM Committee D35 showed that the Poisson's ratio effect can be reduced to an insignificant level if the width-to-length or inverse aspect ratio of the test specimen is equal to or greater than 2.0. Committee D35 has therefore drafted the wide-width strip tensile test, specifying a specimen inverse aspect ratio of 2.0 (200 mm width by 100 mm gauge length).

The strength at failure is normally referred to as the tensile strength. The maximum elongation or strain at failure is normally referred to as elongation. The work done per unit volume before failure is usually taken as the area under the stress–strain curve and is called the toughness. The tensile modulus, or the slope of the initial portion of the stress–strain curve, is referred to as the stiffness. Several choices for defining modulus are available:

(a) initial tangent modulus, used where the initial slope is linear;

(b) offset tangent modulus, sometimes used when the initial slope is very low (the *y* axis is shifted to the right where it meets the downward extension of the linear portion of the response curve); and

(c) secant modulus, obtained by drawing a line from the origin to the curve at a given strain, measuring its slope and reporting this slope as the modulus at that strain.

Many workers have shown that tensile test results of in-soil testing of some geotextiles may be quite different from out-of-soil testing. To obtain realistic results, tensile tests should be conducted on the geotextile confined by the soil in which the geotextile is to be used, or similar soil.

Both tensile strength and elongation at break can vary considerably, depending on the direction in which the force is applied to the fabric. Biaxial tensile tests have been conducted in an attempt to duplicate actual conditions to which a geotextile is exposed. Results may be quite different under various restraints to contraction with such tests compared with uniaxial tests. Uniaxial tests yield much higher values of strain at rupture and a lower stiffness than biaxial tests including tests with prevented cross contraction.

Burst strength can be determined by the Mullen burst test which is described by ASTM Method D3786. In this test, the fabric is subjected to pressure from an inflatable rubber membrane until the fabric bursts.

There are three common tear tests for uncoated fabrics: the trapezoid tear test, the tongue tear test and the Elmendorf tear test. Nonwoven fabrics are tested by ASTM Method D1117, woven fabrics by ASTM D2261, D2262 and D1424. The trapezoid tear test is usually the method of choice.

The impact resistance of uncoated fabrics can be measured by attaching an impacting cone called a Spencer impact attachment to the pendulum arm of an Elmendorf tear apparatus: the energy required for the cone to penetrate through the fabric specimen is measured. Since the limit of the test does not permit the testing of heavy nonwoven fabrics, impact pendulum testers for metals are used according to ASTM Method A370 with the test specimen holder modified to handle fabrics rather than notched metal bars. The impact resistance of a coated fabric or geomembrane can also be measured with a Spencer attachment to the Elmendorf test apparatus. The following three ASTM methods may also be used for geomembranes: ASTM D1709 (free-falling dart), ASTM D3029 (falling weight) and ASTM D3998 (pendulum).

A puncture test to measure resistance of a geotextile to rocks or sticks rather than the resistance to dynamic impact is described under ASTM D3787, with a blunt-ended metal rod substituted for the puncturing ball. For testing a geomembrane, the closest ASTM Method is D2582. An ASTM method for measuring soil-to-fabric friction is in the subcommittee stages.

The effect of sustained loading, that is, the creep of geotextiles, is usually measured by hanging weights on the fabrics and measuring fabric elongation with time. Creep is greater with unconfined geotextiles than with in-soil geotextiles, showing that unconfined creep tests might be misleading. ASTM creep tests, as is the case with friction testing, are only in conceptual stages.

The abrasion testing of geotextiles is described by ASTM D1175, which covers the following six procedures: inflated diaphragm; flexing and abrasion; oscillatory cylinder; rotary platform, double head; uniform abrasion; and impeller tumble. ASTM Method D1175 has been replaced by Methods D4157 and D4158; other abrasion tests are ASTM D3884, D3885 and D3886. The rotary platform, double-head (Taber Tester Model 503) method is the most commonly used abrasion test for geotextiles, although ASTM now seems to be headed toward an oscillating sand paper block method.

9. *Tests for Hydraulic Properties*

Although test methods for hydraulic properties of geotextiles are not standardized to the extent of those for the physical properties, test methods are available.

The size of the openings in geotextiles can be determined by both direct and indirect methods. For knitted and woven geotextiles, analysis of magnified photographs of top view sections can be used, especially in the case of monofilament fabrics, to determine pore size. Another optical means to measure the percentage open area of such fabrics is to shine a light through the fabric and project the light onto a surface; the size of the projected light can be measured with a planimeter or by crosshatching the surface, like graph paper. These test methods are not applicable to nonwoven fabrics, so other methods are used. One method utilizes an image analyzer that can measure the actual distance between fibers. The apparent opening size or equivalent opening size of geotextiles can be determined by dry sieving glass or soil particles of uniform and known sizes. The US Army Corps of Engineers has developed such a test by sieving glass beads of known sizes, using successively finer fractions. Although this test has poor reproducibility, it is widely used, especially with nonwoven fabrics for which accurate optical measurements cannot be made. A more reproducible test carried out in Europe involves a wet sieving test of well-graded soil with alternate flows of water. The test, called hydrodynamic sieving, was developed to determine the filtration opening size of nonwoven geotextiles; it measures the largest soil particles that are able to pass through a textile with the percolation of water. Another test involves filtration of soil particles through the geotextile and collection of these particles downstream on a filter paper. Theoretical methods can also be used to estimate the size of the openings in nonwoven fabrics based on the following four basic geotextile parameters: mass per unit area, thickness, fiber diameter and polymer density.

The test method for permeability is ASTM D4491, which allows the use of either a constant or falling head test apparatus. The transmissivity of a geotextile may be measured with a number of devices for determining in-plane hydraulic properties.

The blocking, clogging and soil retention characteristics of a geotextile can be determined using a crossplane permeability test. Probably the most meaningful test for blocking or clogging of a geotextile is to take a sample of soil at the site and place the soil above the geotextile, which is fixed in a test cylinder, and then measure the rate of water flow under constant head over a long period of time.

For geomembranes, a water vapor transmission test is often carried out using ASTM Method E96. In this test a sample of geomembrane is sealed over an aluminum cup, which is maintained at 100% relative humidity (using water within it) or 0% relative humidity (using a salt desiccant within it). The entire assembly is then placed in a controlled relative humidity chamber, and the weight change is measured over a period of time, usually between 3 d and 30 d.

10. Tests for Environmental Resistance

There are a number of environmental tests for geotextiles. These include tests for resistance to chemicals, temperature, light, weather, bacteria, fungi and aging.

ASTM Method D543 covers the testing of the resistance of plastics to 50 chemical reagents, including provisions for testing the changes in weight, dimensions, appearance and strength properties for various exposure times and temperatures.

ASTM Method D794 describes high-temperature testing for plastics, whereas D746 covers testing for the effects of cold temperatures.

ASTM Method D1435 describes conditions for outdoor weather testing of plastics. Tests for resistance to acclerated weathering using an artificial light source which simulates natural sunshine are described in ASTM Method D3334 for testing polypropylene and polyethylene. The light source usually used is a xenon arc described in ASTM D2565.

Resistance of membranes to ozone can be tested by ASTM Method D518, which is primarily a test for surface cracking of rubber.

Resistance of geotextiles to biodeterioration can be tested by soil burial; American Association of Textile Chemists and Colorists (AATCC) Test Method 30—1981 or Federal Standard No. 191, Method 5762 may be used.

Although there are no standard aging tests, geotextiles could be buried in soil and their long-term aging determined by successive testing of specimens at selected times.

See also: Fibers and Textiles: An Overview; Spunbonding of Polymeric Fibers

Bibliography

Chenard M 1987 A new stitch in time. *Dimensions* 2(3): 52–5

Giroud J P, Noiray L 1981 Design of geotextile-reinforced unpaved roads. *J. Geotech. Eng. Div.* 107: 1233–54

Hoare D J 1986 Geotextiles—Compatibility and use. *Civ. Eng. (London)* 9–48

Koerner R M 1986 *Designing with Geosynthetics*. Prentice-Hall, Englewood Cliffs, NJ

Koerner R M, Bove J A 1983 In-plane hydraulic properties of geotextiles. *Geotech. Test. J.* 6(4): 190–5

Kuhn D J 1986 To leak or not to leak: The role of a geosynthetic lining system. *ASTM Stand. News* 14(5): 36–8

McGown A, Andrawes K Z, Kabir M W 1982 Load–extension testing of geotextiles confined in soil.

Proceedings of the second international conference on geotextiles. *IFAI Publ.* 3: 793–8
Rollin A L 1986 Filtration opening size of geotextiles. *ASTM Stand. News* 14(5): 50–2
Veldhuizen J C, Van Zanten R V (eds.) 1986 *Geotextiles and Geomembranes in Civil Engineering*. Wiley, New York

T. F. Cooke
[Textile Research Institute, Princeton, New Jersey, USA]

Glass-Reinforced Plastics: Thermoplastic Resins

Glass-fiber-reinforced thermoplastics (GFRTP) fall into two distinct categories: injection-moldable compounds in which the reinforcement is in the form of short fibers and thermoformable sheet materials containing either continuous fibers or relatively long chopped strands. The former are well-established materials which may be molded into complex load-bearing components, such as gears and small machine parts, while the latter have only recently been made available and are intended to be shaped, by hot stamping, into parts of similar form to metal pressings.

1. *GFRTP Injection Molding Compounds*

The molding compound is prepared by melt blending the polymer with glass fibers. The most common process is to charge a dry blend of chopped glass strand and polymer granules into a screw extruder, which feeds a "spaghetti" die. The extrudate is chopped, either hot or cold, into pellets, which may then be injection molded. The compounding process disperses the glass strands into individual filaments and breaks them into short lengths, typically less than 0.5 mm. In the injection molding operation, the pellets are remelted and injected at high pressure (~50–200 MPa) into a cold mold. This process may cause further fiber breakage. Alternative compounding processes are available which retain longer fibers (~1–6 mm) in the molding granules; these compounds offer enhanced properties but are not widely exploited.

The molded material consists of an array of short fully dispersed fibers of different lengths, oriented in a pattern determined by the flow of material into the mold. The fiber orientation is complex, generally parallel to the flow direction in convergent passages but normal in divergent regions. Frictional drag at the mold walls also induces a through-thickness variation, especially in the thin platelike sections which are very common in injection moldings.

Glass-fiber reinforcement enhances both stiffness and strength at the expense of ductility and, to some extent, toughness. The in-mold shrinkage and thermal expansion are greatly reduced and the dimensional stability and elevated temperature performance of the material are very greatly improved. The material is anisotropic, however, on account of the flow-induced fiber orientation.

The short fibers can be stressed only by a shear transfer mechanism from the strained matrix. The tensile stress in the fiber builds up, from zero at the fiber ends, over a transfer length l_t, which depends on the average matrix strain ε_c and the effective shear strength of the fiber–matrix interface τ and is given by $l_t = \varepsilon_c E_f r_f / 2\tau$, where E_f is the tensile modulus and r_f the radius of the fiber. If the fiber length l_f exceeds $2l_t$, then the central portion is strained to the same extent as the matrix; where $l_f \gg l_t$, reinforcing efficiency approaches that of a continuous fiber but, where $l_f < 2l_t$, it is less than 50%. Ideally, l_f should approach $10l_t$, when the efficiency is about 90%. Reinforcing efficiency is also reduced if the fibers are misaligned relative to the principal stress direction. A planar random array is only 30% as effective as an aligned array but would, of course, give planar isotropy. In practical terms, τ is limited by the shear yield strength of the matrix (typically ~35 MPa for a polyamide), so that, for fibers of 12 μm diameter, l_t would be of the order of 60 μm at a strain of 1% and 120 μm at 2%. This indicates an ideal fiber length of 1.2 mm for a 90% reinforcement efficiency, assuming perfect alignment. In commercial materials the average fiber length is often only 150–200 μm, so that efficiency is comparatively low. It is desirable therefore that the bond between matrix and fiber should be as effective as possible. This is accomplished by the use of silane-based coupling agents, which are incorporated into the glass-fiber finish, and sometimes by incorporation of comonomers into the thermoplastic (especially in the case of polypropylene) to enhance fiber–matrix adhesion. The properties of some typical GFRTP materials are given in Table 1.

2. *Thermoforming Sheet Materials*

These materials consist of glass-fiber reinforcement impregnated with a thermoplastic resin. The reinforcement may consist of unidirectional continuous fiber strands (UD), woven roving (WR), continuous random mat (CRM), chopped strand mat (CSM) or a combination of these forms. Resins currently being exploited include polypropylene, polyamides, polysulfones, polyimides and, more recently, polyether ether ketone (PEEK), the latter three for higher-temperature applications.

These materials are intended to compete with wet lay-up systems and with the sheet molding compounds (SMC) and "prepreg" based on thermosetting resins. The principle of fabrication is to

Table 1
Typical properties of GFRTP materials (molded test bars)

Material	Density, ρ (10^3 kg m^{-3})	Flexural modulus, E (GPa)	Tensile strength, σ (MPa)	HDT[a] (°C)	In-mold shrinkage, SK (%)
Unfilled polymers					
polypropylene	0.91	1.35	34	60	1.8
polyamide 6.6	1.14	3.2	80	75	2.0
polycarbonate	1.20	2.3	60	140	0.7
Filled with 20 vol.% fiber					
polypropylene	1.14	5.7	45	130	0.35
polypropylcnc (chemically coupled)	1.14	5.7	103	155	0.35
polyamide 6.6	1.46	10.0	210	260	0.50
polycarbonate	1.45	9.0	135	160	0.20

a Heat distortion temperature

preheat the sheet and then to shape it by pressing, or stamping, between cold dies. This offers the possibility of a faster production rate, since there is no chemical curing operation as in the competing processes. This process is comparable with metal pressing and uses similar equipment. The replacement of metal components by reinforced plastics offers the possibility of weight reductions, improved corrosion resistance and lower total energy consumption, especially in automobile applications.

The constitution of the material is determined by the mechanical property and surface finish requirements and by the amount of flow necessary during the shaping operation. Continuous aligned strands and woven cloth give the highest mechanical properties, but limited flow possibilities, so that materials containing this form of reinforcement would be suitable only for panels of shallow curvature. For more complex shapes, especially where stiffening ribs and flanges are to be formed, CSM and/or CRM should be used. The CSM allows the maximum flow, but there is a danger of nonuniform glass distribution, as only the polymer may flow into the extremities of the mold. The CRM type of reinforcement gives much greater uniformity of fiber distribution.

The sheet material may be manufactured by melt-compounding chopped strand and polymer and extruding or calendering into sheet, or by passing a sandwich of reinforcement and thermoplastics film through a hot calender-type mill. The surface finish may be enhanced by placing unreinforced layers of thermoplastic at one or both surfaces of the pack. Currently available products are based on polyamide 6 and polypropylene matrices with 40–50% of chopped or continuous random glass strand. They are less stiff (flexural modulus 5–6 GPa) than SMC due to the lower glass content and matrix stiffness but are generally tougher. A material based on continuous carbon fibers in PEEK resin has recently become commercially available, but no glass fiber version using this matrix system has yet been offered.

See also: Composites: Applications Overview; Glass-Reinforced Plastics: Thermosetting Resins; Thermoplastic Elastomers

Bibliography

Broutman L J, Krock R H (eds.) 1967 *Modern Composite Materials*. Addison-Wesley, Reading, MA
Hull D 1981 *An Introduction to Composite Materials*. Cambridge University Press, Cambridge
Morton-Jones D H 1989 Fiber reinforced plastics. *Polymer Processing*. Chapman and Hall, London, Chap. 12
Ogorkiewicz R M (ed.) 1974 *Thermoplastics: Properties and Design*. Wiley, New York
Parratt N J 1972 *Fiber Reinforced Materials Technology*. Van Nostrand Reinhold, London
Titow W V, Lanham B L 1975 *Reinforced Thermoplastics*. Applied Science, London

M. G. Bader
[University of Surrey, Guildford, UK]

Glass-Reinforced Plastics: Thermosetting Resins

Glass-reinforced plastics (GRPs) consist of fibers of low-alkali E-glass embedded in a thermosetting plastics matrix, such as polyester or epoxide resin. During fabrication, glass fibers are incorporated into the liquid thermosetting resin to which catalyst and hardener have been added. These cause the resin to polymerize and, after curing, give a solid plastic matrix reinforced with glass fibers. The fibers enhance the low stiffness and strength of the resin, whose main purpose is to transmit the load into stiffer, brittle fibers and to protect them from damage.

Single E-glass fibers, typically 8–15 μm in diameter, are easily damaged and difficult to handle. For protection, they are coated with a silane or polyvinyl acetate size, which binds the fibers into strands containing about 200 filaments and acts as a coupling agent to provide a good fiber resin bond. To facilitate fabrication of GRP components, glass strands are incorporated into rovings, mats and fabrics. Glass fiber rovings consist of up to 120 untwisted strands, usually supplied wound together on a spool and suitable for the unidirectional (UD) fiber reinforcement of resins. Woven rovings (WR) are glass fiber rovings woven into a coarse fabric, usually with a balanced square weave. Glass fabrics are woven on textile machinery from twisted glass fibers (glass yarn) and are available in several weaves such as plain, square, twill and satin. Chopped strand mat (CSM) consists of chopped glass strands about 30 mm long held randomly orientated in a mat by a small amount of resin binder.

1. Glass-Fiber-Reinforced Resins

The principal resins used in GRP materials are unsaturated polyester resins, with epoxides being used in more limited quantities for high-technology applications and phenolics, furanes and vinyl esters finding use for specialist applications. Polyesters are less dangerous to handle, easier to catalyze and much cheaper than the other resins. They may be cold cured and are suitable for low-pressure, ambient-temperature fabrication techniques. Epoxide resins have better mechanical properties than polyesters, with higher stiffness and strength, particularly at high temperatures, and lower shrinkage on curing. Phenolics and vinyl esters are used in high-temperature applications, with continuous temperatures up to 260 °C, and furanes are noted for their chemical resistance at high temperatures. These specialist resins must be cured at high temperature and GRP components require hot-press molding. The higher-performance resins are generally used with UD glass fibers and glass fabric reinforcement, while CSM and WR reinforcements are used almost exclusively with polyester resins. Recent surveys show that polyesters accounted for 92% of resin consumption in the USA and 99% in the UK, with the remainder being mainly epoxide. Thus, the important GRP materials in high-volume applications are CSM–polyester, WR–polyester and, to a lesser extent, UD glass–polyester. Glass fabric–epoxide and UD glass–epoxide composites are used in low-volume, high-technology applications.

Glass fiber content is related to reinforcement type and is shown, by weight, in Table 1 for the main GRP materials. A high fraction of aligned fibers (60–80% by weight) can be packed into a UD composite, while more resin is needed to impregnate all the fibers in CSM and WR laminates, giving glass contents as low as 25–35 wt% in CSM–polyester materials. In addition to the fibers, resin, catalyst and hardener, glass–resin composites may contain fire-retardant additives to reduce surface flame spread, ultraviolet absorbers to improve outdoor durability, pigments, dyestuffs and thixotropic additives which thicken resins to aid fabrication on inclined surfaces.

2. GRP Molding Compounds

Several GRP molding compounds are now available in which the resin and glass fibers are premixed before molding. Component fabrication is achieved by placing the uncured compound in a closed mold and hot-press molding. The three main types of molding compound are sheet molding compound (SMC), dough molding compound (DMC), sometimes referred to as bulk molding compound (BMC), and preimpregnated glass fiber sheet (prepreg).

Sheet molding compound consists of E-glass fibers reinforcing a mixture of catalyzed polyester resin and a mineral filler, such as particulate, chalk, limestone or clay. The filler thickens the resin sufficiently for uncured sheets of SMC to be stored, handled and cut to the correct size for molding. SMC usually contains chopped glass strands 20–50 mm in length randomly orientated in the plane of the sheet and is supplied in several nominal glass weights, containing 20 wt%, 25 wt%, 30 wt% or 35 wt% glass fiber, with filler contents of 30–50 wt%. Fillers reduce material costs and modify the handling characteristics of the SMC sheet and the flow properties in the mold. High-performance SMC materials may also contain aligned chopped glass fibers or continuous fibers, at glass contents up to 70 wt%. These high-performance materials, sometimes termed HMC or XMC, are being used in automotive and other demanding high-volume applications. Their further development and a rapid growth in their usage are expected.

DMC contains catalyzed polyester resin, up to 50% mineral filler and 15–25% short glass fibers, 3–12 mm in length. The usual filler is chalk though some resins are chemically thickened with alkali metal oxides. Shorter fibers and less glass than SMC means that DMC flows more easily and is thus used to make small articles of intricate shape by hot-press or injection molding.

Prepregs consist of sheets of glass fiber reinforcement about 1 mm thick preimpregnated with catalyzed resin and ready for hot-press molding. The reinforcement is usually UD fibers or glass fabric with epoxide rather than polyester resin and component thickness is achieved by laminating together several sheets in the mold. The glass fiber content is high, typically 50–80 wt%. Accurate alignment of fibers is possible with prepregs, because there is

Table 1
Typical short-term mechanical property data for GRP materials

	DMC	SMC	CSM–polyester	WR–polyester	UD glass–epoxide longitudinal	UD glass–epoxide transverse
Glass content (wt%)						
average	20	30	30	50	70	
range	15–25	20–40	25–35	45–60	60–80	
Modulus (GPa)						
tensile						
average	9	13	7.7	16	42	12
range	8–11	9–16	6–9	12–22	30–55	8–20
flexural						
average	8	11	6.3	13		
range	7–9	7–14	5–8	10–18		
shear	3	3	3	4	5	
Strength (MPa)						
tensile						
average	45	85	95	250	750	50
range	35–60	50–120	60–150	195–350	600–1000	25–75
flexural						
average	100	180	170	290	1200	
range	85–120	90–240	100–300	160–500	1000–1500	
in-plane shear						
average			80	95	65	
range			60–100	80–120	50–80	
interlaminar shear						
average		15	25	20	40	
range		12–20	20–30	10–30	30–50	
Impact strength[a] (kJ m^{-2})	20–40	50–75	40–80	100–200		

a Measured by Izod-type test, unnotched specimen, according to British Standard BS2782, Sect. 306A

minimal flow during molding and hence they are used mainly in high-technology applications. Prepreg tapes are also available for selective stiffening of complex moldings and for the fabrication of GRP tubes and vessels by a dry filament winding process.

3. Sandwich Materials and Laminates

Different types of GRP material are frequently combined together in laminates or with metals, wood or foamed plastics as sandwich materials. Glass fibers may also be combined with carbon or polyamide fibers in a hybrid composite. A composite structure may thus be built up with material properties tailored to design loads. A widely used GRP laminate consists of CSM combined with WR or UD fibers. The CSM–polyester plies improve the interlaminar shear strength and provide a corrosion barier to the WR or UD fibers which carry the main loads. To simplify fabrication, glass reinforcement may be supplied in a combination mat, consisting of CSM and WR or UD fibers stitched together. With UD reinforcement such as prepreg, or in filament wound structures, fibers may be orientated in different directions through the thickness. A common construction is a crossply laminate, with alternate plies orientated at 0° and 90° to a fixed axis, or an angle-ply laminate with plies at $\pm\alpha°$ for some fixed angle α.

GRP sandwich materials are used widely in panel applications where the main loading is flexural. An ideal sandwich panel should consist of thin stiff skins with a thick low-density or low-cost core. For low and medium-technology applications, GRP skins of CSM–polyester or WR–polyester are combined with cores of end-grain balsa wood or foamed plastic such as PVC or polyurethane. A typical construction for a meter square cladding panel might be CSM–polyester skins 3 mm thick with a balsa core 25 mm thick. For high-technology applications where weight saving is a priority, special low-density honeycomb core materials have been developed. These consist of a hexagonal cell structure arranged in a honeycomb pattern as shown in Fig. 1. The main honeycomb core materials are aluminum foil and a thin resin-impregnated paper. High-performance GRP skin materials are used with honeycomb cores, such as glass fabric–epoxide and crossply or angle-ply UD glass–epoxide laminates. A typical construction for a meter square aircraft floor panel might be crossplied skins 1 mm thick, with an aluminum honeycomb core 10 mm thick.

4. Applications of GRP

GRP materials are used in low-, medium- and high-stress applications in industries as diverse as boat building, chemical plant motor vehicles and aerospace. Important properties of these materials which have led to their wide usage are ease of fabrication into complex shapes, high strength-to-weight ratio, excellent corrosion resistance, good weathering properties, low taint and toxicity for foodstuffs and good thermal and electrical insulation. The selection of a particular material is determined mainly by material properties, costs and the suitability of fabrication and quality-control procedures for use in a particular industry. For large-scale production automated fabrication methods based on filament winding, pultrusion and SMC press molding are used, with hand layup and semiautomated methods such as resin injection with preforms retained for smaller numbers of large moldings.

In low-stress applications such as cladding panels, shower cabinets, small boats and meter boxes, GRP materials act as space-filling panels, supporting their own weight but not subjected to any significant external loads. Stress analysis is not usually carried out and the main design requirements are a good surface finish, good weathering properties and the ability to withstand accidental damage. The most widely used material is CSM–polyester with fabrication by hand layup, spray up or vacuum-bag molding. SMC materials are used for longer production runs of larger moldings, such as business machine cases and cladding panels, and DMC for smaller components, such as electrical switchgear, lampholders and meter boxes. GRP sandwich materials containing a balsa or foamed-plastic core with CSM–polyester skins are used for boats and for some of the larger panel applications.

In medium-stress applications, GRP materials may be subjected to significant loads of short duration or low sustained loads. Stress analysis is usually based on the short-term behavior of the material and product design may be governed by codes of practice or standards. Large design factors are imposed so that stresses in the material are often less than 10% of the short-term strength. Applications in this category include: pressure vessels and storage tanks with associated pipework, for use in chemical plant and the food processing industry; larger marine applications such as workboats, small hovercraft, minesweepers and small submersibles; and land transport applications such as rail coachwork and larger automotive moldings for car bumpers, body panels and truck cabs. Resins are usually polyester, reinforced by WR or UD fibers with CSM–polyester surfaces for improved weather and chemical resistance. Most of the vessels, tanks and the marine applications are fabricated by hand layup, with filament winding becoming increasingly important for chemical plant pipework, sewage pipes and cylindrical tanks and pressure vessels. SMC is used for the volume production of such items as car bumpers, lorry cab panels and rail seat shells. Structural GRP sections are becoming increasingly important, for example box beams, channels and tubes fabricated by pultrusion for use in frame structures, ladders, walkways and ducting, and hollow rotationally molded tapered sections for the volume production of GRP street lighting columns. Where low weight is required, as in storage tank covers, hovercraft body panels and rail carriage doors, GRP–foam sandwich materials are used with fabrication by hand layup or vacuum-bag methods.

The high-stress applications of GRP materials are in the aerospace and defense industries. Here, low weight is important and hence materials are highly stressed and used efficiently. Design factors are kept as low as 1.5 in some applications and detailed stress analysis is thus required. Typical applications include power boats, gliders, which are fabricated by hand layup using WR–polyester or UD glass–polyester, and helicopter rotor blades and rocket motor casings, which are filament wound usually with epoxide resin. Sandwich materials consisting of angle-ply prepreg GRP skins with foam or honeycomb cores find use in lightweight structures such as aircraft flooring, radomes and radar antennas.

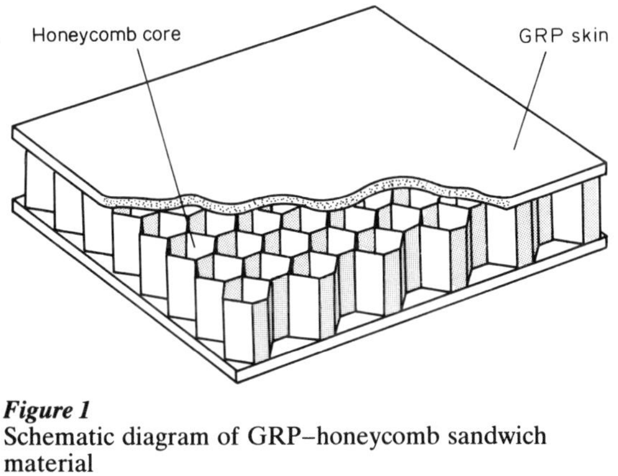

Figure 1
Schematic diagram of GRP–honeycomb sandwich material

5. Fabrication by Hand Laminating

Hand laminating or contact molding is the traditional fabrication technique for GRP materials. It is the method used for about 50% of British and 30% of US consumption of GRP materials. Glass fiber reinforcement, in the form of CSM, WR, fabric or combination mat, is placed on a mold, impregnated with catalyzed resin and consolidated by hand rolling or brushing. The required thickness is built up by adding further layers of reinforcement and resin.

Alternatively, chopped glass rovings are sprayed with a controlled quantity of resin from a special gun onto the mold and then consolidated by hand. The resulting "spray up" material has a random distribution of fibers in the plane of the molding, giving similar mechanical properties to CSM–polyester. After an initial gelation stage, the GRP molding may be removed from the mold and left to cure, during which time the material attains its full mechanical properties. Curing takes about a week at 20 °C, but is accelerated by postcuring at higher temperatures. Typical postcure times vary from 3 h at 80 °C to 30 h at 40 °C and depend on the resin used.

Contact-molded laminates have a single molded surface usually protected by a 0.5 mm thick resin layer termed the gel coat, which is brushed or sprayed onto the mold before the main laminate. For additional protection, the gel coat may be backed by a thin glass surfacing tissue. The main advantage of hand laminating is its versatility. There is no size limitation and metal inserts or additional glass reinforcement can be added where they are needed. Because it is a low-pressure, ambient-temperature process, inexpensive molds of wood, plaster or GRP may be used, which makes it suitable for small production runs. The main disadvantages are that moldings usually only have one good surface and the process is labor intensive with quality dependent on the laminator. Control of glass content, thickness and the cure schedule is poor, which can lead to variability in mechanical properties in a single molding or between moldings in a batch.

6. Automated Fabrication Techniques

To eliminate the variability in hand laminating and to permit longer production runs, several automated and semiautomated fabrication techniques have been developed. The semiautomated methods are based on matched die molding which permits critical control of thickness and glass content and produces a molding with two good surfaces. Cold-press molding makes use of relatively inexpensive tools in a hydraulic press with low pressures up to about 200 kPa. Glass fiber reinforcement such as CSM is cut or preformed to shape and placed by hand in the mold with the required quantity of catalyzed polyester resin. The mold is closed and the resin allowed to cure for about 5 min. Two variations on cold-press molding are vacuum-bag molding and resin injection. In the vacuum-bag method, the matched molds, or an open mold, are sealed inside a rubber bag, which is evacuated so that atmospheric pressure is exerted over the surface of the molding. Resin injection or transfer molding involves the injection of catalyzed resin into the closed mold containing a preformed glass fiber mat. Injection pressures are up to 200 kPa and may be combined with vacuum assistance to improve resin flow. This fabrication method is becoming increasingly important, since it allows accurate placement of reinforcing fibers in the molding and is also suitable for the production of large moldings such as car body shells.

In hot-press molding, matched metal dies are used and a hot curing polyester resin. Pressures are 1 MPa and temperatures are in the range 100–130 °C, with a mold cycle time of 2–5 min. Because of the higher pressures, entrapped air is forced out of the mold giving lower void content than with cold-press molding. In conventional hot-press molding, the operator must still handle the glass fibers and the liquid resin, but the process can be made cleaner by using molding compounds. These have a high viscosity and thus require higher molding pressures, typically 1–7 MPa, with temperatures in the range 120–170 °C. Prepregs are cut accurately and each ply is positioned in the mold to have the required orientation. Because of the short fibers, SMC and, more particularly, DMC are able to flow into the mold. With these materials, a weighed charge is positioned by hand centrally in the mold and on closing flows into the mold causing significant fiber reorientation. SMC has the better mechanical properties and is used in larger panel applications, with DMC retained for smaller, bulkier components. The use of in-mold coatings to improve surface finish and fast-curing resins giving cycle times as low as 1.5 min has meant that SMC is now widely used for the volume production of automotive components such as body panels and bumpers. Because of the good flow characteristics, DMC components may also be fabricated by injection molding.

Fully automated methods have been developed for the large-volume production of GRP pipe, sheet and rod or tube stock. In the filament winding process, glass fiber rovings are fed under tension through a resin bath onto a revolving mandrel. Several layers of reinforcement, with the same or different winding patterns, may be placed on the mandrel in a controlled way. The mandrel is removed after curing, which usually takes place at elevated temperatures. A variation, which eliminates the resin bath, is dry winding using UD prepreg tape. Filament winding is used for GRP pipes and cylindrical tanks and enables high glass contents to be achieved with close control of fiber orientation. In order to improve chemical resistance, the glass fiber reinforcement may be wound over a chemically resistant liner of CSM–polyester or a thermoplastic such as PVC or polypropylene.

Continuous laminating processes have been developed for the manufacture of GRP sheet for applications such as electrical insulating panels, corrugated panels and translucent roof sheets. A moving sheet of CSM or WR reinforcement is impregnated with catalyzed polyester resin and sandwiched between layers of plastic film. This is then passed continuously through a die, to impart the required shape, and a curing oven before being cut into lengths. Pultrusion is a similar process suitable for the production of long lengths of GRP in such forms as rods, tubes or I beams. In pultrusion, glass rovings are impregnated with resin from a bath or by injection and pulled through a shaped, heated die. Other forms of glass reinforcement such as WR or continuous-filament mat may also be pulled through the die with the UD rovings to improve the transverse mechanical properties of the pultruded section. Another recently developed automated process is rotational molding, which is used for hollow lighting and telegraph poles. Here the laminating pressure is derived from centrifugal force in the rotating mold. Methods are also being developed to automate hand lamination, for example, by machine-controlled spray up.

Automated fabrication techniques are suitable for volume production of good quality components through careful control of glass and resin contents, fiber orientation and the cure cycle. Many of these fabrication methods are suitable for control by microprocessor, which can lead to further cost savings and help overcome the styrene emission problem by eliminating manual operations. The main disadvantage is the high capital cost of the processing equipment. Some of the versatility of GRP materials is also lost, since these methods produce GRP stock in the form of pipes or sheets which require bonding or jointing to make into a component, whereas contact or press moldings are often integral GRP structures.

7. *GRP Material Properties*

On loading a GRP material such as CSM–polyester in a short-term tensile test, it is found that the behavior is approximately linearly elastic to failure, which occurs by a brittle fracture at a strain of 1.5–2.5%. The short-term mechanical properties may thus be characterized by an elastic modulus and a failure stress or strength. A more detailed study of failure shows that there is progressive damage to the material at stresses below the ultimate. The first sign of damage under tensile loading is transverse fiber debonding; that is, separation takes place between the resin and those fibers perpendicular to the load direction. In CSM–polyester and WR–polyester, this occurs typically at 0.3% strain and at a stress level of about 30% of the tensile strength. Under increased loading, the debonds initiate resin cracks and, as these cracks spread, more of the load is transferred to the fibers which eventually fracture or pull out at ultimate failure. Resin cracking occurs at approximately 50–70% of the tensile strength and is observed as crazing and by a change of slope in the stress–strain curve. In this article, ultimate failure stress data are given although it should be noted that, for many applications, failure data based on alternative criteria such as resin cracking or fiber debonding may be necessary. Limited data of this type are given, for example by Johnson (1978), or a suitable design factor may be applied to the ultimate strength data.

With a knowledge of the appropriate moduli and strengths, elastic design formulae may be used for stress and deformation analysis of GRP materials under short-term loads. For CSM–polyester and SMC materials, conventional isotropic design formulae may be used while, for WR, fabric or UD reinforcement, account must be taken of the anisotropy in stiffness and strength properties arising from fiber orientation. For GRP materials, it is not usually possible to quote a single value of modulus and strength because of the inhomogeneous structure of the material. Thus, in addition to the effects of anisotropy, mechanical properties also depend on the quantity and type of glass reinforcement and on the fabrication method used.

Comparative short-term mechanical property data are summarized in Table 1 for the principal GRP materials and typical physical properties are given in Table 2. The tables are abstracted from Johnson, who has collated material property data on several types of GRP material from UK resin and glass fiber suppliers, GRP fabricators and research laboratories. Table 1 gives mean values of modulus and strength for DMC, SMC, CSM–polyester, WR–polyester and UD glass–epoxide at a typical glass content for each type of material. Below each mean value approximate bounds are given for the quantities. These bounds are usually wide because they refer to a full range of glass weight fractions, to different resins and glass fibers, fabrication techniques and test methods.

Reference to Table 1 shows tht DMC, SMC and CSM–polyester, which all contain chopped fibers at low glass contents, have corresponding low moduli and strength values. The higher moduli and lower strength values for the molding compounds, compared with CSM–polyester at similar glass contents, are explained by the mineral filler which enhances the modulus but lowers the strength of the molding compound. Higher glass contents are found in WR–polyester laminates with average tensile modulus and strength about twice as high as the corresponding values for CSM–polyester. The data given in Table 1 refer to a fiber direction in a plain weave WR, with an equal number of rovings in the warp and weft directions. The highest glass contents are found in UD glass–epoxide composites and these materials show significant anisotropy in properties. Average tensile moduli range from 42 GPa along the fibers to 12 GPa transverse to them, with corresponding tensile strengths of 750 MPa and 50 MPa.

Table 2
Typical physical property data for GRP materials

	SMC and CSM–polyester	Fabric and WR–polyester	UD glass–epoxide
Density ($kg\,m^{-3}$)	1300–1600 (SMC, 1600–1900)	1500–1900	1800–2000
Linear thermal expansion coefficient ($10^{-6}\,K^{-1}$)	18–35	10–16	5–15
Thermal conductivity ($W\,m^{-1}\,K^{-1}$)	0.16–0.26	0.2–0.3	0.28–0.35
Specific heat ($J\,kg^{-1}\,K^{-1}$)	1200–1400		950
Maximum heat distortion temperature (BS2782, 102 G) (°C)	175	250	300
Dielectric strength, (BS2782, 201 A, C) ($kV\,mm^{-1}$)	9–12	13–16	
Permitivity at 1 MHz (BS2782, 207 B)	4.3–4.7	4.1–5.2	
Power factor at 1 MHz (BS2782, 205 B)	0.015	0.016	
Dry insulation resistance (BS2782, 204 A) (MΩ)	10^6	10^6	

Data on other GRP materials lie within the spectrum of properties shown in the table. Glass-fabric-reinforced resins and UD glass–polyester, fabricated by pultrusion or filament winding, usually have glass contents and mechanical properties intermediate between the WR–polyester and UD glass–epoxide values shown. Other points to note from Table 1 are the low interlaminar shear strength of all GRP materials and the differences observed between tensile and flexural properties, with flexural modulus usually below and flexural strength above the tensile values, as a result of the inhomogeneous structure of the material.

Under long-term loads, fatigue loads and when subjected to high temperature or a wet environment, GRP stiffness and strength properties are below the short-term values given in Table 1.

8. *Calculation of GRP and Sandwich Properties*

A wide range of theoretical formulae are available for predicting the mechanical properties of composite materials in terms of microstructure and fiber and matrix properties. Although these methods give satisfactory predictions of GRP properties, particularly moduli, they are usually only applicable to UD fiber reinforcement. The more rigorous results are also difficult to apply and often require correlation coefficients to be determined. It is therefore of interest to consider a simpler "rule-of-mixtures" approach which accounts for the main trends in the GRP properties reported here. Let E denote the Young's modulus of the composite and E_g and E_r the Young's moduli of the glass fibers and resin matrix, respectively. If v_g and v_r are the volume fractions of glass fiber and resin in the composite, respectively, then the Halpin–Tsai equation may be written in the modified form

$$E = \alpha E_g v_g + E_r v_r \tag{1}$$

where α is a parameter which depends on the efficiency of the reinforcement. Suitable values are $\alpha = 1$ for UD reinforcement, $\alpha = 0.5$ for WR, with E then referring to the modulus along the fiber directions, and $\alpha = 0.3$ for CSM, which has a random fiber distribution. Taking $E_g = 75$ GPa and $E_r = 3$ GPa as typical modulus values for glass fiber and polyester resin and assuming fiber volume fractions v_g of 0.2 for CSM laminates, 0.35 for WR laminates and 0.5 for UD fibers, the following estimates for the GRP modulus are obtained from Eqn. (1): $E = 6.9$ GPa for CSM–polyester, $E = 15.1$ GPa for WR–polyester and $E = 39$ GPa along the fibers in UD glass–polyester. Reference to Table 1 shows that these predicted values agree well with typical GRP data.

An analogous expression to Eqn.(1) may be used for estimating GRP strength properties, in the form

$$\sigma = \alpha \sigma_g v_g + \sigma_r v_r \tag{2}$$

Here, σ is the tensile strength of the composite, σ_g that of the glass fiber and σ_r the tensile stress in the resin at the fiber failure strain. Because of damage, the strength of a glass fiber in a composite may be only half the fresh drawn fiber strength (Parkyn 1970 Chap. 10). Taking as a nominal fiber strength $\sigma_g = 1500$ MPa, with $\sigma_r = 30$ MPa and the v_g and v_r values used previously, Eqn. (2) gives the following estimates of GRP tensile strength properties: $\sigma = 114$ MPa for CSM–polyester, $\sigma = 278$ MPa for WR–polyester and $\sigma = 765$ MPa along the fiber direction in UD glass–polyester. Again, these values are seen to be in reasonable agreement with measured data.

The mechanical property data and the prediction methods described refer to GRP materials containing a single type of reinforcement. In practice, GRP structures often contain several types of reinforcement laminated together or consist of angle-ply or sandwich materials and it is not feasible to provide detailed mechanical property data on each material combination. However, theoretical formulae are available which enable the designer to calculate the stiffness and strength properties of combined materials and these have been found to be in good agreement with measured properties. Laminated plate theory permits the calculation of the full anisotropic stiffness and strength properties of laminated composite materials, consisting of plies of different materials or the same material oriented in different directions. The analysis method is rigorous and requires a complete knowledge of the anisotropic stiffness and strength properties of each lamina. The calculations are lengthy and best carried out by a small computer program. A number of inexpensive microcomputer programs of laminate analysis are now commercially available (e.g., ESDU 1985, 1987, Tsai 1987). A simpler, strength-of-materials method for estimating combined laminate tensile and flexural properties is described by Allen (1969) and by Johnson (1978) for calculating the properties of sandwich structures. When the flexural rigidity of the core material is negligible compared with that of the stiffer skin materials, a simple design formula may be derived for the flexural rigidity per unit width D of a sandwich panel, which is an important quantity for the design of sandwich structures. For a symmetric sandwich with thin skins

$$D=\tfrac{1}{2}Eh(h+c)^2 \tag{3}$$

where E is the Young's modulus of the skin material in the load direction, h is the skin thickness and c the core thickness. Similar design formulae are available for calculating strength properties and the minimum weight design of sandwich panels (e.g., Johnson and Sims 1986).

See also: Composites: Manufacturing Overview; Glass-Reinforced Plastics: Thermoplastic Resins

Bibliography

Allen H G 1969 *Analysis and Design of Structural Sandwich Panels*. Pergamon, Oxford
Anon 1981 Materials survey. *Med. Plast. Int.* 11(1): 34–5
ESDU 1985 *Software for Engineers 2022, Stiffness and Properties of Laminated Plates*. ESDU International, London
ESDU 1987 *Software for Engineers 2033, Failure of Composite Laminates*. ESDU International, London
Holmes M, Just D J 1983 *GRP in Structural Engineering*. Elsevier, London
Hull D 1981 *An Introduction to Composite Materials*. Cambridge University Press, Cambridge
Johnson A F 1978 *Engineering Design Properties of GRP*. British Plastics Federation, London
Johnson A F, Sims G D 1986 Mechanical properties and design of sandwich materials. *Composites* 17(4): 321–8
Melby E G, Castro J M 1989 Glass-reinforced thermosetting polyester molding: Materials and processing. In: Aggarwal S L (ed.) 1989 *Comprehensive Polymer Science*, Vol. 7. Pergamon, Oxford, Chap. 3
Parkyn B (ed.) 1970 *Glass Reinforced Plastics*. Iliffe, London
Rubber and Plastics Research Association 1977 *Survey of Fibre-Reinforced Plastics Markets and Needs*. RAPRA, Shawbury, UK
Tsai S W 1987 *Think Composites Software: User Manual "Composites Design,"* 3rd edn. Think Composites, Dayton, OH
Whitney J M 1987 *Structural Analysis of Laminated Anisotropic Plates*. Technomic, Basel, Switzerland

A. F. Johnson
[National Physical Laboratory, Teddington, UK]

H

Hard-Fiber Elastomers

Traditionally, elastic polymers have been characterized as rubbers and their mechanical properties have been explained by the entropic mechanisms of rubber elasticity theory. Since 1965, growing interest has developed in a new class of crystalline polymer structures that manifest elastic behavior. These hard-fiber elastomers possess the unusual combination of high elastic modulus, high extensibility and high elastic recovery from large extensions. The elasticity of these hard elastic polymers, however, seems to be primarily controlled by energetic rather than entropic mechanisms. Polyoxymethylene, isotactic polypropylene, polyethylene, polybut-1-ene, poly-4-methylpentene, acetal copolymer and polypivalolactone all produce hard elastic structures.

The conversion of a normally nonelastic crystalline polymer to a hard-fiber elastomer is extremely process-sensitive. A row-nucleated crystal morphology, which forms during stress crystallization, is typically observed with these materials. Since stress crystallization is particularly sensitive to such processing parameters as spinning temperature, shear rate, draw-down ratio and annealing temperature, the elasticity or lack of elasticity of the crystalline polymer will depend on the proper choice of these processing parameters.

During room-temperature extension of a hard-fiber elastomer, microcavitation of the row-nucleated structure predominates in the high-extension elastic region, while plastic deformation predominates at still higher extensions. At least five basic structural models have been proposed to explain the elastic properties of these polymers. These are (a) a leaf-spring model involving the elastic bending of lamellae, (b) reversible shear of lamellae between fixed tie points, (c) a general model based on a change in entropy in the intermolecular layer and an increase in surface energy during extension, (d) a combined structural model that attributes the stress for extension to the pulling of fibrils from lamellae and the retractive force resulting in the particular elastic properties to surface energy and entropy effects in the fibrils, and (e) a microparacrystalline model that associates the driving force for elasticity to the surface free energy of destroyed microparacrystals. These models are not consistent and predict different mechanical properties and structures in the strained state. Also, noncrystalline molecular orientation changes have been observed during the crystalline polymer deformation which have not been included in the models. Certainly, both energetic and entropic mechanisms contribute to the observed elasticity and a truly quantitative model has yet to be developed.

The sensitivity of hard-fiber elastomers to morphology also affects the relation between their elasticity and strength. For example, the observed elastic recovery at 50% cyclic extension of isotactic polypropylene hard-fiber elastomer was found to decrease linearly with increasing noncrystalline orientation (and hence increasing tenacity) in the fiber. Thus, isotactic polypropylene fibers with tenacities of 1–2 grams per denier had elastic recoveries of 100% at 50% room-temperature cyclic extension, while a high-strength fiber having a tenacity of 5.5 grams per denier had an elastic recovery of only 70% at 50% room-temperature cyclic extension. This elastic recovery was, however, still far superior to the approximate 40% elastic recovery at 50% room-temperature extension normally observed for non-elastic-drawn fibers of the same polymer.

See also: Elastomers: An Overview

Bibliography

Cannon S L, McKenna G B, Statton W O 1976 Hard-elastic fibers (a review of a novel state for crystalline polymers). *J. Polym. Sci., Macromol. Rev.* 11: 209–75

Celanese Corporation of America 1965 Filamentous material and a process for its manufacture. Belgian Patent 650,890 (January 1965)

Hashimoto T, Nagatoshi K, Todo A, Kawai H 1976 Deformation mechanism of "hard elastic" polyethylene films. *Polymer* 17: 1063–74

Ishikawa H, Numa H, Nagura M 1979 Deformation in hard elastic polypropylene fiber. *Polymer* 20: 516–17

Loboda-Čačković J, Čačković H, Hosemann R 1979 Hard elastic fibers. 1. Four-component analysis of the NMR spectra of isotactic polypropylene. *J. Macromol. Sci., Phys. B* 16: 127–44

Miles M, Petermann J, Gleiter H 1976 Structure and deformation of polyethylene hard elastic fibers. *J. Macromol. Sci., Phys.* 12: 523–34

Noether H D, Brody H 1976 Low modulus "hard" elastic materials. *Text. Res. J.* 46: 467–78

Samuels R J 1979 High strength elastic polypropylene. *J. Polym. Sci., Polym. Phys. Ed.* 17: 535–68

Wool R P, Lohse M I, Rowland T J 1979 Broad-line NMR studies of deformation and recovery in hard elastic polypropylene. *J. Polym. Sci., Polym. Lett. Ed.* 17: 385–9

R. J. Samuels
[Georgia Institute of Technology, Atlanta, Georgia, USA]

Health Hazards in Elastomer Processing

Health hazards in the rubber industry may occur due to physical injury, exposure to noise or exposure to the chemicals and agents used to manufacture the products. The field is very wide and it is not possible in an article of this length to cover all aspects of the different types of hazard that could occur. This article provides a synopsis of the current position in the first two areas and, since chemical hazards have been and are a major concern within elastomer processing, gives a more extended review of the epidemiology of elastomer processing and of the hazards of certain groups of chemicals. For more extensive information on the hazards of individual substances such as isocyanates, rubber chemicals and solvents, reference should be made to the relevant works listed in the bibliography.

1. Physical Hazards

The rubber industry uses a variety of heavy machinery for mixing and processing uncured elastomeric compositions and many severe accidents occured in the early days of production. In the years up to the 1950s, there were, on average, 20 mutilating accidents per year due to trapping at the nip of two-roll mills in the UK. However, the introduction of involuntary safety devices, such as the Lunn bar on two-roll mills and interlocked guarding on other processing machinery, has reduced the incidence of such accidents to negligible proportions, so that the most common cause of accidental injury is now that of damaged backs or muscular strains due to manual handling. The industry is still relatively labor intensive and many of the jobs carried out in factories involve significant physical effort. It is noteworthy, however, that the overall incidence of ischemic heart disease in the rubber industry has generally been found to be lower than in the general population.

The forthcoming introduction of new legislation in the European Community (EC), requiring employers to review all jobs involving manual handling and to assess the hazards involved in them, may well accelerate the more general introduction of mechanical aids (e.g., vacuum lifting equipment) for lifting elastomeric compounds, molds and other items used in rubber production.

Common CEN standards for the guarding of specific machines in the rubber industry, for example two-roll mills and calenders, are now being drafted within the EC and should be available by 1992.

Repetitive strain injuries (RSI), also known as work related upper limb disorders (WRULDS), are also a problem in certain jobs carried out in the industry. Tenosynovitis is an occupational hazard where stressed movements of the wrist and elbows are carried out at frequently repeated intervals. In such situations, a careful review of the ergonomics of the task is necessary in order to find the least-stressed method of working. Attention should be paid to the relative heights of seats and work tables in order to facilitate the operations being carried out and the disposition of other items of equipment may need changing so that the required tasks can be completed with least effort. Where RSI are known to be an occupational problem, good medical supervision must be provided to detect damage at an early stage and prevent progression to a chronic problem.

2. Noise

Noise is generally not a major problem in the rubber industry with the exception of work on two-roll mills or hose braiding machines which can frequently produce noise levels above 90 dB(A). For the mills, this noise is generated partly by the drive train for the mill, but also by the milling action on the rubber, especially where "popping" of entrapped air pockets is a frequent occurrence. For braiding machines, most of the noise is generated mechanically on the braiders. In the UK, control of noise exposure is governed by the 1989 noise at work regulations, which implement the requirements of the EC directive on noise.

In these regulations, three action levels are defined. Where daily personal exposure levels $L_{EP,d}$ are greater than 85 dB(A) (the first action level), ear protectors must be provided to all employees who ask for them. These ear protectors must be maintained and repaired as necessary. Where $L_{EP,d}$ is greater than 90 dB(A) (the second action level), exposure to noise must be reduced as far as reasonably practical by means other than ear protectors. If noise exposure levels cannot be reduced below 90 dB(A), the areas concerned must be clearly marked with warning notices based on the guidance given in BS 5378, Pt I. Ear protection must be provided to all who go into the marked zone and the employer has the duty to ensure that it is worn.

There is also a peak action level of 200 Pa (140 dB are 20 μPa) which is most likely to be important where occasional loud impulsive noises are encountered. The requirements of the peak action level are the same as those of the second action level.

Wherever exposures are over 85 dB(A), there is a duty to provide adequate information, instruction and training to employees about the risks to hearing, what they should do to minimize risk and how they can obtain hearing protection. Employees have a duty to use ear protectors where exposure levels are greater than 90 dB(A).

For two-roll mills, some reduction in overall noise levels may be achieved by provision of a specifically designed noise reduction enclosure around the drive train. However, where most of the noise is produced by the milling action on the rubber, significant reduction in noise levels by such methods is, in practice, likely to be impossible and control by personal protection is the only way of conforming with the regulations.

For hose braiding machines, noise reduction enclosures around the braiders are usually effective at reducing workplace noise levels to acceptable values.

Wherever noise is likely to be a problem, measurements of the hearing performance of all exposed operators (audiometry) should be made at regular intervals to detect any early changes in hearing performance.

3. Chemical Hazards

The rubber industry uses a large number (200–500) of chemicals, many of them specific to the industry. These chemicals often contain active functional groups which are necessary to effect the reactions used in elastomer vulcanization, but which may also result in the chemicals having potential physiological effects. Since many of these chemicals are unique to the rubber industry and contain unusual chemical groupings, the amount of information on their detailed toxicity has often been limited. The position is further complicated by the fact that new chemical species are generated during processing of the elastomer, particularly during the curing operation.

These problems have been tackled from two extremes. Certain groups of chemicals have been the subject of detailed investigations within the industry, both as to the mechanism by which hazards may be caused and in order to investigate the extent of exposure to the chemical. At the other extreme, a number of epidemiological studies have been carried out to determine whether those people who have worked in the industry have suffered damage to their health because of this. Both aspects of these studies will be covered in this article.

3.1 Epidemiology

Whatever speculations are made about the potential effects of the substances used to make polymer-based products, the final conclusions on whether they have produced any harmful effects on health are usually drawn by the epidemiologists. A well-planned epidemiological study should be able to identify and quantify any damage caused to the health of people working in the industry as a result of their work.

Conversely, and often equally importantly, it should be able to rule out the possibility of certain types of damage if these have not occurred.

Of course, things are often not this simple. One well-known problem is that of time; when looking for diseases such as cancer with long induction periods, the study will not be able to produce answers for many years. This means that even recently reported studies reflect conditions in the industry which may have long ceased to exist. A second problem concerns numbers. In order to make the study statistically valid, the population involved must be reasonably large. As numbers employed in the industry decrease due to the introduction of automation, it becomes increasingly difficult to gather numbers to make the treatment statistically valid, especially if the total population is split into subgroups. This has resulted in surveys being carried out on collections of smaller operations, often spanning many different companies rather than a few large factories. Other problems can affect studies that have external controls, where the result obtained may be dependent on the type of control chosen—national figures, figures from the same locality, figures weighted for social class, and so on. These problems have seen attempts being made to use internal comparisons, either of the case/control or regression models and life tables (RMLT) type.

In the UK, the best-known studies of the rubber industry have been made by the Health and Safety Executive (HSE) and the British Rubber Manufacturers Association (BRMA)/Birmingham University.

(*a*) *HSE study*. This large study (41 891 men in 12 tire factories and 369 other rubber factories) was reported in 1980. The study was of the "census" type where all men aged 35 years and over who were employed in the factories concerned on 1 February 1967 and had worked for at least one year in the factory were included. Occupations were recorded for the day of the census; no previous or subsequent occupations were recorded.

The population was followed up over the ten-year period 1967–1976. Of the 98.4% of the population (40 867 men) who were successfully traced over this period, 5733 died.

One of the prime purposes of the study was to find whether occupational bladder cancer had been removed from the industry by the removal of the aldol-α/β-naphthylamine antioxidants (e.g., Nonox S) in 1949. Consequently, the population was divided into three groups.

(i) Group A—the 12 779 men who started work before 1 January 1950 in a factory which had used suspect antioxidants.

(ii) Group B—the 18 118 men who started work on or after 1 January 1950 in a factory which had used suspect antioxidants.

(iii) Group C—the 9970 men who worked in factories which had *never* used the suspect antioxidant.

As far as bladder cancer is concerned, the study provided good evidence that removal of the suspect

Table 1
HSE study of bladder cancer

	Observed deaths	Expected deaths	Standardized mortality rate (SMR)
Total population	73	57.6	127
Group A	36	25.0	144
Group B	24	22.3	108
Group C	13	13.4	97

antioxidants in 1949 had removed the excess of bladder cancer. The overall figures are given in Table 1.

However some evidence was obtained of excess numbers of lung and stomach cancers, as shown in Table 2.

No excesses were recorded for deaths from other diseases of the respiratory system, diseases of the circulatory system (including ischemic heart disease), diseases of the digestive system, cancers of the esophagus, pancreas, prostate, kidney and brain and cancers of the lymphatic and hematopoietic tissue including leukemia.

The authors observed that regional standardization reduced but did not eliminate the excesses of bladder, stomach and lung cancer, as shown in Table 3.

These diseases also have a strong association with social class, with the incidence in unskilled manual workers being higher than in the professional occupations. Allowance for this effect would have eliminated any excesses, as shown in Table 4.

However, standardization of this type must be viewed with some caution. If there are general causative *occupational* factors which may affect a specific social class, for example exposure to high dust levels leading to stomach cancer in unskilled manual workers rather than professional occupations, such a standardization could mask a true occupational problem within the population concerned.

It should be noted that this standardization has also apparently eliminated the excess incidence of bladder cancer, although there is no doubt that an occupational bladder cancer problem was present in this population.

Table 2
HSE study of stomach and lung cancers

	Observed	Expected (by national figures)	SMR
Stomach cancer	216	176.4	122
Lung cancer	822	716.5	115

Table 3
HSE study accounting for regional standardization

	Observed	Expected (by regional Standardization)	SMR
Bladder cancer	73	60.7	120
Stomach cancer	216	191.0	113
Lung cancer	822	764.4	108

Table 4
HSE study accounting for social class

	Observed	Expected (standardized by social class)	SMR
Bladder cancer	73	71.3	102.4
Stomach cancer	216	227.1	95.1
Lung cancer	822	891.0	92.2

The authors of this report discussed the possibility of the lung and stomach cancer excesses being linked with dust and rubber fume exposures and commented that "environmental exposure to these two agents must be reduced generally if the cancer risk found in this study is accepted."

A follow up to this study was undertaken by J. T. Hodgson. The figures for bladder cancer confirm those of the earlier study (see Table 5).

For lung cancer and stomach cancer, the excesses observed are somewhat smaller than in the earlier report, as shown in Table 6.

Of particular interest in this report is an analysis of "latency" (defined here as the period from entry to the industry to death from the disease in question). For occupational cancer, maximal effects can be expected 20–30 years after first exposure. The overall results for the three cancers of most interest are given in Table 7.

As can be seen from these figures, there is no clear pattern of a maximum standardized mortality rate

Table 5
Follow up to HSE study of bladder cancer

	Observed	Expected	SMR
Total population	182	167.7	108
Group A	81	66.8	121
Group B	65	63.4	103
Group C	36	37.5	96

Table 6
Follow up to HSE study of stomach and lung cancers

	Observed	Expected (national figures)	SMR
Stomach cancer	482	419.1	115.0
Lung cancer	2026	1800.2	112.5

(SMR) in the 20–30 year latency although for bladder cancer a raised SMR is present in the 10–39 year period. The high SMRs for the less than ten year latency of lung and stomach cancer are supportive of the hypothesis that this population has a higher baseline rate for these two cancers than would be calculated from the national figures. There is clearly no evidence of a maximum in the figures for stomach cancer and only a weak maximum at 10–19 years in the figures for lung cancer. The figures cast some doubt on the presence of occupational lung and stomach cancer excesses in this study.

(*b*) *BRMA/Birmingham University study*. This study of over 36 000 rubber workers has been reported in 1982 with follow up reports in 1986 and 1989. The population consisted of all male workers entering the industry between 1946 and 1960 and working for at least one year. Thirteen factories were involved in the study, eight of these being tire factories and five being general rubber goods (GRG) factories.

The population was divided into three cohorts according to the date of entry into the industry:

(i) cohort 1 from 1 January 1946 to 31 December 1950,

(ii) cohort 2 from 1 January 1951 to 31 December 1955, and

(iii) cohort 3 from 1 January 1956 to 31 December 1960.

Table 7
Follow up to HSE study showing analysis of latency

	Latency (years)				
	<10	10–19	20–29	30–39	>40
Bladder					
observed	6	46	46	49	35
SMR	75.6	131.9	88.3	127.0	102.5
Stomach					
observed	37	110	148	110	77
SMR	144.5	113.8	113.5	121.3	101.7
Lung					
observed	119	513	680	432	282
SMR	115.4	124.4	119.4	109.0	88.5

Table 8
BRMA/Birmingham University study of various cancers (1982 report)

Site of cancer	Observed	Expected	SMR
All cancers	1359	1221.0	111
Lung	638	517.4	123
Stomach	183	141.8	129
Prostate	30	45.4	66
Thyroid	4	2.2	182
Colon	67	73.7	90
Rectum	50	54.1	92
Esophagus	40	31.8	126
Brain	35	41.1	85
Leukemia	31	28.1	110
Bladder	36	43.0	83
Others	245	242.4	101

The 1982 report (summarizing the position to 1975) found SMRs for deaths from various cancers in this population as shown in Table 8. As can be seen from these figures, there was no evidence of an overall excess of bladder cancer in this population, perhaps due to the small numbers of individuals in the study with significant exposure to antioxidants such as Nonox S. (The longest exposure would have been approximately 3.5 years, from 1 January 1946 until the removal of Nonox S in mid-1949.)

As with the HSE study, there is evidence for an overall excess of lung and stomach cancer and possibly for a small excess of cancer of the esophagus. None of the other cancer sites produced figures that were significantly in excess of expectation.

Table 9 gives the figures for the 1989 report, which took the study to its final end date of 1985. Of the major excesses, lung cancer was significantly different from expectation ($p < 0{\cdot}001$) while stomach

Table 9
BRMA/Birmingham University study of various cancers (1989 report)

Site of cancer	Observed	Expected	SMR
All cancers	3344	2965.6	113
Lung	1592	1219.2	131
Stomach	359	316.5	113
Prostate	91	128.2	71
Testis	4	11.0	36
Thyroid	9	5.4	167
Large intestine	160	177.6	95
Rectum	143	139.9	107
Esophagus	107	87.6	122
Pharynx	30	20.2	149
Skin	17	24.8	69
Brain	67	76.3	88
Leukemia	65	68.1	95
Bladder	106	105.5	100

cancer and cancers of the esophagus were also significantly different, although only at the <0·05 level. Cancers of the prostate and testis were in significant deficit ($p < 0{\cdot}01$ for cancer of the prostate and $p < 0{\cdot}05$ for cancer of the testis). There was again no evidence for an excess of bladder cancer.

The BRMA/Birmingham University study also used the regression models and life tables (RMLT) approach to analyze these figures. In this method, the mean duration of employment in the industry or in certain jobs within the industry for persons dying from causes of interest is compared with that of all matching survivors. In effect, if there is an occupational risk within the industry or within certain parts of it, the longer an employee works within the industry the more likely that employee is to suffer from the disease. On average, therefore, the length of employment for persons with the disease is likely to be greater than for those not affected. In the RMLT method, a "test statistic" is derived to compare the mean durations for the two groups. A positive test statistic of 1·98 indicates significance at the $p < 0{\cdot}05$ level.

The results from this method are revealing and those from the 1989 paper are taken as typical. Values obtained from overall duration of employment in the industry are given in Table 10.

It should be noted that the presence of a "survivor population effect" in the data will tend to produce negative statistics for this analysis of overall duration of employment. In spite of this, three positive test statistics were obtained, for stomach cancer, non-melanoma skin cancer and bladder cancer. Although none of these figures reached formal statistical significance, the value for stomach cancer at +1·69 equates to a probability of $p < 0{\cdot}07$. This situation is in sharp contrast to the figure obtained in this analysis for lung cancer, which was −2·05. With both cancers it should be borne in mind that, in this rather broad analysis, effects in subgroups of the population are likely to be diluted and less easy to recognize. The positive stomach cancer figure must therefore be regarded as worthy of particular attention.

The method was also used to look at the duration of employment in dust exposure and fume and/or solvent exposure jobs, since previous work had hinted at an association between stomach cancer and dust exposure and between lung cancer and fume exposure. These more focused analyses help to remove the survivor population effect observed in the above analysis. Tables 11 and 12 detail the results that were obtained.

In these tables the survivor population effect is minimized by controlling for duration of employment in the industry as well as for age at hire and entry cohort. Probably because of this, more of the results were positive although none reached formal levels of significance. However, it should be noted that stomach cancer gives a positive test statistic in the dust exposure group, but not in the fume and/or solvent exposure group, while lung cancer gives a negative test statistic in the dust exposure group but a positive figure in the fume and/or solvent group. A similar analysis for the "any fume" exposed group (not including exposure to solvents) gave a lower but still positive test statistic of +1·29.

Using the overall duration of employment data, the authors calculated that, for stomach cancer, the maximum likelihood estimate of latent period before development of a cancer was ten years, the "doubling dose" (exposure needed to double the risk of developing the cancer) was 45 years and the percentage of deaths from stomach cancer attributable to occupational exposure was 14% (95% confidence interval 2–25%).

Using the data on employment in fume and/or solvent exposed jobs, the figures for lung cancer were

(i) latent period—maximum likelihood estimate 30 years,

(ii) doubling dose—65 years in fume and/or solvent exposed jobs, and

Table 10
BRMA/Birmingham University, study using the RMLT method (1989 report)

Site	Mean duration of employment	Mean difference between deaths and matching survivors	Test statistic
All causes	9.7	−0.3	−3.85
All cancers	10.0	−0.1	−1.13
Lung	9.8	−0.4	−2.05
Stomach	10.5	+0.7	+1.69
Esophagus	8.7	−2.0	−2.26
Pharynx	6.8	−2.6	−1.66
Melanoma	5.3	−2.9	−0.93
Other skin cancer	14.2	+3.7	+1.43
Bladder	11.2	+0.6	+0.75

Table 11
BRMA/Birmingham University study showing effects of dust exposure

Site	Duration of employment in dust exposure jobs		
	Mean duration of exposed employment	Mean difference between deaths and matching survivors	Test statistic
Lung	1.5	−0.2	−1.62
Stomach	2.2	+0.3	+1.29
Esophagus	1.4	+0.0	+0.03
Pharynx	0.4	−0.8	−1.25
Melanoma	1.0	+0.3	+0.32
Other skin cancer	4.1	+1.7	+1.12
Bladder	2.5	+0.5	+1.04

(iii) percentage of deaths attributable to occupational exposure—6% (95% confidence interval 1–9%).

The authors concluded that, despite the limitations of the method, there was some evidence of occupational causation of stomach and lung cancer with the former being more likely to be connected with exposure to dust and the latter more likely to be connected with exposure to fumes or solvents.

This finding is in contrast to the conclusions of the updated HSE study as described in Sect. 3.1.

However, the HSE study, based as it is on a census structure, is severely limited when attempting to analyze by job or type of exposure. This is because the job descriptions obtained in the study are valid only for the day of the census. Long-term employees within the population (and these are the people most likely to be affected by an occupational hazard) are likely to have had more than one job during their period of employment, making the job descriptions obtained in the study of reduced validity. Occupational hazards that affect only one group of employees will be diluted within the overall population and may be impossible to identify in the overall analysis. In this respect, the BRMA/Birmingham University study is more powerful, since job descriptions were obtained for the full period of each individual's employment. Although the total accuracy of these descriptions may be questioned, the data are more likely to provide a useful base for this more focused type of analysis.

Summarizing the UK studies, care should be taken in interpreting the results from simple SMR analyses, which may be affected by geographical or social class influences. The apparent excesses for lung and stomach cancer in both the HSE and BRMA/Birmingham University studies are likely to be due at least in part to such influences. However, the RMLT analyses undertaken in the BRMA/Birmingham University study indicate that a proportion of these excesses *is* occupationally based, with dust exposure more likely to be associated with stomach cancer and fume and/or solvent exposure more likely to be associated with lung cancer.

(c) *University of North Carolina (UNC) studies.* Two main populations were used in these studies. The first was taken from a single large tire factory in

Table 12
BRMA/Birmingham University study showing effects of fume and/or solvent exposure

Site	Duration of employment in fume and/or solvent exposure jobs		
	Mean duration of exposed employment	Mean difference between deaths and matching survivors	Test statistic
Lung	5.6	+0.3	+1.81
Stomach	5.2	−0.4	−1.31
Esophagus	4.9	+0.4	+0.63
Pharynx	3.1	−0.3	−0.36
Melanoma	2.1	0.0	+0.01
Other skin cancer	8.7	+1.0	+0.46
Bladder	4.8	−0.1	−0.08

Table 13
First UNC study of various cancers

Cause of death	Observed	Expected	SMR
All causes	1983	2033	98
Chronic respiratory disease	61	68.5	89
Ischemic heart disease	867	916	95
Arteriosclerosis	89	60.1	148
Diabetes Mellitus	54	37.2	145
Cancer of lung, etc.	106	115	92
Cancer of stomach	40	23.4	171
Cancer of pancreas	18	21.9	82
Cancer of prostate	53	37.8	140
Cancer of bladder	10	13.3	75
Cancer of brain, CNS	4	6.15	65
Cancer of colon	42	34.4	122
Cancer of lymphatic and hematopoietic system	43	31.6	136
Lymphosarcoma and Hodgkins disease	15	9.15	164
Leukemia	17	13.5	126

Akron, Ohio and consisted of 6678 men who had worked for at least ten years in the factory on 1 January 1964. They were followed for nine years from this date. The factory contained a synthetic elastomer plant and a reclaim plant, in addition to producing tires. The overall results are given in Table 13.

As with the British studies, there is a numerical excess of stomach cancer but, unlike them, there is no overall excess of lung cancer, while there are excesses of prostate cancer and of cancers of the lymphatic and hematopoietic systems which were absent in the British studies.

Some of these causes of death were investigated further using case/control techniques. In these, the job histories of those dying from a specific disease were compared with those of a control group not having the disease. The results are given in Table 14.

Although there was no overall excess of lung cancer, a specific case/control study was carried out on 61 lung cancer cases and 61 matched controls. This showed no correlation between tire building and lung cancer, but a strong correlation between work on tire curing and lung cancer; 25% of the cases worked on tire curing against 15% of the controls. The average duration of work on tire curing was 16 years compared with 7·5 years for the controls.

Although it uses a different methodology, this study shows similar results to the BRMA/ Birmingham University study using the RMLT method.

Table 14
First UNC study showing job histories of diseased patients

Disease	Job	Number of cases	Risk ratio
Stomach cancer	compounding and mixing	5	2.0
	extruding, tread cementing	5	2.3
Colorectal cancer	extruding, tread cementing	7	2.2
	maintenance	5	1.8
Respiratory cancer	receiving and shipping	11	1.9
	compounding and mixing	12	1.4
	mill mixing	5	2.1
	extruding, tread cementing	8	1.4
	reclaim	10	2.3
Prostate cancer	compounding and mixing	6	1.6
	calendering	9	2.4
	janitoring and trucking	16	3.5
Lymphatic and hematopoietic cancers	compounding and mixing	5	1.4
	inspection and finishing	6	2.0
	synthetic plant	6	6.2

Table 15
Second UNC study showing excess incidences of some cancers

Site	Observed	Expected	SMR
Stomach	34	27.7	123
Large intestine	53	45.7	116
Pancreas	34	27.9	122
Prostate	59	45.9	128
Bladder	21	18.1	116
Lymphatic and hematopoietic	52	41.9	124

Table 16
First Harvard study showing excess incidences of some cancers

Site	Observed	Expected	SMR
All cancers	984	1046.4	94
Stomach	98	93.9	104.4
Bladder	48	39.5	121.5
Leukemia	55	43	127.9

The second UNC population was composed of 8418 white male production workers who were working in the industry on 1 January 1964 and were followed to 13 December 1973. Some excess incidences of cancer were identified in the population, as shown in Table 15.

In this study, jobs were assigned to members of the population, based on the most representative department. This produced positive association between stomach cancers and milling (six observed, 1·6 expected), prostate cancer and general service (ten observed, 4·7 expected) and leukemia and general service (six observed, 2·4 expected).

Separate case/control studies were carried out on lymphatic and hematopoietic cancers and on stomach cancer using cases drawn from these Akron populations. For the lymphatic and hematopoietic cancers there was an association between lymphatic leukemia and solvent exposure (risk ratio 3·3) and for stomach cancer there was an association between this and work on batch preparation and cure preparation. In terms of atmospheric contaminants, there was no correlation between stomach cancer and exposure to polycyclic aromatic hydrocarbons, nitrosamines or carbon black, but there was a correlation with exposure to talc.

(*d*) *Harvard studies*. The first study made by Harvard used a population of 13 570 white male workers employed at a rubber plant in Akron between 1940 and 1971. All members of the population had worked for at least five years in the factory and they were initially followed to June 1974. Certain cancers were present in excess over expectation, as shown in Table 16.

A cancer morbidity study was also made at this plant in order to investigate the occurrence of particular diseases and work on specific jobs. Those jobs giving raised SMRs are given in Table 17.

This study was extended in 1982 to include employees who had worked 2–5 years and was followed to an end date of 30 June 1978. The population was split into "front processing" (compounding, mixing, milling) and "back processing" (extrusion, calendering, production of rubber solutions and rubberized fabric). Raised SMRs were found for the groups given in Table 18.

(*e*) *Summary of British and US studies*. There are two main similarities between these studies. Stomach cancer shows an excess incidence in all the four main groups of studies and, wherever an attempt has been made to link it to jobs or exposures, the association is with jobs at the front of the rubber processing operation where exposure to dust can be expected. Lung cancer is also identified in excess of expectation in all four groups of studies, although in

Table 17
First Harvard study showing occurrence of particular diseases in specific jobs

Site	Job	Observed	Expected	SMR
Lung cancer	tire curing	31	14.1	220
	tire molds	10	5.0	200
	fuel cells/deicers	46	28.8	160
Bladder cancer	tire building	16	10.7	150
Skin cancer	tire building	12	1.9	632
Brain cancer	tire building	8	2.0	400
Lymphatic cancer	tire building	8	3.2	250
Leukemia	calendering	8	2.2	364
	tire curing	8	2.6	308
	tire building	12	7.5	160

Table 18
Extended Harvard study showing groups with raised SMRs

	Site	Observed	Expected	SMR
Front processing	all digestive cancers	51	34.0	150
	stomach	15	7.1	211
	large intestine	19	11.2	170
Back processing	biliary and liver cancer	9	5.0	180
	leukemia	14	5.8	241

the US studies it is necessary to focus on a specific group of jobs in order to see the excess. Wherever this is done in the US or in the British studies, the association is with the back stages of rubber processing where rubber fume may be produced. There is little indication in any of these studies of significant trends in these cancer incidences with time, although much of the contribution to the excesses comes from relatively historical times.

Some new studies on later populations are underway, particularly a new BRMA/Birmingham University study taking employees in a mixed group of tire and GRG factories from 1982 onwards. Unfortunately, significant results from this study cannot be expected until the year 2000.

(*f*) *Swiss studies*. One Swiss study compared rates of death from certain diseases in a population working in a rubber factory with the equivalent rates in a population working in an explosives factory between 1 January 1955 and 31 December 1975. Raised SMRs are given in Table 19.

(*g*) *Finnish studies*. In Finland, 1331 male and female workers in a rubber factory were followed from 1 January 1953 to 31 December 1976. Overall, 21 cases of cancer were found and 18·9 expected. Two cases of bladder cancer were found and 0·3 expected.

(*h*) *Swedish studies*. In Sweden, a significant increase in esophageal cancer was identified among workers involved with vulcanization. Eight esophageal cancers were observed and 0·79 expected. An excess incidence of lung cancer was also identified (13 observed, 7·4 expected).

In a separate study, a population of 8734 workers in two Swedish tire factories was investigated from 1952 to 1981. No significant risk excesses were detected for the whole population. The mortality from coronary heart disease and lung cancer was increased when the study period was limited to more than 40 years from first employment. The incidence of bladder cancer was also increased among individuals with heavy and long-term exposure in the weighing and mixing departments.

Table 19
Swiss study showing groups with raised SMRs

	Site	Observed	Expected	SMR
Rubber industry	bladder	4	1.1	364
	stomach	8	4.6	174
Explosives factory	bladder	5	2.6	192

(*i*) *Norwegian studies*. A cohort of 529 men exposed to mineral oils in a Norwegian cable-making company was studied from 1953 to 1984. Statistically significant excesses of deaths were observed in a subcohort of 248 men with known work category and employment for at least one year in oil-exposed work.

(*j*) *Chinese studies*. 1624 employees (957 men, 667 women) in a rubber factory in Shanghai were followed between 1972 and 1984. A significant excess of deaths from lung cancer was found, particularly in those who worked in curing workshops. After allowing for smoking habits, there was a risk ratio of 3·2 for men exposed to curing fumes or inner-tube production and 4·6 for women doing the same jobs. There was little evidence of an interaction between smoking and exposure to curing fumes.

(*k*) *Italian studies*. 4917 male workers who first started work in a large rubber factory between 1962 and 1972 were followed up until 31 January 1983. Excesses were found for stomach cancer and a number of other cancer sites, although observed numbers of deaths were small (six for stomach cancer, less than four for any other site). There was a deficit of lung cancer cases. The authors comment that the average follow-up period of 14 years is too short to produce accurate results for cancers with a latent period of 15–20 years.

3.2 Polycyclic Aromatic Hydrocarbons

Polycylic aromatic hydrocarbons (PAHs) occur in the mineral oils and carbon blacks used by the industry. Many of the five- and six-ring PAHs have been shown to be carcinogenic in animal tests and the aromatic oils, in particular, have produced skin cancer when painted onto animal skin. Amounts of individual PAHs in one series of tests are given in Table 20.

When painted onto mice skin in high amounts, these aromatic oils have given skin cancer in 96% of mice under conditions that would produce a level of approximately 0–4% when using a solvent-refined oil.

For carbon black, the PAHs are bound strongly to the surface of the black so that hot solvent extraction is necessary to release them. The amounts found are given in Table 21.

In animal tests, carbon black has not produced indentifiable cancers whether by skin contact, ingestion or inhalation. Since the solvent extracts from the carbon black have been shown to be carcinoge-

Table 20
Amounts of PAHs found in mineral oils

	Aromatic oil A (ppm by weight)	Aromatic oil B (ppm by weight)	Aromatic oil C (ppm by weight)
Fluoranthene	11.0	0.6	1.4
Pyrene	15.6	6.4	6.9
Benzo[*a*]fluorene	0.9		0.7
Benz[*a*]anthracene	34.2	9.4	5.8
Chrysene	395.3	193.0	63.4
Benzo[*b*]fluoranthene	72.9	43.2	11.7
Benzo[*e*]pyrene	113.2	97.9	43.2
Benzo[*a*]pyrene	13.4	11.5	4.0
Dibenz[*a, j*]anthracene	4.6	5.6	2.8
Dibenz[*a, h*]anthracene	5.7	3.5	1.4
Indeno(1, 2, 3-*c, d*]pyrene	6.2	1.7	1.4
Benzo[*g, h, i*]perylene	17.9	21.1	15.3
Anthanthrene	6.6	9.8	12.3

nic, it can reasonably be concluded that these PAHs are not released into the body in significant amounts after inhalation or ingestion.

Epidemiological studies on carbon black production workers have shown no excess for cancers at any site.

In the rubber industry, skin contact with the carbon blacks may occur at weighing and mixing operations. Direct contact does not occur to any significant extent with the mineral oils since these are usually pumped directly into the mixer. However, indirect skin contact is possible during handling of uncured polymer compounds.

As far as skin cancer is concerned there is little evidence of a problem. One study carried out in the USA using hospital tumor registry records in the Akron area found 12 cases of skin cancer in a group of men who had worked for five or more years in tire building, when only 1·9 was expected. Some of these PAHs might be expected to be released when the polymer compounds are heated. Studies in the UK have shown that the more volatile noncarcinogenic PAHs such as fluoranthene and pyrene are released into the atmosphere under these conditions, but both British and US studies have shown that the higher-molecular-weight carcinogenic PAHs such as benzo[*a*]pyrene are not released into the atmosphere in measurable amounts (greater than the normal urban background levels) from such processes.

It seems unlikely that inhalation of PAHs is a direct cause of any of the cancer excesses identified by epidemiology.

3.3 Nitrosamines

Nitrosamines have the general structure

```
R—N—R′
   |
   NO
```

Table 21
Amounts of PAHs found in various carbon blacks

	Furnace black type[a]								
	N472	N375	N339	N326	N330	N351	N660	N762	LCF4
Fluoranthene	1	59	52	9	10	36	12	32	32
Pyrene	0.5	400	206	58	48	195	52	308	220
Benzo[*g, h, i*]fluoranthene			94	16	20		14		
Cyclopenta[*c, d*]pyrene	<0.2	116	78	<0.8	20	81	13		5
Benzo[*a*]pyrene	<0.05	6.5				5		8	10
			32	1	3		8		
Benzo[*e*]pyrene									12
Indeno[1, 2, 3-*c, d*]pyrene	<0.05	13	35	1	0.3	9	7	9	4
Benzo[*g, h, i*]perylene	<0.05	146	163	16	25	95	41	87	120
Anthanthrene	<0.05	33	42		2	28	7	30	36
Coronene	0.08	219	140	3	11	92	27	80	180

a ASTM designation with amounts in ppm

Many of these substances have been found to be carcinogenic in animals, producing tumors of the liver, kidney, lungs, esophagus and other sites.

Some nitrosamines have been used directly in the elastomer industry, for example the retarder nitrosdiphenylamine (NDPA) and the blowing agent dinitrosopentamethylenetetramine. Both these materials have been found to be noncarcinogenic. However, it has now been shown that some elastomer compounds may generate nitrosamines following the mixing operation. All processes subsequent to the mixing operation may be affected, including storage of finished articles.

Nitrosamines may be produced where the rubber compound generates secondary amines in the presence of a nitrosating source. The three main groups of chemicals that can generate secondary amines are shown in Fig. 1.

The original nitrosating source, recognized by Fajen and co-workers in 1979, was NDPA. The nitroso group from this retarder could be transferred to secondary amines generated by the compound to give different nitrosamines. Production of NDPA has largely ceased following recognition of this phenomenon. However, other materials such as nitrite end stops in certain emulsion polymers, nitrogen oxides dissolved in the polymer, nitrite or nitrate residues on carbon black or nitrite/nitrates in salt bath curing systems may also be effective at producing nitrosamines from secondary amines.

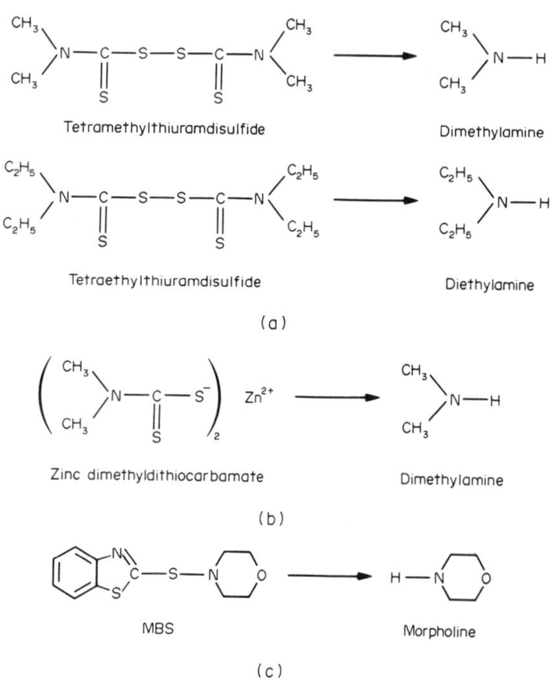

Figure 1
Groups of chemicals that can generate secondary amines: (a) thiuramdisufides, (b) dithiocarbamates and (c) sulfenamides

Production of nitrosamines has been particularly studied in Germany and official limits have been set there for atmospheric nitrosamine levels. The nitroamines concerned, all of which have been shown to be carcinogenic in animal tests, are

(a) *N*-nitrosodi-*n*-butylamine,

(b) *N*-nitrosodiethylamine,

(c) *N*-nitroso diisopropylamine,

(d) *N*-nitrosoethylphenylamine,

(e) *N*-nitrosopiperidine,

(f) *N*-nitrosodiethanolamine,

(g) *N*-nitrosodimethylamine,

(h) *N*-nitrosodi-*n*-propylamine,

(i) *N*-nitrosomethylethylamine,

(j) *N*-nitrosomorpholine, and

(k) *N*-nitrosopyrrolidine.

Of these nitrosamines, types (b), (g) and (j) are common products of elastomer processing, although other members of the list may also be occasionally found.

For general industry in Germany, the limit for total nitrosamines is 1 $\mu g\ m^{-3}$. For most processes in the rubber industry an "orientation value" of 2·5 $\mu g\ m^{-3}$ applies.

Measurement of atmospheric nitrosamine levels in elastomer processing has shown that concentrations may range from well below 1 $\mu g\ m^{-3}$ to greater than 100 $\mu g\ m^{-3}$. There has been a general reduction in atmospheric levels following the removal of NDPA in the early 1980s, although a minority of locations still produces concentrations above the German limit.

Concerns about the generation of nitrosamines from elastomer compounds has prompted a search for accelerators and other chemical systems that will not produce secondary amines. Investigations have also been made of additives such as α-tocopherol which will inhibit nitrosamine formation. These trends have resulted in a decrease in the use of certain common accelerators (e.g., 2-morpholinothio benzothiazole (MBS)) which are known to produce secondary amines (morpholine in this case).

Limits have also been set in some countries for the amounts of nitrosamines or nitrosatable amines in finished products, for example baby teats. Although the methods of extraction vary, current limits are given in Table 22.

It is not clear if nitrosamines play any part in the development of occupational diseases in elastomer processing. It should be noted, however, that at an

Table 22
Limits of nitrosamines and nitrosatable amines in finished products

Country	Volatile nitrosamines (max, ppb)	Nitrosatable amines (max, ppb)
Germany	10–30	200
Switzerland	10	200
Denmark	5	100
The Netherlands	1	100
USA	10[a]	
Canada	10	
Australia	20	400

a For each individual named

atmospheric concentration of $1{\cdot}0\,\mu g\,m^{-3}$, a man working moderately hard would breath in $10\,\mu g$ of nitrosamines over an 8 h shift. This is considerably more than the normal intake of dialkyl nitrosamines from foodstuffs in the UK ($\sim 1\,\mu g/m^{-3}$), although routes of exposure are different.

3.4 Dust Hazards

The rubber industry has historically been prone to high dust concentrations and some operations still struggle to achieve satisfactory control of atmospheric dust levels.

Many chemicals are now supplied in dust-suppressed forms such as pellets or prills or, even better, as wax or polymer-bound granules which are essentially dust free. Use of these forms has markedly reduced exposure to accelerator, antioxidant and other chemical dusts where they have been adopted.

However, it is not usually economical to treat the bulk fillers in this way and powdered material used for antitack, lubricant or support purposes must still be used in forms which may generate dust.

In the early days of elastomer processing, exposure to these dusts must have been grossly excessive by the standards currently set for atmospheric dust levels. There are occasional reports in the early literature of lung fibrosis as a result of exposure to carbon black, talc and other common powdered materials but it is likely that these instances were associated with prolonged exposure to very high dust levels. There is little evidence of radiological or lung function changes at the dust levels set by present legislation, although one US study found minor changes in lung function in men working in the Banbury mixing area of a tire plant.

Talc, which is still widely used in rubber processing, has in the past been associated with a number of cases of lung fibrosis and the term talcosis has been used to describe this syndrome. Some supplies of talc have been shown to contain fibers of some asbestiform minerals (tremolite, anthophyllite, etc.) and it is not clear whether these cases result form exposure to the talc itself or to these fibers. There appears to be a considerable difference between supplies of talc from different sources, with some North American mines producing talc with high fiber levels and other sources being fiber free. Pharmaceutical grades of talc should be free from fiber contamination.

Asbestos itself has been little used in conventional elastomer processing except for specialist areas such as production of sealants or vinyl asbestos flooring compounds.

Carbon black has been studied both for effects on lung function and for carcinogenic effects. As previously stated, no significant lung function changes were found except at concentrations above the current occupational exposure limit of $3{\cdot}5\,mg\,m^{-3}$. No carcinogenic effects have been detected, either in human populations or in animal experiments.

3.5 Monomer Residues in Polymers

Many of the monomers used to make polymers have been shown to be animal carcinogens. Among these must now be counted acrylonitrile, chloroprene and vinyl chloride.

Most solid polymers now contain only small traces of these monomers and atmospheric concentrations during processing are almost always well below the relevant occupational exposure limits, except in very unusual circumstances. However, it should be noted that synthetic latices may contain much higher levels of residual monomer and work with such latices should be carefully checked to ensure that exposures are maintained within existing standards.

3.6 Isocyanates

Diisocyanates, used in the manufacture of polyurethane elastomers, foamed products and so on, can be hazardous to health causing various types of respiratory problems.

The aliphatic diisocyanates have caused problems of skin irritation and sensitization. More detailed information on the hazards of isocyanates and guidance on safe factory working conditions and handling procedures for isocyanates is given by the British Rubber Manufacturers Association (1978).

See also: Elastomers: Additives; Marine Coatings

Bibliography

American Chemical Society 1977 Epidemiological study of chronic illnesses in the rubber industry. Rubber Division Paper 37. ACS, Washington, DC

Assogomma–Cerisie 1984 *Schede Delle Caratteristiche Chemico-Fisiche e Tossicologiche e Delle Misure di Sicurezza Degli Ausiliari dell'Industria Della Gomma*, Vol. 1. Assogomma, Milan, Italy

Baxter P J, Werner J B 1980 *Mortality in the Rubber Industries*. Her Majesty's Stationery Office, London

Bernadinelli L, De Marco R, Tinelli C 1987 Cancer mortality in an Italian rubber factory. *Br. J. Ind. Med.* 44: 187–91

British Rubber Manufacturers Association 1990 *Toxicity and Safe Handling of Rubber Chemicals*, 3rd edn. BRMA, Birmingham, UK

Health and Safety Executive–Rubber National Industries Group 1987 *Clearing the Air—Dust and Fume Control in the Rubber Industry*. HSE-RNIG, Manchester, UK

Holmburg B, Sjostrom B 1977 *A Toxicological Survey of Chemicals used in the Swedish Rubber Industry*. National Board of Occupational Safety and Health, Stockholm

Holmberg B, Westerholm P, Maasing R, Kestrup L, Gumaeliu K, Holmlund L, Englund A 1983 Retrospective cohort study of two plants in the Swedish rubber industry. *Scand J. Work, Environ. Health* 9(S2): 59–68

International Working Group on the Toxicology of Rubber Additives 1984 *Rubber Chemicals Safety Data and Handling Precautions*. WTR, Brussels

Negri E, Piolatto G, Pira E, Decarli A, Kaldo J, La Vecchia 1989 Cancer mortality in a northern Italian cohort of rubber workers. *Br. J. Ind. Med.* 46: 624–8

Nutt A R 1984 *Toxic Hazards of Rubber Chemicals*. Elsevier, London

Parkes, H G, Evans A E J 1984 Epidemiology of aromatic amine cancers. In: Searle C E (ed.) 1984 *Chemical Carcinogens*, 2nd edn., Vol. 1. American Chemical Society, Washington, DC, Chap. 5, pp. 277–301

Parkes H G, Veys C A, Waterhouse J A H, Peters A 1982 Cancer mortality in the British rubber industry. *Br. J. Ind. Med.* 39: 209–20

Rannug A 1984 *Genotoxic Hazards in the Rubber Industry*. University of Stockholm, Stockholm

Schuster R H, Gmünder M, Nabholz F 1989 Inhibition of nitrosamine formation by α-locopherol. *Prep. PRI Symp. Nitrosamines, Challenge and Solutions*. PRI, London

Sorahan T, Parkes H G, Veys C A, Waterhouse J A H 1986 Cancer mortality in the British rubber industry: 1964–80. *Br. J. Ind. Med.* 43: 363–73

Sorahan T, Parkes H G, Veys C A, Waterhouse J A H, Straughan J K, Nutt A 1989 Mortality in the British rubber industry: 1946–85. *Br. J. Ind. Med.* 46: 1–11

Spiegelhalder B, Preussmann R 1982 Nitrosamines and rubber. *N-nitroso Compounds—Occurrence and Biological Effects*. International Agency for Research on Cancer, Lyon, France, pp. 231–43

Spiegelhalder B, Preussmann R 1983 Occupational nitrosamine exposure, 1—Rubber and tyre industry. *Carcinogenesis* 4(9): 1147–52

Vainio H (ed.) 1983 Genotoxic health hazards in the rubber industry. Proceedings of a symposium, November 1982. *Scand J. Work, Environ. Health* 9(S2): 7–93

Veys C A 1973 A study on the incidence of bladder tumours in rubber workers. M.D. thesis, University of Liverpool

Veys C A 1981 Bladder cancer in rubber workers: The story reviewed and updated. *Plast. Rubber Process. Appl.* 1: 207–12

Wang H-W, You X-J, Qu Y-H, Wang W-F, Wang D, Long Y-M, Ni J-A 1984 Investigation of cancer epidemiology and study of carcinogenic agents in the Shanghai rubber industry. *Cancer Res.* 44: 3101–5

Willoughby B G 1984 Vulcanisation fume—Its nature, origin and the options for its control. In: Morrell S H (ed.) 1984 *Progress of Rubber Technology*, The Plastics and Rubber Institute, Vol. 46. Elsevier, London, pp. 143–79

World Health Organization, International Agency for Research on Cancer 1982 *The Rubber Industry*, IARC Monographs on the Evaluation of the Carcinogenic Risk of Chemicals to Humans, Vol. 28. IARC, Lyon, France

Zhang Z-F, Yu S-Z, Li W-X, Choi B C K 1989 Smoking, occupational exposure to rubber and lung cancer. *Br. J. Ind. Med.* 46: 12–15

A. R. Nutt
[Arnold Nutt and Company, Atherstone, UK]

High-Modulus High-Strength Organic Fibers

High-modulus high-strength organic fibers are polymeric fibers characterized by high specific tensile modulus and tensile strength when compared to metals and ceramics, as illustrated graphically in Fig. 1. These fibers, especially the du Pont aramid fiber, Kevlar, have found utility in a broad range of composite and specialty applications from secondary aircraft structure to antiballistic vests. All share a serious deficiency in low compressive strength, which precludes their use in composites for most primary-structure applications and requires well-designed composite assemblies for many others. In applications that require minimum weight and low fiber thermal and electrical conductivity, however, high-modulus organic fibers are often the material of choice.

1. Historical Background

From the late 1960s through the mid-1970s a series of announcements from US industrial laboratories

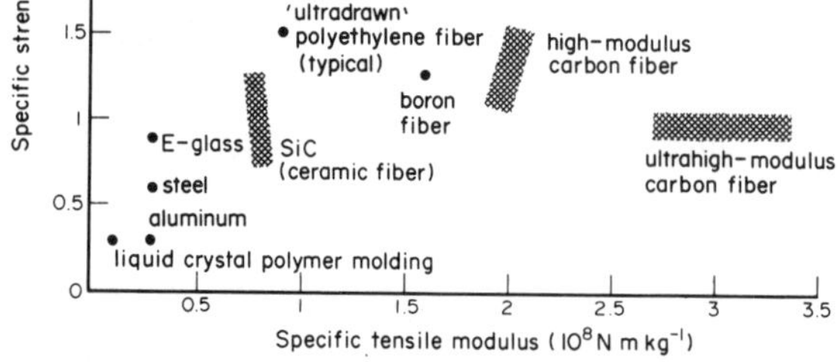

Figure 1
Specific tensile modulus and specific strength for various fibers

(du Pont, Monsanto, Eastman, Celanese) made it apparent that the level of achievable fiber tensile properties had taken a major jump—tensile modulus up by a factor of about 5, tensile strength up by a factor of 2–3—when compared with the commodity industrial fibers such as PET or nylon. All of these commodity fibers were composed of stiff, highly aromatic molecules and were characterized by very high axial molecular orientation in the solid state (orientation function >0.95). These discoveries constituted a major milestone in fiber and polymer technology.

As exciting as the discovery that fibers with tensile properties approaching their theoretical values could be spun was the recognition that many of the stiff molecules exhibited nematic liquid-crystalline behavior in solution or in the melt. The nematic behavior of these stiff polymers leads to low elongational viscosity (ease of orientation) and a solid-state structure similar to a highly oriented assembly of closely packed rods (extended-chain structure). Figure 2 illustrates the processing of these polymers in contrast to a conventional polymer such as PET.

In the mid-1970s it was shown that various essentially paralinked aromatic polyester copolymers exhibit nematic behavior in the melt. These thermotropic copolyesters could be easily melt spun into highly oriented fibers of high tensile modulus and high tensile strength.

The potential for both high modulus and liquid crystallinity in linear synthetic polymers had been

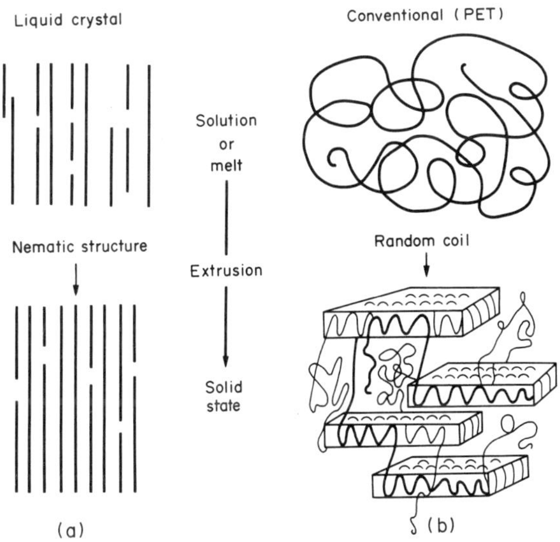

Figure 2
Formation of (a) liquid-crystal fibers with extended chain structure, high chain continuity and high mechanical properties, (b) conventional (PET) fibers with lamellar structure, low chain continuity and low mechanical properties

predicted well ahead of discovery, the former by Mark in 1936 and the latter by Flory in 1956. Mark calulated an upper modulus limit of about 350 GPa. Although it has taken forty years to reach them, values as high as 300 GPa have now been realized in continuous fiber.

Flory, using lattice-model-generated phase diagrams for monodispersed rigid rods in solution, accurately showed the type of phase relationships observed for lyotopic nematic polymers such as poly(*p*-phenylene terephthalamide) (PPT) or poly(*p*-phenylene benzobisthiazole) (PBZT). He also showed that liquid-crystalline behavior was possible for the melt of a stiff-chain polymer.

Polyethylene fibers, carefully processed to minimize internal defects and drawn slowly to very high molecular orientations, are being commercially produced by the Allied Signal Corporation, and others, to mechanical property levels fully competitive with the fibers based on liquid-crystalline polymers. Attempts to extend this technology to other conventional polymers has not proven successful.

It is now evident that two generic routes to high-modulus high-strength fibers exist, namely (a) the spinning of stiff, nematogenic polymers which easily form highly oriented structures in the solid state, and (b) the morphological manipulation of flexible, conventional polymers into highly oriented extended-chain fibers through complex processing.

Two useful subclassifications are the "almost mesogens" such as the aramid–hydrazid copolymers developed by Monsanto and the true molecular rods such as PBZT championed by the USAF. The former polymers are "conventional" in quiescent solution but exhibit nematiclike behavior under elongational flow. The latter are characterized by a persistence length in solution that is equivalent to the fully extended length of the molecular chain.

2. *Chemical Structures and High-Modulus Fiber Formation*

The unifying characteristics of stiff polymers yielding high-modulus fibers are a high degree of aromaticity, high planarity and essential linearity in the chain backbone. Most common are *p*-phenylene or 2,6-naphthylene moieties linked by ester or amide bonds. Even stiffer structures can be produced by incorporating heterocyclic rings in the chain backbone such as bibenzooxazoles or bibenzothiazoles. Examples of such structures are shown in Fig. 3, grouped by mesogenic behavior. The heterocyclic polymers suffer in general from intractability: they are infusible and exceedingly difficult to dissolve, making fabrication an arduous task. Alternatively, only a minor portion of backbone aromatic moieties can have other than linear alignment if high modulus is to be obtained.

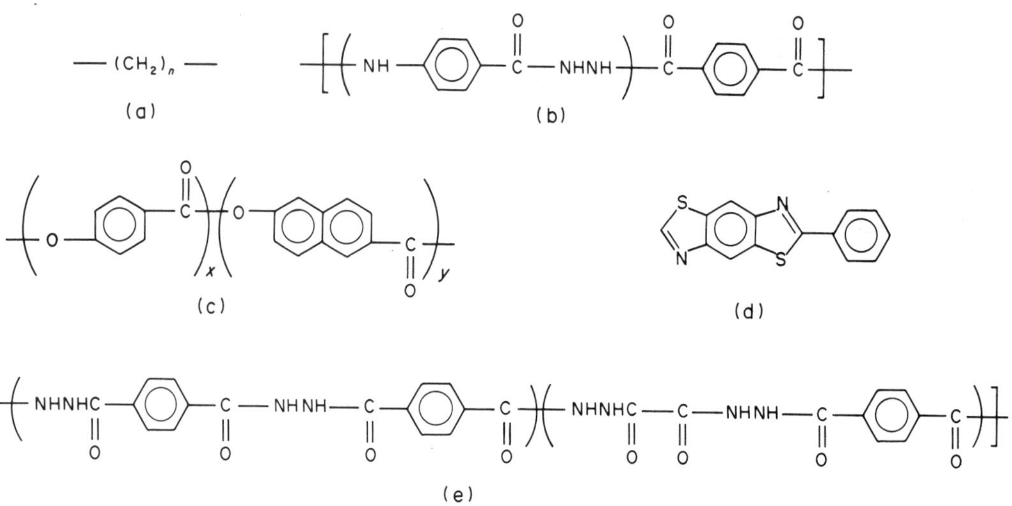

Figure 3
Examples of typical polymers suitable for high-modulus high-strength fibers: (a) polyethylene, (b) aramid: poly(*p*-phenylene terephthalamide), (c) thermotropic copolyester, (d) ordered polymer: poly(*p*-phenylene benzobisthiazole), (e) copolyhydrazide

The rodlike nature of *para*-oriented aromatic polymers, in addition to making the polymer difficult to dissolve or melt, dramatically increases the viscosity of a solution or melt over that of nonrigid structurally isomeric polymers such as poly(*m*-phenylene isophthalamide), the *meta*-analogue of poly(*p*-phenylene terephthalamide) (Kevlar). The solution to this problem is either to spin from very dilute dopes (>6%) or to process from the nematogenic state if possible.

Several important benefits are derived from polymer liquid-crystalline behavior, involving the process and the product. For the lyotropic polymers (those which are liquid crystalline in solution), as the concentration is raised above the critical value the viscosity falls rapidly until a minimum is reached at a substantially higher concentration than the critical value. Thus it is possible to obtain a much higher concentration of polymer in the spinning solution, thereby decreasing the quantity of spinning solvent to be recycled or disposed of. For polymers of useful molecular weight, the concentration of isotropic solutions is limited to 4–6 wt%, whereas concentrations of 15–25 wt% are common for liquid-crystalline solutions. Similar benefits of low viscosity for a given molecular weight are also realized in thermotropic systems.

A second consequence of liquid crystallinity in stiff-chain polymers is the ease of attaining very high mechanical properties directly in spinning. This is a result of the low elongational viscosity associated with nematogens, which leads to the easy formation of highly axially oriented extended-chain structures during the spinning operation. While very useful property sets are obtained as spun, mechanical properties can often be improved through annealing. In the aramid case, annealing leads to a doubling of modulus through an orientation-perfection mechanism. In the case of the thermotropic copolyesters (those which are liquid crystalline when molten), strength improves during annealing through a solid-state polymerization mechanism. Only in the case of the "almost mesogens," such as the hydrazide–amide copolymers, is the processing similar to conventional fiber production, that is, spinning followed by drawing to improve axial orientation. This latter process is also utilized in the production of gel-spun polyethylene where the production of a gel fiber precursor allows drawing to very high draw ratios (20–40) and very high mechanical properties.

3. *Major High-Modulus High-Strength Organic Fibers*

The structural formulae shown in Fig. 3 include the compositions that have received the most developmental attention as high-modulus fibers. Table 1 summarizes and compares important mechanical and other end-use critical properties of these fibers. Note that for all the fibers listed, lateral mechanical properties, as exemplified by the compressive strength, are several orders of magnitude lower than axial tensile properties. This manifests itself in poor compressive performance for this class of fibers. All of these fibers share excellent environmental stability, low coefficient of thermal expansion

Table 1
Properties of high-modulus high-strength organic fibers

Fiber	Typical elastic modulus (GPa)	Typical tensile strength (GPa)	Lateral strength (GPa)	Density ($kg\,m^{-3}$)	T_g (°C)
Gel PE	140	3	0.1	970	−85
Kevlar 49	120	3	0.6	1400	360
Kevlar 29	60	3	0.6	1400	360
Thermotropic copolyester	60	3	0.1	1400	150
Ordered polymer	265	3	0.3	1500	
X-500	100	2.5	N/A	1500	N/A

N/A, not available

($\sim -4\times10^{-6}\,K^{-1}$) and excellent dimensional stability. Not listed in Table 1 is fiber price, which in all cases was over US$10 per pound (US$22 per kilogram) in 1989 and in most cases over US$20 per pound (US$44 per kilogram), a factor that has limited the acceptance of these fibers in many markets. To some extent, the high price reflects the absence of a volume-generating large market such as tire cord, although the combination of specialized monomers and/or process complexity which characterizes the production of high-modulus organic fibers probably precludes low-cost variants. As such, these fibers are finding acceptance in specialty markets such as ballistic protection fabrics, lightweight rigid composites for aerospace applications, and ropes and cables for the anchoring of large structures such as oil-drilling platforms.

3.1 Aramid Fibers

Aramid fibers are commercially produced by du Pont (Kevlar), AKZO (Twarlon) and Teijin (Technora). Within the aramid class there are two basic variants, a low-modulus type (~60 GPa) and a high-modulus type (~120 GPa). Examples of these are Kevlar 29 and Kevlar 49, respectively. All of the aramids possess similar tensile strengths of about 3 GPa. Recently, du Pont has introduced Kevlar 149, a somewhat stronger and stiffer variant. Kevlar 29 is used primarily for such products as tensile members (ropes, cables and webbings) and ballistic cloth, while Kevlar 49 and 149 are reinforcing fibers for high-performance fiber-reinforced composites. To overcome the poor compressive performance of these fibers while utilizing their high specific tensile properties and toughness, they are often used in hybrid constructions with glass or carbon.

The aramid fibers are spun from liquid-crystal dopes through a dry-jet wet-spinning process which yields high tensile property fiber directly. The process is fundamentally a melt-spinning of the dope (~20% solids in 100% sulfuric acid) which is crystalline at room temperature and is extruded at elevated temperature into a coagulation bath (water) to remove the residual solvent. The difference between the high- and low-modulus variants of the aramids is purely structural. Kevlar 29 may be converted to Kevlar 49 through an annealing process which improves overall chain orientation and structural perfection. Advantages of the aramid fibers include high use temperatures (stable up to about 300 °C in the absence of hydrolysis agents) and very low creep. Du Pont scientists have estimated that Kevlar fibers will sustain over half their one-second breaking load for longer than 100 years based on the absolute-rate theory of creep failure for a fixed load.

3.2 Polyethylene Fibers

A number of commercial high-modulus polyethylene fibers entered the market place during 1986–1988, including Spectra (Allied Signal), Dyneema (DSM/Toyobo) and Tekilon (Mitsui Toatsu). Because of the low density of polyethylene ($950\,kg\,m^{-3}$ vs $1400\,kg\,m^{-3}$ for the other fibers) the specific tensile properties of polyethylene are the highest achieved with any organic material (see Fig. 1). The lateral properties of high-modulus polyethylene are the same order of magnitude as those found for the fibers spun from liquid-crystalline polymers, leading to poor compressive performance. Polyethylene fibers also exhibit high levels of creep and are limited, by the low polyethylene melting point, to a use temperature of around 100 °C. Chemically, polyethylene is highly inert and is the best performer of the high-modulus fiber candidates in alkaline environments.

High-modulus polyethylene fibers are produced via a gel-spinning process in which a low-concentration solution of ultrahigh molecular weight ($M>2\times10^6$) polyethylene is extruded to form a gel precursor fiber. This precursor fiber is then hot drawn to draw ratios of 20 or more to produce a very highly oriented fiber with an extended-chain fibrillar microstructure. The high-molecular-weight polymer is necessary to achieve the desired high strength but cannot be conventionally melt-spun because of its high melt viscosity. Lower-molecular-weight-polyethylenes are melt spinnable but do not give the high values of strength after drawing. It is thought that the gel-state precursor fiber has a minimum of internal flaws which allows drawing to the very high draw ratios necessary to obtain the desired properties.

3.3 Thermotropic Copolyester Fibers

The thermotropic copolyesters can be easily melt spun to moduli similar to Kevlar 29 (see Fig. 1), although a range of moduli spanning the aramid axial tensile-property range can be made. Strength can be enhanced through annealing to the 3 GPa range representative of all the high-modulus high-strength organic fibers. The product closest to commercial reality is the fiber Vectran, being evaluated in a joint feasibility study by Hoechst Celanese and the Kuraray company. Vectran is based on the commercial Hoechst Celanese Vectra product line of thermotropic copolyesters for plastics applications. Advantages of this product include very high cut resistance (up to eight times that of the aramids) and excellent hydrolytic stability. Use temperatures are intermediate between Kevlar and polyethylene, in the range 150–200 °C depending on composition.

3.4 Other High-Modulus Organic Fibers

Two other approaches to high-strength high-modulus organic fibers worthy of note are the aromatic heterocyclic ordered polymers developed under the auspices of the USAF Materials Laboratory and the "almost mesogens" represented by the hydrazid copolymers represented by the X-500 products developed by Monsanto in the early 1970s. The former are true molecular rods and have exhibited the highest axial tensile moduli achieved in an organic fiber, values in the region of 300 GPa. Low compressive properties and high inherent cost, however, have slowed commercial development of these fibers. The X-500 fibers were introduced by Monsanto as a competitor to Kevlar. These fibers are conventionally wet-spun and then drawn to high tensile properties. Compared to the aramids, the X-500 fibers offered no advantage and were withdrawn from the market by the mid-1970s.

4. Structure–Property Relationships

A distinctive feature of high-modulus organic fibers is the very high axial molecular orientation, which leads to extreme anisotropy of mechanical properties. In the transverse direction, the strength is only about 20% of the strength along the fiber axis and the modulus is less than 10% of the fiber-axis tensile modulus. In essence, the filaments may be described as a hierarchy composed of fibrillar structures covering a range of several orders of magnitude in diameter. Figure 4 shows a structural model developed for the thermotropic copolyester fibers but which is a reasonable representation of the microstructure of all the fibers discussed. The fibrils are highly aligned along the fiber axis with relatively few connections between the fibrillar entities. In tension, it has been shown that both strength and modulus can be attributed to molecular parameters (chain modulus, chain interactions, molecular weight) and that the microstructure plays a minor role. This suggests that the fibrils are very long compared to the effective testing gauge lengths.

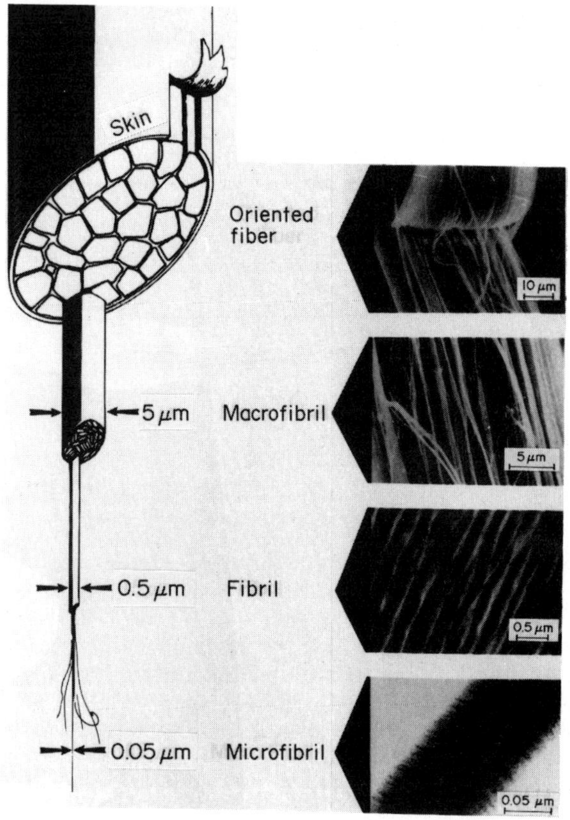

Figure 4
Structural model of a typical thermotropic copolyester fiber

In compression, the situation is much the same as in the transverse direction. The compressive yield as determined in 60 vol.% Kevlar–epoxy composites was less than 20% of the tensile stress for failure. In compression, the analogy to pushing on a steel cable is not a gross simplification. Also related to the low lateral properties are the relatively poor flex fatigue and abrasion resistance exhibited by all these fibers.

5. Applications

High-modulus high-strength organic fibers have found markets in areas where very high performance is required and high costs can be tolerated. Examples of such markets include primary and secondary structures for aerospace applications, ballistics protection for the military, prosthetic devices, sporting goods and ropes and cables for anchoring very large structures. The effectiveness of high-modulus organic fibers in these applications is a function of their unique balance of properties, including low density, high tensile properties, chemical inertness and good thermal stability. This combination allows the production of lightweight, stiff and strong structures whose performance cannot be matched by metals, ceramics or other fiber-reinforced composites.

The increased usage of high-modulus organic fibers is limited by their poor compressive performance as a class and their very high cost. While thermal stability limits the use of polyethylene fibers to a degree, the predominance of applications require thermal stability from about $-20\,°C$ to perhaps $100\,°C$, which all of these fibers possess. To some extent, the compressive problems can be overcome by hybridization with other reinforcing fibers or through composite design (crosslamination, three-dimensional woven structures, etc.). The largest factor restricting the market for high-modulus high-strength organic fibers is cost.

6. Conclusions

The technologies leading to high-modulus high-strength organic fibers are relatively well understood and fibers with tensile moduli of greater than 70 GPa at a strength level of about 3 GPa can be produced from literally hundreds of different polymers. In the limit of the very high molecular orientation inherent in these fibers, all possess similar microstructure and high property anisotropy. Thermal stability varies over a broad range as expected from the range of molecular chemistries being employed, but all these fibers are stable in the neighborhood of room temperature. The limitations to the commercial acceptance, and hence to the replacement of other materials by high-modulus high-strength organic fibers or their composites are (a) the low compressive strength which limits the utility of the fibers to applications where the use is in tension, and (b) the high costs which make less sophisticated approaches to the solution of materials problems attractive on a cost–performance basis.

By 1988, Kevlar usage had grown to greater than 2.5×10^7 kg annually, all in the specialty markets already described. For this market size to increase substantially, the cost–performance of high-modulus organic fibers must be improved. Routes to such an improvement include the increase of lateral strength of these highly oriented fibers, and reduction of monomer costs through new or improved chemical syntheses. In the absence of success in these areas, it is likely that usage of high-modulus high-strength organic fibers will be relatively limited and that few of the fiber chemistries known will survive outside the laboratory.

See also: Polyethylene Processing to Produce Ultimate Properties

Bibliography

Black WB 1979 Stiff-chain aromatic polymer solutions, melts and fibers. In: Miller R L (ed.) 1979 *Flow-Induced Crystallization*. Gordon and Breach, New York, pp. 245–306

Black W B 1980 High modulus/high strength organic fibers. *Annu. Rev. Mater. Sci.* 10: 311–62

Calundann G W, Jaffe M 1982 Anisotropic polymers, their synthesis and properties. *Proc. Robert A. Welch Found. Conf. Chem. Res.* 26: 247

Ciferri A, Ward I M (eds.) 1979 *Ultrahigh Modulus Polymers*. Applied Science, London

du Pont 1978 Kevlar 49 aramid. du Pont Technical Bulletin K-2. E I du Pont de Nemours, Wilmington, DE

Jaffe M 1987 High modulus polymers. In: *The Encyclopedia of Polymer Science and Engineering*. Wiley, New York, pp. 699–722

Jones R S, Jaffe M 1985 High performance aramid fibers. In: Lewin M, Preston J (eds.) 1985 *High Technology Fibers*. Dekker, New York, pp. 349–92

Kavesh S, Prevorsek D C 1983 High tenacity, high modulus polyethylene and polypropylene fibers and intermediates therefore. US Patent No. 4,413,110

Morgan P W 1977 Synthesis and properties of aromatic and extended chain polyamides. *Macromolecules* 10: 1381–90

Sawyer L C, Jaffe M 1986 The structure of thermotropic copolyesters. *J. Mater. Sci.* 21: 1897–913

Smith P, Lemstra P J, Kalb B, Pennings A J 1979 Ultrahigh strength polyethylene filaments by solution spinning and hot drawing. *Polym. Bull.* 1: 733–6

Yang H H 1989 *Aromatic High-Strength Fibers*. Wiley, Chichester, UK

M. Jaffe
[Hoechst Celanese Research Division, Summit, New Jersey, USA]

High-Performance Composites with Thermoplastic Matrices

Throughout the 1970s and 1980s, thermosetting resins were the principal type of matrix used for high-performance fiber composites. The main disadvantages of these resins are a low strain to failure and brittleness, lengthy cure cycles, moisture absorption and its deleterious effects on mechanical properties, and the preclusion of forming operations and certain kinds of repair once the matrix has been fully cured.

Some of these problems can be overcome by using a thermoplastic polymer for the matrix. Compared with thermosets these materials usually have a higher strain to failure and a lower moisture absorption, they can be shaped when heated and repaired, do not require a lengthy cure cycle and have a very long storage life at ambient temperature. Their principal disadvantage has been the difficulty in fabricating fiber-reinforced composites. This arises because, at a given temperature, the viscosity of a thermoplastic is much greater than that of an uncured thermoset. It may not be possible to reduce this sufficiently for fiber impregnation, by raising the temperature, before the polymer degrades. There are also problems of bonding to the fibrous reinforcement. Early composites with thermoplastic matrices contained an excessive void volume and damaged or misaligned fibers, and were inferior in performance to thermoset-based specimens. The approach to the fabrication of thermoplastic-matrix composites that has evolved has been to concentrate on producing fiber-reinforced sheets and tapes by, for instance, solvent and hot-melt impregnation methods, aided by modification of the basic polymer. These feedstocks can then be turned into artifacts by hot pressing, film stacking and hot pressing, tape winding and other techniques. Short-fiber-reinforced molding granules have been developed but, because of the reduced fiber orientation in the final product, composites based on these materials, although having a better degree of isotropy, have poorer properties in a specific direction than unidirectionally reinforced composites.

Because of the difficulties initially encountered in fabricating thermoplastic-matrix composites, the tendency has been, for high-performance continuous-fiber composites, to concentrate on thermoplastics which give maximum resistance to temperature rather than on the use of the commoner, commodity thermoplastics.

1. Types of Thermoplastic Used

Ideally, the thermoplastic polymer matrix should have a high softening point and resistance against pyrolysis and chemical degradation. These are properties that may be shown by crystalline polymers with a rigid backbone chain, regularly spaced substituents and strong interchain forces due to van der Waals or polar interactions. Polymer properties usually show a marked reduction at the glass transition temperature T_g, but crystalline materials have useful property retention until close to the crystalline melting point T_m, as the crystalline phase holds the molecular chains together. Practically, it is necessary to compromise between the desired properties and the ability to process the polymer without causing degradation. Aromatic segments in the main polymer chain can be used to increase chain stiffness and reduce susceptibility to pyrolysis while ether linkages increase chain flexibility. Further details are given by Marks and Atlas (1965), Rose (1984) and Woodhams (1985).

Among the polymers used as matrices are polysulfones, aromatic polyketones, modified polyimides, polysulfides and liquid-crystal systems. To obtain a processible polysulfone, polyether sulfone (PES) or polyaryl sulfone, the backbone polymer chain of the simplest aromatic polysulfone, is modified by the incorporation of ether linkages. Commercial polymers or this type include Udel and Radel (Amoco Chemicals), Ultrason S (BASF) and the polyether sulfones Victrex and Ultrason E, produced by ICI and BASF, respectively. Despite their regular structure, polysulfones are essentially amorphous. These materials are tough, resistant to nuclear but not ultraviolet radiation and are subject to environmental stress cracking. The resistance to most acids, alkalis and aliphatic hydrocarbons is good but the polymers are attacked by concentrated sulfuric acid and aircraft hydraulic fluids. Polysulfones are soluble in dimethyl formamide, dimethyl acetamide, dichloromethane and *N*-methyl pyrrolidone. Temperatures in excess of 300 °C are required for their processing.

Polyketones are represented by ICI's polyether ether ketone, Victrex PEEK, and their polyether ketone, Victrex PEK. BASF and Hoechst also manufacture polyether ketones known as Ultrapek and Hostatec, respectively. Unlike the polysulfones, polyketones are crystalline. Both PEEK and PEK have excellent resistance to nuclear radiation and a wide range of chemicals and solvents, though they are attacked by concentrated sulfuric and hydrofluoric acids. The materials are tough and not susceptible to environmental stress cracking. Processing is more difficult than for polysulfones and a temperature of approximately 390 °C is required. In addition, it may be necessary to control the degree of crystallinity in the final product, to give the desired properties, by cooling at a specified correct rate. There is evidence that reprocessed PEEK may undergo some thermal crosslinking.

Modified polyimides were developed in attempts to retain the heat resistance of polyimides while rendering their processing more tractable. Torlon, produced by Amoco Chemicals, is a polyamide imide (PAI), which is not attacked by aliphatic, aromatic chlorinated or fluorinated hydrocarbons, dilute acids, aldehydes, ketones, esters or ethers, and is very resistant to nuclear and ultraviolet radiation. It is soluble in *N*-methyl pyrrolidone, is attacked by alkalis and is notch sensitive. Its melt temperature is 355 °C and the processed polymer needs a prolonged period of heat treatment to promote chain extension. The time involved can be up to 100 h or more. Polyether imide (PEI) (General Electric Plastics), in which imide groups are interspersed with ether linkages, is sold under the name of Ultem. It has a good resistance to nuclear and ultraviolet radiation and to mineral acids, dilute bases and aliphatic hydrocarbons, but is soluble in halogenated hydrocarbons. It is notch sensitive and subject to environmental stress cracking. When PEI is burnt, the smoke generation is low and the gases produced nontoxic. The processing temperature is in the range 350–425 °C.

Probably the most widely produced high-performance thermoplastic is polyphenylene sulfide (PPS). Phillips Petroleum market the material under the name of Ryton. Bayer also provide the polymer, Ciba Geigy have introduced a variant called Craston and Solvay in conjunction with Tohpren and Celanese together with Kureha Chemicals are entering the market. The Celanese product is more linear and purer than Ryton. Idemitsu have recently developed a very-high-molecular-weight ultrapure product. The linear version of the polymer is highly crystalline and may crosslink oxidatively. The resistance to acids, aqueous bases, halogenated hydrocarbons and alcohols is good but the polymer is attacked by concentrated sulfuric acid and some amines. It is soluble in chloronaphthalene at elevated temperatures. Though brittle, its environmental stress resistance is good, as is its resistance to nuclear, but not ultraviolet, radiation. Smoke generation is low and nontoxic gases are produced on burning. Polyphenylene sulfide is the simplest member of the polyarylene family. Recently two new members known as polyarylene sulfide polymers 1 and 2 (PAS 1 and PAS 2) have been produced by Phillips Petroleum. The first is a semicrystalline material and the second is amorphous in nature. Mechanical properties appear similar to those of polyphenylene sulfide though the strain to failure of PAS 2 is higher, at 8%. The glass transition temperatures of PAS 1 and PAS 2 are 145 °C and 215 °C, respectively, and the crystalline melting point of PAS 1 is 340 °C.

Liquid-crystal polymers are based on aromatic thermoplastic polyesters. Varieties available include Xydar, produced by Dartco, Vectra made by Celanese and Ultrax (BASF). The polymers are self-reinforcing structures in which molecular alignment is frozen in on cooling, giving a significantly higher modulus and strength than those of the other polymers considered here. To ease processing, the molecular backbone chain is kinked by inserting hydroxy naphthoic acid so that the molecules do not fit together too well. The strain to failure is low, presumably because of the fibrous nature of the materials, and behavior is anisotropic. The heat and chemical resistance are excellent and in organic and chlorinated solvents the behavior of Vectra and Xydar is superior to that of the amorphous polymers such as polysulfone, polyether sulfone and polyether imide. In addition, the materials are not attacked by acids, bases, alcohols or esters. The resistance to environmental stress cracking and nuclear radiation is good. The ability of the highly ordered molecular chains to slide over one another in the molten state while retaining their relative orientation enables these materials to be processed relatively easily. The melt temperature of a typical injection or extrusion grade of Vectra is 275–330 °C. For Xydar the figure is about 400 °C and the melt viscosity is approximately two orders of magnitude greater than that of Vectra under similar conditions.

Further details of the various polymers discussed here, except the liquid-crystal systems, are given by Brydson (1982). Liquid-crystal systems are discussed by Collyer and Clegg (1986a). The basic repeat units of many of the systems are illustrated in Fig. 1.

2. *Properties of High-Performance Thermoplastics*

Some typical room-temperature properties of unreinforced grades of various polymers are listed in Table 1. The information has been taken from trade data and the other sources indicated. The values quoted depend on the grade of polymer (e.g., molecular weight, additives to control crystallinity), the nature of the test specimen (e.g., degree of crystallinity, preferred orientation) and the technique used to determine the property. The properties of liquid-crystal polymers are anisotropic and here the data refer to the direction of molecular alignment, except for the coefficient of thermal expansion where the larger value is for the transverse direction. The limiting oxygen index (LOI) indicates the percentage of oxygen in an oxygen–nitrogen mixture necessary to sustain combustion. Since air contains 22% oxygen, all the polymers detailed would be self-extinguishing in air. The continuous service temperature is based on the Underwriters' Laboratories (USA) test method. This gives the temperature at which 50% of the measured property (e.g., tensile strength) will be retained after a period of up to 10^5 h exposure under static conditions. Stress, with or without an aggressive chemical environment, could reduce this figure. It is not possible to condense details of resistance to specific chemicals or, where it is available, fatigue and creep data. The manufacturers' data sheets should be consulted in these cases.

In 1984, the estimated production figures for the West for some of the high-performance polymers listed here were as follows: polyphenylene sulfide 8200 t, polysulfone 6500 t, polyether sulfone 700 t, polyether imide 300 t and polyether ether ketone 100 t. The total market is projected to grow by 460% by 1995. For the year 1984, 8% of all high-performance composites were based on thermoplastic matrices. This represents 400–600 t of

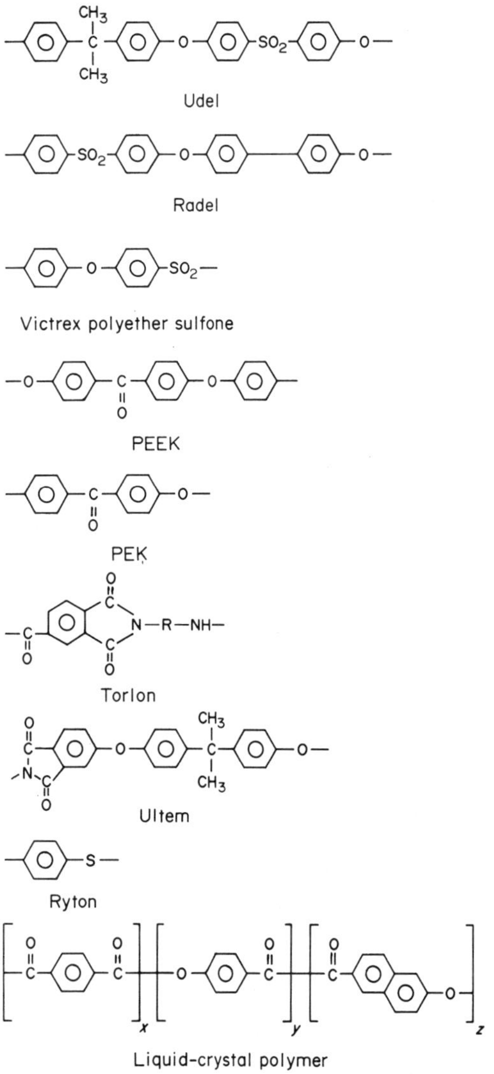

Figure 1
Basic structural repeat units of thermoplastic polymers

thermoplastic. The cost of the unfilled grades (early 1987) ranged from £4000 t^{-1} for polysulfone, through £15 000–£30 000 t^{-1} for liquid-crystal polymers to £30 000 t^{-1} plus for PEEK.

3. *Fabrication of Thermoplastic Preforms*

Mineral-filled, short-glass and carbon-fiber-reinforced thermoplastic materials suitable for injection molding, or bulk materials from which components can be machined, have been available for some years. Though the strength and modulus are increased and thermal expansion and shrinkage reduced, these types of material are not high-performance composites. To qualify for inclusion in this category, reinforcement with continuous, aligned or woven glass, carbon or aramid fibers, or possibly a fiber mat is required. Usually the fabricator buys reinforced thermoplastic preform (analogous to a thermoset prepreg) in the form of a sheet, tape or pultruded section and uses this to produce the final artifact, though some manufacturers make the final, shaped product from separate fibers and polymer. Developments involving liquid-crystal polymers and long, but not continuous, fiber reinforcement are discussed by Cogswell (1987).

In making a preform, the aim is to produce a sheet or tape of constant thickness in which the fibers are aligned and uniformly distributed and individually wetted by the matrix. Since they are amorphous and hence more readily soluble, solvent impregnation was first used with polysulfones, polyether sulfone and later polyamide imide. The sheet material produced must be thin so that the solvent can be removed before any further fabrication operations. Crystalline or semicrystalline polymers tend to have better solvent resistance and melt impregnation is used to fabricate feedstock. Sometimes, to improve alignment, unidirectional fibers are stitched together with thermoplastic fibers based on, for instance, PEEK. Intimate blends of carbon fibers and PEEK, PEK or polyether sulfone fibers can be knitted, woven or combined together as filaments. This leads to better fiber wetout and an absence of polymer-rich areas. Another method of preparing a fabric or sheet is to impregnate with a suspension of fine ($>5\ \mu m$) polymer particles suspended in water or another inert liquid. This method is applicable to polysulfones and polyether ketones as well as polyether imide and polyphenylene sulfide, but not to liquid-crystal polymers since these cannot be cryogenically ground to produce fine particles, because of their inherent toughness and fibrous nature. Pultruded sections or tape may be prepared using a similar approach. Fibers are pulled through powdered resin (30–250 μm) in a coating unit and then shaped and heated in a die. The production rate can be as high as 1 m s^{-1} with the fiber volume loading not deviating by more than ± 1 vol.%. The method has been used with polysulfones, polyphenylene sulfide, PEEK, polyether imide and polyamide imide. Finally, the fibers may be coated with a prepolymer which is then reacted to give a linear thermoplastic. Commercial manufacturers frequently refer to their process as proprietary.

It has been suggested that reinforced thermoplastic sheet and pultruded sections have a lower fiber loading (~ 55 vol.%) and poorer fiber alignment than their thermoset counterparts. The preform products should be stored in clean, dry conditions, since the matrices will pick up small amounts of

Table 1
Properties of unreinforced polymers: * indicates notched specimens, + indicates specimens exposed for 10^5 h

	Polysulfone	Polyether sulfone	Polyether ether ketone	Polyether ketone	Polyether imide	Polyamide imide	Polyphenylene sulfide	Vectra (LCP)	Xydar (LCP)
Crystalline	no	no	semicrystalline		no	no	semicrystalline	liquid crystal	liquid crystal
Density (10^3 kg m^{-3})	1.25[a]	1.37	1.26–1.32		1.27	1.4[d]	1.36[i]	1.37–1.4	1.35–1.4[j]
σ_f(MPa)	106	129	170		145	197	154[i]	169–245	110–143[j]
E_f(GPa)	2.7	2.6	3.6		3.3	4.8[d]	3.5[i]	9–15.2	11–13[j]
σ_t(MPa)	60–75[a]	84	93	110	105	93[d]	84[i]	165–188	80–123[j]
E_t(MPa)	2.5[a]	3.2[e]	3.2[e]		3.0		3.3[e]	9.7–19.3	
ε_f(%)	50–200	40–80	50		60	17[f]	4[i]	1.3–3.0	3.3–4.9[j]
Izod impact J m^{-1})	70	76–84*	83*		50*	53[f]	21*[i]	45–530	75–210[j]
Fracture toughness (MN m$^{-3/2}$)	2.2[b]		7.5[b]						
Critical work of fracture, G_{Ic}(KJ m^{-2})			6.6[l]			1.1	1.4		
LOI	32	36	35		47	43	44[i]	35–50	42–47[j]
Heat distortion temperature (HDT) (1.8 MPa) (°C)	174[a]	200	150	165	200	273[f]	136[e]	222	316–355[j]
T_g(°C)	190[c]	230[c]	143	165	217		93[c]		
T_m(°C)			334	365			285	280	423[j]
Continuous service temp. (°C)	150	180[+]	250[+]		170	230–260[h]	230–240[h]		240[k]
α(°C$^{-1}\times 10^{-6}$)	55.8	55	47		62	63[d]	54	−5 – +75	
Water absorption in 24 h at RT (%)	0.2[d]	0.43	0.1[d]		0.25[d]	0.3[d]	0.2[d]	0.02–0.04	

a Pye 1982 b Gotham and Hough 1984 c Muzzy and Kays 1984 d English 1985 e Specmat (UK) Ltd trade data f Collyer and Clegg 1985a g Collyer and Clegg 1985b h Woodhams 1985 i O'Connor 1987 j Collyer and Clegg 1986a k Collyer and Clegg 1986b l Leach and Moore 1985
All other entries are taken from trade data

water (see Table 1). Unlike thermoset prepregs, which become tacky at room temperature, thermoplastic-based materials are stiff and do not become drapable until above their T_g or, for crystalline polymers, above T_m. If the material is unidirectionally reinforced it will easily split unless handled carefully. A soldering iron is recommended for joining pieces of preform before solidification.

Starting materials based on all the polymers listed in Table 1 except PEK, which is a very recent product, and liquid-crystal polymers are in principle available, though most commercial emphasis appears to be on reinforced PEEK, polyphenylene sulfide and polyether imide products. Phillips Petroleum produce pultruded structural sections and an aramid-fiber tape, both with a polyphenylene sulfide matrix, but the most widely used system appears to be ICI's APC 2 based on carbon fibers in a PEEK matrix. At least one UK manufacturer produces finished articles based on liquid-crystal polymers.

4. *Fabrication of Thermoplastic Composite Artifacts*

Given a supply of preform the final step is to convert the material into the finished article. One method of doing this involves stacking layers of fiber-reinforced preform with the desired orientation, possibly interspersed with very fine layers of polymer film of either the same type as used in the preform or another kind. The stack is processed by pressing in a matched mold at a pressure of 1–10 MPa and at a temperature in excess of T_g, or T_m if appropriate. Processing times vary from 30–300 s while it has been suggested that 30 min may be required. The exact time will depend on the thickness of material used and laying up and cooling may increase the process time very considerably. Care must be taken in molding large, shaped components with shallow curvature because of the long slip path between adjacent plies which can result in fiber wrinkling when a compressive force is applied. In this case, shaping individual preform sheets before fusing them is recommended. As with thermoset fabrication, it is necessary to remove air and any solvents that have been used in preparing the preform before final consolidation. To do this, a properly designed mold with adequate venting or the application of a vacuum is required. Another method is to vacuum form using an autoclave in a manner similar to that used for a thermoset prepreg. A disadvantage of this method is the high temperature that is required.

Certain sheet-metal-forming techniques can be used to form thermoplastic-composite artifacts. In

Table 2
Room-temperature properties of thermoplastic composites based on polysulfone, polyether sulfone and PEEK: * indicates notched specimens

	Polysulfone 40 vol.% glass cloth[a]	Polysulfone 55 vol.% TOHO HTA 7[b]	Polyether sulfone 40 vol.% glass fiber[a]	Polyether sulfone 60 vol.% carbon fiber[c]	PEEK APC 2 61 vol.% AS-4[e]	PEEK APC 2 61 vol.% 1M-6[e]	PEEK 61 vol.% Celion G30–50[f]	PEEK T 300[g]
Density (10^3 kg m^{-3})					1.657			
σ_f(MPa)	172	1560	214	1551	1880	2170	2160	1796
E_f(GPa)	83	112	11	94	121	151	131	139
Interlaminar shear strength (ILSS) (MPa)		75		79			90	107
σ_t(MPa)	131		159	1232	2130	2700	145	
E_t(GPa)	11.7		13.8	153	134	176	145	
ε(%)		1.45			1.45	1.48		1.29
σ_c(MPa)	166		152		1100	1100	1075	
σ_t transverse (MPa)					80			
E_t transverse (GPa)					8.9			
σ_f transverse (MPa)					137	160		171
E_f transverse (GPa)					8.9	9.3		
Izod impact (J m^{-1})	84.6*		79*					
Instrumented impact (KJ m^{-2})		60[c], 62[c]		78[c], 114[c]				75[c](melt)
G_{Ic}(KJ m^{-2})					2.4	2.5		
HDT (°C)	185		216					
α (°C$^{-1}\times 10^{-6}$)	2.34		2.52					

a English 1987 b Weiss and Huttner 1987 c Stori and Magnus 1983 (first reading is for a solution-impregnated specimen, second for a melt impregnated one, both of which have 50 vol.% fiber loading) d McMahon 1984 e Anon 1986 f Clemans et al. 1987 g Owens and Lind 1984

Table 3
Room-temperature properties of thermoplastic composites based on polyether imide and polyamide imide

	Polyether imide 55 vol.% carbon fiber[a]	Polyether imide 54 vol.% glass fiber[a]	Polyether imide 52 vol.% Kevlar fiber[a]	Polyamide imide 0/90 T300[b]	Polyamide imide 0/90 Celion 3000[b]
σ_f(MPa)	711	643	253	1000	1010
E_f (GPa)	51	26	54	63	59
ILSS (MPa)				75	100
σ_t(MPa)	527	355	396		680
E_t(GPa)	57	24	41		62
ε(%)	1.6				
σ_c(MPa)					560
Instrumented impact (KJ m^{-2})	28[c] (sol.)				

a Specmat (UK) Ltd trade data b McMahon 1984 c Stori and Magnus 1983

hydroforming, a preheated, consolidated blank is formed to the shape of the model by using a hydraulically pressurized flexible diaphragm or block of rubber. Because of the hydraulic nature of the loading, an even pressure is applied over the whole surface of the charge, or blank, minimizing fiber damage. A typical forming time of 15 s is claimed. Another method which can be used to produce complex shapes is diaphragm forming. This is analogous to the superplastic forming of metal alloys. A charge of, for instance, APC 2 is laid up in the requisite way and encased between two sheets of Supral aluminum. Air is evacuated from the enclosed charge and the temperature raised to 380 °C when the aluminum sheets become superplastic. Differential pressure and a shaping tool can be used to shape the sandwich structure. The method can presumably be used with other types of thermoplastic, provided that the polymer can withstand the temperature at which the aluminum becomes superplastic. This process can be used to produce complex shapes and is very competitive for short runs. The cycle time can be as low as 20–40 min. One other technique, borrowed from the metal-forming industry, is roll forming. Continuous lengths of preform are heated, consolidated and formed, and further consolidated using shaped rollers. In this way, corrugated sheet and "top hat" sections, among others, can be made. The throughput speed is up to 0.17 m s^{-1}.

Other means of fabrication available are filament winding and tape laying. Filament winding is widely

Table 4
Room-temperature properties of thermoplastic composites based on polyphenylene sulfide

	70 vol.% glass fiber[d]	66 vol.% carbon fiber[a]	55 vol.% pultruded[b] carbon fiber[b]	56 vol.% carbon fiber[c]	61 vol.% Celion G30–50[d]
Density (10^3 kg m^{-3})	2	1.6	1.52		
σ_f(MPa)	1159	1310	1173	1366	1770
E_f(GPa)	44.2	124	96.9	124	131
ILSS (MPa)		70			62
σ_t(MPa)	911	1656	1380	1172	1820
E_t(MPa)	49.7	135	117.3	130	117
ε(%)			1.1	0.8	
σ_c(MPa)	759	655	586		634
Izod impact (J m^{-1})		1586	2078		

a Phillips Petroleum trade data b O'Connor 1987 c Specmat (UK) Ltd trade data d McMahon 1984

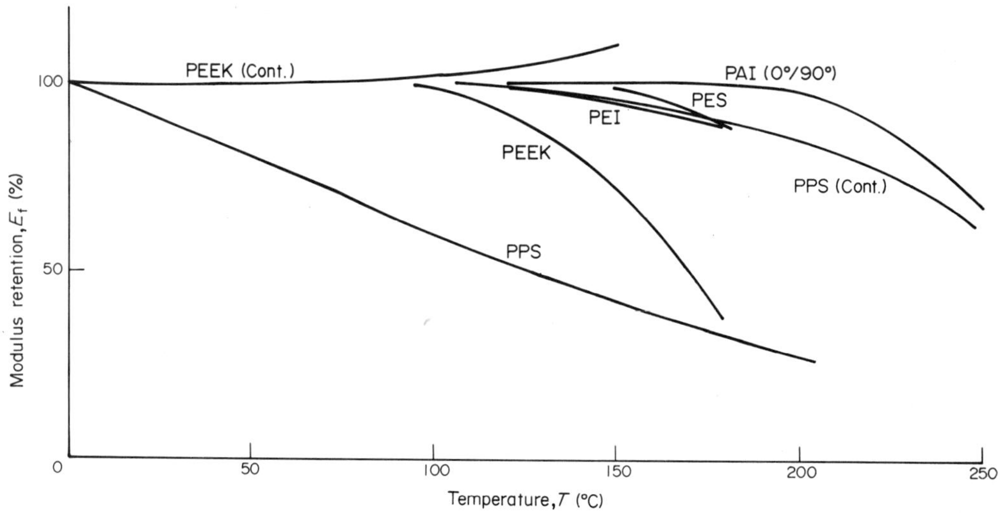

Figure 2
Percentage modulus retention of thermoplastic-matrix composites vs temperature: continuous and 0°/90° reinforcement is carbon fiber; in other cases the reinforcement is 30% short glass fiber

used in thermoset fabrication. The main differences when it is applied to thermoplastics are that a filament already impregnated with the matrix is used and additional heating, possibly of both the mandrel and filament, is required to fuse the filaments together. Methods of heating include infrared, lasers, microwaves, induction, contact, ultrasonic welding, flame and nonoxidising hot gas. Advantages of thermoplastic filament winding include the ability to deviate markedly from geodesic paths and to fabricate structures with reentrant surfaces. Tape laying is used to fabricate large, planar or low-curvature components. Because of the geometry, it is not possible to consolidate by applying tension to the tape. Instead, an external head must be used. In another method, preform tape is spirally wound between two parallel mandrels with continuous butt welding to create a single-ply tube with any fiber orientation between 0° and 90°. The welding speed for PEEK is 0.03 m s^{-1}.

Stamping has been suggested as a means of forming components. The tooling costs are moderate and

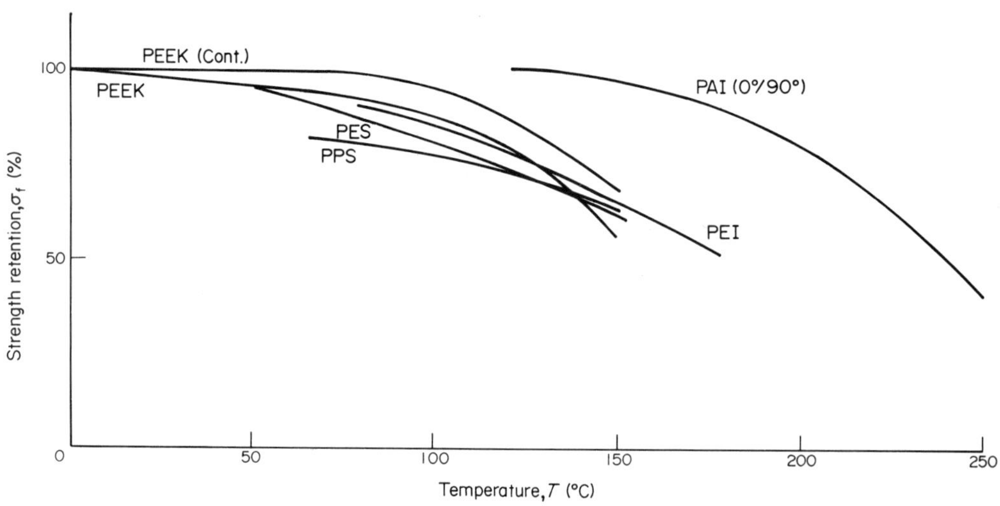

Figure 3
Percentage strength retention of thermoplastic-matrix composites vs temperature: continuous and 0°/90° reinforcement is carbon fiber; in other cases the reinforcement is 30% short glass fiber

the process suitable for long runs and large, simply shaped components. Phillips Petroleum have developed a polyphenylene sulfide reinforced with chopped or swirl mat glass or woven and nonwoven glass fiber. A temperature in the range 315–343 °C and a pressure in the range 14–42 MPa are required for stamping. The cycle time is 20–120 s.

Further details and references on all methods of fabrication, especially when applied to reinforced PEEK (APC 2), are given by Cattanach and Cogswell (1986).

5. *Composite Properties*

Some room-temperature thermomechanical properties of carbon-, glass- and aramid-fiber composites with thermoplastic matrices are given in Tables 2–4. Unless otherwise stated, the specimens are unidirectionally reinforced. Unfortunately, no data covering one type of fiber with all the different matrices is available and thus it is not possible to compare the effectiveness of the various thermoplastic matrices. Information on high-temperature properties and the performance of laminates is available for PEEK and polyphenylene sulfide materials in the trade and open literature. The percentage retention of modulus and strength with temperature for some systems is shown in Figs. 2 and 3.

See also: Carbon-Fiber-Reinforced Plastics; Glass-Reinforced Plastics: Thermoplastic Resins

Bibliography

Anon 1986 Development of thermoplastic composites for airframe construction. *Aerosp. Des. Compon.* October–November 16–21

Brydson J A 1982 *Plastics Materials*, 4th edn. Butterworth, London

Cattanach J B, Cogswell F N 1986 Processing with aromatic polymer composites. In: Pritchard G (ed.) 1986 *Developments in Reinforced Plastics*, 5. Elsevier, London

Clemans S R, Western E D, Handermann A C 1987 *Hybrid Yarns for High Performance Thermoplastic Composites*, Materials Science Monographs 41. Society for the Advancement of Materials and Process Engineering, Azusa, CA

Cogswell F N 1987 The next generation of injection moulding materials. *Plast. Rubber Int.* 12: 36–9

Collyer A A, Clegg D W 1985a Polyimides—Mechanical properties versus processability. *High Performance Plast.* 2(7): 1–5

Collyer A A, Clegg D W 1985b Radiation resistant polymers. *High Performance Plast.* 2(2): 2

Collyer A A, Clegg D W 1986a Self reinforcing polymers. *High Performance Plast.* 4(1): 2–7

Collyer A A, Clegg D W 1986b Xydar electrical component. *High Performance Plast.* 3(5): 8

English L K 1985 Aerospace composites. *Mater. Eng.* 102(4): 32–6

English L K 1987 Fabricating the future with composite materials, part 3: Matrix resins. *Mater. Eng.* 104(2): 33–7

Gotham K V, Hough M C 1984 *Durability of High Temperature Thermoplastics*. RAPRA, Shrewsbury, UK

Leach D C, Moore D R 1985 Toughness of aromatic polymer composites reinforced with carbon fibers. *Compos. Sci. Technol.* 23: 131–61

McMahon P E 1984 Thermoplastic fiber composites. In: Pritchard G (ed.) 1984 *Developments in Reinforced Plastics*, 4. Elsevier, London, Chap. 1

Marks H F, Atlas S H 1965 Principles of polymer stability. *Polym. Eng. Sci.* 5: 204–7

Muzzy J D, Kays A O 1984 Thermoplastic versus thermosetting structural composites. *Polym. Comp.* 5: 169–72

O'Connor J E 1987 Polyphenylene sulfide pultruded type composite structures. *SAMPE Q.* 18: 32–8

Owens G A, Lind D J 1984 *Fibre Reinforced Composites*, Conference Proceedings, Paper 12. Plastics and Rubber Institute, London

Pye A M 1982 High performance engineering plastics. *Mater. Eng.* 3: 407–9

Rose J B 1984 High temperature engineering thermoplastics from aromatic polymers. *Plast. Rubber Int.* 10: 11–15

Stori A, Magnus E 1983 An evaluation of the impact properties of carbon fibre reinforced composites with various matrix materials. In: Marshall I H (ed.) 1983 *Composite Structures*, Paper 24. Elsevier, London

Weiss R, Huttner W 1987 *High Performance Carbon Fiber Reinforced Polysulfone*, Materials Science Monographs 41. Society for the Advancement of Materials and Process Engineering, Azusa, CA

Woodhams R T 1985 History and development of engineering resins. *Polym. Eng. Sci.* 25: 446–52

N. L. Hancox
[UKAEA Harwell Laboratory, Didcot, UK]

Hot-Melt Adhesives

A hot-melt adhesive is a material used to bond surfaces together which functions by changing from a fluid at elevated temperatures during bonding to a strong solid at the use temperature. Hot melts have been used since prehistory when early man first heated a resinous substance obtained from vegetable matter and found that after cooling, it formed a tough material. Sealing wax is an example of a hot-melt adhesive of more recent times. Early hot-melt adhesives were barely functional because of impurities and poor aging resistance of natural resins. Modern hot melts are based on more stable synthetic polymers.

The primary advantage of a hot-melt compound compared with traditional fasteners and other adhesive systems is its rapid setting time. Depending on the adhesive formulation and operating conditions, hot melts can form a strong bond in a fraction of a second. This allows product assembly operations to be run at speeds that are not limited by the adhesive.

Hot melts have a number of other advantages. In the molten state, they have low viscosities and flow readily to fill gaps between the substrates, thus sealing the assembly from vapor transmission. Like other adhesives, a continuous hot-melt film is superior to mechanical fasteners for distributing stresses and damping vibrations. Hot melts can also be designed and used to provide thermal and electrical insulation.

In some respects, hot-melt adhesives are more difficult to use than other assembly methods. They require specialized equipment to transport and deliver the molten mass to the substrate. Heaters, pumps and coating or extrusion equipment are needed. The entire system must be insulated to prevent heat losses. An interruption in the process can result in plugged lines if cooling occurs or degradation of the molten adhesive if it is exposed to the atmosphere for an extended time.

More recently, pressure-sensitive adhesives have been developed which are applied as 100% solids from a hot melt. These products are not traditional hot-melt adhesives and are thus dealt with elsewhere (see *Pressure-Sensitive Adhesives*).

1. Applications

Hot-melt adhesives are used in industries that can benefit from the advantages of fast operation or where bonding of large surfaces is desirable to seal an assembly or to distribute stresses. They are used widely in packaging for bonding paper or cardboard surfaces and for lamination of foil and plastic film. Hot melts are used in the sealing of cartons and cases and for construction of paper bags. Disposable nonwoven products such as diapers, sanitary napkins and incontinence pads are assembled with hot melts, as are books (binding), shoes (counters), furniture (edge veneering and general construction), textile products (seaming, hemming, cuffing) and miscellaneous wood and plastic products. The total hot-melt sales in the USA in 1981 were estimated at about 205 000 t, distributed as follows: 132 000 t (64%) in packaging, 18 000 t (9%) in general product assembly, 14 000 t (7%) in bookbinding, 14 000 t (7%) in disposable nonwoven products, 9000 t (4%) in textiles, 7000 t (3%) in furniture and 11 000 t (6%) miscellaneous. The entire market is estimated to be growing at about 8% per year.

2. Principles of Use

2.1 Variables in Product Assembly

Hot-melt adhesives function simply by cooling from the molten state, during application to the substrate, to a solid that provides strength at the use temperature. The viscosity of the adhesive must be low enough during application to spread and wet the surface of each substrate. The required viscosity depends on the operating temperature and speed, and the shear applied during the coating or wetting step. If the adhesive is too fluid, it may spread excessively on the surface or penetrate and even bleed through porous substrates, leaving too little to form an adequate bond. If the adhesive is too viscous, it may not conform to the microscopic surface roughness of the substrates, resulting in poor contact.

The relationship between the viscosity of the adhesive and its temperature is important. If the viscosity does not increase fast enough after the two substrates are joined (the set time is too long), the bond may deform or be destroyed before setting by the stresses in a continuous manufacturing line. This problem can be recognized in a bonding failure when both substrates have been wet by the adhesive and the hot melt has failed cohesively. If the viscosity increases too fast (the set time is too short), the adhesive may harden before there is contact with the second substrate, preventing wetting and bonding. This condition exhibits itself as adhesive failure between the hot melt and the second substrate.

The rate of increase in viscosity is controlled by both the viscosity–temperature relationship for the adhesive and the cooling rate after bonding. The rate of cooling is governed by the amount of adhesive used, the heat capacity of the substrates and heat losses due to conduction to the processing equipment and the surrounding atmosphere. The time between application of the adhesive to the first substrate and contact with the second substrate (the open time) for a particular piece of equipment must be shorter than the set time. Ideally, the set time should be marginally longer than the open time, so that the substrates bond immediately after contact.

The set time of the adhesive in a specific operation varies somewhat because of uncontrolled variables such as the ambient temperature and variations in the thickness of the substrate and quantity of adhesive applied. These variables can be accommodated to eliminate poor bonding by adjusting the operating speed and the volume of adhesive used. This may not be cost effective with respect to productivity. It may be preferable to accept some level of rejected parts to achieve lower costs through high production rates and reduced consumption of adhesive.

2.2 Adhesive Properties and Performance

An approach can be made to understanding the performance of hot-melt adhesives by considering the bulk properties and rheology of the adhesive. Surface characteristics of the adhesive and substrates are also important for proper wetting and spreading. However, these considerations of surface tension and work of adhesion are the same for all adhesive systems and are therefore not discussed here.

The important performance characteristics of a hot-melt adhesive are its strength and flexibility over its useful temperature range and its viscosity at the temperature of application to the substrates. The useful temperature range is defined by the appearance of brittleness at low temperature and the loss of strength at elevated temperature.

The low-temperature limit of the adhesive is related to the glass transition temperature T_g of the composition. Below T_g, where molecular loss processes become minimal, even small deformations result in brittle failure (fracture). The exact temperature at which brittleness appears depends on the rate of deformation used in the test. If the deformation rate is increased, the brittleness temperature is higher.

The T_g of the adhesive is a function of the values for the individual compatible components. Modifying resins used in hot melts are generally low-molecular-weight (~500–2000) rigid molecules which often have T_g values above room temperature, higher than those of the base polymers. Consequently, addition of resins increases T_g of the adhesive. Addition of compatible oils and waxes of lower T_g, lowers T_g of the adhesive. The T_g of the adhesive can be estimated from the T_g values of the individual compatible components by the following equation:

$$T_{gA} = (W_1 T_{g1}^{-1} + W_2 T_{g2}^{-1} + \ldots + W_n T_{gn}^{-1})^{-1} \quad (1)$$

where T_{gA} is T_g of the adhesive, $T_{g1}, T_{g2}, \ldots, T_{gn}$ are the values for the components and $W_1, W_2, \ldots, W_{gn}$ are the weight fractions of the components; the temperatures are in kelvin.

The stiffness or flexibility of the adhesive within its useful temperature range corresponds to its modulus at the use temperature. Use temperatures are generally in the rubbery plateau region of the system between the transition zone and the melt or terminal zone. The modulus in the plateau is related to the average molecular chain length of the amorphous polymeric system between points of restricted molecular motion. In thermoplastic compositions, polymer mobility can be restricted by chain entanglements or segmental crystallinity. The shorter the amorphous chains, the higher is the modulus of the system in the rubbery plateau. Also, the crystallinity itself increases the modulus by acting as a reinforcing filler. A limit is reached when the crystallinity level is so high and the amorphous chains are so short that the temperature at which brittleness appears is too high for the intended application. Both compatible resins and oils will lower the modulus and increase the flexibility of the adhesive in its normal use range. These additives act as solvents for the amorphous polymeric chains, and the amorphous fraction can be considered to exist as a concentrated solution of polymer. The resins and oils reduce the number of chain entanglements per volume of amorphous polymer and increase the molecular weight between entanglements. The plateau modulus is reduced, according to the classical theory of elasticity, because the modulus is proportional to the concentration of network strands. The number of strands between entanglements and crystalline segments is reduced as the number of entanglements is reduced.

The melt temperature is that temperature at which restrictions to molecular flow disappear. This means that the composition must be above the melting point of the crystalline segments and that entanglements do not impede flow. Each entanglement has a characteristic disentanglement time, depending upon molecular motion, which is related to the temperature of the system. The rate of deformation must be slow enough to allow disentanglement and flow to occur. Therefore, the flow characteristics of a hot-melt adhesive depend on the shear rate as well as the temperature of the system. As indicated earlier, compatible resins and oils reduce the temperature at which flow is observed, because they reduce the number of chain entanglements.

The attainment of the ultimate bond strength of a hot-melt adhesive involves large deformation processes. Cohesive failure depends on the yield or fracture strength of the polymer mixture. Adhesive failure occurs if there is poor contact between adhesive and substrate or if the interfacial work of adhesion is low.

3. Components of Hot-Melt Adhesives

Hot-melt adhesives are composed of a polymer generally combined with modifying ingredients such as low-molecular-weight resins, waxes, oils and fillers. Stabilizers are usually added to combat degradation which can occur both during the molten application stage and as a result of long-term aging. The components and their functions are summarized in Table 1.

3.1 Polymers

The polymer imparts strength to the composition as a result of its molecular weight, crystallinity or hydrogen bonding. Higher-molecular-weight polymers have more chain entanglements per molecule and so have higher melt flow temperatures. If crystallinity is present, the crystalline regions maintain the integrity of the system until the crystalline melting point is reached. Chain entanglements may continue to provide strength for an additional temperature span above the crystalline melting point. Hydrogen bonding in polar polymers produces high

Table 1
Components of hot-melt adhesives

Polymer	
examples:	ethylene–vinyl acetate, ethylene–ethyl acrylate, amorphous polypropylene, polyethylene, polyamides, polyesters
functions:	provides cohesive strength and stiffness, determines melt temperature
Resin	
examples:	petroleum resins, rosin-based resins
functions:	improves specific adhesion and surface wetting, reduces melt viscosity
Wax	
examples:	paraffin and microcrystalline waxes derived from petroleum; synthetic waxes, primarily low-molecular-weight polyethylene
functions:	reduces melt viscosity, eliminates surface tack, reduces cost

melting points, which maintains bond strength and performance at higher temperatures.

(*a*) *Olefin copolymers*. Ethylene–vinyl acetate copolymers (EVA) are the most universally used polymers in hot-melt adhesives. Selection of the proper EVA is based on the vinyl acetate (VA) content and the viscosity at elevated temperatures. A broad spectrum of ethylene–vinyl acetate ratios is available, around 10–40 wt% VA, with the most commonly used concentrations being 25–30 wt%. Vinyl acetate increases the polarity of the polymer and improves wetting and specific adhesion to substrates such as cellulosics. Although the polymers are manufactured in a random process, the polyethylene forms crystalline blocks whose size decreases with increasing vinyl acetate content. The crystalline polyethylene blocks have a melting point of about 75 °C in the polymer, providing strength up to that temperature.

The higher-VA polymers ($\geqslant$28 wt% VA) have less-crystalline polyethylene blocks and are therefore softer and more flexible. They have improved adhesion to nonporous substrates. On the other hand, the lower-VA polymers (<28 wt% VA) have higher polyethylene crystallinity and shorter amorphous segments. This gives a harder, stiffer polymer which has improved compatibility with hydrocarbon additives such as waxes and low-molecular-weight hydrocarbon resins.

The other important factor in choosing the proper EVA is its flow characteristics at elevated temperatures. These are related to the molecular weight and molecular weight distribution of the polymer. Flow is measured by the melt index (ASTM D1238) which is the weight of polymer extruded in a specified time through a designated orifice at a specified temperature and load. The melt index is inversely related to molecular weight. Products having melt indices of around 1–500 are available, but those of around 5–30 are the most used. Generally, the higher-molecular-weight polymers (lower melt index) provide the strongest composition and have more "hot tack" and shorter open time than the lower-molecular-weight polymers. The polymer having the lowest melt index is generally selected, consistent with the handling requirements of the application equipment.

Ethylene–ethyl acrylate copolymers (EEA) are alternatives to the EVA polymers. Poly(ethyl acrylate) has a lower glass transition temperature than poly(vinyl acetate), so EEA copolymers should have a lower T_g than EVA copolymers at the same weight concentration. On the other hand, there is a higher concentration of crystalline polyethylene chains in the EEA copolymers, which results in higher performance temperatures.

(*b*) *Olefin homopolymers*. Amorphous polypropylene (APP), a by-product of the polypropylene process, is used in large volume in hot-melt adhesives. Refined APP is estimated to contain about 15–30% crystallinity by differential scanning calorimetry, much lower than that of polypropylene, but it is still not strictly an amorphous product. The molecular weight is estimated to be in the range 2000–15 000 in the commercially available products used in adhesives.

(*c*) *Polyamides*. These are used in specialty, high-performance hot-melt adhesives. The basic products are made from "dimer acid," which is the dimer of unsaturated C_{18} fatty acids and ethylene diamine. Other modifying acids and amines can be included. The polyamides have high service temperatures because of the polar bonding of the amide links.

(*d*) *Polyesters*. These, like the polyamides, are used in low-volume, specialty adhesives because of their high strength and high service temperatures. They are made from organic dicarboxylic acids and dialcohols.

3.2 Resins

A low-molecular-weight (~500–2000) resin is usually added to the olefinic polymers to reduce the melt viscosity and improve adhesion. The resin must be compatible with the polymer to provide the expected property changes. Resin polarity, molecular weight and molecular weight distribution are important considerations in the functioning of the resin. The softening point of the resin depends on its molecular weight and molecular weight distribution. The viscosity at the softening point is in the range 0.5–1.0 kPa s. The glass transition temperature of the resin is about 40–50 °C below its softening point. For the so-called "hard resins," T_g is well above room temperature, so T_g of the polymer–resin blend is higher than that of the polymer.

There are two major types of resins used in hot-melt adhesives: petroleum derived and resin derived. Polyterpenes and alkylated phenol resins are also employed.

(*a*) *Petroleum resins*. The most useful of these are manufactured from petroleum cracking streams, mostly C_5 hydrocarbons, or pure aromatic monomers such as styrene, α-methylstyrene and vinyltoluene. Mixtures of the pure monomers are often used. The streams are polymerized to low-molecular-weight (500–2000) oligomers by acid catalysts such as acid clays, aluminum chloride or boron trifluoride. The lower-molecular-weight resins from the same hydrocarbon feed are more readily compatible than the higher-molecular-weight products. The broad distribution products are more compatible than products having a narrow molecular weight distribution.

(*b*) *Rosin-based resins*. Rosin, a mixture of C_{20} diterpenoid acids which is produced by pine trees, is the raw material for an important group of resins used in hot-melt adhesives. The most widely used resins are the esters prepared by reaction with polyhydric alcohols such as glycerine and pentaerythritol. The rosins may be stabilized by hydrogenation or disproportionation prior to esterification, to improve retention of viscosity and color of the adhesive during high-temperature application or long-term aging. The residual carboxylic acid and hydroxyl contents are important considerations in compatibility with polymers. The rosin derivatives in general have a broader range of compatibility than the hydrocarbon resins.

The ester resins have a narrow molecular weight distribution because they are mixtures of simple esters. The rosin acids are bulky molecules having molecular weights of about 300. The triester of completely esterified glycerine would have a molecular weight of about 940, while that of completely esterified pentaerythritol would be about 1260. Despite these relatively low molecular weights, rosin-based resins have comparatively high T_g values, well above room temperature for the glycerine and the pentaerythritol esters.

3.3 Wax

A wax may be added to the olefinic polymers to reduce the melt viscosity further, to influence the melting temperature and to eliminate residual tack at room temperature. Wax also reduces the cost of the composition. The wax can be isolated from petroleum or manufactured synthetically, principally from polyethylene by high-pressure polymerization.

Table 2
Typical hot-melt formulations[a]

	Composition (wt%)
General packaging	
EVA (28% VA, MI 6)	30
aromatic copolymer resin (SP 120 °C)	25
paraffin wax (MP 65 °C)	45
Carton sealing	
EVA (28% VA, MI 25)	45
glycerine esters of hydrogenated rosin	25
microcrystalline wax (MP 90 °C)	30
Bookbinding	
EVA (28% VA, MI 6)	35
glycerine ester of hydrogenated rosin	40
microcrystalline wax (MP 70 °C)	25
Carpet seaming tape	
EVA (28% VA, MI 6)	50
pentaerythritol ester of hydrogenated rosin	15
paraffin wax (MP 70 °C)	35

a MI melt index; SP softening point by ring and ball method; MP melting point

3.4 Typical Formulations

Typical examples of hot-melt adhesive formulations are found in the trade literature of suppliers to the adhesives industry. However, finished compositions are considered proprietary by the adhesive manufacturer. A starting point for general EVA-based hot-melt adhesives is equal parts of EVA, resin and wax. Appropriate adjustments to this composition for a specific EVA would be to increase the proportion of polymer for greater strength, increase that of resin for improved adhesion and increase that of wax for reduced viscosity. Some typical formulations are given in Table 2.

4. Testing

Adhesives are tested for quality assurance and to guide the formulator toward performance improvements during development of new compositions. In general, they are tested both for performance as an adhesive and to determine the bulk properties of the composition. The performance tests commonly used are those for shear strength, peel strength and shear adhesion failure temperature. Another important consideration is the temperature at which brittleness occurs. In addition, hot-melt adhesives are often examined for melt viscosity, open time, tensile strength, modulus and flexibility at use temperature. Stability tests are also generally required, stability at application temperatures being particularly important for hot melts. Long-term stability of the assembled bond may or may not be important, depending upon the final use.

4.1 Shear and Peel Tests

The most common test for shear strength of a hot-melt adhesive is the lap shear test, in which the bond is tested to failure. Adhesive peel strength is measured at 180° from a rigid substrate or as a T peel if both substrates are flexible. These two peel tests do not yield the same results, because the peeling angle affects the force required for separation. The rate of peel must be controlled because it affects the measured strength. In the shear and peel tests, the thickness of the adhesive layer must also be carefully controlled.

4.2 Shear Adhesion Failure Temperature

This test measures the temperature at which the viscous resistance of the adhesive in shear can no longer support an applied stress. The test specimen is an overlap shear configuration which is hung from one tab with a weight attached to the other. The assembly is placed in an oven and the temperature is raised either in increments or continuously until at least three of five replicate specimens fail. The temperature at which this occurs is reported as the shear adhesion failure temperature.

4.3 Brittleness

Brittleness or low-temperature flexibility tests measure the temperature at which the adhesive first exhibits brittle characteristics as it is cooled. Various tests have been developed by adhesive manufacturers, but the mandrel bend test is the most common. In this test, a strip of the adhesive is bent around a rod of selected diameter at a set rate at different temperatures. The highest temperature at which the adhesive cracks is recorded as the brittleness temperature. The brittleness temperature is influenced by the thickness of the adhesive, the diameter of the rod and the bending rate. This test is usually done by hand and is therefore carried out by experienced testers, to achieve consistent results.

4.4 Melt Viscosity

The viscosity of hot-melt adhesives at application temperatures is generally measured in continuous shear using a Brookfield Thermosel viscometer. Viscosity–temperature curves are often constructed to cover the temperature range of interest.

4.5 Open Time

Measurement of open time allows the user of hot melts to judge how long an adhesive will remain fluid enough to bond surfaces together. In evaluating adhesives for paper lamination, a molten film of adhesive of controlled thickness is cast onto a paper substrate and strips of a second substrate, such as paper, plastic film or foil, are brought into contact with the molten adhesive after successive 5 s intervals. A non-heat-conductive weight is placed on each bond for 5 s to ensure intimate contact. After the test pieces have cooled, they are lifted to determine the longest time before bond formation that still gives paper tear—in other words, the adhesive is still fluid enough to form a bond when the second substrate is applied to the adhesive surface. This test allows adhesives to be ranked according to open time, but the results do not necessarily relate to specific manufacturing experiences, because of different machine speeds and cooling rates.

4.6 Tensile Strength, Modulus and Flexibility

Hot-melt compositions may be cast into plaques and tested for bulk physical properties by standard test methods. The properties most frequently measured are tensile strength, modulus and elongation at break. Torsional rigidity is sometimes determined at various temperatures to indicate the temperature at which compositions become hard or brittle. There is also a mandrel bend test for flexibility (ASTM D3111-76).

4.7 Dynamic Mechanical Properties

As with any polymeric system, hot-melt adhesives can be examined with a mechanical spectrometer to obtain storage and loss moduli and the loss tangent (tan δ) over a temperature range, or the two moduli and complex viscosity in the melt region. Data also may be obtained at a specified temperature as a function of frequency. This information indicates T_g, the modulus or the stiffness at the use temperature, crystalline melting point and viscosity at application temperature. Compatibility of additives can be assessed by their effects on the dynamic mechanical properties.

4.8 Stability

Hot melt adhesives are especially subject to changes if they are held molten for extended periods before application. The viscosity of the adhesive may change because of chemical reaction, degradation or volatilization of a low-boiling component. If the viscosity changes, the flow characteristics of the adhesive change. In some applications, color and odor are important and these may change if the adhesive is held in the molten state. Testing is straightforward: a sample is held molten at the selected temperature in a suitable container and the viscosity, color and odor as well as the development of a skin on the surface of the melt are monitored.

See also: Adhesion: Fundamental Principles; Adhesive Joining: An Overview; Adhesives: Mechanical Test Methods

Bibliography

Adhesives Age 1972 onwards: issues devoted to Hot-Melt Adhesives

Skeist I (ed.) 1977 *Handbook of Adhesives*, 2nd edn. Van Nostrand Reinhold, New York

Technical Association of the Pulp and Paper Industry 1978 onwards *TAPPI Hot Melt Short Course Notes*. TAPPI, Atlanta, GA

Wake W C 1987 *Synthetic Adhesives and Sealants*. Wiley, Chichester, UK

J. B. Class
[Hercules, Wilmington, Delaware, USA]

Hybrid Fiber–Resin Composites

The concept of mixed fiber composites is an extension of the composites principle of combining materials to optimize their value to the engineer, exploiting their better qualities while mitigating the effects of their less desirable properties. Mixing two or more types of fiber in one matrix allows a closer matching of composite properties to specific requirements than can be achieved with a single type of fiber.

Many hybrids of current interest represent attempts to reduce the cost of expensive composites, which contain reinforcements such as carbon or boron, by incorporating a proportion of cheaper, lower-quality fibers, such as glass, without too seriously reducing the mechanical properties of the original composite. At a lower level, the glass–jute–resin composites reflect the same purpose. Of equal importance is the reverse principle, that of stiffening a glass reinforced plastic (GRP) structure, such as a car body or canoe shell, with a small quantity of judiciously placed carbon or Kevlar 49 fiber, without inflicting too great a cost penalty.

In high-technology fields, the question of cost may be insignificant in comparison with the advantages of optimizing properties, as in the use of carbon–glass–resin hybrids for helicopter rotor blades. In demanding aerospace applications, one of the most important reasons for using hybrids is that the natural toughness of GRP offsets the brittleness of typical carbon or boron composites.

A widely debated issue is whether or not it is possible to obtain greater improvements in some properties than might be predicted from calculations based on properties of the separate components. This so-called hybrid effect has obscured the interpretation of much experimental work. Hybrid properties higher than a mixture-rule prediction have often been hailed as evidence of synergism, regardless of the validity of parallel-connected element models for the composite properties in question.

The elastic properties of hybrid composites can be calculated to within limits acceptable for most design purposes by the application of the principles of composite mechanics. For unidirectional composites, mixture-rule calculations are satisfactory while, for complex laminates, classical laminate theory provides adequate elastic and thermal strain predictions. Elasticity calculations are independent of the nature of fiber distributions. In contrast, predictions of properties such as strength, toughness and failure strain are more difficult because they depend on micromechanisms of damage accumulation which, in turn, are determined by the construction of the laminate and the scale of dispersion of the mixed fibers.

At the grossest level of hybridization, strips of GRP incorporated into a carbon-fiber-reinforced plastic (CFRP) laminate act as efficient crack arresters, provided the width of the strips is sufficient to dissipate the energy of a crack moving rapidly in the CFRP by localized debonding and splitting (Bunsell and Harris 1976). This concept is referred to as the introduction of "softening strips" (Sun and Luo 1985). The commonest type of hybrid, however, and that which is most easily achieved in practice is made by laminating some balanced sequence of plies within each of which there is only a single species of fiber. In such laminates, the failure strain of the higher-elongation (HE) component is usually reduced by the presence of the other, while that of the lower-elongation (LE) component is often increased. This behavior results from the state of elastic constraint in the fabricated laminate (the effect being greater the thinner the plies) and it has been explained by Aveston and Kelly (1980) in terms of energy. An approximate statistical analysis by Zweben (1977) shows that the failure process is affected by the statistical spread of failure strains of the two fiber species and predicts that the addition of HE fibers to a low-failure-strain composite raises the strain level needed to propagate fiber breaks because the HE fibers behave like crack arresters at a micromechanical level. The strength of such composites does not follow a mixture rule. Addition of small quantities of HE glass fiber to an LE CFRP will reduce the strength because there is insufficient glass to carry the load when the carbon breaks. Carbon failure leads to a single fracture failure of the composite. There is a critical level of mixing, however, at which there is sufficient glass to bear the load when the carbon breaks and beyond this level the hybrid strength rises again with further increases in glass content. Multiple fracture of the carbon then occurs and the fracture energy of the hybrid is substantially raised. The question of synergism is thus confused because any measurement of hybrid strength which appears to agree with a mixture-rule prediction is already showing a greater value than expected.

When the fibers are intimately mixed, strength calculations are more difficult because micromechanisms of failure change as the pattern of nearest neighbors around a given fiber alters, and strength is governed either by failure of single fibers or of small bundles of fibers (Parratt and Potter 1980).

In general, toughness or impact strength of brittle composites such as CFRP can indeed by substantially improved by additions of glass or Kevlar 49 fiber, but there is evidence to suggest that the fracture energy is satisfactorily predicted by the mixture rule, as energy arguments would suggest. The fatigue resistance of GRP is also improved by carbon-fiber additions, roughly in proportion to the amount of carbon added (Bunsell and Harris 1974, Harris and Bunsell 1975). Results relating to interlaminar shear strength (ILSS) are not clear cut, but it has been shown that the addition of high-modulus carbon fiber to a high-strength carbon–epoxy composite does not significantly alter the ILSS (Arrington and Harris 1978).

Recent work on the compression of hybrids containing various mixtures of fibers indicates no simple pattern that would permit prediction of compression strengths and moduli, although strengths of most of the hybrids studied fell reasonably close to mixture-rule values (Piggott and Harris 1981). This work emphasizes that changes in local failure modes resulting from hybridization are probably responsible for this uncertainty.

See also: Carbon-Fiber-Reinforced Plastics; Glass-Reinforced Plastics: Thermoplastic Resins; Glass-Reinforced Plastics: Thermosetting Resins

Bibliography

Arrington M, Harris B 1978 Some properties of mixed fibre CFRP. *Composites* 9: 149–52

Aveston J, Kelly A 1980 Tensile first cracking strain and strength of hybrid composites and laminates. In: Watt W, Harris B, Ham A (eds.) 1980 *New Fibres and Their Composites*. The Royal Society, London, pp. 111–26

Bunsell A R, Harris B 1974 Hybrid carbon and glass fibre composites. *Composites* 5: 157–64

Bunsell A R, Harris B 1976 Hybrid carbon/glass fiber composites. In: Scala E, Anderson E, Toth I, Noton B R (eds.) 1976 *Proc. 1st Int. Conf. Composite Materials*, Vol. 2 American Institute of Mining, Metallurgical and Petroleum Engineering, New York, pp. 174–90

Harris B, Bunsell A R 1975 Impact properties of glass fibre/carbon fibre hybrid composites. *Composites* 6: 197–201

Parratt N J, Potter K D 1980 Mechanical behaviour of intimately mixed hybrid composites. In: Bunsell A R, Bathias C, Martrenchar A, Menkes D, Verchery G (eds.) 1980 *Advances in Composite Materials*, Vol. 1. Pergamon, Oxford, pp. 313–26

Piggott M R, Harris B 1981 Compression strength of hybrid fibre-reinforced plastics. *J. Mater. Sci.* 16: 687–93

Sun C T, Luo J 1985 Failure loads for notched graphite epoxy laminates with a softening strip. *Compos. Sci. Technol.* 22: 121–34

Zweben C 1977 Tensile strength of hybrid composites. *J. Mater. Sci.* 12: 1325–37

B. Harris
[University of Bath, Bath, UK]

I

Impact Strength

The impact strength of a polymer is its ability to withstand shock loading (or impact). Since most polymeric components will be subject to impacts of one kind or another during their service life, it is essential that the chosen polymer possesses at least a minimum resistance to impact fracture. Impact fractures generally originate at points of high stress concentration, such as at changes of section, surface flaws or notches. When such polymers are fractured by impact blows, fracture is frequently brittle, even though the same polymers may show considerable ductility before fracture in a standard tension test.

1. Impact Testing

The determination of impact energy (or impact strength) of a solid polymer is somewhat arbitrary, in view of the unknown nature of the impact blows likely to be encountered in handling and in service. Two of the standard test methods are the Izod and Charpy impact tests (see Fig. 1), which are described in BS 2782 and ASTM D256-73. In the Izod apparatus a notched specimen is struck by a pendulum striker at a point 22 mm above the notch. In the Charpy apparatus, the notched specimen is placed between supports 95.3 mm apart and struck at the center directly behind the notch. Neglecting other losses, such as those due to friction, the impact strength of the specimen is determined from the energy given up by the pendulum in striking the test sample per unit of specimen width. Alternatively, the impact strength may be taken as the fracture energy per unit area of fracture surface. With appropriate modification, the Izod and Charpy apparatus

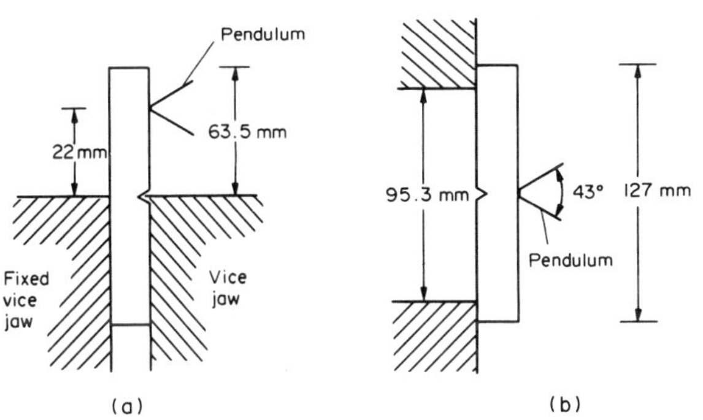

Figure 1
Schematic representation of (a) Izod and (b) Charpy impact test arrangement

can be used for investigating impact behavior under tension rather than in bending. Another type of impact test, used mostly for films, determines impact fracture energy by dropping weights, such as a ball or a dart, onto a clamped specimen (ASTM D1709-75).

2. Effect of Composition and Material Variables

The impact strengths of polymers cover a wide range; various impact strength values can be obtained for a particular polymer, depending on the fabrication conditions, the presence of additives and such external conditions as temperature, the strain rate and the geometry of the notch. Nevertheless, some general conclusions as to the impact strength of different polymers at ambient temperature can be made. One classification, with examples, is

(a) brittle, even unnotched—polystyrene, poly(methyl methacrylate) and cast phenolic;

(b) brittle, if bluntly notched—polypropylene and polyoxymethylene;

(c) brittle, if sharply notched—poly(vinyl chloride), high-density polyethylene, nylon (dry) and poly(2,6-dimethyl-1,4-phenylene ether); and

(d) tough, even sharply notched—low-density polyethylene, polycarbonate and nylon (wet).

The impact strength of many polymers, even brittle ones, can be increased by appropriate modification. Two examples of this are evident from the listed classification; low-density polyethylene has superior impact resistance to the more crystalline high-density polyethylene and nylon saturated with water is superior in impact strength to dry nylon. Impact strength may also be increased by copolymerization (e.g., polypropylene shows increased impact strength on copolymerization with ethylene, owing to reduced crystallinity and increased ductility) and by rubber modification, particularly if the rubber particles are well bonded and well dispersed. The second-phase particles cause increased matrix crazing and plastic deformation and these energy absorbing processes lead to better impact resistance. Thus acrylonitrile–butadiene–styrene (ABS) has much higher impact strength than unfilled styrene–acrylonitrile copolymer (SAN). Rubber modification is used to raise the impact strength of many polymers, including amorphous ones like poly(methyl methacrylate) and poly(vinyl chloride) (PVC), crystalline ones such as polypropylene and nylon, and thermosets such as epoxy resins.

Other types of fillers also affect impact strength. Reinforcement of polystyrene with 15% polyester fibers is reported to increase impact strength by a factor of about four, while addition of 30% glass fibers can raise the impact strength of brittle materials, such as epoxy and phenolic resins, by ten times or more. However, for ductile resins, such as ABS or polycarbonate, the addition of glass fibers tends to lower impact strength. Plasticizers, such as water in nylon, raise ductility and increase impact resistance.

Molecular weight and molecular orientation also play a role in impact performance. With higher molecular weight, there are more molecular entanglements and impact strength rises. Ultrahigh-molecular-weight polyethylene, for example, has exceptionally good impact resistance. Orientation raises impact strength if flow lines are normal to the plane of the notch but lowers impact resistance when flow lines are parallel to the plane of the notch.

Many investigators have reported a correlation between impact strength and damping behavior for a variety of polymers. Figure 2 shows the temperature dependence of impact strength of five different thermoplastics. Of these, polycarbonate has the largest low-temperature mechanical loss, while poly(2,6-dimethyl-1,4-phenylene ether) (PPO) has several overlapping low-temperature damping peaks. PVC and many other polymers show a rapid rise in impact strength over a relatively narrow temperature range; at this temperature T_b, they essentially undergo a transition from a brittle fracture mode to a ductile fracture mode. For polystyrene and poly(methyl methacrylate), T_b is close to the glass-transition temperature, for PVC it is about 40 °C, for PPO about 20 °C and for polycarbonate about −20 °C. The ductile–brittle transition temperature for polycarbonate can be further lowered to −110 °C by forming a block copolymer with 25% silicone.

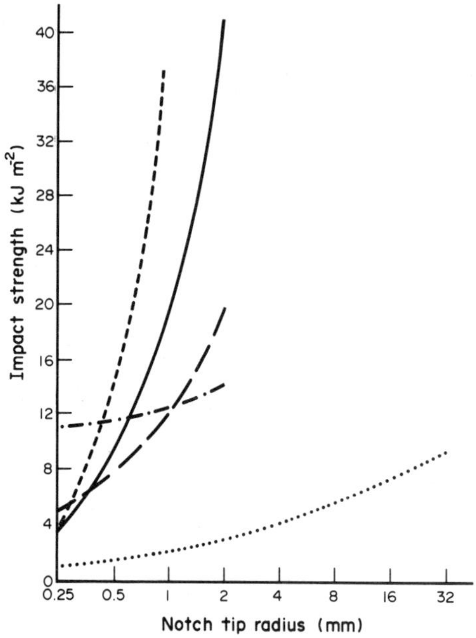

Figure 3
Impact strength against notch tip radius for various polymers: ——, nylon; —–—, acetal; ---, poly(vinyl chloride); -·-, acrylonitrile–butadiene–styrene; · · ·, acrylic (after Vincent 1971)

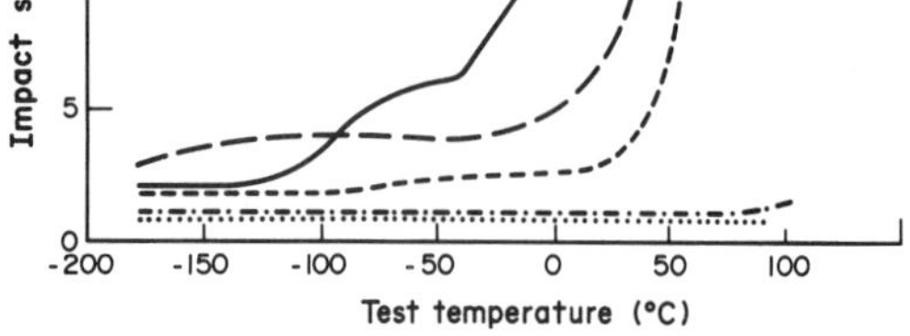

Figure 2
Impact strength against test temperature for five thermoplastics tested at a notch radius of 0.25 mm: ——, polycarbonate; —–—, poly(2.6-dimethyl-1,4-phenylene ether); ---, poly(vinyl chloride); -·-, poly(methyl methacrylate); · · ·, polystyrene (after Vincent 1974. © John Wiley and Sons, Chichester, UK. Reproduced with permission)

3. *Effect of Other Variables*

The geometry of the test piece can greatly alter the recorded impact strength of a given polymer. For example, impact strength will be increased by decreasing the specimen width, as this accentuates plane stress conditions rather than plane strain, decreasing the notch depth, as this lowers the stress concentration factor, and increasing the notch radius, as this blunts the notch and makes it more difficult to initiate a crack.

The influence of notch tip radius on impact strength for a variety of polymers is shown in Fig. 3. For a notch radius of 0.125 mm, the impact strength of the homopolymers is relatively low and ABS is superior in impact strength to the other polymers. As the notch radius is increased, however, the increase in impact strength is more rapid for PVC, nylon and acetal than ABS. Hence these polymers become superior to ABS as the notches become blunter. This behavior results from the fact that

ABS, with its dispersed rubber particles, offers greater resistance to crack propagation than the other polymers, which offer greater resistance to crack initiation.

The impact strength of polymers also depends on fabrication conditions. Rapid cooling frequently leads to a polymer which is more ductile and has higher impact resistance. Annealing near the glass-transition temperature tends to raise the Young's modulus but to reduce ductility and impact toughness. In rubber-modified polymers, control of rubber particle size is important for optimum impact performance. Impact strength of thermoplastics, such as PVC, polycarbonate, PPO and polystyrene, can be raised by subjecting specimens to a thickness reduction by cold rolling. The enhancement that results is attributed to the introduction of beneficial residual compressive stresses at the surface.

See also: Polymers: Mechanical Properties; Polymers: Tests for Mechanical Properties

Bibliography

American Society for Testing and Materials 1970 Standard method of test for tensile-impact energy to break plastics and electrical insulating materials, ASTM D1822-68. *Annual Book of ASTM Standards*, Pt. 27. ASTM, Philadelphia, PA, pp. 526–35

Americal Society for Testing and Materials 1975 Standard methods of test for impact resistance of plastics and electrical insulating materials, ASTM D256-73. *Annual Book of ASTM Standards*, Pt. 35. ASTM, Philadelphia, PA, pp. 73–91

American Society for Testing and Materials 1976 Standard methods of test for impact resistance of polyethylene film by the free-falling dart method, ASTM D1709-75. *Annual Book of ASTM Standards*, Pt. 36. ASTM, Philadelphia, PA, pp. 185–93

British Standards Institution 1970 Mechanical properties, BS 2782. *Methods of Testing Plastics*, Pt. 3. BSI, London

Bucknall C B 1977 *Toughened Plastics*. Applied Science, London

Hall S 1989 Impact performance. *Polymer Materials*, 2nd edn. Macmillan, London, pp. 75–8

Reed P E 1979 Impact performance of polymers. In: Andrews E H (ed.) 1979 *Developments in Polymer Fracture*, Vol. 1. Applied Science, London, pp. 121–53

Sauer J A 1971 The influence of structure and other factors on molecular motions in solid polymers from 4 °K to 300 °K. *J. Polym. Sci., Polym. Symp.* 32: 69–122

Turner S 1983 Impulse tests. *Mechanical Testing of Plastics*, 2nd edn. Longman, New York, pp. 139–54

Vincent P I 1971 *Impact Strength and Service Performance of Thermoplastics*. Plastics Institute, London

Vincent P I 1974 Short-term strength and impact behavior. In: Ogorkiewicz R M (ed.) 1974 *Thermoplastics—Properties and Design*. Wiley, London, pp. 68–83

J. A. Sauer
[Rutgers University, New Brunswick, New Jersey, USA]

Injection Molding of Polymers: Computer Simulation

The design and commissioning of molds for polymers has traditionally been a trial and error procedure. Considerable time is required to test and, if needed, modify the mold before production may begin. A desire to minimize this time and reduce retooling costs had led to the development of computer software for simulating the injection-molding process. Software is commercially available for analyzing both the mold filling and the cooling phases of the injection-molding process. Both types of analysis generally use a three-dimensional representation of the mold cavity. In the case of cooling analysis, a representation of the cooling line locations is also used. These representations or models are typically generated with computer-aided design (CAD) packages. Simulation of the process allows mold and product designers to determine optimal gate positions, feed systems and processing conditions by indicating problem areas so that corrective action may be taken. In addition, the use of analysis software can minimize material costs and cycle times while improving the quality of molded components.

1. Simulation of Mold Filling

Analysis of mold filling is the natural starting point for analysis of the injection-molding process. For modellng purposes it is convenient to consider two stages: the filling phase, during which time a mold is volumetrically filled, and the packing phase, in which volumetric contraction, arising from cooling of the melt, is compensated for by the injection of more material. With an appropriate representation of the cavity geometry and of the rheological and thermal properties of the material, flow analysis can provide insight into the many factors that determine the quality of injection-molded parts.

1.1 Governing Equations

Flow analysis technology is based on the solution of the governing equations of fluid dynamics (Bird et al. 1960):

$$\frac{\partial \rho}{\partial t} = -\nabla \, . \, (\rho \boldsymbol{v}) \tag{1}$$

$$\frac{\partial(\rho \boldsymbol{v})}{\partial t} = -[\nabla \, . \, \rho \boldsymbol{v}\boldsymbol{v}) - [\nabla \, . \, \boldsymbol{\tau}] + \rho \boldsymbol{g} \tag{2}$$

$$\frac{\partial \rho \hat{U}}{\partial t} = -(\nabla \, . \, \rho \hat{U} \boldsymbol{v}) - (\nabla \, . \, \boldsymbol{q}) - (\boldsymbol{\tau} \colon \nabla \boldsymbol{v}) \tag{3}$$

where ρ is the fluid density, $\boldsymbol{v}$ is the velocity vector of a point in the fluid, $\boldsymbol{\tau}$ is the stress tensor, $\boldsymbol{g}$ is the acceleration due to gravity and $\hat{U}$ is the internal energy per unit mass. Equations (1–3) are known as the equations of continuity, motion and energy, respectively. They are valid for any fluid. A constitutive equation is required to relate the stress tensor $\boldsymbol{\tau}$ to the rate of deformation tensor $\nabla \boldsymbol{v}$ through the fluid viscosity. Commonly, a generalized Newtonian fluid constitutive equation is employed in which viscosity varies with shear rate or shear stress. Typical empirical equations (Bird et al. 1977) include the power law model, the Carreau model, the Cross model and polynomial representations. With the exception of the power law model, each of these is capable of representing the transition to Newtonian flow at low shear rates, as observed in polymer melts.

Additional terms in the constitutive equation can account for the temperature dependence of viscosity. Several functional forms are used to represent the dependence, including exponential and reciprocal functions of temperature and relations of the Williams–Landel–Ferry (WLF) type. Another level of complexity in material property representation arises if the pressure dependence of viscosity is considered. Viscosity is most commonly modelled as an exponential function of pressure. In addition, further data about density, heat capacity and thermal diffusivity properties are also required for a complete solution of the governing equations.

Assumptions of creeping flow, no wall slip, incompressibility, zero gravitational body forces and independence of thermal diffusivity from temperature and from cooling rate are widely used to simplify solution of the equations. In addition, the filling phase is often assumed to be adequately represented by steady-state flow at any instant in time whereas transience, which appears through the inclusion of the time derivative in the equation of motion, is essential to the solution in the packing phase.

It is important to note that any flow simulation is only as good as the underlying assumptions and the material data used. Within the guidelines set by simplification of the governing fluid dynamic equations, accurate predictions can only be made with high-quality material data.

1.2 Description of Cavity Geometry

Any flow analysis requires a description of cavity geometry as this comprises one aspect of the resistance to flow of the molten polymer during injection molding. In essence, a mold cavity is modelled by a set of connected points, lines and surfaces. The degree of complexity of the cavity model is generally directly related to the degree of complexity of the equations. Simple one-dimensional surfaces provide a "matchstick" description of the cavity and are only capable of representing simple flows such as rectangular duct flow without edge effects and flow through circular cross-sectional tubes and annuli. Two-dimensional surfaces give a "house of cards" or shell description of a mold cavity and are capable of representing two-dimensional planar flows. This is adequate for the majority of molds, particularly when combined with one-dimensional surfaces where necessary. Three-dimensional cavity descriptions enable analysis of fully three-dimensional flows but modelling and analysis are far more time consuming. Consequently, three-dimensional analyses are rarely used in practical injection-molding simulations.

1.3 Solution of Flow Equations

There are primarily two numerical techniques employed to solve the set of governing flow equations: the finite-difference method and the finite-element method. The finite-difference method reduces a set of differential equations to algebraic equations by substituting for the partial derivatives with approximating finite differences. In the finite-element method (Rao 1982) the solution domain is approximated by a set of piecewise polynomial approximations. In each case, increasing the order of the approximation increases the accuracy of the solution, but the solution cost becomes greater. Ultimately, there is always a trade-off between solution accuracy and computation expense.

The cavity geometry, which defines the solution domain, is mapped into a set of intersecting grid points or connected polyhedral elements, for the finite-difference or finite-element methods, respectively. Both sets are generally referred to as meshes. Solution results in a value of the field variables, pressure and temperature, for example at grid points, element intersections (nodes) or element centroids.

Hybrid finite-element–finite-difference methods are also employed where the cavity is modelled with two-dimensional plates or shells. A finite-element mesh is used as the basis of the calculation of pressure and temperature distribution throughout the mold. Within each finite element a finite-difference mesh is used to account for the variation of velocity, shear rate, viscosity and temperature through the cavity thickness.

Finally, some packages use so-called "fast" algorithms which avoid explicit determination of the variation of parameters through the cavity thickness. These proprietary algorithms effectively increase element viscosity as a function of mold geometry and thermal and pressure history. The net effect is a rapid determination of the cavity fill pressure and of the average temperature distribution.

1.4 Analysis Results and Display

Analysis output is similar between different packages and consists primarily of pressure, temperature, fill time, velocity, shear rate, shear stress and volumetric shrinkage. However, most of these parameters change continuously during both filling and packing phases. It is prohibitively expensive to provide the user with data for each of the major variables at every instant of time during injection, so most commercial packages record results considered to be of particular interest or provide for data storage at various user-defined stages of the analysis. Output is viewed graphically with accompanying postprocessing software or general purpose postprocessors that interface to the analysis output.

One of the most useful outputs is the distribution of fill times at each point in the cavity as such a plot shows how the mold fills. It shows which areas fill first, where opposing melt fronts meet to form weld lines, and leading and lagging sections of the flow front which could lead to air entrapment or weld line formation where it was not anticipated. Consideration of the areas that fill first allows users to avoid overpacking, a common cause of differential shrinkage and warpage. Positioning of weld lines may be of aesthetic or structural importance.

Contour plots of pressure distribution at various stages during filling provide graphic evidence of high-shear-stress areas, flow reversals, areas of overpack and, of course, estimates of fill pressure. When the pressure distribution is integrated over the projected area of the cavity, estimates of clamp force are obtained. Both pressure and clamp force estimates aid in the selection of an injection-molding machine of adequate capacity. Coupled with skillful interpretation of the filling pattern, the molder can use the output to optimize gate positioning so as to avoid overpacking and reduce shrinkage and warpage.

Excessive heat generation, which can lead to material degradation, or premature freezing, which can lead to shrinkage and warpage, can be avoided by examining the distribution of mean temperatures during filling and packing and altering injection conditions accordingly. In addition, plots of flow-front temperature at each point in the cavity give an indication of the quality of weld lines because the hotter the flow fronts at the point of formation of the welds, the more effective the diffusion of material across the interface becomes.

Minimizing shear stress levels during injection is critical to acrylonitrile–butadiene–styrene (ABS) moldings that are to be electroplated. The molder can experiment with molding conditions to ensure maximum stress levels are not exceeded.

Volumetric shrinkage estimates are typical packing phase analysis outputs. These have been related variously to shrinkage and distortion of the entire molding. The formation of unsightly sink marks can be observed through packing analysis and can, therefore, be avoided or minimized.

These examples are just a few of the outputs available from current commercial packages. It is apparent that the advantage of using flow analysis in injection molding is its ability to model multiple injections, varying cavity geometry, molding conditions and material type as appropriate, without the need to build costly molds and run time-consuming mold trials. The expense of component design and manufacture, and of tool making, is reduced because prototyping can be waived. Material savings are possible because runner positioning, diameters and lengths can be optimized for pressure drop and shear heat input. Similarly, local cavity thicknesses can be easily modified to provide optimum filling performance and structural strength without overdesign of the product. The ease with which a user can say "what if?" and then answer the question is certain to spread the use of flow analysis technology through the plastics-injection-molding industry.

2. Cooling Analysis

The cooling phase represents more than two-thirds of the time required to produce an injection molded part. Improperly designed cooling systems can lead to part distortion due to differential cooling and to high levels of residual stress in the molding. Residual stresses may be so high as to cause part cracking and premature failure of the component. The designer of a mold cooling system has two main aims: minimization of the cooling time and uniform cooling.

2.1 Heat Transfer in Moldings

Heat transfer in an injection molding is illustrated in Fig. 1 where Q_p is the heat flow into the mold from the incoming melt, Q_s is the heat flow from the surface of the mold due to convection, Q_r is the

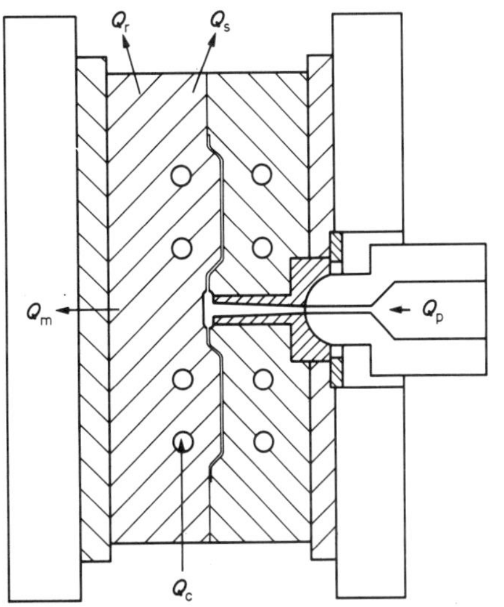

Figure 1
Heat transfer in an injection molding

radiative heat flow from the mold, Q_m is the heat flow due to conduction into the injection molding machine and Q_c is the heat flow into the coolant due to convection.

Conduction of heat into the molding machine is minimized in practice by insulation and may be ignored. Heat loss due to radiation and convection from the outer surface of the mold is significant only if the mold surface temperature is above 80 °C. The major factors are therefore conduction of heat through the plastic occupying the cavity and into the mold and heat flow from the mold into the cooling channels.

2.2 Governing Equations

Heat from the plastic in the cavity is transferred to the mold body by conduction. It is reasonable to assume conduction takes place only in the thickness direction of the plastic and so may be represented by the one-dimensional heat conduction equation:

$$\frac{\partial T}{\partial t}=\kappa\frac{\partial^2 T}{\partial z^2} \tag{4}$$

where $T=T(z,t)$ is the temperature at a point in the cavity, κ is the thermal diffusivity of the material and z is the thickness direction of the cavity.

Heat conduction through the mold body is described by the three-dimensional heat equation:

$$\frac{\partial T}{\partial t}=\kappa\nabla^2 T \tag{5}$$

where ∇^2 is the Laplace operator and $T=T(x, y, z, t)$.

Equations (4) and (5) must be solved subject to suitable boundary conditions. These consist of the prescribed initial temperature field within the plastic or mold and the rate of heat flow across the cavity–mold interface or external surfaces of the mold. The boundary conditions may incorporate the effect of radiation or convection from the mold surface. In particular, the rate of heat flow from the mold into the cooling line is given by

$$q_c=h_c A_c(T_{cs}-\bar{T}_c) \tag{6}$$

where h_c is the heat transfer coefficient of the cooling line, A_c is the area of the cooling line, T_{cs} is the temperature of the cooling line surface and $\bar{T}_c$ is the bulk temperature of the coolant. The value of h_c, and so the ability of the cooling line to extract heat, depends on the Reynolds number and the thermal properties of the coolant. In practice, the dominant factor is the Reynolds number, given by

$$Re=\rho VD/\mu \tag{7}$$

where ρ is the density of the coolant, V is the velocity of the coolant, D is the diameter of the cooling line and μ is the coolant viscosity. Note that although ρ and μ are functions of temperature, if the flow rate is sufficiently high, temperature variation of these properties may be ignored. In order to maximize the heat transfer coefficient, the coolant flow should be turbulent. This is normally easy to achieve with the type of pumps used in the molding industry.

2.3 Description of Cooling Line and Part Geometry

Considerable modelling time is saved if the same model for the cavity can be used for both flow and cooling analyses. Cooling lines must also be described for cooling analysis and are represented as a network of one-dimensional surfaces each of which is defined by two points. Provision to handle special configurations such as bubblers and baffles and lines of different cross-sectional shape is often available in analysis software. The cooling line sections are meshed to form a series of one-dimensional elements. Some analysis software may also require that the outer surface of the mold be described. This will always be necessary for an analysis that takes radiative and convective losses into account.

2.4 Numerical Solution

It is important to note that the solution to the heat transfer problem is transient in nature. As the plastic occupying the cavity cools, the rate of heat transfer into the mold will vary. Despite this fact, several commercial analysis packages use a steady-state analysis. Such analyses are based on the assumption that the heat into the mold may be averaged over the molding cycle.

In order to solve Eqn. (4), it is necessary to specify the initial temperature at the wall of the cavity and the initial temperature of the plastic. These may be given by the user or derived from the results of a flow analysis. Equation (4) may then be solved using finite-difference techniques.

Analysis software invariably includes cooling network calculations. The user specifies an inlet flow rate and temperature. Flow rate and temperature in each branch of the cooling system are then calculated using mass and energy conservation principles.

A very detailed solution of Eqn. (5) may be obtianed with a full finite-element representation of the part and the mold (Karjalainen 1987). Special elements can be used to take into account the thermal resistances between different parts of the mold; molds produced from materials with different conductivities can be handled easily. The major disadvantage is the need to mesh the interior of the mold to produce solid elements. In addition, the computer time needed can be prohibitive.

Boundary integral techniques (Brebbia et al. 1984) are also employed to solve Eqn. (5). One advantage of this approach is that no internal meshing of the mold is required. The process does, however, require the exterior mold surface to be modelled and meshed. One disadvantage is that it is difficult to account for molds constructed with materials of differing conductivity.

Another method involves reducing the problem to a one-dimensional approximation. A three-dimensional finite-element representation of the part is used. The reduced problem consists of a series of one-dimensional analyses where the heat from a plastic element is transferred to a single cooling line element. The distance between the elements is varied according to the heat load. Success or failure of this scheme depends on the quality of the assumptions underlying the approximation. It is necessary to consider the effect of surrounding plastic elements on the effective heat transfer. This manifests itself in the calculations as a change in the distance between the plastic and cooling line elements. For instance, in areas such as corners where the heat load is high the "effective" distance is increased. If the geometry of the part is carefully considered this scheme can rapidly produce very good results. Another advantage is that the calculated one-dimensional distances and effective conductivities can be used by flow analysis software thereby allowing temperature calculations to be based on the effect of the cooling system.

Solution by any of the aforementioned methods is iterative in the sense that after the rate of heat transfer into the cooling lines has been initially estimated, temperatures at the wall of the cavity and cooling lines are calculated, the heat transfer into the cooling lines is then calculated and the process repeats. Final convergence to the correct solution may be determined by an energy balance of heat into and out of the mold.

2.5 Analysis Results and Display

Results from cooling analysis software are displayed in a similar way to those from flow analysis with contours of the calculated variable displayed on the model. The mold temperature at ejection is an important result from cooling analysis and most graphical displays permit the user to display temperatures on either side of the plastic elements. With these results, the uniformity of cooling can be assessed. A particularly useful feature is the ability to plot the difference in temperature across the part. Areas of high temperature differential may then be seen. Minimizing temperature differences across the part will reduce warpage. Often the results from network calculations can be displayed graphically. Typical outputs include flow rate in each branch of the cooling system, the wall temperature of the cooling line and the bulk temperature of the coolant.

3. Future Developments

Flow analysis technology, although now widely accepted by the injection-molding industry and shown to be reasonably successful for a large majority of moldings, is far from being a completed subject. There are a number of areas, particularly in relation to material behavior, where current technology is clearly inadequate.

During the filling phase, for example, material compressibility and density variation with temperature are largely ignored, although it is known that these properties can account for observed differences of 30% between molding machine pumping rate and effective volumetric flow rates within the cavity. Phenomena which reflect the viscoelastic nature of polymer melts, such as extra pressure drops in rapidly converging sections, melt fracture and distortion, orientation development and stress relaxation, are generally not considered by current commercial programs. Polymer melts also exhibit considerable tensile viscous forces, which can dominate the flow field during rapidly converging or diverging flows. This property is particularly important for accurate analysis of fiber-filled resin molding, since it plays a significant part in fiber orientation and subsequent strength and shrinkage. Adequate commercial consideration of these phenomena depends on developments in theoretical modelling of polymer melts, material testing techniques, numerical methods and computer hardware.

Warpage of injection-molded parts remains an area of concern in industry and software for its prediction is currently under development. The main causes of warpage are high temperature differentials at ejection and different amounts of shrinkage throughout the molding. Cooling analysis software can minimize high-temperature-differential problems but shrinkage is a more complex phenomenon (Isayev 1987). It is generally agreed that shrinkage is affected by conditions in both the filling–packing and cooling phases. While existing flow and cooling analysis software can predict the various factors that contribute to shrinkage, the interplay between them is not simple and cannot in general be interpreted by the user. Moreover, there exist no analytical models for shrinkage that may be applied to a wide range of materials.

Current research is centered on correlating values of the variables affecting shrinkage from flow and cooling analyses with measured shrinkages of test samples molded under a range of processing conditions. The results of the correlation are used to calculate material data that relate actual shrinkage to the conditions to which the material is subjected during processing. The problem is further complicated by the fact that material does not always shrink isotropically. In general, high levels of molecular orientation are induced in molded parts and the magnitude of the resulting shrinkage depends on the direction of material orientation. Consequently, it is necessary to calculate shrinkage parallel and perpendicular to the material orientation direction for each plastic element. The effects of temperature differentials are incorporated by calculating the resulting bending moment for each element. In order to predict the warped shape of the part, the elemental shrinkages and bending moments are input to a finite-element stress analysis program. This calculates the deflection of each node. Results are graphically displayed and show the deflected shape of the component and the magnitudes of the deflection.

See also: Processing: Rheological Properties and Fluid Mechanics; Reaction Injection Molding

Bibliography

Bernhardt E C (ed.) 1983 *Computer Aided Engineering for Injection Molding*. Hanser, New York

Bird R B, Armstrong R C, Hassager O 1977 *Dynamics of Polymeric Liquids*, Fluid Mechanics, Vol. 1. Wiley, New York, pp. 442–51

Bird R B, Stewart W E, Lightfoot E N 1960 *Transport Phenomena*. Wiley, New York

Brebbia C A, Telles J C F, Wrobel L C 1984 *Boundary Element Techniques*. Springer, New York

Hill J M, Dewynne J N 1987 *Heat Conduction*. Blackwell Scientific, Melbourne, Australia

Isayev A I 1987 Orientation, residual stresses and volumetric effects in injection molding. In: Isayev A I (ed.) 1987 *Injection and Compression Molding Fundamentals*. Dekker, New York, pp. 227–328

Karjalainen J A 1987 Computer simulation of injection mould cooling. *Acta Univ. Ouluensis Ser. C* 43

Manzione L T (ed.) 1987 *Applications of Computer Aided Engineering in Injection Molding*. Hanser, New York

Menges G, Recker H 1989 Computer-aided engineering in injection molding. In: Aggarwal S L (ed.) *Comprehensive Polymer Science*, Vol. 7. Pergamon, Oxford, Chap. 14

Rao S S 1982 *The Finite Element Method in Engineering*. Pergamon, Oxford

C. Friedl and P. Kennedy
[Moldflow Pty, Victoria, Australia]

Injection Molding of Thermoplastics: Mold Filling

A variety of molding processes are used commercially, including injection, compression, transfer, blow and rotational molding. The particular process which is employed depends largely on the design, final application and requirements of the finished product. Of these molding processes, the most important commercially is injection molding. In this process, raw material is melted during the plastication stage and then injected under pressure into a mold. In order to achieve solidification of the product, the melt is injected into a cold mold cavity in the case of thermoplastics or into a heated mold cavity in the case of thermosets. Cooling or heating continues until the material has solidified sufficiently to permit ejection of the molded article from the mold without damage. The most commonly uscd injection-molding systems are of the reciprocating screw type, although ram injection machines are not uncommon.

1. Injection-Molding Sequence

The injection-molding machine consists of two basic components: the injection unit and the clamping unit. The injection unit heats and melts the raw material and transfers it under pressure into a mold which is held by the clamping unit. Melting is achieved with the help of external heaters and, in most cases, by mechanical heating induced upon shearing and compressing the granular raw material. A typical operating sequence starts with feeding of the material by gravity from a hopper, the screw being in the forward position. The rotation of the screw plasticizes the material and conveys it forward in the screw barrel. Accumulation of the plasticized material at the front of the screw tip forces the screw backwards. A check valve prevents backflow over the screw flights.

When the injection unit is activated, the screw moves forward, as a piston, pushing the melt from the injection chamber through a nozzle into the mold. The hot melt flows into an empty cavity until the mold is filled. More polymer is then packed into the cavity to compensate for shrinkage due to solidification. The cycle continues until enough rigidity is achieved to permit ejection of the part without damage. The cycle is repeated to produce the next part.

2. Melt Delivery Channels: Sprues, Runners and Gates

A diagram of a typical injection-mold arrangement is shown in Fig. 1. After leaving the nozzle, the melt generally flows through the sprue bushing and the runner system and finally enters the cavity through the gate. The opening in the sprue bushing is larger than the adjacent opening in the nozzle and the sprue is tapered to facilitate removal of the plastic after solidification. Moreover, the outlet of the sprue should not be smaller than the runner diameter at the point where the two meet. The runner is usually circular or trapezoidal (if in one plate only). To minimize pressure loss, the runner cross section should be large, but not so large as to require a substantial rise in the cycle time due to cooling requirements.

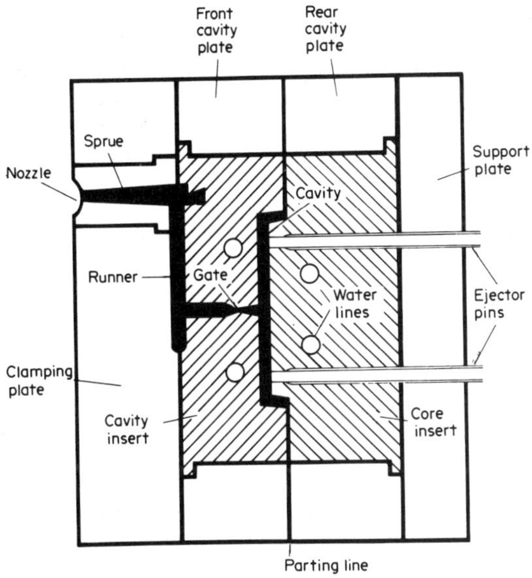

Figure 1
Schematic diagram of injection mold and melt delivery channels

When multiple cavities are used, it is important to ensure that all cavities fill at the same rate and that the material in the different cavities experiences comparable thermal histories. Sometimes this is achieved by a balance between the dimensions of the runner segments and the gates leading to the various cavities. In order to reduce regrinding and other materials handling requirements, hot runner molds have been designed to keep the material in the runner in the molten state.

The gate connects the runner to the mold cavity and controls the flow of the plastic melt into the cavity. It also provides a mechanism for separating the molded article from the runner system. Gates are either restricted (pinpoint gates) or large. Restricted gates are circular in cross section and generally do not exceed 1.5 mm in diameter, although they can be significantly larger for viscous melts. Large gates are square or rectangular and typical side dimensions are around 6 mm. Gates may also be described by location: edge, tab, back, submarine, sprue, diaphragm, ring, fan and flash gates. Ideally the gate should be as short as possible, in order to minimize pressure loss and to avoid the development of flow lines. A long gate could also lead to sink marks and voids in the molding and causes premature freezing. On the other hand, some length is required to prolong the life of the cavity.

Although more accurate estimates may be made by using more complex methods, it is possible to obtain reasonable estimates of flow rates and pressure requirements in delivery channels by making two assumptions: first, that the flow is fully developed, isothermal and incompressible and, second, that the polymer melt behaves as a power law fluid; that is,

$$\tau = m\dot{\gamma}^{n} \tag{1}$$

where τ is the shear stress, $\dot{\gamma}$ is the shear rate and m and n are appropriate power law constants. Thus, the following equations are applicable for the geometries indicated (Tadmor and Gogos 1979).

For a circular tube of radius r and length l,

$$Q = \left(\frac{\pi r^3}{s+3}\right)\left(\frac{r\Delta p}{2ml}\right)^{s} \tag{2}$$

where $s = 1/n$, and Q is the volumetric flow rate.

For a converging conical channel of large radius r_1, small radius r_2 and length l,

$$Q = \left(\frac{n\pi r_1^3}{3n+1}\right)\left(\frac{r_1 \Delta p a_{13}}{2ml}\right)^{5} \tag{3}$$

$$a_{13} = \frac{3n(r_1/r_2 - 1)}{r_1/r_2[(r_1/r_2)^{3n} - 1]} \tag{4}$$

For a half-circle channel (a semicircle with a rectangular extension on the top) of radius r, depth d and length l,

$$Q = \frac{2M_0 r d^3 \Delta p}{12\mu l} \tag{5}$$

where M_0 is shape factor depending on the channel dimensions and μ is a Newtonian viscosity.

For a rectangular channel of length l, breadth b and depth h,

$$Q = \frac{bh^2}{M_1}\left[\frac{bh\Delta p}{2lm(b+h)}\right]^{1/n} \tag{6}$$

where M_1 is a shape factor that depends on the channel dimensions.

3. *The Injection-Molding Cycle*

After the material has traversed the delivery channels and the gate, the process of filling the cavity begins. From this point, the injection-molding cycle may be considered to consist of three stages: filling, packing and cooling. During filling, the polymer melt is introduced into the empty cooled mold. After filling is complete, more material is packed into the cavity at high pressure to compensate for shrinkage as the material cools. Cooling continues after packing is completed. During the cooling state, the material solidifies and the pressure decreases. A typical pressure–time cycle during injection molding is illustrated in Fig. 2. In general, the cooling stage represents the longest segment of the cycle, usually accounting for more than 60% of the overall cycle.

Normally, commercial injection molding involves mold cavities of complex geometry, sometimes incorporating inserts and a variety of obstructions to direct the flow. Gates may be located at various positions and can have a multitude of shapes and dimensions, as mentioned earlier. In some cases, multiple gates are employed to fill a single cavity. All these considerations may have a profound influence on the dynamics of the molding process and on the ultimate characteristics of the molded articles. In particular, some molding arrangements lead to the formation of weld lines in regions where melt fronts intersect inside the cavity (e.g., after flowing around inserts and as a result of filling from multiple gates). Unless carefully manipulated and controlled, weld lines can represent very weak areas in the molding, leading to premature failure.

Ideally, engineering analysis, mathematical modelling or computer simulation of the injection-molding process should recognize these factors and should thus permit the treatment of flow and heat transfer phenomena in complex injection-molding cavities of the types described previously. Modern finite-element numerical techniques make this possible. Even with the finite-element method, the polymers, molds and phenomena involved are so complex that further simplification is usually desirable.

One common approach involves the mold network method. According to this method, the complex mold cavity is divided into a number of units of reasonably simple geometry (e.g., rectangular, circular or cylindrical segments). The solutions for the melt behavior in each of the simpler units are treated separately and are sometimes readily available in graphical, numerical or analytical forms. To obtain the overall solution for the complete cavity, the solutions for the individual components of the network are coupled and dovetailed to ensure the systematic integration of the results from each segment (Hieber 1982).

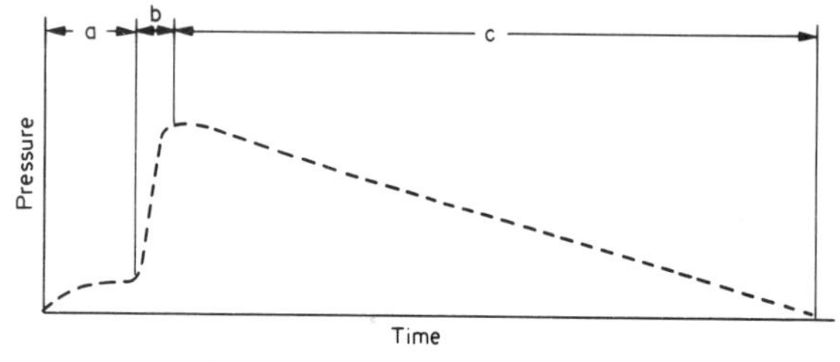

Figure 2
Typical pressure–time cycle for variation of pressure at the gate in injection molding: (a) filling, (b) packing and (c) cooling

4. The Filling Stage

The filling stage is concerned with the unsteady-state flow of a hot, compressible, viscoelastic melt into an empty cavity which is held at a temperature below the solidification temperature of the melt. In this sense, the process can be fully described in terms of the equations of conservation of mass, momentum and energy, coupled with appropriate thermodynamic relationships and a set of constitutive relations which describe the behavior of the material under the influence of stress and thermal fields. In general tensorial formulation, the conservation equations are as follows.

Continuity:

$$\frac{D\rho}{Dt} = -\rho(\nabla \cdot \boldsymbol{v}) \tag{7}$$

Momentum:

$$\rho\frac{D\boldsymbol{v}}{Dt} = -\nabla p - (\nabla \cdot \boldsymbol{\tau}) + \rho \boldsymbol{g} \tag{8}$$

Energy:

$$\rho c_V \frac{DT}{Dt} = -(\nabla \cdot \boldsymbol{q}) - T\left(\frac{\partial p}{\partial T}\right)_\rho (\nabla \cdot \boldsymbol{v}) - (\boldsymbol{\tau} : \nabla \boldsymbol{v}) \tag{9}$$

where ρ is the density, $\boldsymbol{v}$ is the velocity, p is pressure, $\boldsymbol{\tau}$ is the stress tensor, $\boldsymbol{q}$ is the heat flux, T is the temperature and t is time.

To obtain a solution of Eqns. (7–9), each stage of the process is associated with a set of initial and boundary conditions coupled with simplifying assumptions.

There are a variety of constitutive rheological equations that may be employed to describe the viscous or viscoelastic behavior of the polymer melt during the process. The most commonly employed rheological models in the injection-molding process include Newtonian liquid, power law model, second-order fluid (Pilate and Crochet 1977) and Leonov equation (Isayev and Hieber 1980). Whereas some of the models describe only the viscous behavior of the fluid, other models reflect some of the viscoelastic characteristics of the material.

In the filling stage, which starts with an empty cavity, a free surface exists at the advancing melt front. Thus, in a rectangular cavity, the part filled with the melt may be divided into three regions, as

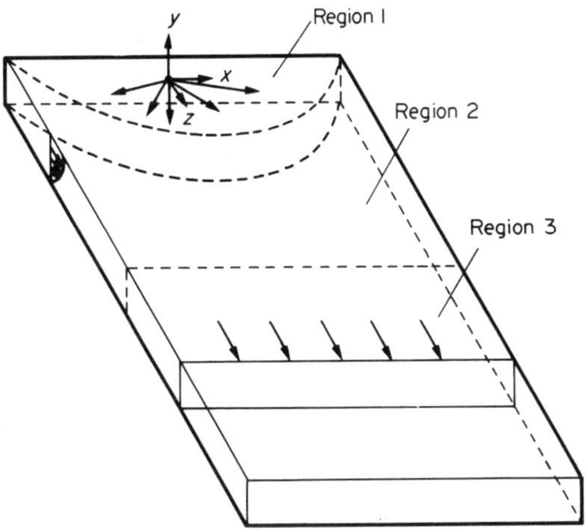

Figure 3
Flow regions during the filling of a rectangular injection-molding cavity

shown in Fig. 3. In region 1, near the gate, the flow is radial until the flow front reaches the wall. Away from the gate, region 2, the flow is characterized by a dominant axial velocity component. In region 3, called the front region, the melt, channelling through the center, tends to move towards the cavity wall in what is referred to as the "fountain flow." Although it is possible to analyze the details of the flow structure in the cavity by dealing with the appropriate three-dimensional flow equations, it is much simpler to treat the regions separately and to solve the corresponding one- or two-dimensional flow equations.

For relatively thick mold cavities or for very short filling times, it is possible to obtain a reasonable analysis of the filling stage by assuming isothermal flow conditions (Berger and Gogos 1973). This is mainly due to the low thermal conductivity of the polymer melt. However, in general, temperature changes due to mold cooling and viscous dissipation must be taken into consideration. Since only moderate pressures are encountered during the filling stage, it is generally reasonable to assume incompressible flow.

Both finite-element and finite-difference numerical techniques may be employed to obtain computer solutions of the relevant equations. The former methods are more complex, but they are well suited for dealing with complex mold cavities and the free surface flow at the melt front. An interesting approach for dealing with the fountain flow near the free surface is based on the finite-difference method in conjunction with the marker-and-cell technique (Kamal and Lafleur 1982).

The analysis of the filling stage yields information regarding the progression of the melt front and the distributions of velocity, pressure and temperature. Moreover, it supplies data regarding the development of the solid skin layer during the filling stage. Typical model predictions relating to the progression of the melt in the cavity and the pressure distribution during mold filling, for a high-density polyethylene resin, are shown in Figs. 4 and 5. By employing models incorporating appropriate rheological constitutive equations and expressions for crystallization kinetics, it is possible to estimate the distributions of crystallinity and frozen stresses in the system. In turn, this leads to reasonable estimates of the distributions of mechanical and optical properties in the molding.

5. *The Packing Stage*

The approach for the simulation of the packing stage is similar to that for the filling stage. For the packing stage, the deceleration motion of the flow reinforces the importance of the unsteady-state terms and they should not be neglected. During packing, the contribution of the unsteady-state terms will tend to be larger than that of the convective terms in the equation.

The dominant feature in the packing stage is the compressibility of the polymer melt. This emphasizes the need for selecting an appropriate equation of state to describe the pressure–volume–temperature relationships for the melt. A number of

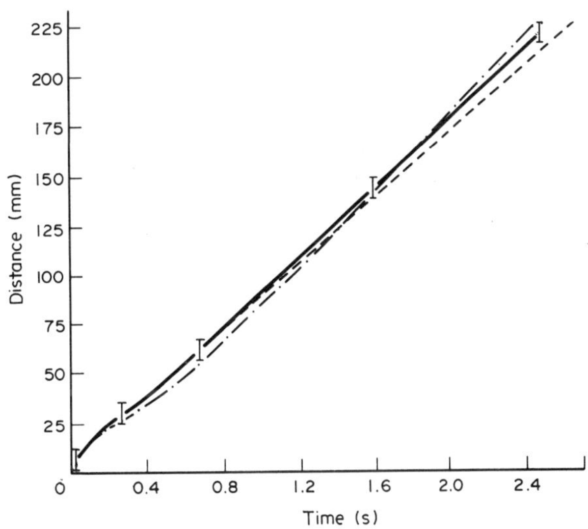

Figure 4
Comparison of model prediction with experimental data for the progression of the melt front during injection molding of high-density polyethylene: ——, experimental transducer; – – –, experimental cinematography; — · —, theoretical (after Doan 1974)

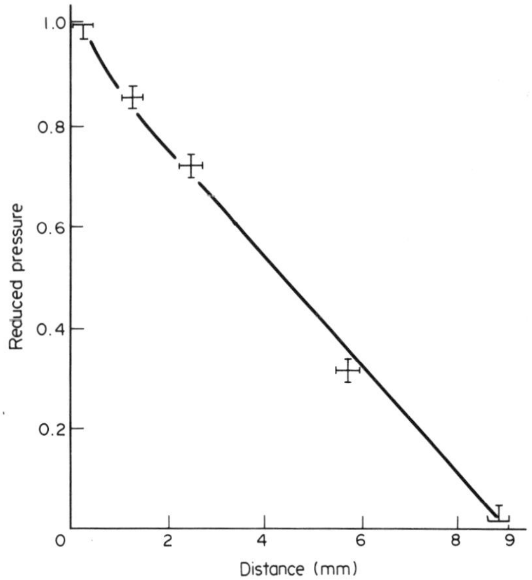

Figure 5
Comparison of model predictions with experimental pressure distribution at the end of filling a rectangular cavity with high-density polyethylene: ⊞, experimental; ——, theoretical (after Doan 1974)

empirical equations of state have been employed in conjunction with the processing of polymer melts (Tadmor and Gogos 1979). The following two equations have been found useful for injection-molding analysis:

$$(p+\pi)(\bar{V}-w)=RT \tag{10}$$

where p is the pressure, T is the temperature, R is the gas constant and π and w are empirical constants (Spencer and Gilmore 1951).

$$\rho=\rho^{\infty}+\left(\frac{\partial\rho}{\partial p}\right)_{p=0} T+(a+bT)p$$
$$+\tfrac{1}{2}(c+dT)p^2 \tag{11}$$

where ρ is the density, ρ^{∞} is the density at 0 K and zero pressure and a, b, c and d are adjustable parameters (Kamal and Levan 1973).

6. The Cooling Stage

After the filling and packing of the cavity, the gate freezes and no additional material enters the cavity. Since fluid motion is negligible or absent during this stage, the problem is reduced to the transient heat transfer by conduction in the polymer melt trapped between the mold walls. For thin cavities, the one-dimensional conduction problem takes the form

$$\rho c_p\frac{\partial T}{\partial t}=\frac{\partial}{dx}\left(k\frac{\partial T}{\partial x}\right)+\frac{T}{V}\left(\frac{608\rangle\beta V}{\partial T}\right)_p\frac{\partial p}{\partial t} \tag{12}$$

where c_p is the specific heat capacity, V is the specific volume and k is the thermal conductivity. All other terms are as defined earlier. The following analysis applies to partly crystalline polymers, which crystallize upon solidification. The treatment of amorphous, noncrystallizing polymers may be considered as a special case of this analysis.

In the cooling process of a crystallizing polymer, Eqn. (12) is applied to both the solid and melt phases, allowing the thermal properties to vary continuously with temperature and pressure. In order to allow for the effect of latent heat of crystallization, the latent heat is included in the specific heat:

$$c_p=c_p'+\lambda\frac{dx}{dt} \tag{13}$$

where c_p is the effective specific heat capacity of the material, c_p' is the specific heat of the material at the same temperature in the absence of crystallization, λ is the latent heat of crystallization and x is the weight fraction crystallinity in the system. In fact, c_p is the property normally measured during a crystallization experiment in the differential scanning calorimeter.

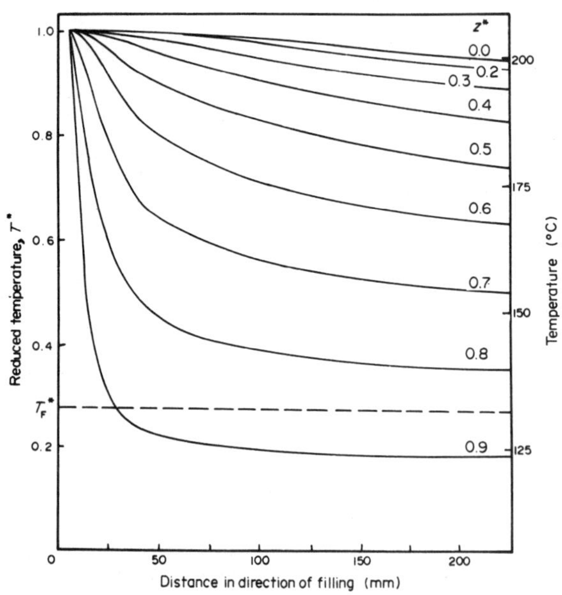

Figure 6
Calculated temperature distribution in a rectangular cavity at the end of filling with high-density polyethylene: z^*, dimensionless transverse distance (center = 0 and wall = 1.0); T_F^*, freezing temperature (after Doan 1974)

The rate of crystallization and the evolution of crystallinity during a nonisothermal crystallization experiment may be based on the extension of Avrami's isothermal treatment to nonisothermal systems (Nakamura et al. 1972):

$$x(t) = 1 - \exp\left[-\left(\int_0^t \kappa^- |T(t)| \, d\tau \right)^n \right] \quad (14)$$

where n is the Avrami index determined from isothermal experiments and $\kappa^-(T)$ is connected with the crystallization rate constant of isothermal crystallization $\kappa(T)$ by the relation

$$\kappa^-(T) = [\kappa(T)]^{1/n} \quad (15)$$

Figure 6 shows typical model predictions of temperature distributions in a rectangular cavity during the cooling stage for a high-density polyethylene resin.

See also: Processing: Classification; Processing: Shear-Controlled Orientation

Bibliography

Berger J L, Gogos C G 1973 A numerical simulation of the cavity filling process with PVC in injection molding. *Polym. Eng. Sci.* 13: 102–12

Doan P H 1974 Injection molding of thermoplastics in rectangular cavities. M.Eng. thesis, McGill University, Montreal

Dym J B 1979 *Injection Molds and Molding—A Practical Manual*. Van Nostrand Reinhold, New York

Gilmore G D, Spencer R S 1951 Photographic study of the polymer cycle in injection molding. *Mod. Plast.* 28(8): 117–85

Hieber C A 1982 Coupled-flow-path method for simulating the injection-molding filling stage. *Soc. Plast. Eng., Tech. Pap.* 28: 356–8

Isayev A I, Hieber C A 1980 Toward a viscoelastic modelling of the injection molding of polymers. *Rheol. Acta* 19: 168–82

Kamal M R, Kenig S 1972 The injection molding of thermoplastics, Pt. I: Theoretical model. *Polym. Eng. Sci.* 12: 294–301

Kamal M R, Lafleur P G 1982 Computer simulation of injection molding. *Polym. Eng. Sci.* 22: 1067–75

Kamal M R, Levan N T 1973 Comparison between some empirical equations of state for polymers. *Polym. Eng. Sci.* 13: 131–8

Kamal M R, Moy F H 1981 Estimation and measurement of the distribution of tensile moduli in injection molded thermoplastics. *Chem. Eng. Commun.* 12: 253–65

Kuo Y, Kamal M R 1976 The fluid mechanics and heat transfer of injection mold filling of thermoplastic materials. *AIChE J.* 22: 661–9

Lafleur P G, Kamal M R 1981 Computer simulation of thermoplastic injection molding. *Adv. Plast. Technol.* 1(4): 8–13

Menges G, Recker H 1989 Computer-aided engineering in injection molding. In: Aggarwal S L (ed.) 1989 *Comprehensive Polymer Science*, Vol. 7. Pergamon, Oxford, pp. 454–64

Mills N J 1986 Injection molding. *Plastics*. Arnold, London, Chap. 4, pp. 100–6

Morton-Jones D H 1989 Injection molding. *Polymer Processing*. Chapman and Hall, London, Chap. 8

Nakamura K, Watanabe T, Katayama K, Amano T 1972 Some aspects of nonisothermal crystallization of polymers, I: Relationship betweeen crystallization temperature, crystallinity, and cooling conditions. *J. Appl. Polym. Sci.* 16: 1077–91

Pilate G, Crochet M J 1977 Plane flow of a second-order fluid past submerged boundaries. *J. Non-Newtonian Fluid Mech.* 2: 323–41

Rubin I I 1973 *Injection Molding: Theory and Practice*. Wiley, New York

Spencer R S, Gilmore G D 1951 Some flow phenomena in the injection molding of polystyrene. *J. Colloid Sci.* 6: 118–32

Tadmor Z, Gogos C G 1979 *Principles of Polymer Processing*. Wiley, New York

Whelan A, Goff J P 1988 Injection molding of engineering thermoplastics—Mold considerations. *Prog. Rubber Plast. Technol.* 4(3): 1–20

M. R. Kamal
[McGill University, Montreal, Canada]

Injection Molding of Thermoplastics: Structural Foams

The thermoplastics structural foam (TSF) process is a derivative plastics injection-molding process for fabricating shapes having integral high-density surface layers or skins, nearly uniform cellular center sections or cores, a gradual transition in density from core to skin, and a high stiffness–weight ratio compared with shapes conventionally molded of the same unfoamed material. A cross section of a TSF product is shown in Fig. 1. Typically, TSF shapes

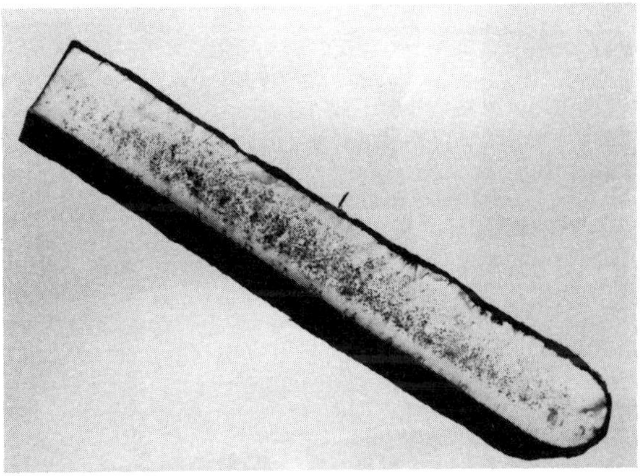

Figure 1
Cross section of a TSF product

have bulk densities of around 50–90% of the bulk density of the unfoamed polymer. The ratio of foamed resin density to unfoamed resin density is known as the reduced bulk density, $\phi = \rho_f/\rho_0$. As a first approximation, the mechanical performance of TSF shapes can be related to the reduced bulk density.

In order to obtain optimum mechanical performance levels, TSF shapes are designed to have wall thickness no less than 3 mm and typically more than about 4 mm. The practical maximum thickness is governed by economics, affected for example by the time required to cool the foamed shape to a temperature where removal from the mold cavity will not cause postprocessing distortion (shrinkage, warpage and reheat blowing). It also depends on the geometry of the article and competitive processing options, such as rotational molding, urethane RIM foam molding and blow molding. If the article is cylindrical or spherical, it can be as much as 100 mm in thickness.

Essentially, TSF processing requires dissolution of a carefully metered amount of a permanent or volatile gas, such as nitrogen, chlorofluorocarbon or carbon dioxide, in a softened or molten polymer under processing conditions approximating those for conventional injection molding. The foamable resin is then transferred from a high-pressure region (such as an accumulator of an injection-molding machine) to a hydraulically clamped two-part injection-mold cavity having the general dimensions of the shape to be fabricated but with a volume somewhat greater than the volume of the foamable resin being transferred. At reduced pressures and excess volume, the gas evolves into microbubbles, the foaming resin expands to fill the mold and the shape thus formed is cooled to a point where it retains the shape of the mold cavity.

1. History

Deliberate attempts to form a cellular thermoplastic composite structure by injection molding a gas-containing plastic apparently began in the mid-1950s with Dow Chemical Corporation's Frostwood process, where polystyrene was foamed with methylene chloride. In the early 1960s, Foster Grant, W. R. Grace, Union Carbide Corporation and Bayer AG worked on various aspects of foam injection molding. The recognition of low mold cavity pressures during filling and expansion of the foaming resin led to the development of an isolatable melt accumulator in which the foamable molten resin could be stored prior to injection. The combination of a fixed plasticating extruder, the separate melt accumulator and a large surface area but relatively low-tonnage hydraulic mold clamp enabled rapid market development of very large fabricated TSF parts in the early 1970s. Computer housings of 45 kg or more have been fabricated, but there is no minimum practical size; 4.5 g paintbrush handles are an example of a small product.

2. Applications

Thermoplastic structural foam shapes have successfully competed with conventional materials in a number of ways: with wood when bulk at reduced weight is required, or when ornateness in carvings must be faithfully repeated; with cast metal when weight reduction, corrosion resistance or ease of assembly is needed; with sheet metal when stiffness is the objective; with sheet molding compound when reduced weight, reduced cost or many parts are needed; with assembled metal parts when labor costs require it; and with conventional injection-molded plastic parts when stiffness, low stress or an absence of sink marks is required. In 1979, about 20% of the TSF shapes fabricated in the USA went into utility conduits and closures and another 10% into each of the following markets: building and construction, furniture, materials handling, retail and janitorial, and business machines and computer housings. High-density polyethylene was used in 40% of these applications, with polystyrene and polypropylene in 20% each. Modified polyphenylene oxide was used in 10% and potentially important TSF materials such as ABS, polycarbonate and nylon in the rest.

3. Processing Equipment

Two TSF processing schools of thought developed in the early 1970s. The European school generated foaming gas primarily through decomposition of a thermally unstable chemical, such as azodicarbonamide, *p, p'*-oxybis(benzenesulfonylhydrazide) or 5-phenyltetrazole. These chemical foaming agents were dry blended with powdered or granular polymer before addition to near-conventional injection-molding machines. The machines were modified to include a positive shutoff nozzle between accumulator and mold cavity and a non-return ring on the reciprocating plasticating screw. These changes were necessary to prevent the gas generated by decomposition of the foaming agent from escaping from the polymer before injection.

The primary emphasis in the USA during this time was focused on the direct addition of nitrogen to the molten polymer through an orifice in the barrel wall at a low pressure zone along the screw. For obvious reasons, it was recognized both in Europe and in the USA that the heavy-walled TSF parts required considerably longer cooling times to achieve removal than did unfoamed injection-molded parts of the same material. This additional time could be used to advantage to plasticate and accumulate large quantities of molten foamable resin and this led to the development of an isolatable foamable-melt accumulator. The primary forces applied against the hydraulic system holding the mold closed are attributable to the gas pressure in the resin. These are typically 1–3 MPa, which are far below conventional packing pressures of 30–70 MPa for nonfoaming injection molding. As a result, the conventional hydraulic forces used to close the mold cavity against the melt pressure can be applied over a much greater mold surface area. This, coupled with the additional time available to accumulate foamable resin, allows for TSF part weights 10–50 times those for nonfoamed parts at the same hydraulic demands on the injection-molding clamp.

In Europe, polymer was introduced to the mold cavity through a single nozzle. In order to spread the polymer throughout the mold cavity before foaming, very high pressures were required. As a result, the isolatable accumulator was fed through a nonreturn valve by a stationary plasticating extruder. Pressures in the region of 100 MPa are developed by applying nitrogen gas pressure of 20 MPa to hydraulic oil and then intensifying the resulting forces. Polymer transfer rates of up to 40 kg s^{-1} have been achieved with materials such as polystyrene. In the USA, the accumulator was the result of the requirements to introduce gas at a specific point along the extruder screw, in effect eliminating the traditional reciprocating motion of the screw. Further, multiple hydraulic nozzles were fed from the accumulator through a heated piping and manifold system. Accumulator pressures of 20–30 MPa were needed primarily to overcome the large pressure losses in the plumbing and this led to relatively slow injection speeds. In recent years, these limitations have for the most part been overcome through redesign of the manifold system.

The processes described above are considered to be low-pressure foaming processes. The lack of control of the foaming process once the accumulator is empty and the mold cavity sealed is considered to be the primary cause of uneven density distribution and varying skin–core thickness ratio across the fabricated shape, and the gas bubble surface splay or streaking so characteristic of low-pressure TSF parts.

As will be discussed, a finite amount of pressure must be applied to the gas-containing resin to control foaming rate. There are many ways to achieve this. One is to essentially injection mold the resin, then slowly expand the mold cavity. In another method, the mold cavity is pressurized with nitrogen or air just before injection, thus minimizing uncontrolled foaming at the entering wavefront of the injected polymer. In a third, the foamable resin is simultaneously injected with unfoamed resin. The latter forms a molten polymer balloon which is subsequently expanded by the foaming polymer in the center to fill the mold cavity. In a fourth, the mold cavity walls are heated to temperatures just below the polymer melt temperature, eliminating bubble freeze off during injection and foam expansion. The walls are then rapidly cooled to remove the sensible heat of the polymer.

4. *Typical Resins and Foaming Agents*

Most conventional injection-molding thermoplastic resins, either "neat" or containing additives, reinforcements and/or fillers, can be foam injection molded. The molecular weight of the resin should be relatively low to permit injection at relatively rapid rates, but high enough to minimize excessive bubble growth and agglomeration to large voids and abnormal in-cavity shrinkage during cooling. The resins should have sufficient melt strengths, as typified by film forming characteristics, thus minimizing bubble agglomeration and blowout during the expansion phase of the process. The resin should also be sufficiently thermally stable to minimize thermal degradation at the high temperatures and long times typical of the TSF process cycle. Materials should be stable for 50 000 K s or longer. These conditions also apply to additives (fire retardants, antioxidants, bubble stabilizers, colorants).

Typical physical foaming agents include nitrogen, carbon dioxide, chlorofluorocarbons (such as refrigerants 11 and 12), chlorinated hydrocarbons (such as methyl chloride) and the volatile hydrocarbons (e.g., propane or butanes). In order to achieve fine cellular products, admixtures such as inorganics (talc, silica) or sodium bicarbonate and citric acid are added as nucleants or cell formers.

Although chemical foaming agents can yield finer cell structures than physical gases, several stringent criteria must be met. Neither the chemical foaming agent itself nor any of its by-products should have an aggressive effect on the polymer, particularly at the long times and high temperatures typical of the TSF process. The agent should be relatively inexpensive at the characteristic dosages (normally <1 wt% and frequently <0.5 wt%) and must be in a form that allows for very fine dispersal in the molten or softened polymer. Normally this implies particle sizes of less than 37 μm. The typical chemical foaming agent decomposes to produce nitrogen and inactive by-products. Regardless of the source, the foaming gas should be sufficiently soluble in the polymer at typical injection-molding conditions to allow for dissolution under typical accumulator conditions. On the other hand, the gas should not be so soluble as to inhibit bubble formation and growth during the expansion phase in the mold cavity.

5. Processing Characteristics

To ensure capture of the foaming gas, the polymer must be molten or softened before decomposition of the chemical foaming agent or introduction of the physical foaming agent. The processing conditions everywhere ahead of the mold cavity must be such that essentially all the gas is dissolved in the polymer. According to Henry's law, which has been shown to be valid up to relatively high pressures for many gases and commodity plastics,

$$S = S_0 p \exp(-E_s/RT)$$

where S is the solubility (cm^3 g^{-1} polymer at STP), p is the pressure (MPa), T is the temperature (K), R is the gas constant, S_0 is a preexponential constant and E_c is the activation energy for solubility. For a chemical foaming agent, the amount of gas generated per gram of polymer is given as V_g(CBA), where V_g is the volume of gas generated per gram of blowing agent and (CBA) is the actual dosage of blowing agent per unit weight of polymer. For physical foaming agents, V_g(CBA) is replaced by V_g', the amount of gas added directly to a fixed amount of polymer melt. The solubility equation can be solved for pressure:

$$P_{\text{equil}} = (V_g(\text{CBA})/S_0)\exp(E_s/RT)$$

Above this pressure, all the gas is soluble in the polymer melt. Similarly, below this pressure, gas begins to form into microbubbles, according to Rayleigh's law.

For low-pressure foam molding, bubbles nucleate at accidental or deliberate sites in the foam and grow inertially until radii of about 100 μm are reached, at which point diffusion of dissolved gas to the growing bubble sites begins to control the growth rate. Initial bubble growth is significantly retarded by extremely small nucleating sites, very low diffusion coefficients and very low foaming agent concentration. Under certain conditions, bubble nucleation and growth can be several orders of magnitude slower than material transfer from accumulator to mold cavity. Typically, under optimal foaming conditions, the gas pressure causing the foaming melt to fill the mold cavity is in the region of 1–3 MPa, rather than the more typical injection pressure of 30–70 MPa found in conventional injection molding.

As the gas expands the polymer against the mold cavity walls, the surface layer begins to cool and the integral skin begins to form. It is thought that the final thickness of the skin depends on the inhibition of bubble growth near the wall owing to the rapidly increasing viscosity, the partial collapse of formed bubbles owing to the temperature difference which allows for increasing solubility of the gas in the plastic near the mold cavity walls, and the internal, centerline gas pressure which allows the centerline bubbles to grow at the expense of the collapsing bubbles near the mold cavity walls. The internal gas pressure is seen to be directly proportional to the concentration of foaming agent in the plastic. Thus the dynamic balance between the internal gas pressure, which slowly falls with time as the polymer cools, and the building skins of solid or hardened polymer near the mold cavity walls determines the minimum time required to cool the shape so that the mold cavity can be opened without excessive shrinkage or postmold expansion. Note that this measure of cycle time differs considerably from the conventional concept of transient heat conduction in the thermoplastic which forms the basis for cycle time determination for unfoamed plastic. The rate of heat removal is dependent on the thermal diffusivity of the material as well as the square of the part thickness. It has been shown that although the thermal conductivity of foam decreases with decreasing density, the thermal diffusivity actually increases with decreasing density, thus allowing foams to cool more quickly per unit thickness than unfoamed resins. On the other hand, TSF shapes are frequently substantially thicker than typical unfoamed shapes, and doubling the thickness increases the time to cool to a given temperature profile by a factor of four. In essence, then, structural foam parts require longer cycle times primarily because of internal gas pressure and only secondarily because of thicker cross sections.

In general, amorphous polymers exhibit thicker skins and more gradual demarcation between skin and foam core than do crystalline polymers. Also, amorphous materials exhibit more uniform cell structure in the foam core than do rapidly crystallizing materials, such as nylons. Polyolefin cell structure is intermediate. Highly filled materials are normally stiffer flowing than the equivalent neat materials. As a result, shapes of filled resins usually have somewhat higher reduced densities for the same foaming agent concentration and processing conditions than shapes of neat resins. However, the cell structure of filled resins is finer, probably due to the nucleating character of the filler. Since the process is characterized as one of high speed but low flow resistance, attrition of glass reinforcement is considerably less than in conventional injection molding. However, in TSF processing, the glass fibers tend to segregate in bands just below the surface in attitudes that are not necessarily parallel to the mold cavity walls. Therefore, the reinforcement of the structure does not occur at the maximum fiber stress in flexure, and the hoped-for enhancement of mechanical properties is not achieved.

6. Mechanical Performance

For typical low-pressure foamed shapes, it has been demonstrated repeatedly that mechanical design properties such as flexural strength and modulus, tensile strength and modulus and, for most materials, fatigue strength and creep resistance can be expressed in terms of the overall bulk density of the material and the property of the unfoamed material as

$$X_f = X_0(\rho_f/\rho_0)^n = X_0\phi^n$$

where X_f is the foamed part property, X_0 is the unfoamed resin property and n is a coefficient with a value between 1 (law of mixtures rule) and 2 (experimentally determined to be a suitable design value for uniform-density foams). Significant deviations from this design equation can occur when the shapes have well-defined thick skins or large voids in the cellular core, or for such design properties as impact strength (tensile, falling weight), deflection temperature and elongation to break. Some ductile polymers exhibit brittle loading behavior and some crystalline polymers exhibit atypical morphological characteristics when foamed. The surface roughness caused by gas bubble ebullition during foaming can lead to serious notch sensitivity in certain polymers and premature failure under impact loading.

See also: Construction Materials, Cellular; Processing: Classification

Bibliography

Benning C J 1969 *Structure, Properties and Applications*, Plastic Foams, Vol. 2. Wiley, New York, pp. 277–80

Berlin A A, Shutov F A 1980 *Strengthened Gas-Filled Plastics*. Khimiya, Moscow

Frados J (ed.) 1976 *Plastics Engineering Handbook*, 4th edn. Van Nostrand Reinhold, New York, pp. 580–99

Morton-Jones D H 1985 Structural foam molding. In: Whelan A, Goff J (eds.) 1985 *Developments in Injection Molding—3*. Elsevier, Amsterdam, Chap. 1

Piechota H, Rohr H 1975 *Integralschaumstoffe*. Hanser, Stuttgart, Germany

Semerdjiev S 1982 *Introduction to Structural Foam*. Society of Plastics Engineers, Brookfield Center, CT

Throne J L 1979 Principles of thermoplastic structural foam molding: A review. In: Suh N P, Sung N H (eds.) 1979 *Science and Technology of Polymer Processing*. MIT Press, Cambridge, MA, pp. 77–131

Throne J L 1982 Structural foam. In: Hilyard N C (ed.) 1982 *Mechanics of Cellular Plastics*. Macmillan, London, Chap. 7

Villamizar C A 1978 A study on thermoplastic structural foam. PhD thesis, Polytechnic Institute of New York

J. L. Throne
[Amoco Chemicals Corporation, Naperville, Illinois, USA]

Injection Molding of Thermoplastics: Structure Development

There are many sources of structure in molded parts. Such structure may vary from the gross scale to the molecular level. If the melt enters the mold in the form of a jet rather than as a uniformly expanding front (Spencer and Gilmore 1951, Yamaguchi et al. 1968, Ballman et al. 1970, Sleeman and West 1974, Oda et al. 1978), the final molded part usually has both surface markings and internal structural discontinuities. This article, however, deals primarily with structure at a smaller scale. On the molecular level, molded parts exhibit polymer chain orientation (Spencer and Gilmore 1950, Ballman and Toor 1960, Wales et al. 1972, Fleissner 1973, White and Dietz 1979). This has been studied extensively using birefringence. Injection-molded parts produced from crystallizing polymer melts exhibit complex distributions of crystalline morphology (Yamaguchi and Oyanagi 1964, Clark 1967, 1973, 1974, Heckmann and Spilgies 1972, Kantz et al. 1972, Heckmann and Johnsen 1974, Matsumoto et al. 1979). Studies involving polarized light microscopy and both wide- and small-angle x-ray diffraction have been carried out.

1. Jetting

A basic study of jetting (see Fig. 1) in injection-mold filling was presented by Oda et al. (1978) (see also White and Dietz 1979). They found that jetting is associated with the swelling melt emerging from the gate and not striking the walls as it enters the mold. Thus, jetting is accentuated by melts exhibiting low extrudate swell and by gate designs which cause emerging melts not to strike the mold walls. Yamaguchi et al. (1968) and Ballman et al. (1970) have discussed characteristics of molded parts formed by jetting.

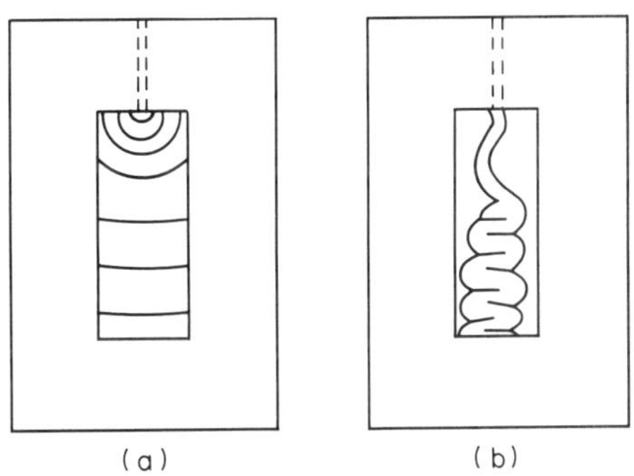

Figure 1
(a) Simple mold filling and (b) jetting

2. Orientation in Molded Glassy Plastics

The existence of orientation at the molecular level in injection-molded parts has been noted by various investigators. Ballman and Toor (1960) made the key basic experimental study of this, using end-gated injection-molded polystyrene bars. The bars were partly abraded away to various extents and birefringence was measured. With "1" taken as the direction of flow, "2" as the direction of the narrow thickness of the molded bar and "3" as the direction of the bar width, the birefringence $\Delta n_{12}(n_1 - n_2)$ exhibited a minimum at the center of the bar and a maximum at about 75% of the distance to the outer surface of the bar (see Fig. 2). The magnitude of the birefringence maximum increased with ram pressure during the molding process and decreased with melt temperature. However, this result could be changed if mold designs with larger gates were used, where sealing occurred in the bar rather than in the gate. The birefringence was found to decrease with increasing mold thickness. Ballman and Toor suggested a mechanism for this behavior, the birefringence in the melt being determined by stresses developed during mold filling and relaxing with the stresses during the cooling process. The intermediate position of the birefringence maximum was due, they suggested, to a vitrified layer existing along the cold mold wall during the filling process, arising from the fountainlike fluid motion from the front (see Fig. 3).

The hypothesis of Ballman and Toor was quantitatively developed in papers by Janeschitz-Kriegl (1977) and Dietz et al. (1978) (see also White and Dietz 1979). The rheooptical law

$$\boldsymbol{n} = \tfrac{1}{3}(\mathrm{tr}\,\boldsymbol{n})\boldsymbol{I} + C\boldsymbol{P} \qquad (1)$$

where $\boldsymbol{n}$ is the refractive index tensor, $\boldsymbol{I}$ is the unit tensor, $\boldsymbol{P}$ is the deviatoric stress tensor and C is the stress optical constant, was applied together with a force balance of form

$$0 = \partial p / \partial x_1 + \partial P_{12} / \partial x_2 \qquad (2)$$

where p is pressure, P_{12} is shear stress and x_1 and x_2 are distances in the 1 and 2 coordinate directions previously defined. A heat balance was utilized to compute the cooling of the melt both during filling and in the filled mold. Birefringence profiles were predicted in molded parts and found to agree semiquantitatively with experiment (Fig. 2).

As the stresses P_{ij} are related through Eqn (2) to the injection pressure, the pressure is predicted to be the primary factor influencing birefringence development. Birefringence levels should also be increased by lower mold wall temperature.

3. Morphology in Molded Crystalline Plastics

Cross sections of injection-molded parts of crystalline polymers were first investigated by Yamaguchi and Oyanagi (1964). Using optical microscopy, they reported that polyethylene molded parts possessed an oriented skin layer and a spherulitic core. Between these two regions an intermediate morphology of ellipsoidal spherulites was observed. A typical cross-sectional morphology of an injection-molded polypropylene part showing similar characteristics is given in Fig. 4. These views were subsequently confirmed and extended in a series of papers by Clark (1967, 1973, 1974) dealing with polyoxymethylene molded products. The observations were

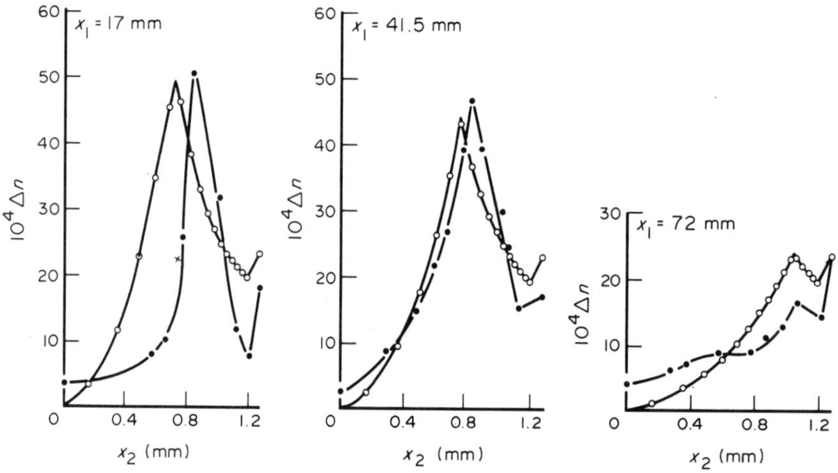

Figure 2
Birefringence distribution across a narrow section of an injection-molded part: •, measured; ○, predicted

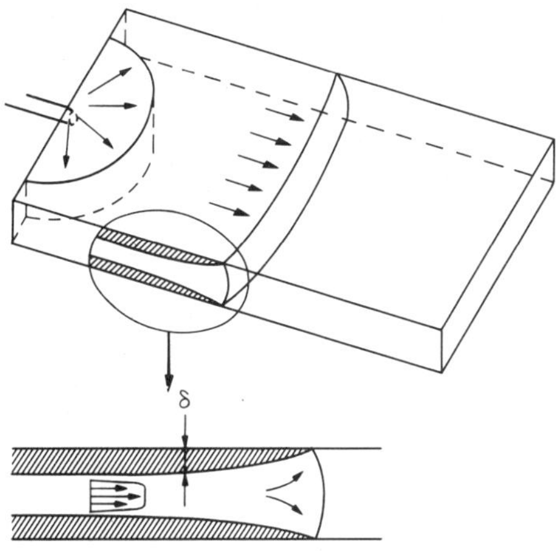

Figure 3
Filling of a mold by a melt front

folded chain lamellae, randomly oriented in the center of the pot and oriented with chains in the flow direction in the outer layers.

Clark (1974) presented detailed optical microscopy observations of injection-molded springs, a cigarette lighter butane reservoir and a container made from acetal resin. In these studies, he noted the importance of factors such as the screw forward time; if the screw was brought back before the solidification was complete, morphology discontinuities were found in the molded parts.

The general observations previously described were confirmed by the studies of Kantz et al. (1972) on polypropylene molded bars using optical microscopy and wide-angle x-ray scattering (WAXS) and by Heckmann and Spilgies (1972) and Heckmann and Johnsen (1974) using WAXS and transmission electron microscopy. Heckmann and Spilgies interpreted their WAXS studies of the structure of injection-molded bars using pole figure analysis.

reminiscent of those for glassy plastics. Decreasing mold wall temperature increased the extent of the oriented regions and decreased the amount of spherulites.

Clark also reported transmission electron microscopy (TEM) and small-angle x-ray scattering (SAXS) examinations of these injection-molded parts. These confirmed the oriented character of the outer layers. He used the results to develop a morphological model of the molded parts in terms of

4. *Molds with Moving Parts*

Cleereman (1967, 1969) has described injection-molding processes to produce plastic cups in which the core of the mold rotates during the mold filling process, inducing a helical orientation in the molded parts. This was noted as preferable to the usual uniaxial orientation resulting from end-gated molds, which suffer from inferior tensile characteristics transverse to the direction of melt flow.

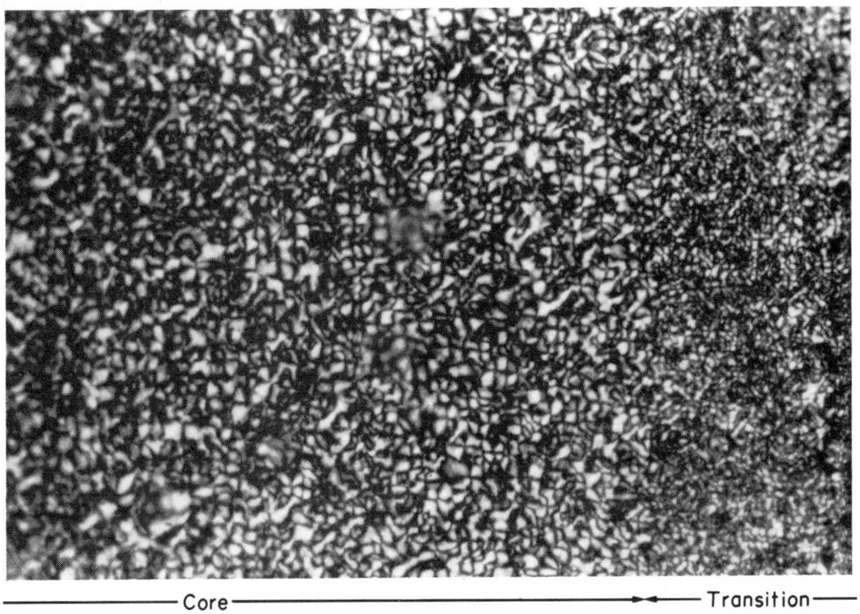

Figure 4
Cross section of molded polypropylene test bar

See also: Processing: Structure Development

Bibliography

Ballman R L, Kruse R L, Taggart W P 1970 Surface fracture in injection molding of filled polymers. *Polym. Eng. Sci.* 10: 154–8

Ballman R L, Toor H L 1960 Orientation in injection molding. *Mod. Plast.* 38(2): 113–24; 205–7

Clark E S 1967 Molecular orientation in injection molding acetal homopolymer. *SPE J.* 23(7): 46–9

Clark E S 1973 Structure development in injection molding acetal homopolymer. *Appl. Polym. Symp.* 20: 325–32.

Clark E S 1974 An in-depth look at molded parts. *Plast. Eng.* 30: 73–8

Cleereman K J 1967 Injection molding of shapes of rotational symmetry with multiaxial orientation. *SPE J.* 23(10): 43–7

Cleereman K J 1969 Injection molding of shapes of rotational symmetry with multiaxial orientation—Part II. *SPE J.* 25(1): 56–61

Dietz W, White J L 1978 Ein einfaches Modell zur Berechnung des Druckverlustes während des Werkzeugfullvorganges und der eingefrorenen Orientierung beim Spritzgiessen amorphol Kunststoffe. *Rheol. Acta* 17: 676–92

Dietz W, White J L, Clarke E S 1978 Orientation development and relaxation in injection molding in amorphous polymers. *Polym. Eng. Sci.* 18: 273–81

Fleissner M 1973 Makroskopische und molekulare Orientierung in Hochpolymeren. *Kolloid-Z. Z. Polym.* 251: 1006–14

Heckmann W, Johnsen U 1974 Scherinduzierte Kristallisation in Spritzgussteilen aus Niederdruck-Polyäthylen. *Colloid Polym. Sci.* 252: 826–35

Heckmann W, Spilgies G 1972 Röntgenographische Bestimmung unterschiedlichen Kristallitorientierungen in Polyäthylen-Spritzgussteilen. *Kolloid-Z. Z. Polym.* 250: 1150–7

Janeschitz-Kriegl H 1977 Injection moulding of plastics: Some ideas about the relationship between mould filling and birefringence. *Rheol. Acta* 16: 327–39

Kantz M R, Newman H D Jr, Stigale F H 1972 The skin-core morphology and structure–property relationships in injection-molded polypropylene. *J. Appl. Polym. Sci.* 16: 1249–60

Matsumoto K, Miura I, Hayashida K 1979 Texture of injection-molded polypropylene. *Kobunshi Ronbushu* 36: 401–6

Oda K, White J L, Clarke E S 1978 Jetting phenomena in injection mold filling. *Polym. Eng. Sci.* 16: 585–92

Sleeman M J, West G H 1974 Rheology and product properties in injection molding. *Br. Polym. J.* 6: 109–18

Spencer R S, Gilmore G D 1950 Residual strains in injection molded polystyrene. *Mod. Plast.* 27: 97–155

Spencer R S, Gilmore G D 1951 Some flow phenomena in the injection molding of polystyrene. *J. Colloid Sci.* 6: 118–32

Wales J L S, van Leeuwen Ir J, van der Vijgh R 1972 Some aspects of orientation in injection molded objects. *Polym. Eng. Sci.* 12: 358–63

White J L, Dietz W 1979 Some relationships between injection molding conditions and the characteristics of vitrified molded parts. *Polym. Eng. Sci.* 19: 1081–91

Yamaguchi Y, Oyanagi K 1964 Study on structural non-uniformity in injection moldings of polyethylene. *Res. Rep. Kogakuin Univ.* 15: 1–9

Yamaguchi Y, Oyanagi K, Sugai Y, Kunitake M 1968 Molding pressure efficiency and gate sealing time in injection molding process of polyethylene. *Res. Rep. Kogakuin Univ.* 24: 71–82

J. L. White
[University of Akron, Akron, Ohio, USA]

J

Joining of Composites

As fiber-reinforced materials, particularly continuous-fiber-reinforced plastics (FRP), become more widely used an important requirement for their full exploitation is the development of suitable attachment methods and general joint design philosophies.

Unlike isotropic materials, problems of joining FRP arise mainly due to the unique characteristics of the material, that is, the inherent weakness of the material in interlaminar shear, transverse tension and transverse compression. Also, since the material remains elastic to failure (in unidirectional material where fibers are oriented in one direction), there is no plastic stress–strain behavior, and therefore the mechanisms of stress relief around geometric discontinuities, such as holes, require special attention. Clearly all of these problems require full consideration before engineers can embark on the design of an efficient joint, efficiency being measured in terms of high strength for low weight and/or cost.

The epoxy-resin-matrix FRP (which make up the bulk of the current high-technology matrices for FRP) cannot be welded, soldered or brazed. With the exception of components made using filament winding techniques, where there is usually no significant end attachment problem, most joints are effected by means of adhesive bonding, mechanical fasteners or a combination of the two.

With the recent introduction of thermoplastics into the field of structural plastics, especially for the aerospace industry, other methods of joining are being explored, and one of particular interest that promises future potential is that of welding.

1. Adhesive Joints

The form of adhesive-bonded joint that has probably received most attention is a simple lap joint. This is attractive for two main reasons: it is simple to fabricate, and many of the problems of load introduction such as are found in mechanical joints are largely eliminated. However, even with the use of advanced adhesive technology only relatively low load-transfer rates can be realized because of the low shear strength of both the adhesive and the composite adherend, and the high thermal strains that can exist within the bonded regions due to the high cure temperature of adhesives. Also degradation of the adhesive, and of the interface between adhesive and composite, can occur after exposure to certain environments. This is discussed in greater depth later in this section.

Failure of FRP adhesive joints can occur in any of four modes: adhesive failure at the interface between the adhesive and adherend, cohesive failure through shearing of the adhesive, peel failure due to out-of-plane loads tending to pull a joint or adherend apart in transverse tension, and adherend failure in axial tension or compression. Axial failure of the adherend occurs at or close to the end of a lap joint with failure initiating in axial tension or compression of the outer plies. Analysis shows that due to a shear lag phenomenon caused by shear deformation in the composite matrix, axial load distribution through a laminate's thickness is a function of axial distance from the joint. Consequently, the most highly loaded plies are those on the face that is bonded, even outside the joint. A solution to the problem is to ensure the outer plies are oriented to give maximum strength, that is, laid parallel to the load direction. Because of this mode of failure, high joint efficiencies are possible only in very thin laminates; for thick laminates, high joint efficiencies are achieved by the use of multistepped or scarf (acute-angled-butt) joints, both of which are more difficult and costly to produce. The configurations of bonded joints available to a designer are numerous; the most common, together with their design limitations, are given in Fig. 1.

One of the problems associated with FRP is that of environmental degradation. This arises because of the readiness of both FRP adherends and adhesive to absorb moisture (water) when exposed to a humid atmosphere. This causes problems during the bonding operation and also after bonding has been successfully achieved. During the bonding operation, which is usually carried out at an elevated temperature to ensure high-temperature structural properties, moisture diffuses out of the adherend into the adhesive producing high voiding in the adhesive and bond line. As a consequence the adhesive and cohesive properties of the adhesive are significantly reduced. Clearly, therefore, the bonding of "wet" adherends must be avoided, either by ensuring the bond surfaces do not absorb moisture or by drying the surfaces prior to bonding. The consequence of environmental exposure of assembled joints is discussed in the following paragraphs.

Central to any successful adhesive joint is joint preparation; this ensures that a maximum strength bond will exist between the adherend and adhesive layer interface. Since both FRP-to-FRP and FRP-to-metal joints will need to be used it is worth

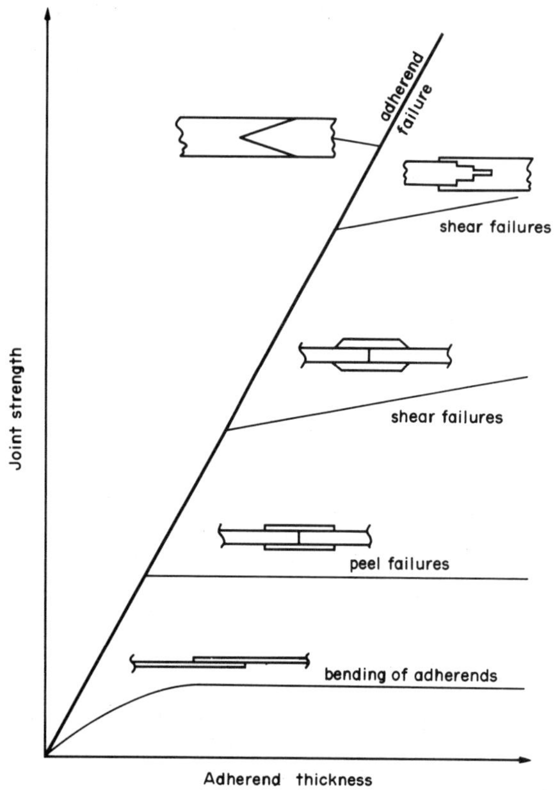

Figure 1
Bonded joint types and their design limitations

commenting on the essential differences between them.

The joint between an FRP adherend and an adhesive layer is, for epoxy-matrix FRP, fairly simple since they are essentially similar materials and usually the interface is not a plane of weakness since joint failure more often occurs in the FRP adherend. A major concern in the surface preparation is that of contamination prior to bonding. Provided this is avoided a good joint should be obtained. During the exposure of the joint to a humid environment moisture is absorbed by the FRP adherend and a plasticization of the matrix and a lowering of the material glass-transition temperature occurs; this directly influences the matrix-dependent strength properties. A similar effect is observed in most adhesives, causing some reduction in compliance and modulus. Moisture absorption also produces swelling of both adherend and adhesive which can result in differential swelling stresses and possible damage of the bond.

Metallic adherends are the most complex to prepare for bonding since it is necessary to etch and/or anodize the bonding surface to produce an oxide layer. Priming of the surface is often necessary to improve the bonding further and to provide a surface protection for production purposes. One of the main difficulties associated with metallic-to-FRP bonding is that of subsequent environmental exposure. Due to moisture ingress through the FRP and through the adhesive edges, hydrolyzing of the metallic bond surface can occur. This is usually brought about when substandard surface treatment has occurred. Unlike the FRP-to-FRP bond the metallic-to-FRP bond can have a major plane of weakness at the metal–adhesive interface.

The adhesive joining of thermoplastics using standard epoxy adhesives is not as simple as joining thermosets, owing to the surface chemistry of the two being so different. The surface pretreatment of a thermoplastic is, therefore, more critical and needs special attention if good adhesion is to be achieved. One method that is currently showing great promise is the corona discharge method. This improves the "wettability" of the surface which gives a correspondingly improved bond over other abrasive type preparations.

2. *Mechanical Joints*

With the increasing interest in the use of mechanical fasteners for joining FRP components a variety of fastening techniques that are used for metals have been successfully applied to many forms of FRP. These range from self-tapping screws which are used for lightly loaded connections, to quality engineering bolted connections used in heavily loaded structures. As with metals, bolted or demountable joints are needed in FRP structures where there is a call for inspection and servicing, modification, repair or replacement of damaged or worn components, or in situations where off-axis loads have to be accounted for and where bonding is uneconomic or impractical. Hence holes or cutouts in FRP laminated composites have to be catered for in design. Like metallic structures, FRP exhibit failure modes in tension, shear and bearing but, because of the complex failure mechanisms of FRP, two further modes are possible, namely cleavage and pullout. Figure 2 shows the location of each of the modes.

It is well known that holes can seriously weaken a laminate, particularly if the laminate is highly anisotropic. This is due in part to the high stress concentrations that occur in the region around such discontinuities, which in tension can be as large as 8 for highly anisotropic carbon-fiber-reinforced plastics (CFRP), compared with a much lower value of 3 normally associated with isotropic materials. Furthermore, isotropic materials yield before failure and the effect of stress concentrations on the net failing stress is small. Since unidirectional FRP do not respond in this way it is not surprising that mechanical fastening of unidirectional-fiber materials is inefficient. However, the problem can be

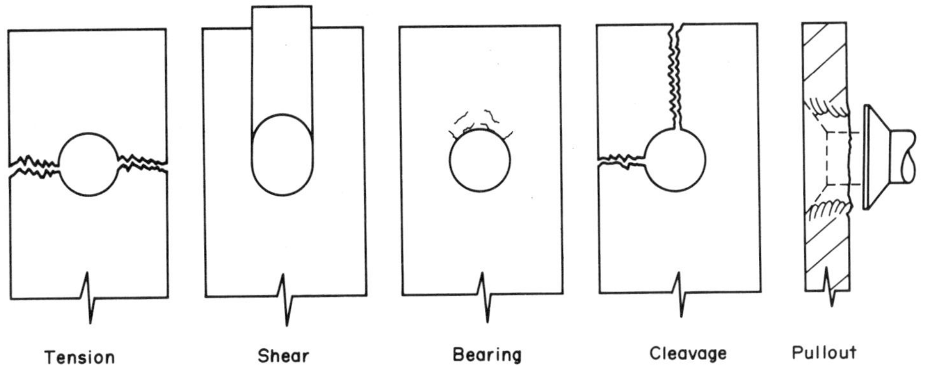

Figure 2
Modes of failure for mechanical joints in FRP

overcome by reducing the degree of anisotropy in the region of a hole and imparting a degree of "plasticity" by the incorporation of fibers oriented in a number of different directions. In fact the failing stress of all three principal modes of failure, that is, tension, bearing and shear pullout, is increased by the inclusion of fibers oriented at ±45°. Figure 3 shows the change in failure mode of a 0°±45° laminate as the contribution of ±45° fibers is varied. If the efficiency of a joint is expressed in terms of its specific strength, then the efficiency of a carbon-fiber-reinforced epoxy joint can be greater than that of an L71 aluminum alloy (Cu $4\frac{3}{4}$%–Si $\frac{3}{4}$%–Mn $\frac{3}{4}$%–Mg $\frac{3}{4}$%) by a factor of 1.32 in tension, 3.7 in bearing and 1.1 in shear pullout.

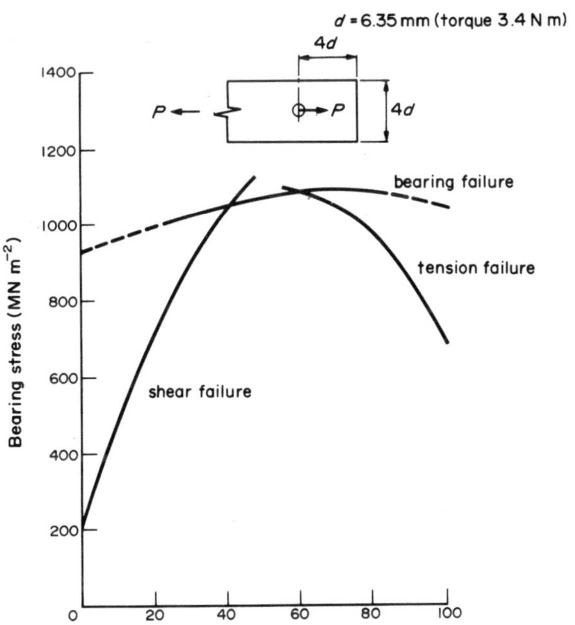

Figure 3
Influence of fiber orientation on failure mode (0°±45° CFRP)

Environmental degradation of a bolted joint, after exposure to a hot, wet environment, is most likely to occur in the shear and bearing strength properties. At present, evidence exists for the bearing property only. The evidence shows for fiber-reinforced epoxies that temperature has a more significant effect than moisture but in the presence of both (at 127 °C) a strength loss of 40% is possible.

The information available for the design of mechanical joints for fiber-reinforced thermoplastics is at present incomplete. The key ingredient still required for establishing a design capability is the characterization of variables such as fiber layup and stacking sequence in terms of joint parameters. Despite the incompleteness of this work sufficient evidence exists to suggest the failure behavior of thermoplastics is much the same as for thermoset FRP and that similar high joint efficiencies can be obtained.

3. *Welding*

The special properties of the currently available thermoplastics, in particular damage tolerance and high resistance to hot wet environments, make them suitable matrices for use with fiber reinforcement and, therefore, a strong contender to thermosets in the engineering and structural plastics field. Thermoplastics are, compared with thermosets, still in their infancy and need further development of fabrication techniques and joining methods before their full potential can be realized.

One of the major features of thermoplastics is their ability to be reformed by heating and for this reason they lend themselves to joining techniques involving remelting that could never be expected of thermosets. Hence, particular attention is now being given to developing welding procedures that are acceptable as efficient load-transfer methods. The experimental work has, so far, examined one particular thermoplastic, a polyether ether ketone (PEEK) manufactured by ICI (UK). Currently several techniques are under review, including the following.

(a) Hot-plate welding—this method gives sound welds and can achieve good lap shear and fracture properties.

(b) Resistance-heating welding—with this method a resistively heated implant technique is needed to produce good properties without disturbance of the fibers. Unfortunately it is limited, subject to further development, to small joint areas.

(c) Induction welding—this technique is capable of continuous welding and produces good mechanical properties. The method shows great promise for the production of large assemblies and potentially of varying joint areas.

All of the three processes show a strong potential for use in production, with induction welding showing the most promise. Further investigation into these techniques is required, together with the development of nondestructive testing techniques to qualify them.

4. Shim Joints

Shim joints use thin metallic interleaved layers to reinforce the composite material in the region of the joint. The action of the shims is to transfer load from the composite material through interlaminar shear between the shim and the composite and then through pin bearing by means of conventional bolts. Failure of this type of joint is much the same as for metals but includes a failure mode due to excessive shear between the composite and metal shims. The multishim joint is a compromise between bonded joints and mechanical joints, and it attempts to make use of the advantages found in each. Unfortunately, because of fabrication difficulties and emphasis on overall cost and weight, shim joints are not always an attractive method of joining.

All of the joining techniques discussed can be used without undue difficulty for FRP, but the behavior of the joints must be properly understood and the limitations realized if structural loads are to be carried safely.

See also: Adhesive Joining: An Overview; Welding of Plastics

Bibliography

Collings T A 1977 The strength of bolted joints in multidirectional CFRP laminates. *Composites* 7: 43–55

Collings T A 1987 Experimentally determined strength of mechanically fastened joints. In: Matthews F L (ed.) 1987 *Joining Fibre-Reinforced Plastics*. Elsevier, London, pp. 9–63

Erdogan F, Ratwani M 1971 Stress distribution in bonded joints. *J. Compos. Mater.* 5: 378–93

Hart-Smith L J 1987 Design of adhesively bonded joints. In: Matthews F L (ed.) 1987 *Joining Fibre-Reinforced Plastics*. Elsevier, London, pp. 271–311

Parker B M 1978 The effect of hot-humid conditions on adhesive-bonded CFRP-CFRP joints. *Symp. on Jointing in Fibre Reinforced Plastics*. IPCS Science and Technology Press, pp. 95–103

Wong J P, Cole B W, Courtney A L 1969 Development of the shim-joint concept for composite structural members. *J. Aircr.* 6(1): 18–24

T. A. Collings
[Royal Aerospace Establishment,
Farnborough, UK]

L

Liquid Crystal Processing

This article deals with the important features and issues involved in the processing of polymer liquid crystal systems into solid-state products. Polymer systems forming liquid crystals are capable of achieving quite high degrees of molecular orientation. This capability lends itself to processing steps which can control and develop orientation. Orientation is of fundamental significance in that it strongly influences the mechanical and physical properties of the solid-state product. With these comments in mind, the various methods for processing polymer liquid crystals will be discussed to analyze those features of the process that influence orientation, fibrillation, microvoiding and, ultimately, the mechanical properties of the solid-state products.

Broadly speaking, polymer liquid crystals arise from lyotropic (solvent-based) systems or are of thermotropic origins. Thus, two basic methods for fabrication exist: wet and melt processing. Melt processing is the more desirable method, as solvent systems are often expensive and/or toxic and lead to very difficult problems on an industrial scale. These problems notwithstanding, industrial wet fabrication processes are often used because a considerable number of the interesting polymers do not melt but thermally degrade before reaching a crystalline melting temperature.

Some of the polymers known to form liquid crystalline phases are *para*-linked aromatic polyamides, cellulose, cellulose derivatives, *para*-linked aromatic polyesters and synthetic polypeptides. From a chemical structural viewpoint, these are all thought of as rigid polymers having strong intermolecular secondary bonding forces. Thus, if they can be melted, it will be at a sufficiently high temperature to produce enough thermokinetic energy to overcome these bonding forces. In the lyotropic cases, the solvents must then be powerful ones for the same reason.

1. Order in Liquid Crystals

The order in liquid crystals is either one or two dimensional as opposed to the three-dimensional order in a typical solid crystal structure. There have been a number of specific structural arrangements suggested for low-molecular-weight liquid crystals that range from two-dimensional smectic cases to one-dimensional nematic systems. In the case of polymers, the majority of systems have been determined to be either nematic or cholesteric. For example, poly(*p*-phenylene terephthalamide) is nematic and hydroxypropylcellulose is cholesteric. Nematic crystals may be thought of as having one-dimensional order with respect to a reference axis. Cholesteric crystals have one-dimensional order within a given layer plane of molecules with a sense of helical pitch for each layer plane normal to the reference axis.

The degree to which order is achieved is a major factor in understanding the behavior of polymeric liquid crystals. One such scale to specify this order is as follows. The alignment of molecules in nematic liquid crystals may be characterized by the value of a factor S defined as

$$S=\tfrac{1}{2}(3\langle\cos^2\theta\rangle-1) \quad (1)$$

where θ is the angle between a molecular and a preferred direction and the angled brackets denote averaging. Thus, S should be seen as essentially the Hermans orientation factor. Complete alignment ($\theta=0$) leads to a value of $S=1$. In polymer liquid crystal processing, it is the degree to which molecular-level order is achieved and maintained up to macroscopic dimensions that sets the basic performance properties of the end product. For fibers, a high degree of unaxial order is important whereas, in the case of films, biaxial orientation should be developed.

2. Theory of Liquid Crystal Formation

The polymers that form liquid crystals are often referred to as rigid macromolecules; there is little if any opportunity for conformer rotation and, in fact, the chain molecule is constrained to approach linearity, especially in the case of *para*-linked aromatic units, as opposed to the three-dimensional random coil associated with the so-called flexible macromolecules.

On this basis, a mechanism of mesophase formation for large rodlike molecules is easy to develop. It may be seen that it is impossible to pack isotropically a concentrated system of rods. In the 1950s Flory proposed a formulation of the problem in terms of a lattice theory and the methods of the Flory–Huggins theory of concentrated polymer solutions, and predicted a liquid crystalline phase separation. Flory has more recently turned again to this problem for the particular cases of polypeptides and aromatic polyamides and has determined the conditions under which the lattice model predicts the formation of an anisotropic phase. A minimum aspect ratio x of approximately 5.4 is required. The critical volume fraction ϕ_c of rigid rods at separation is given approximately by

$$\phi_c = (8/x)[1-(2/x)] \qquad (2)$$

where x is essentially the aspect ratio. This represents the case of zero interaction energy.

These statistcal ideas concerning the space-filling characteristics of molecules with high aspect ratios are seen to be critical to forming an ordered liquid phase. That is to say, it is impossible to have a densely packed liquid possessing one- or two-dimensional order without molecules that are anisometric. In the purest sense these would be athermal. It is also important to keep in mind that the polar and indeed extreme anisotropic character of the electron distributions in polymer structures forming liquid crystals makes them very unlikely to be energetically "noninteracting." Thus Eqn. (2) is qualitatively correct, but is often misused by ignoring other effects of interaction.

3. *Rheology*

From the very first studies of the flow behavior of polymer liquid crystal systems, striking and unusual effects were observed. Instead of the usual monotonic increase in viscosity with concentration, a maximum is consistently observed. This maximum has been related to the transition from the isotropic to the anisotropic phase. Of course, the maximum gives rise to nontypical expressions for the concentration dependence of the viscosity in these different phases. In addition, since molecular weight is intimately related to viscosity, it further complicates the description. Other features of the flow behavior include the observation of yield stresses and nontypical normal stresses.

An approach to organizing the various experimental observations is to categorize the viscosity–shear rate curve into three regions. Region I is associated with the yield stress phenomenon, region II is referred to as the "plateau" range and region III is associated with shear thinning—viscoelastic behavior. Although this provides a framework for a systematic discussion as presented by Wissbrun (1981), it cannot provide fundamental explanations of the behavior. The interpretation of the rheological behavior is at present controversial and far from complete.

The rheooptical behavior of these systems must be briefly mentioned in order to present and understand the significance of certain steps used to process polymer liquid crystals into objects for technical applications. The main point is that an orientation birefringence is introduced coincident with the flow direction. It occurs at low stresses and has a very long relaxation time. The development of flow-induced birefringence means that order becomes a global feature of the system. It is the ability to set the spatial character and extent of this orientation that is crucial to processes for fabricating useful products such as fibers and films.

4. *Fabrication by Wet and Melt Processing*

The spinning of fibers from concentrated liquid crystalline polymer solutions was first carried out using synthetic polypeptides. No commercial significance was attained, however, until the work of Kwolek (1972). She discovered that *para*-linked aromatic polyamide solutions were nematic mesophases and showed that fibers wet spun from such solutions exhibited greatly enhanced mechanical properties. Her 1972 patent describes both wet- and dry-spinning processes for producing *para*-linked aromatic polyamide fibers.

A subsequent patent (Blades 1973) covered modified anisotropic spinning dopes and wet-spinning processes. Of most interest is the dry-jet wet-spinning process, which was shown by Blades to lead to greatly enhanced fiber mechanical properties if applied to liquid crystalline *para*-linked aromatic polyamides. The dry jet, that is the air gap between the spinneret orifice and the coagulation bath, provides for uniaxial extension. This is most beneficial toward developing a high degree of uniaxial orientation.

The total process consists of spinning and washing the coagulated fibers, followed by heat treatment for approximately 1 s under tension at 525 °C. The yarn, however, exhibits little drawing under these conditions. The heat treatment has a tremendously beneficial effect on tensile strength. This is remarkable in view of the fact that tension but little or no drawing is involved in this process step. Thus, here is a process where orientation occurs during spinning and is greatly improved by annealing without drawing. Some perspective on the mechanical capability of these fibers is obtained by comparison with metals and ceramics, on a mass rather than a volume basis. Since the density of such fibers is much lower, their tensile strength is, for example, five times that of steel on this basis.

Flood *et al.* (1982) have investigated the use of wet fabrication to produce biaxially oriented films. Here, a state of biaxial stress is imposed on the liquid crystal by causing it to flow over a lubricated mandrel. This is a step analogous to the air gap in dry-jet wet spinning of fibers. Next the film is coagulated. Postfabrication washing of the film and thermal treatments have a beneficial effect on the mechanical properties in a manner similar to that reported for uniaxially oriented fibers. This study also established the notion that the tendency for fibrillation of a uniaxially oriented solid product disappears as biaxial orientation is achieved.

Among melt fabrication techniques, the melt spinning of fibers using aromatic polyesters is the most notable development. The technology would appear to be very similar to that used for melt spinning of poly(ethylene terephthalate) (PET). The chemical basis of these systems can be seen in the example of a copolymer of PET and *p*-hydroxybenzoic acid (HBA). As the amount of HBA in the copolymer is increased, nematic liquid crystalline behavior is observed at temperatures of 300 °C and above. Here, stresses in the melt spin line lead directly to orientation of the liquid crystalline phase, which is preserved through the solidification step as the filament cools.

5. Conclusions

The processing of liquid crystalline polymers into useful engineering products has achieved considerable success in the case of fibers and films based on poly(*p*-phenylene terephthalamide). The aromatic polyesters also show good promise. To achieve substantial properties of interest in the solid state, various steps in the fabrication process are crucial. The basic features that require understanding and control include knowledge of the phase diagram, setting the flow stress field before transforming to a solid and the proper postfabrication treatments to enhance orientation and strength of the product.

The phase diagram especially needs to be understood in terms of the temperature–composition coordinates. This knowledge is essential to setting the other parameters of the fabrication process so that the material stays in its liquid crystalline state. The flow stress just before solidification must be specified and controlled in order to convert the naturally occurring local order of the liquid crystal to a global order. The solidification step (coagulation for lyotropic systems and cooling for thermotropic ones) must occur in such a way as to preserve the orientation developed by the flow stresses. In wet spinning, care must be taken not to aggravate the intrinsic cavitation that occurs during coagulation. Postfabrication steps to remove liquids from lyotropic systems, to improve material continuity and to enhance the orientation and strength via annealing treatments complete the overall processing scheme. The extent to which these specifications are met determines the ultimate performance characteristics of the product.

As the remaining controversies concerning the behavior of polymer liquid crystals are resolved to provide further understanding of their chemistry and physics, even more interesting engineering performance characteristics are to be expected.

See also: Electrically Conducting Polymers

Bibliography

Blades H 1973 Dry-jet wet-spinning process. US Patent No. 3,767,756

Flood J E, White J L, Fellers J F 1982 Process for the formation of biaxially oriented films of poly(*p*-phenylene terephthalamide) from liquid crystalline solutions. *J. Appl. Polym. Sci.* 27: 2965–85

Kwolek S L 1972 Optically anisotropic aromatic polyamides dopes. US Patent No. 3,671,542

White J L, Fellers J F 1978 Macromolecular liquid crystals and their application to high-modulus and tensile-strength fibers. *J. Appl. Polym. Sci., Appl. Polym. Symp.* 33: 137–73

Wissbrun K F 1981 Rheology of rod-like polymers in the liquid crystalline state. *J. Rheol.* 25: 619–62

J. F. Fellers
[University of Tennessee, Knoxville, Tennessee, USA]

M

Maintenance Coatings

Maintenance coatings are film-forming products used by industry and commerce to give continued protection to capital assets. They are applied to both uncoated new substrates and to previously protected surfaces in order to extend the useful life of the structure.

Substrate protection is required for a number of reasons, including: (a) corrosion protection of metals and alloys; (b) corrosion protection of concrete and masonry; (c) erosion protection of steel and concrete; (d) heat reflection; (e) fire endurance; and (f) maintenance of aesthetics. More recently, aesthetics (or appearance) retention has become a significant factor.

The recoating of previously painted surfaces falls into either the planned or unplanned category. In the former case, generally accepted coating lifetimes are used to judge when the structure will need maintenance coatings. This judgment is virtually always subjective, although some guidelines do exist for both quantitative surveying and determination of recoating need. A major portion of the maintenance-coating activity is of the unplanned variety—either the choice of initial coatings was incorrect, application factors were poor, service conditions have changed from the initial specification or recoating was deferred for economic reasons. The initial step in the recoating process is usually a survey to determine both the required surface preparation and the extent of necessary recoating. This survey must always probe beneath the cosmetics of the surface to determine not only the visible corrosion, but also that hidden beneath apparently intact coating.

Maintenance coatings are normally applied to resist atmospheric and vapor exposures, and, occasionally, splash and spillage. Immersion service coatings are typically applied as tank linings or on the exterior of marine vessels. Tank linings represent the most severe class of service. Any compromise in either surface preparation of choice of lining is unacceptable and usually leads to early failure.

Maintenance protective coatings are typically applied to the following industrial and commercial end-use markets:

(a) chemical processing industry—tanks, structural steel, piping, exterior of process vessels, buildings, machinery, floors;

(b) commercial buildings—walls, floors, structural steel;

(c) food and beverage industry—floors, walls, structural steel, machinery;

(d) fossil-fuelled power plants—structural steel, machinery, floors, tanks;

(e) highways and bridges—structural steel, decking, marking;

(f) metal-finishing industry—tanks, floors, structural steel;

(g) mining and mineral-extraction industry—structural steel, machinery, processing trenches, floors;

(h) nuclear-fuelled power plants—containment building, steel and concrete walls and floors, auxiliary-control buildings, tanks;

(i) offshore structures—piers/decks/pilings, drilling rigs;

(j) oil and gas gathering industry—tanks, piping, machinery;

(k) oil and gas refineries—tanks, structural steel, piping, process vessels, floors, buildings, machinery;

(l) pulp and paper industry—structural steel, tanks, buildings, machinery, floors;

(m) steel industry—buildings, processing and holding tanks, floors;

(n) transportation industry—hopper cars, motive power units, buildings, containers, equipment;

(o) wastewater treatment plants—tanks, floors, walls, structural steel.

1. Surface Treatment

Mechanical surface treatments are by far the most popular for maintenance coatings. This is partly due to their ability to dislodge or uncover contaminants and also because of their portability—enabling them to treat large structures without dismantling or removal.

The US Steel Structures Painting Council, through its published specifications, recognizes nine methods of preparing steel for painting (see Table 1). Of these, four are abrasive blasting procedures which are also recognized by the US National Association of Corrosion Engineers. Photographic standards are included in the publication.

Solvent cleaning is used to remove such contaminants as oil and grease from steel prior to mechanical or acid-pickle treatment. It does nothing to

Table 1
Recognized methods of preparing steel for painting

SSPC	NACE	Description
SP-1		solvent cleaning
SP-2		hand-tool cleaning
SP-3		power-tool cleaning
SP-4		flame cleaning of new steel
SP-8		acid pickling
SP-5	No. 1	white-metal blast cleaning
SP-6	No. 3	commercial blast cleaning
SP-7	No. 4	brush-off blast cleaning
SP-10	No. 2	near-white blast cleaning

remove rust, scale or old coating residues. However, solvent cleaning is often a necessary first step in the preparation of steel, since it is the only effective method of removing heavy deposits of lubricants. Solvent emulsion cleaning followed by water rinsing is a preferred method of solvent cleaning.

Hand-tool cleaning with wire brushes, scrapers, chipping hammers or chisels is a slow operation and is not a reasonable method of cleaning large areas. It is generally employed to remove loose, nonadherent paint, rust or scale. This procedure is satisfactory for indoor spot touchup and repair in less critical service.

Power-tool cleaning with pneumatic or electrically driven rotary-brush, grinding, sanding or impact devices is considerably more efficient than hand-tool cleaning. However, excessive power cleaning may result in a burnished surface to which coatings will not give maximum adhesion. Power-tool cleaning is typically recommended for touchup and repair of coatings.

Flame cleaning by passing a high-velocity oxy-acetylene flame over a steel surface results in a rapid expansion of both the metal and the scale. Since the latter has a different coefficient of expansion from the former, it tends to pop off on cooling. Vaporization of any water under the mill scale aids in dislodging the scale. This method is only practical when the fire hazard can be tolerated. Power wire brushing is a necessary second step. Where the steel contains an old coating, scraping is required to remove the burned residues. Flame cleaning is neither efficient nor particularly effective, and is not often considered a recommended method of preparing steel for painting.

Acid pickling is frequently used to clean steel objects that are small enough to be immersed in a tank of pickling solution. It is often preferred to blast cleaning in a production-line setup, because less labor is required and no dust is created. Obviously, acid pickling is not practical for field-maintenance-painting surface preparation; however, occasionally it is used in the form of a gel which adheres to both vertical and horizontal surfaces. This gel is applied to the rusty steel surface and allowed to remain in place until the rust has been dissolved. It is then rinsed from the surface. This method is not always reliable, due to the great variations in both the type and quantity of rust and it has no effect on loosely adhering coatings. Also, the steel surface tends to remain very active until coated with an inhibitive primer. Therefore, if coating adhesion is critical and exposures are severe, pickling would not normally be considered the preferred means of metal preparation for field surface preparation. In such cases, abrasive blasting would be the better recommendation.

Abrasive-blast cleaning is the preferred method for preparing steel for the application of protective coatings, since it effectively removes rust, scale, old paint and other contaminants. Furthermore, by roughening the surface to a so-called anchor pattern, abrasive blasting increases the area of bond to the primer.

Two elements are needed for effective and efficient abrasive blasting: a hard, clean abrasive and a device that will propel the abrasive against the steel with a speed sufficient to cut into the metal. Manual abrasive blasting, which is typically used for maintenance painting, is accomplished by a system of pots, heavy-duty hose and nozzles, through which the abrasive is propelled at high velocities to the target surface by compressed air. The various types of abrasive that are used perform differently and produce characteristically different types of surface pattern. The pattern, or profile, resulting from blasting is extremely important in its effect on coating performance. If the surface is too smooth, poor adhesion may result with some coatings. On the other hand, if the surface is too rough, sharp ;pinnacles of metal are likely to project through the coating and remain unprotected.

Traditionally, dry abrasive and dry air are used in abrasive blasting to avoid rusting of the newly exposed steel surface. With the enforcement of more stringent environmental laws, wet sandblasting may become the most reasonable approach to cleaning steel surfaces. Injection of water into the sand flow at the nozzle or at the pot to reduce dusting is not a new procedure. Unfortunately, phosphoric acid added to the water to inhibit rusting produces a loose, powdery surface contaminant which has to be removed prior to coating application. Newer corrosion inhibitors with superior properties to phosphoric acid have become available in recent years. A newer technique of wet-abrasive blasting is provided by high-pressure water-blast machines. This apparatus is capable of ejecting a water stream at up to 70 MPa. Such a stream is capable of removing poorly adherent rust, scale and coating from a surface, but for tight scale and for efficient removal of bonded coating and rust, an abrasive is injected into the stream. The precautions to be taken are the same as for conventional wet sandblasting.

There are several grades of abrasive-blast cleaning. The ideal blast cleaning of steel completely removes all rust and scale to produce a surface of uniform light-gray color with no black spots, rust spots or shadows. When a surface meets this standard, it conforms to a white-metal blast (SSPC SP-5 or NACE 1). If 95% of each unit area inspected at random complies to the ideal, it is termed near-white blast (SSPC SP-10 or NACE 2). When only 67% (minimum) of each unit area inspected conforms, the surface has a commercial blast (SSPC SP-6 or NACE 3). In this grade of blast cleaning, pits and depressions often remain not totally clean.

Grading a cleaned surface can result in considerable controversy between the contractor and the inspector. Visual standards, with photographs available from the Steel Structures Painting Council and steel specimens encased in clear plastic from the National Association of Corrosion Engineers, are very useful tools in determining the grade of a sandblasted surface.

Brush blasting (SSPC SP-7 or NACE 4) requires the removal of all oil, grease, dirt, rust scale, loose rust and loose protective coating. Tightly adherent mill scale, rust and coatings are permissible, but the blast pattern must reveal numerous flecks of underlying metal uniformly distributed over the entire surface (salt-and-pepper effect). Visual standards are available also for this grade of blasting.

As previously indicated, blast cleaning is the best method for preparing steel surfaces. There are circumstances, however, where this procedure is not practical or economical and a substitute method is desired. In these cases, the most practical method will depend on the exposure, the type of finish coat to be used and the present condition of the surface. Each case must be analyzed individually. For other substrates, sandblasting is nearly always the best method; however, certain additional special treatments have been recommended.

Cast iron is a porous material and is likely to absorb moisture or other liquids with which it makes contact. To drive out absorbed material from its pores, cast iron should be heated before blasting. This can be done by placing in an oven for 8–12 h at 150 °C, or by heating with torches until this temperature is reached.

New zinc surfaces—galvanized or metal-sprayed—should be prepared by the application of a specially formulated acid treatment. Such solutions clean and etch zinc, thereby providing sufficient adhesion for the subsequent application of coatings. An alternative to this procedure is light sandblasting. However, if zinc-coated surfaces are allowed to weather properly, sufficient roughness is obtained to permit direct coating without prior treating or blasting.

Copper or brass should be blasted lightly to remove oxides and provide adhesion for the coating.

Aluminum surfaces must first be degreased to remove any residual oil. For best results under corrosive environments, a chromate-type conversion treatment should then be applied: where it is impractical to apply a conversion treatment, satisfactory results can be obtained under most conditions by lightly blasting the surface with fine sand.

Concrete surface preparation depends on a number of factors, including whether it is formed, trowelled or gunited. Formed concrete is the most difficult to prepare because of the surface irregularities and surface voids. These irregularities and voids must be minimized. This is achieved by removing fins and protrusions, followed by filling the holes and pits with a mortar mix, allowing to cure and then either blasting the surface or acid etching. Trowelled and gunited surfaces usually only need blasting or acid etching to remove laitance prior to coating application.

Where concrete surfaces have been in use, surface preparation for coating application is often problematical due to oils, grease and other contaminants penetrating beneath the surface. Specific testing on a case-by-case basis is usually the only reasonable alternative.

2. *Major Application Methods*

There are a number of methods of applying protective coatings, ranging from dipping to electrocoating, but maintenance painting is normally accomplished by brush, roller or spray.

2.1 Brush

Although oils and some alkyds apply readily by brush, vinyls, chlorinated rubbers, epoxies, urethanes and inorganics do not easily lend themselves to this type of application, except for touchup. These materials become tacky very rapidly and resist the flowout that is seen in house paint. However, for some systems, brushing of the primer coat is recommended to ensure that it has been worked into the substrate surface. When this is necessary, the primer should be liberally thinned to allow material flow.

2.2 Roller

Rolling works well with some coatings and is impossible with others. A few materials dry on the roller and do not dissolve on dipping; in a short time, the roller becomes too heavy to hold. Some show very poor flow properties. Although some vinyls and chlorinated rubbers roll quite well, application by this method results in pickup of the undercoat. For example, if a white vinyl is applied over a red-vinyl coat, the finish will be pink or rose colored rather than white.

2.3 Spray

Certainly the largest volume of maintenance coatings is applied by spray equipment. This method of application is not only faster than alternative methods, but the resulting films are more uniform in thickness. The object in spray painting is to create a mist of atomized coating particles that will cling to the target in a uniform pattern and then flow into a continuous, even film. In maintenance painting, two types of equipment are used: (a) conventional (air atomization) and (b) airless (hydraulic pressure).

In conventional (air atomization) spray systems, the spray gun, which is the primary component in a spray system, brings the air and paint together. This is accomplished in such a way that the fluid is broken up into a spray, which can then be directed at the surface to be coated.

With airless (hydraulic) spray equipment, there is no air used for atomization. The spray pattern is formed simply by forcing the material under high pressure through a very small orifice in the spray gun. As the material leaves the orifice, it expands and is broken up into fine droplets. Airless spray continues to displace conventional spray systems for maintenance painting, due to the absence of overspray (dry particles of paint hitting the surface) and smoother application to corners and crevices.

Electrostatic application of maintenance coatings has been occasionally recommended for certain difficult-to-coat substrates: these have included wire-mesh fences and expanded steel structures. In these applications, the substrate has been grounded and the coating applied from an electrostatic spray gun. This system has never proved popular, due to equipment limitations in the relatively uncontrolled field environment.

3. Types of Coatings Used

Most maintenance coatings are classified by the binder type that is used in the formulation. Four general types of binders (or film formers) are used to formulate protective maintenance coatings.

3.1 Solvent-Evaporation Types

Solvent-evaporation types (see Table 2) are typically single package and include vinyls, chlorinated rubber and polyacrylics. These coatings represent

Table 2
Solvent evaporation types

	Petroleum hydrocarbon	Vinyl coal tar	Chlorinated rubber	Chlorinated rubber, oil or alkyd modified
Effect of sunlight	minor	slow surface chalk	slow surface chalk	slow surface chalk
Wet or humid environments	excellent	excellent	excellent	fair to good; will yellow
Inustrial atmosphere contaminants				
acid	excellent	excellent	excellent	fair to good
alkali	excellent	excellent	excellent	fair
oxidizing agents	good	limited	excellent	fair
solvents	poor	poor	limited	limited
Spillage and splash of industrial compounds				
acid	good	good	very good	not recommended
alkali	good	good	very good	not recommended
oxidizing agents	fair	not recommended	good	not recommended
solvents	not recommended	not recommended	not recommended	not recommended
Physical properties				
abrasion resistance	poor	good	good	fair
heat stability	softens	limited	limited	fair
hardness	soft	good	good	fair to good
gloss	none	none	SG to matte	wide range
colors	black and aluminum	black/gray	wide range	full range
Additional notes	water resistance and ease of repair result in wide use as ship-bottom systems	use as marine primer	traditionally low solids	

some of the simplest to apply and are relatively insensitive to environmental conditions during application. They can also be formulated to give exceptional longevity with easy touchup and repair. Drawbacks include their relatively low volume solids and sensitivity to solvent environments.

3.2 Oxidation or Drying-Oil Types

Oxidation or drying-oil types (see Table 3) include natural oils (e.g., linseed), alkyds, varnishes, epoxy esters and uralkyds (urethane oils). Coatings of this type have been on the market for the longest period of time and are still used extensively. They offer good wetting of the surface and smooth flow-out when applied by brush. They do not, however, possess exceptional corrosion protection or chemical resistance and often continue to increase in hardness (or brittleness) with exposure to the air and sunlight.

3.3 Polymerization or Coreacting Types

Polymerization or coreacting types (see Table 4) include epoxy polyamide (or amine), polyurethane (isocyanate/polyol), epoxy phenolic, epoxy urethane and coal-tar epoxy. Coatings of this class are of more recent design and continue to gain in popularity. They offer excellent corrosion protection and resistance to chemical attack. The polyurethanes probably give the best combination of aesthetic appeal coupled with longevity; however, they are not typically used to provide corrosion protection alone. The drawbacks of this class of coating include a sensitivity to environmental conditions during application, a two-package product and occasional recoating problems.

3.4 Inorganic Silicate-Binder Types

Inorganic silicate-binder types include inorganic zinc-rich primers and inorganic topcoats. These coatings typically offer the greatest corrosion protection for exposures where pH extremes are minimized. However, they typically require a roughened surface achieved by abrasive blasting, and they are not compatible over organic coatings.

4. Systems for Maintenance Coating

There is no single coating formulation that will satisfy the demands of every environment or set of

Vinyl alkyd	Vinyl chloride copolymers	Poly(vinyl chloride)	Vinyl–acrylic	Acrylic
slow surface chalk	slow surface chalk	slow surface chalk	very slow surface chalk	prolonged chalk resistance
good; slight yellowing	excellent	excellent	very good	good
good	excellent	outstanding	very good	good
fair	excellent	excellent	very good	good
fair	excellent	outstanding	good	good
limited	limited	good	limited	limited
not recommended	very good	excellent	fair	poor to fair
not recommended	very good	excellent	fair	poor to fair
not recommended	good	excellent	fair	poor to fair
not recommended	not recommended	limited	not recommended	not recommended
fair	good	excellent	good	good
fair	limited	limited	limited	limited
fair	good	good	good	good
limited range	SG to matte	gloss to matte	semigloss	gloss, semigloss
full range	full range	most colors	full range	full range
useful as primer or topcoat for more chemically resistant vinyl copolymers	available in high- and low-build formulae; high-build more popular in maintenance applications	low build; Economic limit to tank lining and other severe exposures	wide use as finish coat with vinyl systems	use as finish coat for epoxy systems for improved weathering

application conditions. Usually, a system approach of properly selected primer (intermediate) and finish coats provides the best answer to the requirements of a specific coating problem. A recent innovation has been the introduction of coatings that perform as both primer and topcoat all in one. These are typically epoxy-based high-solids, high-build coatings requring a minimum of surface preparation and application labor.

4.1 Primer Coats

A primer should meet most, if not all, of the following requirements:

(a) good adhesion to the surface to be protected when the latter has been cleaned or prepared according to specification;

(b) a satisfactory bonding surface for the next coat;

(c) the ability to stifle or retard the spread of corrosion from discontinuities such as pinholes, holidays or breaks in the coating film;

(d) enough chemical and weather resistance by itself to protect the surface for a time period in excess of that anticipated before application of the next coating in the system; and

(e) compatibility with any remaining adherent coating so that a continuous primer coat can be uniformly applied to the complete substrate—this includes a lack of lifting the intact, adherent coating.

4.2 Intermediate Coats

Intermediate coats may be required in a system to provide one or more of the following properties:

(a) adequate film thickness of the system (body coat);

(b) a uniform bond between the primer and the topcoat (tie coat); and

(c) a superior barrier with respect to aggressive chemicals in the environment (may be too deficient with respect to appearance or physical properties to be a satisfactory finish coat).

4.3 Finish Coats

Finish coats are the initial barriers to the environment, but are also the surfaces that are seen by management and the public. However, there are situations where the barrier to the environment will primarily be a function of the body or primer coat, while the finish coat serves to provide a pleasing appearance, a nonskid surface, a matrix for antifouling agents or other specialized purposes. Obviously, the chemical resistance of the finish coat in these

Table 3
Oxidation or oil types

	Silicone alkyd	Epoxy ester	Alkyd	Uralkyd
Effect of sunlight	slow surface chalk	rapid surface chalk	slow surface chalk	yellowing and surface chalk
Wet or humid environments	fair to good	fair to good; will yellow	poor to good; will yellow	fair
Industrial atmosphere contaminants				
acid	fair to good	fair	fair to good	poor to fair
alkali	fair	fair	poor	fair
oxidizing agents	fair	poor	fair	poor
solvents	limited	limited	limited	good
Spillage and splash of industrial compounds				
acids	not recommended	not recommended	not recommended	not recommended
alkalis	not recommended	not recommended	not recommended	not recommended
oxidizing agents	not recommended	not recommended	not recommended	not recommended
solvents	not recommended	not recommended	not recommended	not recommended
Physical properties				
abrasion resistance	fair	fair	fair	excellnt
heat stability	good	fair	fair	good
hardness	fair	good	fair	excellent
gloss	good	wide range	wide range	high gloss
colors	limited	full range	full range	limited range

Table 4
Polymerization or coreacting types

	Coal tar epoxy	Epoxy, amine cure	Epoxy, polyamide cure	Urethane moisture cure	Urethane, two package
Effect of sunlight	surface chalking	yellowing and surface chalking	yellowing and surface chalking	see Additional notes	see Additional notes
Wet or humid environments	excellent	very good, may yellow	very good, may yellow	very good	very good
Industrial atmosphere contaminants					
acid	excellent	good	good	good	very good
alkali	excellent	excellent	excellent	good	excellent
oxidizing agents	limited	limited	limited	poor	limited
solvents	limited	excellent	excellent	excellent	excellent
Spillage and splash of industrial compounds					
acids	good	fair	poor to fair	fair	good
alkalis	good	excellent	excellent	fair	good
oxidizing agents	not recommended	not recommended	not recommended	not recommended	not recommended
solvents	not recommended	excellent	very good	good	excellent
Physical properties					
abrasion resistance	limited	good	good	excellent	outstanding
heat stability	excellent	good	good	good	good
hardness	very hard	very hard	hard	excellent	excellent
gloss	none	wide range	wide range	range	range
colors	black, red	full range	full range	limited range	full range
Additional notes	timing between applications critical in obtaining intercoat adhesion	cure rate and pot life affected by temperature (true of all epoxies)		have ability of curing at low temperatures. Urethanes based on aromatic diisocyanates discolor and chalk rapidly in sunlight. Those based on aliphatic diisocyanates have prolonged resistance to yellowing and surface chalking. Care must be taken to avoid moisture contact before cure	

situations must be sufficient to ensure that it remains intact to the environment.

Finish coats represent the first line of defense where the substrate has to be protected from excessive heat or erosion (airborne or suspended solids in liquids). In the former case, the binder is chosen for its heat resistance (typically silicone-based or inorganic) and the pigmentation is designed for maximum reflectivity (aluminum or titanium dioxide). For erosion protection, both the pigmentation and resin type contribute to the protection. Platelet pigments (micaceous iron oxide or mica) are often used to retard erosion damage together with a hard binder (epoxy or aliphatic urethane).

A more recent trend has been to formulate a heavy mastic coating that will act as a primer, intermediate and finish coat, all in one. These high-solids, epoxy-based, heavy-bodied products are often recommended for surfaces that are minimally prepared. Their inhibitive and wetting properties, coupled with a thickness of 125–255 μm, provide the necessary protection. Their performance is, as usual, dependent on the level of surface preparation and the type and nature of any surface contamination.

4.4 Practical Examples of Typical Systems

Maintenance painting in an existing plant can be fraught with many problems, due to the inherent conflicts with plant operations. Often, abrasive blasting is neither safe nor economical unless complete precautions are taken to protect plant equipment, operations and personnel.

When a coating has completely failed, abrasive blasting is the most effective method of providing a base for the high-performance protective coating. In industrial environments, chemical fallout and spillage may render the substrate contaminated and unsuitable for recoating unless it is neutralized, rinsed and mechanically abraded.

Coatings with high solvent content and low flash points may prove dangerous in certain plant locations. Nonflammable, waterborne systems or high-flash-point, high-solids coatings would be the only logical choice. Environmental conditions in the wintertime (or high humidity) may preclude the use of waterborne coatings.

Any coating to be used in an industrial environment should offer fast drying characteristics in order to prevent pickup of detrimental contaminants during the drying process.

Seven typical systems are listed.

(a) Maximum durability in severe industrial environments would typically require a two- or three-coat system: (i) abrasive blast (SSPC SP-10); (ii) 65 μm inorganic zinc-rich primer; (iii) 125 μm high-build epoxy polyamide (or high-build vinyl); and (iv) 50 μm aliphatic urethane or acrylic topcoat.

(b) An alternative recommendation giving slightly less corrosion protection, especially at coating holidays or damaged areas, would include: (i) abrasive blast (SSPC SP-6 or SP-10); (ii) 50 μm inhibitive epoxy primer; (iii) 125 μm high-build epoxy polyamide; and (iv) 50 μm aliphatic urethane or acrylic topcoat.

(c) Where aesthetic considerations are not paramount, the aliphatic urethane or acrylic topcoat may be deleted without adversely affecting the performance of systems (a) and (b).

(d) For previously painted surfaces where abrasive blasting is not permitted, the following system has been used successfully; the performance, while being dependent on the degree and diligence of surface preparation, often provides satisfactory corrosion protection for extended time intervals: (i) power-tool clean and solvent clean (SSPC SP-3 and SP-1), (ii) 75 μm universal synthetic resin "universal" primer and (iii) 125 μm high-build vinyl or epoxy. Alternatively, where a minimum number of steps is required for both surface preparation and coating, a single coat of epoxy mastic can be applied in lieu of primer and high-build topcoat. The mastic epoxy coatings are typically formulated to have excellent wetting properties and can be applied in one application to 200 μm.

(e) Where service conditions are relatively mild (e.g., interior living spaces) or protection is required only for a shortened period, alkyd or latex systems are often specified. A typical system would consist of: (i) power-tool clean and solvent clean (SSPC SP-3 and SP-1); (ii) 40 μm zinc chromate alkyd primer; (iii) 50 μm alkyd topcoat; and (iv) 50 μm alkyd topcoat.

(f) Fire-endurance coatings are applied to metallic or organic substrates in order to protect them from heat damage. In the case of steel subjected to a fire, the goal is to prevent the steel from exceeding 550 °C, where it loses 50% of its mechanical properties. These coatings basically fall into two classes: (i) organic-binder systems, where protection is afforded up to 2 h by their intumescing or ablative properties; and (ii) inorganic coatings offering protection up to 4 h—by insulation or water loss. These coating systems are regulated by various agencies in the USA, including Underwriters Laboratories and Factory Mutual.

(g) For concrete protection, the most commonly specified systems are epoxies and urethane. It must be remembered, however, that concrete is prone to cracking and any adherent coating will also crack. For that reason, a critical appraisal of the concrete structure and expected service conditions is always considered mandatory for continued long-term protection. A typical system is (i) abrasive blast or acid etch, (ii) epoxy primer/sealer and (iii) epoxy monolithic topping (750–5000 μm).

5. *Relative Durability and Economic Considerations*

The economic choice and justification for maintenance painting rest on many factors:

(a) service life of structure to be protected;

(b) cost of surface preparation;

(c) cost of interruption (if any) of plant operations;

(d) coating costs;

(e) cost of coating application; and

(f) accounting practices of the end user.

Expected lifetimes for the first five typical systems listed in Sect. 4.4 will vary dramatically, dependent on relative severity of the service conditions. Some approximate expected lifetimes (time for 5–10% active corrosion of substrate) are listed in Table 5. These should be viewed on the basis of comparative assessment rather than absolute lifetimes, as each set of conditions can give substantial variations in performance.

From a superficial standpoint, the choice is a relatively simple one of low-cost, low-performance coatings applied at more frequent intervals, versus higher-cost, high-performance coatings applied at extended intervals. Unfortunately, the complexities enter when the specifics of each factor listed are considered.

The alkyd types (see Table 3) are usually the least costly systems are are typically applied over surfaces minimally prepared (SSPC SP-1, -2 or -3). Their normal total systems thickness is 75–125 μm applied in three or more coats. It can be seen that they offer the least surface preparation and material cost. Their performance, however, is substantially less than the more sophisticated systems—typically less than 1 year in severe service environments, and 2–4 years in industrial environments. By contrast, an inorganic zinc/high-build epoxy or vinyl system requires abrasive-blast surface preparation and a total thickness of 200 μm in two coats. Expected service lifetimes, however, would be 5–15 years in severe service environments, and up to 20 years in industrial areas.

The cost of surface preparation and coating application is normally three to six times the cost of the coating itself. When comparing relative costs of coating systems, the total applied cost (including surface preparation, coating and coating application) must always be the starting point. Recent studies which included generally accepted accounting practices have compared the relative cost of many different maintenance-coating systems (Brevoort and Roebuck 1980, Roebuck and Brevoort 1983). Only when the expected service lifetimes are compared do the economics favor high-performance protective systems. The initial applied cost is virtually always higher; however, even when the interest expense of borrowing the extra money and its time value are included, the cost per unit area per year is lower for the high-performance protective-coating systems.

Table 5
Approximate comparative lifetimes of typical system types (see Sect. 4.4)

	Sever service[a]	Normal industrial[b] (year)
System (a)	5–15	10–20
System (b)	5–10	10–15
System (c) (from system (a))	5–15	10–20
System (d)	5–10	10–15
System (e)	0.5–2	2–4

a Severe service defined as continuous atmospheric and vapor exposure to high concentrations of industrial and chemical pollutants and occasional splash and spillage of diluted chemicals b Normal industrial defined as intermittent atmospheric and vapor exposure to low concentrations of industrial and chemical pollutants

See also: Bonding: Preparation of Metallic Surfaces; Paints, Varnishes and Lacquers

Bibliography

Brevoort G H, Roebuck A H 1980 Simplified method for calculating cost of an applied paint system. *Mater. Perform.* 19: 24–32

Industrial Maintenance Painting, 4th edn., 1978 National Association of Corrosion Engineers, Katy, TX

Roebuck A H, Brevoort G H 1983 Coating work costs and estimating. *Mater. Perform.* 22: 43–7

Steel Structures Painting Manual, 2nd edn., Vols. I and II, 1982 Steel Structures Painting Council, Pittsburgh, PA

V. P. Simpson
[Ameron, Brea, California, USA]

Marine Coatings

Marine coatings are thin films of polymeric or metallic materials (typically less than 500 μm thick) which are used to protect and modify the surfaces of materials exposed to the marine environment.

Typical coating films are mixtures of polymers, inorganic compounds and additives.

Marine coating films are formed from liquid mixtures of inorganic compounds suspended in a solution of polymers and volatile organic solvents. The liquid coating is applied by brushing, spraying or rolling. The wet film is converted into a dry, hard coating film by one of two mechanisms: solvent evaporation and chemical curing (drying). In the first case, the volatile solvent evaporates leaving behind a dry, hard film. In the second, the solvent evaporates initially leaving a viscous, sticky film, which subsequent chemical reactions, in the film, then convert into a hard, dry film.

Solid plastics are also used to form coating films. The solid plastic is applied in the form of a fine powder. When the powder is melted it fuses into a continuous film.

Metals are also used to form coating films. Metallic coatings are formed either by dipping the object into a bath of molten metal or by spraying molten metal onto its surface.

1. Advantages of Marine Coatings

1.1 The Need for Marine Coatings

Virtually every material exposed to the marine environment (i.e., salt water, sunlight and abrasion) needs to be coated to protect it or to modify its mechanical and chemical properties. Constant exposure to the environment rapidly destroys materials used in the construction of ships and other equipment. Unprotected metals corrode rapidly, lose strength and become more susceptible to damage from wind and waves. Unprotected plastics and wood are damaged by the combined action of sunlight and the penetration of water and salt. Marine coatings are therefore used to

(a) protect metallic structures from damage due to corrosion;

(b) prevent fouling (the attachment and growth of marine organisms);

(c) protect cargoes from contamination;

(d) provide safety warnings and informational markings;

(e) protect from degradation by wind, sunlight and rain;

(f) roughen walking surfaces to prevent uncontrolled slipping and sliding on wet, moving decks;

(g) protect electrical and mechanical equipment, for example antennas and winches, from damage; and

(h) reduce fire hazards.

1.2 Economic Impact

Marine coatings have been important in improving the economics of shipping by prolonging the working life of much of the equipment. For example, in the 1930s and 1940s, no marine coatings were available to help preserve the cargo tanks on ships used to transport petroleum products. The bulkheads of tankers which carried refined products were typically corroded to one-half of the original plate thickness after approximately seven years of service. Further corrosion would have made the ships unsafe and they spent the remainder of their 15–20 year service life carrying less corrosive unrefined products. The cargo tanks of modern tankers are preserved with marine coatings and the service life of a tanker in refined product service is 28–30 years with no allowance for interior corrosion.

2. Corrosion Control Coatings

The two general types of marine corrosion control coatings are those intended for intermittent exposure to moisture and those designed for continuous immersion service. The intermittent-exposure systems are typically used on the exterior portions of structures above the waterline. Continuous-exposure systems are used in ballast and trim tanks, wet bilges, fuel/ballast tanks and on the underwater portions of the structure. The primary difference between these systems is the mechanism used to protect the metal. Continuous-immersion systems use insulating layers coupled with cathodic protection systems to prevent corrosion. Intermittent-exposure systems rely on the formation of an insulating layer and/or chemical inhibition to prevent corrosion.

2.1 Principles of Corrosion Control by Coatings

Ferrous metals, such as steel, corrode because of the formation of anodic and cathodic areas on the same piece of metal. The physical connection between these areas of the steel provides the electrical connection between them. The presence of water (from atmospheric humidity, splashing or immersion) provides the electrolyte which completes the corrosion cell.

The basic steel corrosion reaction is shown in Fig. 1. The oxygen (O_2) is essential for the reaction to occur and the hydroxide ions are created at the cathode (protected area) as a result of the corrosion reaction.

2.2 Barrier Coatings

The most common method used to slow or stop the corrosion reaction is to break the contact with the electrolyte by using a coating to form a barrier between the electrolyte and the metal. Figure 2 illustrates the evolution of coatings used to control

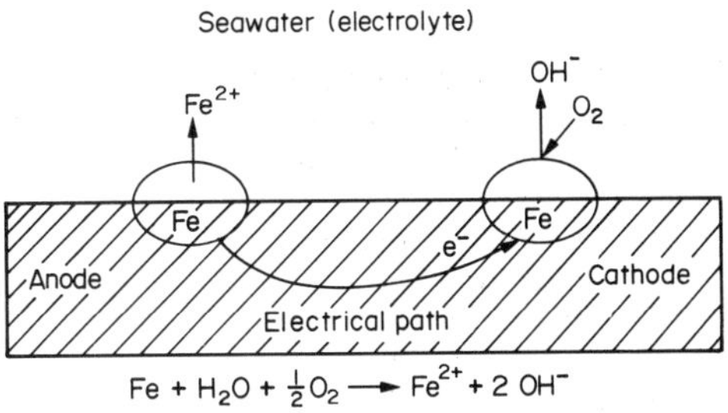

Figure 1
The chemical reaction which produces the local corrosion of steel

corrosion on surfaces exposed to the marine environment. The trend is to increase the thickness of the coating film and to use coating materials with increased chemical resistance and mechanical strength.

2.3 Chemically Active Coatings

Chemically active corrosion control coatings work by being actively involved in the electrochemical reactions that cause corrosion. The two basic mechanisms used are cathodic protection and corrosion inhibition.

(*a*) *Cathodic* (*zinc-rich*) *coatings*. Cathodic protection (sacrificial) coatings contain zinc metal that corrodes in preference to the underlying steel. Inorganic and organic zinc-rich coatings are the most widely used coatings of this type. They form hard, durable films and are particularly useful for protecting areas subject to intermittent immersion. These coatings corrode even in the absence of corrosion of the substrate. They are frequently topcoated with epoxy, vinyl or other barrier coats to prevent their self-destruction. The cathodic protection properties of the topcoated zinc coatings become active only when the topcoat is damaged and the substrate starts to corrode.

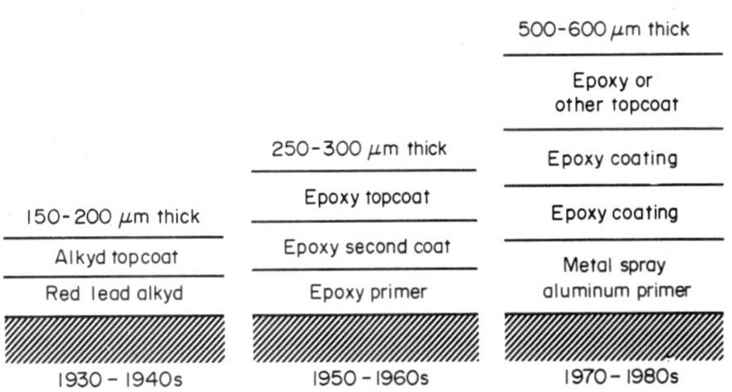

Figure 2
Evolution of barrier-type marine corrosion control coatings

Coatings containing aluminum and metal spray aluminum coatings, while being very effective in marine applications (primarily because of their potency as barrier coatings, rather than their cathodic protection properties), are generally not used as cathodic protection coatings. Aluminum reacts with seawater to form very insoluble oxides that fill voids in the coating and form an effective barrier.

(*b*) *Corrosion-inhibition coatings*. These contain inorganic or organic compounds that are slightly soluble in water and which cause chemical reactions that impede corrosion reaction.

The most common inorganic corrosion inhibitors are slightly soluble chromate, molybdate and borate compounds. These materials act by polarizing the corrosion cells and interfering with the corrosion reactions. Coatings made using these inhibitors are most effective in areas of intermittent exposure to moisture. They are of limited effectiveness in areas that are frequently immersed in seawater. Seawater immersion reduces their effectiveness by diluting the inhibitor concentration at the metal and by chemical action of chloride ions which interferes with the action of the inhibitor.

Amine compounds are effective corrosion inhibitors which increase the pH under the coating and slow down the corrosion reaction. The compounds can be added to the coating or may be part of the binder (e.g., amine- or polyamide-cured epoxy coatings).

3. Antifouling Coatings

Antifouling coatings are unique to the marine environment. They are used to prevent the settlement and growth of marine organisms on structures immersed in the sea. Antifouling coatings act by releasing compounds into the seawater adjacent to the surface. These chemicals then prevent the settlement and growth of fouling organisms by repelling those that approach the surface and by killing those that, having not been successfully deterred, settle on the surface. Many chemicals have been used in antifouling coatings but only copper metal and organic coatings containing cuprous oxide or tributyltin oxide have been shown to have long-term effectiveness and to be suitable for general use.

4. Common Types of Marine Coatings

Liquid marine coatings consist of four categories of ingredients: binders, pigments, additives and solvents. Binders are the materials that form the dry film and without which there would be no coating film. Organic polymers are commonly used as binders. Solvents are used to dissolve the binder and make it liquid so that it can be applied. They are typically volatile organic compounds such as alcohols, ketones and petroleum-based materials. Pigments are inorganic compounds of metals which are added to the binder to provide mechanical, optical and chemical properties. Inert pigments are used to provide physical properties while chemically active pigments react with the environment to produce local changes in the environment. Additives are compounds used to aid in manufacturing coatings or to modify the physical or chemical properties of the applied coating. Additives frequently have a major influence on the properties and performance of the marine coating. They are particularly important ingredients in biocidal coatings and in coatings which require catalysts to cure the applied film.

The single most important ingredient used in marine coatings is the binder. The binder is responsible for formation of the film and is the primary factor in determining the properties of the film. Therefore, coatings are commonly classified on the basis of the generic chemistry of their binder.

4.1 Alkyd

Alkyd resins are polyester compounds that are formed by reactions between polyhydric alcohols (e.g., ethylene glycol or glycerol) and a polybasic acid (e.g., phthalic anhydride) in the presence of a drying oil (e.g., linseed or soybean). The specific oil used determines the curing properties of the resin and its ultimate chemical and physical properties. Alkyds are frequently modified chemically to improve their physical properties or their chemical resistance. Vinyl, silicone and urethane compounds are commonly used to modify alkyd resins.

Coatings based on alkyd resins require chemical catalysts to cure them. Typical catalysts are mixtures of lead, cobalt and manganese salts, for example octoates. Coatings typically become dry to the touch within 1 h by solvent evaporation, but take up to a day before they can be handled without damaging the coating. Depending on the catalysts and the ambient temperature, it will take several days to several weeks before the coating is fully cured.

Alkyd coatings are not suitable for use in immersion service. The alkali generated by the corrosion reactions rapidly attacks the coating and leads to early failure. The presence of cathodic protection, which generates strong alkaline conditions, sharply accelerates this failure.

Alkyd coatings are frequently used as anticorrosive primers and topcoats in interior areas and as cosmetic topcoats over high-performance primers in exterior areas. They have excellent weathering properties, particularly the silicone alkyds, and make durable cosmetic coatings. They are also used to protect wooden and plastic surfaces.

4.2 Vinyl

Vinyl resins are formed by the polymerization of vinyl compounds. The most common resins are based on poly(vinyl chloride) (PVC) copolymers. These resins form coatings by solvent evaporation. Freshly applied coatings are dry to the touch within 1 h and are fully dried within a week. Vinyl coatings are particularly useful when rapid drying is required. The rate at which they dry depends primarily on the ventilation, the solvents used and the ambient temperature.

Coatings based on vinly polymers perform well in immersion situations and are frequently used to protect submerged structures. These coatings have excellent resistance to many chemicals and are good weather-resistant materials.

4.3 Chlorinated Rubber

Chlorinated rubbers are formed by reacting natural rubber with chlorine. They are not suitable for use as coatings by themselves but must be blended with other compounds. Coatings made from chlorinated rubbers that have been blended with highly chlorinated additives provide tough, chemically resistant coatings. These coatings cure by solvent evaporation. Freshly applied films are normally dry to the touch within 1 h and fully dried within a week. Chlorinated rubber coatings are particularly useful where fast drying is required. Like the vinyls, the rate of drying, with respect to both surface and complete dryness, depends on the ambient temperature, ventilation and solvents used in the coating. Chlorinated rubber coatings are noted for their toughness, water impermeability and chemical resistance.

4.4 Epoxy

Epoxy coatings for marine applications are typically formed by the chemical reaction of a Bisphenol-A-type epoxy resin with a curing agent (e.g., amines, amine adducts or polyamide resins). The coatings are packaged with the epoxy portion in one container and the amine portion in a second container. The coatings must be thoroughly mixed prior to use and must be used before the mixture solidifies. The liquid coating forms a film by solvent evaporation and continued chemical reactions between the epoxy resin and the curing agent. The length of time the mixed coating is suitable for use is called its pot life and is different for each formulation. Commonly used coatings have a pot life ranging from 2 h to 8 h at 25 °C.

Epoxy coatings typically dry to touch within 3 h and are fully cured after a week at 25 °C. The time to cure is highly dependent on the ambient and surface temperature during the curing period. The curing reaction slows down rapidly at temperatures below 10 °C.

Epoxy coating films are strongly resistant to most chemicals and make excellent anticorrosion coatings. They chalk when exposed to intense sunlight and are typically used with cosmetic topcoats, for instance silicone alkyds that have a higher resistance to sunlight.

4.5 Coal Tar

Coal tar coatings are made from processed coal tar pitch dissolved in suitable petroleum solvents. They form a film by evaporation of the solvent and the film can be redissolved by solvents. Coal tar films provide very good corrosion protection. The dry film is damaged by direct exposure to sunlight, which causes rapid, severe cracking. Coal tar resins are very dark in color and the coatings made from them are black or silver–black (made by using flake aluminum for pigment). They are normally blended with other resins to improve their light stability, increase their chemical resistance and increase the number of colors available. Common blending resins include vinyl and epoxy materials.

Coal tar epoxy paints are blends of epoxy resins with coal tars. These paints are cured by mixing with a curing agent (either amine or polyamide type). Coal tar epoxy films have high chemical resistance, easily form thick films and have a high dielectric strength. Their high dielectric strength makes them particularly suitable for use near anodes in cathodic protection systems. The high current densities and high alkalinity in these areas can damage other types of coatings.

4.6 Polyurethane

Polyurethane marine coatings are made from resins containing the isocyanate group and thus they react strongly with compounds that contain hydroxyl groups, such as water and alcohols (polyols) which are commonly used for curing agents. Coating films are formed by solvent evaporation followed by chemical reactions between the polyurethane resin and the curing agents.

The most important polyurethane marine coatings are packaged as two- or three-component systems. The first component contains the polyurethane resin and the second component contains an organic polyol. Some systems require the use of a third component containing catalysts, for example metals, soaps or amine compounds, to accelerate curing.

Polyurethane coatings which cure by reaction with water are called moisture-cured polyurethanes. These coatings are packaged in a single can and rely on the moisture in the air to cure the coating. They are commonly used as a cosmetic coating or as a clear varnish.

As in the case of epoxy coatings, polyurethane coatings must be mixed thoroughly with the curing agent and catalyst before use. After mixing, commonly used coatings have a pot life of 2–6 h. The coating film is typically dry to touch within 12 h and fully cured after 14 d at 25 °C. The time taken for curing is highly dependent on the ambient and surface temperature during the curing period as well as the humidity. The curing reaction slows down rapidly at temperatures below 10 °C.

Polyurethane coatings form tough, chemically resistant coatings and make particularly good high-gloss cosmetic finishes. They have good abrasion and impact resistance and are particularly useful in high-wear areas. They have good weather resistance but lose gloss when exposed to intense sunlight. Weathered polyurethane coatings are often difficult to recoat and topcoats will fail to adhere unless care is taken when repainting aged or damaged areas.

4.7 Other Coatings

Other coating types include inorganic zinc, based on powdered zinc metal, organic zinc and waterborne and spray-metallized coatings.

5. Selection of Coating Systems

The selection of coating systems requires consideration of many different factors, such as cost (materials and application), temperature and humidity during application and drying/curing, surface preparation requirements and desired service life. Of particular importance is the accessibility of the area for maintenance. Areas which are easily accessible can be coated with poorer performing coatings and repaired as often as needed, while those which cannot be easily or safely repaired need to be given the best available coating. Sections 5.1–5.8 cover additional considerations for specific problem areas.

5.1 Exterior Underwater Coatings

Exterior underwater areas need protection from corrosion and from the attachment of marine fouling. These portions of ships and structures are generally inaccessible for routine maintenance and the coatings chosen must give reliable performance for extended periods of time. Corrosion control for underwater areas usually includes the use of cathodic protection from sacrificial anodes (zinc or aluminum) or from impressed current cathodic protection systems. The latter generate strongly alkaline environments near the anodes and in areas where damage exposes metal to the water. Both corrosion control and antifouling coatings must be resistant to the environment created by cathodic protection. A coating system consisting of a three-coat (300 μm thick) epoxy polyamide coating with a two-coat (150 μm thick) vinyl cuprous oxide antifouling system is commonly used in this type of service. A longer service life can be obtained by using an ablative, organotin antifouling paint instead of the cuprous oxide paint.

5.2 Fuel Tanks

Coatings for fuel tanks must prevent contamination of the fuel by corrosion products or by materials in the coatings, so they need to be resistant to aliphatic and aromatic petroleum products. A three-coat (300 μm thick) epoxy polyamide coating is frequently used.

5.3 Potable Water Tanks

Coatings for water tanks must prevent contamination of the potable (drinking) water by corrosion products and must not contribute an objectionable smell or taste to the water. The coatings must not react with halogen compounds (e.g., bromine or chlorine) used to disinfect the water. Care must be taken to avoid the use of phenolic compounds in any coating used for potable water tanks (phenolic compounds are sometimes added to epoxy coatings to accelerate curing). Halogenated phenolic compounds in concentrations as low as one part per trillion can make drinking water unfit for use. A three-coat (300 μm thick) epoxy coating is frequently used for potable water tank coating.

5.4 Ballast Tanks

Ballast tanks are exposed to both total immersion and partial immersion in seawater. Marine fouling is typically not a problem. The upper parts of the tank are constantly exposed to high humidity, condensation and salt while the lower portions are constantly immersed. The continually immersed areas can be protected by a combination of cathodic protection and barrier coatings. A three-coat (300 μm thick) epoxy polyamide coating is frequently used as the barrier coating. Alternative systems use inorganic zinc (75 μm thick) either alone or topcoated with three-coat epoxy polyamide coating.

5.5 Weather-Exposed Areas

Weather exposure is characterized by exposure to wind, rain, seawater, sunlight and extremes of temperature. The primer system should give good corrosion protection while the topcoat should give a good, easily maintained appearance. A good low-maintenance system for weather-exposed areas consists of inorganic zinc (75 μm thick), two coats (150 μm thick) of epoxy paint and two coats (75 μm thick) of silicone alkyd or polyurethane paint.

5.6 Identification and Other Markings

It is essential that identification and other markings are durable and visible. Markings are used for a variety of purposes ranging from company logos to identifying danger areas near operating equipment; silicone alkyds and single-component polyurethane coatings are good materials for making markings.

5.7 Nonslip Deck Coatings

Nonslip deck coatings are used to provide a rough surface that prevents the uncontrolled motion of crew and machinery on wet, slippery decks. They are particularly important safety coatings on the weather-exposed decks of ships and oil rigs. Nonslip coatings need to be selected for both their mechanical roughness and their resistance to lubricants and cleaning compounds used on the decks. The most durable nonslip coatings are based on epoxy coatings that contain coarse aluminum oxide grit. A typical epoxy nonslip coating system consists of one coat (100 μm thick) of epoxy primer with one coat (750–1000 μm thick) of epoxy nonslip coating.

5.8 Fire Safety Coatings

Nonflammable and intumescent coatings are the two types of fire safety coatings used on ships and ocean platforms. Nonflammable coatings obviously hinder the spread of fire. Common types are based on chlorinated alkyd resins and on water emulsions of chlorinated polymers. Intumescent coatings contain materials which expand when heated and create a thick insulation film that inhibits damage to surfaces exposed to fire.

6. Trends

6.1 Environmental and Health Considerations

Environmental and industrial health considerations have had a major impact on the trends in materials used in marine coatings. Materials which pose either human health or environmental hazards need to be replaced. These replacements are required to meet statutory requirements or to minimize the legal liabilities associated with the use of hazardous materials. As a result, all components of coatings need to be carefully evaluated for hazards and decisions made to replace or control them, so as to eliminate risks.

Polyurethanes, epoxides and coal tar products have been found to contain residues of hazardous materials used to make them. Changes have been made in raw materials and production processes, in order to minimize the use of hazardous materials. Changes in pigments are particularly important since many anticorrosion coatings have relied upon heavy metal compounds, particularly lead oxides and chromate compounds, as their active ingredients. Lead and chromate compounds are considered hazardous to both human health and the environment and substitutes for these materials must be found. Changes in solvents have had particularly strong effects on the development and marketing of coatings. Environmental regulations controlling air pollution seek to limit the number of solvents used. In many areas, limits have been placed on the percentage of solvent a coating may contain. This has required changes in formulations to meet new compositional requirements. Some of the mandated limits cannot be met with current materials and new coating raw materials are being developed to meet the required composition changes.

Environmental concern about toxins used in antifouling paints, most notably tributyltin compounds, has caused serious restrictions to be placed on their use. These considerations have led to research into coatings which are effective at very low organotin release rates. In addition, alternatives to organotin compounds are being sought by improving existing copper-based coatings and by developing less hazardous toxins.

6.2 Marine Industry Considerations

Requirements for marine coatings vary with the economic climate of the marine industry. Short-term assets are ofen given minimal attention on the assumption that they will be scrapped before they will have to be maintained. Refined-product carriers, oil platforms, ocean liners, warships and other long-term assets are given the best available marine coatings to minimize maintenance costs. The cost of using high-performance marine coatings is usually recovered by reduced operation and maintenance costs. A major challenge for industry is to obtain marine coatings that solve health and environmental problems without reducing the benefits obtained from high-performance marine coatings.

See also: Bonding: Preparation of Metallic Surfaces; Health Hazards in Elastomer Processing; Paints, Varnishes and Lacquers

Bibliography

Fischer E C, Castelli V J, Rodgers S D, Bleile H R 1983 Fouling control technology. In: Costlow J D, Tipper R C (eds.) 1983 *Marine Biodeterioration: An Interdisciplinary Study*. Naval Institute Press, Annapolis, MD, pp. 261–349

Keane J D (ed.) 1982 *Steel Structures Painting Manual*, 3rd edn. Steel Structures Painting Council, Pittsburgh, PA

Ketcham S J (ed.) 1983 *Handbook of Protective Coatings for Military and Aerospace Equipment*. National Association of Corrosion Engineers, Houston, TX

Munger C G 1984 *Corrosion Prevention by Protective Coatings*. National Association of Corrosion Engineers, Houston, TX

H. R. Bleile
[Naval Sea Systems Command, Washington, DC, USA]

Melt Delivery Systems: Corotating Disk Processors

The corotating disk processor or Diskpack is a recent addition to polymer processing machinery. It has the capability to perform the basic functions of polymer processing with efficiency and versatility (Tadmor et al. 1979a). It is, therefore, possible to design Diskpack melt pumps, plasticators, devolatilizers, mixers, compounders and other processors to meet various processing needs.

1. Structure

The Diskpack processor consists of two or more disk-shaped elements attached to a shaft placed within a closely fitted barrel (Fig. 1). Each pair of neighboring disks together with the inner surface of the barrel and shaft surfaces form a single, annular processing chamber. Generally, each processing chamber is designed to perform a specific function such as melting, pumping, mixing or devolatilizing. The barrel disks and shaft are temperature controlled. The disks have a radial thickness profile, which determines the chamber cross section and the radial gap separation profile. The difference between the barrel and shaft radii is generally much larger than the separation between the disks. An inlet or feeding port and a discharge port from the processing chamber are formed by openings through the barrel. The inlet and discharge ports are separated by a channel block which is closely fitted into the processing chamber. Processing chambers may be connected in series or in parallel. Polymer is transferred from chamber to chamber through helical or axial transfer channels machined into the inside surface of the barrel. Diskpacks can be equipped with dynamic seals to separate neighboring chambers and to seal outboard chambers. Diskpacks are fed gravitationally either by molten polymer or by particulate solids.

2. Principles of Operation

The Diskpack operates on the same principles as the single-screw extruder (see *Melt Delivery Systems, Single-Screw*), in which the driving force for executing the five elementary steps of polymer processing (i.e., conveying of solids, melting, mixing, pumping and devolatilizing) is the drag induced by a moving surface (Tadmor and Gogos 1979). In the screw extruder there is a single surface—that of the barrel—moving relative to the screw, whereas in the Diskpack there are two jointly moving surfaces—

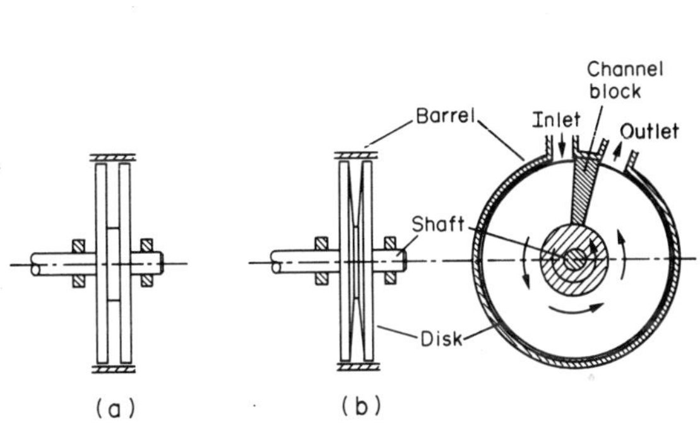

Figure 1
Diagram of a single-stage Diskpack chamber: (a) parallel chamber and (b) wedged chamber

those of the two neighboring disks. Theoretically, all five processing steps can be carried out more efficiently with two jointly moving surfaces than with a single moving surface (Tadmor et al. 1979b).

2.1 Pumping

Material particles in the processing chamber are dragged from the feed port to the discharge port with pressure buildup by the two tangential, jointly moving disk surfaces. The net flow rate achieved by parallel disks for isothermal, laminar, steady flow of a power law model fluid (Tadmor et al. 1979c) is given by

$$Q = \pi NH\, R_d^2(1-\alpha^2) - \frac{H^{2+s}}{2^{s+1}} \times \frac{R_d^{1-s}}{(2+s)m^s}\frac{(1-\alpha^{1-s})}{(1-s)}\left(\frac{dP}{d\theta}\right)^s \quad (1)$$

where Q is the volumetric flow rate, N is the frequency of disk rotation, H is the gap separation between parallel disks, R_d is the disktop radius, $\alpha = R_s/R_d$ with R_s being the shaft radius, $s = 1/n$ with n being the power law exponent, m is the power law parameter Ns^n/H^2 and $dP/d\theta$ is the angular pressure gradient ($=\Delta P/2\pi\varepsilon$, where ε is the effective angular fraction for pumping).

For an optimum wedge-shaped pumping chamber, the volumetric flow rate is given by

$$Q = \tfrac{2}{3}\pi NH_d R_d^2(1-\alpha^3) - \frac{H_d^{s+2}R_d^{1-s}(1-\alpha^3)}{6(s+2)2^s m^s}\left(\frac{dP}{d\theta}\right)^s \quad (2)$$

where H_d is the gap separation between wedge-shaped disks.

2.2 Melting

Particulate polymeric solids in pellet or powder form can be plasticated in the Diskpack chamber. Two melting mechanisms have been observed.

In the first mechanism a molten film is formed on the disk surfaces. The melt is dragged toward the channel block into a "pool," and is then either pumped into a subsequent chamber for further processing or discharged through the discharge port. The drag exerted by the disks through the film to the solids conveys and compresses the loose, solid feed into a solid bed. The mechanism is very similar to the commonly observed melting mechanism in single-screw extruders.

In the second mechanism, known as dissipative mix melting (Tadmor and Gogos 1979), the drag of the surfaces of the disks deforms the solid bed. Mechanical energy is thus converted into heat by mechanisms probably involving solid deformation, solid–solid particle friction and viscous dissipation in the molten regions. This melting mechanism is intensive and efficient and may be triggered by recycling some molten polymer through predetermined clearances over the channel block.

In either mechanism significant mixing has also been observed to occur in feeds consisting of more than one component.

2.3 Mixing

In addition to the mixing obtained in melting chambers, further extensive mixing can be obtained in relatively wide mixing chambers. In such chambers, preferably with parallel disk configuration, the pressure rise is sufficiently low and the residence time is sufficiently high so as to impose large strains on the fluid particle and to achieve composition randomization. For dispersive mixing, mixing blocks of a variety of configurations are employed at different angular positions. Consequently, the polymer melt is forced through narrow clearances between the block and the disks thus imparting a desired shear stress history.

2.4 Devolatilization

In a devolatilizing or stripping chamber, a spreader block is located in the inlet region of the chamber, which spreads the polymer on the surfaces of the disks. Downstream from the spreader block, the deposited films can be exposed to vacuum through a venting port located in the barrel. The deposited films are scraped off the disk by the channel block, collected in a melt pool and transferred to a subsequent chamber. Thus, the polymer melt can be repeatedly spread into thin films and mixed, achieving devolatilization by either a diffusion-controlled mechanism (Biesenberger 1980) or a more characteristic boiling–foaming mechanism (Newman and Simon 1980). Devolatilizing aids such as water and steam can be injected through the spreader block into the melt. The multistage–multichamber characteristic of a Diskpak devolatilizer permits vacuum staging.

3. Diskpack Processor Configurations

Diskpack processors may be tailored to specific applications by assembling the required processing chambers into particular design configurations. Figure 2 shows some typical examples. With the two-stage Diskpack pump (Fig. 2a), melt is fed (gravitationally) into two wide low-pressure outboard chambers which pump the melt into two narrow, high-pressure chambers. A typical Diskpack plasticator design is shown in Fig. 2b. Here, solids are fed in parallel into four melting chambers. The melt from the melting chambers is collected by a single transfer channel, fed into a

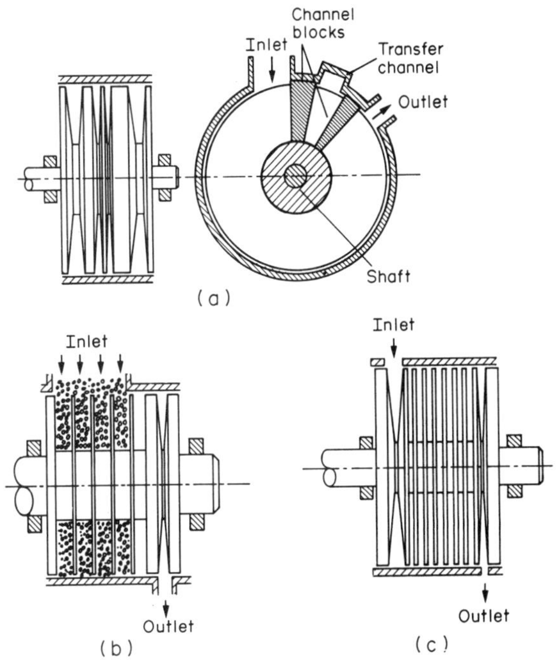

Figure 2
Diskpack processors: (a) Diskpack pump, (b) Diskpack plasticator and (c) Diskpack devolatilizer

single mixing–devolatilizing chamber and then into a high-pressure pumping chamber. Finally, Fig. 2c shows a schematic design of a multichamber devolatilizer with a pumping chamber at the end. Here, polymer melt is fed into a wide chamber and transferred into a set of devolatilizing chambers arranged in series. In the vented devolatilizing chambers, the melt is spread on the disk surfaces. The last devolatilizing chamber transfers the melt into a high-pressure pumping chamber.

By alternative design configurations, Diskpack processors can, in principle, be designed to meet many other processing needs. Hence, the expected fields of application of the processor cover most of the plastics industry from postreactor processing, through mixing and compounding, to precision shape extrusion. Moreover, work is under way to evaluate the Diskpack as a potential polymerization reactor.

See also: Processing: Classification

Bibliography

Biesenberger J A 1980 Polymer devolatilization: Theory of equipment. *Poly. Eng. Sci.* 20: 1015–22

Department of Chemical Engineering, Stevens Institute of Technology 1980 Methods and apparatus for processing polymeric material. US Patent No. 4,194,841 (2 February 1976)

Hold P, Tadmor Z, Valsamis L 1979 Applications and design of a novel polymer processing machine. *Soc. Plast. Eng., Techn. Pap.* 25: 205–11

Newman R E, Simon R H M 1980 A mathematical model of devolatilization promoted by bubble formation. Paper presented at 73rd Annual American Institute of Chemical Engineers Meeting. Chicago, IL

Tadmor Z, Gogos C G 1979 *Principles of Polymer Processing*, Wiley, New York

Tadmor Z, Hold P, Valsamis L 1979a A novel polymer processing machine: Theory and experimental results. *Soc. Plast. Eng., Tech. Pap.* 25: 193–204

Tadmor Z, Hold P, Valsamis L 1979b Radical design that promises an extruder revolution, Pt. 1. *Plast. Eng.* 35(11): 20–5

Tadmor Z, Hold P, Valsamis L 1979c New approach promises to revolutionize extrusion, Pt. 2. *Plast. Eng.* 35(12): 34–8

Tadmor Z, Valsamis L, Yang J C, Mehta P S, Duran O, Hinchcliffe J C 1983 The corotating disk plastics processor. *Poly. Eng. Rev.* 3: 29–62

USM Corporation 1979 Method for processing polymeric materials. US Patent No. 4,142,805 (2 February 1976)

USM Corporation 1980 Rotary processor. US Patent No. 4,213,709 (1 December 1978)

Z. Tadmor
[Israel Institute of Technology, Haifa, Israel]

Melt Delivery Systems: Normal Stress Pumps

A normal stress pump is a device which uses the elastic stored energy in a polymer melt to cause the material to extrude itself. The principle of operation is shown in Fig. 1.

The polymer to be extruded is introduced into the axial gap between a stationary header and end surface of an essentially cylindrical rotor. The polymer is sheared along circular paths, such as A–B, causing an orientation of the molecular segments along circular arcs. The stored elastic energy of orientation tends to pull the circular arcs into straight lines, thereby causing centripetal pumping in the dirction R. This causes the polymer melt to move toward the center of rotation and out through the exit die.

Machines of this type have been built with rotor diameters ranging from 2.5 cm to 75 cm. The throughput is directly related to the rotational speed and the area of the rotor at optimal axial gap. The optimal axial gap is directly related to the rotor diameter. Throughputs of about 450 kg h^{-1} have been demonstrated.

The following are some of the principal advantages of this type of equipment.

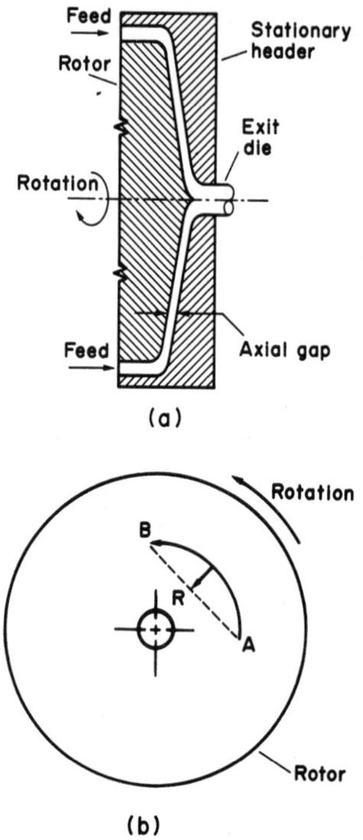

Figure 1
A normal stress pump: (a) side view and (b) front view

(a) One design serves for all materials. In comparison with screw extruders, where the design of the screw depends on the material to be processed, a single design of the normal stress pump works with all thermoplastic materials. Optimum operating conditions are achieved simply by adjusting the temperature and size of the axial gap. Materials ranging from soft rubbers to high-melting crystalline plastics have been extruded in the same apparatus.

(b) Lower extrusion temperatures can be used. Since the polymer is extruding itself there is no need to go to high temperatures to achieve low viscosity to push the polymer through the die. For example, polypropylene has been extruded at only 10 °C above its melting point.

(c) Residence times can be very short. Since the volume of material in the apparatus is very small the polymer is exposed to elevated temperature for only a short time. Residence times of less than 20 s have been measured. This is a great advantage with heat-sensitive materials such as rigid poly(vinyl chloride).

(d) Controlled mixing can be achieved. The amount of work put into the feed material can be controlled by changing both the rotational speed and the size of the axial gap. The mixing characteristics can be adjusted to achieve any desired objective. For example, if the objective is to disperse glass fibers in a thermoplastic, a large axial gap has been found to produce little if any damage to the fibers. On the other hand, if the objective is to break down agglomerates of filler particles such as carbon black, a small axial gap at high rotational speed should be used. The compounding of carbon black into rubber to achieve optimum reinforcing has been demonstrated.

It is easy to produce blends of either compatible or incompatible polymers. Depending on operating conditions, particularly the size of the axial gap, blends can be made where the two components are both continuous phases, or where one component is the continuous phase and the other is a disperse phase.

See also: Processing: Classification

Bibliography

Maxwell B 1970 Scaling up the elastic melt extruder. *Soc. Plast. Eng. J.* 26: 48–50

Maxwell B 1974 A new twist in mixing extruders. *Plast. Eng.* 30: 40–45

Maxwell B, Scalura A J 1959 The elastic melt extruder: Works without screw. *Mod. Plast.* 37: 107–14; 202–10

B. Maxwell
[Princeton University, Princeton, New Jersey, USA]

Melt Delivery Systems, Single-Screw

The single-screw-type melt delivery system is the most important melt delivery system used in the polymer and plastics industry.

The screw–barrel combination (see Fig. 1) is the central component of a whole group of processing machines, which includes the solids-fed plasticating-screw extruder, the melt-fed screw extruder, the reciprocating-screw injection-molding machine and various types of blow-molding machines.

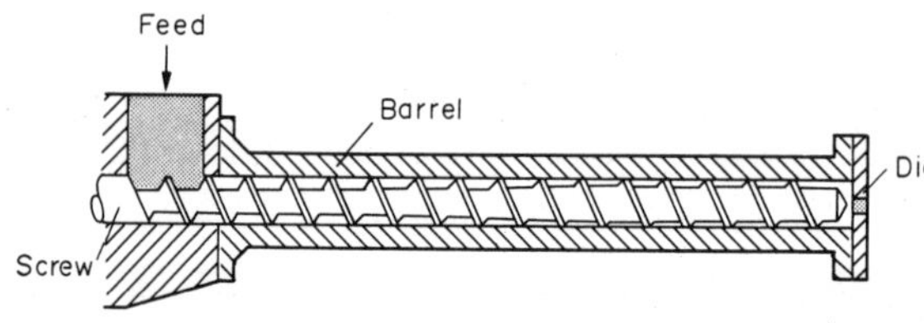

Figure 1
Schematic view of a rotating screw in a cylindrical barrel

1. Historical

The invention of the screw extruder (SE) is linked to the patent of Mathew Gray in England in 1879 (British Patent No. 5,056), which is the first clear and documented exposition of this type of machine (see *Processing: Historical Survey*). Independent of Gray, Francis Shaw in England developed a screw extruder in 1879, as did John Royle in the USA in 1880. Also in the USA, William Kiel and John Prior claimed to have developed screw extruders in 1876 (White 1969). Moreover, circumstantial evidence (Hovey 1961) indicates that DeWolfe, in the USA, may have already developed the first SE in the early 1860s, for insulating wire.

The first SEs were used for rubber compounds, but in the early 1930s extrusion of flexible poly(vinyl chloride) (PVC) compounds marked the start of the modern plastics screw extrusion technology. Since then the short, constant-channel-depth, variable-pitch rubber extruder has evolved into the long plasticating extruder and a host of specialized screw designs. About ten years earlier, the first theoretical analysis of the screw pump was published (Anon 1922, Rowell and Finlayson 1928).

2. Structure and Operation

The screw is driven by a motor through a speed-reducing device (see Fig. 2). It floats in the barrel with a small radial flight clearance δ_f in which the polymer acts as a lubricant, and a thrust bearing at the drive end prevents it from being pushed out by the pressure that develops at the die. The raw material in the form of particulate solids (e.g., pellets, powder) in plasticating extruders and in molten form in melt extruders is fed gravitationally from a hopper through a feed throat to the screw channel. The barrel is temperature controlled. Generally, the feed section (downstream of the hopper) of the plasticating extruder is water cooled; screws are usually neutral (i.e., neither heated nor cooled externally). Either right- or left-handed screws can be used. The barrel and the screw are made of steel and the inner surface of the barrel and the tips of the flights are either hardened or coated with special wear-resistant alloys. At the discharge end of the SE a die is attached for shaping the product. Usually, between the end of the screw and the die there is a breaker plate supporting a screen pack. These prevent contaminants from reaching the die and also aid in homogenizing the melt.

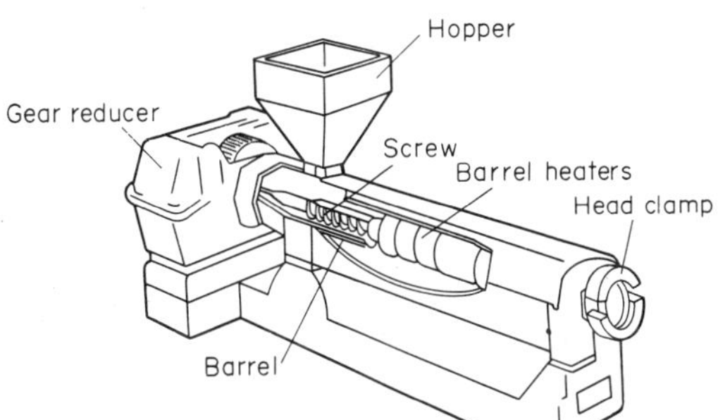

Figure 2
Schematic view of a single-screw extruder

The main operating variables are the frequency of screw rotation (screw speed) and the barrel temperature profile. Occasionally, discharge pressure is controlled independently from the die by an adjustable valve. The main design variables are the screw diameter and length, usually expressed as length-to-diameter (L/D) ratio. These, together with the operating variables, largely determine the throughput, polymer residence time and available barrel surface for heat transfer. Details of screw design, such as the number of parallel flights, helix angle, channel depth profile (i.e., radial distance between the tip of the flights and the root of the screw), and various geometrical modifications largely determine the quality of plastication, mixing and pumping performance. Many screws are designed with a lead (or pitch) equal to the diameter (referred to as square-pitched screws), which results in a helix angle of 17.6°. They generally have a deep channel feed section, to accommodate and properly convey the low bulk density particulate feed (or to accept and convey away molten polymer from under the feed throat), and a relatively shallow channel section at the front end. Connecting these two sections by a constant tapered section leads to the frequently used metering screw design. There are, however, a variety of other screw designs and the trend is deviating more and more from the classical metering-type screw.

Screw extruders are manufactured in a broad range of sizes from 2 cm to 50 cm or more in diameter, delivering molten polymer up to and beyond $10\,t\,h^{-1}$. A typical L/D ratio is 24:1, which has evolved from the short (5:1) rubber extruders. The trend is toward still longer extruders, in particular for two-stage vented and special-purpose extruders. Typical screw speeds range from 20–200 rpm (decreasing with increasing size), although in autogenous extruders screw speeds range from 400–600 rpm and experimental screws have been operated at speeds up to 2000 rpm.

The function of the SE is to deliver homogeneous molten polymer to the die and then to pump it through the die. Thus, the functions of a melt SE are limited to accepting the melt, conveying it toward the die, mixing in additives if needed and pumping (i.e., generating the required die pressure), whereas the functions of the plasticating extruder are more demanding and include the capability to accept and convey particulate solids as well as plasticating or melting them. Some SEs also have to devolatilize (i.e., strip the molten polymer from volatile components). The mixing demands of the SE include mixing in solid, liquid or gaseous additives, as well as homogenizing the mixture both thermally and compositionally. It is because the SE performs all these functions efficiently, that it has retained for many years a central, dominant role in polymer processing.

3. *Applications*

There are numerous applications for SEs in the plastics industry. In fact, most polymers pass at least once through a screw channel on their journey from raw material to finished products. Much of the polymer pelletizing downstream of polymerization reactors is done in SEs. These may be melt extruders or solids-fed (powder-fed) plasticating extruders. Screw extruders are also used for manufacturing a vast number of finished and semifinished plastics products, such as rods, pipes, tubes, hoses, blown and flat films, sheets (as finished or semifinished products for the thermoforming industry), wires, cables, filaments, nets and coatings. In addition to shaping, many SEs with specially designed screws are used for mixing and compounding. They are used for devolatilizing and stripping and even as polymerization reactors. Finally, single-screw melt delivery systems form the central part of reciprocating-screw injection-molding machines for manufacturing a myriad of molded products, and blow-molding machines for manufacturing bottles and other hollow container-type products.

4. *Principles of Operation*

There are five basic functions that a plasticating SE can perform: (a) conveying of particulate solids, (b) melting, (c) pumping and pressurization, (d) mixing, and (e) stripping. These basic functions were defined as the elementary steps of polymer processing (Tadmor and Gogos 1979). Any particular SE may perform one, some, or all of these functions, depending on the particular application it was designed for. The understanding of the principles of operation of the SE requires the understanding of the underlying mechanisms of each of the elementary steps in the SE.

The material fed to the SE passes through a relatively long, shallow helical channel formed by the flight of the screw. The channel is bounded by the inner surface of the barrel, the root of the screw and the flight. It is more convenient to analyze the SE by assuming a stationary screw and a rotating barrel (in the opposite direction). This does not alter

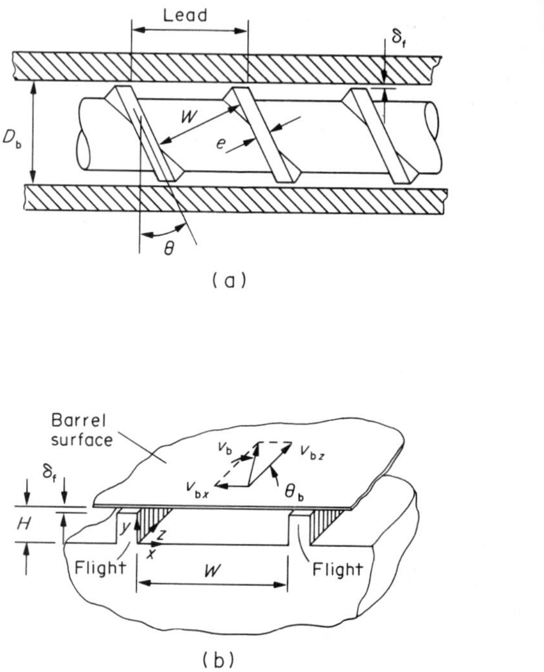

Figure 3
(a) Geometry of a single flighted screw and
(b) geometry of the rectangular screw channel

the analysis, since centrifugal forces are negligible. The velocity of the barrel relative to the screw channel is given by:

$$v_b = \pi N D_b \tag{1}$$

where v_b is the tangential velocity of the barrel surface relative to the screw, N is the frequency of screw rotation and D_b is the inside barrel diameter. This expression may be broken into a down-channel component

$$v_{bz} = v_b \cos \theta_b \tag{2}$$

where θ_b is the helix angle at the barrel surface, and a cross-channel component

$$v_{bx} = v_b \sin \theta_b \tag{3}$$

(see Fig. 3). All mechanisms of the five basic functions result from the relative motion between the surface of the barrel and the screw channel.

4.1 *Solids Conveying*

Solids are conveyed toward the die and compressed by frictional forces and torques exerted by the moving barrel. If these forward dragging forces and torques exceed the opposing retarding ones exerted by the root of the screw and flight, in steady operation (i.e., in absence of acceleration) a positive pressure gradient is generated to balance the forces and torques. Thus pressure is built up in the direction of flow and the low bulk density solids are compressed into a sturdy solid bed or plug which slides down channel. The simplest yet still useful isothermal model (Darnell and Mol 1956) relates the mass flow rate G to the ratio of outlet to inlet pressure P_2/P_1 (over helical length along barrel surface z_b) as follows:

$$G=\pi^2 NHD_b(D_b-H)\rho_b \frac{\tan\phi\tan\theta_b}{\tan\phi+\tan\theta_b}$$

$$\times\left[1-\frac{e}{\pi(D_b-H)\sin\theta}\right] \quad (4)$$

where H is the channel depth, ρ_b is the bulk density, e is the screw flight width, θ is the average helix angle and ϕ is the solids conveying angle (i.e., the angle of the barrel surface which moves relative to a viewer sitting on the solid plug), given by

$$\sin\phi=\frac{(1+K_s^2-M^2)^{1/2}-K_sM}{(1+K_s^2)} \quad (5)$$

where

$$K_s=\frac{\bar{D}}{D_b}\frac{\sin\bar{\theta}+f_s\cos\bar{\theta}}{\cos\bar{\theta}-f_s\sin\bar{\theta}} \quad (6)$$

and

$$M=2\frac{H}{W_b}\frac{f_s}{f_b}\sin\theta_b\left(K_s+\frac{\bar{D}}{D_b}\operatorname{cotan}\bar{\theta}\right)$$

$$+\frac{W_s}{W_b}\frac{f_s}{f_b}\sin\theta_b\left(K_s+\frac{D_s}{D_b}\operatorname{cotan}\bar{\theta}_s\right)$$

$$+\frac{\bar{W}}{W_b}\frac{H}{z_b}\frac{1}{f_b}\sin\bar{\theta}\left(K_s+\frac{\bar{D}}{D_b}\operatorname{cotan}\bar{\theta}\right)\ln\frac{P_2}{P_1} \quad (7)$$

where $\bar{D}$ is the mean diameter at midchannel, D_s is the diameter of the screw, f_s is the coefficient of friction at the screw surface, f_b is the coefficient of friction at the barrel surface, W_b is the channel width at the root of the screw and W is the mean channel width. From Eqns. (4–7) either P_2/P_1 can be calculated from G or vice versa. The maximum practical conveying capacity is obtained by setting $P_2/P_1=1$. A high coefficient of friction on the barrel relative to that on the screw improves conveying capacity. Hence, cutting grooves in the barrel, which brings about an almost mechanical interlocking mechanism (i.e., high apparent coefficient of friction), greatly improves conveying and pressurization capacity. High pressure, however, leads to frictional heat dissipation, hence efficient cooling must be provided to the barrel in the feed section of such designs to avoid premature melting.

The solids-conveying model of Eqns. (4–7) rests on numerous simplifying assumptions and there are improvements and modifications which can be made to it, although these are not discussed here.

4.2 Melting

The solid bed slides down channel in the solids-conveying zone like a nut on a screw. On reaching the heated portion of the barrel (or earlier if the frictional heat at the barrel is not efficiently removed) a thin molten film is formed between the barrel surface and the solid bed. Heat is conducted from the hot barrel to the melt film–solid bed interface. This conduction heat is one source of energy for melting. The film of melt is sheared by the relative motion between the barrel and solid bed. Thus, shaft power is converted into heat by viscous dissipation; this is the second source of energy for melting. The film thickness increases in the down channel direction until it exceeds the flight clearance, at which point the pushing flight (i.e., the flight toward which the barrel moves) scrapes off the melt from the film creating a melt pool at the pushing flight and segregating the slowly deforming solid bed at the trailing flight (see Fig. 4). In this way, at any point in the melting zone the slowly deforming solids flow into the melt–solid interface and the melt from the film is continuously removed by drag induced by the cross-channel component of the barrel velocity and a relatively thin film is maintained. Consequently, relatively high heat conduction rates and viscous dissipation rates are achieved and maintained. Moreover, the melt is quickly removed from the intensively sheared film into a mildly circulating melt pool. The continuous helical

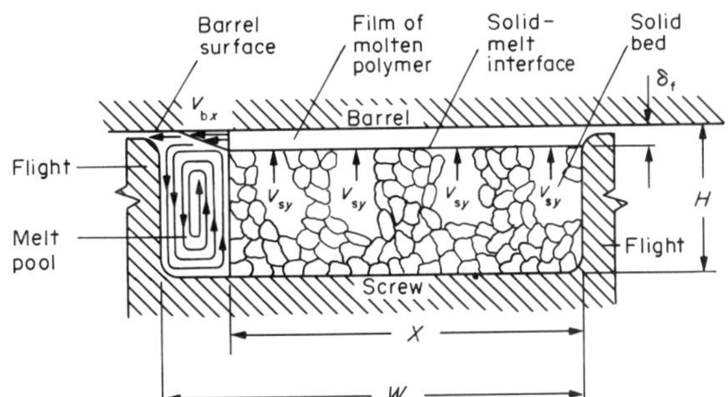

Figure 4
Cross section of screw channel in the melting zone with the solid bed of the "trailing" flight and the melt pool at the "pushing" flight

solid bed slides slowly down channel and as it melts (at the melt–film interface), the width of the solid bed decreases and that of the melt pool increases. In tapered sections, the solid-bed depth is reduced exposing more surface area for melting. Thus, depending on the relative magnitudes of the local rate of melting and taper, the width of the solid bed may continue to decrease, stay constant, or even increase. The latter, if excessive, may lead to surging. The mass flow rate of solids always decreases in the flow direction. Clearly the melting rate in tapered sections is higher than in constant channel depth sections. This two-phase flow and melting mechanism (Tadmor 1966) was proposed on the basis of experimental visual observations (Maddock 1959). Its simplest mathematical formulation, obtained by solving the temperature profiles in the solid bed and film and making the appropriate heat balances assuming constant viscosity Newtonian fluid leads to the local melting rate per unit down-channel distance ω_L:

$$\omega_L = \Phi X^{1/2} \qquad (8)$$

where X is the solid-bed width and

$$\Phi = \left\{ \frac{v_{bx}\rho_m[k_m(T_b - T_m) + (\mu/2)v_j^2]}{2[\lambda + C_s(T_m - T_{so}) + C_m(\bar{T} - T_m)]} \right\}^{1/2} \qquad (9)$$

where ρ_m is the density of the polymer melt, k_m is the thermal conductivity of the polymer melt, T_b is temperature of the barrel inner surface, T_m is polymer melting point, μ is Newtonian viscosity, v_j is $v_b - v_{sz}$ (with v_{sz} being the down-channel velocity of the solids), λ is heat of fusion, C_s is specific heat of polymer solid, C_m is specific heat of polymer melt, T_{so} is temperature of solids fed to the hopper and $\bar{T}$ is the mean temperature of the film (Tadmor and Gogos 1979). If viscous dissipation is not too significant, then $\bar{T} - T \cong \frac{2}{3}(T_b - T_m)$.

The width of the solid bed as a function of down-channel distance z (i.e., the solid-bed profile) is obtained by a down-channel mass balance which, for a constant depth channel, gives

$$\frac{X}{W} = \left(1 - \frac{\psi}{2}\frac{z}{H}\right)^2 \qquad (10)$$

and, for a tapered channel with initial channel depth H, gives

$$\frac{X}{W} = \left[\frac{\psi}{A} - \left(\frac{\psi}{A} - 1\right)\left(\frac{H}{H - Az}\right)^{1/2}\right]^2 \qquad (11)$$

where ψ is a dimensionless melting number given by

$$\psi = (\Phi H W^{1/2}/G) \qquad (12)$$

where W is the channel width and A is the taper in the compression section.

This melting model, termed drag-removal melting (Tadmor and Gogos 1979), was carefully investigated, modified and improved by many workers in the field.

In addition to this, the most common melting mechanism, three other distinct types of mechanisms have been observed. In one, the melt pool segregates at the trailing flight rather than the pushing flight (Menges and Klenk 1967). This was observed with certain powder and granulated PVC compounds and high-molecular-weight polyethylene. In the second mechanism, melt film at the barrel grows in thickness without solid-bed segregation at either side of the channel and at the same time a thick melt film is also observed at the root of the channel (Dekker 1976). This mechanism was analyzed mathematically and appears to be characteristic to large extruders (Lindt 1976). Finally, there are indications that a mechanism termed dissipative mix melting (Tadmor and Gogos 1979) evolves in screw designs and operations that result in very high pressure in the feed section or when molten polymer is recycled to the feed zone. Here, the solid bed undergoes intensive shear deformation and possibly through a mechanism of sintering, solid deformation, interparticle friction and viscous dissipation (of molten regions) shaft power is converted to heat throughout the solid–melt mixture.

4.3 Pumping and Mixing

Viscous drag by the moving barrel surface relative to the screw channel is also the driving force for melt pumping and for laminar mixing (see *Mixing: Single-Screw Extruders*). Specifically, it is the down-channel component of the barrel velocity v_{bz} that drags the melt down channel towards the discharge end, whereas the cross-channel component v_{bx} induces a cross-channel circulatory flow pattern that imposes large shear strains on the fluid and results in good laminar mixing. Moreover, the latter also brings about a desirable narrow residence time distribution function (Pinto and Tadmor 1970). For isothermal steady laminar flow of a Newtonian fluid in rectangular channels (i.e., neglecting curvature effects) the volumetric flow rate is related to the local pressure gradient as follows (Anon 1922, Rowell and Finlayson 1928, Carley et al. 1953):

$$Q = \tfrac{1}{2}v_{bz}WHF_d + \frac{WH^3}{12\mu}\left(-\frac{dP}{dz}\right)F_p \qquad (13)$$

where

$$F_d = \frac{16W}{\pi^3 H}\sum_{i=1,3,5}^{\infty}\frac{1}{i^3}\tanh\left(\frac{i\pi H}{2W}\right) \qquad (14)$$

and

$$F_p = 1 - \frac{192}{\pi^5} \frac{H}{W} \sum_{i=1,3,5}^{\infty} \frac{1}{i^5} \tanh\left(\frac{i\pi W}{2H}\right) \quad (15)$$

The first term on the right hand side of Eqn. (13) is the drag flow rate Q_d and the second term is the pressure flow rate Q_p. Clearly, if Q_d is greater than Q, pressure builds up in the down-channel direction ($dP/dz > 0$) and, if Q_d is less than Q, it drops. In constant channel depth sections the pressure gradient is constant and equal to the pressure rise (or drop). The terms F_d and F_p are the shape factors (which are functions of only H/W) for the drag and pressure flows, respectively. In tapered sections the pressure gradient varies from point to point. Equation (13) further indicates that for any given flow rate the pressure rise is proportional to the viscosity. Hence, the SE is superbly suited for pumping the very viscous polymer melts. Conversely, it is rather impractical to generate very high pressure in SEs with low-viscosity liquids. Since about 1950, a great deal of attention has been focused on flow analysis in SEs. Results of this research effort provide answers to problems concerning flow in SE channels completely filled with melt, as is the case in melt extruders and often in the final stages of plasticating extruders. However, the two-phase flow occurring along most of the length of plasticating extruders, where melting solids and molten polymer flow side by side, still requires much further elucidation.

4.4 Devolatilization

Devolatilization of polymers melts in SEs is achieved in screw sections only partially filled with melt exposed to vacuum through vent ports. Obtaining a partially filled channel is possible by appropriate screw design and operating conditions. These SEs have two- or multiple-stage screw designs; that is, the compression and shallow metering section (terminating frequently with an unflighted "torpedo" section) of the first stage of the screw is followed by a deep constant channel vented section in the second stage. Polymer melt is continuously smeared on the barrel surface through the flight clearance and collected and mixed into the circulating pool at the pushing flight. One possible devolatilization mechanism is molecular diffusion from the film (and the pool itself) into the vapor phase (Latinen 1962, Roberts 1970).

The continuous surface renewal inherent in this mechanism greatly improves the mass transfer rates. However, the stream splitting and back-mixing also inherent in the mechanism limit its efficiency (Biesenberger 1980). Another devolatilization mechanism involves boiling of the volatile component and foaming of the molten polymer–volatile component mixture. Growing experimental evidence seems to indicate that this is the dominant devolatilization mechanism not only at high volatile concentrations, but down to very low volatile concentration ranges where molecular diffusivities drop sharply and a molecular diffusion mechanism cannot explain the separations attained experimentally (Vrentas and Duda 1979).

5. Mathematical Analysis and Design of Screw Extruders

The previous section discussed simple mathematical models of the various subsystems of the plasticating extrusion process. An overall mathematical model of the process can be obtained by assembling the subsystem models. Of course, for a satisfactory representation of the process the simple subsystem models are replaced by greatly improved ones and account must be taken of the complex interactions of the various subsystems.

Such an overall model invariably takes the form of large computer programs (Klein and Tadmor 1968, Tadmor and Klein 1970. Maddock and Smith 1972, Brauner et al. 1976, Shapiro et al. 1976, Halmos et al. 1978, Rao et al. 1979, Losson 1982). These programs are instrumental in simulating SE processes and by repeated simulation designing screws for new processes. The objective of these programs is to predict pressure–temperature and solids-content profiles along the screw, power consumption, surging and mixing levels, as a function of operating conditions (i.e., flow rate, screw speed and barrel and screw temperature profiles) and physical properties of the polymer. At present, practically useful results can be obtained with relatively simple screw design. However, simulation reliability drops with increasing complexity in screw design such as mixing- and barrier-type screws. Theoretical analysis lags behind working practices, although much of the present trend in screw design practices is based on past theoretical research efforts.

5.1 Development Trends in Screw Design

The screw design is the dominant factor in determining product quality as well as production rate. It is not surprising, therefore, that much research and development effort is directed toward improved screw designs. Indeed, screw designs for plasticating extrusion have been undergoing intensive development (Schenkel 1979, Ingen Housz and Meijer 1981). This work is greatly aided by earlier research efforts which helped elucidate the melting and mixing mechanisms in extruders.

The starting point of most screw designs is the classical single-flighted constant-lead, variable-depth, metering-type screw. This evolved from the constant depth, variable lead, short screws used for rubber compounds. It has three sections: a relatively deep feed section, a tapered section and a relatively shallow metering section. The throughput is metered by the whole screw and not only by the last shallow section. The key design variables are the channel depths of the first and the last sections, and the position and length of the tapered section.

A straightforward variation of the metering screw is the two-stage screw used in vented extruders. The two-stage screw consists essentially of two metering-type screw designs in sequence. Following the first metering section, in a properly designed and operated vented extruder, pressure drops to zero and a vacuum is imposed through a vent in the barrel. Entrapped air is thus removed and volatiles are stripped from the melt, which only partially fills the channel. The discharge pressure is generated in the second tapered and metering sections. Venting is generally necessary for a feed in powder form.

From the metering-type screw, two additional groups of screw designs have evolved: (a) mixing-type screws and (b) barrier-type screws. The objectives of all these screw designs are to improve melting, reduce temperature and composition non-uniformity, improve stability and (since throughput is frequently limited by extruder quality) to increase throughput.

From the discussion on the melting mechanism it is evident that the decreasing solid-bed width toward the end of the melting zone is coupled with a diminishing melting rate. Moreover, the unmelted solids in this region are frequently found in the form of chunks and helical segments surrounded by melt and flowing with it toward the die. Thus the melting rate is further impaired by a virtual absence of the drag removal mechanism. Hence, the melting rate can be greatly improved by breaking up the solid into smaller pieces (i.e., creating large solid–melt interfaces) or smearing it into thin sections with the same objective. This is achieved by inserting into the screw channel mixing rings (such as rows of pins or pegs), by multiple flighted screw sections, broken flight sections, reverse flight sections, unflighted torpedo sections and similar geometrical modifications. Most of these mixing sections also bring about variable levels of spatial randomization of composition and temperature homogenization. They may be placed at the end, or elsewhere along the screw. Many screws have more than one mixing section.

The natural melting mechanism in the extruder leads to the segregation of the solids at the trailing flight and the melt at the pushing flight. This two-phase flow configuration has given rise to a host of screw designs in which solids are physically separated from the melt by a barrier wall, such that no solids can escape into the discharged molten stream. In these designs the drag removal melting mechanism is not only retained but frequently maximized,

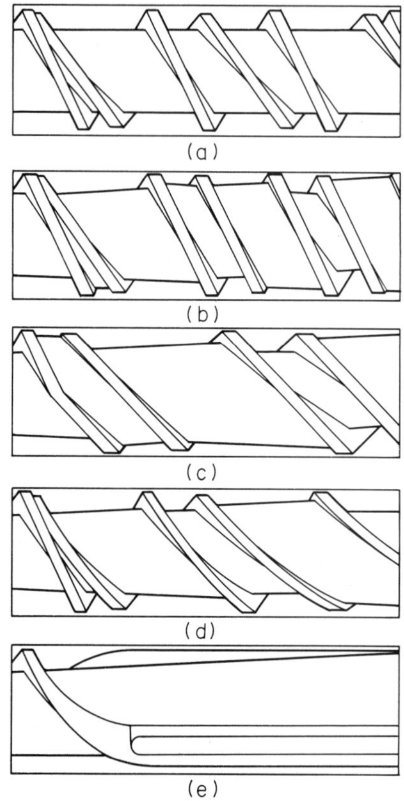

Figure 5
Schematic view of various screw designs: (a) Maillefer screw (after Maillefer C 1963 *Mod. Plast.* 40: 132); (b) Barr screw (after Barr R A 1971 US Patent No. 3,698,541); (c) Dray and Lawrence screw (after Dray R F, Lawrence D L 1970 US Patent No. 3,650,652); (d) Kim screw (after Kim H T 1972 US Patent No. 3,867,079); and (e) Ingen Housz screw (after Ingen Housz and Meijer 1981)

and the drifting of unmelted chunks and pieces of solids toward the discharge is prevented by the barrier wall. The first barrier-type design is the Maillefer screw where an auxiliary flight with variable lead and fairly large clearance generally follows the solid-bed profile. The auxiliary flight creates an inlet and a discharge channel side by side. Most other improved screw designs based on the Maillefer screw have variable channel depth and width profiles of these two channels as well as lead variations of the main flight (see Fig. 5). For example, in the Barr screw, the channel depth of the feeding (solids) channel decreases and that of the discharge (melt) channel increases in the down-channel direction. The Dray and Lawrence screw is similar, but there is a sudden increase in helix angle at the beginning of the melting zone. In the Kim screw, the channel depth of the feeding channel decreases and the width of the discharge channel increases. The latter is accomplished by a gradual increase in helix angle. In the Ingen Housz screw, the channel depth of the feeding channel decreases and is split into a number of parallel channels with helix angle approaching 90°. The Maddock-type fluted mixing section (Maddock 1967) and modifications of this concept also fall within the barrier-type screw of design configuration.

A coupling of screw design with improved homogenizing and mixing, with older concepts of special feed section design led to most effective feed zone oriented screw designs. In these designs the output rate is dominated by the feed section where large pressures are also generated. The rest of the screw downstream from the feed section is designed mostly for melting, mixing and homogenization. The key to these designs is an efficiently cooled barrel in the feed section with axial grooves. It is possible that in certain cases a dissipative mixed-melting process takes place in such screw designs and operations. Finally, the improvements reported by recycling (or feeding) molten material into the feed section may also be associated with the triggering of the dissipative mix-melting mechanism which is potentially more efficient than the drag removal mechanism.

See also: Processing: Relation to Product Properties; Processing: Rheological Properties and Fluid Mechanics

Bibliography

Anon 1922 Screw viscosity pumps. *Engineering* 114: 606

Biesenberger J A 1980 Polymer devolatilization: Theory of equipment. *Polym. Eng. Sci.* 20: 1015–22

Brauner N, Lavie R, Tadmor Z 1976 Control of plasticating extruders. *Proc. 3rd Int. IFAC Conf. Paper, Rubber and Plastics*. International Federation of Automatic Control, Brussels

Carley J F, Mallouk R S , McKelvey J M 1953 Simplified flow theory for screw extruders. *Ind. Eng. Chem.* 45: 974–8

Darnell W H, Mol E A J 1956 Solids conveying in extruders. *Soc. Plast. Eng. J.* 12(4): 20–9

Dekker J 1976 Verbesserte Schneckenkonstruktion für das Extrudieren von Polypropylen. *Kunstsoffe* 66: 130–5

Fenner R T 1979 *Principles of Polymer Processing*. Macmillan, London, pp. 93–144

Halmos A L, Pearson J R A, Trottnow R 1978 Melting in single screw extruders. 3: Solution for a power law temperature dependent viscous melt. *Polymer* 19: 1199–216

Holmes-Walker W A 1975 *Polymer Conversion*. Applied Science, London, pp. 117–71

Hovey V M 1961 History of extrusion equipment for rubber and plastics. *Wire Wire Prod.* 36: 19

Ingen Housz J F, Meijer H E H 1981 The melting performance of single screw extruders. *Polym. Eng. Sci.* 21: 352–9; 1156–61

Klein I, Marshall D I 1968 *Computer Programs for Plastics Engineers*. Reinhold, New York, pp. 238–78

Klein I, Tadmor Z 1968 Operating characteristics of plasticating extruders. In: Klein and Marshall 1968, pp. 254–89

Latinen G 1962 Devolatilization of viscous polymer systems., *Adv. Chem. Ser.* 34: 235–46

Lindt J T 1976 A dynamic melting model for a single screw extruder. *Polym. Eng. Sci.* 16: 284–90

Losson J M 1982 *EXTRAN a Powerful Tool for Extrusion Analysis*. Laboratoire Control-Centre de Calcul, Solvay SA, Brussels

McKelvey J M 1962 *Polymer Processing*. Wiley, New York, pp. 228–98

Maddock B H 1959 A visual analysis of flow and mixing in extruder screws. *SPE J.* 15: 383

Maddock B H 1967 Improved mixing screw design. *Proc. 25th Conf. Society of Plastics Engineers*. Society of Plastics Engineers, Detroit, MI, pp. 835–42

Maddock B H, Smith D J 1972 Extruder design by computer printout. *SPE J.* 28: 12–17

Menges G, Klenk P 1967 Melting and plasticating of unplasticized PVC powder in the single screw extruder. *Kunststoffe* 57: 598–603

Middleman S 1977 *Fundamentals of Polymer Processing*. McGraw-Hill, New York, pp. 123–70

Patton J B, Squires P H, Darnell W H, Cash F M, Carley J F 1959 Extrusion. In: Bernhardt E C (ed.) 1959 *Processing of Thermoplastic Materials*. Reinhold, New York, pp. 154–307

Pearson J R A 1966 *Mechanical Principles of Polymer Melt Processing*. Pergamon, London, pp. 77–91

Pinto G, Tadmor Z 1970 Mixing and residence time distribution melt extruders. *Polym. Eng. Sci.* 10: 279–88

Rao N, Hagen K, Krämer A 1979 Auslegen von Extruderschnecken mit Rechenmodell und EDV. *Kunststoffe* 69: 71–4

Roberts G W 1970 A surface renewal model for the drying of polymers during screw extrusion. *AIChE J.* 16: 878–88

Rowell H S, Finlayson D F 1928 Screw viscosity pumps. *Engineering* 126: 249–50

Schenkel G 1966 *Plastics Extrusion Technology and Theory: The Design and Operation of Screw Extruders for Plastics*. Iliffe, London

Schenkel G 1979 Extruder technology a look back, present state and future prospects. *Kunststoffe* 69: 58–66

Shapiro J, Halmos A L, Pearson J R A 1976 Melting in single screw extruders, Pt. 1, the mathematical model; Pt. 2, solution for a Newtonian fluid. *Polymer* 17: 905–11; 912–18

Tadmor Z 1966 Fundamentals of plasticating extrusion: I. A theoretical model for melting. *Polym. Eng. Sci.* 6: 185–90

Tadmor Z, Gogos C G 1979 *Principles of Polymer Processing*. Wiley, New York

Tadmor Z, Klein I 1970 *Engineering Principles of Plasticating Extrusion*. Van Nostrand Reinhold, New York

Vrentas J S, Duda J L 1979 Molecular diffusion in polymer solutions. *AIChE J.* 25: 1–24

White J L 1969 Elastomer rheology and processing. *Rubber Chem. Technol.* 42: 257–338

Z. Tadmor
[Israel Institute of Technology, Haifa, Israel]

Melt Delivery Systems, Twin-Screw

The basic function of any polymer extrusion operation is to produce and transport a polymer melt. In both single- and twin-screw extruders, a solid is plasticized by frictional and external thermal energy. As the melt is transported through the extruder, it is subjected to shear. Additives may also be incorporated into the melt and volatiles withdrawn. Finally, enough pressure must be generated in the screw to force the polymer melt through a die.

1. Advantages of Twin-Screw Systems

In both single- and twin-screw extruders, the ability of the machine to convey the polymer is a function of the frictional effects between either a melt, solid or a mixture of solid and melt, and the surrounding barrel walls. This frictional relationship also affects the pressures which can be generated at any location in the machine. Single-screw extruders can produce the necessary shear rates to successfully handle polymer viscosities of 1000–100 000 poise. However, adequate control cannot always be maintained over all the unit operations performed in the single-screw extruder. The larger the diameter and length of the single-screw extruder, the more difficult it becomes to control the polymer in the screw channel in terms of shear, conveying forces and stock temperature. Typical problems that occur include surging, poor mixing, overheating of the product and uneven treatment of product particles due to wide variations in stress distribution (see *Melt Delivery Systems, Single-Screw*).

To overcome the limitations of single-screw extruders, industry has found that twin-screw mechanisms are an economical solution for higher line-rate requirements. Use of intermeshing twin screws substantially increases the conveying capability of a twin-screw extruder over that of a single-screw system. It also increases the efficiency of energy input and pressure generation and allows tighter control of the process over a much wider range of polymer viscosities.

2. Twin-Screw Mechanisms

The different types of twin-screw mechanisms are shown in Fig. 1, which illustrates the differences between corotating and counter-rotating as well as between intermeshing and nonintermeshing screws. The counter-rotating screw shown in Fig. 1 (type 1) has been classified (Erdmenger 1964) as an axially closed, single-chamber pumping system, whereas the corotating screw (type 4) can be considered as an axially open mixing system relying on drag forces.

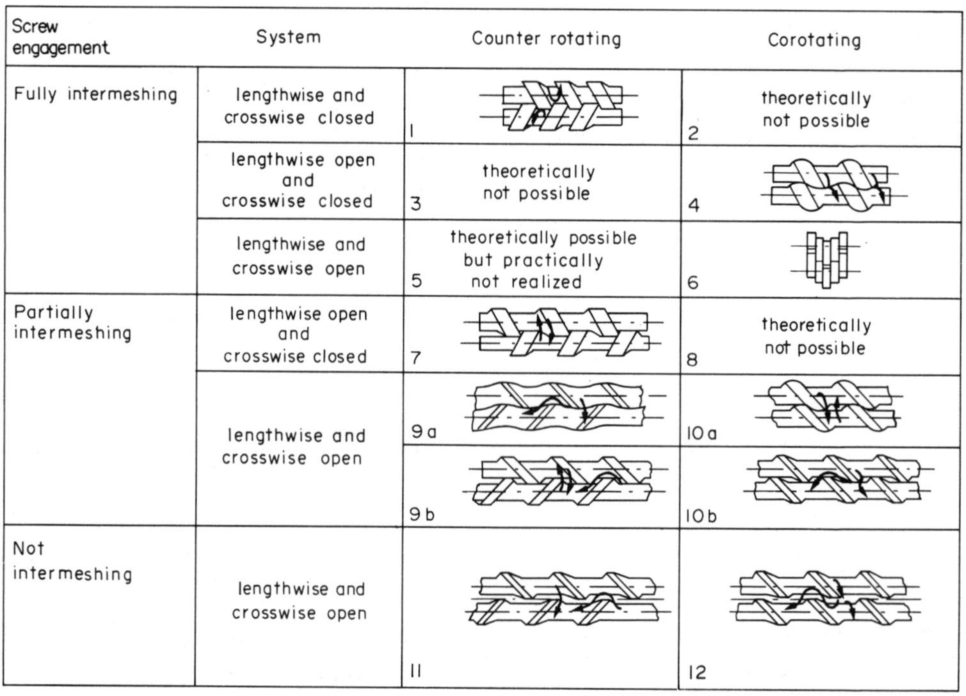

Figure 1
Twin-screw mechanisms

Screw geometry has a direct effect on conveying conditions, mixing and melting action and pressure buildup capacity of the screw system. A screw system open in the axial direction is actually open in the longitudinal direction of the screw channel with material passage from the inlet to the outlet of the machine. Material exchange cannot take place lengthwise in the screw channel. The cross section of the screw channel in the twin-screw mechanism can be either open or closed. An open cross section allows material exchange to take place from one flight to the other in a direction perpendicular to the screw channel. Normal leak flow over the screw crests and through the areas required for mechanical clearances is not considered here.

Only fully intermeshing, counter-rotating systems can be closed lengthwise and crosswise and in these systems closed chambers can be developed regardless of mechanical clearances (Fig. 1, type 1). Fully intermeshing corotating screws are open lengthwise. When normal screw flights (type 4) are used they are closed crosswise but when staggered screw disks (type 6) are used they also open crosswise. Most commercial corotating screw extruders are combinations of systems 4 and 6. Most commercial intermeshing counter-rotating screws normally correspond to system 11 and are used in conjunction with a single-screw discharge extruder.

3. *Flow Principles of Fully Intermeshing Self-Wiping Twin Screws*

Self-cleaning screws are desirable and often necessary to prevent material from adhering to the screw root. This can cause the material to degrade because of long residence time in the extruder. In both counter-rotating and corotating screws, self-cleaning is achieved either by an opposite roll-off motion or by a wiping motion. Self-cleaning of the screw elements is therefore achieved in different ways for each system, with varying degrees of effectiveness.

In counter-rotating screws, both screw shafts turn with an equal angular velocity in a counteracting rotation. In order to take in as much polymer as possible, the screws turn outward on top and inward at the bottom (see Fig. 2a). This means that both screws roll off one another in the intermeshing zone (wedge area) with a gliding velocity such that

$$-2\pi n(R_a - R_i) \leqslant v_{\text{gliding}} \leqslant 2\pi n(R_a - R_i) \qquad (1)$$

where R_a is the outer (crest) diameter, R_i is the root diameter and n is the screw speed (rpm). This gliding velocity is small compared to the circumferential speed of each screw. As a result, the shear velocity required to wipe the material boundary layers is low. The rolling-off mechanisms between the crest of one screw and the root of the other screw, and between the screw flanks, create a calender effect or squeezing of the material. In addition, material is also drawn into the roller gap and then squeezed on the surface of the barrel walls.

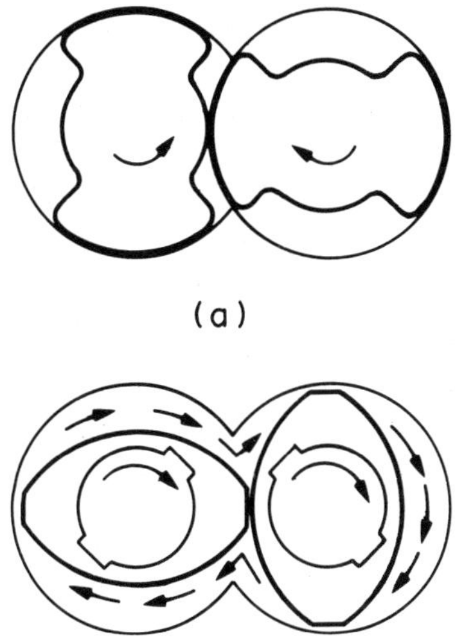

Figure 2
Rotational principles of twin screws: (a) counter-rotating screws and (b) corotating screws

In corotating screws, the screws turn in the same direction (Fig. 2b). The crest of one screw wipes the flank of the other screw with a tangentially oriented, constant relative velocity:

$$v_{\text{rel}} = 2\pi n(R_a + R_i) \qquad (2)$$

This relative velocity is large compared to the circumferential speed of each screw and results in a high shear velocity for wiping the material boundary layers. The calender effect and related squeezing of material to the surface does not occur. In consequence, a more efficient and uniform self-cleaning action results with corotating fully intermeshing screws. A more detailed analysis is contained in Eise et al. (1981).

4. *Melting Processes in Extruders*

After material has been fed into the extruder it is conveyed to a melting section. The melting process in all screw extruders occurs in two phases.

(a) One phase is the compaction of the fluffy polymer in order to minimize air pockets. In most cases the air escapes through a feed opening. Simultaneously, a melt film is formed on the barrel walls. The polymer fuses owing to the application of heat and pressure.

(b) The other phase is generation of melt by shearing the polymer and mixing the unmolten particles with the already formed melt. As a result, a large heat-exchange surface is generated between the solid particles and the melt.

Sections 4.1 and 4.2 describe the differences in the melting processes for twin-screw corotating and counter-rotating mechanisms.

4.1 Fully Intermeshing Self-Wiping Counter-Rotating Screws

Compaction of the fluffy polymer is achieved by several methods: (a) continuously reducing the screw channel, (b) using several melt zones with different screw pitches and number of flights, (c) continuously increasing the crest widths and (d) continuously decreasing outer and/or inner screw diameters.

In counter-rotating screws with tight clearances, a compression over several screw flights cannot be obtained nor will all of the air completely escape out of the individual chambers towards the feed opening. To compensate for this, counter-rotating screws are intentionally constructed with loose clearances. This counteracts the formation of closed chambers by the counter-rotating screws.

The second phase of the melting process is the actual generation of the melt. This involves (a) development of a melt film on the barrel wall, (b) scraping off the melt film by means of a pushing side of the screw flank and (c) collection of melt in front of the pushing flank. The rolling motion of the counter-rotating screw produces the following additional effects: (d) in the wedge intake, a melt whirlpool is created which draws in solid particles and softens them and (e) the plasticized mass is taken in by the calender gap and collected at the trailing flank. The calender effect of the counter-rotating screws does not always produce the desired results. One rotation of the screw causes the chamber volumes to be axially displaced by a distance equal to the screw pitch H. Each time one screw rolls off the other, only a fraction of the chamber volume is drawn into the calender gap. In spite of the high axial velocity, only a relatively low shear action is achieved in the calender gap. This low shear action is caused by the low gliding velocity (Eqn. 1) and the width of the calender gap. The small width of the calender gap results in a high shear velocity, but only allows a small amount of product to enter. A larger calender gap would draw in a greater amount of material, but the shear velocity would be greatly reduced.

Attempts to increase shear by increasing the screw speed result in displacement of the melting zone towards the discharge end of the screw. This can cause poor product homogeneity due to reduced residence time in the melting zone. Additional mixing devices such as interrupted flights result in better mixing and uniform heating of the product; however, the plasticizing process is only marginally improved and unmolten particles can often be observed in the discharge. Klenk (1969, 1977) describes how in a counter-rotating screw during the plasticizing process there is a final stage at which unmelted material is collected in the center of the chamber. The solids are suspended in the surrounding melt, which is available on all boundary walls. Melting of the solids can be achieved by heat transfer or shearing, but this is not without problems. The heat-transfer coefficient of polymer melts in twin-screw extruders is low (450–700 $W\,m^{-2}\,K^{-1}$). In addition, application of shearing stresses can only marginally influence the solids. This is because the zero shear stress point is at the center of the screw channel for axial flow as well as for cross flow. The intermeshing counter screw, therefore, pushes the solids forward on top of a melt pool resulting in severe lack of homogeneity.

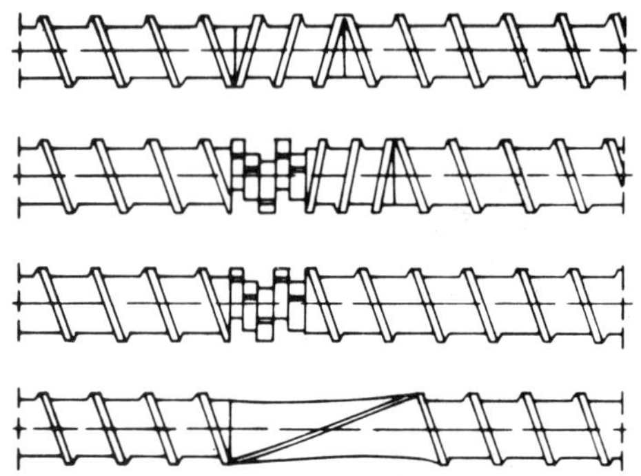

Figure 3
Corotating screws: screw elements for compressing and plasticizing polymer melts

4.2 Fully Intermeshing Self-Wiping Corotating Screws

In corotating screws, compression of the feed is produced by using smaller pitch (tight pitch), forward conveying screw elements or reversed pitch (left-handed) screw elements. These left-handed elements convey material upstream towards the feed section (see Fig. 3). In order to achieve a positive downstream conveyance, the left-handed screw restriction must be overcome by building up pressure in the upstream right-handed screw elements. Plasticizing results from the backup of material generated by either the tight-pitched right-handed elements or by the left-handed elements. The actual melting mechanisms involve (a) generation of the melt film on the barrel wall with sintering of the product, (b) scraping the film from the barrel wall by the pushing flank and (c) collection of a melt pool in front of the pushing flank.

By careful selection of pitch angle and screw speed, it is possible to shift the zero shear strain point of the axial flow components outside the screw channel. Therefore, a greater shear stress can be imposed on suspended solids particles than is possible in counter-rotating screws. In the wedge area, the material is picked up by the other screw and does not undergo a calendering effect. Shear stresses are applied between the flanks moving with constant relative velocity v_{rel}. As the screw pitch and screw speed are varied, the degree of fill in the screw channel can also be adjusted as required. Processing can therefore take place within a large shear speed velocity range.

Plasticizing can be achieved with special mixing and shear elements with large pitch (Fig. 3). With the utilization of these elements, a maximum self-wiping effect is maintained. Kneading disks have the same cross section as the corresponding screw elements. Disks with various widths and staggering angles can be arranged along the shaft to give different plasticizing capabilities. The function of a kneading block can be described by visualizing a single infinitely long disk in a closed-barrel bore. When the barrel is rotated relative to the disk, a wedge-shaped flow develops. Through variation of radial clearances, speed and number of flights (all of which influence the shape of the wedge), leak flow and shear effect can be varied in the shear gap between the kneading disk and barrel wall. A similar wedge effect is not possible in counter-rotating screws because of the slope of each screw flank.

The finite width of each kneading disk results in melt flow in the axial direction on both sides. By varying disk widths, the pressure profile in the cross section can also be influenced. If several disks are staggered at different angles in the axial direction, material flows axially on both sides into the neighboring channels. In this way, an excellent mixing effect is achieved which is not possible with fully intermeshing screws having axially closed channels.

Complete melting can be achieved without the self-handed screw restriction by using only a right-handed kneading block configuration. Left-handed screw elements with large pitch have a processing effect similar to kneading disks. Not only do they restrict the product flow, they also force material back over the shearing edges of the kneading disks. In fully intermeshing, self-wiping, corotating extruders it is possible to localize the plasticizing zone and to assure complete melting in short distances by use of these different elements.

See also: Processing: Classification

Bibliography

Eise K, Werner H, Herrmann H, Burkhardt U 1981 An analysis of twin-screw extruder mechanisms. *Adv. Plast. Technol.* 1(2): 18–32

Erdmenger R 1964 Mehrwellen-Schnecken in der Verfahrenstechnik. *Chemie.-Ing.-Techn.* 36: 175–85

Klenk K P 1969 Beitrag zur werkstoffgerechten Verarbeitung von PVC Hart-Pulver auf Ein-und Zweischnecken-Extrudern, Dissertation TH

Klenk K P 1977 Neue Entwicklungen im Bau von Zweischnecken-Extrudern, Thyssen

Tadmor Z, Gogos C G 1979 *Principles of Polymer Processing*. Wiley, New York, pp. 276–332

K. Eise and M. Abo
[Werner and Pfleiderer, Ramsey, New Jersey, USA]

Melt Spinning: Dynamics

The melt-spinning operation depends on the dynamics of the process and its interaction with heat transport. The development of understanding of the dynamics of polymer melt spinning is largely associated with the efforts of Ziabicki (1960), Kase and Matsuo (1965, 1967) and Kase et al. (1966). Reviews have been published by Ziabicki (1976) and White (1981).

1. Force Balance

A force balance for a fiber spinline will now be developed. Such a balance was first reported by Ziabicki (1960). The forces acting on the spinline include the applied tension F_L, the vertical downward force of gravity F_{grav}, the forces due to frictional air drag F_{drag} and forces due to surface tension F_s. If a balance is made between the takeup roll and an arbitrary position on the spinline (e.g., at position X_1, which has velocity v_1), the rheological force F_{rheo} acting to deform the descending fiber is the resultant of the applied forces and momentum fluxes (compare Ziabicki (1960, 1976) and Matovich and Pearson (1969)):

$$F_L + F_{grav} - F_{drag} - F_s - F_{rheo} = \rho Q(v_L - v_1) \quad (1)$$

where ρ is the density of the fiber, v_L is the takeup-roll velocity, v_1 is the fiber velocity and Q is the volumetric flow rate. These forces may be expressed as

$$\left.\begin{aligned} F_{grav} &= \int_{X_1}^{L} \rho g A \, dz \\ F_{drag} &= \int_{X_1}^{L} \sigma_f(\theta, z) R \, d\theta \, dz \\ F_s &= C\gamma \\ F_{rheo} &= \int_0^A \sigma_{11} \, dA \end{aligned}\right\} \quad (2)$$

where A is the cross-sectional area, z is the distance along the fiber, R is the radius of the fiber, θ is the radial angle from the fiber, σ_f is the air-drag shear stress, σ_{11} is the tensile stress within the filament, ρ is the density, g is the acceleration due to gravity, L is the position along the spinline where it enters the takeup roll, C is a constant determined by the filament shape and γ is surface tension.

The main unknown in the force balances of the fiber spinline is the air drag. This becomes increasingly important as spinline velocity increases. There have been both experimental and theoretical investigations of air drag since the 1960s. Early experimental investigations used blown air over a stationary filament. Sano and Orii (1968) of Mitsubishi Rayon and Hamana et al. (1969) and Matsui (1976) of Teijin determined air drag using a moving threadline. Their results, which show general agreement, indicate lower frictional stresses than the stationary filament results. The air drag is represented, as is traditional in engineering Newtonian fluid mechanics, through a friction factor c_f:

$$\sigma_f = c_f \rho v_1^2/2 \tag{3}$$

which is taken to depend on a Reynolds number Re defined in terms of a local fiber diameter d. The specific expressions are

$$c_f = 0.68\, Re^{-0.8} \tag{4a}$$

$$c_f = 0.37\, Re^{-0.61} \tag{4b}$$

Equation (4a) is derived from Sano and Orii (1968) and Eqn. (4b) is taken from Hamana et al. (1969) and Matsui (1976). In both cases

$$Re = dv_1\rho_a/\eta_a \tag{4c}$$

The subscript a refers to properties of air and η_a is the kinematic viscosity of air. The correlations involve fibers of diameter 30–100 μm and velocities of 200–1200 m min^{-1} in Sano and Orii (1968) and Hamana et al. (1969). This involves an Re 10–100 range. More recently, Matsui (1976) quotes additional data in the 1000–6000 m min^{-1} range which still correlates with Eqn. (4b).

The relative magnitudes of the different forces on the fiber spinline vary considerably with spinning conditions. If no tension is applied to the fiber spinline then gravitational forces F_{grav} alone act. If tension is applied then both F_L and F_{grav} must be considered. As the drawdown is increased, F_L will increasingly dominate F_{grav}. The higher velocities involved in the increased drawdown rates give rise to increased frictional drag and initial forces. Since F_{grav} acts in the same direction as F_L and $F_{drag} + \rho Q(v_L - v_1)$ must be considered, it is possible to identify several regimes depending on drawdown velocity (v denotes fiber velocity).

(a) Regime 1 (gravitational spinning)

$$F_{rheo} = F_{grav} \tag{5a}$$

(b) Regime 2 (low drawdown)

$$F_{rheo} = F_L + F_{grav} \tag{5b}$$

(c) Regime 3 (moderate drawdown)

$$F_{rheo} = F_L \tag{5c}$$

(d) Regime 4 (high drawdown and take up velocities)

$$F_{rheo} = F_L - F_{drag} - \rho Q(v_L - v_1) \tag{5d}$$

2. Heat Transfer

The simplest consideration of energy conservation on a moving threadline involves a one-dimensional balance which presumes a constant temperature across the fiber cross section:

$$\rho Qc\frac{dT}{dx_1} = -C[h(T - T_{surr}) + \varepsilon\kappa(T^4 - T_{surr}^4)] \tag{6}$$

where the first term on the right-hand side involves convective heat transfer and the second term involves radiation. T is the thermodynamic temperature, T_{surr} is the temperature of the surroundings, h is the heat-transfer coefficient, Q is the volumetric flow rate, c is the heat capacity, C is the circumference (which is πd for a cylindrical cross section), ε is the emissivity and κ is the Stefan–Boltzmann coefficient. Expressions of this type were first considered in the context of melt spinning by Kase and Matsuo (1965).

The generally accepted expression for the heat-transfer coefficient to a moving fiber spinline is that developed and used by Kase and Matsuo (1965) of Toyobo:

$$\frac{hd}{k_a} = 0.42\left(\frac{dv_1\rho_a}{\eta_a}\right)^{1/3} \tag{7}$$

where $dv_1\rho_3/\eta_a$ is the Reynolds number of the air relative to a stationary fiber and hd/k_a is known as the Nusselt number (k_a is the heat conductivity of air and ρ_a is the density of air). This result was extended to heat transfer in the presence of crossflows. Kase and Matsuo (1965) express their result as

$$hd/k_a = 0.42(dv_1\rho_a/\eta_a)^{1/3}(1 + K) \tag{8}$$

where K is determined, from the data of Kase and Matsuo, by the expression

$$1+K \simeq [1+8v_y/v_1)^2]^{1/6} \quad (9)$$

where v_y is the crossflow velocity of air.

3. Modelling

The simplest consistent theory of the dynamics of nonisothermal melt spinning is that by Kase and Matsuo (1965). They combine a one-dimensional Newtonian fluid model for the spinline with only applied takeup forces acting with a one-dimensional energy balance with forced convection heat transfer. It is implicit in their work that the temperature dependence of rheological properties masks all complexities of such responses. Basically,

$$A\chi(T)\frac{dv_1}{dx_1}=F_L \quad (10a)$$

$$Qc\frac{dT}{dx}=-h\pi d(T-T_{surr}) \quad (10b)$$

where $\chi(T)$ denotes an apparent tensile viscosity.

Kase and Matsuo (1965, 1967) describe the predictions of this set of equations for various applications. Increasing spinneret diameter with no change in takeup speed or final fiber diameter results in slower drawdown and cooling and increases the drawdown force. Increasing the takeup speed while maintaining a constant fiber diameter has a similar effect but changes F_L very little. If the air temperature is lowered, the filaments thin and cool more rapidly and the tension F_L increases. A similar result is obtained if air is blown perpendicularly over the descending threadline. On the other hand, raising the spinneret temperature T_0 without changes in other spinning conditions makes the spinline hotter and decreases the filament tension.

Hamana et al. (1969) presented a more concise formulation of the nonisothermal melt-spinning problem which includes air drag, gravitation and momentum flux forces. However, they presume, as did Kase and Matsuo, that the melt behaves as a Newtonian fluid. The basic equations of Hamana et al. are

$$A(T)\frac{dv_1}{dx_1}=F_L+g\int_{x_1}^{L}\rho A\,dx_1-\int_{x_1}^{L}\pi d\sigma_f\,dx_1 \quad (11a)$$

$$Gc\frac{dT}{dx_1}=-hd(T-T_{surr}) \quad (11b)$$

where σ_f is given by Eqns. (3) and (4b). They present comparisons between theory and experiment for various cases. While their work is a significant extension, the extent of their calculations is not clear and it is not evident how these differ from the familiar, simpler Kase–Matsuo model.

More recent modelling studies using complex rheological models have been reviewed by White (1981). Most notable are studies by Fisher and Denn (1976) which attempt to include the influence of viscoelastic characterization.

4. Fiber Nonuniformity

Fibers generally exhibit longitudinal variations of diameter. Sometimes these are random small-amplitude fluctuations due to disturbances in the system (Kase and Matsuo 1967). In other cases, large-amplitude periodic diameter variations are observed. These are an instability phenomenon generally termed draw resonance (Ziabicki 1960, 1976, Kase et al. 1966, Sano and Orii 1968, Kase 1974). The phenomenon occurs most readily with isothermal spinlines and with materials which are deformation rate softening in elongational flow and readily exhibit necking. Draw-resonance instability has been analyzed by Kase et al. (1966), Pearson and Matovich (1969) and Fisher and Denn (1976). The influence of polymer structure on this instability has been investigated by White and Ide (1978) and more pertinently by White and Yamane (1985, 1987).

See also: Fibers: Drawing; Fibers: Formation and Processing; Processing: Rheological Properties and Fluid Mechanics

Bibliography

Fisher R J, Denn M M 1976 A theory of isothermal melt spinning and draw resonance. *AIChE J.* 22: 236–46

Hamana I, Matsui M, Kato S 1969 Filament formation in melt spinning. *Melliand Textilber. Int.* 50: 382–8

Kase S 1974 Studies on melt spinning IV. On the stability of melt spinning. *J. Appl. Polym. Sci.* 18: 3279–304

Kase S, Matsuo T 1965 Studies on melt spinning I. Fundamental equations on the dynamics of melt spinning. *J. Polym. Sci. A* 3: 2541–54

Kase S, Matsuo T 1967 Studies on melt spinning II. Steady-state and transient solutions of fundamental equations compared with experimental results. *J. Appl. Polym. Sci.* 11: 251–87

Kase S, Matsuo T, Yoshimoto Y 1966 Theoretical analysis of melt spinning $\rho > 2$ surging phenomena in extrusion casting of plastic films. *Sen'i Kikai Gakkaishi* 19: T63–72

Matovich M A, Pearson J R A 1969 Spinning a molten threadline: Steady-state isothermal viscous flows. *Ind. Eng. Chem., Fundam.* 8: 512–20

Matsui M 1976 Air drag on a continuous filament in melt spinning. *Trans. Soc. Rheol.* 20: 465–73

Pearson J R A, Matovich M A 1969 Spinning a molten threadline: Stability. *Ind. Eng. Chem., Fundam.* 8: 605–9

Sano Y, Orii K 1968 Drag coefficients of filaments in air flow during the spinning. *Sen'i Gakkaishi* 24: 212–18
White J L 1981 Dynamics, heat transfer and rheological aspects of melt spinning: A critical review. *Polym. Eng. Rev.* 1: 297–362
White J L, Ide Y 1978 Instabilities and failure in elongational flow and melt spinning of fibers. *J. Appl. Polym. Sci.* 22: 3057–74
White J L, Yamane H 1985 A collaborative study of the rheological properties of unstable melt spinning characteristics of linear and branded polyethylene terephthalate. *Pure Appl. Chem.* 57: 1441–52
White J L, Yamane H 1987 A collaboraive study of the instability of extrusion, melt spinning and tubular film extrusion of same high-, low and linear low density polyethylene samples. *Pure Appl. Chem.* 59: 193–216
Yamane H, White J L 1983 Extrusion and melt spinning characteristics of thermally degraded polypropylene. *Polym. Eng. Sci.* 23: 516–20
Ziabicki A 1960 Mechanical aspects of fibre spinning process in molten polymers III. Tensile force and stress. *Kolloid-Z. Z. Polym.* 175: 14–27
Ziabicki A 1976 *Fundamentals of Fibre Formation.* Wiley, London

J. L. White
[University of Akron, Akron, Ohio, USA]

Melt Spinning of Fibers: Structure Development

Since the first experiment on melt spinning by Carothers and Hill (1932) it has been known that changes occur in the structure of fibers as a function of melt-spinning conditions. However, it was not until the 1960s that detailed studies of structure development in the melt-spinning process appeared in the literature. These investigations considered both on-line structural changes in the fiber spinline and changes in fiber characteristics with spinning conditions.

1. Theoretical Aspects

The structural changes considerd here involve the development of crystallinity, molecular orientation and crystalline morphology (i.e., the organization or morphology of the crystalline structure). Molecular orientation is often represented in terms of the Hermans orientation factor, which may be expressed as

$$f_H = \tfrac{1}{2}(3\,\overline{\cos^2\phi_{1c}} - 1) \tag{1}$$

where ϕ_{1c} is the angle between the polymer chain axis and the fiber axis (Hermans et al. 1946).

For crystalline polymers this may b e generalized to include all three crystallographic axes by defining Hermans–Stein orientation factors:

$$f_a = \tfrac{1}{2}(3\,\overline{\cos^2\phi_{1a}} - 1) \tag{2a}$$

$$f_b = \tfrac{1}{2}(3\,\overline{\cos^2\phi_{1b}} - 1) \tag{2b}$$

$$f_c = \tfrac{1}{2}(3\,\overline{\cos^2\phi_{1c}} - 1) \tag{2c}$$

where a, b and c represent crystallographic axes and ϕ_{1a}, ϕ_{1b} and ϕ_{1c} represent angles between the fiber axis and these crystallographic axes (Stein 1958). Generally, the c axis is the polymer chain axis. For an isotropic material, the f_j are zero, whereas if the orientation is perpendicular to the fiber axis, f_j is -0.5.

In semicrystalline polymers, $\cos^2\phi_{ij}$ in crystalline regions may be determined by wide-angle x-ray diffraction or scattering (WAXS). However, such polymers possess amorphous as well as crystalline regions and the classical manner of determining amorphous orientation in fibers is to use birefringence combined with WAXS measurements (Hermans et al. 1948, Stein and Norris 1956). The birefringence is given by

$$\Delta n = X f_{\mathrm{Hcryst}} \Delta^{\circ}_{\mathrm{cryst}} + (1 - X) f_{\mathrm{Hamorph}} \Delta^{\circ}_{\mathrm{amorph}} + \Delta n_{\mathrm{form}} \tag{3}$$

where X is the fraction crystallinity, f_{Hcryst} and f_{Hamorph} are the Hermans orientation factors of the crystalline and amorphous phases, $\Delta^{\circ}_{\mathrm{cryst}}$ and $\Delta^{\circ}_{\mathrm{amorph}}$ are the intrinsic birefringences of the crystalline and amorphous regions and Δn_{form} is the form birefringence (which is generally neglected).

2. Polystyrene

There have been numerous studies of melt spinning of fibers of polystyrene that show that it vitrifies into a glass and does not crystallize. High levels of negative birefringence are found in the melt-spun filaments. The magnitude of the birefringence increases with drawdown of the fiber and the level of the birefringence is uniquely and linearly dependent on spinline stress, but independent of melt temperature and polymer molecular weight (Oda et al. 1978). This relationship (see Fig. 1) can be expressed as

$$\Delta n = C\sigma \tag{4}$$

where the coefficient C relates birefringence and stress and is the same as the stress optical constant for molten polystyrene. From Eqns. (3) and (4),

$$f_H = (C/\Delta^{\circ})\sigma \tag{5}$$

Equivalent birefringence–stress correlations are found for rubber-modified polystyrene (Omotoso et al. 1980). Birefringence profiles have also been

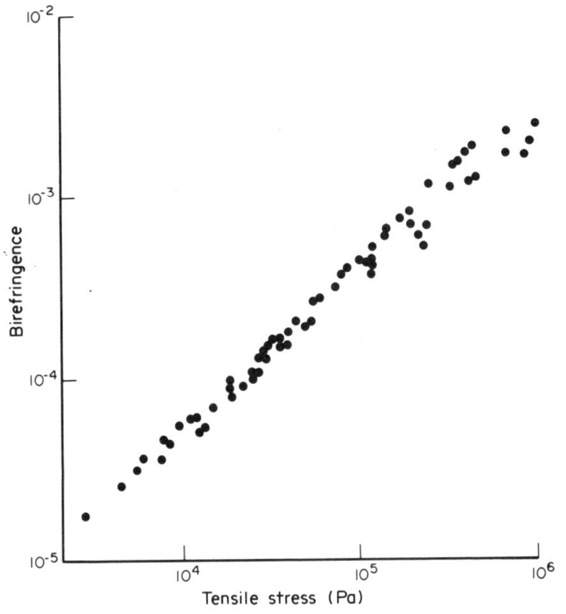

Figure 1
Birefringence of various polystyrene filaments formed in elongational flow as a function of stress

reported in polystyrene filaments, with the maximum values occurring at the outer radii (Andrews 1954).

3. Polyethylene

There have been numerous investigations of structure development in the melt spinning of polyethylene fiber. On-line investigations of x-ray diffraction patterns have shown that polyethylene crystallizes in the spinline to form a well-defined orthorhombic unit cell with the *c* axis being the polyethylene chain axis (Katayama et al. 1968, Dees and Spruiell 1974). The crystallization temperature is depressed below the equilibrium value with the extent of depression increasing with drawdown. The rates of crystallization are significantly enhanced by drawdown.

The crystalline orientation of the melt-spun fibers has been expressed in terms of the Hermans–Stein orientation factors (see Fig. 2) by various investigators (Abbott and White 1973, Kitao et al. 1973, Dees and Spruiell 1974, White et al. 1974). With increasing drawdown f_b rapidly becomes negative, reaching a value of about −0.4. In contrast, f_c gradually increases from zero and f_a increases from zero to a maximum of about 0.25, then decreases and becomes negative. These values indicate that the first step in the development of orientation is the *b* axis becoming perpendicular to the fiber axis. The *c* axis gradually increases with drawdown, and the *a* axis first tends to become parallel and then reverses this tendency and becomes perpendicular to the fiber axis. The crystalline orientation factors have been found to depend on the drawdown stress rather than the kinematics (Dees and Spruiell 1974).

Studies of small-angle x-ray scattering (SAXS) indicate lateral order in melt-spun fibers. This is considered to indicate a row structure consisting of folded-chain lamellae (Dees and Spruiell 1974). The orientation factors are consistent with this morphology if the lamellae are considered to twist and the average level of twisting is considered to decrease with increased take-up stress. The *b* axis corresponds to the growth direction of the lamellae (see Fig. 3).

4. Polypropylene

On-line studies of the crystallization of polypropylene in the fiber spinline have been reported by various investigators (Katayama et al. 1968, Spruiell and White 1975, Ishizuka and Koyama 1976, 1977, Nadella et al. 1977). The crystallization temperature is greatly depressed (as much as 75 °C) below the equilibrium melting temperature and the rate of crystallization is significantly enhanced by the melt-spinning operation. Polypropylene is found to crystallize into a well-defined monoclinic unit cell, whereas quenching of polypropylene fibers in cold water rather than air leads to a poorly defined smectic structure (Sheehan and Cole 1964, Nadella et al. 1977).

The development of crystalline orientation in polypropylene has been investigated by various researchers. It has been found that polypropylene

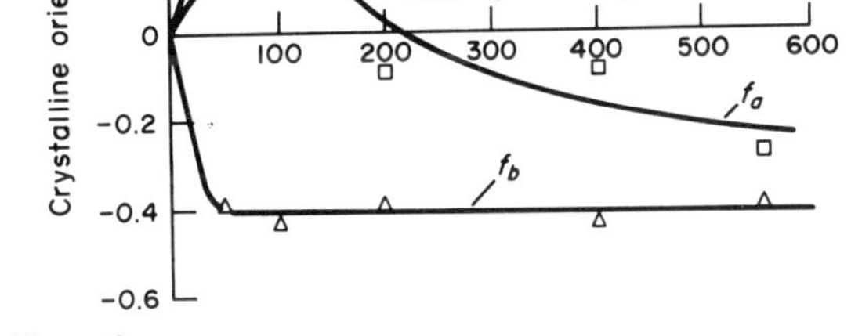

Figure 2
Variation of crystalline orientation factors with take-up velocity for linear polyethylene (measured at 20 ± 3 °C and a polymer flow rate of 1.93 ± 0.02 g min^{-1})

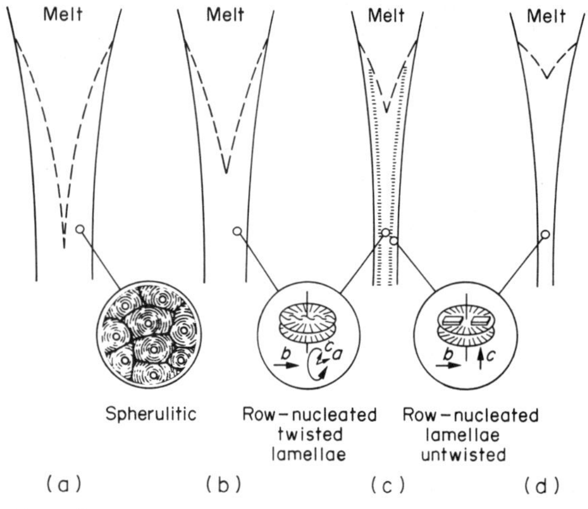

Figure 3
Model for the changes in morphology of linear polyethylene: (a) very low take-up velocity; (b) low take-up velocity; (c) medium take-up velocity; and (d) high take-up velocity

crystallized under stress exhibits bimodal orientation (i.e., regions of two differing molecular orientations, one tending to be parallel to the c axis and the other perpendicular to the stress) (Katayama et al. 1968, Nadella et al. 1977). Interpretation of fiber orientation in terms of Hermans–Stein orientation factors has shown that f_b rapidly becomes negative with increasing drawdown, whereas f_c slowly increases from zero and f_a first becomes positive, goes through a maximum, then decreases and becomes negative (Kitao et al. 1973, Spruiell and White 1975, Nadella et al. 1977, Shimizu et al. 1977). This corresponds to the b axis first rapidly becoming perpendicular to the fiber axis and the c axis increasing slowly from zero. The a-axis orientation tends to become first parallel to the fiber axis and then perpendicular to it. The behavior resembles that found with polyethylene fibers. It has been found that the crystalline orientation factors and birefringence are a unique function of the spinline stress (Shimizu and Shimazaki 1973, Spruiell and White 1975, Nadella et al. 1977).

Most studies of melt-spinning polypropylene have involved take-up velocities of order $\leqslant 1000\ \mathrm{m\ min^{-1}}$. Shimizu et al. (1977) present studies for speeds up to $7000\ \mathrm{m\ min^{-1}}$.

Lateral order is also found in melt-spun polypropylene fibers by SAXS measurements (Nadella et al. 1977). This is also interpreted in terms of lamellar structures. The bimodal orientation may be explained in terms of a secondary crystallization where the chains form crystallites between the primary lamellae.

5. *Nylon 6*

At low take-up velocities (stresses, drawdown ratios) on-line WAXS indicates that Nylon 6 does not appear to crystallize on the spinline (Bankar et al. 1977), though evidence of a poorly defined smectic structure has been hypothesized on the basis of the position of the single broad peak (Hirami and Tanimura 1981). In these cases, the fiber appears to crystallize on the bobbin. Placing a heated tube in the spinline at temperatures over 80 °C was found, using on-line birefringences, to induce crystallization (Ishibashi and Ishii 1976). High-speed melt spinning also seems to induce spinline crystallization (Shimizu et al. 1981a). Crystallization generally occurs into a smectic pseudohexagonal crystalline form related to the γ structure in polyamides (White and Spruiell 1978).

Orientation increases with increasing take-up velocity, drawdown and stress. Bankar et al. (1977) have interpreted this in terms of Hermans–Stein orientation factors (i.e., f_c increases and f_a and f_b decrease with drawdown), and their SAXS measurements on crystalline spun fibers show a meridional two-point pattern suggesting the form of lamellar row structures.

6. *Nylon 66*

There has been relatively little study of the melt spinning of Nylon 66 in the literature. It appears that Nylon 66 crystallizes in the spinline to form an α-triclinic structure (Danford et al. 1978). Rapid quenching can retain the amorphous form or lead to a nematic structure which ages (Chappel et al. 1964, Starkweather et al. 1963).

Crystalline orientation increases with drawdown. Danford et al. (1978) have interpreted this in terms of Hermans–Stein orientation factors; they also showed that a poorly defined two-spot meridional pattern is formed in the melt-spun fiber. This suggests the formation of a folded-chain lamellar structure similar to that presumed to exist in polyethylene fibers.

7. *Poly(Ethylene Terephthalate)*

At low drawdown ratio, poly(ethylene terephthalate) vitrifies into a glass. At higher speeds ($4000\ \mathrm{m\ min^{-1}}$), the polyester crystallizes in the threadline into a triclinic lattice structure. Shimizu et al. (1981b) have detected skin core-structures in high-speed melt-spun fibers. The skin regions have higher orientation than the core and SAXS and density measurements suggest that the fibers also contain small voids.

At low drawdown velocities where a glass is formed, the birefringence is found to be proportional to the stress, as in the case of polystyrene

(Hamana et al. 1969). At higher drawdowns and velocities where the melt crystallizes, high levels of crystalline orientation develop. Hermans crystalline orientation factors of 0.96 have been found in high-speed melt-spun fibers (Heuvel and Huisman 1978).

8. Recent Developments

A new phase in the structure development area of polymer fiber technology has been the production and utilization of high-modulus high-strength organic fibers (see *High-Modulus High-Strength Organic Fibers*). Such fibers, including aramid and polyethylene fibers, have high specific tensile modulus and high tensile strength when compared with metals and ceramics.

See also: Fibers: Formation and Processing; Processing: Classification; Processing: Structure Development

Bibliography

Abbott L E, White J L 1973 Melt spinning of high density and low density polyethylene fibers: Development of orientation and crystallinity and mechanical properties of spun fiber. *Appl. Polym. Symp.* 20: 247–68

Andrews R D 1954 Measurement of orientation in polystyrene filaments by means of double refraction. *J. Appl. Phys.* 25: 1223–31

Bankar V G, Spruiell J E, White J L 1977 Melt spinning of nylon-6: Structure development and mechanical properties of as-spun filaments. *J. Appl. Polym. Sci.* 21: 2341–58

Carothers W H, Hill J W 1932 Studies of polymerisation and ring formation XV: Artificial fibers from synthetic linear condensation superpolymers. *J. Am. Chem. Soc.* 54: 1579–87

Chappel F P, Culpin M F, Gosden R G, Tranter T C 1964 Some crystallization phenomena in rapidly quenched 6.6 nylon. *J. Appl. Chem.* 14: 12–19

Danford M D, Spruiell J E, White J L 1978 Structure development in the melt spinning of nylon-66 fibers and comparison to nylon-6. *J. Appl. Polym. Sci.* 22: 3351–61

Dees J R, Spruiell J E 1974 Structure development during melt spinning of linear polyethylene fibers. *J. Appl. Polym. Sci.* 18: 1053–78

Hamana I, Matsui M, Kato S 1969 Der Verlauf der Fadenbildung beim Schmelzspinnen. *Melliand Textilber.* 4: 382–8; 499–503

Hermans J J, Hermans P H, Vermaas D, Weidinger A 1946 Quantitative evaluation of orientation in cellulose fibers from the x-ray fiber diagram. *Recl. Trav. Chim. Pays-Bas* 65: 427–47

Hermans P H, Hermans J J, Vermaas D, Weidinger A 1948 Deformation mechanism of cellulose gels, IV: General relationship between orientation of the crystalline and that of the amorphous portion. *J. Polym. Sci.* 3: 1–8

Heuvel H M, Huisman R 1978 Effect of winding speed on the physical structure of as-spun poly(ethylene terephthalate) fibers, including orientation induced crystallization. *J. Appl. Polym. Sci.* 22: 2229–43

Hirami M, Tanimura A 1981 Structure development in the melt spinning of nylon-6. *J. Macromol. Sci. Phys.* 19: 205–10

Ishibashi T, Ishii T 1976 Studies on the melt spinning of nylon-6, II: Effect of heating the threadline upon orientation and crystallization. *J. Appl. Polym. Sci.* 20: 335–44

Ishizuka O, Koyama K 1976 Crystal growth in a running filament in melt spinning. *Sen'i Gakkaishi* 32: T43–9

Ishizuka O, Koyama K 1977 Crystallization of running filament in melt spinning of polypropylene. *Polymer* 18: 913–18

Katayama K, Amano T, Nakamura K 1968 Structural formation during melt spinning process. *Kolloid-Z. Z. Polym.* 226: 25–134

Kitao T, Ohya S, Furukawa J, Yamashita S 1973 Orientation of polymer molecules during melt spinning, II: Orientation of crystals in as-spun crystalline fibers. *J. Polym. Sci., Polym. Phys. Ed.* 11: 1091–109

Nadella H P, Henson H M, Spruiell J E, White J L 1977 Melt spinning of isotactic polypropylene: Structure development and relationship to mechanical properties. *J. Appl. Polym. Sci.* 21: 3003–22

Oda K, White J L, Clark E S 1978 Influence of melt deformation history on orientation in vitrified polymers. *Polym. Eng. Sci.* 18: 53–9

Omotoso M A A, White J L, Fellers J F 1980 Melt spinning of high impact polystyrene: Rubber morphology variations, orientation development, and mechanical properties. *J. Appl. Polym. Sci.* 25: 1573–85

Sheehan W C, Cole T B 1964 Production of super-tenacity polypropylene filaments. *J. Appl. Polym. Sci.* 8: 2359–88

Shimizu J, Okui N, Kaneko A, Toriumi K 1978 High speed melt spinning of polyethylene terephthalate (PET): Effects of melt draw ratio and flow rate. *Sen'i Gakkaishi* 34: T54–71

Shimizu J, Okui N, Kikutani T 1981b High speed melt spinning of poly(ethylene terephthate): Radial variation across fibers. *Sen'i Gakkaishi* 37: T135–42

Shimizu J, Okui N, Kikutani T, Ono A, Takaku A 1981a High speed melt spinning of nylon-6. *Sen'i Gakkaishi* 37: T143–52

Shimizu J, Shimazaki K 1973 The estimation of birefringence in polypropylene melt spinning process. *Sen'i Gakkaishi* 29: T205–10

Shimizu J, Toriumi K, Tmai K 1977 High speed melt spinning of polyester filaments—Effect of spinning velocity on the properties and molecular orientation. *Sen'i Gakkaishi* 33: T208–14

Spruiell J E, White J L 1975 Structure development during polymer processing: Studies of the melt spinning of polyethylene and polypropylene fibers. *Polym. Eng. Sci.* 15: 660–7

Starkweather H N, Whitney J F, Johnson D R 1963 Crystalline order in nylon-66. *J. Polym. Sci.* 1: 715–23

Stein R S 1958 The x-ray diffraction birefringence and infrared dichroism of stretched polyethylene, II: Generalized uniaxial crystal orientation. *J. Polym. Sci.* 31: 327–34

Stein R S, Norris F H 1956 The x-ray diffraction, birefringence and infra-red dichroism of stretched polyethylene. *J. Polym. Sci.* 21: 381–95

White J L, Dharod K C, Clark E S 1974 Interaction of melt spinning and drawing variables on the crystalline morphology and mechanical properties of high density and low density polyethylene fiber. *J. Appl. Polym. Sci.* 18: 2359–68

White J L, Spruiell J E 1978 Structure development in fiber formation in polyamides and polypeptides. *Appl. Polym. Symp.* 33: 91–127

J. L. White
[University of Akron, Akron, Ohio, USA]

Mixing: Basic Concepts

Mixing is a limiting step in many polymer processes. In simple plasticating extrusion, mixing is required to achieve thermal uniformity. In systems such as polyurethane reaction injection molding, the difficulty of mixing the components limits the choice of materials. In the compounding of materials, mixing of pigments, fillers, extenders, stabilizers, lubricants and other minor components with polymer is the entire purpose of the operation.

Mixing is the intermingling by mechanical action of differentiated components within a continuum. Two physical phenomena are involved. When one or more of the components has a cohesive nature, such as agglomerated particles or droplets with surface tension, mechanical stress must be applied to reduce the size of the particles or droplets. This is called intensive mixing or dispersion. Intensive mixing is governed by the history of fluid mechanical stresses in a mixture. On the other hand, when no such barriers exist, as in the mixing of hot and cold segments of the same material or in mixing of thermodynamically miscible fluids, mixing is determined simply by the history of deformation imparted to the fluid—the strain. This is called extensive mixing or blending. In systems where intensive mixing is required to break up a phase, extensive mixing is also required to distribute the broken phase throughout the volumes.

1. Measure of Mix

A mixture is a spatial distribution of variation of concentration of components, just as electronic noise is a temporal distribution of variations in voltage or current. The complete analysis of mixture would require three-dimensional analysis of the variation of each component. The data acquisition difficulties associated with mix measurement have precluded the development of sophisticated data analysis.

The degree to which a mixture approaches uniformity has been variously termed (e.g., the degree of mix, the mixedness or the goodness of mix) but the number of measures exceeds the variety of the nomenclature.

The simplest direct method of assessing mix is to look at a mixture through a microscope and measure the layer thickness. Laminar flow inherently produces layered mixtures. Since the thickness of striations is seldom uniform across a sample, distributions of thickness determined by correlation functions are more general expressions of the state of mix.

Normally, mixing is monitored only by its effect on the final properties of a material such as color, solvent resistance or strength. When measured directly, dispersion is determined by direct observation of particle sizes, such as in the American Society for Testing and Materials test for carbon black dispersion (ASTM D2663). Extensive mixing is usually characterized by striation thickness or distributions of thickness as correlograms. These measures have arisen from the historic difficulty of taking the many measurements required for complete descriptions of mixtures. Advances in automated data acquisition and, particularly, image analysis promise to remove this constraint.

2. Principles of Extensive Mixing

The first fundamental consideration of the blending of highly viscous liquids was made by Spencer and Wiley (1951). They used the total interfacial area between the two components as an appropriate measure of degree of mix. They quantitatively related interfacial area growth in plane strain flow with fluid motion and applied this relation to very simpling mixing geometries. The growth of an interface in a fluid undergoing shear was found to follow the relation

$$A/A_0 = s \cos \alpha \tag{1}$$

where s is the shear strain, A_0 is the initial area and α is the angle defining the orientation of the surface to the shear strain. If this expression is averaged over all possible orientations, it becomes

$$A/A_0 = s/2 \tag{2}$$

This large-shear expression has been applied to mixing analysis by Mohr et al. (1957). This expression, however, is applicable only to large simple shear, a very inefficient mixing flow. More recently, Erwin (1978), Cheng (1979) and Ottino et al. (1979) have shown that the effect of deformation on orientation in simple shear aligns mixtures in uniquely unfavorable directions. Interruption of this orientation greatly enhances mixing. The upper bound for plane strain mixing shows an exponential rate with strain when a mixture is aligned with the tensile axis of a plane strain stretching flow. Under these conditions, mixing proceeds as

$$A/A_0 = \exp(s/2) \tag{3}$$

This shows why mixers are nearly always messy devices. Simple devices usually impart simple shear, which is a very inefficient mixing flow. If a mixer is irregular in geometry, the entrance flows give extensional character; also, the orientation imparted to the fluid by shear is disrupted. Both of these effects enhance mixing efficiency. Unfortunately, these irregular geometries are difficult to analyze. This is why no efficient mixer has been subjected to analysis from first principles.

While extensive mixing is always a simple function of strain, strain is not always a simple function of stress. When the components of a mixture have differing constitutive relationships, the fluid mechanics become dependent on the mixing process. One simple model of this process assumes that the stress in the components is equal. This causes the higher-viscosity phase to exhibit a proportionally lower strain rate. The mixing process is assumed to be limited by this lower strain rate. In practice, such systems are dramatically affected by the structure of the mixture. For example, a continuous high-viscosity phase with a discontinuous low-viscosity phase will mix much more readily than the inverse structure. The study of mixing of fluids of radically differing behavior is not yet sufficiently well developed to permit exact analysis.

When one or more of the mixture components is a soluble, low-molecular-weight material, molecular diffusion often plays an important role. Since time is important in diffusion, and diffusion takes place simultaneously with mechanical mixing, the mixing rate must be calculated based on strain rate. This can then be integrated over time to give mixer performance. A theoretical basis for this procedure has been established by Ottino et al. (1979).

3. *Principles of Intensive Mixing*

In contrast to blending, which is caused by strain, dispersion is caused by stress. When a mixture contains agglomerated particles or droplets bound by surface tension, these structures must be broken if the material they contain is to be distributed on a finer scale. The forces required to break the bound phases are provided by fluid mechanical stress. This stress depends on the flow, the fluid and the particle geometry.

For example, if two spherical particles of radii r_1 and r_2 are attached, the maximum force available to break them in steady shear flow of Newtonian liquid is

$$F_{\max} = 3\pi\mu\gamma r_1 r_2 \tag{4}$$

where μ is the viscosity and γ is the strain rate. A similar expression for elongational flow shows that at the same strain rate, elongational flow gives twice the force.

The dispersive force equation can be rewritten in terms of force F as a function of power per unit volume p dissipated in shear flow:

$$F_{\max} = 3\pi(\mu p r_1 r_2)^{1/2} \tag{5}$$

This expression shows that less power is required to mix dispersively mixtures of higher viscosity, a phenomenon commonly observed in practice. This is why bakers knead dough in a stiff form rather than beat a loose batter to break up agglomerations of flour. For either flow, and in general for any flow, dispersive forces are maximized by maximizing the product of the fluid viscosity and the strain rate, the fluid stress.

Since high energy dissipation rates are required to achieve dispersion, requirements for subjecting a large mixture body to a dispersive mechanical stress all at once are prohibitive. Dispersive mixers therefore have small regions of intense deformation, such as the nip between the rolls in a two-roll mill, through which the mixture is passed. In a dispersive mixer the bulk of the fluid is more or less at rest while a small quantity is undergoing dispersion. The irregular geometry required to accomplish this often makes machines designed as dispersive mixers also effective as blending mixers.

The requirement of small regions of intense strain means that the corresponding entrance regions have elongational characteristics. It has been shown that elongational deformations have kinematic advantages in dispersing mixtures. Flow visualization studies show that agglomerates tend to rotate in shearing flows, being kneaded but not separated: whereas in elongational flows, the fractured agglomerates are grossly separated.

Although careful experiments on the relationship between strain and dispersion have been conducted only on isolated droplets or agglomerates, several significant results are known. Stretching flows are much more effective in dispensing droplets than shearing flows. For example, if an isolated droplet has a viscosity more than 3.5 times that of its surroundings, the surface tension forces cannot be overcome in simple shear at any stress. Droplet fracture is much less dependent on droplet viscosity when conducted in stretching flows and is possible throughout the range of visosities. In more realistic concentrated dispersions of droplets, the interactions between droplets play an important but as yet undetermined role.

4. *Types of Mixer*

The discussion thus far shows that the basic requirement for accomplishing mixing is the deformation of the mixture. On this basis, it might be wondered why mixers are necessary at all, since deformation occurs in all flows. Pumps and pipes deform fluids.

However, two considerations of pipe flow reveal the reasons for the existence of mixers:

(a) pipe flow imparts simple shear to fluid, which was shown in Sect. 2 to be a uniquely inefficient flow for extensive mixing; and

(b) the maximum stress in pipe flow is only 1.5 times the volume-averaged shear stress and no small region of high stress is developed, so the maximum dispersive stress is not high.

To create mixing, additional complexity must be incorporated into the fluid handling system. For extensive mixing, the inefficient orientation of the simple shear must be disrupted. Fluid turbulence accomplishes this in lower-viscosity fluids but is not practical in polymer melts. An agitator of irregular geometry can be placed within a pipe, creating a conventional continuous mixer. Deflecting blades, plates or channels can be used, creating a "motionless" mixer. Both modifications disrupt the inefficient orientation and improve mixing by orders of magnitude. They also dissipate more energy in the fluid, which enhances mixing.

If dispersion is required, a small region of intense deformation must exist. Such a small region is created by passing the fluid through a small passage. The passage may have walls which move relative to each other, such as are created by a closely fitting impeller or a single passive orifice which can impart large stresses from the large pressure drop across it.

In the case where the mixing energy is supplied by moving mechanical components rather than the action of an externally applied pressure gradient, it is possible to operate the mixing chamber with otherwise stagnant fluid so that a discontinuous or batch mixer is created. While this is distant from simple pipe flow, it is the most familiar daily experience.

See also: Processing: Classification

Bibliography

Cheng D C-H 1979 The role of extensional flows in fluid mixing and dispersion. *Proc. 3rd European Conf. Mixing,* Vol. 1. British Hydrodynamics Research Association, Cranfield, UK, pp. 73–104

Danckwerts P V 1952 The definition and measurement of some characteristics of mixtures. *Appl. Sci. Res. A* 3: 279–96

Erwin L 1978 Theory of laminar mixing. *Polym. Eng. Sci.* 18: 1044–8

Koa S V, Mason S G 1975 Dispersion of particles by shear. *Nature (London)* 253: 619–21

McKelvey J M 1962 *Polymer Processing*. Wiley, New York

Matthews G 1982 *Polymer Mixing Technology*. Applied Science, London

Mohr W D, Saxton R L, Jepson C H 1957 Mixing in laminar flow systems. *Ind. Eng. Chem.* 49: 1855–6

Ottino J M, Ranz W E, Macosko C W 1979 A lamellar model for analysis of liquid–liquid mixing. *Chem. Eng. Sci.* 14: 877–90

Spencer T S, Wiley R M 1951 The mixing of very viscous liquids. *J. Colloid Sci.* 6: 133–45

Tadmor Z, Gagos C G 1979 *Principles of Polymer Processing*. Wiley, New York

Uhl V W, Gray J B (eds.) 1966, 1967 *Mixing Theory and Practice,* Vols. 1, 2. Academic Press, New York

L. Erwin
[Massachusetts Institute of Technology,
Cambridge, Massachusetts, USA]

Mixing: Internal Mixers

The internal mixer is the oldest of mixing devices used for compounding polymers. As long ago as the 1820s, the English entrepreneur Thomas Hancock developed machines consisting of an annular container with a spiked rotor to mix various ingredients into natural rubber (Hancock 1823, 1837, 1857). A generation later, his younger brother Charles utilized the device with gutta percha to prepare the first polymer blends (with natural rubber) and to introduce ammonium carbonate to produce the first foams (Hancock 1846).

It was only in the twentieth century with the development of rubber that internal mixers came to exist in their current form. Various inventors sought to develop internal mixers based on counter-rotating pairs of rotors for incorporation of fillers, accelerators and sulfur into rubber. The most successful of the early designs was that of Banbury (1916, 1917, 1921, 1924a, b), which came to have a dominant position in succeeding years (see Fig. 1). Banbury developed the design while working with Werner and Pfleiderer in Saginaw, Michigan. Unable to get his invention commercialized by them, he took it to the Birmingham Iron Foundry (later Farrel Corporation), which developed it. This led to the Banbury mixer. Research on detailed designs of mixer rotors for mixing and plastics in an internal mixer has continued to the present time. Later patents have generally involved new rotor designs where the number, shape and angles of the vanes on the rotor are designed in particular manners. Internal mixers generally consist of roughly ∞-shaped cavities in which each section contains a vaned rotor, where vanes make rough screw pumping in opposite directions. Above the cavity is a ram, which also acts as the location for introduction of polymers and fillers. Rotors generally have two or four flights.

1. Mechanisms and Modelling

There have been few basic studies of flow mechanisms in internal mixers. Only in recent years have flow visualization investigations by Min and White (1985, 1987) and Min (1987a, b) and flow modelling

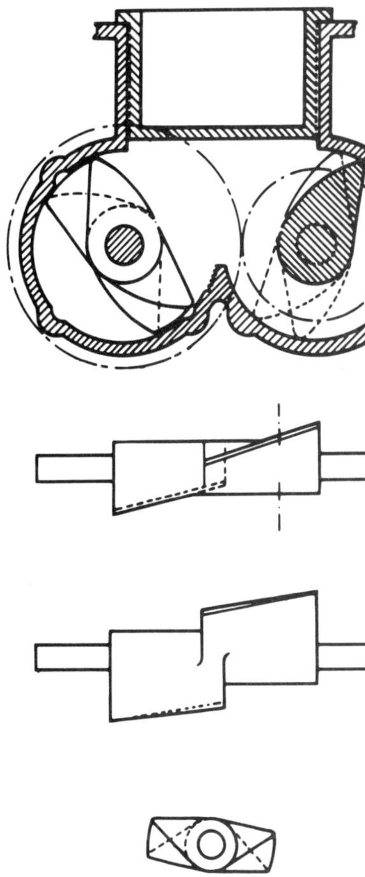

Figure 1
An internal mixer

by Kim et al. (1991) made clear the mechanisms and operations. The internal mixer rotors act as screws, which pump in opposite directions and give rise to circulatory flows (see Fig. 2). Inter-rotor transport occurs at the opposite ends of the mixer and is driven by pressure differences. The longer flights on a two-flighted rotor dominate in pumping. The shorter flights have a secondary role in causing backflows, which aid in distributive mixing.

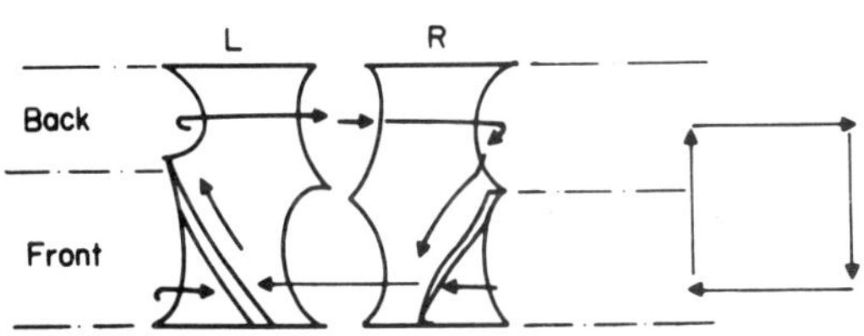

Figure 2
Circulatory flow in an internal mixer

Min (1987a) has examined the behavior of screws in an internal mixer. Putting in two right-handed screws as rotors results in a more perfect circulating motion. However, introduction of right-handed and left-handed screws results in both rotors pumping polymers in the same direction, against one end of the mixing chamber. Under this condition there is no inter-rotor transfer and no circulatory motion. Instead of transferring from rotor to rotor the polymer pumping to the end of the mixing chamber seems to flow back over the tips of the rotor flights, similar to leakage flow in a screw extruder.

Complex inter-rotor flows, including stagnation regions, were observed by Min and White (1985) in studies of elastomers. These authors note that specific material characteristics can lead to tearing or crumbling of elastomers. This interferes with the development of dispersion.

In order to model the mixing and phase morphology development in an internal mixer, it is first necessary to simulate the global motions. There has been little effort in this direction until relatively recently. Kim et al. (1991) at the University of Akron developed a model of the flow based on the observations of Min and White (1985, 1987) and Min (1987a, b).

The method of simulation very much resembles the analysis of flow in single-screw extruders. A coordinate system is placed in one of the rotors which is stretched out and the mixing chamber, and second rotor, are considered to rotate relative to it. The chamber is considered completely full of fluid and hydrodynamic lubrication theory is applied. This involves neglect of normal stresses. A Cartesian coordinate system is used in which 1 is the circumferential direction of the rotor, 3 is the direction along the rotor axis and 2 is the perpendicular direction from the rotor axis to the chamber wall. The shape of the rotor surface is specified by the function

$$\mathrm{H}=\mathrm{H}(x_1, x_3) \tag{1}$$

The equations of motion for the flow between the rotor and chamber wall are of the form

$$0=-\frac{\partial p}{\partial x_1}+\frac{\partial \sigma_{12}}{\partial x_2}=-\frac{\partial p}{\partial x_1}+\frac{\partial}{\partial x_2}\left(\eta\frac{\partial v_1}{\partial x_2}\right) \tag{2a}$$

$$0=-\frac{\partial p}{\partial x_3}+\frac{\partial \sigma_{32}}{\partial x_2}=-\frac{\partial p}{\partial x_3}+\frac{\partial}{\partial x_2}\left(\eta\frac{\partial v_3}{\partial x_2}\right) \tag{2b}$$

with boundary conditions

$$\begin{aligned} v_1(x_1,0,x_3)&=0, \qquad v_1(x_1,\mathrm{H},x_3)=U \\ v_3(x_1,0,x_3)&=0, \qquad v_3(x_1,\mathrm{H},x_3)=0 \end{aligned} \tag{3}$$

where σ_{ij} is stress, p is pressure, v_i is velocity and η is shear viscosity. Kim et al. (1991) have solved Eqn.

(2) with the indicated boundary conditions for the pressure and velocity fields within the apparatus. These indicate the rotor flights act as screws pumping fluid against the walls of the mixer chamber—the longer the flight, the greater the pressure developed. The pressure differences between the two rotors result in an inter-rotor material transfer. The rate of this transport is determined by a set of equations representing flow in the inter-rotor region. These are more complex in form than Eqn. (2) and contain 2 as well as 1 and 3 velocity components.

Computed velocity fields throughout the mixing chamber give rather more information than is wanted. Kim *et al.* have sought to simplify this by computing fractional flow rates representing relative amounts of material flowing longitudinally along the rotors (F_L), circumferentially around them (F_C), transferring from rotor to rotor (F_{TR}) and flowing over the flight tips (F_D). They proceed by calculation of the individual flows Q_L, Q_C, Q_{TR} and Q_D and then forming

$$F_L = Q_L/(Q_L + Q_C + Q_{TR} + Q_D) \tag{4a}$$

$$F_C = Q_C/(Q_L + Q_C + Q_{TR} + Q_D) \tag{4b}$$

$$F_{TR} = Q_{TR}/(Q_L + Q_C + Q_{TR} + Q_D) \tag{4c}$$

$$F_D = Q_D/(Q_L + Q_C + Q_{TR} + Q_D) \tag{4d}$$

The quantities F_L and F_{TR} determine how good the distributive mixing is in the mixer: F_D describes its ability as a dispersive mixer. If F_{TR} is zero, the material does not circulate around the mixing chamber. If F_D is zero, the mixer would have no dispersing ability.

It is possible to compute the factors F_L, F_C, F_{TR} and F_D as a function of the design variables of the mixer. This would include the relative lengths of the flights on the two rotors, the total length of the flights relative to the lengths of the rotors and the angles the flights make with the rotor axis. Calculations of this type have been reported by Kim et al. (1991). Typical results for F_D are shown in Fig. 3.

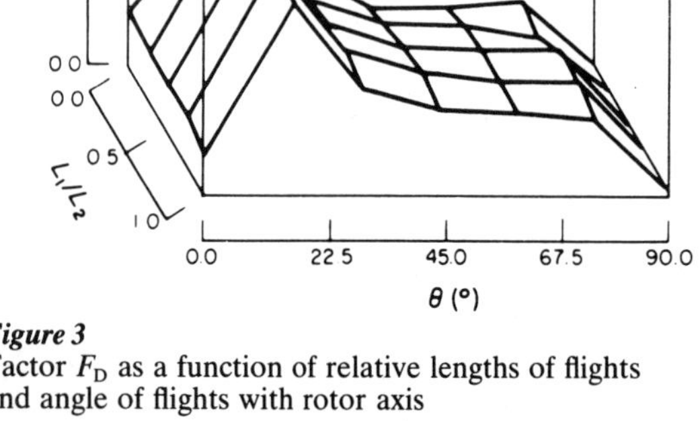

Figure 3
Factor F_D as a function of relative lengths of flights and angle of flights with rotor axis

See also: Rubber Processing

Bibliography

Banbury F H 1916 US Patent No. 1,200,070
Banbury F H 1917 USA Patent No. 1,227,522
Banbury F H 1921 US Patent No. 1,370,398
Banbury F H 1924a US Patent No. 1,513,733
Banbury F H 1924b US Patent No. 1,516,488
Clark W A 1962 US Patent No. 3,641,659
Crixell A R 1972 US Patent No. 3,672,644
Ford E H 1951 US Patent No. 2,559,418
Hancock C 1846 English Patent No. 11,147
Hancock T 1823 English Patent No. 4,768
Hancock T 1837 English Patent No. 7,344
Hancock T 1857 *Origin and Progress of Caoutchouc or India Rubber Manufacture in England.* Longman, London
Kim J, White J L, Min K, Szydlowski W 1991 *Int. Polym. Process.* (in press)
Millauer C 1978 US Patent No. 4,048,263
Min K 1987a *Int. Polym. Process.* 1: 179
Min K 1987b *Adv. Polym. Technol.* 7: 243
Min K, White J L 1985 Flow visualization of the motions of elastomers and molten plastics in an internal mixer. *Rubber Chem. Technol.* 58: 1024–37
Min K, White J L 1987 Flow visualization investigations of the addition of carbon-black and oil to elastomers in an internal mixer. *Rubber Chem. Technol.* 60: 360–80
Pletscher C, Reese L C 1919 US Patent No. 1,324,170
Pointon J E 1915 US Patent No. 1,138,410
Sato N, Miyaoka M, Yamazaki S, Inoue K, Kuriyama A, Fukui T, Asai T, Nakagawa K, Mazaki T 1981 US Patent No. 4,284,358
Schnuck C F, Hooker O W 1951 US Patent No. 2,615,689
Tyson D E, Comper L F 1966 US Patent No. 3,230,581
Wiedemann H, Schmid H M 1980 US Patent No. 4,234,259

J. L. White and K. Min
[University of Akron, Akron, Ohio, USA]

Mixing: Mills for Elastomers

Milling operations play the following roles in today's rubber factories: (a) handling and improving filler dispersion of carbon black masterbatches as they leave the internal mixer; (b) controlling compound feed rates through screw extruders; and (c) laboratory screening to determine the downstream processability of compounds in the factor.

The mill behavior of elastomers is classified into

four different regions as shown in Fig. 1:

(a) elastomer remains in mill bank,

(b) elastomer forms tight elastic band clinging to the mill roll—best region for filler dispersion,

(c) elastomer tears and granulates in mill nip and often bags off the mill roll, and

(d) elastomer forms transparent fluidlike film at high temperature.

The mill behavior of an elastomer depends on its molecular weight distribution (MWD) and its microstructure as well as the mill temperature, nip distance and roll speeds.

In general, elastomers do not exhibit all four regions. Broad-MWD elastomers like natural rubber and emulsion styrene–butadiene rubber (SBR) exhibit region (a) and (b) behavior over a wide temperature range. Narrow-MWD elastomers like butyl-lithium butadiene rubber, SBR and ethylene–propylene copolymers exhibit region (d) behavior at high temperatures. If the temperature is lowered, they exhibit a striking transition to region (c). Since mill behavior is so sensitive to MWD, elastomer producers must maintain careful MWD control to ensure that their materials give consistent finished compounds.

Analytical descriptions of each region have been developed. In region (a), the stress due to the stretching flow of chunks of rubber exceeds the frictional force between rubber and roll surface, slip occurs and rubber does not flow into the nip. The transition between regions (b) and (c) is understood as follows: flow in the nip is predominantly shear flow with an extensional flow component imposed by the contracting channel depth. Although the dispersion of fillers in the rubber is dominated by shear, the extensional flow or extensibility to break plays an important role in maintaining region (b) behavior. The tight band appearance in region (b) is controlled by tear initiation and a balance between the rate of tear propagation and the mill speed, a function of the MWD of the elastomer, the operating temperature, the nip distance and the roll speed. In region (d), the banding appearance is similar to region (b), but the two regions differ in nature. Region (d) is actually a fluid adhering to the roll and shows a negligibly small tension around the roll.

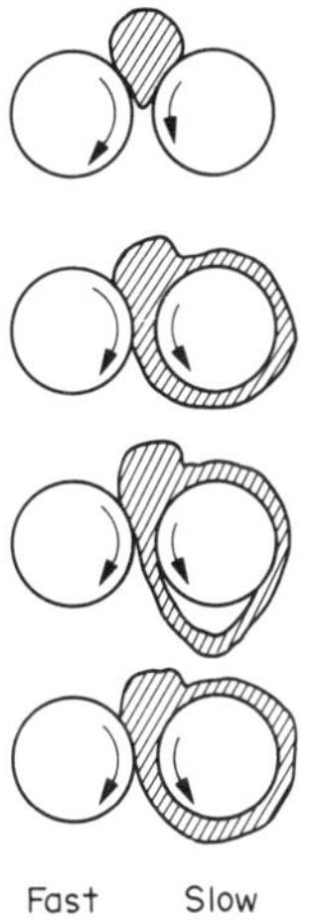

Figure 1
Different types of mill behavior of elastomers

Table 1
Values of N_o and $2H^*$ for a drop mill[a]

Material	Roll temperature (°C)	N_o	$2H^*$ (cm)
Tread	57	0.24	>0.41
	76.7	0.18	>0.23
Footwear	room temp.	0.32	>0.72
	54	0.29	>0.59

a Roll diameter 71 cm, friction ratio 1.25

Efficient mill operation depends on controlling the roll on which the masterbatch bands when it is discharged on the mill bank from the internal mixer. Normally compounds are banded on the slow speed (front) roll, because of safety and ease in stripping and cut–fold operations and of better efficiency of filler dispersion compared with fast-roll banding.

A dimensional analysis for the observed front–back roll transition of several compounds provides a dimensionless number N_o, which is a function of friction ratio (speed ratio between fast and slow rolls) f and nip–roll-radius ratio $2H_0/R$ as described by Eqn. (1):

$$N_o = (f+1)(2H_o/R)^{0.5} \tag{1}$$

This N_o number is a characteristic constant of a material regardless of the type of mill, but it is different for different materials. The typical values of N_o for tread and footwear compounds are given in Table 1.

If the N_o number of a compound does not exceed a critical value, which is defined by friction ratio f^* and nip–radius ratio $2H^*/R^*$ of the operating mill in the same fashion as in Eqn. (1), the compound bands on the slow roll. The rational nip setting of a factory mill to avoid back-roll band of a compound with a given N_o number can be obtained from Eqn. (2):

$$2H^* > N_o^2[R^*/(f^*+1)^2] \tag{2}$$

This analysis provides a technique for utilizing a laboratory mill to determine the nip setting and temperature specification needed to avoid back-roll banding in the factory.

See also: Rubber Processing

Bibliography

Anon 1969 The mixing process, an analysis. *Rubber J.* 10: 48–58

Bergen J T, Benhardt E C (eds.) 1959 *Processing of Thermoplastic Materials.* Reinhold, New York

Bulgin D 1961 The rheological properties of *cis*-polybutadiene. *Rubber Plast. Age* 42: 715–23

Chase D C, Morton M (eds.) 1959 Processing equipment. *Introduction to Rubber Technology.* Reinhold, New York, p. 517

Chong J S 1968 Calendering thermoplastics materials. *J. Appl. Polym. Sci.* 12: 191–211

Cogswell F N, Lenk R S (eds.) 1978 *Polymer Rheology.* Applied Science, London

Cotton G R 1979 Significance of extensional flow in processing rubbers. *Plast. Rubber Process.* 4: 89–95

Freakley P K, Wan Idris W Y 1979 Visualization of flow during the processing of rubber in an internal mixer. *Rubber Chem. Technol.* 52: 134–45

Funt J M 1977 *Rubber Mixing.* RAPRA, Shrewsbury, UK

Funt J M 1980 Rubber mixing. *Rubber Chem. Technol.* 53: 772–9

Gaskell R E 1950 The calendering of plastic materials. *J. Appl. Mech.* 17: 334–6

McKelvey J M 1963 *Polymer Processing.* Wiley, New York

Nakajima N, Collins E A 1976 Linear and nonlinear viscoelastic measurements, ultimate properties and processability of raw elastomers. *Trans. Soc. Rheol.* 20: 1–21

Ng T S, Angerer G 1979 Bestimmung der Scherverformung einer Kautschukmischung beim Durchgang durch den Walzenspalt. *Kautsch. Gummi Kunstst.* 32: 404–6

Pearson J R 1966 *Mechanical Principles of Polymer Melt Processing.* Pergamon, New York

Shih Chi-Kai 1972 Mill processing behavior of ethylene–propylene–hexadiene terpolymers. *Rubber Chem. Technol.* 45: 1412–20

Tadmore Z, Gogos C G 1979 *Principles of Polymer Processing.* Wiley, New York

Tokita N 1979 Analysis of band formation in mill operation. *Rubber Chem. Technol.* 52: 387–99

Tokita N, Pliskin I 1973 The dependence of processability on molecular weight distribution of elastomers. *Rubber Chem. Technol* 46: 1166–87

Tokita N, White J L ;1966 Mill behavior of gum elastomer: Experiment and theory. *J. Appl. Polym. Sci.* 10: 1011–25

Wagner M G 1970 Mill behavior of elastomers: Scale-up criteria. *J. Appl. Polym. Sci.* 14: 757–63

White J L 1969 Elastomer rheology and processing. *Rubber Chem. Technol.* 42: 257–338

White J L 1977 Processability of rubber and rheological behavior. *Rubber Chem. Technol.* 50: 163–85

White J L, Tokita N 1968 Instability and failure phenomena in polymer processing with application to elastomer mill behavior. *J. Appl. Polym. Sci.* 12: 1589–99

N. Tokita
[Uniroyal, Middlebury, Connecticut, USA]

Mixing: Rheological Behavior and Dispersion of Two-Phase Systems

Mechanical blending of rubber and plastics is generally carried out to develop new products with improved mechanical properties (e.g., higher toughness and impact strength), better solvent resistance and increased high-temperature performance. Examples of such products include blends of polypropylene and ethylene—propylene rubber, polycarbonate and acrylonitrile–butadiene–styrene (ABS) resins, poly(*p*-phenylene oxide) and high-impact polystyrene, and poly(vinyl chloride) and butadiene–acrylonitrile rubber.

1. Rheological Properties of Polymer Melt Blends

There have been extensive studies of the rheological behavior of mechanical blends of molten plastics; reviews are given by Han (1981) and White et al. (1981). Many experimental studies have indicated that the viscosity function varies monotonically with composition (Hayashida et al. 1970, Lee and White 1975, Alle and Lyngaae-Joergensen 1980a), whereas other studies have described blend systems which show either minima (Han and Yu 1971, Płochocki 1972, Han et al. 1975) or maxima (Tsebrenko et al. 1974, Iwakura and Fujimura 1975). Some systems exhibit either minima or maxima, depending on the shear rate (Ablazova et al. 1975) but others may exhibit both maxima and minima at the same shear rate or shear stress (Han and Kim 1975).

Although extensive experimental investigations of disperse two-phase flow of polymer melts have been reported, there is little literature attempting to represent quantitatively the dependence of viscosity η on composition. The most widely quoted relation is that given by Hayashida et al. (1970):

$$1/\eta = \phi/\eta_A + (1-\phi)/\eta_B \qquad (1)$$

where ϕ is the volume fraction of phase A and η_A and η_B are the shear viscosities of phases A and B at the same shear stress. The usual derivation assumes that flow takes place in parallel lamellae, which are in a region of constant shear stress. This equation predicts a monotonic variation of the shear viscosity with composition.

Principal normal stress differences for polymer melt blends have been extensively studied. These investigations have used both cone-plate (Han et al. 1973, Liang et al. 1982) and exit pressure (Han and Yu 1971, Han et al., 1975, Han and Kim 1975) measurements. As with shear viscosity–composition plots, these may show maxima or minima.

2. Phenomena Controlling Dispersion

On studying the formation of emulsions from two Newtonian liquids (Taylor 1934, Karam and

Bellinger 1968), it is found that shear stresses disrupt droplets but interfacial tension stabilizes them. Few studies of dispersion in polymer melt blends as a function of mixing conditions exist, but those that do generally show that shear mixing in an internal mixer or mixing with a combined screw extruder–static mixer can yield dispersions with phase sizes of the order of micrometers for nonpolar hydrocarbon thermoplastics (Barentsen et al. 1974, Han et al. 1975, Avgeropoulos et al. 1976, Heikens and Barentsen 1977). Generally the disperse phase is in the form of filaments oriented in the flow direction. Cocontinuous phases are observed when compositions approach 50:50.

Liang et al. (1982) studying Nylon 6/polypropylene found that there are both phase growth and phase size dispersion phenomena in incompatible blend systems. They observed that in quiescent and mildly deformed systems the dispersed polypropylene phases grew in size. Extrusion through a capillary die at high shear rates and shear stresses was found to reduce phase size. It would appear that the growth mechanism is associated with incompatibility, interfacial tension and a low continuous-phase viscosity. Dispersion is caused by shearing stresses and is interpreted in terms of the capillarity number:

$$N_{\mathrm{cap}}=\frac{\text{interfacial forces}}{\text{viscous forces}}=\frac{\gamma/R}{\eta v/R}=\frac{\gamma}{\eta v} \qquad (2)$$

where γ is the surface tension, R is the radius of curvature of the interface, η is the visocity and v is the shear rate (velocity). It was observed that the phase sizes in this system were always an order of magnitude greater than that found in hydrocarbon polymer blends. In earlier studies of polyethylene–polystyrene blends by Heikens and Barentsen (1977), coalescence phenomena for discrete phases were proposed but not clearly observed.

Additives are used to influence the morphology of blends. Maleic anhydride-grafted polypropylene has been used to improve the dispersion of nylon–polypropylene blends (Ide and Hasegawa 1974) and graft and block copolymers of styrene and ethylene have been used in polystyrene–polyethylene blends to reduce the scale of phase morphology (Barentsen et al. 1974, Heikens and Barentsen 1977). This can be interpreted in terms of additives reducing interfacial tensions.

Tsebrenko *et al.* (1980) concluded that the viscosity ratio plays a major role in the phase morphology of extrudates. If the phase viscosities are about the same, a disperse phase in shear flow forms long ligaments or fibrils. If the matrix viscosity is low, the lengths of the ligaments decrease and, if it is high, films form as well as fibers. Han (1981) emphasizes the influence of rheological properties on the development of morphology in processing.

3. Minifibers from Melt-Spun Blends

In the melt spinning of blends, the disperse phase frequently forms high-aspect-ratio, small-diameter fibrils, which may be isolated by extracting the continuous phase (Merriam and Miller 1963, Tsebrenko et al. 1974, 1976, 1980, Liang et al. 1982, Shimizu et al. 1982). Submicrometer minifibers have been produced from a wide range of polymers including polyethylene, polypropylene, nylon and polyoxymethylene. The mechanism of forming the fibrils involves the affine drawdown of cylindrical rods in the dispersed phase. If v_0 is the extrusion velocity, v_{L} the takeup velocity, d_0 the rod diameter and d_{L} the fibril and minifiber diameter then, from continuity,

$$v_0 d_0^2 = v_{\mathrm{L}} d_{\mathrm{L}}^2 \qquad (3)$$

$$d_{\mathrm{L}} = (v_{\mathrm{L}}/v_0)^{1/2} d_0 \qquad (4)$$

Thus if v_{L}/v_0 is 100, d_{L} will be a tenth of d_0.

See also: Polymer Blends; Rheological Characterization of Molten Polymers

Bibliography

Ablazova T I, Tsebrenko M B, Yudin A B V, Vinogradov G V, Yarlykov B V 1975 The rheological properties of molten mixtures of polyoxymethylene and copolyamide: Their fibrillation and microstructure. *J. Appl. Polym. Sci.* 19: 1781–98

Alle N, Lyngaae-Joergensen J 1980a Polypropylene and polyethylene blends. 1. Flow behavior in capillaries. *Rheol. Acta.* 19: 94–103

Alle N, Lyngaae-Joergensen J 1980b Polypropylene–polyethylene blends. 2. The relationship between rheology and morphology in capillary flows. *Rheol. Acta.* 19: 104–10

Avgeropoulos G N, Weissert F C, Biddison P H, Bohm G G A 1976 Heterogeneous blends of polymers: Rheology and morphology. *Rubber Chem. Technol.* 49: 93–104

Barentsen W M, Heikens D, Piet P 1974 Effect of addition of graft copolymer on the microstructure and impact strength of PS/LDPE blends. *Polymer* 15: 119–22

Danesi S, Porter R S 1978 Blends of isotactic polypropylene and ethylene–propylene rubbers: Rheology, morphology and mechanics. *Polymer* 19: 448–57

Han C D 1981 *Multiphase Flow in Polymer Processing.* Academic Press, New York

Han C D, Kim K U, Parker J, Siskovic N, Huang C R 1973 Measurements of the viscoelastic properties of poly(methyl methacrylate) and polystyrene blends. *Appl. Polym. Symp.* 20: 191–8

Han C D, Kim Y W 1975 Dispersed two-phase flow of viscoelastic polymeric melts in a circular tube. *Trans. Soc. Rheol.* 19: 245–69

Han C D, Kim Y W, Chen S J 1975 The effect of mixing on the modes of dispersion and rheological properties of two phase polymer blends in extrusion. *J. Appl. Polym. Sci.* 19: 2831–43

Han C D, Yu T C 1971 Rheological properties of molten polymers, II: Two phase systems. *J. Appl. Polym. Sci.* 15: 1163–80

Hayashida K, Takahashi J, Matsui M 1970 Some contributions to the flow behavior of polymer melts. In: Onogi S (ed.) 1970 *Proc. 5th Int. Rheological Congress,* Vol. 4. University of Tokyo Press, Tokyo, pp. 523–35

Heikens D, Barentsen W 1977 Particle dimensions in polystyrene/polyethylene blends as a function of their melt viscosity and of the concentration of added graft copolymer. *Polymer* 18: 69–72

Ide F, Hasegawa A 1974 Studies on polymer blend of nylon 6 and polypropylene or nylon 6 and polystyrene using the reaction of polymer. *J. Appl. Polym. Sci.* 18: 963–74

Iwakura K, Fujimura T 1975 Rheogoniometry of molten blends of polystyrene and styrene ionomers. *J. Appl. Polym. Sci.* 19: 1427–37

Karam H J, Bellinger J C 1968 Deformation and breakup of liquid droplets in a single shear field. *Ind. Eng. Chem., Fundam.* 7: 576

Lee B L, White J L 1975 Experimental studies of disperse two-phase flow of molten polymers through dies. *Trans. Soc. Rheol.* 19: 481–92

Liang B R, White J L, Spruiell J E, Goswami B C 1982 Polypropylene/nylon 6 blends: Phase distribution morphology, rheological measurements and structure development in melt spinning. *J. Appl. Polym.* 28(6): 2011–32

Merriam C N, Miller W A 1963 Plastic fibers. US Patent No. 3,099, 067

Pľochocki A 1972 On the rheology of molten binary blends of polyolefins: X. Homogeneity and high shear (Poiseuille's) melt flow of polypropylene–polyethylene blends. *J. Appl. Polym. Sci.* 16: 987–1008

Shimizu J, Okui N, Yamamoto T, Ishii M, Takaku A 1982 Melt spinning of polypropylene/polystyrene blends: Morphology and ultra-fine fibril formation. *Sen'i Gakkaishi* 38: T1–9

Taylor G I 1934 The formation of emulsions in definable fields of flow. *Proc. R. Soc. London, Ser. A* 146: 501–23

Tsebrenko M V, Jakob M, Yu M, Kuchinka, Yudin A V, Vinogradov G V 1974 Fibrillation of crystallizable polymers in flow exemplified by melts of mixtures of polyoxymethylene and copolyamides. *Int. J. Polym. Mater.* 3: 99–116

Tsebrenko M V, Rezanova N M, Vinogradov G V 1980 Rheology of molten blends of polyoxymethylene and ethylene–vinyl acetate copolymer and the microstructure of extrudates as a function of their melt viscosities. *Polym. Eng. Sci.* 20: 1023–8

Tsebrenko M V, Yudin A V, Ablazova T I, Vinogradov G V 1976 Mechanism of fibrillation in the flow of molten polymer mixtures. *Polymer* 17: 831–4

White J L, Pľochocki A P, Tanaka H 1981 Shear and elongational flow rheological behavior of disperse two-phase blends of molten polymers. *Polym. Eng. Rev.* 1: 217–48

J. L. White
[University of Akron, Akron, Ohio, USA]

Mixing: Single-Screw Extruders

It is common practice to mix additives (e.g., color concentrates, fillers, reinforcing fibers, stabilizers, lubricants, antioxidants, foaming agents and cross-linking agents) into polymers as well as to blend polymers in single-screw extruders. In such cases, the extruder must deliver to the die a steady melt stream uniform in composition. Moreover, even if the extruder is fed with a single component, the mixing performance of the extruder determines the temperature homogeneity of the melt stream at the die. Hence, in all cases, the mixing capability of the extruder is of great practical importance. Indeed, current trends in screw design are towards such improved mixing capability.

1. Mixing Mechanisms: Extensive and Intensive Mixing

Extensive mixing reduces nonuniformity of composition in a mixture by inducing physical motion of the ingredients. There are three basic types of motion in mixing (Brodkey 1966): molecular diffusion, eddy "diffusion" and convection. Molecular diffusion is a spontaneous process driven by a chemical potential gradient. In turbulent mixing, molecular diffusion is superimposed on the gross random eddy motion, which in turn may occur within a larger-scale convective flow. In mixing polymer melts, because of the high viscosity, eddy diffusion is rarely reached and molecular diffusion is an insignificant contributor to mixing. Convection is therefore the dominant mixing mechanism. It reduces composition nonuniformity if the interfacial area between the components increases (e.g., liquid–liquid mixing) (Spencer and Wiley 1951) or if the minor component is distributed throughout the major component without necessarily increasing the interfacial area (e.g., solid–liquid mixing) (Bergen et al. 1958).

Convective mixing is achieved either by simple bulk—ordered or random—rearrangement of the material by plug-type flow called "distributive mixing" or by imposing a finite deformation on the materials, called laminar mixing. The latter is the main mixing mechanism in screw extruders, with shear strain as the dominant type of deformation (Mohr 1959).

The ratio of interfacial area between two components after a finite deformation to the initial value, in a general homogeneous flow field, is given by Erwin (1978a)

$$A/A_0=\left(\frac{\cos^2\alpha'}{\lambda_x^2}+\frac{\cos^2\beta'}{\lambda_y^2}+\frac{\cos^2\gamma'}{\lambda_z^2}\right)^{1/2} \qquad (1)$$

where $\cos\alpha'$, $\cos\beta'$ and $\cos\gamma'$ are the directional cosines of the area element with respect to the principal axes of the strain tensor in the initial state and λ_x, λ_y and λ_z are the principal elongational ratios. For simple shear flow, Eqn. (1) reduces to

$$A/A_0 = (1 - 2\cos\alpha\cos\beta + \cos^2\alpha\gamma^2)^{1/2} \quad (2)$$

where γ is the total shear strain and the directional cosines of the initial orientation of the area satisfy the relationship

$$\cos^2\alpha + \cos^2\beta + \cos^2\gamma = 1 \quad (3)$$

For large strain, Eqn. (2) further reduces to

$$A/A_0 = |\cos\alpha|\gamma \quad (4)$$

Finally, for an initially randomly oriented interfacial area element distribution the total area ratio is (Tadmor and Gogos 1979)

$$A/A_0 = \gamma/2 \quad (5)$$

indicating that in laminar mixing by shear, A/A_0, which reflects mixedness, is directly proportional to shear strain. In continuous mixers, such as screw extruders, however, exiting fluid particles experience different shear rate histories and residence times and therefore may acquire different strains. In parallel with the definition of residence time distribution (RTD) functions, Lidor and Tadmor (1976) defined the strain distribution functions (SDF) $f(\gamma)d\gamma$ and $F(\gamma)$ for characterizing mixing. The function $f(\gamma)d\gamma$ denotes the fraction of exiting flow rate that has experienced a total strain between γ and $\gamma + d\gamma$ and $F(\gamma)$ denotes the fraction of flow rate that has experienced a strain equal to or less than γ. The mean strain of the exiting stream is $\bar{\gamma} = \int_0^\infty \gamma f(\gamma)d\gamma$.

The SDFs and $\bar{\gamma}$ can be calculated, in principle, from the velocity profiles. Clearly, good mixing requires a large mean strain and a narrow SDF. However, it should be noted that the SDF functions are strictly speaking useful only if little reorientation takes place. A further and independent requirement of "good" mixing is the (random) distribution of the interfacial area elements throughout the system. Depending on initial orientation, this may or may not occur in a given flow field. In screw extruders it is possible to impose on the polymer large strains with a narrow SDF. Furthermore, it is also possible to achieve composition uniformity across a plane perpendicular to the flights but, because of the almost complete lack of backmixing (i.e., the existence of a very narrow RTD), the screw extruder cannot even out temporal nonuniformities in inlet composition. Hence, to ensure an extrudate which has a composition uniform in time, the feed composition must be steady in time.

The foregoing considerations are valid for mixing two rheologically homogeneous liquids or for mixing a particulate solid into a liquid. Mixing considerations for rheologically inhomogeneous liquids are more complex (Bigg and Middleman 1974, Tadmor and Gogos 1979). It can be generally stated that for mixing such systems both phases must undergo finite deformations, that it is more difficult to mix a high-viscosity minor component into a low-viscosity major component than vice versa and that for optimum mixing the two viscosities should not be too far apart.

A different type of mixing of great practical importance is intensive or dispersive mixing, which involves a process of breaking down the minor component into small entities. Dispersion of solid agglomerates (e.g., carbon black) into the polymeric matrix and the dispersion or homogenization of incompatible polymeric melt "blobs" into another polymer can serve as examples. The former involves the rupture of the agglomerates and the latter the breakdown of polymeric blobs. Both involve some finite yield strength. It is believed that in intensive mixing the dominant variables are the stress and rate of stress buildup imposed on the mixture (e.g., under the blade of an internal mixer), although there is some evidence that the rate-controlling step in intensive mixing may be the separation of the particles rather than the rupture process itself (Funt 1980). Single-screw extruders are generally poor intensive mixers. Though high-shear-stress regions exist over the flights, only a portion of the mixture passes over them. Moreover, "good" intensive mixing requires many repeated passes over the high-stress regions (Manas-Zloczower et al. 1982).

2. *Laminar Mixing in Single-Screw Melt Extruders*

Laminar mixing in single-screw extruders is induced by the moving barrel surface relative to the screw channel (see *Melt Delivery Systems, Single-Screw*). The unwound screw channel can be approximated as a shallow rectangular channel. It is covered by an infinite plate—the surface of the barrel—moving at constant velocity $V_b = \pi N D_b$ at an angle θ (the helix angle) to the down-channel direction. The down-channel velocity component of the plate $V_{bz} = V_b \cos\theta$ drags (pumps) the melt towards the exit, whereas the cross channel velocity component $V_{bx} = V_b \sin\theta$ induces crosschannel melt circulation, which is instrumental in imposing on the melt large strains with a narrow SDF as well as a narrow RTD. The velocity profiles, for isothermal flow of a Newtonian fluid, in the crosschannel direction $u_x = v_x/V_{bx}$ and down-channel direction $u_z = v_z/V_{bz}$ are, respectively,

$$u_x = \xi(2 - 3\xi) \quad (6)$$

and

$$u_z = \xi + [3\xi(1-\xi)Q_p/Q_d] \quad (7)$$

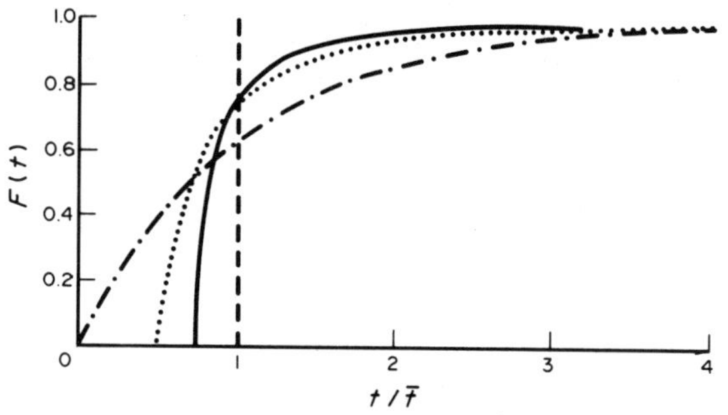

Figure 1
The RTD function for flow in a screw extruder (——) compared with plug flow (– – –), isothermal flow of Newtonian fluids in pipes (· · · ·) and a continuous well-mixed tank (— · —)

where $\xi = y/h$ is the reduced height and Q_p/Q_d is the pressure–drag flow ratio:

$$\frac{Q_p}{Q_d} = -\left(\frac{h^2}{6\mu V_{bz}}\right)\frac{dp}{dz} \qquad (8)$$

which acquires values from 0 to 1, from open discharge to closed discharge conditions, respectively. Equation (6) indicates that crosschannel velocity vanishes at $\xi = \frac{2}{3}$. A fluid particle at ξ ($\xi > \frac{2}{3}$) moves towards the pushing flight, turns at the flight and proceeds in the opposite direction at ξ_c ($\xi_c < \frac{2}{3}$), where

$$\xi_c = \tfrac{1}{2}(1 - \xi + \sqrt{1 + 2\xi - 3\xi^2}) \quad (\tfrac{2}{3} \leqslant \xi \leqslant 1) \qquad (9)$$

Hence, the fluid particles follow a flat helical path in the rectangular channel and the flow path in the screw channel is described as a helical path within the helical channel. The RTD functions can be calculated from the velocity profiles, Eqns. (6, 7), and the relationship in Eqn. (9), to give (Pinto and Tadmor 1970)

$$F(t) = F(\xi) = \tfrac{1}{2}[3\xi^2 - 1 + (\xi - 1)\sqrt{1 + 2\xi - 3\xi^2}] \qquad (10)$$

where ξ is related to residence time by

$$\frac{t(\xi)}{t_0} = \frac{9}{2}\left[\frac{3\xi - 1 + 3\sqrt{1 + 2\xi - 3\xi^2}}{\xi(1 - \xi + \sqrt{1 + 2\xi - 3\xi^2})}\right] \qquad (11)$$

The minimum residence time t_0 is given by

$$t_0 = 3l/2V_b[1 + (Q_p/Q_d)]\sin\theta\cos\theta \qquad (12)$$

where l is the axial length of the screw. The mean residence time $\bar{t} = 4t_0/3$. Figure 1 compares RTD functions in screw channels (verified experimentally by Wolf and White 1976) with those in pipe flow and stirred tanks.

The magnitude of the rate of strain tensor is a function of position ξ:

$$\dot{\gamma}(\xi) = [\dot{\gamma}^2_{yx}(\xi) + \dot{\gamma}^2_{yz}(\xi)]^{1/2} \qquad (13)$$

where the shear rates $\dot{\gamma}_{yx}$ and $\dot{\gamma}_{yz}$ are obtained, respectively, from Eqns. (6) and (7). The total strain as a function of ξ is given (McKelvey 1962, Tadmor and Klein 1970) by

$$\gamma(\xi) = \dot{\gamma}(\xi)t_f(\xi)t(\xi) + \dot{\gamma}(\xi_c)[1 - t_p(\xi)]t(\xi) \qquad (14)$$

where $t_f(\xi)$ is the fraction of time the fluid particle spends in the "upper" portion ($\xi > \frac{2}{3}$) of the channel.

Equation (14) assumes that the strains imposed on the fluid particles in the "upper" and "lower" portions of the channel are additive. This, however, may not be the case and a correct evaluation of $\gamma(\xi)$ requires the exact two-dimensional formulations of the flow field in the cross channel direction. The SDF and the mean strain $\bar{\gamma}$ can now be calculated following a similar procedure used in the RTD function. Figure 2 shows $\bar{\gamma}$ vs Q_p/Q_d with helix angle as a parameter for $l/h = 1$. It supports the experimental observation that, as discharge resistance is increased, mixing is improved and that the longer screws give proportionately better mixng. To the left of this value where $Q_p/Q_d > 0$ (i.e., the pressure drops in the flow direction, as may be the case in certain metering sections), lower strain values are observed, whereas to the right where $Q_p/Q_d < 0$

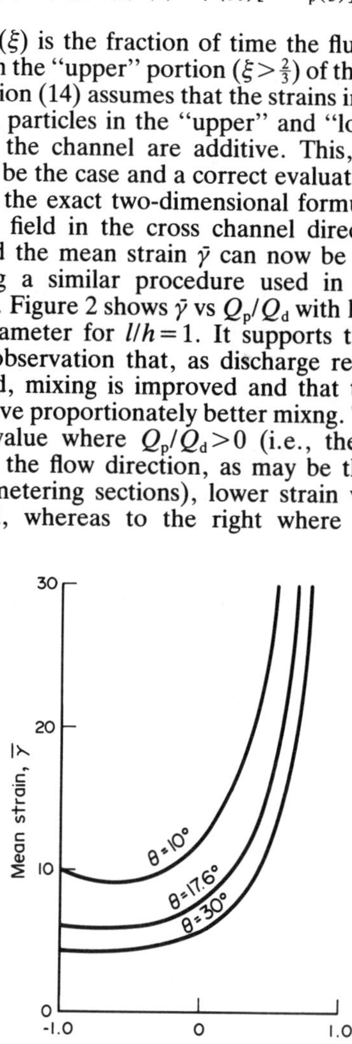

Figure 2
Mean strain as a function of Q_p/Q_d for $l/h = 1$ with helix angle θ as a parameter; $Q_p/Q_d = 0$ corresponds to open discharge conditions

(i.e., the pressure rises in the flow direction), the strain increases sharply, approaching infinity at the closed discharge condition ($Q_p/Q_d=-1$).

In principle, therefore, by manipulating screw design (l/h) and operating conditions (Q_p/Q_d), any desired level of mixing can be imposed on the polymer. However, it can be shown (Erwin 1978b, Tadmor and Gogos 1979) that mixing efficiency is reduced as simple shearing continues, because the orientation of the interfacial area becomes less favorable. Maximum total mixing is obtained by maintaining the latter at an optimum 45° to the direction of shear. The interfacial area ratio in this case is given by

$$A/A_0=\exp(\gamma/2) \tag{15}$$

which corresponds to pure elongation. It is difficult to devise practical elongational mixers; however, Erwin (1978b) showed that, by repeatedly randomizing the shear lines before imposing a given strain γ,

$$A/A_0=\tfrac{1}{2}(\gamma/N)^N \tag{16}$$

where N is the number of randomizing steps. Thus, for example, for $A/A_0=5000$, a total strain of 10 000 is needed for $N=1$, but only 200 strain units per section are needed for $N=2$. It appears reasonable, therefore, to assume that the very beneficial effect of mixing sections (e.g., rings, pins) in screw designs is due to the partial randomization they impose on the fluid, thus reducing the total strains required for a given mixedness.

3. *Laminar Mixing in Single-Screw Plasticating Extruders*

In plasticating extruders, a melt pool or channel is formed side by side with a solid bed (see *Melt Delivery Systems, Single-Screw*). The flow patterns in the melt channel are reasonably well approximated by those presented in Sect. 2. However, throughout the melting zone, newly melted material is fed to the melt channel. This melt has a residence time and strain history in the melt film.

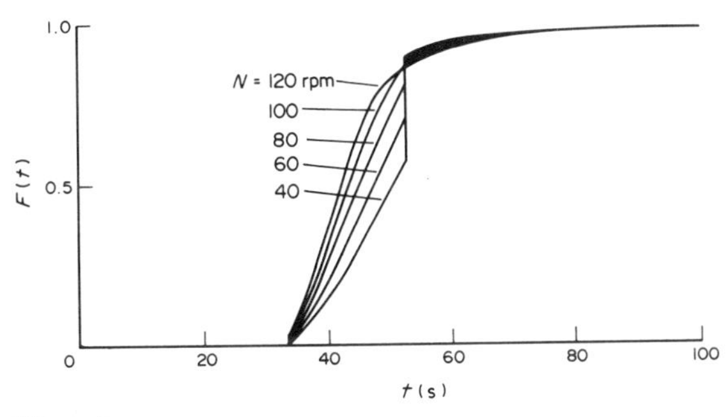

Figure 3
The RTD function in a 64 mm diameter screw extruder of length–diameter ratio 20:1, with 10 mm channel depth and 6 mm channel width at 45 kg h^{-1} flow of low-density polyethylene

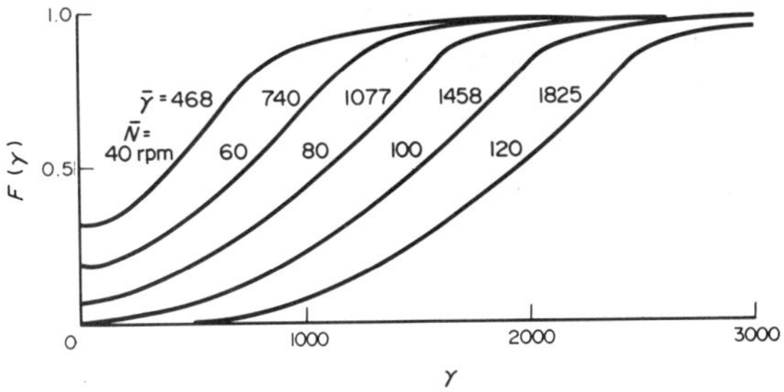

Figure 4
The SDF function for the extruder and conditions in Fig. 3.

Lidor and Tadmor (1976) modelled the RTD and SD functions in plasticating extruders, accounting for the melt film and assuming that the newly melted material is distributed over the melt channel height proportionally to the local flow rates. Clearly, the melting process profoundly affects both distribution functions. Figures 3 and 4 show, respectively, the computer-simulated RTD and SDF in a 64 mm diameter screw extruder, for a constant flow rate with screw speed as a parameter. The straight-line segments in Fig. 3 and the $F(0)$ values in Fig. 4 indicate unmelted solids in the extrudate. Melting is completed in the screw at 100 rpm. Higher screw speeds reduce the melting length, shift the SDF curves to higher γ values and reduce the width of the SDF. The mean residence time is little affected by screw speed, varying only from 45.3 s at 120 rpm to 49.1 s at 40 rpm, owing to changes in the solids content in the screw channel, which affects the mean density. The latter becomes quite narrow for melt extruders. Mixing sections in plasticating extruders, in addition to their randomizing function mentioned in Sect. 2, are also instrumental in greatly improving the melting performance of the screw and consequently improving the mixing performance as well.

See also: Processing: Classification

Bibliography

Bergen J T, Carrier G W, Krumbansh J A 1958 Criteria for mixing and mixing process. *14th Society of Plastics Engineers National Technical Conf.* Society of Plastics Engineers, Detroit, MI

Bigg D, Middleman S 1974 Laminar mixing of a pair of fluids in a rectangular cavity. *Ind. Eng. Chem., Fundam.* 13: 184

Brodkey R S 1966 Fluid motion in mixing. In: Uhl V A, Gray J B (eds.) 1966 *Mixing*, Vol. 1. Academic Press, New York

Erwin L 1978a Theory of laminar mixing. *Polym. Eng. Sci.* 18: 1044
Erwin L 1978b Theory of mixing sections in single screw extruders. *Polym. Eng. Sci.* 18: 572
Funt J M 1980 Rubber mixing. *Rubber Chem. Technol.* 53: 772
Lidor G, Tadmor Z 1976 Theoretical analysis of residence time distribution functions and strain distribution in plasticating screw extruders. *Polym. Eng. Sci.* 16: 450
McKelvey J M 1962 *Polymer Processing*. Wiley, New York
Manas-Zloczower I, Nir A, Tadmor Z 1982 Dispersive mixing in internal mixer: A theoretical model based on agglomerate rupture. *Rubber Chem. Tech.* 55: 1250
Mohr W D 1959 Mixing and dispersing. In: Bernhardt E C (ed.) 1959 *Processing of Thermoplastic Materials*. Reinhold, New York
Pinto G, Tadmor Z 1970 Mixing and residence time distribution in melt extruders. *Polym. Eng. Sci.* 10: 279
Spencer R S, Wiley R N 1951 The mixing of viscous liquids. *J. Colloid Sci.* 6: 133
Tadmor Z, Gogos C G 1979 *Principles of Polymer Processing*. Wiley, New York
Tadmor Z, Klein I 1970 *Engineering Principles of Plasticating Extrusion*. Van Nostrand Reinhold, New York
Wolf D, White D H 1976 Experimental study of the residence time distribution in plasticating screw extruders. *Am. Inst. Chem. Eng. J.* 22: 122

Z. Tadmor
[Israel Institute of Technology, Haifa, Israel]

Mixing: Twin-Screw Extruders

Most plastics are not composed purely of polymer, rather, they are mixtures of a base polymer with a variety of additives. These additives are incorporated into the polymer for a variety of reasons: (a) to improve the properties and appearance of the finished product; (b) to initiate a chemical reaction; (c) to reduce the amount of base polymer needed, thereby making the process more economical; and (d) to facilitate processing. Examples of typical additives include pigments, lubricants, stabilizers, antioxidants, crosslinking agents, fillers, reinforcing agents and plasticizers. In addition, blending of different polymers to achieve improved properties is becoming increasingly significant as the rate of discovery of new polymers levels off. Depending on product requirements, the proportion of additives incorporated can vary from a minute amount to a substantial portion of the mixture.

In recent years, twin-screw extruders have been developed which can handle a wide range of compounding tasks. In twin-screw compounding systems, close control over all variables such as stock temperature, shear and residence-time distribution can be maintained. This gives the twin-screw extruder an advantage over single-screw extruders, in which close process control is not always possible because of their inherent limitations.

Another requirement of many mixing operations is devolatilization—the removal of volatiles such as water, solvents or residual monomer. Devolatilization is generally diffusion controlled and in order to obtain high removal rates, this diffusion is normally coupled with extensive mixing (Tadmor and Gogos 1979). Single- or multiple-port devolatilization is easily accomplished in a twin-screw system, thereby broadening its application for many different compounding operations.

This article deals with the use of twin-screw extruders for mixing plastics and is restricted to fully intermeshing, self-wiping twin-screw extruders. A discussion of the different types of twin-screw mechanisms is included in the article *Melt Delivery Systems, Twin-Screw*.

1. Geometric Differences: Corotating Versus Counter-Rotating Screws

Fully intermeshing counter-rotating screws form lengthwise- and crosswise-closed C-shaped chambers in the intermesh zone (wedge area). Each chamber has only a very small volume. Material is not transferred from one screw to the other, thereby reducing the mixing efficiency of the system. The closed chambers that are formed result in plug flow (similar to that observed in a screw pump) as the polymer is conveyed down the length of the extruder.

Fully intermeshing corotating screws form V-shaped wedge areas which have four to five times the volume of those found in counter-rotating systems. The screw channels are open lengthwise and closed crosswise. Because of the uninterrupted open channel, material exchange can take place lengthwise in the machine. This results in a mixing system based on drag flow within the lengthwise-open screw channel. Polymer is transferred in a figure-eight pattern from one screw to the other, resulting in a renewal of material layers and surfaces. A better degree of mixing is thereby obtained with the corotating mechanism.

Figure 1 illustrates the differences in the wedge areas of corotating and counter-rotating screws.

2. Velocity and Stress Distribution of Down-Channel and Cross-Channel Flow

The geometric differences between corotating and counter-rotating systems result in significant differences in the velocity and stress distribution of the down-channel flow. Although there is no difference between the two systems in cross-channel flow, a distinct difference in resultant shear stress arises

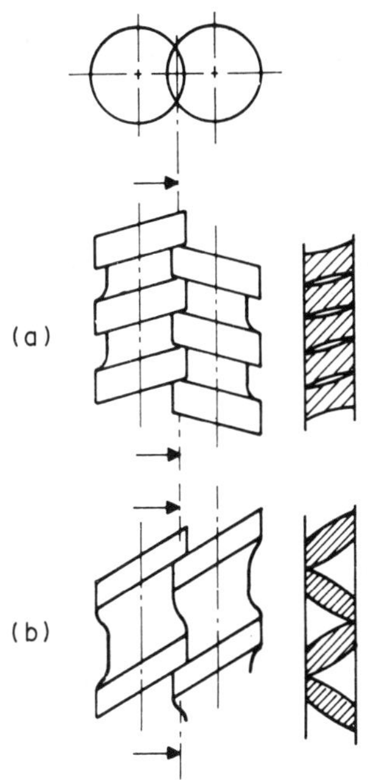

Figure 1
Wedge sections of (a) counter-rotating and (b) corotating twin-screw extruders

when both down- and cross-channel flow are considered simultaneously. These differences have a significant effect on the mixing capabilities of the two twin-screw mechanisms.

Eise et al. (1981) present a detailed analysis of velocity and stress distributions in down-channel, cross-flow and combined flow for both corotating and counter-rotating screws. The coordinate system selected for this analysis is shown in Fig. 2. The x axis was chosen perpendicular to the screw channel, the y axis perpendicular to the barrel surface (pointing towards the rotating screw axis) and the z axis parallel to the screw channel. A summary of the analysis is presented as follows:

(a) The shear stress τ_b produced at the barrel wall is greater for counter-rotating screws than for corotating screws.

(b) The shear stress τ_s produced at the screw root is greater for counter-rotating screws than for corotating screws.

(c) The maximum shear stress τ_{max} in counter-rotating screws (not considering the wedge area and mechanical clearances) is identical to the shear stress at the screw root τ_s. In corotating screws, this maximum stress is identical to the shear stress at the barrel wall.

(d) In counter-rotating screws the minimum shear stress τ_{min} is at $y/h = 0.38$, where h is the channel depth. In corotating screws, $0.66 \leqslant y/h \leqslant 0.9$ depending on throughput.

(e) In counter-rotating screws, the change in shear stress $d\tau/dy$ in the channel depth is considerably larger than with corotating screws. This means that the shear-stress distribution of counter-rotating screws is much broader than in corotating screws.

(f) The shear-stress distribution in counter-rotating screws depends only on the screw pitch H. However, in corotating screws not only does the shear-stress distribution change with the pitch angle ϕ, it also depends on the throughput and the screw speed. Hence, with corotating screws, the shear-stress distribution can be influenced by operating conditions as well as by screw geometry.

Counter-rotating systems have relatively high stresses at the barrel walls and relatively low stresses in the middle of the channel. Corotating screws, operating under the same parameters, have lower stresses at the barrel walls and a much more uniform stress distribution in the screw channel. In counter-rotating screws, the material at the screw root is drawn into the wedge area where it is exposed to a large strain. The minimum shear stress τ_{min} is inside the screw chamber and hence a large fraction of material is not exposed to the shear required for good mixing and dispersion. In corotating machines, however, operating conditions can be adjusted so that the minimum shear stress is closer to the screw root. This allows a greater fraction of the material to be subjected to adequate shear. Material layers at the screw root therefore encounter lower shear strain than in counter-rotating machines. In corotating machines, these layers are wiped by the crest of the oncoming screw in the wedge area and then

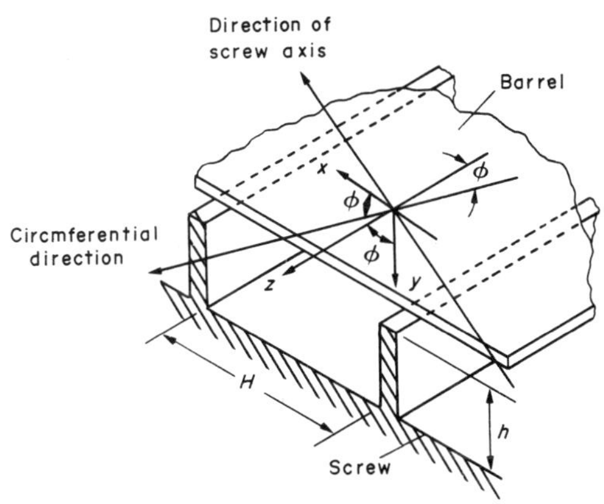

Figure 2
Coordinate system for analysis of velocity and stress distributions of corotating and counter-rotating screw systems

recombined near the barrel wall. Material layers at the barrel wall are in turn forced into the wedge area and transported to the bottom of the channel of the oncoming screw. This rearrangement and transfer of material layers results in a more uniform shear strain and more homogeneous mixing. The use of kneading elements as part of the mixer screws in corotating systems has also added to their effectiveness as mixers. These kneading elements permit the introduction of very high shear energies at specific

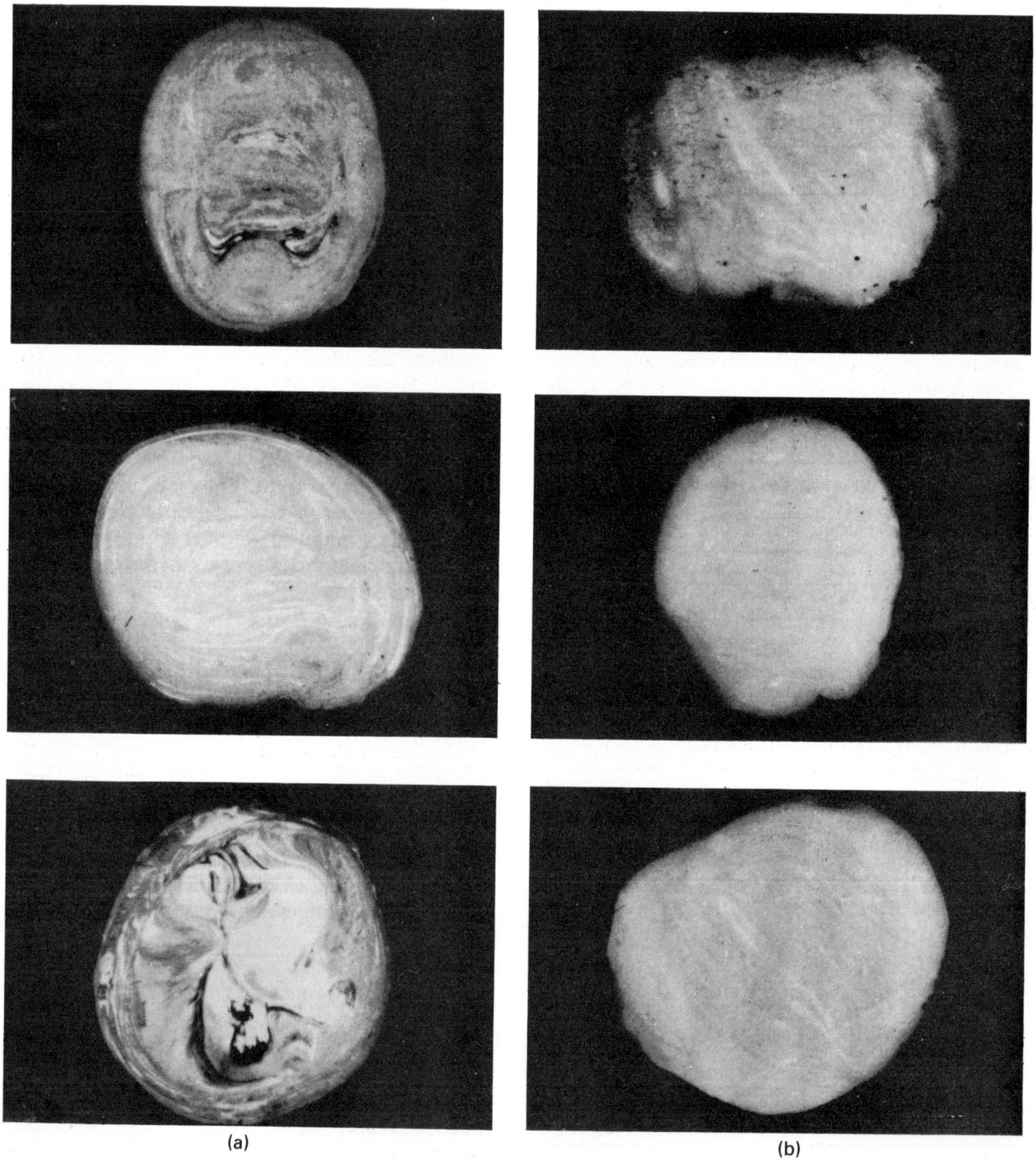

(a) (b)

Figure 3
Mixing and homogenization by (a) 35 mm counter-rotating and (b) 28 mm corotating twin-screw extruders at 25 rpm (top), 50 rpm (center) and 100 rpm (bottom)

locations along the process section. They can also be used to change the direction of shear such that more uniform mixing of the product results. Shear rates up to 10 000 s^{-1} can be obtained using these kneading elements without overworking the product. As a result, products with widely differing viscosities can be successfully mixed in corotating systems.

A comparison of the mixing and homogenization capabilities of similar-size corotating and counter-rotating, fully intermeshing twin-screw extruders is shown in Fig. 3 at various screw speeds. As the screw speed increases, the degree of mixing and homogenization that can be obtained in the corotating machine is superior.

The operating speeds of counter-rotating screws must be kept relatively low because of the shear stresses that are developed at the barrel wall and screw root as previously described in (a) and (b), and the increased tendency for mechanical wear in counter-rotating screws (due to the rolling motion of one screw off the other in the wedge area). Corotating screws can be operated at much higher speeds because of the more uniform overall stress and absence of a calendering effect. When considering throughput, the superior conveying capability of a counter-rotating screw with plug flow is offset by the higher operational speeds of the corotating screw. For example, a 170 mm screw diameter counter-rotating machine used to compound high-density polyethylene has a throughput of up to 1800 kg h^{-1}. A throughput of 5 t h^{-1} could be achieved in the same process using a corotating machine with the same screw diameter (Eise et al. 1981).

3. Residence-Time Distribution

Control of residence time is of great importance in compounding operations, particularly for heat- and shear-sensitive polymers and additives, and for processes that require elevated temperatures. To obtain maximum uniform product quality, the residence time of material in the machine must be short and uniform (i.e., the residence-time distribution must be narrow). Corotating, self-wiping, fully intermeshing twin-screw extruders have such a narrow residence-time curve (see Fig. 4).

Two important elements of residence-time distribution curves are the distance w between the points of inflection and the overall width b of the distribution curve. A small distance between the points of inflection indicates little "back and forth" mixing, whereas a greater distance indicates a larger degree of longitudinal mixing. Width b is influenced by distribution tails, which indicate the cleaning efficiency (self-wiping capability) of the machine.

Figure 4 shows two typical residence-time curves possible for corotating, fully intermeshing twin-screw extruders. The self-wiping capability of these

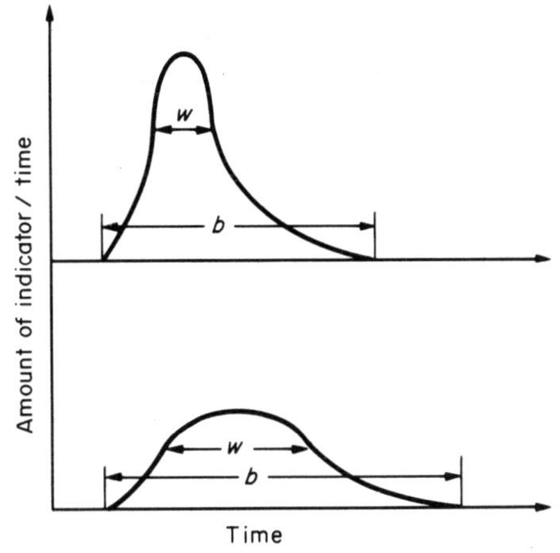

Figure 4
Residence-time curves for corotating twin-screw extruders

machines is evidenced by the absence of the residence-time "tails" discussed previously. This means that no polymer particles remain in the machine for long periods of time, where they could be subjected to severe thermal treatment. There are no dead corners where material can accumulate. The degree of back mixing (width w) is influenced by the screw configuration used in the machine. Therefore, w will be wider. A screw configuration using kneading disks or left-handed pitch-conveying elements will have more back-mixing, and will have more back-mixing, and will therefore provide a wider residence-time curve.

See also: Processing: Classification

Bibliography

Eise K, Werner H, Herrmann H, Burkhardt U 1981 An analysis of twin screw extruder mechanisms. *Adv. Plast. Technol.* 1(2): 18–39

Tadmor Z, Gogos C 1979 *Principles of Polymer Processing*. Wiley, New York, pp. 196–240

K. Eise and M. Abo
[Werner and Pfleiderer, Ramsey, New Jersey, USA]

Molded Fiber Products

Molded fiber is the product of a manufacturing technique in which fibers are dispersed in water and then accreted by filtration on a porous felting die which is in the shape of the desired article. The method has been called "pulp molding" and the product "molded fiber" and the misnamed "molded paper." The US Bureau of the Census uses the term "pressed and molded pulp goods." The products of the industry which are most visible and familiar are flower pots, food trays, meat packaging trays and egg containers. More sophisticated products such as radio speaker diaphragms, automobile components, luggage shells, tropical helmets and combustible cartridge cases are often not recognized as being molded fiber. Williams (1970) has described the process and its many unique advantages.

Tooling for molded fiber involves both expense and expertise but, once the equipment is at hand, the method provides economical mass production. Fibrous raw materials are readily available: strong synthetic fibers are employed when strength is required, refined wood fiber serves for the majority of goods and, for the less costly items, recycled fibers are entirely adequate. Desirable properties are designed into the products. Food trays are porous, will absorb a spill and after use are readily disposable in landfills. Rocket launcher fairings for aircraft are light, rigid and made to break into small pieces as the rockets are launched. Combustible ammunition components, made with a substantial portion of nitrocellulose fibers, are consumed completely on firing, contribute to the propellant charge and leave no unburned residue in the gun tube. As filters, molded fibers form an open structure of graded density which retains the contaminant load while protecting the dense filter bottom from blinding. Shapes which are stable at high temperature, of use in casting metals, can be made by combining ceramic fibers with inorganic binders. Metal fibers can be felted into useful shapes and the fibers bonded by sintering.

The *Annual Survey of Manufacturers* issued by the US Bureau of the Census gives the following values for pressed and molded pulp goods (SIC Code 26460), including bituminous impregnated pipe, in millions of dollars:

1964	94	1980	200
1965	117	1981	240
1966	122	1982	228
1977	226	1983	217
1978	209	1984	236
1979	208	1985	260

1. Manufacture

1.1 Preparation of the Fiber Slurry

Various fiber mixes such as kraft, ground wood, cotton linters, sulfite, rayon, glass fiber, polyester and nitrocellulose are utilized. Thick mats are ordinarily formed, so the stock must be kept open and free draining. The fibers may be cut to a desired length, but heavy beating and "hydration" are almost never required. Addition of size, wet strength resin and resin emulsions are made at the beater. Fibers such as Hercules Pulpex, which cofelt and function as binders, are available. The finished stock is pumped from the beater to the storage chest and diluted to the desired consistency to supply the felting tank.

1.2 Felting Vacuum and Felting Die

Vacuum is ordinarily provided by a liquid piston pump which exhausts a tank preferably placed below the felting tank. The felting die is immersed in the slurry in the felting tank and the fibers drawn on by vacuum-induced flow to the desired weight. With the felting die removed from the slurry, the water collected in the vacuum tank is sent to white water storage or, in some cases, returned to the felting tank. The felting die is made in the shape of the product and must be strong enough to stand up to the considerable force exerted on it when under vacuum. It may be a casting or, for some cylindrical shapes, a metal or plastic pipe. The necessary water channels are provided by drilling $\frac{1}{8}$ in (3.2 mm) holes on $\frac{1}{2}$ in (13 mm) centers. These are interconnected at the outside surface by shallow grooves. The die is then smoothly covered with 40 mesh screen. A male felting die tends to make a thin felt at a sharp radius, while a female felting die thickens the felt at the radius. The latter effect is of value in strengthening corners and radii in luggage shells. The preform can be made thin where desired, as in the hinge of a radio speaker diaphragm, by restricting the drainage at that point in the felting die.

A problem encountered in preforming items of long configuration is unequal distribution of fibers from top to bottom. To correct this, the following procedure is recommended. The inside bottom of the preforming mandrel is closed with a metal disk carrying a metal tube through the center. The inside diameter of the metal tube should carry no more than half the water flow induced by the vacuum system. A second disk is placed near the top of the pipe which divides the mandrel into two chambers, 60% of the volume below and 40% of the volume above. The bottom plate is drilled with a series of holes with an aggregate diameter somewhat less than the diameter of the pipe. These holes are increased or decreased until the felted wall thickness becomes uniform.

1.3 Felting Tank

The fiber concentration of the slurry, the freeness of the stock, the vacuum applied to the felting tool and the time of immersion govern the weight of preform obtained. To maintain consistency at the proper level, stock and water are added in each cycle to supply the slurry which has been removed in the felting operation.

Another method of maintaining felting tank consistency, one which keeps fibers well distributed, is to dilute the stock in the supply tank to the desired concentration, perhaps 0.2%, and pump this continuously at the base of the felting tank, allowing it to overflow symmetrically at the top circumference of the tank and from there return to the supply tank.

The operator periodically weighs the preforms and adjusts the felting time. To assist in holding the assigned weight, the vacuum tank may be equipped with a "sight" gauge in which the water level is marked with a float. The float, at the cutoff volume, optically activates equipment which brings the felting platform out of the slurry.

The felting die is brought out of the slurry with the vacuum still applied to remove water in the preform. A thin rubber blanket may be used at this point, to assist in dewatering. The vacuum is then cut off and air pressure is applied to free and felt. The preform is removed by hand or taken by a vacuum transfer.

1.4 Drying the Preform

The transfer may place the preform on a belt which carries it into an oven for drying. The belt may carry drying forms to receive the felt and to maintain its form during drying. Very open stocks, as used in filtration products, and thin sections, as felted in speaker diaphragms, are dried by pulling heated air through the part. The wet felt may also be dried and pressed in matched metal dies. In this system it is usually the male die which is grooved, and possibly screened, with a vacuum connection to carry away water and steam. There is an unfortunate but natural tendency for the preform to elongate and even tear as the dies are closed. The intrinsic green strength or initial wet strength of the stock is important at this point. The use of the rubber blanket to reduce water content strengthens the preform. Shaped inflatable rubber bags and press-mounted drained dies are used to give firmer, drier felts which do not tear. The female die is given a flared "ironing edge" to reduce the drag on the part as the dies close.

The preform does not flow in the die dry; this is a pressing rather than a molding operation. As a consequence, the weight of the felt must be accurately governed. Too light a felt puts the press pressure on the top of the part, while too heavy a felt puts wedge pressure on the sides. Elongation of the preform as the dies close makes forming the bottom edge difficult, though not impossible. Post-trimming by machine is often required. Drying dies are heated using premixed gas (see Fig. 1) or, where more exact temperature control is required, as in the drying of nitrocellulose, by steam or heated oil.

1.5 Special Forming and Drying Techniques

The fiber slurry may be injected under pressure into split screened and drained molds. After the part is formed, some drying may be carried out by following the slurry with heated air. The dies are then separated, the part is removed and drying is completed in the oven. Duck decoys have been made by this process, with subsequent hardening and water proofing by a varnish impregnation.

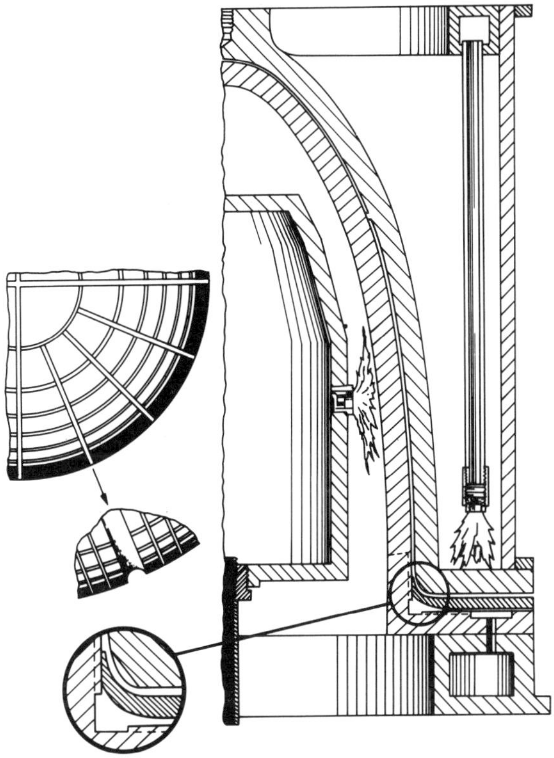

Figure 1
Matched meta driving dies to form break pattern in aircraft rocket launcher fairing (after De Luca 1966)

In the Lass process, large split screened felting dies, which when joined have an aperture at the lower end, are dipped in the fiber slurry and the slurry is drawn in to accrete the part. The dies are then withdrawn and heated air is pulled through the aperture to dry the molding, after which the dies are separated to obtain the part. Large-dimensional advertising signs have been made by this method (Lass 1943).

Preforms which carry from 8% to 20% thermoplastic resin from a beater treatment may be oven dried and then pressed to a dense exact shape in a heated die.

2. *Specific Products*

The radio speaker diaphragm is made in the form of a shallow cone. An outer ring is made thin to act as a hinge, while the center portion is made thicker to carry the voice coil and to reproduce high notes.

Felting is carried out in large tanks with the fiber at low consistency to give optimum formation; diaphragms are held at the drying station to close weight and resonance tolerances. The apex of the cone which receives the voice coil must be molded to an exact diameter. A guided ring is slipped over this section as the drying oven is brought down over the screen-supported wet felt. The dried trimmed cones are dipped in lacquer and may be redipped at the apex to meet waterproofing and sound reproduction specifications.

In a process for felting annular or cylindrical depth-type filter cartridges, the dried cartridges were impregnated with phenolic resin, cured and trimmed (Anderson 1951). Such a cartridge is most open at the outside (upstream) end of the cylinder and most dense at the center-aperture base of the filter. As contaminated liquid flows through the cartridge, particles are removed and stored in the wall according to their size. Thus flow is maintained over a long period of use. The cartridge holds its shape under high pressure drop without a central support. In a modified process the cartridge felt was formed in a small tank fitted close to the felting form (Williams 1970). Slurry was brought into the tank during the felting operation. The composition of the slurry could be changed at will, allowing the filter bottom to be placed at any point in the wall; the rate at which the graded density opened was also brought under control.

Fiber molded shells for luggage are light and strong and have been the basis of a very successful line of carrying cases. If necessary, the shell is strengthened by the incorporation of glass fiber. Such fiber is available as a cut strand coated with an agent which causes it to bond to polyester resin during molding. When cut strand is added to a molding slurry it separates, under agitation, into a multitude of small filaments. These felt to make a bulky preform and a great deal of the reinforcing action is lost. However, if catalyzed polyester resin is added to the slurry along with the cut strand, the resin associates with the glass as a result of the coating applied. Mild agitation is required to bring this about. Cellulose fibers do not make the resin and act as spacers to prevent the glass fibers from sticking together. As the batch is heated, the resin cures and the strands are stabilized. Excellent improvement in impact strength is obtained. If the glass content of the preform is raised to 90% in this method, open preforms are produced which, when dry, can be molded in matched metal trimming dies using the resin pour-on technique.

Dress tropical helmets have been made in large quantities using varnish-impregnated, cloth-covered, molded shells. The edge of the helmet is rolled and adhered, which gives considerable strength to a lightweight article. Combat helmet liners have been made by the same method.

Molded preforms filled with phenolic resin have been dried at low temperatures and then remolded at higher temperatures in polished dies. For a restaurant tray, three preforms have been used; the two outside at 60% resin to give a food and detergent resistant finish and the center preform at 15% resin to give the tray toughness.

Molded fiber frangible nose and tail fairings have been produced in large quantities for airborne rocket launcher systems. The matched metal drying dies press a grenade break pattern into fairings as shown in Fig. 1. The fairings withstand the aerodynamic loading created by high-speed combat aircraft. On firing the first rocket, the fairings disintegrate. The break pattern combined with the phenolic impregnation gives strength, rigidity and small pieces which do not damage the aircraft on breakup (De Luca 1966).

The combustible cartridge case has proved to have advantages. The case is molded of wood and nitrocellulose fibers and contains the propellant: it is itself part of the charge. Use of such a case permits more rapid firing of ammunition rounds. After firing there is no spent case to dispose of, which is a great advantage in the close quarters of a tank turret. Slurry molding and drying of the preform in matched metal dies give the case accurate dimensions. Since nitrocellulose fibers are used, die drying must be done with good control of temperature and time in the die. Dies are shielded to protect operators and safety equipment comes into action in case of a power failure. Preforms containing nitrocellulose fiber have been dried using steam heat at 121 °C without problems (De Luca 1967). The cartridge case may be postimpregnated with resin or the binder may be introduced at the beater. Incorporation of a portion of fibrillated polyacrylic fiber improves the stability of the case (De Luca and Williams 1984).

See also: Papers of Natural and Synthetic Fibers

Bibliography

Anderson L E 1951 Method of making a filter element. US Patent Nos. 2,539,767 and 2,539,768 (30 January 1951)

De Luca P L 1966 Process for making fibrous articles. US Patent No. 3,250,839 (10 May 1966)

De Luca P L 1967 Cartridge case and method for the manufacture thereof. US Patent No 3,320,886 (23 May 1967)

De Luca P L, Williams J C 1984 Fibrillated polyacrylic fiber in combustible cartridge cases. *I & EC Prod. Res. Dev.* 23(3): 438–40

Lass W P 1943 Lass dehydroform. *Pac. Pulp. Pap.* July: 13–16

Williams J C 1970 Pulp molding. Three-dimensional paper products. In: Mosher R H (ed.) 1970 *Industrial and Specialty Papers*, Vol. 4. Chemical Publishing, New York, pp. 198–233

Williams J C 1971 Accretion apparatus. US Patent No. 3,585,107 (15 June 1971)

P. L. De Luca
[Armtec Defense Products, Coachella, California, USA]

J. C. Williams
[Alexandria, Virginia, USA]

Multilayer Polymer Films

There are now many polymer films commercially available, with a wide range of individual properties. However, the complex packaging requirements of today's enormous range of products, coupled with the equally complex requirements of the marketing and distribution functions, means that no single polymer film can be expected to be completely satisfactory in all situations. In general, the answer lies in the use of multilayer polymer films, the deficiencies of one material being compensated for by the advantages of another.

Multilayer polymer films are produced in three main ways: coating, lamination or coextrusion.

1. Coating

1.1 Roller Coating

Lightweight coatings are usually applied by roller-coating methods. Such coatings may be used to confer heat sealability on films such as regenerated cellulose or to improve their barrier properties. Predetermined amounts of coating may be applied by engraved rollers running in a trough of coating material, or by a smooth roller, the excess material being removed by a flexible blade or by a rod that has been spirally wound with a stainless steel wire. Coatings applied by the above methods include vinyl chloride–vinylidene chloride copolymers, acrylic polymers and nitrocellulose.

1.2 Extrusion Coating

Heavier coating weights can be applied by extrusion coating. This is a process in which a molten thermoplastic polymer is forced from an extruder, through a horizontal slit die, onto a moving web of film. As the molten film leaves the die lips, it is pulled down into the nip between two rolls situated close to the die. The substrate, which is moving at a speed faster than the extruded film, draws the molten film to the required thickness, while the pressure between the rolls forces the film onto the substrate. By far the most common polymer used for extrusion coating is low-density polyethylene because of its ease of extrusion, its hot-melt strength and its combination of end-use properties. The most commonly used polymer substrate is regenerated cellulose, but extrusion coating is also used to coat paper, board and aluminum foil.

2. Lamination

Lamination is the process of bringing together two or more films and bonding them with an adhesive or by heating. The main lamination process used for polymer films is known as dry bonding and may incorporate either (a) the use of an aqueous or organic solvent-based adhesive film that is dried prior to laminating, or (b) a hot-melt adhesive, based on wax or on one of a range of polymers. In the first process, the aqueous or solvent-based laminating adhesive is applied in precise amounts to one web. The coated web is then passed through an oven, to remove all water or solvent, and is then combined with the other ply in a pressure nip between two rolls, sometimes with added heat. The second dry-bonding method is based on the use of hot-melt (or 100% solid) adhesives which can be applied by a variety of methods. In one, a slit-die extruder discharges a molten curtain of polyethylene or some other thermoplastic into the nip between the two plies to be laminated. The heated adhesive is then cooled by passing the multi-ply film over a specially designed combining and cooling roll section.

The earliest hot-melt adhesives were waxes of one sort or another and these are still widely used. With more critical performance requirements have come developments in polymer–wax blends and a wide range of modified waxes is available. In general, the addition of polymers to waxes produces tough films, better adhesion characteristics and good moisture vapor barrier properties. Typical modifiers include ethylene–vinyl acetate copolymers and polyisobutylene. There are also hot melts based on polymers, without waxes, including polyethylene, polyamides and polyesters.

3. Coextrusion

The coextrusion process consists in coupling two or more extruders to a single die head. It differs in one important detail from other methods of producing multilayer polymer films in that it can be adapted to produce such films in either lay-flat tubing form or as a flat film. The success of the technique depends mainly on the design of the die and on the way in which the individually extruded polymers are brought together, prior to being extruded as a multilayer film. In essence, there are two different methods of carrying out this operation. In one system, each separate polymer melt is extruded into a separate manifold and then brought together at a common point inside the die. In the other system, known as the laminar now (or combining adaptor) process, the separate melts are brought together in a common manifold. The flow passages are designed to minimize turbulence and mixing of the separate streams is avoided. The main advantages of coextruded films over conventionally produced laminates are lower cost, a lower tendency towards delamination and a greater flexibility in obtaining a wide range of properties. Cost savings are possible (in spite of more expensive equipment) because the finished multilayer film is produced in one operation. In adhesive lamination processes the separate films have to be extruded and then reeled up and put into storage. Later, the reels have to be unreeled and run through the lamination process, thus adding to the labor costs. Cost savings on materials are also possible. The minimum thickness of any particular ply in a conventional laminate is often dictated by its handling characteristics on the laminating equipment rather than the barrier or other properties required by its end use. In coextrusion, however, very thin layers can be produced so that the optimum thickness can be chosen for the job. This is particularly important in the case of high-performance, but expensive, polymers where the thickness used can be kept down to the minimum, consistent with performance.

Disadvantages of coextrusion include the difficulty of utilizing the scrap produced during extrusion. In the extrusion of single-ply films a reasonably high proportion of scrap can be fed back into the extruder. With coextrusion, however, a mixture of polymers is produced without an intervening stage and separation of the plies, even when possible, would not be economic. One solution is to extrude the composite scrap to produce a blended core with other polymers as outer plies to give the required barrier and heat seal characteristics. Another disadvantage is the fact that laminates with "sandwich" printing cannot be produced. This is sometimes important in food packaging where a scuff-proof decoration may be desirable but where the printing cannot be on the inside of the ply in contact with the food.

4. Uses

Multilayer polymer films are used where the performance requirements are such that no single material will supply the requisite properties at an economic cost. For example, low-density polyethylene has excellent heat-seal properties, is a reasonably good barrier to water vapor and is one of the lowest-cost materials. However, it is not a good barrier to oxygen and other gases, neither is it a good barrier to odors. Its softening point, too, is not high enough for uses requiring hot fill or heat-sterilization. Polyamides (nylons), on the other hand, have high softening points and good gas-barrier properties but are not good barriers to water vapor.

4.1 Vacuum and Gas Packaging

Many foodstuffs that are sensitive to oxygen are vacuum-packaged in pouches or sachets. The film must necessarily be an extremely good barrier to oxygen and the pouch seals must also be able to withstand the mechanical stresses entailed. If, in addition, the foodstuff must be protected from moisture pick-up then the performance requirements are quite stringent. One multilayer polymer film that has been used successfully for vacuum packaging is regenerated cellulose, extrusion-coated with low-density polyethylene. The cellulose film supplies the gas barrier properties while the polyethylene gives strong heat seals and water-vapor barrier properties.

Gas packaging is somewhat different. The main factor here is the permeability of the film so that the correct gaseous composition can be maintained inside the pack. One example is that of fresh produce. Fruit and vegetables are still living organisms after they have been harvested, taking in oxygen and giving out carbon dioxide. This respiration depends on the oxidation of sugars to give carbon dioxide, water and energy. If there is not enough energy then the physiological functions depending on that energy cease. Such functions include the ripening of fruits and aging or senescence in leafy vegetables. Until fairly recently the main method of preserving fresh produce has been refrigeration but this is not always a suitable method. Food can lose flavor and texture at temperatures approaching 0 °C and some fruit-loving bacteria thrive in freezing conditions. Refrigeration can also ruin some tropical or sub-tropical crops. Bananas, for example, turn black at temperatures below about 14 °C.

The rate of respiration can also be lowered by limiting the amount of oxygen that enters the pack. However, if a pack contains no oxygen, then the produce inside will start to respire anaerobically and will produce by-products. The balance between starving the fruit of oxygen to retard ripening and providing enough of the gas to prevent the generation of by-products is a subtle one and the plastic film is an important factor. It is not just a matter of picking a film with a particular oxygen permeability. The situation is complicated by the fact that the relative humidity inside the pack affects the way in which the produce modifies its atmosphere, so water vapor permeability is also important. In general, films must be tailor-made to suit each particular application and multilayer films often provide the answer.

4.2 Bag in Box

The bag-in-box concept is probably best known in the form of the 3 l wine boxes, now available in most wine stores and supermarkets. The outer corrugated-board box supplies the rigidity and mechanical protection but the inner plastic film supplies the barrier properties and the flexibility that allows the bag to collapse as the contents are drawn off through a tap. Early bag in boxes were used in the 1960s as bulk packs for milk. The bags were simple affairs, usually made from a single-layer polyethylene film. The development of multilayer polymer films with high barrier properties has opened up the field of use considerably and these containers are now used not only for wines, but also for fruit juices, vinegar, sherry, tomato ketchup, soy sauce and vegetable oils.

Early materials that helped the development of bag-in-box packaging for wine were metallized polyester laminates such as low-density polyethylene/metallized polyester/ethylene–vinyl acetate copolymer. Later developments have been in the field of coextrusions including polyethylene/ionomer/nylon and laminates based on ethylene vinyl alcohol copolymers. Apart from improvements in barrier properties, these coextrusions have the merit of clarity, enabling the air bubble in the bag to be seen, and hence controlled, by the filler. Other advantages are a high interlayer bond strength and improved flexcrack resistance.

4.3 Retort Pouches

The retort pouch is a flexible pack designed to be filled with a food product which is then fully heat processed. The food can then be stored for up to two years or more at ambient temperatures. The performance requirements are therefore protection against oxygen, moisture and light, for around two years, and the ability to maintain seal integrity during heat processing and subsequent storage and distribution. For this application, the barrier properties of a multilayer polymer film are usually reinforced by the addition of an aluminum foil layer. A typical laminate is polyester/aluminum foil/propylene–ethylene copolymer. The foil provides a complete barrier while the polymer layers provide heat seal strength, toughness and a good print surface.

The advantages of the retort pouch are its lightness in weight in comparison with metal cans and glass bottles and its lower storage space requirements. Another factor is its biological stability at ambient temperatures so that it does not require chilled or frozen distribution or storage. Because deep-freeze distribution is expensive, the retort pouch has a distinct advantage over frozen food packs.

4.4 Miscellaneous Uses

Multilayer films are not normally suitable for techniques such as shrink wrapping because of the different degrees of shrinkage possessed by the separate layers. Shrink wrapping would lead to cockling and possibly delamination when heated. Multilayer films are used in stretch wrapping, however, one example being a triple laminate with conventional low-density polyethylene as the outer layers and an inner core of linear low-density polyethylene. The advantages over low-density polyethylene alone are higher tear strength, puncture resistance, tensile strength and elongation. The overall thickness can be reduced, therefore, and the stretchability is greater.

One of the latest innovations in multilayer films is a lamination of two high-density polyethylene films so that the uniaxial orientation of each layer is at right angles to that of the other. The material possesses a very high tear strength by virtue of the cross-lamination and thinner gauges than normal can be used.

See also: Processing: Classification; Tubular Film Extrusion: Dynamics and Heat Transfer

Bibliography

Benning C J 1983 *Plastic Films for Packaging*. Technomic, Lancaster, PA

Briston J H 1983 *Plastics Films*, 2nd edn. Godwin, Harlow, UK

J. H. Briston
[Cheam Sutton, UK]

N

Novel Applications of Polymers

Considered here are a number of specialty polymers developed in the past few years. In many cases the polymers are still in development, while others are being test marketed. They are grouped into five categories.

1. Polymers with Novel Electrical Properties

Among the significant developments reported in the literature are (a) highly conducting and superconducting organic materials, and (b) piezo/pyro/ferroelectric polymers. A number of conjugated and aromatic polymers have been shown to display metallic conductivity when doped with appropriate oxidizing agents such as Br_2 and AsF_5. These polymers include polyacetylene $(CH)_x$, polyparaphenylene and poly(paraphenylene sulfide). The highest conductivity ($10^3\,(\Omega\,cm)^{-1}$) has been achieved with $(CH)_x$ doped with AsF_5. Experimental quantities of $(CH)_x$ are available and there is considerable interest in the development of novel batteries using doped $(CH)_x$ as a cathode. An organic superconductor (tetramethylselenofulvalene perchlorate), with a T_c of 1.3 K at ambient pressure, was reported in 1980. This material closely resembles the behavior of the inorganic polymer SN_x, which was shown to be superconducting in 1975.

Work on piezo/pyroelectric polymers which began in the late 1960s has focused almost completely on poly(vinylidene fluoride) (PVF_2). Companies in both the USA and Japan provide thin films of this material for use as transducers to detect extremely small changes in pressure or temperature. The films have been used as active elements in high-fidelity speakers and as ir detectors. Reports emerged in 1979–1980 that PVF_2 and copolymers of PVF_2 with trifluoroethylene are ferroelectric, which means that the polarization is reversible when the electric field is switched and that a remnant polarization persists when the field is removed. These polymers display very high dielectric constants with values ranging from 10 to 80, an important property for thin-film capacitors.

2. Radiation-Sensitive Polymers for Fine-Line Lithography

The impetus to develop new types of radiation-sensitive polymers (resists) has arisen primarily from the increasing need for higher-density multilevel circuitry for advanced electronic devices. Current lithographic processes depend on near-ultraviolet radiation (404 nm and 436 nm) and are limited to 2.5–3 μm linewidths because of diffraction effects. Thus, polymers are being developed that are sensitive to short-wavelength radiation such as electron beams, x rays and ultraviolet to produce linewidths of 0.5–1.5 μm.

With electron-beam resists, patterns are obtained by focusing the beam sequentially on a series of spots for pulses measured in nanoseconds. This differs from normal lithographic practice, where the resist is exposed to radiation through a mask to obtain the desired pattern. A terpolymer of methyl methacrylate, methacrylic acid and methacrylic anhydride was reported in the late 1970s to undergo rapid chain-scission reactions during E-beam radiation (positive resist), giving the required differential dissolution to solvent developers and thus providing the first useful application of E-beam radiation for direct device fabrication. Recently a need has arisen for more sensitive resists that are also stable to reactive ion etching (RIE), a technique for etching layers between different levels of circuitry on the semiconductor chip. This problem was solved by designing a two-component system consisting of an RIE-resistant Novolac resin and an aliphatic polysulfone which is five times more sensitive to E-beam radiation than the acrylate terpolymer.

X-ray lithography closely resembles E-beam processes, since it depends on formation of secondary electrons to effect chain-scission or crosslinking reactions, but is also similar to optical lithography in that the radiation is passed through a mask pattern. Various x-ray sources can be used, such as a synchrotron, a rotating-anode or a conventional, fixed-anode x-ray source. With synchrotron radiation the flux is so intense that most E-beam-sensitive resists can be used; however, with conventional sources the flux is orders of magnitude lower and very sensitive resists are required to achieve acceptable exposure times. Therefore, researchers have turned to negative resists such as the glycidyl acrylates, which require only a few crosslinking reactions to render them insoluble, permitting selective dissolution of the unexposed regions.

The most promising approach for developing uv resists is to modify the diazo structure (currently used in the near-uv resists) to function at shorter wavelengths. Progress in this area is described primarily in disclosures from the patent literature.

3. Ultrahigh-Strength, Ultrahigh-Modulus Polymers

The announcement in the early 1970s of the high-strength, high-modulus Aramid fibers spun from

H_2SO_4 has prompted considerable effort to develop similar polymers that are melt processible. The most promising candidate is the aromatic copolyester of *p*-hydroxybenzoic acid, biphenol and terephthalic acid. This class of copolymers was shown in the early 1970s to be melt processible at 400 °C, permitting use of standard injection-molding techniques to yield relatively high strength/modulus shapes. Because of the liquid-crystalline nature of the melt (thermotropic), very high mechanical properties can be achieved in the direction of flow, with correspondingly lower properties in the transverse direction. By increasing the content of *p*-hydroxybenzoic acid in the copolymer to 3:1:1, it was possible to prepare a relatively low-molecular-weight copolymer that could be melt spun into filaments and then heated to advance the molecular weight to yield a nonmelting filament whose strength and modulus are comparable to the Aramid fiber. Alternate compositions under development include copolymers of *p*-hydroxybenzoic acid with naphthalene dicarboxylic acid and hydroquinone or *p,p′*-oxydiphenol.

4. Thermally Stable Polymers

Work on thermally stable polymers has been directed at a broad range of needs, such as (a) high-temperature elastomers for use at 250–300 °C under corrosive environments; (b) flame-resistant fabrics that retain dimensional stability in a fire; (c) laminating resins for high-temperature composites; and (d) high-temperature coatings and sintered shapes.

High-temperature corrosion-resistant elastomers stable to 300 °C are required for seals in geothermal-energy-recovery units. Inorganic elastomers such as silicones, polyphosphazenes, polysiloxane–carboranes and fluorinated alkyl silicones have been shown to undergo slow irreversible degradation in air between 200 °C and 250 °C. A system that appears to satisfy these stringent requirements is a perfluoroelastomer consisting of a copolymer of tetrafluoroethylene and (trifluoromethoxy)trifluoroethylene.

There is a critical need for a flame-resistant fiber that is reasonably priced and displays acceptable textile features. The most promising fiber to date is derived from very low cost phenol formaldehyde starting materials. Fabrics do not burn in a flame, but slowly convert to a carbonaceous material which retains excellent dimensional stability providing continued protection to a flame. In addition, little or no smoke is given off, and the primary by-products are water and CO_2. Techniques to dye these fibers to a range of light-stable colors now exist, as well as methods for developing fabrics with necessary wear resistance.

The US Air Force and NASA Laboratories provide much of the impetus for development of high-temperature laminating resins to extend the use of high-strength, high-modulus graphite-fiber composites to temperatures of 250–300 °C. Resins under development include modifications of polyimides which have low viscosities and do not evolve gaseous by-products during cure. For example, low-molecular-weight polyimides with reactive ethynyl end groups can be effectively cured without evolution of any gas.

High-temperature coatings, films and sintered shapes based on polyimides of pyromellitic dianhydride and oxydianiline are available and are stable at 350 °C. Certain aromatic polysulfones and modified polyimides can be injection molded; however, they retain properties only to 250 °C. Coatings and sintered shapes made from the commercially available homopolymer of *p*-hydroxybenzoic acid display significant mechanical properties up to 330 °C.

5. Polymers with Controlled Permeability

A large number of needs exists for polymers that are either highly impervious to, or permit controlled diffusion of, gases or even ions. Illustrative examples are described in the following.

A membrane consisting of a sulfonated tetrafluoroethylene polymer has been developed for use as a solid polyelectrolyte in the electrolytic conversion of water to hydrogen. The membrane is very stable to the KOH environment and provides 85–90% conversion efficiency, compared with 65% for other commercial electrolyzers.

Porous polypropylene films have been prepared by anisotropic crystallization processes which induce micropores extending through the thickness of the film. Such films have been designed to filter particles in the range of several hundred angstroms.

Phenolic fibers have been etched by air at 300–400 °C to produce very fine pores of 10–20 Å and correspondingly high surface areas of 500 m^2 per gram of fiber. Such fabrics absorb trace impurities from gases or liquids in a manner similar to activated carbon granules. These fibers permit the design of greatly simplified air- and water-pollution control devices compared with activated carbon granules, which must be contained and have far lower contact efficiencies.

Coatings that are almost impervious to moisture are required for improved encapsulants. In one approach a prepolymer derived from the reaction of triethynyl benzene with phenyl acetylene was described with this specific goal in mind. The prepolymer was designed to first melt at 110–130 °C to a low viscosity, eliminating any microporosity associated with solvent evaporation, and then the specimen was cured at 150 °C without evolution of volatiles. The water sorption of cured coatings was at least three orders of magnitude lower than values for many solvent processed coatings.

See also: Electrical Properties of Polymers

Bibliography

Bowden M J 1981 Physics and chemistry of the lithographic process. *J. Electrochem. Soc.* 128: 195C

Paul D R 1979 Gas sorption and transport in glassy polymers. *Ber. Bunsenges. Phys. Chem.* 83: 294

Wegner G 1981 Conducting polymers. *Makromol. Chem.* 4 (Suppl.): 155

J. Economy
[University of Illinois, Urbana,
Illinois, USA]

O

Offshore Engineering Applications of Polymers

Many important developments occurred in the production of oil and gas from offshore fields during the 1980s, especially in the North Sea where relatively hostile conditions have stimulated engineering advances. Reductions in the oil price prompted a search for economic solutions to engineering problems. Polymers (and elastomers in particular) now play a vital part in many sophisticated engineering systems and their use can lead to considerable technical and cost advantages. Some components are novel, not found outside the oil industry, making severe demands on knowledge of material properties and require a sound basis of research and development. Most engineering applications have so far involved elastomers rather than plastics or adhesives, but interest in other polymers continues to grow. Applications of polymers in offshore engineering have still to reach their full potential.

In structural applications, there has been little or no use of plastics (fiber reinforced or not) but there is a growing use of elastomers (in structural bearings). Elastomers are unique as engineering materials in their ability to provide designed flexibility in several directions and to sustain large strains without damage in severe service environments. Correct choice can be difficult due to the very wide range of available materials and a general lack of characterization for the engineering properties required for the application. Some costly failures of seals, flexible pipes and other components have occurred, but correctly designed and well-manufactured rubber engineering components can last almost indefinitely in hostile service conditions with no maintenance requirements. The key to future progress will be

(a) greater emphasis on material characterizations, providing engineers with clear performance-related properties rather than standard test data which may be misleading; and

(b) quantitative life prediction.

This article reviews a wide range of different applications of polymers in offshore engineering with an outline of relevant material properties.

1. Structural Design

1.1 Compliant Platforms for Deepwater Oil and Gas Production

In shallow waters, offshore production platforms are usually fixed steel or concrete structures. As exploration and production move to deeper waters, however, the size of a fixed platform required to resist the massive forces of water currents, waves and winds becomes prohibitively expensive. Platform designs have therefore developed which incorporate sufficient compliance to accommodate displacements caused by natural forces without imposing high stresses on the steel or concrete parts of the structure. In several designs this compliance is provided by laminated elastomeric bearings.

The most advanced technology for this use of elastomers has been developed for the tension leg platform (TLP) first installed by Conoco (UK) for the Hutton Field in the North Sea in 1984 (Tetlow and Leece 1982, Wilton et al. 1985). The TLP is a floating production platform held in place by 16 vertical tension legs, which are attached to a foundation templated and piled into the seabed as shown in the model in Fig. 1. The tension legs are held taut because the platform has more buoyancy than weight, being terminated by elastomeric flexjoints at each end which permit substantial angular motions without structural fatigue. The TLP is a compliant structure which responds dynamically to wave excitation. In a free-floating condition, the platform would behave like a semisubmersible and move in the six degrees of freedom associated with all floating bodies, namely surge, sway, yaw, heave, roll and pitch. The tension legs resist the last three but the others still occur and depend on wave excitation and platform response.

The elastomeric flexjoints are primary structural components in this mooring system, located in the crossload bearing assembly and anchor connector with design life requirements equal to that of the platform structure. An extensive design analysis and fatigue test program was conducted to provide detailed information on flexjoint integrity including a fracture mechanics study of the flexjoints, which are composites consisting of alternating layers of nitrile rubber and stainless steel (Leece 1983, Stevenson 1986). The TLP, installed in 1984, has now experienced a number of severe storms without incident. Recent inspections showed that the flexible mooring system was still functioning very well after being in service for five years, in spite of some damage to sealing elements. Some flexjoints have been examined in detail and display no signs of any significant deterioration (McIntosh and Stevenson 1990). This experience gives significant confidence for future use of this class of platform in deeper waters and a number of new TLP structures are now at various stages of design and construction for the USA as well as the North Sea.

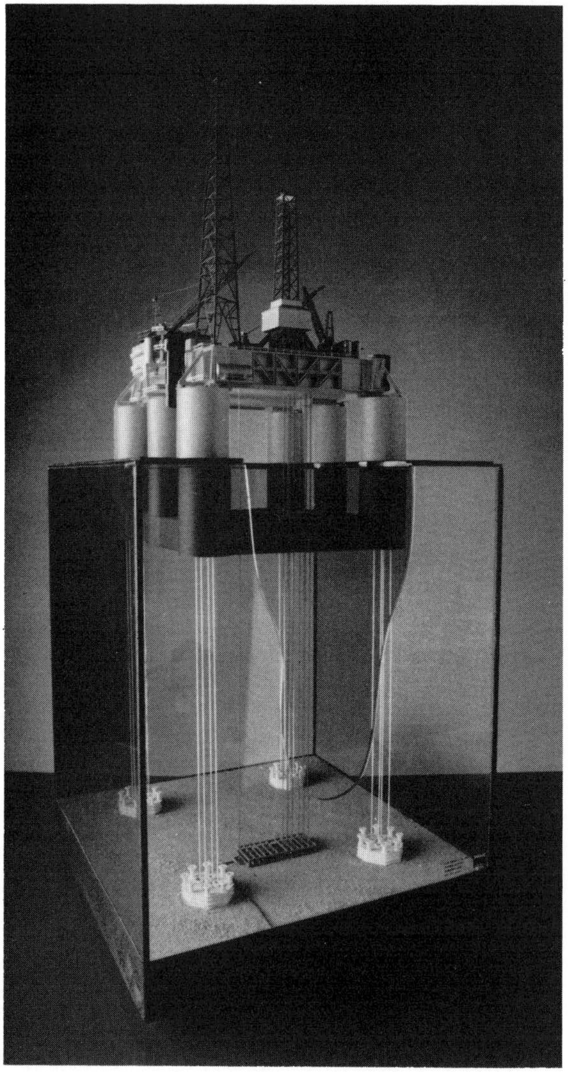

Figure 1
Tension leg platform (courtesy of CONOCO (UK) Ltd)

Detailed design feasibility studies have also been carried out for a deep water gravity tower, designed for water depths of 300–900 m (Sedillot and Stevenson 1983). The tower is supported on a single articulation rubber ball joint at its steel base, which is piled to the seabed. The ball joint comprises 17 individual spherical steel section pads, each laminated with alternate layers of rubber and metal, and is designed for permanent exposure to seawater and to resist the fatigue stressing due to the various complex motions of the tower.

Laminated elastomeric bearings have also been used with success for the huge Oosteschelde Project in the Netherlands, where an estuary has been supplied with a storm surge barrier (van Shaik 1988). The barrier is 3 km long and consists of 65 prefabricated concrete piers between which are 62 sliding steel gates, normally open but able to be closed when storm surges are expected. Large concrete sill beams support the gates and these are fixed and supported by a novel type of bearing which incorporates an inflatable jack, allowing the bearings to be prestressed at installation. Extensive design analysis and environmental testing has been performed to ensure the planned service life of 200 years.

1.2 Design of Structural Bearings and Nonlinear Elastic Properties

The design of structural rubber bearings requires a detailed knowledge of the force–deflection behavior of the material. In contrast to conventional structural materials the working strains are large—up to several hundred percent in some cases—and the behavior is nonlinear. The elastic shear modulus G cannot thus be considered as a material constant as for steel, but will depend on strain. This nonlinearity must be allowed for by the designer.

A second feature, unique to elastomers, is their near incompressibility. In the absence of hydrostatic constraints, they deform at essentially constant volume and are soft materials with shear moduli of typically 1–5 MPa. In the presence of hydrostatic stresses, however, they become extremely stiff with bulk moduli typically 1000–3000 MPa. These effects are exploited in the design of bonded laminated bearings where the degree of hydrostatic stress increases as the rubber layers become thinner (i.e., have higher "shape factor"). This feature, enabling bearings to be designed with stiffnesses differing by two orders of magnitude in different directions, provides enormous potential in design. Further control of directional stiffnesses is made possible by the use of curved laminates, as in the flexjoints for a TLP. Full design equations are to be found in standard texts (Stevenson 1990).

A computer program has been developed by the Materials Engineering Research Laboratory (MERL) to enable nonlinear force–deflection curves of laminated elastomeric bearings to be calculated which can be incorporated into fatigue life analysis. There are also several finite-element (FE) programs available, useful for component designs of complex shape. However, a drawback of existing FE codes for elastomers is that they are based on approximations to linear material behavior which can be unrealistic, and they do not provide estimations of crack growth rate.

1.3 Fracture Mechanics: Fatigue Life Determinations

Structural components are required to have detailed analysis of fatigue life expectations, involving either fracture mechanics analysis, full-scale fatigue testing or both. Fracture mechanics has been used to study crack growth in elastomers with considerable success for many years (Rivlin and Thomas 1953, Lake and Lindley 1966, Sedillot and Stevenson 1983, Stevenson 1986, 1990) and uses tearing energy T defined as the energy required to cause unit area of new crack growth, that is,

$$T = \left(\frac{\partial U}{\partial A}\right)_l$$

where U is the stored elastic energy and A is the area of new crack growth.

The tearing energy T is equivalent to strain energy release rate G or J-integral, the terminology usually used for structural adhesives and plastics, but calculating the tearing energy for complex-shaped engineering components can be difficult. Solutions exist for several different geometries in tension and for bonded elastomeric layers in shear and compression (Stevenson 1986). In shear and in compression, tearing energy is not a direct function of crack length, thus fatigue crack growth will not accelerate as in tension. Thus, elastomers are not normally used in tension for engineering applications, but in compression or shear. This approach has been successful in unifying dynamic fatigue crack growth in shear, compression and tension relevant to the design analysis of flexjoints for the TLP and the deep water gravity tower (Sedillot and Stevenson 1983).

If the tearing energy can be calculated, fatigue lives may be determined by reference to dc/dN curves characteristic of the elastomer, where c is the crack length and N is the number of cycles. The relation between tearing energy and crack growth rate is a material characteristic independent of component geometry. This relation is different for different elastomers, but follows a characteristic form. At very low tearing energies, only chemical failure mechanisms occur, which proceed at an extremely low rate (0.1 mm per year). There is a minimum threshold value of tearing energy T_0 below which no mechanical fatigue crack growth will occur. At higher values of T, growth rate increases with different power law relations. At a high critical value T_c, failure will occur from a single cycle. Consideration of the shape of the T vs dc/dN curve provides a basis for fatigue resistant design—namely, to ensure that the maximum tearing energies of a bearing in service are in a stable region well below T_c.

Different elastomers have different shapes of T vs dc/dN curve and this provides a basis for materials selection. Figure 2 shows the characteristic fatigue behavior of four different engineering elastomers: natural rubber (NR), polychloroprene (CR), ethylene–propylene terpolymer (EPDM) and butadiene–acrylonitrile copolymer (NBR). Across the tearing energy spectrum, different elastomers will exhibit the best fatigue resistance. NBR is more resistant than NR at low tearing energies but less resistant at high tearing energies. Correct materials selection requires identification of the relevant part of the tearing energy spectrum for each application: no simple single-point fatigue test will give meaningful results. Some elastomers will also exhibit time-dependent fracture, which requires separate characterization.

Prediction of the service life of a component requires analysis of the effect of a crack on performance. In compression, even substantial crack growth does not remove load-carrying capability—and a rubber bearing may be considered fail safe in this mode. In tension, much less crack growth may be catastrophic.

1.4 Environmental Durability

At very low temperatures all rubbers undergo a transition to the glassy state, becoming brittle and stiffening by factors of about 1000, when they clearly cannot function in any engineering application. Table 1 summarizes the typical glass-transition temperatures for engineering elastomers. At less severe

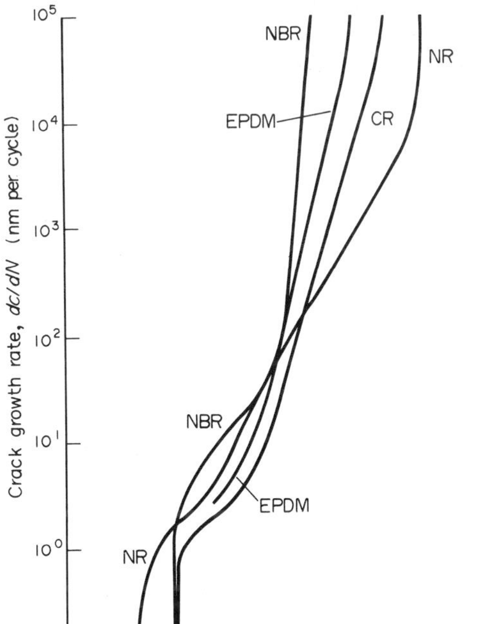

Figure 2
Fatigue behavior of elastomers from fracture mechanics tests (note the logarithmic scale)

Table 1
Typical glass-transition temperatures for engineering elastomers

Elastomer type	Glass-transition temperature, T_g (°C)
Silicone rubber	−80 to −125
Butadiene–rubber (BR)	−75 to −100
Natural rubber (NR)	−70
Butyl rubber	−65
Ethylene–propylene rubber (EPDM)	−40 to −60
Styrene–butadiene rubber (23% styrene) (SBR)	−60
Chlorosulfonated polyethylene (Hypalon)	−57
Polyurethane (polyether type)	−55
Acrylonitrile–butadiene rubber (28% ACN) (NBR)	−50
Polychloroprene (CR)	−40
Epichlorhydrin	−20 to −40
Fluoroelastomer (Viton)	−30
Actylonitrile–butadiene rubber (40% ACN) (NBR)	−15
Acrylonitrile–butadiene rubber (50% ACN) (NBR)	+5
Polyacrylic rubber	−15

low temperatures, strain crystallizing rubbers can undergo crystallization stiffening of 5–100 times depending on the compound. Below a depth of 10 m, seawater temperatures decrease uniformly to a fairly constant value of about 3 °C which could lead to crystallization in some compounds of NR and CR. The ability to strain crystallize is an advantage for fatigue resistance, but low-temperature crystallization can be a disadvantage for long-term flexibility which can be important for structural bearings.

1.5 Seawater Environments

Seawater is not generally a deleterious environment for elastomers. Most studies have shown most elastomer types to be relatively unaffected by seawater. The first British Navy submarine, the Holland I (built in 1901), made use of rubber seals around its torpedo hatch doors. It was lost at sea in heavy weather while being towed to a breakers yard and recovered in 1982. After immersion for 80 years in seawater, the natural rubber seals were still found to be in reasonable condition (Ab-Malek and Tyler 1986). All elastomers will absorb some water over long enough periods of time by diffusion mechanisms, causing increases in mass and volume, but not usually affecting modulus or strength significantly (Ab-Malek and Stevenson 1986, Stevenson 1987). Seawater can, however, be detrimental to rubber–metal adhesive bonds. Laboratory tests and deep-sea trials off the northwest coast of Scotland (Sedillot and Stevenson 1983) suggest that rubber–steel bonds can be highly durable in seawater although, if the steelwork is connected to an anode system, the cathodic potentials may prevent steel corrosion but enhance the rate of adhesive bond failure (Stevenson 1987).

Environmental testing is an important part of the evaluation of any design of structural bearing since its life can be limited as much by environmental effects as by fatigue crack growth.

2. *Antivibration Mounts*

In the development of offshore platforms it was essential to isolate the accommodation modules used by off-duty staff from the noise and vibration generated in working areas. These modules are essentially large multistorey hotels and the isolator units are required to act as structural supports under a wide range of dynamic and static loadings, often in hostile marine environments as well as providing vibration isolation.

The design approach is similar to that for structural bearings, but needing precise control of stiffness to ensure that the natural frequency of the system is correct for the vibration isolation required. Damping characteristics of the material can also be important. Figure 3 shows typical dynamic properties for a polychloroprene compound, illustrating the properties that need to be considered. Dynamic modulus is stable at frequencies up to 200 Hz and the loss angle, representing damping, is relatively low. Combining these results into a single transmissibility curve, as shown, provides a useful indication of the effectiveness of the antivibration mounts over a range of frequencies. The compounds would usually be based on natural rubber or polychloroprene. Antivibration mounts are also necessary for machinery, where similar principles apply, except that the loads and therefore elastomeric mount stiffnesses are usually lower.

An elastomeric isolator designed to support 1000 t maximum load with a natural frequency of about 10 Hz is illustrated in Fig. 4. Elastomeric bushings

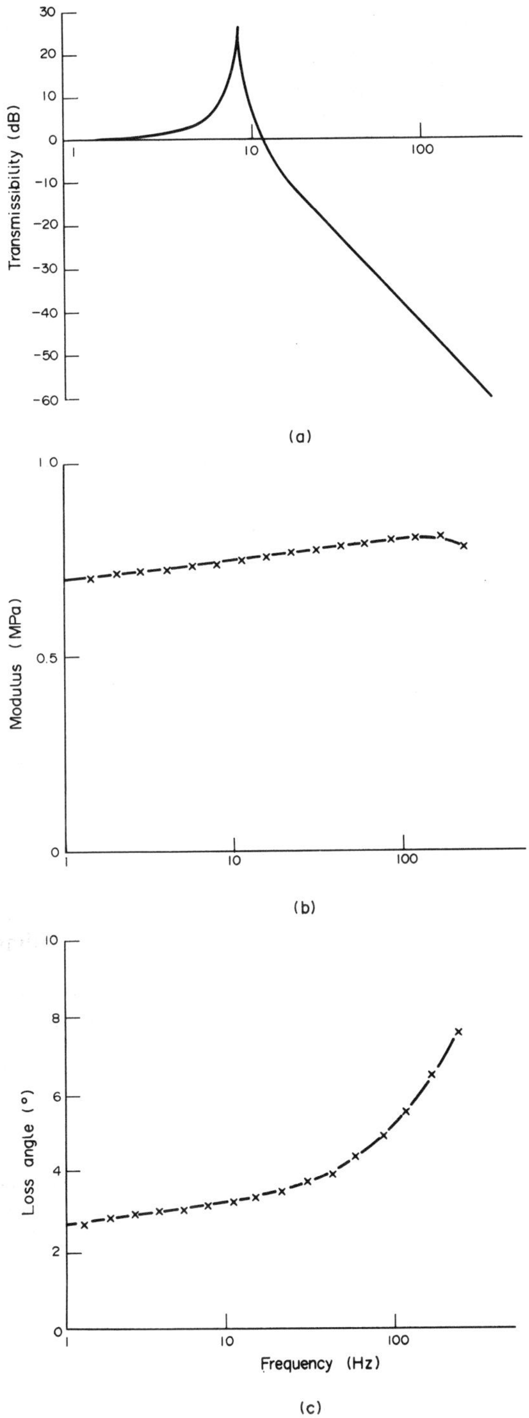

Figure 3
Typical elastomer dynamic properties:
(a) transmissibility vs input frequency, (b) dynamic stiffness (modulus) vs frequency and (c) loss angle vs frequency (data shown are for a sulfur-modified polychloroprene supplied by DuPont (UK) Ltd)

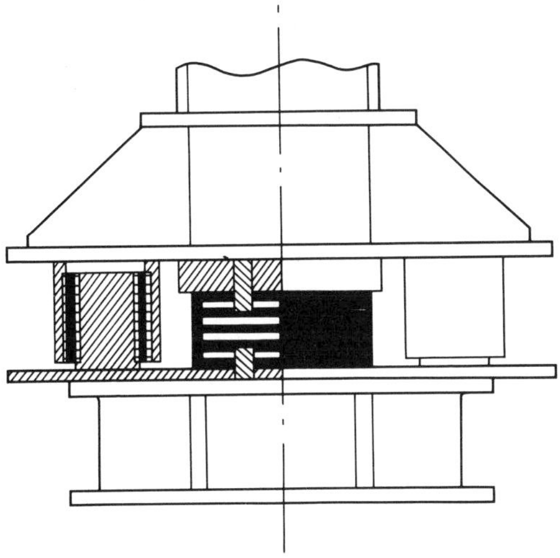

Figure 4
Schematic view of 1000 t load capacity antivibration mount for an accommodation module (courtesy of BTR/Andre Ltd)

are provided in addition to the laminated bearing for lateral constraint against wind loading and so on. Similar antivibration mounts are used to support helicopter landing decks—although usually with lower load capacities (20–50 t). Bridging between different parts of the deck can be supported by elastomeric pot bearings to provide rotational flexibility with high lateral and vertical stiffness.

Introduction of more stringent legislation to restrict noise levels in offshore platform work areas (Vardal 1988) is likely to place further demands on the design of effective polymeric components for noise and vibration attenuation.

3. Deck Mating Systems

It has become increasingly attractive for oil companies to complete structural works on the hull/legs and the deck of an offshore platform separately and then to mate the deck to the hull offshore involving accurate location of connecting components, often under sea state conditions which could cause damage to parts of the structure. To accommodate this, special elastomeric shock cell systems have been designed to absorb the energy, prevent damage and ensure smooth location during installation. This application subjects the elastomer to very severe strains for the few hours it is required to function during the mating operation. This system was used for the Hutton TLP. A further development of this concept is the system developed by Conoco for taking a completed deck, loaded out on a single barge and placing it onto a preinstalled jacket. The load transfer system incorporates a series of elastomeric elements designed especially for this function. An accurate analysis of the nonlinear force–deflection behavior of elements was achieved with the good agreement between analysis and scale test results which is an essential prerequisite for a successful design.

4. Platform Fendering

Energy absorbing elastomeric fenders have been used for over 30 years to protect vessels from damage when berthing against a harbor wall or wharf. There, wave conditions need not generally be considered and the vessel approach velocity is controlled (usually $<0.2\,m\,s^{-1}$). For offshore platforms, this situation is reversed and fenders are required to protect the legs of a platform from damage by supply vessels. Energy absorbing fenders have been developed, particularly for offshore platform conditions as in the North Sea, consisting of a standard shear shock cell with a cylindrical layer of rubber together with an eccentrically mounted loading shaft in a rubber collar inside an outer steel cylindrical can—the eccentric bumper ring (EBR), shown in Fig. 5. This is designed to deflect in a controlled manner under axial and lateral loads. In the open sea, the pitch and roll of the vessel can cause point loads to be applied at almost any point on the contact surface and in almost any direction. The units also need to present as low a surface area as possible to prevent wave damage.

5. Applications in Drilling and Production Riser Systems

5.1 Flexjoints

Drilling and production riser pipes are frequently required to have pivot joints to provide flexible links. Figure 6 shows a typical pivoting flexjoint that uses elastomeric seal and bearing elements to work as a ball-and-socket joint. The total flexjoint assembly is in tension when installed, such that the elastomer elements are loaded in compression. During flexing, the elastomeric bearings and seals are both deflected in a shearing mode. The type of elastomer used, particularly for the sealing elements, will depend on the nature and temperature of the liquid flowing through the pipe, commonly NBR since it

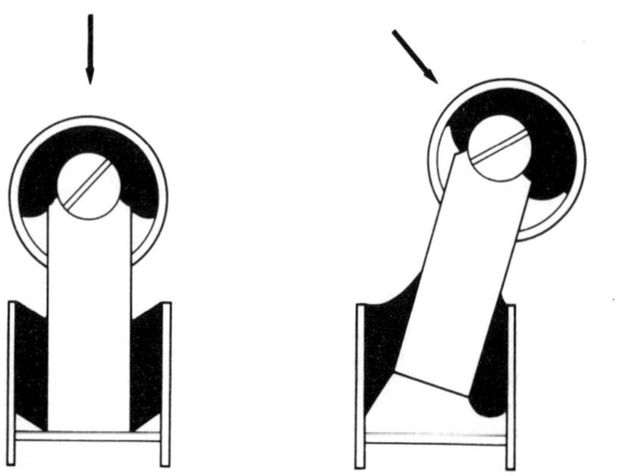

Figure 5
Combined shear-type rubber shock cell and eccentric bumper ring for multidirectional energy absorption (courtesy of Regal Technology Ltd)

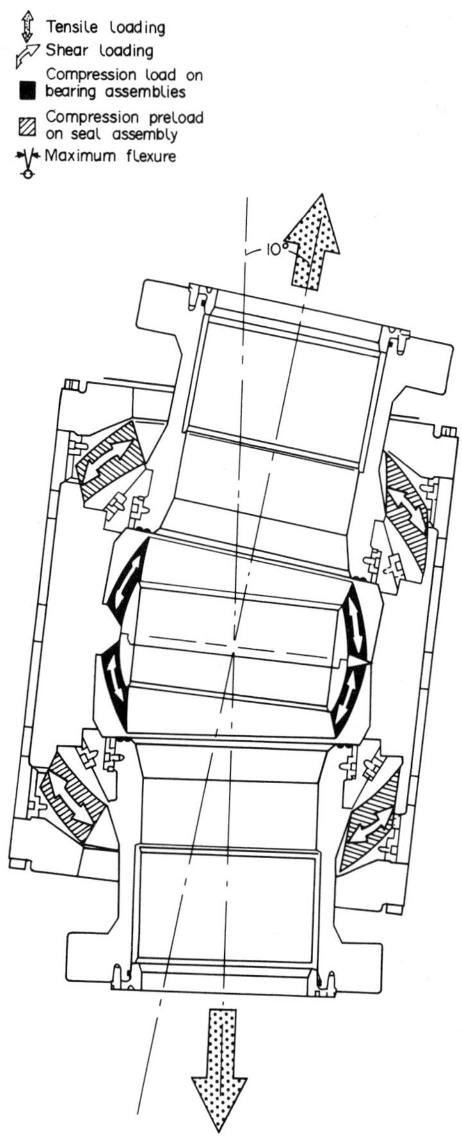

Figure 6
Riser flexjoint arrangement incorporating elastomeric bearings and seals (courtesy of Lord Corporation)

can be compounded to have good mechanical properties while retaining very good resistance to contact with hot oil. This type of flexjoint requires much less maintenance than a nonelastomeric joint since no lubrication is required, the design is fail safe (some elastomer tearing would not impair its function), visual inspection is possible without the need to dismantle a complex assembly and there is no breakway force which occurs with stick–slip condition in sliding joints. Elastomeric flexjoints are finding increasing use and usually have a longer life than other types. The cost of disruption to drilling or production operations caused by bearing failure far outweighs their initial cost.

5.2 Oil Transport Hoses

Many developments have been made in the design of hoses for transporting oil to and from the large tankers used offshore. Hose diameters have increased from 15–20 cm in the early 1960s to a current value of 60 cm. A 1.5 MPa hose type might have to withstand 7.5 MPa pressure for 15 min before bursting. This pressure can create an end loading of over 250 t, a severe test of the bonding systems, the correct winding angle of reinforcements and the integrity of the elastomer itself. Different parts of the hose may utilize different types of elastomer: linear material must be oil resistant requiring use of NBR, fluoroelastomers and so on. The resistance of the hose walls to kinking is also important and thick steel reinforcements are being replaced by polymeric fibers giving structures which have greater recovery.

5.3 High-Pressure Flexible Elastomeric Pipes

High-pressure flexible elastomeric pipes have also been developed for use in riser systems using multiple layers of elastomer bonded to steel wire (Griffiths and Cheetham 1988, Seregely et al. 1988). These are more ruggedly constructed than hoses and may incorporate interlocking steel linings for crush resistance. Problems can be encountered with the linings in particular due to the need for repeated depressurization and the phenomenon of "flow slugging." Some systems are fully elastomeric with braided wire reinforcement. Other types of flexible pipe (less flexible), using unbonded layers of nylon and other plastic reinforcements (Raveau and Simon 1988), have enhanced environmental resistance to oil and gas and particularly to explosive decompression. They can, however, be prone to fretting fatigue between the unbonded layers during bending. The use of flexible pipe is increasing for riser systems and associated systems.

5.4 Rotary Drilling Hose

Rotary drilling hose is similar in construction to bonded elastomeric high-pressure flexible pipe but is required to supply the drilling fluid and survive large numbers of bending motions during typical drilling operations. Bending resistance can become more important than environmental resistance of the lining.

5.5 Umbilicals

Umbilicals have evolved from simple hydraulic lines to sophisticated bundles of control lines needed to control remotely operated valves and other equipment. Oil and gas production in relatively deep waters requires a very high level of reliability for this type of application. Kevlar-type *p*-aramid fibers have found wide usage for the reinforcement of umbilicals—an estimated 10 000 km of such hose have been installed in the North Sea alone. The properties of *p*-aramid fibers of stretch resistance combined with high tensile strength and chemical resistance to hydraulic fluids and seawater have made this polymer an attractive choice (Rothuizen 1988).

6. Sealing Systems

Exploration and/or production of oil and gas are involving increasing well depths with increasingly hostile downhole conditions of high temperatures and pressures with concentrations of wet carbon dioxide, hydrogen sulfide, methane or steam. New materials and seal designs have thus been developed for these environments. Applications generally fall into two categories: drilling and production. Drilling applications include pump valve seats, pump pistons, valve liners and seals and drillhead elastomers. Production applications include packers to tubing seals, packer rings and cups, subsea safety valve seals, O-rings for valve stems, downhole valve seats and low-pressure casing annulus seals, and blowout preventers.

Concentric steel tubular casings are passed from the ground into the oil/gas bearing region below and these are suspended from and sealed off at the wellhead. The casing string is normally run to the bottom of the well and cemented in place, providing a conduit from the producing regions to the well surface. An inner retrievable tubing string is usually run down to protect the casing and to enhance/help control the flow of fluids from the well. At the lowest end of the casing, the tubing string is sealed to it by means of a production packer.

6.1 Wellhead Equipment

The flow out of or into the well is controlled by a series of valves at the wellhead, known as a "Christmas tree." This wellhead equipment may control not only the flow of the hydrocarbon product out of the well, but also the injection of water and compressed gas back into the oil-bearing strata. One of the most critical seals in wellhead equipment isolates one casing anulus from another: failure can result in rupture of the casing and blowout. Seals for this application can be up to 500 mm in diameter and normally use NBR or fluoroelastomer due to the extremes of temperature and aggressive media. They are tested for up to 80 h at fluctuating pressures (up to 70 MPa) and temperatures (up to 90 °C). Wire/fabric reinforcement can be used to inhibit extrusion of the elastomer into unavoidable gaps in the assembly under the high operating pressures.

6.2 Downhole

Below-surface, inside-well (downhole) conditions can be even more aggressive. Elastomeric sealing systems (known as production packers) are used to seal the base of the tubular casing to the inner tubular string. Production packers may be either permanent, being removed from the well by destructive milling, or retrievable, when they may be reconditioned and reused. Figure 7 shows the sequence of operations for a novel design of inflatable packer developed by Nowsco, designed for multiple high-pressure inflations and deflations at full downhole temperatures and pressures. NBRs are most frequently used for downhole seals, although for deep hostile wells a perfluoroelastomer may be required. Elastomeric seals are also used in subsurface safety valves, which are designed to close off and contain a well in an emergency: these valves are usually controlled by hydraulic lines from the surface supplying fluid under pressure to maintain the valve in an open position.

6.3 Drilling

Another type of closure element is the blowout preventer (BOP), designed to seal off gas or oil pressure in the event of a drill breaking into an oil or gas pocket. The elastomeric BOP closure element contains a configuration of bonded steel inserts and is designed to seal around a drill pipe to close off the hole completely. In this application, the mechanical properties and strength of the elastomeric compound are more critical than its long-term thermal or oil resistance.

6.4 Applications at Elevated Temperatures

At high enough temperatures all elastomers will undergo degradation reactions leading to a loss of physical properties. Moderately elevated temperatures may cause the scission and reformation of molecular chains and a net formation of additional crosslinks leading to some hardening of the elastomer. At higher temperatures the scission of crosslinks may outweigh crosslink formation with breakdown in structure and a net softening of the elastomer. At very high temperatures, a breakdown of the main molecular chain will occur with charring and embrittlement. Oxygen concentration is important in determining the effect of temperatures. In the absence of oxygen, most types of elastomer could, for example, survive temperatures of 175 °C whereas, in the presence of oxygen, degradation would occur very rapidly. Hydrogen sulfide gas has also been reported as a cause of chemical degradation. Table 2 summarizes realistic upper service limits of most elastomers in engineering applications. This guide has to be approximate, since the actual suitability will depend on the properties measured, exact compounds used, component size and the length of service required at the temperature. New developments in perfluoroelastomers offer the significant advantages of enhanced chemical and thermal resistance (Weston 1988) but often suffer from the disadvantage of low strength and poorly characterized mechanical properties for engineering applications.

Elastomeric components may be exposed to crude hydrocarbon oils containing gases (CH_4, H_2S or CO_2) at elevated temperatures. In general, the elastomers that have the best mechanical properties and resistance to low temperatures (e.g., NR) have the poorest resistance to gas, oil and elevated temperatures. Thus, Table 2 also provides a rough guide to the suitability of different elastomers to oil and gas environments. A failure problem that adversely affects the functioning of elastomeric sealing rings—particularly in wellhead equipment—is known as explosive decompression. This occurs when the elastomeric material is subjected first to high gas pressure (e.g., CO_2 at 50 MPa) when gas dissolves in the elastomer, followed by a substantial and relatively rapid drop in pressure. This can cause internal rupture of the material as gas comes out of solution to form pockets which expand to cause blisters or splits. Expensive failures have resulted in service; some improved materials have been developed, but no complete answer has yet been found. High resistance to tear and to oil and gas at high temperatures is required, but is difficult to achieve. In practice, improved performance is likely to require a more complete understanding of the interaction of mechanisms of chemical degradation, diffusion and permeation and creep and internal fracture failure (Campion 1988), rather than the invention of new polymers.

7. Corrosion Protection Applications

7.1 Coatings and Linings

Elastomeric linings have been used for many years to protect the steel walls of vessels storing or transporting corrosive fluids. With the development of the offshore oil industry, this application has been extended to many different types of pipework and vessels, used in processes connected with the exploration and production of hydrocarbon oil or gas. The most corrosive media are usually based on seawater, at elevated temperatures. Many methods of protection are available in corrosion engineering, including metallic cladding, cathodic protection and paint films. Thermoplastics and thermoset resins have also been used, specific polymers including polyethlene, PVC, nylon, thermoplastic alloys, epoxy resins and

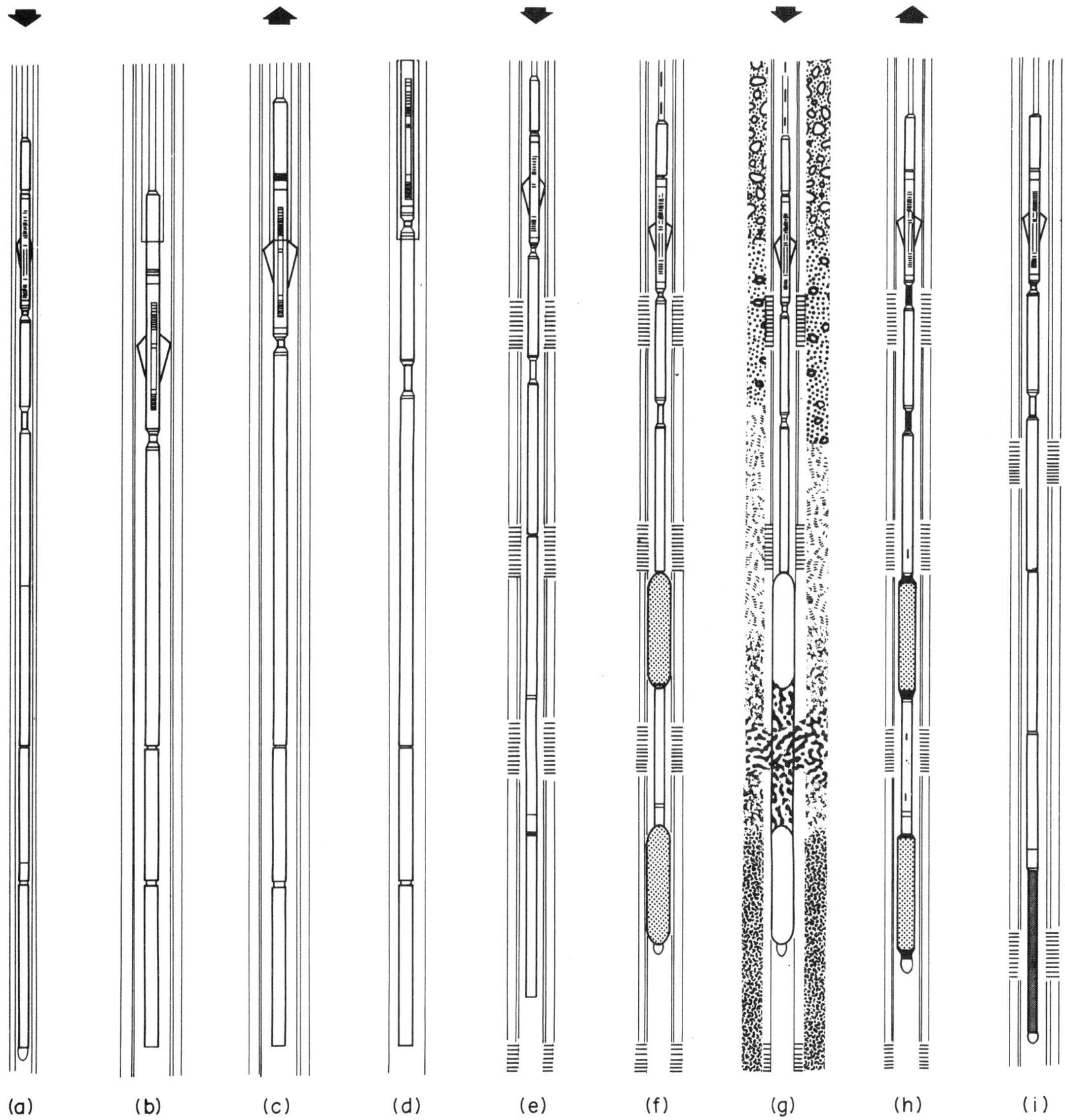

Figure 7
Straddle packer tool sequence of operation: (a) running in hole; (b) run out of tailpipe and stop; (c) pick up so that tubing end locator engages mule shoe or wire-line reentry guide etc., registering additional load on injector weight indicator; (d) increase pull to compress spring in the tubing end locator and collapse arms inside the tailpipe (the additional load on the indicator will disappear and the precise positioning of the tool is now registered, allowing accurate depth counter calibration); (e) run back into well to the required depth of the first treatment interval; (f) once in position increase flow rate to close dump sub, inflating packers; (g) when the correct packer pressure is reached set down weight closing inflation ports and opening annulus treatment ports—start treatment; (h) after treatment pick up weight to reopen packer ports and pull to open dump sub allowing packers to deflate; and (i) in its deenergized condition the tool can be run to the next treatment interval and the entire setting/treatment procedure repeated (courtesy of Nowsco Ltd)

Table 2
Resistance to thermal degradation

Highest service temperature (°C)	Rubber type
75	polyurethane styrene–butadiene natural rubber
100	butadiene rubber polychloroprene
125	nitrile rubber butyl rubber EPDM
150	chlorosulfonated polyethylene polyacrylic rubber epichlorhydrin
200	fluoroelastomer fluorosilicone rubber
250	silicone rubber
300	perfluoroelastomer (Kalrez)

PVDF resin. However, there is a growing use of relatively thick layers of elastomer, bonded to the steel walls of the pipe or vessel. Correctly applied, an elastomeric composition can have an adhesive bond to steel whose durability far exceeds that of plastics or paints. There can, however, be greater difficulties of application and for all coating systems careful metal surface preparation is essential. Underwater pipes are also subject to fouling by marine organisms and some coatings aim to inhibit this. For example, a new type incorporates copper granules, to prevent fouling and at the same time protect the steel from corrosion (Seregely et al. 1988).

In the external coating of risers, flow lines, spools and J-tubes, the corrosive medium is always seawater at ambient temperature. The steel pipe may, however, be at temperatures of 85 °C, due to carrying hot oil. The coating is applied to pipe which has been prepared for bonding by grit blasting, degreasing and application of bonding agents. Uncured elastomeric sheet is then applied by wrapping around the pipe and the assembly is cured in an autoclave or by means of heating blankets. In the internal lining of vessels and pipework, seawater-based fluids can still be the main concern, with up to 5% of oil, plus other fluids. The main property requirements are that the elastomeric coating should prevent the migration of water or chloride ions to the protected steel, possess a durable bond to steel (Sedillot and Stevenson 1983) and be strong enough to resist tearing under normal handling and installation conditions. The elastomer most frequently used is polychloroprene (Stevenson 1990). The Plastics and Rubber Institute (1987) have printed a guide to corrosion control with elastomers. Application techniques, which can be more important than polymer selection, are described in detail.

7.2 Chemical Injection Technology

Continuous downhole injection of inhibitors has been reported as being effective in reducing internal tubular corrosion in both oil and gas wells (Ringler and Howell 1988). A key component of a system to achieve this is the chemical injection line, an encapsulated hydraulic tubing which connects a surface pumping unit with an injection valve at or near the packer. A fluoropolymer has been developed for use in this application with reported success.

8. Structural Composites

There has to date been relatively little use of fiber-reinforced plastics (structural composites) in offshore engineering. However, there have been several extensive studies and a number of applications are under consideration, which offer potential for weight reductions and corrosion resistance.

8.1 Riser Tubes

Research has been carried out on filament-wound fiber-reinforced plastics for riser tubes. These may contain more than one structural layer to resist axial stresses and bursting stresses independently, using carbon fibers for the longitudinal layers and glass fibers for the circumferential layers to form a "hybrid tube." It is claimed that this approach enables axial strains from thermal and pressure effects to be eliminated. Applications proposed for this type of composite tube include kill and choke lines for drilling risers, TLP production risers, satellite lines for production risers with subsurface buoys and production/injection tubings.

8.2 Pipes for Tankers

Glass-reinforced plastic pipe systems have been developed for tankers to replace some metal pipes for problematic applications. Improved fire safety (of pipe contents) has been demonstrated (Grimm 1988).

8.3 Large-Diameter Pipelines

A study has been made of using portable composite pipe manufacturing to construct jointless large-diameter pipes. These could be required to have a 2.5 m diameter and an ability to withstand 6 MPa. The feasibility of this has been studied for the arctic environment between Prudhoe Bay and Edmonton.

8.4 Wellhead Protection Structures

Glass-fiber-reinforced plastics (GRP) have been used in the development of a wellhead protection structure for satellite wells for the Draugen field in Haltenbanken, northwest of Trondheim, Norway. The aim of such structures is to protect wellhead equipment from damage and prevent damage to fishing gear in the event of overtrawling. The development work included a shaping study and also environmental resistance.

8.5 Fire Safety

Concern over fire safety is the single most significant issue concerning the offshore topside use of structural composites. The potential advantages of reduced weight and corrosion, and indeed fire/blast protection, have led to several large feasibility studies to review the whole area.

Prime applications for fire-resistant grades of glass–phenolic composites include fire water-main piping, walkways, accommodation modules and blast walls. The fire advantage such materials offer is due to the low thermal conductivity and heat capacity of the composite (lower than steel by a factor of 100), rather than resistance to flamespread which is inferior to steel. Grades of composite that use reinforcing fibers of high strains to failure can be used for blast-wall design, an application where composites offer preeminent performance as demonstrated by their military use for ballistic and explosive protection.

9. Future Developments

A very wide range of new applications for polymers (and especially elastomers) has been stimulated by the demands of the offshore oil and gas industry. Some of these involve highly critical components which have already made a major contribution to the safe and economic production of oil and gas in increasingly hostile and difficult conditions. The engineering requirements have also stimulated new polymer science and technology advances.

Problems have been encountered, however, where materials have been incorrectly specified or inadequately characterized for the engineering service required. Future developments will require improved communication between engineering end users, manufacturers and raw material producers—plus even more reliance on specialist knowledge of materials. One barrier to engineering confidence is a reluctance by manufacturers to divulge the composition of their materials to end users—or to accept standard composition benchmarks, as in the steel industry.

Another barrier to engineering confidence is a lack of convincing and well-validated methods for determining component life expectancy. Considerable progress has been made (e.g., for TLP flexjoints) but much research remains to be done before engineering design with polymers can be carried out with complete confidence.

See also: Design of Elastomer-Based Engineering Products; Marine Coatings

Bibliography

Ab-Malek K, Stevenson A 1986 The effect of 42 years immersion in seawater on natural rubber. *J. Mater. Sci.* 21: 147–54

Ab-Malek K, Tyler S 1986 *Rubber Dev.* 39: 6

Bowditch M R, Oliver R A 1988 An underwater adhesive method for steel offshore structures. In: Stevenson 1988, Paper 34

Brockbank J W 1988 A flexible neoprene/copper nickel material for long term anti-fouling. In: Stevenson 1988, Paper 21

Campion R P 1988 Failure mechanisms in elastomers in high pressure oil and gas conditions. In: Stevenson 1988, Paper 5

Griffiths A R, Cheetham I 1988 Elastomeric high pressure flexible pipes. In: Stevenson 1988, Paper 16

Grimm G C 1988 Glass reinforced plastics pipe systems for shell fleet marine applications. In: Stevenson 1988, Paper 17

Lake G J, Lindley P B 1966 Fatigue of rubber. *J. Appl. Polym. Sci.* 9: 176–86

Leece M J 1983 Hutton tension leg platform mooring system flexjoint. In: Stevenson A (ed.) 1983 *Rubber in Offshore Engineering*. Hilger, Bristol, UK, Chap. 4

McIntosh W, Stevenson A 1990 Operating experience with elastomer composites on the Hutton tension leg platform. *Proc. 9th Int. Conf. Offshore Mechanics and Arctic Engineering,* Vol. 3A. American Society of Mechanical Engineers, New York

Plastics and Rubber Institute 1987 *Guide to the Elastomeric Coating and Lining of Pipework and Vessels to Provide Corrosion Protection Against Sea Water and Sea Water Based Fluids*. PRI, London.

Raveau G, Simon J P 1988 Polyamide 11 for use in collecting pipes. In: Stevenson 1988, Paper 14

Ringler R, Howell B 1988 Advanced polymer encapsulation for continuous chemical injection technology. In: Stevenson 1988, Paper 13

Rivlin R, Thomas A G 1953 *J. Polym. Sci.* 10: 291

Rothuizen J W 1988 Para-aramid reinforcement for hydraulic umbilicals. In: Stevenson 1988, Paper 18

Sedillot F, Stevenson A 1983 Laminated rubber articulated joint for the deep water gravity tower. *ASME Trans. J. Energy Res. Tech.* 105: 480

Seregely I, Nagy T T, Pfisztner N 1988 Ageing of elastomer for simulated offshore conditions. In: Stevenson 1988, Paper 15

Stevenson A 1986 Fatigue of high load capacity rubber laminates. *Rubber Chem. Technol.* 59: 208

Stevenson A 1987 *J. Adhes.* 21: 313

Stevenson A (ed.) 1988 *Proc. 3rd Int. Conf. Polymers in Offshore Engineering*. Plastics and Rubber Institute, London

Stevenson A 1990 Engineering with rubber. *Kempes Engineering Yearbook*. Morgan Grampian

Tetlow J, Leece M J 1982 Hutton TLP mooring system.

Proc. Offshore Technology Conf., OTC 4428. Offshore Technology Conferences

van Shaik H 1988 Structural rubber bearings for Dutch delta project. In: Stevenson 1988, Paper 29

Vardal O H 1988 Noise survey on Norwegian installations. In: Stevenson 1988, Paper 35

Weston R J 1988 A comparison of perfluoroelastomers and other elastomers tested in oilfield media. In: Stevenson 1988, Paper 6

White G 1988 Rubber energy absorption system for offshore installation of an integrated deck for Miller field. In: Stevenson 1988, Paper 32

Wilton G W et al. 1985 The design substantiation of the anchor connectors and crossload bearings for the Hutton TLP. *Proc. Offshore Technology Conf.*, OTC 4947. Offshore Technology Conferences

A. Stevenson
[Materials Engineering Research Laboratory, Hertford, UK]

P

Paints, Varnishes and Lacquers

A thin film applied to protect or decorate an inanimate surface is known as a surface coating—a term covering paints, varnishes and lacquers. These materials differ not so much in their function as in their mode of drying or curing.

Regardless of the mode of drying, to be effective, coatings must be applied in the proper manner by brushing, rolling, spraying or rubbing, and must go over surfaces that have been suitably prepared. Surface preparation is as important as selection of the coating material. Generally, results are better when a coating of lesser quality is used over a properly prepared surface than when a superb coating material is applied a poorly prepared surface.

Paints are usually pigmented vehicles—derived from petrochemicals or vegetable oils—that dry or cure by oxidation or some other chemical reaction. To facilitate spreading they must be reduced by solvents or some other liquid. They also contain one or several of the following additives: driers, which are catalysts to hasten or complete oxidation; antiskinning agents; ultraviolet inhibitors; microbiocides to prevent mildew and protect against bacterial action in storage; thickening agents; and rheological agents to aid flow and levelling.

Varnishes are transparent or nearly transparent vehicles with additives that aid drying and stabilize color. Lack of pigmentation distinguishes them from paint, although transparent colors, mainly dyes, can be added to produce wood stains.

Lacquers differ from paints and varnishes in that they dry by evaporation of their solvent.

1. *Paints*

Paints are categorized in a number of ways. One categorization is by relative gloss: flat paints, semigloss enamels and high-gloss enamels. Another categorization is by function: architectural, heavy-duty maintenance and factory applied. A third designates paints by the vehicle family on which they are based, which often but not always relates to their functional category: a poly(vinyl acetate), poly(vinyl acrylic), acrylic, linseed oil or alkyd paint is usually an architectural coating; a urethane, epoxy or chlorinated rubber is likely to be maintenance paint for use in a difficult environment; and urea melamines, cellulose acetate butyrates, vinyl chlorides, and certain urethanes and acrylics are used for factory-applied or product finishes.

1.1 Relative Gloss

Flats and enamels are found in all the functional and vehicle categories. Flats are used mostly on the interior and exterior walls of residences, factories or institutions, and also on product finishes; semigloss enamels are used mainly for architectural trim, such as molding, and wherever a dressy appearance is desired without excessive shine. This may include metal, wood, or cementitious surfaces—either painted on-site or in a factory. High-gloss paints are frequenty found on factory-applied finishes such as appliances and furniture, on architectural areas subject to frequent cleaning, and where durability is demanded.

Flat paints differ from the enamels in probable durability as well as gloss, because they usually, but not always, contain less vehicle than enamels and more hiding pigment and extender. For low-cost paints, excessive amounts of extenders—clay, calcium carbonate and talc—are used to reduce raw-material expense.

The more of a given vehicle in a paint, the greater its ability to withstand scrubbing and impact and to avoid cracking and blistering. Therefore, flats are less durable, as a rule, than semigloss and high-gloss enamels. Exceptions are the so-called flat enamels, which are paints with sufficient vehicle content to qualify as enamels but whose gloss is minimized by a flatting agent.

A flat paint, except as noted, has a high pigment volume concentration (or relatively low vehicle content) and an enamel has a low pigment volume concentration (relatively high vehicle content). A semigloss enamel is midway in pigment volume concentration between the two.

1.2 Function

Coatings serve several general functions. As architectural coatings, they usually provide a balance of protection and decoration, mostly in mild environments. Maintenance coatings usually function in harsher environments, some of them extreme, such as in industrial areas where airborne chemicals are present.

Both architectural and maintenance coatings are normally applied in at least two coats: primer and topcoat. Primers differ considerably when used on metal, wood or cementitious surfaces, whereas topcoats may not differ at all if the environment is the same. Where necessary, a third, or intermediate, coat is used to provide extra protection. This is often the same as the topcoat.

2. *Primers and Topcoats*

Primers are basically of two kinds: barrier or reactive. Barrier primers are used to block off environmental influences such as wind, sun, rain and chemicals. They contain the usual vehicles and pigments. Reactive primers contain pigments that combine chemically with materials in the vehicle to provide the substrate with some form of specialized protection.

In the case of steel, this specialized protection reduces corrosion if water reaches the substrate; in the case of natural dye-containing wood, the reactive substance combines with the natural dye and forms an insoluble substance that will not migrate to the surface and stain the paint.

Reactive primers can also serve as barriers. Zinc oxide, the reactive pigment, increases the barrier effect of linseed oil primers by reacting with the oil's fatty acids to enhance the impermeability of the primer to water. Lead pigments act similarly in blocking moisture.

Steel protection, in particular, calls for reactive primers. Finely powdered zinc dust is used in coatings to protect steel substrates. Dispersed in organic vehicles, such as epoxies or phenoxies, or in inorganic vehicles, such as ethyl silicate, zinc protects when water is present by sacrificing itself in a physical reaction known as cathodic protection.

This reaction occurs because zinc in water is more chemically active than iron and is preferentially consumed in the tiny electric cell that results when water reaches two metals. Zinc becomes the anode and is slowly consumed.

Other pigments also change the susceptibility of steel to corrosion. These include zinc phosphate, calcium molydate and barium metaborate. Protection against corrosion is also enhanced by passivating the surface before painting. This is seldom done on structural metal, but is used extensively on home appliances and other devices whose original manufacturer is zealous about providing satisfactory service life.

Passivation is a procedure whereby phosphate or chromate compounds are applied in solution to the metal surface. There they form metal salts that are less active electrically than the original metal in the presence of moisture and, therefore, the resulting surface is less likely to be corroded.

Primers for cementitious surfaces, and for wood surfaces of some types, seal off pores or cells that could absorb too large a portion of vehicle in non-specialized coatings. These are called primer sealers because they block pores so that the vehicle in topcoats will not be excessively absorbed.

Proper formulation of primers to be used with specific topcoats is important for three reasons: first, the topcoat must be able to adhere; second, the primer, when dry, must not degrade a topcoat by absorbing part of its vehicle, thus weakening it and leaving an unsightly surface; and third, the primer must not contain a vehicle which, when dry, can be lifted by the solvent in the topcoat. For these reasons, when buying primer–topcoat systems it is important that one manufacturer's matched products be used and that a dependable manufacturer's products be selected.

Some primer formulations can also be used as topcoats or as intermediate coats. More often, though, coatings with qualifications for particular sets of circumstances are selected for topcoats. Where circumstances are only mildly taxing, relatively inexpensive latex emulsions or acrylics may be used with suitable pigments and additives for decorative effects.

For architectural exteriors, where rain, wind, sun, atmospheric pollutants and microbial attack must be endured, more flexible and durable primers and topcoats are required. These may consist of water-reducible poly(vinyl acetate) copolymers or acrylics, but with a greater percentage of flexible monomers than is used for interior coatings.

The most effective architectural-exterior water-reducible coatings are made of acrylic resins, which use a tough monomer, methyl methacrylate, and soft flexible monomers, such as ethyl or butyl acrylate. Acrylics have the advantage of wet adhesion, which means they are less likely than poly(vinyl acetate) copolymers to be harmed if rain or high humidity is present shortly after application.

Older, solvent-reduced, architectural-exterior paints based on linseed oil or alkyd resins are still used, but they are losing ground to the more convenient water-thinned coatings which most experts believe to be about as good.

For harsh environments, high-performance or heavy-duty topcoats are needed, often with special-purpose primers, as noted. For vehicles, these topcoats use materials derived from somewhat exotic sources, often more difficult to make than those used for mass-produced coatings. Selection should be determined after site conditions are thoroughly understood. Temperature changes, wind velocity, proximity to industrial or marine atmospheres, and resulting contaminants must all be considered when selecting a high-performance coating system.

3. *Vehicle Systems*

Vehicle families used for coating systems—including primers and topcoats and intermediate coats when necessary—are described in the following. Selection of systems rather than choice of individual primers and topcoats is clarified.

3.1 General Systems

(*a*) *Oils and alkyds*. Older vehicles still in use are linseed oil, used exclusively for exteriors, and alkyd resins—both solvent reduced. Alkyds are used for exterior flats, for enamel-trim paints and for some high-quality interior paints. Their main use in architectural coatings is for enamels, but water-reducible versions have been gaining slowly.

Alkyds are really upgraded vegetable oils. Soybean, linseed or tall-oil fatty acids are reacted with a fatty alcohol, usually glycerine or pentaerythritol, and a dicarboxylic acid, usually phthalic anhydride. Drying characteristics and performance of the oil are improved.

Virtually all exterior latex primers require the addition of either a vegetable oil or an alkyd—usually about 20% of the vehicle—to improve adhesion, particularly if it is to go over an old paint with unseen surface breakdown in the form of chalk.

(*b*) *Poly(vinyl acetate) and acrylate monomers*. Water-reduced architectural paints are based mostly on poly(vinyl acetate) copolymerized with an acrylate monomer, usually ethyl or butyl acrylate. They are used extensively in exterior water-reducible formulations and to a lesser extent on interiors, where their higher cost is a disadvantage. Their superior color retention and ability to resist moisture dilution soon after application are advantages over competing vehicles.

The acrylate monomers soften and flexibilize the poly(vinyl acetate) and the acrylic monomers. Because they are more costly than the harder constituents, they are sometimes slighted as to quantity used. While this is not very significant on interior coatings, it may result in cracking and shorter life expectancy in exterior products.

(*c*) *Terpolymers*. Terpolymers which utilize vinyl acetate, an acrylate and a third monomer have been introduced. One version uses vinyl chloride as the third monomer and approaches acrylic emulsions in performance but at a lower cost. A second version uses low-cost ethylene with some success.

3.2 High-Performance Systems

For harsh environments where coatings may have to protect against airborne acids, alkali or salt spray, or may have to withstand rough handling or the impact of sand, rock or battering by shod feet or machines, coatings based on more durable vehicle families are selected. These include the following, which, except as noted, are solvent thinned.

(*a*) *Acrylics*. Acrylics, which are chemically related to the water-reducible versions, are more durable. Factory-applied versions are used or curtainwalls and for automobile enamels. For metal coatings under harsh conditions where solvents cannot be used, for example on delicate machinery, water-reducible acrylic latexes of special kinds are available, usually with acrylonitrile or some other additive that improves durability.

(*b*) *Chlorinated rubber*. Chlorinated rubber is used where low water permeability is required for concrete or steel surfaces. It is also desirable if microorganisms are to be encountered, or if the surface may encounter radioactive materials that must be cleaned off. Effective use of this material in metal primers has been achieved with zinc dust.

(*c*) *Epoxies*. Epoxies are tough chemically-resistant vehicles made by combining various proportions of epichlorohydrin and bisphenol A, and various modifiers. Some liquid epoxies can cure at a temperature as low as 10 °C and some types have been cured in ocean water. Up to 100 mils (2.5 mm) of epoxy have been applied in a single coat; in a suitable solvent, epoxies can be sprayed. Blended with coat tar, epoxies form thick coatings for surface protection underground or underwater. One low-melting epoxy, a solid, liquefies between 65 °C and 75 °C. It is thinned with solvents.

Two-component epoxy systems are used for heavy-duty maintenance and marine and floor coatings. Two subtypes are identified by the resin selected for curing. When cured by a polyamide resin, the system is quite flexible and water resistant; a polyamine curing resin is selected if hardness and chemical resistance are more important.

A third type, epoxy ester, is made by precuring reactive portions of the epoxy with a fatty acid such as those found in soybean oil, dehydrated castor oil or tall-oil fatty acids. They dry and perform somewhat like alkyds, but are more water resistant. Their main use is to fortify acrylics, polyesters or phenolics.

Epoxies in water-reducible forms have shown promise and are used in some cases where complete elimination of moisture is difficult. In recent years, epoxies in powder form have proved effective when near-vitreous, yet moderately flexible, protection is required in environments where salt or corrosive alkali incursions occur. Steel reinforcement bars embedded in concrete bridge decks, for example, and subjected to road salt have been protected by these fused powders, following techniques developed with the support of the US Federal Highway Administration and the National Bureau of Standards.

Although they provide long-term protection, epoxies have limited color stability: they fade and chalk. Some success in stabilizing color has been achieved by using a hydantoin modifier.

Epoxies have been combined with polyesters, acrylics and silicones to yield improved coatings. Epoxy polyesters have added hardness and gloss, and acrylic and silicone modifications improve weatherability and chalk resistance.

(*d*) *Fluorocarbons*. Fluorocarbons, which are among the most expensive coatings, are used where repainting is difficult and costly, such as on metal curtainwalls or skyscrapers and for remote power-line towers, or where chemical exposure is severe, as in pulp mills or water-treatment plants. Shop application, closely controlled and done on meticulously prepared surfaces, is necessary, although field touchup by specially trained applicators is possible. Fluorocarbons are prepared by combining fluorine with hydrocarbons to form a highly stable, weather-resistant polymer with a life expectancy of 40 years.

(*e*) *Formaldehyde resins*. Formaldehyde resins are nitrogen-containing polymers (amines) that form when formaldehyde is reacted with urea or melamine. The resulting urea–formaldehyde or melamine–formaldehyde is added to alkyd baking resins to impart hardness and color stability when used for architectural metals or other factory-applied finishes.

(*f*) *Polyesters in combinations*. Polyesters are often combined with other vehicles to impart characteristics not otherwise obtainable. For example, an epoxy polyester provides a glazelike coating resistant to many chemicals and much abuse; silicone polyesters are chemical resistant and highly durable and heat resistant. Silicones improve chemical and heat resistance when added to epoxies, alkyds or acrylics.

(*g*) *Urethanes*. Urethanes are of two basic types: aliphatic and aromatic. The most effective as to chemical and environmental resistance is the aliphatic, which must be cured by a polyester, an epoxy or an acrylic; hence they are known as two-package systems.

Aromatic urethanes lack the weatherability and color stability of the aliphatic. They are usually cured by a polyester or a polyether component, packaged separately and added just before use.

A third general type is known as moisture cured, a single-package material cured by atmospheric moisture after application. The three types may be applied in the field. In addition, certain elastomeric urethanes are compounded to yield rubbery films for use as tank linings exposed to acids, alkalis or abrasives.

(*h*) *Neoprene*. Neoprene (polychloroprene) is a synthetic rubber applied as a coating if the surface is exposed to ozone, strong chemicals or abrasion. Modified neoprenes are supplied for various circumstances, in solvent-reducible or water-thinnable forms and in cured or uncured sheets. If neoprene is to be exposed to sunlight it is usually coated with Hypalon (chlorosulfonated polyethylene).

(*i*) *Hypalon*. Hypalon provides chemical-resistant, high-build coatings useful in resisting chemicals and ozone. The selection of curing agents must be related to the material to which Hypalon is to be exposed. Highest chemical resistance is imparted by a litharge (lead monoxide) curing agent, whereas magnesium oxide imparts resistance to strong sulfuric acid. Other curing agents enable the coating to function in other circumstances.

Ethyl silicate is used with zinc dust as an effective primer for steel.

(*j*) *Vinyl resins*. Vinyl resins have been effective as intermediate and topcoats in protecting steel in marine environments. They may be applied in very thin coats or in conventional film build. Vinyl resins for coatings consist of a vinyl chloride monomer with a vinyl acetate monomer to provide strength.

(*k*) *Phenoxy resins*. Phenoxy resins are closely related to epoxies, since they are made of the same basic materials, bisphenol A and epichlorohydrin, but they are thermoplastic, whereas epoxies are thermosetting. They have the advantage, however, of being as thermally stable as many thermosetting resins and have a long shelf life. They also have the advantage of requiring no curing agents. Their acceptance has not been great, but they have been used with some success in anticorrosive painting systems as a primer with zinc dust for vinyl systems in marine environments.

4. Varnishes

These transparent or semitransparent coatings consist of any one of several resins, or may consist of one of these resins dissolved in vegetable oil. Varnishes are thinned with hydrocarbon solvents; driers are added to aid curing. Resins used include alkyds, phenolics, coumarone–indene, vinyl toluene, limed rosin, phenol formaldehyde, ester gum, epoxy ester and urethane. Oils used as varnish components are linseed oil, castor oil and tung oil.

The resins in a varnish usually provide the hard portion, whereas oils impart softness and flexibility. The relative amounts of oil and resin are often expressed as oil length, with softer, more flexible varnishes having larger amounts of oil. Those described as 10-gallon lengths or less are known as short-oil varnishes and are hard. Medium-oil varnishes with 20- to 25-gallon lengths (meaning 20–25 gallons of oil per 100 gallons of resin) are not as flexible as spar varnishes, or long-oil versions, with 33- to 50-gallon lengths.

In addition to hardness and flexibility, desirable properties are water resistance, reasonable speed of drying and freedom from wrinkling. Dyes or transparent brown or red pigments may be added to varnishes to provide stains.

5. Lacquers

Advantages of lacquers are rapid drying resulting from solvent evaporation, consistent results since the product does not react or cure, reduced pickup of dirt, and receptivity to polishing that removes imperfections.

Lacquers are used mainly for factory-applied finishes; also, a considerable quantity is used for automobile and furniture refinishing.

Nitrocellulose, modified by plasticizers and other resins, is used extensively for wood coatings, but petrochemically derived synthetics have been growing in popularity because of better abrasion and heat resistance, and higher solids content, which, in some cases, reduces the number of coats to be applied. These synthetics include polyurethanes, cellulose acetate butyrate and melamine–alkyds. Other important lacquer resins include poly(vinyl chloride), poly(vinyl acetate) and acrylics.

See also: Maintenance Coatings; Marine Coatings

Bibliography

Banov A 1979 *Paints and Coatings Handbook for Contractors, Architects, Builders, and Engineers.* Structures, Farmington, MI

Burns R M, Bradley W 1967 *Protective Coatings for Metals,* 3rd edn. Reinhold, New York

Myers R R, Long J S (eds.) 1967–1976 *Treatise on Coatings.* Vols. 1–5. Dekker, New York

Nylén P, Sunderland E 1965 *Modern Surface Coatings: A Textbook of the Chemistry and Technology of Paints, Varnishes, and Lacquers.* Wiley, New York

A. Banov
[American Paint Company, New York, USA]

Papers of Natural and Synthetic Fibers

Papers made of blends of natural and synthetic fibers vary widely in characteristics from the "papery" to the soft nonwoven. Products of such blends may be used in heat-seal teabag tissue, roofing felt, vinyl floor backing, gaskets, vacuum cleaner bags, hospital gowns, filters and wallpapers.

The major justification for making such blends over the years has been the replacement of textiles and very recently the replacement of asbestos.

1. Types of Blend

There are generally three major categories of cellulose–synthetic fiber blends, but these categories overlap.

(a) Blends of relatively small percentages of synthetic fibers to add some special features to a paper. Examples include a filter paper with 1–5% glass fiber to improve the dirt catching efficiency, wallcovering paper with 5–10% nylon to increase resistance to tearing and heat-sealing teabag tissue with 15–25% vinyl chloride fibers over a base structure of manila hemp.

(b) Blends in which the cellulose is present as a diluent for the sake of economy, the special properties having been achieved with less than 100% synthetic fibers. Examples of such products include filter papers with 50% glass and 50% cellulose, and wet-formed nonwoven with 80% polyester and 20% cellulose.

(c) Blends where the cellulose is added primarily to make web handling easier. For example, a diaper lining nonwoven uses 10–20% hardwood fibers to hold the sheet together during manufacture until other binders are added.

2. Manufacture

Pulping may be done conventionally for blends with a low percentage of nonmelting fibers. Generally, the papermaker does not want to refine the cellulose and synthetic fibers together, so each fiber is dispersed separately and mixed after the cellulose has been refined. Agitation to disperse the synthetic fibers in the cellulose fiber slurry should be minimized.

The fiber web may be formed on a conventional cylinder or fourdrinier paper machine if the amount of the synthetic fiber is small or if the synthetic fibers are short or fibrillated. However, with long, free-draining fibers, high levels of stock dilution in the headbox are necessary (0.1% consistency) and inclined fourdriniers or rotoformers are more effective. This is especially true for porous nonwoven products.

Wet pressing is often omitted in blends of high synthetic content, because the synthetic fibers pick out readily on press rolls. Glass fibers can be expected to shear under pressure. In the case of floor backing manufacture, wet pressing is essential for economy.

Bonding with latexes or resins is needed in high-synthetic grades. Spray bonding may be carried out on the forming wire, in the press section or in a special wet-web bonding unit. Saturation of the web may also be performed either before or after web drying. Print bonding may be effected on the dried web.

Whether blends can be dried conventionally depends on the type and amount of synthetic fibers and the binders. Melt fibers such as PVC or polyolefin can be expected to shrink and this shrinkage may require special drives for the dryers, removal of the dryer felts or special heating to help strengthen the sheet. On the other hand, glass fiber papers offer processing problems due to the absence of shrinkage. The wet sheet containing glass fiber may stretch under its own weight, requiring special drying, draw control or removal of felts.

3. Strength Considerations

When the papemaker uses synthetic fibers to build a web, the web is typically not strong enough to cause synthetic fiber failure when the sheet is destructively

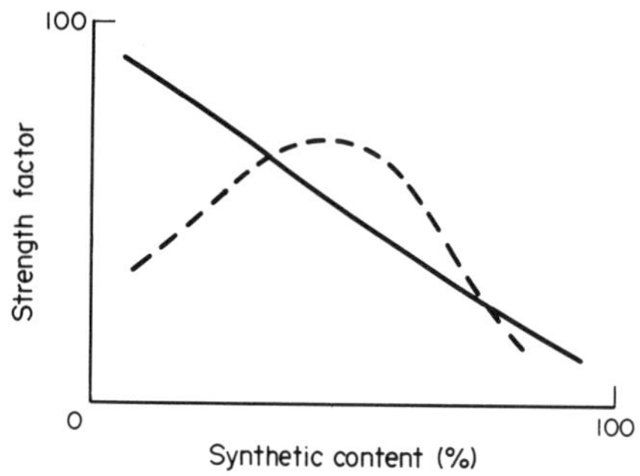

Figure 1
Strength of web of unbonded blends as a function of synthetic content (balance wood pulp): ———, tensile; ---, tear

tested in tensile or tear. One reason for the low strength is the shortness of the fibers, which does not allow fibers to be bonded at enough points. Thus, for applications requiring high strength, the user of synthetic fibers has stayed with woven or dry-formed nonwovens, where enough bonding is available to give high strength by fiber rupture.

When typical 6 mm synthetic fibers are added to well-beaten wood fibers in a web without supplemental bonding, the web tensile strength decreases linearly as the level of synthetic fiber is increased to 100%. As the synthetic fiber level is increased to around 40%, the tear strength increases; beyond this level it decreases. These effects are shown in Fig. 1.

For bonded webs of 100% cut synthetic fiber, the tear strength peaks at 25% binder and the tensile strength is a maximum at 35% binder, as illustrated in Fig. 2. Above these levels of binder, the web behaves more like a film of binder.

4. *Fiber Length Considerations*

Synthetic fibers having a length–diameter ratio of greater than 500 can be expected to be difficult to process, depending on modulus and electrical

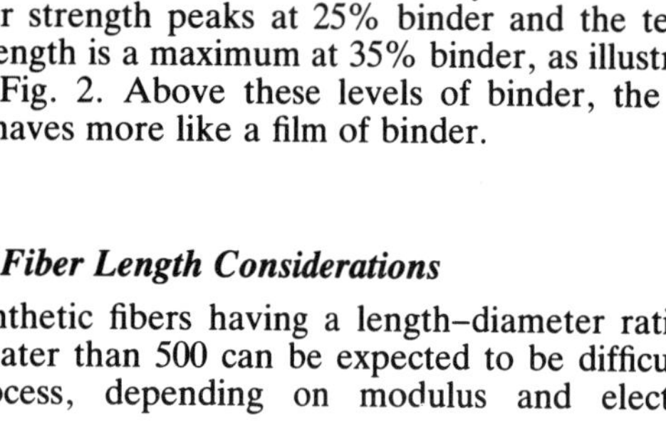

Figure 2
Strength of bonded 100% synthetic webs as a function of binder content: ———, tensile; ---, tear

charge. Glass fibers, which are stiff, form best at an l/d ratio of 650, but commercial practice includes many products at l/d values up to 1800.

The papermaker is always vulnerable to inaccurate cutting of the synthetic fiber. For example, if a rotary cutter allows some $6x$ (where x equals the intended cut length) fibers into the product, the papermaker can expect to have fiber clumps and knots in his product. With strong fibers such as nylon and polyester, some long cuts (multiple-length fibers) are to be expected. With rayon or glass, which are easier to cut, fewer multiple-length cuts generally occur.

5. *Synthetic Fibers Used*

5.1 Polyester

Polyester fibers are the dominant fibers used in wet-laid nonwovens because of their price relative to other fibers. A typical application is interliners for garments using woven or knitted polyester. A binder is required to carry the polyester through the papermaking process. A small amount of wood fiber is the simplest binder for this purpose.

Polyester fibers are widely used in varying lengths and deniers. The level of polyester ranges from papers containing a few percent to nonwovens containing 100% polyester. Strength, porosity, dimensional stability and softness are among the features polyester adds to products.

5.2 Nylon

Nylon was the first synthetic other than cellulose-derived rayon to be made into paper. However, the early interest in paper containing nylon failed to develop major end uses except as blends with wood fibers to make wall coverings and curtain header buckram. Undrawn nylon is also used in nonwoven medical fabrics to improve tear strength.

5.3 Aramid

There are two aramid fibers available to the papermaker: Nomex and Kevlar. Du Pont's Nomex fiber, which will not burn in air, has a high temperature resistance. It is mainly used in one type of product: a 100% Nomex fiber paper. This is made from a blend of cut fibers and fibrids on a conventional paper machine. The primary end use is high-temperature dry electrical insulation. Another application is in honeycomb core structures in airplane panelling.

Du Pont's Kevlar is being developed for papermaking as a replacement for asbestos in markets such as engine head gaskets and frictional products (drum brake linings and disk brake pads). The high tensile strength of the fiber is not fully used, because there are no binders as strong as the fiber. Kevlar fibers can be fibrillated by severe mechanical action like disk refining, developing fibrils and enough surface for a web of these fibers to hold together. Generally, Kevlar usage is in the range 2–10% of the total fiber furnish.

5.4 Polyolefin

There has been much activity in making polyolefin synthetic pulp by several methods. All these methods result in fibers of length, diameter and fibrillation suitable for papermaking. When first developed in the late 1960s, polyolefin pulp gave promise of large market penetration, since the cost was forecast to be only 30% higher than that of wood pulp. Since then, the disproportionate increase in oil prices has resulted in a dwarfed development, with polyolefin pulp price some 300% that of wood pulp.

The major usage of polyolefin fiber has been as a replacement for asbestos fiber in paper for backing of vinyl flooring. In addition, the heat-sealing properties of this fiber have made it useful in teabag tissue. Many other specialty uses are under development, but short staple polyolefin fibers are not widely used in papermaking.

5.5 Rayon

Viscose rayon, which is derived from natural cellulose, was a major factor historically in the development of synthetic fiber papers. At the time it was introduced, it was relatively inexpensive. It was desirable because it was safe for food uses. Currently, the cost of rayon has increased so that it is now more expensive than polyester, which has taken over many of the early end uses for rayon. Rayon has good wet strength in a web and it can be run easily on the paper machine. When dry, rayon papers must have a binder added for web strength. Flat, ribbonlike rayon fibers were developed which had dry strength without a binder. The making of these fibers was discontinued when rayon costs rose.

5.6 Cellulose Acetate

This fiber has been promoted for use in filter papers, where it is added in small quantities.

5.7 Acrylic

Acrylic and modacrylic fibers are not widely used in wet-laid papers. The only useful properties that they impart to papers are increased porosity and strength, and these can be achieved with less expensive fibers. Acrylic gel fibes and other undrawn or semidrawn acrylics have the property of self-bonding by fibrillation or heat.

5.8 Glass Fibers

Major end uses for glass fiber papers are in filtration, roofing, insulation and reinforced plastics. Filter papers of high efficiency are made of 100% glass fibers ranging in diameter from 0.2 μm to 9.0 μm. They may be formed conventionally on a paper machine, and most glass filter papers incorporate 2–5% of organic binders. Glass fibers of diameter less than 1 μm can be made into paper without binder where chemical purity is needed in the paper. Chopped strand glass is used to make roofing shingles. The usage of glass in this application is substantial and growing, since the resulting material requires less asphalt.

5.9 Ceramic Fibers

Various ceramic fibers have been used for making insulating papers for high-temperature applications such as forms for pouring molten metals. Ceramic fibers require a binder for handling during papermaking.

5.10 Other Fibers and Processes

Many other fibers have been used in papermaking. Although, in theory, any fiber can be used in the process, there are applications for which the product may be made better on a loom or cheaper by the dry process. Papermaking as a process has advantages of speed and controlled fiber orientation plus opportunity for excellent blending, but end uses must require sufficient production volume if papermaking is to be chosen instead of another method of fabrication.

Bibliography

Battista O A (ed.) 1964 *Synthetic Fibers in Papermaking*. Wiley, New York

Glomb J W, Mulligan D D 1989 Paper and paperboard. In: Kelly A (ed.) 1989 *Concise Encyclopedia of Composite Materials*. Pergamon, Oxford, pp. 217–27

Hoge W H 1972 *Synpulp Projects and Technologies*. Paperfilm Associates, Flemington, NJ

Merriman E A 1983 Kevlar aramid pulp processing and properties for industrial papers. *Nonwovens Symp*. Technical Association of the Pulp and Paper Industry, Atlanta, GA, pp.13–15

Moncrieff R W 1970 *Man-Made Fibres*. Wiley, New York

Scheffel N B, Cenedella A J 1982 Refractory fibers. *Nonwovens Symp*. Technical Association of the Pulp and Paper Industry, Atlanta, GA, pp. 5–9

Shiffler D A 1984 Characterizing the dispersion kinetics of synthetic fibers in water. *Nonwovens Symp*. Technical Association of the Pulp and Paper Industry, Atlanta, GA, pp.173–7

Stassen W N 1983 Dispersing glass fibers in the wet process. *Nonwovens Symp*. Technical Association of the Pulp and Paper Industry, Atlanta, GA, pp. 7–12

D. G. Magill
[Magill Associates, Hanover, New Hampshire, USA]

Plasticization of Polymers

Plasticization refers to the blending of polymers with relatively low-molecular-weight organic compounds to improve the melt processability or to modify

Table 1
Typical plasticizers

General type	Structure	Examples
Phthalate esters	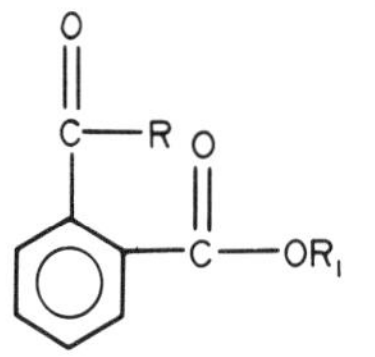	$R = R_1 = $ 2-ethylhexyl, methyl, ethyl, butyl, cyclohexyl; $R =$ butyl, $R_1 =$ benzyl
Phosphate esters	$(RO)_3PO$	$R =$ ethyl, phenyl, cresyl
Adipate esters	$ROC(=O)(CH_2)_4C(=O)OR$	$R =$ butyl, 2-ethylhexyl
Azelate esters	$ROC(=O)(CH_2)_5C(=O)OR$	$R =$ butyl, 2-ethylhexyl
Sebacate esters	$ROC(=O)(CH_2)_8C(=O)OR$	$R =$ butyl
Benzoate esters	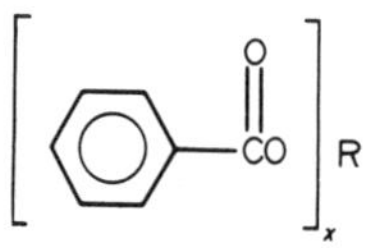	$R =$ neopentyl ($x = 2$), glyceryl ($x = 3$)

certain physical or mechanical properties of the polymer. Through plasticization, polymers that have inherently poor melt processability (due, for example, to only a narrow range between their processing and degradaton temperatures) can be made more useful materials. By variation of the plasticizer level, a given polymer system can be modified to have mechanical properties, particularly stiffness and toughness, that cover a wide range of values. Plasticization has been an important factor in the development of polymer and plastics technology since at least 1868 when camphor was used as a plasticizer in cellulose nitrate to yield celluloid.

1. Plasticizer–Polymer Systems

Some of the common chemical classes of plasticizers are listed in Table 1. Most plasticizers, in contrast to polymers, are low-molecular-weight organic compounds of relatively low volatility. Low-molecular-weight polymers may also be used as plasticizers; typical of these are certain aliphatic polyesters.

Two important aspects of the interaction of plasticizer with polymer are compatibility and efficiency. Compatibility is used here in the sense that the polymer and plasticizer are miscible and form a single phase. Some polymer–plasticizer systems may show limited miscibility, wherein a plasticizer-rich phase separates at a certain concentration. This effect is usually evident because it is accompanied by the development of haze or opacity in the material which can be visually detected.

Efficiency, used here to broadly describe the relative effectiveness of plasticizers in changing a polymer property such as the melt viscosity, modulus or glass-transition temperature, must be defined relative to a specific property and is thus not an absolute concept. It is entirely possible that one plasticizer would be more efficient than another in reducing the glass-transition temperature and at the same time be less efficient in reducing the modulus.

Because plasticizers are fairly low in molecular weight their tendency to migrate from the polymer and be lost to the environment is also of concern. This property is often referred to as "permanence." Where permanence is of primary importance, polymeric plasticizers should be considered, because the higher molecular weights of such plasticizers substantially reduce the migration rate.

Among the polymers most commonly plasticized are poly(vinyl chloride) and the cellulosics (cellulose acetate, cellulose acetate butyrate and cellulose acetate propionate). Acrylics, styrenics and other vinyl polymers are also occasionally plasticized. In poly(vinyl chloride), the phthalates, particularly bis(2-ethylhexyl) phthalate (commonly referred to as dioctyl phthalate), are the most widely used plasticizers. Lower alkyl (such as diethyl and dimethyl) phthalates, phosphates and esters of aliphatic diacids are most common in the cellulosics.

Depending on the polymer, plasticizers may be incorporated by a number of melt-compounding techniques. Roll milling is widely used, but other mixers and compounding extruders may also be used. It is sometimes possible to sorb plasticizer directly into a polymer powder by dry mixing, often by using a brief exposure to slightly elevated temperatures to accelerate the process.

2. *Effects of Plasticizers on Physical and Mechanical Properties*

In general, plasticization has the effect of reducing viscoelastic relaxation times and many of its effects on the physical and mechanical properties of polymers can be understood from this perspective. It is also useful to remember that, at least qualitatively, a reduction in relaxation times is equivalent to an increase in temperature; thus, often the effects of plasticization on mechanical properties are similar to those caused by an increase in temperature.

2.1 The Glass Transition and Melting Point

Reduction of the glass-transition temperature T_g is a major effect accompanying plasticization. Several equations exist to relate the glass-transition temperature of a plasticized polymer system to its composition and to the glass-transition temperatures of the components. The simplest of these is

$$T_g = T_{g_1} - kw_2 \tag{1}$$

where T_g is that of the plasticized system, T_{g_1} is that of the unplasticized polymer, w_2 is the weight fraction of plasticizer and k is an empirical constant. This equation often represents the initial stages of plasticization well but, after a point, the incorporation of additional plasticizer has increasingly smaller effects on the glass-transition temperature and a deviation from the straight line predicted by Eqn. (1) occurs. Other equations, for example the Gordon–Taylor equation, are more successful in predicting the T_g depression over a wider range of plasticizer content:

$$T_g = \frac{w_1 T_{g_1} + c w_2 T_{g_2}}{w_1 + c w_2} \tag{2}$$

Here, subscript "1" again refers to the polymer and "2" refers to the plasticizer. The parameter c is the ratio of the difference in the thermal expansion coefficient of the glass and melt of the polymer to that of the plasticizer.

The decrease in T_g caused by plasticization directly affects the viscoelastic relaxation times and thus the melt viscosity. Reduction of melt viscosity, particularly in polymers with poor thermal stability, is a major reason for the use of plasticizers.

In crystallizable polymers, plasticization lowers the melting point according to the equation developed by Flory for polymer–solvent systems:

$$\frac{1}{T_m} - \frac{1}{T_m^0} = \left(\frac{R}{\Delta H_u}\right)\left(\frac{V_u}{V_2}\right)(v_2 - \chi v_2^2) \tag{3}$$

where T_m is the observed melting point for the polymer–plasticizer system, T_m^0 is the melting point of the pure polymer, ΔH_u is the heat of fusion per mole of polymer repeat unit, V_u is the molar volume of polymer repeat unit, V_2 is the molar volume of the plasticizer, v_2 is the volume fraction of plasticizer, R is the gas constant and χ is the Flory–Huggins interaction parameter.

Figure 1 shows the depression of the melting point and glass-transition temperature of bisphenol-A polycarbonate plasticized with tricresyl phosphate. The strong effect of the first amounts of plasticizer on T_g is evident and it is also clear that the glass transition is affected more by plasticization than the melting point. Because of this increasing divergence of T_m and T_g, crystallization rates are usually enhanced by plasticization.

2.2 The Elastic Modulus

Normally, plasticization of a polymer will reduce its elastic modulus. Anomalies do occur, however, and over a limited temperature range it is possible for a plasticized polymer to have a higher modulus than the unplasticized polymer. This phenomenon is

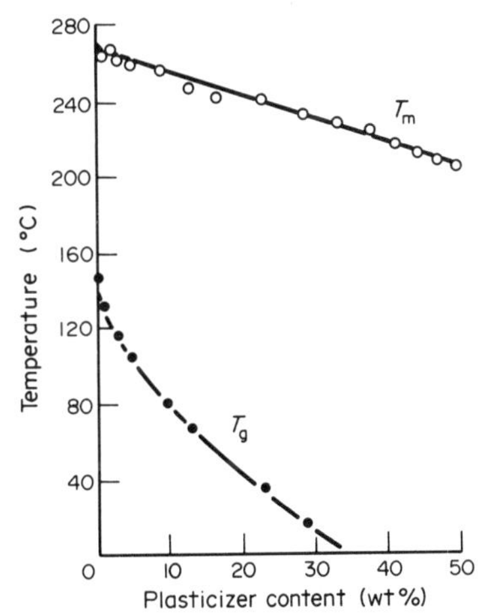

Figure 1
Effect of tricresyl phosphate on the glass-transition temperature and melting point of bisphenol-A polycarbonate (after Conix and Jeurissen 1965. © American Chemical Society, Washington, DC. Reproduced with permission)

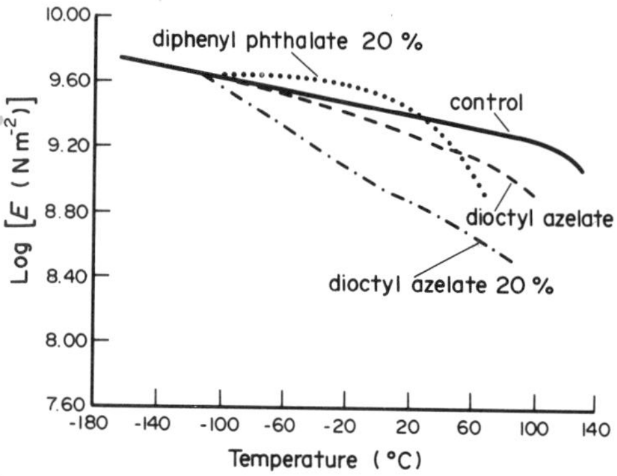

Figure 2
Effect of selected plasticizers on the modulus–temperature curve for cellulose acetate butyrate (after Seymore and Minter 1980. © Society of Plastics Engineers, Brookfield, Connecticut. Reproduced with permission)

often referred to as antiplasticization (Sect. 3). An example for cellulose acetate butyrate is shown in Fig. 2, where diphenyl phthalate is acting as an antiplasticizer.

2.3 Strength Properties

A reduction in tensile strength normally accompanies plasticization. Ductile polymers that show a yield point when unplasticized remain ductile on plasticization, although at high plasticizer contents a distinct yielding may not be observed. Plasticization of brittle polymers (those with ultimate elongations of only a few percent) increases the extensibility but lowers the stress levels that can be supported. Because of increased elongation, impact strengths of polymers are generally increased by plasticization. Again the occurrence of antiplasticization results in opposite trends (increase in tensile strength, decrease in elongation and decrease in toughness) over a restricted range of temperatures.

2.4 Time-Dependent Properties

The behavior in creep and stress-relaxation tests can be predicted from the effect of plasticizer on the viscoelastic relaxation times. Both processes occur at faster rates in plasticized systems. The subject is treated in some detail by Ferry (1970).

3. Polymer–Plasticizer Interactions

Some aspects of the interaction between polymers and plasticizers are similar to polymer–solvent interactions, since the plasticizer acts as a solvent. The degree of solvent activity may be assessed by measuring the intrinsic viscosity of a polymer with the plasticizer as the solvent. Differences in solvating power among plasticizers have sometimes led to their being classified as "primary" and "secondary," primary plasticizers being the better solvents. Such classifications are only qualitative and are often somewhat arbitrary.

The occurrence of specific interactions between plasticizer and polymer molecules has been inferred from a number of studies using dielectric and mechanical relaxation methods. These techniques are sensitive to local motions of molecular segments, such as those that occur below the glass transition. Several changes in such motions can accompany plasticization. The plasticizer–polymer interaction may be such that mobility of some segment of the polymer chain is hindered and a reduction in magnitude of a subsidiary relaxation may occur. Additionally, new motions resulting from the plasticizer–polymer interaction can give rise to a relaxation peak not present in either the polymer or the plasticizer alone.

The glass-transition temperature is invariably lowered by plasticization and, in that sense, plasticization increases the mobility of the polymer chains. This effect is consistent with the general increase in free volume caused by the incorporation of a low-molecular-weight diluent. However, in certain temperature regions below T_g, plasticization can cause a decrease in mobility, as evidenced for example in the suppression of a subsidiary relaxation process as mentioned previously. This decrease in sub-T_g mobility occurs concurrently with the increase in mobility at higher temperatures that causes the reduction in glass-transition temperature. Such a decrease in mobility over a certain temperature range below T_g gives rise to the phenomenon of antiplasticization. Increases in modulus and tensile strength and decreases in elongation and toughness will be observed over temperature regions where molecular mobility is reduced and the polymer is said to be antiplasticized. At higher temperatures, however, the mobility of such a polymer–diluent system will be greater than that of the polymer alone and the normal effects of plasticization (lowered modulus and tensile strength, increased elongation) will be observed. See, for example, Fig. 2, where diphenyl phthalate acts as an antiplasticizer below about 20 °C and as a plasticizer above that temperature. The level of additive used affects whether antiplasticization is observed and also the temperature range over which it occurs. Higher levels of additive restrict or eliminate the region of antiplasticization. The distinction between plasticizers and antiplasticizers is thus largely arbitrary and an identical type of polymer–diluent interaction may be present with either.

4. Plasticization of Elastomers

In the rubber industry, a range of products can be used as plasticizers. Other terms are sometimes used to describe them. Softeners such as hydrocarbon oils are used in hydrocarbon elastomers, reducing hardness and improving processing. Processing aids are additives that reduce unvulcanized compound viscosity and nerve, improving processing (i.e., calendering, extruding and molding). Extenders are hydrocarbon oils used, at substantial percentages, to reduce elastomer costs. Softeners, processing aids and extenders are all types of plasticizer.

In common with the use of plasticizers for thermoplastics, a degree of compatibility between plasticizer and base polymer is needed, involving similar values of solubility parameter. Thus, hydrocarbon oils can be used for hydrocarbon elastomers but more polar elastomers require polar plasticizers. Phosphates, phthalates and ester types are used with butadiene–acrylonitrile copolymers (NBRs), and thioether or thioester plasticizers are used with polychloroprenes. Plasticized NBR compounds exhibit appreciably better low-temperature properties (i.e., less stiffening) due to lowered T_g values.

See also: Elastomers: Additives; Plasticized PVC Elastomers

Bibliography

Bernardo J J, Burrell H 1972 Plasticization. In: Jenkins A D (ed.) 1972 *Polymer Science: A Materials Science Handbook*, Vol. 1. North-Holland, Amsterdam, pp. 537–78

Brydson J A 1988 *Rubbery Materials and Their Compounds*. Elsevier, Amsterdam, Chap. 1, pp. 49–53; Chap. 2, pp. 391–8

Conix A, Jeurissen L 1965 Plasticization of bisphenol A polycarbonate. In: Gould R F (ed.) 1965 *Advances in Chemistry Series*, Vol. 48. American Chemical Society, Washington, DC, pp. 172–84

Darby J R, Sears J. K 1969 Plasticizers. In: Bikales N M (ed.) 1969 *Encyclopedia of Polymer Science and Technology*, Vol. 10. Interscience, New York, pp. 228–306

Ferry J D 1970 *Viscoelastic Properties of Polymers*, 2nd edn. Wiley, New York

Hammer C F 1978 Polymeric plasticizers. In: Paul D R, Newman S (eds.) 1978 *Polymer Blends*, Vol. 2. Academic Press, New York, pp. 219–41

Mellan I 1961 *The Behavior of Plasticizers*. Pergamon, New York

Petrie S E B, Moore R S, Flick J R 1972 Influence of diluent mobility on the plasticization of polymers. *J. Appl. Phys.* 43: 4318–26

Seymour R W, Minter J R 1980 Analysis of plasticizer activity in cellulose esters. *Polym. Eng. Sci.* 20: 1188–91

Seymour E W, Weinhold S, Haynes S K 1979 Mechanical and dielectric relaxation in cellulose esters. *J. Macromol. Sci. Phys.* 16: 337–53

Studebaker M L, Beatty J R 1978 The rubber compound and its composition. In: Eirich F R (ed.) 1978 *Science and Technology of Rubber*. Academic Press, London, Chap. 9, pp. 398–401

R. W. Seymour
[Tennessee Eastman, Kingsport, Tennessee, USA]

Plasticized PVC Elastomers

Plasticized PVC elastomers are poly(vinyl chloride) resins modified by the addition of plasticizers or other polymers, so that the normally hard, rigid PVC resin becomes a flexible, extensible and resilient elastomer. The type and amount of added material, together with the manner of addition and degree of dispersion, determine the properties of the final elastomer.

1. Structure

The additives most commonly used are external plasticizers (e.g., phthalate esters) which do not enter into chemical bonding but are compatible to varying degrees with the PVC polymer through hydrogen bonding and van der Waals intermolecular forces. Molecules of external plasticizers retain their chemical identity as they hold apart a three-dimensional network of PVC polymer chains, linked together at intervals by crystalline regions, to form a three-dimensional gel structure. When stress is applied the increased spacing between PVC molecules allows them to flow over each other into a more relaxed state without breaking chemical bonds or changing basic structure and properties. Large strains deform the network sufficiently to extend the molecules fully between the crystalline regions, which have high cohesive strength. This mechanism gives externally plasticized PVC elastomers their unique combination of high extensibility and toughness.

Some low-molecular-weight rubbery polymers have sufficient compatibility with PVC to form good permanent mechanical blends. These comprise a special class of polymeric external plasticizers.

Other additives are used as internal plasticizers, in which case they are chemically bonded to the PVC polymer chain. Applied stress will eventually result in fracture of the chain, with a consequent change in chemical identity and physical properties. Copolymers with acrylonitrile–butadiene or ethylene–propylene rubbers are examples of internally plasticized PVC elastomers.

2. Manufacture

Externally plasticized PVC elastomers are simply prepared by mechanical mixing of PVC resin with plasticizers, accompanied by heating. Dry blends for direct extrusion, molding, calendering or for pelletizing compound are made in ribbon blenders and high-intensity mixers, the latter providing particularly rapid and even dispersion. Pelletized compounds are made from dry blends in a compounding/pelletizing extruder. Plastisols are made by dispersing very small (usually 1–10 μm) glassy PVC particles in a light plasticizer which solvates the PVC on subsequent heating. Internally plasticized PVC elastomers are made by copolymerization of PVC and the internal plasticizer during manufacture of the resin, or by grafting PVC onto a rubbery polymer and then compounding it into a dry blend, pelletized compound or plastisol for manufacture into products.

3. *Properties*

The wide popularity of plasticized PVC elastomers is due to their extreme versatility. Plasticizer or a mixture of several plasticizers may be mixed with PVC resin to give chosen values of tensile strength, extensibility and flexibility, together with such qualities as low-temperature flexibility, low plasticizer volatility at higher temperatures, resistance to leaching when immersed in solvents or detergents, low flammability, low smoke generation, transparency, uv resistance, electrical resistance, biological compatibility (for medical applications) or incompatibility (to counteract bacterial growth in conditions of high humidity). These qualities may be incorporated singly or in required combinations by a judicious choice of plasticizer formulation.

Plasticized PVC elastomers typically have tensile strengths of 7–28 MPa, elongations of 100–500%, Shore hardnesses of A50–D75 and dielectric strengths of 10–40 $kV\,mm^{-1}$. They may be extruded, molded or calendered as required. Plastisols may be spread, dip coated, slush molded or rotationally cast.

4. *Applications*

Typical applications are flexible coatings for furnishings, clothing, wall coverings and automobile interiors; flexible and cellular molding compounds for shoe soles, toys and sporting goods; flexible film for shower curtains, pool liners, auto tops and packaging; electrical wire and cable insulation, jackets and sleeves; and floor covering, flexible hose and tubing, medical equipment, flexible gaskets and magnetic seals, book coverings and office equipment, baby pants and general profile extrusions.

See also: Plasticization of Polymers; Wire and Cable Systems for Underground Transmission

Bibliography

Burgess R H (ed.) 1982 *Manufacture and Processing of PVC*. Macmillan, New York

Kaufman M (ed.) 1962 *Advances in PVC Compounding and Processing*. Maclaren, London

Nass L I (ed.) 1976 *Enclyclopedia of PVC*, Vols.1–3. Dekker, New York

Penn W S 1972 *PVC Technology*, 3rd edn. Wiley, New York

P. F. V. Lloyd
[B F Goodrich Chemical Company,
Avon Lake, Ohio, USA]

Plastics and Composites as Building Materials

Plastics and composites of plastics combined with other materials are relative newcomers to building but their use is growing rapidly. They possess a number of attributes applicable to building, of which the following are examples.

(a) *Formability*. Plastics have no inherent form, so they can often be shaped efficiently for a given application. Form can also help to overcome limitations such as low stiffness.

(b) *Consolidation of parts*. One-piece forming can often eliminate separate pieces, joints, seams, fasteners and sealants.

(c) *Light weight and good strength-to-weight ratio*. The low specific gravity of most plastics and reinforcing materials results in high ratios of strength to weight (see *Polymers: Mechanical Properties)*. The toughness of many plastics and composites often allows them to be used in thin sections, unlike some conventional hard building materials. By taking advantage of formability, lightweight structural shapes such as shells, folded plates and tension structures can be achieved. The extremely low specific gravity of plastics foams (0.015 and up) contributes to low weight in such structures as foam-cored building sandwich panels (see Fig. 1).

(d) *Corrosion resistance*. Because they are largely immune to electrochemical corrosion, plastics and composites can often be used in such applications as tanks to handle corrosive chemicals, water treatment plants and piping.

(e) *Transparency and light transmission*. Plastics may be highly transparent, translucent or opaque. Examples include dome-shaped skylights, breakage-resistant glazing, lighting fixtures, luminous ceilings, security screens and cover sheets for solar collectors.

(f) *Integral color*. Plastics integrally colored with dyes and pigments can be cut and otherwise worked without loss of color. Examples are flooring and building panels. Permanence is largely a matter of the permanence of the pigments and dyes. Surface colors may be applied as paints or gel coats.

(g) *Wear resistance and smoothness*. Although they may be scratched and indented more easily than

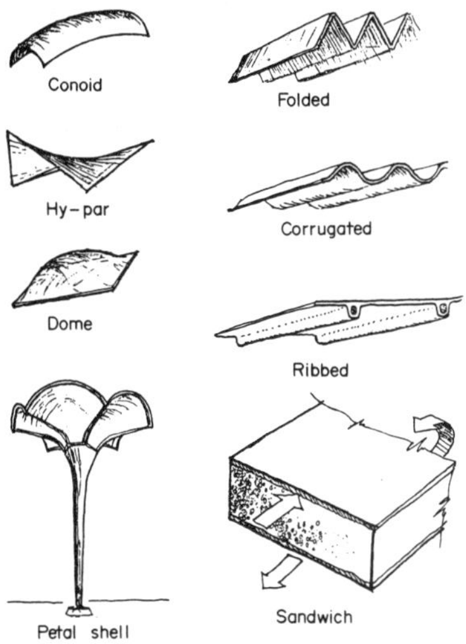

Figure 1
Examples of structural shapes utilizing the formability of plastics. In sandwiches, facings form a moment-resistant couple; the core provides shear resistance and stabilizes facings

glass and other harder materials, plastics often provide good to excellent resistance to wear or abrasion. Nylon hardware parts frequently exhibit outstanding wear resistance and quiet operation. Vinyl flooring is resistant to wear and abrasion. The smooth inner surfaces of extruded pipes result in low resistance to flow of liquids.

(h) *Energy absorption and impact resistance.* Some plastics shatter easily; others are tough and resilient. Examples of tough plastics include acrylic and polycarbonate glazing and lighting fixtures; poly(vinyl chloride), acrylonitrile–butadiene–styrene and polypropylene piping; and glass-fiber-reinforced plastic concrete forms.

(i) *Flexibility.* Some plastics are inherently soft and flexible; others become so by additives and copolymerization. Applications include electrical wire insulation, membranes for air-supported structures, vapor barriers, pond liners and flexible pipe.

(j) *Thermal expansion.* Plastics generally have much higher coefficients of thermal expansion than the other common building materials such as glass, masonry, concrete, metal and wood. Plastic building components must allow for expansion and contraction (see Fig. 2). Incorporation of reinforcement such as glass fiber greatly reduces thermal expansion coefficients but, even so, reinforced plastics typically expand and contract more than conventional structural materials.

1. Major Uses of Plastics in Building

The applications of plastics in building may be classified as (a) auxiliaries to other materials, (b) nonstructural and (c) structural and semistructural.

1.1 Auxiliaries to Other Materials

Among the principal uses of plastics as auxiliaries to other building materials are (a) protective and decorative coatings (see *Paints, Varnishes and Lacquers*); (b) adhesives and resin-bonding agents, especially for wood products (see *Adhesives for Wood*; *Structural Adhesives*); and (c) building sealants in the form of caulking compounds and gaskets. Contemporary sealants augment the older oil- and resin-based varieties and may be two component (mixed on the job) or single component (ready to use, often as cartridges for sealant guns). Polymers employed include polysulfide rubber, acrylics, silicones, chlorosulfonated polyethylene and urethanes. Gaskets, similarly, draw upon various polymers including natural and synthetic rubbers and poly(vinyl chloride).

1.2 Nonstructural Uses

Nonstructural uses constitute by far the most major application in building. In general, they are not called upon to carry significant loads and failure does not entail appreciable danger to occupants of buildings. Among the most important of such uses are the following.

(a) *Pipe, fittings and conduit.* This, the largest single class of uses, includes pipe and fittings mainly for drains, vents, water and underground lines. Poly(vinyl chloride) is the most widely used, followed by high-density and low-density polyethylene, acrylonitrile–butadiene–styrene, reinforced polyester, polypropylene and polystyrene.

(b) *Thermal insulation.* Efficient thermal insulating materials are provided by prefoamed or foamed-in-place plastics. Prefoamed plastics, such as polystyrene and polyurethane, are typically blown as large masses cut into slabs and blocks for use under concrete slabs, in cavity walls and as sheathing and sandwich cores. Two foamed-in-place types are common. In the first, for example polystyrene, pellets containing gas-forming ingredients are placed in the space to be filled and are heated. As gas is formed the pellets soften and expand, filling the space and coalescing. In the second, the liquid components of the foam, for example polyurethane and urea–formaldehyde, are mixed and poured into the space to be filled. Gas is formed as the reaction takes place; the foaming mass rises, fills the space and hardens. Depending upon the foaming agent and density, the thermal insulation achievable is one of the most favorable

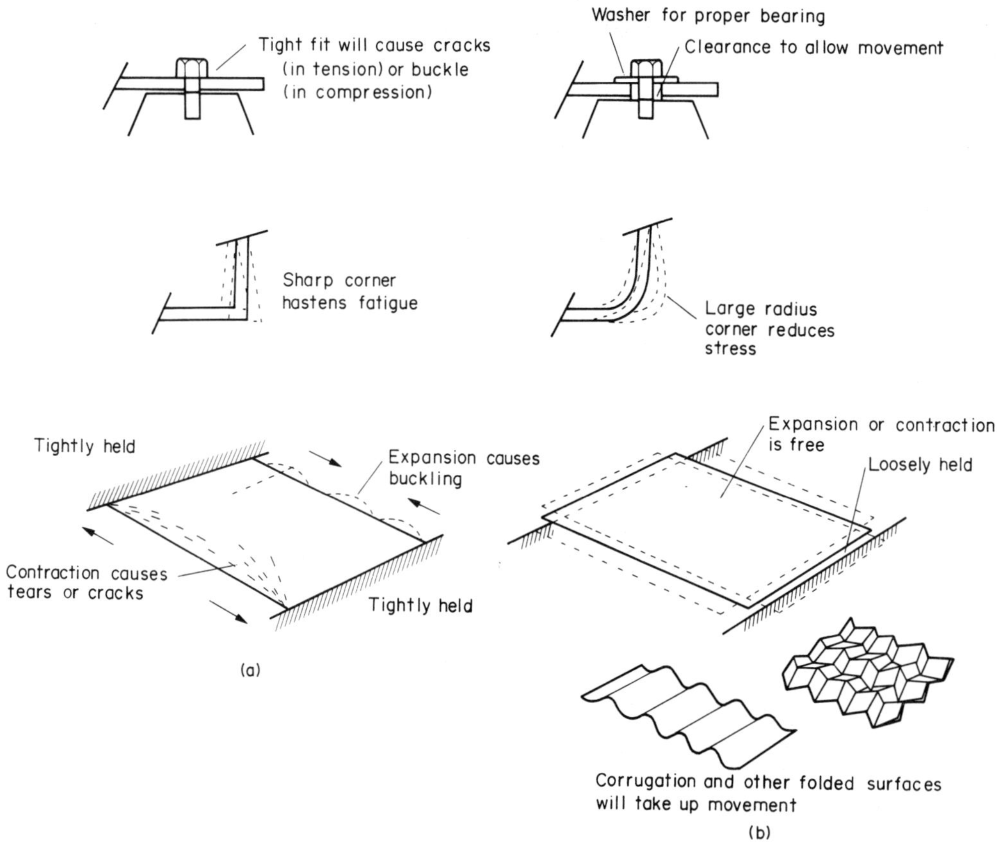

Figure 2
Thermal expansion: (a) bad and (b) good practice

among building materials. Some highly efficient gases, such as fluorocarbon, unless sealed, as in refrigerator walls, may gradually be replaced by air, decreasing the thermal value of the insulation. Cells may be open or closed, that is, interconnected or not. Closed-cell foams may be impervious to the passage of water vapor and constitute vapor seals; open-cell foams are not. In some foams, such as urea–formaldehyde, water is produced as a by-product of the foaming reaction and must be allowed to escape. Some foamed-in-place materials may shrink upon curing and allow heat to escape through the cracks. Precautions may be necessary to avoid residual irritating gases (see Sect. 3).

(c) *Flooring*. Poly(vinyl chloride) is compounded with fillers, fiber and additives to provide resilient sheet and tile flooring. Urethane foams are employed as underlayments for rugs. Epoxies provide paving materials.

(d) *Plumbing*. Bathtubs, shower stalls and other plumbing fixtures utilize large quantities of reinforced polyesters in addition to acrylics, polyacetals, thermoplastic polyesters and polystyrene.

(e) *Glazing and skylights*. Flat, corrugated and thermoformed acrylic and polycarbonate sheets are widely used, as are flat and corrugated reinforced-polyester sheets. Flat reinforced-plastic sheets are bonded to metal grids to form translucent structural wall and roof panels.

(f) *Lighting fixtures*. Acrylics, cellulosics, polycarbonates, polystyrene and poly(vinyl chloride) provide a multiplicity of indoor and outdoor lighting fixtures.

(g) *Moisture barriers*. Low-density polyethylene and poly(vinyl chloride) find extensive use as barriers to prevent moisture migration through concrete slabs on the ground and vapor passage through walls and roofs and as liners for swimming pools.

(h) *Profiles*. These include window parts, rainwater systems, moldings and many others and are chiefly extruded poly(vinyl chloride) and polyethylene, solid or foamed.

(i) *Wall coverings*. Poly(vinyl chloride) and polystyrene account for most wall coverings, including flexible sheets for walls subject to considerable wear.

(j) *Panels and siding*. These are mainly poly(vinyl chloride) and reinforced polysters, with acrylics and cellulose acetate–butyrate also employed.

1.3 Structural and Semistructural Uses

Plain unmodified plastics are moderate in strength, but generally too low in stiffness (having low moduli of elasticity) to be useful for structural and semistructural applications. These may be defined as those in which failure is dangerous to life and property. For such uses, composites are structural; the decorative high-pressure laminates (see Sect. 2.3), for example, are usually not employed structurally. Many fibrous and particulate composite parts (see Sect. 2.1) are nonstructural.

2. *Composites*

Among the most significant uses of plastics in building are the composites, combinations of materials whose properties transcend those of the individual materials acting alone. Plastics-based composites may be classified as particulate, laminar or fibrous.

2.1 Particulate Composites

Portland-cement concrete may be modified with polymers to produce polymer-impregnated concrete, polymer-cement concrete and polymer concrete.

Polymer-impregnated concrete is produced by drying cured standard concrete, evacuating the pores and impregnating with a monomer subsequently polymerized in place by heat, chemical action or high-energy radiation. Compressive strengths may be increased fourfold, creep drastically diminished, and resistance to sulfate attack and freezing and thawing greatly enhanced. Road surfaces have been dried at high temperatures in temporary enclosures, then soaked with monomer which is subsequently cured by heating.

Polymer-cement concrete is standard concrete with a monomer or polymer latex added during mixing. It is subsequently handled like standard concrete. Results have ranged from disappointing to several-fold increases of compressive and tensile strength.

Polymer concrete consists of regular mineral aggregate in a polymer matrix, customarily polyester, instead of Portland-cement paste. The ingredients are mixed and cast into forms or molds where the polymer hardens, in hours at room temperature or minutes with heat. Unlike standard concrete, no curing time is needed. A common application is in wall panels with thin polymer-concrete facings surrounding a foamed plastic core. Polymer concrete is used for rapid repair of holes in concrete roads.

2.2 Fibrous Composites

For building purposes, the most commonly used fibrous composite consists of E-glass fibers in a polyester matrix. Other fibers, natural and synthetic, are used occasionally, as are other matrices such as epoxy, phenolic and acrylic. Glass fibers are employed as continuous-filament and chopped-fiber mats, in various fabrics woven of rovings or twisted yarns and in filament winding. Of these, chopped-fiber is the most common, employed in layup, sprayup, bulk-molding compounds and sheet-molding compounds.

Features of fibrous composites favorable for building applications are increased strength and stiffness relative to plain plastics, toughness, reduced coefficients of expansion and lightness compared to some other building materials. As noted previously, curved shells, folded plates, sandwiches, ribs, corrugations and similar features serve to utilize the material to the fullest extent (Fig. 1).

Membranes are a special case of fibrous composites. These are usually woven fabric, glass or other synthetic, coated with a plastic such as poly(vinyl chloride) or polytetrafluoroethylene. Such membranes are employed in air-supported structures and in tension structures commonly supported and held taut by cables.

2.3 Laminar Composites

For buildings, high-pressure laminates and structural sandwiches are most important.

High-pressure laminates are most commonly decorative. They consist of layers of paper impregnated with phenolic resin, faced with printed paper (sometimes fabric or wood veneer) impregnated with melamine–formaldehyde and overlaid with a melamine–formaldehyde-impregnated veil, all simultaneously pressed into a decorative laminate, usually subsequently bonded to plywood, particle board or other substrates for counter and table tops, door facings, furniture and cabinetwork. Such laminates are resistant to most household solvents and are harder than common varnishes and lacquers, but not as hard as glass.

Structural sandwiches consist of relatively thin facings of dense, strong, stiff materials bonded to relatively thick, less dense, less strong and less stiff cores to provide lightweight combinations that are strong and stiff because of their geometry (Fig. 1). In building panels, the combination is also expected to provide thermal insulation, acoustical properties, fire resistance, weather resistance and resistance to normal wear and tear.

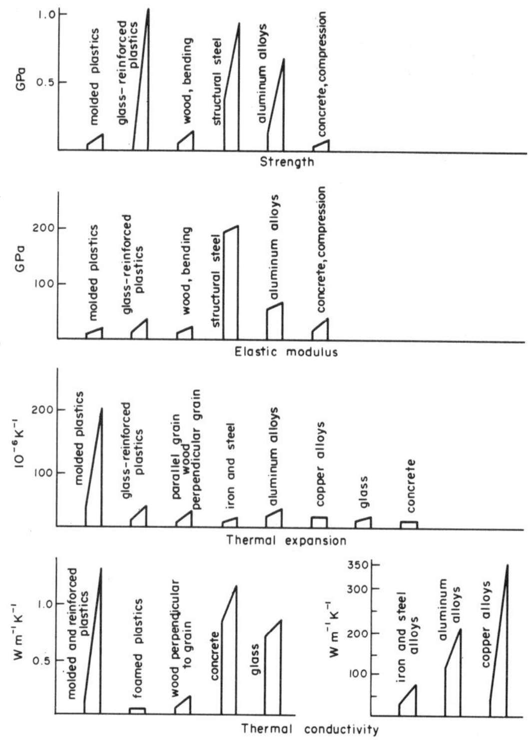

Figure 3
Comparison of the properties of plastics and other materials

Facings most commonly employed include plywood, fiberboard, particle board, cement–asbestos board, gypsum board, metals such as steel and aluminum, high-pressure laminates and glass-fiber-reinforced plastics.

Cores include foamed plastics, honeycombs of paper impregnated with phenolic resin, fiberboards, plywood, particle board, cement–asbestos board, gypsum board, foamed concrete and various wood and metal grids.

Adhesives to bond facings and core are generally based on synthetic resins. Some foamed-in-place plastics, for example polyurethanes, are self-adhering. The adhesive must withstand shear and tensile stresses between facing and core.

3. *Properties of Building Plastics*

Some properties of plastics and composites compared with other building materials are shown in Fig. 3. The higher values of strengths and elastic moduli are achieved by fibrous composites, whereas the lower values are those of plain unreinforced plastics. Although glass has an elastic modulus of about one-third that of steel, when combined with a low-modulus plastic matrix in the usual proportions, the combined elastic modulus is seldom more than half that of glass alone and is frequently less than that of wood parallel to the grain or of unreinforced concrete.

The high values of thermal expansion shown in Fig. 3 are those of plain plastics, whereas the low values are caused by the constraining effects of reinforcements, especially glass with its low thermal expansion and high elastic modulus.

The thermal conductivity of molded and reinforced plastics is comparable to wood, concrete and glass. Foamed plastics, on the other hand, are among the best insulators for buildings.

3.1 Durability and Weathering

Depending on the compositions and on the conditions of exposure, plastics may be highly durable or may disintegrate rapidly. Plain polyethylene deteriorates quickly outdoors but, when compounded with carbon black, provides highly durable cable covering and irrigation pipe. Acrylics and PVC have given many years of good performance. Glass-fiber-reinforced polyesters exposed to weathering have performed well or poorly depending on the type of polyester employed, the care with which the fiber has been thoroughly wetted and embedded in the matrix and the surface treatments such as veils and gel coats. Once exposed by erosion, fibers tend to act as wicks to draw moisture inward, leading to deterioration. Localized stress, as in sharply curved transparent sheets, often results in localized crazing and eventual cracking. Stress may also be caused by ignoring the typically high expansion coefficients of plastics and confining them too rigidly.

Plastics may perform well under conditions such as moisture and polluted atmospheres that cause deterioration of metal, stone and other susceptible materials. They provide some of the best protective coatings for such conditions and may themselves by protected from attack or have eroded surfaces rejuvenated by such coatings.

3.2 Flammability

Like other organic building materials, such as wood, plastics can be destroyed by fire. Some burn readily, some with difficulty and others do not support combustion on removal of flame, but none withstand the temperatures that can be encountered in building fires.

Smoke and gases may be more hazardous than flame. Some gases, such as water and carbon dioxide, may be innocuous, but carbon monoxide is likely to be generated copiously when oxygen is deficient. Depending on the chemical structure, gases may be noxious or toxic and dense smoke may be generated. A choice may therefore have to be made between flame and smoke in a given situation.

Some of the most effective fire retardants diminish the durability of plastics in uses such as outdoor exposure. A choice or balancing of requirements may be necessary.

4. Building Codes and Standards

There are no Federal building codes in the USA. Many states have building codes but they may not be completely binding upon municipalities. Local codes govern. Other countries often have national codes. Codes may be prescriptive (i.e., specify materials and installation in detail) or they may specify the performance to be expected of buildings and the components, but leave the choice of materials and methods open.

Several organizations in the USA write recommended uniform codes that are frequently adopted by municipalities. Their provisions for plastics are similar and may be taken as representative, realizing that the municipal codes, although often based upon the recommended codes, prevail.

In addition to general measures applicable to all materials, the recommended codes have separate sections devoted to light-transmitting plastics and to foamed plastics. Plastics are classified as thermoplastic, thermosetting and reinforced. Both sections refer to standard tests and specifications of the American Society for Testing and Materials (ASTM) respecting strength, durability, sanitary properties, flammability and smoke.

4.1 Light-Transmitting Plastic Construction

For light-transmitting plastic constructions, self-ignition temperatures must be 350 °C or above (ASTM D1929) and smoke density not greater than 450 (ASTM E84). Two classes of plastics are based on burning rates. Types of applications are glazing, plastic wall panels, roof panels, skylights and light-diffusing systems. Glazing of unprotected openings is limited according to occupancy, percentage of wall area, height above ground and presence of absence of automatic fire-suppressant systems such as sprinklers. Exterior wall panels are subject to similar provisions. Roof panels are limited according to occupancy (prohibited in some), requirements for roof coverings, automatic fire-suppressant systems, horizontal spacing and proximity to exterior walls. Skylight assemblies must conform to requirements respecting curbs, edge protection (incombustible strips), maximum area, minimum rise of dome, horizontal separation and proximity of exterior walls. If automatic fire-suppressant systems are present or if skylights are used as fire vents, areas may be more generous. Light-diffusing systems (illuminated ceilings) must comply with interior finish requirements or fall from their supports at specified temperature ranges. Sprinklers may be needed above or below the installation. Areas are limited. Partitions, bathroom accessories, awnings and greenhouses are subject to special provisions.

4.2 Foamed Plastics

Similar language to that used for light-transmitting plastics applies to foamed plastics. Generally, flame-spread ratings are limited to 75 and smoke developed to 450 (ASTM E84) unless specifically exempted.

Foamed plastics are permitted within the cavities of masonry or concrete walls and, generally, on room surfaces or as elements of non-fire-resistive walls if protected by a barrier with at least a 15 min fire rating (ASTM 119), such as 13 mm gypsum board or Portland-cement stucco, 0.81 mm aluminium or 0.45 mm (26 gauge) galvanized steel sheet.

With protective surfaces at least equal to these, foamed plastics are permitted as cores of sandwich panels. When such panels are load bearing, as in roof panels, special requirements may be imposed. Automatic sprinklers may be required, depending on occupancy.

See also: Construction Materials, Cellular; Design with Plastics; Flammability of Polymers; Paints, Varnishes and Lacquers; Structural Adhesives

Bibliography

Agranoff J (ed.) 1984 *Modern Plastics Encyclopedia*. McGraw-Hill, New York

American Society of Civil Engineers 1984 *Structural Plastics Design Manual*. ASCE, New York

Basic Building Code 1975 Building Officials and Code Administrators International, Chicago, IL

Dietz A G H 1969 *Plastics for Architects and Builders*. MIT Press, Cambridge, MA

Saechtling H (ed.) 1973 *Bauen mit Kunststoffe*. Hanser, Munich, Germany

Skeist I (ed.) 1966 *Plastics in Building*. Reinhold, New York

Standard Building Code 1977 Southern Building Code Congress International, Birmingham, AL

Uniform Building Code 1978 International Conference of Building Officials, Whittier, CA

A. G. H. Dietz
[Massachusetts Institute of Technology, Cambridge, Massachusetts, USA]

Plastics for Packaging

Approximately 35% of plastics produced in the developed countries is consumed by packaging, which is the biggest single use of these materials, amounting to US$19.9 billion in 1986. Of this quantity, 71% is accounted for by the three polyolefines: low-density polyethylene (LDPE), high-density polyethylene (HDPE) and polypropylene, which thus dominate. Beyond these, the next two major plastics are polystyrene which accounts for 12% and polyethylene teraphthalate (PET) at 6.4%.

The polymers used in packaging fall into two main price groups, depending on the complexity of their manufacture and the relative cost of the chemical intermediaries used. The simpler and cheaper materials are the polyolefines, followed by polyvinyl chloride (PVC) and polystyrene. Next are the nylons, PET and acrylonitrile–butadiene–styrene copolymer (ABS). The more expensive specialty polymers developed for engineering applications are not generally used for packaging. In some cases, more demanding applications can justify more expensive materials. A survey of the properties and uses of the main plastics used in packaging is given in Table 1.

The prime function of packaging is to retain all the contents of the product packed as near as possible in the state in which they leave the factory and appropriate to the expectations of the final customer. This implies protection from physical damage by external agents such as impact, corrosion, degradation by light and heat, and abuse in transit and storage; it also requires noninteraction chemically between the contents and container materials. There must also be no migration of the constituents of the packaging materials into the contents, nor absorption of the contents into the packaging material. The material used for the packaging must prevent or minimize the passage of gases such as oxygen and carbon dioxide inwards or outwards, and water vapor except in special applications.

During the processing of packaged foods and drinks, temperatures up to 125 °C for a period of up to an hour may be used to sterilize the contents. Alternatively, the pack may be subjected to deep-freeze temperatures as low as −40 °C for periods of months. An important function of packaging is to present an attractive image and to convey information about the contents. This requires the surface to be printable or capable of being labelled. It is against this background that the use of packaging should be viewed and the choice of a specific polymer and its processing selected.

Plastics are preferred to other materials by virtue of their better economics, the freedom with which they can be molded to a wide variety of shapes, and their lightness compared with metal or glass—leading to economies in transport and often savings in space. The moisture- and gas-barrier properties of plastics are not as good as those of metal and glass but are adequate for most applications, or can be enhanced by coating or by the use of special barrier polymers. If dropped or otherwise abused, plastic will perform much better than glass which also represents a safety hazard when broken. Many plastics have high impact strength and resistance to tearing or denting. Metal containers have better barrier and good impact strength but are subject to corrosion and can only be produced in a limited variety of shapes, usually cylindrical or rectangular.

Compared with paper and board containers, plastics again have the advantage of the much greater variety of shapes that can be used. Board containers are usually rectangular or cylindrical and require a lining of plastics to make them resistant to water and to render them sealable.

The polymers used in packaging have properties that are broadly adequate for packing most classes of product and the ease with which different shapes can be produced and linked to specific market images is a major strength compared with other materials. However, they do have a number of weaknesses. Their relatively open structure can lead to significant gas and water vapor transmission, as well as absorption of some molecular species from the contents with a resultant change in the aroma and flavor of foodstuffs and beverages. In addition, some organic solvents can be absorbed and this will lead to a reduction of strength or to swelling. Most polymers are sensitive to degradation in some degree by heat and light, particularly ultraviolet frequencies.

Plastics are inert chemically and therefore resist attack from the most substances. This can make it more difficult to obtain adhesion, and special preparation of the surface of nonpolar polymers is necessary to allow the application of inks or adhesives. Such polymers are satisfactorily treated by the brief application of a gas flame or by corona electrical discharge. Thermal and light degradation have been defeated by the introduction of durable and nonmigrating thermal and ultraviolet stabilizers. Barrier deficiencies have been improved markedly, both by the development of coatings and by the addition of special barrier polymers in multilayer structures.

Nearly all the polymers used, in packaging have softening and melting temperatures in the range 70–270 °C. This allows them to be processed by melting and then extruding from a screw extruder or by injection molding. In both cases, the material is cooled rapidly and at the same time shaped to its desired form, be it sheet, film, container or bottle. In some cases, extra strength is imparted to the material by stretching above the glass-transition temperature in the solid state. This resultant extra degree of molecular orientation and crystallinity of the polymer also improves the gas-barrier properties.

1. Bottles

1.1 Choice of Materials

One of the most common uses of plastics in packaging is the manufacture of bottles in a wide range of

Table 1
Properties of plastic materials used in packaging

Material	Density ($kg\ m^{-3}$)	Material softening (°C)	Oxygen permeability[a] $cm^3\ mm\ m^{-2}\ d^{-1}\ bar^{-1}$ at 20 °C	Water vapor transmission $g\ mm\ m^{-2}\ d^{-1}\ bar^{-1}$ at 20 °C	Oil/grease resistance	Solvent chemical resistance	Impact strength IZOD	Clarity	Uses in packaging
LDPE	917–932	98–115	190	0.6	good	good except nonpolar solvents	very good	fair	films, bags, coatings, moldings, bottles
LLDPE	918–960	122–124	190	0.6	good	good except nonpolar solvents	very good	fair	films, bags
HDPE	950–905	130–137	60	0.2	good	good except nonpolar solvents	very good	poor	bottles, moldings bags, containers
Polypropylene	900–910	165–175	70	0.2	good	good except nonpolar solvents	fair–good	good	films, bags, bottles, containers, caps, tubes
PVC	1300–1580	75–105	4	1.0	good	good for nonpolar solvents	fiar–good	good	films, bottles, containers
Polystyrene GP	1050	74–105	160	2	fair–good	poor	poor	good	containers, expanded foam
High impact	1030–1060	93–105	150		fair	poor	fair–very good	poor	
ABS	1020–1060	88–120	40	5	fair	poor	good–very good	poor	containers
PET	1290–1400	260–270	2.7 (amorphous) 1.4 (crystalline)	1.0	good	good	good	very good	bottles, films
Polycarbonate	1200	306–316	80	5.5	good	good	excellent	good	bottles, films
Nylon 66	1130–1150	255–265	2	2.5	good	good	good	fair	bottles, films
Ionomer	930–960	81–96	190	0.6	good	good	very good	good	adhesive–heatseal layer with nylon, films
			RH 0% 100%						
PVdC Extrusion Grade	1650–1720	160–172	0.06 0.06	0.04	good	fair		good	barrier layer
EVOH 30% Ethylene	1100–1200	140	0.003 0.3	1.0	very good	good		good	barrier layer
Nylon MXD6	1220	243	0.6 0.15	1.0	good	fair		good[b]	barrier layer
Acrylonitrile styrene	1130	110	0.40 0.40	2.0	good	fair		fair	bottles, films, containers

a Standard metric units; standard US units are $cm^3\ (10^{-3}\ in)\ (100\ in^2)^{-1}\ d^{-1}\ atm^{-1}$ at 68 °F with (US units) = 2.54 × (metric units) b Poor when crystalline sizes from 25 ml to 5 l or more. Currently, some four billion plastic bottles are used annually in the UK and about 38 billion in the USA. The products packed cover an enormous range, including household cleaners, industrial chemicals, sauces, beverages, water, milk, pharmaceuticals and cosmetics. Some are dry, some water based and some contain solvents or corrosive liquids; some are toxic.

The polyolefines are the first choice for a large number of products, since for minimum cost they give bottles that are strong, flexible and, in the case of polypropylene, fairly transparent. The water vapor barrier is excellent and they are resistant to all but the most corrosive concentrated chemicals. They are, however, prone to absorb essential oils from some foods, which can result in a loss of flavor and similarly a loss of aroma from cosmetics. Equally, the rather poor gas-barrier properties of the polyolefines mean that they are not used alone for oxygen-sensitive foods, nor for highly carbonated beverages where the useful storage life would be too short. For products with a short shelf life, such as milk, they are widely used.

LDPE was first used in packaging in the 1940s, mainly in the form of film. In the 1950s, bottles made from it became established for a range of chemical and household products, including the first squeeze bottles for washing-up liquids. However, the introduction of HDPE with its better rigidity led to its replacing low-density material for most bottles in the 1960s because considerable weight and thus cost savings could be made. It now accounts for 55% of polymers used in bottles.

Polypropylene came into use in the 1960s and is now used for bottle manufacture on some scale, but has not replaced HDPE to any great extent because it is more difficult to process and is not any cheaper. It is mainly used where its better clarity is an advantage, as for instance in sauce bottles. Polypropylene has a somewhat higher softening temperature than the polyethylenes, which makes it better for applications where hot-fill or in-pack heat sterilization is used. It is increasingly being used for bottles to contain sterile water and other liquids for medical use.

Since these materials are so cost effective over such a wide range of packs, it has been a continual challenge to overcome some of their weaknesses and, in particular, their relatively poor gas barrier. This has been achieved to a considerable degree by the development of coextruded multilayer bottles incorporating an inner layer of a special polymer such as hydrolyzed ethylene vinyl acetate (EVOH) which is bonded to the polyolefine using a special adhesive polymer layer. These coextruded bottles are mainly used for sauces and ketchups which degrade or discolor when oxygen is present.

Where polyolefines do not provide adequate properties the next most commonly used material is PVC. It gives clear bottles with good gas and water vapor barrier and better resistance to absorption of oils. PVC requires the addition of stabilizers and processing aids that must be selected for food contact applications. It is widely used for bottles to contain a range of products including bath foams, sauces, cordials and in France about half the mineral water consumed is in PVC bottles. There are now strict regulations covering the use of PVC in contact with foods.

The packaging of beers and other carbonated beverages in plastic bottles was at a low level before the introduction of PET bottles in the 1970s. PET is a special high-molecular-weight grade of polyethylene terephthalate which is also used for the production of oriented films. It has excellent clarity and water vapor barrier and in its oriented state it has good gas barrier, about 50 times better than the polyolefines. PET bottles have a high degree of molecular orientation and thus a very high tensile modulus. This allows them to contain highly carbonated beverages within which pressures as high as 1 MPa can be experienced if they are left in direct sunlight. The loss of carbonation through the wall of the bottle is such that a useful shelf life of 26 weeks is possible with a 2 l bottle but this is halved when a 0.5 l bottle is used. This reduction is due to the higher surface-to-volume ratio in the smaller bottle.

For packaging of beers the oxygen barrier of PET bottles is not good enough so it is now normal to coat the bottles externally with a layer of polyvinylidene chloride (PVdC) polymer which is applied as an aqueous distribution and then dried in a hot-air oven.

Alternative barrier systems using an internal layer of a high-barrier polymer have recently come into use. These barrier polymers are either of the EVOH type or of special grades of aromatic nylon. They are introduced as the inner layer either by coextrusion or by coninjection with the PET.

Other polymers used to make bottles are polycarbonate and nylons which are used for their clarity and high temperature resistance. Both can be sterilized by boiling and will withstand much higher temperatures in use. They are used for such applications as baby-feeding bottles.

1.2 Production Processes

The production process used for bottle making is highly automated and is usually computer controlled. The output from the largest machines is often as high as 15 000 bottles per hour from up to 28 molds. Small units with one mold are used for low-output specialty bottle production. There are two processes: extrusion blow molding and injection blow molding and these have been extended to include injection stretch–blow molding.

For both processes the polymer in granular form is fed to a screw extruder which has a screw with a helical flight rotating inside a heated barrel. This melts, mixes and compacts the material to the front of the screw. Coloring and processing aids can be added at the input to the extruder. Some moisture-sensitive polymers such as PET and nylons must be well dried before extrusion to avoid degradation.

In extrusion blow molding, which is the process most widely used for bottle manufacture, the molten polymer is forced (usually) downwards through a tube-forming die by the pressure generated by the extruder. Two halves of a cooled bottle mold are then closed around the tube, on parison. It is severed from the emerging tube and air is introduced at pressure through either the top or bottom of the parison. After blowing and cooling, the bottle is removed from the mold and the excess material at the top and bottom is removed and recycled back into the extruder.

The injection blow process first forms the parison by injection molding using an extruder with a reciprocating screw unit to melt and inject the polymer into a multicavity mold. The parisons are then either moved on the mold core to blowing molds and blown while the material is still hot, or the parison is cooled and subsequently reheated before it is placed on a mandrel and introduced to the blowing mold. In both cases, the material is blown at a temperature above the glass-transition point and usually below the melt point.

In injection stretch–blow molding, the parison is first reheated to a temperature just above the glass-transition point and is then stretched longitudinally by a rod inserted through the neck while it is blown against the wall of the cooled mold. This introduces a high degree of molecular orientation into the wall of the bottle or jar. PET bottles for use with carbonated beverages are made this way and have walls with very high tensile and impact strengths. The orientation also improves the barrier of the bottle wall to gases because of the closer packing of the molecular structure. This can be further improved if the material is given increased crystallinity by heating after blowing above the glass-transition temperature. However, the bottle must be restrained from shrinking by internal pressure while this is done.

1.3 Design Considerations

In all packaging applications a prime requirement in designing a pack, once the external shape has been chosen, is to develop a container that will hold, within that shape, the correct volume of contents with the appropriate fill height while minimizing the amount of plastic material used and meeting the criteria for strength, flexibility and storage. Internal pressure, whether positive or negative, which the pack may have to sustain during processing, distribution and in storage must also be considered. Significant savings are made by placing material exactly where it is required in the bottle. In the case of parison extrusion, this is done by moving the inner mandrel of the die vertically during the extrusion of each bottle length to vary the wall thickness.

Generally, a container will be rounded, avoiding flat panels, sharp angles and rectangular shapes.

2. Pots, Tubs and Trays

2.1 General Description

The category of pots, tubs and trays covers open-ended containers with a wide range of sizes in each type. The shapes are either round, rectangular or square, usually with rounded corners. Pots generally range from a capacity of 100–500 ml. Tubs range from 0.5–5 l capacity. These containers are usually closed, with either plastic lids or aluminum foil laminates which are heat sealed to a flange. Trays are relatively shallow, rectangular in shape and are covered with a layer of film or a plastic lid.

Pots and tubs are extensively used for dairy products, soft margarine and a range of moist and dry food products. Larger tubs are used for paint and other similar home-use products as well as catering packs for food products. Trays are used for cooked and uncooked meat and fish, and range of fresh or chilled foods including fruit and vegetables. Trays made from expanded polystyrene are used for fast foods.

2.2 Choice of Materials

The materials most widely used for the manufacture of these containers are polystyrene, polypropylene, ABS, PVC and HDPE. The polystyrene can be either "crystal" or modified with butadiene rubbers to improve impact strength. ABS and PVC have traditionally been used for greasy products because of their good resistance to stress cracking and also because of their very good thermoforming performance. HDPE is used for the manufacture of containers by injection molding. Polypropylene is increasingly replacing the other materials because of its low cost and good physical properties.

None of these polymers has very good gas-barrier properties but the two polyolefines have very good moisture barrier. To improve oxygen barrier, a coextruded inner layer of a high-barrier polymer such as PVdC or EVOH may be incorporated. PVC combines relatively good moisture- and gas-barrier properties and is now widely used for "modified atmosphere packaging" of meats. Its use for margarine packaging is reducing as polypropylene takes over.

Improved oxygen-barrier pots and trays are increasingly being used for shelf-stable food packs. These are either filled under aseptic conditions or are sterilized by heat processing after filling and sealing. The storage life of these packs is often at least 12 months and they are used for both dairy products and ready meals.

Special grades of PET are used for making trays that are crystallized after forming to give them high temperature resistance for use with prepared meals or dishes to be cooked in ovens.

The use of polypropylene for paint containers is increasing, but only at present for water-based paints. The use of plastics for solvent-based paints is so far restricted, due to the relatively high cost of solvent-resistant polymers such as PET and the PVC copolymers specially formulated for this purpose.

2.3 Production Processes

Tubs, pots and trays are manufactured by injection molding or thermoforming. Smaller containers are mainly made by thermoforming; injection molding is used mainly for larger containers and those of higher quality. Containers made by the latter route tend to be thicker and more rigid. The process is the same as that described for the manufacture of injection-molded parisons but the mold walls are much thinner and shorter cooling times can be used. With multicavity molds, outputs as high as 8000 pots per hour are possible.

Thermoforming consists first of continuous extrusion of sheet through a wide-slit die onto a cooled rotating roll which is usually followed by two or three more cooling and polishing rolls. The sheet is reheated to just below or to the melt temperature and then formed into pots, tubs or trays in cooled multicavity molds using either vacuum from below or air at pressure from above to shape it into the separate cavities. It is common practice to use a plug to push the material into the cavity before the air is applied, to improve the distribution in the base and wall. The individual containers are then cut from the web in the forming position or can be conveyed forwards in the web and cut out in a separate trim press. PVC sheet is usually made by calendering between heated rolls which give a high polish to the web and produce very uniform thickness. It is otherwise handled as described above.

Foamed polystyrene sheet is made by extruding with a foaming agent that forms small bubbles in the sheet as the pressure is released after emergence from the die. The sheet is then formed in conventional thermoforming machines. It has been normal to use a chlorinated fluorocarbon as the foaming agent, but since the discovery of the damage to the ozone layer, it is now being replaced by other less harmful gases (e.g., pentane).

2.4 Design Considerations

The range of shapes that can be made by thermoforming is relatively restricted. Circular, square and rectangular containers are all made but the walls must be tapered outwards from the base for ejection from the cavity. It is normal to have a denesting feature below the flange to enable separation from a stack when the pots are placed in the filling machine.

Injection-molded pots are made in a greater variety of shapes (e.g., goblets with integral stems) but usually also require a taper for ejection from the cavity.

3. Film and Flexible Packaging

3.1 General

Film and flexible packaging is based on polymer in the form of film, defined as material of 0.25 mm or less in thickness, either on its own or in combination with paper, aluminum film or metalized with aluminum. It can be used in preformed pouches, bags or as free film for wrapping. It can particularly be applied to other containers such as jars and tubs as a lidding material, or with trays as a covering. It is used for both consumer and industrial products. It forms 40% of the use of thermoplastic polymer in packaging in the developed countries and is the area of most rapid growth.

The term flexible packaging is applied to bags and pouches, made of the materials described, which can be sealed to form a closed container. It accounts for more than half the use of polymer for film.

Cellophane from regenerated cellulose was the first film used for packaging and was introduced in 1929. Subsequently, PVC film appeared in the late 1930s but it was with the introduction of polyethylene film in the late 1940s that the real growth began.

3.2 Choice of Materials

Film for packaging needs to be strong, tough, puncture resistant, usually of reasonable clarity, free from creep under the tension used in wrapping, and show good elastic recovery. Where foodstuffs and beverages are concerned, the consideration of oil absorption, water and gas penetrability apply as in other container forms.

The polymers used in this area are polyethylene in low-density, linear low-density and high-density forms, polypropylene (oriented or unoriented), PVC, nylons, PET, ionomer and PVdC. A wide variety of combinations of these materials in film form are used to match particular requirements. LDPE, especially the linear form (LLDPE), accounts for 79% of film products. PVC film is extensively used as clingfilm.

Polyethylene film was first used on its own for the manufacture of bags and wrappings during World War II. A major advance in its use came with the introduction of coatings to improve surface properties, particularly adhesion, and thus the bonding of one film to another and to metal. This has greatly increased the range of applications by allowing combinations of film or film coating to be tailored to specific package needs. It is still the most widely used polymer in flexible packaging in both low- and high-density forms. To these has recently been added LLDPE, which has much higher impact strength and allows lower thickness and thereby strengthens its position in this field.

Polyethylene is frequently the first-choice material for film for wrapping and bags in noncritical end uses. As with bottles and jars, it provides an adequate performance at low cost for many purposes (e.g., carrier bags and retail takeaway bags). There is a wide range of form-fill bags made and used for magazine and journal transit by post. These are made on special reel-fed machines that form the bag rond the item to be packed and heat seal the back and end seams before separating the unit pack from the reel.

Film is also used for holding a number of items in a collated bundle or on a pallet. The collation is wrapped round with film from one or more reels, usually in an automated process, and the ends heat sealed together. The whole is passed through a hot tunnel where the film shrinks tightly into place round the contents while the natural elasticity of the film coupled with low stress relaxation hold the pack firmly together over a prolonged period. In an alternative mode, a preformed sleeve of film is stretched and the contents inserted. When the tension on the film is relaxed, this allows them to be firmly gripped. These technologies can be usefully applied to cardboard cartons containing goods to provide a good seal, tamper-evident overwrap, to keep water off the vulnerable board containers and to provide a tough outer protection.

Polyethylene is commonly used for these shrink- or stretch-wrap applications, but the elastomeric nature can be enhanced by copolymerization of ethylene vinyl acetate (EVA). Such films will stretch over 100% and still display elastic recovery.

Reel-fed oriented polypropylene film is used in large quantities because of its clarity and strength to make bags and pouches for snack foods (e.g., potato crisps by the form-fill-seal technique). Oriented polypropylene of itself cannot be satisfactorily sealed, so the films are usually made by coextrusion with a lower melting copolymer with polyethylene. Heat seals can then be made without distortion of the polypropylene base film.

For many purposes, particulaly when the product to be packed is food, these relatively simple films cannot provide the combination of properties required. Thus oxygen-sensitive foods require a much better barrier to the gas diffusing in from the atmosphere. Several combination systems have been developed such as polypropylene film and aluminum foil, film coated with a thin layer of PVdC or two layers of film coextruded with an inner layer of EVOH. All have their advantages and disadvantages. Thus, the aluminum foil gives in principle a complete barrier but has to have an inner layer of heat-sealable polymer, usually provided by an inner polyethylene layer. PVdC coating improves the barrier performance by up to 50 times that of polyolefines, but is expensive. The barrier with EVOH is excellent under dry conditions, also having the advantage over aluminum foil of allowing a transparent pack while the metal foil pack is opaque but can have a very attractive appearance. Other factors such as strength and sealability can be similarly manipulated.

The most refined pack of this type is the retort pouch which consists of an outer film of PET laminated to aluminum foil with a further inner layer of polyethylene or polypropylene laminated on the other side of the foil.

A rapidly developing use for film packaging is in "modified atmosphere packaging" for fresh foods (e.g., meats and vegetables. The packs have either to contain the produce in a specified predetermined atmosphere in the pack or to allow the entry of oxygen at a predetermined rate into a pack where its level has been reduced to a chosen degree at packing. In most cases an artificial atmosphere of oxygen, nitrogen and carbon dioxide in various proportions is used to replace the normal atmosphere at packing.

In the UK, 50% of red meat sold in supermarkets is packed in modified-atmosphere-packaging conditions using laminates of PVC and polyethylene to form the tray. The lid is of polyester, coated with PVdC and polyethylene. The polyethylene provides the seals. Combinations of nylon and ionomer or polyethylene are used for cooked meats and cheese and for cooking oil pouches in India.

3.3 Production Processes: Film Products

The majority of films used in packaging are formed by extrusion, either from a single polymer or by combination with one or more materials, using lamination, coextrusion and coating.

As with other plastic packaging products, additives are compounded with the polymer to give ultraviolet and thermal stability, antioxidant properties and, particularly with film slip additives, to allow easy separation and movement of stacked film and its products. Coloring additives are are also incorporated if required.

Whether a single layer or multilayer coextruded film is being produced, the basic unit is a screw extruder, as for other products mentioned earlier. The molten polymer is forced at high pressure, typically 20–30 MPa, by the pumping action of the rotating screw into either a straight-slit flat die or a tubular ring die.

In flat-die extrusion, the molten web is first cooled by laying it on a polished cold casting roll. This is rotated at a greater speed than the extrusion speed, thus stretching and thinning the polymer sheet to a predetermined controlled thickness before winding the filming up a single layer.

In tubular-film manufacture, the molten polymer from the ring die is usually extruded vertically upwards. Air is introduced through the die into the tube to inflate the bubble of film. The tube is hauled off by driven nip rolls and then either wound up as collapsed lay-flat tube or is slit at the edges and wound as two separate rolls.

PVC can be formed into film by extrusion onto calender rolls where its thickness is reduced between the hot polished rolls. PVC film for packaging normally has to be made pliable by the addition of stabilizers and plasticizers. There is also a growing production of blown PVC for clingfilm use.

Orientation of the polymer molecules in the longitudinal and transverse directions is applied to films where increased strength and stability is required. It incidentally provides some improvement in clarity and in gas barrier. The films most often processed thus, in order of importance, are polypropylene, PET, PVC, nylon and polystyrene. Film formed in the normal way is reheated to a temperature above its glass transition and then stretched by between three and eight times in directions at right angles in the plane of the film. The stretch ratios used and their balance depend on the polymer and the properties sought from it.

4. *Other Applications*

There remain a number of other speific uses of plastics in packaging which can be briefly listed. These include medical and pharmaceutical packs, cosmetics and toiletries and aerosols.

Plastics have replaced glass and metal in the packaging of many medial and pharmaceutical items. Among the major uses are small bottles, jars and blister packs for ethical and proprietary pills, larger bottles and pouches for sterile fluids, intravenous feeds and plastic wrapping for disposable instruments and other one-shot items. Polyethylene, polypropylene and PVC are extensively used.

Cosmetics and toiletries are strongly influenced by the requirement to provide a strong and specific image for the product by the style, design, finish and decoration of the container. As many of these products contain oils and perfumes, the plastics used must have good resistance to these and must not absorb components of the scents. PVC, styrene–acrylonitrile and polyacetals are all used.

The main part of aerosol containers are metal or glass but the vital valve parts are plastic. The polymers used are polypropylene for the dip tube and nylon or polyacetal for the actual valve components.

Plastics are widely used for caps, closures, lidding material and overcaps. The polymers used are mainly polyethylene and polypropylene. The latter is particularly used for screw closures since these require higher stiffness to give good thread stability.

5. *Safety and Disposal*

Polymers are relatively inert chemically and, hence, to that degree safe. However, many have added stabilizers and small quantities of low-molecular-weight oligomer which can be leached from the bulk polymer under suitable circumstances. For food contact, all additives require clearance by appropriate authorities and manufacturers that they are safe for this purpose. Oligomers and unreacted monomer are minimized by polymer manufacturing and processing.

Disposal of used plastic packaging is inherently safe and causes minimal environmental damage, whether it is by land fill or incineration. There is increasing pressure for recycling and much effort is being applied to achieving economic reuse of recovered packaging. PET bottles, for example, are estensively recycled in the USA.

See also: Bonding: Preparation of Polymer Surfaces; Bottles and Other Rigid Containers; Degradation and Stabilization, Ultraviolet; Recycling

Bibliography

Juran R (ed.) 1988 *Modern Plastic Encyclopedia.* McGraw-Hill, New York

Rauch R (ed.) 1987 *The Rauch Guide to the U.S. Plastics Industry. Data for 1985, 1986 and Projections to 1991.* Rauch Associates, Bridgewater, NJ

Russo J R (ed.) 1988 *Packaging Encyclopedia 1988.* Cahners, Newton, MA

J. F. E. Adams
[Caversham, UK]

J. McIntosh
[Maidenhead, UK]

Polyethylene Processing to Produce Ultimate Properties

Polyethylene $(CH_2{-}CH_2)_n$ was discovered in 1933 and by the 1950s had become a major commodity thermoplastic. In its usual semicrystalline form, polyethylene has a room temperature Young's modulus of approximately 1 GPa and a tensile strength of 0.5 GPa, although both values depend to some extent on the molecular weight of the polymer. The long polymer chains form thin "chain-folded" platelet lamellar crystals, each of a thickness of about 10 nm. These platelets are usually organized to form centrosymmetric spherulitic crystals of the order of micrometers in diameter which give the material an overall isotropic organization, as shown in Fig. 1.

The unit cell of crystalline polyethylene was established at an early stage by Bunn (1939) and the chain

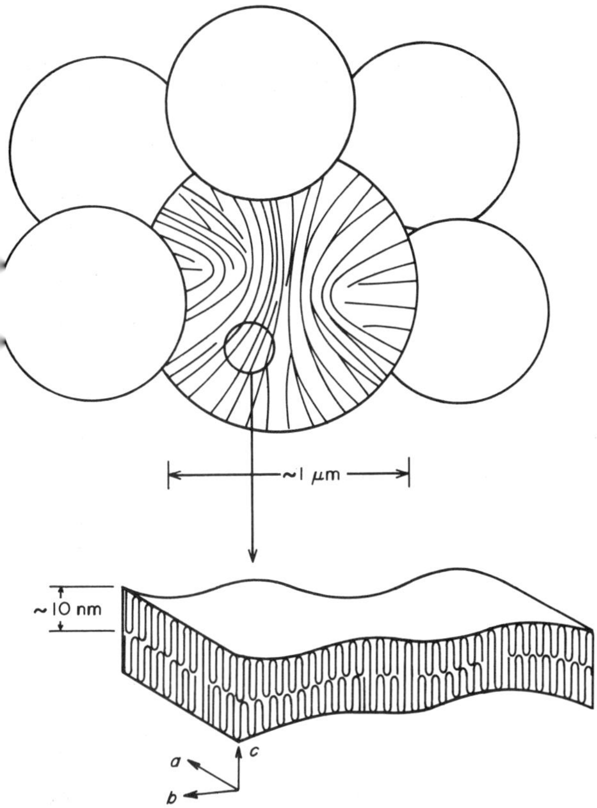

Figure 1
Structure of spherulitic melt-crystallized polyethylene

configuration is shown in Fig. 2. From this diagram it can be seen that within the unit cell, the chain is in an all-*trans* configuration with the *c* axis aligned along the chain axis. Frank (1970) pointed out that the unit cell of crystalline polyethylene was similar to that of diamond so that if polyethylene could be persuaded to crystallize in the all-*trans* configuration with chains extended fully along the *c* axis, a material with significantly improved properties could be expected and, in particular, a room-temperature Young's modulus of the order of 200–300 GPa might be anticipated. The prospect of such a high modulus from a low-cost commodity polymer generated an international effort to realize this objective and this has resulted in a number of process routes being discovered independently.

1. Solid-State Drawing

The classic route for generating anisotropy in polymers is mechanical deformation (Ward 1971). Solid-state drawing is an essential part of most synthetic fiber spinning processes and, after forming a generally isotropic fiber by melt spinning, orientation is induced by mechanically drawing the fiber at room or elevated temperature to draw ratios λ of up to ten. The draw ratio is defined as the ratio of the drawn length to the original length. By mechanically drawing, the isotropically oriented lamellar crystals of polyethylene are disrupted and aligned towards the direction of the draw axis such that the *c* crystallographic axis has a preferred orientation in the draw direction, as shown in Fig. 3

Andrews and Ward (1970) showed that by drawing at an enhanced rate, the modulus of polyethylene could be increased from approximately 1 GPa to approximately 20 GPa. Later, Cappacio and Ward (1975) discovered that by drawing at an elevated temperature (above 80 °C) and by choosing polyethylene of a particular molecular weight distribution and crystallite morphology, it was possible to achieve even higher draw ratios of up to 20, with an associated Young's modulus of 40 GPa. It was found that there appeared to be an essentially linear relationship between the observed tensile modulus and the draw ratio, the modulus varying from 1 GPa at $\lambda = 1$ to 50 GPa at $\lambda = 25$.

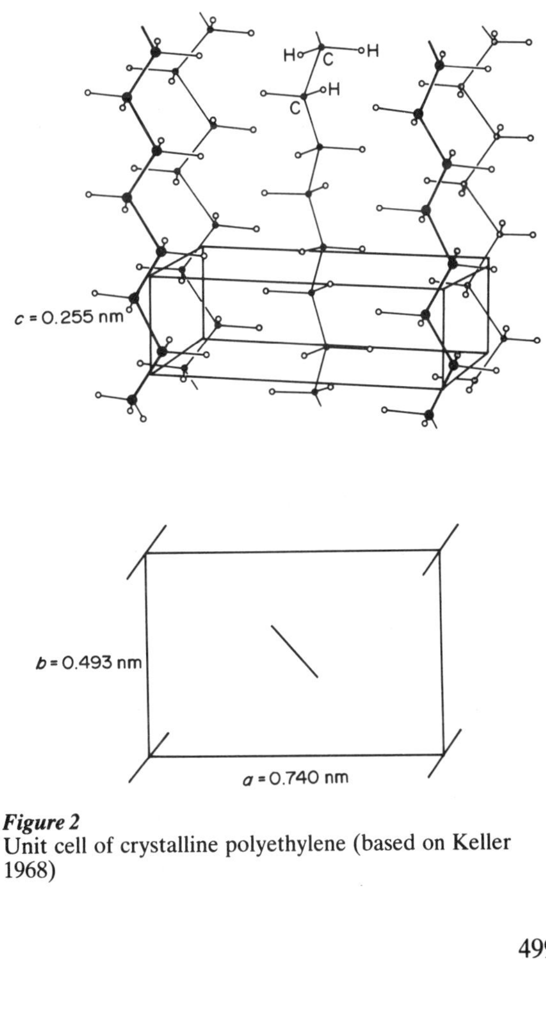

Figure 2
Unit cell of crystalline polyethylene (based on Keller 1968)

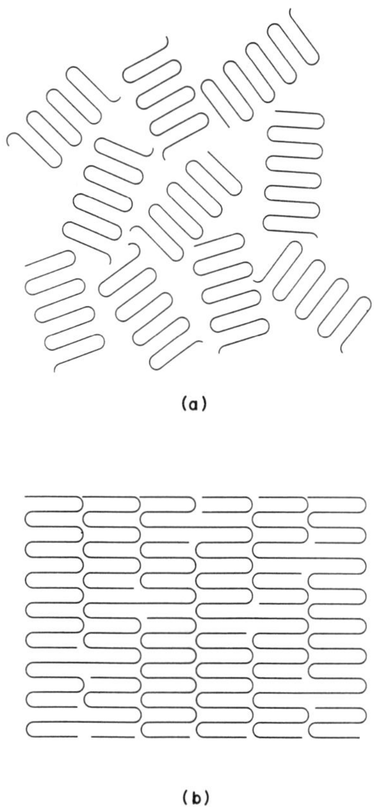

Figure 3
Alignment of crystals of polyethylene: (a) before and (b) after drawing

A variant of the solid-state drawing process is some form of ram extrusion (Ward 1985). In this case a billet of polyethylene is either pushed or pulled through a die of reduced cross section, as shown in Fig. 4. The process has many similarities to tensile drawing but in the extrusion case the dimensional changes of the material are controlled by the configuration of the die rather than by the mechanical behavior of the material.

In the case of solid-state drawing it has generally been found that drawing at high draw ratios is most readily achieved with linear polyethylenes that have an average molecular weight $\bar{M}_w$ below about 10^5. These are the molecular weights associated with normal melt-processable polyethylenes. The detailed mechanisms associated with the ability to generate high draw ratios in this way are not fully understood but the simplicity and elegance of the process has ensured commercial viability for this method in producing polyethylene fibers and monolithic extrusions with a tensile Young's modulus of up to about 50 GPa. This is an increase of 50-fold from the usual melt-crystallized material.

2. Solution Processing

Pioneering studies on the way in which polyethylene crystallizes from solution have been carried out by Keller (1968) for crystallization from quiescent solutions and by Pennings et al. (1970) for crystallization from mechanically stirred solutions. Keller found crystals formed thin chain-folded crystal lamella and Pennings et al. showed that stirring could induce a "shish kebab" morphology. The molecular organization associated with these two morphologies is shown in Fig. 5 and it can be seen that the core of the shish kebab polyethylene crystal has the necessary organization to have the potential for yielding a high-modulus polymer.

The first continuous length of polyethylene fiber produced by solution processing was made using a seeding tecnique developed by Zwijnenburg and Pennings (1975). In this process a seed crystal was placed in a flowing stream of a polyethylene solution consisting of about 0.1% w/v (weight per unit volume) of ultrahigh-molecular-weight polyethylene (UHMWPE) in an organic solvent such as xylene. At a temperature below 113 °C it was found that a continuous length of polyethylene fiber could be produced and, by using the system shown in Fig. 6, growth rates of 2 mm min^{-1} for 20 μm diameter fibers could be achieved. The level of molecular orientation in the fibers was high and tensile moduli of 22 GPa were reported.

This rather slow process was soon superseded by a more rapid and efficient surface growth route discovered by Zwijnenburg and Pennings (1976). In this method, the growing fiber was allowed to migrate towards a rotating surface and, at this point,

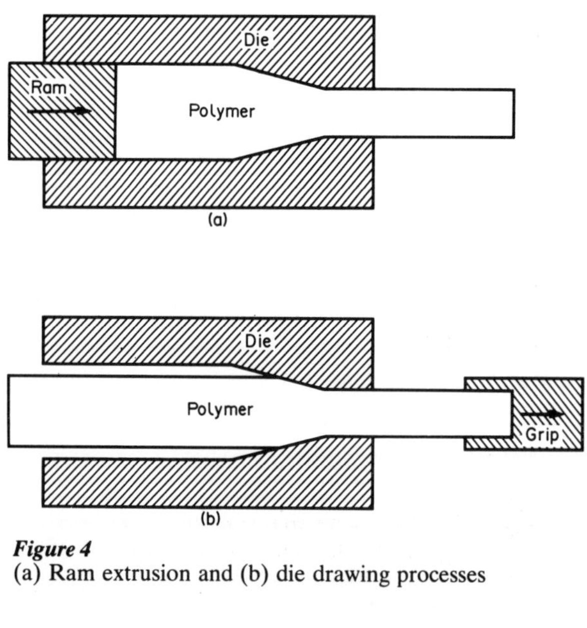

Figure 4
(a) Ram extrusion and (b) die drawing processes

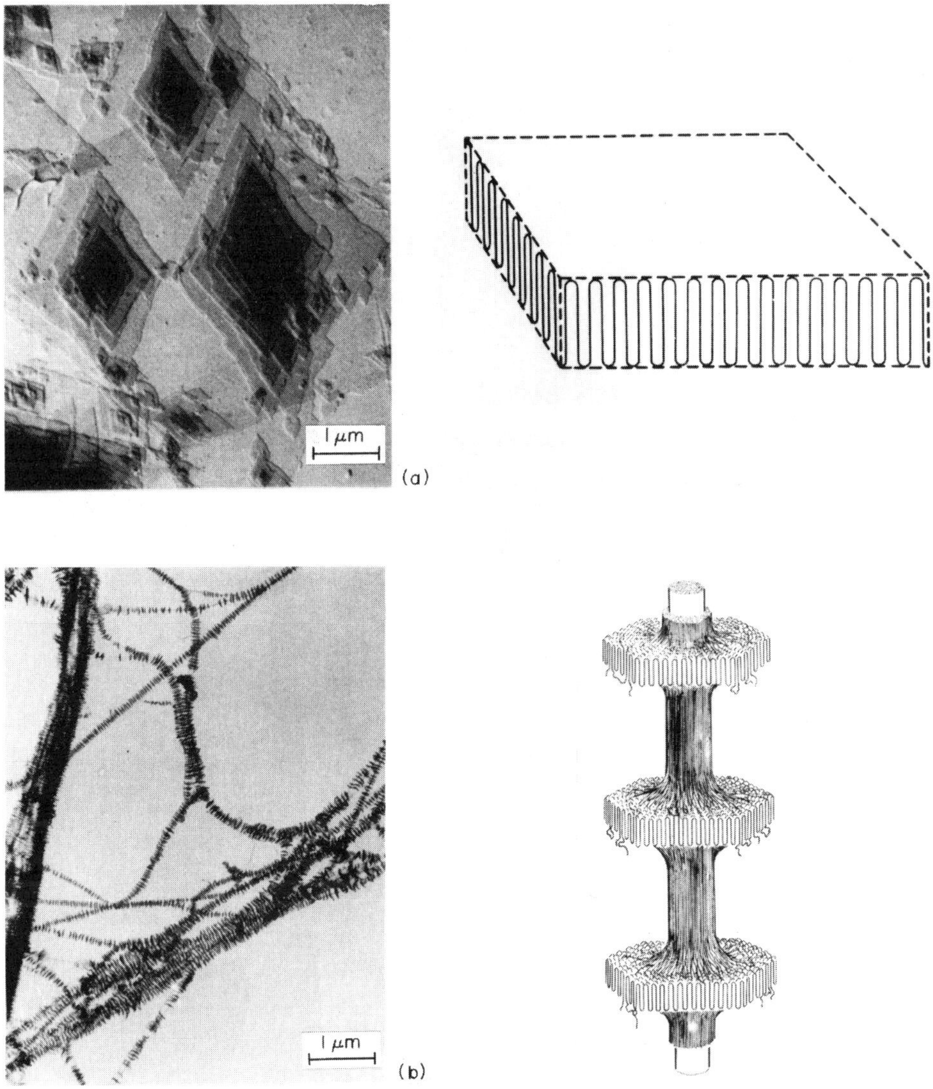

Figure 5
Micrographs and associated diagrams of polyethylene crystallized from (a) quiescent solution (chain-folded lamellar crystals) and (b) mechanically stirred solution (shish kebab crystals)

polymer chains attached to the growing fiber appeared to be caught between the fiber and an entanglement adsorption layer of polymer attached to the moving surface. The relative motion between the fiber and the surfaces causes elongation of the chains and subsequent crystallization occurs if the temperature of the solution is below 118 °C. A diagram of the system is shown in Fig. 7 and, using this type of apparatus, single fiber growth rates of up to 0.2 m min^{-1} could be achieved. Again, fiber orientation was very high and the first polyethylene with a Young's modulus exceeding 100 GPa and a tensile modulus of approximately 2 GPa was achieved in this way.

UHMWPE, with values of $\bar{M}_n$ (number average molecular weight) and $\bar{M}_w$ in excess of 10^6, is extremely difficult to process in either the melt or solid form. Hoever, Zwijnenburg and Pennings (1975) have clearly shown that with the aid of solution processing it is possible to produce polyethylene that is both highly extended and orientated and that yields a tensile modulus that can be compared with other high-modulus fibers such as carbon fibers and Kevlar.

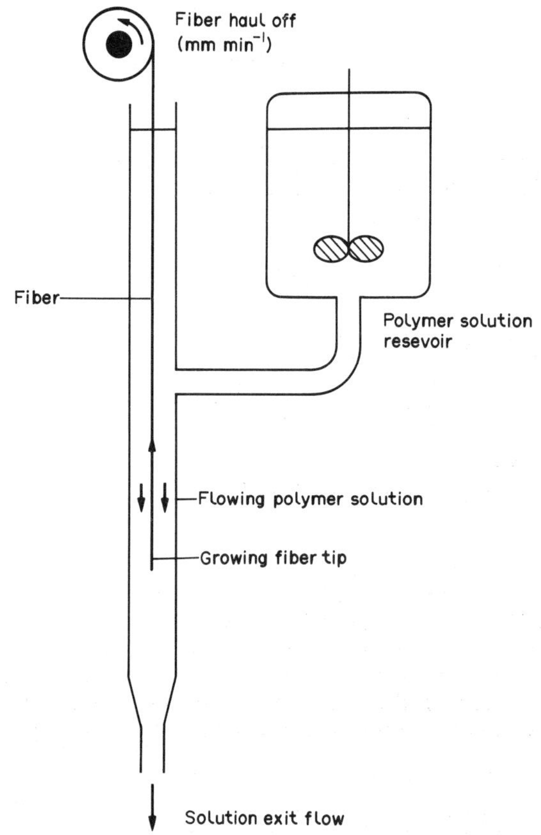

Figure 6
"Free surface" fibrous crystal growth process with 0.1–1.0% w/v UHMWPE in xylene at temperatures of 100–113 °C

3. Gel Spinning

Following the success of the surface growth process, a new gel-spinning route was discovered at Dutch State Mines (DSM). Smith and Lemstra (1980) reported that they were able to gel spin UHMWPE fibers that could subsequently be drawn to a draw ratio λ of 31.7 which resulted in fibers with a Young's modulus of 90 GPa and a tensile strength of 3.0 GPa. Their apparatus is shown in Fig. 8 and the process route is as follows:

(a) a solution of 0.5% w/v UHMWPE in Decalin (decahydronaphthalene) is prepared under agitation at 150 °C;

(b) gel fiber is formed by extrusion of the solution through a nozzle at 130 °C resulting in isotropically oriented fiber (~1 mm diameter);

(c) solidification and crystallization of the fiber is achieved by quenching in a cooling bath; and

(d) the solidified fiber is then heated to remove solvent and drawn at temperatures in excess of 90 °C to draw ratios λ of up to approximately 30.

Processing UHMWPE in this manner causes a dramatic increase in the drawability of the material. Melt-quenched UHMWPE can normally be drawn at 90 °C up to draw ratios λ of 3–4. By dissolving the polymer and subsequently crystallizing in the presence of a solvent the drawability can be increased to values of λ up to 30 with a corresponding increase in Young's modulus to 90 GPa.

Further development of this technology has enabled the polymer solution to be prepared in a continuous process as shown in Fig. 9. In this system, both solvent and polymer powder are fed into a screw extruder in the concentration range of 1–10% w/v and the polymer is dissolved within the flight of the barrel (Kirschbaum and van Dingenen 1988). A metering gear pump controls the volumetric flow rate out of the screw extruder and into a multiorifice spinneret. Downstream of the spinneret the process route is essentially the same as described previously. Fibers manufactured by this or related low-concentration gel-spinning routes are known as Dyneema (manufactured by DSM and Tohobo) and Spectra (manufactured by Allied Chemicals).

Smith et al. (1981) reported that the enhanced drawability of the gel-spun fibers was very dependent on the initial solution concentration and was a consequence of low entanglement levels in solution when compared with the melt. They asserted that these low entanglement levels were quenched into the material on solidification and, in turn, controlled ultimate drawability when hot drawn. This idea is supported in a model developed in Sect. 5; however, a complication to this concept becomes apparent when the third commercial polyethylene fiber

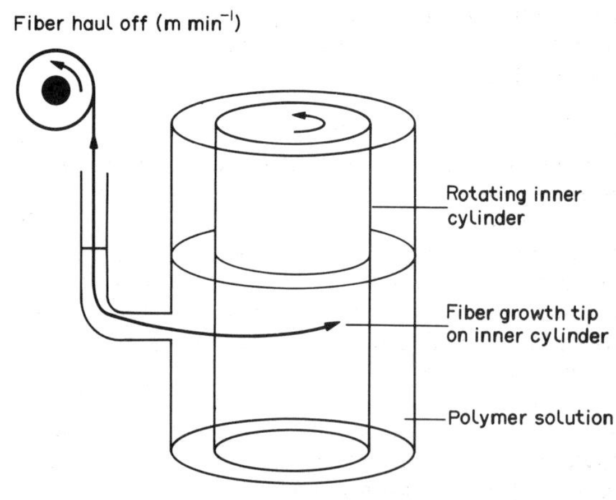

Figure 7
"Surface growth" fibrous crystal process

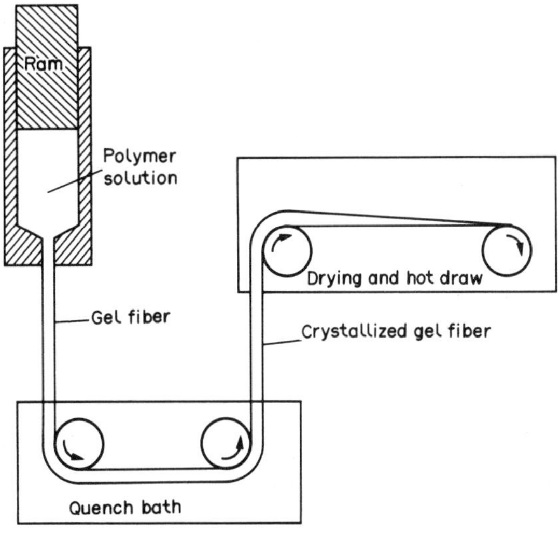

Figure 8
Batch gel-spinning apparatus

known as Teknillon and manufactured by Mitsui is considered. This fiber has similar properties to Dyneema and Spectra but appears to be manufactured by the hot drawing of an isotropic fiber that has been formed by plasticizing UHMWPE with about 20% w/v of paraffin wax. The material can then be melt processed in the normal way to produce a fiber geometry and, after extraction of the paraffin, hot drawing can produce a fiber with a Young's modulus of 100 GPa.

4. *Other Process Routes*

There are a number of variants for producing high-modulus and high-strength polyethylene and some of these are reviewed by Barham and Keller (1985). Most, but not all, of these processes are directly related to the three routes that have been described in Sects. 1–3.

4.1 Two-Stage Superdrawing

In two-stage superdrawing (Kanamoto et al. 1988), solution-crystallized UHMWPE is dried, coextruded in the solid-state at temperatures of approximately 110 °C and subsequently hot drawn at progressively higher temperatures. In this way, a consolidated low-entanglement material is initially obtained and this can then be drawn to draw ratios λ greater than 50, providing the draw temperature is sufficiently high. A gel-crystallized ram extrusion process (Anton et al. 1987) is similar to two-stage superdrawing; an essential feature of both is that deformation of the low-entanglement material always occurs below the melting point T_m, which for UHMWPE is approximately 145 °C.

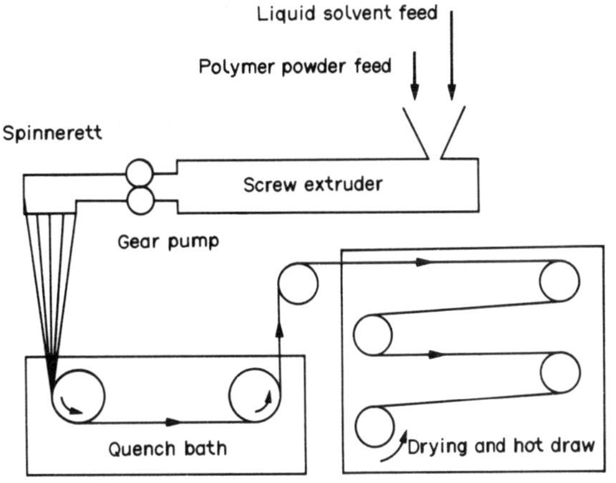

Figure 9
Continuous gel-spinning apparatus

4.2 Swell Drawing

In swell drawing (Mackley and Solbai 1987), the initially solid UHMWPE is swollen with solvent at an elevated temperature and then quenched in the presence of the solvent. The solvent is then removed by mechanically squeezing the material and anisotropy is developed by subsequent hot drawing.

4.3 Processing of Virgin Polymer

A method has been reported (Smith et al. 1985) by which UHMWPE has been successfully drawn to very high draw ratios with corresponding enhanced properties, without the material having previously been subjected to any solvent processing. Working at Du Pont, Smith et al. discovered that it was possible to polymerize polyethylene in such a way that it could be directly hot drawn to very high draw ratios. The polymerization catalyst used allowed UHMWPE to be formed at a temperature well below the melting point of the polymer and in addition it appears that a low-activity catalyst was chosen. Under these polymerization conditions the polymer was unable to become highly entangled during polymerization and, providing subsequent drawing took place below T_m, further entanglement was not possible, thereby allowing high draw ratios to be achieved in the final hot-draw process.

5. *Mechanisms Associated with Ultimate Drawability*

From a knowledge of the crystalline unit cell for polyethylene, modulus enhancement is clearly expected if high levels of uniaxial orientation and continuity along the chain axis are achieved. Modulus enhancement appears to be directly related to levels of chain orientation, while significant strength enhancement seems to only occur with high-molecular-weight variants of polyethylene.

The detailed mechanism associated with the drawing of polyethylene is complex. Drawability above 80 °C is easier for all ranges of molecular weights and this can be related to stess-induced thermal activation of the chains, either within the crystalline or the amorphous phase of the material. The drawing process appears to follow reasonably closely to that predicted by the pseudoaffine deformation assumption.

The limiting drawability of polyethylene and the profound effect of solvent processing before drawing form the basis for the Smith et al. (1981) model for the deformation process. It is well known that polyethylene in the molten state exists as highly entangled chains that are themselves essentially in a random coil configuration. The molecular weight M_e of the polymer segments between these entanglement junctions can be determined from the melt plateau modulus of the polymer G_0', given by

$$G_0' = \frac{\rho RT}{M_e} \tag{1}$$

where ρ is the polymer density, R the universal gas constant and T the absolute temperature. If it is assumed that entanglement junctions are fixed, the distance between these junctions can be thought of in terms of the root mean square end-to-end distance l_0 of the random chain between entanglement points and this is given by $l_0 = ar^{1/2}$ where r equals the number of repeat units between entanglement junctions and a is the link length of each repeat unit. The term r is equal to M_e/M_0 where M_0 is the equivalent molecular weight of a statistical chain segment of polyethylene and the situation is represented in Fig. 10a. Smith et al. (1981) then speculate

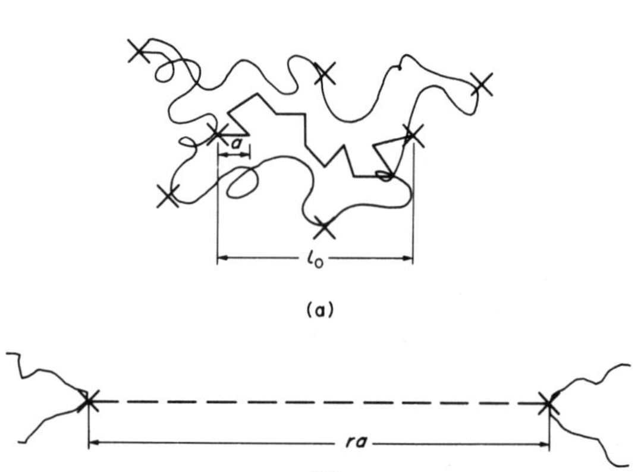

Figure 10
Molecular organization of a network: (a) undeformed and (b) at the point of maximum deformation

that the entanglement level within the melt is trapped in the material during crystallization and that the subsequent drawability of the material is controlled by the drawability of this entanglement network. If the network was fully drawn, as shown in Fig. 10b, the limiting end-to-end distance between entanglement junctions would be ar yielding a maximum draw ratio of

$$\lambda_m = \frac{ar}{ar^{1/2}} = r^{1/2} = \left(\frac{M_e}{M_0}\right)^{1/2} = \left(\frac{\rho RT}{M_0 G_0'}\right)^{1/2} \tag{2}$$

For polyethylene crystallized from the melt, Eqn. (2) gives $\lambda_m \simeq 3$.

If solvent is present, the plateau modulus decreases with increasing solvent concentration and follows an equation of the form

$$G_\phi' = G_0'(1-\phi_s)^{5/3} \tag{3}$$

where ϕ_s is the volume fraction of solvent. Smith et al. use a slightly modified equation to that given in Eqn. (3); however, replacing G_0' in Eqn. (2) by G_ϕ' from Eqn. (3) yields

$$\lambda_\phi = \left(\frac{\rho RT}{M_0 G_0'}\right)^{1/2} (1-\phi_s)^{-5/6} \tag{4}$$

where λ_ϕ is the maximum drawability after prior solvent processing.

If a λ_ϕ is approximately equal to 40 is required, it can be seen from Eqn. (4) that a solvent volume fraction of 0.956 is needed, meaning that a polymer content of 4.4% volume fraction has to be used. For gel-spinning and swell-drawing behavior, Eqn. (4) holds quite well and this implies that processing with low polymer concentration appears to be essential if hig drawability is to be achieved.

To some extent the virgin polymer processing route is consistent with this, because here the material is polymerized well below T_m and in a way this deliberately attempts to produce a low entanglement state. The Mitsui process and some aspects of superdrawing, however, suggest that it is also possible to obtain high drawability using limited plasticization and/or careful control of draw temperature and of molecular weight distribution. A plausible explanation for this is that when drawing occurs close to the melting point, under certain conditions, partial disentanglement of any embedded network within the material can take place simultaneously with the development of anisotropy.

In conclusion, the new processing routes for polyethylene, and particularly UHMWPE, that have been described in this article open the way for a new and exciting chapter in the history of polyethylene. It is now possible for polyethylene to compete in the market place with other high-performance fibers such as carbon and aramid fibers, although the relatively low melting point of polyethylene at 145 °C will restrict material application areas. The potential for manufacturing high-modulus polyethylene structures other than in fiber form has been clearly demonstrated and the scope for the development of novel and more economic processing routes would still appear to be considerable.

See also: Die Extrusion; Fibers: Drawing; Fibers: Formation and Processing; Fibers: Texturing and Draw Texturing; High-Modulus High-Strength Organic Fibers

Bibliography

Andrews J A, Ward I M 1970 The cold-drawing of high density polyethylene. *J. Mater. Sci.* 5: 411–17

Anton C R, Mackley M R, Solbai S 1987 Extrusion processing of ultra high molecular weight polyethylene. *Polym. Bull.* 17: 175–9

Barnham P J, Keller A 1985 Review: High strength polyethylene fibres from solution and gel spinning. *J. Mater. Sci.* 20: 2281–302

Bunn C W 1939 The crystallographic unit cell of polyethylene. *Trans. Faraday Soc.* 35: 482–91

Cappacio G, Ward I M 1975 Ultra high modulus linear polyethylene through controlled molecular weight and drawing. *Polym. Eng. Sci.* 15: 219–24

Frank F C 1970 The strength and stiffness of polymers. *Proc. R. Soc. London, Ser. A* 319: 127–36

Kanamoto T, Tsurata A, Tanaka K, Takeda M, Porter R S 1988 Superdrawing of ultra high molecular weight polyethylene. *Macromolecules* 21: 470–7

Keller A 1968 Polymer crystals. *Rep. Prog. Phys.* 31(2): 623–705

Kirschbaum R, van Dingenen J L J 1988 Advances in gel spinning technology and Dyneema fibre applications. *Integration of Polymer Science and Technology, 3rd Rolduc Conf.* Elsevier, Amsterdam

Mackley M R, Solbai S 1987 Swell drawing: A new process for manufacturing high performance polyethylene. *Polymer* 28: 1115–20

Pennings A J, van der Mark J M A A, Kiel A M 1970 Hydrodynamically induced crystallization of polymers from solution. *Kolloid-Z. Z. Polym.* 237: 336–58

Smith P, Chanzy H D, Rotzinger B P 1985 Drawing of virgin ultrahigh molecular weight polyethylene. *Polym. Commun.* 26: 258–60

Smith P, Lemstra P J 1980 Ultra drawing of high molecular weight polyethylene cast from solution. *J. Mater. Sci.* 15: 505–14

Smith P, Lemstra P J, Booij H C 1981 Ultra drawing of high molecular weight polyethylene: The effect of concentration. *J. Polym. Sci., Polym. Phys. Ed.* 19: 877–88

Ward I M 1971 *Mechanical Properties of Polymers.* Wiley, New York

Ward I M 1985 The preparation, structure and properties of ultra high modulus flexible polymers. *Adv. Polym. Sci.* 70: 1–70

Zwijnenburg A, Pennings A J 1975 Longitudinal growth of crystals from flowing solutions II. Polyethylene crystals in Poiseuille flow. *Colloid Polym. Sci.* 253: 452–61

Zwijnenburg A, Pennings A J 1976 Longitudinal growth of crystals from flowing solutions III. Polyethylene crystals in Conette flow. *Colloid Polym. Sci.* 254: 868–81

M. R. Mackley
[University of Cambridge, Cambridge, UK]

Polymer Blends

In the 1970s, the technology related to polymer blends attained a prominent position as a major area of research and development in polymer science. The interest in polymer blends is due, to the need for a myriad of specific property profiles for the diverse applications requiring polymeric materials. Economic considerations limit the potential number of commercial polymers; thus, only a finite number of property combinations is possible without further modification. Polymer blends greatly extend the potential property profiles and are a significant factor in the science and technology of polymers.

1. Classification

A polymer blend is here considered to be a combination of structurally different polymeric constituents. Such combinations may be either physical blends or covalently bonded polymer chains (e.g., block copolymers), but this article deals mainly with simple polymer blends where no covalent bonding is present. Unique cases of physical polymer blends include polyelectrolyte complexes, where strong ionic bonding exists, and isomorphic polymer blends, where the rare cases of cocrystallization of structurally different polymer chains are observed. Interpenetrating polymer networks are a special case of noncovalently bonded polymer blends where interlocking (catenane) chains can be achieved via crosslinking reactions. Covalently bonded blends include block and graft copolymers.

2. Thermodynamic and Phase Behavior

Although the vast majority of polymer blend combinations exist as phase-separated systems, an increasing number of miscible polymer blends are appearing in the literature. One of the earlier surveys of polymer blends listed twelve miscible pairs, of which several were copolymer mixtures at different monomer contents (Bohn 1968). A more recent comprehensive review of miscible polymer systems listed approximately 180 pairs when simple blends, block and graft copolymers, interpenetrating polymer networks, isomorphic blends and polyelectrolyte complexes were considered (Olabisi et al. 1979). The single feature common to many miscible blends is the potential for specific interaction (e.g., hydrogen bonding).

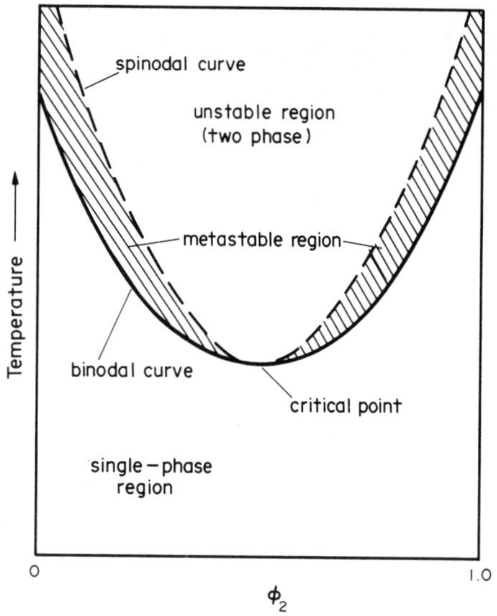

Figure 1
Typical phase diagram for polymer–polymer mixtures

The phase behavior of polymer blends is quite different from that of lower-molecular-weight components in that an increase in temperature makes miscibility less probable. This has been predicted using the equation-of-state thermodynamic treatment of polymer mixtures (McMaster 1973) and has frequently been observed experimentally. The thermodynamic criterion for miscibility requires the free energy of mixing ΔG_m to be negative:

$$\Delta G_m = \Delta H_m - T\Delta S_m < 0 \tag{1}$$

where ΔH_m is the enthalpy of mixing and ΔS_m is the entropy of mixing. For high-molecular-weight polymers, ΔS_m is negligible, so a negative enthalpy of mixing is required for miscibility. Conditions exist where phase separation can occur if $\Delta G_m < 0$; thus, Eqn. (1) is a necessary but not a sufficient condition for miscibility. The necessary and sufficient thermodynamic requirement for miscibility is

$$\left(\frac{\partial^2 \Delta G_m}{\partial \phi_2^2}\right)_{P,T} > 0 \tag{2}$$

where ϕ_2 is the volume fraction of component 2.

The essence of the phase behavior of a polymer blend is depicted in Fig. 1. There are several regions of critical importance in defining the nature of the blend. The spinodal curve defines the position where spontaneous phase separation occurs. This phase separation process, termed spinodal decomposition, originates from composition fluctuations and yields a highly interconnected structure. The spinodal curve is defined by Eqn. (2) with the left-hand side equal to zero.

The binodal curve defines another region of phase separation commonly referred to as nucleation and growth. The area between the binodal and spinodal curves represents the region of temperature and composition where only nucleation and growth occur. The binodal curve is defined by the temperature–composition coordinates where the chemical potential $\Delta\mu_i$ of each component in each phase is equal ($\Delta\mu_1 = \Delta\mu_1'$; $\Delta\mu_2 = \Delta\mu_2'$). The critical point is the position of intersection of the binodal and spinodal curves and can be determined from

$$\left(\frac{\partial^3 \Delta G_m}{\partial \phi_2^3}\right)_{P,T} = 0 \tag{3}$$

Figure 1 depicts a lower critical solution temperature. As the temperature is raised from the miscible region, phase separation occurs and the method of phase separation is generally nucleation and growth unless the critical point is traversed or the nucleation-and-growth region is rapidly crossed into the spinodal region.

3. Characterization

The most important characteristic of a polymer blend is the phase behavior. The property–composition profile depends critically on whether the polymer molecules are phase separated, are mixed to a sufficient degree to exhibit single-phase behavior or exhibit an intermediate level of mixing. In attempting to determine the level of miscibility, the previously stated thermodynamic criterion is usually difficult to satisfy experimentally unless the particular system in question exhibits a lower critical solution temperature and the phase boundary (e.g., cloud point curve) can be defined. More practically, miscibility is defined as the level of molecular mixing required to yield macroscopic properties of the blend indicative of single-phase behavior. The primary property of interest is the glass-transition temperature T_g. A miscible system has a unique T_g, generally intermediate between the values for the constitutents (in cases of strong specific interactions, the T_g for the blend can be higher than that of either component). An immiscible blend retains the T_g values characteristic of the unblended polymers. As T_g is a characteristic polymer property upon which the viscosity, modulus–temperature behavior, crystallization rate, creep rate, toughness and other properties are dependent, the property profile of a miscible polymer blend can be profoundly different from that of an immiscible blend.

Methods capable of measuring the glass-transition temperature are the most common techniques utilized for characterizing the phase behavior of polymer blends. These methods include dynamic mechanical, calorimetric, dielectric and dilatometric techniques.

Scattering techniques are useful in defining the physical structure of polymer blends—specifically, the local order, the chain conformation and the blend morphology. These techniques include x-ray scattering conventional light scattering, pulse-induced critical scattering, Rayleigh–Brillouin scattering and neutron scattering. Neutron scattering experiments on polymer blends are quite new and have been shown to yield important conclusions about the level of molecular mixing for systems defined as miscible by macroscopic (e.g., T_g) property determinations. The cloud point curve method is essentially a scattering technique commonly utilized for experimentally determining the phase diagram (Fig. 1).

Microscopic techniques (e.g., scanning electron, transmission electron) are extremely useful in elucidating the structure of phase-separated polymer blends. Staining techniques and phase contrast microscopic methods are available and widely utilized to define the structure down to 5 nm resolution. Other techniques employed to analyze polymer blends include nuclear magnetic resonance, inverse gas chromatography, excimer fluorescence and infrared spectroscopy.

4. Properties

A specific advantage of achieving miscibility in a polymer blend is the assurance of mechanical compatibility of the constituents, thus attaining a reasonable compromise in ultimate properties (e.g., strength). In many cases, a random combination of polymers yields a phase-separated blend with inferior ultimate properties due to poor adhesion between the phases (interfacial regions representing structural flaws). Block and graft copolymers offer significant property advantages over their physical mixture counterparts, as the covalent bonds connecting the separate phases ensure interfacial adhesion.

One of the major property advantages derived from phase separation in blends is impact strength enhancement. Phase separation *per se* does not promote impact strength improvement but requires the proper combination of materials and microstructural features of the phase-separated blend. Polystyrene and styrene–acrylonitrile copolymers were initially impact-modified by physical blends with butadiene-based elastomer blends. Impact strength efficiency was found to be considerably enhanced through the *in situ* polymerization of styrene or styrene–acrylonitrile mixtures in the presence of a butadiene-based elastomer. The resulting product consists of a matrix of polystyrene or a styrene–acrylonitrile copolymer and a rubber phase of the elastomer grafted with the matrix and lightly crosslinked. Optimum impact efficiency is achieved by proper control of the rubber-phase particle size, graft content and rubber-phase crosslink density. Impact-modified grades of poly(methyl methacrylate) (PMMA) utilize similar technology. Transparent grades of these polymers can be achieved by matching the refractive index of the rubber phase with that of the glassy matrix.

For blends containing a crystalline component, the phase behavior determines the effect on the melting point and the crystallization kinetics. In miscible blends, the crystallizable component exhibits a melting point depression, as with poly(vinylidene fluoride) in PMMA blends and poly(ε-caprolactone) in blends with poly(vinyl chloride). The crystallization rate of a crystallizable component of an immiscible blend is not expected to be significantly different from that in the unblended state unless nucleation occurs. For miscible blends, the crystallization rate either increases or decreases, depending on whether the T_g of the crystalline polymer decreases or increases, respectively. This behavior is predicted by the spherulite growth rate equation, which contains T_g as a variable. Examples of this behavior include the reduced crystallization rate of poly(ε-caprolactone) with poly(vinyl chloride) addition (increasing T_g) and the increased crystallization rate of bisphenol A polycarbonate with poly(ε-caprolactone) addition (decreasing T_g).

Cocrystallinity (isomorphism) in macromolecular blends is extremely rare; only a few cases have been reported (Allegra and Bassi 1969). Isomorphism has been observed where the conformation and size of the polymer chains in the crystalline state are virtually identical. Examples of isomorphic blends include poly(vinyl fluoride)–poly(vinylidene fluoride) and poly(isopropyl vinyl ether)–poly(*s*-butyl vinyl ether).

Blends of immiscible polymers and phase-separated block and graft copolymers generally exhibit additive-specific volume. Densification is often observed in miscible polymer blends and is generally believed to be a consequence of specific interactions.

The gas permeability of a phase-separated blend follows a sigmoidal relationship with composition, dependent on the continuous–discontinuous phase structure of the blend. With miscible blends, a logarithmic relationship versus composition can be expected.

5. Commercial Examples

The primary reasons for the current surge of academic and industrial interest in polymer blends are directly related to their potential for meeting end-use requirements. The property profiles of the unmodified commercial polymers cannot meet all of the application requirements, so further modification is required.

5.1 Miscible Blends

The first commercial miscible blend of synthetic polymers emerged in the early 1940s, involving poly(vinyl chloride) (PVC) and nitrile rubber (butadiene–acrylonitrile copolymers). Plasticized PVC was added to vulcanized nitrile rubber to improve ultraviolet and oxidative stability. These blends are still utilized in sponge, wire and cable, footwear, belt and hose applications. Nitrile rubber (unvulcanized) represented the first permanent plasticizer for PVC, yielding compositions offering resistance to plasticizer migration in adverse environments such as oil and water containment and food contact applications.

Other high-molecular-weight polymers have since been employed as permanent plasticizers for PVC. These include ethylene–vinyl acetate copolymers (65–70 wt% vinyl acetate), ethylene–vinyl acetate–carbon monoxide terpolymers, polyester-based thermoplastic polyurethanes and the poly(butylene terephthalate)–polytetrahydrofuran $\leftarrow\!\!(AB)\!\!\rightarrow_n$ block copolymer.

The upper use temperature (heat distortion temperature) of acrylonitrile–butadiene–styrene (ABS) is typically 90 °C. In order to extend the load-bearing capabilities to higher temperatures, one approach commercially utilized is the addition of an α-methylstyrene–acrylonitrile copolymer to ABS. An α-methylstyrene–acrylonitrile copolymer at the azeotropic composition (69:31 by weight) has a T_g 25 °C higher than that of the styrene–acrylonitrile matrix of ABS. The α-methylstyrene–acrylonitrile copolymers (at the azeotropic composition) are also miscible with PVC and utilized for PVC applications requiring higher heat resistance. Polystyrene, like ABS, has a heat distortion temperature of 90 °C. The addition of poly(2,6-dimethyl-1, 4-phenylene oxide) (PPO) with a T_g of 210 °C can yield a significant increase in the T_g of polystyrene and thus in the upper use temperature. PPO–impactpolystyrene blends are used in appliance and business machine housings, electronic components, pump components and automotive applications.

The coatings industry has utilized miscible polymer principles for years, initially with nitrocellulose and later with cellulosic variants (e.g., cellulose acetate butyrate) modified with natural products (e.g., shellac) as well as synthetic polymers (e.g., phenolics, poly(vinyl acetate)). Specific adhesive formulations employ miscible polymer blends such as nitrile rubber formulations with vinyl chloride copolymers and terpolymers and phenolic resins. Weatherable films of poly(methyl methacrylate)–poly(vinylidene fluoride) have been proposed for commercial applications. Polyelectrolyte complexes, such as poly(vinylbenzyltrimethyl ammonium chloride)–sodium poly(styrene sulfonate), have been utilized or proposed for such diverse applications as filtration of macromolecules, battery separators, hemodialyzers (artifical kidneys), microencapsulation and antistatic coatings and as a matrix for the slow release of implantable drugs.

5.2 Immiscible Blends

The primary commercial application of polymer mixtures involves elastomer addition to rigid polymers for impact modification. Impact polystyrene and ABS are the primary examples, as mentioned in Sect. 4. Additional commercial blends include impact-modified PMMA and PVC, "supertough" nylon and ethylene–propylene rubber-modified polypropylene.

Elastomer blends have been utilized to achieve specific property profiles, to alleviate property deficiencies or both. For the large tire market, blends of polybutadiene with styrene–butadiene copolymers offer a better property profile than either of the constituents. A more complex blend, natural rubber with halobutyl and ethylene–propylene–diene (EPDM), has been employed for white sidewall production. Polyisobutylene has been incorporated in inner tube formulations to reduce air permeability. EPDM blends with the more highly unsaturated elastomers (e.g., polybutadiene, styrene–butadiene copolymers) offer improved thermal, oxidative and ozone resistance. More detailed information relating to elastomer blends is given by Corish and Powell (1974).

Polyolefin blends are also quite common and are utilized to correct specific property deficiencies or to offer unique property combinations. A new series of polypropylene–ethylene–propylene rubber (EPR) blends (>50 wt% EPR) offers flexible compositions for automotive fascia applications, wire and cable insulation and certain plasticized PVC replacements. Many of the commercial blends are simple mixtures, whereas other combinations contain a degree of grafting between the phases.

Immiscible polymer combinations range from translucent to opaque unless the refractive indices of the components are identical. In certain cases where the refractive indices are close, a pearlescent appearance (e.g., polystyrene–PMMA blends) has been utilized for decorative purposes.

Block copolymers of styrene and butadiene derive their strength from dispersed glassy domains of polystyrene and the elastomeric character from the continuous polybutadiene phase. Phase separation is required to achieve the desired thermoplastic elastomer-type behavior.

Other blends of commercial interest include poly(tetrafluoroethylene) addition to various polymers to impart lubricity, the addition of polymers (e.g., poly(vinyl acetate)) termed low-profile additives to unsaturated polyester formulations to improve surface smoothness and the addition of highly brominated polymers to improve the flammability rating without the volatility or sweatout problem of low-molecular-weight flame retardants.

See also: Thermoplastic Elastomers

Bibliography

Allegra G, Bassi I W 1969 Isomorphism in synthetic macromolecular systems. *Adv. Polym. Sci.* 6: 549–74

Baranwal K C, Son P N 1974 Co-curing of EPDM and diene rubbers by grafting accelerators onto EPDM. *Rubber Chem. Technol.* 47: 88–99

Bever M B, Shen M 1974 The morphology or polymeric alloys. *Mater. Sci. Eng.* 15: 145–57

Bohn L 1968 Incompatibility and phase formation in solid polymer mixtures and graft and block copolymers. *Rubber Chem. Technol.* 41: 495

Bucknall C B 1977 *Toughened Plastics*, Material Science Series. Applied Science, London

Callan J E, Hess W M, Scott C E 1967 Carbon black distribution in elastomer blends. *Rubber Chem. Technol.* 40: 371–84

Coran A Y 1988 Thermoplastic rubber–plastic blends. In: Bhowmick A K, Stephens H L (eds.) 1988 *Handbook of Elastomers*. Dekker, New York, Chap. 8

Corish P J 1978 Elastomer blends. In: Eirich F R (ed.) 1978 *Science and Technology of Rubber*. Academic Press, New York, Chap. 12.

Corish P J 1985 Rubber–rubber blends. In: Walsh et al. 1985, Chap. 12

Corish P J, Powell B D W 1974 Elastomer blends. *Rubber Chem. Technol.* 47: 481–510

Gardiner J B 1969 Measurement of curative diffusion between rubbers by microinterferometry. *Rubber Chem. Technol.* 42: 1058–78.

Klempner D, Frisch K C (eds.) 1977 *Polymer Alloys: Blends, Blocks, Grafts, and Interpenetrating Networks*. Plenum, New York

Kohudic M A 1988 *Advances in Polymer Blends and Alloys Technology*, Vol. 1

Krause S 1978 Polymer–polymer compatibility. In: Paul D R, Newman S (eds.) 1978 *Polymer Blends*, Vol. 1. Academic Press, New York, Chap. 2

Livingston D I, Rongone R L 1968 Structure and properties of elastomer blends. *Proc. Int. Rubber Conf. 1967*. Maclaren, London, pp. 337–57

MacKnight W J, Karasz F E 1989 Polymer blends. In: Aggarwal S L (ed.) 1989 *Comprehensive Polymer Science*, Vol. 7. Pergamon, Oxford, Chap. 4.

McMaster L P 1973 Aspects of polymer–polymer thermodynamics. *Macromolecules* 6: 760–73

Manson J A, Sperling L H 1976 *Polymer Blends and Composites*. Plenum, New York

Olabisi O, Robeson L M, Shaw M T 1979 *Polymer–Polymer Miscibility*. Academic Press, New York

Ossefort Z T, Bergstrom E W 1969 Ethylene–propylene rubbers *Rubber Age (N.Y.)* 101(9): 47–60

Paul D R, Newman S (eds.) 1978 *Polymer Blends*. Academic Press, New York

Robeson L M 1984 Applications of polymer blends: Emphasis on recent advances. *Polym. Eng. Sci.* 24: 587

Roland C M 1988 Rubber–rubber blends. In: Bhowmick A K, Stephens H L (eds.) 1988 *Handbook of Elastomers*. Dekker, New York, Chap. 6

Sperling L H (ed.) 1974 *Recent Advances in Polymer Blends, Grafts, and Blocks*. Plenum, New York

Walsh D J, Higgins J S, Maconnachie A (eds.) 1985 *Polymer Blends and Mixtures*. Nijhoff, The Hague, The Netherlands

Yoshimura N, Fujimoto K 1969 Structure of vulcanized and unvulcanized SBR/BR blends. *Rubber Chem. Technol.* 42: 1009–13

L. M. Robeson
[Union Carbide Corporation, Bound Brook, New Jersey, USA]

Polymer Science and Engineering: Historical Development

If one attempts to trace the early concepts of the existence of polymeric systems and very large organic molecules—now called macromolecules—it becomes evident that there are essentially two sequences of events which contributed to the progress and, eventually, to the present state of understanding in this field. One was the continuous intense interest in the composition and structure of such important natural substances as cellulose, starch, proteins and rubber. The other consisted of a long series of occasional independent observations which showed that certain chemical reactions did not lead to the expected liquid or solid crystalline products, but to amorphous resinous substances. In many cases, these were hard and infusible, either completely insoluble or soluble with difficulty and giving sticky and highly viscous solutions. Usually the products of such reactions were only superficially decribed and could not be quantitatively characterized because of the absence of adequate experimental methods.

Both roots of the science of polymers or macromolecules go back to the early decades of the nineteenth century. About 100 years later they began to merge into the intense and systematic study of many different processes of polymerization and polycondensation, which led to an almost confusingly large number of new materials. Some of these became immediately useful for fulfilling existing demands. Others opened up vistas for new applications in a large number of rapidly developing fields, ranging through space exploration, terrestrial transportation, modern communications, energy saving, environmental protection, public health, microbiology and medicine.

1. Natural Rubber

A particularly interesting example of the early research is provided by the continuous efforts to elucidate the composition and structure of rubber.

This occurred not only because of the unusual properties of this substance and its obvious practical applications, but also because it was soluble in available solvents and, very importantly, because it responded favorably to the technique of destructive distillation. In the early decades of the nineteenth century, this was the most popular method for the study of unknown organic substances. Michael Faraday succeeded in 1826 after some effort in isolating a certain quantity of a volatile liquid to which he assigned the formula C_5H_8 as a result of elemental analysis and molecular weight determination. In those days, only volatile substances allowed an estimate of their molecular weight, through the vapor-density method. The deep and sustained interest in this intriguing natural substance was demonstrated during the next decades by such preeminent scientists as Himley, Dumas, Bouchardat, Liebig, Dalton and Gregory, who carried out fractional distillation experiments with many different samples of rubber and essentially confirmed Faraday's original findings. During this period, rubber in solutions and blends was already finding numerous practical applications, which were ultimately crowned by Goodyear's discovery of vulcanization in 1839.

The name of isoprene was given in 1860 to the liquid fraction by Granville Williams, who also noted that this liquid became viscous in the presence of air and, when heated after this apparent oxidation step, was transformed to a "white, spongy, elastic mass." This was the first step in correlating the liquid fragment isoprene with the remarkable properties of its solid "mother substance." In 1879, Bouchardat prepared isoprene by the fractional distillation of rubber and polymerized it with HCl as a catalyst to an "elastic rubber-like solid."

In 1884, Tilden showed that isoprene could also be obtained independently of rubber by the pyrolysis of turpentine and that this product also gave a tough, elastic substance if treated with HCl. He had already assigned to it the (correct) structural formula $CH_2{=}C(CH_3){-}CH{=}CH_2$. Thus the relationship of isoprene as the fragment (monomer) and rubber as the parent substance (polymer) was firmly established. There remained, however, the important and difficult problem of establishing how the individual isoprene units are chemically bonded together in the rubber molecule and what the molecular weight might be. Fortunately, in 1882, Raoult had made the important discovery that molecular weight could be determined in solution and, in 1887, van't Hoff succeeded in developing the rigorous thermodynamics of solutions. Due to the ready solubility of rubber in many organic liquids, it became possible to investigate the parent substance directly. In 1889, Gladstone and Hibbert carried out cryoscopic measurements with rubber solutions and obtained molecular weight values from 6000 to "at least 12 000." Instead of initiating a new era in rubber research, these values were not given serious consideration since they were too high to be acceptable to the organic chemists of the day.

Investigation of the bonding of the isoprene units in the rubber molecule received new impetus only after the turn of the century, through the contribution of Pickles and, particularly, through the momentous work of Harries. In 1906 and again in 1910, Pickles proposed the form of a chain with 1,4-bonds for the rubber molecule. Harries, who was the first to give dependable experimental evidence for this bonding, favored at the end of his studies a cyclic structure comprising four isoprene units. This argument remained unresolved until 1920, when Staudinger, on the basis of new experiments (hydrogenation and dehydrogenation of rubber together with viscosity measurements), came up with convincing evidence in favor of a long-chain structure of the rubber molecule, to which he assigned the 1,4-*trans* bond structure. Soon afterwards (from 1925 to 1928), Katz, Hauser and Mark, and Meyer and Mark provided quantitative figures for the chain skeleton and established definitely the *cis* structure of the consecutive isoprene residues.

Before this level of understanding was attained, numerous empirical attempts had been made to prepare a synthetic rubber by the polymerization of monomers (e.g., butadiene, isoprene and dimethylbutadiene). Between 1900 and 1920, Harries, Hofmann, Kondokov, Kyriakides, Lebedev, Ostromyslensky, Spence and Thiele had prepared elastomers in many different ways and during World War I some 2500 t of an elastic polydimethylbutadiene were produced in Germany.

By the end of the 1920s the molecular weight and the structure of natural rubber were well known and attempts were made to polymerize many different monomers. New polymerization systems and novel powerful catalysts were introduced. Currently, there exist around a hundred different synthetic elastomers, some of which remain soft and resilient even at −100 °C, whereas others may be used over prolonged periods at temperatures as high as 250 °C.

2. *Natural Polysaccharides*

Scientific studies of cellulose began in the first half of the nineteenth century in France with Braconot, Payen, Fremy and Pelouze, who studied the composition of a wide variety of natural cell walls and arrived at the conclusion that a certain well-defined substance was the most important component in all of them. Payen called it cellulose. Elemental analysis showed that cellulose was a polysaccharidic carbohydrate, isomeric in composition with starch, which could be degraded hydrolytically into simple sugars.

These early observations were followed by many

additional studies—some of them controversial—until, in 1910, Ost finally demonstrated that most celluloses can be converted almost quantitatively into β-D-glucose by acid hydrolysis. Thus, it became known that cellulose was a homopolymer of glucose. However, nothing was known of its molecular weight, of the specific structure of the moleucle—ring or chain—or of the bonding of the monomeric units to each other. This last important question, however, was decided in favor of a β-(1→4)-D-glucoside linkage through a series of brilliant experiments by Haworth, Hirst, Irvin, Purves and Freudenberg between 1910 and 1930. Throughout this period, the question of ring or chain structure was still unsettled and Hess in his well-known book *Die Chemie der Zellulose und Ihrer Begleiter* (1928) still listed six different possible structures, three of them with the form of a ring and three of them with the form of a chain. Unfortunately, the direct determination of the molecular weight of cellulose was not possible because of its insolubility in all available solvents. Gradually, however, these two remaining questions found their answers during the period 1925–1935. The insolubility of cellulose itself was bypassed by measurements of viscosity, diffusion, sedimentation and osmotic pressure of soluble derivatives such as cellulose nitrate and cellulose acetate, mainly by Staudinger and his associates during 1925–1935, Hess, Trogus, Sakurada and Akim in 1931 and Dobry in 1937.

Finally, a series of investigations during the 1920s by Staudinger, Freudenberg, Polanyi, Sponsler and Dore, and Meyer and Mark arrived concurrently at the conclusion that the cellulose molecule has the structure of a long chain (500 or more monomers) of β-D-glucose units which are linked together by 1→4-glucose bonds. The remaining important questions concerning the precise conformational character of the chains and their relative positions in the crystalline and amorphous domains of specific cellulose samples are now being attacked and gradually being resolved in several institutes specifically devoted to cellulose research in Finland, Sweden, Western Europe, Canada, the USA and the USSR.

Similar bird's-eye view accounts could be given for the gradual growth of knowledge about starch and proteins. This, however, would take up a considerable amount of space and the reader is referred to the Bibliography for further details.

3. *Synthetic Polymers*

Berzelius was probably the first to publish the word polymer in 1833, when he noticed that certain compounds could have the same empirical formula but different molecular weights, for example acetylene $(CH)_2$, benzene $(CH)_6$ and styrene $(CH)_8$. He considered these materials to represent a special type of isomerism and used the word polymers for the two with the higher molecular weight. Of course, in the present sense of the word polymer, they are not truly polymeric but are oligomers.

The notion of the existence of highly polymeric—amorphous and resinous—compounds gradually developed during the nineteenth century through a series of observations where, in the course of certain reactions, the expected well-defined crystalline end products were not obtained, but instead a mixture of amorphous insoluble and infusible systems were produced which defied the characterization methods of the time.

Thus, in 1839, Simon reported the self-resinification of styrene, which Berthelot called polymerization in 1866. The monomer C_8H_8 was recognized to be unsaturated; that is, it contained an aliphatic double bond. It was found by both observers that this water-clear, mobile liquid formed a hard, brilliantly transparent, colorless and amorphous resin on exposure to heat or light. Similar tendencies were observed during the following years for many other unsaturated substances, for example in 1872 by Baumann for vinyl chloride, later by other authors for vinyl acetate and for several acrylic compounds, and in 1920 by Staudinger for ketenes. This all eventually led to the notion that the aliphatic double bond under certain conditions—heat, light, presence of a catalyst—could act as a vehicle for resinification which was somehow identified with polymerization (i.e., with the formation of synthetic high-molecular-weight substances).

Between the mid-1860s and the early 1910s, large numbers of observations were made which showed that small molecules containing two or more functional groups had a distinct tendency to form polymeric condensates. Flory in his *Principles of Polymer Chemistry* (1953) gave an impressive selection of such cases. In all these numerous investigations the authors did not vigorously follow up on their discoveries because their goal had been to find soluble and, eventually, crystallizable materials which they could readily characterize and use for further synthetic steps of identifiable compounds.

Equivalent observations were made in the course of the years with molecules which were not unsaturated but contained certain reactive—or functional—groups such as OH, NH_2, COH or COOH. Thus, Lourenzo in 1860 reported the preparation of a series of ethylene glycols and ethylene dihalides having the common general formula $Z{+}CH_2{-}CH_2{-}O{+}_nH$ (where Z is OH or a halide) and identified the members with n from 2 to 6. Kraut and Gerhard observed in 1853 and 1869, respectively, that salicylic acid formed high-molecular-weight products by self-condensation under the elimination of water. When Bakeland, around 1900, reacted phenols with formaldehyde, he was also disappointed that his reaction products were insoluble and amorphous, but he did follow-up because he realized that some of his hard, solvent-resistant and high-softening-point resins might have great commercial value in themselves.

Later, in 1930, after the firm establishment of the macromolecular concept. Carothers clearly pointed out that these two historical trial-and-error lines actually represented the two major reactions for the preparation of synthetic polymers: addition and condensation polymerization. A particularly clear case of an early addition polymer was found when von Pechmann in 1898 and Bamberger and Tschirner in 1900 reacted diazomethane with substituted hydroxylamines in order to methylate the free hydroxyl groups of these amines. Instead, they obtained a product in which two phenyl hydroxylamines were connected by a methylene bridge and concluded that diazomethane dissociates into N_2 and CH_2, which in turn can either react with the amine or polymerize to form polymethylene. This material $(CH_2)_n$ was found to be a white, chalklike, fluffy powder, with a melting point of 128 °C. This was the first correct formulation and description of a polyhydrocarbon—namely linear polymethylene—which in its structure and properties is identical with the linear 1,2-polyethylene. However, at that time, because of a lack of interest in the "amorphous by-products" of organic chemical syntheses and, evidently, because of the limited availability of diazomethane, the discovery of polymethylene made no impression on the chemists of 1900.

Summarizing the early decades of the twentieth century, it can be said that (a) for cellulose, rubber, starch and proteins, several ideas and proposals had eventually coined the general name of the "high-molecular-weight natural substances" without any clear quantitative specification of the word "high" and (b) there existed many scattered observations that certain well-known small molecules were able to polymerize and to form amorphous intractable resinous materials.

At this time, new experimental methods became available for the investigation of materials in the solid state (e.g., x-ray scattering and infrared absorption) and in solution or dispersion (e.g., ultramicroscopy, electrodialysis and ultracentrifugation). All of these techniques later became important in the ultimate resolution of the structures of natural and synthetic polymers.

Accelerated development occurred in 1920 as a result of the research of Staudinger on natural rubber and ketenes, a group of unsaturated molecules which exhibited a distinct tendency to polymerize. From a series of new experiments with rubber, polystyrene and polyindene, he concluded that these materials consisted of long chains in which the individual units were connected to each other by normal covalent bonds and postulated for two synthetic polymers and for rubber the following formulae: polystyrene, $-\!\!(CH_2CHPh)\!\!-_n$; polyoxymethylene, $-\!\!(OCH_2)\!\!-_n$; and rubber, $-\!\!(CH_2C(CH_3){=}CHCH_2)\!\!-_n$.

To obtain an estimate for the chain length of these molecules, Staudinger developed a viscometric method in which it was postulated that the relative viscosity η_{sr} (i.e., the relative thickening of the solvent through the presence of the solute, which is equal to $(\eta_e - \eta_0)/\eta_0$, where η_0 is the viscosity of the solvent and η_e is the viscosity of the solution with a concentration c) should be proportional to the molecular weight M and the concentration c:

$$\eta_{sr} = K_m M c$$

where K_m represents the famous Staudinger constant, an empirical number which must be established by an absolute method for the measurement of molecular weight. Methods such as this were only in their infancy in the early 1920s and Staudinger had to use assumptions to arrive at values for K_m. Based on these assumptions, he arrived at values up to the hundreds of thousands for soluble materials such as rubber, polystyrene, polyindene and cellulose acetate. The concept that in polymeric substances, natural or synthetic, a certain number of units (monomers) are bonded together by normal chemical valences was widely accepted, but the postulate that molecular weights could be in the 100 000 range was strongly rejected by the organic chemists of the time. The majority supported another hypothesis, which postulated the existence of small structural units held together by exceptionally strong forces of aggregation or association. These forces were assumed to be of a new and as yet unknown character. The fact that most materials under investigation during this period were products of living matter, plants or animals, was an attractive and probably perfectly legitimate argument in favor of something new—something still to be learned in order to understand the structure and properties of polymeric materials.

Besides arguments of this somewhat romantic character, there existed some experimental evidence for the acceptance of stable, strongly bonded complexes. For example, Werner and his associates were able to prepare a large number of remarkably stable molecular systems which owed their existence not to the classical primary valences but to the action of secondary bonding forces. These bonds produced a wide variety of complex compounds having considerable chemical and thermal stability. However, it was found that all of these Werner complexes had molecular weights within the commonly accepted range.

Another reason for the position that unusually (and "outrageously") high molecular weights were not necessary to explain the special properties of substances such as cellulose, proteins or rubber was derived from the general knowledge and principles of colloid science. This new and rapidly expanding discipline—the "world of the neglected dimensions" —was directed especially towards the study and understanding of many natural substances (e.g., starch, gums and rubber). These materials produce "sols" of extremely high viscosity (dispersions consisting of very stable particles) and gels with great swelling capacity and remarkable elasticity. Many typical colloids were inorganic sols and gels (e.g., clay and numerous heavy metal oxides) while others were soaps or other well-known small organic compounds. The term micelle had been coined for aggregates or agglomerates or ordinary inorganic and organic molecules that were able to exist in the sol and gel state, could be reversibly transferred from one to the other, and covered a wide range of mechanical and thermal behavior.

Many distinguished scientists, among them Bergmann, Harries, Hess, Herzog, Karrer, Pringsheim and Pummerer, could not see the necessity in assuming the presence of exorbitantly large molecules in order to arrive at a satisfactory interpretation of all observed and known phenomena. Furthermore, not a single experimental fact existed in the mid-1920s which would clearly prove the existence of macromolecules. Very high viscosities of very dilute solutions had been frequently observed in colloidal systems (e.g., soaps, gums and metal oxides) which consisted of ordinary molecules. Thus, the explanation which Staudinger offered for the high viscosities of his macromolecular solutions (i.e., the presence of very thin and long rigid rods) was clearly unacceptable. It did not conform to known principles of theoretical hydrodynamics. Moreover, at the same time it was found that the crystallographic elementary cells of cellulose, silk and rubber were so small that they could only accommodate a few—four to eight—building units. From the point of view of classical crystallography, this actually would exclude the existence of very large molecules.

At a special meeting in Düsseldorf in 1926 two factors were presented which started to tip the scales in favor of the macromolecular concept: Staudinger reported a series of experiments on the hydrogenation of rubber, polystyrene and polyindene which demonstrated the remarkable stability of these substances against chemical reactions and Mark pointed out that, on the basis of a new critical analysis by Reis and Weissenberg, a small crystallographic elementary cell could be perfectly compatible with very-long-chain molecules. At the same time, systematic studies by Svedberg and his associates, using ultracentrifugation, made it quite clear that some proteins had molecular weights in the range of 100 000 and that similar values were obtained with cellulose derivatives and polystyrene.

Furthermore, the systematic application of x-ray diffraction not only showed the possibility of long-chain structures, but actually presented quantitative models for the arrangement of the macromolecules in the crystalline domains of cellulose (native and regenerated), silk, ruber, chitin and several cellulose derivatives. Soon afterwards, Carothers extended his investigation of polyesters and polyamides into the truly macromolecular range and followed the gradual extension of the growing chains by rigorous quantitative end-group measurements. Results from a series of experiments left no doubt that normal chemical bonds between the dicarboxylic acids and diamines or glycols extended into the molecular weight range of several tens of thousands. These studies were accompanied by systematic x-ray diffraction measurements, which resulted in the quantitative establishment of chain structures for these materials. The body of this new evidence—all of it favorable to the macromolecular concept—was rounded out by osmotic, electrophoretic and diffusion measurements which refuted prior claims for the existence of small units in cellulose derivatives and rubber. All of this evidence added weight to the breakthrough of Staudinger's macromolecule concept.

Beginning in the 1930s, emphasis shifted away from the controversy on basic concepts to vigorous and sustained research in other directions. Vinyl, diene and acrylic addition polymers and copolymers were studied and scaled up to commercial products in Europe, the USA and the USSR. Notable among these were polystyrene, polyvinyl chloride, polyvinyl esters and polyacrylic esters. Butadiene polymers and copolymers were developed at the end of this period (1937) and, after long and hazardous efforts, ICI succeeded in polymerizing ethylene at high pressures and temperatures. During this same period, several condensation polymers, particularly polyesters and polyamides, were developed and made commercially available. Nylon was a key product developed in this period (1937).

The mechanisms of all polymerization processes were studied intensely and comprehensively classified in many academic and industrial research laboratories. This not only gave important clues, leading to the improvement and simplification of existing processes, but also provided valuable guidelines for new processes. Gradually, a general pattern emerged for commercially viable polymerization processes. Scientifically, large areas for the discovery of new fundamental relationships opened up. From the practical point of view, wide vistas for commercially attractive materials provided incentives for enormous industrial developments.

It would go far beyond the scope of this brief chronology to give any kind of detailed description of all of the developments following World War II, but it must be added that all facets of human life and activities have been modified and enriched by the impact of polymeric materials. The influence of polymers extends into medicine, public health and safety, housing, transportation, communication, education and an appreciation of energy saving, and the protection of the environment.

See also: Elastomers: An Overview; Polymers: Classification by Use and Properties; Processing: Historical Survey

Bibliography

(a) Reviews of early aspects of natural polymeric organic substances

(i) RUBBER

Davies C C, Blake J T (eds.) 1937 *Chemistry and Technology of Rubber*, American Chemical Society Monograph Series No. 74. Reinhold, New York

Dunbrook R F 1954 Historical review. In: Whitby G S (ed.) 1954 *Synthetic Rubber*. Wiley, New York, pp. 32–55

Meyer K H 1942 *Natural and Synthetic High Polymers*, High Polymers, 1st edn, Vol. 4. Interscience, New York, pp. 119–122

Priesner C 1980 *H. Staudinger, H. Mark and K. H. Meyer: Thesen zur Grösse und Struktur der Makromoleküle*. Verlag Chemie, Basel, Switzerland, Pts. 1–2, pp. 1–31

(ii) CELLULOSE

Bikales N M, Segal L 1971 *Cellulose and Cellulose Derivatives*. Interscience, New York

Hess K 1928 *Die Chemie der Zellulose und Ihrer Begleiter*. Akademische Verlagsgesellschaft, Leipzig, Germany

Heuser E 1944 *The Chemistry of Cellulose*. Wiley, New York

Mark H 1932 *Physik und Chemie der Zellulose*. Springer, Berlin

Meyer K H 1942 *Natural and Synthetic High Polymers*, High Polymers, 1st edn, Vol. 4. Interscience, New York

Meyer K H, Mark H 1930 *Aufbau der Hochmolekularen Organischen Substanzen*. Akademische Verlagsgesellschaft, Leipzig, Germany

Ott E, Spurlin H M, Grafflin M W 1954 *Cellulose and Cellulose Derivatives*, High Polymers, 2nd edn, Vol. 5. Interscience, New York

Staudinger H 1932 *Aufbau der Hochmolekularen Organischen Substanzen*. Springer, Berlin

Tollens B C G 1914 *Kurzes Handbuch der Kohlenhydrate*. Barth, Leipzig, Germany

(iii) STARCH

Meyer A 1895 *Untersuchungen über Stärke Körner*. Fischer, Jena, Germany

Samec M 1927 *Kolloidchemie der Stärke*, Steinkopff, Leipzig, Germany

Waltur R P 1928 *The Chemistry of Starch*. Reinhold, New York

Whistler R L, Paschall E F (eds.) 1965 *Starch: Chemistry and Technology*. Academic Press, New York

(iv) PROTEINS

Liebig J 1838 *Animal Chemistry*; reprinted by Johnsons Reprints, New York (1964)

Linderstroem-Lang K U 1952 *Proteins and Enzymes*. Stanford University Press, Stanford, CA

Neurath H, Hill R L 1963–1970 *The Proteins*, 2nd edn, Vols. 1–5. Academic Press, New York

Stryer L 1974 *Biochemistry*. Freeman, San Francisco, CA

(b) Reviews of early work on synthetic polymers

Flory P J 1953 *Principles of Polymer Chemistry*. Cornell University Press, Ithaca, NY, pp. 5–40

Fredga A 1954 *Les Prix Nobel 1953*. The Nobel Foundation, Stockholm, p. 27

Milligan W O (ed.) 1967 *Polymers, Proc. Robert A Welch Foundation Conf. Chemical Research*, Vol. 10. Robert A Welch Foundation, Houston, TX, pp. 19–58

Olby R 1974 *The Path to the Double Helix*. University of Washington Press, Seattle, WA, pp. 6–37

(c) Reviews of contemporary work on synthetic polymers

Mark H F 1976 Polymer chemistry: The past 100 years. *Chem. Eng. News* 54(15): 176–89

Sailors H R, Hogan J P 1981 History of polyolefins. *J. Macromol. Sci. Chem. A* 15: 1377–1402

Stahl G A 1981 *Polymer Science Overview*, American Chemical Society Symposium Series 175. ACS, Washington, DC

H. Mark
[Polytechnic Institute of New York,
New York, USA]

Polymer Sheet: Manufacture and Applications

A generally accepted definition of sheet is that it is an artifact with two dimensions very much greater than the third; indeed, one of the two may be very long, for example the extrusion direction for extruded sheet. The thickness of sheet is accepted as being greater than that of foil, which in turn is greater than that of film. A lower limit of 0.25 mm is generally accepted for sheet material.

The methods of manufacturing thermoplastics sheet include the following techniques.

(a) Casting into sheet from a monomer or monomer–polymer syrup and polymerizing in this form (e.g., cast acrylic sheet).

(b) Extruding through an appropriate sheet die, the polymer being made either outside the extruder (e.g., high-impact polystyrene (HIPS)) or by polymerizing in the extruder, utilizing the high-powered stirring action and the good heat transfer inherent in the equipment. The latter is again exemplified by acrylic sheet manufacture but, in this case, the molecular weight must be reduced to enable the extrusion of sheet to take place through the die. The molecular weight is lower that that of cast sheet.

(c) Compression molding of laminated foils; often the foils are prepared by calendering (e.g., unplasticized poly(vinyl chloride) (UPVC) sheet).

(d) Calendering (occasionally used for making flexible sheets).

In addition to these methods, sheet products with patterned surfaces, specific textural features or with a specific construction can be manufactured by variants of these processes.

Crosslinked or crosslinkable sheet can be used in the manufacture of the following.

(a) Sheet material that has not been crosslinked, but is shaped before crosslinking and is finally shape stabilized by crosslinking.

(b) Panels that are subsequently joined for a variety of applications; such panels may be homogeneous or heterogenous, the latter exemplified by sandwich panels (e.g., GRP–foam–GRP (GRP denotes glass-reinforced plastic); sheet may be generated by hand layup, spray techniques or by machine impregnation of chopped strand mat.

More detailed discussion of these individual matters is given in the following sections, together with some consideration of the shaping or fabrication of the sheet for a variety of applications. Not included in the present review are some very specific items of equipment, for example the topformer, which first establishes a small sheet at the head of the extruder, then subjects it to a thermoforming process producing, typically, disposable drinking beakers of HIPS.

1. Manufacture of Thermoplastics Sheet

1.1 Casting from the Monomer

Casting from the monomer is, in principle, the simplest method of polymerizing/shaping. To illustrate the method, consider methyl methacrylate, a liquid monomer which is polymerized by free-radical mechanisms. Common sense indicates that the polymer will be obtained if a suitable source of free radicals is supplied, which initiates the reaction in a temperature region where the rate of reaction is satisfactory. This simple process is not operated industrially for two reasons: (a) reacting monomer to polymer is accompanied by high shrinkage, and (b) the polymerization is highly exothermic.

The first problem is solved by employing a syrup of preformed, low-molecular-weight polymer in monomer (in which it is soluble); also present is a free-radical initiator which is activated at high temperatures. Problems associated with the heat generated by the polymerization reaction are also aided by the use of syrup. Even so, the heat generated still remains the controlling factor in the manufacture of all but the thinnest sheet. Polymerization temperatures are maintained at lower-than-optimum values so that there is little risk of runaway reactions. As a consequence, sheet (better described as blocks) 150 mm in thickness may take up to three weeks to polymerize completely.

Cast acrylic sheet is characterized by near-perfect optical transparency and very high molecular weight. The latter affects the shaping process as entanglements restrict the movement of the polymer chains, even when they are rendered relatively mobile by operating at a temperature above the softening point. The system responds to stress or strain by deforming, as does crosslinked natural rubber. In other words, the strain imparted to the cast poly(methyl methacrylate) (PMMA) in the thermoforming process is largely recoverable. Whenever the temperature of an artifact, thermoformed from PMMA cast sheet, approaches the softening temperature, some reversion occurs. The additional problem is that the residual stress decreases the effective softening temperature by up to 10 °C.

1.2 Sheet Die Extrusion

Extrusion through an appropriately shaped die is the most apt way of manufacturing thermoplastics sheet; the simple analysis

$$\text{polymer melt} \rightarrow \text{rectangular slot die} \rightarrow \text{cool, maintaining section}$$

is, however, too naive an approach. Generally the polymer is available as granules or powder which is "stirred" into a homogeneous mass by the screw of the extruder. There is the possibility, as previously mentioned, of employing the extruder barrel as a reaction vessel, stirred by the screw, and where the excellent heat transfer serves to keep the reaction under temperature control while the polymerization proceeds rapidly. Of course, it is necessary that the polymer should pass through the die, requiring control of the molecular weight by addition of a chain transfer agent, frequently dodecyl mercaptan.

The extrusion stage, whether it be of polymer fed to the extruder or of polymer made *in situ*, introduces two problems (Tadmor and Gogos 1979):

(a) the transformation of flow in a circular geometry to that of a flat sheet while at the same time ensuring, as far as possible, that all elements undergo the same shear history; and

(b) the occurrence at the cooling stage of warping, if the cooling is unbalanced, or voiding, particularly for thick sheets of crystallizing plastics, unless remedial measures are adopted.

The first problem can be tackled either empirically or theoretically, although the theoretical treatment does not, as yet, enable a final die design to be established without fine tuning. Both theoretical and experimental approaches lead to the coathanger design. Pearson (1966) has examined the problem by modelling the polymer as a power law fluid and assuming the flow to be isothermal (which it is not; temperature gradients will be present in the melt and across the die).

The problem of extruding and cooling, without voiding, a thick sheet of a crytallizing polymer has been solved by careful control and matching of the cooling and crystallization processes.

1.3 Coextrusion

Two or more polymer melts can be coextruded from flat sheet or film dies, fed by the appropriate number of extruders. Many of the properties of such systems are additive; for example, the gas permeability can be calculated by "adding the resistances in series." In other cases, synergism is encountered, as when alternating tough and brittle layers leads to a tough, strong composite.

1.4 Compression Molding

Compression molding is used only rarely for thermoplastics sheet manufacture since, with conventional compression molding presses, cycle times are unacceptably long as, inevitably, the press platens are heated and cooled for the shaping and shape stabilization stages. Only one major product, UPVC, is manufactured by compression molding. An appropriate number of calendered foils are loaded into the press at 165 °C; after molding, the platens are water cooled until the temperature of the sheet is at 50 °C or less, at which stage the sheet is form stable.

1.5 Calendering

A calender consists essentially of four rolls, usually assembled into an inverted "L" or an inverted "Z" configuration. The four rolls give three gaps: the first controls the feed rate to the calender, the second defines the final thickness of the sheet and the third further refines this thickness (see Fenner 1979). Desirable features of this process include its economy, its suitability for heat-sensitive plastics and the capability of handling high-viscosity plastics.

1.6 Shaping and Applications of Thermoplastics Sheet

The shaping of a plastics sheet is to be regarded as a secondary shaping, the primary operation being the manufacture of the sheet. Thus, whereas shear forces usually dominate the primary shaping stage, for example shaping by passing through an extrusion die or calendering, it is frequently tensile forces that dominate secondary shaping and this is almost inevitably the case for the shaping of sheet.

The process by which thermoplastics sheet is shaped is generally known as thermoforming and comprises a range of related processes involving vacuum or air-pressure forming, sometimes with the aid of a pneumatically or hydraulically powered plug (Throne 1987). The thermoforming process has three stages:

(a) heating the sheet to a uniform temperature;

(b) shaping, with or without plug assist; and

(c) cooling to stabilize the shape.

As noted, shaping is frequently accomplished by the application of tensile or compressive forces. The response of the material may be predominantly viscous flow at the high-temperature limit and mainly high elastic deformation at the lowest tempertures. These limits define the operating window within which thermoforming can be applied successfully. For some plastics, the situation is complicated by the occurrence of crystallinity, which modifies the shaping behavior. The crystallinity may be inevitable as the sheet is cooled, as occurs for polyethylene and polypropylene, or it may only appear following orientation and thermal conditioning, which are part of the thermoforming process, exemplified by poly(ethylene terephthalate).

Since thermoforming is a secondary shaping process depending on the supply of preformed sheet, it is a process which is suited to prototyping and to small production runs. The equipment is relatively cheap (for example the molds do not have to withstand high pressure) but in its simplest form the process is relatively slow and often labor intensive. Commercial thermoforming equipment utilizes radiant heaters, almost without exception, with the lack of uniformity which is inevitable with that mode of heating. The alternative static or circulating air ovens are considered too slow for economic use.

2. Crosslinked Plastics Sheet

2.1 Manufacture of Sheet Material for Subsequent Shaping and Crosslinking

The fundamental difference between crosslinkable sheet material and the thermoplastic sheet discussed in Sect. 1 is that, in this case, the shaping is completed before the chemical reaction is invoked. The benefits from this include the following:

(a) shaping may be possible at much lower pressures than for thermoplastics sheet;

(b) the shape is truly stable after the completion of the crosslinking reaction; and

(c) these sheet materials often contain fillers or reinforcements which enhance the properties and improve the economics.

Shapings based on these materials are suitable for high-temperature use without the problems of softening and demolding; they will be illustrated by the manufacture and application of sheet molding compound (SMC), which is frequently based on unsaturated polyester resin.

The uncured resin, containing a suitable high-temperature initiator and filled with cheap, particulate filler, is spread onto continuous polyethylene (or polyamide) film. Chopped glass fibers are spread evenly on the uncured resin, followed by a further layer of resin and the containing polyethylene or polyamide film. By treatment with profiled rollers, the glass fibers are dispersed in the polyester resin, within the confines of the containing film, imparting an appreciable amount of planar orientation to the glass fibers. The system is known as SMC, but requires a further stage before it can be applied. This is known as maturing and involves a viscosity increase of two orders of magnitude, achieved by adding magnesium oxide or hydroxide to the unsaturated polyester. The base is presumed to react with terminal carboxyl groups, linking the chains and thereby increasing the viscosity of the matured resin system, allowing the polyethylene or polyamide film to be stripped. It should be noted that since the SMC contains an initiator to effect the curing reaction the shelf life of the system is limited. A typical formulation is 30 wt% resin, 0.5 wt% peroxide initiator, 33.5 wt% calcium crbonate, 6 wt% low-shrink additive, 2 wt% release agent, 2 wt% pigment, 1 wt% thickener (MgO) and 25 wt% chopped glass fiber. The resin is an unsaturated polyester dissolved in styrene, with which it copolymerizes.

2.2 Shaping and Application of SMC

Shaping of SMC is achieved basically by compression molding: analysis of the process is associated with squeezing flows (Leider and Bird 1974), which is geometry suited to the characterization of polymer melts. SMC is generally specified for comparatively large areas of high stiffness, typified by the hood cover of the Citröen BX series of cars.

2.3 Manufacture of Panels

Composite panels comprising GRP–foam–GRP can be fabricated into lightweight structures, which are applied where stiffness and light weight are desirable features. Examples are trailers and the workshop module of British Telecom transport. The panels are assembled essentially by sticking GRP faces to a slab of foam; performance is critically dependent on the density of the foam and on the integrity of the interface.

2.4 Applications of Composite Panels

As noted in Sect. 2.3, composite panels are used extensively by trailer makers; the thermal insulation provided by the foam is an added advantage over solid structures.

See also: Processing: Classification

Bibliography

Fenner R T 1979 *Principles of Polymer Processing.* Macmillan, London

Leider P J, Bird R B 1974 Squeezing flow between parallel disks, I. Theoretical analysis. *Ind. Eng. Chem., Fundam.* 13: 336–41

Pearson J R A 1966 *Mechanical Principles of Polymer Melt Processing.* Pergamon, Oxford

Tadmor Z, Gogos C G 1979 *Principles of Polymer Processing.* Wiley, New York

Throne J L 1987 *Thermoforming.* Hanser, Munich, Germany

A. W. Birley

[University of Technology, Loughborough, UK]

Polymers: Classification by Use and Properties

Polymer science and technology have been developed to a level of understanding and engineering capability which permits a wide variety of properties to be conferred on products fabricated from polymers. The high level of polymer usage in the USA (~25 Mt in 1984) is due substantially to the ability to tailor properties to suit a very broad spectrum of applications. Polymeric materials may be produced as stable, tough, solid plastics, fibers, foams, films, adhesives or coatings. They may be hard or soft, rigid or flexible, transparent or opaque, rubbery, leathery or glassy. They can be made resistant to the effects of temperatures up to several hundred degrees and they can also be useful at very low temperatures such as those encountered in space applications.

The control of polymer properties depends not only on chemical composition but, equally important, on the structure and arrangement of the atoms within the molecules and the structural arrangement of the molecules or segments thereof with respect to one another (morphology).

The classification of polymers may be undertaken using various concepts as criteria; those selected here attempt to provide a useful basis for understanding the relations between properties, structure and indications for end-use applications. This basic classification depends on a pragmatic consideration of gross mechanical properties (stress–strain relations) at ambient temperatures. Thus, the following classes of polymeric materials may be recognized:

(a) class I materials, which are soft, flexible, extensible and possibly elastomeric (e.g., polybutadiene, polyacrylates, ethylene–propylene rubber);

(b) class II materials, which hard, stiff, tough and pliable (e.g., polyethylene, polypropylene, Nylon 66); and

(c) class III materials, which are hard, rigid, dimensionally stable, often brittle (e.g., polystyrene, poly(methyl methacrylate) and thermosets).

A detailed interpretation of the stress–strain relations for each of these classes is given in Sect. 2. The structure–property relations necessary to this understanding are discussed in Sect. 1.

Table 1
Glass-transition temperatures for several amorphous polymers

Polymer	T_g (°C)
Silicone rubber	−123
Polyisobutylene	−75
Natural rubber	−72
Poly(*n*-butyl acrylate)	−55
Poly(ethyl acrylate)	−20
Poly(methyl acrylate)	10
Poly(vinyl acetate)	30
Poly(vinyl chloride)	65
Polystyrene	100
Poly(methyl methacrylate)	105
Poly(*M*-vinyl carbazole)	211

1. Structure–Property Relations

The special characteristics associated with the polymers in each of the classes defined above depend very strongly on two basic parameters, the glass-transition temperature and the degree of crystallinity. These, in turn, are influenced by a number of properties which provide a basis for further classification. They include molecular weight and degree of crosslinking, chain flexibility or stiffness and polar effects.

1.1 Glass-Transition Temperature

The glass-transition temperature T_g is defined as the temperature below which thermally induced molecular motion is frozen in. Above this temperature there is sufficient thermal energy to permit motion and undulations in the chain. In amorphous, morphologically disordered polymers this chain motion is associated with softness and plastic or elastic characteristics. In crystalline polymers (see Sect 1.2) chain motion is hindered by the crystalline forces. Below T_g, amorphous materials are stiff, hard and often brittle. Here too, the presence of crystallinity modifies the behavior, especially embrittlement. Generally, the level of T_g is associated with the relative flexibility of the polymer chain: a lower T_g gives a more flexible chain (see Sect. 1.4). Inclusion of large atoms or groups and especially rings in the molecules increaes the stiffness and causes an increase in T_g. Examples of amorphous (noncrystallizable) polymers and their glass transition temperatures are given in Table 1.

1.2 Crystallinity

Crystalline polymers are characterized by the capability of their molecules, or segments of these molecules, to form three-dimensionally ordered arrays. These exhibit characteristic intermolecular or intramolecular spacings and melt at a specific temperature, T_m. Crystalline polymers generally have a variable degree of crystallinity depending on average chain length and thermal treatment. Shorter chains at elevated temperatures, above T_g, can more readily move into crystalline arrays, but longer chains are more likely to be involved in extensive entanglements, crossovers and random kinks and folds which cause disordered regions. The longer chains are usually associated with several crystalline regions and they serve as "tie" molecules interconnecting the various reinforcing crystalline volume elements which act as physical crosslinks. Thus, as crystallinity increases, tensile strength increases and elongation tends to decrease. Usually, increasing crystallinity in a polymer system is associated with decreasing transparency, and most crystalline polymers are quite opaque.

Not all polymers can be crystallized. The basic requirement for forming three-dimensional, close-packed molecular arrangements are: (a) linear, regularly shaped symmetrical molecules that have relatively free vibrational and rotational motions so that they can pack efficiently; and (b) the absence of bulky, irregularly spaced side chains or groups which prevent close packing. Examples of crystalline polymers and their melting temperatures are shown in Table 2.

1.3 Molecular Weight

The unique mechanical properties of polymers (i.e., strength, flexibility and toughness) depend on the long-chain-like nature of polymer molecules and the entanglements and interactions between them. Mechanical properties increase almost linearly with molecular weight until an average degree of polymerization (DP, the number of monomer units per polymer molecule) of about 600 or more is reached. Above this level there is a very slow increase in properties, with maximum toughness achieved in practice at DP levels of about 100 000 as found in ultrahigh-molecular-weight polyethylene used for

Table 2
Glass-transition temperatures and melting points for several partially crystalline polymers

Polymer	T_g (°C)	T_m (°C)
Polyethylene (branched, low density)	−125	110
Polyethylene (linear, high density)	−125	137
Polyoxymethylene	−70	181
Poly(1-butene)	−25	125
Poly(vinylidene chloride)	−15	198
Polypropylene	−10	176
Poly(4-methyl-l-pentene) (isotactic)	20	250
Nylon 610	40	227
Nylon 66	50	265
Poly(ethylene terephthalate)	70	267
Poly(vinyl chloride)	87	212

bumpers, bearings and other high-stress applications. At these high DP levels, the chain entanglements are very numerous and the materials effectively behave as though there were many physical crosslinks which impede the flow. Another feature of molecular weight M and associated chain entanglements is its very important effect on polymer melt viscosity N_{melt}, as expressed by

$$N_{melt} = KM^{3.4}$$

where K is a constant characteristic of the polymer. This relation has a very strong influence on melt processibility. Doubling the molecular weight increases the melt viscosity by a factor of about ten. Low-molecular-weight polymers tend to serve as waxes, lubricants or coatings while medium- to high-molecular-weight materials make up the bulk of useful molded or extruded polymeric products. The ultrahigh-molecular-weight materials (polytetrafluoroethylene and polyethylene) require special fabrication techniques.

Generally, because extruded products must have form stability (hot melt strength) as they emerge from the hot extrusion die, it is necessary to use higher-molecular-weight polymers (having more reinforcing entanglements) for fabrication by extrusion as compared with injection molding, where the relatively cool mold causes rapid quenching and solidification of the polymer in the desired shape.

The special character of thermoplastic polymers which causes them to soften with increased temperature is due to increasing chain mobility as the temperature is raised. Introduction of chemical links between the molecular chains (by irradiation or chemical means) has the effect of raising the average chain molecular weight to very high levels and strongly hinders chain mobility. Thus soft, irreversibly extensible materials are made elastomeric by low levels of crosslinking. This is the basis of the large number of rubbers used in many applications. Increasing crosslink density produces harder, less elastomeric rubbers which are stable at high temperatures. Similarly, crystalline polyethylene (normally melting at 135 °C) on being crosslinked does not show the typical abrupt change in mechanical properties usually observed when heating the material above T_m.

1.4 Chain Flexibility

An important molecular parameter which affects polymer properties is the relative flexibility of the polymer chains. This is related to the ease of thermal activation of vibrational and rotational motions involving the covalent bonds. Rotation is inhibited by stiff bonds and by bulky side groups. Thus, polymers essentially made up of C—C, C—O—C or C═C bonds in the *cis* form have flexible chains because the energy barrier to rotation is quite low.

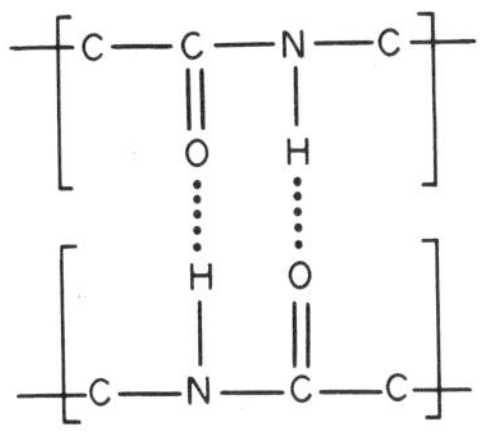

Figure 1
Hydrogen bonding in nylon

These flexible chains form polymers with low glass-transition temperatures (<−50 °C). When the chains are in a relatively disordered arrangement (amorphous), the polymer mass tends to be soft, extensible and readily soluble at room temperature and above. Introduction of order into the arrangement of the chains results in crystallization, which causes a reinforcement of the total system and results in harder, stronger, tougher, less soluble polymers.

Inclusion of cyclic structures (phenyl rings) in the polymer backbone or bulky side chains (methyl, phenyl, acetate or acrylate) increases the chain stiffness and results in high-T_g material which tend to be hard, rigid and have a high modulus. Still higher T_g values are attained in "ladder" polymers, which have regularly spaced chemical links between two parallel carbon backbone chains (e.g., polyimides).

Chain stiffness can also have an important effect on the melting temperature of crystallized polymers. Rigid, regular chain molecules can close pack in a three-dimensional array and the resulting polymer mass tends to have a high melting temperature. Thus, poly(4-methylpentane) forms a helical, relatively rigid chain and even without polar effects it has a melting point of 260 °C.

1.5 Polar Effects

Polymer molecules with polar groups along the chain may tend to produce electrical interactions which cause a reinforcing or stiffening effect on the polymer system. Thus Nylon 66 in a dry state tends to be rigid and stiff owing to the hydrogen bonding between the negative C═O groups and the positive N—H groups (see Fig. 1). The regularity of the spacings of the polar groups tends to produce a reinforcing effect along the length of the aligned chains. Thus, T_m is quite high at 265 °C, while that of polyethylene, which has no interactive groups, is only 135 °C.

Polyamide polymer systems tend to be sensitive to water, a polar molecule, and readily absorb it. As water is absorbed it tends to diminish the electrical interaction and the polymer becomes somewhat more flexible, has a lower modulus and higher elongation. The water acts as a plasticizer.

2. Interpretation of Stress–Strain Curves

The stress–strain curves shown in Fig. 2 provide a visual expression of the basic characteristics of the three general classes of polymers. The soft, flexible materials (class I) show a shallow slope in the low-elongation part of the curve. This is indicative of a low modulus. Further, as the stress is increased the material stretches readily and elongates by several hundred percent. As this takes place, the molecules stretch out and orient causing the sample to increase in strength. If the chains are long enough (high molecular weight) there are sufficient entanglements to permit spontaneous retraction when the force is removed. This behavior is associated with relatively free motion of the molecules and involves low-T_g ($<-20\,°C$) flexible materials in relatively amorphous (disordered) arrangements.

Thus, in terms of these property generalizations, the relatively flexible polymer molecules (low T_g) with low crystallization tendency, with or without polar interactions, fall into this class of soft, flexible materials. If the molecular weights are sufficiently high or some crosslinking is present, the materials tend to be elastomeric.

Materials of higher T_g such as poly(vinyl chloride) ($T_g = 87\,°C$) or Nylon 66 ($T_g = 50\,°C$) can be made soft and flexible by use of suitable plasticizers or by copolymerization. In both cases the effective T_g is lowered. Similarly, crystalline, low-T_g polymers such as polyethylene, which is normally rigid (class II) can be copolymerized with ethyl acrylate or vinyl acetate to make soft, flexible, rubbery products of class I behavior which are used in film, hose and gasket applications.

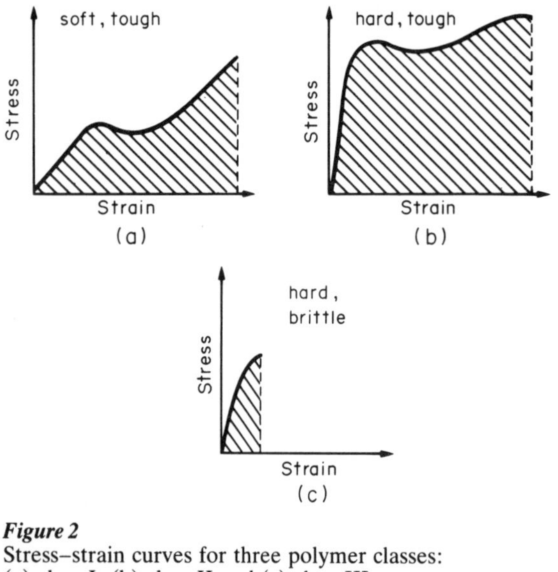

Figure 2
Stress–strain curves for three polymer classes: (a) class I, (b) class II and (c) class III

The curve for class II materials shows a steep initial slope (at low elongations) followed by a high elongation and high tensile strength. This is usually associated with crystalline structure where the intermolecular and intramolecular forces reinforce the entire system and thereby result in higher-strength, stiff, leathery materials. The high elongation noted is especially associated with high-molecular-weight, long-chain-like molecules which tend to be more physically associated and entangled. The T_g range may be quite broad (about $-100\,°C$ to $100\,°C$). This means that molecules may be quite flexible or moderately stiff, but most important they must be able to pack in a crystalline array, so that as the polymer is stretched, the system of chains orients and the strength increases.

The polymers in this class make up the largest portion of materials used in synthetic fibers (crystallinity and orientation enhance properties), films and general molding applications. Many of these materials, especially when reinforced with glass or other fibers are used as engineering resins.

The curve for class III hard, rigid materials shows a rapid initial rise in strength at low elongation (high modulus) and a relatively low elongation to break (usually associated with brittleness). This behavior is associated with relatively stiff, high-T_g polymer molecules which form an amorphous structure. Generally, these materials are hard, transparent and dimensionally stable up to their T_g. Examples inlcude polystyrene, poly(methyl methacrylate) and thermoplastic copolyesters. These materials are used in place of glass (e.g., for glazing, drinking cups and bottles).

Thermosets (crosslinked network polymers) also fall into this class. They are not generally transparent and are usually filled with various reinforcing materials to improve the impact strength. Since they are thermosets, they can be used at elevated temperatures and maintain their dimensions and properties. They are used in electrical applications, dishwasher-proof dinnerware and as metal part replacements (engineering resins).

The term engineering resin refers to both engineering thermoplastics and thermosets which can be used as replacements for metals in automobile and appliance parts and in industrial applications where rigid engineering specifications are required. These materials are generally characterized by the following key properties:

(a) high strength, comparable with that of metals on a specific weight basis;

(b) low creep and retention of mechanical properties at elevated temperatures:

(c) retention of physical dimensions at elevated temperature (i.e., low coefficient of expansion); and

(d) good electrical properties and chemical resistance.

The following are some commonly used polymers that are associated with this category.

(a) *Crystalline polymers* (class II). Among these are acetal polymers and copolymers (formaldehyde polymer), fluoropolymers (e.g., chlorotrifluoroethylene polymer, ethylene–tetrafluoroethylene copolymer, vinylidene fluoride polymer, tetrafluoroethylene polymer), polyamides (e.g., Nylon 6, Nylon 66, Nylon 612), polycarbonate, poly(phenylene sulfide) and thermoplastic polyesters.

(b) *Amorphous polymers* (high T_g, class III). Among these are ABS resins (acrylonitrile–butadiene–styrene copolymer), epoxy resins suitably crosslinked, phenol, urea- and melamine-based thermosets, polyimides and thermoset polyesters.

See also: Design with Plastics

Bibliography

Bikales N M (ed.) 1971 *Mechanical Properties of Polymers*. Wiley, New York

Billmeyer F W Jr 1971 *Textbook Polymer Science*, 2nd end. Wiley, New York

Hall C 1981 *Polymer Materials: An Introduction for Technologists and Scientists*. Wiley, New York

Kaufman H S, Falcetta J J (eds.) 1977 *Introduction to Polymer Science and Technology: An SPE Textbook*. Wiley, New York

Nielsen L E 1974 *Mechanical Properties of Polymers and Composites*. Dekker, New York

Odian G C 1981 *Principles of Polymerization*, 2nd edn. Wiley, New York

Rodriguez F 1982 *Principles of Polymer Systems*, 2nd edn. McGraw-Hill, New York

Rudin A 1982 *The Elements of Polymer Science and Engineering: An Introductory Text for Engineers and Chemists*. Academic Press, New York

Williams D J 1971 *Polymer Science and Engineering*. Prentice-Hall, Englewood Clifs, NJ

H. S. Kaufman
[Ramapo College, Mahwah, New Jersey, USA]

Polymers: Mechanical Properties

Most applications of polymers require load-bearing capacity. The geometrical responses to loading lead to a wide range of mechanical properties. However, nearly all can be grouped under stress–strain properties, viscoelasticity and failure properties. These properties are in turn largely determined by the polymer structure (i.e., molecular weight, crosslinking, branching, segmental structure and morphology) and the nature and extent of compounding. In addition, external conditions such as temperature, loading rate and environment must be considered when characterizing the mechanical behavior of a polymeric system.

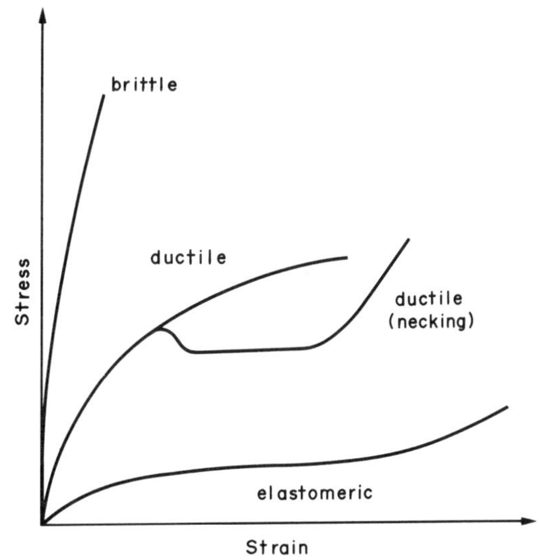

Figure 1
Typical stress–strain curves of brittle, ductile and elastomeric polymers

1. Stress–Strain Properties

The most common technique for measuring mechanical properties is tensile or compression testing. The results of such tests are stress–strain diagrams. Polymers exhibit three basic stress–strain behaviors: brittle, ductile and elastomeric (see Fig. 1). There are a number of factors that affect the stress–strain curves of a particular polymer, such as strain rate, compression or tension, and orientaton. Strain rate is important because polymers are generally viscoelastic and thus their behavior is time dependent. Figure 2 illustrates the effect of strain rate on tensile properties. Note that increasing temperature has the same effect as decreasing the strain rate. In glassy polymers, compression differs from tension because surface flaws are detrimental to tensile properties (see Fig. 3). Finally, orientation of the chains in the direction of stress substantially increases the properties in that direction, but properties in other directions decrease for both crystalline and amorphous polymers (see Fig. 4).

Modulus is a function of temperature and is generally divided into three regions—rigid, leathery and rubbery—as temperature increases. The actual shape of the modulus vs temperature curves is dependent on molecular weight and morphology.

Increasing molecular weight increases the length of the rubbery region because of increased molecular entanglements (see Fig. 5). Crosslinking broadens the leathery region because the random crosslinks hinder molecular motion, increasing and broadening the glass-transition temperature range. Also, the rubbery region remains constant and increases in modulus with increased crosslinking (see Fig. 6). Crystallinity raises the melting point and increases the modulus of the rubbery region because crystallites act as crosslinks in the amorphous region. The negative slope in the rubbery region is due to the melting of imperfect crystallites and thermal expansion (see Fig. 7). Finally, plasticizers, copolymerization and pressure also affect the modulus vs temperature curves.

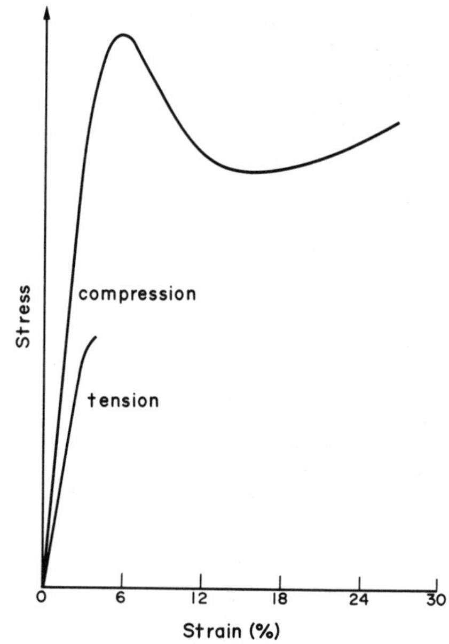

Figure 3
The stress–strain behavior of a glassy polymer showing the difference between tension and compression

2. *Viscoelasticy*

2.1 Creep and Stress Relaxation

Polymers are generally viscoelastic and hence exhibit creep and stress relaxation behavior. Creep and stress relaxation tests measure the viscoelastic behavior by monitoring elongation changes at a constant stress and by measuring stress reduction at constant elongation, respectively. Both tests permit calculation of the same viscoelastic constants. Most of the examples given here involve creep phenomena but can be applied to stress relaxation.

Several models have been developed to describe the viscoelastic nature of polymers. The models consist of Hookean springs and viscous Newtonian dashpots (see Fig. 8). A Hookean spring elongates

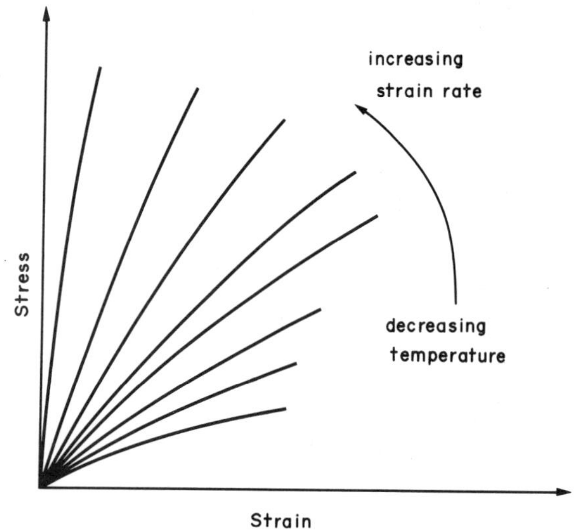

Figure 2
Stress–strain curves as affected by strain rate and temperature

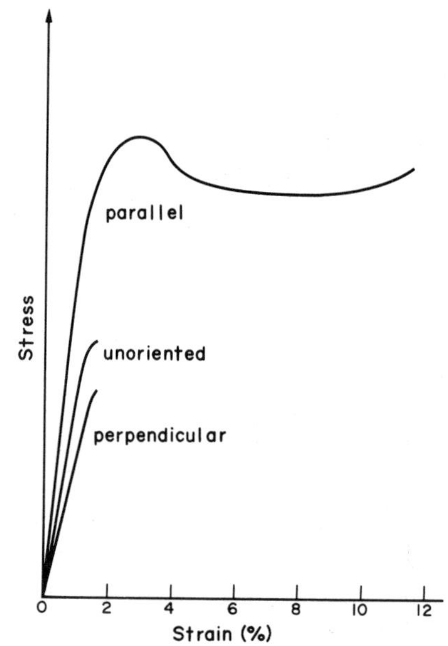

Figure 4
The stress–strain behavior of an oriented, glassy polymer tested in tension parallel and perpendicular to the orientation compared with that of an unoriented polymer

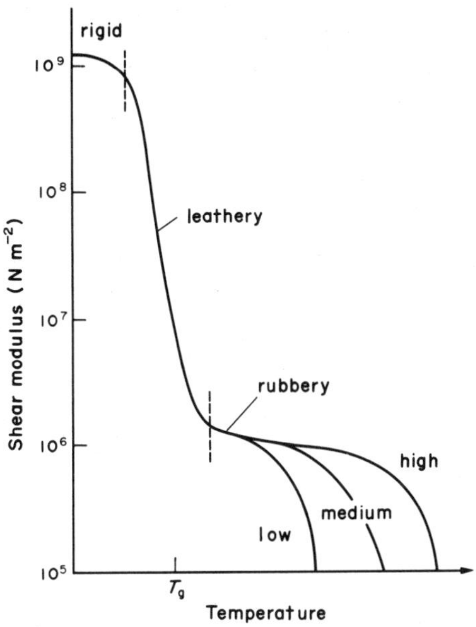

Figure 5
Shear modulus vs temperature for an amorphous polymer of relatively low, medium and high molecular weight

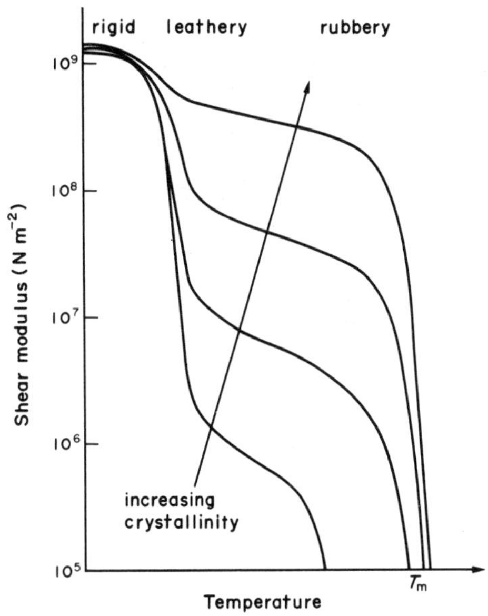

Figure 7
Shear modulus vs temperature as a function of increasing crystallinity

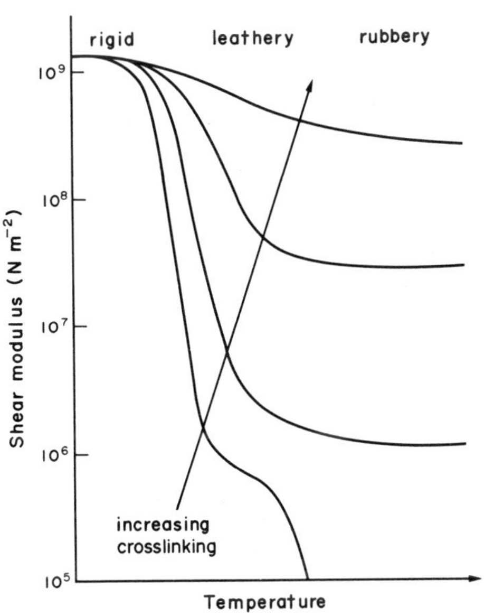

Figure 6
Shear modulus vs temperature of an amorphous polymer as the amount of crosslinking increases

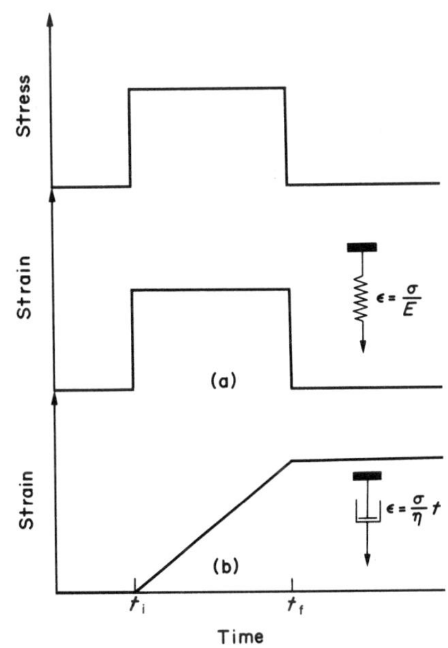

Figure 8
The creep behavior of (a) a spring and (b) a dashpot when a stress is instantaneously applied and removed

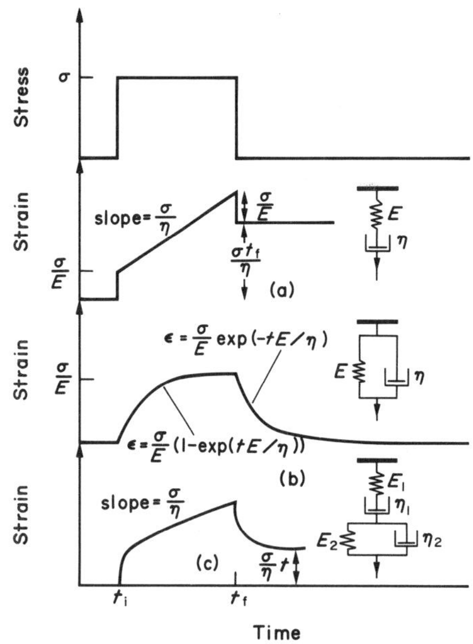

Figure 9
The creep behavior of (a) a Maxwell element, (b) a Voigt–Kelvin element and (c) a four-parameter model when stress is instantaneously applied and removed

instantaneously to an equilibrium elongation ε which is directly proportional to the applied stress σ (i.e., $\varepsilon=\sigma/E$ where E is Young's modulus) and instantaneously recovers when the stress is removed. A dashpot is similar to a shock absorber in that an equilibrium elongation is never reached and it does not recover elongation already incurred. Mathematically, the applied stress σ is equal to the rate of elongation through a proportionality constant η known as the viscosity (i.e., $\varepsilon=(\sigma/\eta)t$) effect of an applied and removed stress as a function of time.

Three relatively simple models (Maxwell, Voigt–Kelvin and the four-parameter) and their creep behavior are shown in Fig. 9. The creep behavior of the Maxwell model adequately represents the instantaneous elongation or recovery and permanent deformation due to stress of an actual polymer. However, it does not characterize the exponential decrease in elongation rate or recovery demonstrated by most polymers. The Voigt–Kelvin model demonstrates the exponential creep and recovery of polymers, but is deficient in near instantaneous elongation or recovery. Also, this model reaches an equilibrium elongation not indicative of a true polymer and shows no permanent set.

The best features of both models are combined into the four-parameter model and thus the creep response of this model adequately matches the creep behavior of polymers. However, this model cannot fully represent polymeric creep behavior because of the distribution in molecular weight and other complex morphological structures associated with polymers. Thus, several models consisting of many Maxwell or Voigt–Kelvin elements with a distribution of relaxation times have been introduced, although the equations are complicated, as shown for a Voigt–Kelvin series:

$$J(t)=\int_0^\infty L(\lambda)[1-\exp(-t/\lambda_i)]\,d\lambda \qquad (1)$$

where J is the compliance and $L(\lambda)$ is the continuous relaxation time distribution.

2.2 *Superposition Principles*

In order to describe the creep response as a function of loading history, the Boltzmann superposition principle states that the sum of the creep responses due to each stress is the total creep response. Thus, the individual creep response from each stress is independent of other creep responses (see Fig. 10) and is mathematically represented by

$$E(t)=J(t-t_1)(\sigma_1-\sigma_0)+J(t-t_2)(\sigma_2-\sigma_1) \\ +J(t-t_3)(\sigma_3-\sigma_2)+\ldots \qquad (2)$$

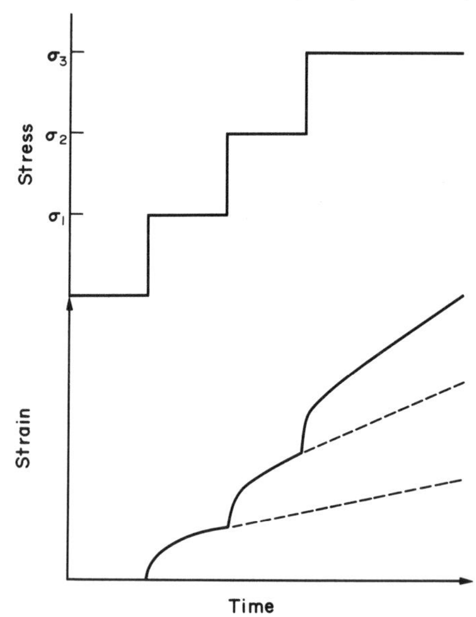

Figure 10
The creep behavior due to the subsequent addition of stresses as predicted by the Boltzmann superposition principle; dotted lines indicate creep behavior boundaries for each stress applied

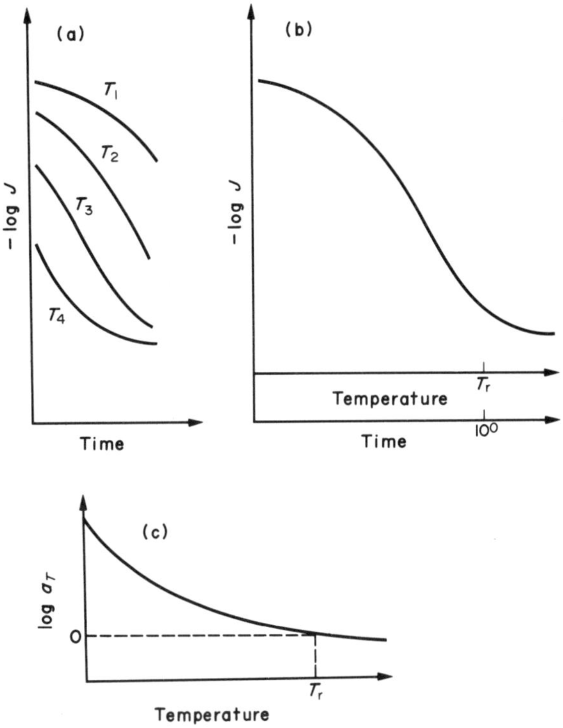

Figure 11
(a) Stress relaxation data for different temperatures, (b) the resulting master curve at reference temperature T_r and (c) shift factor vs temperature

Unfortunately, the Boltzmann superposition principle is only valid for small strains (<2%) and is thus extremely limited. The Nutting equation takes into account the nonlinear creep behavior when large stresses are applied:

$$E = K\sigma^{\beta}t^{n} \tag{3}$$

Nonlinearity occurs when the stress exponent β exceeds unity.

The time–temperature superposition principle establishes an equivalence between time and temperature which allows the prediction of long-term properties from short-term, high-temperature tests. This is accomplished by smoothly shifting the log of the creep compliance versus the log of the time curves (see Fig. 11a) along the log of the time axis to form a master curve (Fig. 11b) in reference to a curve conducted atthe desired temperature T_r. Unfortunately, the "shift factor" itself is dependent on temperature (Fig. 11c). The WLF equation relates the shift factor a_T to the glass-transition temperature T_g:

$$\log a_T = -17.4\,(T - T_g)/(51.6 + T - T_g) \tag{4}$$

However, this equation is only applicable to amorphous polymers and to the temperature range from T_g to 100 °C above T_g.

2.3 Dynamic Mechanical Properties

Dynamic mechanical properties indicate the ability of a polymer to dissipate energy. Owing to their viscoelastic nature, dynamic mechanical properties measure the in-phase and out-of-phase moduli with an applied oscillatory stress function. More importantly, tan δ, the "damping factor" (a measure of the phase angle between the applied stress and subsequent strain), is also obtained. This allows a qualitative estimation of the ability to dissipate energy. Figure 12 shows shear storage modulus (G', in phase), shear loss modulus (G'', out of phase) and tan δ for a typical polymer as a function of temperature, where

$$G''/G' = \tan\delta \tag{5}$$

3. Deformation and Failure

There are many types of mechanical failure. At the microscopic level, most of these processes involve bond stretching and bond rotation leading to a reduction in secondary bonding and, to a lesser extent, primary bonding. Other influences such as

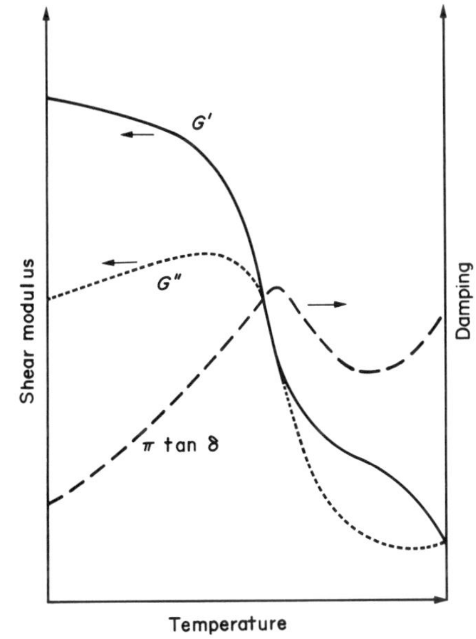

Figure 12
Dynamic mechanical properties as a function of temperature

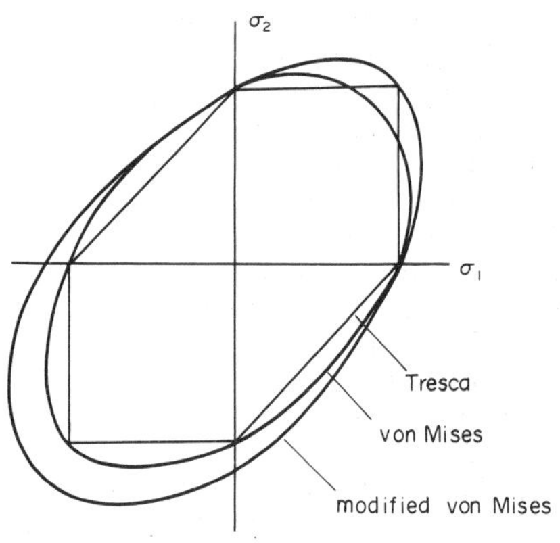

Figure 13
Yield criteria for Tresca, von Mises and pressure-modified von Mises models for biaxial stress conditions

chemical reactions or radiation usually sever primary bonds, thereby affecting the material properties.

Since most products are designed so as to avoid plastic deformation and failure, it is necessary to establish the yield point, where plastic deformation begins. This may be done by taking the value from a stress–strain curve. However, this assumes uniaxial tension and is insufficient for most cases where three-dimensional stresses need to be considered. In these cases a yield criterion is used. This is a general condition that must be satisfied for yield to occur. It may be defined in terms of strain or stress and is represented by a surface in space surrounding the origin. As stress is increased from the origin, yield does not occur until a point on the surface is reached.

The Tresca yield criterion is based on a simple concept which states that yield will occur when the resolved shear stress on any plane reaches a critical value. Figure 13 shows the yield criteria in a plane. The actual yield surface in three dimensions is a hexagonal prism.

A further developed model which works better for polymers is the von Mises yield criterion, which was originally developed because the discontinuities at the corners of the Tresca model were difficult to deal with mathematically. In this model, yield begins when the elastic shear strain energy density reaches a critical value. The three-dimensional yield surface in this case is cylindrical.

There are a number of more fully developed, more complex yield criteria. A model that fits real polymer data reasonably well is the modified von Mises model, which includes a correction for hydrostatic pressure.

The ability of the polymeric material to yield is an indication of ductility and usually implies that the polymer is tough. A brittle polymer will generally fail before the onset of yielding while a ductile polymer will yield and then elongate with little or no increase in stress. The amount of elongation is related to the ductility of the polymer (see Table 1). The interaction of bulk yielding and brittle fracture processes determines whether or not brittle fracture occurs. Figure 14 shows how each of these modes varies with temperature for a glassy polymer. Although the yield strength generally decreases with increasing temperature, the toughness increases significantly above the ductile–brittle transition.

Many polymers exhibit a transition from ductile or high-energy failure to brittle or low-energy failure as the temperature is lowered. Generally, this transition occurs just below T_g when T_g is measured at the frequency of testing, although some polymers

Table 1
Typical mechanical properties of polymers

Material	Tensile modulus (10^9 N m^{-2})	Tensile strength (10^6 N m^{-2}) Yield	Tensile strength (10^6 N m^{-2}) Ultimate	Elongation (%) Yield	Elongation (%) Break	Izo impact strength (kJ m^{-1} notch)
Acrylonitrile–butadiene–styrene	2.4	50	55	10	15	17.7
Acetal	3.2	55	68	10	50	10.2
Nylon 66	2.0	59	80	18	150	7.6
Phenolic	3.5		56		1	1.6
Polycarbonate	2.6	66	64	25	100	76
Polypropylene	1.35	32	34	12	400	7.6
Polyethylene						
high density	1.0	29	30	10	500	5–100
low density	0.2	7.5	10	20	1000	>100
Poly(methyl methacrylate)	3.1	65	65	10	10	2.1
Polystyrene	3.4	51	51	2	2	1.8
Poly(vinyl chloride)	2.5	49	50	3	29	5.1
Rubber (natural)			8		>1000	no break

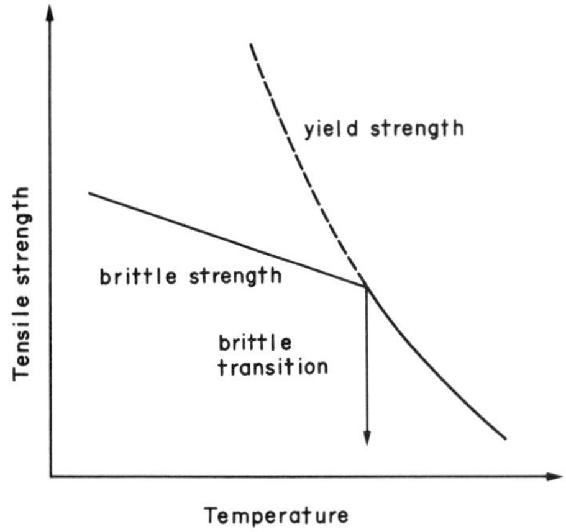

Figure 14
Resultant tensile strength from interaction of the brittle strength and the yield strength

such as polycarbonate have ductile–brittle transitions as much as 200 °C below T_g. Although there is some disagreement, this behavior is attributed to secondary relaxations below T_g. These secondary transitions are associated with very localized chain motions compared with the long-range segmental motions associated with T_g. Generally, in order to affect the ductile–brittle transition temperature, a secondary transition must be associated with a main chain motion.

The brittle nature of a polymeric material may be characterized by the use of fracture mechanics. Fracture mechanics theory presupposes the existence of a crack of length $2c$ subjected to a stress σ. Failure will occur when

$$\sigma^2 \pi c = K_c^2 \tag{6}$$

where K_c is the fracture toughness. An additional parameter is the energy release rate G_c necessary to initiate a crack:

$$G_c = K_c^2 / E \tag{7}$$

where E is Young's modulus. The values of K_c and G_c are dependent on sample thickness as the state of stress at the crack tip varies from plane stress in very thin sections to plane strain near the center of thick sections.

3.1 Mechanisms of Deformation

There are two different types of deformation in amorphous polymers. The deformation of brittle amorphous polymers is characterized by crazing, a voiding sprocess; ductile deformation is characterized by shear yielding.

Crazes are narrow zones of highly deformed and voided polymer resembling a crack. They are initiated in localized regions where inhomogeneities exist. As the stress near the craze tip increases, coiled molecules align in the direction of the applied stress and form fibrils connecting the sides of the craze. This allows an increase in thickness without lateral contraction of the material. As stress increases the craze grows perpendicular to the stress with constant thickness (usually 1–3 μm) until the fibrils break resulting in crack formation and fracture (see Fig. 15).

Crazing occurs to some extent in ductile amorphous plastics but deformation here is dominated by shear yielding. Shear yielding includes both diffuse shear yielding and localized shear band formation. These shear bands are thin planar regions of high shear strain formed by a cooperative movement of molecular segments of the chains. They allow the material to deform for some time without loss of intermolecular cohesion. The bands form at about 45° to the stress and are composed of highly oriented material.

The deformation of semicrystalline polymers containing inhomogeneous spherulites involves a more complex mode of deformation. The spherulite in uniaxial tension undergoes a transformation to a fibrillar structure. As the material is strained the crystalline lamellae undergo small-scale changes (phase changes, twinning, tilt) which allow the parallel packed chains to slip apart near the macroscopic yield point. The breakup of these lamellae and subsequent fibril formation then takes place as blocks of folded chains from individual lamellae are drawn into microfibrils. The fibrils have higher strength than the spherulitic material and contain folded chain blocks with chain segments and tie molecules oriented in the axial direction. The yielding phenomena in ductile amorphous and semicrystalline polymers absorb a considerable amount of energy, which accounts for the higher impact strengths of these materials.

3.2 Types of Failure

It is convenient to examine the mechanical behavior of polymers by observing stress–strain relationships.

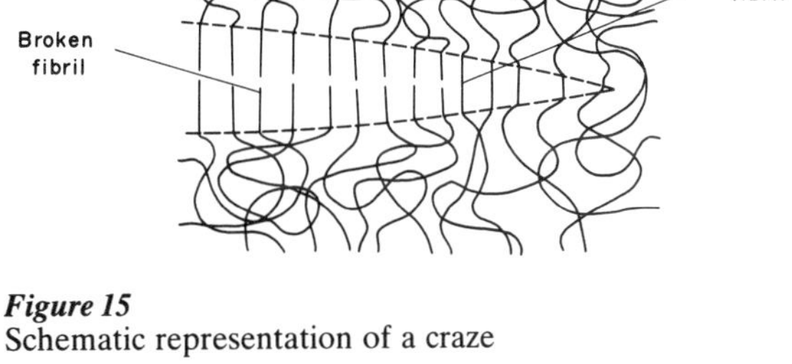

Figure 15
Schematic representation of a craze

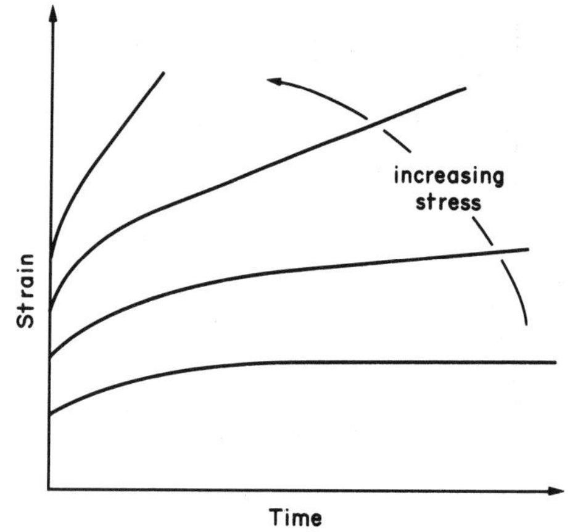

Figure 16
The effect of increasing stress on the creep properties

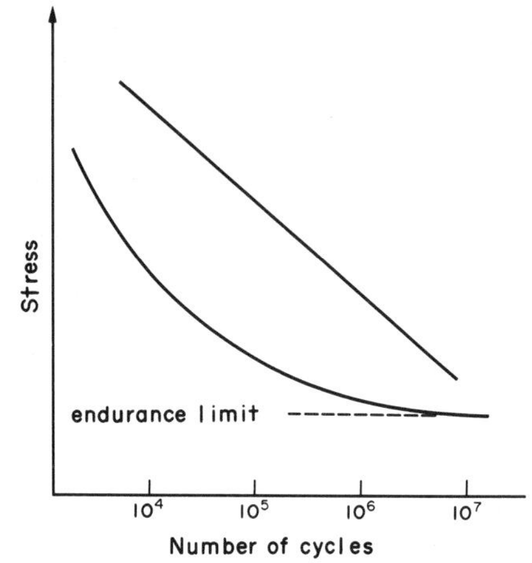

Figure 17
Fatigue life curves of typical polymers

Even though most mechanical failures of polymer products involve more than simple tension and compression, understanding the mechanisms involved in tension and compression contributes to the understanding of these other, more complex modes of failure. However, factors such as time, impact loading, repetitious loading and environment must also be considered.

The effect of a constant load over a long period of time (creep) is a very important design consideration. Creep basically involves the strain varying as a function of time for a given load. Initially the strain increases instantaneously by an amount determined by the modulus of the material. Then it gradually increases with time owing to the viscoelastic nature of the polymer, with a strain rate which decreases with time. The rate of increase is highly dependent on temperature and markedly increases as the temperature nears T_g. The effect of stress on the creep properties of a typical polymer can be seen in Fig. 16.

The response of polymeric materials to high rates of loading is related to the toughness of the polymer. The amount of energy needed to rupture the polymer in an impact loading is the impact energy, which is used as a measure of the toughness. As noted in the previous section, ductile yielding in response to such a loading absorbs substantial energy and is indicative of a tough material. Temperature is very important as many polymers undergo a ductile–brittle transtion near T_g. Sample geometry, if and how the sample is notched and the test method itself are all significant factors. Most impact tests use some form of swinging pendulum (as in the Izod impact strength) or falling weight to supply the impact energy.

Cyclic loading can often lead to fatigue failure at lower stresses than expected. Samples are usually tested while cycling the stress between zero and some fixed maximum value, although other cycles may be used. The number of cycles before failure is plotted for several stresses to obtain fatigue curves. The lower the applied stress, the greater the number of cycles that are needed to cause failure. Many polymers exhibit an endurance or fatigue limit: an applied stress below which failure will not occur in a measurable number of cycles. Temperature and cycle frequency have important effects on the fatigue properties. High cycle freqencies may cause a significant temperature rise in the polymer sample. A sufficient increase in temperature may lead to T_g or the melting point being reached resulting in softening of the material and rapid failure. Fatigue life curves for a polymer with and a polymer without an endurance limit are given in Fig. 17.

Other factors such as machining, abrasion and environmental effects (e.g., solvents, oxidation and radiation) also affect mechanical properties.

See also: Impact Strength

Bibliography

Andrews E H 1973 Cracking and crazing in polymeric glasses. In: Haward R N (ed.) 1973 *The Physics of Glassy Polymers*. Wiley, New York, pp. 394–453

Bowden P B 1973 The yield behaviour of glassy polymers. In: Haward R N (ed.) 1973 *The Physics of Glassy Polymers*. Wiley, New York, pp. 279–339

Bucknall C B 1977 *Toughened Plastics*. Applied Science, Barking, UK

Hartmann B, Lee G F 1979 Impact resistance and secondary transitions. *J. Appl. Polym. Sci.* 23: 3639–56

Hertzberg R W 1976 *Deformations and Fracture Mechanics of Engineering Materials*. Wiley, New York

Hertzberg R W, Manson J A 1980 *Failure of Engineering Plastics*. Academic Press, New York

Kaufman H S, Falcetta J A (eds.) 1977 *Introduction to Polymer Science and Technology*. Wiley, New York
Kramer O, Ferry J D 1978 Dynamic mechanical properties. In: Eirich F R (ed.) 1978 *Science and Technology of Rubber*. Academic Press, London, Chap. 5
Nielsen L E 1962 *Mechanical Properties of Polymers*. Reinhold, New York
Peterlin A 1977 Plastic deformation of crystalline polymers. *Polym. Eng. Sci.* 17: 183–93
Rodriguez F 1970 *Principles of Polymer Systems*. McGraw-Hill, New York
Ward I M 1971 *Mechanical Properties of Solid Polymers*. Wiley, New York
Williams J G 1977 Fracture mechanics of polymers. *Polym. Eng. Sci.* 17: 144–9

R. D. Corneliussen, T. Frederick and R. J. Gordon
[Drexel University, Philadelphia, Pennsylvania, USA]

Polymers: Technical Properties and Applications

The number and variety of applications in which polymers are used have increased at a phenomenal rate since the first totally synthetic polymer (bakelite) became available commercially at around the beginning of the twentieth century. Applications of the naturally occurring polymers, cellulose and rubber, can be traced back to the earliest history of mankind. Coupling the many applications of synthetic polymers with the historical use of the naturally occurring polymers, it is not surprising that these complex macromolecules now constitute an important class of materials which have become a major factor in improving the standard of living.

The remarkable versatility of polymers allows them to compete in many applications including transportation, construction, wiring, agriculture and medical devices. As decorative materials, polymers have a distinct advantage over competitive materials as a result of their low density, moderate cost, ease of pigmentation and the low energy required for their fabrication into finished products.

1. *Economic Considerations*

The growth of the synthetic polymer industry is illustrated in Fig. 1. Spurred by the need to develop synthetic rubber during World War II, the industry expanded rapidly and continues to grow. Table 1 shows production figures for the USA. Much of the growth can be attributed to the introduction of a wide variety of new polymers. Polymers are now classified as commercial, engineering and specialty types. Polymers produced in large quantity are classified as commercial or commodity products. Engineering polymers are those having exceptional mechanical strength which are therefore suitable for

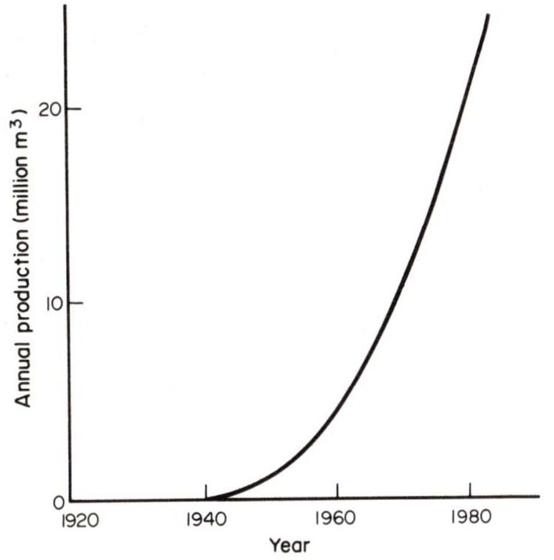

Figure 1
Growth of the synthetic polymer industry

applications in which strength is an essential requirement. Several engineering polymers are now produced in quantities comparable with the commodity polymers. Although not strictly an engineering application, packaging is a major application for synthetic polymers, consuming about half the annual production.

Since almost all synthetic polymers are made from fossil fuel, costs of these materials have increased sharply with the rise in cost of crude oil and natural gas. However, the use of inexpensive fillers as extenders has helped to hold down the costs of products made from polymers.

It is difficult to predict the point at which production of synthetic polymers will level off or when prices will stabilize. The industry is still in its infancy

Table 1
Polymer sales in the USA (10^3 t)

	1981	1990
Low-density polyethylene	3416	5404
High-density polyethylene	2129	3870
Polypropylene and copolymers	1766	3700
Poly(vinyl chloride) and copolymers	2548	4230
Polystyrene	1638	2337
ABS resins	441	551
Phenolic resins	1048	1286
Saturated polyesters	294	1063
Unsaturated polyesters	450	558
Polyurethane foams	785	1486
Polycarbonate	109	282
Epoxy resins	150	211
Acrylics	255	342

Source: *Modern Plastics*, January 1983; *Modern Plastics International*, January 1991

and further development of new polymers can be expected. In engineering applications, the introduction of reinforcing fillers has enabled polymers to compete effectively with metals in certain areas. Composites using carbon fibers are a typical example (see *Carbon-Fiber-Reinforced Plastics*). Advances in reinforcement and induced orientation of fibrous fillers may be expected to widen still further the scope of applications for polymer composites.

Interest is growing in the use of polymer blends and laminate structures. Blending of two or more different polymers aims to take advantage of the unique properties of each component. For example, poly(methyl methacrylate) (PMMA) is used as a glazing material despite its vulnerability to abrasion. Lamination of an abrasion-resistant polymer over the surface would allow PMMA to compete more effectively with conventional glazing materials.

As the technology for compatibilization of polymer blends advances, these composites may find much broader application. Combinations of polymers with other materials will also contribute to growth of the synthetic polymer industry. The lightweight but exceptionally strong polymer concretes are an example. Glass-reinforced polymers are widely used in the automotive and other industries.

2. Properties Influencing Engineering Applications

Several key properties are important to the suitability of polymers for engineering applications. A brief summary is presented here with emphasis on comparison of corresponding properties with other, competitive materials.

2.1 Density and Ultimate Strength

The low density of polymers provides an important advantage over metals and ceramics in many applications (see *Polymers: Mechanical Properties*). The density of most engineering polymers lies between 0.9 and 1.9. These values compare with a range of 2.7–3.0 for the lightweight metals and 7.8–8.0 for steel. Therefore, polymers are very useful in those applications in which the weight–volume ratio is critical. The ultimate strength of polymers, on the other hand, does not compare favorably with that of metals. For example, polyesters have a Young's modulus in tension in the region of 3 GPa, in contrast to 200 GPa for steel and 10 GPa for wood. It should be noted, however, that some of the new polymer composites, such as carbon-fiber-reinforced polyimides, have strengths well within the competitive range. Because of the trade-off between low density and high strength, polymers are usually materials of choice when very high strength is not required but where advantage can be taken of their low density and ease of fabrication.

2.2 Morphology

Many polymers exist in a semicrystalline state. This complex morphology has the advantage of strength in the ordered crystalline regions and flexibility in the disordered amorphous regions. Thus, polymers find application when strength must be combined with flexibility, as in fibers, electrical insulation and some pipe applications. Orientation at the macro level has been used effectively to maximize strength in polymers. Monofilaments owe their great strength per unit cross-sectional area to orientation induced in the drawing process. The single thread spun by a silkworm has a tensile strength comparable with that of metals of the same cross-sectional area. In general, polymers exhibit much higher strength in tension than in compression.

2.3 Dielectric Properties

Polymers as a class have little competition in nonrigid dielectric applications. Their strength and flexibility provide an ideal combination of properties for wire insulation and cable jacketing. Being less fragile than ceramics, polymers also have the advantage of resistance to breakage when used as rigid insulators. However, the electrical breakdown strength of polymers does not compare favorably with that of ceramics, so they are less widely used in high-voltage applications. Polymers are ideal materials for the manufacture of printed circuit boards. Their ease of fabrication and flexibility are important factors favoring the use of polymers in these applications.

Polymers have excellent dielectric properties. Polyethylene has a dielectric constant about twice that of dry air at 1000 Hz. Polytetrafluoroethylene has an even lower value, while poly(vinyl chloride) represents a compromise with respect to resistance to burning; it has a dielectric constant between 4 and 8, depending on the level of plasticization. Both the dielectric constant and the surface resistivity of polymers are influenced by moisture. Since most polymers are hydrophobic, moisture absorption must be attributed to additives such as the fillers used in thermosetting resins. Large amounts of fillers such as wood fibers or cotton are responsible for dielectric losses in filled phenol resins. For use in dielectric applications, a water-resistant filler such as mica is preferred.

2.4 Processability

Polymers are fabricated into final products by a variety of processes including calendering, extrusion and a number of molding techniques. Because of their low melting temperatures, polymers can be formed into useful shapes at temperatures in the region of 200 °C. In contrast, if melting is required, metals must be fabricated at much higher temperatures.

Despite the fact that almost all synthetic polymers use fossil fuel as feedstock, the overall energy requirement for fabricating the same article from a polymer is still much less than from metals. It has been estimated that 360 kJ would be required to fabricate a 0.35 kg container from polyethylene, whereas 10 800 kJ would be needed to form a similar container from aluminum.

Polymers have a further advantage in fabrication, as a result of the ease with which their melts flow. Thus, it is relatively easy to fill complex and intricate molds to form products with a high degree of detail.

Reaction injection molding is a new fabrication technique for polymers in which monomers are polymerized within the mold. Relatively lightweight molds can be used. Since monomers flow quite freely, considerable detail can be obtained in the final polymeric product. This process is expected to lead to an extended use of polymers in automobiles, where they contribute to weight reduction and increased mileage.

2.5 Resistance to Deterioration

All polymers degrade when exposed to high temperatures or ultraviolet radiation, although there is considerable variation between individual polymers. However, major advances have been made in the stabilization of polymers, so the required service life is now attainable in many applications. Polyethylene, for example, can be stabilized to resist outdoor weathering for more than 50 years by addition of carbon black and an effective thermal antioxidant.

Those polymers that are formed by addition reactions are very resistant to hydrolytic breakdown. In contrast to many metals which corrode on exposure to moisture, this class of polymers is ideally suited for applications in which high humidity or moisture is present. Synthetic polymers are also very resistant to biodegradation.

2.6 Reuse of Scrap Polymeric Material

The difficulty of reusing polymer scrap is a serious restriction on the expanding application of these materials. There are well-established processes for recycling metals and ceramics, but the recycling of polymer scrap has not developed to a comparable extent. The higher densities of metals and ceramics makes it easier to separate these materials from other components of an industrial or municipal waste stream. Polymer scrap, on the other hand, is difficult to recover once it has been mixed into a waste stream. There are, however, several successful processes for the recycling of source-separated polymeric scrap.

High temperatures can be used in recycling metals and ceramics. Because of their lower decomposition temperatures, polymers are likely to decompose under the severe conditions required for recycling. In many instances, clean polymer scrap is recycled in fabrication processes. However, the danger of decomposition must always be taken into account.

3. Applications for Engineering Polymers

Although the use of polymers in packaging is not generally considered to be an engineering application, protective coating and encapsulation are important in certain engineering designs or processes. Furthermore, the volume of polymeric materials used for packaging is large.

3.1 Polymers in Packaging

Lower cost and ease of fabrication account for the increased use of polymer protective coatings in engineering applications. Foamed cushioning to protect instruments during shipping and encapsulation of solar cells are important applications. The preservation of fresh and frozen food products by polymeric materials has been responsible for extending the food supply by avoiding contamination and spoilage. The impermeable nature of polymer films tends to exclude both moisture and microorganisms.

Polymers are finding increasing application in the manufacutre of beverage bottles, where they have an advantage over glass in resistance to breakage as well as lighter weight. When polymeric materials are used for packaging food or medicine, it is very important that no additives or residual monomer are present which could be absorbed by the contents of the container. A further problem with polymers in beverage containers is the difficulty encountered in washing and sterilizing plastic containers. The opacity of many plastic bottles prevents adequate visual inspection after washing. Moreover, most polymeric materials would distort at sterilization temperatures.

Many household chemicals are packaged in containers made from polymers. In these applications, resistance to breakage is an important advantage when the contents are corrosive or toxic.

Disposal of waste products in plastic bags is another important application of polymers in packaging. Plastic garbage bags have the advantage of light weight, strength and the capability to retain odors as a deterrent to rodents. Plastic garbage cans are not only lighter in weight than those fabricated from metals but they also resist damage effectively, when the correct polymer is selected. Leaf bags of thin polymeric films are also popular because of the ease of disposal.

3.2 Polymers in Construction

Polymers are used in a wide variety of construction applications where advantage can be taken of their excellent insulation capability (both thermal and electrical), low weight, design flexibility and ease of pigmentation. Even in applications where moderate strength is required, polymers are finding wide use. Such items as extruded gutters, siding, imitation wood beams, room dividers and window or door frames are growing in popularity. Since colorants can be dispersed throughout the polymer mass, repeated painting is not required. Resistance to biodegradation also favors polymers in those applications where attack by organisms is likely to occur.

Poly(vinyl chloride) (PVC) is used widely as siding for houses, competing with wood and aluminum. Color permanence, resistance to impact, insulating value and relatively low cost are all in favor of polymers in this application. PVC can be extruded as siding in long, uniform lengths as required and either applied directly over sheathing in new construction or over deteriorated wood siding. Installation is at least as simple as for competitive materials. Resilience of PVC siding minimizes damage by impact, a distinct advantage over aluminum siding. Stability to biodegradation, especially in humid areas, is a further advantage of PVC siding.

Pipe fabricated from PVC, ABS molding resin, high-density polyethylene and other polymers is popular with "do-it-yourself" homeowners. Perforated drainage pipe is not as fragile as ceramic tile used for this purpose and long lengths can be extruded easily. Cutting into desired lengths is easy and joining is relatively simple. However, in those applications in which the pipe must withstand high pressure, metal pipe is superior.

Polyolefins, paper and rubber are used as wiring insulation, but PVC is favored because of its greater resistance to burning. Codes covering inside wiring often restrict the use of polymers, but new polymers have been developed which provide outstanding resistance to burning. These new materials belong to the fluorocarbon class. Because of their improved flame resistance, these materials compete effectively as insulation for inside wiring, particularly in aircraft and marine construction, where weight is an important factor.

Decorative items constitute a major application area for polymers in construction. Decorative panels, room dividers, light diffusers, imitation beams, doors, moldings, various hardware items, bathroom fixtures and other components are easily fabricated from polymers. Only moderate strength is required in most of these applications and attractive colors can be obtained. Ceiling panels are fabricated from a variety of polymers or a moisture-resistant polymer can be used as a protective film over conventional ceiling tiles. Bathroom wall tiles made from polymers have the advantage that they are not cold to the touch.

3.3 Polymers in Transportation

Several factors have contributed to the expanded use of polymers in transportation. Their basic property of low density provides the main reason for most of these applications. This property has always been very important in aircraft construction. More recently, the use of light polymers in automobile design has attracted considerable attention as one method for reducing fuel consumption. Marine applications, particularly in hull construction, take advantage of the resistance of polymers to attack by marine organisms.

Polymers are well adapted to the fabrication of complex decorative parts in automobiles, both exterior and interior. Polymer foams are used extensively as seat cushions and in impact-resistant dashboards. Ductwork for heaters and air-conditioning units is fabricated from polymers, as are most under-the-hood containers. Battery cases are now made almost entirely of polypropylene, which has adequate strength for this application while providing a significant weight advantage over bakelite or hard rubber. As reaction-injection-molding technology expands in the automotive and other industries, thermosetting polymers are expected to become a major fraction of the polymers used in automobile construction. Complete body sections are envisioned and longer-range predictions include drive shafts and even engine blocks with metal inserts.

Polymers as adhesives are widely used in aircraft construction. Laminated structural units are especially adaptable to small aircraft. Thermoforming of polymers is ideally suited to the fabrication of hulls for small boats. Larger hulls have been made using glass-reinforced thermosetting polymers. Though barnacles may attach themselves to plastic hulls, they do not degrade the material and are easily removed.

3.4 Polymers in the Communications Industry

The introduction of polyethylene in 1942 revolutionized the construction of wire and cable. Before the discovery of this synthetic polymer, communication cable used lead as the outer protective cover. Replacement of lead by polyethylene greatly reduced the weight while still providing adequate strength and abrasion resistance. Cost savings were substantial and sheath fabrication was greatly simplified. Although the original polyethylene was rapidly degraded by heat and ultraviolet radiation, modern formulations of this polymer provide more than adequate service life.

The introduction of polyethylene and other synthetic polymers as a replacement for paper insulation on primary conductors has provided many improvements in cable construction. All of the synthetics have very good insulating properties and in addition they are hydrophobic. Several synthetic polymers are now used as expanded insulation to reduce the weight still further while providing even better dielectric quality. The semicrystalline morphology of most polymers used in this application provides the desired properties of strength and flexibility. Even in the advanced technology of transmission through optical fibers, a polymeric coating must be applied immediately to the glass fibers, to prevent development of surface cracks which would lead to rapid embrittlement of the glass.

3.5 Polymers in Biomedical Applications

Although the volume used is small, polymers have assumed an important role in medical applications. In most of these applications, polymers have little or no competition from other types of materials. Their unique properties of flexibility, resistance to biochemical attack and low weight make polymers the primary, if not the only, choice for the following applications.

From the familiar use of plastic tubing for intravenous infusion, biomedical applications of polymers have now been extended into complex and sophisticated techniques. A wide variety of products fabricated from polymers are used as surgical implants. The inert nature of these materials reduces the possibility of rejection by the host which is often encountered with implants of living tissue. The artificial hip joint has been used successfully for many years to restore mobility to arthritic patients.

Synthetic polymers are also fabricated into thin-wall tubing as replacements in the circulatory system. The value of breakage-resistant containers for medicine and whole blood is very important in these applications. The transport of blood plasma to soldiers in combat zones by poly(vinyl chloride) bags has been responsible for the saving of many lives. The artificial heart, which is receiving much attention, can be expected to make use of synthetic polymers in its final design. The list of biomedical products fabricated from polymers is certain to expand as medicine advances.

See also: Automotive Applications of Polymers; Degradation and Stabilization, Ultraviolet; Design with Plastics; Plastics and Composites as Building Materials; Plastics for Packaging; Processing: Classification; Wire and Cable Systems for Underground Transmission

Bibliography

Hall C 1981 *Polymer Materials: An Introduction for Technologists and Scientists.* Wiley, New York

Kaufman H S, Falcetta J J 1977 *Introduction to Polymer Science and Technology*. Wiley, New York

Kaufman M 1968 *Giant Molecules: The Technology of Plastics, Fibers and Rubber*. Doubleday, Garden City, New York

Patton W J 1976 *Plastics Technology*. Reston Publishing, Reston, VA

Rogers D C 1964 Applications. In: Raff R A V, Doak K W (eds.) 1964 *Crystalline Olefin Polymers*. Interscience, New York

Seymour R B 1975 *Modern Plastics Technology*. Reston Publishing, Reston, VA

Williams H L 1975 *Polymer Engineering*. Elsevier, Amsterdam

W. L. Hawkins
[Montclair, New Jersey, USA]

Polymers: Tests for Mechanical Properties

Polymers are used to an increasing extent for applications in which they replace traditional materials of construction, such as metals and wood, where their lower weight, resistance to corrosion, ease of fabriction and potentially lower cost offer significant advantages. In addition, newer polymers are replacing older polymeric systems; for example, engineering thermoplastics are replacing thermosetting polymers, in applications where economy of manufacture and design latitude favor thermoplastic polymers.

Since there are many areas where polymers are being considered and there are literally hundreds of polymers to choose from, including blends, alloys and composites, it is necessary to understand the mechanical properties of polymers to be able to design functional items which can withstand the expected stresses in use. Typical properties of interst to the designer or engineer faced with selection and specification of polymers are tensile, flexural, impact, creep and fatigue stress relaxation as well as compressive and shear properties (see *Polymers: Mechanical Properties*).

Most technical data available to the engineer to assist in making a choice of materials are found in technical literature supplied by vendors. These are generally useful for making preliminary selections of potential materials but do not help when several different materials are under consideration. This is because the test conditions and sample geometry often vary from one material to another. Also, basic material property data of this type do not represent conditions of use.

This article presents the basic steps in determining the mechanical properties of polymers and suggested techniques for evaluating the properties in relation to the application. In the latter case, as much information as possible regarding the service conditions of the fabricated part must be considered when selecting the material. Conditions of loading, temperature, chemical exposure (if any) and expected service life must all be included.

Basic to any determination of mechanical properties is a review of the design parameters to establish the required properties. It must also be established whether specific test procedures are prescribed in the purchase specifications. If so, the procedures will most probably be those laid down by the various standards organizations. These procedures define the test conditions, preferred sample geometry and sample preparation and conditioning methods, and also specify the rcommended procedures to be used when samples are taken from fabricated parts.

The test procedures described here relate to the basic properties employed in design and use of polymers.

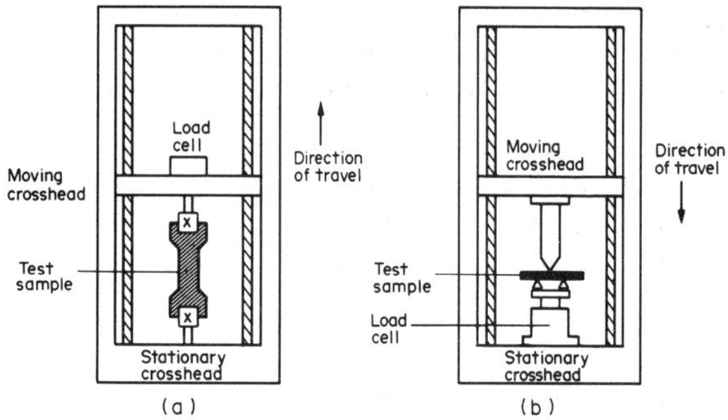

Figure 1
Schematic arrangements for determination of mechanical properties of polymers: (a) tensile and (b) flexural

1. Tensile Properties

The tensile behavior of a polymer is probably the most fundamental mechanical property used in the evaluation of polymers. The basic method is employed to determine yield strength, ultimate strength and elongation properties as well as various moduli which are important in designing with polymers. The stress–strain diagram obtained can also be used to measure the ability of a polymer to absorb energy. This is particularly useful when testing vulcanized elastomers.

The tensile strength of a material is the ratio of the force applied to the material at rupture to its original cross-sectional area. This property is typically called the ultimate tensile strength, that is, the strength at break, and is usually expressed in megapascals. The corresponding extension is called the ultimate strain or ultimate elongation and is typically given as a percentage.

The equipment required consists of a machine having a moving and a fixed crosshead to which clamping fixtures are affixed to hold the specimen. One of the clamps is attached to a load cell which may be mounted on either crosshead. The movable crosshead can be driven at a constant rate which can be varied from 0.25 mm min^{-1} to 1250 mm min^{-1} or sometimes up to 25 000 mm min^{-1}. Other types of equipment are able to apply a constant rate of loading to a specimen.

The load or resistance of the specimen to deformation is continuously recorded and the extension determined by use of an extensometer connected directly to the specimen or by measuring the distance between gauge marks on the specimen during the progress of the test. To prevent failure of the sample in the grips, the test specimens usually have a reduced cross section and are sometimes referred to as "dumb-bell" or "dog-bone" specimens. Figure 1a illustrates the specimen loaded in tension. The force is recorded up to the point of rupture. These readings are converted to stresses and are plotted against strain. Figure 2 illustrates typical stress–strain diagrams for the major types of polymer.

As stated, the tensile strength is the ratio of force to unit area and is expressed as follows:

$$\sigma = F/A \tag{1}$$

where force F is expressed in newtons and the cross-sectional area A is in square meters. Strain is given by the following expression:

$$\varepsilon = (l - l_0)/l_0 \tag{2}$$

where l is the final length and l_0 is the original length. Typically, strain is reported as percent strain or percent elongation, which is 100ε. The properties calculated as shown are referred to as the engineering stress and strain or elongation. These values are shown as the ultimate properties or breaking stress and strain.

Features to be noted are yielding, necking and strain hardening. Semicrystalline polymers such as polyethylene and polypropylene show yielding, followed by necking, then an increase in strength before rupture, as in Fig. 2a. Figure 2b shows a

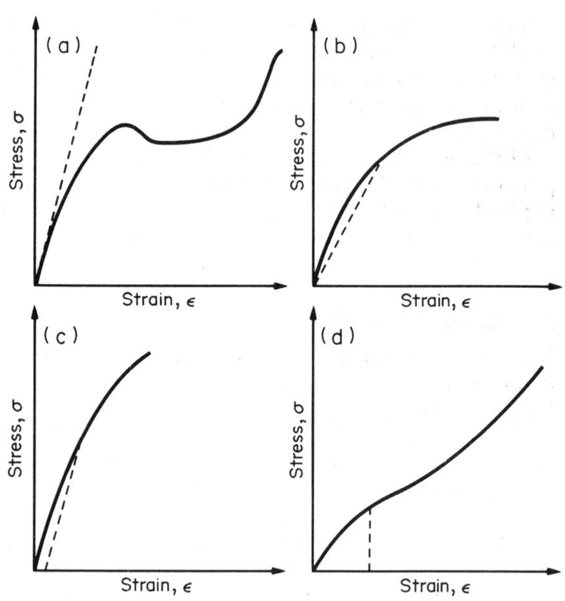

Figure 2
Tensile curves for polymeric materials: (a) semicrystalline polymer (—— stress–strain curve, - - - tangent modulus of elasticity); (b) amorphous ductile polymer (—— stress–strain curve, - - - secant modulus of elasticity); (c) amorphous brittle polymer (—— stress–strain curve, - - - offset modulus); and (d) crosslinked rubber (—— stress–strain curve, - - - 300% modulus)

ductile polymer such as plasticized poly(vinyl chloride) which shows yielding and elongation before failure. Figure 2c shows the behavior of a brittle polymer such as polystyrene; the stress–strain curve shows no yield point and low elongation before failure. Crosslinked rubbers behave behave as in Fig. 2d.

The modulus may also be obtained from the stress–strain or load–deformation curve. The modulus value most frequently used in design is Young's modulus E, which is determined from the initial slope of the stress–strain curve:

$$E=\sigma/\varepsilon \tag{3}$$

The dashed line in Fig. 2a is an extrapolation of the initial linear portion of the curve for the calculation of E.

Other modulus values commonly reported are the secant modulus, tangent modulus and, in the case of elastomers, the 300% modulus; the stress at 300% elongation is a commonly reported property. These are shown by a dashed line in Figs. 2b, 2c and 2d, respectively. The relationship between the shear modulus and tensile (Young's) modulus is given by

$$E=2G(1+\nu) \tag{4}$$

where E is the tensile modulus, G is the shear modulus and ν is Poisson's ratio. This relation can be used to determine the shear modulus from tensile measurements when Poisson's ratio is known for the material under test.

Modulus values are useful when designing with polymers, but in using the data obtained by these methods, caution must be exercised that they are not extrapolated to extreme conditions of service.

2. *Flexural Properties*

Many uses of polymers involve their subjection to bending or flexural stresses. In these cases, knowledge of the response of the material to such conditions is required. One of the most useful methods of determining the flexural properties of polymers employs the same basic equipment as used to determine the tensile properties. This technique makes use of a specially constructed fixture which substitutes the flexural test fixture for the grips used in tensile testing. Another convenient method is to use the tensile machine in the compression mode: the tensile load cell is replaced by a compressive load cell and the movable crosshead is moved towards the stationary crosshead. A schematic diagram of this test arragement is shown in Fig. 1b in the three-point bending configuration.

A typical load–deformation curve for a polymer tested in flexure is similar to the initial portion of a tensile test, as shown in Fig. 2. These curves can be used to determine flexural yield strength as well as the tangent or secant modulus, as illustrated in Figs. 2a, b.

The rate of straining the outer fibers is usually $0.01\ s^{-1}$. To obtain this strain rate, the crosshead speed varies with the support span L and the depth or thickness of the sample d as given by

$$R=\dot{\varepsilon}L^2/6d \tag{5}$$

where $\dot{\varepsilon}$ is the strain rate and R is the crosshead speed. The load is applied until the sample fails or the strain reaches 0.05 in the outer fibers. The deflection at this latter point is calculated as

$$D=0.05L^2/6d \tag{6}$$

Higher strain rates may be used to cause rupture before the 0.05 limit or to determine the flexural properties under loading conditions approximating those expected in use.

The maximum stress under three-point loading conditions occurs at the midpoint and can be calculated as

$$\sigma_B=3PL/2bd^2 \tag{7}$$

where σ_B is the bending stress (Pa), P is the force (N), L is the span length (m), b is the sample width (m) and d is the thickness (m).

The modulus of elasticity in bending E_B is calculated, as stated, from the load deformation curve by constructing the tangent to the initial straight portion of the curve and determining the slope m of the tangent ($N\ m^{-1}$):

$$E_B=L^3m/4bd^3 \tag{8}$$

The secant modulus of elasticity in bending may also be calculated for any specified strain by constructing a secant from the origin to the selected strain, determining the slope m and applying Eqn. (8).

In the case of high strain rates, some slippage may occur at the supports, requiring modified test procedures. Similarly, large span–depth ratios require correction factors.

Flexural properties are often useful in the design and selection of materials, particularly when the fabricated part is subjected to bending stresses. It should be noted here that the modulus of elasticity E determined in tension is not equivalent to the modulus of elasticity E_B determined in flexure.

3. *Compressive Strength*

Compressive strength data are useful for predicting the service life of polymeric devices which are used in load-bearing applications. The compressive yield stress, strain and modulus of elasticity as well as ultimate compressive strength may be determined for materials subject to brittle failure.

The machine used is a tensile testing machine set up in the compression mode. Special test fixtures are required to provide vertical support for thin specimens. The testing arrangement is similar to that used for flexural properties, as shown in Fig. 1b, with the three-point bending fixture replaced by a fixture to support the specimen for compressive loading. The load–deformation curve from which the stress–strain curve is obtained is similar to that for tensile and flexural tests at low levels of strain (Fig. 2). The compressive stress, strain and modulus are calculated as for the tensile properties, by Eqns. (1), (2) and (4), respectively.

4. Time-Dependent Properties

Creep and stress relaxation are important properties of polymers, since they must be considered in the design of any fabricated part subjected to constant load or deformation. The influence of the environment to which the part is exposed must also be taken into consideration, including not only the variations in temperature but also potential exposure to chemical agents which might accelerate fracture in service.

Creep occurs when the polymer is subjected to a constant stress below the yield stress or fracture stress of a material. After an initial deformation ε_0 which is proportional to the applied stress σ, the material continues to elongate or deform under the applied load until fracture occurs. Fracture may be either ductile or brittle. The strain at any time is a function of the original strain which occurs upon loading plus the time-dependent component of strain. The strain depends on the level of stress and the temperature at which the test is carried out. A general expression for time-dependent strain is

$$\varepsilon(t) = \varepsilon_0(\sigma) + F[t^n g(\sigma)] \tag{9}$$

where $\varepsilon(t)$ is the strain at time t resulting from applied stress σ, ε_0 is the initial strain resulting from application of σ, and $F[t^n g(\sigma)]$ is the time-dependent component of strain which is the incremental increase in strain at time t resulting from the applied stress σ. This concept is discussed by Levy (1979).

Creep measurements are made by applying a static load to the sample and recording the deformation as a function of time. The data are then used to construct a creep curve. It is useful to conduct the test under several different loads and plot the data for each test condition. A typical series of creep curves is shown in Fig. 3a ($\sigma_5 > \sigma_4 > \sigma_3 > \sigma_2 > \sigma_1$). These determinations are made at constant temperature so that the strain is a function of stress σ and time h (hours) at the test temperature T.

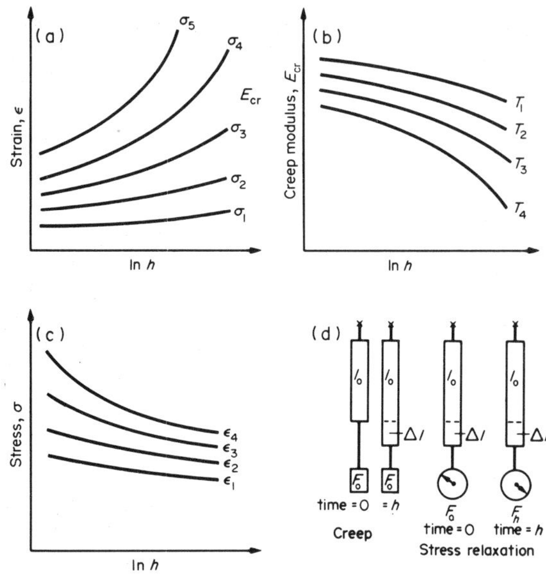

Figure 3
Time-dependent properties: (a) creep curves, strain vs time (log hours) under load σ_i; (b) creep modulus time (log hours) at temperatures T_i; (c) stress relaxation, stress time (log hours) under deformation ε_i; and (d) schematic representations of creep and stress relaxation

Measurements made at several temperatures, including those expected in practice, allow service life to be predicted, taking into account the variation of creep with temperature as well as load.

For design purposes the modulus is used to predict deformation under applied stress; however, since it has been shown that the strain increases with time, it follows that the modulus also changes with time. This change is determined by dividing the applied stress σ by the strain at time h and plotting the modulus E_{cr} against time. The modulus value is called the creep modulus. Figure 3b illustrates this property for a series of temperatures, where $T_4 > T_3 > T_2 > T_1$.

Stress relaxation is the decrease in stress with time for a material subjected to constant deformation. It is related to creep. In this case the sample is strained to a predetermined level and the reduction in force is measured as a function of time. A stress relaxation curve is shown in Fig. 3c. In this case the stress σ is plotted against time (ln h) for a series of strain levels ε, where $\varepsilon_4 > \varepsilon_3 > \varepsilon_2 > \varepsilon_1$.

A schematic arrangement for the determination of creep and stress relaxation is shown in Fig. 3d. In this example the creep $\Delta\varepsilon$ is the increase in length at time h caused by force F_0. The stress relaxation indicates a decrese in stress, $F_0 > F_h$, at time h resulting from a constant deflection $\Delta\varepsilon$.

Further discussion on time-dependent properties is to be found in Nelson (1974a, b), Turner (1974), Levy (1979) and Ward (1983).

5. Impact Properties

The impact behavior of polymers is a cause for concern among designers because the performance of fabricated parts subjected to impact loads or to deformation at high strain rates is often difficult to predict. It is also a complicated matter to develop tests which can accurately model the actual use conditions for the material under consideration. In many cases failure occurs at impact loads far below the inherent strength or toughness of the material. These problems have led to a large number of methods for determining the impact properties of polymers and likewise a large number of machines of increasing sophistication to measure the behavior of polymers subjected to impact loading conditions. Instrumented impact testing machines have become popular in recent years because more information is generated regarding the mode of failure. These techniques have been applied to tensile impact, Charpy impact and falling weight impact tests, among others.

In the instrumented impact tests, a load–time (deformation) curve is generated as well as an indication of the energy required to cause failure. A typical curve from an instrumented Charpy test is shown in Fig. 4a. The maximum force to failure, which is shown as P_m, can be used to calculate the fracture toughness K_{Ic} for the material:

$$K_{Ic} = 1.5\, Y P_m L a^{1/2} / B W^2 \tag{10}$$

where L, a, W and B are the dimensions (m) of the specimen shown in fig. 4c and Y is a geometrical factor related to the notch depth–thickness ratio a/W. The units of K_{Ic} are MPa m$^{1/2}$. The impact strength (kJ m^{-2}) is calculated from the failure energy E_a by

$$I_s = E_a / B(W-a) \tag{11}$$

Figure 4d shows a tensile impact sample in a typical test fixture. Tensile impact strength is also calculated from the energy of failure and cross-sectional area of the specimen

$$I_t = E_a / Wt \tag{12}$$

where W is the width and t the thickness of the specimen. This quantity also has units of kJ m^{-2}.

Tensile impact and Charpy impact strengths are determined on pendulum-type impact machines with test fixtures appropriate for the specific method. A typical arrangement for the Charpy test is shown in Fig. 4b. Many types of falling weight impact tests are also available, depending on the type of specimen and test data required. Data may be reported in various forms, such as kilojoules per meter, kilojoules for 50% failure and failure height in meters for a specified drop weight.

Other properties of interest to designers are the behavior of polymers in friction, wear and fatigue, as well as the dynamic mechanical behavior. Test methods used to determine these properties are described in the references cited in the bibliography.

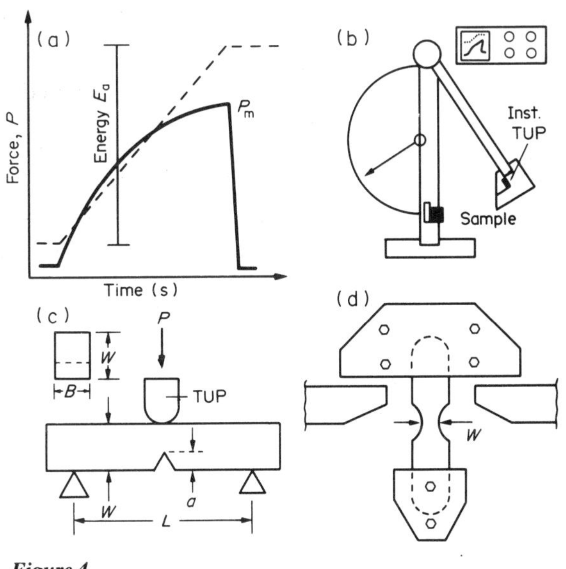

Figure 4
Impact properties and testing: (a) curves for force (——) and energy (- - -) from instrumented impact test; (b) schematic of pendulum-type impact tester; (c) specimen Charpy impact test; and (d) specimen configuration and loading arrangement for tensile impact test

6. Data Handling

Advances in computer technology, such as the widespread availability of minicomputers and microcomputers, have led to a rapid growth in automated data handling in the laboratory. It is now possible to intereface many pieces of equipment to a single computer so that data is collected from each simultaneously. Software is available to calculate such properties as tensile strength, elastic modulus, flexural modulus and impact properties, among others. The data can be analyzed statistically and curves generated through the use of printer–plotter equipment.

These systems have improved data handling and have reduced the materials characterization required. Further discussions of basic systems may be found in Chudzicki and Peake (1984) and Holland (1985).

See also: Impact Strength

Bibliography

American Society for Testing and Materials 1977 *Standard Test Method for Tensile, Compressive and Flexural Creep and Creep-Rupture of Plastics*, ASTM Designation D-2990-77. ASTM, Philadelphia, PA

American Society for Testing and Materials 1979 *Standard Test Method for Tensile-Impact Energy to Break Plastics and Electrical Insulating Materials,* ASTM Designation D-1822-79. ASTM, Philadelphia, PA

American Society for Testing and Materials 1980a *Standard Test Method for Compressive Properties of Rigid Plastics,* ASTM Designation D-695-80. ASTM, Philadelphia, PA

American Society for Testing and Materials 1980b *Standard Test Method for Rubber Properties in Tension*, ASTM Designation A-412-80. ASTM, Philadelphia, PA

American Society for Testing and Materials 1981a *Standard Test Methods for Flexural Properties of Unreinforced and Reinforced Plastics and Electrical Insulating Materials,* ASTM Designation D-790-81. ASTM, Philadelphia, PA

American Society for Testing and Materials 1981b *Standard Test Methods for Impact Resistance of Plastics and Electrical Insulating Materials,* ASTM Designation D-256-81. ASTM, Philadelphia, PA

American Society for Testing and Materials 1982 *Standard Test Method for Tensile Properties of Plastics,* ASTM, Designation D-638-82a. ASTM, Philadelphia, PA

Brown R P 1986 *Physical Testing of Rubber,* 2nd edn. Elsevier, Amsterdam

Carlowtz B 1972 *Tabellarische Ubersicht über die Prüfung von Kunststoffen*, 4th edn. Umschau, Frankfurt, Germany, pp. 6–23; 26–31

Chudzicki M J, Peake R J 1984 Unified approach to computer-aided polymer characterization. *Plastics in a World Economy, Proc. SPE 42nd Annual Technical Conf.* Society of Plastics Engineers, Greenwich, CT, p. 342

Gent A N 1978 Strength of elastomers. In: Eirich F R (ed.) 1978 *Science and Technology of Rubber.* Academic Press, New York, Chap. 10

Holland L B 1985 Materials evaluation laboratory automation. *Am. Lab.* January: 42

Levy S 1979 Projecting short-term data into long-term creep. *Plastics Design Forum.* Industry Media, Denver, CO, pp. 38–45

Nelson L E 1974a *Mechanical Properties of Polymers and Composites*. Dekker, New York, Chaps. 1, 3

Nelson L E 1974b *Mechanical Properties of Polymers and Composites*. Dekker, New York, Chap. 5

Turner S 1973 *Mechanical Testing of Plastics,* 2nd edn. Longman, New York

Turner S 1974 Deformational behaviour. In: Ogorkiewicz R M (ed.) 1974 *Thermoplastics: Properties and Design*. Wiley, London, Chap. 3

Vincent P I 1974 Short term strength and impact behaviour. In: Ogorkiewicz R M (ed.) 1974 *Thermoplastics: Properties and Design*. Wiley, London, Chap. 5

Ward I M 1983 *Mechanical Properties of Solid Polymers,* 2nd edn. Wiley, Chichester, UK, Chaps. 5–7

E. V. Lind
[Container Corporation of America,
Wilmington, Delaware, USA]

Powdered and Particulate Elastomer Processing

In powdered and particulate elastomer processing, elastomers and all compounding ingredients are in powder form and liquids are directly incorporated. Powder implies that the maximum particle size of the elastomer is 1 mm. Processing of elastomers with larger particle sizes (1–12 mm) is termed particulate technology and is very similar in principle to the powder methods. Some exceptions exist, notably in the compactor route, where powder-size particles (<1 mm) are necessary. The chosen route determines the particle size of the elastomer and is very important in obtaining maximum benefits, including price advantage.

Many of the high-volume elastomers are available in powder/particulate form, including natural rubber (NR), styrene–butadiene (SBR) copolymers, acrylonitrile–butadiene (NBR) copolymers and polybutadienes (BR). However, usage of other elastomers may also be considered as it is comparatively easy to grind them "in-house" with the correct equipment.

Initially, all the compounding ingredients are blended with the powdered/particulate elastomer in powder-blending equipment. Final mixing follows in internal mixers, on open mills or by direct feed into extruders. Obvious savings in mixing time and greater cleanliness are obtained with the latter process. The advantages of powdered elastomers are outlined later in relation to hose, cables and extruded products but are applicable to most branches of the industry.

1. Preparation of Powdered Elastomers

Thermoplastics polymers have been available in powder form for many years; the first elastomer to be converted to powder form, NBR copolymer, was used for blending with poly(vinyl chloride) as a solid polymeric plasticizer. Many major elastomer manufacturers now incorporate a powdering operation as part of their finishing processes for a range of elastomers.

It is also possible to convert baled elastomers to powder form by a process involving primary granulation, followed by reduction to powder size in attrition mills, adding minimum quantities of antitack/partitioning agents to prevent sticking during storage. Other routes involve the use of freeze/spray drying, freeze grinding, precipitation and coagulation in line.

Powdered elastomers need to be free flowing and capable of being transported and stored without compacting. This requires the use of a partitioning agent, its amount depending on the elastomer type and on the method of powder production. Studies reported by Evans (1988) showed that siliceous materials of particle size less than 200 mesh gave free-flowing powders when used in a range of elastomers after considerable periods of storage/stacking and transportation. At the optimum concentration (5%), the physical properties of the powder-mixed compounds were better than those of the bale-mixed control.

Later discussion will confirm these results which are attributed to better dispersion of all compounding ingredients and to less elastomer chain scission. A greater consistency of results is also demonstrated. Care should be taken to avoid levels of partitioning agent that are too high, as property levels will be lower.

2. *Mixing and Blending the Recipe*

Processes widely used for preparing a powdered elastomer preblend are similar to domestic mixing with a high-speed blender. Much early laboratory work used such equipment. Powdered elastomer compounds are ideal for automatic powder-handling systems. Thus, all ingredients of the mix must be in powder, crumb or flake form before weighing and preblending commences, that is, up to 2 mm particle size. Oils or other plasticizes are added directly to the mixer.

For some final mixing operations, such as the internal mixer route, particulate-sized granulated elastomer (up to 12 mm) may be used with satisfactory results.

The compounding process for powdered elastomer mixing is similar to that for thermoplastics and involves blending all the compounding ingredients in an intensive rapid mixer. Suitable equipment is manufactured by Werner and Pfleiderer, T. K. Fielder, Papenmeier, Henschel, and Lodige and Drosna. These machines have been designed to enable intensive mixing to be carried out, gaining the full effect of the mixing action. The equipment is jacketed so that additional heating or cooling may be used as necessary, depending on the elastomer and formulation used. However, unlike thermoplastics (e.g., poly(vinyl chloride)) mixes, an after-cooling mixer is not required with elastomer powders.

A typical powder mixing cycle is given in Table 1. The resulting free-flowing, powdered preblend is ready for final mixing and processing.

Table 1
Typical powder mixing cycle (300 l mixer, 45 kg batch weight)

Minutes	Mixer speed	Mixing sequence
0	start—slow speed	add main powders, elastomer on top
1	switch to high speed	
2.5	revert to slow speed	
3		add oils, etc.
3.5	continue at slow speed	
4	stop	

3. *Powder Mastication Principles*

The basic steps in the conversion of powdered elastomer preblends into usable compounds are

(a) compaction or compression;

(b) deformation to establish elastomer–elastomer contacts;

(c) dispersion, especially of fillers; and

(d) ultimate mixing and homogenization.

These basic steps for powdered elastomers can apply to open-mill mixing or internal mixer sequences, or within other ultimate processes such as direct extrusion or injection molding.

3.1 Open-Mill Processing

With a tight nip setting, it is possible to "trickle feed" the preblend from the storage bin onto a mill, producing satisfactorily mixed, uniform material in a short time. Such a system is used in industry, but the operation is dusty and efficient extraction is essential. Also, due to the tight nip setting, throughput, while greater than with conventional open-mill mixing, is not as high as when using the compactor route.

The compactor route involves the use of a compactor unit that produces continuous strips of compressed mix from the preblend storage container, delivering this material to a mill with normal nip settings. Very satisfactory rapid mixing takes place with output rates up to four times those of conventional open-mill mixing techniques. Dispersion is excellent, leading to greater reproducibility of results and improved vulcanizate properties.

3.2 Internal Mixer Process

If the powder preblend is fed directly to an internal mixer, mixing speed is substantially increased, cycle times being reduced by factors of up to four at normal rotor speeds. Additionally, two-stage mixing can usually be eliminated, with reductions in energy usage and lower dump temperatures.

Elastomer particles of up to 1 mm are not essential in the preblend if the internal mixer route is subsequently used. Particulate elastomers can give satisfactory results even when mixed by an "upside-down" process, in which compounding ingredients are loaded first into the internal mixer followed by the elastomer—without preblending.

Excellent dispersion, greater reproducibility of results and, usually, improved physical properties are obtained by the internal mixing process.

3.3 Continuous Mixing and Processing Equipment

The ultimate aim of powdered elastomer technology is continuous mixing/processing of preblended material and this has recently been achieved to a large extent.

The Mixtrumat machine, developed by Stewart-Belling Inc. in the USA, was first demonstrated at the ACS Rubber Division Meeting in 1973, where it mixed powdered elastomer compounds. Farrel-Bridge in the UK have introduced their MVX machine, which continuously mixes and extrudes powdered and particulate preblend. Four stages are involved, namely feed, mix, vent and extrude; all stages are automatically synchronized and controlled by the speed of the extruder screw. Werner and Pfleiderer in Germany have developed their EVK mixer, a continuous machine capable of mixing powder preblends, either direct or compacted, plus extruding, if required. Other extruder manufacturers including Troester and Francis Shaw have also developed machines capable of direct processing. The feeding may be via feed hoppers with or without vibration depending on polymer and formulation, by suction (Meyer Mashinen, Howe-Richardson) or by tube conveyor (Floveyer), depending on conditions.

Direct mixing and extruding of hose or cable jacket is now possible, continuously vulcanizing the product in one pass using suitably sized equipment (Evans 1983). Considerable cost reduction and, for hydraulic hose, greatly increased impulse and flex performance are obtained.

4. *Disadvantages*

The present disadvantages of powder/particulate processes are minimal and consist of:

(a) powder premium charge, and

(b) limited availability of elastomer in powder/particulate form.

The premium charge for powdered elastomers has been appreciably reduced since the late 1970s due to higher-volume production and changes in finishing operations. Final grinding can now take place at the coagulation/crumb stage of the latex rather than after baling. Alternatively, in-house grinding eliminates both the premium charge for grinding and the availability drawback.

5. *Advantages*

The main advantages of powder/particulate processes are

(a) elimination of bale cutting;

(b) shorter mixing cycles than conventional internal mixer or open-mill techniques;

(c) lower power consumption;

(d) reduced plant maintenance;

(e) the possibility of a less capital-intensive plant;

(f) better ultimate dispersion;

(g) possible use of more highly accelerated compounds;

(h) less heat memory (retention of heat from processing conditions) for difficult elastomers (e.g., polychloroprene (CR)), thus less scorch tendency;

(i) direct feed possibility to extruders and molding presses, thus eliminating internal mixer and/or milling operations;

(j) elimination of premastications (reduction of elastomer molecular weight by premixing, often chemically activated) and masterbatching (masterbatches are concentrates of appropriate materials in base polymers: they enable small amounts of critical components such as accelerators to be added more accurately and improve cleanliness and distribution of filler-loaded compounds);

(k) easier factory control;

Table 2
Mixing cycle comparisons for the internal mixer

	Control two-stage mix 12 min		Powder preblend 2 min		Powder no preblend 3 min		Bale straight mix 10 min	
Elastomer	TS[a] (kg cm^{-2})	EB[b] (%)	TS (kg cm^{-2})	EB (%)	TS (kg cm^{-2})	EB (%)	TS (kg cm^{-2})	EB (%)
NBR	109	160	118	180	116	175		
CR	112	235	113	240	114	240	112	235
NR/SBR			95	590	98	585	87	560

a Tensile strength b Elongation at break

Table 3
Power consumption and dump temperature comparisons

Elastomer	Control (bale mixing) Average dump temperature (°C)	Control (bale mixing) Maximum power (A)	Powder mix (same recipe) Average dump temperature (°C)	Powder mix (same recipe) Maximum power (A)
NBR	140	220	115	145
CR	145	170	115	135
NR/SBR	150	165	110	125

(l) considerably lower dump temperatures (temperatures at which a solid mass of compound is released from an internal mixer or from mill rolls); and

(m) better environmental conditions.

These advantages are highlighted by data given in the following tables.

Table 2 shows that internal mixer outputs can be increased by factors of three to five by the direct mixing of powdered elastomer compounds compared with normal bale mixing. It also demonstrates slight improvements in physical properties for the powder mixes.

Lower power consumption and lower dump temperatures for some powder mixes compared with control bale mixing are listed in Table 3.

Typical energy usages for hose compounds extruded by bale routes and powder/particulate routes are given in Tables 4 and 5, respectively, and demonstrate appreciable savings for the powder/particulate routes, whether using a Banbury-type internal mixer or the MVX machine in either mode.

6. Economics

At present, the premium cost of powdered/particulate elastomer makes it essential to examine all aspects of the operating process to ascertain whether existing stages can be eliminated or speeded up. Comparative flow charts of the current methods and powder routes are helpful. A useful formula for cost calculation is

$$\text{conversion cost per kilogram } g = \frac{M+F+V}{R}$$

where M is manpower (wage rate × number of personnel), F is fixed costs (overheads + depreciation + maintenance), V is variable costs (energy, etc.) and R is mixing rate (kg h^{-1}).

Calculations for various processes will establish savings for powder/particulate processes which can be compared with the premium for the powdered/particulate elastomer. In many cases, the powder/particulate processing routes will be favored.

7. Other Usage Areas

The usual methods for producing adhesives and doughs involve mastication of elastomers and/or compounds to a lower plasticity and then sheeting out thinly to improve their solubility in organic solvents. An appreciable "laying down" time is necessary for the solvent to swell and then dissolve the elastomer; thus, preparation of satisfactory adhesives and doughs can take days to complete. Use of powdered or particulate elastomers can significantly reduce or eliminate the laying down period, enabling completion of adhesive manufacture in hours, thus allowing more economic production schedules.

Table 4
Energy usages for bale route extrusions (kWh kg^{-1})

Bale cutting	0.32
Banbury—stage 1	0.97
Banbury—stage 2	0.97
Mill 1	0.52
Mill 2—strip	0.52
Cold feed extruder	0.64
Total	3.94

Table 5
Energy usages for powder/particulate extrusions (kWh kg^{-1})

	Banbury	MVX strip	MVX direct
Intensive mixing	0.06	0.06	0.06
Banbury	0.52		
Mill 1	0.52		
Mill 2	0.52		
MVX		0.64	0.64
Cold feed extruder	0.64	0.64	
Total	2.26	1.34	0.70

See also: Elastomers: Additives; Fillers in Elastomers; Rubber Processing

Bibliography

Ellwood H 1981 MVX machine. Farrel-Bridge Technical Leaflet MVX R79. Farrel-Bridge, Rochdale, UK

Evans C W 1978 *Powdered and Particulate Rubber Technology*. Applied Science, London

Evans C W 1983 PLCV system. In: Evans C W (ed.) 1983 *Developments in Rubber and Rubber Composites-Two*. Applied Science, London

Evans C W 1988 Powdered rubber. In: Bhowmick A K, Stephens H L (eds.) 1988 *Handbook of Elastomers*. Dekker, New York, pp. 141–82

Smith L P 1978 European powdered rubber technology. *Elastomerics* 110(11): 32–6

C. W. Evans†
[Gateshead, UK]

P. J. Corish
[Patrick J. Corish & Associates, Stratford-on-Avon, UK]

Pressure-Sensitive Adhesives

A pressure-sensitive adhesive is a material which forms a bond of measurable and useful strength between two adherends when the adherends are brought together with only light pressure. Pressure-sensitive adhesives are most familiar in everyday life as transparent household tapes, masking tapes and labels. The earliest use for pressure-sensitive adhesives was on medicinal plasters and surgical tapes. One of the first pressure-sensitive compositions is credited to Shecut and Day, who in 1845 patented a fabric-backed construction for use as a medicinal plaster. This patent describes a pressure-sensitive formulation consisting of a blend of rubber and resin not dissimilar to formulations currently in use. For many years, surgical tape remained the major use for pressure-sensitive adhesives. When spray painting of automobiles was developed, workers found it convenient to use surgical tapes to protect surfaces which were not to be painted. As this application grew, the limitations of the adhesive used on the surgical tape became troublesome. There was difficulty in removing the tape because the adhesive tended to transfer and stick to the unpainted surface. The development of new adhesive formulations for this masking tape application marked the birth of the pressure-sensitive adhesives industry.

1. Uses

The major applications for pressure-sensitive adhesives are in tapes and labels. General-purpose pressure-sensitive tapes are widely used in the developed countries of the world for wrapping packages, sealing cartons, repairing paper tears and many other low-stress applications. Transparent tape, masking tape and reinforced carton sealing tape can be found in many homes. Other consumer applications include double-sided pressure-sensitive carpet tape, automobile bumper stickers, identification tags and easy-installation floor tiles. In industrial applications, pressure-sensitive adhesives are used to apply decorative laminates, to place trim onto automobile bodies and to construct lightweight assemblies. In addition, tapes are used to seal heating ducts. In this application, the tape must maintain its bond without slipping as the duct cycles from room temperature to hot air temperatures. Refrigerator tapes must adhere at freezer temperatures and must maintain their bond to food packages after low-temperature storage.

The primary advantages of pressure-sensitive adhesives are convenience and fast application. Pressure-sensitive adhesives generally do not require heating, drying or removal of solvent to form a bond. While other adhesives must undergo a phase change (curing) for bonding, a pressure-sensitive adhesive requires only that it be pressed lightly onto a surface. The main deficiency of pressure-sensitive adhesives is that they form relatively weak bonds. They do not have the strength of structural adhesives. However, for some applications the pressure-sensitive adhesive may have a slow-curing mechanism built into it which will strengthen the composition during aging.

The use of pressure-sensitive adhesives has grown rapidly since the early 1960s. The market was approximately 180 kt in 1980, about half being used for tapes. The current growth rate of the market is about 5% per year.

2. Method of Application to Substrate

Pressure-sensitive adhesives have traditionally been applied to the carrier from solution in organic solvents. The solvent must then be evaporated. The carrier may be cloth, paper, plastic film or metal foil. The adhesive may also be deposited onto a release surface for later transfer to another substrate. In recent years, the pressure-sensitive tape and label industries have begun to convert from organic solvents because of increasing costs and potential toxicological problems. Two approaches presently being considered are to apply the adhesive as an aqueous emulsion or as a hot melt. The aqueous approach involves the use of polymer lattices and aqueous resin dispersions which can be blended to formulate a waterborne system. As with a solvent system, the water must be evaporated to deposit the adhesive. Hot-melt adhesives are applied as molten liquids but exhibit pressure-sensitive performance at room temperature. They are primarily based on styrenic block copolymers (styrene–isoprene–styrene, styrene–butadiene–styrene or styrene–ethylene butylene–styrene), although formulations have been designed using ethylene–vinyl acetate and acrylic polymers.

3. *Requirements for Pressure-Sensitive Performance*

Pressure-sensitive adhesives have the ability to adhere to a surface, forming a bond having measurable strength, when applied under a light deforming stress. Finger pressure is often all that is required and, with some compositions, the weight of the tape or label itself is sufficient to form a bond. The commonly used term to describe this property is tack.

Tack is not a material property such as modulus, tensile strength, elongation or viscosity (though all these properties contribute to it). It is very much a function of the test used to determine it. The surface being bonded and the conditions under which tack is measured are important considerations. Tack involves both a bonding and a debonding step. Each must be considered separately, because they are distinct operations in testing for tack. Bonding is a low-deformation process carried out at a low rate of deformation. On the other hand, the debonding step takes place at a comparatively high rate relative to the thickness of the adhesive layer and involves large deformations. Thus, poor tack can result from either deficiencies in the initial bonding or low holding power after the bond has been formed.

If the flow characteristics of the adhesive mass are inadequate under the application conditions of pressure and temperature, the adhesive will not flow to conform to the microscopically rough surface of the substrate. Consequently, this results in poor wetting and contact. In 1969, Dahlquist proposed a criterion for tack, based upon creep measurements, which defines the response required for the adhesive to flow under the deforming stress of a typical tack test. According to Dahlquist, an adhesive must have a 1 s compressive creep compliance of greater than $10^{-6}\,m^2\,N^{-1}$ to wet a surface properly and exhibit tack. If the adhesive is less compliant (i.e., it has less flow when subjected to a deforming stress), then the adhesive does not flow sufficiently and cannot form an adequate bond. Of course, if surface energies or surface roughness inhibit intimate contact, no bonding or poor bonding will occur. Thus, the Dahlquist criterion, which is concerned with the bonding step, is a necessary but not a sufficient condition for tack.

The adhesive must also demonstrate satisfactory cohesive and adhesive strengths under the conditions of debonding. Aside from surface energy and adhesion consideration, a functional pressure-sensitive adhesive composition must have both a low modulus at the rate of bonding at low deformation as defined by Dahlquist and a high modulus and cohesive strength at the rate of debonding at high deformation. A difference in modulus with respect to rate of deformation indicates that the adhesive has a significant mechanical loss component at the temperature of use. Since the time constant for a loss process is temperature dependent, the modulus of the adhesive composition is also dependent upon temperature in the range embracing the temperature of use. In all amorphous polymeric systems, including pressure-sensitive adhesives, the greatest change in modulus and compliance with temperature, excluding melt flow, is observed in the transition zone where the composition passes from the glassy state into the rubbery plateau. Figure 1 represents a plot of modulus vs temperature for a typical amorphous polymeric composition. Scale values are not indicated, because they depend upon the composition of the system and the method used to determine the modulus.

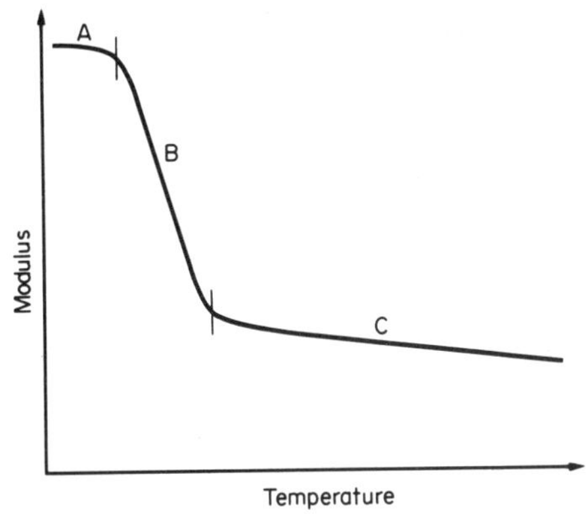

Figure 1
Modulus–temperature curve for a typical amorphous polymeric composition: A, glassy state; B, transition zone; C, rubbery plateau

To demonstrate maximum performance, the adhesive should be designed so that its use temperature will be near the lower-modulus end of the transition zone, at the entry to the rubbery plateau. Since the dependence of modulus on temperature in the transition zone is simliar for all amorphous systems, the modulus vs deformation rate requirement for maximum pressure sensitivity suggests that the composition should have a glass-transition temperature T_g about 10–30 °C below the use temperature. The requirement that the slope of the plot of modulus vs deformation rate must be greater than some minimum value, in conjunction with the Dahlquist criterion, describes the viscoelastic requirements for a pressure-sensitive adhesive.

The viscoelastic requirements should be considered in designing formulations. A high-molecular-weight, low-T_g elastomer is generally used to impart strength to the adhesive. A compatible low-molecular-weight resin (MW≈500–2000), generally known as a tackifying resin, is added to provide tack. The resin, which has a T_g above the use temperature of the adhesive, modifies the viscoelastic properties of the elastomer in two ways: (a) it raises the glass-transition temperature of the composition and (b) it reduces the modulus, as well as increasing the slope of the plot of modulus vs temperature, in the rubbery plateau. This is shown in Fig. 2. During removal of the tape, the higher strain rate corresponds rheologically to a lower effective temperature; Fig. 2 shows that the response of the adhesive is to exhibit a higher modulus and thus a higher bond strength.

An estimate of the T_g of the composition can be calculated from the T_g of the compatible components using the following equation proposed by Fox:

$$T_{gA}=(W_1T_{g1}^{-1}+W_2T_{g2}^{-1}+\cdots+W_nT_{gn}^{-1})^{-1} \quad (1)$$

where T_{gA} is the glass-transition temperature of the adhesive (in K), T_{g1}, T_{g2}, . . . are the glass-transition temperatures of the components (K) and W_1, W_2, . . . are the weight fractions of the components. The selection of a resin and its concentration in the composition are governed by the T_g desired for the adhesive.

The modulus of the composition must be adjusted so that its creep compliance meets the Dahlquist criterion at the use temperature. As shown in Fig. 2, the plateau modulus of the unmodified elastomer may be too high to flow and wet the substrate, according to the Dahlquist criterion. The modulus can be reduced by adding the appropriate compatible resin at the proper concentration. A compatible

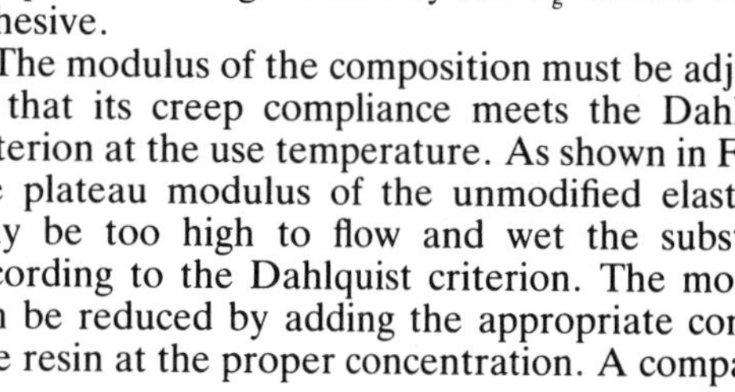

Figure 2
Effect of resin addition on modulus–temperature curve: ——, elastomer alone; - - -, elastomer–resin blend

oil can also be used to reduce the modulus to meet the Dahlquist criterion. Care must be taken in using an oil, because it reduces the T_g value of the composition, possibly to a temperature below the useful range. In addition, if the oil is not completely compatible, it may exude at the surface of the composition and destroy the bond.

Thus, guidance in formulating a pressure-sensitive adhesive is provided by the glass-transition temperatures of the polymer and compatible resin, the modulus of the polymer at the use temperature, and the glass-transition temperature and creep compliance of candidate compoisitions.

4. Components of Pressure-Sensitive Adhesives

As described in Sect. 3, pressure-sensitive adhesives primarily consist of a polymer to provide strength and a low-molecular-weight resin to tackify the composition. Oil and fillers may be added to adjust performance and to reduce cost. Stabilizers are included to protect the products during both storage and long-term use.

4.1 Polymers

A high-molecuar-weight, low-crystallinity, low-T_g polymer is preferred. The high molecular weight imparts strength and higher-temperature performance to the composition. Low crystallinity is required because crystallinity will reinforce the composition. This may increase its modulus above the limiting level which allows the composition to be readily deformed and wet the substrate. T_g obviously must be below the use temperature, for the same modulus considerations. Below T_g, the polymer will be too hard to conform under light deforming stresses to the microscopic surface irregularities of the substrate.

The commonly used polymers are natural rubber, styrene–butadiene rubber, acrylic polymers and the styrenic block copolymers. Polyurethanes have received some attention, as described in the patent literature.

(a) Natural rubber (NR) and styrene–butadiene rubber (SBR). These are the polymers traditionally used for solvent-applied pressure-sensitive adhesives. With the recent consideration of waterborne systems by manufacturers, rubber lattices of both NR and SBR are being evaluated. The polymers supplied as lattices may have much higher average molecular weights than are possible in organic solvents at reasonable concentrations. These higher-molecuar-weight polymers provide adhesive performance characteristics which differ from those of the traditional polymers applied from solvent. Therefore, formulations usually require adjustment to achieve similar end-use performance.

(b) Acrylic polymers. Before the current interest in

NR and SBR lattices for pressure-sensitive adhesives, a reference to waterborne pressure-sensitive adhesives was usually understood to imply acrylic polymers. The acrylics were designed as single-component copolymers to provide both tack and holding power. Consideration is being given to formulating acrylic adhesives with resins, to provide formulating flexibility. A higher-molecular-weight acrylic polymer is used to provide strength, combined with a resin to provide tack. Hot-melt-applied pressure-sensitive adhesives based on acrylic polymers have also been described.

(c) Styrenic block copolymers. The block copolymers of styrene–isoprene–styrene and styrene–ethylene butylene–styrene are the major candidates for hot-melt-applied pressure-sensitive adhesives. The polymer blocks segregate into separate phases, each exhibiting its own T_g. The rubbery midblock imparts tack to the composition, while the styrene end block provides cohesive strength below its T_g of about 100 °C. The resin selected to tackify the midblock must be completely incompatible with the end block, to prevent softening of the end block and loss of strength. An appropriate resin compatible with the styrene blocks may be added to increase the temperature at which bond failure may occur. Its T_g should be no lower than that of polystyrene. If a resin is used with the end block, it must be incompatible with the midblock so that it does not raise the T_g of the midblock and destroy tack.

4.2 Resins

A low-molecular-weight resin (MW ≈ 500–2000) is usually added to impart tack to the composition. The resin must be compatible with the polymer to increase T_g and to reduce the modulus of the composition at the use temperature to the level necessary to exhibit sufficient tack. The polarity, molecular weight and molecular weight distribution of the resin are important considerations. The softening point of the resin depends on its molecular weight and molecular weight distribution. The T_g of the resin is about 40–50 °C below its softening point. For the so-called hard resins, T_g is above room temperature.

There are two major types of resins: those based on petroleum streams and those based on rosin. Polyterpenes and alkylated phenol resins are also used (see *Hot-Melt Adhesives*).

5. Formulations

Formulations are designed to meet the requirements of the various applications. For example, masking tapes must be easy to remove and must not leave a residue. Freezer tapes must have tack at freezer temperatures. High strength is not necessary for label adhesives. The ratio of polymer to resin depends on their T_g values. Typical formulations (wt%) are

Component	wt%
Natural rubber	40
Poly(α-pinene) resin (SP 115 °C)	60
Styrene–butadiene rubber	45
Glycerine ester of hydrogenated rosin	55
Acrylic polymer	70
Aromatic copolymer resin (SP 75 °C)	30
Styrene–isoprene–styrene block copolymer	45
Aliphatic hydrocarbon resin (SP 95 °C)	45
Paraffinic oil	10

where SP is the softening point by the ring and ball method. All compositions should include appropriate levels of antioxidants.

6. Testing

Pressure-sensitive adhesive products are tested to assure that they meet product specifications as well as the requirements of the user. In developing new compositions, tests are run to guide the formulator towards performance improvements. The performance tests commonly used measure tack or quick stick, peel strength and shear strength or holding power. Hot-melt pressure-sensitive adhesives also require testing for melt viscosity, shear adhesion failure temperature and stability of the adhesive at melt temperatures. Long-term stability of the assembled bond may or may not be important, depending on the final application. Recommended test procedures for pressure-sensitive tapes have been developed and published in the USA by the Pressure Sensitive Tape Council (1976). The more routinely used methods are described here.

6.1 Tack

Tack values depend very much on the test method. Both bonding and debonding conditions influence the tack values obtained by the standard tests. Greater bonding forces and longer bonding times result in higher values in the debonding process. The usual tests are the probe, quick stick and rolling ball tack tests.

In the probe tack test (ASTM D2979-79), a small polished stainless steel probe is brought into contact with the surface of the pressure-sensitive adhesive at a selected rate and held for a set time under a specified load. The probe is then withdrawn at the same rate at which contact was made and the force required to separate the probe is recorded. The surface of the probe must contact the adhesive surface uniformly to obtain meaningful data. Surface impurities such as dust or fingerprints obviously have a significant and unwanted effect on the tack values obtained. Probes other than stainless steel can be used to obtain tack to various surfaces.

In the quick stick test (Pressure Sensitive Tape Council 1976, Method 5), a bond is formed by placing a pressure-sensitive tape of measured width on a clean steel plate, using only the weight of the tape itself as the bonding force. The tape is then peeled at a 90° angle to the plate surface. The force required is the quick stick value. A test jig is used to maintain the 90° angle during the peel. The jig moves the test plate horizontally as the testing instrument pulls the tape vertically. Data are reported as force per unit width of tape.

The rolling ball tack test (Pressure Sensitive Tape Council 1976, Method 6) operates at the force and bonding–debonding rate set by a small steel ball rolling on the adhesive surface. The ball is allowed to roll down an inclined channel, of specified length and incline angle, onto the surface of a pressure-sensitive tape. The distance the ball rolls before stopping on the adhesive surface is inversely related to tack.

Although the foregoing tests measure the same phenomenon, ranking of adhesives can differ, depending upon the test used. This reflects the differences in bonding and debonding conditions. For example, the rolling ball tack test allows a very short contact time to wet the ball. Therefore, the adhesive must be more compliant to obtain good tack in this test than in the other tests. Adhesives manufacturers and industrial consumers may favor one test over the others because in their experience the results are more meaningful to their operation.

6.2 Peel Strength

The peel test is conducted after a tape has been applied to a surface. In the standard test (Pressure Sensitive Tape Council 1976, Method 1), the tape is rolled onto a stainless steel panel using a 2 kg rubber-covered roller. After an appropriate contact time, 24 h for example, it is peeled from the surface of the panel at an angle of 180° and the peeling force is measured. The peeling force is not uniform as the tape is pulled, because of imperfections and irregularities in the bond or the properties of the adhesive. The average peel strength per unit width of tape is calculated from a recorder chart of the peeling force.

6.3 Shear Strength or Holding Power

The shear test measures the ability of a pressure-sensitive adhesive to withstand a shearing force without exhibiting cold flow or creep (Pressure Sensitive Tape Council 1976, Method 7). A measured area of tape is applied to a steel panel using a 2 kg rubber-covered roller, as in the peel strength test. A specified weight is attached to the end of the tape to apply a shearing force to the bond. The panel is suspended at an angle of 178° from the adhesive side of the extended piece of tape. The time to failure under a standard load is usually reported. Alternatively, the displacement of the tape due to creep after a specified time may be reported.

6.4 Shear Adhesion Failure Temperature

The shear adhesion failure temperature test is a measure of the temperature at which the adhesive in shear can no longer support an applied stress. It indicates the temperature where there is a significant viscous flow of the adhesive. The test specimen is an overlap shear configuration which is hung by one unbonded tab with a weight attached to the other. The assembly is placed in an oven and the temperature is raised either in increments or continuously until at least three of five replicate samples fail. The temperature at which this occurs is reported as the shear adhesion failure temperature.

6.5 Melt Viscosity

The viscosity of hot-melt pressure-sensitive adhesives at application temperatures is generally measured in continuous shear using a Brookfield Thermosel viscometer. Viscosity–temperature curves are often constructed to indicate viscosities at the temperatures which may be required to apply the composition.

6.6 Melt Stability Tests

Hot-melt pressure-sensitive adhesives based on styrenic block copolymers are particularly subject to changes if they are held molten before application to the substrate. The viscosity of the adhesive may change because of degradation, chemical reaction or partial volatilization. Development of color may also be undesirable.

Testing the stability of the hot melt is straightforward. A sample is held molten at the selected temperature in a suitable container. Changes in viscosity and color, as well as the development of a skin on the surface of the adhesive, are monitored as a function of time. A control adhesive may be included to provide comparative information.

See also: Adhesive Joining: An Overview; Elastomer–Elastomer Adhesion Processes; Hot-melt Adhesives

Bibliography

Aubrey D W 1977 Pressure sensitive adhesives—Principles of formulation. In: Wake W C (ed.) 1977 *Developments in Adhesives,* Series 1. Applied Science, Englewood, NJ, pp. 127–56

Aubrey D W Sherriff M 1980 Peel adhesion and viscoelasticity of rubber resins blends. *J. Polym. Sci., Polym. Chem. Ed.* 18: 2597–608

Counsell P J C, Whitehouse R S 1977 Tack and morphology of pressure sensitive adhesives. In: Wake W C (ed.)

1977 *Developments in Adhesives*, Series 1. Applied Science, Englewood, NJ, pp. 99–126
Dahlquist C A 1959 An investigation into the nature of tack. *Adhes. Age* 2: 25–9
Dahlquist C A 1969 Pressure sensitive adhesives. In: Patrick R L (ed.) 1969 *Treatise on Adhesives and Adhesion*, Vol. 2. Dekker, New York, pp. 219–59
Hodgson M E 1979 Pressure sensitive adhesives and their applications. In: Allen K W (ed.) 1979 *Adhesion*, Series 3. Applied Science, London, pp. 207–20
Kraus G, Jones F B, Marrs O L, Rollmann K W 1977 Morphology and viscoelastic behavior of styrene-diene block copolymers in pressure sensitive adhesives. *J. Adhes.* 8: 235–58
Kraus G, Rollmann K W 1977 The entanglement plateau in the dynamic modulus of rubbery styrene-diene block copolymers. Significance to pressure sensitive adhesive formulations. *J. Appl. Polym. Sci.* 21: 3311–18
Kraus G, Rollmann K W, Gray R A 1979 Tack and viscoelasticity of block copolymer based adhesives. *J. Adhes.* 10: 221–36
New England Research Application Center 1980 Pressure sensitive adhesives (Jun 70–Jun 80): Citations from the engineering index data base. National Technical Information Service, Springfield, VA
Pressure Sensitive Tape Council 1976 *Test Methods for Pressure Sensitive Adhesives*. Specifications and Technical Committee, Pressure Sensitive Tape Council, Glenview, IL
Sherriff M, Knibbs R W, Langley P G 1973 Mechanism for the action of tackifying resins in pressure-sensitive adhesives. *J. Appl. Polym. Sci.* 17: 3423–38
Toyama M, Ito T 1973 Pressure sensitive adhesives. *Polym. Plast. Technol. Eng.* 2: 161–230

J. B. Class
[Hercules, Wilmington, Delaware, USA]

Processing: Classification

The procedures used to fabricate parts from polymers are highly varied, both in type and in the number and character of the steps. Many manufacturing procedures involve only a single operation whereas others involve numerous interacting steps. In this article, the more important examples of the manufacture of polymeric products will be described.

First, a major division of processing into (a) pumping, (b) mixing and (c) shaping will be considered. The most important of these steps is the final fabrication step, the shaping of parts. A more specific inclusive outline of polymer processing operations may be developed based on the details of shaping given in the following.

1. Simple Machine Fabrication Operations Based on Screw Extrusion/Injection Devices

1.1 Extrusion Through Dies

In the simplest shaping step, the melt pumped by the screw extruder is acted on only by a shaping device known as a die. A simple example of a shaping die is the "coathanger" die in which the melt emerging from the extruder is formed into a sheet (see *Die Extrusion*). Wire coating is another example of this operation.

Several screw extruders may be combined to produce multilayer sheet. This is called coextrusion (see *Coextrusion*).

1.2 Extrusion with Postdie Shaping—Tubular Film Extrusion

Postdie devices which draw down the dimensions of emerging simple shapes are widely used in plastics processing. Take-up rolls are often used to produce thin film from melt emerging from a slit die. More commonly, melt is extruded through an annular die to form a tube into which air is blown (tubular film extrusion). The air pressure expands the tube to several times its diameter, which together with the drawdown rolls acts to decrease the film thickness. It is, of course, possible to produce multilayer films using these methods by combining melts emerging from two or more extruders (see *Tubular Film Extrusion: Dynamics and Heat Transfer*).

1.3 Injection Molding

In injection molding, the emerging melt is molded into a desired shape. This may be accomplished with a screw extrusion device which injects the polymer melt into a cold mold (see *Injection Molding of Thermoplastics: Mold Filling*).

If gas is dissolved in the melt and allowed to come out of solution in the mold, very large parts may be produced. In the sandwich molding process, skin–core (sandwich) parts involving different polymers may be produced using two screw-injection-molding machines which sequentially (or sometimes simultaneously) inject two melts into a mold (see *Sandwich Molding of Plastics*).

1.4 Blow Molding

In blow molding, melt is extruded out of an annular die to form a parison. Air is then injected into the center of the annulus and a cold mold closes around it. The top of the parison is "snipped off" by the closing of the mold and the bottom of the mold, through which air is injected, becomes the mouth of the bottle. The molten inflated parison solidifies in the mold, which then opens and ejects the bottle. The extruder then begins operation again (or the reciprocating screw moves forward) and a new parison is produced. The cycle is repeated (see *Blow Molding of Thermoplastics*).

2. Two-Machine Manufacture of Polymer Products

There are many cases in which polymer products are made using two sequentially operating processing machines. The first of these is an extrusion device which makes a preform. In the second step, the solidified preform is heated, sometimes above its softening temperature, and then further deformed to produce the final product. There is some arbitrariness in this classification since tubular film extrusion and traditional blow molding were included in Sect. 1 under single-machine manufacture rather than here.

There are two major reasons for carrying out two-stage manufacturing operations. One is to obtain a desired set of mechanical properties not achievable in processing operations involving single-machine manufacturing operations. For example, filaments, film and bottles can be shaped in a single-machine process, but the desired levels of molecular orientation and properties cannot be attained in this way. The second reason is the formation of intricate shapes that are difficult to produce through injection molding (see *Processing: Structure Development*).

2.1 Spinning/Drawing of Fibers

High-modulus fibers require a two-step manufacturing process. In the first step, the polymer (generally in the form of a melt) is extruded using a screw extruder and/or a gear pump through a multihole die called a spinneret. The emerging extrudates descend vertically to a rotating roll where they are taken up at a much higher velocity than they are extruded from the die. This results in both the production of smaller diameters and enhanced anisotropy. This is called melt spinning (see *Melt Spinning of Fibers: Structure Development*).

It is also possible to form fibers from solutions with the solvent being removed by either evaporation (dry spinning) or extraction (wet spinning).

In the second processing stage, filaments are run from slow rolls over a hot plate or through a hot tube, which softens but does not melt the filament, and onto a fast roll. This further draws down the diameter and enhances the modulus and anisotropy. This process is known as drawing (see *Fibers: Drawing*). In some drawing operations, a rotating spindle twists the filament over a hot plate or in a hot tube and gives the filament helical as well as transverse anisotropy. This is called draw texturing (see *Fibers: Texturing and Draw Texturing*).

A novel method for management of fiber orientation and microstructures in molded and extruded polymer composites has recently been reported (see *Processing: Shear-Controlled Orientation*).

2.2 Two-Stage Film Formation

Films are often manufactured in two-stage manufacturing operations. Pellets are melted and pumped through a slit die to form a film which vertically descends onto a rotating roll. The film is then taken from a slow roll through a region of elevated temperature from where it moves onto a faster roll. The film is clamped in the transverse direction and is simultaneously stretched in this as well as the machine direction. This type of film manufacturing process yields products which have high moduli and equal properties in all directions in the plane of the film (see *Tubular Film Extrusion: Structure Development*).

2.3 Stretch Blow Molding

A two-stage process may be applied to produce bottles. Preforms are extruded or molded before being placed in a blow molding press. The inflation is carried out in such a manner as to stretch and orient the material biaxially as it forms a bottle (see *Bottles and Other Rigid Containers*).

3. Multistage Processing Incluling Compounding and Curing

There are many manufacturing operations for polymer products which involve only a single processing machine. Typically, they involve single-screw extruders coupled with a shaping apparatus (see *Mixing: Single-Screw Extruders*). In these operations, the screw extruder serves as both a melting device and a pump, and often as a mixer for additives. In some cases, special static mixing devices are added following the extruder screw to aid in dispersing these added ingredients. Bags of pelletized plastics are added to the hopper and the melts produced are pumped into the shaping apparatus (see *Melt Delivery Systems, Single-Screw*).

Multistage processing including compounding and shaping steps is often found in thermoplastic fabrication. Compounding of reinforcing fillers such as fibers is often carried out in twin-screw extruders. The compounds so produced are subsequently fabricated using extrusion through dies, injection molding or similar processing methods.

In the rubber industry, several independent processing steps and machines are generally required (see *Rubber Processing*). One begins with elastomers (generally solid bales) and compounding ingredients (e.g., carbon black, oils, curatives) and produces a shaped, cured (crosslinked) product. Compounding is usually carried out in an internal mixer combined with roll mills or an extruder. The product is then shaped, usually with a screw extruder or with the rolls of a calender and sometimes a screw-injection-molding machine. Calenders are generally used as a step in multistage processing of compounds (rubber, poly(vinyl chloride)) (see *Calendering*). The extruded or calendered product is then vulcanized in an oven or by dielectric heating. Injection-molded products are generally cured during the molding step.

The most complex fabricated rubber product is probably the pneumatic tire. This involves serveral components including a tread, carcass (rubber-coated fabric), sidewall and liner, each of which has a different composition involving distinct polymers and compounding ingredients in different proportions. These are separately compounded. The various components are then built together on a drum and the annular product is placed on a deformable membrane in a mold. When the mold closes, the membrane is inflated by steam pressure to fill the mold and form a tire; this is subsequently vulcanized by heat (see *Automobile Tires*).

Bibliography

Fenner R T 1979 *Principles of Polymer Processing*. Macmillan, London

Han C D 1976 *Rheology in Polymer Processing*. Academic Press, New York

McKelvey J M 1962 *Polymer Processing*. Wiley, New York

Middleman S 1977 *Fundamentals of Polymer Processing*. McGraw-Hill, New York

Tadmor Z, Gogos C G 1979 *Principles of Polymer Processing*. Wiley, New York

J. L. White
[University of Akron, Akron, Ohio, USA]

Processing: Historical Survey

This article considers the historical development of methods used to fabricate parts from polymers including plastics, elastomers and fibers.

1. Rubber Processing

The development of modern polymer processing technology is associated with the fabrication of products from naturally occurring polymers. It is not with cotton, wool and leather that the origins of fabrication methods are seen, but with natural rubber and gutta-percha, in particular the former. There is an important but uncertain early history of rubber processing. This occurred in Central and South America before the coming of the Europeans. Columbus found and took back rubber balls to Spain. Cortez and the conquistadors found rubber balls and figures in Mexico and described games similar to basketball played with the balls. This was in the period 1492–1530. In the 1740s, the French explorer Charles de la Condamine described a more advanced technology among the Indians of the Amazon. They produced rubber bottles and boots by dipping molds into latex and coagulating them over a fire. This was repeated several times to produce multiple layers. Patterns were then incised on the rubber with a knife.

The polymer industry in the modern sense begins with the commercial development of rubber in Europe. The first pioneer of the rubber industry was the English entrepreneur Thomas Hancock, who began manufacturing rubber products in London in about 1820. The first descriptions of mastication degradation and mixing of elastomers are found in Hancock's patents and autobiography. Hancock used an internal mixer which he called a "pickle." It was originally kept secret at the time to give Hancock advantage over his competitors, but was later described in a patent of 1837 (English Patent No. 7,344). The machine consisted of horizontal shaft with spikes which rotated in a cylindrical box. The rubber would be torn apart by the spikes.

The second major innovation in modern polymer processing was the coating of rubber solutions on fabrics. This was developed in connection with the 1823 patent of the Macintosh sandwich fabric (English Patent No. 4,804), a waterproof material consisting of two layers of fabric with an intermediate rubber layer. A detailed coating process is described in Hancock's 1837 patent (Hancock and Macintosh were then partners).

The origins of modern rubber latex technology were also pioneered by Thomas Hancock (English Patent No. 5,045) in 1824. Hancock imported the same latex used by the South American Indians to England and impregnated it into fabrics to produce leatherlike materials and waterproof fabrics competitive with those of Macintosh. Unfortunately, he could not stabilize the latex during transport and industry had to wait for a century to exploit this technology.

The use of pairs of rotating steel rolls for processing rubber appears to date from the 1820s and is discussed by Thomas Hancock. However, the use of a pair of horizontal steam-heated iron rolls as a mastication and mixing device was conceived by the American entrepreneur Edwin M. Chaffee of the Roxbury India Rubber Factory and is cited in his patent of 1836 (US Patent No. 16). Chaffee goes beyond this in his patent to describe the use of several rolls to coat solid rubber onto fabric. This was the first calender. This machine was developed to be competitive with the solution coating procedure described above. A fabric would be fed horizontally through the small gap between two vertically separated horizontal steel rolls rotating in opposite directions. The rubber would be metered to the fabric by the upper of the two rolls and a third roll. The roll mill was also used for preparing the compounds for the calender. Chaffee built a 30 ton calender of this type which was called the "Monster."

One of Chaffee's motivations in developing the calender was to eliminate the disastrous adhesiveness of rubber products during the summer months. This he associated with the use of solvents such as turpentine which could not be completely removed. It came to be realized in following years when calendered articles proved to be adhesive that this was not the case. This led in the years around 1840 to the invention of curing with sulfur at elevated temperatures. This development can be followed in successive patents by Nathaniel Hayward (US Patent No. 1,090) and Charles Goodyear (US Patent No. 3,633) in the USA and Thomas Hancock (English Patent No. 9,952) in England, all of whom claimed independent invention. Posterity has generally given the credit to Goodyear. The term vulcanization which describes this process was coined by Hancock's colleague, William Brockedon. The development of vulcanization rapidly led to the addition of sulfur to rubber compounds and the introduction of ovens and/or curing presses in rubber factories throughout the world. (An 1840 patent of Chaffee (US Patent 1,939) describes an early prevulcanization compression molding process.) A few years later in 1846, Alexander Parkes (English Patent No. 11,147) invented a cold vulcanization process using sulfur dichloride dissolved in carbon disulfide which could rapidly vulcanize rubber sheets dipped into it.

2. *Plastics Processing*

The beginning of plastics processing and extrusion processes is associated with the introduction of gutta-percha into England during the 1840s and its commercial development as insulation for electrical wire. One of the early pioneers of the new industry was Thomas Hancock's younger brother, Charles Hancock, one of the founders of the Gutta Percha Company. In his patents of 1846–1847 Charles Hancock described fabrication of gutta-percha using a processing technology similar to that developed in the rubber industry largely by his brother. He used a pickle-type masticator for compounding gutta-percha with additives including sulfur and softeners. He also sheeted with rollers and vulcanized the products with sulfur.

The first foamed plastic and rubber products were developed in 1846 in separate patents by the Hancock brothers. Charles Hancock (English Patent No. 11,032) foamed gutta-percha using ammonium carbonate and similar compounds. William Brockedon and Thomas Hancock (English Patent No. 11,455) produced foamed products using sulfur chloride dissolved in a rubber or gutta-percha solution.

The first ram extrusion devices are described in the patents of 1845 by Richard A. Brooman (English Patent No. 10,582) and Henry Bewley (English Patent No. 10,825), which discuss the manufacture of gutta-percha thread, tubes and hose. Brooman's patent uses a five-hole die that produces five simultaneous continuous filaments which are extruded into a bath and taken up on a roll. Bewley's patent extruded tubes and hose. Charles Hancock, who was a partner of Bewley in the Gutta Percha Company, is said to have developed insulation coating for wire using Bewley's extrusion methods. Methods of coating wires are described in patents by Barlow and Forster (English Patent No. 12,136) and by Siemens (English Patent No. 13,062) in 1848–1850. The first great successes of gutta-percha were its application to electrical insulation of the Dover–Calais and trans-Atlantic cables.

The development of continuous extrusion of plastics using screw extruders began with gutta-percha and natural rubber and dates from the 1870s. The concept of screw pumping seems to be attributable to Archimedes. The earlier use of screw pumps in the soap industry is described in the patent literature. The first patent for screw extrusion is that of Matthew Gray of London in 1879 (English Patent No. 5,056). Interestingly, the reason for the invention as cited by Gray is the existence of defects in coatings placed on wires. The extruder was fed from a two-roll mill or calendering device. There appear to have been independent developments of the screw extruder in Germany and the USA at about the same time, but Gray's patent is the first clear statement.

The next stage in the development of polymer processing methods came with the commercial development of cellulose nitrate as a plastic. The first moves in this direction during the 1860s by Alexander Parkes and Daniel Spill in England met with only limited success. Cellulose nitrate could not be melted and they used a range of volatile solvents that evaporated from their products. These left high levels of residual stresses which caused shrinking and cracking. Parkes and Spill had rubber-processing backgrounds and apparently used rubber-processing machinery. In the USA, John Wesley Hyatt and his brother Isaiah Smith Hyatt found that compounds or solutions of cellulose nitrate in nonvolatile camphor produced more desirable products. This was called Celluloid. The Celluloid Manufacturing Company was formed in the 1870s in Newark, New Jersey, to exploit this product and proved to be a great success. The Hyatts and their associates developed many important industrial processing operations to exploit Celluloid.

An 1872 patent by the Hyatt brothers (US Patent No. 133,229) contains both the reinvention of the ram extruder and the first ram injection-molding machine. They called this a stuffing machine. John Wesley Hyatt later described the use of complex multiple-cavity molds to be used in conjunction with the stuffing machine. This would either mold objects or coat cores of objects in the mold.

In an 1878 patent, John Wesley Hyatt (US Patent No. 204,228) described the extrusion of Celluloid from the stuffing machine over a mandrel coated with a lubricant. This mandrel could be programmable and expand to produce complex hollow shapes. This led to the development of blow molding in 1881 by the Hyatts' colleague William B. Carpenter (US Patent No. 237,168). Here, a preformed extruded tube is placed in a mold and is then expanded to fill the mold by pumping a heated fluid into the tube. These inventions were largely employed to produce a range of products including components of dolls and liners for pipes.

The 1880s saw the development of the synthetic fiber industry. Brooman's 1845 patent for the formation of gutta-percha thread sets out clear procedures for producing fibers from the melt. The synthetic fibers sold commercially in this period were produced from cellulose nitrate which could not be melted. A method of producing fibers by extruding acetic acid solutions of cellulose nitrate into a water or alcohol coagulation bath was described by Joseph Wilson Swan (English Patent No. 5,978) in 1883. Swan's patent describes the later carbonization of the fibers with heat and thus represents the beginning of the carbon-fiber industry. Swan's application was filaments for incandescent lights. Shortly thereafter in France, the Count de Chardonnet (US Patent No. 304,559) described a process for forming fibers from ether alcohol solutions by extruding the solutions into a water coagulation bath. De Chardonnet produced much finer fibers than Swan; he formed a company and commercialized them as an artificial silk. Later, de Chardonnet (US Patent No. 531,158) described a dry spinning process in which the filaments were extruded into the air where the solvent was evaporated. Also, during the 1890s, using the system of and collaborating with Cross, Bevan and Beadle, Stearn invented a reactive spinning method in which cellulose is dissolved in a mixture of sodium hydroxide and carbon disulfide to form cellulose xanthate, which is extruded into an acid coagulating bath that regenerates the cellulose. This material became known as rayon.

The first truly synthetic plastics, phenol formaldehyde resins, were developed commercially by Leo Hendrik Baekeland, a Belgian immigrant to the USA, just before 1910. These were poured as low- or intermediate-molecular-weight liquids into molds where they were polymerized into three-dimensional networks. Bakelite products were compression molded.

Large internal mixers were introduced into the rubber industry in the second decade of the twentieth century by Fernley H. Banbury, an English immigrant to the USA (US Patent No. 1,200,070). These were made necessary by increased production needs, the toxic accelerators used to hasten vulcanization and the increasing use of large quantities of finely divided particles, especially in the new pneumatic-tire industry. To some extent, these represented a new generation of Hancock's pickle mixers suitable for the modern period of rubber processing.

By 1920, the essential features of the processing of rubber, plastics and fibers had been established. The years since then have seen additional major innovations. These are associated with the development of new synthetic thermoplastics which were mostly introduced from 1920 to 1960. Screw extrusion devices using two corotating or counter-rotating screws were developed and applied as mixing devices. Reciprocating screws were introduced and replaced rams in injection-molding machines.

Numerous innovations in forming technology have been developed. In the late 1930s, Du Pont, notably through the efforts of Carothers, developed a commercial process for continuously forming highly oriented synthetic fibers from aliphatic polyamides (US Patent No. 2,130,948). This included the melt spinning and drawing processes.

Since the late 1940s, films have been produced by extruding melt from an annular die and expanding the melt with air. The basic patent of tubular film extrusion seems to be that of Fuller of the Visking Corporation issued in 1949 (US Patent No. 2,461,975).

Early process technologies mixed, pumped and shaped products. New process technologies have been developed since the mid-1960s which go beyond this to produce products with desirable mechanical properties inducing certain types of molecular orientation. The manufacture of melt-spun and drawn fibers is a forerunner of these technologies. Slit-film extrusion, notably of poly(ethylene terephthalate), followed by the subsequent continuous development of biaxial orientation on a tentering frame was developed by Du Pont investigators in the mid-1950s (US Patient No. 2,728,941). A generation later, Wyeth and Roseveare of Du Pont (US Patent No. 3,849,530) devised stretch blow molding of bottles where an inner mandrel vertically deforms as well as inflates a parison preform to produce a biaxially oriented bottle.

Most spectacular has been the invention of liquid crystalline process technology for aromatic polyamides by Kwolek, Morgan and their co-workers at Du Pont which has been applied to the manufacture of super high modulus and tensile strength fibers (US Patent No. 3,671,542).

Other developments have been the invention of processes for the formation of large foamed parts manufacture of multilayer film and sheet, and skin–core (sandwich) injection-molded parts.

See also: Processing: Classification

Bibliography

Adams R 1964 Wallace H. Carothers (1896–1937). *Rubber Chem. Technol.* 37(1): 27–34

Baekeland L 1909 The synthesis, constitution and uses of bakelite. *J. Ind. Eng. Chem.* 1: 149–61

Baekeland L 1916 Perkin medal address. *J. Ind. Eng. Chem.* 8: 184–90
Bolton E K 1942 Development of nylon. *Ind. Eng. Chem.* 34: 53–8
Carothers W H, Hill J W 1932 Artificial fibers from synthetic linear condensation superpolymers. *J. Am. Chem. Soc.* 54: 1579–87
Chardonnet H de (Count) 1914 Perkin medal address. *J. Soc. Dyers Colour* 176–9
Cross C F, Bevan E J, Beadle C 1893 New cellulose derivatives and their industrial applications. *J. Soc. Chem. Ind. London* 12: 498–501
Doree C 1935 Charles Frederick Cross (1855–1935). *J. Chem. Soc.* 91: 337–1340
Goodyear C 1855 Gum elastic. Privately printed, New Haven, CT
Hancock T 1857 *Personal Narrative of the Origin and Progress of Caoutchouc or India Rubber Manufacture in England.* Longmans, London
Hyatt J W 1914 Address of acceptance of Perkin medal. *J. Ind. Eng. Chem.* 6: 158–61
Kaufman M 1963 *The First Century of Plastics.* Plastics Institute, London
Killefer D H 1962 *Banbury, the Master Mixer.* Palmerton, New York
Tadmor Z, Gogos C G 1979 *Principles of Polymer Processesing.* Wiley, New York
White J L 1974 Charles Macintosh of Glasgow, his enterprises and the foundation of the rubber industry. *Rubber Ind. London* 8(4): 148–55
Wolf R F 1939 *India Rubber Man.* Caxton, Caldwell, ID
Woodruff W 1958 *The Rise of the British Rubber Industry During the Nineteenth Century.* Liverpool University Press, Liverpool, UK
Worden E C 1911 *Nitrocellulose Industry.* Van Nostrand, New York

J. L. White
[University of Akron, Akron, Ohio, USA]

Processing: Relation to Product Properties

The methods used to fabricate polymer products have a strong influence on the properties and performance of the product. Of special concern are mechanical and optical properties. The variations in properties are largely associated with the differences in structural order in these products, the most important influence probably being that of molecular orientation. However, other physical phenomena may also have a signficant effect, such as the thermodynamic state (including crystallinity and physical aging of glassy plastics) and the development of residual thermal stresses by rapid cooling. Also, optical characteristics are determined by structures within the polymer which have dimensions approximately equal to the wavelength of visible light and by defects on surfaces.

The general approach in this article is to relate properties to structural characteristics which are influenced by processing. Detailed relations between processing and such structural characteristics are given in the article *Processing: Structure Development*.

1. Uniaxial Orientation Effects on Mechanical Properties

Uniaxial orientation is a cylindrically symmetric orientation relative to a specific axis in the material. Such orientation occurs in fibers but arises more generally in products solidified under the action of a uniaxial tension.

The influence of orientation on the mechanical properties of polymers was noted in the literature by Carothers and Hill (1932) on melt spinning of aliphatic polyesters:

> "... if sufficient tension is applied during the spinning operation the filaments are very different from the initial mass in their physical properties.... They also have a very much higher breaking strength than the initial mass, and they are sufficiently pliable to be tied into hand knots, whereas the opaque filaments produced without stress are so fragile that they can hardly be bent without breaking. The opaque polyester and the transparent filaments produced under tension furnish very different x-ray diffraction patterns. The pattern for the opaque ester indicates that it is crystalline, but there is no sign of general orientation. The transparent filaments on the other hand furnish a fiber pattern such as one obtains from certain natural silk and cellulose fibers and rayon filaments that have been spun under tension."

Alfrey (1948) notes, citing studies of polystyrene and cellulose acetate, that uniaxial orientation increases tensile strength.

The general view of these early studies has been extended by numerous later investigators. It has been found that the modulus and tensile strength of melt spun and drawn fibers of amorphous and crystalline polymers increase with molecular orientation (Samuels 1974, Nadella et al. 1977, 1978). Figure 1 relates the tensile strength of polypropylene filaments to birefringence, which represents orientation. In glassy polystyrene, uniaxial orientation induces a brittle–ductile transition. Elongation to break first increases, reaches a maximum at small strain levels and then decreases (Kanetsuna and Tanabe 1978, Carey et al. 1980) (see Fig. 2). In crystalline polymers which have a two-phase structure, the variation of mechanical properties with uniaxial orientation may be correlated with the orientation of the amorphous rather than the crystalline regions (Samuels 1974).

The transverse ultimate properties of uniaxially oriented products frequently deteriorate. Transverse elongation to break in polystyrene decreases with uniaxial orientation (see Fig. 3). This is usually more noticeable in films (Matsumoto et al. 1981), but is also found in fibers. This effect is also seen in the fibrillation of drawn polyolefin fibers (Sze et al. 1976).

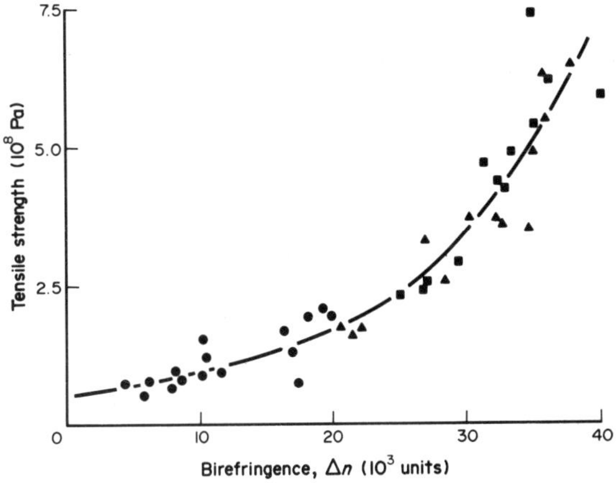

Figure 1
Variation of tensile strength with orientation for polypropylene filaments as measured by birefringence: ●, as spun; ▲, cold drawn (annealed 140 °C); ■, hot drawn (140 °C)

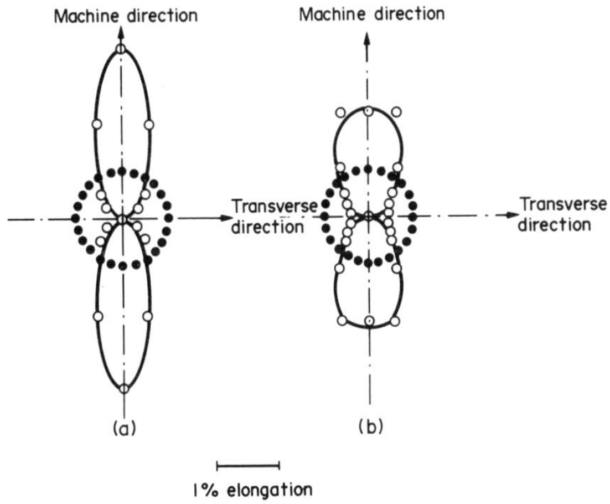

Figure 3
Elongation to break as a function of angle from machine direction at stretch temperature of (a) 110 °C and (b) 130 °C for a uniaxially oriented film of polystyrene

General formulations of the influence of orientation on mechanical properties have been developed by Raumann and Saunders (1961) and more recently by Ward (1971). They have interpreted the influence of orientation on mechanical properties in terms of anisotropic elastic theory. Basically, the infinitesimal strain γ_{ij} is written in terms of the stress tensor $\boldsymbol{\sigma}_{ij}$:

$$\gamma_{ij} = \boldsymbol{S}_{ijkl}\boldsymbol{\sigma}_{kl} \tag{1}$$

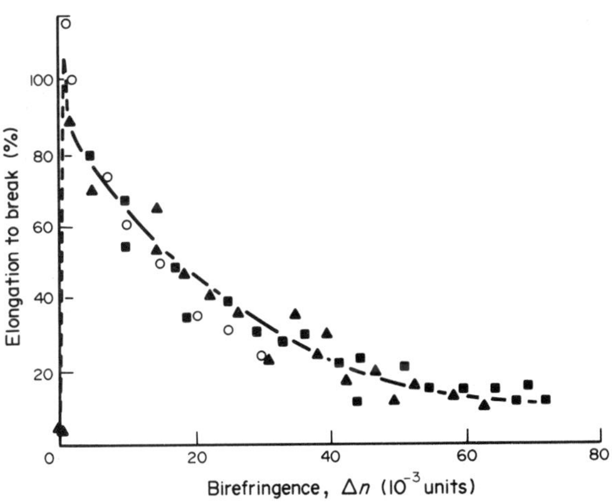

Figure 2
Elongation to break as a function of orientation for polystyrene as measured by birefringence. All results at room temperature: ■, nonisothermal; ▲, isothermal; ○, data from Kanetsuna and Tanabe (1978)

where $\boldsymbol{S}_{ijkl}$ is the compliance tensor and summation notation is used. This has in general 81 constants but, because of the symmetry of the stress and strain tensors, this simplifies to 36. Equation (1) is usually written

$$\gamma_i = \boldsymbol{S}_{ij}\boldsymbol{\sigma}_j \tag{2}$$

where the indices run from 1 to 6 (Raumann and Saunders 1961, Ward 1971). Energy arguments show that $\boldsymbol{S}_{ij}$ is symmetric ($\boldsymbol{S}_{ij} = \boldsymbol{S}_{ji}$), reducing the number of independent constants from 36 to 21. For uniaxial orientation, there are five independent compliances: S_{11}, S_{33}, S_{44}, S_{66} (or S_{12}) and S_{13}. S_{11} and S_{33} correspond to compliances in the plane of a sheet (filament) perpendicular and parallel to the axis of symmetry. S_{44} is a shear modulus in a plane containing the axis of symmetry. S_{66} is a shear modulus in the plane perpendicular to the symmetry axis. $-S_{12}/S_{11}$ and $-S_{13}/S_{33}$ corresponds to Poisson's ratios. Clearly S_{11}, S_{33}, S_{44} and S_{66} are all positive and S_{12} and S_{13} are probably negative.

Raumann and Saunders (1961), Raumann (1963), and Pinnock and Ward (1964, 1966) have determined the influence of uniaxial orientation, as specified by birefringence, on the mechanical properties of filaments. They found that in uniaxially oriented poly(ethylene terephthalate) and polypropylene the extensional compliance S_{33} and the torsional compliance S_{44} decrease with increasing birefringence. (The corresponding moduli, of course, increase.) The effect is much stronger for the extensional compliance. Little variation of the properties perpendicular to the symmetry axis (S_{11}) has been found.

Sophisticated studies of the effects of polymer chain orientation on plastic yielding have also been reported. The influence of orientation on the yield surface in principal stress space has been investigated by Bridle et al. (1968), Brown et al. (1968) and Caddell et al. (1973) among others. Hill's anisotropic yield criterion which seeks to generalize the von Mises yield criterion to an orthorhombic crystal is used as the basis for these formulations (Ward 1971).

2. Biaxial Orientation Effects on Mechanical Properties

Uniaxial orientation in films and injection-molded products is often found to be undesirable as the improvement of mechanical properties in the machine direction is sometimes more than offset by the deterioration of the transverse properties. The effects of biaxial orientation as opposed to uniaxial orientation on films/sheets of polystyrene have been described by Thomas and Cleereman (1972) and by Matsumoto et al. (1981). They have found that the material becomes tough in the transverse direction as well as in the machine direction (see Fig. 4).

In any case, it is desirable to manufacture products with uniform properties in the plane of the film or molded product. Processes have been described to develop such orientation using: (a) tentering frames (Alles 1956) and mandrels (Carlson 1971) to produce biaxial films, (b) translating (Wyeth and Roseveare 1973) or rotating (Cleereman 1967) mold components to produce biaxial injection or blow-molded containers or (c) special compression molding (scrapless forming) (Ayres et al. 1973). Various biaxially oriented products including polyester film and bottles are commercially available.

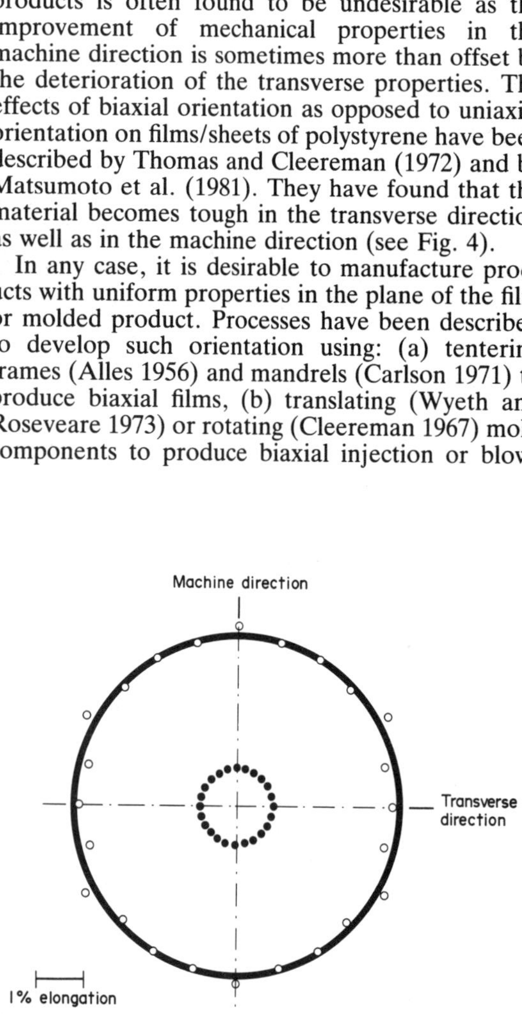

Figure 4
Elongation to break as a function of angle from one machine direction for a biaxially oriented film of polystyrene at 115 °C

3. Thermodynamic State, Physical Aging, Residual Stresses and Mechanical Properties

The state of matter of many polymers depends on cooling, heat treatment and/or environment melt flow conditions in processing. The most important of such possible changes is the occurrence and level of crystallinity. Thus, poly(ethylene terephthalate) filaments may be crystalline or amorphous depending on the state of quench and applied stresses.

Perhaps a more common problem is the quenching of a vitrifying melt into a nonequilibrium glassy state (Struik 1977). Generally, amorphous polymers are not in thermodynamic equilibrium below their glass-transition temperature T_g. As supercooled liquids, they have a higher volume and entropy than they would have in the equilibrium state. Molecular mobility below T_g is not quite zero and there is a slow approach to equilibrium which acts as a gradual continuation of the vitrification process. This depends on $T_g - T$, that is, how far room temperature is below the glass transition. Many physical properties of the polymer are changed during the aging process as the material becomes more glasslike and less rubberlike. The polymer becomes stiffer and brittle and its damping capacity decreases. This aging phenomenon has been found in poly(methyl methacrylate), polystyrene, polycarbonate, poly(ethylene terephthalate) and polycarbonate. In past research into physical aging, care has not been taken with the variation of cooling rates across the thickness of parts and the resulting heterogeneity of the specimens, and studies of heat conduction and temperature profiles are needed.

The effects of physical aging on the properties of fabricated glassy polymers should be clear. The mechanical behavior of parts varies depending on cooling conditions, being more rubbery for quenched samples and stiff and brittle for slow cooling. Furthermore, the aging of mechanical properties creates difficulty in supplying reliable products to customers. The longer the product is stored, the more it changes.

Another factor is residual thermal stress. When a polymer melt is cooled to induce solidification, the material first solidifies at its periphery. The volume of the product at this stage exceeds that which would exist if the material were in equilibrium at room temperature. When the product reaches room temperature, it has an abnormally large volume, causing volumetric–dilational strains in the inner portions. The stresses set up are referred to as residual thermal stresses (Knappe 1961, Wubken 1975,

Broutman and Khrishnakumar 1976, Struik 1978). The stress distribution is extensional or dilational in the inner part of cross sections and compressive in the outer layers. Thermal stresses may be determined by layer removal or by birefringence in polymers (if there is little contribution from orientation). The former method is recommended.

The influence of thermal stresses as opposed to physical aging on the properties of fabricated thermoplastics is unclear. Thermal stresses have been claimed to increase impact strength in glassy plastics (Broutman and Khrishnakumar 1976), though Wubken (1975) suggests that they propagate cracks. It is not clear in such studies if physical aging is distinguished from residual thermal stresses.

4. Optical Properties

Another characteristic of importance is the optical behavior of polymers, especially films. The state of orientation itself determines the optical isotropy of films. Indeed, birefringence is used to measure orientation (Matsumoto et al. 1981).

The haze of films is caused by both scattering from the surface and from the internal structure of films (Clampitt et al. 1969, Stehling et al. 1981). Surface roughness seems to be the dominant problem in commercial tubular film processes for low-density polyethylene film. This surface roughness is associated both with die exit flows and disturbances during crystallizastion. This may be reduced by mechanical deformation before extrusion (Stehling et al. 1981).

The variation in transparency of high-density polyethylene films with processing has been attributed primarily to variations in crystalline morphology and not simply the size of scattering units (Stein and Prud'homme 1971). Rodlike morphologies due to row structures give less scattering.

See also: Polymers: Mechanical Properties; Processing: Shear-Controlled Orientation

Bibliography

Alfrey T 1948 *Mechanical Behavior of High Polymers*. Interscience, New York

Alles F P 1956 Process for longitudinally stretching polymeric film. US Patent No. 2,767,435

Ayres R C, Cleereman K J, Schrenk W J 1973 Scrapless forming of plastic articles. US Patent No. 3,739,052

Bridle C, Buckley A, Scanlan J 1968 Mechanical anisotropy of oriented polymers—I. A yield criterion for uniaxially drawn polyethylene teephthalate. *J. Mater. Sci.* 3: 622–8

Broutman L J, Khrishnakumar S M 1976 Impact strength of polymers—I. The effect of thermal treatment and residual stress. *Polym. Eng. Sci.* 16: 74–81

Brown N, Duckett R A, Ward I M 1968 The yield behavior of oriented polyethylene terephthalate. *Philos. Mag.* 18: 483–502

Caddell R M, Raghava R S, Atkins A G 1973 A yield criterion for anisotropic and pressure dependent solids such as oriented polymers. *J. Mater. Sci.* 8: 1641–6

Carey D A, Wust C J, Bogue D C 1980 Studies in nonisothermal rheology: Behavior near the glass transition temperature and in the oriented glassy state. *J. Appl. Polym. Sci.* 25: 575–88

Carlson F A 1971 Method for producing biaxially oriented polystyrene heavy gauge sheet. US Patent No. 3,619,545

Carothers W H, Hill J W 1932 Artificial fibers from synthetic linear condensation superpolymers. *J. Am. Chem. Soc.* 54: 1579–87

Clampitt B H, German D E, Anspon H D 1969 Techniques for measuring surface and bulk properties of polyethylene films. *Anal. Chem.* 41: 1306–9

Cleereman K J 1967 Injection molding of shapes of rotational symmetry with multiaxial orientation. *Soc. Plast. Eng. J.* 25: 43–7

Cleereman K J 1967 Multilayered thermoplastic article. US Patent No. 3,307,726

Jones T T 1976 The effect of molecular orientation on the mechanical properties of polystyrene. *Proc. Appl. Chem.* 45: 39–57

Kanetsuna H, Tanabe Y 1978 Brittle-to-ductile transition based upon amorphous orientation of polystyrene monofilaments. *J. Appl. Polym. Sci.* 22: 2707–11

Knappe W 1961 Die Festigkeit thermoplastichen Kunststoffe in abhangigkeit von den Verarbeitungsbedingungen. *Kunststoffe* 51: 562–9

Matsumoto K, Fellers J F, White J L 1981 Uniaxial orientation development and mechanical properties of polystyrene film. *J. Appl. Polym. Sci.* 26: 85–96

Nadella H P, Henson H M, Spruiell J E, White J L 1977 Melt spinning of isotactic polypropylene: Structure development and relationship to mechanical properties. *J. Appl. Polym. Sci.* 21: 3003–22

Nadella H P, Spruiell J E, White J L 1978 Drawing and annealing of polypropylene fibers: Structural changes and mechanical properties. *J. Appl. Polym. Sci.* 22: 3121–33

Pinnock P R, Ward I M 1964 Mechanical and optical anisotropy in polyethylene terephthalate fibres. *Br. J. Appl. Phys.* 15: 1559–68

Pinnock P R, Ward I M 1966 Mechanical and optical anisotropy in polypropylene fibres. *Br. J. Appl. Phys.* 17: 575–86

Raumann G 1963 The anisotropy of the elastic constants of transversely isotropic polyethylene terephthalate. *Br. J. Appl. Phys.* 14: 795–9

Raumann G, Saunders D W 1961 The anisotropy of Young's modulus in drawn polyethylene. *Proc. Phys. Soc.* 77: 1028–37

Samuels R J 1974 *Structured Polymer Properties: The Identification, Interpretation, and Application of Crystalline Polymer Structure*. Wiley, New York

Stehling F C, Speed C S, Westerman L 1981 Causes of haze of low-density polyethylene blown films. *Macromolecules* 14: 698–708

Stein R S, Prud'homme R 1971 Origin of polyethylene transparency. *J. Polym. Sci. B* 9: 595–8

Struik L C E 1977 Physical aging in plastics and other glassy materials. *Polym. Eng. Sci.* 17: 165–73

Struik L C E 1978 Orientation effects and cooling stresses in amorphous polymers. *Polym. Eng. Sci.* 18: 799–811

Sze G M, Spruiell J E, White J L 1976 The influence of

drawing, twisting, heat setting and untwisting on the structure and mechanical properties of melt-spun high-density polyethylene fiber. *J. Appl. Polym. Sci.* 20: 1823–47

Thomas L S, Cleereman K J 1972 Gaging the benefits of biaxial orientation—I. General-purpose polystyrene. *Soc. Plast. Eng. J.* 28: 61–7

Treloar L R G 1971 Molecular orientation in polymers. *Plast. Polym.* 39: 29–37

Ward I M 1971 *Mechanical Properties of Solid Polymers*. Wiley, New York

Ward I M (ed.) 1975 *Structure and Properties of Oriented Polymers*. Wiley, New York

Wubken G 1975 Eigenspannungen in Spritzgussteilen. *Plastverarbeiter* 26: 17–23

Wyeth N C, Roseveare R N 1973 Biaxially oriented poly(ethylene terephthalate) bottle. US Patent No. 3,733,309

J. L. White
[University of Akron, Akron, Ohio, USA]

Processing: Rheological Properties and Fluid Mechanics

In polymer processing, a supply of polymer, typically in the form of granules or pellets, is subjected to heating or mechanical working or (usually) both to produce a homogeneous melt, which is shaped in some way and then cooled, with the possibility of further shaping during cooling. The rheology of the melt, over a temperature range that involves substantial changes in the ease with which it is deformed, influences the design and operation of polymer processing equipment and the properties of finished articles made from the polymer.

The two most significant features of polymer melt rheology, after the temperature dependence of properties, are shear-thinning behavior and elastic (or viscoelastic) behavior. Effects of pressure are commonly ignored (by theorists at least) and are not discussed here though perhaps more should be said (see Cogswell 1981 pp. 43–5). Shear thinning is exaggerated in practical situations by viscous heat generation, so that the most rapidly sheared material tends to be hottest and, hence, least viscous, but even for isothermal flow the viscosity at shear stresses occurring in polymer processing operations is (for some polymers) less than one-hundredth of the viscosity at very low stresses. (For other polymers, notably those used in artificial fiber production, the reduction may be much less, and viscosity may be nearly independent of shear rate.) Apart from the fact that it makes calculations more difficult, shear-thinning behavior is a great asset in polymer processing since it cuts down the energy requirements for extrusion and for flow in dies and molds.

Melt elasticity manifest itself in a number of ways, is less obviously advantageous and raises much greater problems of mathematical description. However, it does seem to be associated with the ability of some polymer melts to form stable fibers and films, and with the fundamental differences in response of polymer melts to shear and elongation. Other phenomena associated with the elastic nature of polymer melts include the existence of normal stress differences in shear (which may be larger than the shear stress at the high shear rates typical of many polymer processing operations), the swelling of an extrudate after emerging from a die and large recoverable deformations (often frozen in by solidification). Much has been written on polymer rheology as a science and in relation to polymer processing, and viewpoints differ widely. Hence, it is necessary to offer a large list of reference texts (Brydson 1970, Astarita and Marrucci 1974, Han 1976, Bird et al. 1977, Middleman 1977, Schowalter 1978, Fenner 1979, Tadmor and Gogos 1979, Vinogradov and Malkin 1980, Cogswell 1981, Pearson 1985). It is also worth drawing attention to computational investigations both in non-Newtonian fluid mechanics (Crochet et al. 1984) and in the analysis of polymer processing operations (Pearson and Richardson 1983). Both finite-difference and finite-element methods are used and most of the models discussed below have been investigated for some flow fields.

In discussing polymer rheology in the context of modelling the mechanics of melt processing operations it is convenient to divide flows into two categories. Internal flows (in extruders, dies and mold channels) are predominantly shear flows and the basic concern is with the melt viscosity and its dependence on shear rate and temperature. External flows (where there is a free surface, as in melt spinning, film blowing, blow molding and thermoforming) are predominantly elongational flows and, moreover, flows in which the stress experienced by the melt changes with time (if we follow a particular particle of the melt through the process). Hence, the response to changing stress, which brings in the viscoelastic nature of the melt, ought to be considered, as well as the elongational viscosity.

There are also situations, of which converging flow is the most significant, where the responses of the melt both to shear and to elongation are important and it involves a very great assumption to treat this by adding the two contributions—although admittedly the state of the art is such that we cannot do much better than make this approximation (Cogswell 1978, Petrie 1979 Chap. 7).

Another aspect of the mechanics of polymer processing which cannot be ignored is that of instability. The rate of flow and, hence, of production is often limited by instabilities such as surging in extruders, melt fracture and sharkskin in extrudates, and draw resonance in melt spinning (Pearson 1976, Petrie and Denn 1976, White and Ide 1978). This is also an aspect of processability—instability may prevent the production of an acceptable product with a particular polymer or grade of polymer—and rheological tests are not always useful in distinguishing between processable and nonprocessable polymers (Meissner 1975).

1. Shear Flows: Viscometric Functions

Viscometric functions are defined for steady simple shear, and for a class of flows that are geometrically equivalent to steady simple shear. In a steady simple shear using the notation recommended by the Society of Rheology (Sieglaff 1976), we have a velocity v_1 in the x_1 direction which is a function of x_2 only, where (x_1, x_2, x_3) are Cartesian coordinates of a point in the liquid. The shear stress is $\sigma = \sigma_{21}$, and the normal stresses are σ_{11}, σ_{22} and σ_{33}. If the shear rate is $\dot{\gamma}$, defined to be $\partial v_1/\partial x_2$, then the viscosity ($\eta$) and normal stress coefficients (Ψ_1, Ψ_2) are even functions of $\dot{\gamma}$ and are defined by

$$\eta(\dot{\gamma}) = \sigma(\dot{\gamma})/\dot{\gamma} \tag{1}$$

$$\Psi_1(\dot{\gamma}) = N_1(\dot{\gamma})/\dot{\gamma}^2 = [\sigma_{11}(\dot{\gamma}) - \sigma_{22}(\dot{\gamma})]/\dot{\gamma}^2 \tag{2}$$

$$\Psi_2(\dot{\gamma}) = N_2(\dot{\gamma})/\dot{\gamma}^2 = [\sigma_{22}(\dot{\gamma}) - \sigma_{33}(\dot{\gamma})]/\dot{\gamma}^2 \tag{3}$$

where N_1 and N_2 are known as the primary and secondary normal stress differences, respectively.

These functions may be obtained experimentally; $\eta(\dot{\gamma})$ should be available for any polymer, $\Psi_1(\dot{\gamma})$ may be available (at least for a similar polymer) but $\Psi_2(\dot{\gamma})$ is not easily measured for polymer melts.

There are a number of expressions for the viscosity function which are used to fit empirical data and which may be used in calculations. The commonest of these is the power law model:

$$\sigma(\dot{\gamma}) = K\dot{\gamma}^n \tag{4a}$$

or

$$\eta(\dot{\gamma}) = K\dot{\gamma}^{n-1} \tag{4b}$$

where K and (to a lesser extent) n are temperature-dependent parameters. For polymer melts, $0<n<1$ is generally observed (shear-thinning behavior); n may be obtained from the slope of a graph of $\log \sigma$ against $\log \dot{\gamma}$. Generally, this will be adequate over restricted ranges of the variables and, in particular, will not be good for very small values of $\dot{\gamma}$ or σ where Newtonian behavior ($n=1$) is observed. An improvement which shows power law behavior for large $\dot{\gamma}$ and Newtonian behavior for small $\dot{\gamma}$ is

$$\eta(\dot{\gamma}) = \eta_0/(1 + \eta_0/K\dot{\gamma}^{n-1}) \tag{5}$$

It is often asserted that there is also an "upper Newtonian" region, that is, that at very high shear rates the viscosity again becomes constant, at a value η_∞ which is less than η_0. Experimental evidence for this is difficult to obtain (because of flow instabilities), but the viscosity function proposed by Carreau,

$$\eta(\dot{\gamma}) = \eta_\infty + (\eta_0 - \eta_\infty)[1 + (\lambda\dot{\gamma})^2]^{(n-1)/2} \tag{6}$$

has this additional feature and is quite successful in fitting experimental data by choice of its four parameters.

It is possible to construct a constitutive equation that describes any observed behavior in a shear flow, that is, can fit any observed set of viscometric function, Eqns. (1–3). This is the Criminale–Ericksen–Filbey equation, which may be written

$$\boldsymbol{\sigma} = 2\eta \boldsymbol{D} + 4\Psi_2 \boldsymbol{D}^2 - \Psi_1 \boldsymbol{D}^\triangledown \tag{7}$$

where $\boldsymbol{\sigma}$ is the extra-stress tensor, $\boldsymbol{D}$ is the rate of strain tensor, $\boldsymbol{D}^\triangledown$ is the upper convected derivative of $\boldsymbol{D}$ and η, Ψ_1 and Ψ_2 are functions of the second invariant of $\boldsymbol{D}$, which reduces to $\frac{1}{2}\dot{\gamma}^2$ in simple shear. If the velocity has Cartesian components $v_i(\boldsymbol{x})$ then $\boldsymbol{D}$ has components

$$d_{ij} = \frac{1}{2}\left(\frac{\partial v_i}{\partial x_j} + \frac{\partial v_j}{\partial x_i}\right) \tag{8}$$

and if $\boldsymbol{W}$ is the vorticity tensor, with components

$$w_{ij} = \frac{1}{2}\left(\frac{\partial v_i}{\partial x_j} + \frac{\partial v_j}{\partial x_i}\right) \tag{9}$$

then the upper convected derivative of a tensor $\boldsymbol{T}$ is defined by

$$\boldsymbol{T}^\triangledown = \frac{\partial \boldsymbol{T}}{\partial t} + \boldsymbol{v}\,.\,\nabla \boldsymbol{T} - \boldsymbol{W}\,.\,\boldsymbol{T} + \boldsymbol{T}\,.\,\boldsymbol{W} - (\boldsymbol{D}\,.\,\boldsymbol{T} + \boldsymbol{T}\,.\,\boldsymbol{D}) \tag{10}$$

(also known as the Oldroyd contravariant derivative). Other convected derivatives commonly used are the corotational (or Jaumann) derivative $\boldsymbol{T}^\circ = \boldsymbol{T}^\circ + \boldsymbol{D}\,.\,\boldsymbol{T} + \boldsymbol{T}\,.\,\boldsymbol{D}$ and the lower convected (or Oldroyd covariant) derivative $\boldsymbol{T}^\triangle = \boldsymbol{T}^\circ + \boldsymbol{D}\,.\,\boldsymbol{T} + \boldsymbol{T}\,.\,\boldsymbol{D}$. An alternative development of Eqn. (7) uses the Rivlin–Ericksen tensors $\boldsymbol{A}_1 = 2\boldsymbol{D}$ and $\boldsymbol{A}_2 = 2\boldsymbol{D}^\triangle$. Once Eqn. (7), which is exactly true in all steady shear flows, has been obtained, it may be used in flows close to steady shear as an approximation.

Observations show quite clearly that Ψ_1 is positive and the limited data available indicate that, at least at low shear rates, Ψ_2 is negative. Note that Eqn. (7) will not describe stress growth (or decay) when motion starts (or ceases), which is a feature of the viscoelastic behavior or polymer melts. Another important feature of practical processing is temperature variation and it is essential, in modelling most processing operations, to take account of this, at least through an emprical law (e.g., an Arrhenius equation) describing viscosity dependence on temperature.

2. The Lubrication Approximation

If the polymer melt flows in a channel described by two surfaces which are close together, with a gap $h(x, z)$ in the y direction which does not change significantly with x and z in distances comparable to h, we have a geometry similar to that occurring in lubrication by thin films of oil. The theory developed by Reynolds for this situation may be extended to deal with non-Newtonian liquids and has application to screw extrusion, injection molding, flow through sheet and film dies, calendering and coating (Middleman 1977 pp. 172, 192, 199, Fenner 1979 p. 45, Tadmor and Gogos 1979 p. 126, Pearson 1985 Chap. 8).

The basic approximation consists in assuming that the velocity profile at any position in the flow channel is the same as that in a gap of width $h(x, z)$ between parallel planes under the action of a pressure gradient or motion of the two planes (in the case of screw extrusion, for example). The basic equations are

$$\frac{\partial p}{\partial x}=\frac{\partial(\eta\, \partial v_x/\partial y)}{\partial y} \qquad (11)$$

$$\frac{\partial p}{\partial z}=\frac{\partial(\eta\, \partial v_z/\partial y)}{\partial y} \qquad (12)$$

$$\frac{\partial(k\, \partial T/\partial y)}{\partial y}=-\eta\dot{\gamma}^2 \qquad (13)$$

where

$$\dot{\gamma}^2=\left(\frac{\partial v_x}{\partial y}\right)^2+\left(\frac{\partial v_z}{\partial y}\right)^2 \qquad (14)$$

and η is a function of $\dot{\gamma}$, given by Eqn. (4) for example. The thermal conductivity k and viscosity η both depend on the temperature T. Here it is assumed in Eqn. (13) that the temperature profile is fully developed, which may not be a good approximation unless the flow channel is long (Winter 1975), and is certainly more likely to be in error than the assumption of a fully developed velocity profile (which is also implicit in the lubrication approximation).

Appropriate boundary conditions for these partial differential equations are that v_x and v_z are zero (or are prescribed, when one surface moves as in screw extrusion of coating) at the surfaces [$y=0$ and $y=h(x, z)$], and either T or $\partial T/\partial y$ or some combination is given at the surfaces, depending on the heat capacity of the channel walls and the heat transfer characteristics of the surfaces. Further boundary conditions are needed for values of x and z describing the edges of the flow region.

3. Elongation Flows

In flows that involve stretching of a thread or sheet of molten or soft solid polymer, the deformation is fundamentally different from simple shear (Petrie 1979); indeed such flows are sometimes called shear-free flows. Moreover, processing operations involving these flows are not carried out with constant rates of strain, although these are sought in elongational viscometry. Hence, information on transient rheological behavior, for example stress growth at the start of flow, is needed for any realistic modelling of the processes. With the present state of the art it has to be said that the influence of extrusion conditions (upstream of the elongational flow) is known to be important but has not been modelled with complete success.

A starting point for attempts to describe time-dependent elongational behavior is the Maxwell model, which may be written as a rate equation, for example the upper convected Maxwell model with relaxation time λ,

$$\boldsymbol{\sigma}+\lambda\boldsymbol{\sigma}^{\nabla}=2\eta\boldsymbol{D} \qquad (15)$$

or as an integral equation. As an illustration, in uniaxial elongation, if the rate of strain in the direction of stretching is $\dot{\varepsilon}$, for axisymmetric flow the rate of strain tensor is

$$\boldsymbol{D}=\begin{bmatrix}\dot{\varepsilon} & 0 & 0\\ 0 & \frac{1}{2}\dot{\varepsilon} & 0\\ 0 & 0 & \frac{1}{2}\dot{\varepsilon}\end{bmatrix} \qquad (16)$$

in rectangular Cartesian coordinates, with stretching in the x_1 direction. Then the upper convected Maxwell model gives

$$\sigma_{11}+\lambda(\dot{\sigma}_{11}-2\dot{\varepsilon}\sigma_{11})=2\eta\dot{\varepsilon} \qquad (17)$$

$$\sigma_{11}+\lambda(\dot{\sigma}_{11}-\dot{\varepsilon}\sigma_{11})=-\eta\dot{\varepsilon} \qquad (18)$$

with $\sigma_{33}=\sigma_{22}$ and the dot denoting the usual material derivative,

$$\dot{\sigma}_{ii} = \frac{\partial \sigma_{ii}}{\partial t} + v_1 \frac{\partial \sigma_{ii}}{\partial x_1} \tag{19}$$

For the elongation of a uniform specimen of length L, as in a tensile test, quantities are independent of x_1 and $\dot{\sigma}_{ii} = \partial \sigma_{ii}/\partial t$. In this case $\varepsilon = \ln(L/L_0)$, the Hencky strain relative to length L_0 at time t_0, and then $\dot{\varepsilon} = L^{-1}\, dL/dt$. Another simplification is obtained in steady spinning, when quantities are independent of t, and then $\dot{\sigma}_{ii} = v_1\, \partial \sigma_{ii}/\partial v_1$, $\varepsilon = \ln(v_1/v_0)$, where v_0 is the value of the velocity at some cross section $x_1 = x_0$, and $\dot{\varepsilon} = \partial v_1/\partial x_1$.

The corresponding equations for a Newtonian liquid are

$$\sigma_{11} = 2\eta\dot{\varepsilon}, \qquad \sigma_{22} = -\eta\dot{\varepsilon} \tag{20}$$

and, using the boundary conditions of no net radial stress ($\sigma_{22} - p = 0$, where p is the isotropic pressure), the applied stress in the direction of stretching is $\sigma_{11} - \sigma_{22}$; hence the classical result of Trouton,

$$\eta_E = (\sigma_{11} - \sigma_{22})/\dot{\varepsilon} = 3\eta \tag{21}$$

where the nomenclature of the Society of Rheology (Dealy 1984) is used. For the corotational Maxwell model, that is,

$$\boldsymbol{\sigma} + \lambda\overset{\circ}{\boldsymbol{\sigma}} = 2\eta\boldsymbol{D} \tag{22}$$

we have

$$\sigma_{11} + \lambda\dot{\sigma}_{11} = 2\eta\dot{\varepsilon} \tag{23}$$

$$\sigma_{22} + \lambda\dot{\sigma}_{22} = -\eta\dot{\varepsilon} \tag{24}$$

Note that this is exactly the same as would be predicted by linear viscoelasticity, valid here for large deformation. The Trouton viscosity here is 3η, exceeding three times the shear viscosity at corresponding nonzero rates of strain, since for this model $\eta(\dot{\gamma}) = \eta/(1 + \lambda^2\dot{\gamma}^2)$. From Eqns. (17) and (18) for uniform steady flow of the upper convected Maxwell model, we have

$$\eta_E = 3\eta/[(1 - 2\lambda\dot{\varepsilon})(1 + \lambda\dot{\varepsilon})] \tag{25}$$

This has two important features: first, the Trouton viscosity increases with increasing rate of strain, while in shear the model predicts a constant viscosity; and second, for large rates of strain ($\dot{\varepsilon} > 1/2\lambda$) the stress never reaches a steady value and so Eqn. (25) is not valid (since it is obtained by assuming that the time derivatives of $\boldsymbol{\sigma}$ are zero). This is interpreted by calling the response at low rates of strain liquidlike and the response at high rates of strain solidlike, since the stress increases with the strain (and does so without limit as the model contains no rupture criterion).

Neither of the convected Maxwell models discussed, nor any model of similar simplicity, gives an adequate quantitative description of polymer melt rheology. For some polymers (e.g., polystyrene), Eqn. (15) with strain-rate-dependent parameters seems to be quite successful, even with a single relaxation time. Other polymers (e.g., low-density polyethylene) require a spectrum of relaxation times, and also strain- or stress-dependent parameters. The latter are typified by the Phan-Thien–Tanner model:

$$\boldsymbol{\sigma} = \sum_{i=1}^{n} \boldsymbol{\sigma}^{(i)} \tag{26}$$

$$\boldsymbol{\sigma}^{(i)} + \lambda^{(i)}[\boldsymbol{\sigma}^{\nabla(i)} + \xi(\boldsymbol{\sigma}^{(i)} . \boldsymbol{D} + \boldsymbol{D} . \boldsymbol{\sigma}^{(i)})] = 2\mathrm{G}^{(i)}\lambda^{(i)}\boldsymbol{D} \tag{27}$$

$$\lambda^{(i)} = \tau^{(i)} \exp[-\mathrm{b}\, \mathrm{tr}(\boldsymbol{\sigma}^{(i)})] \tag{28}$$

where $\mathrm{tr}(\boldsymbol{\sigma})$ denotes the trace of $\boldsymbol{\sigma}$ and where ξ, b, $G^{(i)}$ and $\tau^{(i)}$ are constants (Phan-Thien 1978). The idea of strain-dependent parameters is more easily accommodated in an integral model of the Kaye–BKZ (Bernstein–Kearsley–Zapas) type:

$$\boldsymbol{\sigma} = \int_0^\infty [m_1(s, I, II)\boldsymbol{C}_t^{-1}(t-s) + m_2(s, I, II)\boldsymbol{C}_t(t-s)]\, ds \tag{29}$$

where I and II are invariants of the deformation tensor, $\mathrm{tr}[\boldsymbol{C}_t^{-1}(t-s)]$ and $\mathrm{tr}[\boldsymbol{C}_t(t-s)]$. In uniaxial elongation (Eqn. (16)) if we define the strain (at time $t-s$, relative to time t),

$$\varepsilon_t(t-s) = \int_t^{t-s} \dot{\varepsilon}(t')\, dt' \tag{30}$$

we have principal stretch ratios $\exp[-\varepsilon_t(t-s)]$ and $\exp[\varepsilon_t(t-s)/2]$, each being the ratio of lengths of a material line at times t and $t-s$, and the relative Finger deformation tensor has components

$$\boldsymbol{C}_t^{-1} \begin{bmatrix} \exp[-2\varepsilon_t(t-s)] & 0 & 0 \\ 0 & \exp \varepsilon_t(t-s) & 0 \\ 0 & 0 & \exp \varepsilon_t(t-s) \end{bmatrix} \tag{31}$$

There are empirical models of this type and similar models based on molecular modelling (Petrie and Dealy 1980).

Some recent work on elongational flows has been concerned with nonisothermal rheology, and this is of fundamental importance in polymer processing (Matsumoto and Bogue 1977, Gupta and Metzner 1980). Instability has also been mentioned already, but one final area of theoretical uncertainty needs to be mentioned. The transition from shear to elongational flow in a die is extremely difficult to model and it is accompanied by the phenomenon of extrudate swell (sometimes called die swell). For an emerging extrudate of polymer melt which is not under tension, the diameter may increase by a factor of two or three. This cannot be predicted reliably, although some correlations are available, nor is the relative importance of various possible influences certain. Furthermore, the effect of an applied tension is significant and would need to be considered in any realistic model of melt spinning or film blowing. In parison formation for blow molding the time-dependent nature of extrudate swell is also highly relevant.

See also: Processing: Classification; Rheological Characterization of Molten Polymers

Bibliography

Astarita G, Marrucci G 1974 *Principles of Non-Newtonian Fluid Mechanics.* McGraw-Hill, London

Bird R B, Armstrong R C, Hassager O 1977 *Fluid Mechanics,* Dymanics of Polymeric Liquids, Vol. 1. Wiley, New York

Brydson J A 1970 *Flow Properties of Polymer Melts.* Plastics and Rubber Institute, London

Cogswell F N 1978 Converging flow and stretching flow: A compilation. *J. Non-Newtonian Fluid Mech.* 4: 23–38

Cogswell F N 1981 *Polymer Melt Rheology.* Wiley, New York

Crochet M J, Davies A R, Walters K 1984 *Numerical Simulation of Non-Newtonian Flow.* Elsevier, Amsterdam

Dealy J M 1984 Official nomenclature for material functions describing the response of a viscoelastic fluid to various shearing and extensional deformations. *J. Rheol.* 28: 181–95

Fenner R T 1979 *Principles of Polymer Processing.* Macmillan, London

Gupta R K, Metzner A B 1980 Non-isothermal flows of viscoelastic fluids. In: Astarita G, Marrucci G, Nicolais L (eds.) 1980 *Applications*, Rheology, Vol. 3. Plenum, New York, pp. 3–8

Han C D 1976 *Rheology in Polymer Processing.* Academic Press, New York

Matsumoto T, Bogue D C 1977 Non-isothermal rheological response during elongational flow. *Trans. Soc. Rheol.* 21: 453–68

Meissner J 1975 Basic parameters, melt rheology and end-use properties of three similar low-density polyethylene samples. *Pure Appl. Chem.* 42: 553–612

Middleman S 1977 *Fundamentals of Polymer Processing.* McGraw-Hill, New York

Pearson J R A 1976 Instability of non-Newtonian flow. *Annu. Rev. Fluid Mech.* 8: 163–81

Pearson J R A 1985 *Mechanics of Polymer Processing.* Elsevier, London

Pearson J R A, Richardson S M (eds.) 1983 *Computational Analysis of Polymer Processing.* Applied Science, London

Petrie C J S 1979 *Elongational Flows*, Research Notes in Mathematics, 29. Pitman, London

Petrie C J S, Dealy J M 1980 Uniform elongational flow of molten polymers. In: Astarita G, Marrucci G, Nicolais L (eds.) 1980 *Principles*, Rheology, Vol. 1. Plenum, New York, pp. 171–94

Petrie C J S, Denn M M 1976 Instabilities in polymer processing. *AIChE J.* 22: 209–36

Phan-Thien N 1978 A nonlinear network viscoelastic model. *J. Rheol.* 22: 259–83

Schowalter W R 1978 *Mechanics of Non-Newtonian Fluids.* Pergamon, Oxford

Sieglaff C L 1976 Proposed nomenclature for steady shear flow and linear viscoelastic behaviour. *Trans. Soc. Rheol.* 20: 311–17

Tadmor Z, Gogos C G 1979 *Principles of Polymer Processing.* Wiley, New York

Vinogradov G V, Malkin A Ya 1980 *Rheology of Polymers.* Mir, Moscow

White J L 1990 *Principles of Polymer Engineering Rheology.* Wiley, Chichester, UK

White J L, Ide Y 1978 Instabilities and failure in elongational flow and melt spinning of fibers. *J. Appl. Polym. Sci.* 22: 3057–74

Winter H H 1975 Temperature fields in extruder dies with circular, annular or slit cross-sections *Polym. Eng. Sci.* 15: 84–9

C. J. S. Petrie
[University of Newcastle-upon-Tyne,
Newcastle-upon-Tyne, UK]

Processing: Shear-Controlled Orientation

Shear-controlled orientation technology (SCORTEC) provides routes for the management of fiber orientation and microstructure in molded and extruded polymer-matrix composite materials, including ceramic- and metal-matrix composites processed with the aid of sacrificial polymer binders.

The basis of SCORTEC is the control of the preferred orientation of constituents susceptible to alignment, by the application of specific macroscopic shears to solidifying melts. The process is completed when the orientation is "locked in" by the influence of carefully controlled cooling of the polymer matrix. Shear-controlled orientation of reinforcement results in;

(a) enhancement of strength and stiffness;

(b) elimination of internal weld lines;

(c) molding to closer tolerances;

(d) *in situ* formation and management of fibers;

(e) elimination of microcracks, microvoids and surface sinkage; and

(f) control of thermal expansion and other physical properties influenced by preferred orientation of microstructure.

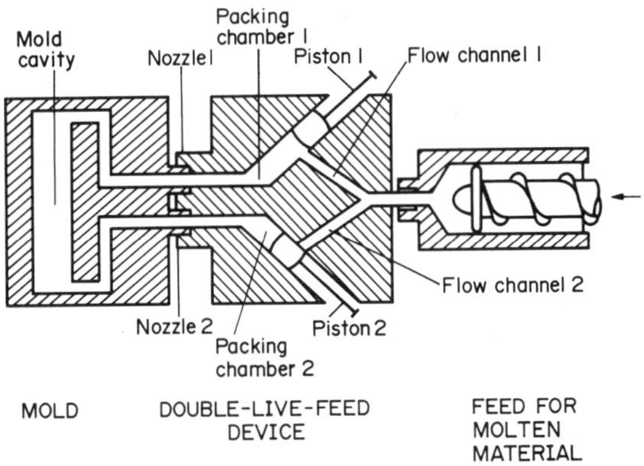

Figure 1
Schematic diagram of one embodiment of a double-live-feed process (after Allan and Bevis 1987)

The application of SCORTEC in injection molding (SCORIM) (Allan and Bevis 1987) and in extrusion (SCOREX) (Allan and Bevis 1991) will be used to illustrate the principles and levels of physical property enhancement associated with the application of SCORTEC. Short-glass-fiber-reinforced polypropylene will be the principal thermoplastic-matrix compound used to illustrate the resultant microstructure and physical property modifications attainable.

1. SCORIM

Injection molding provides an automatic mass-production route for the conversion of short-fiber-reinforced polymers into complex shaped components. The management of the orientation of fibers in injection-molded components is imprecise and the optimum physical properties are seldom realized in practice. In thick-sectioned moldings, the lack of control over the orientation of fibers is pronounced, as is the formation of microvoids and microcracks. SCORIM, alternatively referred to as multiple live-feed molding (MLFM) (Allan and Bevis 1987), was introduced to provide for the better management of fibers and control of microporosity in injection moldings. The basis of the process is to split the single feed from an injection-molding machine into a plurality of feeds in which each one is capable of supplying pressure to the cavity independently of the others. Two-live-feed and four-live-feed arrangements will be used to illustrate SCORIM and the management of short fibers that is possible using the process.

Figure 1 is a schematic diagram of a double-live-feed process (Allan and Bevis 1987). The diagram consists of three parts: the injection-molding machine screw and barrel, the double-live-feed mold packing device and the mold. The two packing pistons are driven by a hydraulic pump and valve systems. The hydraulic feeds to the pistons are controlled by microprocessor via electronically activated pressure valves.

A typical packing cycle would take the following form. Piston 1 is set in the back position and piston 2 in the forward position. Molten material is injected from the barrel of the injection molder through the nozzle and into the flow channels. Piston 2 obstructs the flow in channel 2 so the material enters the mold cavity via channel 1, packing chamber 1 and nozzle 1. On completion of mold filling, there are three principal options for packing with the device.

(a) *Mode A*. The pistons can be pumped back and forth at the same frequency but with a phase difference of 180°.

(b) *Mode B*. The pistons can be pumped back and forth at the same frequency and in phase.

(c) *Mode C*. The pistons can be held down under a static pressure.

With mode A, the melt in the cavity is repetitively injected alternately from the two packing chambers at the set frequency as illustrated in Fig. 2a. The effect of this is to shear the melt within the cavity while it is solidifying. This results in the alignment of fibers and other constituents susceptible to alignment at the melt–solid interface, which moves progressively from the surface to the center of the molding during solidification.

Mode B causes the two packing chambers to work as two separate oscillating pressure packing heads (Allan and Bevis 1983a, b, Allan et al. 1990), as illustrated in Fig. 2b. The effect of mode B is to compress and decompress the remaining melt in the mold cavity.

Mode C applies a static packing pressure to the cavity via the two sprues, as shown in Fig. 2c.

The operation of mode A provides a route for the elimination of processing defects in molded parts as in the following three examples.

(a) Internal weld lines formed by the meeting of flow fronts can be erased, leading typically to a 200–300% increase in the weld line strength of molded thermoplastic-matrix composites (Allan and Bevis 1987) and thermoset-matrix composites (Gibson et al. 1990). However, the most remarkable results are in thermotropic liquid crystal polymers where 1000% increases in weld line strength are possible.

(b) Knit lines associated with jetting or uncontrolled filling of melt into a thick-section cavity can be erased.

(c) The loss of control over fiber orientation and the tendency for voids and/or cracks to form in thick sections remote from the gate can be rectified.

The appropriate sequencing of modes A, B and C can lead to substantial enhancement of modulus through the control of fiber orientation and this is illustrated by the production of uniaxial alignment in 30 wt% short-glass-fiber-reinforced polypropylene bars with a 20 mm × 20 mm cross section. Figure 3 is a schematic diagram of the bar mold cavity used for the production of two series of moldings. The first series of moldings was made by injecting the glass-fiber-reinforced polypropylene at both A and B, followed by an oscillating packing pressure. The packing pressure was achieved by synchronizing the two packing pistons, as defined by mode B in Fig. 2b. This resulted in a fiber orientation similar to that which occurs in conventional injection molding, but without the gross internal voiding that would apply in a thick-section molding. The second series of moldings was produced by filling the cavity from A and B as with the first series. The cavity was then packed by the action of the pistons operating for short periods of time at 180° out of phase (mode A), in between longer periods when the pistons were made to pack the molding in phase (mode B). The effect of this sequence of operations was to sequentially align the fibers in the mold cavity as the material froze and to produce moldings free from voids.

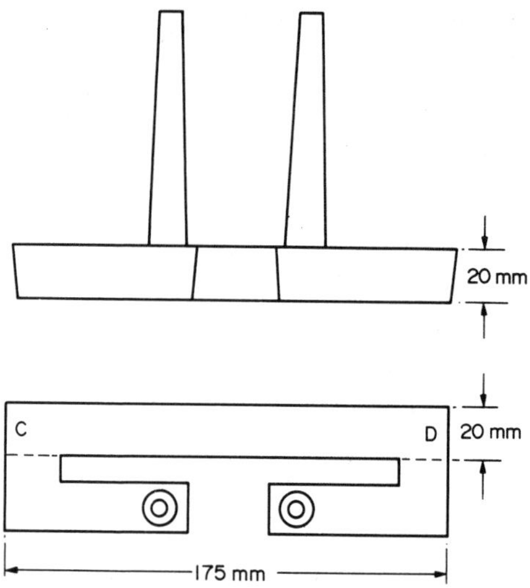

Figure 3
20 mm × 20 mm bar mold cavity showing the twin sprues (after Allan and Bevis 1987)

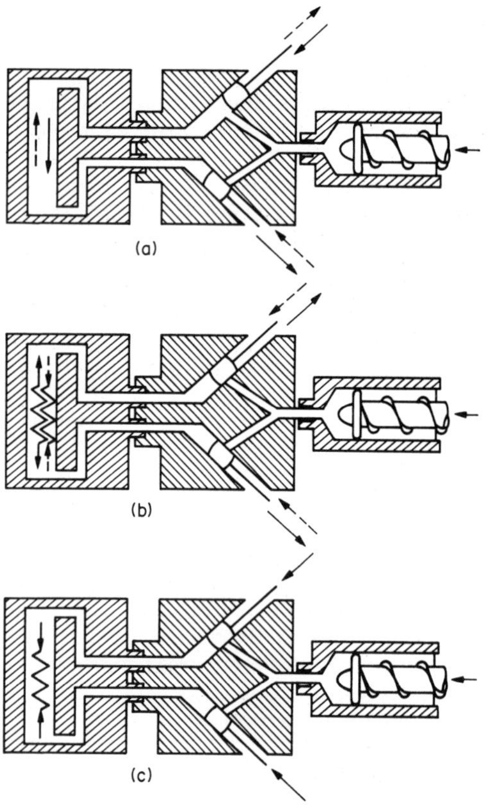

Figure 2
Schematic diagrams of the three principal modes of operation of the double-live-feed molding device: (a) mode A, (b) mode B and (c) mode C (after Allan and Bevis 1987)

Samples were cut from moldings from each of the two series so that the effect of applying macroscopic shears to solidifying melts could be assessed. Five strips were cut from the moldings as shown in Fig. 4. Samples 1 and 5 were from the outer surfaces of the moldings, sample 3 was from the center and samples 2 and 4 were taken halfway between the center and the respective outside surface. The results of measurement of tensile modulus through the thickness bars produced in the two series of moldings are summarized in Fig. 5. From this, it is evident that SCORIM results in a significantly higher modulus than that measured in moldings made with a symmetric mold packing pressure. In particular, at the center of the molding, there is an approximately fourfold increase in the modulus of the second series when compared with the first series. The reason for

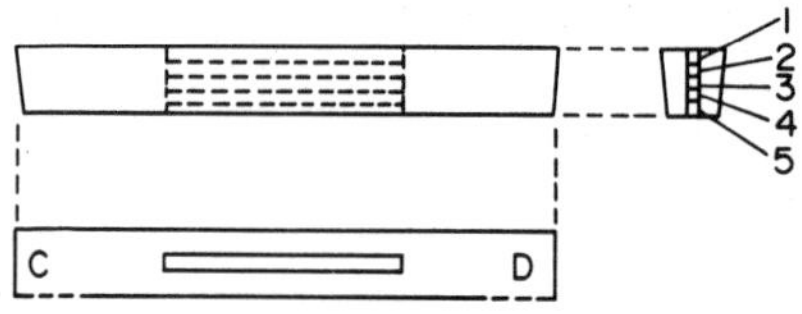

Figure 4
Location of test bars (−5) used to measure modulus in 20 mm × 20 mm 30% glass-fiber-reinforced polypropylene (GFRPP) injection moldings (after Allan and Bevis 1987)

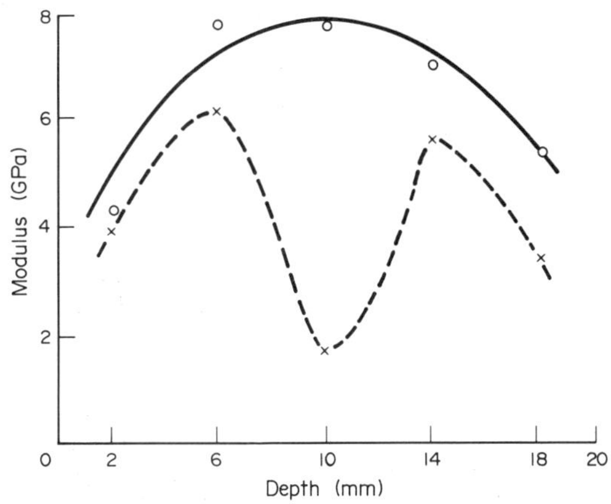

Figure 5
Modulus measurements taken from 20 mm × 20 mm molded bars of 30% GFRPP: ×, bars molded using symmetric packing pressures; ○, bars in which mode A was incorporated into the packing pressure profile (after Allan and Bevis 1987)

this substantial enhancement of modulus is attributed to the more pronounced fiber alignment along the length of the bar in the molding produced by SCORIM. Figures 6a and 6b are contact x-ray microradiographs from sections taken parallel to the length of bars produced without and with the application of shear during solidification, respectively. The second series of moldings exhibits almost total uniaxial fiber alignment along the length of moldings for a large proportion of the section thickness.

SCORIM in the form of a two-live-feed molding has been used in this article to illustrate the effect of applying a macroscopic shear to a solidifying thermoplastic composite melt. A more general example of composite microstructure control is illustrated by four-live-feed molding (Allan and Bevis 1990). Figure 7 shows a four-live-feed arrangement based on twin injection units and two double-live-feed devices. On completion of mold filling of the mold cavity through one or more of gates 1–4 as required and in place of the conventional mold packing stage, the four pistons are caused to oscillate according to a specific sequence and thereby influence fiber orientation during solidification of the melt.

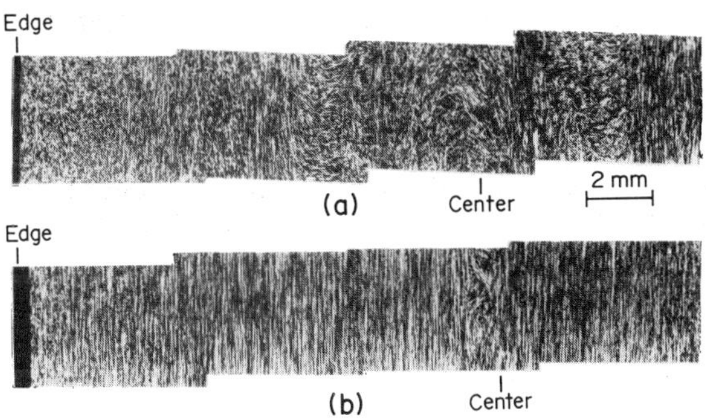

Figure 6
Photographs of parts of the contact microradiographs taken parallel to the process direction in moldings produced when (a) mode B and (b) mode A were incorporated into the packing profile; the edge and the center of the molding are indicated (after Allan and Bevis 1987)

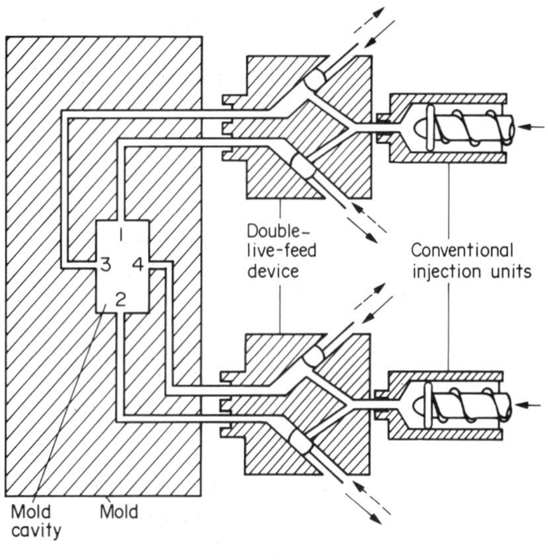

Figure 7
Schematic diagram of a four-live-feed arrangement based on twin injection units and two double-live-feed devices: the gates are labelled 1–4 (after Allan and Bevis 1990)

The four-live-feed arrangement in the Wolfson Centre for Materials Processing at Brunel University is based on a Negri Bossi 130-90-90 two-color injection-molding machine. The control signals that have to be provided by the injection-molding machine to effect the operation of SCORIM, together with other requirements likely to feature in selecting or adapting a machine for use, have been outlined by Allan and Bevis (1990).

A 9 mm thick, 85 mm square plaque molding featuring four independently controlled fan gates 1–4 (see Fig. 8) will be used to illustrate the control of microstructure that is possible with four live feeds. The out-of-phase operation (mode A) of the live feeds at gates 1 and 2 after mold filling tends to cause alignment of the fibers in the direction indicated in Fig. 8a. After a preset time, the pistons serving gates 3 and 4 are caused to operate in mode A, causing a layer of fibers to form with the preferred orientation direction shown in Fig. 8b. A multilayer laminated structure may be formed through the thickness of the plaque according to the sequencing and duration of the operation of pairs of live feeds, as the melt

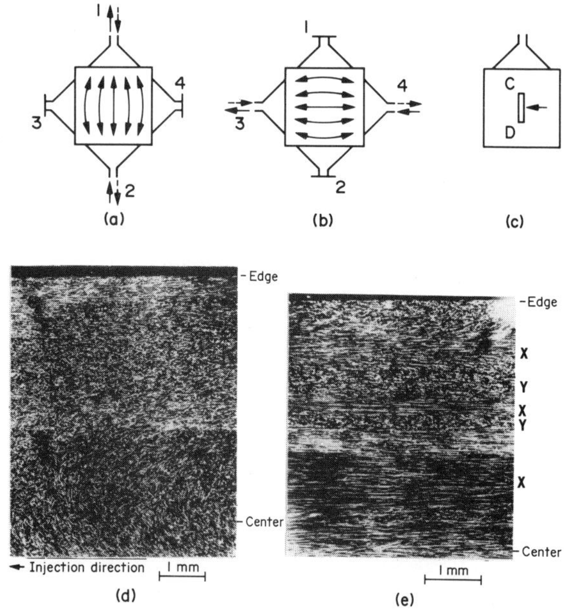

Figure 8
(a) Operation of the live feeds at gates 1 and 2 tends to cause alignment in the direction of the arrows; (b) operation of the live feeds at gates 3 and 4 tends to cause alignment in the direction of the arrows; (c) the orientation of micrographs in Figs. 8d, e; (d) microradiograph showing through the thickness orientation of glass fibers in molding produced using conventional injection-molding route; and (e) microradiograph showing the laminate structure X, Y, X, Y, . . . produced by alternate operation of live feeds 1–2 and 3–4 (the preferred uniaxial fiber orientation at the center of the molding was realized by maintaining operation of live feeds 3–4 during the later part of the molding cycle) (after Allan and Bevis 1990)

solidifies from the outside to the center of the plaque cavity. Figure 8c indicates the position of sections used for contact x-ray microradiography of short-glass-fiber-reinforced polypropylene moldings produced using

(i) a single gate, packed with a static pressure; and

(ii) four gates, packed using a four-live-feed arrangement according to Figs. 8a and 8b.

Figures 8d and 8e are microradiographs corresponding to the conventional molding and the SCORIM molding produced by four live feeds, respectively. The latter clearly shows the layered structures exhibiting different preferred orientations (X and Y) of fibers, produced by alternate operation of live feeds 1–2 and 3–4 and concluding with prolonged operation of live feeds 3–4, to give the uniaxially aligned fibers in the center of the molding.

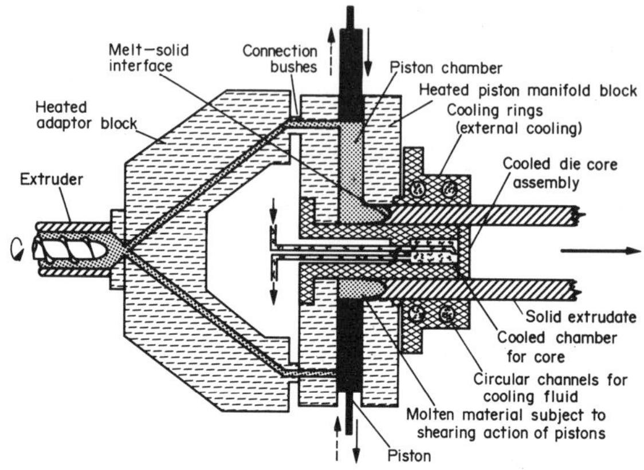

Figure 9
A schematic diagram of the center section through the SCOREX die assembly used for the production of tubes (after Allan and Bevis 1991)

2. SCOREX

The concept of shear-controlled orientation in solidifying polymer melts by the application of macroscopic shears may also be used to manage fiber orientation in extruded profile. The SCOREX process will be illustrated by an arrangement used to produce circumferential alignment in short-fiber-reinforced thermoplastic tubes (Allan and Bevis 1991).

The main functions of the SCOREX die illustrated in Fig. 9 and designed to produce a tubular extrudate are

(a) to transmit the feed of molten material from the extruder through one or more channels in an adaptor manifold to the main die block;

(b) to pass the molten material into one or more chambers, each containing a reciprocating piston, within the die block;

(c) to solidify the material within the die block;

(d) to shear the solidifying material within the die block by the controlled reciprocating displacements of the pistons within the die block; and

(e) to expel the solid extrudate from the die.

The molten material from the extruder enters an adaptor block. The melt stream is split into two equal channels which carry the material to the piston manifold block. The material enters the piston manifold block through two channels positioned at the top end of two piston chambers and progresses down to the annular die. The die consists of a cooled core section which is surrounded by a set of annular cooling rings. The temperatures of the piston manifold block and the die sections are set so that a solid–plastic interface is created in close proximity to the bottom of the piston chambers. (The position of the interface defines the cross-sectional area of the molten material being sheared.) The volume of molten material within the die section which is enclosed by the die walls and the solid–plastic extrudate is repetitively sheared by the reciprocating action of the pistons. For the tube die there are four such pistons which are mounted at right angles to each other.

The movement of the pistons, which are driven by a hydraulic power unit with four separate pumps, is controlled by a microprocessor.

A typical sequence of movements of the four pistons A, B, C and D for the production of a tube section is illustrated in Fig. 10a and can be described as follows.

(a) Stage 1—A down, C back; B and D fixed as shown.

(b) Stage 2—D down, B back; C down to shut off the feed channel. A fixed as shown.

(c) Stage 3—C down, A back; B and D fixed as shown.

(d) Stage 4—B down, D back; A down to shut off the feed channel. C fixed as shown.

The four stages are repeated in this sequence for the duration of the extrusion run. Stage 5, shown in Fig. 10b, uses one or more of the pistons to compress the melt in the die to cause the expulsion of a section of the solid extrudate from the die. This stage can be incorporated into the main sequence at regular intervals if required.

The SCOREX die and the operating conditions described have been applied to the extrusion of short-glass-fiber-reinforced tubes (Allan and Bevis 1991). They provide for the production of shear-controlled orientation for short fibers in extrusion and give the following advantages over the conventional extrusion process:

(a) enhanced hoop orientation of fibers, resulting in substantial enhancement of failure hoop stress and hoop stiffness; and

(b) production of a uniform fiber orientation around the hoop and along the axis of the extrusion, resulting in significant reduction in the variation of axial failure stress and axial stiffness around the hoop direction.

3. *More General Application of SCORTEC*

The use of shear-controlled orientation technology for the control of fiber orientation in relatively simple mold and die geometries has been described. This illustrates the extent to which the orientation of reinforcing fibers can be managed in molded and extruded components, coupled with control of porosity in thick-sectioned composite components. In more complex geometries, the orientation of fibers can be managed by the strategic placing of live feeds to induce preferred orientation in specified directions and regions of moldings or extrusions (Allan and Bevis 1990). Hot-runner systems can be used for the routine production of moldings using SCORIM with multicavity mold tools.

SCORTEC can also be used for the management of fibers in molded and extruded ceramic- or metal-matrix composites. The conversion of these composites may be realized with the aid of moldable formulations containing typically 30 vol.% sacrificial polymer binder and 70 vol.% inorganic filler comprising a combination of ceramic or metal powders and reinforcing fibers. The polymer binder is the vehicle for conversion of the ceramic- or metal-matrix composite into complex shapes, imparting

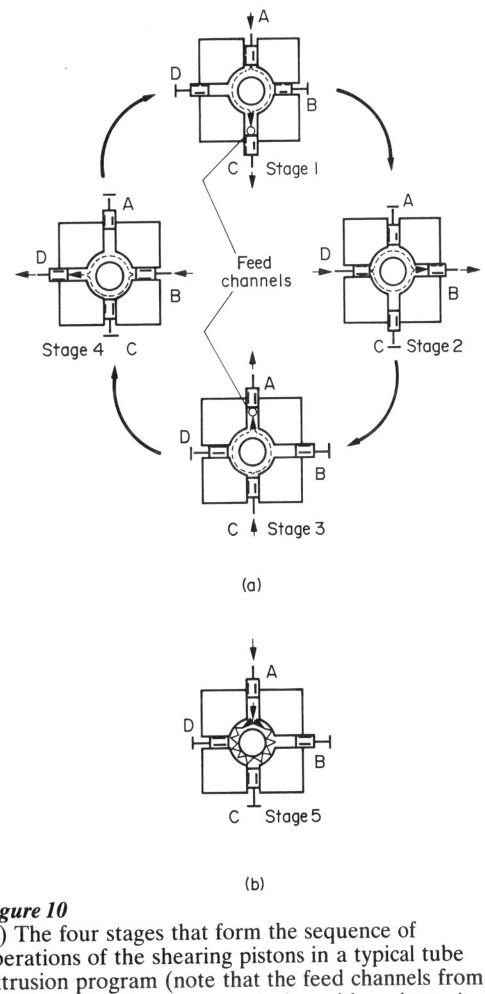

Figure 10
(a) The four stages that form the sequence of operations of the shearing pistons in a typical tube extrusion program (note that the feed channels from the extruder are only opened when either piston A or piston C are fully retracted in stages 1 and 3 only), (b) compression stage (stage 5) which can be introduced into the cycle shown in Fig. 10a if a high pressure is required to expel the solid extrudate (after Allan and Bevis 1991)

preferred orientation of reinforcing fibers by the action of controlled shear of the solidifying melt and elimination of voids and cracks in thick-section green components by the action of oscillating packing pressures—this provides for controlled solidification and compensation for shrinkage defects, all prior to removal of the polymer binder and subsequent sintering to consolidate the extrudate or molding.

See also: Computer Control in Polymer Processing; Extrusion of Fiber-Reinforced Polymer Systems; Injection Molding of Thermoplastics: Mold Filling; Processing: Classification; Processing: Relation to Product Properties; Residual Stresses

Bibliography

Allan P S, Bevis M J 1983a The production of void-free thick-section mouldings I: Shot-weight and dimensional reproducibility. *Plast. Rubber Process. Appl.* 3:85–91

Allan P S, Bevis M J 1983b The production of void-free thick-section mouldings II: Preferred orientation and residual stress measurements. *Plast. Rubber Process. Appl.* 3: 331–6

Allan P S, Bevis M J 1987 Multiple live-feed injection moulding. *Plast. Rubber Process. Appl.* 7: 3–10

Allan P S, Bevis M J 1990 Development and application of multiple live-feed moulding for the management of fibres in moulded parts. *Compos. Manuf.* 1: 79–84

Allan P S, Bevis M J 1991 Shear controlled orientation in extrusion. *Plast. Rubber Compos. Process. Appl.* (in press)

Allan P S, Bevis M J, Hardwick S T, Murphy M, Sculley D J 1990 *Plast. Rubber Proc. Appl.* 13: 15–27

Gibson J R, Allan P S, Bevis M J 1990 The multiple live-feed moulding of DMC's. *Compos. Manuf.* 1: 183–90

P. S. Allan and M. J. Bevis
[Brunel University, Uxbridge, UK]

Processing: Structure Development

Structural order in polymer systems can exist on various scales. At a level of 0.1 nm, polymer chains can order into crystalline regions. On a slightly larger scale, the orientation of the polymer chains and associated crystallographic axes along characteristic directions in the object—for example, the axis of a fiber or the machine and transverse directions in a film—become significant. At dimensions in the region of 7–20 nm, a lamellar structure is found to exist in crystalline polymers. On a still larger scale, 0.1–10 μm, spherulitic structures are observed in isotropic crystalline polymer systems and prolate (and probably oblate) spheroidal and cylindrical structures in anisotropic materials. Specification of the structure of a polymer includes all these features and all depend on the processing conditions used in fabrication.

Various characterization techniques must be used to specify polymer structure. Most important is x-ray diffraction, which can not only detect the existence of crystallinity but also specify the parameters of the crystalline lattice and the orientation of this lattice. This involves the use of wide-angle x-ray diffraction or scattering (WAXS). From Bragg's law for diffraction of radiation of wavelength λ,

$$n\lambda = 2d \sin\theta \tag{1}$$

which relates the d spacing to the Bragg angle θ of the diffracted beam, n being an integer. The Bragg angles for the d spacing of the unit cell are large; hence the term wide-angle x-ray scattering. The lamellar structure at 7–20 nm involves much smaller Bragg angles and the appropriate technique is hence referred to as small-angle x-ray scattering (SAXS).

Other methods are used in addition to WAXS to determine the amount of crystallinity in polymer systems. These include calorimetry, in which the amount of heat evolved in melting a sample is measured and compared with the heat of fusion of a hypothetical completely crystalline polymer. Another method is to measure the density of the polymer and compare it with the hypothetical densities of completely crystalline and amorphous polymers. Generally, results of this type are interpreted in terms of a two-phase model and are expressed as the fraction X of the crystalline phase.

WAXS represents a major method of investigating orientation in polymers. There are other methods; most notably, birefringence, representing the difference in refractive index with direction, is widely determined. Another method is based on infrared dichroism, the anisotropy of absorption of infrared radiation. Birefringence and dichroism may be used to measure orientation in amorphous polymers, unlike WAXS, which is based on the existence of crystalline structural order. Other methods used to study orientation are nuclear magnetic resonance and Raman scattering.

Electron microscopy, especially transmission electron microscopy (TEM), has proved to be valuable in determining qualitative morphological features at the 10–100 nm level.

1. Orientation

Orientation is most conveniently expressed in terms of orientation factors which characterize second moments of the orientation distribution relative to specified directions in the materials, such as the axis of a fiber in the machine and transverse directions in a film. The first orientation representations of this type were introduced by Hermans in the 1930s, who specified polymer chain orientation in fibers by an expression of the form

$$f_H = \tfrac{1}{2}(3\,\overline{\cos\phi_{zc}} - 1) \tag{2}$$

where ϕ_{zc} is the angle between the polymer chains and the fiber axis. The superposed bar indicates averaging over all the polymer chains. The arguments leading to Eqn. (2) involve the ratio of the difference of the measured anisotropy of the polarizability along the perpendicular to the fiber axis to the difference in the principal values of the polarizability tensor. Note that for isotropy f_H is zero, for perfect orientation it is unity for orientation perpendicular to the fiber axis it is -0.5.

A generation later, Stein (1958) noted that the definitions of Eqn. (2) may be generalized for crystalline fibers to all three crystallographic axes:

$$f_a = \tfrac{1}{2}(3\,\overline{\cos^2\phi_{za}} - 1) \quad (3a)$$
$$f_b = \tfrac{1}{2}(3\,\overline{\cos^2\phi_{zb}} - 1) \quad (3b)$$
$$f_c = \tfrac{1}{2}(3\,\overline{\cos^2\phi_{zc}} - 1) \quad (3c)$$

Here ϕ_{za}, ϕ_{zb} and ϕ_{zc} are angles between the crystallographic axes and the polymer chain axis. If the axes of the unit cell are orthogonal, as in tetragonal or orthorhombic crystals, then

$$\cos^2\phi_{za} + \cos^2\phi_{zb} + \cos^2\phi_{zc} = 1 \quad (4a)$$

or

$$f_a + f_b + f_c = 0 \quad (4b)$$

Various researchers have sought to generalize this formulation to biaxial orientation. The most successful of such attempts seems to be that of White and Spruiell (1981) using Hermans' polarizability analysis and the angles ϕ_{cz} and ϕ_{cv} between the polymer chain axis and two orthogonal axes set in the surface, such as a machine and a transverse direction (see Fig. 1). These orientation factors are

$$f_z^B = \overline{\cos^2\phi_{cz}} - \overline{\cos^2\phi_{cx}} = 2\,\overline{\cos^2\phi_{cz}} + \overline{\cos^2\phi_{cy}} - 1 \quad (5a)$$
$$f_y^B = \overline{\cos^2\phi_{cy}} - \overline{\cos^2\phi_{cx}} = 2\,\overline{\cos^2\phi_{cy}} + \overline{\cos^2\phi_{cz}} - 1 \quad (5b)$$

where f_z^B and f_y^B may vary from -1 to $+1$, with zero representing isotropy. These orientation factors reduce to the Hermans' orientation factor when there is fiber symmetry. For symmetry with respect to the z axis, $\cos^2\phi_{cy}$ equals $\cos^2\phi_{cx}$, f_y^B is zero and f_z^B is f_H. When there is equal biaxial orientation, $\cos^2\phi_{cz}$ and $\cos^2\phi_{cy}$ are equal, with $\cos^2\phi_{cx}$ zero. This makes f_z^B and f_y^B both equal to $+0.5$.

The orientation factors of Eqn. (5) may readily be generalized to apply to all three crystallographic axes; that is,

$$f_{zj}^B = 2\,\overline{\cos^2\phi_{jz}} + \overline{\cos^2\phi_{jy}} - 1 \quad (6a)$$
$$f_{yl}^B = 2\,\overline{\cos^2\phi_{jy}} + \overline{\cos^2\phi_{jz}} - 1 \quad (6b)$$

where the subscript j represents the jth crystallographic axis. The six f_{zj}^B and f_{yj}^B are not independent

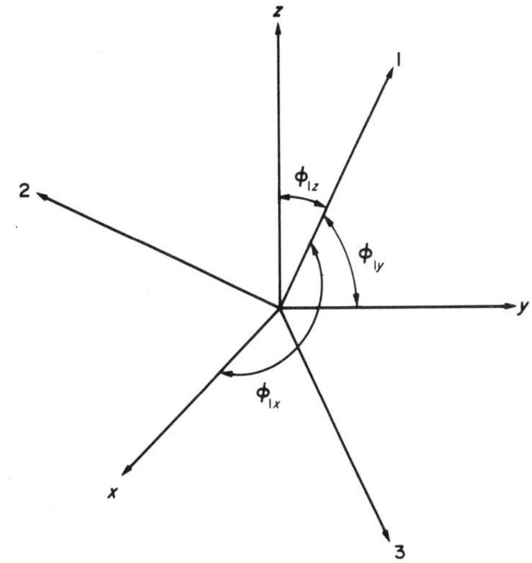

Figure 1
Diagram showing angles for biaxial orientation analysis: z is the machine direction and y the transverse direction

but are interrelated through relationships between $\cos^2\phi_{zj}$ and $\cos^2\phi_{yj}$.

The orientation factors in homogeneous materials such as melts and glasses may be determined by methods such as birefringence and dichroism. The polymer chain orientation factors f_z^B and f_y^B may be shown to be related to the birefringences Δn_{zx} and Δn_{yx} through

$$\Delta n_{zx} = n_z - n_x = f_z^B \Delta^\circ \quad (7a)$$
$$\Delta n_{yx} = n_y - n_x = f_y^B \Delta^\circ \quad (7b)$$

where n is the refractive index and Δ° is the intrinsic birefringence. The f_{zj}^B, f_{yj}^B in crystalline regions in polymers may be determined by WAXS.

In crystalline polymers, the total orientation may be measured using birefringence and the crystalline orientation by WAXS. Amorphous orientation may be determined from the difference. In the case of uniaxial orientation, if X is the fraction of crystalline polymer, then

$$\Delta n = X f_{\mathrm{H\,cryst}} \Delta^\circ_{\mathrm{cryst}} + (1-X) f_{\mathrm{H\,amorph}} \Delta^\circ_{\mathrm{amorph}} \quad (8)$$

where $\Delta^\circ_{\mathrm{cryst}}$ and $\Delta^\circ_{\mathrm{amorph}}$ are the maximum (intrinsic) birefringences of the crystalline and amorphous phases; $f_{\mathrm{H\,cryst}}$ is determined by WAXS.

2. Solidification: Crystallization and Vitrification

In processing operations, the polymer is generally reduced to a molten state in a screw pumping device and shaped in a die or in a postdie processing operation such as inflation of film or in a molding step. Solidification in the case of profile extrusion, melt spinning of fibers and tubular film occurs in quiescent air or a blowing air stream. In some cases, extrudates emerge into a water bath. Such air and water cooling of melt streams gives a very rapid quench with a temperature drop of 50–2000 K s^{-1} in a melt spinning operation, as reported by Dees and Spruiell (1974). On the other hand, cooling of a polymer mass in a mold is much slower. This is due to the thickness of molded parts, coupled with the low thermal conductivity of polymers, as much as to the lower heat-transfer coefficients at the mold walls.

When melts solidify, either they undergo a first-order phase transition and crystallize or they vitrify into a glass. In polymers, as already mentioned, crystallization is generally imperfect and the products are only semicrystalline. Some polymers that possess inherently regular structures such as atactic vinyl polymers (e.g., commercial polystyrene) vitrify into glasses. On the other hand, "linear" polyethylene crystallizes rapidly. Many polymers ranging from polyolefins to polyesters and polyamides are intermediate between these extremes. The character of the solid-state structure formed depends on the rate and environment of cooling. Poly(ethylene terephthalate) generally vitrifies into a glass, but crystallizes if melt spun at high drawdowns (stresses) (Shimizu et al., 1977). Isotactic polypropylene cooled in a mold or in air generally forms a well-defined monoclinic structure, but if quenched in cold water forms a poorly defined smectic structure which ages towards the monoclinic from (Nadella et al. 1977). Nylon 66 cooled in a mold or in air forms a well-defined triclinic crystalline structure, but when quenched in cold water forms a nematic structure (Starkweather et al. 1963). Nylon 6 quenched in air forms a glass but ages in moist air, forming a poorly defined crystalline structure of hexagonal symmetry (γ form). If this product is annealed or not drawn, it transforms to a more defined monoclinic structure (α form) (Bankar et al. 1977).

3. Orientation Development in Melts and Solidification of Glasses

Most thermoplastic fabrication operations stress the polymer while shaping it. This leads to the development of polymer chain orientation in the molten state which is usually partially or totally frozen in during solidification. Such orientation has usually been determined by WAXS and birefringence. The type of orientation developed generally varies in character with the processing operation involved. Melt-spun fibers exhibit orientation only relative to the fiber axis. Continuously fabricated films as a rule exhibit orientation relative to two directions, the machine direction and a perpendicular transverse direction. Fabricated bottles usually similarly exhibit orientation relative to both the hoop and lengthwise directions.

It has been realized by various investigators since perhaps the 1930s (Carothers and Hill 1932) that orientation in fabricated polymers is due to the deformation of the polymer melts during processing. Uniaxial stretching or stress application leads to uniaxial orientation. Biaxial deformations and stress fields lead to biaxial orientation. Films may be made biaxially with tentering frames (Alles 1956) or with a mandrel (Carlson 1971). Bottles may be produced with biaxial orientation by stretching the bottle as it inflates in a mold (Wyeth and Roseveare 1973). Injection-molded products may be given biaxial orientation through rotating the core of the mold (Cleereman 1967).

Efforts have been made to relate the level of orientation quantitatively to the details of processing conditions. These begin with the studies of the optical retardation of flowing polymer melts possessing flexible chains, such as polyolefins and polystyrene, which are isotropic in a quiescent state but become increasingly birefringent with applied stress. It has been found that these melts obey the rheo-optical law for photoelastic stress analysis:

$$n_i - n_j = C(\sigma_i - \sigma_j) \tag{9}$$

where the σ_i are the principal stresses. The refractive index and stress ellipsoids have parallel principal axes whose magnitudes are linearly related to each other. It is found from Eqn. (7) that Eqn. (9) is equivalent to

$$f_z^{\mathrm{B}} = \frac{\Delta n_{zx}}{\Delta^{\circ}} = \frac{C}{\Delta^{\circ}}(\sigma_z - \sigma_x) \tag{10a}$$

$$f_y^{\mathrm{B}} = \frac{\Delta n_{yx}}{\Delta^{\circ}} = \frac{C}{\Delta^{\circ}}(\sigma_y - \sigma_x) \tag{10b}$$

The state of orientation in the melt is determined by, and is an increasing function of, the differences in principal stresses applied.

In solidification by vitrification, polymer chain orientation would be expected to be frozen in. This would make Eqns. (9) and (10) valid where birefringence and orientation would be measured in the glass and stresses determined at the position of vitrification. This hypothesis has been tested for polystyrene in a series of experiments by Oda et al. (1978) and by Choi et al. (1980), who found it to be valid for vitrified uniaxially stretched filaments, sheared sheets, melt-spun fibers and tubular extruded film (see Fig. 2).

In molding operations, polymer melt is injected or "blown" with fluid pressure to fill the mold. The melt subsequently cools and solidifies. During this period of cooling, the orientation developed in the

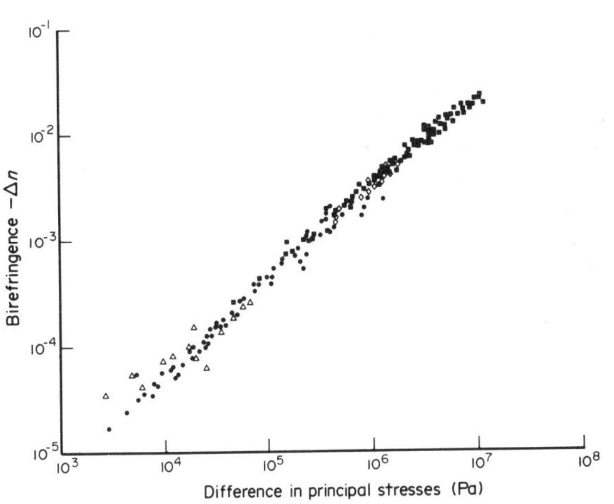

Figure 2
Birefringence of tensile and shear specimens of polystyrene as a function of difference in principal stresses: ■, melt spinning; ●, elongational flow; △, simple shear flow; ◇, melt spining (on-line measurement)

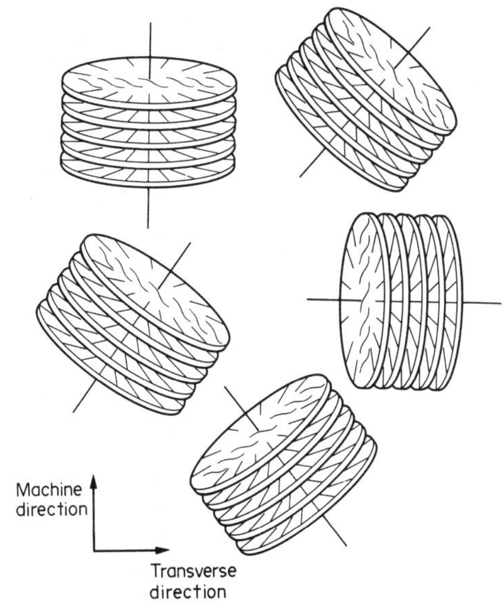

Figure 3
Diagrammatic representation of lamellar structures formed under the action of uniaxial stress: arrows denote the machine and transverse directions

injection or blowing step gradually decays. Various research groups have developed modelling of this hypothesis for injection molding and testing it on polystyrene (White and Dietz 1979). It has also been applied to blow molding (White and Agarwal 1981).

4. Crystalline Morphology and Superstructure Development

Crystalline polymers exhibit a complex morphology and superstructure at varying levels of dimensions. This structure is strongly influenced by process conditions. The variations encountered are closely associated with both molecular orientation and states of cooling.

The basic investigations and generally accepted views of morphology development in crystalline polymers during processing were developed during the 1960s by Keller and Machin (1967). Under the action of uniaxial streses, it has been shown using SAXS, TEM and WAXS that row structures are formed. These are lamellae of thickness 7–20 nm perpendicular to the direction of stretch. It has been surmised that these are essentially the "shish kebab" structures observed in crystallization from stirred polymer solutions. The lamellae are considered to consist of a folded chain structure in which the polymer chains are perpendicular to the lamella surface and are thought to represent the growth of crystalline material from a filamentary nucleus oriented in the stress direction (see Fig. 3). Choi, Spruiell and White, who produced and characterized tubular film under a range of conditions, introducing biaxial as well as uniaxial orientation, concluded that these lamellae distribute themselves in space according to the stress field.

At a higher level, the spherulitic structures observed in compression-molded isotropic parts are transformed to prolate spheroids and cylindrites under the action of stress. The elongated axis of the spheroid or cylindrites is oriented in the direction of a uniaxial stress.

The crystalline orientation of polymer products has been investigated by WAXS and the amorphous orientation described by combining WAXS and birefringence measurements. For polymers with multiaxial orientation, pole figures of the pertinent crystallographic planes have generally been used to express orientation distributions in space. The basic studies are those of Heffelfinger and Burton (1960) and Lindenmeyer and Lustig (1965). These have been related to specific process conditions by later investigators (Desper 1969, Maddams and Preedy 1978, Choi et al. 1982).

During crystallization, the level of molecular orientation increases substantially. However, the final state of crystalline orientation should still be expected to depend on the orientation in the melt but to have a different functionality for each material. This leads to a dependence of orientation, represented by the orientation factors of Eqns. (3) and (6), on stress because of Eqn. (10). This view has been tested and verified experimentally in studies of melt spinning and tubular film extrusion of polyethylene and polypropylene (Dees and Spruiell 1974, Nadella et al. 1977, Choi et al. 1982).

See also: Processing: Classification

Bibliography

Alles F P 1956 Process for longitudinally stretching polymeric film. US Patent No. 2,767,435

Ayres R E, Cleereman K J, Schrenk W J 1973 Scrapless forming of plastic articles. US Patent No. 3,739,052

Bankar V G, Spruiell J E, White J L 1977 Melt spinning of nylon 6: Structure development and mechanical properties of as-spun filaments *J. Appl. Polym. Sci.* 21: 2341–58

Carlson F A Jr 1971 Producing biaxially oriented polystyrene heavy-gauge sheet. US Patent No. 3,619,445

Carothers W H, Hill J W 1932 Studies of polymerization and ring-formation XV. Artificial fibers from synthetic linear condensation superpolymers. *J. Am. Chem. Soc.* 54: 1579–87

Choi K J, Spruiell J E, White J L 1982 Orientation and morphology of high density polyethylene films produced by the tubular blowing method and its relation to process conditons. *J. Polym. Sci., Polym. Phys. Ed.* 20: 27–47

Choi K J, White J L Spruiell J E 1980 Orientation development in tubular film extrusion of polystyrene. *J. Appl. Polym. Sci.* 25: 2777–88

Cleereman K J 1967 Multilayered thermoplastic article. US Patent No. 3,307,726

Dees J R, Spruiell J E 1974 Structure development during melt spinning of linear polyethylene fibers. *J. Appl. Polym. Sci.* 18: 1053–78

Desper C R 1969 Structure and properties of extruded polyethylene film. *J. Appl. Polym. Sci.* 13: 169–91

Heffelfinger C J, Burton R L 1960 X-ray determination of the crystallite orientation distributions of polyethylene terephthalate films. *J. Polym. Sci.* 47: 289–306

Katayama K, Amono T, Nakamura K 1968 Structural formation during melt spinning process. *Kolloid-Z. Z. Polym.* 226: 125–34

Keller A, Machin M J 1967 Oriented crystallization in polymers. *J. Macromol. Sci.* [ ] *Phys.* 1: 41–91

Kitao T, Ohya S, Furukawa J, Yamashita S 1973 Orientation of polymer molecules during melt spinning, II. Orientation of crystals in as-spun polyolefin fibers. *J. Polym. Sci., Polym. Phys. Ed.* 11; 1091–109

Lindenmeyer P H, Lustig S 1965 Crystalline orientation in extruded polyethylene film. *J. Appl. Polym. Sci.* 9: 227–40

Maddams W F, Preedy J E 1978 X-ray diffraction orientation studies on blown polyethylene films. I. Preliminary measurements. II. Measurements on films from commercial blowing units. III. High stress crystallization orientation *J. Appl. Polym. Sci.* 22; 2721–59

Nadella H P, Henson H M, Spruiell J E, White J L 1977 Melt spinning of isotactic polypropylene: Structure development and relationship to mechanical properties. *J. Appl. Polym. Sci.* 21: 3003–22

Nagasawa T, Matsumura T, Hoshino S, Kobayashi K 1973 Film forming process of crystalline polymer. I. Factors inducing a molecular orientation in tubular blown film. II. Microstructure. *Appl. Polym. Symp.* 20: 275–313

Oda K, White J L, Clark E S 1978 Influence of melt deformation history on orientation in vitrified polymers. *Polym. Eng. Sci.* 18: 53–9

Shimizu J, Toriumi K, Tamai K 1977 High speed melt spinning of polyester filaments—Effect of spinning velocity on the properties of molecular orientation. *Sen'i Gakkaishi* 33: T208–14

Spruiell J E, White J L 1975 Structure development during polymer processing: Studies of melt spinning of polyethylene and polypropylene fibers. *Polym. Eng. Sci.* 15: 660–7

Starkweather H W Jr, Whitney J F, Johnson D R 1963 Crystalline order in nylon 66. *J. Polym. Sci.* Al: 715–23

Stein R S 1958 The x-ray diffraction, birefringence, and infrared dichroism of stretched polyethylene, II. Generalized uniaxial crystal orientation. *J. Polym. Sci.* 31: 327–34

Stein R S, Wilkes G L 1975 Physico-chemical approaches to the measurement of anisotropy. In: Ward I M (ed.) 1975 *Structure and Properties of Oriented Polymers.* Applied Science, London, Chap. 3, pp. 57–149

White J L, Agarwal A 1981 Orientation development in blow molding: Polystyrene. *Polym. Eng. Rev.* 1: 267–93

White J L, Deitz W 1979 Some relationships between injection molding conditions and the characteristics of vitrified molded parts. *Polym. Eng. Sci.* 19: 1081–91

White J L, Spruiell J E 1981 Specification of biaxial orientation in amorphous and crystalline polymers. *Polym. Eng. Sci.* 21: 859–68

Wyeth N C, Roseveare R N 1973 Biaxially oriented poly(ethylene-terephthalate) bottle. US Patent No. 3,733,309

J. L. White
[University of Akron, Akron, Ohio, USA]

Protective Tape Coatings

Corrosion of steel substrates, particularly pipelines, occurs when they come into direct contact with their underground or underwater environment. All soils contain water and oxygen, which are the basic elements that cause corrosion of steel underground. The purpose of a coating is to isolate the steel from its environment so that the corrosive chemical and electrolytic reactions cannot take place.

A pipeline coating must have the following properties in order to prevent active corrosion on the substrate:

(a) high resistance to moisture absorption and water vapor transmission;

(b) high electrical resistance;

(c) long-term chemical, physical and adhesive stability;

(d) thermal stability; and

(e) flexibility and ductility over a broad temperature range.

One known and proven effective method for mitigation of the corrosion of steel pipelines is a two-part system: (a) a protective coatings applied to the surface of a pipeline and (b) supplemental cathodic protection applied after the pipeline has been installed and during its operation. This concept is now enforced by US Federal regulations (US Federal OPS 49CFR Part 192) which clearly state that all pipelines in the USA must be coated and cathodically protected. Similar methods are recommended by the National Association of Corrosion Engineers (NACE).

Extensive laboratory and field tests, plus massive worldwide use of plastic tape, have shown that this type of coating clearly meets the long-term requirements for pipeline coatings. However, it should be noted that PVC tapes, which once dominated the pipeline coatings industry, have been phased out. This change, which occurred in the late 1970s, came about because PVC lacked physical, chemical and electrical stability with time and temperature. Hence, the plastic tape in current use for coatings is polyethylene.

1. Compositions of Primer and Tape Systems

A tape system consists of a primer, an inner corrosion-protection tape and an outer wrap, either with or without adhesive. In some instances the outer wrap, which is a mechanical protection layer for the inner corrosion protection tape, can be eliminated, but the combination is most widely used.

The primer consists of a rubber elastomer in a solvent which evaporates prior to tape application. The tape and outer wrap consist of a polymer film backing and a rubber-modified adhesive. The primer, which wets and flows onto the steel surface, functions as a bond coat or tie coat between the steel substrate and the adhesive layer of the tape. Tenacious bonds are generated between steel and primer, and primer and adhesive.

For manual applications such as joint and patch tapes, the system usually consists of a primer and two layers of joint wrap tape. The joint wrap usually consists of a thin layer of backing film (0.15 mm) and a very thick layer of adhesive ($\geq$0.5 mm). The tape is designed to give good conformability at odd shapes and/or girth welds and good conformability to itself at the overlap. It is usually applied in two layers to provide a total thickness nearly equal to that of the line tape and to give adequate chemical, mechanical and corrosion protection.

In the "line-travel" mechanical wrapping system or manual application "over the ditch," the tape system usually consists of a primer, an inner corrosion primary tape and a mechanical outer wrap. This construction is different from that used in joint wrapping and other hand applications in that the primary tape has a different construction and composition and the outer wrap does not have a high impact resistance. Likewise, the outer wrap differs in construction and composition from that used in plant coating, the essential difference being in impact properties.

The line-travel system usually has the same adhesive system for the inner corrosion-protection layer, but the film to adhesive thickness ratio differs. The inner corrosion tape has, initially and over the long term, a very high bond strength which allows for immediately lowering the sytem into the ditch and, if necessary, directly into water. The optimum bond strength is attained as soon as the tape comes into contact with the primer and requires no bonding or setup time to generate a good bond. When the outer wrap is applied, it generates a good bond to the inner tape, which also permits immediate usage.

For the plant coating or yard coating system, the same primer is used as for hand or line-travel wrapping. Also, the adhesive used is the same as for the other systems. However, the tape construction differs, usually having greater adhesive thicknesses. The outer wrap is usually a very tough and highly abrasion-resistant tape which requires much higher tape tensions for proper application. The outer wrap must be highly resistant to damage because it must prevent damage to the inner corrosion-protection tape during transportation, bending and yard storage. Because there is much handling of the coated system from coating to final welding into a pipeline at the right-of-way, the outer wrap must possess high impact strength and have high abrasion resistance, good flexibility and strong adhesion to the inner corrosion tape.

The tape system may be viewed as a viscoelastic material which behaves both as an elastomer and as a liquid. At low temperatures, it has the flexibility and ductility to be functional for pipeline coating application. Requirements include bendability, transportability and multiple handling resistance while stil retaining ductility at −30 °C.

2. Tape Manufacturing Methods

Tapes, either polyethylene or poly(vinyl chloride) (PVC), are manufactured by several different polymeric film processes. The most common are extrusion sheet line, coextrusion and calendering, using either low-density polyethylene or a blend of high- and low-density polyethylene, plus color masterbatches (e.g., carbon black, yellow, white or orange pigment) and antioxidants. Plasticizers are compounded in the film manufacturing process in the case of PVC tapes. The adhesive is manufactured using two-roll mills, Banburying or twin-screw extruders. In all processes, the elastomer is compounded with filler, tackifier, processing oils and antioxidants. The tape is made in master drums, which are then slit to the specified width.

3. Methods of Application of Tape Systems to Pipelines

There are three accepted methods used in wrapping pipelines: hand wrapping, over the ditch by line-travel machine and plant coatings. The hand wrapping method is the direct application of tape to pipe using hand tension. This method is used most often by utilities and gas companies for the coating of field joints, girth welds, service lines, tees, valves and other odd-shaped piping structures. Line-travel wrapping over the ditch uses a machine which follows the welded pipe and applies the tape under tension. The plant coating method is done in coating yards where a constant tension dispenser ("let-off stand") applies the tape to individual lengths (usually 40 ft (12 m) or 80 ft (24 m)) of pipe. The coated pipe is stored in a yard and at the appropriate time is shipped to the pipeline right-of-way where the coated pipe is bent, welded and laid into the ditch. A factor affecting the long-term performance of a pipeline coating is the quality of its application; that is, the more carefully the coating is applied, the better will be the protection and in-service performance of the coating.

3.1 Manual and Over-the-Ditch Application

The application steps are simple.

(a) The metallic surface to be coated is cleaned by wire brushing.

(b) The cleaned surface and any coating adjacent to it is primed with an appropriate primer.

(c) The tape is applied over the primed surfaces with sufficient tension and overlap to ensure that the bare metal and any adjacent coatings are completely covered with the tape; either a spiral or cigarette wrap may be used.

The advantages of manual or over-the-ditch tapes in these types of applications are as follows.

(a) No heat or open flames are required.

(b) A wide application temperature range is possible.

(c) No air-polluting components or by-products are present in or evolved during the application.

(d) The tapes are compatible with other types of coatings.

Tapes are applied by power-driven or hand-operated machines. The use of hand-operated wrapping machines is limited to short runs of pipe and almost always to small diameters. The tapes are often overwrapped with a mechanical outer wrap. The most effective and compatible outer wraps used over tapes are based on polyethylene resins similar to those used in the corrosion-protection tapes.

Power-driven tape-wrapping equipment is the most common method of applying polyethylene tapes. It is estimated that 80–90% of all plastic tape coatings are applied over the ditch by power-driven wrapping equipment. This equipment is designed, built and operated to (a) clean the pipe, (b) apply the primer and (c) apply the corrosion-protective tape and, if specified, the outer wrap in one single operation.

Over-the-ditch wrapping equipment of this type applies tape coatings at very consistent and fast rates. There are no limitations on pipe diameter and the only limiting weather factor in over-the-ditch application is moisture. If the pipe surface is dry at the time of tape application, power equipment can be used to apply polyethylene tapes throughout the temperature range of −34 °C to 81 °C. Over-the-ditch coating production rates of 3–4 km per day are easily and consistently attained with modern tape-wrapping equipment.

3.2 Plant Coating

Improvements in tape coatings and the methods for their application are continuously being made by users as well as manufacturers of tape coatings and application equipment. The latest developments are in plant coatings. The new plant coating system is based on a novel, multilayered combination of pipeline products. The requirements for plant coatings are very rigorous compared with those for over-the-ditch coatings. The major additional requirements of a plant coating are bendability of precoated pipe over a broad temperature range, ability to withstand multiple handling and transportation and sunlight (ultraviolet) resistance. All these factors must be satisfied in establishing an acceptable and functional plant coating system. The coating system consists of the following components starting from the pipe surface:

(a) pipeline primer applied to cleaned pipe;

(b) a polyolefin film laminated to a pressure-sensitive adhesive, in which the resultant pipeline coating tape acts as the basic corrosion-protective layer; and

(c) an outer wrap whose purpose is to provide mechanical protection.

These pipeline coating materials are applied in a single pass operation which includes the cleaning and priming, as well as the application of the tape and the outer wrap.

No significant chemical change has been observed in tape after 11 years in the ground. Tape coatings applied via plant coatings require 10.8 mA m^{-2} for protection, a value which is considered acceptable for a functional coating.

It is possible that mulitiple layers of tape and outer wrap, rather than just one of each, can be applied by the plant coating process if more than two layers are needed to meet the requirements of any construction or operating condition.

The multilayer polyethylene tape processes and systems offer important advantages in the area of product performance. These product advantages are as follows.

(a) The corrosion-protection layer is completely protected during outdoor storage. The outer wrap layer completely covers and shields the corrosion-protective layer from ultraviolet light, cold, heat and moisture. With all other types of plant coatings, incuding epoxy and extruded or sintered polyethylene, the protective layer is subject to deterioration during outdoor storage. Any deterioration of the corrosion-protective layer which occurs to these coatings during storage reduces the level of corrosion protection after the coated pipe is installed underground.

(b) The polyethylene-based tape and outer wrap remain ductile and flexible in temperatures down to −40 °C. Thus, pipe coated with the multilayer polyethylene tape coating systems can be transported and handled in extremely cold weather without damage. Other plant coatings become brittle at temperatures considerably above −40 °C.

(c) Because of good ductility and flexibility, pipe coated with the multilayer polyethylene tape systems can be bent in the field without damage on the inside or outside of bends throughout a wide temperature range.

(d) Pipe coated with the multilayer polyethylene tape coating systems can be coated with concrete by all of the commonly used concrete coating techniques. The outer wraps are tough enough to protect the corrosion-protection layer during the concrete coating operation.

(e) The tough polyethylene outer wrap absorbs the mechanical forces which cause damage during handling and transportation and protects the anticorrosion tape layer.

3.3 Other Coating Methods

Another accepted method of applying polyethylene coatings to pipes in plant coating is by direct extrusion onto the pipe. The coating consists of a butyl extrusion layer (~0.25 mm) and a polyethylene layer (~1 mm). Each layer is applied to the pipe using separate extruders operating in tandem.

Other new tape coating systems use polyethylene backing films with hot-melt adhesives, such as polyolefins, polyamides and butyl. These adhesive systems require heat to bond the adhesive to the pipe, whereas present systems are applied cold and are not energy dependent and/or intensive. Hot-melt adhesive systems are limited in their use at this time, but will find a place in the pipeline industry with rising operating temperatures, pressures and more aggressive environments. One of their outstanding features is their very high modulus and high softening temperature. Therefore, their shear-resistant properties are good. In situations where shear stress and temperature are significant factors, these coatings will find a use.

A proposed new system combines polyethylene tape over a fusion-bonded epoxy coating. This system adds to the advantages of epoxy by increasing the barrier properties of the total system as well as improving the impact strength. Another coating system for pipeline applications makes use of foams to maintain and prevent temperature changes and heat fluxes in the medium being transported through the pipeline. These foams can be applied directly onto the pipe as a premolded material. Each method has its advantages, limitations and economic considerations. However, all perform the same basic function, which is to maintain a constant temperature and heat profile in the medium being transported through the pipe.

4. In-Service Performance

The fourth coating requirement listed by the US Federal Pipeline Safety Code is that a coating must have properties compatible with cathodic protection.

Proof of the compatibility of polyethylene tape coatings with cathodic protection is available through coating conductance surveys performed on pipelines in service for many years. In one case, after 12 years of service, the coating conductance of the polyethylene tape increased less than five times. This illustrates the remarkable ability of polyethylene to resist moisture penetration and absorption, as well as the cathodic-protection-produced stresses of electroosmosis (see Table 1). Coatings with coating conductance values in the range of 53.8–215.2 $\mu\Omega^{-1}\,m^{-2}$ are rated as excellent.

Tape pipeline coatings have been used on underground pipelines since the late 1940s. At the present time, more than 200 000 km of pipe are being protected from corrosion by tape systems with pipe diameters up to 1400 mm. That these tape coatings, in conjunction with cathodic protection, have

Table 1
Coating conductance performance of a polyethylene tape

Time underground (years)	Coating conductance ($\mu\Omega^{-1}\,m^{-2}$)
1	
2	45.9
3	70
4	76.5
5	78.3
6	70
7	106.4
8	98.3
9	108.9
10	188.1
11	239.0
12	212.4
16	520.0
19	700.0

Table 2
Current densities required for cathodic protection of a hydrocarbon pipeline with polyethylene tape coating

Year	Current density for protection ($\mu A\ m^{-2}$)	Pipe-to-soil potential (mV)
1958	7.5	1270–2000
1963	11.4	1275–1875
1964	10.5	1350–1700
1965	12.6	1200–1675
1968	13.6	1175–1600
1969	14.8	1125–1600
1970	16.1	1050–1325
1971	18.1	950–1650[a]
1972	14.9	950–1650[a]
1974	16.9	1000–1650[a]

a Low PS potential reading was taken at point of bond to a crossing pipeline

provided corrosion protection is shown by tests using leakage conductance and cathodic protection current requirements cited in the NACE Standard RP-01-69.

The performance of the polyethylene tape pipeline coating on a pipeline owned by a major US oil company has been evaluted at intervals in terms of current requirements, since the pipeline was built in 1958. The current requirements of this 48 km of 200 mm pipe, as shown in Table 2, ranged from 7.5 $\mu A\ m^{-2}$ in 1958 (year of construction) to 16.9 $\mu A\ m^{-2}$ in 1974. This represents an increase of 0.4 A in total current to protect the entire 48 km. This shows that the electrical, chemical and moisture resistant properties of the polyethylene tape coatings remain unchanged during long-term underground exposure and that the coating has the strength to resist physical damage.

In summary, tape systems have the following advantages.

(a) Application is cold; no heat is necessary to bond or apply the coating to the pipe.

(b) Capital investment is low for plant coating applicators.

(c) The system is nonpolluting and meets Occupational Safety and Health Administration regulations.

(d) The thickness of the total coating system is changed simply by the addition of individual tape layers.

(e) Tapes are flexible and ductile down to −35 °C.

(f) Tape systems have high impact and abrasion resistance and can be bent to conform to the required geometry of pipeline without affecting integrity of the system.

(g) For plant coating systems, tapes can be concrete coated.

(h) There exists a broad temperature range for both application to the pipe and in-service functional performance.

(i) There is high resistance to soil stresses imposed by the environment.

(j) The system has proven in-ground service performance with low conductance and current density.

(k) The inner corrosion tape is isolated and protected from all environments by the outer wrap, which is not the case with fusion-bonded and extrusion systems.

(l) Tapes have high dielectric strength, high electrical resistance, low water absorption and low water vapor transmission. Also, tapes achieve high bond strengths.

(m) There is no limitation on pipe diameter, as is the case with fusion-bonded and extrusion systems.

There are, however, some disadvantages.

(a) They are not operational at temperatures above 85 °C, due to softening of backing films.

(b) For vinyl tapes, there is a migration of plasticizer from the film backing of the tape which causes a loss in physical, mechanical and functional properties.

(c) The tensions must be correct in order to obtain good conformability of the tape at the overlap. This requires properly engineered tension-dispensing apparatus.

(d) At elevated temperatures, adhesive shear strength decreases in accordance with the Arrhenius equation.

(e) Oxygen permeability through polyethylene is high, but for butyl adhesive there is good oxygen permeability resistance.

See also: Maintenance Coatings

Bibliography

Chapman B N, Anderson J C 1979 *Science and Technology of Surface Coating*. Academic Press, London
Harris G M 1979 A review of recent developments and the performance of tape coatings for underground pipelines. *Mater. Perform.* 18(9): 17–21
McKelvey J M 1962 *Polymer Processing*. Wiley, New York
Samour C M, Jackson E G, Thomas S J, Davidson L E 1980 Coated pipe and process for making same. US Patent No. 4,213,486 (22 July 1980)
Tadmor Z, Gogos C 1979 *Principles of Polymer Processing*. Wiley, New York

C. M. Samour and S. J. Thomas
[Kendall Company, Lexington, Massachusetts, USA]

Q

Quality Requirements in Elastomer Processing

Raw elastomers, whether natural rubber or one of the many varieties of synthetic rubber, have to undergo a series of processing steps, such as mixing, milling, extrusion and molding, to convert them into usable products such as tires, hoses or seals. In each of these steps there are quality requirements and specifications to be met.

There are three contributory factors which, between them, determine both the efficiency of each process and the quality of the resultant product. These are: (a) the quality of the material entering the process, (b) the precision and accuracy of the test methods used to assess both the feedstock and the product and (c) the degree of the control of each process.

Elastomers, and their compounds, are more difficult to mix and process than thermoplastics. One reason for this is that elastomers are viscoelastic and so their deformation and flow behaviors are more complex than those of thermoplastics. A second reason is that in order for the end product to have adequate physical properties, compounds require high levels of reinforcing fillers such as carbon black. In addition, and most importantly, in the final, forming step, rubber compounds have to be vulcanized; that is, crosslinks have to be formed between the separate carbon backbone chains, in order to give the required elasticity in the end product.

1. *Quality of Raw Materials*

The major raw materials of the rubber industry are, in addition to the elastomer itself, carbon black, other fillers, oils and various chemicals. For each of these categories, the processor has to be establish minimum requirements or specifications, whose aim is to ensure the quality of the raw materials entering the process. For raw elastomers, such specifications usually include minimum and/or maximum property limits, both for the elastomer itself and for specified compounds prepared from the elastomer in the supplier's testing laboratory. There are also certain implicit, but unstated, requirements, such as that the material should be free of contamination by foreign materials. The specifications, in themselves, do not guarantee that the elastomer will mix, extrude mold, cure or perform satisfactionaly in the customer's process or product. In many cases, the test recipe bears little or no resemblance to the compound that the customer produces from raw elastomer. Equally, the tests that are performed on the raw elastomer or compound, and which feature in the specification, often bear little resemblance to the actual processing or service requirements. The object, or purpose, of specifications, as applied to elastomers, is to guarantee to customers that they receive the same product shipment by shipment. Thus, the specification is, in effect, a benchmark or template. If the raw elastomer meets the agreed specification then it can be expected to perform in processing and application in a consistent manner. Meeting the specification is thus a guarantee of the consistency of the product.

The width of a specification range for a particular property is also important. It must reflect both the customer's needs and the manufacturer's capability to produce and supply. The midpoint should express the most desirable value for a parameter critical to the customer. Any deviation from this value that is not test error, represents a loss. This loss is a continuous function, although specifications imply, illogically, that there is a precipitous loss at the specification limits. Thus, the producer's objective should be to reduce variability continuously within the specification. To measure this, it is now common to express the ratio of the specification range to the actual process variability as the "process capability." In the rubber industry, this concept is complicated by the relatively large error inherent in most test methods.

2. *Reproducibility of Test Methods*

Determining the processability of elastomers and their compounds, whether in mixing, extrusion, molding or curing, is not easy. There is a lack of simple tests suitable for quality control in factory conditions. In fact, if the test is simple it usually does not correlate directly with processability and, if it does correlate, it is rarely a simple test to perform.

The basic problems of process control in the manufacture of elastomeric products are threefold:

(a) there is a need for material characterization at each stage of the process, from incoming raw material to finished product;

(b) the property measured must relate to a significant response of the material to processing conditions; and

(c) the method of measurement must be consistent with the practical constraints of manufacturing operations—this means that the test procedure must be rapid and amenable to execution and interpretation by semiskilled or nontechnical staff.

Processability of a raw elastomer in an internal

mixer, and of the resulting compound in later stages of processing, depends on polymer type, molecular weight distribution, degree and type of branching, catalyst and emulsifier residues, stabilizer and other additives. In addition, filler type, filler concentration, polymer–filler interaction and the levels and types of oils, plasticizers and process aids also affect the process.

When all the above have been accounted for there are three residual sources of test error:

(a) calibration of the equipment,

(b) the care taken in performing the test, and

(c) the quality of the written standard test method.

3. *Control of the Processes*

There are four major steps in converting a raw elastomer into a finished product. These are mixing, milling, forming and curing.

Mixing involves the preparation of a homogeneous, well-dispersed compound from the initially separate ingredients. Milling involves converting the product dumped from the mixer, into sheets, strips or pellets suitable for feeding to the forming stage. This latter step involves converting the mixed compound into the shape required for the end use, by extrusion or molding. Curing or vulcanization requires a period at an elevated temperature so that the chemical reactions involved in developing cross-links (the process of vulcanization) can occur. This article will deal only with the physical processes; it will not consider the chemical process of vulcanization. Nor will it deal in any detail with the procedures or equipment used in the industry. Details of these can be found by consulting the sources listed in the bibliography.

At one time, it was usual in the rubber industry to consider all aspects of processing as more of an art than a science. As such, expertise had to be transmitted on the factory floor from master to pupil by a sort of apprenticeship system. Although this is still true, a much clearer understanding of the process involved has now been achieved. This has allowed better control and, therefore, optimization of the various processes.

3.1 Mixing

When mixing fillers, plasticizers and chemicals into an elastomer, the aim is to produce a product that has the ingredients incorporated and dispersed sufficiently thoroughly, that it will shape easily, cure efficiently and give the necessary end-use properties—all with the minimum expenditure of machine time and energy. Once the compound formulation has been decided on, converting the separate ingredients into the mixed compound involves problems of rheology, heat exchange, engineering and economics. Consideration has to be given to the effect of machine variables on the microscopic properties of the material and, in turn, how these affect the properties of the end product.

It is important to distinguish between the two important but distinct actions which take place—distributive and dispersive mixing.

Distributive mixing is the increase in the randomness of the spatial distribution of the minor constituents within the major base without further change in the size of that minor constituent; for example, distributing black masterbatch or accelerator masterbatch throughout a compound. (Masterbatches are concentrates of appropriate materials in base polymers: they enable small amounts of critical components such as accelerators to be added more accurately or improve the cleanliness and distribution of filler-loaded compounds.)

Dispersive mixing is the reduction in agglomerate size of the minor constituent to its ultimate particle size (e.g., compounding carbon black into a polymer).

In practice, both actions occur simultaneously, until the required degrees of homogeneity and particle size have been reached. However, before either distributive or dispersive mixing can effectively take place, there is an initial step in which the originally separate ingredients form a coherent mass. This process is known as incorporation or welting.

(*a*) *Incorporation*. In most processing operations, elastomers do not deform smoothly or flow steadily, but rather yield and rupture. The mixing process imposes a large deformation, which often exceeds the breaking strain and so material failure is important in mixing, especially in the incorporation stage. The incorporation stage has two mechanisms. In the first, the elastomer undergoes a large deformation, increasing the surface area for accepting filler agglomerates and then sealing them inside. In the second mechanism, the elastomer breaks down into small pieces, mixes with the filler agglomerates and, once again, seals them inside. In an internal mixer, the material is forced through narrow passages between the rotor blades and between the rotor blades and the chamber wall and, in the process, undergoes extensional deformation. This extensional flow is believed to be more efficient in incorporation and in breaking up rigid aggregates (dispersive mixing) than the concomitant shear flow. At the same time as incorporation, compaction or massing has to occur. This is the displacement of entrapped air in the material and requires deformation of elastomer domains to match the shape of the filler.

(*b*) *Dispersion*. In the early stages of mixing, carbon black forms relatively large agglomerates (up to 100 μm); during dispersion they are broken down to a size of less than 1 μm. The dispersion stage of mixing requires higher shear stress and energy input than the incorporation stage and the physical properties of the mix alter as the dispersion proceeds. The changes in power consumption in a typical elastomeric mix are indicative of stages in the process and they can be related to the development of end-product properties.

Properties such as viscosity and die swell, which are related to the effective volume fraction of elastomer in the mix, approach a steady-state value at the end of the dispersion stage. Thus, a well-dispersed mixture always has a lower viscosity and a higher die swell than a comparable, less homogeneous, mixture.

(*c*) *Process variables*. A major concern in most mixing operations is to control the temperature at the end of the mixing cycle. Modern internal mixers are very efficient as heat exchangers, but a major problem in cooling elastomers, in any process, is its inherently poor conductivity, which means that it is only cooled where it touches a cool surface. Thus, total heat removal depends on the area of the cooling surface and also on the efficiency with which the rotors cause fresh elastomeric surfaces to come into contact with that cooling surface.

The most commonly used criteria to determine when to terminate the mixing process have always been time and temperature, based on previous experience with the same or similar mixes. The aim is to achieve sufficient dispersion, avoid overmixing, and to minimize variation between batches. For many compounds, monitoring the instantaneous power usage and total energy input per unit volume of compound has been found to give a more precise and reproducible control of the mixing cycle. Several microprocessor control systems are now available which monitor and control temperture, energy input, mixing time and so on. Some of these also control weighing, charging and inventory control of the raw materials.

3.2 Milling

Milling was the original method used to mix elastomeric compounds but, because mill mixing is slow and environmentally potentially hazardous, mills are normally used today only to prepare the batch, dumped from the mill as a solid mass, for further processing by producing sheets and strips. In addition to changing the physical shape of the batch, the mill also cools the elastomer by exposing it to a large cool metal surface.

Depending on the base elastomer and the composition of the compound, varying degrees of difficulty may be encountered in forming a band on the mill rolls. Some materials may sag, fail to adhere to either the fast or the slow roll or crumble on passage through the nip. The behavior of a particular compound on the dump mill depends on the balance between its viscous and viscoelastic properties. If the stresses caused by deformation in the nip exceed a certain critical value then, in the time of one revolution of the roll (normally 2–4 s), the material can relax and sag. The stresses (or stored energy) may be high enough to initiate tears. If the speed of the loaded roll is too high, the reverse can happen; the stresses don't relax enough and therefore build up with further passes through the nip, resulting in tearing and crumbling. The correct temperature, nip gap and roll speeds for any particular compound have to be determined by practical experiment.

3.3 Extrusion

The extrusion process consists of two distinct parts: a delivery system that pumps the elastomeric compound and a die system that forms the material into the required cross-sectional shape. The phenomena observed during the extrusion of elastomers are different from those observed for thermoplastic materials, which exhibit three consecutive stages—solid conveying, melting and melt conveying. With elastomers, the cold elastomer feed is above its glass-transition temperature, its yield value is low and the transiton from solid to melt flow takes place in the entrance region as soon as the elastomer is subjected to the influence of the screw. As a result, the only zone which is clearly distinguishable is the metering zone.

A well-designed extruder screw provides high output rates at low extrudate temperatures, high homogeneity at both high and low speed, good self-cleaning properties to facilitate compound changes and the ability to handle compounds with a wide range of viscoelastic properties. It is not easy to achieve all these in a single screw and so most manufacturers provide interchangeable sections to enable changes from a standard to either a high-shear or high-output configuration. Extruder manufacturers have developed mixing elements and screws whose aim is to introduce turbulent flow patterns in the melt, preferably without increasing pressure drop. However, it must be clearly understood that mixing sections can only perform local homogenization or distributive mixing; macrohomogenization and dispersive mixing are not achieved by passage through an extruder. This local homogenization can be optimized by use of a cavity transfer mixer, which comprises a rotor and a stator, both of which have rows of hemispherical cavities, and which fits at the end of the screw and produces the low shear, cutting and blending action required.

The purpose of the extruder head is to equalize pressure across the profile, to guide compound into the die and to hold other items such as screenpack, pressure and temperature sensors and "spider." The flow of compound through the extruder head has to be smooth and without deadspots or temperature inhomogeneties, right up to the die entrance.

For continuous vulcanization applications, a shear extrusion head is a useful device. This fits between the extruder screw and the die and raises the compound temperature to the required prevulcanization temperature. The material is sheared in an annular shearing gap between a rotating mandrel and a stationary cylinder, as it is flowing past the mandrel.

There are three phenomena of importance in the flow of elastomers through a die; these are (a) pressure drop, (b) die swell and (c) extrudate distortion. Much of the pressure loss in extrusion through a die occurs externally, as entrance and exit losses, due to changes in extensional flow patterns. After a polymer is forced through a restriction, such as a die, it expands in cross section (die swell) and, correspondingly, shrinks in length (extrusion shrinkage). For a given compound, extrusion shrinkage at a constant shear stress is independent of die geometry and temperature of extrusion, but rather depends on the molecular orientation imposed on the compound during extrusion. The rate of molecular relaxation is extremely rapid, 80% occurring in less than 0.1 s after exit from the die.

Developments in electronics in recent years have made microprocessor control of extrusion processes relatively simple and inexpensive. However, their use depends on the experimental development of extruder operation curves, which relate output to head pressure, because accurate theoretical predictions of this relationship are not yet possible. Where dimensional tolerance of the extrudate is vital, they can be monitored continuously with laser triangulation gauges, which require feedback control to screwspeed or takeoff conveyors.

3.4 Molding

The major molding processes for elastomers are: (a) compression molding, in which elastomeric blanks, previously shaped and weighed, are placed in the mold manually; (b) transfer molding, in which blanks, previously shaped, weighed and preheated, are placed manually in a transfer chamber and the cavities are fed automatically; and (c) injection molding, in which the injection chamber is automatically fed the exact amount of the plasticized elastomeric compound, which is then automatically fed to the mold.

Injection molding is the preferred technique on economic grounds, except for the largest moldings. However, in injection molding, elastomeric compounds are subjected to more severe processing conditions than in compression or transfer molding. Temperature, pressure and shear stresses are higher, although cure times are shorter. Control over process variables is more precise and the cycle time can be minimized by independently controlling barrel temperature, screw speed, mold temperature, cure time and injection pressure. To a large extent, controlling the injection-molding process reduces to a question of homogeneity of temperature and heat history. The temperature of the mix in the injection chamber, prior to injection, is determined by the temperature of the extruder and chamber, by the screw speed and design and by the applied back pressure. The injection, or mold filling, time and temperature depend on the mix temperature, as determined by the above factors, the injection pressure and the dimensions of nozzle, runners and gates. The cure time depends on the mold temperature and the injection temperature. Computer packages, both to control the injection process and to simulate the filling and curing phases of the injection-molding process, are now available.

4. Future Developments in Elastomer Processing

There are a number of areas in which improvements are possible. There will continue to be developments in our understanding of the processes of mixing, milling, extrusion and injection molding of elastomers. Parallel with these improvements in understanding will be improvements in understanding will be improvements in microprocessor modelling and control and also in design and operation of equipment. Elastomer processing is gradually evolving from an art into a controlled, science-based technology.

Some recent publications, whose preparation was coordinated by the UK Plastics and Rubber Institute, have given guides to the development of total quality management in the polymer industry, aimed at improvements in products, processes and services (West 1987, Pimblott 1988, Coulson 1989, Buist and Coulson 1990).

See also: Elastomers: Additives; Fillers in Elastomers; Mixing: Basic Concepts; Mixing: Internal Mixers; Mixing: Mills for Elastomers; Rheological Characterization of Molten Polymers; Rubber Processing; Vulcanization

Bibliography

Allee W A 1985 Calendering equipment and process. In: Long 1985, Chap. 5

Buist J M, Coulson S H (eds.) 1990 *Total Quality Management*. Plastics and Rubber Institute–Dunholm, Ashton-Under-Lyne, UK

Coulson S H (ed.) 1989 *Managing Improvement: Where to Start*. Plastics and Rubber Institute–Dunholm, Ashton-Under-Lyne, UK

Ellwood H 1987 Using modern mill room equipment. In: Whelan and Lee 1987, Chap. 6

Freakley P K 1985 *Rubber Processing and Production Organisation*. Plenum, New York

Johson P S 1983 Developments in injection-molding science and technology. *Elastomerics* 115(5): 22–32

Johnson P S 1985 Mixing equipment and the mixing process. In: Long 1985, Chap 3

Johnson P S 1987 Quality requirements and rubber mixing. In: Whelan and Lee 1987, Chap. 7

Johnson P S 1989 *Extrusion of Rubber*. Nova Rubber Division, Sarnia, Canada

Johnson P S 1990 Rubber processing in the 1990's. *Elastomerics, Annu. Rev.*

Lambright A J 1985 Extrusion equipment and process. In: Long 1985, Chap. 4

Leblanc J R 1981 Die swell and melt fracture in rubber extrusion. *Rubber Chem. Technol.* 54: 905

Long H (ed.) 1985 *Basic Compounding and Processing of Rubber*. American Chemical Society Rubber Division, Akron, OH

Norman R H, Johnson P S 1981 Processability testing of elastomers: Rubber reviews for 1981. *Rubber Chem. Technol.* 54: 493

Pimblott J G (ed.) 1988 *Profit and Growth Through Quality*. Plastics and Rubber Institute–Dunholm, Ashton-Under-Lyne, UK

Reed D, Jagit K (eds.) 1988 *Rubber Injection Molding—Improving on Sophistication*. Rapra Technology Conference Papers

Sezna J A 1988 Processability testing, I. *Polym. Test.* 8(3): 161–90

Sezna J A 1988 Processability testing, II. *Rubber World* 199(4): 21

Sommer J G 1985 Molding equipment and methods. In: Long 1985, Chap. 6

Stiehler R D 1985 Testing of rubber. In: Long 1985, Chap. 8

Vieth A G 1987 Precision and accuracy in rubber testing. *Polym. Test.* 7(4): 239

West G H 1987 *Quality Control for the Plastics Industry*. Kaiser–Plastics and Rubber Institute

Whelan A, Lee K S 1987 *Developments in Rubber Technology*, Vol. 4. Elsevier, London

P. S. Johnson
[Sarnia, Canada]

R

Radiation Effects on Polymers

The interaction of high-energy radiation with matter can have three main effects: changes in nuclear structure with the creation of new isotopes or elements; displacement of the nucleus, for example, in a crystal; and disturbance of electronic structure, leading to ionization, excitation, trapped electrons and hence to chemical changes. In considering radiation effects in polymers, it is the last group of effects that is involved.

The study of radiation-induced chemical reactions forms the subject matter of radiation chemistry, which is now largely concerned, at academic level, with reaction kinetics and processes in liquids. The importance of irradiation in polymers is that although the chemical changes involved are essentially similar to those observed in the irradiation of low-molecular-weight organic compounds, the physical and biological changes can be considerably greater. This is because a single modification in the chemical structure of a macromolecule can affect the behavior of the very large remainder. Examples include major changes in solubility and in mechanical properties of a polymer, and in the biological behaviour of a biopolymer.

Although the changes described here can be induced by many forms of high-energy radiation—fast neutrons, protons, alpha particles and so on—most of the research on the effect of radiation on polymers has been carried out with gamma radiation from ^{60}Co sources or electron radiation from high-voltage accelerators, with energy per electron from about 250 keV up to about 4 MeV, insufficient to produce nuclear changes or displacements of significance compared with the chemical changes. Apart from the very different radiation intensities involved (and hence the different times available for competitive reaction rates and for oxygen diffusion into the irradiated specimen), as well as the very different penetrations, there does not appear to be any fundamental difference between the effects of these two very different forms of high-energy radiation. A gamma photon will eject energetic electrons during its passage through a material and this will produce the same reactions as those resulting from electrons whose high energy is imparted by electrical accelerating equipment outside the specimen.

Differences do however appear when these two forms of radiation are compared with that involving heavier particles of the same energy. The rate of energy loss of these in passing through matter is much greater, resulting in a much shorter range, very much higher energy density of ionization and excitation, so that there is a very real difference between the effects of a fast proton, an alpha particle or an electron. This difference does not often arise in the irradiation of polymers, since in practical terms this is almost entirely concerned with the effects of the more penetrating gamma photons, fast electrons or high-energy x rays.

1. Dose and Radiation Sources

In penetrating a polymer (solid or liquid) of about unit specific gravity, an electron of initial energy 2 MeV will lose energy by a series of collisions and will eventually be "thermalized," that is, reach an energy of kT, at a penetration of about 6–8 mm. In so doing it will undergo some 60 000 collisions, with resultant excitations and ionizations, spaced about 100 nm apart. If each of these collisions causes chemical change in a different polymer molecule, of average molecular weight 10^5, the net effect could be to modify the physical properties of a total molecular weight of 3×10^{10}. Since a biological molecule may have a molecular weight of about 10^9, the net effect here could be correspondingly greater.

Radiation doses are measured in terms of the energy absorbed per unit mass and the unit used is the gray ($1\ Gy=1\ J\ kg^{-1}$). For polymer work, the doses involved are usually a few tens of kilograys. Until recently, the unit in widespread use was the rad ($10\ nJ\ g^{-1}$). Much earlier, the radiation unit was the röntgen, based on ionization of air.

If fully absorbed, 3.6 MJ of high-energy radiation corresponds to a dose of 1 kGy (0.1 Mrad) delivered to 3600 kg, 10 kGy (1 Mrad) to 360 kg or 25 kGy (2.5 Mrad) to 144 kg (the usual dose for such applications as sterilization). It will therefore be appreciated that a relatively low amount of high-energy radiation power can achieve a considerable effect in biological, medical and polymer applications.

The yield of specific chemical change (denoted here by c.ch.) is expressed as a G(c.ch.) value, where G represents the number of such changes per hundred electron volts absorbed. 3.6 MJ (1 kWh) equals 2.25×10^{25} eV and therefore promotes $2.25\times10^{23}\,G$ (c.ch.) chemical changes of this nature. In polymers, G (crosslinked units) is about 2 in many cases so that 3.6 MJ produces about 0.7 mol of reactive groups of type c.ch. If the polymer molecular weight averages M, the mass affected may be as high as $0.7\times M$ g, that is, 210 kg if $M=3\times10^5$. In biological systems, where M may be as high as 10^9, the mass involved could be correspondingly higher.

In biological systems, account must be taken of the different biological effect of the same energy input, but from different types of radiation, notably alpha particles and fast neutrons. This is dealt with by using units that take in certain factors, including relative biological effectiveness (RBE). The rem (röntgen-equivalent man) is numerically equal to the absorbed dose in rads multiplied by a quality factor, and the sievert is the present biological unit equivalent to the gray or 100 rad times this quality factor. In comparing radiation effects in polymers and biological systems, the former are almost invariably assessed in Gy or Mrad with electron or gamma radiation, so that for many biopolymers the chemical changes with the same radiation are directly comparable.

Gamma-radiation sources for polymer research and applications are almost invariably based on rods of ^{60}Co, emitting two photons, of 1.33 MeV and 1.17 MeV, per decomposition. Equipment can be in the form of a hollow cylinder formed of such rods, with space inside and outside the cylinder, in a shielded laboratory. The source intensity is typically 10^3–10^4 Ci with a dose rate of the order of 10 kGy h^{-1} (1 Mrad h^{-1}). For large-scale industrial use, gamma radiation is less frequently used for thin materials because of their high transparency to gamma radiation and consequent waste of effective irradiation power. Very large cobalt sources are used, however, for sterilization of packages on a conveyor.

Electron radiation can be based on a number of different types of accelerator, ranging from linear accelerators (10 MeV), through resonant transformers (2–3 MeV) to equipment of much lower voltage (0.25–0.5 MeV). The power available can extend up to 200 kW for large-scale industrial applications, or a few hundred watts for research purposes. The electron beam is highly concentrated, so that extremely high doses can be accumulated in very small volumes and in very short time. The main disadvantage of electron sources is their low penetration, around 3 mm MeV^{-1}.

2. *Radiation Processes*

One of the most important observations made when high-energy radiation is absorbed in polymers, or in other organic systems, is that although the energy is presumably absorbed at random, the chemical changes produced, primarily the scission of primary bonds, is not random and the radicals produced are largely specific in nature. Nor are the weakest bonds necessarily those most readily fractured—polyethylene and polyisobutylene both have C—H and C—C bonds, but in the former it is the C—H bond which fractures while in the latter it is the C—C bond. The nature of radiation damage must also depend on polymer morphology, mobility and so on.

In many gases, the G value for ionization is about three, whereas for organic solids and liquids it is usually only about 0.1. This is because in gases, the ejected electron can move considerable distances before it loses most of its energy by collision and is thermalized, whereas in the condensed phase, the distance for thermalization is much smaller (more collisions for the same distance) and the electron no longer has the energy to escape the electric field of the parent atom. The electron is recaptured by the latter to give a neutral but excited molecule, producing a radical. Most of the reactions described here are due to such radicals. At sufficiently low temperatures, their nature and location can be studied by electron-spin resonance and their subsequent reactions followed. The long lifetime of such radicals indicates that one of the explanations proposed for the degree of bond selectivity, that is, energy transfer, cannot often be accepted for such systems. Alternative explanations include the hopping of radicals between adjacent sites by a hydrogen transfer reaction, or a series of addition and subtraction reactions with H and H_2. Molecular rearrangements at sites of fracture must also occur.

The major changes in linear polymers as a result of radiation may be grouped in the following categories.

(*a*) *Main-chain scission*. This results in a reduction in mean molecular weight, a lowered viscosity if in the molten state, reduced mechanical strength and a rearrangement of the chemical groups near the point of scission. Such reactions occur, for example, in cellulose, polytetrafluoroethylene (PTFE), polyisobutylene (PIB) and poly(methyl methacrylate) (PMMA, lucite or Perspex).

(*b*) *Side-chain scission*. In this case the main chain is left unaltered. As a result, radicals are left attached to the main chain, enabling adjacent molecules to become linked together by new permanent bonds. As a result, the molecular weight is increased, but more important is the fact that above a certain density of such "crosslinking," a network is formed with properties quite different from those of the initial polymer. This change corresponds to the vulcanization of rubber, which in the tree has no true elastic properties. Such crosslinking occurs in a wide variety of polymers, including natural rubber, synthetic rubber, polyethylene (PE), polydimethylsiloxane (PDMS), poly(methyl acrylate) (PMA) and so on. Polypropylene (PP), halfway between PE and PIB in the location and number of side chains, is one of the few polymers that shows approximately equal amounts of crosslinking and scission.

(*c*) *Other chemical reactions*. These may occur within the polymer main chain and include, notably, the formation of double bonds (a convenient basis for dosimetry, using infrared spectra of polyethylene), or their destruction, for example, vinyl groups at the end of a PE chain.

(*d*) *Destruction of crystallinity in partially crystalline polymers*. Such an example is PE at room temperature. The rate of loss of crystallinity is, in fact, very low and doses of up to a hundredfold the amount for significant crosslinking may be needed to produce marked destruction of crystallinity.

(*e*) *Competitive reactions*. The radicals produced by radiation may engage in competitive reactions, notably combination with oxygen diffused into the specimen. A destructive oxidative degradation can then ensue by a chain reaction and the effect may continue over many months, owing to slow diffusion of oxygen, for example, and result in a polymer of weaker strength. To avoid this damaging effect, antioxidants are sometimes included. These can continue their protection, or be themselves effectively eliminated by reacting with the radicals produced by radiation. In this latter case, radiation is used to produce radicals (which might otherwise give useful crosslinks) and then additives are used to remove them. Such behavior may nevertheless be necessary if the prime objective is to irradiate objects packaged in polymeric films (such as medical equipment for sterilization), whose physical properties should not be damaged by the sterilization process.

2.1 Polymer Scission

The number of main-chain scissions occurring in the first group of polymers is directly proportional to the radiation dose *r*, so that the molecular weight can be adjusted over very wide limits by choosing the correct radiation dose.

There are a number of ways of measuring average molecular weight, depending on how much importance is attached to shape as well as numbers and also on the degree of dispersion about the mean. For polymers that crosslink (see Sect. 2.2), the most important is the weight average M_w. For scission, where the average weight distribution is of no importance, it is the number average M_n. With A (Avogadro's number$=6.02\times10^{23}$) each gram of polymer contains $A/M_n(0)$ molecules and this will increase to $A/M_n(r)$ after a dose of r kGy (0.1 r Mrad). Then

$$A/M_n(r)=A/M_n(0)+(0.624\times10^{17}\ G(\text{scission})r)$$

or

$$1/M_n(r)=1/M_n(0)+(1.04\times10^{-7}\ G(\text{scission})r)$$

If the initial distribution of molecular weights is random, it will remain so when scissions occur at random. If it is not random, then it will rapidly become so. However, many polymers show a lack of homogeneity, such as cellulose with amorphous and crystalline structure, and it is the former, with its far more mobile structure, which will be more readily affected. This has been suggested as a means of degrading cellulose, to allow more rapid attack later by bacteria. Although very considerable damage can be caused by radiation alone, the doses required to achieve an edible form of some cellulose structures are too high to be economically feasible.

Polyisobutylene also suffers main-chain scission and the G(scission) value, which appears to be independent of dose, varies with temperature, being greater at higher temperatures. This might be explained by the higher mobility of the polymer chains, so that molecular chain separation is more readily achieved. Of considerable interest is the fact that this temperature dependence is identical with that observed in certain bacteria. This is a strong argument in favor of the idea that the temperature effect on radiation damage in such biological systems has no purely biological origins, but arises primarily from enhanced molecular mobility and is of a purely physicochemical origin.

Polytetrafluoroethylene is another, highly crystalline polymer that suffers irradiation-induced main-chain scission, with a consequent reduction in mechanical strength. It is of interest to note that this increases the degree of crystallinity to only a minor extent, but subsequent heating greatly enhances it. This would indicate that it is the chains linking adjacent crystalline regions, and only partly accommodated in either, that are in the most sensitive position and that scission of such chain lengths permits subsequent recrystallization into larger and more perfect crystals which were otherwise restrained by sections of molecules linking adjacent smaller crystals.

2.2 Polymer Crosslinking

Polymer crosslinking involves essentially linear molecular chains, initially separate and only linked together by entanglement or weak hydrogen bonds. On irradiation, the radicals on adjacent chains can link to form a primary, covalent bond, as strong as the original main-chain bond, for example:

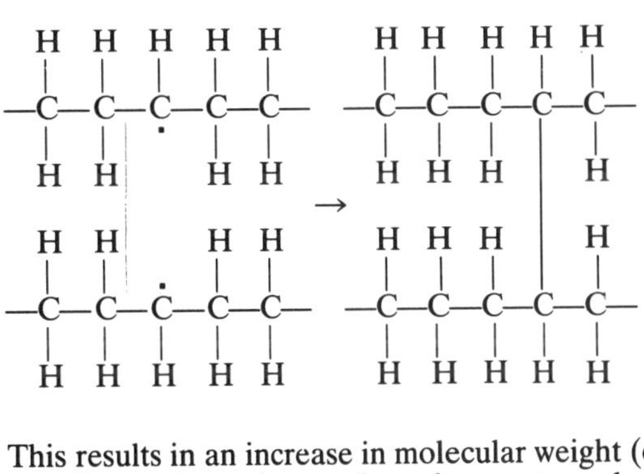

This results in an increase in molecular weight (each crosslink reduces the number of separate molecules until many such links become internal).

A point is soon reached at which a three-dimensional network begins to form. This is essentially a single molecule with bonds that are all covalent and which is therefore insoluble in the same sense as any single molecule cannot be dissolved—it always remains a single molecule which cannot be split without primary bond fracture. The irradiated specimen then separates into (a) a soluble fraction s (single molecules, or molecules comprising two or more simple links), referred to as the sol, and (b) an insoluble fraction or gel, essentially a single giant molecule or network. The properties of (a) and (b) are completely different and the gelation dose r_g at which this network first appears is a very important value.

The theory of random crosslinking leads to the evaluation of the gel/crosslink density. The network first starts to appear at a crosslink density equivalent to an average of one crosslinked unit per molecule with weight-average molecular weight. Since each crosslink requires two crosslinked units (one per molecule) this means that 0.5 crosslinks per molecule with weight-average molecular weight is required.

The number of crosslinked units per molecule with weight-average molecular weight is denoted by the crosslink coefficient δ, and the gel fraction g begins to increase above zero only when $\delta > 1$. The rate of increase beyond this point depends on the molecular weight distribution. If all molecules are originally the same size ($M_w = M_n$), then

$$s = \exp(-\delta g) \quad (s = 1 - g)$$

while for an initially random distribution, with $M_w = 2M_n$

$$s + s^{1/2} = 2/\delta$$

Other formulas for other distributions can be derived from the general formula (Charlesby 1954, 1960). Furthermore, since crosslink density is proportional to dose r, and $\delta = 1$ at the gelation dose r_g, then $\delta = r/r_g$.

At low temperatures, when the molecular chains are still rigid, below the crystallization or glass temperature, the existence of crosslinks (except at very high densities) has very little influence on the physical properties of the system. At higher temperatures, when the polymer chains have adequate mobility, they attempt to acquire the random configuration predicted for thermodynamic reasons. (This is somewhat akin to the behavior of gases for the same thermodynamic reasons.) Under an external stress, the network takes up a less random, more oriented arrangement, but attempts to return to a purely random configuration, thus exerting a retractive force. In this respect, each length of chain between two successive links attempts to behave as though it were a separate chain, so that the retractive force is proportional to the number of such chains, of average molecular weight M_c, which is inversely proportional to radiation dose. In line with the somewhat analogous formula for the pressure inside a gas, $pV = RT$ per mole of gas where p is the pressure, V is the volume, T is the temperature and R is the universal gas constant, or $p = \rho RT/M$ per gram of gas, we have

$$E = 3\rho RT/M_c \quad \text{per gram}$$
$$\propto 3\rho RTr$$

where ρ is the density.

By increasing the radiation dose we can therefore pass from a separate, fluid system above the melting or glass-transition temperature to an elastic network whose modulus is proportional to the dose r (apart from corrections for free ends, soluble fractions, etc.). The modulus increases steadily although the tensile strength reaches a maximum, at about 400–500 kGy for rubber. It then decreases as the overcured network takes on the properties of a cheeselike structure, a leathery character and eventually the nature of a glass as the value of M_c diminishes. This behavior can be traced in many crosslinking polymers and follows the behavior of rubber, crosslinked by a vulcanizing agent such as sulfur. However, it has the advantage of being quantitatively proportional to the dose over an extremely wide range and of leaving no residue. The physical behavior as a function of structure can therefore be analyzed in detail.

A second important property is the existence of a so-called memory effect, best demonstrated in the case of polyethylene. This polymer is largely crystalline at room temperature and it is the crystals that are largely responsible for giving the polymer its solid behavior. Above the melting point, however, the crystals disappear and the polymer is turned into a highly viscous liquid, in which it is only the presence of temporary physical chain entanglements which holds the specimen together for a short period. It is these chain entanglements that give natural rubber its "green" strength, allowing the rubber to be manipulated, at least for a short time.

For irradiated polyethylene (or rubber, etc.) well above the melting point, the crosslinks have converted the polymer into an elastic system. If, while it is under stress and therefore oriented, it is cooled down and allowed to crystallize, it will maintain this shape (on removal of the stress) because of the formation of these crystallites for an indefinite period. If however the specimen is subsequently reheated above the melting temperature, the crystals disappear again and the elastic forces, due to thermodynamic influences, will attempt to restore the molecules to their original, unstrained state. The irradiated specimen has therefore acquired a memory of its initial shape and will attempt to return to this whenever it is given sufficient freedom of motion. It should, however, be added that this memory is in fact of the random configuration and not (at least for small doses) of the shape it held during irradiation if the latter is different. The behavior may be different at high crosslink densities.

The crosslinked (network) structures of polymer chains will, at sufficiently high temperatures, be flexible and therefore capable of swelling in a suitable solvent. Being effectively a single molecule, it cannot dissolve and the degree of swelling depends on the pore size, that is, on the crosslink density and the radiation dose. Apart from initial effects such as the soluble fraction s or the end-lengths in the network, the average molecular weight between crosslinks is represented by M_c ($\propto r^{-1}$ approximately) and the degree of swelling, if large, can be represented by

$$V^{5/3} = (0.5 - \mu)M_c/\rho v$$

where V is the swollen volume/initial volume, μ is the parameter associated with the heat of mixing, ρ is the density and v is the molar volume of the solvent. Of course, this relation will not apply at low doses when there is still a significant sol (non-network) fraction, or when M_c is not very much smaller than the initial molecular weight M.

Another method of measuring crosslink density is by the use of pulsed nuclear magnetic resonance, when the T_2 relaxation curve provides not only a measure of sol and gel fractions, but also one of M_c. However, at the very short times involved in such measurements, the presence of chain entanglements of adequate lifetime may also appear as (temporary) crosslinks, so that the true measure offered by T_2 relaxation is M_{ce} (crosslinks and entanglements). The distinction between such entanglements and permanent crosslinks is in their temperature dependence. At higher temperatures, the number (or lifetime) of the former decreases, while the latter are independent of temperature. The method, therefore, offers a new method of assessing entanglements and their effect on mechanical properties and this can be established quantitatively by direct comparison with crosslinks, whose numbers can be determined directly from radiation experiments.

3. *Water-Soluble Polymers*

Many polymers can be readily dissolved, in water for example, and they then crosslink or scission under radiation in the usual manner. However, when a polymer is irradiated in aqueous solution, the degree of crosslinking can be greatly enhanced for the same radiation dose, despite the fact that at lower concentration the polymer molecules are further apart and have more difficulty in finding a partner with which to form a crosslink.

The reason for this behavior is that in irradiating an aqueous polymer solution one is also irradiating the water itself, and the radiolytic products of the water can then attack the polymer molecules to provide additional radicals to promote more crosslinking. At polymer concentrations as low as 1%, a crosslinked network can be formed at very low doses, well below 10 kGy (1 Mrad). The crosslinked network will of course be greatly swollen; but at higher doses, and thus higher crosslink densities, M_c diminishes, as does the degree of swelling.

At very low initial concentrations ($\leqslant 1\%$), this behavior is not always shown. On the contrary, the individual molecules tend to shrink as the radiation dose increases and no network is formed. This is because two radicals on the same chain are apt to find each other and bind two parts of the molecule together before meeting a radical on another polymer chain. Such internal links reduce the effective size of the individual molecule, which no longer has a linear flexible configuration.

Another aspect which is worthy of further attention is the irradiation of aqueous polymeric systems with the addition of very low concentrations of suitable additives capable of acting as radiation protectors, for instance, by reacting with the aqueous radiation products before these are able to react with the polymer. This type of information could be of value not only in the study of polymeric systems, but also as a quantitative model for biological systems. Some additives may indeed act as means of enhancing the effect of radiation.

4. *Polymerization*

From a different perspective, the use of high-energy radiation to promote polymer formation from the original, low-molecular-weight monomer can be considered. This process is also the standard method of producing polymeric materials, using chemical initiation techniques, but radiation has certain advantages for this reaction. It requires no additional catalysts and leaves no residues (except possibly some by-product of the irradiated monomer or polymer, and this is generally negligible). Furthermore, it can be used over an extremely wide range of intensities and, hence, rates of initiation, far beyond that possible with chemical catalysts. Another disadvantage with chemical initiators is that the temperature must be raised to initiate the reaction and this will intervene in the subsequent reactions of propagation and termination. This may result in an uncontrollable rate of reaction, as heat engendered by these reactions raises the temperature and the rate of initiation. With radiation, no catalyst is present and the temperature change has an insignificant effect on the initiation.

Since the early 1950s, much attention has been devoted to the study of radiation-induced polymerization for a range of monomers, under a variety of conditions. Unlike the reaction of scission or crosslinking, where dose rate has no direct effect, polymerization shows a considerable dependence on dose rate since it depends on initiation and termination by another growing chain:

$$M \rightsquigarrow M\cdot + M \rightarrow MM\cdot \rightarrow M_n\cdot$$

$$M_n\cdot + M_p\cdot \rightarrow M_{n+p}$$

Other mechanisms of termination (e.g., transfer) are possible.

At higher radiation intensities (and therefore at a higher concentration of radiation-induced initiated chains) the termination reaction, which does not depend on radiation intensity but rather on the concentration of growing chains, will also increase and, in general, the yield of polymer per unit time varies as $I^{0.5}$ (where I is the radiation intensity), while the average molecular weight decreases at the higher intensities owing to increased termination. Transfer from one growing chain to initiate another from a monomer complicates the issue. Detailed studies of these reactions have provided values of the reaction parameters for a series of monomers under different conditions. The doses involved are usually small, but the economic advantages do not seem to have promoted radiation as a large-scale commercial process.

Several features of radiation-induced polymerization deserve special attention. At first it appeared that only radical polymerization reactions were possible, but later experiments at low temperatures and under dry conditions showed that ionic polymerization can be induced by radiation. The basic reactions are quite different and the electron ejected by radiation leaves a positively charged species to propagate, with the electron retained in the vicinity. The latter condition appears necessary since, unlike radical polymerization, the yield varies with the radiation intensity I and the growing chains do not interfere with each other.

In the presence of an inorganic solid (fine silica powder) a much higher yield of polymer is obtained; this can be ascribed to the removal of traces of water on the surface of the powder so that the irradiation takes place in an ultradry environment.

Somewhat surprisingly, radiation can also initiate polymerization of a monomer in the solid state, often with a high yield. The arrangement of polymer molecules might be expected to be quite different from that of the originating monomer and the constraints imposed by the regular crystalline structure to inhibit this reaction, but this is not always true.

4.1 Radiation-Induced Grafting

By irradiating a monomer in the presence of a polymer, homopolymer and crosslinking/scission may result but, in addition, a graft may be the major product. This can be viewed as a derivation from the crosslinking reaction, whereby the side-radicals, normally used for the formation of a polymer–polymer link, are now used to initiate the growth of side polymer chains from the monomer which is present.

The monomer may be the precursor of the original polymer, in which case a highly branched form of the original polymer is obtained. A far more important situation arises when the original polymer and the monomer are of distinct chemical structures. For example, the polymer may be hydrophobic whereas the side branches created from the monomer may be hydrophilic. Very advantageous combinations might be expected—or indeed the converse.

Several methods of producing such polymer grafts are available. One is to irradiate the polymer, which may retain some of the radicals produced by radiation, and then introduce the monomer. The monomer itself is then never subject to radiation, so no homopolymer is theoretically produced from it. A second version is to irradiate the polymer in the presence of the monomer, with homopolymer formation from the monomer; in some cases this can be prevented by use of an additive. The best example is polyethylene film irradiated in an aqueous solution of acrylic acid as monomer, so that, in addition to the grafted material, a considerable amount of polyacrylic acid is formed, dissolved in the water or coating the surface of the polyethylene. The formation of this undesirable homopolymer may be almost completely suppressed by the presence of a small concentration of Moore's salt (ferrous sulfate). One possible explanation is that the salt suppresses initiation from the monomer, but not the propagation step, so that only chains initiated by polymer radicals can be effective. It is difficult to see why, if the monomer of acrylic acid can penetrate into the polymer, the salt cannot do likewise. A third grafting process involves the peroxide reaction on irradiated polymer. On heating, the peroxide groups can decompose, producing radicals capable of initiating polymerization. The ability to exploit such grafting reactions on the surface of thin films, or in membranes, can be of great practical value.

4.2 Enhanced Crosslinking

The theoretical calculations on the formation of a network by random crosslinking have been amply verified in a wide range of polymers, radiation doses, intensities and so on. The basic relationship governing even an incipient insoluble network is readily given:

$$r_g M_w G(R) = 0.96 \times 10^7$$

where $G(R)$ is typically about 3, M_w is about (1–5) × 10^5, and r_g is about 10–30 kGy (1–3 Mrad). However, in some systems, far smaller molecules are involved, yet the dose to form a highly crosslinked system can be considerably lower than this figure. In fact, such a system is not a random distribution of crosslinks, formed by pairs of radicals on adjacent chains. Instead, a system of unsaturated molecules is involved, such that a radical is produced on a polymer A by radiation; this can link A to another polymer B (through the unsaturated groups) and this creates a second radical on B, which then links it to a polymer C, and in its turn this propagates further:

$$\mathrm{A} \rightsquigarrow \mathrm{A\cdot(B)} \rightarrow \mathrm{AB\cdot(C)} \rightarrow \mathrm{ABC\cdot(D)} \rightarrow \cdots$$

Only A is involved in the primary radiation event, yet a number of polymer chains are linked together.

To achieve this high yield of links, it is necessary to make use of unsaturated groups, either within the polymer A itself, or in monomer M (= B, C, D) with the appropriate degree of unsaturation. In this respect, the reaction is a form of direct polymerization of polymer chains which are themselves unsaturated or, more probably, through a crosslinking polyunsaturated monomer.

In another version of this reaction, a polyfunctional monomer may react to give a three-dimensional network of small pore size, trapping the polymer itself with little or no crosslinking. This may account for the apparent formation of a crosslinked poly(vinyl chloride).

A very frequently observed example is the irradiation of an unsaturated polyester in a vinyl monomer and, as in simple radical polymerization, the degree of cure varies as the root of radiation intensity, $I^{1/2}$. An additive such as benzoquinone can greatly increase the dose required for gel formation, and, moreover, by an amount which is proportional to the concentration of additive (which can either react with the primary radicals initiating the reaction, or with growing chains). Different reaction rates are involved.

In addition to the initiating reaction, due to radiation, a termination reaction must also be envisaged and this is best ascribed to the reaction of two growing chains, just as in polymerization. The mathematics of this form of "enhanced crosslinking" has been analyzed and confirmed. The use of such unsaturated systems has the advantage of very low dose requirements and the possible disadvantage of higher cost of unsaturated polyester or polyfunctional monomer.

5. *Electrical and Optical Behavior*

Although most electrons formed by ionization during the irradiation of insulating polymers are recaptured to form an unstable system, a certain small fraction escape, either providing an increase in electrical conductivity during or after irradiation, or alternatively being trapped, for example, in a defect center from which they can escape after a lapse of time or by a temperature rise. Apart from the induced conductivity there is also the emission of light (thermally induced phosphorescence when the temperature is raised, or isothermal luminescence with lapse of time). The temperatures at which the light appears are characteristic of the polymer, showing that the trap is within the polymer, but the spectrum is that of an additive such as anthracene, indicating that this is ultimately responsible for the recapture of the electron, or at least the energy it emits on neutralization.

The isothermal luminescence intensity decays with time, but may be enhanced for a short time when an external electric field is applied. This spurt in light emission can be repeated, but with a very rapid decay of the additional light, by a series of pulsed electric fields. However, if the electric field is reversed, a large spurt is again observed. This may be taken to show that the external field helps to free the trapped electrons in one direction only, presumably those electrons which lie on the appropriate side of a positive charge. When these have been largely eliminated, it is only the electrons on the opposite side which can be freed by the opposite external field. This kind of experiment yields an estimate of trap depth.

A full and satisfactory explanation of the steady current induced by a steady radiation intensity and electric field does not yet appear to have been reached, although such an explanation might well help towards an understanding of electrical breakdown in insulators.

6. *Polymers and Biopolymers*

Sufficient research has been carried out to emphasize the many analogous reactions induced by radiation in simple polymers and in biopolymers. The G values for radical formation are generally similar; the effects of temperature are similar in polyisobutylene and in dry bacteria; the indirect effects seen when biological systems are irradiated in water are also seen in aqueous polymer solutions. Many additives acting as radiation protectors are effective in both kinds of system, and the important oxygen effect is also found in both. Hopefully, these analogies will be analyzed more fully, since the effects in simple polymers are relatively easy to follow quantitatively and results could be transferred to important biological systems subject to irradiation, as in cancer.

See also: Radiation Effects on Polytetrafluoroethylene; Radiation Processing; Vulcanization

Bibliography

American Chemical Society 1967 *Irradiation of Polymers, Symposium*, Advances in Chemistry, Series 66. American Chemical Society, Washington, DC

Brown W L 1986 Ion bombardment effects in polymers. *Radiat. Eff.* 98: 115–30

Chapiro A 1962 *Radiation Chemistry of Polymeric Systems*. Interscience, Chichester, UK

Charlesby A 1954 Gel formation and molecular weight distribution in long-chain polymers. *Proc. R. Soc. London, Ser. A* 222: 542–57

Charlesby A 1957 A theory of network formation in irradiated polyesters. *Proc. R. Soc. London, Ser. A* 241: 495–507

Charlesby A 1960 *Atomic Radiation and Polymers*, Appendix. Pergamon, Oxford

Dole M (ed.) 1972 *The Radiation Chemistry of Macromolecules*, Vols. 1 and 2. Academic Press, London

Spinks J W, Woods R J (eds.) 1976 *An Introduction to Radiation Chemistry*, 2nd edn. Wiley, New York

A. Charlesby
[Swindon, UK]

Radiation Effects on Polytetrafluoroethylene

Because of the many technological applications of polytetrafluoroethylene (PTFE), there is a long-standing and widespread interest in bonding this material to other substrates, particularly polymeric adhesives and metal thin films. This entails modifying the chemical and/or physical properties of the surface region without altering bulk properties. The current standard industrial method for preparing PTFE for bonding is to chemically etch the surface with an organic solution containing sodium metal. There are many situations, however, where such a method is undesirable or impractical because of processing or environmental concerns. For this reason, numerous investigations have been made concerning modification of PTFE surface properties by ionizing radiation, particularly electron and photon bombardment, ion bombardment and exposure to a radio-frequency-induced gas plasma. This article reviews work in all three of these areas and, in the case of ion bombardment, bombardment at moderate or low energies (⩾10 keV) is considered; that is, where nuclear momentum transfer processes are significant (Watson and Tombrello 1985).

Since the late 1960s, significant progress has been made in the development of techniques for the analysis of the chemical and physical structure of polymer surfaces. As a result, investigation of radiation-induced changes in surface structure is often straightforward. However, the relationship between a particular set of surface structural or chemical features and adhesion to that surface is incompletely understood in most cases. This is largely due to the fact that accepted methods for measuring adhesion between two materials are neither universally applicable nor quantitative. Therefore, changes in the surface structure of PTFE wrought by exposure to various types of ionizing radiation, and the resulting relative changes in adhesion between PTFE and other materials, are reviewed here. A precise description of the relative importance of chemical versus morphological changes in promoting the adhesion of PTFE to another surface is in most cases not yet possible.

1. Ion Bombardment

While some studies have been reported concerning the interaction of PTFE with reactive ions, mainly N_2^+ (Dillon et al. 1968, Taylor et al. 1978a), most work has concentrated on bombardment by noble-gas ions and corresponding fast atoms. Such ion or fast-atom bombardment induces both morphological and chemical changes in the polymer surface.

1.1 Morphological Changes

At low ion-beam current densities (<1 mA cm^{-2}, 750 eV Ar^+), a morphology of closely spaced filaments is observed on the surfaces of PTFE and other fluoropolymers, while at higher densities more widely spaced truncated cones are formed (Sovey 1979). Similar surface texturing of PTFE occurs under bombardment by 6 keV xenon atoms (Michael and Stulik 1986). A detailed model for this surface texturing process has yet to be formulated, but a number of facts indicate that thermal effects are important. These include both dose rate effects as well as the observed sputtering yields which are significantly higher than predicted by sputtering theory for nonmolecular solids (Sovey 1979, Michael and Stulik 1986). Filament or cone formation in PTFE and other fluoropolymers with poor thermal conductivity occurs at ion doses which are up to several orders of magnitude below the doses necessary to induce similar effects in nickel, stainless steel and other metals, suggesting that local heating of the sample occurs (Mirtich and Sovey 1979, Ghose et al. 1983), or that ion-induced or thermally induced chemical processes are occurring. However, the observation of secondary-ion energy distributions which are strongly uncharacteristic of pyrolysis (Dillon et al. 1968) indicate that thermal degradation is not the primary method of bond scission.

1.2 Chemical Changes

Ion- or fast-atom-induced morphological changes have been correlated with changes in PTFE surface composition as observed by x-ray photoelectron spectroscopy (XPS). Unexposed PTFE exhibits a single peak in the C(1s) region due to photoemission from CF_2 groups (Clark and Feast 1975). Bombardment by Ar^+ or Xe^+ ions leads to some defluorination (Taylor et al. 1978b, Dwight and Beck 1979, Sovey 1979). The amount of defluorination observed, however, is rather small because the peak due to photoemission from CF_2 groups still dominates the C(1s) region. The amount of defluorination increases with dose up to a saturation point (Taylor et al. 1978b) but is independent of dose rate, with the CF_2 feature still dominating the C(1s) photoemission spectrum at saturation doses (Sovey 1979). Under noble gas or ion bombardment, no incorporation of surface species other than carbon or fluorine has been reported, but bombardment by N_2^+ does lead to some nitrogen incorporation (Taylor et al. 1978a). The studies reported here have generally been carried out under controlled vacuum environments and the effects of exposure of a freshly bombarded surface to the atmosphere have not been investigated.

1.3 *Effect on Adhesive Properties*

The effects of ion bombardment on the adhesive properties of PTFE and related fluoropolymers have been explored by Mirtich and Sovey (1979). Ion-beam texturing of PTFE results in a significant improvement in the tensile and shear strengths of PTFE–epoxy joints compared with either untreated PTFE or PTFE etched by a sodium–naphthalene complex in an organic solvent (Mirtich and Sovey 1979). The strength of the adhesive bond increases with the beam exposure time up to a certain point (750 eV Ar^+, 0.5 mA cm^{-2}, 30 min) but possible degradation effects at longer times have not been explored. For sodium-etched PTFE, the bond strength to epoxy decreases as the length of time between etching and bonding is increased (Mirtich and Sovey 1979). In contrast, ion-textured PTFE bonded to epoxy showed no decrease in joint strength with elapsed time between treatment and bonding. Samples bonded 20 d after etching still exhibited tensile and shear strengths 46% and 100% greater, respectively, than for sodium–naphthalene-etched PTFE bonded immediately after etching. In summary, ion-beam texturing of PTFE and related fluoropolymers results in PTFE–epoxy bond strengths comparable or superior to those achieved by sodium–naphthalene etching, with no observed degradation of adhesive properties due to atmospheric exposure prior to bonding. The extent to which the improved adhesive qualities of the ion-textured surface are due to mechanical interlocking of the epoxy with the roughened surface, a greatly enhanced interfacial area, or increased chemical interaction has not yet been determined.

2. *Plasma Treatment*

Early studies of the exposure of polymers to gas plasmas often involved plasmas excited by direct current (dc) discharges (Schonhorn et al. 1970). Corona treatment (dc discharge in air) has been used extensively (Rossman 1956). Much of the more recent work has involved plasmas excited by radiofrequency (rf) power, particularly because of the increased flexibility in reactor design and improved control of processing variables which this arrangement permits (Clark et al. 1978). Since most of the work on fluoropolymers involves rf plasmas, this article concentrates on this type of plasma. A typical rf-activated gas plasma (often referred to as a "glow discharge") contains electrons (energies 0–20 eV), ions (0–2 eV), metastables (0–20 eV) and ultraviolet (uv)/visible radiation (3–40 eV) (Clark et al. 1978). Bell (1974) discusses the concentrations of these species under various plasma conditions. Exposure of polymer surfaces (including but not limited to PTFE) to either noble or reactive gas plasmas is becoming a popular industrial method for the alteration of surface properties to improve adhesion. Plasma treatment of PTFE and related materials, such as the well-studied ethylene–tetrafluoroethylene copolymer (ETFE), results in morphological and chemical changes which are a sensitive function of the type of gas used.

2.1 *Noble Gas Plasma Treatment of PTFE*

Exposure of PTFE to an argon gas plasma results in the formation of needles and cones similar to those observed for argon ion bombardment (Yamamoto et al. 1987). Such treatment of PTFE or ETFE also results in defluorination and crosslinking, with a small amount of surface oxidation (Yamamoto et al. 1987, Clark and Dilks 1977). A major source of oxygen appears to be chemisorbed oxygen-containing species on the walls of the reaction vessels (Clark and Dilks 1977). In the case of ETFE, the kinetics of these chemical reactions have been monitored by XPS (Clark and Dilks 1978). These studies indicate that argon ions and metastables induce crosslinking and defluorination in the outer monolayer or so of the material by direct energy transfer, while uv radiation induces such changes both at the surface and in the bulk of the material. Similar effects are observed for helium, neon and krypton plasmas, with the rates for the ion- and metastable-induced surface reactions roughly an order of magnitude larger than for uv radiation-induced bulk reactions. Since uv radiation can penetrate hundreds of nanometers or more into solid materials (Clark and Dilks 1977), the commonly accepted notion that plasma bombardment induces strictly surface modification of the material is not true. The reaction rates and the extent of the reaction are also influenced by the various plasma parameters such as gas pressure, power levels, chamber geometry and so on (Clark and Dilks 1978).

2.2 Adhesion Enhancement by Noble Gas Plasma Treatment

Two particularly pertinent issues are the extent to which adhesion of PTFE to other polymers can be improved by noble gas plasma treatment and, if adhesion is improved, whether the observed improvement in bond strengths is attributable to mechanical or chemical factors. Adhesion of argon plasma-treated PTFE to epoxy has been examined under conditions which induce cone formation and some surface chemical modification (Yamamoto et al. 1984). Adhesive bond strengths, as determined by 180° peel tests, were achieved equal to those induced by the sodium-etch treatment. While the latter method generally results in cohesive failure within the sodium-etched layer, the mode of failure in this study was not reported. However, sodium-etched surfaces are known to be unstable under uv radiation and prolonged exposure leads to drastic degradation of adhesive properties (Dwight and Riggs 1974). At uv exposure levels which resulted in almost a complete loss of adhesion for sodium-etched samples (i.e., practically no peel strength as observed for untreated PTFE), the argon plasma-treated samples exhibited peel strengths within 80% of the unirradiated samples. XPS and scanning electron microscopy (SEM) studies of argon plasma-treated samples subsequently exposed to uv irradiation showed that such treatment induces a surface chemical structure similar to virgin PTFE without affecting the morphology. These results conclusively demonstrate that for argon plasma treatment of PTFE, the improvement of adhesion to epoxy is due to mechanical interlocking of the plasma-treated surface with the epoxy (Yamamoto et al. 1984).

2.3 Reactive Gas Plasma Treatment of PTFE

The treatment of PTFE surfaces with reactive gas plasmas induces changes in surface morphology and chemistry significantly different from those obtained with noble gas plasmas. Moreover, the type of reactive gas used has a pronounced effect on the types of changes induced. Exposure of PTFE to an ammonia plasma results in defluorination of the surface and the adsorption of significant amounts of oxygen and nitrogen, as observed by XPS (Collins et al. 1973). The samples were exposed to atmosphere between plasma treatment and examination by XPS but the source of oxygen contamination in this study was not determined. The cone formation which results from noble gas plasma treatment was not observed. SEM photographs revealed pinhole formation and the destruction of any filaments and tendrils existing at the untreated film surface. A comparison of XPS spectra and attenuated total reflectance (ATR) infrared measurements, with respective sampling depths of about 3 nm and 1 μm, indicated that prolonged exposure to the ammonia discharge results in an increasing modification depth, with no significant removal of material from the surface (Collins et al. 1973). By contrast, a plasma discharge in air continuously removes material from the surface (Collins et al. 1973) and sufficiently long exposure leads to sample disintegration. The air plasma also results in defluorination of the sample surface, but no significant oxidation (Collins et al. 1973). These contrasting results underline the variety of chemical reactions which can be induced by different reactive gas plasmas at the polymer surface. The chemical reactions induced by hydrogen and oxygen at the PTFE surface have been investigated in some detail by Smolinsky and Vasile (1979), who examined the effects of noble gas, hydrogen and oxygen discharges on plasma-polymerized tetrafluoroethylene films. It was found that the addition of as little as 0.1 mol.% hydrogen or oxygen to a noble gas discharge produces significant changes in the ion-etching chemistry. Hydrogen was found to preferentially attack carbon–fluorine bonds while oxygen preferentially breaks carbon–carbon bonds. Thus, the presence of even small levels of impurity gases in a noble gas plasma can lead to noticeable changes in plasma-induced surface morphology and chemical structure.

2.4 Adhesion After Reactive Gas Plasma Treatment

In view of the significant influences that low-level gas impurities or variations in processing parameters can have on surface modifications, comparisons of adhesion results from different sets of experiments must be made only with caution. However, one general trend is that exposure of PTFE or other fluorinated polymers to either a reactive or noble gas discharge results in an initial increase in surface wettability and bond strength (to epoxy), but that prolonged exposure results in no further increases and occasionally in some degradation of these properties (Hall et al. 1972, Hirotsu and Ohnishi 1980). A careful examination of the mode of failure between the treated surface and an epoxy adhesive during peel tests was made for PTFE exposed to an ammonia plasma (Collins et al. 1973). Plasma treatment resulted in the formation of a brown surface layer. Adhesive failure occurred either within this layer or between this layer and the unmodified PTFE. The modified surface layer was found to be susceptible to attack by hot nitric/perchloric acid.

2.5 Plasma Treatment Effects

In summary, exposure of PTFE and related fluoropolymers to noble gas plasmas results in defluorination, crosslinking and formation of cones or filaments at the textured surface. Most of the improved adhesion of the treated material to epoxy is due to the mechanical properties of the surface. Exposure to a reactive gas plasma, or to a noble gas plasma with reactive gas impurities, results in radically different plasma-surface chemistry, the type of chemical and mechanical surface changes induced being highly dependent on the type of gas used. The types of chemical and mechanical interactions which occur between such surfaces and polymeric adhesives (or other materials) have yet to be explored.

3. *Electrons and X Rays*

Electron radiation damage (modification) in PTFE has been known and characterized since the late 1960s. Since 1980, use has been made of electron and x-ray irradiation to enhance adhesion of metals and epoxies to PTFE. Bulk damage to PTFE by high-energy electrons (>3 keV) has been investigated by Lerner (1969) and De Wilde (1974). In each case the predominant mode of damage appears to be electron-induced defluorination and crosslinking by nonthermal mechanisms.

3.1 Electron Bombardment of PTFE

Lerner (1969) used electron-spin-resonance (ESR) spectroscopy to characterize free-radical formation in PTFE caused by high-energy electrons (2 MeV). In the presence of air or oxygen, including oxygen dissolved in the bulk, both $—CF_2(CFO\dot{O})CF_2—$ and $—CF_2O\dot{O}$ are formed whereas in vacuum, $—CF_2\dot{C}FCF_2—$ and $—CF_2\dot{C}F_2$ radicals are formed. From the decay and line shape of the ESR signals from these species he postulated that the main mechanism of damage caused by the electron beam is chain scission and defluorination. In addition, these radical species were shown to be stable at room temperature for weeks and the oxygen-containing radicals are thermally stable to 150 °C and 250 °C, respectively.

De Wilde (1974) used a 4 keV electron beam to decompose bulk PTFE from a crucible and deposit thin fluorocarbon films on a substrate in vacuum. Residual gas analysis indicated that only CF, CF_2, CF_4, and C_2F_2 species were produced in the gas phase as opposed to C_2F_4 which is produced by thermal decomposition. Again, these results suggest that the electron beam is directly causing chain scission in bulk PTFE. He suggested that the polymerization of fluorine-containing fragments on the substrate is caused by stray 4 keV electrons in the chamber.

The interactions of low-energy electrons (~100–3000 eV) with the surface regions of PTFE have been investigated by Pepper (1974), Kelber et al. (1986), and Clark and Brennan (1986). Again, the major effects appear to be defluorination and crosslinking by nonthermal mechanisms. Pepper (1973) used dynamic Auger electron spectroscopy (AES) to characterize electron-beam damage to pure polymers (including PTFE) as well as thin polymer films transferred to clean metal surfaces by friction in vacuum. Kelber et al. (1986) and Clark and Brennan (1986) have utilized the chemical state information available from XPS to characterize electron damage in PTFE.

Pepper found that the electron beam (2 kV, 2 μA) caused electron-stimulated desorption (ESD) of fluorine from the thin PTFE film, with measured cross sections for fluorine desorption ranging from 4.4×10^{-16} cm^2 initially, to 1.0×10^{-18} cm^2 after prolonged bombardment. For comparison, typical gas-phase ionization cross sections are of the order of 10^{-16} cm^2 (Pepper 1973). In general, he also found that the electrons could rupture chemical bonds allowing crosslinking or degradation of the polymer and a variety of changes in the polymer properties.

For low-energy electron bombardment (250–350 eV, 35–50 nA cm^{-2}), Kelber et al. (1986) found defluorination and crosslinking of the near surface region as evidenced by changes in the C(1s) and F(1s) photoemission spectra with electron-beam exposure time. Initially, both defluorination and crosslinking proceed rapidly, followed by a decrease in the rates with increasing bombardment time. Measured defluorination cross sections are in qualitative agreement with the findings of Pepper (1974). The depth of surface modification, as determined by attenuation of the XPS signals and angle-resolved XPS measurements, is confined to the top 3 nm of the surface, that is, of the order of the distance travelled by electrons of these energies.

Clark and Brennan (1986) found that exposure of PTFE to higher-energy electrons (1–2.5 keV and up to 7.5 μA cm^{-2}) also leads to defluorination and crosslinking of the near-surface region. Both effects lead to an increase in the density of carbon in the analysis volume. The first-order fluorine desorption cross sections were found to vary with dose rate and ranged from 1×10^{-15} cm^2 at 0.15 μA to 1×10^{-16} cm^2 at 6 μA. The depth of damage was found to increase with exposure time but was not found to be strongly dependent on beam energy in the range 1–2.5 keV. They estimated that electrons of these energies travel 72–277 nm into PTFE and, over these distances, the depth of modification increases with irradiation time. The depth of modification appears to be controlled by electron-stimulated diffusion of CF_x moieties from the bottom of the modified region to the surface.

3.2 X-Ray Bombardment of PTFE

The bombardment of PTFE by soft x rays (~1.2–1.5 keV) causes similar damage to that seen in the case of electron bombardment, defluorination and crosslinking, but to much greater depths. Wheeler and Pepper (1982a) used Mg(Kα) x rays ($h\nu = 1253.6$ eV, 400 W) to irradiate PTFE and PTFE transfer films for periods of up to 20 h. The changes induced in the near-surface region were then monitored by following changes in the C(1s) and F(1s) XPS transitions as a function of exposure time. The maximum dose was estimated at 5.8×10^{11} rad (5.8×10^{9} Gy). The changes in the C(1s) and F(1s) regions indicate that crosslinking and chain scission are caused by interaction of the x rays with PTFE. Gases leaving the surface during x-ray exposure were monitored with a residual gas analyzer. Saturated fluorine-containing fragments were observed to leave the surface during irradiation, but no C_2F_4 species were observed. Because C_2F_4 is the major product of thermal degradation in a vacuum, the observed damage is not due to thermal effects.

Chaney and Barth (1987) looked at the changes induced in the near-surface region of PTFE and other polymers by both nonmonochromatic and monochromatic Al(Kα) radiation ($h\nu = 1486.6$ eV) from a small-spot x-ray source. Use of both types of radiation induces damage similar to that observed by Wheeler and Pepper (1982a). Damage was confined to the area exposed to the small-spot x-ray source indicating that secondary electrons from the x-ray source impinging on the sample were not responsible for the observed damage. Angle-resolved XPS measurements indicate the damage is greatest at the surface and is nonuniform within the sampling depth of XPS (~6 nm).

3.3 Morphological and Chemical Changes Caused by Bombardment

In contrast to the chemical changes caused by the interaction of photons and electrons with PTFE, comparatively little work has been done to examine morphological changes produced by these interactions. SEM studies performed by Wheeler and Pepper (1983) on PTFE surfaces exposed to x-ray and electron bombardment indicate negligible changes in surface morphology. However, noticeable changes were observed in the contact angle from wetting experiments with hexadecane indicating a higher energy surface following irradiation. A low-temperature thermal desorption study (Wheeler and Pepper 1987) of the interaction of xenon with virgin and irradiated (1 keV electrons) PTFE surfaces also indicates that the irradiated surfaces have a higher surface energy.

Irradiation of PTFE with electrons or x rays leads to similar chemical changes in the near-surface region. These changes include defluorination, crosslinking and a greater density of carbon at the surface. The high-energy surfaces which result might be expected to show improved bonding characteristics to metals and other polymers. This has been shown to be the case in both qualitative and quantitative measurements. Wheeler and Pepper (1982b) evaporated nickel and gold in vacuum onto PTFE surfaces, both virgin and irradiated (by electrons or soft x rays), until no substrate XPS features could be seen (>10 nm). The initial rate of uptake for both metals, as monitored by XPS, is greater for the irradiated compared with the nonirradiated surface, again indicating a higher energy surface after irradiation. After removal of the samples from the vacuum, cellophane tape was applied to the metal overlayer and a qualitative peel test was performed. Both gold and nickel were found to adhere better to an irradiated surface, nickel adhering better than gold. For nickel on the irradiated surface, failure always occurred at the tape–metal interface, whereas the nickel film was always removed by the tape on the unirradiated PTFE. For gold, the results were not as clearcut; tape with a weaker adhesive had to be used to avoid removing all the gold from the PTFE but irradiated surfaces did show enhanced adhesion as compared with virgin PTFE.

The fact that nickel, which is more reactive than gold, forms a stronger bond following irradiation indicates that chemical reaction is at least partially responsible for the enhanced adhesion rather than just physical interactions. XPS examination of the PTFE–nickel interface during the early stages of metal deposition confirm a chemical interaction between the metal and the irradiated polymer. NiF_2 and, possibly, —F—Ni—C— were found at the interface but interestingly, no —Ni—C— was detected. No such interactions were detected for unirradiated samples. Although NiF_2 *per se* is not expected to contribute to enhanced bonding between the metal and the PTFE, since it does not involve nickel directly bonded to the polymer backbone, its presence does demonstrate that chemical interactions occur at the interface for irradiated surfaces.

Quantitative adhesion tests have also been performed with nickel and gold films on PTFE and other polymers (Yamada et al. 1984). The PTFE was irradiated with 2 keV electrons (6 μA cm^2, 30 min) and nickel or gold was then evaporated onto the substrate. Tensile pull tests were then performed. For nickel, the bond strength increased from $1.9\ (\pm 0.5) \times 10^6$ Pa to $4.9(\pm 0.9) \times 10^6$ Pa after irradiation. In the case of gold, the bond strength increased from $1.1(\pm 0.5) \times 10^6$ Pa to $1.9(\pm 0.5) \times 10^6$ Pa after irradiation. Again, the fact that the bond strength is higher for the nickel–PTFE system as compared with the gold–PTFE system (for both virgin and irradiated surfaces) suggests a chemical interaction in addition to any physical interactions responsible for enhanced bonding.

Sessler et al. (1973) also investigated improving adhesion of PTFE to gold by irradiating the PTFE with 5–20 keV electron beams. No surface-analytical work was reported but they did find that adhesion increased with primary beam energy up to 10 keV. Higher primary beam energies did not improve adhesion relative to the maximum for 10 keV and this observation was attributed to the fact that the depth of penetration of electrons whose energy is greater than 10 keV is too deep to influence bonding at the metal–PTFE interface. In addition, at a constant beam energy, they observed that adhesion also increased with delivered current density to the PTFE. With a sample irradiated with a 10 keV beam for a delivered charge density of around 5×10^{-5} C cm^{-2}, the tensile–shear joint strength goes from 1×10^6 Pa (unirradiated) to 6×10^6 Pa.

Kelber et al. (1986) examined electron-beam enhanced bonding of PTFE to epoxy (Epon 828/Versamid 140). With exposure to a 300 eV electron beam at a current density of 0.1 μA cm^{-2} for 60 min, the tensile bond strength increased from zero (the epoxy will not wet the PTFE surface) to $1.9–2.5\times10^6$ Pa. This is to be compared with the cohesive failure of PTFE at around 1.4×10^7 Pa. Because the epoxy was not applied *in vacuo*, the possibility that exposure of irradiated PTFE to air results in partial oxidation or hydroxylation cannot be discounted. However, the attachment of NH_x groups to the surface under electron bombardment was examined. It was found that attachment occurs only during bombardment in the presence of NH_3 gas. Sequential exposure of the bombarded surface to NH_3 yields no significant NH_x adsorption, indicating that attack of free-radical surface sites by gas-phase species competes with a more rapid crosslinking process. Interactions of the surface with O_2, which reacts more readily with free radicals, were not examined.

4. Future Developments

The body of work reviewed in this article shows that much understanding of the chemical and morphological changes induced at PTFE surfaces by ion, plasma, electron and photon bombardment has been obtained. Relatively less understanding has been gained on how these changes affect adhesion to metal and polymer surfaces. This is due in part to the difficulty in measuring interfacial adhesion in a quantitative and consistent manner. Another factor is that most studies reported to date typically involve detailed examination of the treated surface prior to bonding, but no analysis of interfacial interactions occurring as the metal or polymeric film is deposited. While such experiments are difficult, increasing sophistication in the application of modern surface-analytical methods and in the handling of materials in ultrahigh vacuum environments should lead to progress in this area.

See also: Bonding: Preparation of Polymer Surfaces; Elastomer–Textile Adhesion Processes

Bibliography

Bell A T 1974 Fundamentals of plasma chemistry. In: Hollahan J R, Bell A T (eds.) 1974 *Techniques and Applications of Plasma Chemistry*. Wiley, New York, pp. 1–56

Chaney R, Barth G 1987 X-ray induced surface damage of PTFE. *Anal. Chem.* 329: 143

Clark D T, Brennan W J 1986 An ESCA investigation of low energy electron beam interactions with polymers, I. PTFE. *J. Electron Spectrosc. Relat. Phenom.* 41: 399–410

Clark D T, Dilks A 1977 ESCA applied to polymers, XV. RF glow-discharge modification of polymers, studied by means of ESCA in terms of a direct and radiative energy-transfer model. *J. Polym. Sci., Polym. Chem. Ed.* 15: 2321–45

Clark D T, Dilks A 1978 ESCA applied to polymers, XVIII. RF glow discharge modification of polymers in helium, neon, argon and krypton. *J. Polym. Sci., Polym. Chem. Ed.* 16: 911–36

Clark D T, Dilks A, Shuttleworth D 1978 The application of plasmas to the synthesis and surface modification of polymers. In: Clark D T, Feast W J (eds.) 1978 *Polymer Surfaces*. Wiley, New York, pp. 185–211

Clark D T, Feast W J 1975 Application of electron spectroscopy for chemical applications (ESCA) to studies of structure and bonding in polymeric systems. *J. Macromol. Sci., Rev. Macromol. Chem.* 12: 191–286

Collins G C S, Lowe A C, Nicholas D 1973 An analysis of PTFE surfaces modified by exposure to glow discharges. *Eur. Polym. J.* 9: 1173–85

De Wilde W 1974 Evaporation of polytetrafluoroethylene by electron bombardment of the bulk material. *Thin Solid Films* 24: 101–11

Dillon A F, Lehrle R S, Robb J C, Thomas D W 1968 Positive-ion emission from polymers under ion impact. *Adv. Mass Spectrom.* 4: 477–90

Dwight D W, Beck B R 1979 Ion beam polymer surface modification: Characterization by ESCA and SEM. *Org. Coat. Plast. Chem.* 40: 494–9

Dwight D W, Riggs W M 1974 Fluoropolymer surface studies. *J. Colloid Interface Sci.* 47: 650–60

Ghose D, Basu D, Karmohapatro S B 1983 Cone formation on argon-bombarded copper. *J. Appl. Phys.* 54: 1169–71

Hall J R, Westerdahl C A L, Bodnar M J, Levi D W 1972 Effect of activated gas plasma treatment time on adhesive bondability of polymers. *J. Appl. Polym. Sci.* 16: 1465–77

Hirotsu T, Ohnishi S 1980 Surface modification of some fluorine polymer films by glow discharges. *J. Adhes.* 11: 57–67

Kelber J A, Rogers J W Jr, Ward S J 1986 Effects of low-energy electron bombardment on the surface chemical structure and adhesive properties of polytetrafluoroethylene (PTFE). *J. Mater. Res.* 1: 717–23

Lerner N R 1969 ESR spectra of orientated polytetrafluoroethylene subjected to 20-nsec pulses of electrons. *J. Chem. Phys.* 50: 2902–10

Michael R, Stulik D 1986 Surface topography development on Teflon® under keV Xe atom bombardment. *J. Vac. Sci. Technol. A* 16: 809–12

Mirtich M J, Sovey J S 1979 Adhesive bonding of ion-beam-textured metals and fluoropolymers. *J. Vac. Sci. Technol.* 16: 809–12

Pepper S V 1973 A new technique for Auger analysis of surface species subject to electron-induced desorption. *Rev. Sci. Instrum.* 44: 828–9
Pepper S V 1974 Auger analysis of films formed on metals in sliding contact with halogenated polymers. *J. Appl. Phys.* 45: 2947–56
Rossmann K 1956 Improvement of bonding properties of polyethylene. *J. Polym. Sci.* 19: 141–4
Schonhorn H, Ryan F W, Hansen R H 1970 Surface treatment of polypropylene for adhesive bonding. *J. Adhes.* 2: 93–9
Sessler G M, West J E, Ryan F W, Schonhorn H 1973 Increase of gold-teflon FEP joint strength by electron bombardment. *J. Appl. Polym. Sci.* 17: 3199–209
Smolinsky G, Vasile M J 1979 Ion-etching of discharge-produced tetrafluoroethylene film with hydrogen, oxygen and noble gases. A mass spectrometric analysis. *Eur. Polym. J.* 15: 87–91
Sovey J S 1979 Ion beam sputtering of fluoropolymers on PTFE surface morphology and composition. *J. Vac. Sci. Technol.* 16: 813–16
Taylor J A, Lancaster G M, Rabalais J W 1978a Interactions of N_2^+ and NO^+ ions with surfaces of graphite, diamond, teflon, and graphite monofluoride. *J. Am. Chem. Soc.* 100: 4441–7
Taylor J A, Lancaster G M, Rabalais J W 1978b Surface alteration of graphite, graphite monofluoride and teflon by interaction with Ar^+Xe^+ beams. *Appl. Surf. Sci.* 1: 503–14
Watson C C, Tombrello T A 1985 A modified lattice potential model of electronically mediated sputtering. *Radiat. Eff.* 89: 263–83
Wheeler D R, Pepper S V 1982a Effects of x-ray flux on polytetrafluoroethylene in x-ray photoelectron spectroscopy. *J. Vac. Sci. Technol.* 20: 226–32
Wheeler D R, Pepper S V 1982b Summary abstract: Improved adhesion of Ni films on x-ray damaged polytetrafluoroethylene. *J. Vac. Sci. Technol.* 20: 442–3
Wheeler D R, Pepper S V 1983 XPS, mass spectroscopy and SEM studies of x-ray and electron irradiated PTFE. *NASA Tech. Memo* 83413
Wheeler D R, Pepper S V 1987 Thermal desorption of Xe probe of PTFE surface forces before and after electron bombardment. In: Mittal K L (ed.) 1987 *Surface and Colloid Science in Computer Technology*. Plenum, New York, pp. 309–20
Yamada Y, Wheeler D R, Buckley D H 1984 Quantitative measurements of the effects of electron irradiation on adhesion of PTFE to evaporated Ni and Au films. *NASA Tech. Pap.* 2360
Yamamoto S, Tabata H, Ezoe M, Uemori K, Oya Y, Moriuchi T 1984 Effects of Ar plasma on morphology and adhesive properties of PTFE. *Oyo Butsuri* 53: 639–43

J. A. Kelber and J. W. Rogers Jr
[Sandia National Laboratories, Albuquerque, New Mexico, USA]

Radiation Processing

The early applications of high-energy radiation were limited by the low powers available, usually x-ray sources associated with low-voltage electron accelerators, and were therefore largely limited to use in the examination of microstructures of materials and also in cancer therapy, where only low powers are permitted for medical reasons. In view of this limitation of dose output, it is not surprising that the earliest application, other than for materials microstructures, was to biological structures, where the molecular weights are extremely high and the sensitivity equally great.

The first two major applications of high-energy radiation appeared at the beginning of the 1950s and were associated, indirectly, with research for the nuclear energy industry. This industry was in a position to provide much larger radiation sources in the form of gamma radiation from ^{60}Co (made by irradiating cobalt in a nuclear reactor) and fast electrons from accelerators, initially used in the examination of materials and structures intended for use in reactors. Early basic research on polymers was directed at the polymerization of monomers and the stability of materials in a radiation environment, but two lines of investigation showed considerable promise for industrial and medical purposes—the crosslinking of polymers, especially polyethylene and rubber, and the sterilization of medical disposables. These two areas have developed very rapidly and each must now have a turnover of over US\$$10^9$ per annum. Other potential fields began development somewhat later and are rapidly entering industrial processes. Two major directions can be discerned: the application of radiation in fields where no alternatives are readily available and those where the products have to compete with well-established conventional industrial processes.

The major fields of application of large-scale radiation treatment are primarily those in which radiation ruptures chemical bonds and produces modified chemical structures. This is a major activity of the chemical industry but, in the radiation processing of polymers, elastomers and natural products, it is primarily the improvement of the physical or biological properties which has been the objective. This is because high-energy radiation is an expensive form of energy, costing up to US\$10 kWh^{-1} from a large source, so that the major fields of commercial exploitation have been those in which a small chemical change produces a large physical or biological improvement. This applies largely to macromolecules.

1. Irradiation of Polyethylene

Irradiation of polyethylene was found to produce valuable changes in chemical structure, linking the molecules together into a network system. Above the melting point of the crystallites, the irradiated polyethylene, instead of flowing, behaves as the equivalent of a vulcanized rubber, insoluble but with high resilience (see *Radiation Effects on Polymers*). The elastic modulus is proportional to the radiation dose over a very wide range. In this respect, also, irradiated polyethylene behaves as the equivalent of a vulcanized rubber. It differs from it in one important respect, however, in that, on cooling to room temperature, the crystallites reform and lock the structure into any desired shape, which the material retains indefinitely until it is reheated above the usual melting point. At this point, the crystallites unlock, allowing the crosslinked polyethylene to revert to its original shape.

This "memory" effect has proved of very great industrial value, for example in electrical jointing components. An irradiated polyethylene tube is heated, extended or otherwise modified and then cooled in an expanded or strained condition. Sections of the tube can be slipped over a cable or wire joint and then heated above the usual melting point whereupon the locking crystallites disappear and the tube shrinks back over the joint, ensuring a neat watertight closure. An improvement on this device incorporates a form of solder, so that heating both will seal the two metal cables together and will provide an excellent insulator.

Another major application consists of irradiating thin polyethylene foil and then stretching it to provide the thin film needed in food or other packaging. Subsequent heating, for example in cooking, merely contracts the film and provides a tight cover to the contents.

With an appropriate loading of carbon black, polyethylene sheet, film or tubing can act as an electrical conductor and heating device. When heated above a certain temperature, however, the polyethylene expands and the consequent separation of the carbon particles greatly increases the electrical resistance. The material can therefore serve as a self-limiting heating device.

The simplest application of irradiated polyethylene is, however, in polyethylene-insulated electrical cables. Under normal conditions of design such cables run the risk of losing their insulating characteristics, by melting if the current is excessive, as in a short circuit, and a fire can result. If, however, the cable is irradiated after manufacture, overheating merely converts the largely crystalline polyethylene insulating cover to a very flexible one, rather similar to rubber insulation, and very high temperatures can be supported, at least for a limited time. As a very primitive demonstration of the possible commercial exploitation of irradiated polyethylene, an electrical cable could be safely run, with good insulating properties, in a bath of molten lead. A major restriction is due to the oxidation of the polymer at high temperatures and this may impose a serious restriction on its use in situations where temperatures well in excess of the usual melting temperature will be maintained over long periods. To some extent this is overcome by the incorporation of antioxidants which may, however, act in part as radiation protectors, reducing the crosslinking power of the radiation.

The formulae set out in the article *Radiation Effects on Polymers* allow the dose required to create an incipient network to be calculated. However, this procedure is not entirely satisfactory, since it assumes a knowledge of molecular weight and ignores the reduction in effectiveness of radiation because of the presence of crystallites and the degree of crosslinking beyond the incipient needed for the requisite properties. For many purposes, a dose in the range 100–250 kGy (10–25 Mrad) is found to be adequate and only a small fraction of the polymer remains not crosslinked. Since the polyethylene specimens are usually a few millimeters in thickness, electron beams of several million volts energy are very adequate and economically advantageous.

2. *Irradiation of Rubber*

The irradiation of rubber to promote its crosslinking parallels the use of sulfur as a vulcanizing agent, but has certain advantages over this process: any temperature rise is scarcely significant, the degree of crosslinking can be carefully and reproducibly controlled by the radiation dose and no extra chemicals or indeed sulfur need be introduced. To achieve a degree of crosslinking equivalent to that usually deemed necessary with chemical vulcanization, a dose of about 400–500 kGy (40–50 Mrad) is adequate, but this radiation dose is relatively high and expensive. (The dose to achieve the minimum degree of crosslinking to initiate a network is about one-tenth of this dose.) The reason for requiring the higher dose is that it is the maximum tensile strength of a highly elastic system that is desired, which requires a much higher crosslink density. Another possible disadvantage is that radiation, in comparison with chemical crosslinking, is stated to give a lower maximum tensile strength; this may be due to the chemical nature of the bridges formed, which for relatively long sulfur chains may be self-regenerating in the case of scission under excessive strain. Apart from specialist applications, full radiation crosslinking of rubber does not appear to have been very successful.

A much more widespread use of radiation in the rubber industry is in the improvement of natural green strength. In the manufacture of rubber components such as car tyres, some inherent strength is needed in shaping the rubber component before it is finally vulcanized with sulfur and this has been provided by dynamic chain entanglements which maintain the elastic properties of the rubber, at least for a short time. However, this may be inadequate and a small radiation dose may be used to maintain a more permanent, even if local, structure while the correct shape is prepared prior to vulcanization.

A further application is in the direct vulcanization of rubber latex. The latex contains natural rubber in a fluid medium and this may be irradiated and then subsequently dried to form a vulcanized (i.e., crosslinked and entangled) sheet, film and so on. This is an alternative to the usual vulcanization of dried rubber and for many purposes has some advantages. By the use of additives such as carbon tetrachloride at low concentrations, the dose requirements can be reduced to only a few tens of kGy or less (see *Elastomers: Additives*).

3. *Miscellaneous Applications of Irradiation*

The production of foamed polyethylene by means of radiation is a very valuable industrial application. It parallels the production of foamed rubber by conventional chemical techniques, but the radiation approach has certain advantages in the case of polyethylene. The polymer is formed containing a chemical foaming agent which is active at a high temperature, well above the melting point of polyethylene. The polymer is then irradiated, heated to transform it into a flexible network structure and then heated further to activate the foaming agent. The bubbles trapped in the flexible polyethylene matrix produce the required foam. When the polyethylene has cooled, the crystals that are formed lock the polymer in the foamed configuration.

Polytetrafluoroethylene (PTFE) is one of the large group of polymers which degrade under irradiation, that is, the main chain is scissioned. This reduction in molecular weight allows the polymer to be used for various purposes such as the nonstick surface coating on pans (owing to its resistance to high temperatures) or on mechanical devices (owing to its very low surface friction properties). Irradiation causes scission primarily in the amorphous regions between crystallites and subsequent melting and recrystallization allows a greatly increased degree of crystallinity, previously inhibited by the chains between adjacent crystallites, which after scissioning no longer prevent further crystallization (see *Radiation Effects on Polytetrafluoroethylene*).

Poly(vinyl chloride) has also been used in an irradiated form for cable insulation at higher temperatures but, in this polymer, additives must be included to allow the crosslinking reaction to proceed. It may even be argued that the additives form a network in which the polymer is held.

4. *Sterilization*

The use of radiation for sterilizing medical disposables was studied at Harwell (the UK nuclear research center) and other laboratories in the early 1950s and was taken up on an industrial scale in the USA some years later. The disposables, such as syringes, placed in sealed plastic packages, are irradiated on a conveyor system when already in their external packaging cases—a very convenient process thanks to the high penetration of ^{60}Co gamma radiation. A number of large radiation centers provide this service to companies which do not have the necessary radiation sources, which may amount to several megacuries.

The syringes and packages often consist of polypropylene and this polymer suffers both scission and crosslinking while irradiation proceeds. This has to be overcome by modification of the polymer through the use of radiation-protecting additives which reduce the long-term deterioration in physical properties due to oxidative degradation.

5. *Applications Under Development*

A number of potential radiation applications have been reported in scientific publications and many are now approaching, or may indeed have reached, industrial or medical applications, although not on a large scale.

Exposure to electron sources of relatively low voltage (0.5 MeV or less), with a corresponding lower cost, is being used for surface treatment of coatings involving a highly reactive monomer. This polymerizes and crosslinks by a chain reaction, so that the radiation is only used as a catalyst to initiate the process. With the very high intensities available, the reaction rate is extremely rapid. This approach envisages applications in the graphics industry, for example in web-offset printing or paint curing. The curing of a surface paint layer has already been undertaken in a number of industrial concerns and has a number of advantages over a thermal process, including excellent energy economy, less danger from contamination by dust and so forth and extreme rapidity. One unfortunate feature is that the paint used for radiation treatment differs from that needed for conventional surface painting. The process is also envisaged for the curing of polymer blends for magnetic media, where it has the advantages of the fullest control on rate and degree of cure, lack of catalytic impurities and accurate temperature control over a very limited range.

A major improvement is possible in the size and response time of microelectronic components, whose present size is limited by the wavelength of light. By using a high-voltage electron beam of much shorter wavelength to crosslink or scission a thin polymer coating on a silica or similar device and then removing the irradiated (or nonirradiated) polymer, devices with far finer resolution and faster response can be prepared.

A further potential application is the radiation-induced polymerization of suitable glass-forming monomers as a means of casting for optical uses such as lenses and prisms. This is one of the few instances of practical uses for radiation-induced polymerization.

On the medical and biological side, the immobilization of biologically active materials inside a polymer matrix may be envisaged. One approach is to graft onto a backbone polymer a monomer containing a functional group to which an enzyme can be bonded. Biocompatible polymers are developed for such uses as artificial organs. Another approach is the physical entrapment of a biocomponent solution or dispersion within a polymer matrix. Drug-delivery systems for controlled slow release have also been developed.

Polymeric films, tubes and other structures can be prepared for use in medical prostheses and have the requisite physical properties, but fail when implanted owing to thrombus formation. To avoid this, a different surface grafted onto the base polymer is prepared by radiation and this can impart the necessary nonthrombogenic properties to the prostheses.

See also: Elastomers: Additives; Radiation Effects on Polymers; Radiation Effects on Polytetrafluoroethylene; Vulcanization

Bibliography

Baird W G, Soonase P, Ross A B, Helman W P 1982 *Radiat. Phys. Chem.* 19: 339–60
Chapiro A 1962 *Radiation Chemistry of Polymeric Systems*. Interscience, Chichester, UK
Charlesby A 1960 *Atomic Radiation and Polymers*. Pergamon, Oxford
Dole M (ed.) 1972 *The Radiation Chemistry of Macromolecules*, Vols. 1, 2. Academic Press, London
Spinks J W, Woods R J (eds.) 1976 *An Introduction to Radiation Chemistry*, 2nd edn. Wiley, New York

A. Charlesby
[Swindon, UK]

Reaction Injection Molding

Reaction injection molding (RIM) is one of several reactive processing techniques for the production of polymeric solids or polymer composites, the final article and the polymer being produced simultaneously in a closed mold. It is characterized by the need for rapid intensive mixing of the polymer precursors immediately prior to the mold. Reactive processing demands a careful balance between the processing technology and the polymerization chemistry or chain-extension reactions. Some of the key reactive processing technologies are shown schematically in Fig. 1. The simplest technology is that of resin injection, where reactants with relatively long shelf life are introduced into a mold without mixing. This will not be dealt with further.

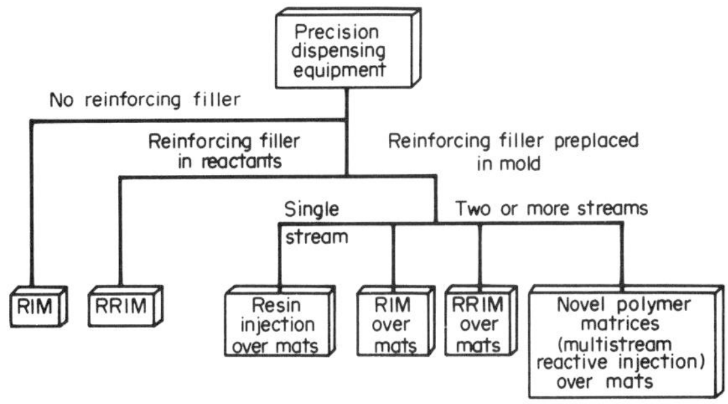

Figure 1
Reactive injection routes for polymers and polymer composites

The forerunner to the current high-pressure RIM technology was a low-pressure process developed specifically for polyurethane (PU) chemistry, largely for low-modulus products such as low-density foams and shoe-soiling materials, which employed static or dynamic mechanical mixers. In the high-pressure process, two or more reactants are mixed intensively by high-pressure jet impingement at a precise ratio. The relatively low-viscosity mixture then passes at low pressures into a closed mold, generally of large surface area, where polymerization is completed.

The attractions of low process-energy requirements and relatively low capital costs have ensured an expansion of interest in this process which has extended to other polymeric materials and also to the incorporation of particulate or continuous fiber reinforcement for the enhancement of mechanical properties. This article will focus on RIM processes, including reinforced RIM and RIM over reinforcing fiber mats.

1. Materials for RIM

Key variables in RIM processes may be conveniently categorized as those related to materials and those which relate to the process technology (see Fig. 2).

1.1 Polymerization and Chain-Extension Processes

The requirement for rapid reactive injection cycle times has intensified with improvements in the competitive thermoplastics injection-molding process: polymerization must be essentially completed in the mold in less than 2 min, ideally with no postcure or finishing requirements. Such rapid liquid-to-solid transformations can be achieved in polymerization reactions which are not intrinsically fast through the use of catalysts, initiators and by adjustment of mold temperatures.

The major problems in achieving rapid formation of polymeric materials with good physical properties

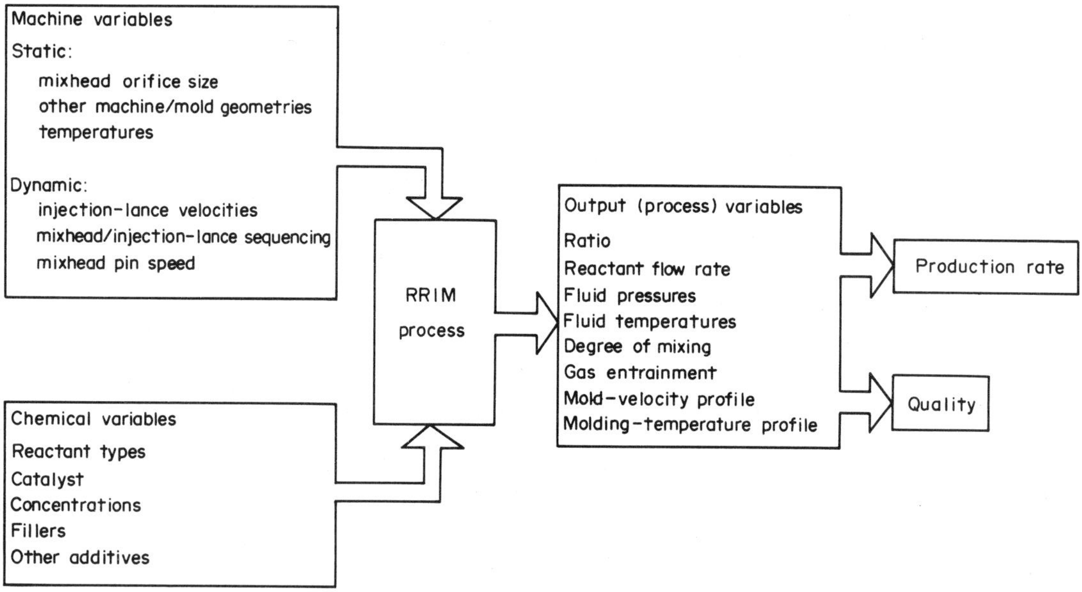

Figure 2
RIM process input and output variables

is that most reactions are exothermic and necessitate rapid heat removal from the mold. Physical factors such as crystallization and phase separation are also significant features of the materials in relation to their properties, and these are controlled by the thermal history of the product, which is dependent on both the exothermic reaction heat and heat-transfer characteristics in the mold.

(*a*) *Polyurethanes and polyureas*. Polyurethanes are the most extensively studied materials for RIM and dominate the RIM market. The reactants (isocyanate stream and polyol stream) are dispensed at conveniently low temperatures (~15–40 °C) and mold temperatures are also moderate (~65 °C). However, the reactants also need to be dispensed accurately to maintain the appropriate stoichiometry and unless adequate mixing is achieved, physical properties of the product will be impaired.

Polyurethanes derive their properties from the chemical constitution and composition of the polymer chains, crosslinking and morphology, which is frequently two phase. The polymer precursors are polymeric materials of low molecular weight, designed to have relatively low viscosities at processing temperatures for ease of dispensing and mixing. The density of the final product is readily controlled by the addition of water which, on reaction with isocyanates, liberates carbon dioxide.

The rates of the chain-extension reactions which give rise to the polymer and the liberation of carbon dioxide can be controlled through the use of organometallic and amine catalysts. The precise balance of the relative rates of these reactions is crucial to the production of well-developed foams and the overall rate of production of polymer.

For many RIM products, for example for the automotive industry, high-density materials are required and the diol stream may need to be dried to ensure that only a microcellular foam is produced.

High temperatures can be reached in the RIM process, the center of the product commonly being 30–80 °C higher than the mold surface. At these temperatures side reactions can occur, for example the formation of additional crosslinks or, in the extreme, polymer degradation. It is common to employ postcuring of products where the solid polymer is formed prior to complete reaction, leading to diffusion-controlled reaction rates.

Typical RIM polyurethane formulations will include the following essential components:

(i) a multifunctional isocyanate,

(ii) an oligomeric diol,

(iii) multifunctional chain extenders, and

(iv) appropriate catalysts.

The precise formulation has to be carefully tailored to achieve the necessary low viscosities for the RIM technology as well as for the appropriate physical properties in the final product.

The isocyanate reactant frequently comprises a multifunctional prepolymer with a functionality of the order of 2.1–2.4 and is often based on 4,4-diphenyl methane diisocyanate (MDI) or its derivatives.

The most commonly used oligomeric diols for PU soft segments are based on polyesters or polyethers (Noshay and McGrath 1977, Speckhard and Cooper 1986). Polyesters give good physical (e.g., impact and tensile) properties but can be hydrolytically unstable. For conditions which might lead to hydrolytic degradation, polyethers are employed.

Typical chain extenders are ethylene oxide, 1,4-butane diol or multifunctional low-molecular-weight diols which react with the isocyanates to give the hard-block component. Multifunctional chain extenders can be used to control the extent of crosslinking. Differences in final properties will result from the type of isocyanate employed, through the differing hard segments obtained (Gillis et al. 1983).

Often a synergistic mixture of tertiary amine and tin catalysts is used to achieve the correct balance of reaction rates.

Polyureas also comprise hard- and soft-block components, the hard block being formed from polymeric isocyanates as with PUs, but the chain-extension reaction involving either difunctional or trifunctionalized low-molecular-weight amines. Amine-terminated polyesters or polyethers are used for the soft-block component. Amine–isocyanate reactions are intrinsically fast and do not require catalysts. Indeed, the reactions can be so fast that problems can be encountered in the completion of mold-filling and in achieving adequate mixing. Reaction rates can be controlled by selecting the appropriate amine functionality or through blocking agents.

Polyureas give good physical properties through strong hydrogen bonding and are less dependent on phase separation of the hard- and soft-block components for the development of good properties. Interesting properties can be achieved by using a mixture of alcohol and amine-terminated chain extenders or polyols, to produce poly(urea–urethanes).

Polyureas offer the combination of good properties with fast reactions, for example gel times of around 1 s are typical of commercial polyurea RIM, and these materials have been finding increasing application, particularly in automotive RIM production.

Recent commercial developments in polyurethane RIM for specialist applications include internal mold-release agents, based, for example, on modified zinc stearates for PUs or acidic siloxanes for polyureas (Dominguez et al. 1983, Grigsby and Dominguez 1985) incorporated into the PUs to decrease process cycle times, and PU paint layers sprayed onto the mold surface prior to injection, for "in-mold painted" articles.

Despite the wide range of properties available in polyurethane RIM products, there remains considerable interest in the development of RIM materials for structural applications which imply the need for a higher modulus than PUs, yet with inherent toughness and an increased high-temperature range. There is a limited choice of possible RIM systems which can hope to satisfy the practical processing requirements for RIM and which have sufficiently rapid reactions.

(*b*) *Amide copolymers*. The reactive casting of caprolactam (Nylon 6) using anionic initiators has been known for many years. Significant developments have taken place in improving the rate of polymerization and the impact resistance of these materials to make them acceptable for RIM. Commercial formulations are available (e.g., DSM Nyrim and Starim). The main attraction of these materials is the high-temperature elastic modulus and this has been sufficient incentive for the development of the required high-temperature process technology which is necessary to ensure melting of the caprolactam reactant (melting point 67 °C). Molding temperatures are also high (~140 °C) to achieve acceptable reaction rates. Care has to be taken to ensure that molding temperatures are not so high as to result in high concentrations of caprolactam, since the polymerization is reversible.

The initiation reaction is crucial in this polymerization and acceptable reaction rates can only be achieved through the use of acyllactam intermediates (Hedrick et al. 1985).

The products are crystalline and the good high-temperature modulus stems from the high melting point of the polyamide (~256 °C). As with many linear segmented copolymers, a compromise has to be achieved between mechanical properties and impact character; increasing the polyether soft-block content increases impact character but reduces the high-temperature working capability of the material (Kurz 1985).

Phase separation is an important process in polyamide RIM. Polymerization commences at the mold surface (unlike PUs which polymerize from the center of the molding outwards) and crystallization occurs concurrent with polymerization. In addition, soft-block phase separation occurs, producing complex solid-state morphology, with the soft blocks separating into fine lamellae (van der Loos and van Geenen 1985).

Typical two-stream formulations for polyamide RIM include significant proportions of caprolactam in both reactants. There is consequently no specific demand for precise stoichiometric ratio of the reactants, the only requirement being to achieve precise mold filling. Mixing requirements are also less stringent than for PUs because of the common content of the two feed streams. However, because of the relatively slow reaction rates, the polymerization occupies a greater proportion of the RIM cycle time than is the case for PUs, but approximately equal overall cycle times may be achieved for both polymers because the highly crystalline amide copolymers offer easy mold release.

Polymerization should be more than 96% complete on demolding to avoid monomer release—residual monomer can cause surface blemishes and coating or painting problems.

Overall, given the attractive end properties of Nylon 6 block copolymers, these materials have considerable potential in RIM and RIM-related processes (Fig. 1).

Other nonpolyurethane RIM polymerizations include the commercially available dicyclopentadiene (Geer 1983, Klosiewicz 1983) and acrylamates (Wilkinson and Eckler 1986). Polyester, acrylic, epoxy and phenolic formulations are under development for RIM. All of these materials require moderate temperature dispensing (<50 °C) with relatively high-temperature molds, typically 120 °C to speed up the polymerization reaction. The phenolic system differs from the others in being a condensation reaction, with water as a by-product.

An attractive route for the control of physical properties is to produce blends, alloys or interpenetrating polymer networks (IPNs) from polymers of different types. These materials often have a two-phase morphology, one polymer component giving mechanical strength to the material and the second giving the impact character. There have been numerous accounts of blends, alloys and IPNs; most studies have been carried out under conditions far removed from those encountered in RIM, particularly with respect to the required short cycle times. Coates and Johnson (1986) and Armitage et al. (1987a) have shown the feasibility of producing polyurethane–acrylic alloys from commercially available polyurethane and acrylic systems in a cycle time of around 3 min, in a mold at 50 °C. Two to four reactant streams are required for RIM alloy or IPN production. In the reported work, the following four reactant streams were used:

(i) acrylic prepolymer plus peroxide initiator,

(ii) acrylic prepolymer plus amine accelerator,

(iii) diisocyanate, and

(iv) poly(ether polyol), chain extender and dibutyltin dilaurate catalyst.

The polyurethane had a low flexural modulus and the toughened acrylic component comprised a mixture of an oligourethane methacrylate with methyl methacrylate. The observed morphology was more complex than expected from a simple consideration of network interpretation. At 20 wt% of PU in the blend, there is evidence of well-distributed spherical domains of diameter 1–2 μm separating out, with some signs of fine structure in the continuous phase. When the PU content in the blend reached 40 wt%, a distinct two-phase morphology could be observed, with domain sizes around 0.02 μm in the matrix and the larger spherical domains had increased in number and grown in diameter to around 3–4 μm. It is believed that these domains are attributable to poly(methyl methacrylate) and that the matrix contains some interpenetration of the PU and acrylic networks.

1.2 Fillers

Low-pressure RIM was originally developed for PU shoe-sole manufacture; the automotive industry provided the motivation for the development of high-pressure PU reinforced RIM (RRIM), in which reinforcing fillers were incorporated into one or both reactant streams, in order to enhance the values of product flexural modulus (and hence heat-sag resistance at elevated temperatures) and reduce the linear coefficient of thermal expansion (Coates and Johnson 1981). Typical reinforcing fillers which have been investigated include hammer-milled glass (the most used commercially), flake glass, chopped-strand glass, wollastonite and mica.

Commercial hammer-milled glasses typically exhibit a broad fiber length distribution, from tens of micrometers to a millimeter (rather lower than estimated critical fiber lengths), for fibers of diameter approximately 10–20 μm. Up to around 40 wt% loadings have been employed in polyurethane RRIM. The fibers are normally coated with a silane size, to enhance the bond with the eventual polymer matrix; the size may also include dispersing agents to help the fibers to distribute homogeneously in the polyol, seeking to avoid aggregation of fibers which are a cause of RIM mixhead orifice blockage. Chopped glass is more monodisperse, typically around 1.6 mm in length, 10–20 μm in diameter for RIM and, hence, offers greater reinforcing efficiency. It is, however, relatively expensive and even at low loadings causes rapid viscosity buildup at low shear rates. Flake glass platelets are typically up to 1 mm^2 in area and 10–20 μm in thickness. Other fillers may be incorporated for special applications, for example various fire retardants and barium sulfate for sound insulation.

Fibrous reinforcing fillers introduce problems in handling and health, settlement and machine wear, but provide a method for enhancing the Young's modulus of PU products and decreasing the linear coefficient of thermal expansion to a value nearer that of steel (which is important for steel–polymer car-body construction). However, the reinforcing efficiency of short fiber fillers tends to be low, particularly after attrition of the filler during high flow-rate recirculation in standard RRIM processing equipment. Fibrous fillers also lead to anisotropy of the product (much reduced in the case of platelet-type fillers) which may be exploited in some cases but in most cases is not desirable. Platelet-type materials (such as flake glass) help reduce anisotropy.

Consequently, there has been growing interest in various forms of continuous reinforcing filler, for example random glass mats, woven glass or carbon fiber cloths or knitted structures. These provide a theoretically more efficient reinforcement owing to the effectively infinite length of fiber. Packings up to about 80 wt% can be achieved, increasing further the stiffness of the product, and these fillers also lead to a more isotropic product. However, knitted structures, in particular, are not inherently mechanically stiff and, hence, the achievable product stiffness is reduced. Also, at high loadings the flowpaths in a mold may demand relatively high pressure drops, leading to loss of dimensional accuracy in the molding through mold/press deflection or opening. Technological problems relating to the manufacture, handling and placement of suitable preformed mats, mold design and filling and chemical formulations are currently being addressed.

Nevertheless, the RIM over-mats technique shows great promise in achieving far higher property levels than by RRIM alone (see Sect. 4). It is also feasible that reinforcing (or other) fillers will be required in the reactants in a combined RRIM over-mats process.

2. *Rheology*

A notable feature of RIM formulations is the relatively low viscosities of the reactants and the reacting mixture before significant reaction occurs. Polyamide and acrylic RIM reactants may have apparent viscosities as low as 10^{-3} Pa s while those of PU reactants may be up to about 0.5 Pa s. The addition of reinforcing filler to the reactants causes a rise in this initial viscosity, so that at the low shear rates encountered in stirred storage tanks it is not possible to incorporate more than around 40 wt% hammer-milled glass or 16 wt% chopped-strand glass in PU reactants (air inclusion also becomes a significant problem at these loadings). The behavior of PU reactants, even those containing chopped-strand-glass reinforcing filler, has been observed by Coates et al. (1987a) to be essentially Newtonian over the range of shear rate expected in RIM mix-heads (10^4–$10^6\,s^{-1}$). As reaction and mold filling proceed there may be an initial fall in the viscosity of the reacting mixture owing to heat transfer from the mold and exothermic heat.

It is desirable to fill a mold completely before the occurrence of the increasingly rapid rise in viscosity owing to polymerization. The form of the viscosity–time curve is discussed by Castro et al. (1984)—simple modelling, combining reaction kinetics with structural relationships, can be used to predict the buildup in molecular weight and, hence, the rise in viscosity with time and temperature in the mold. The true chemorheology of polymerizing mixtures in reactive injection molding is particularly complex, with the superposition of heat transfer, polymerization, phase separation and crosslinking all possibly in the presence of reinforcing fillers in suspension or flow through reinforcing fiber mats (such fillers may act as heat sinks and barriers to diffusion of the reacting species).

Castro et al. (1984) also report laboratory measurements of PU reacting-system rheology and suggest an empirical relationship between viscosity and the degree of conversion of isocyanate groups. The critical parameter for fitting the observed viscosity rise as polymerization progresses is the fractional conversion of isocyanate groups at physical gelation (at which point the reacting mixture viscosity rises very rapidly).

Laboratory measurements are unrealistic for very fast reacting systems, such as those commercially in use. Hence, attempting to measure shear viscosity in the mold has considerable significance. It is possible to use the initial pressure rise during mold filling to predict initial viscosity (Castro and Macosko 1982, Vespoli and Alberino 1985).

3. *Processing*

A schematic diagram of an RIM machine is shown in Fig. 3. Two key elements are the dispensing system and the mixing head. Good control of dispensing and adequate mixing are fundamental to successful RIM production.

3.1 Machine Control

In the case of RIM of PUs or related formulations, accurate dispensing at the stoichiometric ratio is

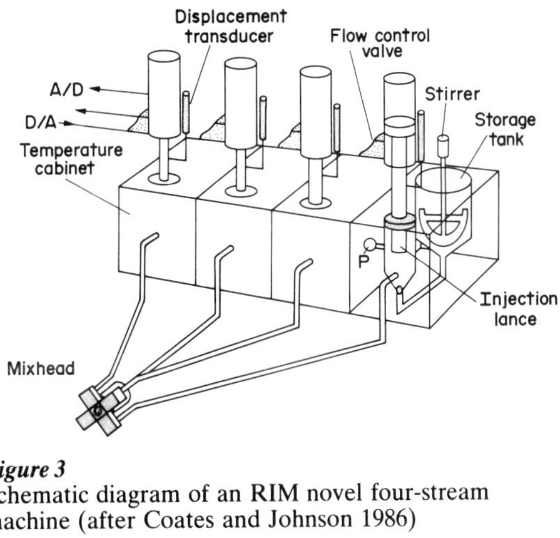

Figure 3
Schematic diagram of an RIM novel four-stream machine (after Coates and Johnson 1986)

necessary to achieve the desired product properties. A suitable measure of machine dispense accuracy is the root-mean-square (rms) deviation of the reactant ratio from the set ratio. However, this may typically be up to 5% on commercial dispensing equipment and, hence, commercial RIM formulations often have to be designed to provide a tolerance in ratio. Accurate dispensing control and the effects of lack of control on dispense ratio have been discussed by Coates et al. (1986) who describe research equipment that is consistently capable of less than 1% rms deviation in ratio.

Commercial mixheads were designed essentially empirically and act as directional flow valves for the reactants—when the mixhead is actuated reactants are diverted from flowing back to their storage tanks to flowing into the small-volume (of the order of milliliters) mixing chamber through small-diameter orifices, typically at velocities of around $100\,\mathrm{m\,s^{-1}}$. The orifices may be adjustable (by means of flow restrictors). Two orifices are generally used, but larger mixheads can have three or four inlets, allowing multicomponent mixing. In commercial practice this allows the incorporation of colorants or foaming agents. Multicomponent injection has been used in a novel manner by Coates and Johnson (1986) and Coates et al. (1987b) in a three-stream dispensing of PU precursors, effectively blending and mixing in one stage at the mixhead, and in simultaneous injection of four streams to produce a polymer blend in a single mold cavity.

Coates et al. (1987c) further indicate the effects of poor sequence control at the mixhead in causing uncontrolled compressible flows of reactants, particularly at the start of injection, leading to the well-known phenomenon of "lead-lag," that is, off-ratio production at the commencement of injection (see Fig. 4). Although accurate ratio control is less critical for nonpolyurethane RIM systems, accurate dispensing is a prerequisite for consistent production and properties.

3.2 Mixing

Good mixing of reactants is critical to the completion of fast urethane polymerization. High-pressure jet impingement has become the dominant route to the required intensive mixing for PUs (essentially the rapid creation of adequate contact area between the reactants in a timescale which is short compared with total injection time). Since the technology for impingement mixheads with self-cleaning valving (reviewed by Oertel 1985) is attractive from a manufacturing viewpoint, impingement mixing is also the major mixing method for nonpolyurethane RIM.

The assessment of mixing quality has been the subject of much study and some controversy. The lowest value of Reynolds number Re in an impingement-mixing nozzle has become a favored criterion for mixing quality. Reynolds number is given by

$$Re = \rho v d / \eta$$

where ρ is the reactant density, v is the average stream velocity, d is the orifice minimum opening and η is the reactant apparent viscosity. In the case of polyurethane RIM, the lowest values of Re will be obtained for the polyol blend, which normally exhibits apparent viscosities around four times that of the isocyanate reactant.

Attempts to quantify the degree of mixing will be categorized here as nonreacting-systems and reacting-systems measurements.

(*a*) *Nonreacting systems*. Early work (e.g., Malguarnera and Suh 1977, Tucker and Suh 1980) indicated that a value of Re in excess of about 200 was required to obtain "good mixing," with no apparent improvement in mixing for values of Re greater than 500. The assessment was based on the variance of concentration of tracers in samples of impingement-mixed model fluids. However, the size scale resolution of these techniques is greater than about 200 μm. Begemann et al. (1986) also worked with nonreacting systems, impinging polyol with glycerine and then measuring the percentage of sample emulsified. It appears from this that emulsifi-

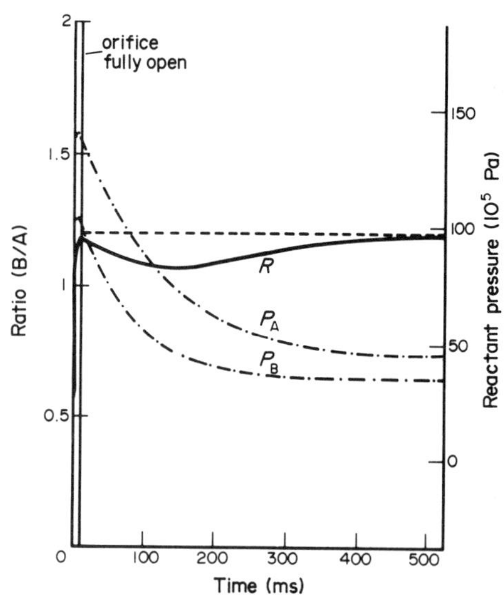

Figure 4
Modelling results for grossly nonideal sequence control at the mixhead, showing off-ratio R production at the commencement of injection, due to compressible flows of the isocyanate (A) and polyol (B) reactants

cation increases with Re for values of $Re \leqslant 300$. Visualization techniques have been used to study the flow patterns in mixheads. Early work of this kind using injected dies in water and water–glycerine mixtures indicated the onset of flow instabilities (as required for mixing) to be at $Re > 60$ (Lee et al. 1980, Tucker and Suh 1980). Sandell et al. (1985) injected a rapidly fading dye to visualize flows of $Re \leqslant 700$; mixing improvements were observed for $Re \leqslant 400$. The various visualization studies indicate that the instabilities (perhaps best described as disturbed laminar flows) produced in the mixing chamber on impingement are reduced to simple laminar flows, which incur less mixing, within around three diameters of the impingement point.

(*b*) *Reacting systems*. A wide range of techniques has been employed in an attempt to infer or measure mixing quality in reacting systems. Fruzzetti et al. (1977) used product flexural modulus as a measure of mixing efficiency and found increases in modulus with Re for $Re \leqslant 300$. Richter and Macosko (1978) found that the adiabatic temperature rise of mixed reactants for a crosslinking polyurethane increases with Re, but only for a limited range of Re. The adiabatic temperature rise caused by polymerization exothermic heat would be expected to be a maximum for a perfectly mixed system. In practice, impingement mixing is thought to create lamellae or striations, probably with a distribution of thicknesses. Where the lamellae are thin, the polymerization reaction can proceed by diffusion of reactants into each other; where lamellae are thicker, unreacted material is likely to be trapped in the lamellae. Thus, efficient mixing would create sufficiently thin lamellae to allow diffusion to complete the mixing process. This would lead to a higher conversion and, hence, a greater adiabatic temperature rise. Chella and Ottino (1983) analyzed these results and concluded that for gel times of the order of 10 s, striations must be smaller than 10 μm for the temperature rise to be controlled by reaction kinetics.

Kolodziej et al. (1982) studied a phase-separating formulation, which indicated the same trend with Re. Molecular-weight distributions in their samples showed distinct differences with Re. As Re increased (up to 210), molecular weight increased and the distribution narrowed, indicating a more homogeneous polymerization and, hence, by inference, better mixing. These workers also employed polymerization to freeze mixing patterns. Their polyol contained 0.1% carbon black and samples were collected from the polymerized material from the center of a tapered tubular mold. Measurements of striation thickness in these samples showed that this thickness decreases as Re increases up to about 250, and average striation thicknesses were around 100 μm (see Fig. 5). Baldyga and Bourne (1983)

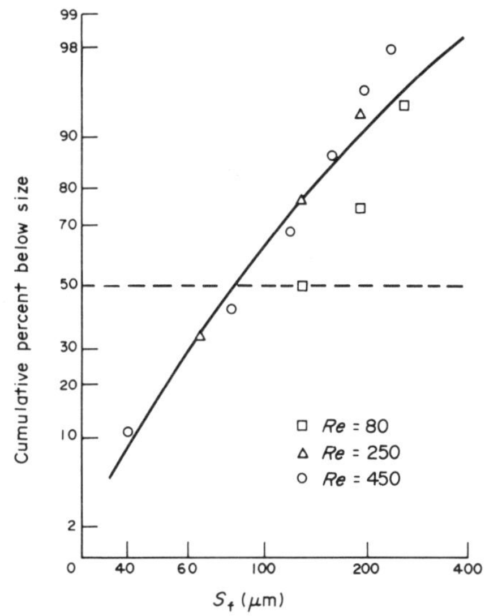

Figure 5
Cumulative size distribution of striations at various Re (after Kolodziej et al. 1982), with the line from the isotropic turbulence model of Baldyga and Bourne (1983)

proposed a theoretical model for impingement mixing which assumes disturbed flow in a region of two chamber diameters from the impingement point and an exponential residence-time distribution in this zone. They obtained a fit to the experimental results of Kolodziej et al. (1982), which is shown in Fig. 5.

The flow visualization experiments clearly indicate a larger-scale striation than the 1 μm which would be needed for molecular diffusion to allow completion of the polymerization reaction in a striation. However, the RIM process does produce well-reacted polymers and consequently the "fluid mechanics" approaches above cannot represent a full understanding of mixing processes for reacting systems. Various workers (Ranz 1979, Fields et al. 1986, Wickert et al. 1987) tentatively suggest a micromixing mechanism related to surface effects as follows. The interface between two reactants may become unstable and droplets with diameters of the order of a micrometer may be formed (each drop being encapsulated in a newly formed urethane structure) which then disperse into one of the reactant lamellae. Molecular diffusion could then account for completion of the reaction. Such micromixing, with diffusion, would thus be the final stage of an intensive mixing process in which reactant jets of the order of 2 mm diameter are impinged to produce lamellae of thickness of the order of 100 μm. This process requires critical values of Re of the order of 300 at the mixhead orifice—a condition which can be readily achieved in RIM dispensing equipment.

3.3 Mold Filling

Generally, low-viscosity mixtures passing into a mold imply that fill pressures will be low and that wetting of the surface (to pick up surface detail) will be good. However, the danger of air entrainment at the flow front is also increased. A major processing requirement is to fill molds in a laminar-flow mode, but in times somewhat shorter than the gelation time for the reacting system.

Both experimental (Coyle et al. 1987, working on model fluids) and modelling studies (e.g., Domine and Gogos 1980, Castro and Macosko 1982, Lekakou and Richardson 1986) have indicated a parabolic velocity profile in the major part of the advancing fluid, with a "fountain flow" at the flow front. If such a flow genuinely occurs in a reacting system, it is important that it be quantified since it determines the location of each element of the reacting mixture in the mold.

Flow-front instabilities begin to occur when too rapid a mold filling is attempted. At typical polyurethane RIM mold-filling viscosities, it is suggested that a critical Reynolds number of between 3 and 30 in the cavity should not be exceeded to avoid instabilities leading to air entrainment, gate design also being important in this matter.

Of further practical importance is whether the mold can be filled in a steady laminar manner at *Re* below some critical value, before gelation occurs or before any significant rise in viscosity causes too great a demand in filling pressure. Castro and Macosko (1982) suggest a finite-difference analysis of the filling problem, which requires several major assumptions. Their work indicates that reasonable pressure–time predictions can be made and that if an approximation for the fountain flow is not included, pressures are predicted considerably under their actual value.

Adiabatic filling is a further possibility in rapid mold filling of rapidly reacting systems, as in the work of Vespoli and Alberino (1985) and Vespoli et al. (1986). These results suggest a simple criterion for mold filling—if the time to fill a mold is greater than the adiabatic gel time then the mold will not be filled. Predictions of the adiabatic gel time for a specific RIM system have been made by Manas-Zloczower and Macosko (1987), who also discuss the concept of an RIM "moldability diagram" (i.e., a designation of the conditions under which a molding can be successfully made) similar to those used in thermoplastic injection molding.

Gate designs and the filling of more complex geometry molds, modelling commercial designs, have been studied by Müller et al. (1985), Begemann et al. (1986) and Maier and Menges (1986), with indication of specific gate and mold design parameters to avoid air entrainment at gates and inserts which obstruct flow.

Mold filling when continuous fillers are preplaced in the mold has been modelled by Gonzalez (1983) who used a simple analysis for flow through glass mats, employing D'Arcy's law.

3.4 Curing

Curing is often the rate-limiting step in the RIM molding cycle. Sufficient "green strength" must be built during cure for demolding. In rapid reacting RIM systems the inmold time is generally fixed by heat-transfer limitations, that is, the time taken to conduct away exothermic heat to the mold. The conversion and temperature profiles in the molding will determine the development of properties and, hence, strongly influence demolding. The Damkohler number *Da* is an important parameter in assessing cure times. For a mold of typical thickness H,

$$Da = \frac{\text{heat production by chemical reaction}}{\text{heat transport by conduction}} = \frac{C_0 k(T_0)}{4a/H^2}$$

where $k(T_0)$ is the overall rate constant evaluated at initial material temperature, a is the thermal diffusivity of the reacting mixture and C_0 is the initial concentration of reactive groups. Large values of *Da* clearly suggest that adiabatic conditions apply; when *Da* is small, reaction heat may be neglected and the problem reduces to one of transient heat transfer through the mold thickness.

4. Properties

The final properties of an RIM polymer matrix will depend on formulation variables and the process history of all the elements making up the final product. The occurrence and degree of crosslinking, phase separation and alloying will have large effects. A wide range of physical properties may therefore be achieved in a controlled manner in an RIM polymer matrix by choice of formulation (e.g., amount and type of chain extender in a PU formulation) and there will also possibly be further less-controlled variation through process history. An indication of some physical properties of RIM polymers is given in Table 1.

However, the amount of enhancement of mechanical properties which can be achieved by manipulation of chemistry and process history is limited compared with the demands for structural engineering materials, that is, those with adequate levels of stiffness. Incorporation of reinforcing fillers of the type discussed in Sect. 1.2 appears to be the only satisfactory way of attaining relatively high-modulus materials with adequate impact properties and

Table 1
Selected results from a polyurethane RRIM program with mixed reinforcing fillers (ambient temperature testing)

Reinforcing agent used	Product density (kg m^{-3})	Product hardness Shore D	Heat sag[a] (mm) $\parallel/\perp$[b]	Impact energy (J)	Tensile strength (MPa) $\parallel/\perp$[b]	Flexural modulus (MPa) $\parallel/\perp$[b]
Unfilled	1100	58.0	35	55	20	300
5% CSG[c]	1180	59.5	n/a	54.2	24/24	576/390
5% CSG + 5% Woll[d]	1150	60.5	15/28	40.3	21/20	525/375
5% CSG + 10% Woll	1170	61.0	20/28	37.2	19/19	470/415
5% CSG + 5% HMG[e]	1230	60.0	n/a	36.1	25/25	555/380
5% Flake[f]	1200	60.0	32/40	21.0	17/15	490/490
5% Flake + 10% HMG	1230	62.5	22/33	21.2	23/21	535/470
2% Flake + 4% CSG	1130	60.0	10/26	24.5	23/20	730/455
4% Flake + 3% CSG	1200	62.0	10/25	24.5	20/19	690/510
4% Flake + 8% HMG + 3% CSG	1180	63.0	9/20	24.4	21/19	740/530

a 125 mm overhang at 120 °C for 1 h b Parallel/perpendicular to the injection direction c Pilkington Brothers WX6450 1.5 mm chopped-strand glass d Wollastonite (Colin McNeal Ltd) G-Wollastikup 1100/4 10 u e Hammer-milled glass f Owens Corning 3 mm flake glass 74 × 59639 no coupling agent

reduced coefficient of thermal expansion. Indeed, addition of only moderate levels of reinforcing fibrous fillers has a marked effect on modulus, but leads to a decrease in elongation-to-break and impact strength. Mixed reinforcing fillers have been used to modify the often unwanted anisotropy of mechanical properties caused by flow-aligned fibers. Table 1 shows some selected results from the studies by Armitage et al. (1987b). The effect of a two-dimensional-platelet filler, flake glass, in reducing anisotropy is clear.

By far the most significant enhancement of modulus and improvement in other physical properties such as impact strength, high-temperature creep behavior and linear thermal-expansion coefficient (Gonzalez and Macosko 1983, Nelson 1986, Wilkinson and Eckler 1986) is obtained by RIM (or resin injection) over continuous mats. Hence, there is increasing interest in this process. Table 2 shows the large increases obtained in flexural modulus, to levels which may be considered structural, by incorporation of high weight percentages of continuous glass mats. Figure 6 shows storage modulus E', measured at 0.3 Hz, as a function of temperature (Armitage et al. 1987b) for a PU produced by RIM and the same polymer containing mixed discrete glass reinforcement, produced by RRIM, with two examples of the same polymer containing about 20 wt% and about 40 wt%, respectively, of continuous random glass mat, produced by RIM over these mats.

Although the continuous reinforcement clearly has a major effect in determining the overall modulus level, the polymer-matrix temperature performance cannot be ignored. From a structural engineering viewpoint it is preferable to have as flat a modulus–temperature response as possible. This suggests that there is room for manipulation of the

Table 2
Some comparative details for reactive injection over mats (ambient temperature properties except heat distortion temperature)

Material	Tensile strength (MPa)	Elongation (%)	Izod impact (J m^{-1})	Flexural modulus (MPa)	Heat distortion temperature (°C)
Acrylic unreinforced[a]	56	6.5		2860	83
Acrylic + 100 wt% $CaMgCO_3$				9000	
Acrylic + 40 wt% woven glass mat[b]	200			18000	
PU structural foam + 18 wt% glass mat[c]	69			3000	130
Isocyanurate + 33 wt% glass mat[d]	180	7.1	650	11200	203
Acrylamate + 33 wt% glass mat[e]	103	1.7	598	7500	
Acrylamate + 50 wt% glass mat[f]	276	2.5	939	14000	210

a Modar series 80, ICI (Mond division) b Leadbitter et al., University of Bradford c Nelson 1986 d Carleton et al. 1985 e Wilkinson and Eckler 1986 f Arimax 2100, Ashland Chemical Company

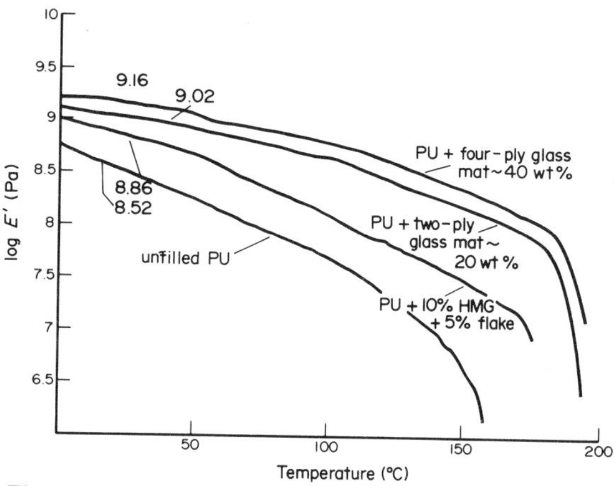

Figure 6
Storage modulus E' as a function of temperature for an RIM polyurethane and various RIM polyurethane–glass composites (after Armitage et al. 1987b)

polymer matrix to reduce the fall in composite properties with temperature, while maintaining polyamides over mats, and RIM of alloys or IPNs over mats would seem to offer attractive routes to the desired property levels for structural products.

RIM and associated reactive-injection processes and products are consequently challenging traditional sheet-metal-forming processes and products for automotive bodywork applications. They are themselves, however, in competition with other polymer-forming techniques, in particular with the well-established large scale injection molding of thermoplastics blends and alloys for the same automotive applications, as well as furniture, appliance housings and leisure goods. The potential for development of structural materials for high-volume applications remains a strong incentive for developments in RIM and, in particular, RIM over long fiber mats.

See also: Injection Molding of Thermoplastics: Mold Filling; Injection Molding of Thermoplastics: Structural Foams; Injection Molding of Thermoplastics: Structure Development; Processing: Classification

Bibliography

Armitage P D, Coates P D, Hynds J, Johnson A F, Leadbitter J 1987a Reaction injection moulding of polymer blends: An investigation of the morphology of an acrylic–urethane polymer. *Proc. Int. Conf. Reinforced Injection Processing*. Plastics and Rubber Institute, London, Chap. 8, pp. 1–10

Armitage P D, Coates P D, Hynds J, Johnson A F, Leadbitter J 1987b Rapid novel composite formation by reactive injection. *Proc. 2nd Int. Conf. Polymer Processing Machinery*, London, Chap. 16, pp. 1–13

Baldyga J, Bourne J 1983 Distribution of striation thickness from impingement mixers in reaction injection molding. *Polym. Eng. Sci.* 23: 556–9

Begemann M, Maier U, Muller H, Pierkes L 1986 RIM improved technology of the machines and selective mould design enhance moulding quality. *Proc. 13th Kolloquium*. Institut für Kunststoffverarbeitung (IKV), Aachen, Germany, pp. 465–96

Carleton P D, Waszeciak D P, Alberino L M 1985 MM/RIM: RIM composites using preplaced reinforcement. *J. Cell. Plast.* November: 409–16

Castro J M, Macosko C W 1982 Studies of mold filling and curing in the reaction injection molding process. *AIChE J.* 28: 250–60

Castro J M, Macosko C W, Perry S J 1984 Viscosity changes during urethane polymerization with phase separation. *Polym. Commun. (Peking)* 25(3): 82–7

Chella R, Ottino J M 1983 Modelling of rapidly-mixed, fast-crosslinking exothermic polymerizations. Part 1: Adiabatic temperature rise. *AIChE J.* 29: 373–82

Coates P D, Johnson A F 1981 An introduction to reinforced reaction injection moulding. *Plast. Rubber Process. Appl.* 1: 223–38

Coates P D, Johnson A F 1986 Reinforced reaction injection moulding: Exploring the machine–materials interface. *Proc. UTech 86.* Crain Communications, London, pp. 49–54

Coates P D, Johnson A F, Armitage P D, Hynds J, Leadbitter J 1987b Novel reinforced reaction injection molding (RRIM) processing: Computer-controlled multiple-stream RRIM production of interpenetrating polymer networks and polyurethanes. *Polym. Eng. Sci.* 27: 1209–15

Coates P D, Sivakumar A I, Johnson A F 1986 RRIM: Computer control of injection. *Plast. Rubber Process. Appl.* 6: 141–51

Coates P D, Sivakumar A I, Johnson A F 1987c RRIM: Experimental studies of flow transients in mixheads. *Plast. Rubber Process. Appl.* 8: 203–13

Coates P D, Sivakumar A I, Johnson A F 1987a RRIM: Rheology of glass-filled polyol slurries and polymeric isocyanate reagents. *Plast. Rubber Process. Appl.* 7: 19–28

Coyle D J, Blake J W, Macosko C W 1987 The kinematics of fountain flow in mold-filling. *AIChE J.* 33: 1168–77

Domine J D, Gogos C G 1980 Simulation of reactive injection molding. *Polym. Eng. Sci.* 20: 847–58

Dominguez R J G, Rice D M, Lloyd R F 1983 Reaction injection molded elastomer containing an internal mold release made by a two-stream system. US Patent No. 4,396,729

Fields S D, Thomas E L, Ottino J M 1986 Visualization of interfacial urethane polymerization by means of a new microstage reactor. *Polymer* 27: 1423–32

Fruzzetti R E, Hogan J M, Murray F J, White J R 1977 Factors affecting the quality of impingement mixed RIM urethanes. *Soc. Automot. Eng. Tech. Pap. Ser.* 770839

Geer R P 1983 Polydicyclopentadiene: A new RIM thermoset. *Soc. Plast. Eng., Nat. Tech. Conf. 1983.* Society of Plastics Engineers, Brookfield, CT, pp. 104–5

Gillis H R, Hurst A T, Ferrarini L J, Watts A 1983 Liquid MDI isocyanates: Effects of structure on properties of derived polyurethane elastomers. *SPI Polyurethane Division 6th Int. Tech/Marketing Conf.* Technomic, Lancaster, PA, pp. 375–80

Gonzalez V M 1983 Studies of reactive polymer processing with fiberglass reinforcement. Ph.D. thesis, University of Minnesota
Gonzalez V M, Macosko C W 1983 Properties of mat reinforced reaction injection molded materials. *Polym. Compos.* 4(3): 190–5
Grigsby R A, Dominguez R J G 1985 Polyurea RIM—A versatile high performance material. *Soc. Plast. Ind.* November: 248–53
Hedrick R M, Gabbert J D, Wohl M H 1985 Nylon 6 RIM. *ACS Symp. Ser.* 270: 135–62
Klosiewicz D W 1983 Plasticized polydicyclopentadiene. US Patent No. 4,400,340
Kolodziej P, Macosko C W, Ranz W E 1982 The influence of impingement mixing on striation thickness distribution and properties in fast polyurethane polymerization. *Polym. Eng. Sci.* 22: 388–92
Kurz J E 1985 Block size and distribution in nylon-6 RIM. *Polym. Process. Eng.* 3: 7–24
Lee L J, Ottino J M, Ranz W E, Macosko C W 1980 Impingement mixing in reaction injection molding. *Polym. Eng. Sci.* 20: 868–74
Lekakou C N, Richardson S M 1986 Simulation of reacting flow during filling in reaction injection molding (RIM). *Polym. Eng. Sci.* 26: 1264–75
Maier U, Menges G 1986 Design of RIM moulds. *Soc. Plast. Eng. (Tech. Pap.)* 44: 1350–5
Malguarnera S C, Suh N P 1977 Liquid injection moulding, I. An investigation of impingement mixing. *Polym. Eng. Sci.* 17: 111–15
Manas-Zloczower I, Macosko C W 1987 Moldability diagrams for reaction injection moulding. *Polym. Process Eng.* 4: 173–84
Müller H, Mrotzek W, Menges G 1985 Mold filling with polyurethane. *ACS Symp. Ser.* 270: 237–58
Nelson D 1986 Reinforcing structural foam RIM polymers with continuous fiberglass mat. *J. Cell. Plast.* 22: 104–21
Noshay A, McGrath J E 1977 *Block Copolymers: Overview and Critical Survey*. Academic Press, New York
Oertel G (ed.) 1985 *Polyurethane Handbook*. Hanser, Munich, Germany
Ranz W E 1979 Applications of a stretch model to mixing, diffusion, and reaction in laminar and turbulent flows. *AIChE J.* 25: 41–7
Richter E B, Macosko C W 1978 Kinetics of fast (RIM) urethane polymerization. *Polym. Eng. Sci.* 18: 1012–18
Sandell D J, Macosko C W, Ranz W E 1985 Visualization technique for studying impingement mixing at representative Reynolds numbers. *Polym. Process Eng.* 3: 57–70
Speckhard T A, Cooper S L 1986 Ultimate tensile properties of segmented polyurethane elastomers: Factors leading to reduced properties for polyurethanes based on nonpolar soft segments. *Rubber. Chem. Technol.* 59: 405–31
Tucker C L III, Suh N P 1980 Mixing for reaction injection moulding, I. Impingement mixing of liquids. *Polym. Eng. Sci.* 20: 875–86
van der Loos J L M, van Geenen A A 1985 Properties and morphology of impact modified RIM nylon. *ACS Symp. Ser.* 270: 181–91
Vespoli N P, Alberino L M 1985 Comparison of rheological measurements during mold filling for glycol and amine extended urethane RIM systems. *J. Elastomers Plast.* 17: 173–82
Vespoli N P, Alberino L M, Peterson A A, Ewen J H 1986 Mold filling studies of polyurea RIM systems. *J. Elastomers Plast.* 18: 159–76
Wickert P D, Ranz W E, Macosko C W 1987 Small scale mixing phenomena during reaction injection moulding. *Polymer* 28: 1105–10
Wilkinson T C, Eckler J H 1986 Advances in structural RIM. *J. Elastomers Plast.* 18: 128–35

P. D. Coates and A. F. Johnson
[University of Bradford, Bradford, UK]

Reclaiming of Elastomers

The economic and social value of recycling elastomers for reuse was recognized if not practiced to a great extent even before vulcanization was discovered. Goodyear's famous invention was not only the real start of the rubber industry but also of the rubber reclaiming industry.

Recycled crumb rubber and reclaimed rubber make up the bulk of recovered elastomeric materials that are presently available to the materials engineer. Unvulcanized and thermoplastic elastomers present no special reclaiming problems since they can be simply chopped, milled and blended with new materials for recycling directly back into the same or equivalent products as those from which they came.

Many elastomers have only limited utility in the virgin state and must be compounded with reinforcing materials and vulcanized to develop necessary physical properties. Vulcanization of an elastomer serves to crosslink chemically the very long chains of a polymer so that their movement is restricted. This changes the polymer's character from a millable, plasticlike material to a dry, nontacky thermoset which, in most cases, cannot be replasticized by simple milling.

Most vulcanized elastomers can easily be mechanically ground to a fine particle size with minimal agglomeration. Mechanical grinding, accompanied by the removal of contaminants such as wire and fiber, was one of the first methods developed for recycling elastomers. It is still widely used today. None of the compounding ingredients contained in the scrap elastomer are removed during mechanical grinding, so the physical properties remain essentially unchanged. The quantity and gradation of the crumb to be used in a new compound is dictated by the properties of the desired finished product. Although the function of recycled crumb rubber in a compound is essentially that of a resilient filler, it does impart the properties of the original scrap to the new compound and is often used to impart elastic properties to plastics and thermoplastics such as asphalt road materials.

Commercial roll mills are widely used in mechanical grinding for economically reducing vulcanized rubbers down to 50 mesh (250 μm) (LaGrone

Table 1
Applications of recycled crumb rubber

Application	Quantity	Purpose
Rubber compounding	3–10%[a]	low cost; improve molding
Brake pads		prevent fade
Plastics extender	3–50%	filler; increase elasticity
Asphalts:		
athletic surfaces		
hot	8–12%	impart resilience
cold emulsion	20–80%	impart resilience
road mixes		
hot plant mix	2%	improve adhesion, reduce rutting and bleeding
chip seal	20–25%	improve adhesion, reduce brittle point
SAMI	20–25%	strain absorbing membrane
joint sealers	20–25%	increase softening point; reduce brittle point

a Up to 30 parts at 20 μm fine crumb is possible

1980). The cryogenic system of Zolin et al. (1977) is capable of producing a fineness of 100 mesh (150 μm) but, due to its higher cost, has generally been limited to recycling very expensive elastomers. A proprietary process that will produce a 500 mesh (20 μm) grind has also been developed (Swore et al. 1980) which, if economically feasible, could expand considerably the possibilities for using vulcanized crumb rubber. Table 1 shows some current applications of recycled crumb rubber.

Automobile and truck tires represent the largest and most easily gathered source of scrap elastomer. It is estimated that over 2×10^8 tires are scrapped each year in the USA alone. Several billion kilograms of other types of rubber scrap also become available for recycling each year.

The vulcanization process is not considered to be truly reversible, although several laboratories have claimed to have achieved "devulcanization" (Nicholas 1979). However, no such process has ever been commercialized. Nonetheless, a rubber reclaiming industry has existed for over 150 years and has produced many billions of kilograms of product commonly referred to as "reclaim" rubber. Smith (1978) has defined reclaimed rubber as the product resulting from the treatment of ground vulcanized scrap rubber tires, tubes and miscellaneous waste rubber by the application of heat and chemical agents, followed by intense mechanical working, whereby a substantial "devulcanization" or regeneration of the rubber component to its original plastic state is effected, thus permitting the product to be compounded, processed and revulcanized. More scientifically, devulcanization is a change in vulcanized rubber which results in a decreased resistance to deformation at ordinary temperatures (Williams 1947). This definition requires only that the vulcanized rubber lose its elastic properties and become less resistant to compression, stretching or swelling.

The chemistry of reclaiming rubber is complex and not fully understood. It is believed to involve depolymerization by chain scission involving oxygen in which none of the combined sulfur nor any of the other compounding chemicals are removed.

A detailed review of the chemistry, processing and compounding of reclaimed rubber has been given by Sverdrup and Wendrow (1970).

In contrast to recycled crumb rubber, the physical properties of most reclaimed elastomers are considerably degraded from those of the original scrap. While the average tensile properties of compounds contained in a passenger or truck tire are in the 17–20 MPa range, reclaimed rubber produced from this scrap would typically only develop ultimate tensile properties in the range of 7–8 MPa. The degradation of properties is not nearly as great when butyl, EPDM or silicone scrap is reclaimed.

Notwithstanding this reduction in physical properties, the utility of reclaimed rubber is unquestioned and restricted only by the requirements of the physical properties of the final products. Thus, reclaimed rubber may constitute all the rubber in certain articles or it can be blended with new rubbers to give a whole range of acceptable properties. Specialty reclaims such as those made from butyl and silicone scrap are generally used in the same type of new compound. However, reclaims produced from compatible scraps can be used for other new products. For example, reclaim produced from a high natural rubber scrap can be used as an extender and processing aid for SBR compounds.

Although reclaimed rubber offers the materials engineer a low-cost source of elastomer, its most outstanding contribution is as a processing aid. Processing advantages of reclaimed rubber are:

(a) a shorter breakdown and mixing time;

(b) low power consumption during breakdown and mixing;

(c) low mixing, calendering and extrusion temperatures;

(d) fast, uniform calendering and extrusion;
(e) better strike-through during calendering;
(f) improved building tack;
(g) less susceptibility to temperature variations;
(h) improved green strength and the firming of uncured stocks;
(i) reduced swelling and shrinkage during extrusion and calendering;
(j) the imparting of low thermoplasticity characteristics; and
(k) a fast cure rate.

Probably in excess of 4×10^9 kg of scrap elastomeric compounds become available in the USA each year. Various reclaiming processes have been developed over the years to accommodate the polymers available and new ones are still being suggested.

Since the inception of the rubber industry, it has been considered prudent to reclaim and reuse scrap elastomers. With the rapid depletion of world sources of raw materials and increasing prices, recycling will make even more sense in the future.

See also: Elastomers: An Overview; Elastomers: Applications Overview

Bibliography

Dufton P W 1987 *The Value and Use of Scrap Tires*. Rubber and Plastics Research Association, Shrewsbury, UK
Dufton P W (ed.) 1988 Waste tyres, problems and opportunities. *Proc. Conf. Rubber and Plastics Research Association* (26 November 1987)
Fix S 1980 Microwave devulcanization of rubber. *Elastomerics* 112(6): 38–40
LaGrone B D 1980 *Proc. 1st Asphalt-Rubber User-Producer Workshop*. Arizona Department of Transportation, Scottsdale, AZ, pp. 39–48
Nicholas P P 1979 Devulcanized rubber composition. US Patent No. 4,161,464 (17 July 1979)
Sittig M 1975 *Resource Recovery and Recycling Handbook of Industrial Waste*. Noyes Data Corporation, Park Ridge, NJ
Smith F G 1978 Reclaimed rubber. In: Babbit R O (ed.) 1978 *The Vanderbilt Rubber Handbook*. Vanderbilt, Norwalk, CT
Sverdrup E F, Wendrow B R 1970 Rubber reclaiming. In: Bikales N M (ed.) 1970 *Encyclopedia of Polymer Science and Technology*, Vol. 12. Interscience, New York, pp. 341–55
Swore R A, Jensen L W, Budzol M 1980 Ultrafine recycled rubber. *Rubber Chem. Technol.* 53: 1215–25
Williams I 1947 Goodyear metal lecture. *Rubber Chem. Technol.* 21: 1
Zolin D J, Frable N B, Savage R W 1977 Cryogenic grinding of cured rubber scrap for reuse in molded parts. *Elastomerics* 109(9): 38–9

B. D. LaGrone
[US Rubber Reclaiming Company, Vicksburg, Mississippi, USA]

Recycling

It is not possible to evaluate with accuracy the total amount of polymer scrap which is discarded each year. However, statistics on the production of plastics are reasonably reliable. In 1990, approximately 100 Mt of plastics were produced worldwide. Various types of packaging account for nearly half the annual production in the industrialized nations and plastics used in these applications are discarded as scrap within a year or less of production. Other plastics, used in applications where a longer service lifetime is expected, eventually add to the accumulation of scrap. Taking all these factors into account, it is quite likely that the worldwide accumulation of polymer scrap each year is roughly half the reported production.

Since plastics are derived from fossil fuel (either oil or natural gas), replacement of scrap with new material represents an additional, and perhaps unnecessary, drain on an important energy resource. Environmental problems associated with the disposal of polymer scrap provide further impetus to recycling efforts. In this article, polymer scrap is considered to be any material which is rejected in the manufacturing process or discarded after performing its original function.

1. Classification of Scrap

It is convenient to classify polymer scrap into preapplication and postapplication scrap. Preapplication scrap includes only material generated at an industrial site—either a polymerization plant or plant where the polymer is fabricated or compounded into a final product. This class of scrap is sometimes referred to as preconsumer scrap, but this term has led to some confusion. Significant amounts of polymer accumulate as scrap from products which, though used by a consumer, still remain the property of the supplying industry. Since products that generate this type of scrap have been in service, this material is here included in the postapplication category.

Preapplication scrap generally consists of a single material. Since it is generated in plant, identity, cleanliness and homogeneity can be controlled. In contrast, postapplication scrap is usually a mixture of different plastics and is generally contaminated with dirt and other waste products. Recycling of postapplication scrap is therefore much more difficult. Postapplication scrap usually ends up in the solid waste stream where it is discarded. Much of the scrap in this category is generated from packaging applications.

For successful recycling of postapplication scrap, polymer articles should be separated from the waste stream before the collection of refuse. Once incorporated into the waste stream, the polymer component is extremely difficult to separate. The amount of polymer in municipal waste varies between 3% and 6% among countries. Only a few commercial processes are in operation for the recycling of postapplication scrap and each of these is based on a source separation of selected polymer items.

Postapplication scrap is further classified into source-separated scrap and scrap already in the solid waste stream. For the latter category, burning as a means of recovering energy is the most likely method of disposal.

2. *Recycling of Preapplication Scrap*

Preapplication scrap originates from a variety of in-plant operations. For example, polymerization processes often yield product batches which are classified as off grade due to failure to meet specifications. Off-color lots are fairly common and other lots may be rejected because they do not meet the specifications for molecular-weight distribution or some other important characteristic. It has been estimated that as much as 14% of polymers produced are rejected as off grade. However, much of this scrap can be used in less demanding applications and so finds its way into useful products.

Another major source of preapplication scrap is fabrication and compounding operations. For example, in filling molds during injection molding, a considerable amount of molten polymer is required to fill channels leading to the mold cavity. This is particularly true for multicavity molds. At the end of the mold cycle, runners, gates and sprues remain attached to the final product and must be removed in a later operation. Most fabricators recover this type of scrap, which is ground or repelletized and then returned to the molding machine, often mixed with various amounts of virgin polymer.

Carefully controlled scrap which originates in plant as imperfect product, as for example from an incompletely filled mold or an off-dimension extrudate, can be recycled in the same way as trimmings. It is now common practice to recycle many types of scrap from fabrication or compounding operations. Those fabricators who do not have equipment to recycle their scrap may sell the product to companies which specialize in the purchase of scrap. These scrap dealers may fabricate the scrap into finished articles or they may convert it into pellets for resale.

Purging of processing equipment provides another source of preapplication scrap. Molten polymer is often passed through the equipment at the start of an operation to clean the machine or when changing from one polymer to another. Scrap collected in this way, however, is likely to be contaminated with partly degraded material or it may be a mixture of more than one polymer. Therefore, machine purgings are not usually recycled.

Scrap which is derived from thermosetting resins cannot be recycled by thermoforming, since these materials are infusible after the final step of polymerization. However, thermosetting scrap can be reground and used as filler for new thermosetting resins. Only preapplication scrap which is derived from thermoplastics can be recycled by thermoforming.

3. *Recycling of Source-Separated, Postapplication Scrap*

Several processes have been developed for the recycling of polymer scrap that has been separated before reaching the industrial or municipal waste stream. Scrap in this category is generated in manufacturing operations or from programs designed to source separate the polymer. However, only those articles which are easily identified are likely candidates for the latter approach. Polymer beverage bottles are an example of a product that lends itself to source separation.

3.1 Recycling of Plastics from Scrap Wire and Cable

Several technologies are available for the recycling of insulation and jacketing from scrap wire and cable products. The primary purpose of these processes is to recover copper or aluminum used as the conductor. Environmental constraints now restrict burning of the polymer components to recover and reuse the metal conductor material. As a result, insulation and jacketing are removed by nondestructive processes. Polymer recovered in this way can be further purified for reuse in secondary applications.

Poly(vinyl chloride) (PVC) from scrap wire and cable can be dissolved in appropriate solvents and the residual, metallic component then refined for further use. Plasticizers, dyes and other organic additives also go into solution with the PVC. Evaporation of the solvent then yields a crude mixture of PVC contaminated with these additives. After minimum further cleaning, this recovered polymer is suitable for less demanding products such as floor tile or garden hose. Alternatively, the PVC can be precipitated from solution, filtered off and dried. This method yields a relatively pure PVC that can be reformulated with plasticizers, fillers, stabilizers and colorants for recycling into a variety of useful products.

Insulation formed as a tubular sleeve over the conductor, that is, without adhesion to the metal, can be separated by shaking short sections of the wire on a vibrating table or in a fluidized bed. The polymer must be further treated to remove the last traces of metal before it can be reformulated and processed into new products.

3.2 Recycling of Discarded Beverage Bottles

Considerable attention has been directed to the possible recycling of discarded packaging materials.

Although many of these can be identified and perhaps isolated from the municipal waste stream, only beverage bottles made from poly(ethylene terephthalate) (PET) have been successfully recycled. In 1982, approximately 11×10^6 kg of PET were recovered and recycled in the USA. The process starts with accumulation of discarded bottles, usually from the areas where a deposit law stimulates return of specific types of bottle. Densification to 270–320 kg m^{-3} is advisable for economical shipment to collection centers. Bottles may be compacted or roughly ground before shipment. Since many bottles used to package household chemicals are made from other polymers, it is important that these should not be collected with PET bottles. Public-sponsored collection systems usually result in a mixture of PET with other polymers, making recycling much more difficult.

Scrap PET must first be washed thoroughly, since residual contents can interfere with the recycling process. Bottles may be washed intact, but preferably after crushing or grinding. After washing and drying, the clean PET scrap can be reprocessed to supply established markets.

The major outlet for recovered PET is fiberfill or batting, which utilizes almost half the scrap. PET scrap, often mixed with virgin material, is spun into a fiber which is used as a filler in pillows and similar articles. It is marketed in various grades, depending on the quality or color of the scrap used. PET scrap from industrial sources is often used to supplement the scrap obtained from recovered bottles.

The second important outlet for clean PET scrap is in the production of strapping. In this application, scrap is normally blended with virgin PET to obtain the high strength required. Another small, though increasing, use for PET scrap is in the production of engineering resins for molding. Here, the scrap may be blended with other molding compounds including nylon and poly(butylene terephthalate). PET scrap may also find an outlet in the unsaturated polyester industry. For this application, the scrap would be broken down by hydrolysis or alcoholysis into intermediates such as phthalic anhydride or isophthalic acid.

Several large companies are now engaged in the recycling of PET from discarded beverage bottles and others are investigating the process and may soon enter the recycling field.

3.3 Recycling of Plastics from Junked Automobiles

Scrap iron and steel are recovered from junked automobiles and there is also some recovery of nonferrous metals. Discarded, usually crushed automobiles, after removal of gas tanks, radiators and tires, are fed into machines which shear them into fist-sized pieces. The ferrous metal fraction is separated by magnetic devices and nonferrous metals are recovered from the residue by heavy medium separation or other techniques. There is also a light or fluff fraction which consists of thermoplastics, thermosets, glass, cellulosics and a large quantity of dirt. This fluff fraction is an important potential resource for recycling. Although its heterogeneity poses a difficult challenge, the quantity of polymer scrap from this source is potentially enormous: in the USA alone some 10^7 automobiles are discarded annually and it has been predicted that as much as 135 kg of polymers may be used in future cars. An annual production of greater than 1 Mt of scrap from this source would be a reasonable prediction. As a result, considerable interest is being shown in recovering some value from this material.

Polyurethane foam, which constitutes a large part of the fluff fraction, has been isolated by an aeration process. The recovered polyurethane can be recycled by hydrolysis or alcoholysis into monomers from which new polymer could be prepared. Recovery of other plastics from the fluff fraction has shown little promise so far. Attempts are underway, however, to form composites from the whole fluff fraction through the use of inexpensive bonding agents. Such composites might be used as wood or cardboard substitutes. Burning of the fluff fraction for its fuel value is also a viable alternative.

4. Recycling of Postapplication Scrap from the Municipal Waste Stream

Polymer scrap which has become part of the solid waste stream is extremely difficult to recover for recycling. When it is recognized that plastics represent only 3–6% of the total waste, it is difficult to conceive of a process for its removal. However, a few waste treatment facilities in Europe and in Japan, which are capable of isolating a polymer-enriched fraction from municipal waste, have been reported, but these are still experimental and very little polymer is being recovered. Furthermore, contamination poses difficult problems for reuse of the mixed plastics obtained from municipal waste.

The most likely outlet for postapplication scrap which cannot be source separated is as an alternative or supplemental fuel. Since synthetic polymers are derived from fossil fuel, their heat content is very high. Polyolefins, for example, have about the same heating value as crude oil. Other plastics, however, may present serious problems in a combustion process. Poly(vinyl chloride) liberates large volumes of hydrogen chloride when burned, which can cause extensive corrosion of grates and flues. Other plastics release toxic products during combustion. Nonetheless, the polymer component of municipal solid waste provides an effective fuel to assist in the burning of other waste components in the generation of electricity.

5. Conclusion

Considerable progress has been made in the recycling of polymer scrap. Preapplication scrap is almost completely utilized. Source-separated post-application scrap also lends itself to recycling processes and there are several notable examples of successful recycling of this type of scrap. There are in addition several minor processes in operation for the recycling of various plastics, but these add very little to the overall volume of scrap being reused.

For a polymer recycling process to become successful, four important requirements must be met.

(a) A continuous and reasonably consistent source of scrap must be available.

(b) Technology must exist for cleaning and refining the scrap.

(c) There must be market outlets for the recycled product.

(d) The use of recycled scrap must provide an economic advantage over virgin material for each selected application.

See also: Reclaiming of Elastomers

Bibliography

Alpher J, Nelson G L 1989 The promise of recycling. *Polymeric Materials*. American Chemical Society, Washington, DC
Kaminsky W, Menzel J, Sinn H 1976 Recycling of plastics. *Conserv. Recycling* 1: 91–110
Kaufman F S 1975 Opportunities for plastics in resource recovery. *Resource Recovery and Utilization*, STP 592. American Society for Testing and Materials, Philadelphia, PA, pp. 122–31
Leidner J 1981 *Plastics Waste—Recovery of Economic Value*. Dekker, New York
Menzel J, Perkow H, Sinn H 1973 Recycling plastics. *Chem. Ind. (London)* 1973 (12): 570–3
Milgrom J 1972 Incentives for recycling and reuse of plastics. *Report for US Environmental Protection Agency*. Arthur D Little, Cambridge, MA
Scott G Some chemical problems in the recycling of plastics. *Resour. Recovery Conserv.* 1: 381–95
Tillman D A 1975 Fuels from recycling systems. *Environ. Sci. Technol.* 9(5): 418–20

W. L. Hawkins
[Plastics Institute of America, Hoboken, New Jersey, USA]

Residual Stresses

Residual stresses develop in polymers during fabrication as the result of differential shrinkage. This is generally caused by differential cooling rates during molding operations and with thermosets may be associated with the curing process as well. Residual stresses cause distortion when they are not balanced. They may influence the fracture behavior of a molded artifact; when the stresses are compressive near the surface their effect may be beneficial. In fiber-reinforced polymers, shrinkage stresses increase the frictional force between fiber and matrix and enhance the toughness of the composite.

Residual stresses can be modified significantly by postmolding conditioning, including annealing and outdoor weathering and even aging indoors at ambient temperature. A temperature gradient can produce significant unwelcome changes. Other property-controlling characteristics also depend on the thermomechanical history of the polymer, and the relative importance of the residual stresses is sometimes difficult to establish.

1. Origin of Residual Stresses

1.1 Thermoplastics

Thermoplastics are shaped at elevated temperature in the melt state and then cooled. In molding operations, the tool is held below the solidification temperature and the material begins to solidify as soon as it comes into contact with the mold wall. A solidified skin quickly forms, lining the mold cavity. Polymers are poor thermal conductors so a steep temperature gradient develops and the solidification front moves slowly into the material. In injection molding, the molten material in the center of the mold cavity cools slowly and contracts. The injection system continues to force more material into the mold to compensate for the contraction until the gate at the entrance to the mold freezes off. Once this has happened no more topping up can occur. The material in the interior ("core") continues to cool, but shrinkage is opposed by the solidified skin and the core goes into a state of hydrostatic tension, causing the skin to compress to maintain internal force equilibrium. Sometimes the shrinkage stress in the core is sufficient to cause voiding or the production of sinking, partially relieving the stresses locally. Thus, in an injection molding the stresses are normally tensile in the interior and compressive near the surface.

Similar temperature gradients develop during other molding operations, although detailed differences may occur. With blow molding, cooling takes place inside the mold (as in injection molding) and the thermal transfer characteristics at the outer surface of the molding are similar. The inner surface of the hollow article is likely to cool more slowly, causing a nonsymmetric stress distribution to form. The cooling rate at the inner surface can be increased by forced cooling, using a cold vapor spray. The residual stress distribution then becomes more nearly symmetric and a faster production rate can be achieved because the overall cooling rate is increased, permitting more rapid demolding. With profile extrusion, cooling occurs in a shaping die, in air or in a water bath. As with injection molding, cooling proceeds from the outside inwards and so the outer skin sets before the interior has cooled below the solidification temperature. The interior develops hydrostatic tensile stresses, causing compressive stresses to form near the surface. These stresses have both axial and hoop components. In the case of pipe production, the material near the outer surface will normally cool more quickly than that near the inner surface where tensile stresses may form. This can be altered by applying forced cooling in the bore of the pipe, whence tensile stresses form in the wall interior and compressive stresses are produced near both inner and outer surfaces.

Temperature gradients also occur during the formation and drawing of fibers, producing residual stresses similar in sense to those obtained by profile extrusion. Fibers spun from solution contain stresses of a similar sense, though a different mechanism is responsible. A skin forms first and the subsequent loss of solvent from the interior produces shrinkage, leading to tensile stresses in the core and compressive stresses near to the surface. The stresses are largely relieved by buckling of the skin.

The shrinkage obtained when a noncrystalline polymer cools through a large temperature range is significant, but even greater shrinkage may be observed with semicrystalline polymers because of the greater molecular packing efficiency normally obtained in crystals. This does not usually result in higher stresses, probably because of the ease of relaxation in the amorphous phase at room temperature.

1.2 Thermosets

Considerable shrinkage accompanies curing in thermosets. Shrinkage alone does not generate residual stress; there must be differential shrinkage. If solidification proceeds from the surface inwards, as with thermoplastics, then a similar stress distribution will develop. The curing process is normally too slow for the temperature gradient present initially to have much influence, however, except perhaps with larger section thicknesses. Thus solidification usually proceeds fairly uniformly throughout the material and no significant contribution to residual stress is derived from this source. A temperature gradient often develops during curing as a result of the exothermic nature of the reaction. This may lead to residual stresses when the material is restored to room temperature equilibrium, as a result of differential shrinkage. The highest temperature is attained in the center of the molding, which will therefore shrink the most; again the result is a stress distribution that is tensile in the interior and compressive near to the surface.

So far, consideration has been given only to the self-stressed state developed in an isolated body. Thermosets are often used in contact with other materials, for example, as coatings and adhesives. When this is so, shrinkage of the resin during curing is restrained by the substrate to which it adheres, so tensile stresses develop within the resin. Changes in service temperature lead to changes in stress as a consequence of differential thermal expansion between resin and substrate. Similar circumstances can arise with thermoplastics, although they are less frequently associated with service failure.

1.3 Polymer Composites

Polymer composites based on thermoplastics are generally processed in similar ways to unfilled grades. The most important type is the injection-molding resin reinforced with short fibers (normally glass or carbon). Solidification inwards from the surface occurs as before, leading to residual stresses that are tensile in the interior and compressive near the surface. The fibers become oriented during mold filling and cause considerable anisotropy in the molding. Strong anisotropy in residual stress has been measured in glass-fiber-reinforced polypropylene moldings. The origin of this anisotropy is not certain, but it can be explained if an entropy-elastic contribution is present. This will be directional and additional to the energy-elastic effect produced by thermal shrinkage. The implication is that some molecule segments remain extended after solidification, exerting a stress along the extension axis. The unfilled polymer does not show the same degree of residual stress anisotropy even though molecular orientation occurs during mold filling. During cooling the molecule segments probably become pinned, either by direct adhesion to the filler particles or through connection with the crystals. These noncrystallized segments are likely to be held in oriented configurations.

In addition to the long-range stresses already dealt with, there are local stresses that form when the composite cools as a result of differential thermal contraction between fiber and matrix. The polymer has a much greater thermal expansion coefficient than the fiber so that the fiber is in compression. Thus the residual stresses will enhance the opposing friction force during pull-out and so increase toughness.

In a similar way the shrinkage stress that develops in a thermosetting matrix provides toughening in fiber composites. Thus, short-range stresses surrounding the fibers are generally of much greater importance than the overall long-range stresses, which are small.

2. Prediction of Residual Stresses

A review of the attempts to calculate the distribution of residual stress in quenched flat plates has been presented by Isayev and Crouthamel (1984). Much of the original work was conducted with reference to inorganic glasses but has been adopted to model the solidification of a thermoplastic in a cold mold. Each calculation contains simplifying assumptions to make the computation manageable and most include the assumption, or an equivalent assumption, that the viscosity is infinite below the grass-transition temperature and zero above it. If no account is taken of the viscoelastic nature of the material, the through-thickness profile is predicted to be parabolic:

$$\sigma_i(z_1) = A(\tfrac{1}{3}z_0^2 - z_1^2) \qquad (1)$$

where $\sigma_i(z_1)$ is the residual stress in the xy plane at depth $z = z_1$ from the center of the plate, thickness $2z_0$, and A depends on the thermoelastic properties of the material and the cooling conditions. Several theoretical studies have attempted to take account of the viscoelastic properties of the material and have indicated how the measured departures from the parabolic form may arise. The calculated profiles cannot be used confidently with polymers, however, especially in the case of crystallizing polymers. No satisfactory treatment of the effect of the volumetric changes that accompany crystallization has yet been presented. Furthermore, no theories have yet been developed that account adequately for the stress profiles measured in injection moldings made from thermosets or from reinforced thermoplastics.

After the flat plate, the shape that has received most attention for analysis of this kind is the cylindrical pipe. A large volume of polymer is processed into pipe for the transmission of a variety of liquids and gases and it is believed that residual stresses can have a significant influence on service lifetime. This geometry is almost as amenable to treatment as the flat plate, although the same difficulties regarding the introduction of material properties are present. Williams (1981) has presented a treatment in which time-dependent properties are included and predicts that substantial tensile stresses form near to the inner surface when conventional extrusion conditions are used.

3. Measurement

3.1 The Layer Removal Procedure

The most widely used method for measuring residual stresses in polymer moldings is the layer removal procedure. This technique provides the stress distribution through the thickness of the test piece, but is suitable only for parallelepipeds and artifacts with cylindrical symmetry. With parallelepipeds, thin layers are removed using a high-speed milling machine, preferably with a fly-cutting action. After removal of a layer, and following the removal of the molding from the machine bed and release of all external forces, the residual stresses redistribute to restore force equilibrium. This results in changes in the linear dimensions and also produces bending because material is removed from one surface only. It is generally possible to measure the curvature with much more accuracy than the length change. Thus, the preferred method consists of removing thin layers (say 0.05–0.1 mm thick) and measuring the curvature at each removal. If it is known that the stresses are symmetrical it is sufficient to terminate this process on reaching the center of the molding. If the stresses are likely to be biaxial the curvature should be measured in the two principal directions, that is, along and perpendicular to flow in an injection-molded plaque. Methods of measuring curvature have been described by Haworth et al. (1982). With flat moldings possessing biaxial stresses, pairs of bars can be extracted, one with its axis parallel to the flow direction and the other with its axis perpendicular to it. Curvature measurements are then made on both bars and the results used in a full biaxial stress analysis.

The data obtained in this way are used to produce plots of the curvature $\rho(= 1/\text{radius of curvature})$ against the thickness of material removed $(z_0 - z_1)$. When a biaxial analysis is required, two curvature plots are made, represented by $\rho_x(z_1)$ and $\rho_y(z_1)$. These plots are then converted to a stress–depth profile using a formula presented by Treuting and Read (1951) for residual stress analysis in metals:

$$\sigma_{i,x}(z_1) = \frac{-E}{6(1-\nu^2)} \times \left\{ (z_0 + z_1)^2 \left[\frac{d\rho_x(z_1)}{dz_1} + \frac{\nu\, d\rho_y(z_1)}{dz_1} \right] + 4(z_0 - z_1)[\rho_x(z_1) + \nu\rho_y(z_1)] - 2\int_{z_1}^{z_0} [\rho_x(z) + \nu\rho_y(z)]\, dz \right\} \qquad (2)$$

where $\sigma_{i,x}(z_1)$ is the residual stress in the x direction in the plane z_1 from the midplane of the bar prior to layer removal, E is Young's modulus and ν is Poisson's ratio, both assumed to be isotropic in the xy plane and uniform throughout the thickness.

If the stresses are equibiaxial ($\sigma_{i,x} = \sigma_{i,y}$) as in a carefully made compression-molded plaque, then the curvature is the same in all directions (equal to ρ) and Eqn. (2) becomes

$$\sigma_{i,x}(z_1) = -B\left[(z_0+z_1)^2\frac{d\rho}{dz_1} + 4(z_0+z_1)\rho - 2\int_{z_1}^{z_0}\rho\, dz\right] \quad (3)$$

where $B = E/[6(1-\nu)]$.

If the residual stresses are uniaxial (parallel to x) and ρ is measured in this direction then Eqn. (3) is still applicable, but with $B = E/6$. Finally, in some cases the curvature in one direction is very small (e.g., $\rho_y = 0$), so Eqn. (2) again reduces to Eqn. (3), this time with $B = E/[6(1-\nu^2)]$.

There are two major problems with the layer removal procedure. The first is that Young's modulus is not usually uniform, contrary to one of the assumptions upon which Eqn. (2) is derived. Significant variations in Young's modulus are present in injection moldings and can result from nonuniformity in molecular orientation, crystallinity and fiber orientation. It has been shown by White (1985) that the general conclusions drawn from layer removal studies are valid, although some of the stress magnitudes are in error. The second problem is that stress measurements cannot be made within a layer removal increment. Thus, no accurate measurement can be made within the first 0.1 mm or so from the surface, which is unfortunate since the stress often changes very steeply there. A further related problem exists if the molded surface is not truly planar, as when sinking is present. Then the first layer removed is not uniform and its average thickness is not known, adding further uncertainty to the stress measurement near the surface.

Layer removal can also be used to assess residual stresses in cylindrical bars or pipes. Layers can be removed from the outside or from the bore and measurements of the consequent length change made. Better accuracy is possible than with length measurement in bars because of the absence of bending, but the changes in length are generally quite small and the large thermal expansion coefficient of polymers leads to large errors unless careful temperature control is exercised. Inaccuracy is also introduced if the molding possesses any ellipticity or if it is not accurately centered during machining. This method reveals the axial stresses but for a complete analysis, including hoop stresses, it should be combined with the split ring method.

3.2 Split Ring Method for Pipe

A useful method for assessing the residual stress levels in pipe is the split ring method described by Williams et al. (1981). This involves slicing a ring-shaped section from the pipe, then making an axial cut through the wall and allowing the split ring to bend so as to equilibrate the hoop stresses. The sense of bending indicates the sense of the stresses and the amount of bending, conveniently measured in terms of the displacement of the cut surfaces, depends on the stress magnitude. To obtain the through-thickness stress distribution, sections must be taken for various depth removals, again using removals either from the outside or from the bore. A continuous stress distribution can then be derived in a manner analogous to that described in Sect. 3.1.

3.3 Hole Drilling

A popular technique for measuring residual stress in metals is to drill a hole and measure the strain produced at several adjacent locations when the material adjusts to its new stress equilibrium state. The strain is measured using a standard array of strain gauges cemented to the surface and using a centrally positioned hole of standard size. This method has been less widely used with polymers, probably because of the difficulty of finding a suitable adhesive with which to attach the strain gauges and because of the difficulty in drilling a reproducible hole without introducing stresses during the machining process. Some attempts at using this method are described in the review by Ito (1981).

3.4 Other Methods

The polarizability of a chemical bond changes when it is stressed, giving rise to birefringence that is related to the stress field. If favored bond orientations are present this also produces birefringence. In an injection molding the orientation effect is normally dominant even when quite high residual stresses are present. It is difficult to separate the contributions of stress and orientation and even a most careful (and time-consuming) procedure developed by Saffell and Windle (1980) has been criticized as inadequate.

Barrett and Predecki (1976) have described a method of residual-stress determination based on x-ray lattice parameter measurements on small crystalline particles embedded in the molding resin. This technique appears to be best suited to thermosets used in casting and hand-layup artifacts.

4. Effect of Processing Conditions on Residual Stress

When a polymer melt is quenched, the residual stress magnitudes are increased by raising the initial temperature or by lowering the quench temperature. The sensitivity is much greater to changes in the quench temperature than to changes in the initial temperature (Isayev and Crouthamel 1984). If quenching is sufficiently rapid, stresses as high as $10\,MN\,m^{-2}$ tensile (at the center) and $40\,MN\,m^{-2}$ compressive (near the surface) can be obtained with unfilled polymers, including poly(methyl methacrylate). The magnitudes are usually less than those calculated from theory and the stress distribution profile is not identical to that predicted. The departures are attributed to nonlinear effects not taken account of in the theories or to stress relaxation after solidification. The thermal transfer conditions obtained in injection molding lead to much lower stress levels, usually less than $5\,MN\,m^{-2}$ near the center (more usually $0.5–2.0\,MN\,m^{-2}$) and compressive stresses as high as two or three times these values near the surface. Higher tensile stresses have been observed in filled polymers, values of $20\,MN\,m^{-2}$ being common with glass-fiber-filled polypropylene (White 1984). Residual stress anisotropy is often observed with injection moldings and is much greater when fillers are present (see Sect. 1.3). For example, in glass-fiber-filled polypropylene injection-molded plaques for which a maximum tensile residual stress of over $20\,MN\,m^{-2}$ was measured in the flow direction, the maximum was approximately $7\,MN\,m^{-2}$ perpendicular to flow (White 1984). Thus, in injection molding, the tool temperature and the melt temperature are important, controlling the thermal residual stress, but in addition other factors, such as injection and flow-path geometry, which influence molecular orientation, may influence the orientation-related contribution. The orientation component is likely to be significant in polymers only above their glass-transition temperature, where the entropy-elastic stresses are active. In glassy polymers the orientation does not produce an active stress, though the presence of orientation just prior to freezing may cause directional shrinkage and, as a consequence, modest residual stress anisotropy.

5. *Postmolding Conditioning*

The residual stress distribution present at the end of the molding operation is often subject to modification. Prolonged room temperature aging causes substantial reduction in the stress levels in polypropylene (falling to about 50% of the as-molded values) but less reduction in glassy polymers and in reinforced polypropylene (falling to 80–90% of the as-molded values). Annealing at elevated temperature accelerates the process and may cause further reductions, but the stresses are rarely removed altogether. A temperature gradient causes the stress to become imbalanced and when uniform temperature is restored the compressive stresses near the surface that was formerly at the higher temperature are reduced and may even reverse, becoming tensile (Thompson and White 1984). Similar reversal has been observed in polystyrene and poly(vinyl chloride) on outdoor weathering in an extreme climate (Qayyum and White 1985). The absorption of a liquid may modify residual stresses as a result of nonuniform swelling or plasticization (permitting relaxation of stress) and, in the case of Nylon 6, Russell and Beaumont (1980) suggest that at elevated temperatures water promotes crystallization which in turn causes the residual stress distribution to change.

6. *Effect of Residual Stresses on Service Behavior*

Distortion, especially warping of flat sections, occurs when residual stresses are not balanced. This happens if cooling is nonuniform, as when opposite faces of an injection-molding tool are at different temperatures, or may develop if a temperature gradient is subsequently applied. Balanced stresses do not cause distortion, even if they are large, but relaxation over a prolonged period is more likely to result in imbalance if the as-molded stresses are high. Balanced relaxation may cause changes in dimensions.

Tensile residual stresses are normally located in the interior and are usually small compared to the (average) breaking strength. Thus, they rarely increase the likelihood of fracture substantially. The presence of compressive stresses (often in the range $20–40\,MN\,m^{-2}$) near the surface will inhibit crack growth from a surface flaw. On the other hand, if tensile stresses are present near the surface, as at the inner surface in a conventionally extruded pipe or in a molding subjected to a temperature gradient or an extreme weathering treatment, a significant loss in strength can be expected. Experiments by So and Broutman (1976) and Broutman and Krishnakumar (1976) claim to have shown that residual stresses influence impact behavior. The different thermal treatments used to produce different residual stress profiles, however, must have created very different states of physical aging and this may be an alternative explanation for the observations. Toughening of composites as a result of enhancement by shrinkage stresses of the friction force opposing fiber pull-out is, however, generally accepted to occur.

Environmental failure occurs when an aggressive liquid (e.g., an alcohol for many polymers) is in contact with a polymer under stress. Residual stresses may provide the driving force. For example, holes drilled into injection-molded crash helmets to provide chin-strap fixtures expose regions of tensile stress in which crazes grow readily when contact is made with a crazing agent even without application of an external load.

Shrinkage of coatings provides tensile forces that may be sufficient to cause peeling from the substrate.

Thus, residual stresses have an important

influence over properties and serviceability of polymer artifacts, though their importance relative to that of physical aging has not been established in all circumstances. Some measure of control over their magnitude can be exercised during fabrication. The stresses, however, may become modified in storage or in service and so an assessment of the long-term effect of the environment (thermal, mechanical and chemical) should always be made with any critical component.

See also: Injection Molding of Thermoplastics: Mold Filling; Injection Molding of Thermoplastics: Structural Foams; Injection Molding of Thermoplastics: Structure Development; Processing: Relation to Product Properties; Processing: Shear-Controlled Orientation; Processing: Structure Development

Bibliography

Barrett C S, Predecki P 1976 Stress measurement in polymeric materials by x-ray diffraction. *Polym. Eng. Sci.* 16: 602–8

Broutman L J, Krishnakumar S M 1976 Impact strength of polymers: 1. The effect of thermal treatment and residual stress. *Polym. Eng. Sci.* 16: 74–81

Haworth B, Hindle C S, Sandilands G J, White J R 1982 Assessment of internal stresses in injection-molded thermoplastics. *Plast. Rubber Proc. Appl.* 2: 59–71

Isayev A I, Crouthamel D L 1984 Residual stress development in the injection molding of polymers. *Polym. Plast. Technol. Eng.* 22: 177–232

Ito K 1981 Residual stress and strain in moulded plastic products with dimensional stability and environmental cracking. *Rep. College Eng. Hosei Univ.* 21: 1–87

Qayyum M M, White J R 1985 Weathering of injection-moulded glassy polymers: Changes in residual stress and fracture behaviour. *J. Mater. Sci.* 20: 2557–74

Russell D P, Beaumont P W R 1980 Structure and properties of injection-moulded nylon 6, Part 2. Residual stresses in injection-moulded nylon 6. *J. Mater. Sci.* 15: 208–15

Saffell J R, Windle A H 1980 The influence of thermal history on internal stress distributions in sheets of PMMA and polycarbonate. *J. Appl. Polym. Sci.* 25: 1117–29

So P, Broutman L J 1976 Residual stresses in polymers and their effect on mechanical behavior. *Polym. Eng. Sci.* 16: 785–91

Thompson M, White J R 1984 The effects of a temperature gradient on residual stresses and distortion in injection moldings. *Polym. Eng. Sci.* 24: 227–41

Treuting R G, Read W T Jr 1951 A mechanical determination of biaxial residual stress in sheet materials. *J. Appl. Phys.* 22: 130–4

White J R 1984 Origins and measurement of internal stress in plastics. *Polym. Test.* 4: 165–91

White J R 1985 On the layer removal analysis of residual stress, Part 1. Polymer mouldings with depth-varying Young's modulus. *J. Mater. Sci.* 20: 2377–87

Williams J G 1981 On the prediction of residual stresses in polymers. *Plast. Rubber Proc. Appl.* 1: 369–77

Williams J G, Gray A, Hodgkinson J M 1981 The determination of residual stresses in plastic pipe and their role in fracture. *Polym. Eng. Sci.* 21: 822–8

J. R. White
[University of Newcastle-upon-Tyne, Newcastle-upon-Tyne, UK]

Rheological Characterization of Molten Polymers

To understand polymer processing operations, an understanding of the flow or rheological properties of the material is needed. Rheological measurements on molten plastics and elastomers are, in general, subject to great difficulties. Experiments must be carried out at elevated temperatures and sometimes at elevated pressures. In many cases, special precautions with the environment or with removal of impurities must be made. Not only rheological principles but also the peculiar characteristics of the systems to be investigated must be grasped.

The instruments used to characterize these materials are of two general types, as is the case with most classes of industrial materials: scientific instruments intended to make basic measurements of rheological properties and rougher simpler apparatus which is primarily used for industrial quality control. Both of these are of concern here.

1. General Observations of Kinematics and Processing

Various types of rheological instruments may be used to investigate the response of polymer melts during flow. These may be classified according to the kinematics of the flow. Generally, the flows of interest may be divided into shear and elongational flows.

A "123" cartesian coordinate system is first established. A shear flow may be specified by

$$v_1 = \dot{\gamma} x_2, \qquad v_2 = v_3 = 0 \tag{1}$$

where $\dot{\gamma}$ is a velocity gradient or shear rate. The primary stress response to this flow is the shear stress σ_{12}, which is related to the shear rate $\dot{\gamma}$ by the viscosity η:

$$\sigma_{12} = \eta \dot{\gamma} \tag{2a}$$

though nonisotropic normal stresses σ_{11}, σ_{22} and σ_{33}

usually develop. Two independent normal stress differences may be defined:

$$N_1 = \sigma_{11} - \sigma_{22}, \qquad N_2 = \sigma_{22} - \sigma_{33} \qquad \text{(2b)}$$

In elongational flows,

$$v_1 = E_1 x_1, \qquad v_2 = E_2 x_2, \qquad v_3 = E_3 x_3 \qquad \text{(3a)}$$

where the E are elongation rates, which are restricted by

$$E_1 + E_2 + E_3 = 0 \qquad \text{(3b)}$$

Different classes of elongational flow may be distinguished. First, there is uniaxial extension, for which

$$E_2 = E_3 = -\tfrac{1}{2}E_1 \qquad \text{(4a)}$$

Here stresses σ_{ii} develop, but σ_{11} is the primary stress of interest. σ_{22} and σ_{33} are generally equal to the pressure of the environment. σ_{11} is related to E_1 through an elongational viscosity χ:

$$\sigma_{11} = \chi E_1 \qquad \text{(4b)}$$

Multiaxial stretching and flows may also be considered. Equal biaxial extension may be defined:

$$E_1 = E_2, \qquad E_3 = -2E_1 \qquad \text{(5a)}$$

The stresses developed are σ_{ii} with σ_{11} equal to σ_{22}. In planar extension, there is no deformation in one direction:

$$E_1 = -E_2, \qquad E_3 = 0 \qquad \text{(5b)}$$

The major purpose of industrial rheological measurements is the interpretation of flow in polymer processing operations and the forming of products from polymer systems. The application of measurements of the type described previously to these operations is now considered.

In most processing operations, the polymer is pumped through the action of a rotating screw, that is, a screw pump. The transport in the fully melted polymer in the metering region of the screw extruder, which is furthest from the hopper, is basically a shearing flow. Analysis of transport in a screw involves an understanding of solid transport, melting and rheological behavior. However, only shear flow behavior need be understood and applied. The output of the extruder screw generally proceeds through a die. Flow of the melt in the die as well as in the metering region of the screw is dominated by the contact and adhesion of the melt with the solid walls. The motion of the melt is determined by the application of pressure or the relative motion of the walls. This again causes the motions to be basically shearing in character. An exception is in the region where the melt converges into the narrow die channel. Here there are both shearing and uniaxial stretching motions. Near the centerline of the die, the shearing behavior disappears because of symmetry and uniaxial stretching dominates.

Beyond the die exit, different types of subsequent history may occur. In injection molding, the melt is injected into a mold. Again the walls dominate and shearing flow occurs. As in the extruder screw, the flow may be highly nonisothermal. This is due to cooling of the melt by cold mold walls and heating by viscous dissipation.

In profile and pipe extrusion processes, the melt emerges into the air, where it undergoes a "recovery" process due to the release of the applied stress along the die wall. The resulting swelling can result in changes of the profile shape.

In many processing operations, tensions are applied to the emerging melt in order to change its shape. The drawdown in melt spinning of fibers may be classified as such a process, as well as the combined drawdown and blowup of the annular melt sheet during tubular film extrusion. In some processes such as blow molding and thermoforming, the emerging melt is allowed to undergo an unstressed "recovery" period as in profile extrusion, but within a short time interval is subjected to applied tensions which stretch it into a new shape. Melt spinning involves a uniaxial extension process, while tubular film extrusion, blow molding and thermoforming involve multiaxial stretching. Understanding of the rheological response in these flows requires the study of elongational flow behavior. It is also to be noted that all these elongational flow processing operations are highly nonisothermal because of cooling of the emerging melt by the environment.

Processing problems vary with the type of process. For design of "internal flows," as in extrusion and injection molding, the shear viscosity and its shear rate and temperature dependence are required. In extrusion, there are also problems with varying extrudate swell levels and extrudate distortion, which seem related to melt elasticity and "memory." Fiber and film extrusion moreover have fluctuations in diameter or thickness related to system response to input disturbances (and sometimes the "draw resonance instability"). This may be traced to the elongational flow properties of polymer melts and the cooling conditions during drawdown.

2. *Shearing Flow Instruments*

The simplest of viscometers used for polymer melts are the parallel plate and sandwich instruments (see Fig. 1). In this type of instrument, a very viscous

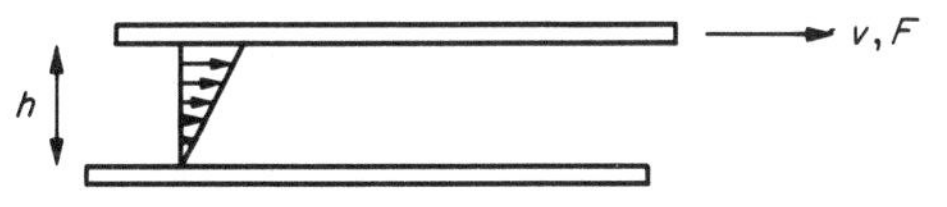

Figure 1
Parallel plate viscometer

fluid, generally an elastomer, is sheared between two parallel steel plates or in a sandwich between three plates. The shear rate is $\dot{\gamma} = v/h$ where v is the velocity of the moving member and h is the interplate distance. If the moving member is programmed at a constant velocity, a constant shear rate is obtained. The applied force F may be determined as a function of time. This yields, after normalization with area A, the shear stress as a function of time. The advantage of this type of instrument is the ability to achieve very low shear rates for materials which generally exhibit slippage in nonpressurized rotational instruments such as those based on the cone–plate geometry.

The coaxial cylinder viscometer (see Fig. 2) is one of the most important of rheological characterization instruments. It is one of the oldest of viscometers, having been developed and widely used in the late nineteenth century. The coaxial cylinder can involve either a rotating inner cylinder (bob) or an outer cylinder (cup). This instrument was first applied to polymer systems by Mooney in the 1930s, who reported the first true viscosity shear rate data obtained on such a material (natural rubber) to appear in the literature. The shear stress σ_{12} at a surface of the sensing member with radius r is simply related to the couple or torque T through

$$T = 2\pi r^2 l \sigma_{12}(r) \tag{6}$$

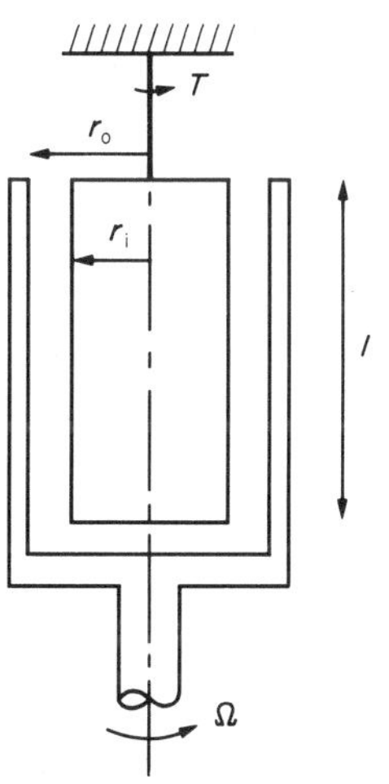

Figure 2
Coaxial cylinder viscometer

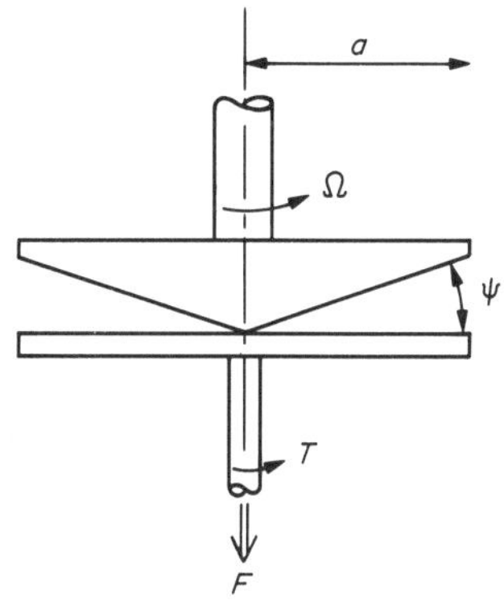

Figure 3
Cone-and-plate viscometer

The shear rate $\dot{\gamma}$ is more difficult to determine. For the case of a rotating outer cup and stationary inner bob, a good approximation for the shear rate at the bob is

$$\dot{\gamma}(r_i) \simeq \frac{2N\Omega}{1 - \left(\dfrac{r_i}{r_0}\right)^{-2N}} \tag{7}$$

where N is the exponent of the empirical power law fit of shear stress–shear rate data ($\dot{\gamma} = K\sigma_{12}^N$).

A widely used instrument for measurement of shear flow properties is the cone-and-plate system (see Fig. 3). Weissenberg used this geometry as the basis of his rheogoniometer to measure both shear stresses and normal stresses in steady shear flow. This was first applied to polymer melts by Pollett and Cross in 1950. The cone–plate instrument has been widely used since that time in many laboratories to measure both viscosity and normal stresses. Generally, the cone angles in this instrument are small. With this geometry, it is possible to develop a constant shear rate throughout which is given by

$$\dot{\gamma} = \Omega/\psi \tag{8}$$

where Ω is the angular velocity of the moving member and ψ the angle between the cone and the plate. The shear stress and principal normal stress difference are given by the torque T and normal thrust F exerted on the instrument through

$$\sigma_{12} = 3M/2\pi a^3, \qquad \sigma_{11} - \sigma_{22} = N_1 = 2F/\pi a^2 \tag{9}$$

where a is the cone radius. Use of this instrument is generally limited to low shear rates. A flow instability in polymer melts occurs at approximately 1–10 s^{-1} in which the melt emerges from the gap between the cone and the plate.

It is also possible to measure viscosity in torsional flow between a stationary and a rotating disk. Instruments of this type have been reported in the literature for more than a century. The first instrument of this type to be used on bulk polymers was Mooney's 1934 "shearing disk" viscometer (see Fig. 4). In this instrument a serrated disk is rotated in a sample fixed in a pressurized cavity. It was developed for application to rubber and its compounds and both pressurization and serrations were meant to prevent slippage. This has become the standard quality control instrument of the rubber industry and will be mentioned again in Sect. 5.

Since the 1960s, various investigators have used unpressurized smooth parallel disks to determine viscosity and normal stresses in shear flow of molten plastics. Rheological properties in the parallel disk instrument are generally determined at the value of the shear rate at the outer radius of the disk which is

$$\dot{\gamma} = \frac{a\Omega}{h} \tag{10}$$

The shear stress and normal stress differences are given by

$$\sigma_{12}\Big|_a = \frac{3M}{2\pi a^3}\left(1 + \frac{1}{3}\frac{d\ln M}{d\ln \dot{\gamma}_a}\right) \tag{11a}$$

$$\sigma_{11} - 2\sigma_{22} + \sigma_{33}\Big|_a = N_1 - N_2$$

$$= \frac{2F}{\pi a^2}\left(1 + \frac{1}{2}\frac{d\ln F}{d\ln \dot{\gamma}_a}\right) \tag{11b}$$

It is to be noted that Eqn. (11b) gives a different combination of normal stresses from Eqn. (9).

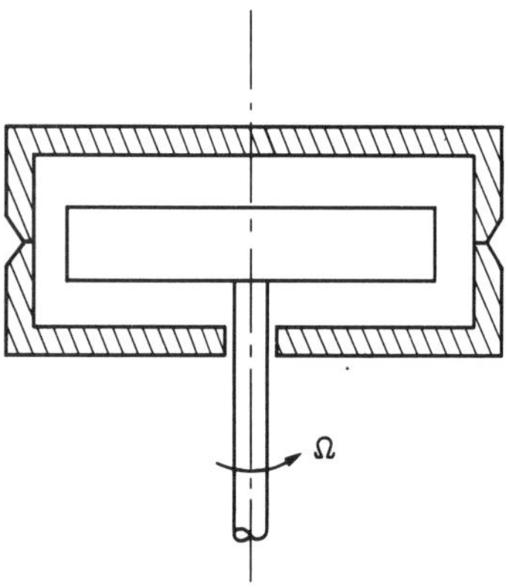

Figure 4
Mooney viscometer

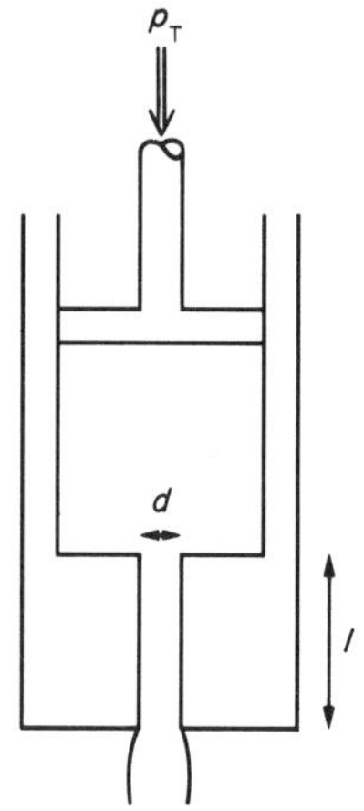

Figure 5
Capillary rheometer

Equations (9, 11b) may be used to obtain the two independent normal stress differences for polymer melts if normal thrust measurements are made for cone–plate and parallel disk instruments. As in the case of the cone–plate instrument, the simple parallel disk apparatus is limited to low shear rates. At higher shear rates, there is a flow instability in which the melt emerges from the gap between the disks. In Mooney's shearing disk viscometer, the shear stress at the outer radius of the disk must be corrected because of the contributions of the Couette flow at the outer periphery.

The most widely used of all rheological instruments for polymer melts is the capillary rheometer (see Fig. 5). It consists of a reservoir to which is connected a capillary die with a diameter generally in the range 0.05–0.20 cm. The melt is generally extruded at either a constant volumetric throughput or at a constant pressure. Capillary viscometric instruments have been applied to Newtonian fluids for more than a century. They have been widely used for polymer melts since 1930. The shear stress along the wall of a capillary die is closely related to the pressure drop p_T required to extrude melt through the die. A force balance gives

$$p_T = 4(\sigma_{12})_w \frac{l}{d} + \Delta p_{ends} \tag{12}$$

Large "ends" pressure losses are associated with the process and must in general be corrected for. To determine $(\sigma_{12})_w$, either data must be obtained at several l/d ratios and the slope of a p_T vs l/d plot measured or pressure transducers must be used along the length of the die. The shear rate at the die wall is given by the Weissenberg equation:

$$\dot{\gamma}_w = \frac{4Q}{\pi r^3}\left[\frac{3}{4}+\frac{1}{4}\frac{d\ln 4Q/\pi r^3}{d\ln(0\sigma_{12})_w}\right] \tag{13}$$

and requires knowledge of both $4Q/\pi r^3$ and its slope as a function of shear stress.

Generally, capillary viscometers may be used at higher shear rate levels than rotational viscometers for polymer melts. Flow instabilities, variously called simply extrudate distortion or melt fracture, represent the limitation at higher extrusion rates. This flow instability generally occurs at about die wall shear stress levels of 0.1 MPa. The capillary instrument has also been used to measure normal stresses, although such methods have never gained general agreement.

In polymer solutions, both the swell and backward thrust forces of jets have been used. In polymer melts, the total ends pressure loss and the exit pressure loss have been applied. The efforts of Han (1976) in developing such an instrument should be noted. Of the techniques used for melts, the exit pressure loss technique has the more rigorous basis. It depends, however, on the presumption of fully developed flow extending to the exit of the tube. This, however, appears not to be the case, especially in flow at very low Reynolds numbers, as occurs in polymer extrusion.

The slit rheometer is similar in concept to the capillary apparatus, differing only in the orifice cross section. This type of instrument has been used for polymer melts by various investigators, but has most notably been exploited by Han. Generally, a series of flush-mounted transducers is placed along the flow direction of the instrument to measure pressure gradients and exit pressure. The slit rheometer operates in a similar region of high shear rates as does the capillary instrument. It is generally limited at low shear rates by instrument sensitivity and at high shear rates by extrudate distortion. Han and his coworkers have used exit pressure measurements to determine normal stresses with a slit rheometer.

3. *Elongation Flow Instruments*

Many polymer processing operations involve uniaxial elongational flow rather than shear flow. The importance of such processes has led to the extensive development of instruments to carry out elongational measurements. The earlier investigation of instruments of this type was for viscous Newtonian liquids by Trouton as early as 1906. The first studies on polymer melts were by Ballman in 1965 and later by Cogswell, Meissner and Vinogradov, and Fikham and Radushkevich in the late 1960s. Since that time, many investigators have reported experimental work on elongational viscosity of melts.

Measurements of elongational viscosity are generally more difficult to carry out than shear flow determinations. The problems involve the difficulty of maintaining a constant elongation rate or stress and achieving a steady state. All data reported to date on polymer melts have been on high-molecular-weight polyolefins, elastomers and styrenics and their compounds.

The major method of measurement of elongational viscosity is to stretch a filament. This may be done in a variety of ways. However, a general limitation in all these techniques is that the fluid must be sufficiently viscous to be maintained in a controllable deformation field. This is the case for melts with viscosities of the order of 10 kPa s and above, such as those of polyolefins and elastomers. It is not possible to carry out this type of deformation with lower-viscosity melts, such as those of polyesters and aliphatic polyamides, which are often in the 100 Pa s region. The method is also limited to deformation rates of less than 5 s^{-1}.

There are alternative filament stretching methods of measurement of elongational viscosity. In the original method of Ballman, a vertical thermostated filament is clamped at both ends and stretched at a rate which varies $l(t)$ in such a way as to maintain a constant rate of deformation E. In the method of Meissner, a horizontal filament of length l is supported on the surface of a hot (thermostating) immiscible oil and held at both ends between pairs of toothed wheels with linear velocity $v/2$. The filament is drawn at both ends through the pair of wheels, one of which contains a tension-sensing instrument. The deformation rate E is v/l.

Others describe methods in which filaments are clamped at one end and taken up on a rotating roll which reduces the amount of filament being stretched, as in the Meissner apparatus, and attempts to produce a constant elongation rate. Constant-stress measurements of elongational viscosity have also been reported in the literature. A creep experiment using a constant force coupled with a cam to make it respond as a constant stress is utilized. The apparatus of Cogswell uses a filament extended vertically on top of a bath, while that of Munstedt has a vertical sample immersed in a bath. The latter instrument allows only a limited extension, determined by the apparatus design.

Existing methods of measurement of biaxial and planar extensional flow are based on the work of Denson and involve inflation of clamped circular or high-aspect-ratio rectangular sheets to form "bubbles." The material near the upper tip of the bubble undergoes a biaxial elongational flow for circular sheets and this can be programmed to be constant in rate if the movement of fiducial marks or growth of the bubble is monitored and the inflation pressure is varied to maintain this constant. A limitation of this apparatus lies in the total strain that can be applied before the sheet becomes too distorted (beyond a hemisphere or half cylinder). Beyond this, a constant elongation rate cannot be maintained. Thus, a steady-state elongational viscosity may not often be achievable. Another limitation is that the materials studied by such an apparatus must obviously have very high elongational viscosities.

4. *Small-Strain Transient Experiments*

An important class of rheological experiments performed on bulk polymer systems involves studying the response of these materials to small strains. The strain is sometimes applied as a step function and the relaxation of stress following its application is measured. In other experiments, a sinusoidal oscillatory deformation is applied and the stress response to it is measured. Other variants have been investigated. Experiments of this type are best classified together because their results are usually given a uniform interpretation in terms of the theory of linear viscoelasticity.

Stress relaxation is an experiment which has been widely carried out on elastomers and other highly viscous polymer systems, involving application of a strain to a sample in extension or shear. The force response $F(t)$ to the applied strain is measured. The data are interpreted in terms of a relaxation modulus. For a sample of initial cross section A_0 and thickness d which is sheared by an amount Δ_l, the relaxation modulus $G(t)$ is defined as

$$G(t) = \lim_{\Delta l \to 0} \left[\frac{F(t)/A_0}{\Delta l / d} \right] \tag{14}$$

Extension measurements have also been reported.

A common experiment performed to polymer melts is the determination of the stress response to an imposed sinusoidal oscillation. Measurements of this type on polymer solutions were first carried out in the 1930s and became common on bulk polymers in the 1950s. For a sinusoidal stress,

$$\gamma(t) = \gamma_0 \sin \omega t \tag{15}$$

The stress (generally shear stress) response is out of phase with $\gamma(t)$ by an angle δ. Considering a shear stress $\sigma_{12}(t)$, we have

$$\sigma_{12}(t) = G^* \gamma_0 \sin(\omega t + \delta) \tag{16}$$

where G^* is the complex modulus. Equation (16) may be rewritten as

$$\sigma_{12}(t) = G'(\omega)\gamma(t)\eta'(\omega)\frac{d\gamma}{dt} \tag{17}$$

where $G'(\omega)$ is the storage modulus and $\eta'(\omega)$ the dynamic viscosity. To carry out such measurements, the melt may be placed in the gap between a cone and plate and one member oscillated relative to the other with a shear strain $\gamma(t)$.

More recently, a technique has been proposed by Maxwell and Chartoff which involves the rotation of two parallel eccentric disks. Fluid elements in this geometry undergo a periodic sinusoidal deformation and the forces exerted on the disk may be interpreted as $G'(\omega)$ and $G''(\omega)$.

5. *Quality Control Measurements*

This section deals with quality control instrumentation in the polymer industry and, specifically, instruments which are used by polymer manufacturers and often by fabricators either to specify and control the polymer sold to customers or to determine whether it is suitable for production by a specific process. There are two widely used instruments of this type, dominating all other procedures used in the trade. These are the Mooney viscometer in the rubber industry and the melt indexer in the plastics industry. These instruments yield rapid single-point measurements of quantities related to shear viscosity.

It has long been realized that such procedures give only rough estimates of the rheological behavior of elastomers and polymer melts and are especially inadequate as production rates approach conditions where flow instabilities occur or where it is necessary to control closely some aspect of a product such as extrudate dimensions. Various "multiple-point" modifications of the Mooney viscometer and melt indexer have been proposed for this purpose.

As mentioned in Sect. 2, the Mooney viscometer consists of a rotating disk in a cylindrical cavity. According to the American Society for Testing and Materials (ASTM) specification, the disk has a diameter of 19.05 mm and a thickness of 5.54 mm. A single-point viscosity measurement is made at a rotor speed of 2 rpm after a period of 4 min. The measurements are reported in terms of a calibrated torque measurement on a standard dial and referred to as the ML-4 reading. Here M refers to the Mooney viscometer, L to a large disk size (distinguished from a smaller diameter disk used for vulcanization scorch tests) and 4 to the time of test. Measurements are carried out at 100 °C. Typical ML-4 values for commercial synthetic rubbers are in the range 45–55.

The melt indexer developed by Du Pont is the standard quality control device used in the plastics industry. Polyethylene, polypropylene and polystyrene are generally specified by melt index (MI). The instrument is basically an extrusion rheometer of the type described in Sect. 2. It uses a single capillary of diameter 0.655 mm and a length–diameter ratio of 12. Extrusion is effected by pressure applied through a deadweight. The melt index is the mass (g) extruded in a 10 min test. The ASTM specifies a series of testing conditions A–K which involves changes in temperature (125–275 °C) and applied load (0.325–21.6 kg) or pressure (0.5–29.5 bar). The specifications are chosen to give melt indices between 0.15 and 25. Condition E is the most widely used: 190 °C and 2.95 bar. The melt index is related to the reciprocal of the viscosity.

See also: Mixing: Rheological Behavior and Dispersion of Two-Phase Systems; Processing: Rheological Properties and Fluid Mechanics

Bibliography

Coleman B D, Markovitz H, Noll W 1965 *Viscometric Flows of Non-Newtonian Fluids*. Springer, Berlin

Dealy J M 1978 Extensional rheometers for molten polymers. *J. Non-Newt. Fluid Mech.* 4: 9

Han C D 1976 *Rheology in Polymer Processing*. Academic Press, New York

Middleman S 1967 *The Flow of High Polymers*. Wiley, New York

Van Wazer J R, Lyons J W, Kim K Y, Cogswell R E 1963 *Viscosity and Flow Measurement*. Interscience, New York

Walters K 1975 *Rheometry*. Chapman and Hall, London

White J L 1978 Rheological behavior of unvulcanized rubber. In: Eirich F R (ed.) 1978 *Science and Technology of Rubber*. Academic Press, New York

White J L 1980 Molten polymers. In: Walters K (ed.) 1980 *Rheometry*. Wiley, New York

J. L. White
[University of Akron, Akron, Ohio, USA]

Roll and Coil Coatings

The term roll coating as used in this article describes a process whereby a liquid chemical (paint) is applied to a substrate by a combination of synthetic rubber and steel rolls. Roll coating is employed in the coil coating process, which consists of several continuous operations combined into one coating line. Coil coating is one of the fastest growing methods of coating application.

1. Roll Coating

Roll coaters can be set up to operate in either "direct" or "reverse" modes and usually employ either two or three rollers, depending upon the application (see Fig. 1). Direct roll coating describes a setup in which the applicator roll is moving in the same direction as the substrate at the point of contact. Reverse roll coating describes a setup in which the applicator roll is moving in the opposite direction to the substrate at the point of contact.

Direct roll coating is used mainly for deposition of very thin (<25 μm) wet films of coatings at normal roll coating solids contents (30–50 vol.%) or in applying unusually high-solids coatings. Direct roll coating also has the advantage of extended roll life when painting rough substrates or those with wavy or burred edges. When setting up for direct roll coating, the peripheral speed of the applicator roll is usually set 2–5% below line speed, resulting in a slight "smearing" of the paint from the applicator roll onto the substrate. However, there are also circumstances which necessitate applicator roll speeds slightly faster than line speed. The pickup roll is usually set at about 50% of line speed. Direct roll coating setups are also employed where offset gravure or flexographic printing of the substrate are desired. For printing, all rolls are set equal to line speed.

Reverse roll coating has the advantage of permitting deposition of much thicker wet films than direct roll coating and of providing much better flow and levelling properties than direct roll coating. When setting up to coat in the reverse roll coat mode, the peripheral speed of the applicator roll is normally set at 110–120% of line speed. The pickup roll is normally set at 50% of line speed. Experienced coater

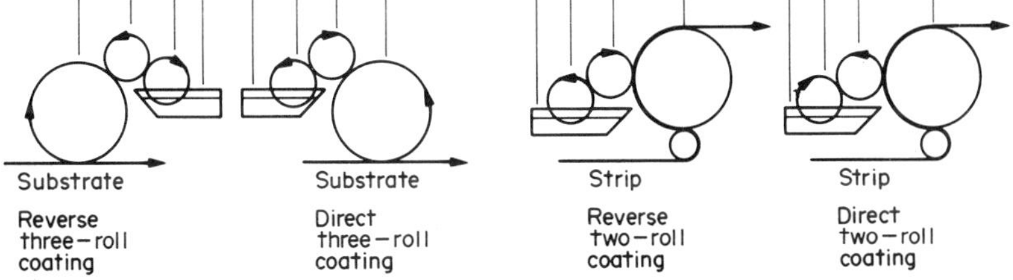

Figure 1
Roll coating operations: A, applicator roll (rubber); B, metering roll (steel); C, pickup roll (rubber on steel); D, paint pan; E, pickup roll (steel); F, bed or backup roll (steel)

operators then finely adjust the applicator and pickup roll speeds to achieve the best application of a given coating on a given substrate.

The roll coating process has the advantage over other application methods, such as spray and dip coating, of being able to apply uniformly very thin wet films with precise control of film thickness and with excellent deposition efficiency. For example, in spray application, a considerable amount of the coating is lost as overspray. Although electrostatic spray application is considerably more efficient than conventional spray application, a significant amount of coating is still lost as overspray. Roll coating results in virtually 100% deposition efficiency. Precise film thickness control and film thickness uniformity are very difficult with spray techniques. Dip coating also results in uneven film deposition and precise control of film thickness is not possible. The powder coating process is useful only for thick-film deposition. Coatings with the excellent exterior durability available in roll coat and spray finishes are not yet commercially available in powder form. Electrodeposited coatings are generally used as interior finishes or as primers, because they lack the exterior durability available in roll coat and spray finishes, but they do provide excellent corrosion resistance. They are also limited to thin films.

Roll coating has the capability of applying different colors to the substrate as stripes and of printing various designs such as wood grain, brick and stone patterns and the many designs necessary for various container labels. Typical roll coating applications include wood panelling, container sheet stock and two-piece cans, which represent applications to separate individual substrates as opposed to continuous coating of metal coils for subsequent slitting, cut-to-length and forming operations.

2. Coil Coating

2.1 The Process

Large modern coil coating lines can be some 0.5 km long and have some 2 km of strip in the line. A typical flowsheet is shown in Fig. 2. Large coils of metal are fed into the line at the uncoilers. The ends are trimmed by a shear and then spliced together by crimping, stitching, stapling or welding equipment. The edges are flattened (if necessary) by deburring rolls to prevent cutting of the rubber rolls throughout the rest of the line. The strip then enters the entry accumulator, which stores metal so that the line can continue operating at normal speed while coil splices are being made, after which it passes through the "wet section," which is usually a five-stage operation consisting of alkaline cleaner, water rinse, metal treatment, water rinse and acid rinse tanks. The acid rinse tank is kept at 65–70 °C to help dry the strip in preparation for coating.

The strip then enters the primer coating station, where the primer is applied to one or both sides of the strip by a roll coater. The suspended strip then passes, after a 2–5 s flash-off, into the prime oven, where it is cured at air temperatures of 95–425 °C, resulting in peak metal temperatures of 65–260 °C depending on the coating and metal properties

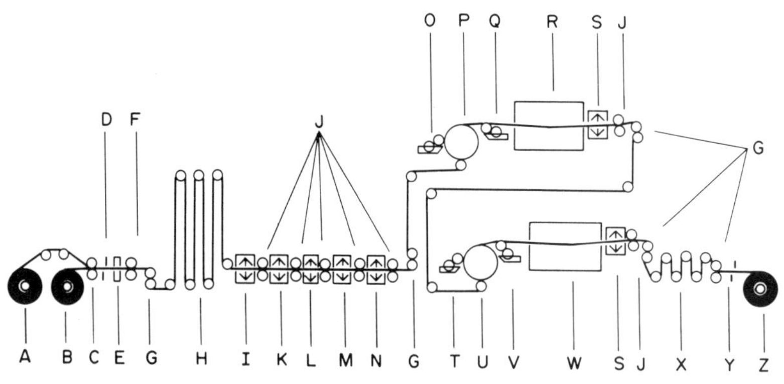

A Uncoiler "A"
B Uncoiler "B"
C Feed rolls
D Shear
E Splicer
F Deburring rolls
G Tension and steering rolls
H Entry accumulator
I Cleaner tank
J Squeegee rolls
K Water rinse tank
L Metal treatment tank
M Water rinse tank
N Acid rinse tank
O Top primer coater
P Primer coater bed roll
Q Bottom primer coater
R Primer oven
S Water quench
T Top finish coater
U Finish coater bed roll
V Bottom finish coater
W Finish oven
X Exit accumulator
Y Shear
Z Recoiler

Figure 2
A typical coil coating line

required. Most modern coil coating lines operate at speeds of 90–240 m min^{-1} and have ovens 45–75 m long. Some ovens have dwell times as short as 8 s. After the primer is cured, it is cooled by high-velocity air followed by a water spray. Excess water is squeezed from the strip by rubber rollers.

The strip then passes into the finish coater, where the topcoat is applied to one or both sides of the strip. If a finish coat is specified for one side of the strip only, a backer coating is usually applied to the reverse side of the strip to prevent blocking (the coil sticking together after being rewound) and burnishing in the coil, to facilitate subsequent forming operations and to provide corrosion protection to the substrate. The curing and quenching operations are similar to those used for the primer. The strip finally passes into the exit accumulator section which can expand to store metal so that the line can continue to operate at normal speed while a coil is being removed from the exit end of the line.

Subsequent operations performed on coil-coated metal can include slitting to the desired width, cutting to the desired length, roll forming, stamping, brake forming or drawing. Most modern coil coating lines also have the capability of printing and wet inking, laminating, in-line embossing, waxing or other operations.

The coil coating process has several advantages over other methods of coating application. The most important of these is probably the low labor cost per unit area of coated substrate. Most coil lines operate with a crew of six to eight people. These include one entry operator, one or two coater operators, one exit operator, one paint handler, one or two fork lift or crane operators at entry and exit and one person functioning as a combination line foreman, wet-section operator and quality control tester. At line speeds of ≥90 m min^{-1} and strip widths of 1.2–1.5 m being painted on both sides, the labor cost per unit area is extremely low compared with that of other surface coating operations.

Another very important advantage of the coil coating process is energy efficiency. Even though coil coating ovens sometimes operate at temperatures as high as 425 °C, large quantities of material are being processed through these ovens with very short dwell times compared with other coating and curing operations. Also, the quantities of makeup air are extremely small compared with other coating and curing operations.

Another very important advantage of the coil coating process is its efficiency of pollution control. Since the coater room, flash-off section, curing ovens and quench section are almost totally enclosed, it is relatively simple to operate the ovens at a slight negative pressure which pulls all fumes and solvent vapors from these areas into the cure oven. The exhaust from the oven is then relatively easy to incinerate so as to remove >90% of the volatile organic compounds from the exhaust stream and thus reduce solvent emissions to a much lower level than is possible with other coating and curing methods. Most modern coil coating lines are designed to recover the heat generated during incineration and recycle it to the curing oven, the wet section and other areas of the plant requiring heat. This contributes greatly to the energy efficiency of the entire operation and some lines have reduced their overall gas consumption by as much as 70% by reclaiming the heat generated in burning the solvents from the paint.

2.2 Substrates and Coating Materials

Substrates painted on coil coating lines include aluminum, cold-rolled steel, electrogalvanized steel and hot-dip-galvanized steel. The substrate choice depends on the end use for the coated sheet. Metal pretreatments or conversion coatings used are primarily chromate types for aluminum and phosphate or complex oxide types for steel substrates. These conversion coatings are usually applied by spray, although several lines employ roll coating or dip coating procedures. Most current commercial coil coating pretreatments are of the type that require water and/or acid rinses. However, increasingly stringent pollution regulations have forced some coaters to change to "nonrinse" or "dry-in-place" metal treatments even though corrosion resistance and physical properties obtained with these types are somewhat poorer than with those metal pretreatments requiring rinses.

There are many generic types of coating used on coil coating lines, but the discussion here is limited to the largest volume coatings currently in commercial use.

The type of coating selected depends on the applied cost and the physical properties required for a given end use. All coatings formulated for roll coating application must have desirable rheological properties to permit good flow, levelling, film build and liquid transfer of the coating from the pan to the pickup roll, between the rolls and onto the substrate. Normally, coatings for roll coating application are formulated to have a rheology as near Newtonian as possible. However, there are a few coatings, such as vinyl plastisols designed for striated application, which have pseudoplastic properties. Proper control of surface tension is also necessary. Coatings for high-speed application must have a surface tension low enough to enable proper substrate wetting but not so low as to create problems with paint pickup and transfer. Stability under high shear is also an important requirement of high-speed coatings. Color or gloss development or excessive shear thinning under the high shear conditions encountered on modern high-speed roll coaters cannot be tolerated.

A wide range of physical properties is available in modern coil coatings. Some of these include a good balance of hardness and flexibility, exterior durability, solvent resistance, stain and dirt resistance, adhesion, corrosion resistance, lubricity and precise gloss and color control. Proper cure response for the short dwell times encountered on modern high-speed coating lines and blister resistance when subjected to the high curing temperatures are also essential.

Most of the primers currently in use are either epoxy–urea–formaldehyde or epoxy–melamine–formaldehyde based. These are usually formulated with chromate and/or phosphate inhibitive pigmentation for optimum corrosion resistance. The adhesion and toughness of these polymer systems, as well as their chemical and corrosion resistance, make them highly desirable for primer applications. However, the extremely poor chalk and fade resistance of epoxy coatings prohibits their use as exterior topcoat finishes.

Alkyd–melamine coatings are used where cost is the primary consideration. They are relatively brittle and have only moderate exterior durability compared with other commercially available coil coating systems. Their principal uses are for backer coatings and interior decorative applications which do not require good flexibility.

Solution vinyls, organosols and plastisols are moderately expensive coatings but are still preferred for some specialty applications. The solution vinyls have excellent flexibility and are used where severe forming is encountered. Problems with adequate film build because of their low solids content, poor solvent resistance and high applied cost limit their usage. Vinyl organosols and plastisols are used for higher film build applications where very good flexibility is required. The organosols are applied at dry films up to 25 μm. Application problems, cost, low solvent resistance and moderate exterior durability have limited the demand for these coatings. Vinyl plastisols are employed for very high film build coatings. Dry films obtained with these coatings are normally 75–100 μm although, for some applications such as culvert stock, thicknesses of 250 μm are not uncommon. Factors limiting growth of these coatings are their high cost, only moderate chalk and fade resistance, low line speeds and lack of on-line gloss and color control.

Thermosetting acrylics and polyesters are the most widely used coil coatings because of their relatively low cost and good balance of hardness and flexibility. Their exterior durability is also good. Siliconized thermosetting acrylics and polyesters have very good exterior durability with only a slight reduction in their hardness and flexibility.

Fluorocarbon finishes based on polyvinylidene fluoride are used where outstanding flexibility, exterior durability and chemical resistance are required. They are very high in cost but provide the properties necessary for monumental buildings, power plants and comparable structures.

The coatings previously mentioned are all solvent borne and, due to modern incineration and heat recovery techniques, will continue to be employed in large volume. Thermosetting acrylic emulsion coatings are the only waterborne finishes that have achieved large volume usage on coil coating lines. They have excellent exterior durability and flexibility and have a very low solvent content, making them very desirable for those coil coating lines not fitted with incineration and heat recovery equipment. However, they do not have the "toughness" of most solvent-borne finishes. Therefore, their usage is currently limited primarily to residential building products, which receive much better packaging and handling in the field than most other coil coated products.

Newer technologies such as electron beam and ultraviolet cure coatings and high-solids (>80 vol.%) coatings have had experimental runs during the past few years but have not achieved commercial acceptance, due to problems with application and physical properties, exterior durability and cost. Electrodeposition coatings have been commercially used for several years as primers on coil coating lines. Installation costs and on-line technical expertise requirements have limited the expansion of this technology on coil coating lines because similar properties can be obtained from solvent-borne and waterborne roll coating finishes at lower equipment and technical costs.

While generalizations have been made concerning properties of various generic types of coating, it must be emphasized that wide ranges of properties, including flexibility, hardness, exterior durability and cost, can be achieved with different polymers and formulating techniques within each generic class.

2.3 Applications

Several major market areas employ metal prepainted by the coil coating process. The residential building products market includes siding and accessories, rainware, awnings and carports and simulated cedar shake shingles. The majority of this market is prepainted aluminum, although steel siding and accessories continue to be a significant portion. Acrylics and polyesters, siliconized acrylics and polyesters and acrylic emulsion coatings are used for these products. The mobile home, recreational vehicle and truck trailer markets are also primarily prepainted aluminum. However, usage of prepainted light gauge hot-dip-galvanized steel is increasing in these market areas. Acrylic and polyester finishes dominate these markets.

The agricultural and preengineered building panel

market is primarily hot-dip-galvanized steel. Small quantities of prepainted aluminum are used, but substrate cost is a limiting factor. Acrylic and polyester and siliconized acrylic and polyester finishes are used. The architectural building panel market is also dominated by prepainted hot-dip-galvanized steel. The cost of prepainted aluminum keeps its market share small. Fluorocarbon enamels dominate this market area, due to their outstanding exterior durability and chemical resistance. The use of prepainted sheet for the (major) appliance industry is currently limited to refrigerator/freezer wrappers and liners. Hard, flexible, stain-resistant polyester finishes are used on cold-rolled steel for wrappers and on aluminum or cold-rolled steel for liners. Smaller appliances such as air conditioners and humidifier–dehumidifiers employ polyester-prepainted cold-rolled steel for cabinets and covers.

The automotive industry consumes large quantities of cold-rolled steel painted with zinc-rich primers for their weldability and corrosion resistance. Penetration of other areas of the automotive original equipment market with prepainted metal has been extremely slow. The container market for prepainted coil stock is limited primarily to aluminum can end stock, shallow draw "margarine" and "pudding tub" applications and draw/redraw food cans. There are many products in the miscellaneous category which account for large volumes of prepainted metal. These include office furniture, toys, tool boxes, file boxes, pipe jacketing, sign blanks, office partitions and small appliance cabinets.

The coil coating industry has grown rapidly since the mid-1950s, from lines 50 mm wide running at 12 m min^{-1} painting venetian blind stock to lines 1.85 m wide painting at speeds up to 250 m min^{-1} for a vast number of end uses. Since the mid-1960s, the coil coating industry has averaged a 15% per year growth rate. With its labor, energy and pollution control efficiencies becoming increasingly important, the dynamic growth rate of the coil coating industry is expected to continue for many years.

See also: Maintenance Coatings; Paints, Varnishes and Lacquers

Bibliography

Annual Book of ASTM Standards, Pt. 27: Paint—Tests for formulated products and applied coatings. American Society for Testing and Materials, Philadelphia, PA

National Coil Coaters' Association Technical Manual. Fernley and Fernley, Philadelphia, PA

J. A. Knuttila
[PPG Industries, Delaware, Ohio, USA]

Roofing Materials

The function of roofing is to provide shelter against the effects of the sun, rain, wind and other climatic conditions. An increasingly wide range of materials is used as roofing in two general categories of roof structures, depending on the manner in which penetration of water is prevented. Sloped roofs, conventionally used to shed water rapidly from structures, are covered with unit roofing materials arranged in an overlapping fashion to prevent water penetration through the joints. Flat or nearly flat roofs, a more recent development, contain rain and snow for longer periods of time than sloped roofs and are covered with continuous roofing materials which depend on durable waterproofing properties for their effectiveness. All roofing materials must be used in a compatible manner with supporting roof structures, and with adjoining insulation and retardant materials to control heat flow and water vapor flow between building interior and exterior environments.

1. Historical Development

The first sloping roofs developed to provide shelter from the elements made use of the watershedding properties of sloping overlapping branches, leaves and crude forms of thatch. Later, other locally available materials, such as slate and stone slabs, were found to be effective roof coverings when laid in an overlapping fashion on sloped structures. Where slate and stone were scarce, roofing tiles of clay and, later, wood shakes, shingles and metal roof coverings were developed. All of these materials continue to be used for covering sloped roof construction. When laid, they act to prevent water penetration, despite voids or holes in the roofing at each lap joint, by sufficiently overlapping the unit materials and making the roof slopes steep enough. Thus, the gravitational forces which draw the water out of the joints and down the slopes exceed all other forces which would carry the water inward.

The modern flat roof covered with a continuous waterproof membrane was not introduced until the mid-nineteenth century. The first membranes were based on tar-saturated felts, initially pine tars and later coal tars, coal-tar pitches and asphalts. The felts were attached directly to wood and concrete decks using fasteners and bituminous materials between felt layers to achieve a continuous built-up roofing membrane capable of performing satisfactorily for 20 years or more. Following World War II, changes took place in building design and construction which affected the development of roofing technology. Larger industrial buildings with greater column spans and larger roof areas were constructed, resulting in greater thermal and structural stresses to the roofing membranes. Steel decks were used instead of concrete and wood decks, air conditioning of buildings was developed and insulation was added between the decks and the membranes to control heat flows. The need developed in many buildings for vapor-retarding materials on the warm side of the insulation to control the occurrence of condensation on the underside of the membranes. Earlier installation practices of applying heavy quantities of asphalt to adhere the roofing membranes and the insulation and vapor-retardant materials were found to result in increased fire spread, while subsequent reductions in asphalt applications increased the exposure to wind uplift.

In the 1960s, modifications to existing materials and the development of new plastic and rubber roofing materials resulted in an increasing number of attempts to improve the performance of roofs. To distinguish them from bituminous built-up roofing systems which are often referred to as conventional systems, the new systems were termed nonconventional systems. These newly developed polymeric systems were designed to be lighter, to have fewer reinforcing fabrics or felts to provide strength, to be more elastic over a wider range of temperatures and to be resistant to a wider variety of weather conditions, physical abuse and chemical attack. Covering a broad spectrum of materials with almost infinite variations, these materials were categorized as elastomeric systems which, after substantial deformation and release of a weak stress, quickly return to their approximate initial dimensions. The materials also were categorized as single-ply roofing because they were developed either as preformed sheet systems manufactured in factories or as liquid applied systems which solidify as a film in place by solvent or water evaporation or by chemical curing upon mixing. The wide spectrum of systems has had varying degrees of success and continues to be developed and modified.

2. *Sloped Roofing Materials*

Sloped roofing materials include thatch, slate, clay tiles, wood shakes and shingles, asphalt shingles and metal roof coverings.

3. *Bituminous Built-Up Roofing*

The largest category of roofing systems in use on modern flat roofs is hot process built-up roofing, in which plies/layers of roofing felts are bonded together with hot bitumen to constitute a waterproof membrane.

Another variant is cold process built-up roofing, which uses solvent-thinned cutbacks and asphalt emulsions as adhesives, which do not require heat to melt.

4. *Elastomeric Roofing*

Elastomeric roofing includes a wide variety of thermosetting and thermoplastic materials possessing different chemical, physical and mechanical properties. Thermosetting materials solidify when first heated under pressure and cannot be remelted or remolded without destroying their original characteristics. Thermoplastic materials, which will repeatedly soften when heated and harden when cooled, are not truly elastomeric materials and are separately categorized here, as are groupings for composite roofing materials and other nonconventional systems.

The elastomers consist of natural and synthetic rubbers and rubberlike materials. They are resistant to creep because of their elasticity and are tougher than thermoplastics. The vulcanized elastomers are the least affected by physical and chemical stress, are flexible at 10 °C and are virtually insoluble in petroleum distillates. In the factory, sheets are seamed together under pressure before vulcanization, to minimize the amount of field sealing required. The membranes can usually be adhered to the substrate or can be installed loose-laid with a covering of stone or other ballast weighing around $60\,kg\,m^{-2}$. They cannot be solvent or heat welded and field sealing typically is accomplished by contact adhesive.

The most commonly used elastomeric roofing materials are neoprene, ethylene–propylene–diene monomer (EPDM) and polyisobutylene.

Neoprene is the generic name for the polymers of chloroprene and exhibits good resistance to petroleum oils, solvents and weathering. It is available in liquid-applied systems and preformed sheet systems. Some formulations require a protective coating of chlorosulfonated polyethylene to retard weathering. Thickness varies from 30 mils to 120 mils (0.76–3.05 mm).

EPDM is a synthesized elastomer which has good resistance to ozone and weathering, and has low water vapor and gas permeability. Although resistant to corrosive chemicals and dilute mineral acids, it is not resistant to petroleum oils and gasoline. Thickness varies from 30 mils to 60 mils (0.76–1.52 mm).

Polyisobutylene (PIB) is usually blended with EPDM and other synthetic rubbers and has good resistance to ozone and weathering. Although its resistance to dilute mineral acids and dilute solutions of alkalis is good, its resistance to oils and solvents is low. Thickness varies from 30 mils to 60 mils (0.76–1.52 mm).

5. *Thermoplastic Roofing*

Thermoplastic materials melt when heated and resolidify when cooled. At normal temperatures they are brittle unless blended with plasticizers to improve their flexibility. Chlorinated polymers improve fire resistance, solubility and control of creep. Thermoplastic sheets are sealed by heat welding or by solvent welding which fuses the seams of the somewhat soluble resin sheets in solvent cement.

Thermoplastic materials include polyethylene and polypropylene, both of which are difficult to bond and are not used as a principal material in roofing. The most commonly used thermoplastic roofing material is poly(vinyl chloride) (PVC). In addition, chlorinated polyethylene and chlorosulfonated polyethylene are produced as thermoplastics but these materials can be vulcanized and therefore are sometimes considered as semielastomerics.

PVC, a thermoplastic polymer, exhibits good resistance to acids, alkalis and many chemicals. Through plasticizing and proper formulation, elastomeric properties and improved resistance to weathering are obtained. The loss of plasticizers by leaching, aging, solar exposure or other factors causes embrittlement and possible shrinkage. Ultraviolet inhibitors are added if the installed membrane is not to be protected from the sun by ballast. Thickness varies from 32 mils to 60 mils (0.81–1.52 mm) and reinforcement with nylon, fiberglass or polyester fabrics is added in some membranes to form composite sheets. Application can be loose-laid with ballast, fully adhered or mechanically fastened.

Chlorosulfonated polyethylene (CSPE) is a chlorinated polyethylene containing a small number of chlorosulfonyl groups and exhibits good resistance to ozone, heat and weathering. It also has good resistance to mineral oils and chemical attack but will swell in chlorinated solvents. The liquid-applied system is used as a coating on concrete and plywood decks and on foamed-in-place roofing, and the sheet material contains an asbestos backing to facilitate handling. Thickness is approximately 35 mils (0.89 mm) and application is by hot asphalt or contact adhesive.

6. Modified Bituminous Roofing

Modified bituminous roofing materials are composite systems which, in general, have more elongation than conventional bituminous membranes and use fewer plies. The modifications render the bitumen more flexible and less brittle at cold temperatures and also improve elasticity at high temperatures. The modifiers are generally macromolecules such as natural rubber or synthetic rubber. Because of their greater elongation, the membranes are reinforced with polyethylene films, nonwoven polypropylene and polyester mats, or similar materials with greater stretch properties than reinforcing mats used in conventional bituminous systems. Surfacing may be a thin embossed metal, special protective coating or gravel adhesive with mineral granules. The systems are adhered with adhesives or hot bitumen or in some applications are loose-laid and ballasted. Joint laps and adhesive bonding are accomplished with propane torches or with solvents or solvent-based adhesives. Thicknesses vary from 60 mils to 160 mils (1.52–4.06 mm), depending on the manufacturer.

7. Other Liquid-Applied Roofing

Liquid-applied roof coatings are used to provide a weathering skin on concrete and plywood decks, and on foamed-in-place roofing such as sprayed urethane foam. The coatings are generally applied by roller or airless spray equipment in two or three coats to give a minimum dry-film thickness of 20 mils to 30 mils (0.51–0.76 mm). In addition to the liquid-applied roofing materials already described, several other materials are commonly used.

Acrylics are single-component, water-based coatings that cure by evaporation of water and are the most widely used coatings.

Silicone coatings are available as single- and two-component types. The single-component silicone cures by reacting with moisture vapor in the air following solvent evaporation. The two-component silicone system cures by rapid chemical reaction.

Urethane coatings are also available as single- and two-component types. Again, the single-component types cure by reaction with moisture vapor in the air while the two-component types cure by chemical reaction.

8. Relation of Roofing to the Completed Roof Structure

The wide range of available roofing materials must be used in a compatible manner with a variety of other components to complete a weathertight and durable roof structure. Flashings must be used at all roofing membrane terminations and penetrations to intercept and lead the water to drains. The manner and location of placement of components to control heat and moisture vapor flows between the outside and inside of roof structures impose different requirements on the component materials and affect the ultimate durability of the roofing.

Flashings are required to withstand differential structural and thermal movements which occur between roof membranes and independent adjoining building elements. Flashing materials include bituminous felts, glass-fiber fabrics, various metals and the newer polymer materials. Base flashings are made part of the roofing membrane to bridge the joints to the adjoining building elements. They are required to be flexible, compatible with the membrane and sag resistant where they turn up vertically. Metal counterflashings are attached to the adjoining building elements and overlap and remain free oif the base flashings. Control or movement joint flashings are deliberate discontinuities in membrane installations to allow for movements resulting from expansion and contraction.

Insulations are built into many roof structures to limit heat flow. Insulation materials include a variety of open-fibrous and closed-cell materials which are produced in board form in a factory or are mixed or foamed-in-place at the site. Their location in the completed roof system has an effect on the thermal and moisture conditions in other components. The usual location of insulation in flat roof construction is under the roofing membrane on top of the structural deck. This position subjects the membrane to wider extremes of daily and seasonal temperatures with consequent material stressing than would be experienced if no insulation were used. A new type of roofing assembly has become available which rearranges the normal elements of the roof system. Termed the protected membrane system, inverted roof or upside-down roof, the system places the roofing membrane directly on the structural deck and the preformed insulation boards outside the membrane. The extreme temperature variations experienced by the membrane are avoided in this configuration. The system requires that the insulation (usually polystyrene foam insulation boards) be nonabsorptive of water and that it be held in place against wind uplift and flotation by adhesive bonds or ballast.

Vapor retarders are incorporated into some roof structures of buildings experiencing moist air conditions or located in colder climates. Their purpose is to prevent moist inside air from reaching and condensing on cold roof surfaces, including insulation placed under the membrane. Vapor retarders are normally placed on the warm side of the insulation and are installed to be continuous, without gaps, and sealed at all joints and edges. Vapor retarders have low permeance to moisture vapor and include such materials as asphalt-saturated and coated roofing felts and kraft-paper laminates. The use of hot-applied bituminous felts is not recommended as vapor retarders for steel roof decks because of poor resistance to internal fires.

The roof structure and deck must be designed to support all applied loads without failure. Typical modern flat roof decks, classified as nailable or nonnailable, are constructed of wood, metal, cast-in-place materials and preformed units. When selecting roofing materials, it is necessary to consider roof structure deflections and differential movements arising from loads, shrinkage, creep, moisture changes and thermal variations. Cracks and deflections, even if not structurally significant, may induce sufficient stress to tear bonded membranes and permit increased heat flow, and moisture and vapor migration. Slopes of all roofs should be designed to a minimum slope of 1.875° (1 in 48). Excessive deflections may allow ponding of water which can adversely affect the durability of the roofing materials.

See also: Plastics and Composites as Building Materials

Bibliography

American Society for Testing and Materials 1966 *Engineering Properties of Roofing Systems*, Special Publication No. 409. ASTM, Philadelphia, PA

Baker M C 1980 *Roofs: Design, Application, Maintenance*. National Research Council of Canada (Multiscience), Montreal, Canada

Building Research Establishment 1970 *Built-Up Felt Roofs*, BRE Digest No. 8. BRE, Watford, UK

Building Research Establishment 1979 *Flat Roof Design: The Technical Options*, BRE Digest No. 221. BRE, Watford, UK

Bushing H W, Mathey R G, Rossiter W J T, Cullen W 1978 *Effects of Moisture in Built-Up Roofing—A State-of-the-Art Literature Survey*, NBS Technical Note 965. National Bureau of Standards, Washington, DC

Griffin C W Jr 1982 *Manual of Built-Up Roof Systems*, 2nd edn. McGraw-Hill, New York

Handegord G O 1966 *Moisture Considerations in Roof Design*, Document No. CBD 73. National Research Council of Canada (Division of Building Research), Ottawa

Mathey R G, Cullen W C 1974 *Preliminary Performance Criteria for Bituminous Membrane Roofing*, NBS Building Science Series 55. National Bureau of Standards, Washington, DC

National Academy of Sciences/National Research Council 1974 *Roofing in Developing Countries*. NAS/NRC, Washington, DC

National Roofing Contractors Association 1981 *NRCA Roofing and Waterproofing Manual*. NRCA, Oak Park, IL

Rossiter W Jr, Mathey R 1979 *Elastomeric Roofing: A Survey*, Technical Note NBS 972. National Bureau of Standards, Washington, DC

Watson J A 1979 *Roofing Systems: Materials and Applications*. Reston Publishing Company, Reston, VA

J. D. Stelling
[Thompson and Lichtner Company,
Cambridge, Massachusetts, USA]

Rubber Processing

The processing of rubber usually consists of three stages: mixing, shaping and crosslinking (Pearson 1968, White 1969, Rapetski 1978, Freakley 1985). However, for many applications the building of a composite structure is required before crosslinking is performed. Specialty rubber products, such as rubber cements, latexes, thermoplastic elastomers and low-molecular-weight rubbers, may require different processing steps.

1. Mixing

The elastomer is seldom used in its pure form; it is usually mixed with various additives, which either help to reduce the cost, improve processability or impart a desired property to the finished product (Studebaker 1978).

1.1 Formulation

The formulation (compounding ingredients) may typically contain the elastomer, filler, process oil, crosslinking system and protective agents.

The choice of elastomer depends primarily on its intended application. Often, more than one elastomer is blended to achieve specific properties. For a given type of elastomer the molecular weight is an important parameter: a high-molecular-weight elastomer provides better product properties, whereas a low-molecular-weight elastomer is used for its ease of processing. The molecular weight distribution, branching and differences in microstructure are other variables affecting the properties and processability. The relationship between microstructure and macrostructure of elastomers and compound properties and processing has been reviewed by Corish (1980).

Most rubber products contain either reinforcing or nonreinforcing fillers. The latter are primarily for cost reduction and include calcium carbonate, nonreinforcing carbon black, clay and other finely divided minerals. The largest volume use of the reinforcing fillers is carbon black, but silica is also important, particularly for nonblack applications. Fibrous fillers have been developed for special applications. The properties of carbon black, which influence the properties of elastomer products, are particle size, surface area, hydrogen content, oxygen content and structure (Studebaker 1965).

The choice of crosslinking system (Ambelang 1964) depends not only on the chemical structure of the polymer with respect to its cure site, but also on the intended end use of the product—low sulfur or sulfurless cure systems promote heat resistance and peroxide systems promote low compression set. Ultrafast systems are desirable for injection molding and slower curing systems for thick articles such as conveyor belts.

Compounding ingredients which modify the rate and temperature of the cure are called accelerators or retarders, depending on their effect on cure.

Various types of oil are added for economic reasons (extender oil) and for ease of processing, particularly for the high-molecular-weight rubber. Lubricants and plasticizers are added for particular requirements in processing or in the product properties. Stabilizers, such as antioxidants and antiozonants, are added for protection against heat, fatigue, light and weathering.

1.2 Equipment

Basically, only two types of equipment are used for mixing elastomer compounds (Stephens 1973): mills and internal mixers. The former is a batch mixer whereas the latter may be either batch or continuous (White 1969, Palmgren 1975).

Among the additives carbon black is the most difficult to mix; for this reason, mixing is often described as that of the carbon black and the elastomer. It consists of three processes: incorporation, dispersion and macroscopic homogenization.

During incorporation, the carbon black and elastomer are brought together, with the former inside the latter. The dispersion process improves microscopic homogeneity of carbon black and elastomer domains. Macroscopic homogenization is achieved manually on the mill by cutting, folding and recharging of the stock, whereas in the internal mixer it is achieved by axial and lateral transport of the stock between two rotors. Although the three processes occur simultaneously, incorporation is dominant at the onset and is completed earlier. The other two processes continue until the desired mixing is achieved (Nakajima and Harrell 1984).

Present industrial processes favor single-step mixing, with curatives added during the last stage of the mixing. However, the batch temperature must not exceed 125 °C and higher filler loadings with high-molecular-weight polymers develop too much heat. Therefore, in such processing, curatives must be added in a second step or on a mill.

2. Shaping

The mixed compound is either shaped into the finished product and vulcanized, or shaped into a form used to build a composite structure. Compression molding, transfer molding and injection molding are the common processes for the former (Thompson 1955) and extrusion and calendering for either.

2.1 Molding

In compression molding, transfer molding and injection molding, the compound is cured in the mold with the application of heat and pressure after the mold cavity is filled. Gaskets and seals can be made in this way and they may be molded together with inserted metal parts if desired. In compression molding, the compound is placed in the mold cavity and compressed to make it conform to the shape of the cavity. Compression molded items, such as O-rings, seals and heels, are usually preformed to size by the blanking out of extruded or calendered sheet, or by extruding a shape and knife cutting to preform size.

In transfer molding, the compound is loaded into a pot which is placed adjacent to the mold cavity. A plunger forces the compound to flow into the mold and fill the cavity.

Most of the injection-molding equipment presently used is of the screw or screw–ram type (Wheelans 1974). With the screw–ram type, the screw plasticates the compound and feeds a prescribed amount through a check valve into an accumulation chamber. The plunger of the accumulator injects this compound into the mold cavity (Wolff 1979). Injection-molded items may use granular, crumb or strip-fed compound.

An in-depth review of injection molding discussing various controlling factors such as machine/mold design and compounding variables such as rheology and scorch is given by Isayev (1989).

2.2 Extrusion Processing

Extrusion (Boguslawski 1972) is used for preparing a continuous profile, which may be cured as a finished product or used as a part of a composite material. For example, extruded hose may be braided with a fabric, a filament or a metal wire and then have a cover extruded over it; wire coating and cable jacketing are extruder products; tire tread is extruded and then cut to size for building a tire; roller-die extruders can be used for the manufacture of conveyor belt covers or sheet rubber goods.

The extrusion process may be subdivided into hot- and cold-feed extrusions (Christy 1979). Hot-feed extrusion requires prewarming and plastication of the rubber stock, usually with a mill. Hot-feed extrusion has lower tolerance to scorch and therefore the feed screw generally has a deeper channel and a smaller length–diameter (L/D) ratio in order to minimize shear heating; the L/D ratio ranges typically from 3:1 to 8:1. The extrusion pressure is generally lower than that for cold-feed extrusion which requires plastication to be accomplished in the extruder. This is achieved with a shallower channel and a larger L/D ratio of the feed screw; the L/D ratio typically ranges from 8:1 to 17:1. The screw is capable of building up a high extrusion pressure against a small die opening.

In general, cold-feed extrusion has several advantages: it is less sensitive to the rheological properties of the feed compound, the rate diminishes less when the die restriction is increased, the process is more receptive to external temperature control and results in lower heat history. In addition, cold-feed extrusion has a cost advantage because it does not require a warming mill; labor costs are also decreased considerably.

On the other hand, hot-feed extrusion is sometimes advantageous; for example, when the compound requires extensive plastication, a mill is more versatile for this purpose. In recent years the trend has been towards cold-feed extrusion, but the pros and cons of the two processes depend upon the given situation.

The critical requirements of extrusion are the output rate, the smoothness of the surface, the sharpness of the edges and the dimensional accuracy. The last is affected by the die swell, which tends to increase the cross-sectional area and decrease the length.

2.3 Calendering

Calenders can be used for forming sheet rubber, coating fabric or wire, laminating plies together and stock preparation for molded items.

For flooring applications, the calendered sheets are subsequently vulcanized; in this case the surface finish is important. When the sheet is used for building a tire, it is cut to the required size. The sheet may also be slit to a narrow strip for feeding into an extruder or a screw-injection molding machine. For this purpose, the calendered sheets are generally thick and the surface is not necessarily smooth.

A critical requirement of the calendering operation is that the compound should stay on one roll and separate cleanly from another roll. This behavior is affected by the balance of viscosity and tack of the compound (Tokita 1979).

3. Crosslinking

Crosslinking is a process essential to most rubber products. It is also called curing or vulcanization. In the absence of crosslinking, the shapes of the articles may change permanently, particularly when they are under load. The crosslinking process ties the molecular chains into a network, thus limiting deformability. The crosslink points also provide "markers" to restore the original shape when the rubber articles recover from deformation.

Historically, the discovery of sulfur as a crosslinking agent for natural rubber opened a vast number of applications for rubber products. Currently, sulfur and sulfur-donor compounds are still the major crosslinking agents. Peroxides which yield free radicals are also used for crosslinking. In principle, any difunctional or multifunctional chemical which is capable of reacting with a part of the rubber molecule can yield the crosslinked network. In practice, a variety of chemicals are used, depending on the nature of the reactive group in the rubber molecule. Also, a physical means such as high-energy radiation can be used. The chemistry of vulcanization is reviewed in the article *Vulcanization* and by Coran (1978), Kirkham (1978), Baker (1988) and Chapman and Porter (1988).

Chemical crosslinking is effected by the application of heat. In compression molding, heat is externally supplied; in injection molding, shear heating precedes the external heating from the mold wall. Calendered and extruded products as well as composite products can be vulcanized in an autoclave using steam under pressure or in a hot-air oven. Another process, known as the continuous vulcanization process, utilizes molten salts, certain metal alloys, fluids or glass beads as the heat transfer media (see *Vulcanization, Continuous*; Stephens 1973). Recently, microwave ovens have been introduced for preheating rubber compounds; this results in faster cures (Meredith 1975).

See also: Automobile Tires; Elastomers: Additives; Fillers in Elastomers; Mixing: Basic Concepts; Mixing: Internal Mixers; Mixing: Mills for Elastomers; Vulcanization

Bibliography

Ambelang J C 1964 Curatives for general-purpose elastomers. In: Alliger G, Sjothum I J (eds.) 1964 *Vulcanization of Elastomers*. Reinhold, New York, pp. 126–52

Baker C S L 1988 Non-sulfur vulcanization. In: Roberts A D (ed.) 1988 *Natural Rubber Science and Technology*. Oxford University Press, Oxford, Chap. 11

Boguslawski J J 1972 Basic fundamentals of extrusion and mixing theory. *Rubber Chem. Technol.* 45: 421–33

Chapman A V, Porter M 1988 Sulfur vulcanization chemistry. In: Roberts A D (ed.) 1988 *Natural Rubber Science and Technology*. Oxford University Press, Oxford, Chap. 12

Christy R L 1979 Hot vs cold feed extrusion. *Rubber World* 180(4): 100–1; 180(6): 30–2

Coran A Y 1978 Vulcanization. In: Eirich F R (ed.) 1978 *Science and Technology of Rubber*. Academic Press, London, Chap. 7

Corish P J 1980 Elastomer architecture: The relation of elastomer micro- and macrostructure to compound properties. *Plast. Rubber Mater. Appl.* 5(3): 109–21

Freakley P K 1985 *Rubber Processing and Production Organization*. Plenum, New York

Isayev A I 1989 Injection molding of rubbers. In: Aggarwal S L (ed.) 1989 *Comprehensive Polymer Science*, Vol. 7. Pergamon, Oxford, Chap. 11

Kirkham M C 1978 Current status of elastomer vulcanization. *Prog. Rubber Technol.* 41: 61–95

Meredith R J 1975 Microwave energy for high-speed, efficient vulcanization of extruded rubber. *Rubber Chem. Technol.* 48: 1100

Nakajima N, Harrell E R 1984 Contribution of elastomer behavior to mechanisms of carbon black dispersion. *Rubber Chem. Technol.* 57: 153–67

Palmgren H 1975 Processing conditions in the batch-operated internal mixer. *Rubber Chem. Technol.* 48: 462–94

Pearson J R A 1968 Application of rheology to rubber processing. *Rubber Chem. Technol.* 41: 23–46

Rapetski W A 1978 Rubber processing and equipment. In: Babbit R O (ed.) 1973 *The Vanderbilt Rubber Handbook*. Vanderbilt, Norwalk, CT, pp. 450–61

Stephens H L 1973 The compounding and vulcanization of rubber. In: Morton M (ed.) 1973 *Rubber Technology*, 2nd edn. Van Nostrand Reinhold, New York, pp. 19–50

Studebaker M L 1965 Compounding with carbon black. In: Krause G (ed.) 1965 *Reinforcement of Elastomers*. Interscience, New York, pp. 319–403

Studebaker M L, Beatty J R 1978 The rubber compound and its composition. In: Eirich F R (ed.) 1978 *Science and Technology of Rubber*. Academic Press, London, Chap. 9

Thompson D C 1955 *Mechanical Molded Goods, Neoprene and Hypalon®*. Du Pont, Wilmington, DE

Tokita N 1979 Analysis of band formation in mill operation. *Rubber Chem. Technol.* 52: 387–99

Wheelans M A 1974 *Injection Molding of Rubber*. Halsted, New York

White J L 1969 Elastomer rheology and processing. *Rubber Chem. Technol.* 42: 257–338

Wolff T J 1979 Injection molding equipment, tooling, processes. *Rubber World* 180(4): 91–4

N Nakajima, E R Harrell and D A Seil
[B F Goodrich Chemical Company,
Avon Lake, Ohio, USA]

S

Sandwich Molding of Plastics

Sandwich molding is a process by which a skin core structure, usually of two different polymers, is produced in a twin injection process. Such structures are often produced by cleaning out injection-molding machines with a second polymer. It was introduced as a new commercial process by ICI (Oxley and Sandiford 1971, Garner 1973, Oxley 1973) in the late 1960s. The intention was to produce parts with high-quality, smooth thin-skinned surfaces with perhaps expensive additives and a foamed or perhaps reclaimed plastic core. Much of the early literature treats this as a structural foam process (Throne 1976).

In the first technologies, sequential injection of two melts was used. Later patents and papers discussed processes with simultaneous injection as well (Fillmann and Eckhardt 1975, White and Lee 1975).

Basic studies of the influence of melt rheological variables on mold filling have been reported (White and Dee 1974, White and Lee 1975, Young et al. 1980). Most studies have focused on sequential injection molding. The most uniform sandwich-molded parts have come from matching shear viscosities in the two melt phases. If the first melt injected has a much higher viscosity, the second melt will penetrate and burst through it. If the second melt injected has the higher viscosity, it will not penetrate to the far reaches of the mold. The molded parts in this region will have a very thick skin.

See also: Injection Molding of Thermoplastics: Structural Foams; Processing: Classification

Bibliography

Fillmann W, Eckhardt H 1975 Verfarhren zum diskontinuierlichen Herstellen von Mehrchicht Formeteimen aus thermoplastiche Kunststoff. German Offenlenlgungsschrift 23,42,957

Garner P J 1973 Injection molding machines. US Patent No. 3,733,156

Oxley D F 1973 Process for production of laminar cellular articles. US Patent No. 3,751,534

Oxley D F, Sandiford D J H 1971 Sandwich molding. *Plast. Polym.* 39: 288–92

Throne J L 1976 The newer structural foam techniques. *J. Cell. Plast.* 12: 264–83

White J L, Dee H B 1974 Flow visualization for injection molding of polyethylene and polystyrene melts. *Polym. Eng. Sci.* 14: 212–22

White J L, Lee B L 1975 An experimental study of sandwich injection molding of two melts using simultaneous injection. *Polym. Eng. Sci.* 15: 481–5

Young S S, White J L, Clark E S, Oyanagi Y 1980 A basic experimental study of sandwich injection molding with sequential injection. *Polym. Eng. Sci.* 20: 748–804

J. L. White
[University of Akron, Akron, Ohio, USA]

Sealants as Building Materials

Sealants represent a fraction of 1% of the cost of a building but, after the roof, represent one of the most troublesome areas of the exterior skin of a building. Air, dust and water infiltration can damage or destroy interior walls and contents. Leaking seals waste energy and are an annoyance to the occupants. However, in the last few decades the science of building sealants and the art of joint design have advanced considerably, so that failures of the sealant need not happen. No single sealant is ideal in all applications, so a knowledge of what type of sealant to use in each application and how to use it is fundamental to solving the sealing problems.

1. Factors Affecting Selection

When two or three sealants have all the necessary performance characteristics for a specific sealing job, a cost comparison will determine which one is ideal. Provision of the required performance at the lowest cost determines the ideal sealant. However, performance is the first consideration, since a failed sealant has a high labor replacement cost. Increased costs for building and from deterioration of contents due to seal failures often exceed the initial cost of the proper sealant.

The selection of a sealant, as with any building material starts with an estimation of what kind of performance is needed. The following checklist is a guide to sealant selection, with items listed in order of descending importance.

(a) *Dynamic movement expected.* How much movement will the joint encounter due to expansion and contraction of adjacent materials, vibration, wind forces, settling of the structure and other sources? The amount of movement should be estimated as a percentage of the width of the joint to be sealed. Sealants are generally classed for use with movements of less than ±5%, ±5% to ±12.5%, ±12.5% to 25% and ±25% to ±50%. The sealing joint width is decided on the basis of the total movement expected and the movement capability of the sealant chosen, or the sealant is chosen on the basis of the desired joint width and the movement expected.

(b) *Weatherability*. What environment will the sealant encounter? Outdoors requires a sunlight- and rainwater-resistant seal; indoors, less durability is generally required. Below grade or continuous water immersion provides different conditions again. The length of time that the sealant is required to perform in the environment is also a key consideration: a month, a year or many years? Every sealant performs for some time in most environments.

(c) *Extremes in environment*. Many construction seals are exposed to temperature extremes. Solar collectors can reach temperatures of 95 °C, 140 °C or 185 °C when in static conditions. Sealing around hot-water pipes, heating ducts and the like also presents special problems. Extreme cold causes many sealants to stiffen, but many building materials such as metals and plastics shrink or contract by large amounts as they cool down. Thus, when expansion is most needed, as the temperature decreases, the sealant is least able to expand.

(d) *Adhesive and cohesive strength*. Sealants are required to adhere to the substrate; thus, specifying the substrate is fundamental in all sealing operations. Many sealants require a primer for some substrates. If the sealant also serves as a glue to hold materials in place, the strength requirement for the adhesive seal should also be noted in sealant selection.

(e) *Abuse and abrasion*. Some walkways and factory floors are sealed flush with the surface. The sealant chosen for high-traffic areas, with the flush joint design, needs to be abrasion resistant.

(f) *Other factors*. Many seals need to be resistant to attack by oils and solvents. Perhaps radiation resistance or electrical insulation will be required. All such special features must be considered before selection.

If more than one sealant satisfies all the performance requirements, there are two further factors to be considered.

(g) *Ease of installation*. The amount of joint preparation with a given substrate can vary with the type of sealant. The temperature of installation can also be a factor. If mixing and bulk dispensing are problems, a premixed, one-package sealant is to be preferred to a two-package product that requires mixing.

(h) *Cost*. Only when all the foregoing requirements are specified is the price of the sealant a consideration. When two or more materials satisfy all requirements, then the sealant of lowest price is the obvious choice.

2. *Types of Sealant*

Building sealants currently available include oil-extended mastics, oil-based caulks, butyl sealants, latex acrylic sealants, solvent-based acrylic sealants and specialty sealants made of Neoprene–Hypalon, ethylene–propylene and polybutene–polyisobutylene copolymers and SBR. There is also a group generally referred to as high-performance sealants in the broad generic classes of polysulfides, silicones and polyurethanes. For each of these types, especially the high-performance classes, there are several variations such as one-package ready-to-use versus two-package or three-package mix-on-the-job products; there are products with high, medium and low movement capability.

Most wet sealing work employs one or other of the high-performance sealants. They weather well, with outdoor lifetimes of more than ten years (sometimes >20 years) and have joint movement tolerances of ±25% or more (occasionally exceeding ±50%). Their major market share is reflected in the greater detail in which they are considered in this article (see Sects. 2.5–2.7).

2.1 Oil-Extended or Oil-Based Caulks

Until 1950 oil-extended or oil-based caulks dominated the market. They are composed of filler(s), generally calcium carbonate and/or a clay, and silica together with one or more oils such as linseed oil, fish oil, soybean oil or tung oil. Some of the oils are of the drying type; others are not, and serve as long-term plasticizers. The most important attribute of this type of sealant is its low cost; it uses the least expensive materials. These sealants are most suitable for temporary structures that undergo no movement or very limited movement. Most materials in this class have a useful life expectancy of 2–10 years and a movement capability, based on joint width, of 5% or less. Failure occurs by hardening, shrinkage and cracking. The best of these materials comply with ASTM specification C570, which details initial performance and has no durability requirement.

2.2 Butyls

The butyls are generally copolymers of isobutylene and isoprene which form a cured rubber which is then swelled in solvents and filled with clay, chalk or similar filler. The butyls use a higher-molecular-weight "permanent" plasticizer for greater elasticity, together with volatile solvents which are temporary plasticizers that impart gunnability and allow for easy extrusion. The ratio of permanent to temporary plasticizer is critical, because the permanent plasticizer causes the principal problems of tackiness and staining often encountered with these sealants. These sealants cure by solvent evaporation and form a rather long-lasting seal. The expected lifetime is 10–15 years. These sealants can typically tolerate movements up to ±10% of the joint width.

The relatively low cost, coupled with rather good unprimed adhesion, has made butyls the most popular sealants in the mobile-home industry and also popular with the homeowner.

This type of sealant is often used in inappropriate places: where there is more movement than the sealant can tolerate; on porous substrates where staining is most likely to occur; in highly visible positions where the tacky surface can become unsightly; or in structures whose lifetime is much greater than the expected useful lifetime of the sealant. Historically, the butyls represent an early advance in technology over the oil-based caulks. In many markets they are being replaced by sealants with greater movement capability and fewer staining problems.

2.3 Acrylics

The latex and solvent acrylics are copolymers of acrylates and substituted acrylates mixed with typical extending fillers and pigments. Both have remarkably good unprimed adhesion and good weathering characteristics. The solvent acrylics contain significant monomer, so that when extruded to cure, they exude a strong, obnoxious odor. They are not to be used in occupied dwellings but are quite acceptable outdoors. The monomer that causes the odor is also responsible for the ability to stick to most substrates, even those coated with dust or light oils. Solvent acrylics weather quite well, with a 10–15 year life expectancy outdoors. The best of the solvent acrylics tolerate movements up to ±12.5% of the joint width. They are rather soft, pick up appreciable dirt and are called self-healing, because the polymer relaxes and reorients under strain. This last property is also called compression set and is one of the reasons why these sealants tolerate only ±12.5% joint movement.

The largest use of solvent acrylics is in commercial exterior construction applications in joints with little or no movement. They are generally of intermediate cost, higher than the oil-based caulks but lower than the high-performance products. They enjoy some popularity based on their adhesion, weathering and relative cheapness, but are receiving an ever-decreasing market share as durability and performance are becoming more sought after.

The latex acrylics differ from the solvent type in that they have almost no odor. They are used extensively in some parts of the sealing trade. Since the carrier is water, installation and subsequent cleanup are easy. The elastomeric properties are usually a bit better than those of solvent acrylics because of prepolymerization to a higher molecular weight.

The ease of cleanup, good adhesion even to damp substrates, lack of odor and moderate cost make the latex acrylics the most popular sealant with the do-it-yourself (DIY) trade. This type of sealant has a 15–20 year life expectancy indoors and 10–15 years outdoors. Poor low-temperature flexibility, high shrinkage and slow cure below 3 °C are the major problems. The relatively low movement capability keeps these products from taking an important share in most commercial and industrial sealant markets. Even in the DIY market the push for longer lifetimes and more water resistance after cure is starting to erode the share of the acrylics in favor of more durable products.

2.4 Specialties

The specialty systems made with materials such as polychloroprene, ethylene–propylene copolymer systems, styrene–butadiene copolymer systems, polybutene and polyisobutylene copolymer systems all offer one or more material advantages that keep them out of the high-performance sealant class. Together, these specialty products represent a very small fraction of the sealing market.

2.5 Polysulfides

The polysulfides were the first of the high-performance materials to achieve broad introduction and use in the early 1950s. The polymeric backbone of the system is generally of the type $(SSCH_2CH_2OCH_2OCH_2CH_2)_n$. The system commonly cures according to the general equation:

$$2R\text{—}SH \xrightarrow{\text{catalyst}} R\text{—}SS\text{—}R + H_2O$$

where the catalyst is a peroxide. The sealant always contains one or more of the common fillers and additives as needed.

The advantage of the polysulfides over the urethanes and silicones is primarily that of cost, because the selling price averages 15–20% less than that of either of the others. The difference in total job costs is smaller since the labor, joint preparation and installation costs are similar for all three types of sealants and many times exceed the material cost. The historical factor is also important. Sealant specifiers and applicators are more familiar with this product than with the newer silicones and urethanes and, thus, the polysulfides retain some of their popularity due to this familiarity and performance history.

The first polysulfides were two-package systems mixed on the job and this is still the most popular package system for polysulfides. With the two-package system, coloring agents can be added at the site, making an almost infinite variety of colors and shades possible. With the one-package, premixed sealant, 5–10 standard colors are available. The two-package system cures to a dry surface in several hours, but the one-package system can take several days. Unfortunately, this slow cure allows dust and dirt to be embedded in the sealant surface. There is therefore a trade-off between the convenience of the one-package, no-mix system and the color flexibility and cure rate advantages of the two-package system. However, the inconvenience of mixing should not be passed over lightly. Many sealant experts judge that inadequate mixing of the two-package systems is the greatest single cause of problems (with both polysulfides and polyurethane two-package systems), after adhesion and surface preparation problems.

Adhesion of the polysulfides generally requires priming on most substrates other than steel, glass and aluminum and there are no less than ten different primers available and recommended for various substrates. This is one of the factors in the marginal acceptance of this product in the DIY market. Priming is acceptable for large jobs with trained crews that are accustomed to the task; the sealant distributor who sells the job will have the primer needed and will give advice in the primer selection process. Small stores almost never carry primers and the DIY user is often not aware of the need for them.

Polysulfides do almost everything the sealants mentioned earlier do and they have the ability to seal a joint which moves up to ±25% and to do so successfully for many years.

Ongoing markets for polysulfides are the expansion joint and perimeter sealing market and the glazing market. Another large market is the insulated glass market. Two-package systems that are easily mixed in automated equipment can cure quickly with heat acceleration. Such systems are used to seal insulated glass units. The key features here are good strength, ease of application and reasonable cost. These features, combined with acceptable weathering properties and fairly low moisture vapor transmission rate (midway between silicones and butyls), are the reason for their popularity. Polysulfides allow for the production of sealed glass windows with expected trouble-free lifetimes of ten years and often more.

2.6 Silicones

Silicone sealants did not penetrate the sealing market until the mid-1960s. They offer some unique features over the polysulfides and other sealants available. Their backbone polymer is as follows, where X represents a reactive group:

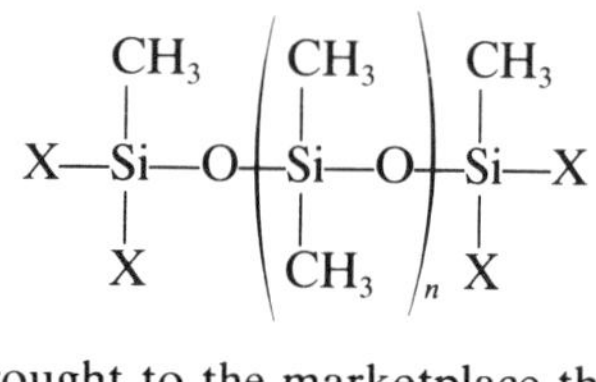

Silicones brought to the marketplace thermal stability and the ability to perform at 140 °C. (Previous sealants could tolerate ~85 °C and this is still the operating limit of the nonsilicone sealants.) It also brought moisture stability so that glass could be bonded and sealed to glass for frameless aquariums. Also, for the first time a clear sealant was available that was more aesthetically pleasing in some applications.

Along with these properties, the silicones featured better weatherability and better resistance to ultraviolet degradation than any previous sealing material, for an expected lifetime of more than 20 years. They also cured faster than the polysulfides or the newly emerging urethanes, giving tack-free surfaces in less than 1 h. In addition, the first silicones had a joint movement capability equal to the best, able to take movement of ±25%.

One of the key limitations of the first silicones was that they released acetic acid during the curing process and thus were not acceptable against concrete, marble and some metal substances. After the initial introduction and acceptance of silicones, modifications soon produced amine cures, caustic cures and neutral cures so that silicone sealants now available can be used on virtually any substrate. The initial silicones had unprimed adhesion to glass and many household-type materials, which allowed their placement into the consumer and DIY areas, but many substrates still needed priming.

Further developments in silicone sealants produced materials that adhere unprimed to most substrates. The newer developments also produced sealants with increased capability for joint movement—silicone sealants with joint movement tolerances up to ±50% of the joint width are available. In fact, a very wide variety is now available and care must be used in choosing the proper silicone for each application.

One important distinction in high-performance sealants and especially within the generic class of silicone sealants is the modulus, or the force required to stretch the sealant a given distance. The lower the modulus, the less stress on the bond line and the substrate. The very high movement capability mentioned, ±50% of the joint width, is possible because some silicones have a very low modulus and also very high recovery values after stretching. The low-modulus sealants are ideal for expansion joints and sealing applications where high movement occurs (where sidewalks or decks meet side walls or in highways, in airfield runway joints or for sealing large precast panels).

The low modulus is not ideal for some applications where the sealant also performs as an adhesive. A classic case of this is in structural glazing, where the glass lite is literally glued on the side of the building with the sealant. The sealant is the glue and also the weather seal. The sealant here must not be so soft or flexible that it permits excess movement of the glass lite. Only some of the higher-modulus sealants can be used in this application and then only a silicone with its ultraviolet stability and resistance to oxidation is permitted, since the bond is in full exposure to sunlight and must not deteriorate.

Thus, silicones of one type or another are being used in virtually every segment of the construction industry, from pure glazing to adhesive seals, insulated glass, expansion joints and highway seals. While the majority are of the one-package ready-to-use type, some two-package mix-on-site materials are being used.

2.7 *Urethanes*

While the silicones were being introduced, yet a third generic type of high-performance sealant was being developed. The urethanes and modified urethanes were introduced in the 1960s. The urethanes generally contain polyethers and urethanes in the base polymer. The polymer is generally terminated with isocyanate groups and cures by the addition of a polyol. The basic chemical structure is

$$\left[O(CH_2)_a-O-\overset{\overset{O}{\|}}{C}-\overset{\overset{H}{|}}{N}-R-\overset{\overset{H}{|}}{N}-\overset{\overset{O}{\|}}{C}-O \right]_n$$

Many commercial variations on this exist, including the termination of the base polymer with epoxy functional groups and curing by the addition of a polyamine. Thus, some differences exist in the urethanes available. Also, some come as the one-package ready-to-use system and others as the two-package mix-on-site variety. However, the urethane sealants as a generic class have less variety and more common features than the generic class of silicones. They accept fillers well and some lower-cost, low-performance materials can be made by extensive filler addition, but only the high-performance products are considered here.

Urethanes show very little creep or flow. They generally recover from stress better than the polysulfides and not as well as the silicones, but all three types are better than any of the other common sealants on the market. The urethanes weather well enough to be acceptable in most sealing jobs, although some are not recommended for glazing and no urethane is used in structural glazing. They generally cure faster than similar polysulfides and slower than similar silicones. They cure better in the cold than polysulfides but not nearly as fast as silicones. They adhere well to glass and some metals, but every urethane manufacturer recommends primed surfaces for most substrates. Thus, in adhesion too they are similar to the polysulfides and many, but not all, of the silicones.

In joint preparation, all three major sealant groups require a clean and dry surface, the urethanes being generally more susceptible to problems if any dampness remains in the surface. The urethanes, like the polysulfides, accept paint quite well. The silicones, except for specially formulated products, do not allow paint adhesion. The paintability feature, however, is not of high value, since urethanes and polysulfides are generally used in external applications where the sealing joint is not usually painted.

In heat stability, the urethanes (up to ~85 °C) are comparable with the polysulfides and less stable than the silicones. Thus, active solar collectors and other hot areas are more likely to use silicone seals.

In the key feature of movement capability, the urethanes are quite good, with most able to tolerate ±25% and some of the lower-modulus products accepting movement ±30% or ±35% of the joint width. They are therefore generally near the middle of the range in this respect also, with the polysulfides and the high-modulus silicones on one side and the very low-modulus silicones on the other.

In another key feature, cost, urethanes again started out as intermediate between polysulfides (cheaper) and silicones (more expensive). This position has changed somewhat with the rising cost of hydrocarbons, because premium urethanes are now equal to or only a few percent lower than silicones in cost and the latter are 15–20% more expensive than polysulfides (for materials only).

None of the premium sealants are recommended underwater on concrete unless special barrier coat primers are used. Underwater applications of all types are generally tough on urethanes and care should be taken in the selection of the sealant for such situations.

In toughness and tear resistance the urethanes are generally better than both polysulfides and silicones. This property makes urethanes very acceptable for sealing joints in high-abrasion areas and situations where cutting, picking and vandalism are likely to occur. Areas such as sidewalks and industrial floors where the sealant is often flush with the wearing surface are suitable for urethanes.

2.8 *Combinations*

Many sealants are available which use urethanes, polysulfides and silicones as a small part of their polymeric backbone or as an additive to enhance one property or another. These sealants do not closely resemble the premium products already described. Coal-tar-extended urethanes and polysulfides are closer to coal tar than to urethanes or polysulfides in performance. Butyl sealants with an additive of silicone or siliconized acrylic latex are really butyl or acrylic sealants with all the properties of a butyl or acrylic except for adhesion, cure rate, extrudability or some other single property or couple of properties that are enhanced slightly by the use of the silicone additive. If high performance is needed, a high-performance sealant is required. The manufacturer's data sheets are the best source of data for finding the sealant that has the properties required for a given application.

3. Installation

With any sealant, for any application, proper joint design is fundamental. The manufacturer's data sheets are also the best source of instruction as to the proper joint preparation procedure and as to the necessity of priming for particular substrates. However, the following are certain general procedures that are usually appropriate for most sealants.

(a) Choose the sealant on the basis of the properties desired.

(b) Plan the joint dimensions on the basis of the total expected movement and the movement capability of the sealant.

(c) Test the adhesion, either primed or unprimed, by putting down a bead of sealant on a few centimeters of a typical joint (prepared as expected to be done on the job), let it cure for 7–14 d and then try pulling it off. All this implies, correctly, that any important sealing job should be planned and investigated several weeks before sealing is actually to be done. When the sealing is actually started, attention must be paid to the joint preparation. Even if no priming is needed, the joint must be clean and dry and free of any oil or detergent residue.

(d) Install a backing material that will control the depth of the sealant and provide a surface to which the sealant will not stick. Adhesion at the joint sides is fundamental, but adhesion to the back restricts the ability of the sealant to expand and contract.

(e) Now install the sealant with a pushing motion, forcing it into the joint. Immediately after installation, tool the sealant with a concave instrument. This forces the sealant into contact with the joint faces—a step that should never be skipped.

If all these steps are followed, the performance advertised by the sealant manufacturer can be expected and the job will be well done. A good sealant, well installed, will then be more than a building material—it will also be a protection for the structure that will help to prevent deterioration of other building elements.

See also: Adhesives in Building; Elastomeric Products from Liquid Polymers; Structural Sealant Glazing

Bibliography

Anon 1978 Weather tester "climate" doesn't faze silicone sealant. *Mater. News* November/December: 5

Baldwin B 1976 Selecting high performance building sealants. *Plant Eng.* 30(1): 58–64

Bruner L B 1963 Room temperature-vulcanizable polysiloxane elastomers. US Patent No. 3,077,465

Cahill J M 1966 Weathering of one-part silicone building materials. *Build. Res.* 3(3): 38–9

Ceyzeriat L, Dumont P 1959 Elastomers based on diorganopolysiloxanes. French Patent No. 1,188,495

Crossan I D 1977 Proper selection and use of silicone adhesives insures extended life. *J. Appl. Polym. Sci., Appl. Polym. Symp.* 32: 421–42

General Electric Company 1967 Platinum catalyst. British Patent No. 1,054,658

Gibb J F 1980 Hidden, but essential—A technical review of backer rod. *Constr. Specifier* 3(3): 40–5

Hyde J F 1951 Siloxane elastomers. US Patent No. 2,571,039

Kasperski M G, Klosowski J M 1978 Construction sealants. *Constr. Specifier* 31(10): 21–6

Klosowski J M 1981 "Premium" sealants of the 1980s. *Adhes. Age* 24(11): 32–9

Klosowski J M 1982 Byword for sealants: Durability. *Adhes. Age* 25(1): 30–2

Matherly J E 1972 Selection and use of RTV silicones as structural adhesives. *Appl. Polym. Symp.* 19: 231–55

Noll W 1968 *Chemistry and Technology of Silicones*, 2nd edn. Academic Press, New York

Rochow E G, LeClair H G 1955 On the molecular structure of methyl silicone. *J. Inorg. Nucl. Chem.* 1: 92–111

Roth W L 1947 The possibility of free rotation in the silicones. *J. Am. Chem. Soc.* 69: 474–5

Smith A H Jr 1972 Silicon dioxide-containing fillers for hardenable silicones. US Patent No. 3,635,743

Smith A L, Angelotti N C 1959 Correlation of the SiH stretching frequency with molecular structure. *Spectrochim. Acta* 15: 412–20

J. M. Klosowski
[Dow Corning Corporation, Midland, Michigan, USA]

Sports Equipment Manufacture

Polymeric materials are now used in large volumes for sports and games equipment which was inevitably first based on naturally occurring materials, particularly wood, leather and natural rubber (NR). In general terms, equipment for sports and games can be divided into three broad categories:

(a) bats, sticks, rackets, rods, skis and so on, each of which might be regarded as an extension of a person's limb;

(b) balls, shuttlecocks, pucks and so on, which are basic to many games; and

(c) injury protection devices for players of games and sports.

Taking these in sequence, very different mechanical properties are demanded for such diverse applications:

(a) high specific stiffness and strength for bats, sticks, rods and so on;

(b) high resilience and elasticity for balls; and

(c) high energy-absorbing properties and toughness for protectors.

Fiber-reinforced polymers (polymer composites) are now used extensively for bats, sticks, rods and so on and have replaced the wood and sometimes metal used previously.

NR is still preferred for many balls, particularly for tennis balls, but modern synthetic elastomers have replaced NR in other cases, particularly where the ultimate in resilience is required.

For protection devices, expanded thermoplastic materials contained in a pouch or shell have significantly replaced the devices of cane in padded fabric or leather used for many years.

While a wide range of polymeric materials is used in specialized footwear developed for sports and games, this subject will not be discussed here.

1. *Historical Note*

The first serious application of natural polymers to games equipment occurred around 1850 when gutta-percha was first used to fashion (solid) golf balls. Around 1900, golf balls were first made of tightly stretched NR thread wound to form a core which was molded inside a shell of gutta-percha to protect the windings. The game of tennis as it is known today was developed in the 1870s and was possible only because rubber balls could be made by the developing rubber industry. Another important early use for NR was as an envelope for pressurized air as a "bladder" for leather footballs (where it replaced pigs bladders used previously) where the low gas permeability of NR could be used to advantage. The high surface friction of NR meant that it also found application in enhancing the gripping of hand-held artifacts.

It was not until after World War II that plastic materials began to be applied to sports equipment and this did not gather pace until after 1960 when, for example, glass-reinforced and, in 1970, carbon-fiber-reinforced epoxy resins began to challenge wood for structural purposes. Wood ceased to be used for rackets in the mid-1980s.

Golf balls remained unchanged in general format for over 70 years but then changed significantly after 1970 due to new materials and manufacturing methods, leading to the reinvention of the solid golf ball.

Protection devices took advantage of the many plastic materials produced in expanded form for shock absorption purposes from 1960 onwards.

2. *Specifications and Other Requirements*

The development of sports and games equipment and the application of new materials have to take place within the confines of specifications designed to preserve the "nature" of each sport or game and to allow appropriate standardization in terms of both geography and time. Such specifications vary considerably in their rigor from one controlling authority to another.

There are also certain necessary but intangible qualities in sports equipment which are difficult to specify and yet are extremely important to the sportsperson or player for thc full and satisfying appreciation of the sport or game being played. Examples are the "click" of a golf ball on club contact, the "response" of a tennis ball off the racket and the "action" of a fishing rod. Such qualities can determine whether or not an otherwise satisfactory product is successful commercially and so manufacturers spend considerable time and energy in developing their own particular expertise and know-how.

3. *Polymer Composites*

3.1 General

Tubular format is usually necessary for hand-held artifacts because polymer composites have a higher density than wood which they usually replace and, in any case, tubular structures are the most efficient in terms of specific stiffness and strength.

Usually, the fiber reinforcement is glass or carbon and, in some cases, aramid and the polymer-matrix epoxy resin. The fiber–matrix ratio is normally 60:40 by weight and the fibers may be in the form of a woven fabric or longitudinally aligned yarns (warp sheet) located and held together by the resin itself in preimpregnated form.

Special attention is paid to fiber alignment in relation to the required properties of the product, with maximum alignment in the direction of required stiffness and strength. Generally, considerable hand fabrication is necessary and automatic methods are rarely applicable. After fabrication, molding takes place under heat and pressure by a variety of methods.

3.2 Fishing Rods, Golf Club Shafts and Ski Sticks

Fishing rods, golf club shafts and ski sticks are of straight, tapered tubular format and manufacture is by the simple expedient of wrapping resin-impregnated glass or carbon fiber fabric onto a removable mandrel and overwrapping with shrink-wrap tape, before curing in an autoclave. Subsequent surface finishing is by centerless grinding.

In the case of golf shafts, torsional properties are important and special consideration has to be given to optimum fiber alignment for both good longitudinal and torsional stiffness. Composite shafts are lighter than steel shafts (wooden shafts were replaced by steel before 1920), but they are considerably more expensive and their advantage in playing terms is debatable. For this reason, they have only had limited commercial success.

3.3 Tennis Racket Frames

Tennis racket frame manufacture from polymer composites is altogether more complicated due to the shaped hollow structure that is necessary. Glass and/or carbon fiber fabric impregnated with epoxy resin is wrapped onto a parallel mandrel but, in this case, a nylon (polyamide (PA)) tube for subsequent pressurization is interposed between the mandrel and the wrapping. The mandrel is removed and the wrapped inflation tube is inserted into a keyhole-shaped mold to form the recognized racket shape. Extra reinforcement is added to the central throat area and a "bridge piece" to complete the oval of the head is fitted into place—this being in the form of resin-impregnated fabric wrapped around a shaped former made from expanded polyurethane rubber (PUR). The mold is closed, heated and the inflation tube pressurized. Consolidation and curing take place and the throat piece core also expands under the action of head so that it becomes integrated into the structure of the frame.

After removal from the mold and deflashing, surface finishing is necessary by repetitive filling and sanding to achieve a high finish for subsequent painting. This is costly in terms of labor and is one reason why racket manufacture has moved substantially to the low-labor-cost areas of the world.

String holes have to be drilled and PA grommets fitted to insulate the sharp edges from the strings. The frames are customized for weight, balance and grip size by molding on a handle from a rigid PU foam. This is done by clamping a mold of appropriate size around the lower shaft of the frame and adding lead weights as appropriate, prior to the introduction of the PU mix from a dispenser.

Racket frame quality is largely related to overall and local stiffness and to vibration suppression and to this end different types, amounts and orientations of fibers are incorporated at different positions around the frame. The frame cross section is also varied for this purpose.

Racket frames have evolved with larger and larger heads (unrestricted by specification) since the mid-1970s, such that the strung area is now up to 40% greater than the previous standard and, even so, frames can now be made 10% lighter. It is here that the justification for polymer composites becomes apparent, because both factors lead to better playing properties in terms of greater energy transference to the ball and improved wieldability.

3.4 Unique Racket Frames

While the tennis racket frame manufacturing method described in Sect. 3.3 is used by the majority of manufacturers, one manufacturer has developed a significantly different method which avoids problems of surface finishing and hole drilling and has other significant advantages.

This method involves injection molding the frame from a chopped-carbon-fiber-reinforced thermoplastic material composed of 40 wt% carbon fiber in PA. To produce the necessary hollow frame, the material is injected around a fusible core which is subsequently melted out. Although the flexural modulus of the chopped-fiber thermoplastic material is significantly less than that of continuous filament/thermosetting material, frames of appropriate stiffness and strength can be produced by increasing the frame cross section. Superior vibration damping properties are also obtained using this technique.

3.5 Other Racket Frames

The remarks made previously for tennis rackets apply generally to squash and badminton rackets, although the latter cannot be made by the fusible core process due to the fragility of the necessary core.

4. Racket Strings

Racket strings were originally based on the intestines of sheep which were appropriately processed and then slit and twisted longitudinally. The greater proportion are now made from extruded/drawn thermoplastic filaments which are then either used in monofilament or multifilament form. In multifilament form, the separate filaments are wrapped around and bonded to a central core, which itself may be either monofilament or multifilament. In the latter case, the core may be a high-strength yarn, for example carbon or aramid fiber, and an outer protective skin may be incorporated by an extrusion coating process. Polyester or, more usually, PA filaments are the main constituents, but a recent innovation is a string based on polyether ether ketone (PEEK). Properties sought in a string are high strength at relatively low modulus with low energy absorption (i.e., high resilience). Strings are tensioned to up to 30 kg and stress relaxation must be minimal.

5. Skis

Skis were originally made from wood due to the required combination of low weight and relatively high stiffness. Modern skis developed since the 1960s are a complex combination of metal and plastics materials and sometimes wood. The design requirements are for controlled longitudinal and torsional stiffness, high damping in flexure and, of course, low weight. To achieve these properties a central "torsion box" is made from a core of wood, PU foam or a honeycomb structure sheathed in fiber-reinforced epoxy resin. The upper surface and sides of the ski are usually of acrylonitrile–

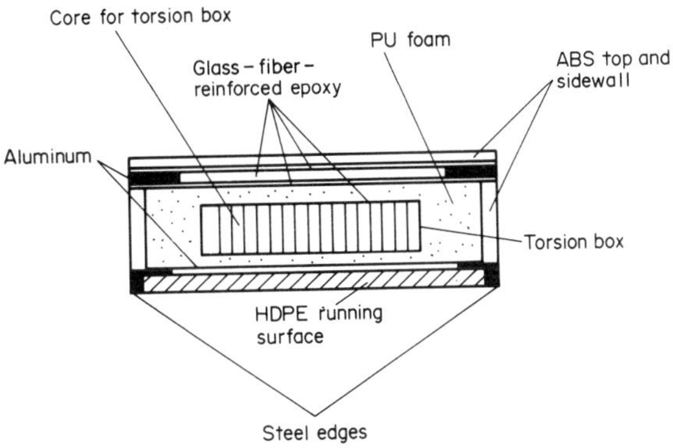

Figure 1
Ski construction

butadiene–styrene (ABS) and the running surface of high-density polyethylene (HDPE) to give low friction with good wear.

The running surface is edged with steel and the technology of ski manufacture is primarily that of satisfactorily bonding together the many dissimilar materials. This may be carried out by using epoxy or other adhesives followed by heat curing, but an elegant modern method is to appropriately assemble the components in a mold and inject an integral skin PUR foam into the hollow box section, so that the components become bonded together (see Fig. 1).

In 1986, the market for skis amounted to nearly 10^7 pairs and this market is serviced by many manufacturers who develop special constructions and manufacturing methods in a highly competitive environment.

6. Balls

Balls are of many and varied types, but most incorporate polymeric materials of some kind. Where the ball is required to bounce and have high elasticity, natural or synthetic elastomers are invariably used and the elastomer may form an enclosure for pressurized air contained in an outer casing, which can be separately optimized in properties according to the duties the ball has to perform. In certain types of ball (such as for cricket or baseball), a hard core may be molded, for example, from expanded PUR.

6.1 Tennis Balls

Tennis balls are made by a process which has changed little over the years and consists of producing a pressurized core based on NR-containing mineral fillers, designed to increase modulus and reduce gas permeation, while detracting little from resilience.

The core is produced by first compression molding hemispherical shells which are then joined adhesively at their rims in a second molding operation in which compressed air is introduced prior to the core being "closed." The pressurized core is then covered with a special cloth which is stamped into 8-shaped blanks, two of which can be made to fit over the spherical surface of the core. NR adhesive has previously been applied to the edges and reverse side of the blanks and to the core, so that the core and cover can be vulcanized together in a third molding operation.

The cloth is specially woven from a wool and nylon blend in a "sateen" pattern so that a nap can be produced on one side by a "raising and fulling" operation and this becomes the wearing surface.

Some 10% of tennis balls are of the nonpressurized type in which the core itself, unsupported by internal pressurization, must provide the required properties of rebound. The core is normally made from polybutadiene (BR) compounded with low-density fillers so that a thicker core wall can be incorporated, but with no increase in weight. Nonpressurized or pressureless tennis balls have the advantage that they do not need pressurized packaging, necessary with pressurized balls, to prevent loss of pressure in storage.

Tennis balls must meet stringent specifications for weight, size, rebound and compression (i.e., deformation under load).

6.2 Squash Balls

Squash balls constitute perhaps the simplest balls in format, but the most complex in terms of material formulation. While they are formed solely from two hemispherical moldings, as in the case of a tennis ball core, a complex combination of properties is required. These are related to playing "speed" (which depends on ball rebound), surface friction (which determines rebound angles) and durability, together with non-wall-marking properties. Due to their required low resilience (about 20% rebound), they absorb considerable energy in play, heat up and so become more resilient and hence faster in play. Speed at temperature equilibrium (about 40 °C) is, therefore, an important factor.

Formulations for squash balls are jealously guarded by manufacturers, but are based mainly on butyl rubber (IIR). Squash balls must meet specifications for size, weight, rebound and compression.

6.3 Golf Balls

Golf balls, in contrast to tennis balls, have changed markedly in construction over the years. As mentioned in Sect. 1, the original polymer-based golf ball of 1850 (the "gutty") was made solely from molded gutta-percha. The more complex rubber-thread-wound ball had greatly improved resilience and hence flight distance.

This ball has been developed significantly to date, but since 1980 has been progressively superseded by a ball with a molded core made from the latest polymeric materials, such that, in 1990, 75% of the annual demand for some 5×10^7 dozen golf balls was of this type.

At its inception, the wound golf ball was made from NR thread highly stretched and tightly wound around a "center" which was initially a liquid in a rubber container. This allowed maximum distortion of the windings and so the generation of high resilience and hence ball speed. The windings were protected from damage by the club head by a molded cover of gutta-percha or balata, which would be molded at a temperature (90 °C) sufficiently low so that degradation of the NR thread was avoided. In addition, the aerodynamically important dimple pattern could be molded into the cover.

Liquid center manufacture was by a variety of methods, but one way was to encapsulate with rubber a frozen liquid based on a suspension of barytes in glycerine and water. This was then vulcanized to produce a liquid-filled rubber sphere, which was again frozen prior to the winding operation. Liquid centers were superseded by solid molded centers made from BR, which enhanced the resilience of the core, but this did not occur until the 1960s. Figure 2 illustrates the components of a modern wound ball.

While gutta-percha and balata compounds were used for golf ball covers for many years, synthetic *trans*-polyisoprene (TPI) was used for some years until the development of the so-called ionomer materials in the 1960s.

Ionomers are thermoplastic copolymers incorporating ionic crosslinks which are thermolabile. The materials are tough and can be molded at relatively low temperatures (160 °C) and thread degradation can be avoided by special molding procedures. Ionomers have now replaced TPI in all but about 5% of golf balls produced in the world, although TPI balls are still preferred by top tournament golfers for whom golf ball durability is not a problem and the "feel" of the TPI-covered wound ball is preferred.

Molded core technology was developed in the 1960s to a level such that, for the first time, the performance of wound balls could be challenged. The core material is again based on BR compounded with certain crosslinking monomers of which zinc methacrylate has been found to be most suitable. Zinc oxide is used as a weight adjusting filler and the cores are made by a simple compression-molding process.

The molded core is hard and highly resilient, but is somewhat brittle and, like the wound core, needs the protection of a separate cover. However, the molded core is not heat sensitive and so the cover can be applied by a direct injection-molding process in which the core is centralized in a mold and direct injection of hot cover material takes place. This allows rapid molding cycles with robotic core placement and ball removal. A significant reduction in expenditure on molds is possible compared with the conventional compression-molding process.

Golf ball specifications cover minimum size, maximum weight, maximum resilience and the necessity for "spherical symmetry" in all its connotations.

6.4 Table Tennis Balls

Balls for table tennis are made mainly from cellulose nitrate (CN) by molding separate hemispherical shells in a precision compression-molding process. These are then joined together with an overlapping joint which is adhesively bonded. Table tennis balls have also been made from acrylonitrile–styrene–acrylate (ASA) terpolymer using a vacuum forming process for the hemispherical shells and joining them by a hot-plate welding process in an automatic

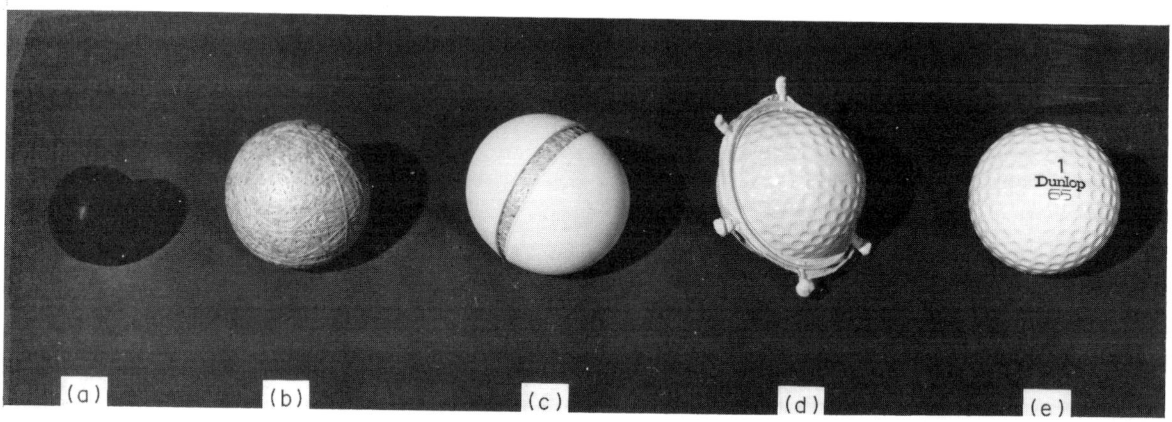

Figure 2
Components of a wound golf ball: (a) center, (b) wound core, (c) molding assembly, (d) molded ball and (e) finished ball

sequence of operations. Centerless grinding is necessary in both cases to produce a perfectly spherical surface and to provide appropriate surface roughness.

6.5 Games Balls

Various types of games balls (e.g., basketballs) were originally made by fabricating layers of rubber-impregnated cloth onto a rubber bladder with the aid of a rubber-based adhesive. This construction has since been superseded by a process in which a rubber-impregnated textile yarn is machine wound around the bladder in a continual fashion so that a reinforcing casing is formed. Compounded rubber sheet is then applied to the surface of the casing and the structure is molded under heat and pressure with an appropriate surface profile.

7. Shuttlecocks

Shuttlecocks for badminton provide another example where modern polymeric materials compete with natural materials—this time feathers and cork. The feathered structure of a conventional shuttlecock can be imitated by a lacelike molding in conical form designed to reproduce the desirable guiding and aerodynamic effect of overlapping feathers. PA molding compound is used specifically because of its good mold flow, but special high-pressure injection techniques are necessary to ensure that the delicate and intricate mold profiles are filled. The qualities of cork have been reproduced by an expanded PVC compound molded to the appropriate shape which is then fixed to the conical skirt to form the complete shuttlecock. Shuttlecocks made in this way are considerably more durable than the feathered variety and can be made semiautomatically to a high degree of consistency.

8. Protection Devices

Protection devices for use by hockey and ice hockey players and other sportspeople are composed of various types of expanded plastic materials which include PVC, polyester, PE and PPE, fitted into an outer flexible pouch or rigid shell. The shell may be of PE for limb protection or PC if in the form of a helmet. Materials and designs are many and various and injection molding or thermoforming are particularly suitable methods for producing the often complex outer casings.

Both open- and closed-cell plastic foams may be used. In one design for a shoulder protector for hockey, an open-cell foam holds air which is trapped in a pouch of restricted air permeability, which itself is fitted into a PE outer shell. On impact, air flow from the structure takes place in a controlled way thus improving on the shock absorbing properties of the materials alone.

9. Table Tennis Bat Rubbers

The rubber covering for table tennis bats has required significant technical innovation to meet critical requirements. The rubber may be in one of two forms:

(a) "pimpled" sheet, that is, sheet with molded projections or pimples; and

(b) pimpled sheet with a sponge rubber underlayer termed sandwich rubber.

The pimpling on the surface was originally designed to improve friction between bat and ball but this effect can be alternatively obtained by using "reversed" pimples (i.e., the plain back of the pimpled sheet) and a rubber compound which produces a "tacky" surface. In the sandwich constructions, sponge rubber (NR and/or BR) may either be made to have high resilience (for attacking play) or low resilience (for defensive play) and these qualities combined with different types and orientations of pimpled sheet lead to a wide available range of playing properties for table tennis bats.

Strict specifications are laid down for the thickness of the pimpled sheet and sandwich layers, together with the length and density of the pimples.

See also: Carbon-Fiber-Reinforced Plastics; Design with Plastics; Elastomers: Applications Overview

Bibliography

Dunlop Ltd 1957 Golf ball core winding machine. GB Patent No. 777,959

Dunlop Ltd 1964 Cellulosic fillers in tennis ball core compounds. GB Patent No. 1,108,556

Dunlop Ltd 1965 Method of vacuum forming and assembling table tennis balls. GB Patent No. 1,131,323

Dunlop Ltd 1968 Carbon fibre reinforced golf shaft. GB Patent No. 1,261,541

Haines R C 1983 Natural and synthetic rubber in sports goods. In: Evans C W (ed.) 1983 *Developments in Rubber and Rubber Composites 2*. Applied Science, London, UK

Haines R C, Curtis M E, Mullaney F M, Ramsden G 1983 Design development and manufacture of a new and unique tennis racket. *Proc., Inst. Mech. Eng., London B* 197

Hull D 1981 *Introduction to Composite Materials*. Cambridge University Press, Cambridge

Spalding 1939 Method for injection molding cover onto golfball core. US Patent No. 2,361,348

R. C. Haines
[Dunlop Slazenger International,
Wakefield, UK]

Spunbonding of Polymeric Fibers

A spunbonded fabric has been defined as a fabric formed by filaments that have been extruded, drawn, laid on a continuous belt and then immediately bonded. Spunbonded fabrics can also be defined

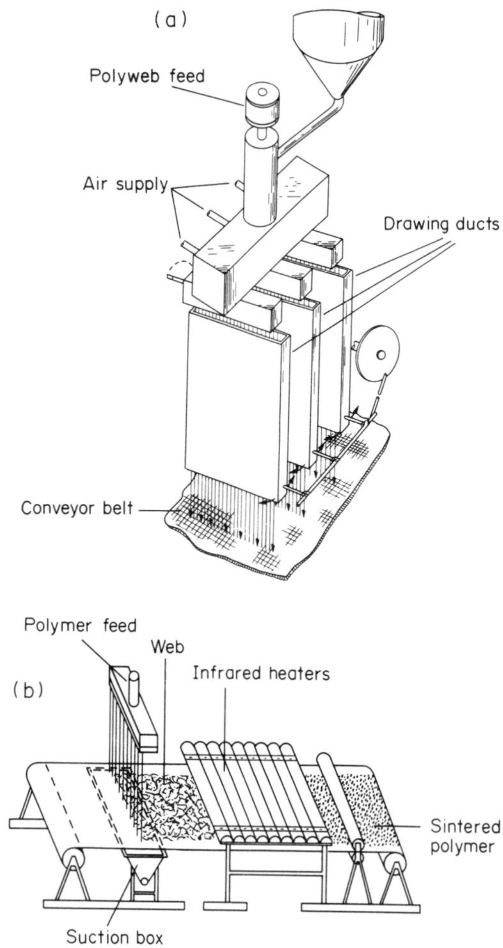

Figure 1
Freudenberg spunbonding: (a) web formation, (b) fabric formation

generically as continuous-filament fibrous structures which can be made in the form of fabrics, sheets and tapes and are prepared from synthetic polymers in a process integrated with fiber manufacture. The origins of the spunbonding process appear to date at least from 1940, when the processes of the spunbonding type were patented by Corning Company to produce glass wool (Slaytor and Thomas 1940). Similar processes for the manufacture of mineral wool from molten siliceous compounds were patented shortly thereafter (Callander 1945). Commercially produced spunbonded fabrics made from synthetic polymers are generally based on the methods patented in the 1950s and 1960s by Freudenberg (1963) of West Germany (see Fig. 1) and by Du Pont of the USA. There is a growing body of patent literature related to spunbonding (e.g., Till and Smallman 1957, Dobo et al. 1967), and summaries can be found in *Textile Progress* (1970, 1973), Casper (1975) and Gillies (1979). Most of the literature, however, pertains to the manufacturing processes and the end uses of spunbonded nonwoven structures. Since the spunbonding process begins with the filament production, the most convenient fiber materials that have been commercially successful are those that can be melt extruded (polyamides, polyesters, polyethylene and polypropylene). However, spunbonded fabrics made from viscose rayon, which were initially uneconomical to produce due to relatively low spinning speeds ($125\ m\ min^{-1}$), have now appeared on the market.

1. Fabric Formation

The process of making a spunbonded fabric involves web formation (extrusion, attenuation and collection of filament bundle) and bonding of web to impart strength, cohesiveness and integrity to the structure. The attenuation of the web may be achieved either by the use of draw-off rollers and subsequently forcing the filament bundle through an aspirator jet with a spreading operation that involves charging the filament bundle with an electrostatic field to deposit the filaments in the form of a web on a conveyor belt, or by drawing filaments into a number of separated sheets with the help of air ducts and then laying the sheets sequentially, such as in the Freudenberg process (Freudenberg 1963). In recent years, the use of supersonic-velocity air streams and the movement of air jets from side to side with some mechanical devices have been used to prepare webs. In other processes, the filaments are blown radially outward in a plane normal to the axis of their flow and into a sidewall of a cylindrical collector to form a tubular web or the polymer may be melt blown onto a moving screen (Buntin and Lohkamp 1973).

2. Bonding of Webs

Bonding of webs is carried out in a variety of ways that include reacting the web with gases such as $HC1$, HBr, BF_3, $BC1_2$, $C1_2$, SO_3, N_2O_3, NO_2 or an SO_2–$C1_2$ mixture (Dong 1970), or by a self-bonding process. In the latter process, the web may be produced from a mixture of two systems of filaments, the one having the lower softening point bonding the other on application of heat and pressure (see Fig. 2). In yet another system, the bonding polymer (i.e., the polymer with the lower softening point) is an integral part of the melt-spinnable fiber-forming polymer, such as a sheath core or a side-by-side spun polymer running along the length of the filament as shown in Fig. 3. The fusion at the crossover point in the web occurs on application of heat and pressure.

Figure 2
Scanning electron micrograph of polyethylene TYVEK style No. 1073 spunbonding fabric 500

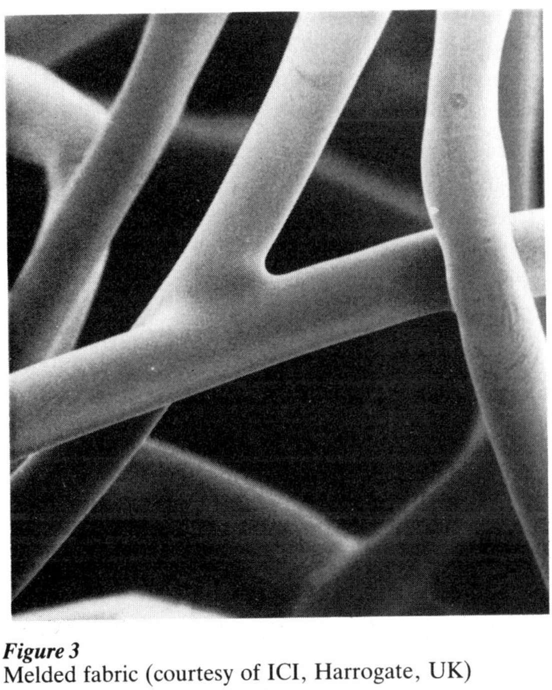

Figure 3
Melded fabric (courtesy of ICI, Harrogate, UK)

3. *Mechanics of Web Formation*

There have been numerous empirical studies dealing with the mechanical and physical properties of commercial products and the suitability and performance of these products as compared with already existing textile structures in various industrial applications (e.g., carpet backing, reinforcement in pavements, packaging, hospital use, furniture industry, shoe and luggage industry and filtration applications). Spunbonded fabrics exhibit extremely high tensile and tear strengths in comparison with conventional woven or adhesive-bonded nonwoven fabrics. Properties of some commercial fabrics are given in Table 1. The mechanical and aesthetic properties of the spunbonded materials depend on the fiber and web structural factors listed in Table 2.

Analysis of the mechanics of spunbonded structures involves the study of (a) the dynamics of the threadline and the effect of air drag on the spinline, (b) fine-structure development and (c) the formation, bonding and mechanics of fabric structure. However, it is only recently that the mechanics of formation and the structural mechanics of the spunbonded fabrics have received attention in the literature. The understanding and prediction of the behavior of filaments during melt spinning and the pattern of spun laying down during fabric formation require the consideration of aerodynamics and the electrostatic and mechanical forces acting on the spinline.

One of the first studies that attempts to simulate the formation of web (Hearle et al. 1976a, b) suggests that the mechanism of buckling (i.e., the form taken by threads being fed too rapidly onto a moving belt) is governed by the geometry of feeding and by the linear density and the bending and torsional rigidities of the threads. The form taken by the threads laid down can be represented by the equation of a modified cycloid, provided the speed ratio between the feed and the belt is greater than 2. However, the experimental conditions are far from realistic.

The mechanics of structure formation in the melt-spun filament have been extensively reported in the literature, but the information on the threadline dynamics in the spunbonding-type systems is scanty. Gould and Smith (1980) point out that since the speed at which filaments or yarns flow can be extremely high, the air drag will have an effect on the development of the fine structure and consequently on the mechanical properties of the filaments. They suggest that the drag coefficient C_D on fine textile filaments less than 0.3 mm in diameter in asymmetric flow in a wind tunnel can be expressed as a function of the radius Reynolds number Re_a: $C_D = 0.27\,Re_a - 0.61$ for $20 < Re_a < 200$ at air velocities below $100\ m\ s^{-1}$, where the length Reynolds number $Re_x = Ux/v$, the radius Reynolds number $Re_a = Ua/v$

Table 1
Some typical physical properties of commercial spunbonded nonwoven fabrics

Property	Cerex	Fibertex 200	Reemay 2470	Terram 1000	Tybar 3351	Tyvek 1085D
Polymer type	Nylon 6	polypropylene	polyester	low-density polyethylene/ polypropylene	polypropylene	high-density polyethylene
Type of bonding	autogenous area bonded	needle punched	thermally bonded at filament crossovers	area thermally bonded bicomponent	area thermally bonded	area thermally bonded under pressure
Nominal basis weight range ($g\,m^{-2}$)	12–17	20–100	20–200	70–300	50–140	54–110
Thickness[a] (mm)	0.10–1.15	1.27	0.13–0.75	0.5–1.5	0.15–0.25	0.17–0.25
Breaking strength[b,c] ($N\,cm^{-1}$)	182/116	570[d]	565/500[d]	850[e]	512[d]	119/140
Breaking elongation[b] (%)	45	125[c]	100[f]	80		26/33
Tongue tear strength[c] (N)	36/30	200	54/63	250[g]	255[h]	5.3/5.5[j]
Bursting strength (kPa)	276[k]	1724[h]	445[l]	1100[m]		1840[l]
MIT-Flex test (cycle)						7100[m]
Air permeability ($dm^2\,min^{-1}$)	1094[n]	3040[n]			306[n]	
Water-vapor permeability ($gm^{-2}\,d^{-1}$)				24[p]		647
Opacity	moderate	high	high	high		90[q]

a Static load b Strip test c Machine direction/transverse direction d Grab strength e Grab strength ($N\ 50\,mm^{-1}$) f Crimped-type polyester g BS 4303 (1968) h Trapezoid test j Elmendorf tear test (impact tear) k ASTM-D2265 l Mullen test m BS 4768 (1972) n ASTM-D-737-75 p Manufacturers test method q Eddy opacity test

and the drag coefficient $C_D = D/(\rho U^2 \pi a L)$; U is the relative velocity between filament and air, a is the filament radius, υ is the kinetic viscosity of air ($1.57\times10^{-5}\,m^2\,s^{-1}$), ρ is the density of air ($1.18\,kg\,m^{-3}$), L is the length of filament exposed to air flow, D is the total drag force and x is the axial distance along the filament.

In this experiment, the test filament is suspended freely from a sensitive strain gauge and a small weight is attached to the other end to keep the filament taut. The midsection of the filament passes through the wind tunnel. Alternatively, Chen et al. (1980) have shown that the air drag coefficient on a spinline (polypropylene filament) being drawn down through an aspirator (an arrangement similar to a spunbonding process) may be expressed by the following relationship which shows a much better agreement with the experimental results: $C_D = 0.91\,Re_D^{-0.61}$. The difference in the results is due to the fact that in the latter case the filament is falling freely on the conveyor belt.

Theoretical treatment of the mechanics of bonded fabrics has received considerable attention (Hearle and Newton 1968). Hearle and Ozsanlav (1970), in a theoretical treatment of the mechanics of bonded nonwoven fabrics, discuss the contribution of the binder deformation in addition to the component fibers. They suggest that the total energy expended in deforming a bonded fiber network can be expressed by

$$E = \sum_{\text{all } i}\left(m_i\int_0^{\varepsilon_i'}\sigma\varepsilon_i'\,d\varepsilon_i' + m_i\int_0^{\varepsilon_i'}\sigma\varepsilon_i''\,d\varepsilon_i''\right) \qquad (1)$$

where σ is the stress, ε_i' is the fiber strain, ε_i'' is the binder strain and m_i is the mass of the segments. Also, the contribution of orientation angle distribution and curl factor is taken into account in predicting the overall mechanical behavior of the structure, but the contribution of the buckling of transverse filaments is ignored. The filament bonding and bond rotation should be taken into account for the theory to be effective. Dent (1980) suggests that this is important in computing the initial modulus of fila-

Table 2
Factors affecting the properties of spunbonded fabric

Filament properties	linear density, tenacity, modulus, elongation, cross section, crimp, fiber morphology
Bonding parameters	binder type, binder concentration, binder distribution, self-bonding
Filament arrangement	fabric weight uniformity, filament separation, orientation

ment webs and has found good agreement with the theory.

The classical nonwoven theory among several deficiencies ignores one major parameter, that is, the bonding level. Bonding level and the geometry of the bonding polymer are quite different from the latex-bonded fiber web. The level of bonding usually creates large discrepancies between the predicted ($\simeq$0.5) and the measured ($\simeq$3) Poisson's ratio. The phenomenon of lateral contraction is modified by compressive forces generated during uniaxial extension as has been suggested by several authors (Goswami et al. 1984). The mechanics of deformation of spunbonded fabrics are further complicated by the lack of estimation of free length of fibers between the bonds. A comprehensive theoretical treatment of the mechanics of tensile deformation of spunbonded fabric is still not available.

See also: Fibers and Textiles: An Overview; Fibers: Formation and Processing

Bibliography

Buntin R R, Lohkamp D T 1973 Melt-blowing: One-step web process for new nonwoven products. *Tappi* 56: 74–7

Casper M S 1975 *Nonwoven Textiles*. Noyes Data Corporation, Park Ridge, NJ

Chen C H, Goswami B C, White J L, Spruiell J E 1980 Mechanics of spunbonded fabrics. *ACA Cellulose, Paper and Textile Division Conf.*

Chen C H, White J L, Spruiell J E, Goswami B C 1983 Mechanics of spunbonded fabrics. *Text. Res. J.* 53: 44–51

Dent R W 1980 The initial modulus of filament webs. *J. Eng. Ind.* 102(4): 360–5

Gillies M T 1979 *Nonwoven Materials: Recent Developments*. Noyes Data Corporation, Park Ridge, NJ

Goswami B C, Suryaderara, Vigo T L 1984 Poisson's ratio in the thermally bonded non woven fabrics. *Text. Res. J.* 54: 391–6

Gould J, Smith F S 1980 Air-drag on synthetic-fibre textile monofilaments and yarns in axial flow at speeds of up to 100 meters per second. *J. Text. Inst.* 71: 38–49

Hearle J W S, Newton A 1968 Non woven fabric studies. 15. Application of fiber network theory. *Text. Res. J.* 38: 343–68

Hearle J W S, Ozsanlav V 1970 Studies of adhesive-bonded non-woven fabrics, 1: A theoretical model of tensile response incorporating binder deformation. *J. Text. Inst.* 70: 19–28

Hearle J W S, Sultan M I, Govender S 1976a The form taken by threads laid on a moving belt, 1: Experimental study. *J. Text. Inst.* 67: 373–6

Hearle J W S, Sultan M I, Govender S 1976b The form taken by thread laid on a moving belt, 2: Mechanisms and theory. *J. Text. Inst.* 67: 377

Textile Progress (September 1970, September 1973) issues devoted to Spunbonding

There is also an extensive patent literature covering all aspects of spunbonding:

Callander M E 1945 US Patent No. 2,382,290

Dobo E J, Dong W K, Mallonee W C 1967 US Patent No. 3,542,615

Dong W K 1970 US Patent No. 3,676,244

Freudenberg C K G 1963 BD 1,064,847

Levy M R 1966 US Patent No. 3,276,944

Parrish E, Katz M 1964 US Patent No. 3,117,055

Slaytor G, Thomas J H 1940 US Patent No. 2,206,058

Till D E, Smallman C R 1957 US Patent No. 2,810,426

Vogt C M 1968 US Patent No. 3,772,417

Wagner W S, Roberts J D 1970 US Patent No. 3,543,332

B. C. Goswami

[University of Tennessee, Knoxville, Tennessee, USA]

Structural Adhesives

The first important structural adhesive, a vinyl phenolic resin composition, was developed for the bonding of aircraft fuselage structures during World War II (de Bruyne 1950). Shortly after, the adhesive qualities and processing advantages of epoxy resins were recognized and the foundations for structural bonding formed.

For the purpose of this article structural adhesives are materials used to join rigid load-bearing adherends. They must be able to distribute stresses uniformly throughout the bonded area, have sufficient cohesive strength to sustain applied loads, and be able to withstand service environments for a period of time much greater than the expected lifetime of the structure. The principal materials bonded are metals, although in recent years a need has developed to bond fiber-reinforced composite structures.

The bonding of rigid substrates is of importance to a wide variety of manufactured products. Although most adhesives were originally designed for aerospace applications, their use has gradually expanded to other industries. This article deals primarily with metal-to-metal bonding as applied to the aerospace, appliance, automotive, electrical and electronics industries. However, even in the construction field some combinations of materials, such as the epoxy–polysulfides and epoxy–polyamides, originally developed for aircraft applications, have become important commercial products.

The physical characteristics of the substrate dictate the types of materials (chemically) best suited for bonding. Proper substrate preparation, interfacial wetting and interfacial attractions are prerequisite to good bonding but beyond the scope of this article; an excellent review of the theories of bonding, however, is given by Zisman (1977). Much of the thinking encompassing these phenomena can be expressed in a simple rule of adhesion published by de Bruyne (1939): "Provided we use pure or simple substances as adhesives then there is a good deal of evidence that strong joints can never be made by polar adherends with nonpolar adhesives or by nonpolar adherends with polar adhesives."

Since most structural bonds involve polar substrates (metal oxide surfaces), it follows that polar resins are the basis for most adhesive formulations. Thermoset resins are favored over thermoplastics. They resist creep under sustained loads and are highly fluid at some point in the bond formation process. The latter condition provides better wetting of substrates and superior interfacial bonding. In keeping with these concepts, the preferred adhesive resins are epoxies, phenolics and blends of these products with other polymers. Vinyl-based polymers including cyanoacrylates are used where rapid bond formation is important and polyimides or other special polymers are employed where elevated temperature performance is a requirement. Urea and melamine formaldehyde resins are not widely used as structural adhesives as ureas have poor water resistance and melamine resins are outperformed by the phenolics on metal substrates. Isocyanates or polyurethanes likewise find only limited use in structural bonding. They tend to depolymerize above 120 °C, are subject to hydrolysis and autoxidize on thermal or ultraviolet light exposure (Frisch and Reegen 1972).

The sections that follow include a description of the more popular resins and their chemistry, together with a brief discussion of the important bond test methods, formulation, bond properties, end-use patterns and future trends. Throughout the text various resins and curing agents will be referred to by chemical name and all are commercially available from chemical suppliers.

1. Polymer Chemistry

Undoubtedly the most versatile materials used in the formulation of structural adhesives are epoxy resins. Besides metal-to-metal bonding, they are excellent adhesives for fiber-reinforced plastics, paper, wood, ceramics and rubber. They cure without the evolution of volatile by-products and are highly resistant to moisture, a variety of solvents, alkalis and acids and natural environments. Service temperatures range from the cryogenic regions to beyond 200 °C in specialized formulations. In combination with phenolics, polyamides, nitrile elastomers, polysulfides, vinyl polymers and other structural resins, properties can be tailored to fit a broad spectrum of applications.

Historically, the first mention of the basic epoxy resin, the diglycidyl ether of bisphenol A (**1**) was by Schlack (1938, 1939). He disclosed that this resin could be hardened with equivalent amounts of organic amines and showed strong adhesion. Parallel development activities started a few years later: patents were issued to Castan (1944) in Switzerland on the use of the resin as a denture material and to Greenlee (1944) in the USA as an intermediate for surface coatings. The development in adhesives started around 1948 and has since moved forward to the wide range of products available today.

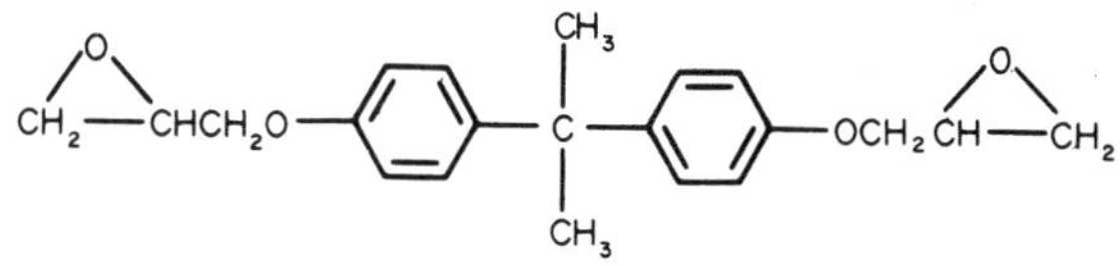

(**1**)

The versatility of epoxy resins stems from the many variations in chemical structures available from both the resins and the curing agents. Although a majority of the epoxy adhesives are based on the diglycidyl ether of bisphenol A (**1**), if resins with higher-temperature performance are required the bisphenol part of the molecule can be replaced with phenolic aldehyde condensation products which increase the number of functional groups per molecule. Typical products of this class are glycidyl ethers based on novolacs (**2**, $n=2,3$), tetraphenol ethane (**3**) and bisphenol A novolacs (**4**, $n=3,4$).

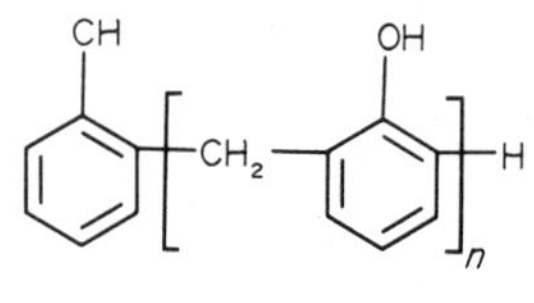

(**2**)

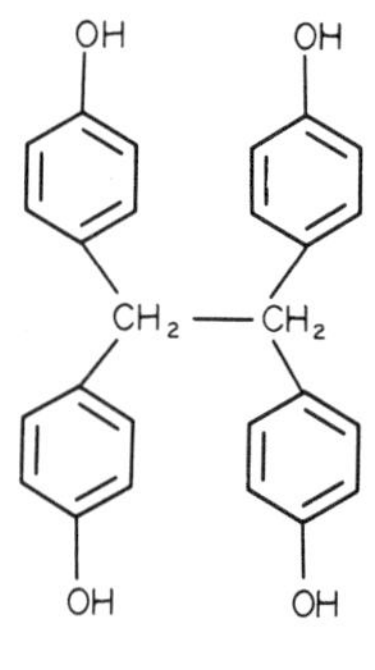

(**3**)

There are a number of low-viscosity epoxy resins which can be used to control the viscosity and tack of epoxy formulations. Most common are the monofunctional diluents such as butyl glycidyl ether and

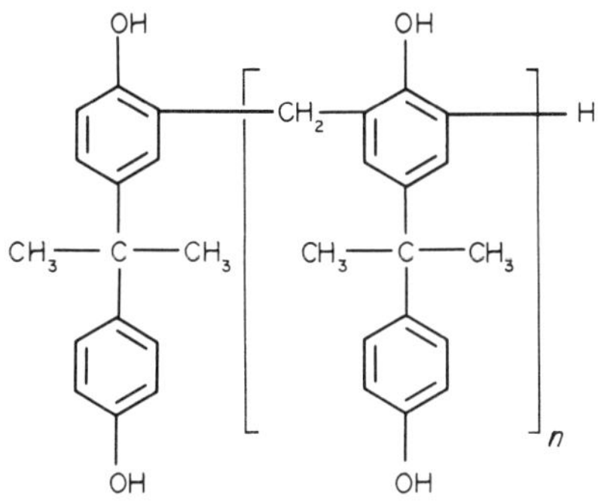

(**4**)

phenyl glycidyl ether. However, these materials reduce elevated-temperature performance, and vinylcyclohexene diepoxide, which has a minimal effect on the glass-transition temperature, is often used as an alternative. A cycloaliphatic diepoxy carboxylate (**5**) although primarily used in electrical castings, has also been demonstrated to be an excellent diluent for use in high-temperature epoxy adhesives (May and Nixon 1961).

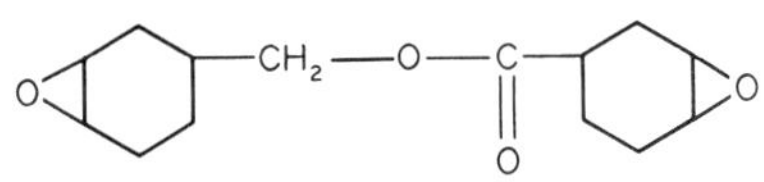

(**5**)

More recently, a number of adhesive formulations based on glycidyl amines have become available. Bonds prepared from products such as tetraglycidyl methylenedianiline (**6**) and triglycidyl *p*-aminophenol (**7**) display excellent bond strengths at both ambient and elevated temperatures. Also available to the adhesive formulator are flexibilizing resins where greater toughness or peel strength is required, halogenated resins, which afford fire retardancy, and a variety of specialty resins designed for specific applications.

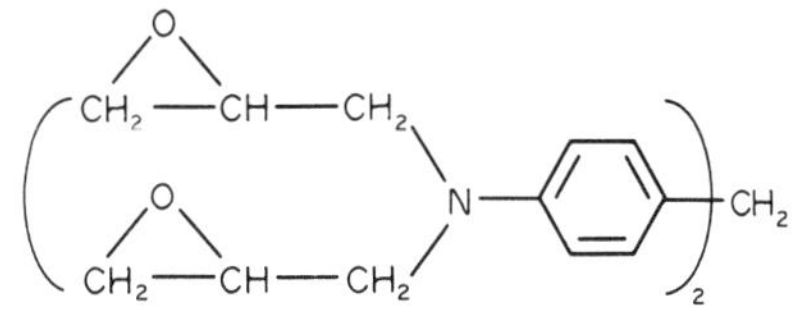

(**6**)

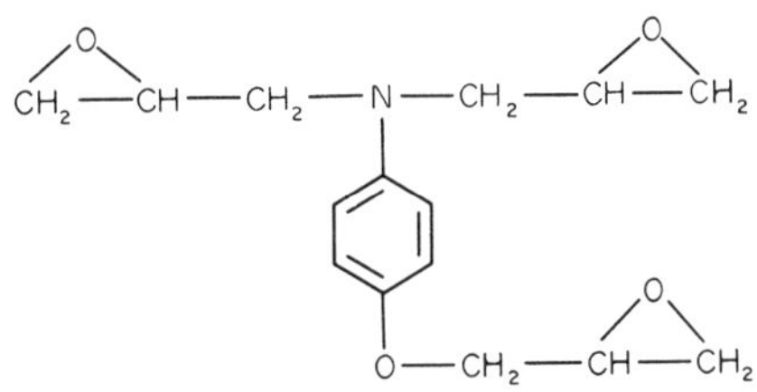

(**7**)

Epoxy resins developed their three-dimensional crosslinked structure by chemical reactions with curing agents. Table 1 shows the bond strengths that might be expected when the major curing agents are used in combination with compound (**1**) (Dannenberg and May 1969). With a wide variety of resins and curing agents available, it is readily apparent that the adhesive formulator has an almost infinite number of material combinations to choose from.

The most important curing agents are the aromatic and aliphatic amines. Thus, the curing reactions of greatest importance are as illustrated in Scheme 1. It is also possible to cure epoxy resins with organic carboxylic acids or anhydrides and by catalytic homopolymerization of the epoxide group using Lewis acids or bases. However, these types of curing

Table 1
Bond strengths expected from major curing agents

Curing agent	Typical cure temperature (°C)	Expected tensile shear strength on aluminum (MPa)					
		(−57 °C)	(24 °C)	(82 °C)	(121 °C)	(149 °C)	(260 °C)
Polyamides, polysulfides	24	20.7	27.6	6.9			
Aliphatic amines	24	13.8	20.7	17.2			
	80–99	20.7	24.1	17.2	3.4		
Mixed aliphatic tertiary amines	80–99	20.7	24.1	24.1	6.9		
Aromatic amines	135–163	20.7	24.1	27.6	24.1	13.8	
Insoluble latent amines	149–177	17.2	24.1	24.1	20.7	10.3	
Carboxylic acid anhydrides	93–149	13.8	17.2	20.7	17.2	10.3	
Phenolic resole resins	166–177	17.2	17.2	13.8	13.8	13.8	10.3

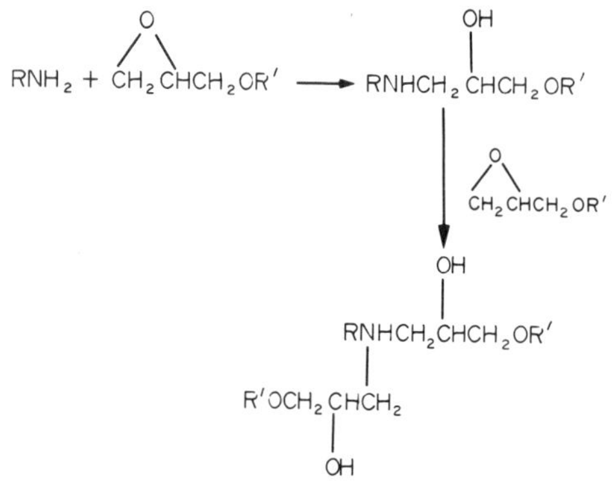

Scheme 1

agents are not widely used in epoxy adhesive formulations for a variety of reasons.

Phenol–formaldehyde resins are useful in structural adhesives because they are highly resistant to hostile chemical environments and form strong bonds at elevated temperatures. Service temperatures exceed those of the epoxies. The main disadvantages are the evolution of volatiles during cure and susceptibility towards atmospheric oxidation, which limit the useful life of the resins at elevated temperatures. Most structural adhesive formulations involve combinations with high-molecular-weight thermoplastics, elastomers or epoxy resins. The major use of phenolic resins is in the manufacture of plywood; major structural applications are in the aircraft and automotive fields.

There are two types of phenolic resins used in structural adhesives—the resoles and the novolacs. The resoles are generally made using an alkaline catalyst and an excess of formaldehyde. Because of the excess formaldehyde, the resole structure (**8**) is dominated by oxymethyl groups as illustrated. The predominance of oxymethyl groups dictates a highly reactive and chemically complex cure mechanism. Curing involves the elimination of water and formaldehyde. The ratio of the benzyl ether linkages x to the methylene linkages y is dependent on the phenol–formaldehyde ratio and the reaction conditions under which the resin is prepared and cured. Formaldehyde is eliminated during the cure, but not in proportion to the number of benzyl ether linkages (Zinke et al. 1939). This suggests that the formaldehyde reacts further with the phenolic nucleus (both *meta* and *para* positions are active sites) or the methylene bridges (Waterman and Veldman 1936).

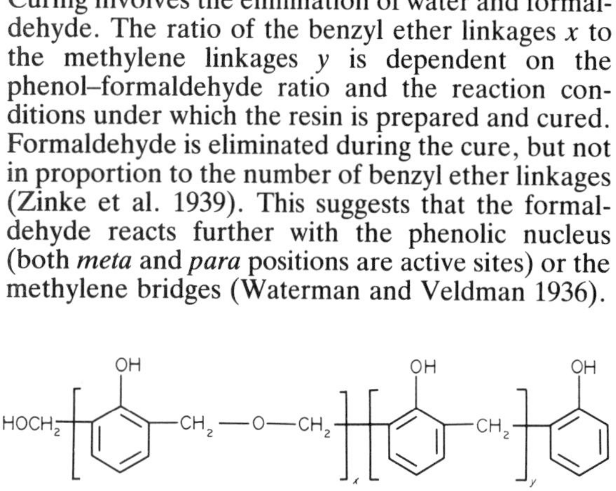

(**8**)

The novolac resins (**2**, $n=2,3$) are prepared using a molar deficiency of formaldehyde. An acidic catalyst is generally preferred. Since these resins are less reactive, a catalyst or curing agent is generally added to promote crosslinking. The most common is hexa methylenetetraamine or "hexa." The curing agent serves as a latent source of formaldehyde. This combination has the advantage that minimal water of reaction is evolved during the cure. As in the case of the resoles, the chemistry of the novolac cure is highly complex and beyond the scope of this article. The chemistry of phenolic resins is discussed further by Martin (1956) and Megson (1958).

The isocyanates or urethanes, in spite of the previously mentioned drawbacks, find use in the bonding of fiber-reinforced plastics for the automotive industry (Kimball 1981) and in the bonding of elastomers to themselves or metals (Schollenberger 1977). The reactions of isocyanates with alcohols and amines form urethanes (**9**) and ureas (**10**).

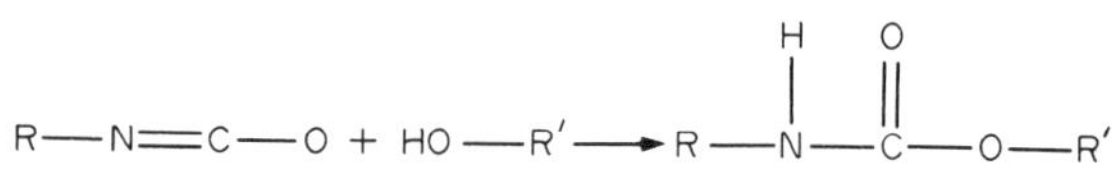

(**9**)

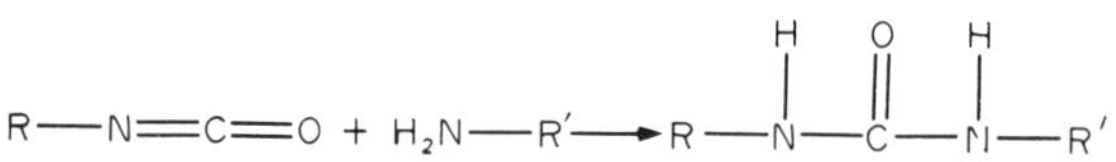

(**10**)

There are two types of urethane adhesives. The two-component systems utilize the chemistry described earlier. One component contains the isocyanate, while the other is an amine, an alcohol-containing compound or a polymer. The one-component urethanes rely on atmospheric moisture for cure. Water reacts with the isocyanate group to form a carboxylate group, which splits off CO_2 and NH_2^-; the latter catalyzes the polymerization of the isocyanate. The rapid cure of the urethanes is of importance for assembly line applications.

Two different monomeric species are of importance to the chemistry of the vinyl structural adhesives—the glycol diacrylates or methacrylates (**11**, R=H, CH_3) and the alkyl 2-cyanoacrylates (**12**). Glycol acrylates are used to formulate products known as anaerobic adhesives. They are single-component liquids that can be stored at room temperature in the presence of oxygen and will polymerize rapidly to a solid when confined between two surfaces which exclude air. They also cure rapidly when exposed to ultraviolet light. Because of the rapid cure, products of this type are also readily adaptable to assembly line operations in the automotive and electronics industries. Second-generation acrylics are now appearing, which are less affected by surface contamination and are capable of bonding a greater variety of surfaces.

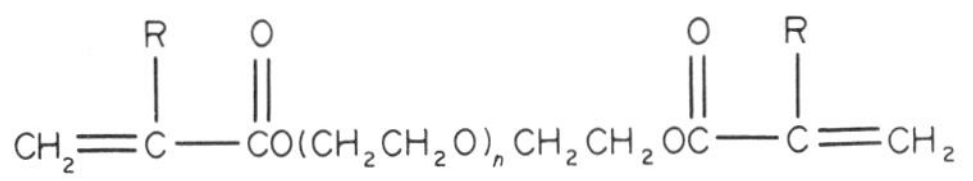

(11)

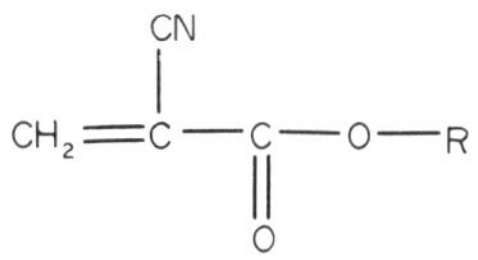

(12)

Cyanoacrylate adhesives display the same room-temperature storage and cure characteristics as the anaerobics. However, in this case the cure mechanism is an anionic polymerization catalyzed by a weak base. The formulated products thus contain inhibitors such as sulfur dioxide for stability against anionic polymerization and quinone derivatives as free-radical inhibitors. These products adhere well to a variety of substrates and are also candidates for assembly line manufacturing. Excellent reviews, with numerous references, can be found for both the anaerobic (Murray et al. 1977) and cyanoacrylate (Cover and McIntire 1977) types of products.

Another important class of structural adhesives are those which yield useful bonds above 260 °C. The classic high-temperature adhesive is a phenolic–epoxy combination (Naps 1953) which displays useful bond properties up to 425 °C. Aromatic polyimides, made primarily by the reactions of aromatic diamines with tetrafunctional aromatic carboxylic acid diahydrides, have useful bond strengths at about 370 °C and have better high-temperature aging characteristics than the epoxy–phenolics. Other polymers which appear of interest for this purpose are the polybenzimidazoles, polybenzothiazoles, polyquinoxalines, polyphenylquinoxalines and the polybenzoxazoles. Useful bond strengths have been reported up to 540 °C; however, the useful lifetime of the bonds drops to 1–10 h above 425 °C (Aponyi 1977).

2. *Testing*

Important adhesive-bond properties to be considered are strength in tension, shear and peel, toughness, environmental resistance and fatigue strength.

The most common test is the lap or tensile shear test. The specimens are inexpensive and easy to make and the procedure is readily reproducible. The test is used for preliminary evaluation of adhesives and for quality assurance. The forces involved in bond rupture are a combination of the tensile and shear strengths of the adhesive and peel forces acting primarily at the adhesive–adherend interface. Testing in pure tension merely reflects the tensile properties of the polymer involved and pure-shear specimens are difficult to prepare.

The toughness of adhesive bonds can be measured in a number of ways. Most common are the T-peel strength and floating roller peel tests. In peel testing at least one of the adherends is flexible and is pulled away from the bondline in a manner which concentrates the stress at the peel front. Another measure of toughness is the impact or block-shear test, whereby the bond between two blocks of adherend is sheared by a single, calibrated blow of a test hammer. The difficulty with this test is the need for perfect alignment between the test specimen and the hammer. Recently, toughness of adhesive bonds has been evaluated using a double-cantilever-beam specimen. This is widely recognized as an appropriate measure of bond toughness. The measurement is called the fracture energy (G_{Ic}) and is the energy required to initiate the propagation of a crack in the bondline.

Another important characteristic of a bonded structure is its ability to resist hostile environments. The most common method of evaluation is to expose lap-shear specimens, either under tension or in the relaxed state, to the environment in question. The environmental effects most commonly studied are those of moisture, salt spray, selected organic solvents and temperature. One of the more important tests developed is the wedge-crack test, which serves to evaluate the entire bonding process. The substrate is bonded in a manner whereby a wedge of fixed dimension can be forced into the end of the bondline to initiate a crack. The rate at which the crack growth progresses with time is then measured at 49 °C in an atmosphere of 95% relative humidity. The rate is not only dependent on the adhesive used, but also on the quality of the substrate preparation (cleaning) prior to bonding.

A further bond property tested is the resistance to fatigue. The test requires a machine that can supply a cyclic load to bonded specimens in tension, compression or both. Since the test involves expensive equipment and quality adhesives usually require numerous test cycles, fatigue data are time consuming and expensive to obtain. Consequently, only a few commercial adhesives have fully characterized fatigue behavior.

Finally, testing is also important for quality assurance. Until recently, the acceptability of purchased products was based on mechanical testing of bonds using the methods described. However, because of

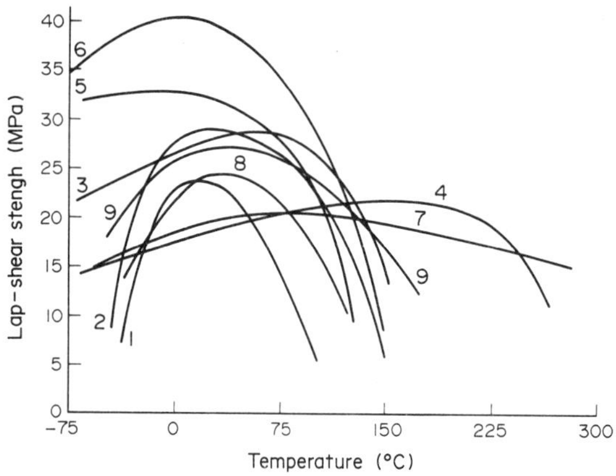

Figure 1
Lap-shear strengths of typical structural adhesives: 1, general-purpose epoxy polyamide; 2, one-part general-purpose epoxy; 3, aromatic amine-cured epoxy; 4, epoxy phenolic; 5, rubber-modified epoxy; 6, nylon epoxy; 7, polyimide; 8, vinyl phenolic; and 9, nitrile phenolic

the wide range of formulative components available to the manufacturer, changes can occur in formulations, but go undetected by mechanical tests. This is a serious problem. Whereas a formulation change may not affect short-term mechanical strengths, long-term effects, such as reduced environmental or fatigue resistance, could result. Fortunately, recent efforts have done much to alleviate this problem; for example, procedures have been evolved for the chemical characterization of epoxy formulations. Both analytical procedures (May et al. 1976) and characterization methods (May et al. 1979) are documented in the literature.

3. Formulation

The lap-shear strengths of some typical structural adhesives are shown as a function of temperature in Fig. 1. Adhesive formulations consist of a matrix–resin system and, in general, a filler or mixtures thereof. The resin system can be highly complex consisting of thermoset resins, curing agents, catalysts, polymers added to achieve special properties, and other additives such as heat stabilizers. The technology of adhesive formulation is the art of compromise; the formulator must add the correct concentration of a component to achieve the desired effect without unwittingly destroying some other needed property. For example, a thermoplastic may be added to a thermoset for toughening, but too much could reduce the elevated-temperature performance of the bond to undesirable levels.

Representative formulations of structural adhesives are given in Table 2. The general-purpose epoxy polyamide formulation is typical of the twin-tube kits for home use. Satisfactory cures can be obtained at room temperature; however, mild heat cures (up to 100 °C) will result in better bond performance. Fumed silica is added to one-part general-purpose epoxy adhesives to give thixotropy. This reduces sag of the adhesive on vertical surfaces. The curing agent is highly stable until heated. The normal curing temperature is 177 °C; however, accelerators can be added to the formulation which permit cures in the 115 °C range.

Aromatic amine-cured epoxy formulations are versatile products with good elevated-temperature performance up to 150 °C. There is a wide variety of aromatic amines available which will give roughly the same strengths when used in an amount chemically equivalent to the epoxide content of the resin.

Table 2
Representative formulations of structural adhesives

Adhesive	Formulation	Parts by weight
General-purpose epoxy polyamide	liquid diglycidyl BPA[a] resin	100
	polyamide curing agent	75
One-part general-purpose epoxy	liquid diglycidyl BPA resin	100
	alumina	20
	fumed silica	10
	dicyandiamide curing agent	6
Aromatic amine-cured epoxy	liquid diglycidyl BPA resin	100
	silica flour	80
	liquid aromatic amine	20
Epoxy phenolic	solid diglycidyl BPA	33
	phenolic resole	67
	aluminum dust	100
	dicyandiamide curing agent	6
	Cu 8-hydroxyquinoline	1
Rubber-modified epoxy	rubber-modified resin	100
	melamine resin	2
	fumed silica	5
	dicyandiamide	8
	3-(3,4-dichlorophenyl)-1,1-dimethylurea (Diuron)	1.5
	aluminum dust	50
Nylon epoxy	liquid diglycidyl PBA resin	25
	nylon	75
	dicyandiamide	10
Polyimide	linear condensation polyimide	100
	aluminum dust	80
	fumed silica	5
	arsenic thioarsenate	2
Nitrile phenolic film	nitrile rubber	100
	phenolic resin	100
	sulfur	2
	zinc oxide	5
	benzothiazyl disulfide	1.5
	carbon black	20

a BPA, bisphenol A

Copper 8-hydroxyquinoline is added to epoxy phenolic formulations to improve the high-temperature aging of bonded structures. Because of the presence of the phenolic resole, water is evolved during the cure. This results in a foamy bondline, but the consistency of the bond performance is unaffected.

The rubber-modified resin, used in formulation of rubber-modified epoxy adhesives, a 10:1 mixture of liquid diglycidyl bisphenol A resin and carboxy-terminated nitrile rubber with an epoxide–carbonyl reaction catalyst at about 100 °C. The exact reaction conditions determine the morphology of the rubber as a dispersed phase and accordingly govern bond performance. Addition of Diuron permits cures in the range 110–120 °C.

Of the other formulations given in Table 2, nylon epoxy has a very high peel strength and fracture resistance. It is fabricated as a film cast from a solution of the components in an alcohol–water or an alcohol–chlorinated-hydrocarbon solvent.

Vinyl phenolic was the first structural adhesive. It was used in the UK during World War II to assemble the DeHavilland aircraft. Bonds were made by spreading a solution of a phenolic resole on the parts to be bonded and sprinkling it with a powdered polyvinyl formal resin. The solvent was allowed to evaporate and the resultant assembly was subsequently cured with heat and pressure. An exact formulation is thus not possible. Although the technique was crude, metal-to-metal bonds proved to be consistently strong and durable.

Formulations for isocyanate and vinyl structural adhesives cover a wide variety of combinations of materials and are usually tailored for very specific applications. The main factors to consider in selection of a particular formulation are ease of application and speed of cure.

4. Current and Future Uses

Presently, the major use of structural adhesives is in the fabrication of aircraft. Adhesively bonded primary structures have been in service in Europe for many years. In the USA, the approach has been more conservative—there are essentially no bonded primary aircraft structures in service today. However, in the US Air Force PABST (primary adhesively bonded structure technology) program, tests proved that bonding technology is ready for use on the next aircraft to be designed (Thrall 1979). That is not to say that adhesive bonding is not a part of aircraft manufacture in the US: many aircraft use adhesives to bond secondary structures. The Lockheed L-1011 TriStar, for example, uses adhesives for the bonding of skin panels, riveted and bonded doublers, triplers and failsafe straps (Petrino 1972). The US Air Force makes extensive use of adhesive bonding. The C-5 aircraft, for example, has a total of over 2300 m^2 bonded honeycomb structure and about 900 m^2 metal–metal bonded structure (Reinhart 1981). Suffice to say that under the impetus of programs such as PABST, structural bonding in the aerospace industry still represents a major growth area for the future.

For other industrial uses of adhesive bonding, it is difficult to define single operations which use large volumes of structural adhesives. In the automotive field, although there is a considerable amount of adhesive bonding, the material types and their use vary widely. Urethanes find widespread application in bonding fiberglass-reinforced-plastic body components (Kimball 1981). Epoxies are used for bonding numerous small parts, such as quartz headlamps, underhood electronic devices, automatic chokes, voltage regulators and ignition control modules. Weld and rivet bonding of sheet metal structures is increasing in importance. Epoxies are replacing the traditional lead body solders on trucks, buses and taxicabs (Bolger 1980). As the emphasis on fuel economy through reduced weight continues to become more important, adhesive bonding in the automotive industry will continue to grow.

Structural adhesives are also found in refrigerators, freezers, food disposers, air conditioners, dishwashers, washing machines, driers, ovens and numerous other small appliances. At one factory, over 40 000 electric motor adhesive assembly operations take place each day (Huntress 1979). The electrical and electronic industries use adhesive bonding extensively in the manufacture of motors, generators, transformers, coils and microcircuitry and for assembling the various components into products such as computers, radios, television sets and microprocessors. For these latter purposes, ease of application and speed of cure are important and, hence, cyanoacrylates and acrylic-type adhesives find widespread use.

As structural adhesives continue to broaden their range of physical properties and methods of application, the use of these products will continue to rise. The primary areas in need of improvement are higher-temperature performance and increased toughness. Analytical methods for assuring the proper chemical composition of the adhesive and the proper cure chemistry are realizable goals. The future depends heavily on the reliability and durability of bonded structures.

See also: Adhesive Joining: An Overview; Adhesives in Building; Adhesives in Civil Engineering

Bibliography

Aponyi T J 1977 Other high temperature adhesives. In: Skeist 1977, Chap. 38, pp. 619–27

Bolger J C 1980 Epoxies for manufacturing cars, buses and

trucks. *Adhes. Age* 23(12): 14–20
Borstell H 1974 *Chemical Composition and Processing Specifications for Air Force Advanced Composite Matrix Materials*. Grumman Aerospace, Bethpage, NY
de Bruyne N A 1939 The nature of adhesion. *Aircraft Eng.* 18(12): 52–4
de Bruyne N A 1950 (to Rohm and Haas Co) Adhesion between surfaces. US Patent No. 2,499,134
Carpenter J F 1978 *Test Program Evaluation of Hercules 3501-6 Resin*. Report No. N00019-77-6-0155. McDonnell Douglas Aircraft, St Louis, MO
Castan P 1944 Synthetic resins. US Patent No. 2,324,954
Cover H W Jr, McIntire J M 1977 Cyanoacrylate adhesives. In: Skeist 1977, Chap. 34, pp. 569–80
Dannenberg H, May C A 1969 Epoxide adhesives. In: Patrick R L (ed.) 1969 *Treatise on Adhesion and Adhesives*, Vol. 2. Dekker, New York, Chap. 2, pp. 3–76
Flickinger J H 1974 Elastomer modified epoxy resins. *Repr. Adhes. Seal. Counc. Spring Meeting* 3: 84–102
Floyd D E 1966 *Polyamide Resins*, 2nd edn. Van Nostrand Reinhold, New York
Frisch K C, Reegen S L (eds.) 1972 *Urethane Science and Technology*, Vol. 1. Technomic, Westport, CT
Greenlee S O 1944 (to Devoe and Raynolds Co) Synthetic surface coatings. US Patent No. 2,456,408
Huntress E A 1979 Adhesive assembly. *Am. Mach.* 123(10): 145–60
Kimball M E 1981 Polymethane adhesives: Properties and bonding procedures. *Adhes. Age* 24(6): 21–6
Kinlich A J (ed.) 1983 *Durability of Structural Adhesives*. Elsevier, Amsterdam
Martin R W 1956 *The Chemistry of Phenolic Resins*. Wiley, New York
May C A 1977 Composite quality assurance—An art becomes a science. *Natl. SAMPE Symp. Exhib.* 24: 390–403
May C A, Fritzen J S, Wearty D K 1976 *Exploratory Development of Chemical Quality Assurance and Composition of Epoxy Formulations*. AFML Report TR-76-112. US Air Force Materials Laboratory, Dayton, OH
May C A, Hadad D K, Browning C E 1979 Physiochemical quality assurance methods for composite matrix resins. *Polym. Eng. Sci.* 19: 545–51
May C A, Helminiak T E, Newey H A 1976 Chemical characterization plan for advanced composite prepregs. *Natl. SAMPE Tech. Conf.* 8: 274–94
May C A, Nixon A C 1961 Reactive diluents in aromatic amine cured epoxy adhesives. *J. Chem. Eng. Data* 6(2): 290–3
Megson N J L 1958 *Phenolic Resin Chemistry*. Academic Press, New York
Murray B D, Hauser M, Elliott J B 1977 Anaerobic adhesives. In: Skeist 1977, Chap. 33, pp. 560–9
Naps M 1953 *Elevated Temperature Resistant Modified Epoxide Resin Adhesives for Metals*. AFML Report TR 53-126. Air Force Materials Laboratory, Dayton, OH
Petrino D A 1972 L-1011 Tristar places new demands on aircraft sealants and adhesives. *Adhes. Age* 15(2): 15–19
Reinhart T J 1981 Use of structural adhesives by the US Air Force. *Adhes. Age* 24(8): 20–3
Schlack P 1938 Resinous amines rich in nitrogen and suitable for coating paper, cloth, etc. US Patent No. 2,136,928
Schlack P 1939 (to I G F Farbenindustrie) Nitrogenous condensation products. German Patent No. 676,117
Schollenberger C S 1977 Polyurethane and isocyanate based adhesives. In: Skeist 1977, Chap. 27, pp. 446–64
Skeist I (ed.) 1977 *Handbook of Adhesives*, 2nd edn. Van Nostrand Reinhold, New York
Thrall E W Jr 1979 PABST program test results. *Adhes. Age* 22(10): 22–33
Waterman H I, Veldman A R 1936 Destructive hydrogenization of resins from phenol and formaldehyde I, II. *Br. Plast.* 8: 125–8; 182–4
Zinke A, Hanus F, Ziegler E 1939 Beitrag zur Kenntnis des Härtungsprozessen von Phenol-Formaldehyd-Harzen. *J. Prakt. Chem.* 152: 126–44
Zisman W A 1977 Influence of constitution on adhesion. In: Skeist 1977, Chap. 3, pp. 33–71

C. A. May
[Lockheed Missiles and Space Company,
Sunnyvale, California, USA]

Structural Sealant Glazing

Structural sealant glazing is the architectural concept in which glass panels are glued to the side of a building. This type of construction allows for "glass cube" buildings and continuous runs of glass as in strip windows.

This concept of structural glazing was tested extensively in sealant-development laboratories in the late 1960s and the early 1970s. A few buildings were constructed in this manner each year from 1968. In 1978, after 10 years of testing both in the laboratory and in mockups, the concept had become accepted and it has since grown rapidly in popularity. Routine inspections of early structural glazing projects continue to provide confidence in the performance of this technique. The sealant that glues the glass to the building also performs the task of weathersealing the building against the elements. Besides being aesthetically pleasing and giving new freedom to the architect, glazing of this design is easy to clean and maintain due to the uncomplicated exterior.

The heart of the structural system is the sealant that transfers the wind load from the glass panel to the support mullions. The sealant, or glue, must be sufficiently strong to counteract the negative wind load yet remain flexible enough to accommodate the movement from differential thermal expansion experienced in metal curtainwall construction.

Since this adhesive or sealant must not stiffen and lose its elastic properties in extreme cold (−35 °C to −40 °C is experienced at some installations) or soften and lose its adhesion strength in the heat (dark glass and dark metal systems can experience 80–85 °C on a southern exposure in southern climates), only silicone adhesives or sealants are at present allowed in this application. Also the bond line sees the full radiation of the sun either directly (at the edge) or through the glass (back-sealed system) and must not deteriorate. At present only certain silicones can perform in this high-stress environment. Structural glazing is in fact a concept made possible only by the existence of a durable, elastomeric glue.

There are two different types of structural glazing design. The first, and the one with the longest history, can be classified as "two-sided" silicone glazing. This design concept incorporates silicone sealant as the structural adhesive on only two opposite sides (edges) of the glass panel. The other two sides are typically held in a metal glazing channel at the head and sill. The glazing channel supports the weight of the glass and provides outside capture of the glass.

The surface contact of the silicone sealant on the verticals is calculated on the basis of both the specified wind load and the glass width. This calculation utilizes the trapezoidal distribution of stress on the glass to approximate the force transmitted through the silicone sealant.

The sealant most often specified for this application is a high- or medium-modulus silicone. The design stress for acetoxy sealant is generally accepted as 140 kPa. Based on this design stress, the calculated silicone sealant contact surface allows for a two and a half- to seven-fold safety factor based on perfect laboratory specimens.

The second and newer type of structural silicone glazing is referred to as "four-sided" sealant glazing. This type incorporates silicone sealant as the only structural link between the glass panel and the metal curtainwall. Some designs specify a metal fin at the horizontals. These fins accept the weight of the glass through the setting blocks placed at the quarter points. The silicone sealant provides both the structural fastening and the weathersealing of the panels.

The most innovative utilization of "four-sided" structural sealant glazing eliminates the fin from the design. The metal framing is literally glued to the back side of the glass with the silicone sealant. In this design, the sealant constantly bears the weight of the glass. The sealant also counteracts the negative wind load as well as weathersealing the building.

The successful utilization of silicone sealant glazing demands that all accessory spacers, setting blocks and gaskets be totally compatible with the sealant. Since the ultraviolet exposure of the spacers and setting blocks is much higher than in typical wet or dry glazing in a glazing channel, newer, more stable compounds must be used in the fabrication of these rubber extrusions. Use of silicone rubber extrusions for all the accessory materials assures complete compatibility. Selected organic accessory materials can also be incorporated into the design, provided that they have been evaluated by the sealant manufacturer.

The laboratory test procedure for determining compatibility involves a 21-day exposure period under concentrated ultraviolet exposure. The criterion for determining compatibility depends on the sealant manufacturer. To determine the compatibility of spacers, a clear silicone, attached to the spacer, is exposed to accelerated ultraviolet weathering. No color change should be observed after 21 days. The result of extreme incompatibility could be total loss of adhesion of the silicone sealant.

The ultimate success of structural sealant glazing lies in the long-term durability of the adhesive bond of the sealant to both the glass and the metal mullion. The industry presently accepts ASTM C794-80 (*Standard Test for Adhesion-in-Peel of Elastomeric Joint Sealants*) as a method for determining the adhesion of structural sealants. This test method makes use of a metal screen embedded in a strip of sealant. The test also includes a 7-day water immersion cycle.

ASTM Committee C-24 on Building Seals and Sealants is currently developing a more realistic method for testing adhesion. This test method utilizes a sealant specimen 13 × 13 × 51 mm that more closely approximates the typical structural sealant joint.

Thus, structural sealant glazing is a systems concept in which compatible glass (or other panels), metal mullions, setting blocks, spacers and high-strength, high-modulus silicone adhesives or sealants all work together. To make such a system successful demands competent workmanship during installation. In particular, the sealant needs to be applied to a clean, dry surface according to the manufacturer's recommendations.

See also: Adhesives in Building; Sealants as Building Materials

Bibliography

Brower J R 1980 Considerations and advantages of silicone sealants in curtainwall applications. *US Glass, Met. Glazing* July/August: 45–53

Hilliard J R, Parise C J, Peterson C O Jr 1977 *Structural Sealant Glazing: Sealant Technology in Glazing Systems*, ASTM STP-6-1977. American Society for Testing and Materials, Philadelphia, PA, pp. 67–99

Kasperski M G, Klosowski J M 1978 Construction sealants. *Constr. Specifier* 31 (10): 21–6

Klosowski J M, Gant G A L 1980 *The Chemistry of Silicone Room Temperature Vulcanizing Sealants*. Reprinted from ACS Symposium Series No. 113, *Plastic Mortar, Sealants and Caulking Compounds*, pp. 113–27. American Chemical Society, Washington, DC

Peterson C O Jr 1980 Structural silicone glazing. *US Glass, Met. Glazing* July/August: 38

Romig C A, Klosowski J M 1977 Sealants. *Glass Dig.* 56: 27; 33

J. M. Klosowski and C. M. Schmidt
[Dow Corning Corporation, Midland, Michigan, USA]

Synthetic Adhesives

The American Society for Testing and Materials (ASTM) defines an adhesive as a substance capable of holding materials together by surface attachment. Synthetic polymer adhesives are those based on man-made polymers.

For a good bond between an adhesive and a substrate, the adhesive must be fluid enough to wet the surface completely. This is achieved when the surface free energy of the adhesive is less than that of the substrate. After flowing and wetting the surface of the substrate, the adhesive must harden into a solid that has cohesive strength.

1. Hardening Mechanisms

The hardening of fluid adhesives into solids is accomplished in several ways, including chemical reaction, evaporation of water or solvent, and cooling.

Chemical reaction generally involves polymerization and/or crosslinking. This may be obtained with or without heat or by electromagnetic irradiation. The most common examples are thermosetting adhesives based on epoxies, phenolics, amino resins and polyurethanes. These are usually two-component systems. Chemically reactive one-component adhesives include cyanoacrylates, which polymerize anionically upon contact with a trace amount of moisture, and anaerobics, which polymerize by a redox mechanism when oxygen is excluded.

Evaporation of water or solvent is the setting mechanism for adhesives based on solutions, dispersions or latices. Water-based synthetic adhesives include such polymers as poly(vinyl acetate), poly(vinyl alcohol) and acrylics. Solvent cements are usually based on elastomers such as styrene–butadiene rubber, neoprene and nitrile rubber. Some thermoplastic polymers such as linear polyurethanes, saturated polyesters and vinyl chloride copolymers also find use in solution form. Contact cements are exemplified by solutions of neoprene rubber plus phenolic resin. They are coated on both of the substrates to be bonded; then the solvent is evaporated and the substrates are brought together.

Pressure-sensitive adhesives adhere tenaciously upon application with only slight finger pressure. They are widely used for tapes and labels made by coating the adhesives on such substrates as plastic, paper and cloth. The adhesives are most commonly solutions or latices; however, hot-melt types are gaining acceptance. Solutions and waterborne adhesives are based on a variety of elastomers: natural, reclaimed, styrene–butadiene or acrylic. The formulations may also include resin tackifiers, plasticizers, fillers, antioxidants and other ingredients.

Hot-melt adhesives are thermoplastic solid compositions which become fluid when heated and solidify upon cooling. The oldest examples of such adhesives include bitumen and beeswax. Modern hot-melt adhesives are based on polyethylene, polypropylene, ethylene–vinyl acetate copolymers, saturated polyesters and polyamides. These polymers may be utilized by themselves as hot-melt adhesives. However, those based on polyethylene and ethylene–vinyl acetate also incorporate resin tackifiers and waxes.

Hot-melt pressure-sensitive adhesives are applied to substrates in a molten state. They are formulated from thermoplastic elastomers such as styrene–butadiene–styrene or styrene–isoprene–styrene block copolymers. Tackifiers, plasticizers and fillers may be present as additives.

2. Types of Synthetic Polymers

The advent of synthetic polymers early in the twentieth century has revolutionized many manufacturing industries. The older glues and pastes derived from natural materials have been supplemented or superseded by adhesives based on synthetic polymers, making it possible to bond almost any substrate. Important synthetic polymers for adhesives are discussed in the following sections.

2.1 Elastomers

Butene polymers are elastomeric polymers employed in adhesives and sealants, both as a primary binder and as a tackifier and a modifier. Butyl rubber is a copolymer of isobutylene, with 0.5–2.5 mol.% of isoprene. Polyisobutylene is a high-purity homopolymer of isobutylene, available in a range of molecular weights. Polybutene is a low-molecular-weight copolymer of butene isomers made from a cracked C_4-olefin refinery stream.

While butyl rubber is generally employed as a primary binder in sealants, polybutene and polyisobutylene are more often used as tackifiers and modifiers in sealants, pressure-sensitive adhesives and other solvent-based adhesives. Because double bonds are sparse (butyl rubber) or absent altogether, these hydrocarbon polymers offer excellent aging characteristics. Although their tack is high, absence of polar groups results in weak adhesion to the substrate. One of the major uses of these polymers is in removable pressure-sensitive labels and tapes.

Nitrile rubbers are copolymers of butadiene and acrylonitrile, in weight ratios ranging from 82:18 to 50:50. High acrylonitrile content results in strong and hard rubbers with good resistance to oil, fuel, abrasion and heat. Lower acrylonitrile content results in rubbers with more flexibility at low temperatures. A third comonomer such as acrylic acid is often included to improve adhesion and compatibility with other polymers. Nitrile rubbers have been coformulated with phenolic resins, epoxy resins, vinyl chloride polymers and others.

Nitrile–phenolic adhesives find use in building construction, attachment of linings to automotive brakes, bonded abrasives, such as grinding wheels, and aerospace structural bonding. Epoxy–nitrile combinations have aerospace applications. Solvent cements of nitrile rubber contribute to film-to-film laminations, appliances and mobile homes. Nitrile latices are among the binders for nonwoven fabrics.

Neoprene or polychloroprene, the first synthetic elastomer to become a significant adhesive, was developed by polymerization of 2-chlorobutadiene. Special grades are available for adhesives. Most neoprene adhesives are contact cements. They are applied from solution to both the surfaces to be bonded. After evaporation of the solvent, a strong bond is formed upon contact. Contact cements usually contain, besides neoprene, a heat-reactive phenolic resin, magnesium oxide crosslinker and other additives. They are popular in construction, in the manufacture of automobiles and shoes and for use in the home.

Styrene–butadiene rubber, known initially as GR-S and later as SBR, was developed in the 1940s as a general-purpose rubber to replace natural rubber. As an adhesive, it is generally modified with resins to improve such properties as tack, cohesive strength, heat resistance and adhesion. Among the resins are rosins, rosin esters and phenolics.

In the 1960s, new block copolymers were introduced based on styrene and butadiene or styrene and isoprene. At room temperature, these polymers form a two-phase structure in which the styrene end blocks cluster together to form hard domains, while the diene midblocks constitute a rubbery phase. These physically crosslinked systems are known as thermoplastic rubbers. They provide hot-melt adhesives and hot-melt pressure-sensitive adhesives.

Conventional styrene–butadiene rubber is employed in solvent-based pressure-sensitive adhesives as well as construction adhesives. Carboxylated SB latices incorporating acrylic acid as a comonomer are utilized in carpet-backing adhesives. A terpolymer latex with vinyl pyridine as the third monomer is combined with resorcinol–formaldehyde resins for bonding tire cord to rubber.

2.2 *Predominantly Thermosetting Resins*

Phenolic resins, which were the first synthetic polymers, are the reaction products of phenol and/or substituted phenols with formaldehyde. Novolacs are formed when phenol is reacted under acidic conditions with less than an equimolar amount of formaldehyde. Resole phenolic resins are obtained under alkaline conditions using an excess of formaldehyde.

The largest requirement for phenolic adhesives is the bonding of wood, especially plywood. Here, resoles are employed. They can be extended with such additives as bark flour and wheat flour. Phenolics offer excellent resistance to moisture and weathering.

Phenolics are the leading binder for glass-wool insulation. In foundries, these resins are involved in shell molding, hot-box processes and no-bake as well as the cold-box process. The heat resistance of phenolics makes them suitable for attaching abrasives to paper and cloth and for the bonding of grinding wheels. They are also favored for automotive brake linings and clutch facings, with asbestos as the friction material.

Phenolics are also incorporated into other adhesive systems as modifiers to impart heat, chemical and moisture resistance.

Resorcinol–formaldehyde resins have two phenolic hydroxyls per aromatic ring and are consequently more active than phenolics. They cure at room temperature and offer excellent weathering and chemical resistance. They are employed in tire-cord bonding, along with butadiene–styrene–vinyl pyridine latex. They are also added to phenolics to facilitate curing at lower temperatures.

Amino resins are condensation products of urea and/or melamine with formaldehyde. Urea resins offer low cost, a colorless glue line, room-temperature curing and resistance to molds as well as fungus. They are deficient in resistance to moisture and water, however. Their most important application is the production of particleboard. They are also used in other wood bonding and furniture production.

Melamine resins are more expensive but provide better resistance to heat and water. They are used to upgrade urea resin adhesives. They also form the surface layers of high-pressure decorative laminates.

Epoxy resin adhesives are among the most versatile of adhesives, adhering well to metals, glass fiber and other substrates and providing high strength when cured. They offer low shrinkage, 100%-solids system, room-temperature cure and resistance to heat and chemicals.

The resins are most commonly the condensation products of epichlorohydrin and bisphenol A (p,p'-isopropylidenediphenol). The viscous liquid epoxy resins that have epoxide equivalent weights of 170–300 are the most useful as adhesives. They are cured with compounds containing active hydrogen, especially amines, acids and mercaptans. The most important curing agents in adhesives are the polyamidoamines or reactive polyamides, derived by reacting diethylenetriamine or triethylenetetramine with dimer acid, obtained by Diels–Alder dimerization of unsaturated fatty acids. One-component systems require latent curing agents such as dicyandiamide or BF_3–amine complexes.

In construction, epoxy adhesives are essential for vinyl film laminates, metal doors, flooring and ceramic tiles, traffic markets and other uses. In aircraft and aerospace, they are employed principally as one-component adhesives in the form of film adhesives. Here they may be modified with such additives as phenolics, nitrile rubber and nylon. The electrical and electronic industries utilize them in such areas as television tubes, printed-circuit laminates, batteries and telecommunications. They are involved in the assembly of such products as paint brushes, sports equipment and glassware and are used as household adhesives.

Polyurethanes are the reaction products of isocyanates and polyols. They can be either thermosetting or thermoplastic and are available as one- or two-component systems. Isocyanates react readily with many functional groups and form chemical bonds with a variety of substrates. The adhesives provide excellent strength and chemical resistance.

One-component polyurethanes usually cure with moisture and find uses in flocking and carpet installation, as well as in sealants for automotive windshields. Two-component adhesives offer maximum versatility. They are employed as structural adhesives, especially for automotive plastics. Other uses include film-to-film laminations and plywood-patching compounds. These adhesives are usually accelerated with amines. Thermoplastic polyurethanes are employed essentially in textile laminating and footwear applications, including decals and emblems providing resistance to dry-cleaning solvents.

2.3 Thermoplastics and Chemically Related Thermosets

Poly(vinyl acetate) and vinyl acetate copolymers are the leading thermoplastic adhesives. Most of these polymers are prepared by emulsion polymerization. They are applied mainly to cellulosic products such as paper, wood and textiles. In addition, they are the most popular household adhesives.

Formulated poly(vinyl acetate) adhesive may include poly(vinyl alcohol) as emulsifier, thickener and adhesion promoter, starch or clay as extenders and dibutyl phthalate as plasticizer. Small amounts of solvents and proprietary additives are incorporated in special-purpose formulations.

The vinyl acetate polymer may be internally plasticized by copolymerizing the monomer with ethylene. Emulsion copolymers with ethylene content ranging from 5 wt% to 40 wt% are extensively employed for such applications as film-to-wood laminations and packaging products.

Poly(vinyl alcohol) is obtained by hydrolysis of poly(vinyl acetate). It is soluble in water. In the common grades for adhesives, about 88% of the acetate groups have been converted to hydroxyl. Main applications are in paper products, construction and as protective colloid or additive for poly(vinyl acetate).

Poly(vinyl butyral) is made by reaction of 98%-hydrolyzed poly(vinyl alcohol) with butyraldehyde. The degree of acetalization is usually about 80%. Its major application is in the form of a 750 μm film to bond the glass layer in the construction of laminated automotive safety windshields. For this, the polymer is plasticized with triethylene glycol di(2-ethyl butyrate), in a resin–plasticizer ratio of 70:30.

Acrylics are proving to be another versatile family of polymers for adhesives applications. They are based on acrylate as well as methacrylate ester monomers. The adhesives may be thermoplastic or thermosetting; water based, solvent based or 100% solids; one or two component; and self-curing, heat cured or radiation cured. Weather resistance and aging characteristics are excellent.

Most polymers for adhesive applications utilize one or more of the following: ethyl acrylate, butyl acrylate, 2-ethylhexyl acrylate, acrylic acid and methacrylic acid. The major outlet for acrylic adhesives is textile bonding (e.g., nonwovens, flocking and fiberfill). Here, most of the adhesives are employed in emulsion form and are crosslinkable. The crosslinking sites are introduced into the backbone polymers through functional comonomers such as *N*-methylol acrylamide. Pressure-sensitive acrylics include higher alkyl esters for tack. Electron beams or ultraviolet radiation can serve to cure mixtures of acrylic monomers. Two-component acrylics are emerging as structural adhesives for bonding of metals and plastics.

Anaerobics are single-component adhesives which cure at ambient temperature when air is excluded. They are usually formulated from dimethacrylate esters of low-molecular-weight diols. They are utilized for locking nuts and bolts, sealing fittings and flanges and bonding a variety of mechanical parts. They are ideal for machine assembly, because they do not cure until the parts are joined together. They also prevent loosening of a mechanical assembly subject to constant vibration.

Cyanoacrylates have been termed wonder glue or super glue. Their success lies in being able to make a bond of good strength in as little as 10 s. Almost any nonporous material can be bonded. The methyl cyanoacrylate or ethyl cyanoacrylate monomer polymerizes rapidly in thin films in the presence of minute amounts of moisture such as are present on most surfaces. Resistance to aging and heat is only mediocre. They are employed in a variety of product assembly uses, in mechanical repair and are also popular as household adhesives.

Polyolefins and olefin copolymers such as polyethylene, polypropylene and ethylene–vinyl acetate copolymers are the basis for inexpensive hot-melt adhesives. Major outlets are in packaging, bookbinding and the textile and automotive industries.

Table 1
Some 100%-solids adhesives based on recent technology

Ingredients	Parts by weight
A hot-melt pressure-sensitive adhesive[a]	
styrene–isoprene–styrene block copolymer	12.5
styrene–butadiene–styrene block copolymer	37.5
stabilized pentaerythritol rosin ester	37.5
poly(α-pinene) resin	12
antioxidant	0.5
An ultraviolet-light-curable dental adhesive[b]	
bisphenol A–glycidyl methacrylate oligomer	67
methyl methacrylate	33
5% benzoyl peroxide in methyl methacrylate	40
5% Michler's ketone in chloroform	20
An anaerobic adhesive[c]	
polyethyleneglycol dimethacrylate	94
saccharin	0.4
cumene hydroperoxide	4.0
N,N-Dimethyl-*p*-toluidine	0.4
hydroquinone monomethyl ether	0.2
triallyl phosphate	1.0

a Russel T E 1973 US Patent No. 3,736,281, assigned to The Flintkote Company b Reaville E T, Streicher G M 1978 US Patent No. 4,097,994, assigned to Monsanto Company c O'Sullivan D J 1978 US Patent No. 4,100,141, assigned to Loctite (Ireland) Ltd

Polyethylene and some polypropylene is usually modified with 20 wt% of a tackifying resin, but most polypropylene hot melts are uncompounded low-molecular-weight atactic polymers. EVA hot-melt adhesives, containing 18–28% or more of vinyl acetate, are usually formulated with a tackifier and microcrystalline wax.

Polyamides are also employed in hot-melt adhesives. They are generally derived from condensation of dimer acid with diamines and higher functional amines. They find applications in packaging, footwear, textiles and the automotive and electrical industries.

Saturated polyesters are thermoplastics derived from such diacids and diols as terephthalic acid, isophthalic acid, adipic acid and 1,4-butanediol. They are employed as hot melts, film adhesives and in solution form. Hot melts are utilized in footwear and various product assembly uses. Polyester-film adhesive is used to attach labels and emblems to a variety of textiles. Solution adhesives are employed in film laminating applications. They may be cross-linked with small amounts of isocyanates.

Unsaturated polyesters compounded with fillers find application in patching compounds for auto-body repair. The resins are derived from phthalic anhydride, maleic anhydride and a glycol. They are cured with peroxides.

Poly(vinyl chloride), when formulated into plastisols, can be used as an adhesive. The plastisol adhesives have outlets in automotive and carpet-backing uses. Vinyl chloride copolymers and terpolymers in solution form find applications in pipe cements and film laminations.

Polyimides are suitable for aerospace applications requiring adhesives with higher temperature resistance than afforded by epoxies or phenolics. They are obtained by either a diamine–dianhydride condensation or by addition polymerization of an unsaturated precursor containing imide linkages.

3. Formulations

Selection and formulation of an adhesive involves a variety of factors, including type of substrate, form desired, application method, properties required and cost. It should be pointed out that because of energy cost and ecological considerations, a general trend in the adhesives industry is towards solventless, high-solids systems. Legislation introduced in the USA, the UK and Europe is accelerating this trend. Some representative 100%-solids adhesive formulations based on recent technologies are given in Table 1.

See also: Adhesive Joining: An Overview; Adhesives for Wood; Adhesives in Building; Adhesives in Civil Engineering; Elastomer–Textile Adhesion Processes

Bibliography

Adhesives 1979–80 Desk-Top Data Bank. Cordura Publications, La Jolla, CA
Cagle C V (ed.) 1973 *Handbook of Adhesive Bonding*. McGraw-Hill, New York
Fabris H J, Knauss W G 1989 Synthetic polymer adhesives. In: Aggarwal S L (ed.) 1989 *Comprehensive Polymer Science*, Vol. 7. Pergamon, Oxford, Chap. 5
Shields J 1984 *Adhesives Handbook*, 3rd edn. Butterworth, Sevenoaks, UK
Skeist I (ed.) 1977 *Handbook of Adhesives*, 2nd edn. Van Nostrand Reinhold, New York
Wake W C 1976 *Adhesion and the Formulation of Adhesives*. Applied Science, London
Wake W C (ed.) 1987 *Synthetic Adhesives and Sealants*. Wiley, New York
White L 1990 Elastomeric adhesives. *Eur. Rubber J.* 172(9): 28–32
White L 1990 Rubber/metal bonding. *Eur. Rubber J.* 172(9): 38–40

M. Savla
[Savla Technical Services,
Baroda, India]

T

Textiles: Structures and Processes

Garments, upholstery, tapestry, carpets, drapes, blankets, bed sheets and towels are ubiquitous consumer products made of textile fabrics comprising fiber-based textile yarns. In recent years the use of fibrous materials has been extended to many engineering (Textile Institute 1977) and bioengineering applications, such as inflated buildings, stadium roofs, geotechnical construction, composites reinforcement, aerodynamic decelerators, aircraft emergency slide chutes, arterial prostheses, sutures and artificial skin.

Fibrous sheet materials may be assembled directly from fiber masses using the technology of the paper or the textile industries. The resulting products are named nonwovens; their mechanical properties are derived from the strength of their fibers and of the bonds between fibers. Nonwoven textiles vary widely in physical properties (e.g., strength, thickness, compressibility and flexibility) and in areas of application (e.g., for garment reinforcement or stiffening, subsoil construction and material interface for roads and railways) (Koerner and Welsh 1980). The term nonwoven materials is, to say the least, somewhat ambiguous, suggesting a division of the materials universe into two categories—one woven fabrics and the other encompassing all other materials as nonwovens. Suffice it to say that nonwovens are generally defined as fibrous sheet materials which share one attribute, namely the absence of interwoven or otherwise interlaced yarns as their principal structural component (Textile Institute 1977).

1. Yarn Structures and Properties

The more common textile fabrics are constructed of visually identifiable, compacted bundles of fibers, that is, yarns. Yarns may comprise continuous (infinitely long) filaments (such as silk or extruded synthetic fibers) or they may be made of relatively short, staple fibers, such as cotton or wool or equivalent cut lengths of synthetic fibers such as nylon or polyester. In the case of short (staple) fibers, a cohesive structure is obtained by twisting the fibers together in a strand or yarns.

1.1 Strength

The strength of a yarn determines its performance in the yarn preparation steps and during fabric formation, as well as in fabric usage. Accordingly, this is the property of key importance.

Tensile strength in the case of a low-twist continuous filament yarn is approximately equal to the sum of its filament strengths (statistics aside). As the twist of filament yarns is increased, their strength decreases, roughly as the cosine squared of the outer fiber helix angle, as is well documented (Hearle et al. 1969, Zurek 1975).

Strength prediction for staple fiber yarns is more complex, for at very low twists the yarn is extremely weak due to slippage between fibers. At higher twists (as characterized by outer helix angle or turns per unit length) the staple yarn strength increases, reaches a maximum, then falls. The strength peak corresponds to the point where the lateral tightening action of the helically twisted filaments develops sufficient interfiber friction to preclude fiber slippage. In the postpeak region, obliquity of the filaments causes a cosine-squared strength reduction, such as occurs in twisted continuous filament yarns. The mechanics of staple yarns, although the subject of many papers, is still a problem area and warrants considerable research, particularly for the entire class of yarns composed of blends of fibers which differ in stress–strain properties (Hearle et al. 1969, Zurek 1975).

1.2 Flexibility

Besides strength and resistance to deterioration in use, a primary property of textile yarns is their unusual flexibility. This flexibility is often translated into the textile fabric, which is essentially an interlaced network of yarns. Considering the range of consumer textile products (e.g., in clothing or household articles), the importance of sheet flexibility, limpness and drape is easily perceived.

The flexibility of textile materials is partly due to the fact that apparel textile fibers have tensile moduli one to two orders of magnitude below those of engineering metals. An even more important contribution to yarn flexibility is the fineness of the fibers, in the region of 0.025 mm. The bending rigidity of an isotropic cylinder is proportional to the first power of its modulus, but to the fourth power of its diameter. Thus, while a specified requirement of tensile strength for a given fiber–yarn combination determines its total material cross-sectional area, its flexibility can still be varied significantly by altering the diameter and, correspondingly, the number of its filaments. If yarn bending occurs with free slippage between fibers, that is, with no frictional restraint, it can be easily shown that a square yarn bundle with n fibers has n times the flexibility of a square rod possessing a cross-sectional material area equal to that of the yarn. In other words, the monofilament is approximately n times as rigid as the equivalent yarn of n fibers. Many apparel yarns contain from 30 to 80 fibers, which explains why an application of starch to a shirt collar stiffens it severely—the starch converts the multifiber yarns into pseudo-monofilaments by binding their fibers together and preventing their slippage over one another.

1.3 Structure

As has been indicated, twist is necessary to ensure strength in staple fiber (spun) yarns, yet twist is often inserted in continuous filament yarns where it is not needed as such and actually results in some reduction of strength. On the other hand, increasing the twist in a yarn modifies its cross-sectional shape, increases its lateral bundle cohesion (to prevent damage in high-speed processing) and alters its torsional rigidity and residual torque. All these properties affect the mechanical response of the yarn when incorporated into a textile fabric.

A few words about the geometric structure of yarns are necessary. Traditionally, the singles yarn is considered as a set of concentric and helically twisted fibers. A ply yarn consists of multiple singles yarns plied and twisted together in an opposite sense to the original singles twist direction—with a view towards balancing out the residual torque. There are twistless stretch yarns comprising continuous filaments which have been previously deformed and heat set (textured) to introduce filament crimp, sometimes planar, sometimes three dimensional (Piller 1973). There are staple yarns with filament cores and yarns with staple fiber cores and filament wrappings. There are fasciated yarns with parallel (untwisted) staple fiber cores and occasional short fiber wrappings designed to compress the core fibers and to enhance frictional transfer of axial stresses (Morihashi 1976). Depending on the dictionary or thesaurus used, as many as 40 yarn types may be found listed, differentiated for the most part according to process but occasionally according to usage (Backer and Valko 1969).

While most of the yarns encountered in traditional textile applications are of the usual shingles or ply twisted types, a marked increase of late may be noted in the number and volume of novel yarn structures employed in commerce. This reflects active worldwide development of radically new methods of yarn formation, such as the air torque jet spinning process. While such methods were designed originally to enhance productivity, their resulting products are often significantly different in appearance and properties. Similar development is occurring in processes which are intended, in some cases, to produce traditional yarns (as in rotor spinning), although careful study of the mechanics of such processes and of the geometry of their resulting yarns indicates that there are still considerable differences in their yarn behavior compared with that of the traditional ring-spun yarn.

In summary, it is important to note that yarn properties vary not only with the properties of their constituent fibers but also with the type of manufacturing process employed for their production and with the settings of the particular operation or machine. Photographs of staple yarns produced by different spinning methods are shown in Fig. 1.

2. *Yarn Manufacturing Processes*

The first known form of yarn processing technology goes back about 15 000 years (in Asia) or 12 000 years (in North Africa). It consisted of hand spinning, which is still practised in parts of the world. In hand spinning, a loose mass of fibers is held in one hand and the other hand partly pulls out a small bundle and draws it to yarn size while rotating the free end, thereby twisting the attenuated fibers into a strand. Rotation of the free end is facilitated by attaching it to a spindle which, in turn, is attached to a small weight or whorl. The spindle–whorl combination is then sharply rotated by hand; acting as a flywheel, it continues to rotate for a time, twisting the yarn in the process. After a short length of yarn is thus produced, it is wound on the spindle. The process consists of four essential parts: (a) preparation of the loose (disentangled) mass of fibers; (b) continuously drawing down a subgroup of fibers to yarn size; (c) twisting the yarn; and (d) yarn windup.

2.1 Opening and Carding

The first of these steps, fiber preparation, has always been a separate process and was originally accomplished by hand carding and combing of cotton or wool masses. Current production methods provide for high-production opening equipment, as well as feeds and continuous monitoring of the linear density of the output sliver. However, the new processes are similar in principle to the hand processes, although flat carding boards have been replaced by wire-clothed or sawtooth-covered cylinders rotating at high speeds, often with clearances of less than 0.25 mm. Their function is to shear tear fiber aggregates, thereby detaching or disentangling the fiber clumps and separating dirt and other foreign matter present in the fiber stock. In the case of staple synthetic fibers which are received at the mill in high-pressure bales prepared by the fiber producer, the opening and carding processes have the same functions, less that of dirt removal.

When the fibers are doffed from the very last cylinder of the card, they form a thin web possessing very little strength, yet enough for web conveyance over short distances without support. At this point, operations differ, depending on whether the final

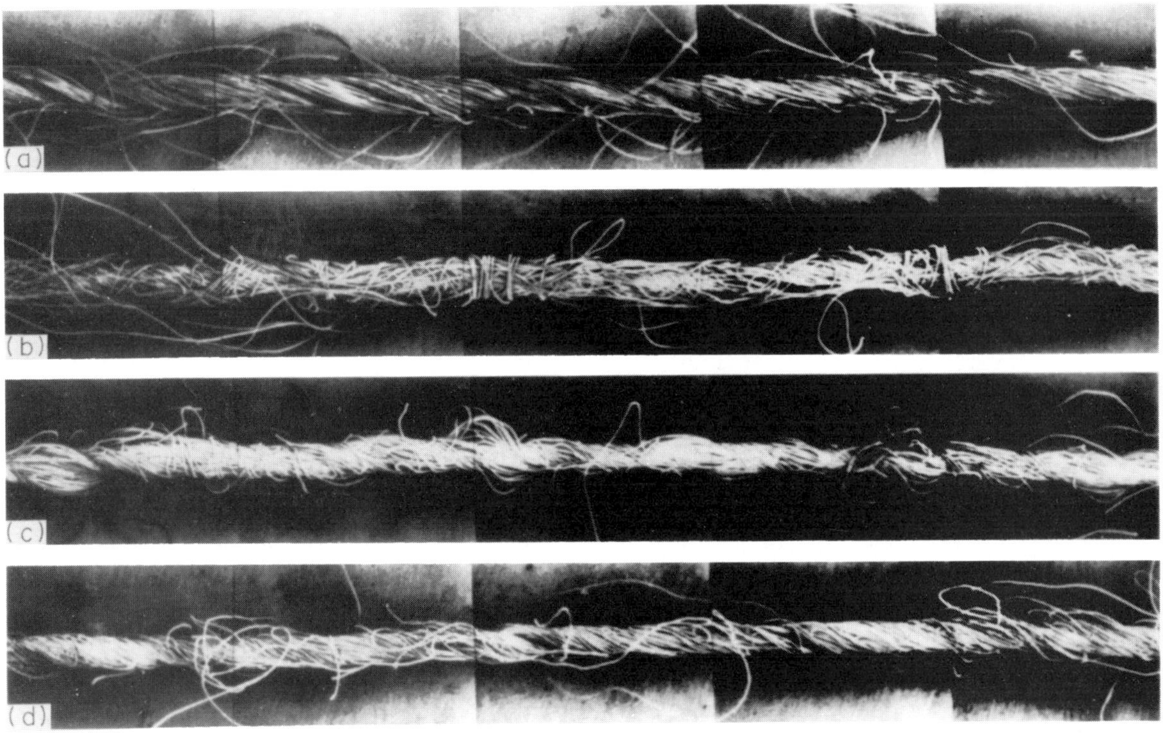

Figure 1
Dependence of yarn structure on spinning system: (a) ring spun; (b) rotor spun; (c) air jet spun; and (d) friction spun (courtesy of Dr F. Looney, E. I. Du Pont de Nemours and Co.)

product is a nonwoven or a spun yarn fabric. Webs for nonwovens may be lapped with other webs and sprayed by or dipped in latex emulsions and heat dried. The resulting latex bond provides for stress transfer capacity between fibers and thus for fabric strength.

The card web derives its coherence from residual entanglements or interleaving of individual crimped fibers, or from some adherence relating to fiber finish added by the fiber producer. While natural fibers, such as cotton or wool, have an inherent crimped or convoluted geometry, synthetic fibers are generally extruded in a straight configuration and must be subjected to a separate crimping process (generally of stuffer box form). Such crimp is essential to provide processing coherence in flat card webs and in funnel-gathered webs, called card slivers.

At first glance, a general machine-direction alignment and a straightening of fibers which have been subjected to the combing action of a carding machine might be expected. Instead, however, it is observed that the high-speed clump-shearing action and fiber transfer from cylinder to cylinder of the card more often results in fiber snapback, buckling, bending, hooking and general disorientation. Disorientation introduced by these actions must be overcome in subsequent drawing operations.

2.2 Drawing and Drafting

In the ancient hand spinning method, the second and third steps of drawdown and of twisting were generally done simultaneously, but intermittently, and were then followed, also intermittently, by the windup on the spindle. These spinning actions were made sequential and continuous at the onset of the industrial revolution in the mid-eighteenth century by development of a roller drawing process operating at constant velocity to serve as a continuous input to a combined twisting and winding process. As a result of this development, the final spinning process became capable of continuous operation and this is the mode of yarn production by the flyer method, as on roving frames, and by the widely used ring spinning system depicted in Fig. 2.

In modern spinning systems, the drawing-down function is performed initially on separate machines. Then, in later stages of processing, it is combined with the twisting operation. The first step reduction

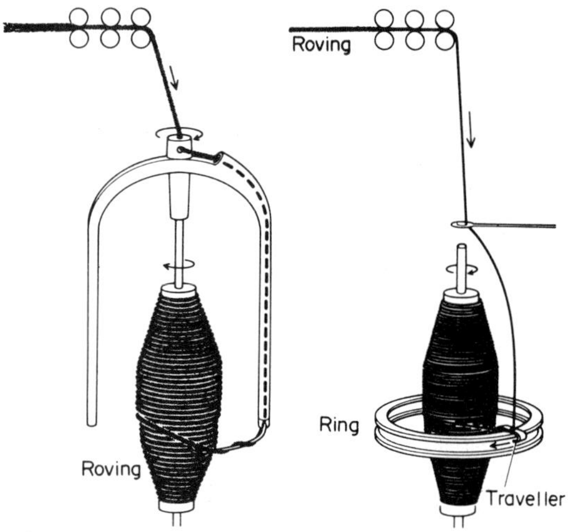

Figure 2
(a) Roving draft–twisting and (b) yarn draft–spinning (after Backer 1972. © Scientific American, Inc., New York. Reproduced with permission)

of sliver size is usually accomplished by roller drawing in a drawframe. In this operation, the degree of reduction, or draft, is simply the ratio of the linear mass of the input sliver to that of the output sliver.

The roller drawing process uses two or more sets of rollers, each set faster than the previous one and each set spaced apart slightly more than the maximum fiber lengths of the assembly. Subjected to such a process, the sliver, a ropelike fiber assembly, is elongated and attenuated, due to interfiber slippage accompanied by increased flow-direction fiber orientation and temporary decrimping.

The fiber interactions in this simple roller drawing device are such that the amount of drafting (or the draft ratio) which can be effected at one stage is limited by the presence of entanglement and hooked and disoriented fibers. The disoriented fiber arrangements in the typical card sliver preclude processing with high draft ratios in early drawing stages: thus, card slivers are given one or two drawing steps at low ratios so as to improve the parallelism, straightness and disentanglement of their constituent fibers. Further, the card sliver input and the draw sliver output of the early drawings are generally "doubled," that is, combined in parallel with other slivers, so as to balance off statistically any weak (fiber-starved) zone which might have developed in any one of the parallel fiber assembly components. Doublings are required because weak zones prevailing entirely across a local section of the sliver would invariably result in disruption of the assembly flow, leading to "ends down" in subsequent drafting. The drafting process is, after all, a form of continuous tensile elongation process of a fiber assembly chain, in which the weakest link always fails first. Not only that, but also it always draws first.

Doubling also serves as a mixing scheme to promote product uniformity. Likewise, doubling can be done with different fiber varieties; the resulting product is a blended-fiber yarn reflecting the desirable properties of each of its fiber components.

After such preliminary drawing and doubling, the sliver may now be drafted down to an intermediate "rope" size, then twisted slightly and packaged in the form of roving. The final processing of the fiber assembly takes place in the ring spinning machine, in which the roving is drafted down to actual yarn size and promptly twisted and bobbin wound as finished yarn. The linear density and the twist of this yarn is dictated by its intended application and by the nature of subsequent processes to which it is to be subjected, whether weaving, knitting, braiding, tufting or even sewing. The earlier stages of yarn preparation may likewise be varied; for example, the opening and carding processes present in virtually all staple yarn systems may be supplemented with an extra combing process which may be introduced to eliminate shorter fibers and to increase the parallelism of the fibers in the sliver assembly, hence in the subsequently drafted and twisted yarn. The term carded yarn implies yarns made without the extra uniformity and parallelism implied in the term combed yarn.

2.3 Spinning Processes

In summary, yarns which are the basic components of interlaced (e.g., woven, knitted) textile fabrics are themselves structures whose principal components are fibers. Yarn properties are directly dependent on fiber properties, whether singly or in blends, but they are also affected by the type of preparatory process employed and by the machine settings, which influence fiber parallelism, packing factor, residual crimp, radial migration, hairiness and twist. Finally, they are strongly influenced by the mechanism of the spinning system process involved. The above descriptions focus on the ring spinning system which dominated the industry until the early 1970s. Since that time, several major new spinning and twisting systems have been commercialized and are now spreading across the textile industries, including open-end (or rotor) spinning, air jet spinning, friction spinning, self-twisting and jet interlacing. These methods will now be compared with the conventional ring spinning process.

The ring–traveller frame is seen in Fig. 2 to provide rollers moving at increasing speeds to draft the loose input roving into a thinner strand. This strand or yarn is immediately twisted upon exit from the last pair of rollers and then wound around the vertical rotating bobbin. Twist is inserted as a result of movement of the small metal clip, or traveller, which is pulled around the nonrotating ring by the

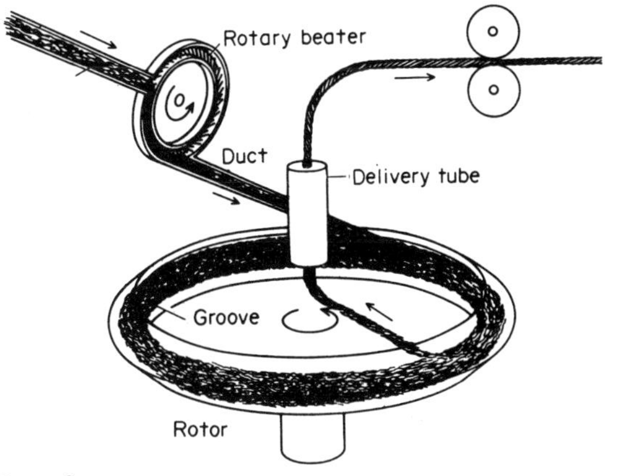

Figure 3
Rotor (open-end) spinning (after Backer 1972. © Scientific American, Inc., New York. Reproduced with permission)

relative motion and friction of the yarn in its passage to the rotating bobbin. The ring is raised and lowered continuously to provide even winding. When the bobbin is filled, the process is stopped, the full bobbin is "doffed" and a fresh bobbin is inserted. Use of larger packages permits longer machine runs between doffs, but at the same time, larger bobbin sizes tend to limit machine speeds due to requirements of dynamic balance. A major goal in spinning development of the past 15 years has been to insert real twist without rotating the entire package except for the purpose of windup. Two approaches to this goal have been taken thus far: open-end spinning (see Fig. 3) and the self-twist method of ply yarn formation (see Fig. 4). Note that both of these methods involve real or alternating twists. Other yarn process developments take alternative routes.

The open-end spinning (OES) method was the subject of intense development around 1970 and several open-end frames of European and Japanese design are now available for routine production. This process, which is also called break spinning or rotor spinning, differs from ring spinning in that the flow or sequencing of fibers is interrupted at a point just ahead of spin twisting. The interruption makes it possible to rotate the downstream end of the strand just below the point of interruption and thus to insert twist into the fiber assembly as it approaches and moves into the delivery tube. Then new fibers are plucked from the upstream sliver by a rotating beater and are transported by air to the groove of the rotor. Here they are entangled with and twisted onto the end of the exiting, twisting strand. Production rates, in terms of twisting revolutions per minute, are several times faster in rotor spinning than in ring–traveller spinning. However, rotor yarns cannot be produced efficiently in fine counts and their mechanical properties differ significantly from ring-spun yarns. Consequently, ring-spun yarns still dominate the fine-goods end of the industry.

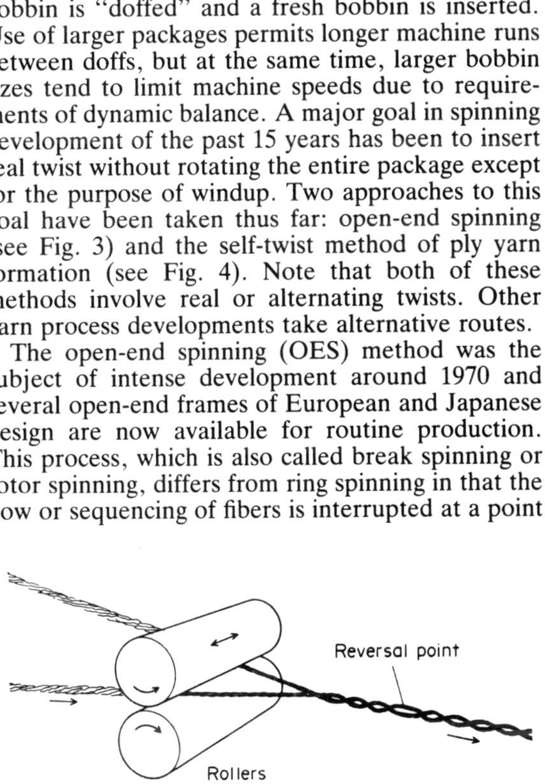

Figure 4
Self-twist spinning (after Backer 1972. © Scientific American, Inc., New York. Reproduced with permission)

Self-twist spinning, depicted in Fig. 4, was developed in Australia in the 1960s. In this process, designed originally for wool yarns, two lines of roving are carried between two rollers that have a double motion: a continuous rotational motion, as with conventional rollers, and a lateral shuffling motion. The latter motion compacts and false twists the rovings as they are drafted and carried forward, giving each line of roving an alternating twist upstream and downstream of the contact point. When adjacent pairs of rovings are brought into contact with each other, they tend to intermesh to some degree and become entangled. When the contact pressure is released, the two rovings try to untwist, but because of their entanglement they ply over each other until the combined torsional and bending energies in the system are minimized and the now plied yarn reaches equilibrium. In one step this system converts loose slivers or rovings into a two-ply yarn without the need for twist rotating the entire package.

Fasciated yarns, or air jet yarns, represent a new kind of yarn structure, as seen in Fig. 1c. They differ from the other yarns described previously in that most of their fibers are not configured in concentric helices. Instead, a fasciated yarn consists of a core of parallel staple fibers wrapped at intervals by a small fraction of the fiber components of the yarn. An aspirating jet pulls a staple fiber roving into a torque jet, which strikes the periphery of the strand tangentially, thereby introducing a false twist. Since the strand is discontinuous, in contrast to the continuous filament on which the usual false-twist spindle operates, some of the upstream fibers are for the moment not twisted as much as the core fibers. On passing the jet, however, they are back twisted more than the core fibers to achieve a localized wrapping over the core fibers, which the airstream has by now made parallel. In current models of the air jet spinner, a second torque jet is sometimes provided with an opposite sense of rotation to that developed in the first jet, as shown in Fig. 5.

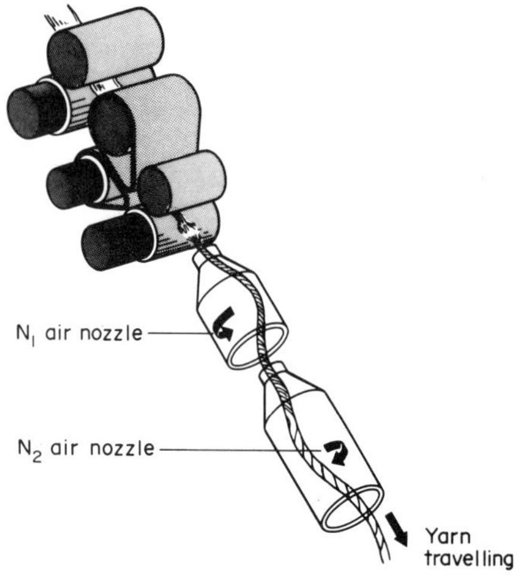

Figure 5
Air jet spinning (Murata, MJS system)

The structure of the fasciated yarn provides for unusual properties in a fabric, since the yarn has a filamentous character because of its parallel core fibers and a spun-yarn character because of its wrappings. Fasciated yarns are cleaner, less fuzzy and more lustrous than ring-spun yarns, but they are also less flexible in the context of woven fabrics than are traditionally ring-spun yarns.

Bonded yarn, in which staple fibers are temporarily held together by an adhesive, is a form of twistless structure. The adhesive is removed after the yarn has been made into a fabric and has thus achieved strength from the interlacing of the weave. Fabrics produced of bonded yarn have good strength, luster and softness and have performed well in wear and washing.

Whereas staple yarns require a substantial degree of twist to provide strength, filament yarns need only a low level of twist (one turn per 50–75 mm), enough to hold the bundle of filaments together during weaving or knitting. With the ring frame, these low levels of twist could be introduced no faster than at rates of 700–800 m min^{-1}, so manufacturers were anxious to find other ways of providing cohesion in the bundle. One successful method relies on air nozzles that pass a stream of air through a bundle of filaments (see Fig. 6), separating filaments and groups of filaments from one another and then randomly intermingling them along the length of the yarn. Cohesion is thus imparted to the bundle and a strand of "interlaced" yarn results. Production rates of more than 3000 min^{-1} have been reported (Bunting and Nelson 1963) and the large, complex twisting frames have been replaced by small, specially designed air nozzles.

2.4 Production Rates

Because of their structural need for substantial twist, staple yarns are twisted (or spun) at much lower linear speeds than are continuous filament yarns and orders of magnitude less than the continuous filament interlacing process. Also, there are significant speed differences among the various staple yarn spinning systems.

Finer (small diameter) staple yarns need higher twist levels than do coarser yarns to achieve equivalent conversion of fiber strength to yarn strength; therefore, it is to be expected that linear production rates decline for finer yarns, as shown in Fig. 7. It is worth noting also that current open-end (OES) rotor spinning methods have production speeds 8–10 times those of ring–traveller spinning frames, while the air jet systems operate at speeds 1.5 times those of the OES. However, the range of fineness counts produced by each system differs considerably, the ring–traveller system furnishing the broadest spectrum of yarn fineness. To some degree this partly explains the persistence of the ring frame in the textile industry despite the availability of much faster systems. Investment inertia no doubt accounts for much of the resistance to equipment change, but the difference in properties and behavior of the different yarns in some fabrics is often the deciding factor.

3. Fabric Structures and Processes

The term textile fabric generally pertains to fibrous sheets comprising yarns which have been interlaced to form a cloth structure. The mode of interlacing employed in forming the fabric serves to identify its structure as, for example, woven, knitted, braided or tufted.

Textile fabric structures are often viewed as two-dimensional structures, in consideration of their small dimensions normal to the plane relative to in-plane dimensions. They are usually flat sheet structures whose thickness is in the region of two or three

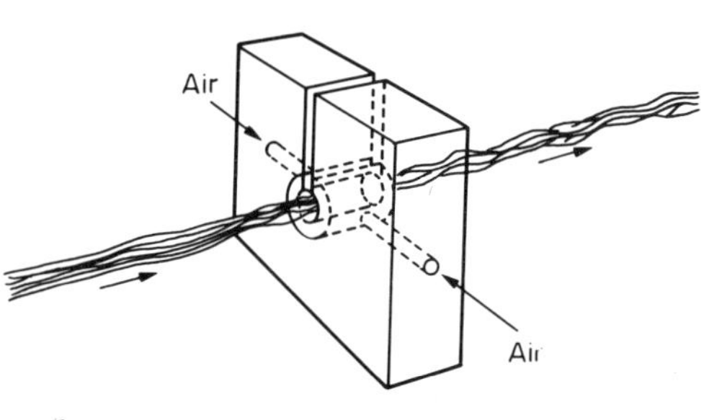

Figure 6
Air jet interlacing (entanglement) (after Backer 1972. © Scientific American, Inc., New York. Reproduced with permission)

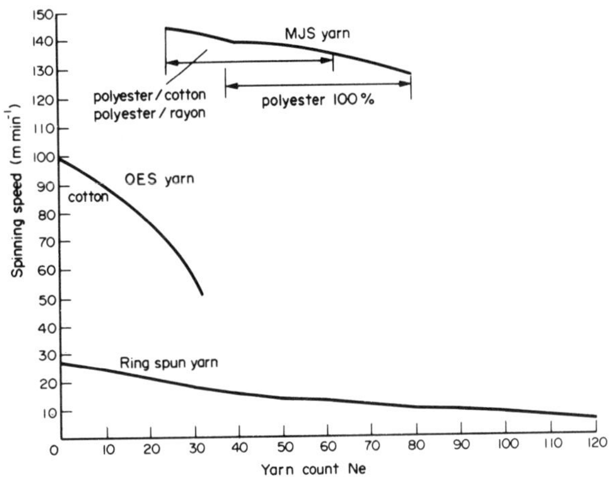

Figure 7
Production rate vs yarn count for various spinning systems

times their yarn diameters. In some cases, however, textile materials are much thicker three-dimensional structures due to periodic out-of-plane orientation of yarn components. Examples of such materials include pile carpets, fleece linings and even double and multiple layer fabrics.

The appearance of textile fabrics, as well as their mechanical and aesthetic properties, depends on the way in which their yarn components are formed and interlaced. The traditional woven fabric is orthogonally interlaced, with yarns undulating above and below yarns lying at right angles. The weft-knit fabric is constructed by laying in each new yarn length across the width of the fabric and interlooping it with the loops of the previously inserted yarn. In the warp-knit fabric, numerous parallel yarns running along the fabric length are shifted and interlooped from side to side to various degrees. The design variations in each of these fabric structures are numerous and extensive, yet each of the three fabric classes cited maintains a distinctive appearance (requiring slight magnification to detect in the case of fine yarn cloths) and different mechanical properties.

Generally, the flat woven (nonpile type) fabric is the thinnest of the three classes, given the same yarn system. In addition, woven fabrics have the highest tensile modulus in both machine and cross machine directions and the lowest extensibility. The weft-knit (or filling-knit) fabric usually has the lowest modulus in both directions and the highest extensibility of the three types. Thus it finds widespread use in sweaters and stockings. Finally, the warp-knit fabric has an intermediate modulus in the machine direction but often a very low modulus in the cross-machine direction. Such variations in mechanical behavior are a reflection of the differences in interlacing methods. This will be made evident in the following discussion of the manufacturing process and the structural geometry of each class of material.

4. Woven Fabrics and Weaving Processes

Woven fabric is one of the oldest textile structures, yet it is currently the most widely used textile material. Although the earliest hand weaver is lost in antiquity, hand weaving is still being practised in developing countries using the same basic techniques as used by the earliest weavers. In fact, the essential mechanics of weaving in a high-speed automatic loom are not much different from those employed in hand weaving, as can be seen in Fig. 8a.

4.1 Weaving Mechanisms

Weaving is accomplished by first arranging numerous longitudinal warp yarns in parallel, stretching them between two cylindrical rolls or beams. At given moments, harness frames are activated to raise selected yarns in the parallel array at a point approximately midway between the rolls to form a tent or shed of raised yarns, with a floor of unraised (or lowered) yarns. A single filling yarn or weft is then inserted under the roof of the shed, perpendicular to the warp array. The raised yarns are then lowered and other warp yarns are raised. Meanwhile, a comb (reed) is moved downstream, pressing the freshly inserted filling yarn (pick) against the newly formed edge (fell) of the fabric. As selected warp yarns are raised and lowered by the harnesses after each pick insertion, they are bent over and above successive picks. At the instant of pick insertion, the filling yarn is essentially straight and the warp yarns are then immediately bent, or crimped around it, forming a fresh increment of woven material at the cloth fell.

As a result of such bending, normal forces develop at each warp–filling yarn intersection. Such forces induce alternate bending or crimping of the filling yarn and promote its longitudinal contraction. However, a "temple" device is provided on most looms, just downstream of the cloth fell, to prevent such filling yarn crimping and fabric narrowing. Accordingly, a biaxial state of stress develops between the fell and the temple but, as the fabric moves by the temple, the cross-machine tension which it induces is dissipated and the cloth experiences only a uniaxial warpwise tension. Major alterations occur in these early moments of cloth formation, in response to the indicated changes in planar stress conditions. Figure 9 provides a schematic representation of four stages of alteration of fabric structure.

Stage 1 represents a cross-sectional view of the warp yarns and a longitudinal view of the most recently placed filling yarn an instant before the warp shed is changed. Stage 2 shows the sectional

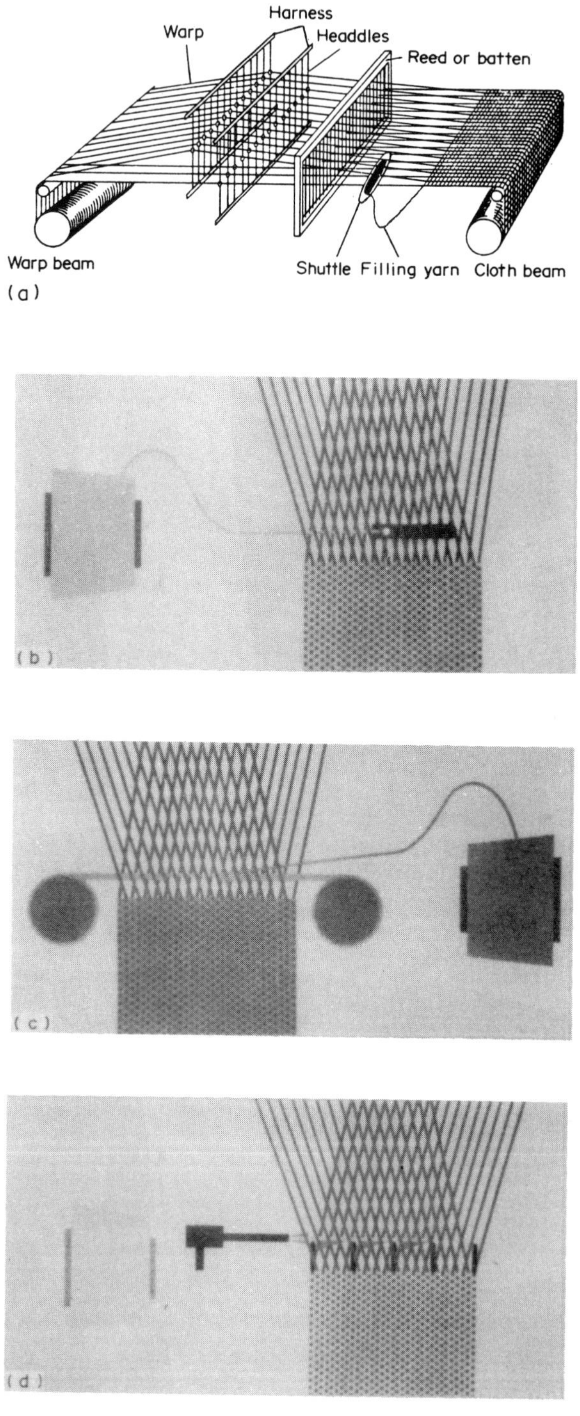

Figure 8
Weaving systems: (a) shuttle, (b) projectile, (c) rapier and (d) with compressed air (Part (a) after Hollen et al. 1979 © Macmillan, New York. Parts (b–d) after Kessels K H 1983 *Sulzer Tech. Rev.* 4: 3. © Sulzer Bros. Ltd., Switzerland. All figures reproduced with permission)

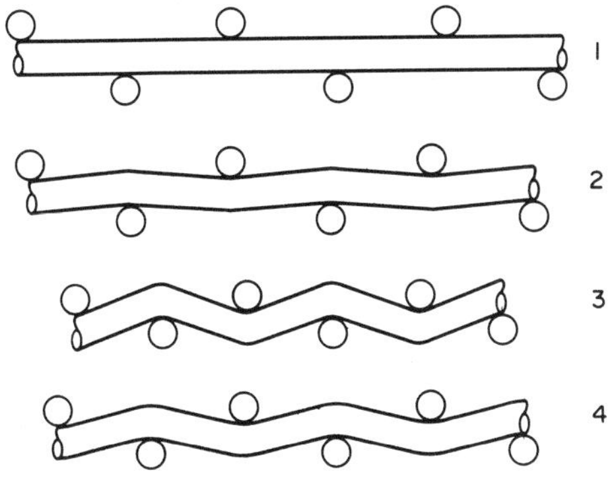

Figure 9
Changes in crimp balance during weaving

view of the fabric in the biaxial stress regime just upstream of the temple. Stage 3 is the sectional view of the filling in the uniaxial stress regime well downstream of the temple and stage 4 represents the fabric in the unstressed off-loom state.

During stages 2 and 3, the yarn blending and crimp interchange present are dictated primarily by the ratio of stresses warpwise and fillingwise, by yarn diameter and by yarn planar spacing. The mechanical requirement is for normal force equality, that is, equilibrium at each intersection. Neglecting bending rigidities, it is easily shown that such force is determined by the product of horizontal tension in each yarn and the tangent of its inclination angle at the fabric center plane.

When the weaving tensions are released in the off-loom state, the fabric assumes still another geometry—that of stage 4. Here, the mechanical requirement is again for force equilibrium between warp and filling, but forces are now dictated by bending rigidities as well as yarn diameter and spacing. Clearly, other geometric states exist in real fabrics comprising viscoelastic, thermoplastic and hydrophilic fibers. For, in addition to the linear elastic bending effects on normal forces, cold stress relaxation and recovery play a role, as well as stress relaxation due to fabric heat setting, or yarn swelling in aqueous finishing treatments. Analytical considerations of these changes are well documented in the textile mechanics literature (Hearle et al. 1969).

4.2 Loom Types

It has been said that the history of weaving from the time of the industrial revolution until World War II was marked by technological advances in shedding, that is, in raising and lowering harnesses to create the yarn shed for pick insertion. Actually, there was a step increase in loom output in the mid-1920s when automatic looms were developed capable of automatic stopping for yarn breakage and with automatic quill (or filling bobbin) resupply. The next step increase took place in the mid-1950s with the introduction of shuttleless weaving and, since then, the emphasis has been on various techniques of rapid pick insertion, as illustrated in Figs. 8b–d.

In shuttleless weaving, the quill-bearing shuttle (of considerable mass) was at first replaced by a small projectile designed to take the filling yarn from an externally placed package and to draw it through the shed across the warp yarns. Not only was the small projectile pick insertion faster than full shuttle projection, but it also required lower shed heights, facilitating rapid shedding motions. In addition, both rigid and flexible band rapiers were developed to pull the yarn across the loom from an outside filling supply. Eventually, both water jets and air jets were employed to impel the pick across the loom. The result of such developments was a rapid increase in pick rate or, just as important, in filling length insertion rate. For example, weaving machines are available with air or water jet weft insertion rates exceeding 600 picks per minute and 2000 m min^{-1}, while current rigid or flexible rapier systems perform at somewhat lower production rates.

From the viewpoint of most users there is a general equivalence of fabrics woven on various loom types, except that cloth woven on a classical shuttle loom comprises a continuous filling yarn which interlaces back and forth across the warp, making 180° turns at each selvage. In newer looms with modified weft propulsion systems, the cloth produced comprises filling yarn metered according to cloth width either in single lengths or in hairpin form.

On the classical fly shuttle loom, multiple shuttles have been possible for some time, permitting the weaving of differently colored yarns or different yarn structures in alternate picks or according to design schedule (e.g., in familiar checks and plaids). Rapier weaving systems handle six- or eight-color systems, while air jet weaving machines, which at first could handle only one color, are now available with two-color capacity and even four colors.

Finally, note should be taken of a revolutionary development in weaving, in which warp yarns are moved into shedding positions in stages, rather than all at once, so that several filling yarns can be inserted at once. The multiphase loom possesses numerous weft carriers (or small shuttles) operating at velocities about one-tenth those of other modern weaving machines. However, because of the simultaneous action of many such carriers, the multiphase weft insertion rate is two to three times higher, at more than 2000 m min^{-1}. The necessity to beat up multiple filling threads simultaneously results in a somewhat different fabric structure from that achieved with other weaving systems.

To the fabric consumer, however, the use of the newer types of weaving machines for fabric production does not significantly affect cloth properties compared with those woven on older loom types (given equivalent yarns and cloth structural design). This is contrary to experience in yarn production, where the mechanisms and actions of the various spinning systems are quite different and, as a result, the ensuing yarn structures and properties are significantly different (despite the use of equivalent fibers). It should be noted, however, that the multiphase loom cannot make certain tight constructions such as broadcloths.

4.3 Woven Fabric Structures

The most common fabric structure is the plain weave, in which each yarn rises to the cloth surface, then falls to the back at each cross yarn intersection. Every unit cell of the plain weave is of type I shown in Fig. 10. However, three other interlacing modes are possible, as shown by types II, III and IV. Fabrics structured with a mixture of all interlacing types are other than plain weave, for example twill, satin or basket weaves, with appearances and characteristics well known to the average consumer.

The sectional view of a plain weave is seen in Fig. 11 and its labelled parameters for warp and filling yarn sets include yarn diameter $d_{w,f}$, yarn length $l_{w,f}$, spacing $p_{w,f}$, wave height $h_{w,f}$, inclination angle $\theta_{w,f}$ and thickness G. Crimp is defined as the fractional difference between actual yarn length and projected length in the unit cell (Hearle et al. 1969). If values of any or all of these structural parameters are changed, given the same fiber composition, the resulting cloth structure will likewise change, along with its mechanical properties, while still being classified as a plain weave. There have been many such plain weave modifications and they are given different generic names in commerce, such as broadcloth, duck, awning cloth, cheesecloth, poplin and taffeta, each being identified with area of usage and often possessing significantly different properties (Backer and Valko 1969, Textile Institute 1975, 1977).

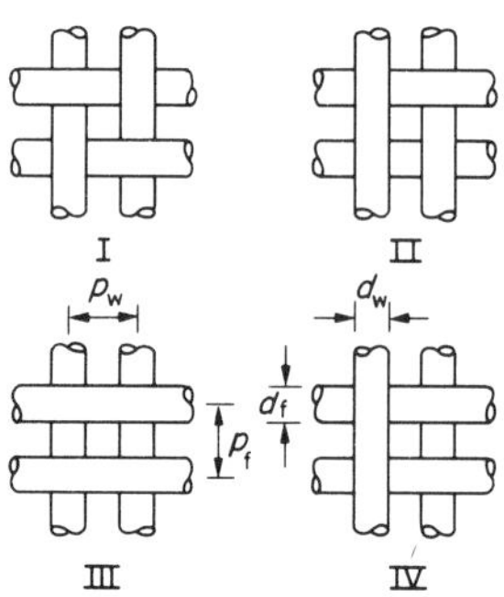

Figure 10
Interlacing types in woven fabrics (after Backer 1951 *Text. Res. J.* 21: 704. © Textile Research Institute, New Jersey. Reproduced with permission)

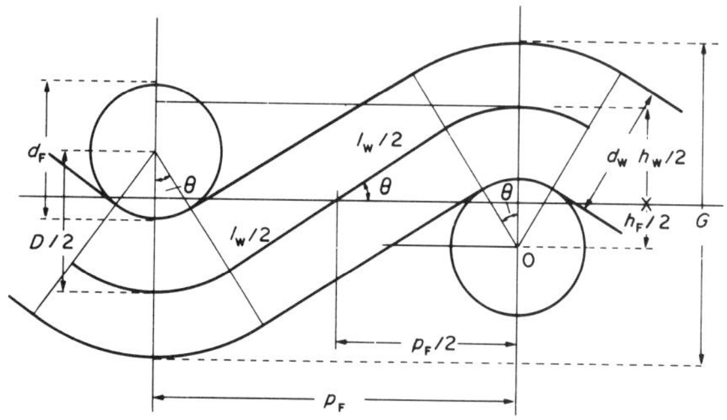

Figure 11
Plain weave geometric model

As to properties of interest, the user of fabrics may be concerned with aesthetic, biochemical, chemical, mechanical, physical or physical/chemical properties. Each such classification of properties possesses an extended list of narrower terms (Backer and Valko 1969), some of which relate to all types of material; others pertain more directly to textile materials. For example, there are more than 40 mechanical properties of textile fabrics, including dimensional stability, air or water permeability, puncture resistance, snag resistance, softness, tear resistance and bending rigidity.

These properties vary directly with changes in the geometric parameters labelled in Fig. 11. The resulting relationships are easily deduced in some cases but not in others. The all-important property of flexibility is clearly related to yarn diameter (or fibers per yarn), to yarn structure or spinning system and to the number of yarns per unit width being bent. It is less clear that warp-direction fabric bending rigidity depends also on relationships between the pattern of fabric interlacing, as well as on the number of filling yarns per unit width, the underlying mechanism having to do with structural jamming and a buildup of normal forces between yarn systems serving to restrict fiber movement during bending (Hearle et al. 1969).

Permeability is yet another property influenced by yarn spacing and by interlacing pattern. The effect of yarn spacing is obvious, but it can also be shown that gas or water permeability is reduced as the proportion of type I cells is increased in the fabric. This has to do with the convoluted geometry of the fabric pores, which are greatly affected by interlace pattern.

A further example of a fabric structure–property relationship is that between tear resistance and yarn spacing and interlacing pattern. It is recognized that yarn mobility is the critical factor influencing tear resistance of woven fabrics, and clearly reduced yarn spacing, as well as maximum interlacing of type I, reduces mobility. With minimum yarn freedom to slip longitudinally and laterally in the weave structure, tearing forces easily cause local failure one yarn at a time. In an open structure, however, yarns may slip against each other and "rope" together at the apex of the tear crack, thus increasing the threshold level of tear force required to break though the rope and to propagate the tear. Such behavior is not encountered in homogeneous solids, but crack bridging by reinforcing fibers in a composite does resemble it to some extent.

The basic elements of the woven fabric construction shown in Figs. 10 and 11 suggest the possibility of weaving a variety of cloth structures, limited only by the number of loom harnesses and harness raising controls. Carried to an extreme in the Jacquard loom, each yarn crossover (warp versus weft) can be individually controlled, making possible an infinite variety of colored yarn designs, even to the point of scenic or portrait patterns—all of the above being related to flat woven fabrics of ordinary commerce.

The same weaving process, however, can be employed in the production of "backed" cloths, double and triple cloths, pile fabrics and a range of three-dimensional materials. The forms of instructions supplied to the loom for weaving such complex structures are the same as for weaving ordinary flat fabrics, namely the square design (or pattern) paper. Its vertical columns represent warp ends and its rows represent the filling picks—the marked square indicating that the end is raised above the pick and a blank square representing an end placed under the pick. The plain weave pattern of Fig. 12 or the twill weave of Fig. 13 can be combined in double weaving, that is, backed with the same or other weaves. To do so, a given repeat pattern of the face fabric (Fig. 14a) is merely opened up and the repeat pattern of the back fabric (Fig. 14b) is superimposed, as shown in Fig. 14c. This provides separate instructions to the face fabric warp and filling and to the back fabric warp and filling. It remains to require the face fabric warp and filling always to rise above the back fabric filling and warp, respectively, as per the pattern instruction of Fig. 14d. The direct combination of Figs. 14c and 14d (not shown) then

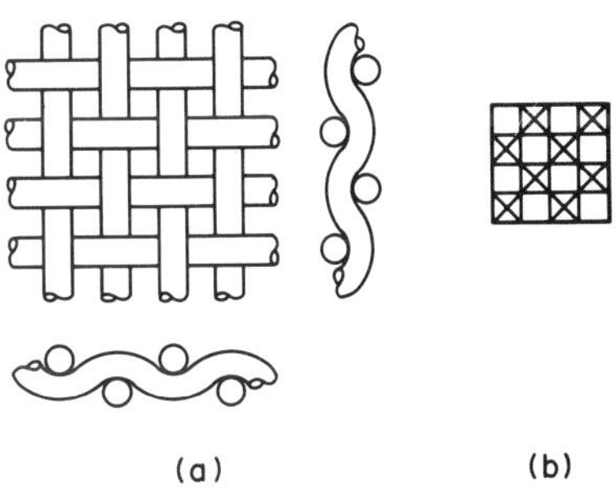

Figure 12
Plain weave structure

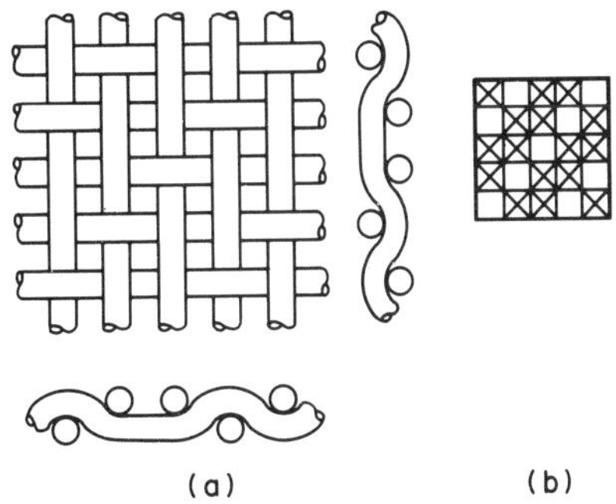

Figure 13
Plain weave structure (2/1 twill)

provides the necessary loom instruction. Only the instruction for a binder yarn is necessary to bond the two fabrics together and this can be provided by selectively raising a back fabric warp above a face fabric filling.

Examples of double and triple fabrics are shown in Fig. 15a, b, of circular weaves in Fig. 15c and of double velvet in Fig. 15d. The purpose of double weaving and so on is twofold: first, to provide thick fabrics with fine surface texture and with pattern differences face and back; and second, to make circular (i.e., cylindrical) fabrics or fabrics capable of unfolding to dimensions far in excess of the loom width (as seen in Fig. 15c). Suffice it to say that weave patterns and structural varieties are boundless, even on conventional looms. However, as to structural mechanics treatments in the literature, only the simplest weave patterns have been dealt with so far, and the user of more complex woven cloth structures in engineering applications must allow for significant differences in initial material behavior and in product durability between the plain

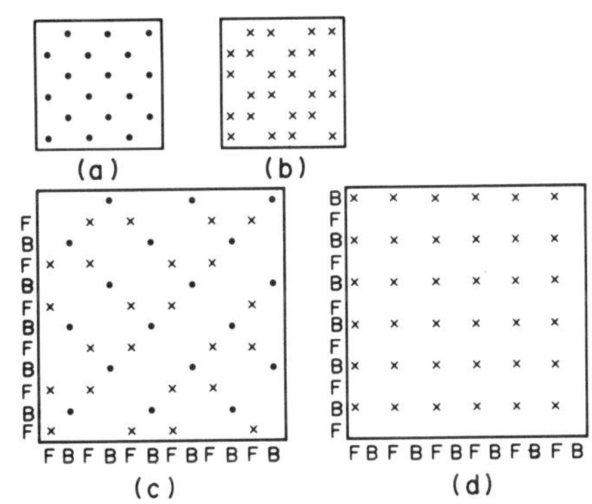

Figure 14
Double weave: (a) face fabric pattern, (b) back fabric pattern, (c) superimposition of face (F) and back (B) patterns and (d) pattern instruction for face fabric to rise above back fabric

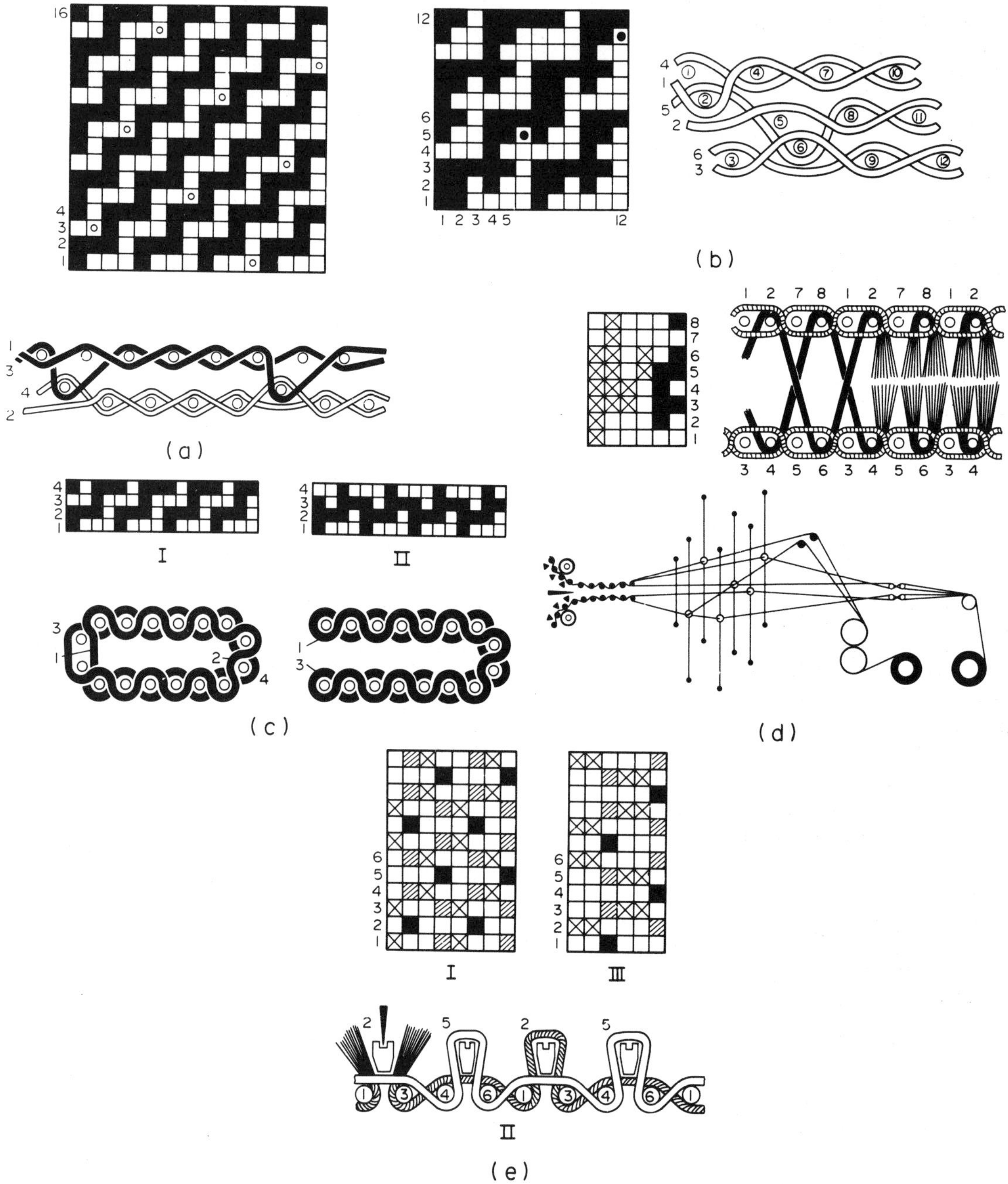

Figure 15
Compound weaves: (a) double, (b) triple, (c) circular, (d) velvet and (e) pile (after Takematsu S 1982 *J. Text. Mach. Soc. Jpn.* 28: 292–5. © Textile, Machinery Society of Japan, Osaka. Reproduced with permission)

weave and the more complex flat weaves, or three-dimensional constructions.

While novelty and fancy patterns are of little interest to the industrial user of woven fabrics, the successful application of complex weaving patterns in the construction of engineered reinforcement for composite systems should be noted. Examples of such applications include three-dimensional I beams, channel beams and conical and spherical nose cones for rockets and radomes.

To the consumer, the most familiar "thick" or three-dimensional fabrics are floor coverings or carpets, fleece garment linings, pile fabrics and synthetic furs. In current technology, most of these structures are produced on machines other than the traditional loom. Still, a limited output of carpets is woven in the traditional way, using special filling elements (as seen in Fig. 15e) to ensure local elevations of the warp yarn, which in turn become the uncut, or subsequently cut, pile tuft of the carpet. The dominant production method for wall-to-wall carpets is an entirely different branch of textile technology, that is, tufting. Briefly stated, the tufting process employs a separate preformed fabric (either woven or nonwoven) and a separate pile yarn. The latter yarn is needle tufted through the backing fabric, the resulting loop being hooked and held below while the needle is withdrawn, then reinserted a stitch length away to form another loop below. As shown in Fig. 16, the loops may be cut after the fabric has moved a few stitch lengths through the machine. The structure pictured is, of course, dimensionally unstable and is generally reinforced afterwards with a coating of adhesive at the base of the loops and with an additional backing fabric.

As has been indicated above, addition of adhesive to fibrous systems increases bending rigidity by one or two orders of magnitude. Thus, most tufted carpets are relatively stiff in bending, while truly woven (and unbacked) carpets are uniquely flexible.

5. Knitted Fabrics and Knitting Processes

The art of hand knitting has been traced back to 250 AD, but the first knitting machine on record, a flat bed unit, was constructed towards the end of the sixteenth century. Two centuries later, the circular knitter and the warp knitting machines were introduced. Both the flat bed and the circular knitting machines produce weft-knit fabrics and, of course, the warp knitting process provides warp-knit structures. As indicated above, these terms derive from the direction in which the yarns are introduced into the fabric and interlooped at the newly formed edge of the cloth.

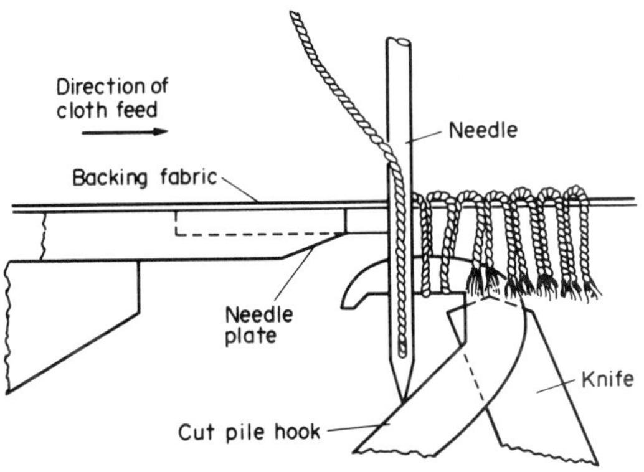

Figure 16
Standard cut pile machine

5.1 Weft-Knitting Mechanisms

Hand knitting is a weft-knitting operation based on a two-needle system. Weft machine knitting, however, is performed on multiple needles, each of which goes through standard knitting actions in sequence. The knitting needles are of several types and are either laid in parallel grooves machined in the surface of a cylinder or in a flat bed or in radial grooves on the surface of a circular disk. In the conventional circular knitting machine, a section of which is depicted in Fig. 17, the cylinder is rotated about its vertical axis, carrying the needles in its parallel vertical surface grooves. The lower end of each needle shank is enlarged to form a butt which projects radically out of the groove and permits cam activation of the upward and downward motion of the individual needle at the appropriate moment. With the cams fixed in space, the rotation of the cylinder thus causes its needles to rise and fall at various points around the fixed periphery of the machine.

Generally, at each cam-directed rise point, a weft yarn is fed under the hook of the highest-positioned needle. As the cylinder progresses rotationally, that needle is then cam-motivated to move downward as it catches the new yarn and initiates the knitting action described below. Thus, each needle goes through a rapid undulating motion at the surface of the rapidly rotating cylinder, with each rise point in its locus representing a yarn feeding or insertion station. As many as 108 feed stations are present on a modern circular jersey machine, indicating that 108 yarns are being simultaneously inserted at the forming edge of the fabric, although at different positions equally spaced around the machine periphery. Thus, the method of fabric formation in circular knitting is highly efficient and its rate of production expressed in terms of fabric area per unit time is several times that of the fastest looms. (Interlock machines have 144 feeds.)

The knitting action referred to can best be shown with a single needle pictured in several vertical positions, as in Fig. 18. There are several different needles employed in knitting: the latch needle shown in Fig. 18, the bearded needle and the compound needle. Since the pertinent knitting actions of all three needles are similar, consideration is given here only to the latch needle shown in Fig. 18. Here, the same needle is seen in different vertical positions, as well as in different circumferential positions on the circular machine. Since both the

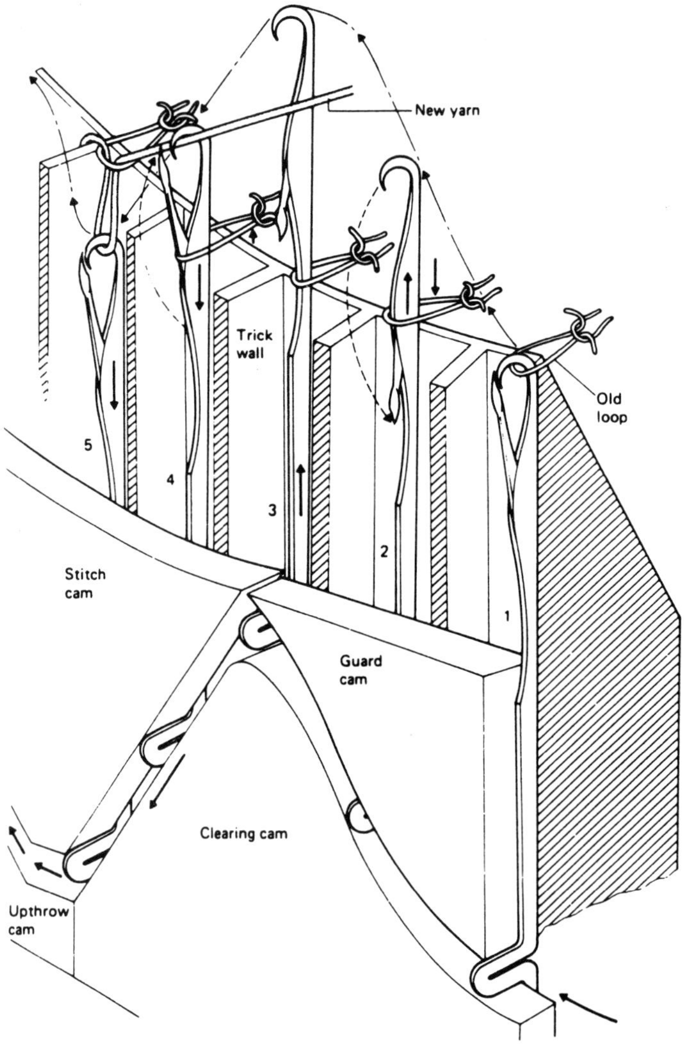

Figure 17
Circular weft knitting machine (after Spencer 1983)

newly knitted fabric and the needle move circumferentially with the rotation of the cylinder, the relative motion between needles and fabric loops occurs in the vertical direction and thus provides the knitting action. This motion may be cam activated, dial activated or electronically triggered.

Each phase of knitting is represented in Fig. 18: (a) the needle is closed on the new loop, after the previous loop has been cast off; (b) the needle is raised so that the new loop slips down the shank, opening the latch; (c) the needle is raised further so that the new loop slips down the shank below the tip of the latch and the needle head just passes its high point and catches the feed yarn at an input station; (d) the needle descends, causing the new loop to ride up its shank, closing the latch and capturing the new segment of feed yarn; and (e) the needle descends still further as the new loop rides up and over the closed latch needle end and pulls the feed yarn segment down through the new loop to form another "new" loop. The process repeats with Fig. 18a.

As the needle of Fig. 18 goes through one such undulation cycle at each feed station (in production of the simple jersey fabric), the needle immediately to its left or right is slightly retarded or slightly advanced in its own undulation cycle. Thus, in the region of each feed station, a wave form can be pictured of vertical positions of adjacent needles, some capturing the feed yarn, others drawing it downward through the previous loop to form a new loop, others slipping the previous loop up the shank to close the latch and others casting off the previous loop, then rising to open the latch for approach to the next feed station along the machine circumference.

It is obvious that the needle sequences shown in Fig. 18 are for a single vertical line of loops called a wale and that all such wales must be horizontally connected to form the knitted fabric. Not so evident is the fact that sequential forming of a new horizontal row of stitches or loops, called a course, and casting of the previous row of stitches at each feed station are accompanied by relative slippage motion of the feed yarn between needles and other machine components. As the stitch is formed by the needle pulling the loop of yarn, some of the yarn length requirement is taken from straight feed yarn upstream and some from already formed stitches downstream. In the last stage of stitch formation, the stitch loop gives back some of its yarn to the upstream stitches which are following it. The yarn slippage or capstanlike snubbing over multiple needles and other machine components leads to tension buildup which can cause filament rupture or even total yarn failure. Numerous analyses have been proposed in the past to relate such yarn stress buildup to process conditions, to machine component geometry and surface properties and to yarn tensile and frictional properties (Knapton and Munden 1966).

So far the production of a single-knit fabric has been considered, that is, the single jersey structure or plain knit, analogous to the plain weave (Fig. 19a). The plain knit is widely used, as it allows for the highest rate of production on the least complicated machinery. However, numerous modifications of the plain knit are possible by appropriate programming of the stitching cycles on different needles. Selected modifications of the weft-knit structures are outlined below.

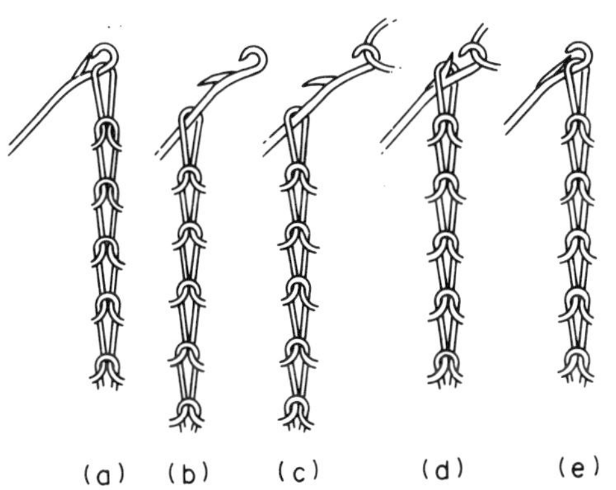

Figure 18
(a–e) Successive knitting steps with latch needle

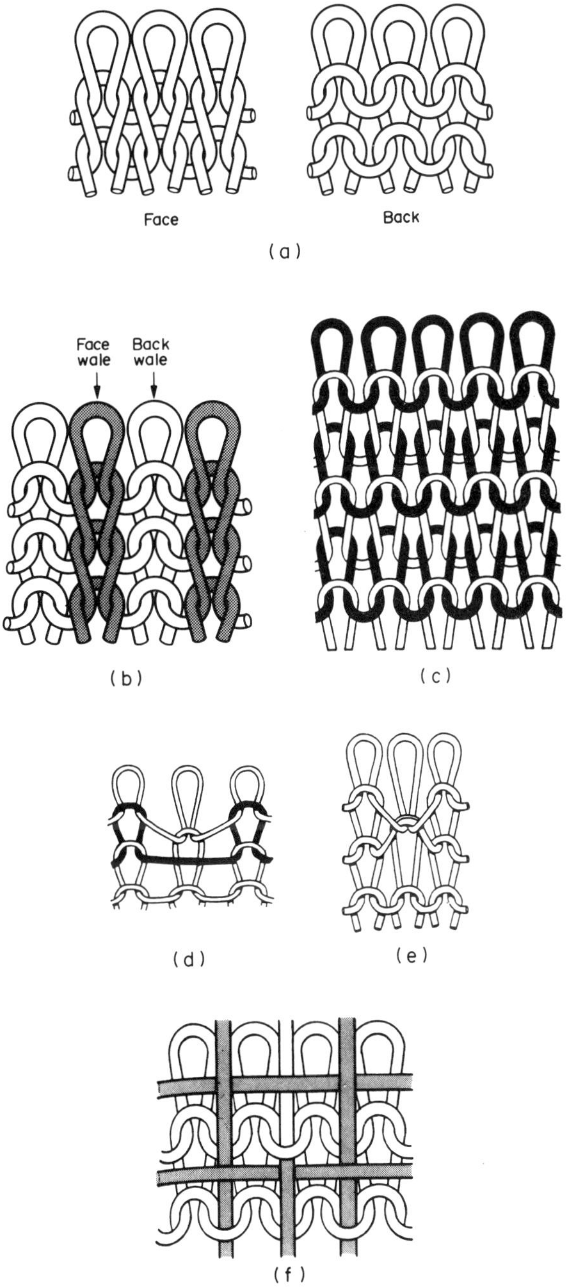

Figure 19
Weft-knit structures: (a) single jersey, (b) 1×1 rib, (c) purl fabric, (d) float stitch, (e) tuck stitch, (f) warp-and-weft insertion structure (Parts (a) and (c) from *Knitting Encyclopedia*, 1972. © National Knitwear and Sportswear Association, USA. Part (b) after Schwartz et al. 1979. © North Carolina State University Press, North Carolina. Parts (d) and (e) after Spencer 1983. Part (f) after Hollen et al. 1979. © Macmillan, New York. All figures reproduced with permission)

5.2 Weft-Knit Structures

Face and back views of a single jersey fabric are shown in Fig. 19a, the face being identified by the fact that in this simplest of knit structures all loops are drawn to the face of the fabric. Double sets of needles (either in a flat bed or cylinder–dial machines) produce double-knit fabrics. They can be knitted with alternating stitch columns, or wales can be constructed first with stitches drawn to the fabric face, then to the fabric back, as shown in Fig. 19b. Such a construction is termed rib knit; a 1 × 1 rib is seen here, but the alteration may be 2 × 2 or other. With the reversal of loop draw direction in successive wales, the structure undergoes a change in both appearance and in horizontal extensibility, due to the corrugation at the lines of loop reversal.

Reversal of the loop draw direction can also be accomplished in successive horizontal lines or courses. The resulting structure pictured in Fig. 19c is a purl fabric. Other stitch variations can be introduced at specific fabric locations by controlling the knitting action of a single needle at a given instant. For example, the needle may be held down in a nonworking position so that it fails to capture the feed yarn at a given input station; this results in a float stitch or miss stitch, as shown in Fig. 19d. Alternatively, the needle may be raised halfway so that the last loop remains on the open latch while a new yarn segment is captured in the needle hook; then both yarns are drawn down through the prior loop to create the structural modification called a tuck stitch, as shown in Fig. 19e.

Another double-knit structure is the interlock fabric, based on an intermeshing of two 1 × 1 rib fabrics. Double-knit fabrics with tuck and float loops, in addition to knit loops, provide for the greatest dimensional stability of the weft-knit class of fabrics and find considerable application in apparel markets usually dominated by woven fabrics.

A final modification of the weft-knit structure introduced in recent years is the insertion of straight (i.e., unlooped) warp yarn and straight filling yarn against the backdrop of the conventional jersey structure. Figure 19f illustrates such a fabric with both warp and weft inserts. Clearly, such a weft-knit fabric has considerable dimensional stability, due primarily to the stiffness and elasticity of the insert yarns.

5.3 Warp-Knitting Processes and Fabrics

The novice observer of warp knitting machines, for example the tricot or raschel machines, is likely to conclude that they are in no way related to the circular knitting equipment employed in production of simple jersey fabrics. Such a conclusion may be motivated by absence of any resemblance in the structure of the equipment, by differences in feed yarn configuration and by mode of yarn input at the freshly formed fabric edge. However, the basic warp knitting action illustrated in Fig. 20a is essentially the same as that shown for weft knitting in Fig. 18, that is, for a single needle. The real difference lies in the mode and sequence of feeding the yarn to multiple needles, for, as implied in the process name, the input is fed as multiple yarns from a warp beam, namely in sheet form. The multiple needle action is simultaneous, rather than sequential as in circular knitting. Instead of the feeding of a single yarn through each fixed inlet station, as in circular knitting, the multiple warp ends are fed and guided through a series of parallel eyes which are mounted on one or more bars running across the machine width.

The sequential motion of these bars across and along the needles, as pictured in Fig. 21, serves to bring the appropriate yarn(s) across given needles to be subsequently formed into a loop. In addition, the bar motion can move the yarn guides along the line of needles, skipping one or several wales so that a single warp end may alternately focus its vertical line of loops in one fabric wale, then in another wale displaced by one or more needles. As a result, the path of a given warp end is vertical on average, with short inclined or horizontal excursions, as seen in Fig. 20b.

There are four basic warp knitting units, including the tricot machine which uses spring-beard, or compound, needles (often with two to four bars,

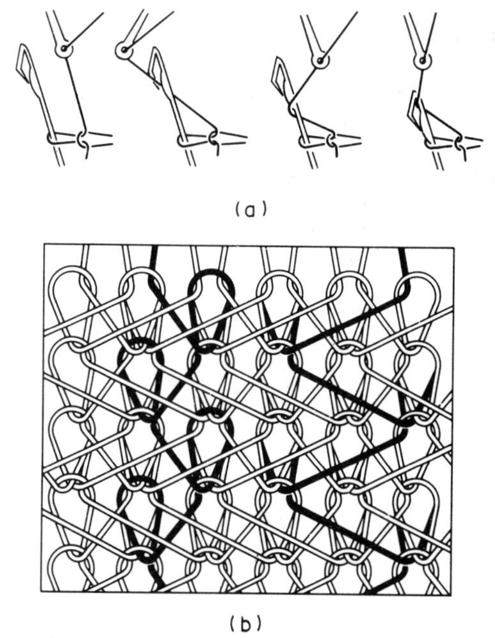

Figure 20
Warp knitting; (a) stitch formation with spring-beard needle and (b) warp-knit fabric

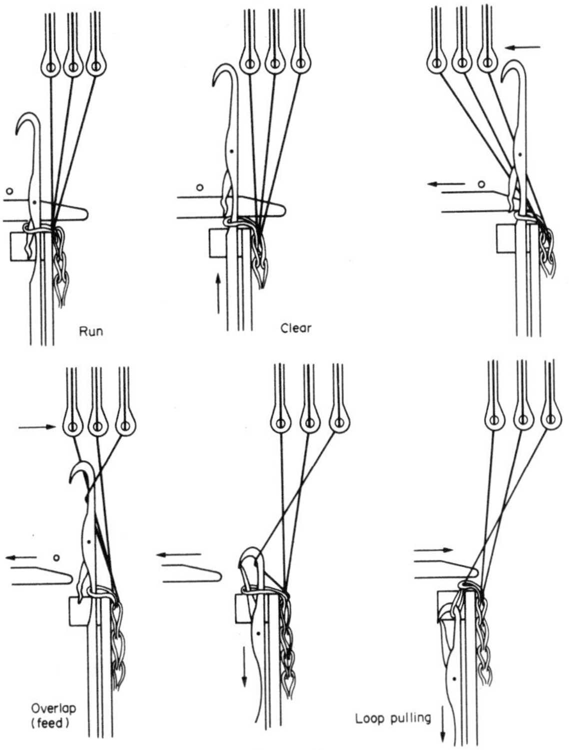

Figure 21
Warp-knitting steps (after Schwartz et al. 1979. © North Carolina State University Press, North Carolina. Reproduced with permission)

depending on the fabric being produced), and the raschel machine with as many as 56 bars and capable of knitting a wide range of complex fabrics, as well as lacelike or netlike structures for apparel, home furnishing or industrial uses.

Finally, there is a relatively new class of processes called stitch bonding, based on the warp knitting system. In stitch bonding, a relatively unstable fibrous sheet is run through a warp-knitting (or stitching) system with its multiple needles running back and forth through the sheet and thus knitting a series of chain stitches, or a simple warp-knit structure, which form a mechanical bond to stabilize the sheet. The original sheet can comprise two sets of parallel yarns orthogonally laid, a fibrous nonwoven web or a composite of a web and a fabric. In some instances, a fleecy web can be used with the bonding chain stitches formed of the actual fibers of the web rather than requiring a separate binding thread. The range of materials which can be produced on these systems is extensive, the productivity level is an order of magnitude greater than that of weaving and the product often approaches the woven cloth in appearance. The process facilitates manufacture of a planar, stiff, orthogonally laid sheet of noncrimped high-modulus yarns bound together in a chain stitching mode. Thus, the stiffness of such a sheet for a given set of yarns can exceed that of a conventional fabric interlaced in conventional plain or twill weaves.

In conclusion, emphasis must be laid on the dominant structural contribution of fabric geometry and processing methods to the mechanical properties of fibrous sheet materials. Since the mid-1940s, great strides have been made in developing improved properties of synthetic fibers. In particular, since the mid-1970s, high-performance synthetics have become commonplace, although they are still very costly. The point to be underscored is that advances in structural mechanics of fibrous systems must move in parallel with advances in materials science, if fibrous structures are to perform in engineering applications at a high level of efficiency. The role dominance of structural mechanics is aptly summarized in Table 1 showing property ratios reflecting the effect of various textile configurations on the translation of selected fiber properties into the performance and properties of certain textile products.

See also: Fibers and Textiles: An Overview; Fibers: Texturing and Draw Texturing

Table 1
Fabric–fiber property ratios

Type of fabric	Ratio of initial extension modulus	Ratio of maximum extension	Ratio of bending modulus
Random-laid bonded fiber	0.1	1.7	100
Crosslaid bonded fiber	0.6	2.0	100
Heavy cotton duck	0.04	2.0	10
Lightweight worsted suiting	0.02	2.0	2
Stretch fabric with elastomer fiber yarn	0.8	1.3	1.3
Warp knit	0.001	3.0	1.0
Weft knit	0.0002	5.0	0.6

Bibliography

Backer S 1972 Yarn. *Sci. Am.* 227(6): 46–56

Backer S, Valko E I 1969 *Thesaurus of Textile Terms Covering Fibrous Materials and Processes*, 2nd edn. MIT Press, Cambridge, MA

Bunting W W, Nelson T L 1963 Process for producing compact interlaced yarn. US Patent No. 3,110,151

Hearle J W S, Grosberg P, Backer S 1969 *Structural Mechanics of Fibers Yarns and Fabrics*, Vol. 1. Wiley, New York

Hollen N, Saddler J, Langford A L 1979 *Textiles*, 5th edn. Macmillan, New York

Knapton J F, Munden D L 1966 A study of the mechanics of loop formation on weft knitting machinery, Pts. I, II. *Text. Res. J.* 36: 1072–80; 1081–91

Koerner R M, Welsh J P 1980 *Construction and Geotechnical Engineering Using Synthetics*. Wiley, New York

Krčma R 1969 *Nonwoven Textiles*. Textile Free Press, Manchester

Lord P R 1970 *Spinning in the '70's*. Merrow Publishing, Watford, UK

Morihashi T 1976 Spinning apparatus for spun yarn. US Patent No. 4,112,658

Nakahara T 1979 Direct spinning apparatus. US Patent No. 4,142,354

Piller B 1973 *Bulked Yarns, Production Processing and Applications*. Textile Free Press, Manchester

Schwartz P, Rhodes T, Mohamed M 1979 *An Introduction to Fabric Forming Systems*. North Carolina State University Press, Raleigh, NC

Spencer D J 1983 *Knitting Technology*. Pergamon, Oxford

Summerscales J 1987 High performance reinforcement fabrics. *Prog. Rubber Plast. Technol.* 3(3): 20–32

Textile Institute 1975 *Textile Terms and Definitions*, 7th edn. Textile Institute, Manchester

Textile Institute 1976 The yarn revolution, new developments in the production of spun and textured yarns and their exploitation in fabric form. *Pap. 60th Annu. Conf. Textile Institute*. Textile Institute, Manchester

Textile Institute 1977 The design of textiles for industrial applications. *Pap. 61st Annu. Conf. Textile Institute*. Textile Institute, Manchester

Zurek W 1975 *The Structure of Yarn*. US Department of Agriculture and The National Science Foundation. Washington, DC

S. Backer
[Massachusetts Institute of Technology, Cambridge, Massachusetts, USA]

Thermoplastic Elastomers

Conventional elastomers are part of the larger category of materials known as thermosets. During the fabrication process of a conventional elastomeric product, long flexible chain molecules are chemically linked together to form a single interconnected macromolecular network. Most of the unusual physical properties of elastomers can be related directly to this covalently crosslinked network structure. A major disadvantage is the lack of thermal or solvent-induced reversibility of this structure; once formed into a finished article, it is not possible to reprocess the material. Furthermore, many of the economical high-rate processing methods developed for the engineering thermoplastics are not readily adaptable for use with conventional thermosetting elastomers.

Partly in response to some of the foregoing problems, a new variety of elastomeric materials became available from a number of sources around 1960. These materials combined the unique mechanical properties (large recoverable deformation) of conventional crosslinked elastomers with the advantageous melt-processing characteristics of uncrosslinked thermoplastics. These thermoplastic elastomers can be formed into complex shapes without requiring long cycle times for vulcanization; scrap material and articles not meeting specifications can be recycled. Because most thermoplastic elastomers are also soluble in a wide range of organic solvents, various solution-casting methods are also possible. This unusual combination of properties has led to continued growth in the marketplace and to a considerable amount of research and development on thermoplastic elastomers.

To explain this unique behavior of thermoplastic elastomers, it is necessary to consider the fact that such materials generally exhibit phase-separated morphologies in bulk. In thermoplastic elastomers the long flexible chains of the rubbery network are held together as multifunctional junction points which act as physical, rather than chemical, linkages. The source of these physical linkages in one major category of thermoplastic elastomer is provided by small discrete domains (e.g., spheres of approximately 20 nm diameter) of rigid glassy polymer embedded in the rubbery matrix. Because many flexible rubbery chains emanate from each domain and terminate in neighboring domains, a continuous network structure is obtained. At temperatures above the softening point, or glass transition (T_g) of the glassy domains, the material can be made to flow like a thermoplastic polymer melt. On cooling below the T_g of the glassy domains, the network structure is reestablished and elastomeric behavior is observed. The lower bound on the temperature of utilization of these materials is governed by the value of T_g of the continuous rubbery phase.

Variations of the domain structure described are also found in commercial thermoplastic elastomers. In addition to dispersed glassy domains, multifunctional physical linkages can be formed by regions of crystallinity, ionic attraction or strong hydrogen bonding. In essentially all cases the characteristic dimensions of the domains are smaller than the wavelength of visible light, so that, even though a phase-separated morphology exists, optical transparency is observed. The fact that these domains are, however, much larger than the typical tetrafunctional chemical linkage of a conventional elastomeric network means that some account must be made of the filler effect imparted by the domains. Often this filler effect has a favorable, reinforcing influence on the large-strain mechanical behavior.

The microphase-separated morphologies which give rise to thermoplastic elastomer behavior are in turn made possible by a certain type of macromolecular architecture known as block copolymerization. In a block copolymer, the chain molecule is divided into long sequences or blocks, each of which is composed of a different type of repeat unit. In one example, the ABA triblock copolymer, each linear macromolecule is comprised of three sequences, a relatively short glassy polymer sequence at each end (A blocks) and a longer rubbery B block between. Owing to the usual thermodynamic incompatibility between dissimilar polymeric species, the minor component glassy endblocks segregate into domains embedded in the rubbery continuous phase; restrictions on the placement of A and B blocks which arise from the sequential structure of the ABA block copolymer molecules lead to microphases, rather than macrophases, in the phase-separated morphologies of thermoplastic elastomers (Meier 1969, Helfand 1975). In addition to ABA block copolymers, various linear and star-branched multiblock copolymers, certain blends of amorphous and semicrystalline polymers and copolymers, and homopolymers with carefully controlled tacticity have been used as alternative molecular bases for thermoplastic elastomers.

Currently, the largest share of the marketplace is held by the linear and star-branched polystyrene–polydiene block copolymer elastomers. Hydrogenation of the unsaturated diene blocks yields derivative products with ethylene–butylene rubber moieties which have greater solvent and oxygen/ozone resistance. Major applications for the styrene-based thermoplastic elastomers are injection-molded parts for footwear and automobiles, wire coating, impact modifiers for rigid thermoplastics, and pressure-sensitive adhesives.

A second major category of thermoplastic elastomer is based on a class of linear multiblock copolymers known as segmented polyurethanes. In these materials, long flexible polyether or polyester segments are connected by short isocyanate–diol segments which in the bulk polymer cluster together to form small rigid domains. Applications for segmented polyurethanes include those previously mentioned as well as in coating applications and as belts and hoses.

The remainder of the thermoplastic elastomer market is essentially filled by materials that have regions of crystalline order embedded in an amorphous rubbery phase. Included in this category are polyester copolymers, copolymers containing syndiotactic 1,2-polybutadiene and blends or copolymers of polypropylene, polyethylene and polybutene. These materials are competitive with styrene-based copolymers and the segmented polyurethanes in the variety of applications mentioned.

The methods of synthesis of thermoplastic elastomers vary widely, as may be expected, owing to the wide variety of molecular structures that can be used to provide desired morphologies and properties (Dreyfus et al. 1980). One particularly versatile synthetic method is homogeneous anionic polymerization (Fetters 1969), which can be used to obtain an essentially unlimited variety of block copolymers. Using this technique, it is possible to control essential details of the chain structure, including tacticity, microstructure, molecular weight and its distribution, in each block sequence of the copolymer. With such powerful synthetic methods available it appears certain that new and improved varieties of thermoplastic elastomers will be developed in the future and that further understanding of the relationships between the molecular structure of these materials and their properties will be obtained. In view of the very great interest in this developing area, several recent reviews have given detailed data plus cross-comparisons between the various types of thermoplastic elastomers available (Legge et al. 1987, Bhowmick and Stephens 1988a, Coran 1988, Quirk et al. 1988, Schollenberger 1988, Wilder 1988, Rader and Stemper 1990).

See also: Elastomers: An Overview

Bibliography

Bhowmick A K, Stephens H L 1988a Additional types of thermoplastic elastomers. In: Bhowmick and Stephens 1988b, Chap. 12

Bhowmick A K, Stephens H L (eds.) 1988b *Handbook of Elastomers*. Dekker, New York

Coran A Y 1988 Thermoplastic rubber–plastic blends. In: Bhowmick and Stephens 1988b, Chap. 8

Dreyfus P, Fetters L J, Hansen D R 1980 Elastomeric block polymers. *Rubber Chem. Technol.* 53: 728–71

Fetters L J 1969 Synthesis of block polymers by homogeneous anionic polymerization. *J. Polym. Sci.* C26: 1–35

Helfand E 1975 Block copolymers, polymer–polymer interfaces, and the theory of inhomogeneous polymers. *Acc. Chem. Res.* 8: 295–9

Legge N R, Holder G, Schroeder H E 1987 *Thermoplastic Elastomers*. Hanser, Munich, Germany

Meier D J 1969 Theory of block copolymers. I. Domain formation in A–B block copolymers. *J. Polym. Sci.* C26: 81–98

Noshay A, McGrath J E 1977 *Block Copolymers: Overview and Critical Survey*. Academic Press, New York

Quirk R P, Alsamarraie M A, Schoenmaker W 1988 Polyester thermoplastic elastomers. In: Bhowmick and Stephens 1988b, Chap. 10

Rader C P, Stemper J 1990 Thermoplastic elastomers—A major innovation in rubber. *Prog. Rubber Plast. Technol.* 6(1): 50–99
Schollenberger C S 1988 Thermoplastic polyurethane elastomers. In: Bhowmick and Stephens 1988b, Chap. 11
Wilder C R 1988 Thermoplastic styrene block copolymers. In: Bhowmick and Stephens 1988b, Chap. 9

R. E. Cohen
[Massachusetts Institute of Technology, Cambridge, Massachusetts, USA]

Tubular Film Extrusion: Dynamics and Heat Transfer

The process of tubular film extrusion, or film blowing, converts a polymer into a flattened tube of thin sheet or film, typically of thickness 20–150 μm for low-density polyethylene. The polymer to be processed is delivered by an extruder through the annular gap in a crosshead or spider die (see *Die Extrusion*) as a tube of molten material which undergoes a combined process of stretching and cooling to form the film. When it is sufficiently cool to prevent mutual adhesion of its surfaces ("blocking"), the stretched tube is collapsed by guides to a "layflat" which passes through driven nip rolls and then is conveyed over further rollers either to be wound up for storage and transport or for subsequent processing, such as printing and bagmaking.

The part of the process under discussion here, in which the tubular film is stretched and cooled, is shown schematically in Fig. 1 (Han 1976 Chap. 9, Middleman 1977 Chap. 10, Tadmor and Gogos 1979 pp. 637–41). The processes which take place between the die and the freeze line are stretching in the machine and transverse directions and cooling (with crystallization) to below the normal temperature for solidification. The dynamics of the process govern the effect on the stretching of the forces applied to the tubular film bubble. Heat transfer (mainly by air blown from the air ring) influences the rate of cooling and, hence, the distance above the die where solidification is complete (the freeze line height). Because molten film is much more easily deformed than solid film, the stresses in the film are not large enough to stretch the solid film and deformation effectively ceases at the freeze line.

Since the haul-off force applied at the nip rolls is thus in effect transmitted to the upper boundary of the molten polymer at the freeze line, a two-point boundary-value problem describes the bubble dynamics and, because the major influence on the freeze line height is the cooling rate, which is largely determined by cooling-air flow rate and air-ring design, there is close coupling between the dynamics and the heat-transfer problems. Conceptually it seems easier to start by estimating how rapidly the temperature of the film decreases, thus obtaining a

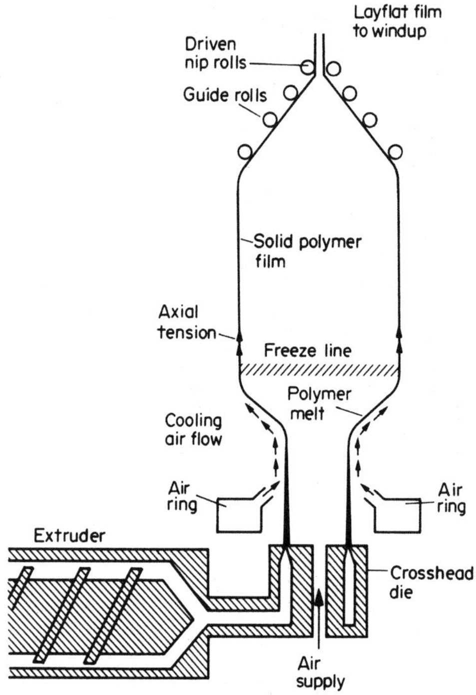

Figure 1
Tubular film extrusion (after Petrie 1975b. Reproduced by permission of the Society of Plastics Engineers, Inc.)

measure of the freeze line height. Then the dynamics can be analyzed in detail, after which the heat transfer can be considered, to show how bubble shape and film thickness modify the estimates of temperature and freeze line height. As well as being conceptually useful, this approach offers a computational strategy that works (Petrie 1975a, b).

1. Dynamics

If viscoelastic effects are neglected, the stresses in the film, on the assumption of a locally plane film of constant density undergoing unequal biaxial stretching, are

$$L=2N_v[dv/dz)-(v/h)(dh/dz)]\cos\theta \quad (1)$$

$$C=2N_v[(v/r)(dr/dz)-(v/h)(dh/dz)]\cos\theta \quad (2)$$

where L and C are dimensionless stresses in the machine (longitudinal) and transverse (circumferential) directions, respectively; $N_v=\eta/\eta_0$, where the local viscosity η depends on temperature and possibly on the strain rate; the dimensionless quantities v, h, r and z correspond to the film velocity, film thickness, bubble radius and distance from the die, respectively; and $\tan\theta = dr/dz$. Equilibrium of forces, using equations of membrane theory, gives dimensionless equations:

$$H = r(hL + N_s)\cos\theta - r^2 B - N_i v \cos\theta \tag{3}$$

$$dH/dz = r(N_d + N_g h/\cos\theta) \tag{4}$$

$$2B = (hC + N_s)\cos\theta/r - (hL + N_s - N_i v/r) \times \cos\theta\, d\theta/dz - N_g h \sin\theta \tag{5}$$

where the dimensionless parameters are inflation pressure $B = r_0^2 P/2\eta_0 h_0 v_0$; axial force at any cross section $H = F/2\pi\eta_0 h_0 v_0$; force due to gravity $N_g = \rho g r^2{}_0/\eta_0 v_0$, which varies with temperature through the density ρ; inertia $N_i = \rho_0 r_0 v_0/\eta_0$; surface tension $N_s = 2\gamma r_0/\eta_0 h_0 v_0$, which varies with temperature through the surface tension coefficient γ; and air drag $N_d = r_0^2 \tau_d/\eta_0 h_0 v_0$, which varies with the shear stress τ_d exerted on the film by the cooling airflow. The subscript 0 denotes a reference variable, probably taken to be the value at the die and used to define the dimensionless variables. Conservation of mass gives a final equation:

$$\rho r v h = \rho_0 \tag{6}$$

and if the air drag and the temperature along the bubble $T(z)$ and the dependence on T (and on strain rate) of the material properties (η, γ, ρ) are known, the differential equations can be solved.

Correct boundary conditions are important. The system is fourth order and in principle the prescribed quantities are the film dimensions at the die exit (taking account of extrudate swell perhaps) and the haul-off force or velocity (both in magnitude and direction). This gives dimensionless values

$$r = 1 \text{ and } h = 1 \text{ at the die, } z = 0 \tag{7}$$

$\theta = 0$ and $v = v_1$ or $H = H_1$ at the freeze line,

$$z = z_1 \tag{8}$$

and a computer program to solve this problem is not hard to write (Petrie 1983). The fact that the inflation pressure is not usually measured can be dealt with by choosing a value for computation to give a prescribed blow ratio (which is how the bubble inflation is done in practice). Computations have been carried out for a number of cases, with the indication that effects of surface tension, inertia and variation of density are not important. Gravity is likely to be important for large-scale processes and has a small but noticeable effect in calculations for small-scale processes. No calculations have been done on the effect of air drag. A non-Newtonian viscosity has been used (Han 1976), but the effect is secondary to the effect of the temperature change along the bubble on the viscosity.

Other constitutive equations have been used to a limited extent, but there are difficulties, both in the choice of an adequate rheological model (Petrie 1979) and in the computation (Petrie 1983). Results for a rubber-elastic model have been obtained (Petrie 1975a) and the rubberlike liquid model has been used (Wagner 1976). The nonisothermal rheology of viscoelastic liquids in elongational flows is not a subject where definitive answers can be given (see *Processing: Rheological Properties and Fluid Mechanics*).

It is worth noting that the dynamical equations for processes such as parison inflation (blow molding) and thermoforming are very similar, so this work has application outside the one particular process. The rapid inflation in blow molding lends plausibility to a rubber-elastic model in this case (Petrie and Ito 1980).

2. *Heat Transfer*

The dominant mechanism for heat transfer is forced convection, that is, the cooling of the tubular bubble of molten polymer by means of air blown onto and along its surface. This air is supplied through an annular gap in the air ring, as shown schematically in Fig. 1, and details of air-ring design remain part of the art and commercially valuable know-how associated with the process. It is also necessary to take account of radiative cooling, since estimates show that this may be responsible for 20% of the heat transfer. If other mechanisms are ignored, the appropriate equation is

$$M\, d(cT)/dz = -[\alpha(T - T_c) + \varepsilon\sigma(T^4 - T_a^4)]2\pi r/\cos\theta \tag{9}$$

where the polymer melt has specific heat c and emissivity ε, α is the heat-transfer coefficient for forced convection and T_c and T_a are the temperatures of cooling and ambient air, respectively.

Both α and ε have to be estimated first (Petrie 1974). A value of 0.4–0.6 is suggested for ε (which is dimensionless). There is some doubt about the correct form of dependence of α on the cooling-air velocity, but the correlation $\alpha = k_1(w_{max})^n$, $w_{max} = k_2(z)^m$, seems preferable, where w_{max} is the maximum velocity of cooling air at any distance from the die. Typical values of the exponents are $n = 1.5$, $m = -0.75$ and also $k_1 = 3.5$–$4\ \mathrm{W\,m^{-2}\,K^{-1}}$. The parameter k_2 clearly depends on the cooling-air flow and could be taken to be proportional to the volume flow rate of cooling air supplied. Use of correlations like this is necessarily limited to similar geometries and may not be accurate over very large ranges of values of variables (such as cooling-air flow rate).

If Eqn. (9) is to be incorporated in a nonisothermal model of the process, the position of the freeze line needs to be specified in some way. Clearly this depends on temperature, but also on the rate of cooling, since in the present model the freeze line is defined as the place where deformation of the film stops. The simplest condition to use in the model is a specified temperature (which can allow for some supercooling).

See also: Processing: Classification; Processing: Relation to Product Properties

Bibliography

Han C D 1976 *Rheology in Polymer Processing*. Academic Press, New York

Middleman S 1977 *Fundamentals of Polymer Processing*. McGraw-Hill, New York

Morton-Jones D H 1989 Extrusion-based processes. *Polymer Processing*. Chapman and Hall, London, Chap. 5

Petrie C J S 1974 Mathematical modelling of heat transfer in film blowing: A case study. *Plast. Polym.* 42: 259–64

Petrie C J S 1975a A comparison of theoretical predictions with published experimental measurements on the blown film process. *AIChE J.* 21: 275–82

Petrie C J S 1975b Mathematical modelling and the systems approach in plastics processing: The blown film process. *Polym. Eng. Sci.* 15: 708–24

Petrie C J S 1979 *Elongational Flows: Aspects of the Behaviour of Model Elasticoviscous Fluids*. Pitman, London

Petrie C J S 1983 Film blowing, blow moulding and thermoforming. In: Pearson J R A, Richardson S M (eds.) 1983 *Computational Analysis of Polymer Processing*. Applied Science, London, pp. 217–41

Petrie C J S, Ito K 1980 Predictions of wall thickness of blow moulded containers. *Plast. Rubber Process.* 5: 68–72

Tadmor Z, Gogos C G 1979 *Principles of Polymer Processing*. Wiley, New York

Wagner M H 1976 A combined rheological and thermal analysis of the film blowing process. In: Klason C, Kubat J (eds.) 1976 *Proc. 7th Int. Congress on Rheology*. Swedish Society of Rheology, Gothenburg, Sweden, pp. 217–19

C. J. S. Petrie
[University of Newcastle-upon-Tyne, Newcastle-upon-Tyne, UK]

Tubular Film Extrusion: Structure Development

The characteristics of tubular film depend on process conditions. Obviously, the film thickness is set by the die geometry and the kinematics (extrusion rate, take-up velocity, bubble diameter). This article, however, is concerned more with the polymer chain orientation, the crystalline morphology and the surface characteristics. The characterization of biaxial orientation in films, the influence of processing conditions on orientation and morphology will be successively discussed and the article will conclude with considerations of surface roughness.

1. Orientation Representation

Beginning with Holmes and Palmer (1958), orientation in tubular film has been specified by wide-angle x-ray diffraction measurements. Three-dimensional orientation has been expressed through pole figures which represent stereographic projections of polar distributions of normals to crystallographic planes. These have been utilized by various investigators of orientation in tubular film (Lindenmeyer and Lustig 1965, Desper 1969, Maddams and Preedy 1978, Choi et al. 1982, Shimomura et al. 1982).

White and Spruiell (1981) and their co-workers (Choi et al. 1980, 1982, Shimomura et al. 1982) have specified orientation in terms of biaxial orientation factors defined by

$$f_{1j}^{B}=2\,\overline{\cos^2\phi_{1j}}+\overline{\cos^2\phi_{2j}}-1 \tag{1a}$$

$$f_{2j}^{B}=2\,\overline{\cos^2\phi_{2j}}+\overline{\cos^2\phi_{1j}}-1 \tag{1b}$$

where ϕ_{1j} is the angle between the jth crystallographic axis and the machine direction, and ϕ_{2j} is the angle with the transverse direction. The c axis is generally the polymer chain axis.

Equations of this type may be derived for the polymer chain axis from the anisotropy of the polarizability tensor (White and Spruiell 1981). This leads to the suggestion of birefringence measurements of f_1^B and f_2^B in *amorphous* polymers. Specifically,

$$f_1^B=(n_1-n_3)/\Delta^\circ \tag{2a}$$

$$f_2^B=(n_2-n_3)/\Delta^\circ \tag{2b}$$

where Δ° is the intrinsic birefringence and n_1, n_2 and n_3 are the refractive indices in the 1, 2 and 3 directions. Birefringence measurements for partly crystalline polymers require consideration of both crystalline and amorphous regions.

2. Criteria for Orientation Development in Tubular Film Extrusion

2.1 General

It has been argued by Choi et al. (1980, 1982) that orientation directions in tubular film are related to the kinematics and stress fields occurring in the tubular film process. We take "1" as the machine direction and "2" as the transverse direction. Uniaxial orientation in the machine directions should develop from

$$d_{22} = d_{33}, \qquad \sigma_{22} = \sigma_{33} \tag{3a, b}$$

Equation (3a) may be shown to be equivalent to the global form:

$$v_L/v_0 = 1/B^2 \tag{4a}$$

and Eqn. (3b) implies that

$$\Delta p = 0 \tag{4b}$$

where B is the blowup ratio (bubble diameter/annular die diameter), v_0 the extrusion velocity, v_L the take-up velocity and Δ_p the inflation pressure. Equal biaxial orientation should result from

$$d_{11} = d_{22}, \qquad \sigma_{11} = \sigma_{22} \tag{5a, b}$$

Equations (5a, b) are equivalent to the global:

$$v_L/v_0 = B, \qquad F_L/2\pi r_L h_L = r_L \Delta p/h_L \tag{6a, b}$$

where r_L is the bubble radius, h_L is the film thickness at solidification and F_L is the take-up tension.

2.2 *Amorphous Vitrifying Polymers*

The tubular film extrusion of atactic vitrifying polystyrene has been investigated by Choi et al. (1980), who determined the birefringences of the film at various drawdown and blowup ratios. Experiments carried out according to Eqns. (3) and (4) lead to uniaxial orientation, while those carried out according to Eqns.(5) and (6) lead to equal biaxial orientation.

Choi et al.(1980) found that the birefringences of their film were uniquely related to the differences in principal stresses. Indeed, they were found to be directly proportional to these stresses through

$$n_i - n_j = C(\sigma_i - \sigma_j) \tag{7}$$

where σ_i represents a principal stress. The value of C determined is essentially the value of the stress-optical constant in the melt above the glass-transition temperature. This leads to the interpretation that orientation developed in the flowing melt is frozen in when the melt vitrifies at the freeze line. This view is confirmed by independent experiments on uniaxial extension, simple shear and melt spinning (Oda et al. 1978).

2.3 *Crystalline Orientation and Morphology*

The crystalline orientation of polyethylene tubular film has been investigated by various researchers (Holmes and Palmer 1958, Lindenmeyer and Lustig 1965, Nagasawa et al. 1973a, Maddams and Preedy 1978, Choi et al. 1982, Ashizawa et al. 1984). Polypropylene has been investigated by Shimomura et al. (1982).

In uniaxial polyethylene tubular film, orientation of the *a* crystallographic axis in the machine direction is observed, with the *b* axis perpendicular to the machine direction. With increasing blowup ratio, the *b* axis first remains uniformly distributed in the plane perpendicular to the machine direction and then concentrates in the plane normal to the film surface. The *c* axis tends to orient in the plane of the film, rather more equally biaxially than might be expected from the stress field and experience with polystyrene.

Studies with polypropylene also show *b*-axis orientation perpendicular to the machine direction. With increasing blowup ratio it becomes normal to the film surface. The tendency for *a*-axis orientation in the machine direction is much less. The *c* axis tends again to exhibit biaxial orientation in the plane of the film.

Choi et al. (1982) have concluded that the crystalline orientation factors (Eqn. (1)) of polyethylene tubular film are uniquely determined by the stress field at crystallization. The dependence of the biaxial orientation factors on the stress field at solidification is the same as that found for melt-spun polyethylene fibers. Observations by Shimomura et al. (1982) for polypropylene are consistent with this.

Small-angle x-ray scattering (SAXS) studies indicate lamellae which orient perpendicular to the machine direction in uniaxial film and at all directions in the plane in biaxial film (Choi et al. 1982). The lamellae presumably contain folded chain structures. Structures of this type have been detected by transmission electron microscopy.

3. *Optical Clarity and Surface Roughness*

The optical characteristics of polyethylene tubular film have been investigated by various researchers (Webber 1957, Huck and Clegg 1961, Clampitt et al. 1969, Stehling et al. 1981, Ashizawa et al. 1984). Measurements of the specular transmittance T of films in air and films coated with liquids of equal refractive index may distinguish the relative importance of scattering of light from the surface and bulk of films. The transmission of light I/I_0 through a film in air (Clampitt et al. 1969) may be written

$$I/I_0 = T = T_{si} T_B T_{so} \tag{8a}$$

$$= T_{si} \exp(-\tau h) T_{so} \tag{8b}$$

where subscripts si, so and B indicate inner surface, outer surface and bulk, respectively, τ is the turbidity and h is the film thickness. Coating the surface of films with liquids of equal refractive index allows determination of T_B directly.

It is generally found that the predominant source of scattering by tubular film is the surfaces (Stehling et al. 1981, Ashizawa et al. 1984). Ashizawa et al. have used a surface profiler to measure the roughness of film surfaces. The protuberances are of the order of 0.01–0.1 μm. They have correlated surface specular transmittance with asperity height. The greater the asperity height, the smaller is the value of T_s.

The intensities of the bulk and surface scattering depend on process conditions. In low-density (and linear low-density) polyethylene, both have been found to increase with frostline height. This indicates that slower cooling, lower drawdown stresses and axial velocity gradients promote haze (Huck and Clegg 1961, Ashizawa et al. 1984). T_s increases with decreasing v_L/v_0 at constant blowup ratio and frostline height, but τ remains the same (Huck and Clegg 1961, Ashizawa et al. 1984). Huck and Clegg (1961) discuss the influence of die design on extrusion haze in low-density polyethylene.

See also: Melt Delivery Systems: Corotating Disk Processors; Processing: Classification; Processing: Structure Development

Bibliography

Ashizawa H, Spruiell J E, White J L 1984 An investigation of optical clarity and crystalline orientation in polyethylene tubular film. *Polym. Eng. Sci.* 24: 1035–42

Choi K J, Spruiell J E, White J L 1982 Orientation and morphology of high density polyethylene film produced by the tubular blowing method and its relationship to process conditions. *J. Polym. Sci., Polym. Phys. Ed.* 20: 27–47

Choi K J, White J L, Spruiell J E 1980 Orientation development in tubular film extrusion of polystyrene. *J. Appl. Polym. Sci.* 25: 2777–88

Clampitt B H, German D E, Anspon H D 1969 Techniques for measuring surface and bulk optical properties of polyethylene films. *Anal. Chem.* 41: 1306–9

Desper C R 1969 Structure and properties of extruded polyethylene film. *J. Appl. Polym. Sci.* 13: 169–91

Hashimoto T, Todo A, Murakami Y, Kawai H 1977 Light scattering from crystalline superstructure in tubular-extruded polybutene-1 films. *J. Polym. Sci., Polym. Phys. Ed.* 15: 501–21

Holmes D R, Palmer R P 1958 The orientation of the crystalline and amorphous regions in polyethylene film. *J. Polym. Sci.* 31: 345–58

Huck N D, Clegg P L 1961 The effect of extrusion on the fundamental properties of tubular polyethylene film. *Soc. Plast. Eng. Trans.* 1: 121–32

Lindenmeyer P H, Lustig S 1965 Crystallite orientation in extruded polyethylene film. *J. Appl. Polym. Sci.* 9: 227–40

Maddams W F, Preedy J E 1978 X-ray diffraction orientation studies on blown polyethylene films, 1: Preliminary measurements. *J. Appl. Polym. Sci.* 22: 2721–37

Nagasawa T, Matsumura T, Hoshino S 1973b Film forming process of crystalline polymer, 2: Microstructure. *Appl. Polym. Symp.* 20: 295–313

Nagasawa T, Matsumura T, Hoshino S, Kobayashi K 1973a Film forming process of crystalline polymer, 1: Factors inducing a molecular orientation in tubular blown film. *Appl. Polym. Symp.* 20: 275–93

Oda K, White J L, Clark E S 1978 Influence of melt deformation history on orientation in vitrified polymers. *Polym. Eng. Sci.* 18: 53–9

Shimomura Y, Spruiell J E, White J L 1982 Orientation development in the tubular film extrusion of polypropylene. *J. Appl. Polym. Sci.* 27: 2663–74

Stehling F C, Speed C S, Westerman L 1981 Causes of haze of low-density polyethylene blown films. *Macromol.* 14: 698–708

Webber A C 1957 Method for the transparency of sheet materials. *J. Opt. Soc. Am.* 47: 785–9

White J L, Spruiell J E 1981 Specification of biaxial orientation in amorphous and crystalline polymers. *Polym. Eng. Sci.* 21: 859–68

J. L. White
[University of Akron, Akron, Ohio, USA]

V

Vulcanization

Vulcanization is an important industrial process. It is required to convert raw or unvulcanized natural or synthetic rubber into useful items such as tires, hose, shoe soles and mechanical goods. The process is generally carried out after the rubber is in its final shape, frequently while it is in a mold at temperatures between 135 °C and 180 °C. The first example of commercial vulcanization has been attributed to Charles Goodyear who carried out the process by heating natural rubber with sulfur in Springfield, Massachusetts in 1841. Thomas Hancock did the same thing in England about a year later. In this article, the effects of vulcanization on rubber molecules and the changes in rubber properties due to vulcanization are considered.

1. Definition of Vulcanization

Vulcanization is usually applied to rubbery or elastomeric materials. Such materials forcibly retract to approximately their original shapes after rather large deformations. Vulcanization is a process that increases the retractive force but reduces the amount of permanent deformation remaining after removal of the deforming force. The process increases elasticity while it decreases plasticity. This is generally the result of the formation of a molecular network during the vulcanization process.

According to theory (Flory 1953), the retractive force to resist deformation is proportional to the number of supporting network chains per unit volume of elastomer. A supporting chain is a linear polymer segment between network junctures; thus, an increase in the number of junctures gives an increase in the number of supporting chains.

During vulcanization, junctures are produced by the insertion of crosslinks between polymer chains. A crosslink can be a group of several sulfur atoms in a chain, a single sulfur atom, a carbon-to-carbon bond, a divalent organic radical or a polyvalent metal ion.

2. Effect of Vulcanization on Vulcanizate Properties

Effects of vulcanization (Hofmann 1967, Coran 1989) are illustrated in Fig. 1. The static modulus increases with vulcanization to a greater extent than does the dynamic modulus. Dynamic modulus is a resultant of both viscous and elastic behavior, whereas static (or equilibrium) modulus measures only the elastic behavior. Elasticity is gained at the expense of plasticity. Hysteresis is a measure of deformation energy which is not stored or borne by the network chains, but which is instead converted to heat. Energy-to-break properties increase with increases in both the number of network chains and hysteresis. Since hysteresis decreases as more network chains are developed, the energy-to-break-related properties (tear strength, fatigue life, toughness) are maximized at some intermediate crosslink density.

Reversion is a term applied to the loss of network structures by nonoxidative thermal aging. It is usually associated with sulfur-cured isoprene rubbers that have been overcured or aged in hot, thick sections. It occurs in vulcanizates containing polysulfidic crosslinks. Bateman et al. (1963) have suggested that crosslinks break and reform as cyclic chain modifications; in addition, Chapman and Porter (1988) have discussed reversion in great detail.

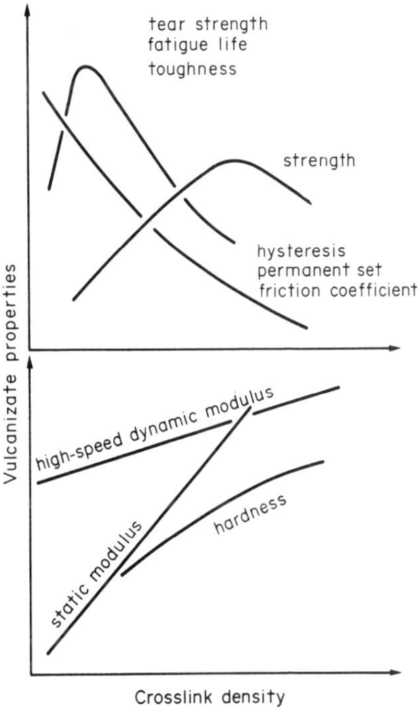

Figure 1
Effect of crosslink density (extent of vulcanization) on vulcanizate properties

3. Parameters of the Vulcanization Process

Critical parameters related to the process of vulcanization are the time elapsed before it starts, the rate at which it occurs and the extent to which it occurs. There must be sufficient delay or scorch resistance to permit mixing, forming and molding (see Fig. 2). The formation of crosslinks should then be rapid and its extent must be controlled.

Resistance to scorch (premature vulcanization) is usually measured by the time required for the onset of crosslink formation at a given temperature as measured by an abrupt increase in viscosity. The Mooney viscosimeter is usually used. It consists of a rotating disk in a cavity which contains the test rubber. A typical or average processing temperature is selected.

The extent and rate of vulcanization have been measured by determinations of mechanical properties after various cure periods. More recently, specially designed rheometers, or "cure meters," have been used. The test sample of rubber is cured automatically in a cavity while a disk embedded in the sample oscillates through a small arc. Thus, the device is called an oscillating-disk rheometer or ODR. The resistance to oscillation is measured and recorded as a function of time. Thus, the so-called rheometer charts (similar to Fig. 2) are generated. It should be noted that there are now cure meters in which the cavity or die moves (rotationally) around an embedded stator. These are called moving-die rheometers or MDRs.

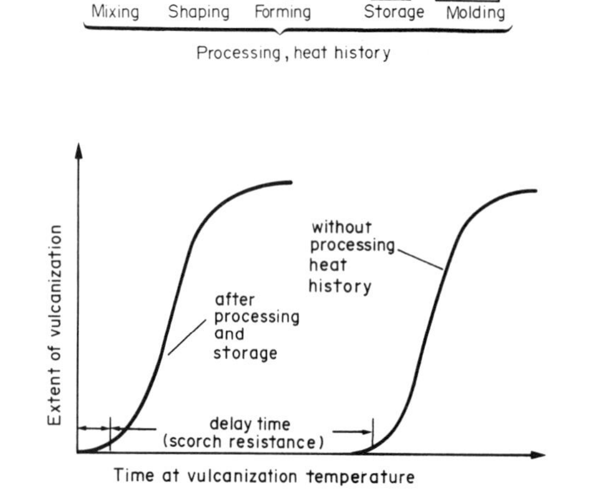

Figure 2
Effect of processing heat history on scorch resistance: a rubber article must not begin to cure until it is in its final shape (e.g., in the heated mold)

4. Vulcanization by Sulfur

Sulfur was the first agent to be used to vulcanize the first commercial elastomer, natural rubber. However, vulcanization by sulfur with no accelerator is no longer of commercial significance. Accelerated sulfur vulcanization is suitable for natural rubber (NR), synthetic isoprene rubber (IR), butadiene rubber (BR), styrene–butadiene rubber (SBR), nitrile rubber (NBR), butyl rubber (IIR) and ethylene–propylene–diene terpolymer rubber (EPDM). The reactive portion of each of these elastomers can be simply represented by >CH—CH=CH—.

A typical recipe for the vulcanization system contains 2–10 parts of zinc oxide, 1–4 parts of fatty acid, 0.5–4 parts of sulfur and 0.5–2 parts of accelerator per 100 parts by weight of elastomer. The zinc oxide and fatty acid constitute the activator system where zinc ion is made soluble by salt formation between the acid and oxide.

Types of accelerators commonly used are thiazoles (2-mercaptobenzothiazole and 2,2′-dithiobisbenzothiazole), thiazolesulfenamides (e.g., *N-t*-butylbenzothiazole-2-sulfenamide and 2-morpholinothiobenzothiazole), dithiocarbamate types (e.g., tetramethyl thiuram monosulfide or disulfide and zinc diethyldithiocarbamate) and amine types (e.g., diphenylguanidine).

It should be noted that there is now a sense of urgency to reduce the usage of accelerators that are based on secondary amines. These amines can react with environmental NO_x to form *N*-nitroso derivatives which are potentially carcinogenic. Such accelerators, during vulcanization or otherwise, could liberate secondary amines (e.g., those from which they were derived) into the atmosphere where they could become nitrosated. These accelerators include the dithiocarbamates (thiurams, etc.) and sulfenamides derived from morpholine.

The different types of accelerators give vulcanization characteristics which differ in both resistance to scorch or premature vulcanization and the rate of vulcanization after crosslink formation starts. These differences are illustrated by Fig. 3.

5. Chemistry of Accelerated Sulfur Vulcanization

Accelerated sulfur vulcanization is thought to proceed as follows. The accelerator reacts with sulfur to give monomeric polysulfidic sulfurating agents of the type Ac—S_x—Ac where Ac is an organic radical

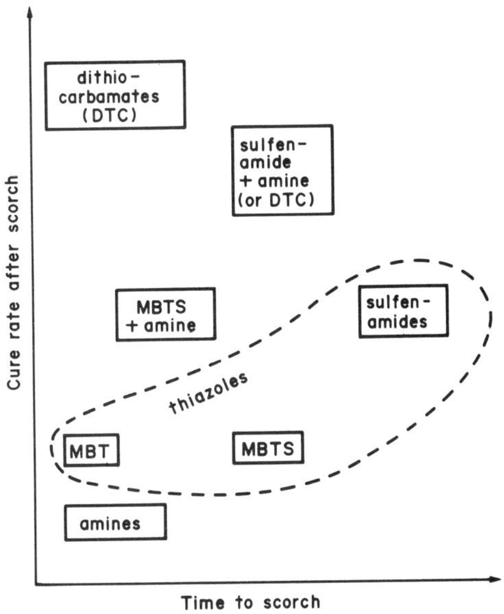

Figure 3
Vulcanization characteristics of various accelerators

derived from the accelerator. The polysulfides can sulfurate rubber to give polymeric polysulfides of the type rubber—S_x—Ac. During this reaction, 2-mercaptobenzothiazole (MBT) forms when an accelerator derived from MBT has been used. When MBT itself is used, it first disappears, then reappears during the formation of rubber polysulfides. The rubber polysulfides then undergo reaction to give crosslinks or rubber polysulfides of the type rubber—S_x—rubber.

In the presence of an accelerator, rubber molecules are efficiently sulfurated in allylic positions. A mechanism for this could be illustrated as in Fig. 4. The rubber polysulfide could cleave to give a polymeric polythiyl free radical. If crosslink formation is by a free-radical mechanism, through the polymeric polythiyl radical (rubber—S_x—S·), delayed action could be the result of a quenching action by monomeric polysulfidic sulfurating agent (Ac—S_x—Ac). Crosslink formation would be impeded until substantial depletion of the monomeric polysulfides (Coran 1983).

When benzothiazolesulfenamides are used, they react with MBT to form 2,2′-bisbenzothiazole disulfide (BTSSBT) which reacts with sulfur to form polysulfides of the type BT—S—S_x—S—BT. These polysulfides react with rubber to form sulfurated rubber (rubber—S_x—BT) with the liberation of MBT. The higher the concentration of MBT, the faster is the depletion of sulfenamide accelerator and the formation of sulfurated rubber. If MBT could be taken out of the system as fast as it forms, substantial increases in processing safety would result. When the premature vulcanization inhibitor (PVI) *N*-cyclohexylthiophthalimide (CTP) is used it reacts rapidly with MBT and converts it to less reactive compounds. There is also the possibility that the PVI reacts in the system to give increased amounts of monomeric polysulfides which could react with crosslink precursor free radicals (rubber—S_x·).

Increases in the concentration of soluble Zn^{2+} (made soluble by fatty acid) cause an increased overall rate in the early reactions which lead to the formation of rubber—S_x—Ac, an increase in the extent of crosslink formation, but a decrease in the rate of crosslinking, once it starts (Coran 1989).

Though this discussion is concerned with accelerators derived from 2-mercaptobenzothiazole, many of the concepts can be applied to dithiocarbamate-type accelerators. Both types contain the molecular moiety, >N—C(—S—)$_2$.

Most of the work done relating to network structure has been with natural rubber. Increases in sulfur and accelerator concentrations give rise to higher crosslink densities. However, as the ratio of the concentration of accelerator to the concentration of sulfur increases, the proportion of monosulfidic

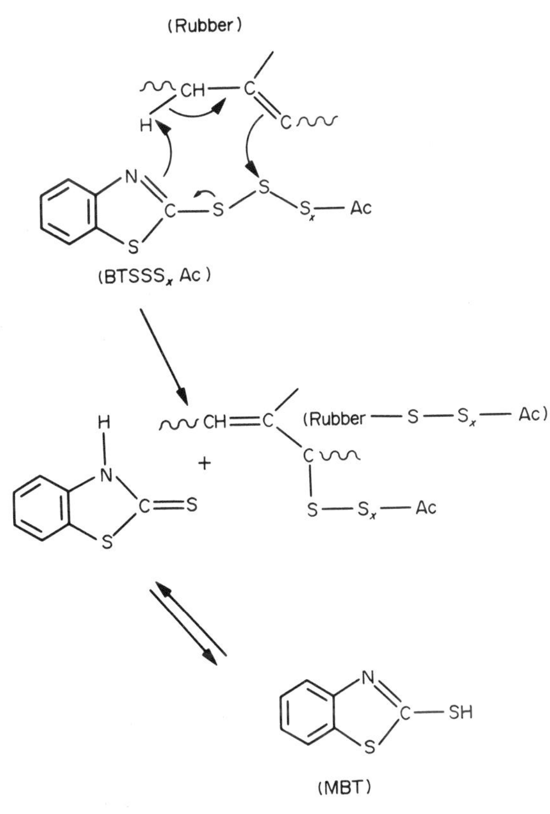

Figure 4
Sulfuration of rubber molecules

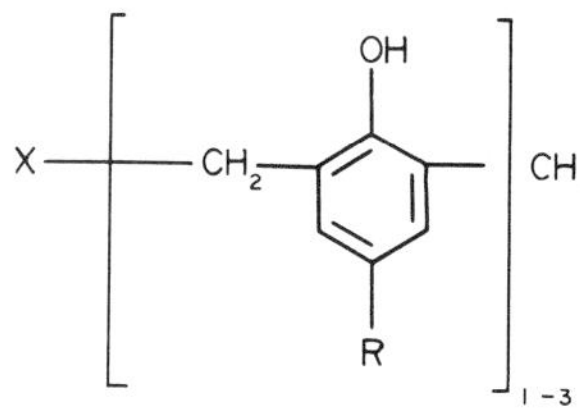

Figure 5
Structure of phenolic compounds which can crosslink diene rubbers; X is OH or a halogen atom and R is an alkyl radical

crosslinks increases. Higher accelerator concentrations also lead to an abundance of pendent groups of the type rubber—S_x—Ac. Higher concentrations of sulfur (in respect to accelerator concentrations) give both more polysulfide crosslinks and more sulfur combined with the rubber chains as cyclic polymer chain modifications. A loss of fatigue life at low levels of sulfur but high levels of accelerator has been attributed to a high concentration of bound accelerator and a high concentration of monosulfidic crosslinks. Hence, where moderately high concentrations of sulfur (in respect to accelerator) are used, flex life improves. However, when even higher concentrations of sulfur are used, flex life decreases.

6. *Other Vulcanization Systems for Diene Rubbers*

Diene rubbers can be crosslinked by compounds of the general structure shown in Fig. 5. These materials are called phenolic resin curatives. Elastomers which are appropriate for this type of vulcanization are the same as for accelerated sulfur. The vulcanization mechanism is believed to require the loss of HX to give quinone methide structures.

Benzoquinone dioxime can be used to vulcanize the same diene rubbers curable by accelerated sulfur. The mechanism is thought to involve the formation of nitroso groups by oxidation (by lead peroxide which is also incorporated into the rubber).

The attack of the vulcanization system on rubber molecules can be similarly visualized in the case of accelerated sulfur, phenolic resin curatives or quinone dioxime (Fig. 6).

7. *Vulcanization by Metal Oxides*

Chlorobutadiene rubber (chloroprene or neoprene rubber) is generally vulcanized by the action of metal oxides. The combination of zinc oxide and magnesium oxide is generally used. Zinc oxide alone tends to be too scorchy while magnesium oxide alone is inefficient. An accelerator widely used with metal oxide cures is ethylenethiourea (ETU). However, chloroprene rubber is sometimes vulcanized by a mixed accelerated sulfur–metal oxide curing system.

Types of crosslinks proposed for the vulcanization of chloroprene are —Mg—, —Zn— and —O— for metal oxide vulcanization, and —S— for ethylenethiourea-accelerated metal oxide vulcanization. In addition to these types, polysulfidic crosslinks (—S_x—) are formed when the mixed metal oxide–accelerated sulfur vulcanization system is used.

8. *Vulcanization by Peroxide*

Most types of elastomers can be vulcanized by the action of organic peroxides. Those which have been considered for use with elastomers are diacyl

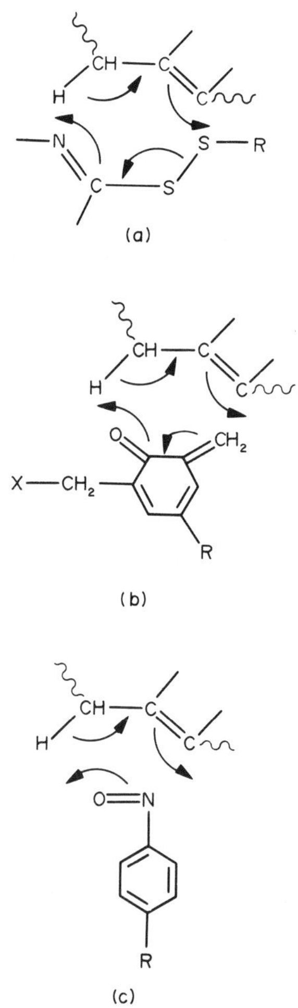

Figure 6
Curing system reactions with rubber: (a) accelerated sulfur, (b) phenolic resin curative and (c) quinone-dioxime-derived curative

peroxides, dialkyl peroxides and peresters. Di-*t*-alkylperoxides and dicumylperoxides give good vulcanizates in compounds that contain reinforcing carbon blacks.

Peroxides are of importance mainly because of their ability to crosslink elastomers containing no sites for attack by other types of vulcanizing agents. They are particularly useful for ethylene propylene rubber, certain millable urethane rubbers and silicone rubbers. They are not generally applicable to polyisobutylene elastomers (butyl rubber) because of a tendency to produce too many polymer-chain scissions.

The first step in peroxide-induced vulcanization is the decomposition of the peroxide to give free radicals. The next step is either the abstraction of a hydrogen atom from the elastomer molecule or the addition of the peroxide-derived radical to a double bond of the elastomer to form a polymeric free radical. The linkage or coupling of two polymeric free radicals or the addition of a polymeric free radical to a polymeric double bond forms a crosslink. In the latter case, addition to the double bond also gives a new free radical adjacent to the crosslink. This free radical can abstract a hydrogen atom from another polymer molecule, for example. Silicone rubbers (mainly polydimethylsiloxanes) are also cured by organic peroxides. Frequently, a high-temperature (~180 °C) postcure is carried out to give additional crosslinks of the type Si—O—Si. Silicone rubbers containing vinyl methyl siloxane groups can be cured by dialkyl peroxides such as dicumylperoxide. Others, such as the purely dimethyl siloxane types, require diacyl peroxides such as bis(2,4-dichlorobenzoyl) peroxide.

9. Radiation Crosslinking

Physical methods of vulcanization can be used including the use of high-energy radiation (see *Radiation Processing*). Industrial usage includes both elastomers (saturated and unsaturated types) and plastics, especially polyethylene. In hydrocarbon polymers, carbon–carbon crosslinks are produced, similar to those resulting from peroxide-induced vulcanization.

10. Urethane Vulcanization

Another nonsulfur vulcanization system is a system based on a nitroso additive reaction to provide pendent groups on elastomer chains, which can subsequently be used for crosslinking. This MRPRA (Malaysian Rubber Producers Research Association) development is described by Baker (1988). Natural rubber vulcanizates with a high degree of reversion resistance are obtained.

11. Vulcanization by Silanes

Vulcanization by silanes is another method that involves production of active groups on elastomer chains for subsequent crosslinking and is detailed elsewhere (see *Vulcanization by Silanes*). It is fairly widely used in the manufacture of polymeric cables.

See also: Elastomers: Additives; Radiation Processing; Rubber Processing; Vulcanization by Silanes; Vulcanization, Continuous

Bibliography

Baker C S L 1988 Non-sulphur vulcanization. In: Roberts A D (ed.) 1988 *Natural Rubber Science and Technology*. Oxford University Press, Oxford, Chap. 11

Bateman L, Moore C G, Porter M, Seville B 1963 Chemistry of vulcanization. In: Bateman L (ed.) 1963 *The Chemistry and Physics of Rubber-Like Substances*. Maclaren, London, Chap. 15

Chapman A V, Porter M 1988 Sulphur vulcanization chemistry. In: Roberts A D (ed.) 1988 *Natural Rubber Science and Technology*. Oxford University Press, Oxford, Chap. 12

Coran A Y 1983 The art of sulfur vulcanization. *Chemtech.* 13(2): 106–16

Coran A Y 1989 Vulcanization. In: Kroschwitz J I (ed.) 1989 *Encyclopedia of Polymer Science and Engineering*, 2nd edn., Vol. 17. Wiley, New York

Flory P J 1953 *Principles of Polymer Chemistry*. Cornell University Press, Ithaca, New York

Hofmann W 1967 *Vulcanization and Vulcanizing Agents*. Palmerton, New York

A. Y. Coran
[Monsanto Company, Akron, Ohio, USA]

Vulcanization by Silanes

Vulcanization by silanes, or silane crosslinking, is a method used primarily to crosslink homopolymers and copolymers of ethylene. The development derives from silicone rubber technology and was invented by the Dow Corning Company in the late 1960s. Originally, the method was conceived for injection molding but the main commercial development of the process has been in the field of wire and cable and pipe extrusion. The main advantages of the silane process which is achieved by the introduction of siloxane bonds between the molecules (i.e., Si—O—Si) is the freedom from melt temperature limitations imposed by the conventional peroxide curing process and the simplicity of the curing technique which involves exposure to hot water or to low-pressure steam. High-density polyethylenes can be conveniently crosslinked by this method but the major application has been for the large-scale production of low- and medium-voltage cables in Europe.

Originally, the process as developed by Dow Corning was conceived in two stages and this has become known in the industry as the two-stage process:

(a) preparation of a reactive graft polymer by grafting alkoxy silane groups onto the polymer molecule; and

(b) extrusion and crosslinking in a separate stage involving catalyzed activation by water and simultaneous combination of reactive sites.

1. Preparation of the Graft Polymer

The graft polymer is prepared by reacting the polyolefin with an unsaturated silane, usually vinyltrimethoxysilane initiated by a small amount of peroxide (e.g., dicumylperoxide). The silane and peroxide together with some antioxidant are dispersed, preferably using continuous compounding plant; Buss Kokneaders or twin-screw extruders are suitable for this purpose. After dispersion is achieved the mixer temperature profile is raised to effect decomposition of the peroxide and cause the grafting reaction to proceed. The graft polymer is then cooled, granulated and stored for further use.

2. Crosslinking

In the crosslinking part of the process the graft polymer is extruded together with a very small amount of catalyst added as a masterbatch in polyethylene using conventional extrusion equipment. (A masterbatch is a concentrate of an appropriate material in a base polymer; in this case it enables small amounts of a critical component to be added more accurately.) The product is crosslinked in a separate stage, off line from the extrusion process, usually by immersing the product in hot water or exposing it to low-pressure steam (e.g., a sauna bath). Under these conditions, activation of the grafted silane by reaction with water (i.e., hydrolysis of the alkoxy side groups to give silanol) takes place followed by a condensation reaction between an adjacent silane and silanol group leading to crosslinking via a siloxane bond formation. The presence of the catalyst accelerates this sequence of reactions. Many acids and bases will do this, but the preferred catalyst has been found to be an organotin compound, dibutyltindilaurate.

The two-stage process is now well established and a large number of graft polymer compounds are now available from various suppliers. There are however a number of limitations with this two-stage technology.

(a) The reactive graft polymer has to be stored under conditions that prevent premature crosslinking occurring during the second extrusion stage. This is particularly a problem if the graft polymer contains carbon black (for semiconducting purposes) or mineral fillers.

(b) A decrease in melt flow index occurs on grafting and, because the graft polymer must be capable of reextrusion at the second stage, this limits the choice of molecular weight when selecting polymers for grafting.

(c) The addition of catalyst via a masterbatch technique requires that approximately 5% ungrafted polymer be added to the mixture at the extrusion stage. This will lead to a slight reduction in crosslink density.

(d) Materials cannot be dried by conventional techniques prior to the second stage.

The recognition of the problems that these limitations might present in the manufacture of wire and cable led the BICC Cables and Maillefer Companies to introduce a single-stage process in the early 1970s. This process is sometimes referred to as Monosil.

3. Single-Stage Process

In the single-stage process, a single-screw thermoplastic extruder is employed with a special long screw designed to perform the three consecutive processing functions of efficient dispersion of the reactants, silane grafting and metering at high throughput speeds. The polymer is fed to the extruder in the normal way, antioxidant is added and a mixture of the reactants (e.g., peroxide, catalyst and silane), dissolved in the silane, is injected. A facility for accurately metering the liquids is an essential feature of the plant. The rest of the extrusion line is of standard design. The crosslinking process is carried out off line in the normal way. Since the grafting and extrusion is carried out in one operation and in the presence of the catalyst, careful selection of the formulation ingredients is necessary to avoid adverse chemical interactions which may, for example, inhibit grafting or lead to premature crosslinking. Major advantages of this process derive from the ability to thoroughly dry all ingredients before use, particularly in the case of filled materials and the unlimited choice of the molecular weight of the polymers that can be used.

4. Other Single-Stage Developments

A third system has been proposed by the Union Carbide Corporation and RAPRA Technology which makes use of the RAPRA cavity transfer mixer (CTM) development (Hindmarch and Gale 1985). The CTM is a dynamic mixer that can be fitted to the output end of a conventional extruder barrel. The mixer is claimed by its inventor to give good distributive mixing. The development work showed that conventional pipe extrusion lines could be converted to produce silane crosslinked polyethylene pipe by the addition of the CTM to the extruder and injecting the liquid cocktail of peroxide, silane and catalyst directly into the mixer.

5. *Materials Crosslinkable by the Silane Process*

Originally, the silane process was applied to polyethylene, but it was quickly found that copolymers of ethylene with vinyl acetate, vinyl acrylate and ethylene–propylene rubbers could be crosslinked in this way. Claims have also been made for chlorinated polyethylene and poly(vinyl chloride). The poly(vinyl chloride) case is slightly different in that an aminosilane is used to couple to the polymer through the amino group, although the crosslinking mechanism is the same.

6. *Use of Additives*

Because the grafting process is a free-radical reaction and the grafted product is extremely sensitive to moisture, extreme care is necessary in the choice of additives. The choice of antioxidant is important. The quantity of peroxide required to initiate the graft is small compared with the amounts used for normal crosslinking purposes, therefore antioxidants present during the grafting process should not be of the type to "scavenge" the peroxide. It is, however, possible to add these types of stabilizers after grafting has been affected. Alkalinity of the system is important—if it is too acidic then the free-radical formation will be inhibited. If halogenated materials are present in the composition, then the most temperature-stable materials should be used and adequate stabilization provided. The presence of certain mineral fillers in the composition may compete for the silanol groups, which may surface treat the filler rather than crosslink or may liberate water.

7. *Ethylene–Silane Copolymers*

In 1981, the Mitsubishi Petroleum Company introduced a reactive copolymer of ethylene and vinylsilane. Because the vinylsilane is directly incorporated as a comonomer rather than grafted, the final chemical structure is slightly different and may account, in part, for the longer shelf life claimed by the manufacturers. The reduction in free silane often present in the graft polymer is also a contributing factor. In all other respects processing is identical with the graft polymer. The same catalyst masterbatch (organotin) is used and added at the extrusion stage and the final product exposed to hot water or low-pressure steam for final crosslinking. During the 1980s, other raw material suppliers (Union Carbide and Neste Chemicals) have produced similar materials and the products are now becoming well established. The next stage in this development will be to produce filled materials. To do this a terpolymer will be necessary. It is not possible to produce a terpolymer with vinyl acetate as the third monomer since problems occur in the reactor with the silane in the presence of acetic acid. However, experimental materials have been made based on ethyl acrylate but these have yet to be commercialized.

8. *Other Developments*

Attempts have been made to apply this technology to rubber compounds. A major obstacle to this has been the difficulties experienced with fillers due to competition for the silanol groups. Nevertheless, blends of ethylene–propylene rubbers with polyethylene have been successfully crosslinked by this method and have found some application in the wire and cable industry.

The latest material developments have been in the area of the silane cocktails. All of the ingredients necessary to perform the silane crosslinking process are now available as single mixtures in liquid form or absorbed as masterbatches on a polyolefin base. These new developments considerably simplify handling procedures and improve safety in the processing plant.

The latest ideas for a single-stage system have come from the Swiss company Buss who have combined the use of their mixing plant with gear pump technology to produce a machine capable of grafting and extruding in a single operation. The excellent dispersion of the ingredients achieved by the combination has produced a product with superior mechanical properties.

See also: Elastomers: Additives; Rubber Processing; Vulcanization; Wire and Cable Systems for Underground Transmission

Bibliography

Gross L H 1988 Polyethylene silane copolymers as new low voltage insulation systems. *Wire J. Int.* 38: 59–66

Hindmarch R S, Gale G M 1985 Silane moisture curing: A one shot extrusion process using CTM. *ACS Rubber Division, Cleveland Meeting*. American Chemical Society Rubber Division, Akron, OH, Paper 25

Narkis M, Tzur R, Vauxman A, Fritz H G 1985 Some properties of grafted moisture crosslinked polyethylene. *Polym. Eng. Sci.* 25(13): 857–62

Swarbrick P 1977 Developments in the manufacture of crosslinked polyethylene cables. *Electr. Rev.* 200(4): 23–5

B. G. Howell
[BICC Cables, Wrexham, UK]

Vulcanization, Continuous

The term continuous vulcanization is generally applied to methods of effecting crosslinking in line with the product manufacture. Products treated in this manner are those produced in long lengths such as electric cables and extruded profiles. The methods used to effect continuous vulcanization can be divided into two categories: pressurized and unpressurized systems. Unpressurized systems tend to be favored where short-length production is a requirement—pressurization of the curing system prevents inspection of the product until the curing process is complete, thus its use for short-length production would result in excessive scrap rates on start-up and finish. There are a large number of different systems in use and a variety of heat transfer media are employed. The reasons for this are that many techniques are variations of the original steam curing methods which have been introduced to give improvements in economy or technical performance of the finished products.

1. Steam Continuous Vulcanization

In its simplest form, steam continuous vulcanization is often referred to as horizontal continuous vulcanization (HCV). A long tube (~65 m) is connected directly to the extruder die. On emerging from the die the extruded product first passes through a zone of high-pressure steam (14 bar, ~200 °C) and then into high-pressure water to cool. As the product cools, the pressure gradually reduces until the fully crosslinked product emerges from the end of the tube. Although referred to as horizontal, the tube is slightly inclined to allow draining of any condensed steam. During the passage through the tube it is necessary to support the extrudate, thus this system is popular in the wire and cable industry where the conductor supports the weight of the catenary. By careful control of the catenary, the product can be prevented from touching down on the tube until crosslinking is sufficient to withstand any deformation. Given that the extruder has sufficient output, then the factors affecting the line speed are the steam temperature and the ratio of steam to water length.

As the product diameter increases there is a tendency to flow and give an eccentric profile. To prevent this the continuous vulcanization tube may be arranged vertically (VCV). When used in this configuration, the extruder is usually situated at the top of a tower and the product extruded downwards. Because of the cost of the tower and the more complex installation, VCV is expensive to operate. More popular is to let the tube take up the shape of the catenary (CCV). The extruder is mounted on a raised platform and the initial part of the tube is curved to the catenary profile. Automatic sensing devices are fitted inside the tube, adjusting the line tension to maintain the catenary in the correct position.

In theory, it is possible to increase production speed by progressively lengthening the steam tube. In practice, this is not possible because of the excessive tensions required to maintain the catenary and the increase in scrap on start-up and shutdown. An alternative solution is to increase the tube temperatures. Unfortunately, the nature of saturated steam is such that for large increases in pressure there is only a modest gain in temperature (i.e., 14 bar and 200 °C, 28 bar and 230 °C). The solution used is to separate heating from pressure. This has led to the development of supersteam systems and other methods based on the use of heated inert gas. Such systems are used in the wire and cable industry for crosslinking high-voltage cables. In the case of supersteam, in addition to the tube being supplied with high-pressure steam up to 25 bar, external electric heating is applied to the tube so that the steam becomes superheated, the temperature reaching as high as 400 °C, the actual temperature used depending on the material to be crosslinked. In all other respects the system is similar to the standard steam continuous vulcanization, cooling being by pressurized cold water.

2. Hot-Gas Systems

In hot-gas systems the tube is heated externally and is pressurized internally with nitrogen gas at about 10 bar. The normal temperature of operation of the tube is 400 °C. Heat transfer to the cable is by radiation and convection. In some cases, cooling is effected by pressurized cold water, but in the more sophisticated systems the cable is cooled by circulating nitrogen gas at high velocity, thus giving a completely dry curing system (CDCS).

3. Unpressurized Systems

A number of unpressurized systems for unsupported profiles and short-length production are available. These include hot-air tunnels, fluidized beds, salt baths and microwave curing techniques. In the case of fluidized beds and salt baths, pressurized systems are also available.

In all of these techniques, the extrudate is supported during cure either on a moving belt as in the case of microwave or hot-air cure, by the fluidized medium forming the bed or, in the case of liquid salt cure, the buoyancy of the curing medium. Whenever unpressurized techniques are employed, special precautions in extrusion and compounding are necessary to avoid porosity in the finished product. The following practices are followed:

(a) a vacuum vented extruder is often used to remove moisture and other volatiles from the extrudate;

(b) calcium oxide is often added to the compound to act as a desiccant;

(c) fillers and other compounding ingredients with volatile components are to be avoided, as are oils with low flash points; and

(d) in the case of wire and cable jacketing (salt bath cure is a popular method for short lengths), vacuum is often applied to the extruder core point to remove air and moisture between the laid-up insulated cores.

4. *Hot Air*

Vulcanization in hot air is one of the oldest forms of curing technique used in the rubber industry. It is often used where a very smooth surface finish is required (steam and other contact heat transfer mediums tend to mark the surface). For continuous operation, the apparatus used is in the form of a hot-air tunnel containing a moving belt on which the extrudate is supported. The tunnel may contain infrared heaters or the hot air may be circulated from an external heat exchanger. Often, this method of crosslinking is used in conjunction with a microwave unit to maintain the vulcanization temperature after the initial heating has been supplied by the microwave energy.

5. *Microwave Heating*

In microwave heating, the extrudate is exposed to an ultrahigh-frequency electromagnetic field where rapid heating occurs through dielectric losses. By this method, very high rates of rise in temperature can be achieved which are uniform throughout the cross section of the extrudate. Most microwave units operate in the 2450 MHz band.

A typical microwave unit consists of a waveguide (hollow conductor) through which the extrudate is conveyed on a moving glass-fiber belt. The microwave energy is generated by magnetrons and is fed into the waveguide where energy is absorbed by the extrudate passing through; nonabsorbed energy is reduced by special tuning elements. To prevent any heat losses due to radiation from the extrudate, preheated air is blown through the tunnel. Having used the microwave energy to quickly heat up the extrudate to vulcanizing temperature, heating is maintained for the duration of the curing cycle by arranging an infrared or hot-air tunnel in tandem with the microwave unit. The rate at which the temperature can be raised by this method is very dependent on the power factor of the rubber compound. Polar materials such as polychloroprene and chlorinated polyethylene present few problems and the presence of fillers such as carbon black and, to a lesser extent, mineral fillers enhances the heating process. Hydrocarbon rubbers and polyethylene do not respond well to microwave heating methods and it is necessary to carefully select the filler systems and to add polar materials such as glycols and/or blend with some polar rubber to achieve the best results. Materials used for insulants for wire and cable are unsuitable for this process. Pressurized microwave tunnels and microwave-sensitive peroxide curing systems have been developed for this purpose but have not been commercialized. The principal application for microwave curing is in the profile extrusion industry where it is used extensively for the vulcanization of automotive weatherstrip products. Originally, problems were encountered when the extruded sections contained metal inserts, but this has been overcome by the use of coaxial waveguides within the normal waveguide.

6. *Salt Bath Curing*

In salt bath curing, the extrudate is passed through a heated bath of a eutectic mixture of water-soluble inorganic salts (usually a mixture of 53% potassium nitrate, 40% sodium nitrite and 7% sodium nitrate). The melting point of the eutectic mixture is between 145 °C and 150 °C and is claimed to be stable to 540 °C. For unpressurized systems, the equipment is very simple, consisting of a closed heated bath of the mixed salts, often some 12 m in length. On emerging from the bath the extrudate is sprayed with cold water to remove the excess salt and then cooled and further washed in a cold-water trough. The density of the salt mixture is such that most extrudates float on the surface of the bath and it is therefore necessary to arrange for the extrudate to be held below the surface by rollers or a continuously moving belt. Good ventilation of the bath is necessary since a by-product of curing in salt can be nitrosamines. Throughout industry this method is known as the LCM process, standing for liquid curing medium. A pressurized version of this method has been developed and this is referred to as PLCV (pressurized liquid continuous vulcanization). As with most continuous vulcanization pressurized systems, the entry end of the vulcanizing tube is fixed directly to the extruder head. The tube is in three sections. The extrudate first enters the hot-salt section and then passes through a separation chamber before entering the cooling section. The tube section containing the salt is double walled to allow for hot-oil circulation which provides the heating. The salt is circulated around this heated section from a salt reservoir by means of a pump. In a similar way, cold water is circulated around the cooling section. The separation chamber separates the hot salt from the cold water. The whole system is pressurized from an external source of compressed gas.

7. Fluidized Beds

Small glass spheres (ballotini) can be heated and fluidized by hot-air or dry-steam circulation to give the effect of a gently boiling liquid.

As the extrudate passes through the bath of heated ballotini it is supported in the bed. It is claimed that this method of heating is very safe at temperatures over 300 °C, but the maximum temperature used depends on the type of polymer to be processed. After transport through the fluidized bed, excess ballotini are removed by high-pressure waterjets or by brushing. In the unpressurized system the ballotini are supported on a metal gauze inside a covered, heated metal trough. Fluidization is effected by blowing air into the system through the bottom of the trough. A pressurized version originally developed by RAPRA Technology (UK) works under pressure of about 5 bar. In this version, the ballotini are supported on a porous ceramic through which steam is blown to effect both fluidization and pressurization. The inlet seal is effected by bolting the inlet end of the tube to the extruder head. The exit seal consists of a head of water maintained in a down tube at a pressure equal to that of the bed but leaking through the annular space surrounding the extrudate into a tank from where it is pumped back into the down tube. It is essential that all surplus ballotini should be removed to prevent ground glass getting into the mechanism of ancillary plant.

8. Du Pont Helicure System

In the Du Pont helicure system, the extruder is fitted with a long tube in the shape of a helix. The tube is heated and filled with polyalkene glycol which acts both as a heat transfer medium and lubricates the passage of the extrudate through the tube. A drawback to this process is that the heat transfer medium can be absorbed onto the surface of the extrudate imparting a silght odor.

9. Rotocure

The rotocure process is used for the vulcanization of flat articles such as conveyor belting and sheeting. It was originally developed by the Boston Woven Hose Company. In this process, the article to be vulcanized is held against a large steam-filled rotating cylindrical drum by an endless belt, the time in contact with the drum being equal to the cure cycle.

10. MDCV Process

The MDCV (Mitsibishi–Dainichi continuous vulcanization) process was developed jointly by the two Japanese companies Mitsibishi Petroleum Company and Dainichi-Nippon Cables for the crosslinking of large power cables. Essentially, it is an extruder fitted with a long land die, the land being many meters in length. Lubricant is injected into the die to prevent the extrudate sticking to the die wall. A disadvantage of this process is the need to change the die for the different sizes of cable.

See also: Processing: Classification; Wire and Cable Systems for Underground Transmission

Bibliography

Anon 1972 Hertz-Four introduce UHF cable curing line. *Rubber J.* 154(11): 18; 57

Anon 1976 A new approach to the production of vulcanised cables. *Wire Ind.* 43: 841

Herlaut E 1988 Continuous vulcanization processes. *Rubber World* 198(4): 20–2

Hofmann 1989 *Rubber Technology Handbook*. Hanser, Munich, Germany, pp. 394–407

Izod D A W, Meredith R J 1972 The magnacure. *Rubber J.* 154(5): 55–62

Masashi I, Takada A, Shimizu M, Nanjo S 1976 Development of a new infrared vulcanising system for rubber insulated cables. *Wire J.* 9(10): 102–5

Mitzukami T, Takahashi K, Ikeda C, Kato N, Yoda B 1975 Insulation properties of crosslinked polyethylene cables cured in inert gas. *IEEE Trans. Power Appar. Syst.* 94(2): 467–72

Oettner C 1972 Continuous vulcanization by the microwave method. *Bayer Tech. Notes Rubber Ind.* 44: 41–56

Ohashi N 1988 Continuous annealing of steel sheet. In: Cahn R W (ed.) 1988 *Encyclopedia of Materials Science and Engineering,* Supplementary Volume 1. Pergamon, Oxford, pp. 85–92

Soden A L 1972 Pressurized fluid beds—A report on hose curing trials at RAPRA. *Rubber J.* 154(1): 44–6

B. G. Howell
[BICC Cables, Wrexham, UK]

W

Welding of Plastics

Complex shapes can be made easily and cheaply in plastics by molding or extrusion, but many applications require joints to be made. Joining can be achieved by mechanical fastening, adhesives or welding, but welding is usually preferred because it is simple, fast and requires no additional material. However, welding can be applied only to thermoplastic materials. Thermoset plastics (e.g., phenol formaldehyde and epoxies) are formed by a chemical reaction during molding and so the resultant molding is rigid when cool and cannot be softened again by heat. Thermoplastic materials, on the other hand, can be softened and melted by heat to allow welding. Typical examples are poly(vinyl chloride) (PVC), polyethylene and nylon. Heat can be applied directly, as in hot gas or heated tool welding, or generated mechanically as in ultrasonic, friction or vibration welding.

1. Ultrasonic Welding

In ultrasonic welding, high-frequency (ultrasonic) mechanical vibrations are used for joining thermoplastic materials. The parts are held together under pressure and are then subjected to ultrasonic vibrations, usually of 20 kHz or 40 kHz frequency. The consequent alternating stresses generate heat in the plastic and if the components are properly designed this heat can be localized at the joint interface. The ability to weld a component is governed by the design of the equipment, the mechanical properties of the material and the design of the components.

Ultrasonic welding is a very widely used technique because it is easily automated and very fast. Weld times of less than one second are typical. However, careful design of components and jigs and fixtures is required, so it is most suited to mass-production applications. The power available from current equipment is limited (<3000 W) so that maximum joint lengths are about 200 mm. Ultrasonic welding only works well on rigid materials which will transmit the vibrations (e.g., polystyrene).

The ultrasonic vibrations are generated by a transducer. This is usually piezoelectric as high efficiencies (95%) can be obtained, although magnetostrictive transducers are also used. The commonest piezoelectric material used is lead zirconate titanate. Disks of the piezoelectric material are sandwiched between two metal sections clamped tightly together to ensure the whole assembly is always in compression. The metal blocks are dimensioned to make the assembly resonate at the required frequency.

The welding head consists of the horn and booster horn. The horn amplifies and focuses the mechanical energy produced by the transducers and imparts it to the workpiece. It also transmits the clamping force to the workpiece. Horns are made from materials that are wear resistant and which transmit ultrasonics easily, usually titanium, although aluminum can be used for prototypes.

The design of the horn is critical. At one end it must be shaped to focus energy into the workpiece while at the other it must match the transducer closely. The horn must be resonant at the operating frequency, that is, it must be half a wavelength long (~130 mm at 20 kHz). Each horn must be tuned individually to achieve this. In addition, the horn must act as an amplitude transformer since the vibration amplitude required by plastics is typically in the range 15–60 μm but the typical vibration amplitude of the transducer is 15 μm.

Amplitude transformation is achieved by reducing the cross section along the length of the horn. Often, it is not possible to design a horn with the required gain because of the limitations imposed by the shape of the components to be welded. In this case, an additional half-wavelength section, called the booster horn, is inserted between the transducer and the welding horn to effect an additional amplitude transformation. Amplification factors in the range 0.5–2.5 are usually available as standard items and other ratios can be achieved if necessary.

For most practical purposes the weldability of a material is determined empirically and it should be noted that many other factors such as joint and component design are also important. For each material, there is a critical thickness beyond which welding is not possible. This varies from 10 mm for a material with poor weldability, such as some nylons, to 120 mm for a good material such as polystyrene.

Parts made of different thermoplastics can be welded together. However, experimental trials are usually necessary as it is not known which factors govern compatibility of materials, although it is known that amorphous and semicrystalline plastics are not mutually weldable.

The design of the component, particularly the joint area, is critical for ultrasonic welding. The joint should be as close as possible to the horn face, particularly for semicrystalline material. In addition, the entire horn face must be in uniform contact with the part to be welded.

The design of the joint area is governed by three factors:

(a) the type of plastics to be welded;

(b) joint requirements, such as liquid- or gas-tight mechanical properties; and

(c) desired outward appearance of the final joint.

Whichever criteria finally dictate the choice, a loose fit is essential since the welding process depends on movement between the two parts as well as friction and pressure. A tight fit or too much clamping pressure prevents relative motion so that no frictional heat is produced.

The basic requirement of joint design is a small, uniform initial contact area. In the most commonly used joint design, this is provided by a triangular section called an energy director. This gives a high concentration of vibratory energy at its apex resulting in a rapid heat buildup. Molten material from the energy director flows across the joint interface and fuses with the two components to form a weld. If cosmetic joints are required, they can be achieved by the use of a step or tongue and groove design. A minimum energy-director height of 0.25 mm is used for materials that are easy to weld such as rigid, amorphous plastics. For semicrystalline or amorphous materials with low stiffness, higher levels of energy are required and welding is more difficult. There is some disagreement as to the best joint design to use. The shear joint is usually recommended to provide a small initial contact area followed by a controlled interference along the joint as the parts collapse together, but sometimes for semicrystalline plastics a projection joint with a large energy director, minimum height 0.4 mm, is used.

The applications of ultrasonic welding to plastics assemblies are diverse but because of the nature of the process tend to be found mostly in the mass-production industries such as automotive fittings, domestic appliances, medical products and packaging. Specific examples include automotive dash assemblies, electric motor housings, bearing cages, cassette bodies, lamp housings, toys, motor vehicle lamps, cigarette lighters, fishing floats and parts for domestic appliances.

2. *Friction Welding*

Friction welding of thermoplastics, also called spin welding, was first used in the 1930s for the manufacture of plastic compasses. The principles are the same as with the more commonly encountered friction-welding methods used with metal components except that, for metals, welding is achieved without melting, while for plastics, melting occurs at the joint interface, presumably because of the lower thermal conductivity of plastics. However, this molten material is substantially displaced from the joint interface into the weld flash. Heat is generated at the joint interface by the rapid rotation of one component against the other, under an applied pressure.

More complex arrangements involve the rotation of an intermediate plastic part, an angular reciprocating movement or the use of an orbital rotation. The first arrangement has the advantage that neither of the two components being welded is rotated, while the last two are capable of welding noncircular components and can maintain any required angular orientation between the components when welded. Furthermore, orbital welding can be employed for welding several components during one welding sequence. A technique using a rotation metal disk to heat, for example, the ends of pipes to be butt welded has also been proposed.

The major welding parameters are: rotational speed, friction pressure, forge pressure, displacement and duration. All of these are interrelated and require optimization for each application and material combination. In general, however, a rotational speed of 1–20 m s^{-1} (as measured at the mid-radius of the component) and friction and forge pressures of 0.8–1.5 bar and 1–3 bar, respectively, are employed, giving a weld within a few seconds.

The advantages of friction welding are the high weld quality achievable and the speed, simplicity and reproducibility of the process. For many applications, little end preparation is necessary, a rough sawn finish often being sufficient, although moisture can reduce joint strength and quality and, consequently, those plastics susceptible to humidity (e.g., nylon) should be dried prior to welding. Another problem that can be encountered is that of cracking in the weld due to residual stresses when components are exposed to certain solvents. Postweld annealing may be necessary if this is to be avoided. The major disadvantage of the process is that, in its simplest form, it is only suitable for welding circular components requiring no angular alignment. Although such welds can be achieved by orbital friction welding the process is then more complex.

Both amorphous and semicrystalline materials can be friction welded including: polycarbonate, polyethylene, polyamide and polyoxymethylene. Although 50–95% of parent material strength is achievable when welding joints between components of the same thermoplastic, the majority of dissimilar material combinations give negligible joint strengths (<10%).

In order to achieve good strengths, the joint line of circular seams can be enlarged by using cylindrical or conical mating surfaces. These designs also have the advantage of being self centering. For the more complex orbital motions square, flat end preparations must be employed.

Direct rotational friction welding has been used for welding circular solid or hollow-section components, for example, spherical polyethylene floats up to 500 mm in diameter, aerosol bottles, transmission shafts and PVC pipes and fittings up to 100 mm in diameter. One particular advantage is the ability to weld beneath the surface of a liquid, so enabling the rapid sealing of liquid containers without entrapment of any gas bubbles. (This capacity was exploited in the manufacture of the previously mentioned plastic compasses.)

Little information is available on the use of the more complex friction-welding arrangements, although the angular reciprocating movement technique has been used for welding moped petrol tanks, because it enables maintenance of orientation of branch connections, and orbital friction welding has been used to weld floats.

3. *Vibration Welding*

Vibration welding, also called linear friction welding, was first used in the mid-1970s and has since become increasingly used for the welding of various mass-produced thermoplastic components. As with conventional friction (spin) welding, frictional heat is generated by relative movement between the two parts to be welded which are held under pressure. The difference between the processes is that, with vibration welding, the movement applied is linear. Once molten material has been generated at the joint interface, vibration is stopped and the parts are aligned to allow the weld to consolidate on cooling. The process is rapid (weld times of 1–5 s are generally used) the vibrations applied being typically 100–300 Hz and 1–5 mm amplitude, while the pressures employed are similar to those used in friction (spin) and ultrasonic welding. Consequently, the main advantage of vibration welding is its ability to weld complex linear joints at high production rates. Other advantages include the capability of welding a number of components simultaneously, simplicity of tooling and the ability to weld almost all thermoplastic materials. Noise is not generally a problem but special sound enclosures may be required.

A vibration-welding machine consists of three components: vibrator assembly, power supply and clamping mechanisms. Three types of machines have been constructed: those with mechanical vibrator assemblies, those with electromagnetic assemblies and those with hydraulic assemblies.

Early mechanical designs involved the generation of vibrations by complex eccentrically rotating bearings. Such machines have not been commercially developed.

Commercial machines comprise a set of flat plate springs, the drive unit, a vibrating element, a stationary element and a clamping mechanism. The springs have three functions: they resonate, support the vibrating element against the welding force and return the head to the correct alignment at the end of the weld cycle. Electromagnetic machines use a fixed frequency: usually 120 Hz or 240 Hz (60 Hz supply), or 100 Hz or 200 Hz (50 Hz supply). Hydraulic machines can be operated at variable frequencies in the range 100–300 Hz. High-frequency, low-amplitude equipment is recommended when joint clearances are restricted to below 1.5 mm, while lower frequencies are recommended when welding long thin unsupported sidewalls, since the large amplitudes employed give greater tolerance to flexing. Lower frequencies allow more flexibility when welding larger components because of greater tolerance to slippage in the tooling.

Vibration welding can be applied to almost all thermoplastic materials. The technique is also believed to be particularly suitable for crystalline plastics such as acetal, nylon, polyethylene and polypropylene, which are not easily joined by ultrasonic welding or by solvents. Hygroscopic resins, such as nylon, do not need to be dry as for other processes, provided storage is at a relative humidity of less than 50%. With regard to dissimilar combinations, these are also possible. The incorporation of coloring can, however, lower the weld strength, as do fillers and glass fibers, for example, a 30 wt% glass fiber content can reduce joint strength by a factor of two.

The material consumption due to welding gives a reduction in length of 0.4–0.8 mm. For specific welding conditions, a tolerance on this reduction of ±0.1 mm should be achievable, with a horizontal alignment (in the direction of welding) of ±0.25 mm and perpendicular alignment of ±0.125 mm.

The application of vibration welding depends on its advantages relative to the competing friction (spin), ultrasonic and hot plate welding techniques. For circular parts, which do not require the maintenance of a radial position, friction welding is cheaper and faster. Similarly, ultrasonic welding will remain most competitive for smaller, simpler geometries because of its shorter cycle time. For larger, linear joints, vibration and hot plate welding are the main competitors. Although compared to hot plate welding, vibration welding has the advantages of shorter cycle times, less tendency to warpage and lower energy consumption, hot plate welding has advantages when the parts have three-dimensional joints or when the walls to be welded are long and thin and cannot tolerate application of vibration during welding.

Vibration welding has found most use in the automotive and domestic appliance industries but limited use has also been reported in aircraft and some other applications. A wide range of automotive parts is welded by this technique, including front and rear light assemblies, fuel filler doors, spoilers, instrument panels, fan surrounds and reservoirs for brake, power-steering and vacuum systems. Perhaps the most interesting automotive application is in the manufacture of the two-part plastic bumper. The basic construction comprises a long, curved U-section thermoplastic bumper welded to a rear reinforcement. This offers weight reduction, lower manufacturing costs and good aerodynamics. The joint is 1.5 m long and parts are vibration welded at a rate of 70 h^{-1}.

Applications in the domestic appliance industries include the welding of dishwasher conduits and pumps. An aerospace application involves the welding of passenger-aircraft air ducts. Injection-molded solar panels are welded because of their large size and because the cellulose acetate butyrate materials used are not readily welded by ultrasonics, while adhesives lead to crazing and leaky joints. These components can be vibration welded in 3 s at a rate of 3000 components per day.

4. Hot Gas Welding

Hot gas welding of plastics closely resembles the oxygas welding of metals. It was first described in 1937. Filler material is fed into a prepared joint and a stream of hot gas (rather than a flame) is used to heat filler and parent material. The equipment can be manual, semiautomatic or fully automatic. Normally, the filler material is a solid rod which is forced under light pressure into the joint. The filler rod is not melted in the gas stream but is sufficiently softened to allow it to fuse to the parent material. The filler rod can be circular in section but recently triangular-section rods have been used as they allow multirun welds and fillet welds to be made more easily. In some cases, particularly where semiautomatic or automatic welding of large assemblies, such as pipe flanges to pipes, is required, a variant of the technique called extrusion welding is employed. In this case, instead of a solid rod, the filler material is extruded into the joint. Hot gas is still needed to soften the parent material and occasionally additional radiant heating is used to preheat the parent material.

The hot-gas-welding tool consists of a nozzle and heater unit to heat the gas and to direct it to the workpiece. If the gas used is air, then either a remote or integral blower can be used or, in the case of a gas, connection to a gas supply must be provided. In most applications hot air is used but, for materials sensitive to oxidation, such as polyethylene and PVC, nitrogen or another inert gas is employed.

The shape of the gas nozzle can be chosen to suit the weld preparation, standard nozzles being supplied for tacking, manual welding or "speed welding" in which the filler rod is fed through an opening in the nozzle. Speed welding is easier and faster than fully-manual welding but joint properties are low. Nozzle designs can vary depending on the material to be welded.

In extrusion welding the equipment is too bulky to be easily portable so is often mounted on a trolley, although handheld systems are available.

It is believed that most thermoplastics are weldable by hot gas welding, provided that care is taken to optimize the welding conditions. However, only six materials are routinely welded: polyethylene, polypropylene, poly(vinyl chloride), poly(methyl methacrylate), polycarbonate and polyamide. If procedures are correct, joint properties close to those of the parent material can be achieved.

Hot gas welding is mostly used for butt and fillet welding but is also used for lap welding of thin sheet. In lap and fillet welding, no joint preparation is used. The filler is fed into the 90° angle of the fillet in fillet welding but no filler is used in lap welding of thin sheet.

For butt welding, the edges of the parent material are bevelled to give an included angle of 60–70°, depending on thickness, and a root gap of 0.5–1 mm. The parent material can be machined by scraping, filling, turning or milling. The parent material and filler rod should be clean and free from oil and grease. Cleaning fluids that attack plastics, such as acetone, are not suitable: petroleum spirit is often used.

In extrusion welding, identical joint preparations and cleaning procedures are recommended, although a root gap of 2 mm is used since extrusion welding is used on thicker sections (5–50 mm) than hot gas welding.

As hot gas welding is a manual process, its success depends entirely on the skill of the operator. The only machine-settable parameters are the gas-flow rate and temperature. Typically, temperatures are 200–300 °C and flow rates are 15–160 l min^{-1}. The technique usually employed in manual operation is first to heat the tip of the filler rod which is then pressed into the root of the weld. Welding then proceeds by heating both the joint and the filler at a uniform rate using an oscillating motion of the torch. Overheating the weld results in charring, blistering or foaming. Underheating results in the rod and parent material becoming insufficiently plastic so that notches, cavities and bad adhesion occur.

The principal advantages of hot gas welding are that it can be used manually and it is fully portable. It is thus suitable for large, complex fabrication and for repair operations. Most of the applications reflect this. The most important applications are in the fabrication of sheet and tube into vessels and pipework for use in chemical and other industrial plant, agriculture and buildings. In these applications, wall thicknesses of up to 50 mm are welded, although less than 15 mm is usual. Hot gas welding of thin sheeting can be accomplished without filler rod and is important in the lining of reservoirs. The most important disadvantages of the process are that it is slow and the weld quality is entirely dependent on the skill of the operator.

5. Heated Tool Welding

Heated tool welding, more commonly called hot plate welding, is perhaps the simplest welding technique with plastics. It is very widely used both in mass production and for large structures such as pipelines. Equipment for shop-floor and on-site use is available, capable of welding components from a few centimeters to 1.5 m in diameter.

The principle of the process is simple. The parts are held in fixtures which press them against a heated tool. In its simplest form this is a flat plate but more complex forms are used in, for example, socket welding of pipe, and three-dimensional joint profiles can be welded. The heating takes place in two stages. First, the tool melts the surface and, as this material is displaced, a smooth surface is obtained. Second, stops in the equipment or a reduction in pressure prevent too much displacement of material but the parts continue to be held against the heated tool until they are softened for some distance away from the tool. The welding fixtures then open, the heated tool is withdrawn and the fixtures reclose so forcing the two parts together. Again, stops in the equipment or pressure control regulate the amount of deformation that occurs. The parts are held together until the weld is cool. This sequence can be achieved completely automatically, although some parts of the operation such as insertion and removal of the heating tool may be done manually.

Heated tool welding is capable of reliable, high-quality joints and can easily be incorporated into a production line. However, the welding cycle time is relatively long, ranging from about 15 s for small components to more than 60 min for butt joints in large-diameter thick-wall pipes.

The heat is usually supplied by resistance heaters, although hot gas or gas burners have been used, to obtain an even temperature across the heated tool (the temperature must not vary by more than about 10 °C over the surface). Typical temperatures are 180–230 °C depending on the material to be welded, although sometimes temperatures as high as 500 °C are used. The surface of the tool is usually coated with PTFE to prevent the workpiece material from sticking to it but this cannot be used at temperatures greater than about 270 °C so, for higher temperatures, aluminum-bronze platens are used. For small machines, the welding pressure is usually obtained pneumatically.

For site welding of pipelines, the principle of operation is the same but generators must be provided to supply the electricity for the heater and hydraulic power for the welding force. On smaller site-welding machines, the welding force may be provided manually through a mechanical or hydraulic system. Site-welding machines also incorporate planing cutters to make square butt ends on the pipe prior to welding.

Most thermoplastic materials can be welded by hot plate welding and joint strengths equal to the parent material can usually be obtained if correct procedures are followed. The weldability of dissimilar material combinations is important in some applications, for example, car rear lights in PMMA (poly(methyl methacrylate)) and ABS (acrylonitrile–butadiene–styrene). Usually, properties equal to those expected from the weaker material can be obtained but only if the temperature and possibly the heating time are different for the different materials. This requires a more complex machine.

As even contact with the heating tool is necessary, the only joint design that can be used is the square butt joint, although the tool can be shaped to allow three-dimensional joints to be welded, as in socket and saddle joints of pipes. Flash traps and recesses can be incorporated into the design to hide the welding flash if this is considered unsightly. In a few applications, the flash is removed mechanically after welding but this is not preferred as it involves another operation. As heated tool welding is a relatively slow process it is not suited to high-volume mass production of small components for which ultrasonic or friction welding would be preferred. However, its simplicity and ability to weld all thermoplastics make it attractive where time is not a restriction and it is widely used in applications where high-integrity joints are required, for instance for fluid containment. Examples are automotive header tanks and hydraulic reservoirs. Other important automotive applications are welding fittings and filler pipes for polyethylene petrol tanks, and polypropylene copolymer battery cases.

The dimensional accuracy and good cosmetic appearance that can be achieved with hot plate welding have made it important in the building industry where it is used in the manufacture of unplasticized PVC door and window frames which are made from extruded sections and welded at each corner. Heated tool welding is also used for lamp assemblies, domestic appliance components such as washing machine agitators, soap dispensers and pump bodies, and for packaging containers such as pallets and crates.

Another important application of the process is the on-site joining of polyethylene pipes for gas distribution, water mains, sewers and sea outfalls. Welded pipe is also used in industrial plant and mining. The high integrity of the joints and the ability to weld pipes of diameters up to 1500 mm have made heated tool welding the only suitable process, although for small pipes (<180 mm diameter) alternatives are now being used.

6. *Implant Welding*

Implant welding is based on the principle of trapping a metal insert between the two parts to be joined and then heating the insert, by induction or resistance heating, so that the plastics material around the implant fuses to form the joint.

With induction heating, a high-frequency magnetic field (2–30 MHz) is used to induce electric current and hence, heat, in the metal implant. A development of this uses submicrometer-size ferromagnetic particles (usually iron oxide) dispersed in a thermoplastic matrix, which is preplaced along the joint in the form of paste or tape.

In resistance-implant welding, conductive wire or braid is again trapped between the components. However, in this case the wire is directly resistance-heated by the passage of an electric current. Currents as high as 150 A may be used.

Welding times are short: up to 20 s for the largest components. The process can be used to join almost all thermoplastics but the presence of the insert limits the joint strength.

Implant welding is a relatively new process and, as such, few details are available on welding conditions or joint properties. It is capable, at least in the case of resistively heated wire or braid, of welding large components of complex shape. The cost of the consumable can be outweighed by savings made because the process requires only simple equipment compared, for instance, with vibration welding. Thus, implant welding is a suitable technique for small production runs and is used in the manufacture of sailing dinghy hulls and electrically driven vehicles.

An important recent application of implant welding is in the production of the two-part plastic bumper for automobiles. The material used is poly(butylene terephthalate) (PBT) thermoplastic polyester. The final bumper assembly is strong enough to pass through the conventional paint-stoving cycle of 140 °C for 30 min.

A modification of resistance-heated implant welding is becoming widely used for joining small diameter (<180 mm) pipes. The process, known as electrofusion, employs a fitting into which the ends of the pipe to be welded are inserted. An electric current is then passed through a heating element in the fitting to melt material in the fitting and pipe and so to form the joint. Control of the process is by a timer and the applied voltage. Voltages in the range 20–40 V are used. The heating coil heats both the fitting and the surface of the pipe.

7. *Other Processes*

A number of other welding processes are employed with plastics, mostly for sealing thin film, for example, in the manufacture of plastic bags. These include HF welding, band sealing, flame sealing and hot bar sealing.

High-frequency (HF), or dielectric, welding is the most widely used of these other processes. It works well on plastics having a high dielectric loss factor such as ABS, PVC and cellulose acetate. Materials such as polyethylene and polystyrene have poor weldability. The technique uses a radio transmitter, usually operating at 27.120 MHz, connected to electrodes which hold the sheets together. The equipment must be shielded to prevent transmission of its signal. Typical applications are toys, raincoats and other mass-production sheet items. The process is well suited to mass production as it is fast and gives reliable results, but the cost of the equipment is high.

8. *Joint Properties and Quality Assurance*

As plastics become more widely used in structural and other critical applications, the mechanical properties and integrity of the welded joints become important. It is notable that in almost all service failures, fracture starts in a weld and is either due to fatigue or is brittle in nature.

Apart from tensile testing, almost no attention has been given to the properties of welded joints in plastics, although since the early 1970s some understanding of the properties and fracture behavior of the materials themselves has been developed. The lack of data on weld properties is partly because the need for such data has arisen only recently and partly because of distrust of fundamental test data in the plastics industry and a reliance on component testing under pseudoservice conditions. The complex behavior of plastics makes service performance difficult to predict from laboratory test data. A need exists for development of a better understanding of the relationship between service performance and test data and to devise more realistic laboratory tests for welds in plastics.

Quality assurance of welds in plastics is limited to destructive testing of batch samples in most applications. This is adequate in the majority of mass-production applications but, in critical and structural welds, techniques for nondestructive testing (NDT) are required. Some progress with a number of techniques has been made, particularly in the pipeline industry. The most successfully used technique at present is radiography with soft x rays. In welds in pipe, x rays can find lack of fusion, porosity, shrinkage voids, cracks, inclusions (e.g., sand or rust particles) and pipe misalignment. Depending on orientation and type of defect, defect sizes as small as 50 μm can be detected. Most specifications require zero detectable defects but, as advances in detection techniques are made to resolve flaws of even smaller size, this requirement will be increasingly hard to meet, so definition of the maximum tolerable defect sizes will be necessary.

Ultrasonic NDT has been tried on welds in plastic, although the technique has met with little success. It is used, however, to detect defects in fiber-reinforced composites and for monitoring the thickness of extruded plastics products.

The importance of obtaining sound joints is being

highlighted by the growing use of plastics in structural and load-bearing applications. Current joining technology for plastics is approached in an empirical way and is not based on an understanding of the underlying physical processes. Neither the process mechanics nor the effects that process parameters have on joint properties and quality are well understood. This situation needs to be improved to allow plastics to be used more widely with confidence in demanding, load-bearing applications.

See also: Adhesive Joining: An Overview; Elastomer–Thermoplastics Adhesion Processes; Joining of Composites

Bibliography

Bucknall C B et al. 1980 Hot plate welding of plastics: Factors affecting weld strength. *Polym. Eng. Sci.* 20: 432–40

Crawford R J, Tam Y 1981 Friction welding of plastics. *J. Mater. Sci.* 16: 3275–82

Frankel E J, Wang K K 1980 Energy transfer and bond strength in ultrasonic welding of plastics. *Polym. Eng. Sci.* 20: 396–401

Neumann J A, Bockholff R J 1959 *Welding of Plastics*. Reinhold, New York

Potente H 1980 Current state of technology and development in welding and joining techniques for plastics. *Kunststoffe* 70: 193–8

Stokes V J 1989 Joining methods for plastics and plastic composites: An overview. *Polym. Eng. Sci.* 29: 1310–24

Watson M N 1988 *Joining Plastics in Production*. The Welding Institute, Abington, UK

M. N. Watson
[The Welding Institute, Abington, UK]

Wire and Cable Systems for Submarine Transmission

Many kilometers of power and communication cables have been laid in rivers, lakes, bays, estuaries, seas and open oceans. Since the mid-1970s a number of extremely long, high-powered submarine cables have been placed in successful operation. The Skagerrak crossing between Norway and Denmark was accomplished with two 110 km long cables, each operating at 250 kV dc and carrying 250 MW, laid at depths of over 550 m. The Cook Strait crossing in New Zealand operates at ±250 kV dc and the 39 km long pressure cables carry some 600 MW. Present prospects and plans indicate that submarine power cables will be used even more extensively in the future.

This article concentrates on submarine power cables and more particularly on their component materials. Submarine cables can differ from land cables in design, materials, manufacture, transportation and laying, but once laid their operation is similar. Almost every type of insulation has been used for submarine transmission, but extruded crosslinked polyethylene or ethylene–propylene rubber insulation is favored for 5–69 kV cables, and impregnated paper for 69 kV and above. For submarine cables special design and materials considerations result from the requirements for: (a) single-pass, nontwisted lay of very long lengths of cable, even under rough sea bottom conditions; and (b) protection of the cable from fouling danger and from seawater corrosion. In addition, it is essential that all cable components be compatible, under all service conditions.

1. Conductor Materials

Although aluminum conductors are being used increasingly for land cables, submarine cable conductors are exclusively copper. Conductors are in general concentrically stranded and of circular cross section. For one-core filled cables, the conductor has a hollow core made by stranding over an open spiralled stainless steel or bronze flat spring of suitable diameter. Conversely, the conductor may be formed into keystones which support an oil passage. In large conductors, skin effects are reduced by longitudinally insulating the strand groups or segments from one another with paper or fiber inserts. Sector-shaped conductors for three-core cable are rarely used. Conductors over 2×10^6 circular mils ($1014\ mm^2$) have their segments bound together by a thin stainless-steel taping. Tinning of the conductor is not required for sulfurless insulations.

2. Conductor Shielding

For extruded insulation, such as crosslinked polyethylene and ethylene–propylene rubber, the conductor shield is extruded in tandem with the insulation and consists of the same polymer loaded (20–25%) with suitable carbon black (<25 nm particles) to provide good conductance ($>0.005\ \Omega^{-1}\ cm^{-1}$ at 25 °C). For impregnated paper insulation several conductive paper tapes form the shield, the first two intercalated and a third taped over. Carbon black paper (kraft paper impregnated with carbon black to secure good conductance) is used extensively for this purpose. Other conductive tape combinations are paper with one surface aluminized and a duplex tape with one surface carbon blacked. These must be wound to provide almost no potential drop from conductor to insulation and to present a smooth outer surface.

3. Insulation

Although extruded insulation is commonly used only up to 69 kV, there are a few installations with crosslinked polyethylene at 115 kV, 138 kV and even 230 kV. Beyond overcoming manufacturing defects such as bubbles, shield skips or intrusions and contaminant particles, extruded insulation suffers an almost inevitable growth of "water trees." In the presence of water, and accelerated by electric stress, these "trees" can grow across the insulation and ultimately cause its destruction. Compounds are now available that provide resistance to tree growth but their long-term effectiveness is unknown. At high voltages, it appears necessary to use a lead or aluminum sheath and even to block the conductor strands, to prevent moisture ingression.

Impregnated paper insulation with some form of ionization control is regularly used at 69 kV and higher voltages. The paper is made of kraft sulfate wood pulp from slow-growing conifers with extremely low mineral content. The pulp is extracted, beaten, and then formed and calendered into uniform, dense, low-loss electrical paper. The product, in the form of super-density, high-density and medium-density papers of thickness 0.003–0.012 in (0.08–0.3 mm) is slit into tape widths and stored under controlled ambient conditions prior to cable taping. After taping, the cable is thoroughly dried under high vacuum and temperature and then impregnated with a dried and degassed low-loss insulating liquid.

The most common impregnants are mineral oils. Paraffinic oils are highly oxidation resistant but gas under stress; aromatic oils do not gas but oxidize easily. Viscosity and viscosity index are also important considerations. Oil-filled cables require impregnants of very low viscosity, whose viscosity changes little with temperature. Pipe-cable cores require impregnants with high viscosity under ambient conditions and low viscosity at impregnating temperature. Mineral oils made to required viscosities from a 70:30 naphthenic/benzene base appear to be the best compromise. Some synthetic impregnants are being used successfully. Polyisobutene has shown excellent performance in pipe-cable cores and several compositions of dodecylbenzene are being used widely in oil-filled cables. These impregnants have lower viscosities and lower dielectric losses than mineral oils.

4. *Insulation Shielding*

The insulation shield over extruded insulations is preferably extruded in tandem, but more often is applied separately. It consists of the same base polymer loaded with carbon black to obtain the required low resistivity and has substantial thickness (0.030–0.050 in, 0.76–1.27 mm). For distribution circuits, this shield is applied tightly but not bonded, so that field stripping by hand is possible. For transmission circuits the shield must be bonded. This poses no hardship since submarine cable joints are usually factory made.

Impregnated paper insulation shields consist of the same types of conductive papers described, with the outer two layers always intercalated. Around these (as well as over the extruded shields) is wound a 0.005 in or 0.006 in (0.13 mm or 0.15 mm) soft copper tape or a copper-woven fabric tape, intercalated with a conductive paper tape.

5. *Sheath Materials*

Extruded cables may or may not have a coverall sheath. Below 35 kV, some "wet" designs apply the bedding and armoring directly over the cable core. For three-core cables, interstices are filled out with fibrous fillers (e.g., jute), and a copper binder tape (0.006–0.008 in, 0.15–0.20 mm) is applied before armoring. However, for higher voltages, either a thick (0.09–0.125 in, 2.3–3.2 mm) sheath made of polyethylene loaded with 4–5% carbon black, or better still a lead or aluminum sheath, is used.

Impregnated paper insulation must always have a hermetically sealed sheath. Lead alloy sheaths are most common, but aluminum has also found substantial usage. The sheaths are extruded directly over the completed cable cores. Lead, melting at 300 °C, is extruded at 200 °C or less, and does no harm to the insulation. Successful lead alloys combine strength and flexibility with resistance to fatigue and corrosion. Copper-bearing lead (0.06% Cu) is used extensively, but "half C" (Pb–0.2% Sn–0.075% Cd), F-3 alloy (Pb–0.15% As–0.10% Sn–0.10% Bi) and alloy "E" (Pb–0.4% Sn–0.2% Sb) all have use in various countries.

Aluminum sheaths are much stronger than lead ones, their permitted hoop stress being 41.4 MPa as opposed to 1.2 MPa. There are, however, some disadvantages: much lower flexibility, higher sheath losses, high corrosion susceptibility and sealing problems at joints and terminals.

Lead sheaths require reinforcement to increase hoop strength. This generally consists of two or three layers of stainless steel or copper–cadmium tapes (of thickness 0.004 in (0.1 mm)) over a conductive fabric-tape bedding.

6. *Protective Covering and Armoring*

Corrosion protection and some mechanical protection for lead and aluminum sheaths is provided by covering with either a vulcanized neoprene base jacket or, more often, an extruded black polymer jacket (poly(vinyl chloride) or polyethylene) of substantial thickness.

Submarine cables (other than pipe type) must be armored, since they are subjected to high tensile and transverse stresses during laying, and physical abuse under varying bottom and water movement conditions. One or two layers of heavy galvanized steel wire wound with long lay in good coverage has had most success.

Armor wires are generally laid over asphalted jute, compounded fabric or hessian-tape bedding. For protection, another layer of asphalted jute or compounded fabric is tightly wound over the armor wires and finished with an asphalt slush and an antistick talc or soapstone dusting.

The usual methods for reducing sheath–armor losses (e.g., insulated joints and crossbonding) are not applicable to submarine cables. For large cables, such losses can reach 1.75 times conductor loss. Nonmagnetic aluminum armor wire covered with 0.030 in (0.75 mm) of polyethylene has been used, but if the covering becomes abraded severe "holiday" corrosion results. Development of fiber-reinforced plastic wires appears to have more promise.

Difficult laying and service conditions may require two layers of armor wire. If so, they are wound in opposite directions with a heavy-duty bedding between. This is an excellent antitwist design. Another less heavy and bulky antitwist design is to use a flat galvanized steel tape (0.5 mm) wound over the jacket of the sheath.

Pipe-type cable for submarine service requires no armoring, since the heavy steel pipe provides adequate protection. However, the pipe must be protected from corrosion. For this, the pipe is covered with a dense somastic hot-applied coating (0.5–0.75 in, 12.7–19.1 mm) or with a woven glass tape–asbestos felt–tar enamel coating. The underwater section is then gunnited with concrete.

See also: Electrical Insulation: Solid Polymeric Materials; Electrical Properties of Polymers; Vulcanization, Continuous

Bibliography

Black R M Recent developments in cable technology. *Prog. Rubber Plast. Technol.* 2(4): 25–39

Mildner R C, Humphries E D 1958 Plastic-insulated cables for ac and dc transmission. *Proc. Conf. Int. Grande Réseaux Electrique à Haute Tension.* CIGRE, Paris

Oudin I M, Belle J 1958 Experimental studies of submarine cables. *Proc. Conf. Int. Grande Réseaux Electrique à Haute Tension.* CIGRE, Paris

Piscioneri J H, Luther R 1970 *The Long Island Sound Submarine Interconnection.* Electric Council of New England, South Egremont, MA

Takenouchi T, Minemuru S 1977 250 kV dc submarine cable for Hokkaido–Honshu link. *Proc. Conf. Int. Grande Réseaux Electrique à Haute Tension.* CIGRE, Paris

E. W. Greenfield
[Pullman, Washington, USA]

Wire and Cable Systems for Underground Transmission

Underground power cables may be buried (trenched) directly in the ground or pulled through ducts usually made of rectangular concrete or round metal or plastic pipes. The cable consists of the conductor surrounded by a primary electrical insulation and usually a protective shield. Well over 100 years ago, the first cables were made of copper wire surrounded by insulation made of rubber strips, gutta-percha or jute soaked in pitch, asphalt or paraffin. The earliest cables had limited flexibility and sometimes were laid in creosoted wooden troughs. At that time underground cables were much more expensive than overhead lines and were used only in congested areas or where obstacles limited overhead use. Over the years, however, the tremendous developments in plastics and other insulating materials have made it possible to produce cables for a wide range of voltages at ever decreasing cost. At the same time, increased consideration of aesthetic and other social factors made it mandatory to eliminate overhead lines so far as was feasible. Around 1960, underground residential distribution (URD) cable, made of relatively inexpensive, extruded plastic and laid directly in the ground, completely eliminated the use of overhead lines in many residential areas. Nevertheless, cable is still more expensive and has replaced overhead transmission lines only in cities and where obstacles such as rivers must be crossed.

1. Conductors

Good electrical conductivity and mechanical flexibility are the primary requirements for the conductors used in cable. Table 1 lists the electrical resistivity (the inverse of conductivity) for several metals. Only copper and aluminum are ordinarily used. (The use of the lightweight sodium metal in glass or extruded plastic tubing has been proposed periodically for years. Cable made with sodium encased in polyethylene was sold commercially for a short while in the USA. Unfortunately, sodium reacts explosively in contact with liquid water and even slowly

Table 1
Electrical resistivities at 20 °C for several metals

Metal	Resistivity (μΩ cm)
Silver	1.629
Copper	1.724
Aluminum	2.828
Sodium	4.3
Nickel	6.844

with water vapor, which permeates the polyethylene insulation.) Copper is the preferred metal because it is of smaller dimension for equivalent conductivity and is easier to join at terminations and splices. Aluminum can be justified only when it is sufficiently less expensive than copper to make up for the increased cost of the insulation on the larger diameter.

To achieve flexibility, annealed copper is used. The degree of anneal may be limited by the strength and elongation needed in manufacture and installation. Large conductors are made of stranded, smaller wires in numerous configurations: concentric, bunched or compressed to meet required flexibility and compactness.

In some liquid-impregnated paper-insulated cable for use at high voltages, the central part of the multiple-stranded cable is replaced by a slightly open spiral of steel wire. This provides a duct or channel to distribute oil in the cable system.

2. Primary Insulation

Insulating materials may be extruded over the conductor or applied in thin layers wrapped around the conductor. In the past, low-voltage cable was often insulated with varnished cotton cloth coated with a viscous liquid, which sealed the completed structure. Additional sealant was sometimes applied during the taping process. For higher voltages, multiple wraps of cellulosic paper impregnated with mineral oil were used. Varnished cloth has been almost entirely replaced by extruded polymers except for silicone resin or rubber-coated glass cloth in Navy cable, which is not truly an underground application.

The primary insulation serves to prevent electrical breakdown but must also contribute as little as possible to charging current (electrical capacitance) and dielectric loss. These characteristics of an insulating material can be best described by the vector diagram in Fig. 1. Dielectric loss, the power lost in the insulating material, is given by the relationship

$$W = 2\pi f V^2 C \tan\delta$$

where W is the power loss in watts, V is the voltage in volts, f is the frequency in hertz, C is the capacitance in farads and $\tan\delta$ is the dissipation factor (for values <0.01, $\tan\delta \simeq \cos\theta$, the power factor). Since capacitance depends on geometry, for materials, reference is made to the dimensionless unit of permittivity ε', often called the dielectric constant and, in cable terminology, the specific inductive capacitance (SIC). Permittivity is the ratio of the capacitance of a specific geometry of the insulation to that of the same geometry in a vacuum. Voltage breakdown, $\tan\delta$ and ε', may be greatly influenced by

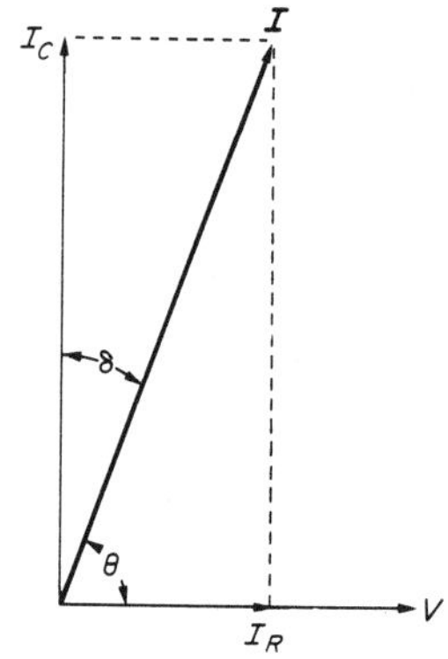

Figure 1
Vector diagram for parallel representation of dielectric loss: I_C, capacitive current; I_R, resistive current

temperature, aging in storage and use and environmental factors such as moisture and other contaminants.

Although voltage breakdown in cable is the most important factor, it is also the most difficult property to deal with from the material point of view. Cable is exposed to operating voltage, to overvoltage in short-time factory and field test and to switching surges which may be three or more times the operating voltage. Some cable also may be exposed to very high-voltage lightning surges, which may rise to a peak value in less than a microsecond. For materials, voltage-breakdown measurements are made on slabs, very often in the form of short lengths of cable with appropriate terminations. The geometry of the specimen greatly influences the breakdown stress. Some typical breakdown values are given in Table 2 and compared with operating stresses in cable. It is obvious that the conventional short-time breakdown voltage for materials may bear little relationship to the operating capability. However, comparative voltage-breakdown tests on materials are useful in screening, in determining the influence of operating and ambient factors such as temperature and moisture and for quality control especially as related to defects and foreign inclusions.

Table 2
Typical voltage-breakdown values and operating stresses in cable

Materials	Short-time breakdown ($kV\,mm^{-1}$)	Operating voltage ($kV\,mm^{-1}$)
Unimpregnated dry paper	4	not used
Oil-impregnated paper	88	up to 14
Polyethylene	92	1.6–2.6
Ethylene–propylene rubber	24	1.6–2.6
Poly(vinyl chloride)	14	0.8–1.6

The long-term voltage endurance of insulating materials is especially important but often very difficult to determine. For example, serious problems have arisen with direct burial of polyethylene cables. Especially in wet locations, the phenomenon of "water treeing" develops, in which slow progressive electrical failure takes place. Dry treeing may also take place starting at voids or conducting inclusions or asperities on the conductor (or conductor shield) and the outer conducting shield. (Wet and dry treeing sometimes occur together.) Considerable controversy exists about the relative treeing resistance of thermoplastic polyethylene, crosslinked polyethylene, filled polyethylenes and ethylene–propylene rubbers of various types. Special formulations and additives, which appear to be effective, have been developed to provide resistance to both wet and dry treeing, but these have not yet been fully proven in service.

Fortunately, tan δ and ε' measurements on materials are easily made and can be functionally significant. Typical curves of tan δ versus temperature are shown for solid materials in Fig. 2a and for liquid-impregnated paper in Fig. 2b. In Fig. 2a it is apparent that poly(vinyl chloride) (PVC) has a relatively high value of tan δ which increases markedly as the temperature increases. If such lossy materials are used at high voltage, it becomes obvious that an increase in temperature will increase the loss still further until either the outward heat flow can balance the heat generated or a thermal runaway will take place. For this reason, PVC is seldom used in cable operating above 600 V and never above 2300 V in underground application. Since the dielectric loss is proportional to the square of the voltage, tan δ becomes especially important at very high voltages. The data in Fig. 2b illustrate steps which have been taken to decrease the loss in oil–paper cable. With the use of synthetic liquids and improved papers, the dielectric loss has been decreased so that 500 kV, and perhaps 750 kV, cable become technically feasible. It is obvious to ask why polyethylene with its very low value of tan δ is not a suitable material in very-high-voltage cable. Polyethylene is used up to 69 kV and in a rather limited experimental way up to 150 kV. Unfortunately, it is extremely difficult to eliminate voids or inclusions, which cause dry treeing especially in thick sections. Moreover, such extruded cable probably should have a metallic sheath to prevent ingress of water and resultant water treeing. The metallic sheath increases costs and makes cable splicing and termination much more difficult, so that some of the potential competitive advantage of extruded cable insulation over oil–paper insulation is thereby lost. Before the use of extruded insulation up to 69 kV, lead-sheathed paper wrapping impregnated with a viscous mineral oil was extensively used and is still used where great reliability is desired. Unfortunately, the oil impregnant in such cable tends to migrate, especially in long vertical runs. To prevent oil migration and to improve impregnation, two types of pressurized oil-filled cables were developed. In one a low-viscosity oil is fed to the cable through a hollow section in the center of the conductor and contained on the outside with a lead or aluminum sheath. In the other, one or three paper-insulated cables are drawn into a metal pipe which is then filled under pressure with mineral oil. ASTM D1818 gives specification values for oils to be used in low-pressure cable systems. ASTM D1819 provides similar data for high-pressure systems. In addition to viscosity, very important requirements are given for water content and specific optical dispersion. The optical characteristics (see ASTM D1807) indirectly measure the amount of unsaturation and aromaticity in the oil. Unsaturation is important because it enables the oil to combine chemically with the hydrogen produced by corona discharge in voids or poorly impregnated areas of the paper. Under such conditions the oil polymerizes to form a wax (commonly called x wax), which fills the void and prevents further discharge. This self-healing characteristic of oil-impregnated paper cable is one major reason for its ability to operate so well at high voltages and to provide the great reliability needed in utility transmission systems. Moisture in the paper and oil can greatly decrease the voltage capability. Great care must be used to thoroughly dry both the oil and the paper without embrittling the paper.

Synthetic liquids including polybutenes and alkylbenzenes, and even diaromatic or polyaromatic compounds, are being used or developed either alone or in combination with mineral oil for use in

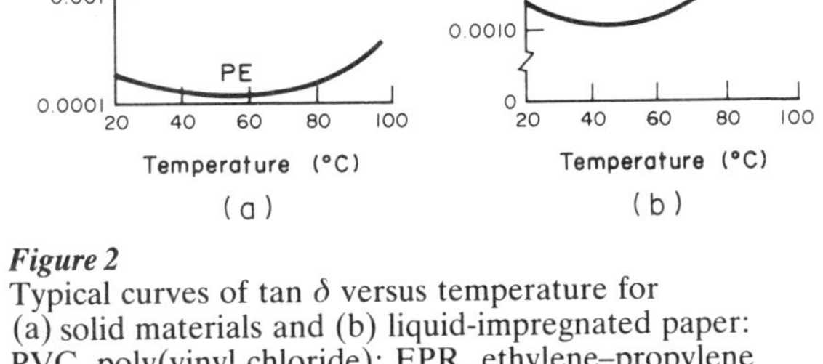

Figure 2
Typical curves of tan δ versus temperature for (a) solid materials and (b) liquid-impregnated paper: PVC, poly(vinyl chloride); EPR, ethylene–propylene rubber; PE, polyethylene; A, paper and mineral oil; B, polypropylene–paper laminate and synthetic liquid

Table 3
Typical performance characteristics of commercial sheat materials

Materials	Relative resistance to:				
	Moisture permeability	Heat	Flame	Aging	Oil
Thermoplastics					
Poly(vinyl chloride)	good		good	poor	very good
Polyethylene	excellent		poor	fair	poor to fair
Chlorinated polyethylene	excellent		fair	fair	fair
Thermoset (vulcanized)					
Neoprene	good	fair	excellent	fair	good
Chlorosulfonated polyethylene	very good	very good	excellent	good	very good
Nitrile rubber	fair	poor	fair	poor	excellent
Nitrile–butadiene/ poly(vinyl chloride)	good	fair	very good	good	good
Ethylene–propylene	excellent	very good	fair	excellent	poor
Silicone rubber	poor	excellent	good	excellent	poor
Crosslinked polyethylene	excellent	fair	poor	good	poor

high-voltage cable. Laminates of paper and polypropylene film are in use to decrease dielectric loss and improve the electrical strength. Plastic films do not provide the wicking action for oil, as paper does, but embossed and textured films are under development to provide the necessary wicking for adequate impregnation.

For cables operating above 2.3 kV it is necessary to provide a means for preventing electrical discharges at the interface between the conductor and insulation. At somewhat higher voltages it is usual to provide similar protection around the outside of the primary insulation. In liquid-impregnated paper cables, a layer of vapor-deposited, metal-coated or carbon-impregnated paper is used for this purpose. For extruded insulation, a semiconducting layer of plastic is usually used. Triple-head extruders provide the best manufacturing techniques to apply first the semiconducting layer over the conductor, then the primary insulation and finally a semiconducting layer around the outside. To make joints or splices it is essential to remove a length of the outer semiconducting layer completely and cleanly. A special composition is used which gives some adhesion but can be easily removed from the primary insulation. For cables above 34 kV, however, it is desirable to use a tightly bonded outer semiconducting layer.

The superior electrical properties of compressed gases, including sulfur hexafluoride (SF_6) and some fluorocarbons, have been recognized for a long time. Compressed-gas-insulated (CGI) cable is now under development with obvious advantages for use in combination with other totally enclosed electrical equipment, which utilizes the same compressed gas. Most of the CGI cables are spacer types with a conductor supported in the center of a rigid or semiflexible metal pipe by post or disk insulators made of various insulating materials with a number of different geometries. CGI cable is limited by the tendency of the spacers to flashover and by electrostatic levitation and movement of foreign metallic particles which reduce the voltage-breakdown strength. The larger diameter of the exterior pipe and its limited flexibility restrict application even more than the somewhat similar disadvantages with oil-filled pipe-type cable. Nevertheless, CGI cable may find use in river crossings and similar applications.

3. *Sheaths and Coverings*

The exterior covering of underground cable must provide protection against a wide variety of ambient conditions and protect the cable during installation. Continuous lead and aluminum sheaths can be applied over a variety of primary insulations to provide maximum protection against the ingress of water. However, metals are subject to corrosion and lead especially is subject to damage by squirrels and other rodents. Metallic sheathed cables are also difficult and expensive to splice and terminate. Metal coatings are fire resistant and, when in good condition, are for all practical purposes impervious to moisture.

In contrast, all plastic coverings for cable are more or less permeable to moisture. Fillers, antioxidants, flame-retardants and stabilizers of various types are added in different amounts to many basic materials. In this way a tremendous variety of products is produced in which balance of performance characteristics is achieved in different ways to different degrees. Table 3 gives rather typical performance characteristics of commercially used sheath materials. Up to 600 V some of the sheath materials, especially PVC and polyethylene, are also used as primary insulation utilizing a single-pass manufacturing operation.

The sheath may not entirely overcome poor characteristics of the underlying primary insulation. For example, PVC shield may not prevent an underlying polyethylene primary insulation from contributing to fire damage. Even worse, high temperatures from fire may decompose PVC or neoprene to produce very corrosive degradation products. Sheaths may be selected to meet the special needs of a specific application. For example, special formulations of PVC and especially ethylene–propylene–diene provide low-temperature flexibility to allow installation in cold weather. Neoprene has outstanding ability to resist mechanical abuse. Crosslinked polyethylene with added flame retardant can attain good flame resistance but with much poorer resistance to moisture.

Metallic braids and spirally wrapped skid wires are sometimes used to aid in installation or to provide protection in use.

See also: Electrical Insulation: Solid Polymeric Materials; Electrical Properties of Polymers; Vulcanization by Silanes; Vulcanization, Continuous

Bibliography

American Society for Testing and Materials 1984 *Annual Book of ASTM Standards*, Sect. 10, Pts. 10.1–10.3, *Electrical Insulation Specifications: Solids, Liquids and Gases*. ASTM, Philadelphia, PA

Baer E (ed.) 1964 *Engineering Design for Plastics*. Reinhold, New York

Eichhorn R M 1977 Treeing in solid extruded electrical insulation. *IEEE Trans. Electr. Insul.* 12:2–18

Friedlander G D 1972 Is "power to the people" going underground? *IEEE Spectrum* 9: 62–71

Robinson D M 1936 *Dielectric Phenomena in High Voltage Cables*. Chapman and Hall, London

Tonaka T, Greenwood A 1983 *Advanced Power Cable Technology*. CRC Press, Boca Raton, FL

K. N. Mathes
[Schenectady, New York, USA]

LIST OF CONTRIBUTORS

Contributors are listed in alphabetical order together with their addresses. Titles of articles that they have authored follow in alphabetical order. Where articles are coauthored, this has been indicated by an asterisk preceding the title.

Abo, M.
Werner and Pfleiderer Corporation
663 East Crescent Avenue
Ramsey, NJ 07446
USA
**Melt Delivery Systems, Twin-Screw*
**Mixing: Twin-Screw Extruders*

Adams, J. F. E.
"St Margarets"
Upper Wadham Avenue
Caversham
Reading
RG4 7EB
UK
**Plastics for Packaging*

Adams, R. D.
Department of Mechanical Engineering
Queen's Building
University of Bristol
Bristol
BS8 1TR
UK
Adhesives in Civil Engineering

Allan, P. S.
Department of Materials Technology
The Wolfson Centre for Materials Processing
Brunel University
Uxbridge
UB8 3PH
UK
**Processing: Shear-Controlled Orientation*

Baccei, L. J.
Loctite Corporation
705 North Mountain Road
Newington, CT 06111
USA
**Acrylate Adhesives*

Backer, S.
Fiber and Polymer Division
Room 3-338
Massachusetts Institute of Technology
Cambridge, MA 02139
USA
Textiles: Structures and Processes

Bader, M. G.
Department of Metallurgy and Materials Technology
University of Surrey
Guildford
GU2 5XH
UK
Glass-Reinforced Plastics: Thermoplastic Resins

Baier, R. E.
Health Care Instruments and Devices Institute
State University of New York
105 Parker Hall
Buffalo, NY 14214
USA
Adhesives in Medicine

Banov, A.
American Paint Company
Room 813
370 Lexington Avenue
New York, NY 10017
USA
Paints, Varnishes and Lacquers

Barito, R. W.
c/o Mr Gary Mellinger
General Electric Company
Building 35-1010
Appliance Park
Louisville, KY 40225
USA
Appliances: Applications of Engineered Plastics

Bascom, W. D.
Hercules Aerospace Division
Hercules Inc.
Bacchus Works
Magna, UT 84044
USA
Adhesive Joining: An Overview
**Bond Durability*
Bond Failure: Fracture

Beardmore, P.
Ford Motor Company
PO Box 2053
Dearborn, MI 48121
USA
Automotive Composite Components: Fabrication
**Design of Composite Structures*

Beck, F.
Universität Gesamthochschule Duisburg
Biologie-Chemie-Geographie
Postfach 10 16 29
D-4100 Duisburg 1
Germany
Electrodeposited Polymer Coatings

Bevis, M. J.
Department of Materials Technology
The Wolfson Centre for Materials Processing
Brunel University
Uxbridge
UB8 3PH
UK
**Processing: Shear-Controlled Orientation*

Birley, A. W.
Polymer Institute of Technology
University of Technology
Loughborough
Leicestershire
LE11 3TU
UK
Polymer Sheet: Manufacture and Applications

Bleile, H. R.
12508 Starlight Lane
Bowie, MD 20715
USA
Marine Coatings

Bowen, D. H.
Materials Development Division
Building 47
United Kingdom Atomic Energy Authority Harwell Laboratory
Didcot
Oxfordshire
OX11 0RA
UK
Composites: Applications Overview
Composites: Manufacturing Overview

Brewis, D. M.
School of Chemistry
Leicester Polytechnic
Leicester
LE1 9BH
UK
Bonding: Preparation of Polymer Surfaces

Briston, J. H.
30 Ranmore Road
Cheam Sutton
Surrey
SM2 7LT
UK
Multilayer Polymer Films

Brody, H.
ICI Fibres
Hornbeam Park
Hookstone Road
Harrogate
HG2 8QN
UK
**Fibers: Drawing*

Brookstein, D. S.
Albany International Research Company
1000 Providence Highway
Dedham, MA 02026
USA
Fibers: Texturing and Draw Texturing

Burlant, W.
General Electric Company
500 West Wilson Bridge Road
Columbus, OH 43285
USA
Automotive Applications of Polymers

Campion, R. P.
Materials Engineering Research Laboratory Ltd
Tamworth Road
Hertford
Hertfordshire
SG13 7DG
UK
**Elastomer–Elastomer Adhesion Processes*
**Elastomer–Metal Adhesion Processes*
**Elastomer–Textile Adhesion Processes*
**Elastomer–Thermoplastics Adhesion Processes*
**Elastomer–Wire Adhesion Processes*

Charlesby, A.
Silverspring
Eagle Lane
Watchfield
Swindon
SN6 8LA
UK
Radiation Effects on Polymers
Radiation Processing

Class, J. B.
Research Center
Hercules Inc.
RTE 48
8 Hercules Road
Wilmington, DE 19899
USA
Hot-Melt Adhesives
Pressure-Sensitive Adhesives

Coates, P. D.
Polymer Research Unit

School of Mechanical Engineering
University of Bradford
Bradford
BD7 1DP
UK
**Reaction Injection Molding*

Cohen, R. E.
Department of Chemical Engineering
Room 66-513
Massachusetts Institute of Technology
Cambridge, MA 02139
USA
Thermoplastic Elastomers

Collings, T. A.
Materials and Structures Department
Royal Aerospace Establishment
Farnborough
Hampshire
GU14 6TD
UK
Joining of Composites

Conant, F. S.
143 Gunarh Drive
Akron, OH 44139
USA
Elastomers: Tests for Mechanical Properties

Cooke, T. F.
Textile Research Institute
PO Box 625
Princeton, NJ 08542
USA
Fabrics: Finishing
Fibers and Textiles: An Overview
Geotextiles

Coran, A. Y.
Monsanto Company
260 Springside Drive
PO Box 5444
Akron, OH 44313
USA
**Elastomers: Additives*
Vulcanization

Corish, P. J.
Patrick J. Corish & Associates
"Mayberry"
Church Road
Snitterfield
Stratford-on-Avon
Warwickshire
CV37 0LN
UK
**Elastomer–Elastomer Adhesion Processes*
**Elastomer–Metal Adhesion Processes*
**Elastomer–Textile Adhesion Processes*
**Elastomer–Thermoplastics Adhesion Processes*
**Elastomer–Wire Adhesion Processes*
**Powdered and Particulate Elastomer Processing*

Corneliussen, R. D.
Drexel University
Department of Materials Engineering
32nd and Chestnut Streets
Philadelphia, PA 19104
USA
**Polymers: Mechanical Properties*

Coveney, V. A.
Senior Lecturer in Mechanical Engineering
Department of Engineering
Bristol Polytechnic
Coldharbour Lane
Frenchay
Bristol
BS16 1QY
UK
**Design of Elastomer-Based Engineering Products*

De Luca, P. L.
Armtec Defense Products Inc.
85-901 Avenue 53
PO Box 848
Coachella, CA 92236
USA
**Molded Fiber Products*

Dietz, A. G. H.
19 Cambridge Street
Winchester, MA 01890
USA
Plastics and Composites as Building Materials

Dunn, J. R.
Research and Development Department
Polysar Limited
Sarnia
Ontario
N7T 7MZ
Canada
Aging of Elastomers

Economy, J.
Department of Materials Science and Engineering
University of Illinois
1304 West Green Street
Urbana, IL 61801
USA
Novel Applications of Polymers

Eise, K.
Werner and Pfleiderer Corporation
663 East Crescent Avenue
Ramsey, NJ 07446
USA
**Melt Delivery Systems, Twin-Screw*
**Mixing: Twin-Screw Extruders*

Erwin, L.
Room 35-008
Massachusetts Institute of Technology
Cambridge, MA 02139
USA
Mixing: Basic Concepts

Evans, C. W.
[deceased; late of Gateshead, UK]
**Powdered and Particulate Elastomer Processing*

Fabris, H. J.
Research Division
General Tire and Rubber Company
Akron, OH 44329
USA
Flammability of Elastomers

Fellers, J. F.
Polymer Engineering
Dougherty Engineering Building
University of Tennesse
Knoxville, TN 37916
USA
Liquid Crystal Processing

Ferry, J. D.
Department of Chemistry
University of Wisconsin
Madison, WI 53706
USA
Elastomers: Dynamic Mechanical Properties

Finelli, A. F.
Goodyear Tire and Rubber Company
Akron
Ohio
USA
**Elastomeric Foams*

Frederick, T.
Drexel University
32nd and Chestnut Streets
Philadelphia, PA 19104
USA
**Polymers: Mechanical Properties*

Friedl, C.
Leader Flow Analysis Group
Moldflow Pty Ltd
Colchester Road
Kilsyth 3137
Victoria
Australia
**Injection Molding of Polymers: Computer Simulation*

Friend, R. H.
Cavendish Laboratory
University of Cambridge
Madingley Road
Cambridge
CB3 0HE
UK
Electrically Conducting Polymers

Gall, H.
c/o Bayer A G
FS-A PV3
Postfach 1140
D-4047 Dormagen
Germany
**Elastomeric Fibers*

Gent, A. N.
Professor of Polymer Physics
University of Akron
Akron, OH 44325
USA
**Automobile Tires*
**Bond Strength: Peel*

Gibson, L. J.
Massachusetts Institute of Technology
Cambridge, MA 02139
USA
Construction Materials, Cellular

Goettler, L. A.
Monsanto Company
PO Box 5444
Akron, OH 44313
USA
Extrusion of Fiber-Reinforced Polymer Systems

Goldstein, I. S.
Department of Wood and Paper Science
North Carolina State University
PO Box 5488
Raleigh, NC 27650
USA
Cellulose: Nature and Applications

Gordon, R. J.
Drexel University
32nd and Chestnut Streets
Philadelphia, PA 19104

USA
**Polymers: Mechanical Properties*

Goswami, B. C.
Department of Textile Merchandising and Design
239 Home Economics Building
University of Tennessee
Knoxville, TN 37916
USA
Spunbonding of Polymeric Fibers

Gray, W.
Southwest Research Institute
6220 Culebra Road
PO Drawer 28510
San Antonio, TX 78284
USA
**Armor, Composite*

Greenfield, E. W.
425 Dack Street North West
Pullman, WA 99163
USA
Wire and Cable Systems for Submarine Transmission

Grey, P. J.
BTR Industries Ltd
Horninglow Road
Burton-on-Trent
Staffordshire
DE13 0SN
UK
Dipped Polymer Products

Haines, R. C.
Racket Sports Division
Dunlop Slazenger International Ltd
PO Box 8
Carr Gate
Wakefield
WF2 0XB
UK
Sports Equipment Manufacture

Hamed, G. R.
University of Akron
Akron, OH 44325
USA
**Bond Strength: Peel*

Hancox, N. L.
Materials Development Division
Building 47
United Kingdom Atomic Energy Authority Harwell Laboratory
Didcot
Oxfordshire
OX11 0RA
UK
High-Performance Composites with Thermoplastic Matrices

Haresceugh, R. I.
Structural Engineering Division
British Aerospace plc
Military Aircraft Division
Warton Aerodrome
Preston
PR4 1AX
UK
Aircraft and Aerospace Applications of Composites

Harper, C. A.
Westinghouse Electric Corporation
Baltimore, MD 21203
USA
Electrical Properties of Elastomers

Harrell, E. R.
B. F. Goodrich Chemical Company
Avon Lake Technical Center
Moore and Walker Road
PO Box 122
Avon Lake, OH 44012
USA
**Rubber Processing*

Harris, B.
School of Materials Science
University of Bath
Claverton Down
Bath
BA2 7AY
UK
Hybrid Fiber–Resin Composites

Haskins, J. F.
31 Belmont Street
Brunswick, ME 04011
USA
Decorative Laminates

Hauser, M.
Loctite Corporation
705 North Mountain Road
Newington, CT 06111
USA
**Acrylate Adhesives*

Hawkins, W. L.
26 High Street
Montclair, NJ 07042
USA
Degradation and Stabilization: Test Methods
Polymers: Technical Properties and Applications
Recycling

Hilado, C. J.
Product Safety Corporation
1457 Firebird Way
Sunnyvale, CA 94087
USA
**Flammability of Structural Materials and Furnishings*

Howell, B. G.
Polymer Processes Department
BICC Cables Ltd
Construction and Wiring Division
Wrexham Technology Centre
Wrexham
Clwyd
LL13 9XP
UK
Vulcanization by Silanes
Vulcanization, Continuous

Huttlinger, P. A.
Product Safety Corporation
1457 Firebird Way
Sunnyvale, CA 94087
USA
**Flammability of Structural Materials and Furnishings*

Jaffe, M.
Hoechst Celanese Research Division
86 Morris Avenue
Summit, NJ 07901
USA
High-Modulus High-Strength Organic Fibers

Johnson, A. F.
Polymer Research Unit
School of Mechanical Engineering
University of Bradford
Bradford
BD7 1DP
UK
Glass-Reinforced Plastics: Thermosetting Resins
**Reaction Injection Molding*

Johnson, G. E.
Bell Laboratories
Room 7D214
600 Mountain Avenue
Murray Hill, NJ 07974
USA
**Electrical Properties of Polymers*

Johnson, P. S.
1098 Hagle Street
Sarnia
Ontario
N7V 4B1
Canada
Quality Requirements in Elastomer Processing

Kamal, M. R.
Department of Chemical Engineering
McGill University
PO Box 6070, Station A
Montreal
Quebec
H3A 2T5
Canada
Blow Molding of Thermoplastics
Injection Molding of Thermoplastics: Mold Filling

Kaufman, H. S.
Ramapo College
505 Ramapo Valley Road
Mahwah, NJ 07430
USA
Polymers: Classification by Use and Properties

Kaufman, S.
AT&T Bell Laboratories
2000 Northeast Expressway
Norcross, GA 30071
USA
Flammability of Polymers: Test Methods

Kelber, J. A.
Physical Chemistry and Mechanical Properties of Polymers
Division 1812
Sandia National Laboratories
Albuquerque, NM 87185
USA
**Radiation Effects on Polytetrafluoroethylene*

Kennedy, P.
Technology Transfer Manager
Moldflow Pty Ltd
Colchester Road
Kilsyth 3137
Victoria
Australia
**Injection Molding of Polymers: Computer Simulation*

Khanna, Y. P.
Allied Chemical Corporation
PO Box 1021R
Morristown, NJ 07960
USA
Flammability of Polymers

Kinloch, A. J.
Department of Mechanical Engineering
Imperial College of Science, Technology and Medicine
Exhibition Road
London SW7 2BX
UK
Adhesion: Fundamental Principles
**Bond Durability*

Klemchuk, P. P.
Additives Department
Ciba Geigy Corporation
Ardsley, NY 10502
USA
Degradation and Stabilization, Ultraviolet

Klosowski, J. M.
Dow Corning Corporation
Construction Specialist Division
3901 South Saginaw Road
Midland, MI 48640
USA
Sealants as Building Materials
**Structural Sealant Glazing*

Knuttila, J. A.
PPG Industries
760 Pittsburgh Drive
Delaware, OH 43015
USA
Roll and Coil Coatings

Kraus, G.
Research and Development
Phillips Petroleum Company
Bartlesville, OK 74004
USA
Fillers in Elastomers

Kulkarni, H. T.
Ford Motor Company
PO Box 2053
Dearborn, MI 48121
USA
**Design of Composite Structures*

LaGrone, B. D.
US Rubber Reclaiming Company, Inc.
Vicksburg, MS 39180
USA
Reclaiming of Elastomers

Lankford, J.
Southwest Research Institute
6220 Culebra Road
PO Drawer 28510
San Antonio, TX 78284
USA
**Armor, Composite*

Lederer, D. A.
Monsanto Company
260 Springside Drive
PO Box 5444
Akron, OH 44313
USA
**Elastomers: Additives*

Lee, T. C. P.
Morton International Ltd
Specialty Chemicals Group
University of Warwick Science Park
Sir William Lyons Road
Coventry
CV4 7EZ
UK
Elastomeric Products from Liquid Polymers

Lind, E. V.
Container Corporation of America
PO Box 1648
Wilmington, DE 19899
USA
Polymers: Tests for Mechanical Properties

Link, G. L.
AT&T Bell Laboratories
Room 7E207
600 Mountain Avenue
Murray Hill, NJ 07974
USA
**Electrical Properties of Polymers*

Livingston, D. J.
Livingston Associates
731 Frank Boulevard
Akron, OH 44320
USA
**Automobile Tires*

Lloyd, P. F. V.
B. F. Goodrich Chemical Company
Avon Lake Technical Center
Moore and Walker Road
PO Box 122
Avon Lake, OH 44012
USA
Plasticized PVC Elastomers

McIntosh, J.
26 Harvest Hill Road
Maidenhead
Berkshire
SL6 2QQ
UK
**Plastics for Packaging*

Mackley, M. R.
Department of Chemical Engineering
University of Cambridge
Pembroke Street
Cambridge
CB2 3RA
UK
Polyethylene Processing to Produce Ultimate Properties

Magill, D. G.
7 Wyeth Road
Hanover, NH 03755
USA
Papers of Natural and Synthetic Fibers

Mark, H.
Polytechnic Institute of New York
333 Jay Street
Brooklyn, NY 11201
USA
Polymer Science and Engineering: Historical Development

Mark, J. E.
Department of Chemistry
University of Cincinnati
Cincinnati, OH 45221
USA
Elastomers: An Overview

Mathes, K. N.
2052 Baker Avenue
Schenectady, NY 12308
USA
Wire and Cable Systems for Underground Transmission

Maxwell, B.
Department of Chemical Engineering
Princeton University
Princeton, NJ 08544
USA
Melt Delivery Systems: Normal Stress Pumps

May, C. A.
Manufacturing Research
Lockheed Missiles and Space Company
PO Box 504
B/170, 0/86-54
Sunnyvale, CA 94086
USA
Structural Adhesives

Menges, G.
Institut für Kunststoffverarbeitung
Technische Hochschule Aachen
Ponstrasse 49
D-5100 Aachen
Germany
**Computer Control in Polymer Processing*

Mills, P. J.
Department of Materials Science and Engineering
University of Surrey
Guildford
GU2 5XH
UK
**Carbon-Fiber-Reinforced Plastics*

Min, K.
Department of Polymer Engineering
College of Polymer Science and Polymer Engineering
University of Akron
Akron, OH 44325
USA
**Mixing: Internal Mixers*

Mock, J. A.
220 Florida Road
Ridgefield, CT 06877
USA
Elastomeric Coatings

Muhr, A. H.
Malaysian Rubber Producers' Association
Brickendonbury
Hertfordshire
SG13 8NL
UK
**Design of Elastomer-Based Engineering Products*

Nakajima, N.
B. F. Goodrich Chemical Company
Avon Lake Technical Center
Moore and Walker Road
PO Box 122
Avon Lake, OH 44012
USA
**Rubber Processing*

Nelson, J. K.
Rensselaer Polytechnic Institute
Troy, NY 12181
USA
Capacitor Materials: Solid Dielectrics

Nutt, A. R.
Arnold Nutt and Company
"The Mount"
Flax Lane
Twycross
Atherstone
Warwickshire
CV9 3SB
UK
Health Hazards in Elastomer Processing

Pendleton, W. W.
1542 Clinton Street
Muskegon, MI 49442
USA
Electrical Insulation: Solid Polymeric Materials

Petrie, C. J. S.
Department of Engineering Mathematics
University of Newcastle-upon-Tyne

Claremeont Road
Newcastle-upon-Tyne
NE1 7RU
UK
Processing: Rheological Properties and Fluid Mechanics
Tubular Film Extrusion: Dynamics and Heat Transfer

Recker, H.
Institut für Kunststoffverarbeitung
Technische Hochschule Aachen
Ponstrasse 49
D-5100 Aachen
Germany
**Computer Control in Polymer Processing*

Reinhart, T. J. Jr
Air Force Materials Laboratory (MEC)
Wright-Patterson IAFB
Dayton, OH 45433
USA
Bonding: Preparation of Metallic Surfaces

Renton, W. J.
Advanced Composites Division
Vought Corporation
PO Box 225907
Dallas, TX 75265
USA
Adhesives: Mechanical Test Methods

Robeson, L. M.
Union Carbide Corporation
Coatings Materials Division
River Road
Bound Brook, NJ 08805
USA
Polymer Blends

Rogers, J. W. Jr
Physical Chemistry and Mechanical Properties of Polymers
Division 1812
Sandia National Laboratories
Albuquerque, NM 87185
USA
**Radiation Effects on Polytetrafluoroethylene*

Rogers, T. H.
Apartment 26
2426 Ecuodorian Way
Clearwater, FL 33515
USA
**Elastomeric Foams*

Samour, C. M.
Laboratory Director
The Kendall Company
17 Hartwell Avenue
Lexington, MA 02173
USA
**Protective Tape Coatings*

Samuels, R. J.
Department of Chemical Engineering
Georgia Institute of Technology
Atlanta, GA 30332
USA
Hard-Fiber Elastomers

Sauer, J. A.
College of Engineering
Department of Mechanics and Material Science
Rutgers University
Busch Campus
PO Box 909
Piscataway, NJ 08854
USA
Fatigue
Impact Strength

Saunders, J. H.
Monsanto Textiles Company
PO Box 12830
Pensacola, FL 32575
USA
Fibers: Formation and Processing

Savla, M.
Savla Technical Services
77 Matri Mandir SCC
Gotri Road
Baroda
390015
India
Synthetic Adhesives

Schmidt, C. M.
Dow Corning Corporation
Sealant Division
3901 South Saginaw Road
Midland, MI 48640
USA
**Structural Sealant Glazing*

Seil, D. A.
B. F. Goodrich Chemical Company
Avon Lake Technical Center
Moore and Walker Road
PO Box 122
Avon Lake, OH 44012
USA
**Rubber Processing*

Seymour, R. W.
Research Laboratories

Tennessee Eastman Company
PO Box 511
Kingsport, TN 37662
USA
Plasticization of Polymers

Shelton, J. R.
32742 Washington Court
Solon, OH 44139
USA
Degradation and Stabilization, Thermal

Siadat, B.
W R Grace and Company
7379 Route 32
Columbia, MD 21044
USA
Dipped Organic Coatings

Simpson, V. P.
Protective Coatings Division
Ameron
201 North Berry Street
Brea, CA 92621
USA
Maintenance Coatings

Smith, P. A.
Department of Materials Science and Engineering
University of Surrey
Guildford
GU2 5XH
UK
**Carbon-Fiber-Reinforced Plastics*

Smith, T. L.
Research Laboratories
IBM
Cottle Road
San Jose, CA 95193
USA
Elastomers: Strength and Extensibility in Simple Tension

Spaak, A.
Plastics Institute of America
Stevens Institute of Technology
Hoboken, NJ 07030
USA
Design with Plastics

Steiner, P. R.
Faculty of Forestry
University of British Columbia
2357 Main Hall
Vancouver
British Columbia
V6T 1W5
Canada
Adhesives for Wood

Stelling, J. D.
Thompson and Lichtner Company
111 First Street
Cambridge, MA 02141
USA
Roofing Materials

Stevenson, A.
Materials Engineering Research Laboratory Ltd
Tamworth Road
Hertford
Hertfordshire
SG13 7DG
UK
Offshore Engineering Applications of Polymers

Sudekum, J. H.
B. F. Goodrich Chemical Company
Research and Development Center
Breckesville, OH 44141
USA
**Elastomers: Applications Overview*

Tadmor, Z.
Department of Chemical Engineering
Israel Institute of Technology
Technion City
Haifa 32000
Israel
Melt Delivery Systems: Corotating Disk Processors
Melt Delivery Systems, Single-Screw
Mixing: Single-Screw Extruders

Temple, E. J.
Monsanto Chemical Company
101 Granby Street
Bloomfield, CT 06002
USA
Bottles and Other Rigid Containers

Thomas, S. J.
The Kendall Company
Lexington Laboratory
17 Hartwell Avenue
Lexington, MA 02173
USA
**Protective Tape Coatings*

Throne, J. L.
Amoco Chemicals Corporation
Research and Development Department
PO Box 400
Naperville, IL 60540
USA
Injection Molding of Thermoplastics: Structural Foams

Tokita, N.
Uniroyal

Research and Development
Middlebury, CT 06762
USA
Mixing: Mills for Elastomers

Trewin, E. M.
Carbon Fibres Division
Courtaulds Ltd
PO Box 16
Coventry
CV6 5AE
UK
**Carbon-Fiber-Reinforced Plastics*

Varga, R. S.
B. F. Goodrich Chemical Company
Industrial Products Division
500 South Main Street
D/1753, B/24B
Akron, OH 44318
USA
**Elastomers: Applications Overview*

Vick, C. B.
US Department of Agriculture Forest Service
Forestry Sciences Laboratory
Carlton Street
Athens, GA 30601
USA
Adhesives in Building

Vlachopoulos, J.
Department of Chemical Engineering
McMaster University
Hamilton
Ontario
L8S 4L8
Canada
Calendering

Wagner, J. R.
2996 Runnymeade Drive
Norristown, PA 19401
USA
Bonding Nonwovens

Ward, I. M.
Department of Physics
University of Leeds
Leeds
LS2 9JT
UK
**Fibers: Drawing*

Watson, M. N.
The Welding Institute
Abington Hall
Abington
Cambridge
CB1 6AL
UK
Welding of Plastics

White, J. L.
Department of Polymer Engineering
College of Polymer Science and Polymer Engineering
University of Akron
Akron, OH 44325
USA
Coextrusion
Die Extrusion
Injection Molding of Thermoplastics: Structure Development
Melt Spinning: Dynamics
Melt Spinning of Fibers: Structure Development
**Mixing: Internal Mixers*
Mixing: Rheological Behavior and Dispersion of Two-Phase Systems
Processing: Classification
Processing: Historical Survey
Processing: Relation to Product Properties
Processing: Structure Development
Rheological Characterization of Molten Polymers
Sandwich Molding of Plastics
Tubular Film Extrusion: Structure Development

White, J. R.
University of Newcastle-upon-Tyne
Claremont Road
Newcastle-upon-Tyne
NE1 7RU
UK
Residual Stresses

Williams, J. C.
Technical Consultant
512 Braxton Place
Alexandria, VA 22301
USA
**Molded Fiber Products*

Wolf, K. H.
c/o Bayer A G
FS-T Dorlastan
Gebaude E 4
Postfach 1140
D-4047 Dormagen
Germany
**Elastomeric Fibers*

SUBJECT INDEX

The Subject Index has been compiled to assist the reader in locating all references to a particular topic in the Encyclopedia. Entries may have up to three levels of heading. Where there is a substantive discussion of the topic, the page numbers appear in ***bold italic*** type. As a further aid to the reader, cross-references have also been given to terms of related interest. These can be found at the bottom of the entry for the first-level term to which they apply. Every effort has been made to make the index as comprehensive as possible and to standardize the terms used.